AF104896

MATRICES
AND
DETERMINOIDS

University of Calcutta
Readership Lectures

MATRICES

AND

DETERMINOIDS

BY

C. E. CULLIS, M.A., Ph.D., D.Sc.

EMERITUS PROFESSOR OF MATHEMATICS, UNIVERSITY OF CALCUTTA;
FORMERLY PROFESSOR OF MATHEMATICS IN THE PRESIDENCY COLLEGE, CALCUTTA,
AND SOMETIME FELLOW OF GONVILLE AND CAIUS COLLEGE, CAMBRIDGE

VOLUME III, PART I

Cambridge:
at the University Press
1925

CAMBRIDGE UNIVERSITY PRESS
Cambridge, New York, Melbourne, Madrid, Cape Town,
Singapore, São Paulo, Delhi, Mexico City

Cambridge University Press
The Edinburgh Building, Cambridge CB2 8RU, UK

Published in the United States of America by Cambridge University Press, New York

www.cambridge.org
Information on this title: www.cambridge.org/9781107414266

© Cambridge University Press 1925

This publication is in copyright. Subject to statutory exception
and to the provisions of relevant collective licensing agreements,
no reproduction of any part may take place without the written
permission of Cambridge University Press.

First published 1925
First paperback edition 2013

A catalogue record for this publication is available from the British Library

ISBN 978-0-521-04760-9 Hardback
ISBN 978-1-107-41426-6 Paperback

Cambridge University Press has no responsibility for the persistence or
accuracy of URLs for external or third-party internet websites referred to in
this publication, and does not guarantee that any content on such websites is,
or will remain, accurate or appropriate.

PREFACE

THE printing of Vol. III of this work was suspended for several years during the completion of some investigations which have necessitated the division of it into two parts and the carrying forward of some of the matter designed for it to a later volume. The only material change in Part I is the expansion of a single chapter on commutants and invariant transformands into three chapters involving references to Part II, which will deal chiefly with structural matrices. Because Part I is issued in advance of Part II, it has been rendered complete in itself by the addition of § 241 and the appendices, except for some references to easily proved properties of ruled simple slopes which occur in § 248. The author has been much indebted to the writings of E. Netto (*Vorlesungen über Algebra*) and P. Muth (*Theorie und Anwendung der Elementartheiler*); and to the investigations of *Weierstrass*, *Kronecker* and *Frobenius* relating chiefly to algebraic forms and therefore necessarily to matrices. His indebtedness to *Sylvester* and *Cayley*, the pioneers of the Calculus of Matrices,—especially to the former,—is always to be understood. Some portions of Chapter XXVI are derived from N. Whitehead's *Universal Algebra*, and some portions of the following three chapters from H. Hilton's *Homogeneous Linear Substitutions*. Where abbreviated proofs are provisionally used to replace longer trains of reasoning, they can usually be attributed to Frobenius. Bibliographical references are few because it was not originally intended to tarry over the historical development of the subject, which has to be traced through a very large number of more or less relevant and more or less inter-related contributions by the investigators mentioned above and others such as *H. J. S. Smith*, *Hermite*, *Stickelberger*, *S. Cantor* and *K. Hensel*. Writings which deal at all *directly* with matrices in general are few in number, the list of them being fairly complete when we have added Kronecker and Hensel's *Theorie der Determinanten*, some papers by Frobenius, and Bôcher's *Introduction to Higher Algebra* to the papers of Sylvester and Cayley. On the other hand writings which deal professedly with determinants, and therefore actually with *square* matrices in a less general way, are very numerous; and of these a very complete account is contained in Muir's *History of the Theory of Determinants*. The terms 'matrix,' 'rank,' 'latent root' and 'canonical square matrix' can most fitly be attributed to Cayley, Frobenius, Sylvester and Weierstrass respectively, whilst the notations D_i and E_i for a maximum and potent factor are reminiscent of

Kronecker's 'Determinantenteiler' and 'Elementarteiler.' Sylvester's 'nullity' has become the author's 'degeneracy'; and his 'vacuity,' or total number of zero latent roots, is the excess of the order of a square matrix over that of its final transmutes. In fact much of Sylvester's work is included in portions of Chapters XIII, XV and XXVI.

The conception of a matrix as a *single* multiple or complex entity has been maintained throughout, and is the distinctive feature of this book. According to the notations which have been adopted, the first of the expressions

$$\begin{bmatrix} a_1, & b_1, & c_1 \\ a_2, & b_2, & c_2 \end{bmatrix}, \qquad \begin{vmatrix} a_1, & b_1, & c_1 \\ a_2, & b_2, & c_2 \end{vmatrix}, \qquad \begin{Vmatrix} a_1, & b_1, & c_1 \\ a_2, & b_2, & c_2 \end{Vmatrix}$$

denotes a single matrix A having six elements which (ordinarily) are scalar numbers, whilst the second denotes a certain function of those elements, viz. the determinoid of A. The third expression does not occur because it has not been precisely defined, and is not adapted for manipulation; as ordinarily used it means some complete matrix of the simple minor determinants of A such as the 3-element matrix $[b_1c_2 - b_2c_1,\ c_1a_2 - c_2a_1,\ a_1b_2 - a_2b_1]$. In the *pure* calculus with which this book deals, the elements of a simple matrix are always scalar numbers. But it is always possible to introduce hyper-numbers satisfying special laws into the elements. For example if i, j, k are hyper-numbers satisfying the relations

$$i^2 = j^2 = k^2 = ijk = -1,$$

whilst x, y, z, w and x', y', z', w' are scalar numbers, then the equation

$$[x, y, z, w]\begin{bmatrix} i \\ j \\ k \\ 1 \end{bmatrix} . [i, j, k, 1]\begin{bmatrix} x' \\ y' \\ z' \\ w' \end{bmatrix} = [x, y, z, w]\begin{bmatrix} -1, & k, & -j, & i \\ -k, & -1, & i, & j \\ j, & -i, & -1, & k \\ i, & j, & k, & 1 \end{bmatrix}\begin{bmatrix} x' \\ y' \\ z' \\ w' \end{bmatrix}$$

represents a development of the product qq' of the two quaternions

$$q = ix + jy + kz + w, \qquad q' = ix' + jy' + kz' + w'.$$

Matrices whose elements are themselves matrices have already occurred, and will be discussed in Part II.

Every physical entity, however complicated, can be represented mathematically by a set of two-dimensional tables giving the values of the parameters which determine its characters; and no other ways of completely representing it are possible to us. Consequently every physical entity can be represented by a matrix (which may be a sequence), and every mathematical theorem can be reduced to a property of matrices. The author is further of opinion that every special mathematical discipline (e.g. Analytical Solid Geometry, the Theory of Elasticity, the Theory of Groups) assumes its simplest aspect when it is treated as a branch of the Calculus of Matrices; ordinary Algebra in particular being the theory of one-element matrices, i.e. matrices of complexity 0. The general principle underlying applications of the calculus is the separation of variables from constants by means of matrix products, so as to reduce each problem to the determination of a matrix or the solution of a matrix equation. In dealing with very simple entities or very simple aspects of entities the calculus can be dispensed with, though not advantageously. When highly complicated entities become the subjects of investigation, the need for it becomes imperative, though at the same time a more precise acquaintance with it becomes necessary. Abundant evidence in support of these views will be forthcoming if it should be the author's good fortune to carry through this work from its present chrysalis state to its appointed end. The programme is substantially that of Sylvester, with enlargements which have been rendered possible by the labours of his successors.

A few explanatory remarks concerning the ten chapters which are now issued may serve to make clear the order in which the subject is being developed.

Chapters XX—XXII are purely algebraic and deal with rational integral functions of any number of variables, irresoluble and irreducible functions and factors, highest common factors, resultants, eliminants, discriminants and common roots. They are to a large extent based on Netto's invaluable *Vorlesungen über Algebra*; but the resultants and eliminants of general (non-homogeneous) functions have been derived from those of *homogeneous* functions, the latter being regarded as the more fundamental. These chapters include most of the algebraical theorems needed in the treatment of rational integral functional matrices, and will also serve as an introduction to the elegant applications of matrices to Algebra, some first indications of which will be found at the end of Chapter XXI. They formed the subject-matter of a course of lectures delivered in Calcutta early in 1920, and were printed during the same year for the use of students under the title of 'Chapters on Algebra' with a descriptive preface, some extracts from which are given below.

The distinction drawn between resultants and eliminants is convenient, but not essential. The eliminant of n functions (of n variables) is a resultant of $n+1$ functions of which one is linear, but it is a resultant of pre-eminent importance; it serves to determine the exact number of the common roots of the n functions and the values of all symmetric functions of the common roots, and it serves the same purpose for the finite roots only and for the infinite roots only.... The distinction arises naturally in the determination of the resultant of n functions (of $n-1$ variables) for successively increasing values of n. Each stage of the process commences with the special case in which one of the n functions is linear, i.e. with the determination of an eliminant; and from this special case we can pass at once to the general case by using the properties of symmetric functions.... The evaluation of symmetric functions of common roots has been simplified by the use of *homogeneous* monotypic symmetric functions, which enables us to equate any monotypic symmetric function S of the common roots of a system of n *homogeneous* functions (of $n+1$ variables) to a *rational integral* function H of the coefficients.

The operations of Algebra have been regarded as derived from practical operations facilitating the process of counting. If we put $a^p = a \setminus p$, making $\setminus$ the symbol for potentation, and if all letters denote natural numbers, we can define a counting operation of the fourth stage as a chain of potentations (or counting operations of the third stage) by the equation

$$a \setminus a \setminus a \setminus a \setminus \ldots \setminus a = a \setminus\!\!\setminus p,$$

where there are p equal operands on the left, and where

$$a \setminus b \setminus c = a \setminus (b \setminus c), \quad \text{whilst} \quad (a \setminus b) \setminus c = a \setminus (b \times c).$$

Whereas chains of additions and chains of multiplications are both commutative and associative, chains of potentations are in general not commutative and only associative with respect to final operands. It was estimated by Krönig (see H. J. Klein's *Kosmologische Briefe*) that if N is the fourth of the integers

$$1 \setminus\!\!\setminus 1 = 1, \quad 2 \setminus\!\!\setminus 2 = 4, \quad 3 \setminus\!\!\setminus 3 = 7,\ 625\ 597,\ 484\ 987, \quad 4 \setminus\!\!\setminus 4 = N$$

expressed in the ordinary decimal system of numeration, the printing of N in the smallest legible type would require more printing ink than would fill a sphere having a radius equal to a quintillion of light-years. This estimate is inadequate ; for if a sphere having a radius equal to 10^x light-years could contain sufficient ink for printing N, and if a cubic inch of ink cannot print more than 10^y digits, where y could be taken to be 8, we should have $3x + y > 8 \times 10^{153}$, therefore certainly $x > 10^{153}$, and not merely $x > 30$. The fourth counting operation has not been introduced (except speculatively) because we have not yet had any practical dealings with such large numbers. It should be observed however that chains of potentations do occur in Algebra, and that the direct operation $a \setminus\!\!\setminus p$ of the fourth stage is definable in Algebra whenever p is a natural number, and therefore has a place in the Theory of Numbers.

Chapters XXIII and XXIV deal with *potent divisors* and *equipotent transformations*. The general discussion of rational integral functional matrices is commenced in these chapters; but successive interpolations have created a wide gap between them and the chapter on primitive matrices (see Appendix C) in which the discussion will be resumed. The need for precision has led the author to define separately the 'irresoluble and irreducible divisors,' the 'maximum and potent factors,' and the 'maximum and potent divisors' of a rational integral functional matrix. We shall chiefly be concerned with the potent divisors; and it is shown that the potent divisors of a compartite matrix are the potent divisors of its several parts. The term 'potent divisor' replaces the terms 'invariant factor' and 'Elementarteiler,' neither of which is used. For the sake of brevity the potent divisors of the characteristic matrix of a square matrix A (see § 222) are called the 'characteristic potent divisors' of A. When the domain of rationality is not specified, the potent divisors of a matrix are ordinarily understood to be those corresponding to its irresoluble divisors. Various reductions by equipotent transformations are described in Chapter XXIV, the most important of them being the reduction of an x-matrix A to a matrix whose non-zero elements are the potent factors of A.

Chapter XXV on rational integral functions of a square matrix contains one of Cayley's most important contributions to the Calculus of Matrices together with developments of it due chiefly to Frobenius. It introduces the ***latent roots*** and the ***characteristic matrix*** (or associated arbitrary linear function) of a square matrix ϕ, and includes the determination of all *rational integral equations satisfied by* ϕ. It ends with a description of Frobenius's solutions of the matrix equation $\psi^2 = \phi$ when ϕ is a given undegenerate square matrix with constant elements, these being rational integral functions of ϕ. Appendix A and Ex. iv of § 228 are addenda to this chapter. The index will give references to other treatments of the subjects discussed.

Chapter XXVI deals with *equimutant* and *isomorphic transformations*, i.e. equigradent transformations by mutually inverse undegenerate square matrices, and also with Weierstrass's special reduction of a square matrix with constant elements to a *canonical square matrix* by isomorphic transformations. Necessary and sufficient conditions are first obtained for the existence of an equimutant transformation between two square matrices with constant elements, and these lead to (being in fact equivalent to) the reduction mentioned. This special reduction, which can be obtained in many ways (see Note 2 of § 240) besides those indicated in this chapter, is a focus of many converging and diverging investigations. It is included in the more general reductions described in Appendix C, which will be established in due course. The determination of the rank of any rational integral function $f(\phi)$ of a square matrix ϕ generalises results given in Chapter XXV, where we were concerned with rational integral functions of zero rank; and the investigations of § 235 are equivalent to the determination of all matrices normal to $f(\phi)$. The introduction of the *transmutes* of a square matrix connects the results of this chapter with those obtained in §§ 153 and 154, and explains the origin of the term 'equimutant.' It is shown that two square matrices with constant elements are connected by an equimutant transformation when and only when they have a first transmute in common, i.e. when and only when they have the same successive transmutes. Appendix C and Note 1 of § 246 are addenda to this chapter.

Chapter XXVII deals with *commutants* and true *commutantal transformations*, a commutant $X = \{A, B\}$ being defined to be a solution of the matrix equation $AX = XB$ (in which A and B are necessarily square matrices). The general commutant $X = \{A, B\}$ is either a zero matrix or a matrix having a determinate structure and a determinate rank, its elements being homogeneous linear functions of a determinate number of independent arbitrary parameters; when the elements of A and B are constants, it is a non-zero matrix if and only if A and B have a latent root in common; when A and B are unilatent canonical square matrices having the same latent root, it is a general ruled compound slope. In ordinary Algebra, where every matrix is a single scalar number, the general commutant $x = \{a, b\}$ is obviously 0 or an arbitrary parameter according as $a \neq b$ or $a = b$. Special attention is paid to symmetric, skew-symmetric and undegenerate commutants; to the cases in which $B = \pm A$ or $\pm A'$, where A' is the conjugate of A; and to commutants

$\{A, A\}$ which are rational integral functions of A. The chapter concludes with a discussion of the reduction and determination of commutantal transformations converting one commutant into another, with applications to symmetric and skew-symmetric matrices. Appendix B and § 253 are addenda to this chapter.

In Chapter XXVIII a matrix which is a commutant of every commutant $\{A, A\}$ of a square matrix A is called a *commutant of the commutants* (or commutant of the commutant) of A. Methods of constructing such matrices are described, and it is shown that they are identifiable with rational integral functions of A. This chapter makes clear the structure of Frobenius's solutions of the matrix equation $\psi^2 = \phi$.

Chapter XXIX deals with *invariant transformands* and the corresponding *bilinear* and *quadratic scalar invariants*, an invariant transformand $X = \text{inv}\ \{A, B\}$ being defined to be a solution of the matrix equation $AXB = X$ (in which A and B are necessarily square matrices). The general invariant transformand $X = \text{inv}\ \{A, B\}$ is either a zero matrix or a matrix having a determinate structure and a determinate rank, its elements being homogeneous linear functions of a determinate number of independent arbitrary parameters; when the elements of A and B are constants, it is a non-zero matrix if and only if there exist latent roots α, β of A, B such that $\alpha\beta = 1$; when A and B are unilatent bi-canonical square matrices whose latent roots satisfy this condition, it is a compound slope whose constituents have a certain common law of construction. In ordinary Algebra the general invariant transformand $x = \text{inv}\ \{a, b\}$ is obviously 0 or an arbitrary parameter according as $ab \neq 1$ or $ab = 1$. Special attention is paid to symmetric, skew-symmetric and undegenerate invariant transformands, and to the cases in which $B = \pm A$ or $\pm A'$, where A' is the conjugate of A.

It is my duty and pleasure to offer my sincere thanks to the authorities of the University of Calcutta for their generous help both in undertaking the publication of this volume and in furnishing facilities for the preparation of it; also to thank the officials and staff of the Cambridge University Press for their patience and careful attention to the printing. The appreciative and suggestive criticisms of such competent reviewers of the preceding volumes as Sir T. Muir, J. B. Shaw and Rai Bahadur A. C. Bose should receive acknowledgement. I can only put on record my special gratitude to the late Sir Asutosh Mookerjee whose sympathetic interest and helpful advice will be sorely missed.

C. E. CULLIS.

KEMPSFORD HOUSE, GLOUCESTER.
May, 1925.

CONTENTS

CHAPTER XX

THE IRRESOLUBLE AND IRREDUCIBLE FACTORS OF RATIONAL INTEGRAL FUNCTIONS

CHAPTER XXI

RESULTANTS AND ELIMINANTS OF RATIONAL INTEGRAL FUNCTIONS AND EQUATIONS

§§ PAGES

192. General, special and particular rational integral functions; weights of the coefficients in the variables; ordinary values of variables and coefficients; divisibility deduced from divisibility for ordinary values of the variables 67–70

193. Resultants of rational integral functions: properties of the resultant R of n general homogeneous (or general) functions of n (or $n-1$) variables; the resultant is homogeneous in the coefficients of each function, isobaric with respect to each variable, and irresoluble; relations of R to the partial resultants and the common roots of $n-1$ of the functions; resultants of functions which are products of general functions. Resultant of 2 functions, of n linear functions. Effect of a linear transformation of the variables on the resultant . 70–88

194. Unreduced and reduced resultants of specialised rational integral functions; unreduced and reduced actual resultants; the discriminant of a single rational integral function 88–96

195. Existence of common roots of rational integral functions: limits to be number of common roots for ordinary values of the coefficients; partial resultants and the information derivable from them . . 96–105

196. Eliminants of rational integral functions, complete and partial eliminants; properties of the X-eliminant E of n general homogeneous (or general) functions of $n+1$ (or n) variables; it is the resultant of the n functions and X; it is a homogeneous rational integral function of the auxiliary parameters or the coefficients of X; terms of highest degree in the individual parameters; the eliminant E is a product of linear factors for all ordinary values of the coefficients of the n functions; interpretation of the identical vanishing of E for particular values of the coefficients; eliminants of functions which are products of general functions. Exact number of common roots of the n functions, exact number of infinite common roots; repeated common roots; the discriminant of the system of n functions. Eliminants of special sets of n functions; of two functions of degrees 1 and 2; of two functions of degrees 2 and 2 105–129

CHAPTER XXII

SYMMETRIC FUNCTIONS OF THE ELEMENTS OF SIMILAR SEQUENCES

197. Symmetric functions of the elements of r similar sequences of n variables: definitions of any monotypic symmetric function S, its weights, degree and order, monotypic symmetric functions of order 1 (the σ's) and of degree 1 (the ϖ's or elementary symmetric functions); relations between the σ's and ϖ's; total number of independent ϖ's; standard relations between the ϖ's; interpretation of the relations between the ϖ's; expressions for S as a rational integral function of the σ's or the ϖ's 130–150

198. Homogeneous symmetric functions of the elements of r similar sequences of n variables; homogenisation of any monotypic symmetric function; relations between the homogeneous ϖ's; interpretations of these relations; expressions for any homogeneous monotypic symmetric function as a rational function of the homogeneous ϖ's . . . 150–155

§§ PAGES

199. Monotypic symmetric functions of common roots: any homogeneous monotypic symmetric function S of the common roots of n general homogeneous rational integral functions of $n+1$ variables can be equated to an equivalent rational integral function H of the coefficients of the n functions; properties of H; any monotypic symmetric function of the common roots of n general rational integral functions of n variables can be equated to an equivalent rational function of the coefficients of the n functions 155–165

200. Determination of resultants and eliminants: determination of the resultant of n general homogeneous (or general) rational integral functions of n (or $n-1$) variables; determination of the eliminant of n general homogeneous (or general) rational integral functions of $n+1$ (or n) variables; final proofs of the theorems of Chapter XXI . 165–179

CHAPTER XXIII

THE POTENT DIVISORS OF A RATIONAL INTEGRAL FUNCTIONAL MATRIX

201. Rational integral functional matrices. Rational integral x-matrices; analogues to the quotient and remainder in the division of one rational integral function by another 180–184

202. The irreducible, irresoluble and linear divisors of a rational integral functional matrix. Properties of the indices of the highest powers of such a divisor which are factors of all minor determinants of orders $1, 2, \dots i-1, i, i+1$ 184–195

203. Regular minors of a rational integral functional matrix. Every regular square minor of order $i+1$ contains a regular square minor of order i, and every regular square minor of order i is contained in a regular square minor of order $i+1$ 195–199

204. Regular minors of a rational integral functional matrix which is symmetric. Every regular diagonal square minor of order $i+1$ contains a regular diagonal square minor of order i or $i-1$, and every regular diagonal square minor of order $i-1$ is contained in a regular diagonal square minor of order i or $i+1$ 199–204

205. The potent divisors and potent factors of a rational integral functional matrix. Maximum and potent indices; maximum and potent divisors; maximum and potent factors 204–212

206. Potent factors and potent divisors of a product of two or more matrices 212–216

207. The irreducible divisors and the potent divisors of a compartite matrix. They are those of its several parts 216–226

208. Irreducible divisors and maximum factors of a complete matrix of minor determinants 226–228

209. Reduction of a rational integral functional matrix to one whose successive leading minors are all regular (with respect to any number of given divisors) by equigradent transformations 228–236

210. Reduction of a symmetric matrix to one whose successive leading diagonal minors are all regular (with respect to any number of given divisors) by symmetric equigradent transformations 236–240

CHAPTER XXIV

EQUIPOTENT TRANSFORMATIONS OF RATIONAL INTEGRAL FUNCTIONAL MATRICES

CHAPTER XXV

RATIONAL INTEGRAL FUNCTIONS OF A SQUARE MATRIX

§§ PAGES

224. Rational integral equations satisfied by a given square matrix ϕ of order m. Cayley's Equation, $-D_m(\phi)=0$; the equation of lowest degree, $-E_m(\phi)=0$ 316–322

225. Frobenius's solutions of the matrix equation $\psi^2=\phi$, when ϕ is a given undegenerate square matrix 323–329

CHAPTER XXVI

EQUIMUTANT TRANSFORMATIONS OF A SQUARE MATRIX WHOSE ELEMENTS ARE CONSTANTS

226. Definition of an equimutant transformation. Isomorphic transformations; symmetric equimutant (or semi-unit) transformations; composition of equimutant transformations 330–336

227. First form of the necessary and sufficient conditions for the existence of an equimutant transformation between two given square matrices ϕ and ψ. The potent divisors of the characteristic matrices $\phi(\lambda)$ and $\psi(\lambda)$ can only differ by some which are equal to λ 337–341

228. Reduction of a square matrix to a canonical square matrix of the same order by isomorphic (equimutant) transformations. Construction of canonical square matrices having given characteristic potent divisors. Equicanonical square matrices; unilatent and unipotent square matrices. Simple, unilatent and standard canonical square matrices. The unilatent super-parts of a standard (or any) canonical square matrix. The characteristic symbol of a square matrix whose elements are constants. Construction of all square matrices satisfying a given rational integral equation 341–349

229. Some properties of a simple canonical square matrix ϕ. Linear functions of ϕ; powers of ϕ; the inverse of ϕ when it is undegenerate; potent divisors of the characteristic matrices of the powers of ϕ or any linear function of ϕ. Potent divisors of the characteristic matrices of the powers of any given square matrix 349–354

230. Rank of any given rational integral function of a given square matrix . 355–363

231. Non-extravagant rational integral functions of a given square matrix . 363–365

232. The first transmutes of a square matrix whose elements are constants . 365–375

233. The successive transmutes of a given square matrix 375–383

234. Second form of the necessary and sufficient conditions for the existence of an equimutant transformation between two given square matrices. They have the same transmutes 383–385

235. Solution of any equation of the form $f(\phi)\,.\,\underline{\overline{x}}_{m}=0$, where ϕ is a given square matrix of order m. The normals to the matrix $f(\phi)$ . . . 385–401

236. The conjugate reciprocal and inverse of the characteristic matrix of a given square matrix ϕ. They are rational integral functions of ϕ . 401–409

CHAPTER XXVII

COMMUTANTS

CHAPTER XXVIII

COMMUTANTS OF COMMUTANTS

CHAPTER XXIX

INVARIANT TRANSFORMANDS

CHAPTER XX

THE IRRESOLUBLE AND IRREDUCIBLE FACTORS OF RATIONAL INTEGRAL FUNCTIONS

[The first three chapters of this volume deal with rational integral functions of several scalar variables, and are introductory to the succeeding chapters on functional matrices. We begin in § 184 with a brief survey of the numbers and operations and the domains of rationality of Algebra. In §§ 185—6 we define rational integral functions, irresoluble and irreducible factors, and highest common factors. The usual methods of regularising and homogenising rational integral functions are described in § 187; which deals generally with ordinary linear transformations of their variables. In § 188 we prove the existence of highest common factors, and the uniqueness of the resolution of a rational integral function into irresoluble or irreducible factors. The last two articles of this chapter contain some useful miscellaneous theorems, and a description of the extension of a domain of rationality by adjunction.]

§ 184. The numbers and operations of Algebra.

1. *Classification of scalar numbers.*

The word 'number' usually means a 'scalar number,' which is tacitly assumed to be finite. It will always be clear from the context when it means a natural number or a positive integer.

Scalar numbers are entities, including the natural numbers, with which all the fundamental operations of Algebra can be performed in accordance with the fundamental laws of Algebra, without any violation of those laws, and without the introduction of any new entities. Any entities which satisfy this definition are to be considered as scalar numbers in whatever way they are obtained. The most general conception of a scalar number has been evolved from that of a natural number by successive introductions of new numbers required for the performance of the operations of Algebra.

Stages	*Direct operations*	*Inverse operations*
I	Addition: $a+b$	Subtraction: $a-b$
II	Multiplication: $a\times b$, ab	Division: $a\div b$, $\frac{a}{b}$
III	Potentation: a^b	Radication: $\sqrt[b]{a}$ Exponentation: $\log_b a$

The *fundamental operations* of Algebra performable with any two ordinary scalar numbers a and b may be considered to be the three direct operations of the first, second and third stages and their inverses as shown in the table above.

Here potentation is the determination of a power, and has two operations inverse to it, viz. radication, by which we mean root-extraction, and exponentation, by which we mean the determination of a logarithm or exponent. The direct operations are derived from the processes of counting, and the inverse operations are those of solving the equations $b+x=a$, $b \times x=a$, $x^b=a$, $b^x=a$. Each of the first three inverse operations is reducible to the corresponding direct operation after the introduction of the negatives and reciprocals of numbers.

The *rational operations* of Algebra are addition, subtraction, multiplication, and division by a number which is not 0; i.e. they are the first two direct operations and their inverses, division by 0 being excluded. They are the operations which lead to results which are capable of only one interpretation.

The *fundamental laws* of Algebra are the laws of the fundamental operations for ordinary numbers together with the law of cancellation, the other special laws for the number 0, and the modified laws for infinite numbers. The most important laws are the commutative, associative and distributive laws for addition and multiplication.

NOTE 1. *The strictly algebraic operations.* We could define the fundamental operations of Algebra to be the rational operations together with root-extractions of the form $\sqrt[n]{a}$, where n is a natural number. We may call these the strictly algebraic operations, regarding the other operations of the text as belonging to the wider field of Algebraic Analysis.

NOTE 2. *The algebraic law of cancellation.* This law states that if a and b are numbers, and if $ab=0$, then at least one of the two numbers a and b must be 0. Thus the equation $ab=0$ leads to $b=0$ when $a \neq 0$, and to $a=0$ when $b \neq 0$. It follows that if $\rho a=\rho b$, where $\rho \neq 0$, we must have $a=b$, even when ρ is infinite.

The *natural numbers* are the numbers 1, 2, 3, ... of which no one is the greatest. They are derivable from the number 1 by repeated additions.

By *integers* we mean the numbers ± 1, ± 2, ± 3, ... together with the special number 0; the numbers 0, 1, 2, 3, ... being *positive integers*, and the numbers 0, -1, -2, -3, ... being *negative integers*. Integers can be formed from the natural numbers by the first fundamental operation and its inverse, i.e. by addition and subtraction.

By *rational numbers* we mean all numbers expressible in the form $\frac{p}{q}$ (i.e. all solutions of equations of the form $qx=p$), where p and q are integers, and $q \neq 0$. A rational number is one which it is possible to derive from the natural numbers by a finite number of rational operations.

An *infinite number* is one which is expressible in the form $\frac{p}{0}$ (i.e. a solution of an equation of the form $0 \, . \, x = p$), where $p \neq 0$. The ratio of two infinite numbers is that of any two finite numbers proportional to them. Accordingly two infinite numbers whose difference is a finite number have a ratio of equality, and (in this sense) are regarded as equal.

An *ordinary number* is one which is neither 0 nor infinite.

We may interpret *real numbers* to be all scalar numbers which can be derived from the natural numbers by rational operations, the rational operations being not restricted to be finite in number.

By *algebraic numbers* we mean the roots of all equations of the form

$$a_0 x^n + a_1 x^{n-1} + \dots + a_{n-1} x + a_n = 0, \quad \dots\dots\dots\dots\dots\dots(1)$$

where n is a natural number, the coefficients $a_0, a_1, \dots a_{n-1}, a_n$ are rational numbers, and $a_0 \neq 0$. The coefficients may without loss of generality be supposed to be integers. By a root of such an equation we here mean a scalar number which satisfies it.

Every *scalar number* z can be expressed in the form $z = x + y\sqrt{-1}$, where x and y are real numbers, and is then called a complex number. All numbers expressible in this form are necessarily scalar numbers, and it can be shown to be a consequence of the law of cancellation that there are no other scalar numbers. In the particular case when z is algebraic, the real numbers x and y are also algebraic. A non-zero number expressed in this form is real only when $y = 0$, imaginary when $y \neq 0$, and purely imaginary when $x = 0$. The number 0 can be regarded as real, purely imaginary or complex.

A scalar number which is not rational is called *irrational*, and a scalar number which is not algebraic is called *transcendental*.

All integers are rational numbers. All rational numbers are both real and algebraic. An irrational number may be either real or not real, and at the same time either algebraic or transcendental. A transcendental number is necessarily irrational, but it may be either real or not real.

Ex. i. The number $\sqrt{-1}$ is algebraic; the numbers e and π are transcendental.

Ex. ii. If x is a *real algebraic* number, it is known that:

(1) e^x is transcendental when $x \neq 0$;

(2) $\log_e x$ is transcendental when $x \neq 1$;

(3) $\sin x$, $\cos x$, $\tan x$ are transcendental when $x \neq 0$.

Ex. iii. If a and b are real and different, there always exist real transcendental numbers lying between a and b.

NOTE 3. *Super-rational and super-algebraic numbers.* If the numbers $a_0, a_1, \ldots a_{n-1}, a_n$ are all rational but not all 0, a root of an equation of the form (1) which is not rational will be called *super-rational*; and if they are all algebraic but not all 0, a root of an equation of the same form which is not algebraic will be called *super-algebraic.* Thus algebraic numbers consist of all rational and super-rational numbers.

We could (see Note in § 190) give wider meanings to these terms.

NOTE 4. *Numbers which are not scalar.* The word 'number' can be used in a wider sense, as when it denotes a quaternion, or when it denotes a quantity of the form $x_1e_1+x_2e_2+\ldots+x_ne_n$, where $x_1, x_2, \ldots x_n$ are any scalar numbers, and $e_1, e_2, \ldots e_n$ are given quantities subject to special laws, but not to all the laws of Algebra.

2. *Domains of rationality.*

An aggregate Ω of scalar numbers, not consisting of 0 only, is said to form a *domain of rationality,* or simply a *domain,* when the performance of a finite number of rational operations with any of them always produces a number belonging to the same

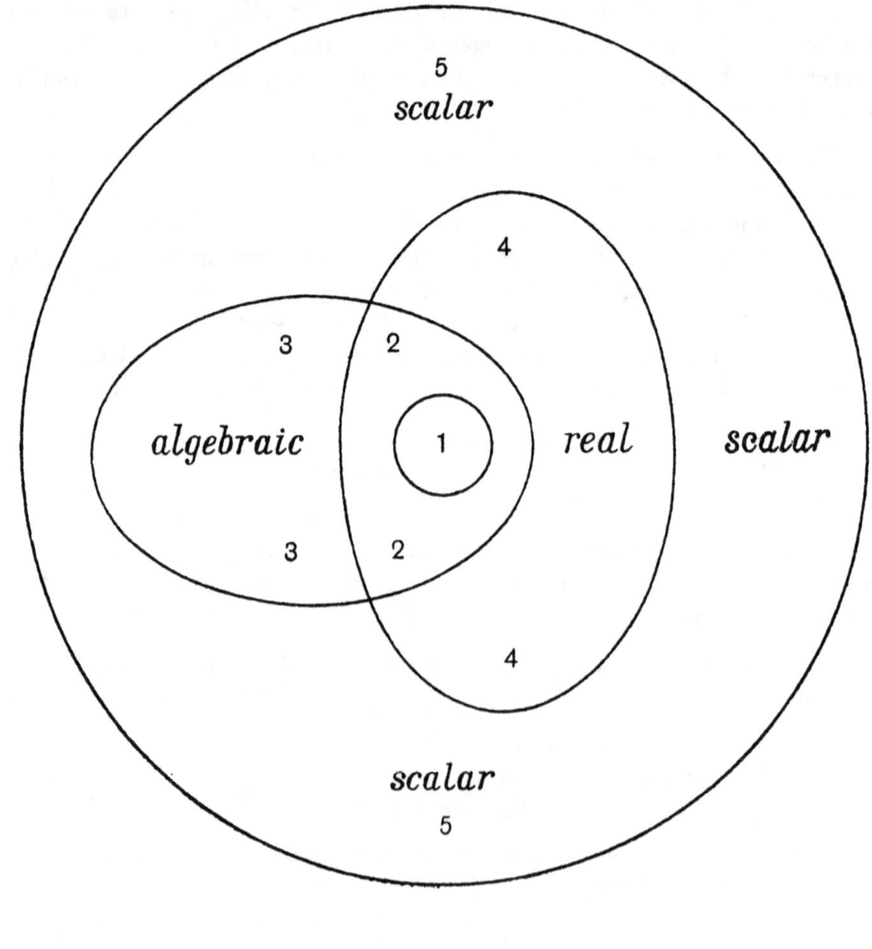

aggregate Ω. The numbers which constitute a domain Ω are called the elements of Ω. As illustrations of such domains we may cite:

(1) The domain of all rational numbers; which will be denoted by $\Omega(1)$ or Ω_1.

(2) The domain of all real algebraic numbers.

(3) The domain of all algebraic numbers.

(4) The domain of all real numbers.

(5) The unrestricted domain, or the domain of all scalar numbers.

In the figure the area of the large circle represents the domain (5); the areas of the two ellipses represent the domains (4) and (3); the area common to the two ellipses represents the domain (2); and the area of the small circle represents the domain (1).

The smallest domain which contains given numbers $(a, b, c, \ldots)$ is denoted by $\Omega(a, b, c, \ldots)$ or $\Omega_{a, b, c, \ldots}$. A domain which contains any non-zero number a necessarily contains the number 1, and a domain which contains 1 must contain all rational numbers. Consequently Ω_1 is the most restricted domain, and every domain must contain Ω_1.

A domain is *algebraic* or *transcendental* according as it does or does not lie in the domain of all algebraic numbers; and it is *real* or *not real* according as it does or does not lie in the domain of all real numbers.

Ex. iv. If $\sqrt{q}$ is a given irrational number whose square is rational, and if x and y are any rational numbers, all numbers of the form $x+y\sqrt{q}$ constitute a domain of rationality.

Ex. v. If x and y are any real numbers, all numbers of the form $x+y\sqrt{-1}$ constitute the unrestricted domain.

NOTE 5. *The zero domain.* The number 0 can be regarded as forming by itself a domain of rationality which is called the zero domain. In the definition of a domain given above the zero domain is excluded.

NOTE 6. *Domains whose elements are algebraic functions.* There exist domains of rationality whose elements are algebraic functions. For example we have:

(1) The domain of all rational functions of x.

(2) The domain of all functions of the form $A+B\sqrt{Q}$, where A and B are any rational functions of x, and Q is a given rational integral function of x which is not a perfect square.

In this volume a domain is always one whose elements are scalar numbers.

§ 185. Rational integral functions of scalar variables.

1. *Rational integral functions of a single variable.*

A rational integral function of a single scalar variable x is an expression of the form

$$f = a_0 + a_1 x + a_2 x^2 + \ldots + a_r x^r, \quad \ldots\ldots\ldots\ldots\ldots\ldots(1)$$

where r is a positive integer; where the coefficients $a_0, a_1, a_2, \ldots a_r$ are constant scalar numbers, and are all finite; and where x is a scalar number to which any value can be ascribed. If the coefficients $a_0, a_1, a_2, \ldots a_r$ all lie in a domain of rationality Ω, then the function f will be said to *lie in* or to *belong to* Ω, or to be *a function in* Ω; and we call $f=0$ the corresponding *rational integral equation in* Ω.

Any particular value of x for which f vanishes is called a *root* of the function f or of the corresponding rational integral equation $f=0$.

The function f is said to *vanish identically* when the coefficients $a_0, a_1, a_2, \ldots a_r$ all have the value 0. This is the case when and only when f vanishes for all finite values of x.

Two rational integral functions of x are *identically equal* when their difference vanishes identically. If we secure the existence of corresponding terms in the two functions by supplying terms with zero coefficients where necessary, this is so when and only when the coefficients of the same powers of x in the two functions are equal. When f vanishes identically, it is identically equal to 0, and may be said simply to be 0.

When f does not vanish identically, and is given in the form (1), the index r of the highest power of x which occurs in f is the *assigned degree* of f in x, even when $a_r=0$; and the index s of the highest power of x which occurs in (1) with a non-zero coefficient a_s is the *actual degree* of f in x. Both r and s are integers not less than 0, and we must have $s \not> r$. When terms with zero coefficients can be omitted or supplied as we please, a function whose actual degree is s can have any positive integral assigned degree which is not less than s. The function f is completely known when we know all its terms, including those with zero coefficients; and it is also known when we know the terms with non-zero coefficients and the assigned degree.

Ordinarily and throughout this chapter, when the terms of a non-zero function f are not shown, the assigned degree is supposed to be the same as the actual degree, and the *degree* of f means its 'actual degree,' but at the same time has the natural meaning of 'assigned degree.' In this case a given rational integral function f of the single variable x which does not vanish identically and has degree r is one which can be expressed in the form (1), where $a_r \neq 0$. When however the terms of f are shown, the 'degree' of f means its 'assigned degree'; in particular the degree of any term of f is always its assigned degree, i.e. the index of x in that term.

Ex. i. A rational integral function of x which does not vanish identically has degree 0 (actual or assigned) when and only when it is a non-vanishing constant.

Ex. ii. *Degrees of a function which vanishes identically.*

When f vanishes identically, its assigned degree can be any positive integer, including 0; and it is usually considered to have no actual degree. When we give to it the assigned degree 0, we are regarding it as a zero constant.

If we make the conventional assertion (see Note 2) that a rational integral function having a negative degree (actual or assigned) is one which vanishes identically, we can say that:

The rational integral function f vanishes identically if and only if its actual degree is less than 0.

If F and f are two given rational integral functions of x which do not vanish identically, the degree of f being not greater than the degree of F, there always exists an identical equation of the form

$$F = fQ + R,$$

where Q and R are rational integral functions of x, and where R is either 0, or does not vanish identically and has a lower degree in x than f. The functions Q and R are uniquely determinate and are called respectively the *quotient* and *remainder* in the division of F by f. When F and f both lie in a restricted domain of rationality Ω, the quotient Q and the remainder R also lie in Ω. When the remainder R is 0, the function F is said to be *divisible* by f. In the particular case when f is a non-vanishing constant, we always have $R = 0$, and F is necessarily divisible by f.

Again if F is a rational integral function of x which does not vanish identically, and if there exists an identical equation of the form

$$F = fg,$$

where f and g are rational integral functions of x, then F is said to be *divisible* by f, f is called a *factor* of F, and g is called the *quotient* in the division of F by f. At the same time F is divisible by g, g is a factor of F, and f is the quotient in the division of F by g. Neither of the functions f and g can vanish identically, and neither of them can have a greater degree in x than F; in fact the sum of the degrees of f and g is equal to the degree of F. Hence this new definition of divisibility agrees with the former, and the term *quotient* is used in the same sense as before.

According to this definition of divisibility we see that:

When the rational integral function f does not vanish identically, it is not divisible by 0.

NOTE 1. *Divisibility of F by f when F vanishes identically.*

When F vanishes identically, we may still retain the second of the definitions of divisibility given above. Then in this special case F is divisible by every rational integral function f of x, including 0; when f does not vanish identically, the quotient, by which F

must also be divisible, is 0; and when f vanishes identically, the quotient may be any rational integral function of x whatever. Accordingly we may say that:

The rational integral function F is divisible by 0, or has 0 as a factor, when and only when it vanishes identically.

2. *Rational integral functions of several variables.*

A rational integral function of several scalar variables $x, y, z, \ldots$ is an expression of the form

$$f = \Sigma a_{ijk\ldots} x^i y^j z^k \ldots, \quad \ldots\ldots\ldots\ldots\ldots\ldots(2)$$

where $i, j, k, \ldots$ are positive integers, each of which may be 0, whose values are not all the same in any two terms of the sum; where the coefficients $a_{ijk\ldots}$ are constant scalar numbers, and are all finite; and where $x, y, z, \ldots$ are scalar numbers to which any values may be assigned. If the coefficients $a_{ijk\ldots}$ all lie in a domain of rationality Ω, then the function f will be said to *lie in* or to *belong to* Ω, or to be *a function in* Ω; and we call $f = 0$ the corresponding *rational integral equation in* Ω.

Any set of values of the variables $x, y, z, \ldots$ for which f vanishes is called a *root* of the function f or of the corresponding rational integral equation $f = 0$.

The function f is said to *vanish identically* when the coefficients $a_{ijk\ldots}$ all have the value 0. This is the case (see Ex. vii) when and only when f vanishes for all sets of finite values of the variables $x, y, z, \ldots$.

Two rational integral functions f and g of the same variables $x, y, z, \ldots$ are identically equal when their difference vanishes identically. If we write

$$f = \Sigma a_{ijk\ldots} x^i y^j z^k \ldots, \quad g = \Sigma b_{ijk\ldots} x^i y^j z^k \ldots,$$

supplying terms with zero coefficients where necessary in order to ensure the existence of corresponding terms in both functions, this is the case when and only when $a_{ijk\ldots} = b_{ijk\ldots}$ for all sets of values of $i, j, k, \ldots$, i.e. when and only when the coefficients of corresponding terms in f and g are equal. When f vanishes identically, it is identically equal to 0, and may be said simply to be 0.

When the function f does not vanish identically, and is given in the form (2), where terms with zero coefficients may or may not occur, let the sum $i + j + k + \ldots$ of the indices of all the variables $x, y, z, \ldots$ be formed for each separate term. Then the greatest value of this sum which occurs amongst all the terms of f is the *assigned degree of f in all the variables $x, y, z, \ldots$*, or the *assigned total degree of f*; and the greatest value of this sum which occurs amongst all those terms of f which have non-zero coefficients is the *actual degree of f in all the variables $x, y, z, \ldots$*, or the *actual total degree of f*. Also the greatest values of the index i of any one variable x

which occur amongst all the terms of f and amongst those terms only of f which have non-zero coefficients are respectively the *assigned degree* and the *actual degree of f in the single variable x.* The assigned and actual degrees of f in any number of selected variables are similarly defined. If r and s are the assigned and actual total degrees of f, both r and s are integers not less than 0, and we must have $s \ngtr r$. When terms with zero coefficients can be omitted or supplied as we please, a function whose actual total degree is s can have any positive integral assigned degree which is not less than s. The function f is completely known when we know all its terms, including those with zero coefficients; and it is also known when we know the terms with non-zero coefficients and the assigned total degree.

Ordinarily and throughout this chapter, when the terms of a non-zero function f are not shown, the assigned degrees are supposed to be the same as the actual degrees, and the *degrees* of f mean its 'actual degrees,' but at the same time have the natural meanings of 'assigned degrees.' In this case a given rational integral function f of the variables $x, y, z, \ldots$ which does not vanish identically and has the total degree r in all the variables is one which can be expressed in the form (2), where there is at least one term with a non-zero coefficient in which $i+j+k+\ldots=r$, but no term in which $i+j+k+\ldots>r$. When however the terms of f are shown, as in the general theorems of the next chapter, the 'degrees' of f mean its 'assigned degrees'; in particular the degrees of any term of f are always its assigned degrees as shown by the indices of the variables in that term.

The most general rational integral function of n variables $x_1, x_2, \ldots x_n$ whose assigned degree in all the variables is r contains $\binom{n+r}{r}$ terms.

A *rational function* of scalar variables is one of the form $\frac{F}{G}$, where F and G are rational integral functions of the same variables, and G does not vanish identically.

Ex. iii. If $f=ax^2+bx^3y+3x^3y^2+0.x^4y^3$, $g=ax^2+bx^3y+3x^3y^2$, then f and g have the same actual total degree 5, but their assigned total degrees are respectively 7 and 5; also f and g have the same actual degree 3 in x, but their assigned degrees in x are respectively 4 and 3.

Ex. iv. A rational integral function of several variables which does not vanish identically has degree 0 (actual or assigned) in all the variables, or in every separate variable, when and only when it is a non-vanishing constant.

Ex. v. *Degrees of a function which vanishes identically.*

When f vanishes identically, its assigned total degree can be any positive integer, including 0; and it is usually considered to have no actual degree. When we give to it the assigned total degree 0, we are regarding it as a zero constant.

If we adopt the interpretations of Note 2, we can say that:

A rational integral function f vanishes identically when and only when its actual total degree is less than 0.

Moreover we can then speak simply of the total 'degree' of a function which vanishes identically, interpreting this to be an integer r which is subject solely to the condition $r<0$.

NOTE 2. *Rational integral functions with negative degrees.*

Since a rational integral function which vanishes identically cannot have any positive actual degrees, it is sometimes convenient, in order to extend the generality of our theorems, to regard it as having negative degrees. This will be legitimate if we interpret a rational integral function whose total degree (actual or assigned) in all the variables is negative to be one which vanishes identically, and a rational integral function whose total degree (actual or assigned) in any selected variables is negative to be one in which the terms involving those selected variables all vanish identically; for these interpretations simply mean that such functions do not exist.

When the rational integral function f does not vanish identically, it will be said to be *regular* in any one variable x when its actual degree in x is the same as its assigned total degree in all the variables. This is the case when and only when f contains a term of the form ax^r, where r is the assigned total degree of f in all the variables, and a is a non-vanishing constant.

Ex. vi. In the special case when f vanishes identically, it can be regarded as regular in any one or all the variables if and only if its assigned degree in all the variables is less than 0.

A rational integral function F of several variables which does not vanish identically is said to be *divisible* by another rational integral function f when there exists an identical equation of the form

$$F=fg,$$

where g also is a rational integral function; and g is then the *quotient* in the division of F by f. More generally when a rational integral function F which does not vanish identically can be expressed as a product of several rational integral functions $f_1, f_2, \dots f_m$, so that the equation

$$F=f_1f_2\dots f_m$$

is an identity in the variables, then each of the functions $f_1, f_2, \dots f_m$ is called a *factor* of F, and F is *divisible* by each of the functions $f_1, f_2, \dots f_m$.

According to this definition of divisibility we see that:

A rational integral function F which does not vanish identically is not divisible by 0.

NOTE 3. *Divisibility of a function which vanishes identically.*

If we still retain the definition of divisibility given above when F vanishes identically, then F in this special case is divisible by every rational integral function f, including 0; when f does not vanish identically, the quotient, by which F must also be divisible, is 0;

and when f vanishes identically, the quotient may be any rational integral function whatever. Accordingly we can make the following statement, which is equivalent to the interpretation which must be given to 'divisibility by 0':

A rational integral function F is divisible by 0, or has 0 as a factor, when and only when it vanishes identically.

We have illustrations of this principle in the theorems of § 187.3.

A rational integral function of several variables which does not vanish identically is said to be *homogeneous* in those variables when all its terms have the same total degree in all those variables. The actual total degree of such a function is necessarily the same as its assigned total degree, and cannot be less than 0. A rational integral function which vanishes identically can be regarded either as homogeneous or as non-homogeneous. The most general homogeneous rational integral function of n variables $x_1, x_2, \ldots x_n$ which has degree r in all the variables contains $\binom{n+r-1}{r}$ terms.

In the examples which follow we regard properties of rational integral functions of a single variable as being known.

Ex. vii. *If f is a rational integral function of the n variables $x_1, x_2, \ldots x_n$ which does not vanish identically, we can determine sets of finite values of $x_1, x_2, \ldots x_n$ for which f does not vanish.*

Knowing that this theorem is true for functions of a single variable, we can show by induction that it is true generally. We will make the hypothesis that the theorem is true for functions of $n-1$ variables, and show that it must then be true for every function of n variables such as f.

If f is a non-vanishing constant, the theorem is obviously true. We may therefore suppose that f is not merely a constant. In this case the degree of f in some one x_1 of the variables must be greater than 0, and we may suppose that

$$f = u_0 x_1^r + u_1 x_1^{r-1} + \ldots + u_{r-1} x_1 + u_r, \quad \ldots\ldots\ldots(1)$$

where $r \nless 1$; where $u_0, u_1, \ldots u_r$ are rational integral functions of $x_2, x_3, \ldots x_n$ only; and where u_0 does not vanish identically. By hypothesis we can assign such finite values $a_2, a_3, \ldots a_n$ to $x_2, x_3, \ldots x_n$ that $u_0 \neq 0$; and x_1 remaining arbitrary, the function f then becomes

$$f(x_1) = A_0 x_1^r + A_1 x_1^{r-1} + \ldots + A_{r-1} x_1 + A_r \quad \ldots\ldots\ldots(2)$$

where $A_0, A_1, \ldots A_r$ are constants, and where $A_0 \neq 0$. Since $f(x_1)$ is a rational integral function of the single variable x_1 which does not vanish identically, we can further assign such a finite value a_1 to x_1 that $f(x_1) \neq 0$. Then $a_1, a_2, \ldots a_n$ are finite values of $x_1, x_2, \ldots x_n$ for which the function f does not vanish.

We conclude that:

A rational integral function f of the n variables $x_1, x_2, \ldots x_n$ vanishes identically when and only when it vanishes for all sets of finite values of the variables $x_1, x_2, \ldots x_n$.

Ex. viii. *If f is a rational integral function of the n variables $x_1, x_2, \ldots x_n$ which is not a non-vanishing constant, we can determine sets of finite values of the variables for which f vanishes, i.e. there exist finite roots of the function f and of the equation $f=0$.*

Knowing that this theorem is true for functions of a single variable, we can show that it is true generally.

If f vanishes identically, the theorem is obviously true. We may therefore suppose that f is not merely a constant. In this case the degree of f in some one x_1 of the variables must be greater than 0, and we may suppose that f has the form (1) of Ex. vii. Then by Ex. vii we can assign such finite values $a_2, a_3, \ldots a_n$ to $x_2, x_3, \ldots x_n$ that u_0 does not vanish; and when x_1 remains arbitrary the function f becomes the function $f(x_1)$ shown in (2) of Ex. vii. Since the rational integral function $f(x_1)$ of the single variable x_1, which is not merely a constant, has finite roots, we can further ascribe such a finite value a_1 to x_1 that $f(x_1)=0$. Then $a_1, a_2, \ldots a_n$ are a set of finite values of $x_1, x_2, \ldots x_n$ for which f vanishes.

Ex. ix. *A rational integral function f of the n variables $x_1, x_2, \ldots x_n$ which does not vanish for any set of values of the variables must be a non-vanishing constant.*

This follows immediately from Ex. viii.

Ex. x. *Let f and g be two rational integral functions of the n variables $x_1, x_2, \ldots x_n$. Then if the product fg vanishes identically and f does not vanish identically, g must vanish identically.*

This is true when f and g are constants or functions of a single variable. We will make the hypothesis that it is true for functions of the $n-1$ variables $x_2, x_3, \ldots x_n$, and show that it must then be true for functions of the n variables $x_1, x_2 \ldots x_n$.

Let $$f=u_0x_1^r+u_1x_1^{r-1}+\ldots+u_r, \quad g=v_0x_1^s+v_1x_1^{s-1}+\ldots+v_s,$$

where $u_0, u_1, \ldots u_r, v_0, v_1, \ldots v_s$ are rational integral functions of $x_2, x_3, \ldots x_n$ only, and where u_0 does not vanish identically.

Then from the identity $fg=0$ we deduce the $s+1$ identical equations

$$u_0v_0=0, \quad u_0v_1+u_1v_0=0, \quad u_0v_2+u_1v_1+u_2v_0=0, \ \ldots \ u_0v_s+u_1v_{s-1}+\ldots=0,$$

the $(k+1)$th of these equations being

$$u_0v_k+u_1v_{k-1}+\ldots+u_kv_0=0 \quad \text{or} \quad u_0v_k+u_1v_{k-1}+\ldots+u_rv_{k-r}=0$$

according as $k \not> r$ or $k>r$.

Applying the hypothesis it follows from these equations in succession that $v_0, v_1, \ldots v_s$ vanish identically, and that therefore g vanishes identically.

Thus if the product fg vanishes identically, one at least of the functions f and g must vanish identically.

Ex. xi. *Let $f_1, f_2, \ldots f_m$ be m rational integral functions of the n variables $x_1, x_2, \ldots x_n$. Then the product $F=f_1f_2\ldots f_m$ vanishes identically when and only when at least one of the functions $f_1, f_2, \ldots f_m$ vanishes identically.*

This is immediately deducible from Ex. x.

Ex. xii. *Let f be a rational integral function of the n variables $x_1, x_2, \ldots x_n$ which does not vanish identically, and let g_1 and g_2 be any two rational integral functions of the same variables.*

Then if $fg_1=fg_2$ identically, we must have $g_1=g_2$ identically.

For if $f(g_1-g_2)=0$ identically, it follows by Ex. x that $g_1-g_2=0$ identically.

Hence if a rational integral function F is divisible by a rational integral function f which does not vanish identically, the quotient is a uniquely determinate function.

Ex. xiii. *If f and g are rational integral functions of the n variables $x_1, x_2, \dots x_n$ neither of which vanishes identically, and if $F=fg$, then the degree of F in all the variables is the sum of the degrees of f and g in all the variables.*

Let the degrees of f and g in all the variables be r and s, and let f' and g' be respectively the sums of all the terms of f and g which have degrees r and s in all the variables; also let F' be the sum of all the terms of F which have degree $r+s$ in all the variables. Then since the equation $F=fg$ is an identity, we have $F'=f'g'$ identically. By hypothesis neither f' nor g' vanishes identically. Therefore by Ex. x the function F' does not vanish identically, i.e. F has terms of degree $r+s$ in all the variables whose coefficients do not vanish.

Further the degree of F in any one variable x_1 is the sum of the degrees of f and g in x_1.

Let the degrees of f and g in x_1 be r and s, and let

$$f=u_0x_1^r+u_1x_1^{r-1}+\dots+u_r, \quad g=v_0x_1^s+v_1x_1^{s-1}+\dots+v_s,$$

where $u_0, u_1, \dots u_r, v_0, v_1, \dots v_s$ are rational integral functions of $x_2, x_3, \dots x_n$ only, and where neither u_0 nor v_0 vanishes identically. Then we have

$$F=U_0x_1^{r+s}+U_1x_1^{r+s-1}+\dots+U_{r+s},$$

where $U_0, U_1, \dots U_{r+s}$ are rational integral functions of $x_2, x_3, \dots x_n$ only, and where $U_0=u_0v_0$. By Ex. x the function U_0 does not vanish identically, i.e. F has degree $r+s$ in x_1.

Ex. xiv. *If $f_1, f_2, \dots f_m$ are m rational integral functions of the n variables $x_1, x_2, \dots x_n$ no one of which vanishes identically, and if $F=f_1f_2\dots f_m$, then the degree of F in all the variables is the sum of the degrees of $f_1, f_2, \dots f_m$ in all the variables, also the degree of F in any one variable x_1 is the sum of the degrees of $f_1, f_2, \dots f_m$ in x_1.*

We can prove this by repeated applications of Ex. xiii.

Further if $f_1f_2\dots f_m=c$, where c is a non-vanishing constant, then each of the functions $f_1, f_2, \dots f_m$ is a non-vanishing constant.

For no one of the functions $f_1, f_2, \dots f_m$ can vanish identically, and if $f_1, f_2, \dots f_m$ were not all constants, the product $f_1f_2\dots f_m$ could not have degree 0 in all the variables.

Ex. xv. *Let $f_1, f_2, \dots f_m$ be any m given rational integral functions of the n variables $x_1, x_2, \dots x_n$ which do not vanish identically; and let F be any rational integral function of the same variables. Then F vanishes identically when and only when it vanishes for every set of finite values of the variables which is not a root of any one of the functions $f_1, f_2, \dots f_m$.*

If F does not vanish identically, then by Ex. xi the product $Ff_1f_2\dots f_m$ does not vanish identically. Therefore by Ex. vii we can determine sets of finite values of the variables for which $Ff_1f_2\dots f_m\neq 0$. Therefore F does not vanish for every set of finite values of the variables which is not a root of any one of the functions $f_1, f_2, \dots f_m$.

Conversely if F vanishes identically, then it vanishes for every set of finite values of the variables, and therefore vanishes for every set of finite values of the variables which is not a root of any one of the functions $f_1, f_2, \dots f_m$.

Hence two rational integral functions F and G of the n variables $x_1, x_2, \dots x_n$ are identically equal when and only when they have equal values for every set of finite values of the variables which is not a root of any one of the functions $f_1, f_2, \dots f_m$.

Ex. xvi. *If the product of two rational integral functions f and g of the variables $x, y, z, \dots$ does not vanish identically and is homogeneous in the variables, then both the functions f and g must be homogeneous in the variables.*

Let f and g have degrees r and s in all the variables, where $r \nless s$. Then we can write

$$f = U_r + U_{r-1} + \dots + U_1 + U_0, \quad g = V_s + V_{s-1} + \dots + V_1 + V_0,$$

where U_i and V_j are homogeneous rational integral functions of degrees i and j in all the variables; and we have an identity of the form

$$(U_r + U_{r-1} + \dots + U_1 + U_0)(V_s + V_{s-1} + \dots + V_1 + V_0) = H_{r+s},$$

where H_{r+s} is a homogeneous rational integral function of degree $r+s$ in all the variables. No one of the functions U_r, V_s, H_{r+s} vanishes identically. From this identity we deduce the $r+s+1$ identical equations

$$U_0 V_0 = 0, \quad U_0 V_1 + U_1 V_0 = 0, \ \dots \ U_0 V_s + U_1 V_{s-1} + \dots + U_{s-1} V_1 + U_s V_0 = 0,$$

$$\dots \ U_k V_s + U_{k+1} V_{s-1} + \dots + U_{s+k-1} V_1 + U_{s+k} V_0 = 0, \ \dots$$

$$U_{r-s+1} V_s + U_{r-s+2} V_{s-1} + \dots + U_{r-1} V_2 + U_r V_1 = 0, \ \dots \ U_{r-1} V_s + U_r V_{s-1} = 0, \quad U_r V_s = H_{r+s},$$

where k receives the values $s+1, s+2, \dots r-s$.

We can write the $(i+1)$th of these equations in the form

$$\Sigma U_{i-\mu} V_\mu = 0, \quad \text{where } \mu \nless 0, \ \mu \nless i-r, \ \mu \ngtr s, \ \mu \ngtr i,$$

i.e. where μ receives all integral values consistent with the conditions $i-\mu \nless 0$, $i-\mu \ngtr r$, $\mu \nless 0$, $\mu \ngtr s$.

Or we can write the $(j+1)$th of these equations in the form

$$\Sigma U_\lambda V_{j-\lambda} = 0, \quad \text{where } \lambda \nless 0, \ \lambda \nless j-s, \ \lambda \ngtr r, \ \lambda \ngtr j,$$

i.e. where λ receives all integral values consistent with the conditions $\lambda \nless 0$, $\lambda \ngtr r$, $j-\lambda \nless 0$, $j-\lambda \ngtr s$.

If $U_0, U_1, \dots U_{r-1}$ do not all vanish identically, let U_i be the first of them which does not vanish identically. Then the $(i+1)$th, $(i+2)$th, ... $(i+s+1)$th of the above equations show in succession that $V_0, V_1, \dots V_s$ vanish identically. Since V_s does not vanish identically, we conclude that $U_0, U_1, \dots U_{r-1}$ all vanish identically.

Again if $V_0, V_1, \dots V_{s-1}$ do not all vanish identically, let V_j be the first of them which does not vanish identically. Then the $(j+1)$th, $(j+2)$th, ... $(j+r+1)$th of the above equations show in succession that $U_0, U_1, \dots U_r$ vanish identically. Since U_r does not vanish identically, we conclude that $V_0, V_1, \dots V_{s-1}$ all vanish identically.

Thus we have $f = U_r$, $g = V_s$; i.e. f and g are homogeneous in the variables.

Ex. xvii. *If H is a homogeneous rational integral function of the variables $x, y, z, \dots$ which does not vanish identically, then every factor of H must be homogeneous in all the variables.*

This follows immediately from Ex. xvi.

Ex. xviii. *Let $F = \phi(x_1, x_2, \dots x_n, x_{n+1})$ be a homogeneous rational integral function of the $n+1$ variables $x_1, x_2, \dots x_{n+1}$, and let $f = \phi(x_1, x_2, \dots x_n, c)$ be the rational integral function of the n variables $x_1, x_2, \dots x_n$ derived from it by putting $x_{n+1} = c$, where c is any finite non-vanishing constant. Then F vanishes identically when and only when f vanishes identically.*

For there is a one-one correspondence between the terms of F and f, any two corresponding terms T and t having the forms

$$T = a x_1^{p_1} x_2^{p_2} \dots x_n^{p_n} x_{n+1}^{p_{n+1}}, \quad t = a c^{p_{n+1}} x_1^{p_1} x_2^{p_2} \dots x_n^{p_n};$$

and the coefficient a vanishes when and only when the coefficient $ac^{p_{n+1}}$ vanishes. Thus all the coefficients in F vanish when and only when all the coefficients in f vanish.

It follows that F vanishes identically if and only if it vanishes for all finite values of $x_1, x_2, \ldots x_n$ when $x_{n+1}=c$.

This result is generalised in Ex. x of § 187 and Ex. vi of § 189.

In particular F vanishes identically when and only when the rational integral function $f'=\phi(x_1, x_2, \ldots x_n, 1)$ derived from it by putting $x_{n+1}=1$ vanishes identically; and therefore F vanishes identically if and only if it vanishes for all finite values of $x_1, x_2, \ldots x_n$ when $x_{n+1}=1$.

Ex. xix. Two homogeneous rational integral functions F and G of the $n+1$ variables $x_1, x_2, \ldots x_{n+1}$ are identically equal when and only when the following two conditions are satisfied:

(1) The functions F and G have the same degree in all the variables.

(2) The rational integral functions of $x_1, x_2, \ldots x_n$ derived from F and G by putting $x_{n+1}=c$, where c is any given finite non-vanishing constant, are identically equal.

The second condition is satisfied if and only if F and G are equal for all sets of finite values of the variables when $x_{n+1}=c$.

Ex. xx. *If f is a rational integral function of the n variables $x_1, x_2, \ldots x_n$, and if f vanishes for all sets of finite values of $x_1, x_2, \ldots x_n$ lying in any assigned domain of rationality Ω, then it vanishes identically.*

This theorem is true for functions of a single variable x; for if any such function has degree r in x and vanishes for more than r different values of x, then it must vanish identically.

We make the hypothesis that the theorem is true for all rational integral functions of the $n-1$ variables $x_2, x_3, \ldots x_n$, and show that it is then also true for all rational integral functions of the n variables $x_1, x_2, \ldots x_n$.

Let
$$f=U_0x_1^r+U_1x_1^{r-1}+\ldots+U_{r-1}x_1+U_r,$$
where $U_0, U_1, \ldots U_{r-1}, U_r$ are rational integral functions of $x_2, x_3, \ldots x_n$ only.

When we assign to $x_2, x_3, \ldots x_n$ any particular values $a_2, a_3, \ldots a_n$ lying in Ω, the function f becomes
$$f(x_1)=A_0x_1^r+A_1x_1^{r-1}+\ldots+A_{r-1}x_1+A_r,$$
where A_i is the value assumed by U_i when $x_2, x_3, \ldots x_n$ have the values $a_2, a_3, \ldots a_n$.

If f vanishes for all finite values of $x_1, x_2, \ldots x_n$ lying in Ω, then $f(x_1)$ vanishes for more than r different values of x_1. Therefore $f(x_1)$ is a rational integral function of x_1 which vanishes identically, and $A_0, A_1, \ldots A_{r-1}, A_r$ all vanish. This shows that $U_0, U_1, \ldots U_{r-1}, U_r$ vanish for all finite values of $x_2, x_3, \ldots x_n$ lying in Ω. Therefore by the hypothesis the functions $U_0, U_1, \ldots U_{r-1}, U_r$ all vanish identically, i.e. f vanishes identically.

Hence if f does not vanish identically, there must exist sets of finite values of $x_1, x_2, \ldots x_n$ lying in any assigned domain Ω for which f does not vanish.

Ex. xxi. *Let $f_1, f_2, \ldots f_m$ be m rational integral functions of the n variables $x_1, x_2, \ldots x_n$, no one of which vanishes identically. Then we can determine sets of finite values of $x_1, x_2, \ldots x_n$ lying in any assigned domain of rationality Ω for which no one of the functions $f_1, f_2, \ldots f_m$ vanishes.*

For by Ex. xi the product $F=f_1 f_2 \ldots f_m$ does not vanish identically, and therefore by Ex. xx there exist sets of finite values of $x_1, x_2, \ldots x_n$ lying in Ω for which F does not vanish.

Hence as in Ex. xv *any rational integral function F of the n variables $x_1, x_2, \ldots x_n$ vanishes identically when and only when it vanishes for every set of finite values of $x_1, x_2, \ldots x_n$ lying in Ω which is not a root of any one of the functions $f_1, f_2, \ldots f_m$.*

Ex. xxii. *Let F and f be rational integral functions of the n variables $x_1, x_2, \ldots x_n$ which lie in any domain of rationality Ω and do not vanish identically. Then if F is divisible by f, the quotient must lie in the domain Ω.*

This theorem is true when F and f are constants and when they are rational integral functions of a single variable. We will make the hypothesis that it is true for rational integral functions of the $n-1$ variables $x_2, x_3, \ldots x_n$, and show that it is then also true for rational integral functions of the n variables $x_1, x_2, \ldots x_n$.

If F is divisible by f, and if g is the quotient, then by Ex. xiii the sum of the degrees of f and g in x_1 is equal to the degree of F in x_1; and if we denote the degrees of f, g and F in x_1 by r, s and $r+s$, we can write

$$f=u_0 x_1^r+u_1 x_1^{r-1}+\ldots+u_{r-1}x_1+u_r, \quad g=v_0 x_1^s+v_1 x_1^{s-1}+\ldots+v_{s-1}x_1+v_s,$$

$$F=U_0 x_1^{r+s}+U_1 x_1^{r+s-1}+\ldots+U_{r+s-1}x_1+U_{r+s},$$

where $u_0, u_1, \ldots u_r, v_0, v_1, \ldots v_s, U_0, U_1, \ldots U_{r+s}$ are rational integral functions of $x_2, x_3, \ldots x_n$ only, and where u_0, v_0, U_0 do not vanish identically. Then from the identity $F=fg$ we deduce the $r+s+1$ identical equations

$$u_0 v_0=U_0, \quad u_0 v_1+u_1 v_0=U_1, \quad u_0 v_2+u_1 v_1+u_2 v_0=U_2, \ldots u_r v_s=U_{r+s}.$$

The $(k+1)$th of these equations is

$$\Sigma u_{k-\mu} v_\mu=U_k, \quad \text{where} \quad \mu \nless 0, \ \mu \nless k-r, \ \mu \ngtr s, \ \mu \ngtr k,$$

i.e. where μ receives all integral values consistent with the conditions $k-\mu \nless 0$, $k-\mu \ngtr r$, $\mu \nless 0$, $\mu \ngtr s$.

Since u_0 does not vanish identically, it follows by the hypothesis from the first $s+1$ of these equations taken in succession that v_0 lies in Ω and that those of the functions $v_1, v_2, \ldots v_s$ which do not vanish identically lie in Ω.

The theorem obviously remains true when F, but not f, vanishes identically.

Ex. xxiii. *Let F be a rational integral function of the n variables $x_1, x_2, \ldots x_n$, and let it be written in the form*

$$F=U_0 x_1^r+U_1 x_1^{r-1}+\ldots+U_{r-1}x_1+U_r,$$

where $U_0, U_1, \ldots U_r$ are rational integral functions of $x_2, x_3, \ldots x_n$ only; also let ϕ be a rational integral function of the $n-1$ variables $x_2, x_3, \ldots x_n$ only which does not vanish identically. Then if F is divisible by ϕ, each of the coefficients $U_0, U_1, \ldots U_r$ must be divisible by ϕ.

For if F is divisible by ϕ, there exists an identical equation of the form

$$U_0 x_1^r+U_1 x_1^{r-1}+\ldots+U_r=\phi\,(V_0 x_1^r+V_1 x_1^{r-1}+\ldots+V_r),$$

where $V_0, V_1, \dots V_r$ are rational integral functions of $x_2, x_3, \dots x_n$ only; and from this equation we deduce the identities

$$U_0 = \phi V_0, \quad U_1 = \phi V_1, \ \dots \ U_{r-1} = \phi V_{r-1}, \quad U_r = \phi V_r.$$

Thus ϕ is a factor of every one of the coefficients $U_0, U_1, \dots U_r$.

If F does not vanish identically and is regular in x_1 (i.e. has the same degree in x_1 as in all the variables), then every rational integral function ϕ of $x_2, x_3, \dots x_n$ only by which F is divisible is merely a constant.

For if F has degree r in x_1 and in all the variables, ϕ must be a factor of U_0 which is a non-vanishing constant.

3. *The highest common factor of a number of rational integral functions.*

Let $f_1, f_2, \dots f_m$ be any m rational integral functions of the n variables $x_1, x_2, \dots x_n$ which do not all vanish identically. Then any rational integral function of $x_1, x_2, \dots x_n$ which is a common factor of all the functions $f_1, f_2, \dots f_m$ (and therefore does not vanish identically), and which is divisible by every common factor of these functions is called a highest common factor of $f_1, f_2, \dots f_m$. That such functions exist is shown in § 188.

If h is a highest common factor of $f_1, f_2, \dots f_m$, then any function which differs from h only by a non-vanishing constant factor is clearly also a highest common factor of them. Conversely any two highest common factors of these functions must differ from one another only by a non-vanishing constant factor. For if h and k are both highest common factors of $f_1, f_2, \dots f_m$, then h must be divisible by k, and k must be divisible by h. Accordingly we have

$$h = uk, \quad k = vh, \quad h = uvh, \quad uv = 1,$$

where u and v are rational integral functions of $x_1, x_2, \dots x_n$ which do not vanish identically; and from the equation $uv = 1$ it follows (see Ex. xiv) that u and v are non-vanishing constants.

Thus a highest common factor of the functions $f_1, f_2, \dots f_m$ is uniquely determinate save for a non-vanishing constant factor to which any value may be ascribed; and on this account we may speak of *the* highest common factor of $f_1, f_2, \dots f_m$, this being a rational integral function of $x_1, x_2, \dots x_n$ containing an arbitrary non-vanishing constant factor.

For 'highest common factor' the abbreviation H.C.F. will be used.

Ex. xxiv. If any one of the functions $f_1, f_2, \dots f_m$ is a non-vanishing constant, their H.C.F. is a non-vanishing constant, and may be taken to be 1.

Ex. xxv. The H.C.F. of the functions $f_1, f_2, \dots f_m$ is the same as the H.C.F. of those of the functions which do not vanish identically.

Ex. xxvi. *If no one of the functions $f_1, f_2, \dots f_m$ vanishes identically, and if k is the H.C.F. of $f_2, f_3, \dots f_m$, then the H.C.F. of $f_1, f_2, \dots f_m$ is the H.C.F. of f_1 and k.*

For the common factors of $f_1, f_2, \dots f_m$ are the same as the common factors of f_1 and k. Hence if h is the highest common factor of $f_1, f_2, \dots f_m$, it must be a common factor of f_1 and k which is divisible by every common factor of f_1 and k, i.e. it must be a highest common factor of f_1 and k.

Ex. xxvii. *When the rational integral functions $f_1, f_2, \dots f_m$ all vanish identically, we can define their* H.C.F. *to be* 0.

This is the only possible interpretation of the H.C.F. if we consider that the definition given in the text still holds good in this special case, and if moreover we regard 0 as a common factor of $f_1, f_2, \dots f_m$.

For then 0 is a common factor of $f_1, f_2, \dots f_m$; it is divisible by every rational integral function which does not vanish identically, and therefore by every common factor of $f_1, f_2, \dots f_m$ other than 0; and it is the only common factor of $f_1, f_2, \dots f_m$ which can be regarded as divisible by their common factor 0.

§ 186. Irresoluble and irreducible functions.

A rational integral function of a single scalar variable or of several scalar variables, which is not merely a constant, is said to be *resoluble* or *irresoluble* according as it can or cannot be expressed as a product of factors each of which is a rational integral function of the same variable or variables, and is not merely a constant.

Two irresoluble functions are *distinct* from one another when and only when they do not differ only by a constant factor, i.e. when neither of them is equal to the other multiplied by a non-vanishing constant.

If f is a rational integral function which is not merely a constant, any irresoluble rational integral function (not being merely a constant) which is a factor of f will be called an *irresoluble factor* of f. If g is an irresoluble factor of f, and if g^α is the highest power of g which is a factor of f, then the irresoluble factor g will be said to be *repeated* α times in f. When in particular $\alpha = 1$, g is an *unrepeated* irresoluble factor of f.

Clearly every rational integral function f which is not merely a constant can be expressed as a product of distinct irresoluble factors, unrepeated and repeated; and it will be shown in Corollary 3 to Theorem III in § 188 that it can be so expressed in only one way, save for a non-vanishing constant factor.

Two rational integral functions of the same variables which do not vanish identically will be said to be *prime to one another* when they have no irresoluble factor in common, i.e. when they have no factor in common other than a non-vanishing constant.

Ex. i. A linear function, i.e. a rational integral function of the first degree in all the variables, is necessarily irresoluble.

Ex. ii. *Irresoluble factors of a function of a single variable.*

All irresoluble factors of a rational integral function of a single variable x are linear.

Ex. iii. *Irresoluble factors of a homogeneous function of two variables.*

All irresoluble factors of a homogeneous rational integral function of two variables x and y are homogeneous and linear.

Ex. iv. *A determinoid with arbitrary elements is an irresoluble function of its elements.*

This follows from the fact that a determinoid contains no term in which two elements belonging to the same horizontal row or the same vertical row are multiplied together.

Let $A=[a]_m^n$ be a matrix with arbitrary elements, and let $\Delta=(a)_m^n=uv$, where u and v are rational integral functions of the elements of A.

The element a_{ij} of A must occur (with a non-vanishing coefficient) in one of the factors u and v. Suppose that it occurs in u. Then no element of A belonging to the same horizontal row or the same vertical row as a_{ij} can occur (with a non-vanishing coefficient) in v. Therefore every element of A belonging to the same horizontal row or the same vertical row as a_{ij} must occur in u. Therefore no element of A can occur in v, i.e. v must be a non-vanishing constant.

A rational integral function of a single scalar variable or of several scalar variables, which is not merely a constant, is said to be *reducible* or *irreducible* in any domain of rationality Ω in which it lies according as it can or cannot be expressed as a product of factors each of which is a rational integral function of the same variable or variables lying in the domain Ω, and is not merely a constant.

Two irreducible functions belonging to the same domain of rationality Ω are *distinct* from one another when and only when they do not differ only by a constant factor, i.e. when neither of them is equal to the other multiplied by a non-vanishing constant (lying in Ω).

If f is a rational integral function which lies in any domain of rationality Ω, and is not merely a constant, any rational integral function (not being merely a constant) which lies in Ω, is irreducible in Ω, and is a factor of f, will be called an *irreducible* factor of f, or more precisely a factor of f irreducible in Ω. If g is an irreducible factor of f, and if g^β is the highest power of g which is a factor of f, then the irreducible factor g will be said to be *repeated* β times in f. When in particular $\beta=1$, g is an unrepeated irreducible factor of f.

Clearly every rational integral function f lying in a domain of rationality Ω, which is not merely a constant, can be expressed as a product of distinct irreducible factors in Ω, unrepeated and repeated; and it will be shown in Corollary 2 to Theorem III in § 188 that it can be so expressed in only one way, save for a non-vanishing constant factor (lying in Ω).

Two rational integral functions of the same variables which lie in the same domain of rationality Ω and do not vanish identically will be said to be *prime to one another in* Ω when they have no factor in common which is irreducible in Ω. It will be shown in Corollary 3 to Theorem I in § 188 that this is the case when and only when they have no irresoluble factor in common, i.e. are prime to one another, i.e. have no factor in common other than a non-vanishing constant.

An irresoluble function is one which is irreducible in the domain of all scalar numbers.

A rational integral equation $f=0$ is or is not resoluble or reducible according as the function f is or is not resoluble or reducible.

Ex. v. The function $(x-y)^2-2$ is resoluble. It is irreducible in the domain of all rational numbers; but it is reducible in the domain of all real numbers, in the domain of all algebraic numbers, and in the domain of all scalar numbers.

Ex. vi. The function $(x+y)^2+2$ is resoluble. It is irreducible in the domain of all rational numbers and in the domain of all real numbers; but it is reducible in the domain of all algebraic numbers, and in the domain of all scalar numbers.

Ex. vii. The function x^2+y^2+2 is irresoluble. It is irreducible even in the domain of all scalar numbers.

Ex. viii. *Irreducible factors of a function of a single variable.*

In the domain of all scalar numbers all irreducible factors of a rational integral function of a single variable x are linear.

In the domain of all real numbers every irreducible factor of such a function is either linear or of the second degree.

Ex. ix. The function x^3-2 is irreducible in the domain $\Omega(1)$ of all rational numbers.

Ex. x. *In the domain $\Omega(1)$ of all rational numbers there exist irreducible rational integral functions of x of degree n when n is any positive integer.*

In fact if p is any prime number, and if $a_0, a_1, \ldots a_n$ are integers such that the first is not divisible by p, the rest are all divisible by p, and the last is not divisible by p^2, then the rational integral function

$$f(x)=a_0x^n+a_1x^{n-1}+\ldots+a_n$$

is irreducible in $\Omega(1)$. This is *Eisenstein's Theorem.*

Ex. xi. If p is a prime number, and if $q=p^{\pi-1}$, where π is any non-zero positive integer, then the functions

$$x^{p-1}+x^{p-2}+\ldots+x+1 \quad \text{and} \quad x^{q(p-1)}+x^{q(p-2)}+\ldots+x^q+1$$

are irreducible in the domain $\Omega(1)$ of all rational numbers.

Ex. xii. More generally if m is any non-zero positive integer, and if $\phi(m)$ is the number of integers which are less than m and prime to it, then the primitive roots of the equation $x^m-1=0$ are the roots of an irreducible equation of degree $\phi(m)$ in the domain $\Omega(1)$.

For example the primitive roots of the equation $x^{15}-1=0$ are the roots of the rational integral equation

$$x^8-x^7+x^5-x^4+x^3-x+1=0,$$

which is irreducible in $\Omega(1)$.

Ex. xiii. *An irreducible function of the single variable x cannot have a repeated linear (or irresoluble) factor.*

Let $f(x)$ be any rational integral function of x in any domain Ω which is not merely a constant, and let $f'(x)$ be its first derived function. Then if $h(x)$ is the H.C.F. of $f(x)$ and $f'(x)$, we have a rational integral identity in Ω of the form

$$f(x) = g(x) \,.\, h(x).$$

If $f(x)$ has a linear factor t repeated p times, p being greater than 1, then $h(x)$ is divisible by t^{p-1} but not by t^p; therefore $g(x)$ is divisible by t (but not by t^2), and cannot be merely a constant. Thus if $f(x)$ has a repeated linear factor, it must be reducible in Ω.

This theorem is generalised in Ex. xi of § 189.

Ex. xiv. *Irreducible factors of a homogeneous function of two variables.*

In the domain of all scalar numbers all irreducible factors of a homogeneous rational integral function of two variables x and y are homogeneous and linear.

In the domain of all real numbers every irreducible factor of such a function is either homogeneous and linear or homogeneous and of the second degree.

Ex. xv. *The irreducible and irresoluble factors of any homogeneous rational integral function are all homogeneous.*

This follows from Ex. xvii of § 185.

§ 187. Ordinary linear transformations of the variables in a rational integral function.

1. *Homogeneous linear transformations of the variables.*

Let $[l]_n^n$ and $\overline{\underline{L}}{}_n^n$ be two mutually inverse undegenerate square matrices with constant elements.

Then the two mutually inverse transformations or substitutions

$$\overline{\underline{x}}_n = [l]_n^n \overline{\underline{y}}_n, \quad \overline{\underline{y}}_n = \overline{\underline{L}}{}_n^n \overline{\underline{x}}_n \qquad \text{.................(A)}$$

establish a one-one correspondence between all sets of finite values of the variables $x_1, x_2, \ldots x_n$ and all sets of finite values of the variables $y_1, y_2, \ldots y_n$.

Any two corresponding sets of finite values of the x's and y's satisfy both the equations (A), and each set of values is uniquely determinate when the other is given.

Let $X_1, X_2, \ldots X_n$ be the homogeneous linear functions of $x_1, x_2, \ldots x_n$ defined by the equations

$$X_i = L_{1i}x_1 + L_{2i}x_2 + \ldots + L_{ni}x_n, \quad (i = 1, 2, \ldots n),$$

so that the equations

$$\overline{\underline{X}}_n = \overline{\underline{L}}^n_n \overline{\underline{x}}_n, \quad \overline{\underline{x}}_n = [l]^n_n \overline{\underline{X}}_n, \quad \text{..................(A}'\text{)}$$

and $\qquad x_i = l_{i1} X_1 + l_{i2} X_2 + \ldots + l_{in} X_n, \quad (i = 1, 2, \ldots n),$

are identities in the x's.

Also let $Y_1, Y_2, \ldots Y_n$ be the homogeneous linear functions of $y_1, y_2, \ldots y_n$ defined by the equations

$$Y_i = l_{i1} y_1 + l_{i2} y_2 + \ldots + l_{in} y_n, \quad (i = 1, 2, \ldots n),$$

so that the equations

$$\overline{\underline{Y}}_n = [l]^n_n \overline{\underline{y}}_n, \quad \overline{\underline{y}}_n = \overline{\underline{L}}^n_n \overline{\underline{Y}}_n, \quad \text{..................(A}''\text{)}$$

and $\qquad y_i = L_{1i} Y_1 + L_{2i} Y_2 + \ldots + L_{ni} Y_n, \quad (i = 1, 2, \ldots n),$

are identities in the y's.

Then every rational integral function of $x_1, x_2, \ldots x_n$ can be expressed in terms of the homogeneous linear functions $X_1, X_2, \ldots X_n$; and every rational integral function of $y_1, y_2, \ldots y_n$ can be expressed in terms of the homogeneous linear functions $Y_1, Y_2, \ldots Y_n$.

Now let $\qquad F = \Sigma a x_1^{p_1} x_2^{p_2} \ldots x_n^{p_n} = \Sigma b X_1^{q_1} X_2^{q_2} \ldots X_n^{q_n}$ ....(a)

be any rational integral function of the n variables $x_1, x_2, \ldots x_n$. We suppose that the indices $p_1, p_2, \ldots p_n$ have not all the same value in any two terms of the first sum, and that the indices $q_1, q_2, \ldots q_n$ have not all the same value in any two terms of the second sum.

If we substitute for $x_1, x_2, \ldots x_n$ in every term of F their values given by the first of the equations (A), we convert F into the rational integral function G of $y_1, y_2, \ldots y_n$ given by

$$G = \Sigma b y_1^{q_1} y_2^{q_2} \ldots y_n^{q_n} = \Sigma a Y_1^{p_1} Y_2^{p_2} \ldots Y_n^{p_n}; \quad \text{...........(a}'\text{)}$$

for the effect of the transformation is to replace x_i by Y_i and X_i by y_i.

Conversely if G is any rational integral function of the n variables $y_1, y_2, \ldots y_n$ given by (a′), and if we substitute for $y_1, y_2, \ldots y_n$ in every term of G their values given by the second of the equations (A), we convert G into the rational integral function F of $x_1, x_2, \ldots x_n$ given by (a).

Thus the two mutually inverse transformations (A) *establish a one-one correspondence between all rational integral functions F of $x_1, x_2, \ldots x_n$ and all rational integral functions G of $y_1, y_2, \ldots y_n$; two such functions F and G corresponding when and only when they are convertible into one another by the transformations* (A).

Any two corresponding functions F and G can be expressed in the forms (a) and (a′); the first of the transformations (A) converts F into G; the second of the transformations (A) converts G into F; and each of the functions F and G is uniquely determinate when the other is given. The equation (a′), being an identity in the y's, determines the coefficients b uniquely when the coefficients a are given; and the equation (a), being an identity in the x's, determines the coefficients a uniquely when the coefficients b are given. Or to determine one set of constants when the other is given, we can regard (a) both as an identity in $x_1, x_2, \ldots x_n$, and as an identity in $X_1, X_2, \ldots X_n$; or we can regard (a′) both as an identity in $y_1, y_2, \ldots y_n$, and as an identity in $Y_1, Y_2, \ldots Y_n$. From the forms of F and G we see that:

If one of two corresponding functions F and G is homogeneous in all the variables, the other also is homogeneous in all the variables.

Ex. i. *One of two corresponding functions F and G vanishes identically when and only when the other vanishes identically.*

For if F vanishes identically, all the coefficients a vanish, and therefore G vanishes identically; and if G vanishes identically, all the coefficients b vanish, and therefore F vanishes identically.

Ex. ii. *Two corresponding functions F and G which do not vanish identically have the same degree in all their variables.*

Let r be the degree of F in *all* the variables $x_1, x_2, \ldots x_n$; and let s be the degree of G in *all* the variables $y_1, y_2, \ldots y_n$.

From (a′) we see that G is the sum of a number of rational integral functions of $y_1, y_2, \ldots y_n$ such as $g' = aY_1^{p_1}Y_2^{p_2}\ldots Y_n^{p_n}$, the function g' corresponding to the term $f = ax_1^{p_1}x_2^{p_2}\ldots x_n^{p_n}$ of F. Since no row of $[l]_n^n$ is a row of 0's, the degree of g' in $y_1, y_2, \ldots y_n$ is $p_1+p_2+\ldots+p_n$, i.e. it is equal to the degree in $x_1, x_2, \ldots x_n$ of the term f of F which cannot exceed r. Thus G is the sum of a number of rational integral functions of $y_1, y_2, \ldots y_n$, no one of which has degree greater than r in all those variables. Consequently the degree of G in all the variables $y_1, y_2, \ldots y_n$ cannot exceed r; i.e. we have $s \not> r$. In the same way we see from (a) that $r \not> s$; and it follows that $s = r$.

Although a homogeneous linear transformation cannot raise (and therefore cannot depress) the degree of a rational integral function in *all* the variables, it can raise (and therefore can depress) the degrees of the function in individual variables; for it is clearly possible for the degree of G in y_1 to exceed the degree of F in x_1.

Ex. iii. *One of two corresponding functions F and G is a non-vanishing constant c when and only when*

$$F = G = c.$$

Ex. iv. *Two corresponding functions F and G have equal values for corresponding sets of values of the variables.*

For if $x_1, x_2, \ldots x_n$ and $y_1, y_2, \ldots y_n$ satisfy the equations (A), we have

$$\overline{\underline{X}}_n = \overline{\underline{y}}_n, \quad \overline{\underline{Y}}_n = \overline{\underline{x}}_n,$$

and therefore

$$F(x_1, x_2, \ldots x_n) = G(y_1, y_2, \ldots y_n).$$

Ex. v. *The transformations* (A) *convert every root of F into a root of the corresponding function G, and every root of G into a root of the corresponding function F.*

For if $x_1, x_2, \dots x_n$ and $y_1, y_2, \dots y_n$ satisfy the equations (A), we have

$$F(x_1, x_2, \dots x_n)=0$$

when and only when $G(y_1, y_2, \dots y_n)=0$.

Ex. vi. If $f_1, f_2, f_3, \dots$ and $g_1, g_2, g_3, \dots$ are two sets of corresponding rational integral functions of $x_1, x_2, \dots x_n$ and $y_1, y_2, \dots y_n$, then $f_1^{u_1}f_2^{u_2}f_3^{u_3}\dots$ and $g_1^{u_1}g_2^{u_2}g_3^{u_3}\dots$ are corresponding functions.

Ex. vii. If $F, f, f_1, f_2, f_3, \dots$ *are rational integral functions of* $x_1, x_2, \dots x_n$, *and if* $G, g, g_1, g_2, g_3, \dots$ *are the corresponding rational integral functions of* $y_1, y_2, \dots y_n$, *we have*

$$F=f_1^{u_1}f_2^{u_2}f_3^{u_3}\dots \quad \textit{when and only when} \quad G=g_1^{u_1}g_2^{u_2}g_3^{u_3}\dots.$$

This follows from Ex. vi; for when F (or G) is given, there is only one function G (or F) which corresponds to it.

Hence f is a factor of F when and only when g is a factor of G; and f is a factor of F repeated u times when and only when g is a factor of G repeated u times.

Ex. viii. *Two corresponding functions F and G which do not vanish identically are either both resoluble or both irresoluble.*

This follows from Ex. vii.

Ex. ix. If $f_1, f_2, \dots f_m$ and $g_1, g_2, \dots g_m$ are two corresponding sets of rational integral functions of $x_1, x_2, \dots x_n$ and $y_1, y_2, \dots y_n$, then the transformations (A) convert every highest common factor of either set of functions into a highest common factor of the other set of functions.

Ex. x. If $x_1, x_2, \dots x_n$ and $y_1, y_2, \dots y_n$ are corresponding sets of finite values of the x's and y's satisfying the equations (A), and if c is any given finite non-vanishing constant, we have

$$L_{1n}x_1+L_{2n}x_2+\dots+L_{nn}x_n=c, \quad\dots\dots(1)$$

when and only when $y_n=c$.

Now let F and G be two corresponding *homogeneous* functions of the x's and y's. Then if F vanishes for all finite values of $x_1, x_2, \dots x_n$ satisfying the equation (1), G vanishes for all finite values of $y_1, y_2, \dots y_{n-1}$ when $y_n=c$. Therefore by Ex. xviii of § 185 the function G vanishes identically, and it follows that the function F vanishes identically.

We deduce the following result, which is a generalisation of Ex. xviii of § 185:

Let F be a homogeneous rational integral function of the variables $x_1, x_2, \dots x_n$; let $k_1, k_2, \dots k_n$ be finite constants which are not all zero; and let c be a finite non-vanishing constant. Then if F vanishes for all finite values of the variables which satisfy the equation

$$k_1x_1+k_2x_2+\dots+k_nx_n=c, \quad\dots\dots(2)$$

it must vanish identically.

This result is further generalised in Ex. vi of § 189.

Hence two homogeneous rational integral functions F and G of the variables $x_1, x_2, \dots x_n$ are identically equal when and only when they satisfy the following two conditions:

(1) The functions F and G have the same degree in all the variables.

(2) The functions F and G are equal for all finite values of $x_1, x_2, \dots x_n$ which satisfy the equation (2).

NOTE 1. *Transformations in Ω.*

If the matrices $[l]_n^n$ and $\overline{\underline{L}}_n^n$ lie in any restricted domain of rationality Ω, i.e. if all the elements of one (and therefore of both) lie in Ω, the transformations (A) are called *transformations in Ω.*

In this case they establish in the same way a one-one correspondence between all sets of finite values of the variables $x_1, x_2, \ldots x_n$ lying in Ω and all sets of finite values of the variables $y_1, y_2, \ldots y_n$ lying in Ω. Also they establish a one-one correspondence of the same character as before between all rational integral functions in Ω of $x_1, x_2, \ldots x_n$ and all rational integral functions in Ω of $y_1, y_2, \ldots y_n$. Two corresponding functions F and G which lie in Ω and do not vanish identically are either both reducible or both irreducible in Ω.

2. *Regularisation of a number of rational integral functions.*

Let $f_1, f_2, \ldots f_m$ be any m given rational integral functions of the n variables $x_1, x_2, \ldots x_n$ which do not vanish identically; and let $g_1, g_2, \ldots g_m$ be the rational integral functions of the n variables $y_1, y_2, \ldots y_n$ into which they are converted by the homogeneous linear transformation $\overline{\underline{x}}_n = [l]_n^n \overline{\underline{y}}_n$, where $[l]_n^n$ is an undegenerate square matrix with constant elements. Then we will prove the following result:

It is possible to choose the undegenerate matrix $[l]_n^n$ so that it lies in any assigned domain of rationality Ω, and so that the functions $g_1, g_2, \ldots g_m$ are regular in each one of the variables $y_1, y_2, \ldots y_k$, where k has any one of the values 1, 2, ... *n.*

Let the degrees of $f_1, f_2, \ldots f_m$ in all the variables $x_1, x_2, \ldots x_n$ (which are also the degrees of $g_1, g_2, \ldots g_m$ in all the variables $y_1, y_2, \ldots y_n$) be $r_1, r_2, \ldots r_m$, and let

$$r_1 + r_2 + \ldots + r_m = r, \quad f_1 f_2 \ldots f_m = F, \quad g_1 g_2 \ldots g_m = G.$$

Also let $$F' = a x_1^{\alpha_1} x_2^{\alpha_2} \ldots x_n^{\alpha_n} + b x_1^{\beta_1} x_2^{\beta_2} \ldots x_n^{\beta_n} + \ldots$$

be the sum of those terms of F which have degree r in all the variables $x_1, x_2, \ldots x_n$, so that

$$\alpha_1 + \alpha_2 + \ldots + \alpha_n = \beta_1 + \beta_2 + \ldots + \beta_n = \ldots = r;$$

and let G' be the sum of those terms of G which have degree r in all the variables $y_1, y_2, \ldots y_n$, so that

$$G' = a Y_1^{\alpha_1} Y_2^{\alpha_2} \ldots Y_n^{\alpha_n} + b Y_1^{\beta_1} Y_2^{\beta_2} \ldots Y_n^{\beta_n} + \ldots,$$

where $$Y_i = l_{i1} y_1 + l_{i2} y_2 + \ldots + l_{in} y_n, \quad (i = 1, 2, \ldots n).$$

Further let $A_1, A_2, \ldots A_k$ be the coefficients of $y_1^r, y_2^r, \ldots y_k^r$ in G. Because these are also the coefficients of $y_1^r, y_2^r, \ldots y_k^r$ in G', we have

$$A_k = a l_{1k}^{\alpha_1} l_{2k}^{\alpha_2} \ldots l_{nk}^{\alpha_n} + b l_{1k}^{\beta_1} l_{2k}^{\beta_2} \ldots l_{nk}^{\beta_n} + \ldots;$$

and because $a, b, \dots$ do not all vanish, it follows that $A_1, A_2, \dots A_k$ as well as $(l)^n_n$ are rational integral functions of the elements of $[l]^n_n$ which do not vanish identically. Hence by Ex. xx of § 185 it is possible to choose the elements of $[l]^n_n$ so that they all lie in Ω, and so that the product

$$(l)^n_n A_1 A_2 \dots A_k \neq 0.$$

When this is done the square matrix $[l]^n_n$ is undegenerate and lies in Ω, and the functions $g_1, g_2, \dots g_m$ are all regular in each one of the variables $y_1, y_2, \dots y_k$, which may be considered to be any k of the variables $y_1, y_2, \dots y_n$.

If $f_1, f_2, \dots f_m$ all lie in any domain of rationality Ω, we can choose $[l]^n_n$ so as to lie in the same domain Ω; and then $g_1, g_2, \dots g_m$ also lie in the same domain Ω.

The one-one correspondence between the functions $f_1, f_2, \dots f_m$ and the regularised functions $g_1, g_2, \dots g_m$ has the character described in sub-article 1.

Ex. xi. The properties of the factors of m rational integral functions lying in any domain of rationality Ω can be deduced from the properties of m corresponding rational integral functions which lie in Ω and are regular in all or any number of the variables.

3. *Homogenisation of a rational integral function by a linear transformation of the variables.*

Let $[l]^{n+1}_{n+1}$ and $\overline{\underline{L}}^{n+1}_{n+1}$ be two mutually inverse undegenerate square matrices with constant elements.

Then the two mutually inverse transformations or substitutions

$$\begin{bmatrix} x \\ 1 \end{bmatrix}_{n,1} = [l]^{n+1}_{n+1} \overline{\underline{y}}_{n+1}, \quad \overline{\underline{y}}_{n+1} = \overline{\underline{L}}^{n+1}_{n+1} \begin{bmatrix} x \\ 1 \end{bmatrix}_{n,1} \quad \dots\dots\dots\dots\dots(B)$$

establish a one-one correspondence between all sets of finite values of the n variables $x_1, x_2, \dots x_n$ and all sets of finite values of the $n+1$ variables $y_1, y_2, \dots y_n, y_{n+1}$ satisfying the equation

$$l_{n+1,1}y_1 + l_{n+1,2}y_2 + \dots + l_{n+1,n}y_n + l_{n+1,n+1}y_{n+1} = 1. \quad \dots\dots\dots(3)$$

Any two corresponding sets of finite values of the x's and y's satisfy both the equations (B), and each set of values is uniquely determinate when the other set is given.

Let $X_1, X_2, \dots X_{n+1}$ be the linear functions of $x_1, x_2, \dots x_n$ defined by the equations

$$X_i = L_{1i}x_1 + L_{2i}x_2 + \dots + L_{ni}x_n + L_{n+1,i}, \quad (i = 1, 2, \dots n+1),$$

so that the equations

$$\underline{\overline{X}}_{n+1} = \underline{\overline{L}}_{n+1}^{n+1} \underline{\overline{\begin{matrix} x \\ 1 \end{matrix}}}_{n,1}, \quad \underline{\overline{\begin{matrix} x \\ 1 \end{matrix}}}_{n,1} = [l]_{n+1}^{n+1} \underline{\overline{X}}_{n+1}, \quad \text{............(B}')$$

$$x_i = l_{i1}X_1 + l_{i2}X_2 + \ldots + l_{i,n+1}X_{n+1}, \quad (i = 1, 2, \ldots n),$$

and
$$l_{n+1,1}X_1 + l_{n+1,2}X_2 + \ldots + l_{n+1,n+1}X_{n+1} = 1, \quad \text{..............(4)}$$

are identities in $x_1, x_2, \ldots x_n$.

Also let $Y_1, Y_2, \ldots Y_{n+1}$ be the homogeneous linear functions of $y_1, y_2, \ldots y_{n+1}$ defined by the equations

$$Y_i = l_{i1}y_1 + l_{i2}y_2 + \ldots + l_{i,n+1}y_{n+1}, \quad (i = 1, 2, \ldots n+1),$$

so that the equations

$$\underline{\overline{Y}}_{n+1} = [l]_{n+1}^{n+1} \underline{\overline{y}}_{n+1}, \quad \underline{\overline{y}}_{n+1} = \underline{\overline{L}}_{n+1}^{n+1} \underline{\overline{Y}}_{n+1}, \quad \text{............(B}'')$$

and
$$y_i = L_{1i}Y_1 + L_{2i}Y_2 + \ldots + L_{n+1,i}Y_{n+1}, \quad (i = 1, 2, \ldots n+1),$$

are identities in $y_1, y_2, \ldots y_{n+1}$.

Then every rational integral function of the x's which does not vanish identically and has degree s in all the variables can with the aid of (4) be expressed as a homogeneous rational integral function of the functions $X_1, X_2, \ldots X_{n+1}$ of degree s; and every homogeneous rational integral function of the y's which does not vanish identically and has degree s in all the variables can be expressed as a homogeneous rational integral function of the functions $Y_1, Y_2, \ldots Y_{n+1}$ of degree s.

First let F be any rational integral function of the n variables $x_1, x_2, \ldots x_n$ which does not vanish identically and has degree s in all the variables. Then we can write

$$F = \Sigma a x_1^{p_1} x_2^{p_2} \ldots x_n^{p_n} = \Sigma a x_1^{p_1} x_2^{p_2} \ldots x_n^{p_n} . 1^{p_{n+1}}$$
$$= \Sigma b X_1^{q_1} X_2^{q_2} \ldots X_{n+1}^{q_{n+1}}, \quad \text{...............(b)}$$

where the coefficients a and b in the various terms of these sums are non-vanishing constants, and where in the various terms $p_1, p_2, \ldots p_n, p_{n+1}$ and $q_1, q_2, \ldots q_{n+1}$ are positive integers such that

$$p_1 + p_2 + \ldots + p_n \not> s, \quad p_1 + p_2 + \ldots + p_n + p_{n+1} = s, \quad q_1 + q_2 + \ldots + q_{n+1} = s.$$

There are terms of the first sum in which $p_1 + p_2 + \ldots + p_n = s$, and therefore there are terms of the second sum in which $p_{n+1} = 0$. We assume as usual that $p_1, p_2, \ldots p_n$ have not all the same values in any two terms of the first sum, and that $q_1, q_2, \ldots q_{n+1}$ have not all the same values in any two terms of the third sum. We can obtain the third sum in (b) from the second by substituting for $x_1, x_2, \ldots x_n, 1$ in each term the expressions identically equal to them which are given by the second of the equations (B′). There cannot be two different expressions for F which are homogeneous rational integral functions of the same degree of the functions $X_1, X_2, \dots X_{n+1}$. For any two such expressions would necessarily be equal for all finite values of $X_1, X_2, \dots X_{n+1}$ satisfying the equation (4), and therefore by Ex. x would be identically equal functions of $X_1, X_2, \dots X_{n+1}$. Accordingly the coefficients b are uniquely determinate when the coefficients a are known; and clearly the coefficients a are uniquely determinate when the coefficients b are known.

By substituting for $x_1, x_2, \dots x_n, 1$ the values given by the first of the equations (B), i.e. by making the substitutions

$$x_i = l_{i1}y_1 + l_{i2}y_2 + \dots + l_{i,n+1}y_{n+1}, \quad (i = 1, 2, \dots n),$$

$$1 = l_{n+1,1}y_1 + l_{n+1,2}y_2 + \dots + l_{n+1,n+1}y_{n+1},$$

we can derive from F the homogeneous rational integral function G of $y_1, y_2, \dots y_{n+1}$ given by

$$G = \Sigma b y_1^{q_1} y_2^{q_2} \dots y_{n+1}^{q_{n+1}} = \Sigma a Y_1^{p_1} Y_2^{p_2} \dots Y_{n+1}^{p_{n+1}}, \quad \text{............(b′)}$$

which does not vanish identically and has degree s in all the variables. The second expression in (b′) is derived in this way from the second expression in (b). The first expression in (b′) is derived in this way from the third expression in (b) when we write X_i in the form

$$X_i = L_{1i}x_1 + L_{2i}x_2 + \dots + L_{ni}x_n + L_{n+1,i} \cdot 1,$$

and $X_1, X_2, \dots X_{n+1}$ in similar forms.

There cannot be two different homogeneous rational integral functions of $y_1, y_2, \dots y_{n+1}$ into which F can be converted by these substitutions. For if ϕ and ψ were any two such functions, then both ϕ and F, and ψ and F, would have equal values for corresponding sets of finite values of the y's and x's; therefore ϕ and ψ would be equal for all finite values of $y_1, y_2, \dots y_{n+1}$ satisfying the equation (3); therefore by Ex. x they would be identically equal functions of $y_1, y_2, \dots y_{n+1}$. In particular the two expressions for G in (b′) are identically equal functions of $y_1, y_2, \dots y_{n+1}$. Accordingly the function G given by (b′) is the only homogeneous rational integral function of $y_1, y_2, \dots y_{n+1}$ of degree s into which F can be converted by the first of the substitutions (B).

If for a moment we regard G as a function G' of $Y_1, Y_2, \dots Y_{n+1}$, we see from the second sum in (b′), which contains terms in which $p_{n+1} = 0$, that G' is not divisible by Y_{n+1}. Since G and G are convertible into one another by the two mutually inverse homogeneous linear transformations (B″), it follows by sub-article 1 that the function G of $y_1, y_2, \dots y_{n+1}$ is not divisible by the homogeneous linear (and irresoluble) function

$$Y_{n+1} = l_{n+1,1}y_1 + l_{n+1,2}y_2 + \dots + l_{n+1,n+1}y_{n+1}. \quad \text{............(5)}$$

Next let G be any homogeneous rational integral function of the $n+1$ variables $y_1, y_2, \dots y_{n+1}$ which does not vanish identically, has degree s in all the variables, and is not divisible by the linear (or irresoluble) function Y_{n+1}. Then we can express G in the forms (b′), where the coefficients a and b in the various terms are non-vanishing constants, and where

$$q_1 + q_2 + \dots + q_{n+1} = s, \quad p_1 + p_2 + \dots + p_n + p_{n+1} = s. \quad \text{.........(6)}$$

The coefficients b are uniquely determinate when the coefficients a are known. Also by sub-article 1 the coefficients a are uniquely determinate when the coefficients b are known, and the second expression, regarded as a function G' of $Y_1, Y_2, \dots Y_{n+1}$, is not divisible by Y_{n+1}. Consequently there are terms of the second sum in which $p_{n+1} = 0$, $p_1 + p_2 + \dots + p_n = s$.

By substituting for $y_1, y_2, \dots y_{n+1}$ the values given by the second of the equations (B), i.e. by making the substitutions

$$y_i = L_{1i}x_1 + L_{2i}x_2 + \dots + L_{ni}x_n + L_{n+1,i}, \quad (i = 1, 2, \dots n+1),$$

we can convert G into the rational integral function F of $x_1, x_2, \dots x_n$ given by (b), which does not vanish identically, the first and third expressions for F being obtained respectively from the second and first expressions for G.

There cannot be two different rational integral functions of $x_1, x_2, \dots x_n$ into which G can be converted by these substitutions; for any two such functions would necessarily be equal for all finite values of $x_1, x_2, \dots x_n$, and must therefore be identically equal. In particular the first and third expressions for F are identically equal functions of $x_1, x_2, \dots x_n$. Accordingly the function F given by (b) is the only rational integral function of $x_1, x_2, \dots x_n$ into which G can be converted by the second of the substitutions (B).

Since there are terms of the first expression for F in which

$$p_1 + p_2 + \dots + p_n = s,$$

we see that F has degree s in all the variables $x_1, x_2, \dots x_n$.

In the particular case when F vanishes identically, any function into which it is converted by the first of the substitutions (B) must vanish identically; and conversely when G vanishes identically, any function into which it is converted by the second of the substitutions (B) must vanish identically.

Thus the two mutually inverse transformations (B) *establish a one-one correspondence between*

(1) *all rational integral functions F of the n variables $x_1, x_2, \dots x_n$;*

(2) *all those homogeneous rational integral functions G of the $n+1$ variables $y_1, y_2, \dots y_{n+1}$ which are not divisible by the homogeneous linear function Y_{n+1};*

two corresponding functions F and G having the same degree in all their variables, and being convertible into one another by the substitutions (B).

Any two corresponding functions F and G can be expressed in the forms (b) and (b′); the first of the transformations (B) converts F into G; the second of the transformations (B) converts G into F; and each of the functions F and G is uniquely determinate when the other is given. In particular one of the two functions vanishes identically when and only when the other vanishes identically.

We call G the homogenised function (of the same degree as F) corresponding to F.

Ex. xii. Two corresponding functions F and G have equal values for corresponding sets of finite values of the variables.

Ex. xiii. The transformations (B) convert every finite root of F into a finite root of G satisfying the equation (3), and every finite root of G satisfying the equation (3) into a finite root of F.

Ex. xiv. If $F, f, f_1, f_2, \ldots f_m$ and $G, g, g_1, g_2, \ldots g_m$ are two sets of corresponding functions which do not vanish identically (no function of the second set being divisible by the homogeneous linear function Y_{n+1}), then:

(1) The products $f_1^{u_1} f_2^{u_2} \ldots f_m^{u_m}$ and $g_1^{u_1} g_2^{u_2} \ldots g_m^{u_m}$ are mutually corresponding functions of $x_1, x_2, \ldots x_n$ and $y_1, y_2, \ldots y_{n+1}$.

(2) We have $F = f_1^{u_1} f_2^{u_2} \ldots f_m^{u_m}$ when and only when $G = g_1^{u_1} g_2^{u_2} \ldots g_m^{u_m}$.

(3) The function f is a factor of F (repeated u times) when and only when the function g is a factor of G (repeated u times).

(4) The two corresponding functions F and G are either both resoluble or both irresoluble.

(5) The first of the transformations (B) converts every highest common factor of $f_1, f_2, \ldots f_m$ into a highest common factor of $g_1, g_2, \ldots g_m$; and the second of the transformations (B) converts every highest common factor of $g_1, g_2, \ldots g_m$ into a highest common factor of $f_1, f_2, \ldots f_m$.

Ex. xv. *Corresponding sets of values of the variables which are not finite.*

If the equations (B) are satisfied and at least one of the x's is infinite, then at least one of the y's is infinite; also if at least one of the y's is infinite, then at least one of the x's is infinite. In such cases we can write

$$[x_1 x_2 \ldots x_n] = \rho [\xi_1 \xi_2 \ldots \xi_n], \quad [y_1 y_2 \ldots y_{n+1}] = \sigma [\eta_1 \eta_2 \ldots \eta_{n+1}],$$

where ρ and σ are infinite scalar quantities, and where the ξ's are finite quantities which do not all vanish and also the η's are finite quantities which do not all vanish. The equations (B) are then satisfied when and only when

$$\rho \begin{bmatrix} \xi \\ 0 \end{bmatrix}_{n,1} = \sigma [l]_{n+1}^{n+1} \begin{bmatrix} \eta \end{bmatrix}_{n+1}, \quad \sigma \begin{bmatrix} \eta \end{bmatrix}_{n+1} = \rho \begin{bmatrix} L \end{bmatrix}_{n+1}^{n+1} \begin{bmatrix} \xi \\ 0 \end{bmatrix}_{n,1}. \quad \ldots\ldots\ldots\ldots\ldots(7)$$

Since the ξ's do not all vanish and the η's do not all vanish, the equations (7) cannot be satisfied when either of the ratios $\rho : \sigma$ and $\sigma : \rho$ vanishes. Hence we must have $\rho = \tau h$, $\sigma = \tau k$, where τ is an infinite scalar quantity, and h and k are finite scalar quantities neither of which vanishes; and in (7) we can replace ρ and σ by the finite non-zero quantities h and k which are proportional to them.

We observe that the η's must always satisfy the equation

$$l_{n+1,1}\eta_1+l_{n+1,2}\eta_2+\dots+l_{n+1,n+1}\eta_{n+1}=0, \quad\text{..........................(3')}$$

which is the limiting form assumed by the equation (3) when some of the y's are infinite.

NOTE 2. *Transformations in Ω.*

When the matrices $[l]_{n+1}^{n+1}$ and $\overline{\underline{L}}_{n+1}^{n+1}$ lie in any restricted domain of rationality Ω, the transformations (B), which are then transformations in Ω, also establish a one-one correspondence between all sets of finite values of the n variables $x_1, x_2, \dots x_n$ lying in Ω and all sets of finite values of the $n+1$ variables $y_1, y_2, \dots y_{n+1}$ which lie in Ω and satisfy the equation (3).

Moreover they also establish a one-one correspondence between

(1) all rational integral functions in Ω of the n variables $x_1, x_2, \dots x_n$;

(2) all those homogeneous rational integral functions in Ω of the $n+1$ variables $y_1, y_2, \dots y_{n+1}$ which are not divisible by the homogeneous linear function Y_{n+1};

two corresponding functions F and G having the same degree in all their variables, and being convertible into one another by the substitutions (B).

Two corresponding functions F and G of the same degree lying in Ω have the same properties as before. One of them vanishes identically when and only when the other vanishes identically. When they do not vanish identically, they are either both reducible in Ω or both irreducible in Ω.

NOTE 3. *Homogenisation with change of degree.*

If $[l]_{n+1}^{n+1}$ and $\overline{\underline{L}}_{n+1}^{n+1}$ are two given mutually inverse undegenerate square matrices in the domain of rationality Ω, and if r and s are any two given positive integers such that $r \nless s$, we see in the same way that:

The two mutually inverse transformations (B) *establish a one-one correspondence between*

(1) *all rational integral functions in Ω of degree s of the n variables $x_1, x_2, \dots x_n$;*

(2) *all those homogeneous rational integral functions in Ω of degree r of the $n+1$ variables $y_1, y_2, \dots y_{n+1}$ which are divisible by Y_{n+1}^{r-s} and are not divisible by any higher power of Y_{n+1};*

two such functions F and G' of degrees s and r of the x's and y's corresponding when and only when they are convertible into one another by the transformations (B).

In this case any two corresponding functions F and G' of degrees s and r have the forms

$$F=\Sigma a x_1^{p_1}x_2^{p_2}\dots x_n^{p_n}=\Sigma a x_1^{p_1}x_2^{p_2}\dots x_n^{p_n}\,.\,1^{p_{n+1}}$$

$$=\Sigma b X_1^{q_1}X_2^{q_2}\dots X_{n+1}^{q_{n+1}}, \quad\text{..........................(b}_1\text{)}$$

$$G'=\Sigma b y_1^{q_1}y_2^{q_2}\dots y_{n+1}^{q_{n+1}}=\Sigma a Y_1^{p_1}Y_2^{p_2}\dots Y_{n+1}^{p_{n+1}}, \quad\text{..........................(b}_1'\text{)}$$

where the coefficients a and b are constants in Ω, and where

$$p_1+p_2+\dots+p_n \ngtr s,\quad p_1+p_2+\dots+p_n+p_{n+1}=r,\quad q_1+q_2+\dots+q_{n+1}=r.$$

One of the functions F and G' vanishes identically when and only when the other vanishes identically. When F and G' do not vanish identically, there are terms in the first expression for F with non-vanishing coefficients in which $p_1+p_2+\ldots+p_n=s$, and there are terms in the second expression for G' with non-vanishing coefficients in which $p_{n+1}=r-s$, but no such terms in which $p_{n+1}>r-s$. The first of the transformations (B) converts F into G'; and the second of the transformations (B) converts G' into F. When F is given, there cannot be two different homogeneous rational integral functions of $y_1, y_2, \ldots y_{n+1}$ of the same degree which are derived from F by the first of the transformations (B); and when G' is given, there cannot be two different rational integral functions of $x_1, x_2, \ldots x_n$ which are derived from G' by the second of the transformations (B). Consequently each of the two corresponding functions F and G' is uniquely determinate when the other is given.

We call G' the homogenised function of degree r corresponding to F.

Two corresponding functions F and G' of degrees s and r have equal values for corresponding sets of finite values of the x's and y's satisfying the equations (B).

If G is the homogenised function of the same degree s as F which corresponds to F, and is obtained as in the text, so that G is not divisible by Y_{n+1}, we have

$$G'=Y_{n+1}^{r-s}G;$$

and G is irreducible in Ω (or irresoluble) when and only when F is irreducible in Ω (or irresoluble).

If $f_1, f_2, \ldots f_m$ are rational integral functions of $x_1, x_2, \ldots x_n$, and if $g_1, g_2, \ldots g_m$ are the corresponding homogenised rational integral functions of $y_1, y_2, \ldots y_{n+1}$ of the same degrees, obtained as in the text, which are not divisible by Y_{n+1}, we have

$$F=f_1^{u_1}f_2^{u_2}\ldots f_m^{u_m} \text{ when and only when } G'=Y_{n+1}^{r-s}g_1^{u_1}g_2^{u_2}\ldots g_m^{u_m};$$

and here $g_1, g_2, \ldots g_m$ are irreducible functions in Ω distinct from one another and distinct from Y_{n+1} (or irresoluble functions distinct from one another and distinct from Y_{n+1}) when and only when $f_1, f_2, \ldots f_m$ are distinct irreducible functions in Ω (or distinct irresoluble functions).

If $G=\Sigma c y_1^{s_1} y_2^{s_2}\ldots y_{n+1}^{s_{n+1}}$, where $s_1+s_2+\ldots+s_{n+1}=s$, we have

$$G'=(l_{n+1,1}\,y_1+l_{n+1,2}\,y_2+\ldots+l_{n+1,n+1}\,y_{n+1})^{r-s}.\ \Sigma c y_1^{s_1}\, y_2^{s_2}\ldots y_{n+1}^{s_{n+1}},$$

$$F=(l_{n+1,1}X_1+l_{n+1,2}X_2+\ldots+l_{n+1,n+1}X_{n+1})^{r-s}.\,\Sigma c X_1^{s_1}X_2^{s_2}\ldots X_{n+1}^{s_{n+1}}.$$

4. *Homogenisation of a rational integral function by the introduction of a new variable.*

In the particular case when $[l]_{n+1}^{n+1}=[1]_{n+1}^{n+1}$, the two mutually inverse transformations (B) of sub-article 3 become respectively

$$[x_1\,x_2\ldots x_n\,1]=[y_1\,y_2\ldots y_n\,y_{n+1}];\quad [y_1\,y_2\ldots y_n\,y_{n+1}]=[x_1\,x_2\ldots x_n\,1].$$

In this case we have

$$[X_1\,X_2\ldots X_n\,X_{n+1}]=[x_1\,x_2\ldots x_n\,1];\quad [Y_1\,Y_2\ldots Y_n\,Y_{n+1}]=[y_1\,y_2\ldots y_n\,y_{n+1}],$$

and the coefficients b are the same as the coefficients a.

If in the homogenised function G we replace $y_1, y_2, \dots y_{n+1}$ by $x_1, x_2, \dots x_{n+1}$, the substitutions which convert F into G and G into F are respectively

$$1 = x_{n+1}; \quad x_{n+1} = 1. \quad \dots\dots\dots\dots\dots\dots\text{(C)}$$

Let F be any rational integral function of the n variables $x_1, x_2, \dots x_n$ which does not vanish identically and has degree s in all the variables, so that

$$F = \Sigma a x_1^{p_1} x_2^{p_2} \dots x_n^{p_n} = \Sigma a x_1^{p_1} x_2^{p_2} \dots x_n^{p_n} . 1^{p_{n+1}}, \quad \dots\dots\dots\text{(c)}$$

where $p_1 + p_2 + \dots + p_n \not> s$, $p_1 + p_2 + \dots + p_n + p_{n+1} = s$, there being terms in F with non-vanishing coefficients in which $p_1 + p_2 + \dots + p_n = s$, and therefore $p_{n+1} = 0$. Then the first of the substitutions (C) converts F into the homogeneous rational integral function G of degree s of the $n+1$ variables $x_1, x_2, \dots x_n, x_{n+1}$ given by

$$G = \Sigma a x_1^{p_1} x_2^{p_2} \dots x_n^{p_n} x_{n+1}^{p_{n+1}}; \quad \dots\dots\dots\dots\dots\dots\text{(c}'\text{)}$$

and this function G does not vanish identically, and is not divisible by x_{n+1}. Also the second of the substitutions (C) re-converts G into F. Again the function F given by (c) vanishes identically when and only when the function G given by (c′) vanishes identically.

Thus the two mutually inverse transformations (C) *establish a one-one correspondence between*

(1) *all rational integral functions F of the n variables $x_1, x_2, \dots x_n$;*

(2) *all those homogeneous rational integral functions G of the $n+1$ variables $x_1, x_2, \dots x_n, x_{n+1}$ which are not divisible by x_{n+1};*

two corresponding functions F and G having the same degree in all the variables, and being convertible into one another by the substitutions (C).

Any two corresponding functions F and G of the same degree have the forms shown in (c) and (c′); the first of the substitutions (C) converts F into G; and the second of the substitutions (C) converts G into F. One of the two corresponding functions F and G vanishes identically when and only when the other vanishes identically; and each of them is uniquely determinate when the other is given.

We call G the homogenised function of the same degree as F formed by the introduction of the new variable x_{n+1}.

Ex. xvi. To each set of values $a_1, a_2, \dots a_n$ of the n variables $x_1, x_2, \dots x_n$ there corresponds the set of values $a_1, a_2, \dots a_n, 1$ of the $n+1$ variables $x_1, x_2, \dots x_n, x_{n+1}$; and two corresponding functions F and G have equal values for corresponding sets of finite values of the variables.

Ex. xvii. If one of two corresponding functions F and G lies in a restricted domain of rationality Ω, the other also lies in Ω.

Ex. xviii. If $F, f, f_1, f_2, \dots f_m$ and $G, g, g_1, g_2, \dots g_m$ are two sets of corresponding functions which do not vanish identically (no function of the second set being divisible by x_{n+1}), then:

(1) The products $f_1^{u_1} f_2^{u_2} \dots f_m^{u_m}$ and $g_1^{u_1} g_2^{u_2} \dots g_m^{u_m}$ are corresponding functions.

(2) We have $F = f_1^{u_1} f_2^{u_2} \dots f_m^{u_m}$ when and only when $G = g_1^{u_1} g_2^{u_2} \dots g_m^{u_m}$.

(3) The function f is a factor of F (repeated u times) when and only when the function g is a factor of G (repeated u times).

(4) Two corresponding functions F and G are either both resoluble or both irresoluble; also they are either both reducible in Ω or both irreducible in Ω.

(5) The highest common factors of $f_1, f_2, \dots f_m$ and of $g_1, g_2, \dots g_m$ are corresponding functions of the x's and y's of the same degrees convertible into one another by the substitutions (C).

NOTE 4. *Homogenisation with change of degree.*

If r and s are any two given positive integers such that $r \nless s$, we see in the same way that:

The two mutually inverse transformations (C) *establish a one-one correspondence between*

(1) *all rational integral functions F of degree s of the n variables $x_1, x_2, \dots x_n$;*

(2) *all those homogeneous rational integral functions G' of degree r of the $n+1$ variables $x_1, x_2, \dots x_n, x_{n+1}$ which are divisible by x_{n+1}^{r-s}, and are not divisible by any higher power of x_{n+1};*

two corresponding functions F and G' of degrees s and r being convertible into one another by the transformations (C).

In this case any two corresponding functions F and G' of degrees s and r have the forms

$$F = \Sigma a x_1^{p_1} x_2^{p_2} \dots x_n^{p_n} = \Sigma a x_1^{p_1} x_2^{p_2} \dots x_n^{p_n} . 1^{p_{n+1}}, \quad \dots\dots\dots (c_1)$$

$$G' = \Sigma a x_1^{p_1} x_2^{p_2} \dots x_n^{p_n} x_{n+1}^{p_{n+1}}, \quad \dots\dots\dots (c_1')$$

where $p_1 + p_2 + \dots + p_n \ngtr s$, $p_1 + p_2 + \dots + p_n + p_{n+1} = r$. Each of the two functions F and G' is uniquely determinate when the other is given, and each of them vanishes identically when and only when the other vanishes identically. When F and G' do not vanish identically, there are terms in F with non-vanishing coefficients in which

$$p_1 + p_2 + \dots + p_n = s;$$

and there are terms in G' with non-vanishing coefficients in which $p_{n+1} = r - s$, but no such terms in which $p_{n+1} > r - s$.

We call G' the homogenised function of degree r corresponding to F and formed by the introduction of the new variable x_{n+1}; and if G is the homogenised function of degree s corresponding to F, which is not divisible by x_{n+1}, we have

$$G' = x_{n+1}^{r-s} G.$$

If $f_1, f_2, \dots f_m$ are rational integral functions of the n variables $x_1, x_2, \dots x_n$ which do not vanish identically, and if $g_1, g_2, \dots g_m$ are the corresponding homogenised rational integral functions of the same degrees of the $n+1$ variables $x_1, x_2, \dots x_n, x_{n+1}$, no one of which is divisible by x_{n+1}, we have

$$F = f_1^{u_1} f_2^{u_2} \dots f_m^{u_m} \text{ when and only when } G' = x_{n+1}^{r-s} g_1^{u_1} g_2^{u_2} \dots g_m^{u_m};$$

and here $g_1, g_2, \dots g_m$ are irreducible functions in Ω distinct from one another and distinct from x_{n+1} when and only when $f_1, f_2, \dots f_m$ are distinct irreducible functions in Ω.

5. *Non-homogeneous linear transformations of the variables.*

An *ordinary* linear transformation from one set of n variables to another is given by n equations of the form

$$\begin{aligned} x_1 &= l_{11}y_1 + l_{12}y_2 + \dots + l_{1n}y_n + e_{11}, \\ x_2 &= l_{21}y_1 + l_{22}y_2 + \dots + l_{2n}y_n + e_{21}, \\ &\dots\dots\dots\dots\dots\dots\dots\dots\dots\dots \\ x_n &= l_{n1}y_1 + l_{n2}y_2 + \dots + l_{nn}y_n + e_{n1}, \end{aligned}$$

where $[l]_n^n$ is an undegenerate square matrix with constant elements, and where the e's are any constants. If the e's all vanish, it is the homogeneous transformation considered in sub-article 1. When the e's do not all vanish, it is non-homogeneous. The matrix equations of this transformation and its inverse are

$$\begin{bmatrix} x \\ 1 \end{bmatrix}_{n,1} = \begin{bmatrix} l, & e \\ 0, & 1 \end{bmatrix}_{n,1}^{n,1} \begin{bmatrix} y \\ 1 \end{bmatrix}_{n,1}, \quad \begin{bmatrix} y \\ 1 \end{bmatrix}_{n,1} = \begin{bmatrix} L, & E \\ 0, & 1 \end{bmatrix}_{n,1}^{n,1} \begin{bmatrix} x \\ 1 \end{bmatrix}_{n,1}, \quad \dots\dots\dots\text{(D)}$$

where $\overline{\underline{L}}_n^n$ is the undegenerate square matrix inverse to $[l]_n^n$, and where

$$\overline{\underline{E}}_n^1 = -\overline{\underline{L}}_n^n [e]_n^1, \quad [e]_n^1 = -[l]_n^n \overline{\underline{E}}_n^1 .$$

The two mutually inverse transformations or substitutions (D) *establish a one-one correspondence between all sets of finite values of the n variables $x_1, x_2, \dots x_n$ and all sets of finite values of the n variables $y_1, y_2, \dots y_n$. Any two corresponding sets of finite values of the x's and y's satisfy both the equations* (D), *and each set is uniquely determinate when the other is given.*

Let $X_1, X_2, \dots X_n$ and $Y_1, Y_2, \dots Y_n$ be the linear functions of the x's and y's respectively defined by the equations

$$\begin{aligned} X_i &= L_{1i}x_1 + L_{2i}x_2 + \dots + L_{ni}x_n + E_{1i}, \\ Y_i &= l_{i1}y_1 + l_{i2}y_2 + \dots + l_{in}y_n + e_{i1}, \end{aligned}$$

so that the two pairs of equations

$$\begin{bmatrix} X \\ 1 \end{bmatrix}_{n,1} = \begin{bmatrix} L, & E \\ 0, & 1 \end{bmatrix}_{n,1}^{n,1} \begin{bmatrix} x \\ 1 \end{bmatrix}_{n,1}, \quad \begin{bmatrix} x \\ 1 \end{bmatrix}_{n,1} = \begin{bmatrix} l, & e \\ 0, & 1 \end{bmatrix}_{n,1}^{n,1} \begin{bmatrix} X \\ 1 \end{bmatrix}_{n,1}, \quad \dots\dots\dots\text{(D}')$$

$$\begin{bmatrix} Y \\ 1 \end{bmatrix}_{n,1} = \begin{bmatrix} l, & e \\ 0, & 1 \end{bmatrix}_{n,1}^{n,1} \begin{bmatrix} y \\ 1 \end{bmatrix}_{n,1}, \quad \begin{bmatrix} y \\ 1 \end{bmatrix}_{n,1} = \begin{bmatrix} L, & E \\ 0, & 1 \end{bmatrix}_{n,1}^{n,1} \begin{bmatrix} Y \\ 1 \end{bmatrix}_{n,1}, \quad \dots\dots\dots\text{(D}'')$$

are identities in the x's and y's respectively, and therefore the equations

$$x_i = l_{i1}X_1 + l_{i2}X_2 + \dots + l_{in}X_n + e_{i1},$$
$$y_i = L_{1i}Y_1 + L_{2i}Y_2 + \dots + L_{ni}Y_n + E_{1i}$$

are identities in the x's and y's respectively. Then every rational integral function F of $x_1, x_2, \dots x_n$ can be expressed as a rational integral function of the linear functions $X_1, X_2, \dots X_n$; and this can be done in only one way; for if $G_1(X_1, X_2, \dots X_n)$ and $G_2(X_1, X_2, \dots X_n)$ were two such expressions for F, then G_1 and G_2 would be identically equal functions of the x's; therefore they would be equal for all finite values of $x_1, x_2, \dots x_n$; therefore they would be equal for all finite values of $X_1, X_2 \dots X_n$ when these are regarded as variables; i.e. they would be identically equal functions of $X_1, X_2, \dots X_n$. Similarly every rational integral function G of $y_1, y_2, \dots y_n$ can be expressed in one and only one way as a rational integral function of the linear functions $Y_1, Y_2, \dots Y_n$.

Proceeding now as in sub-article 1 we can show that:

The two mutually inverse transformations (D) *establish a one-one correspondence between all rational integral functions F of the n variables $x_1, x_2, \dots x_n$ and all rational integral functions G of the n variables $y_1, y_2, \dots y_n$; two such functions corresponding when and only when they are convertible into one another by the transformations* (D); *and any two such corresponding functions F and G can be expressed in the forms*

$$F = \Sigma\, a\, x_1^{p_1} x_2^{p_2} \dots x_n^{p_n} = \Sigma\, b\, X_1^{q_1} X_2^{q_2} \dots X_n^{q_n}, \quad \dots\dots\dots\dots\text{(d)}$$

$$G = \Sigma\, b\, y_1^{q_1} y_2^{q_2} \dots y_n^{q_n} = \Sigma\, a\, Y_1^{p_1} Y_2^{p_2} \dots Y_n^{p_n}. \quad \dots\dots\dots\dots\text{(d')}$$

For if F is any given rational integral function of the x's, it can be expressed as in (d), the coefficients b being uniquely determinate; the first of the substitutions (D) then converts it into the function G given in (d'); and the second of the substitutions (D) re-converts G into F. Again if G is any given rational integral function of the y's, it can be expressed as in (d'), the coefficients a being uniquely determinate; the second of the substitutions (D) then converts it into the function F given in (d); and the first of the substitutions (D) re-converts F into G.

Most of the properties of two corresponding functions obtained in sub-article 1 remain true for the more general transformations (D), the proofs being similar. The essential properties which remain true are summarised in the examples which follow.

Ex. xix. If F and G are two corresponding rational integral functions of the x's and y's, then:

(1) Each of them has the same degree in all its variables as the other; but their degrees in the individual variables need not be the same. In particular each of them vanishes identically when and only when the other vanishes identically.

(2) They have equal values for corresponding sets of finite values of the variables. Consequently the transformations (D) convert every finite root of either of them into a finite root of the other.

Ex. xx. If $F, f, f_1, f_2, \ldots f_m$ and $G, g, g_1, g_2, \ldots g_m$ are two sets of corresponding rational integral functions of the x's and y's, then:

(1) The products $f_1^{u_1} f_2^{u_2} \ldots f_m^{u_m}$ and $g_1^{u_1} g_2^{u_2} \ldots g_m^{u_m}$ are corresponding functions of the x's and y's.

(2) We have $F = f_1^{u_1} f_2^{u_2} \ldots f_m^{u_m}$ when and only when $G = g_1^{u_1} g_2^{u_2} \ldots g_m^{u_m}$.

(3) The function f is a highest common factor of $f_1, f_2, \ldots f_m$ when and only when the corresponding function g is a highest common factor of $g_1, g_2, \ldots g_m$.

Ex. xxi. If $x_1, x_2, \ldots x_n$ and $y_1, y_2, \ldots y_n$ are values of the variables which satisfy the equations (D) and are not all finite, we can write

$$[x_1 x_2 \ldots x_n] = \rho\, [\xi_1 \xi_2 \ldots \xi_n], \quad [y_1, y_2 \ldots y_n] = \sigma [\eta_1 \eta_2 \ldots \eta_n],$$

where ρ and σ are infinite, the ξ's are all finite and do not all vanish, and the η's are all finite and do not all vanish; and we can then replace the equations (D) by

$$h \lceil \xi \rfloor_n = [l]_n^n \lceil \eta \rfloor_n, \quad k \lceil \eta \rfloor_n = \lceil L \rfloor_n^n \lceil \xi \rfloor_n,$$

where h and k are finite non-zero numbers, and $hk = 1$.

§ 188. General properties of the factors of rational integral functions.

We will now enunciate three important theorems and deduce from them certain other theorems which will be regarded as corollaries. The general proofs of the three fundamental theorems will be deferred to the latter portion of the present article. We shall there prove the three theorems by induction. We shall assume it to be known that they are true for functions of a single variable; and we shall show that if they are true generally for functions of $n-1$ variables, then they must be true generally for functions of n variables. It will then follow that they and their corollaries are true whatever the number of variables may be.

We use the definition of a highest common factor given in § 185. 3.

Theorem I. *Let f_1 and f_2 be two rational integral functions of the n variables $x_1, x_2, \ldots x_n$ lying in any domain of rationality Ω, neither of which vanishes identically. Then f_1 and f_2 have a highest common factor h which has the following properties:*

(1) *It is a rational integral function of $x_1, x_2, \ldots x_n$ lying in Ω.*

(2) *It differs by a non-vanishing constant factor only from the product of all the distinct irreducible factors (unrepeated and repeated) common to f_1 and f_2.*

(3) *It differs by a non-vanishing constant factor only from the product of all the distinct irresoluble factors (unrepeated and repeated) common to f_1 and f_2.*

It is to be understood that in the products mentioned in (2) and (3) each distinct irreducible (or irresoluble) factor occurs just as many times as it is repeated in both f_1 and f_2. If it is repeated r times in f_1 and s times in f_2, then in the product mentioned in (2) or (3) it is repeated r times or s times according as $r \not> s$ or $s \not> r$. In the particular case when one of the functions f_1 and f_2 is a non-vanishing constant, each of the products mentioned in (2) and (3) is interpreted to be a non-vanishing constant or to be 1. Thus each of those two products is determinate save for a non-vanishing constant factor.

Let $t_1, t_2, \ldots t_r$ be all the distinct irresoluble factors common to f_1 and f_2; and let $T_1, T_2, \ldots T_s$ be all the distinct factors common to f_1 and f_2 which lie in Ω and are irreducible in Ω. Let $t_1^{\alpha_1}, t_2^{\alpha_2}, \ldots t_r^{\alpha_r}$ be the highest powers of $t_1, t_2, \ldots t_r$ respectively which are factors of both f_1 and f_2; and let $T_1^{\beta_1}, T_2^{\beta_2}, \ldots T_s^{\beta_s}$ be the highest powers of $T_1, T_2, \ldots T_s$ respectively which are factors of both f_1 and f_2.

Then by the theorem we have

$$h = ct_1^{\alpha_1} t_2^{\alpha_2} \ldots t_r^{\alpha_r} = CT_1^{\beta_1} T_2^{\beta_2} \ldots T_s^{\beta_s},$$

where c and C are non-vanishing constants.

From Theorem I we can deduce the corollaries which follow.

COROLLARY 1. *If f_2 is irreducible in Ω, then either f_2 or a constant is a highest common factor of f_1 and f_2.*

For the highest common factor h of Theorem I lies in Ω and is a factor of the irreducible function f_2. It must therefore be either f_2 or a constant.

COROLLARY 2. *Let f_2 be irreducible in Ω but resoluble. Then if f_1 (which lies in Ω) is divisible by any factor of f_2 which is not merely a constant, it must be divisible by f_2.*

Let ϕ be a factor of f_2 which is not merely a constant. Then if ϕ is also a factor of f_1, h must be divisible by ϕ and cannot be a constant. Therefore by Corollary 1 the function f_2 is a highest common factor of f_1 and f_2. Consequently f_2 is a factor of f_1.

COROLLARY 3. *If f_1 and f_2 have no irreducible factor in common, then they have no irresoluble factor in common.*

For in this case h must be a constant, and therefore every common factor of f_1 and f_2 must be a constant.

COROLLARY 4. *Let $f_1, f_2, \ldots f_m$ be rational integral functions of the variables $x_1, x_2, \ldots x_n$ lying in any domain Ω, no one of which vanishes identically; and let $t_1, t_2, \ldots t_r$ be all the distinct irresoluble factors and $T_1, T_2, \ldots T_s$ be all the distinct irreducible factors which are common to all the functions $f_1, f_2, \ldots f_m$. Then $f_1, f_2, \ldots f_m$ have a highest common factor h which lies in Ω and is such that*

$$h = ct_1^{\alpha_1} t_2^{\alpha_2} \ldots t_r^{\alpha_r} = CT_1^{\beta_1} T_2^{\beta_2} \ldots T_s^{\beta_s},$$

where c and C are non-vanishing constants, $t_i^{\alpha_i}$ is the highest power of t_i which is a factor of every one of the functions $f_1, f_2, \ldots f_m$, and $T_j^{\beta_j}$ is the highest power of T_j which is a factor of every one of the functions $f_1, f_2, \ldots f_m$.

We can prove this corollary by induction. Assuming it to be true for the $m-1$ functions $f_2, f_3, \dots f_m$, it follows from Ex. xxvi of § 185 that it is true for the m functions $f_1, f_2, \dots f_m$.

Theorem II. *Let f_1, f_2 and g be rational integral functions of the n variables $x_1, x_2, \dots x_n$ lying in any domain of rationality Ω, no one of which vanishes identically. Then if the product $f_1 f_2$ is divisible by g, and if g and f_1 are prime to one another in Ω (i.e. have no irreducible factor in common), f_2 must be divisible by g.*

From Theorem II we can deduce the corollaries which follow.

COROLLARY 1. *If $f_1 f_2$ is divisible by g, and if g and f_1 are prime to one another (i.e. have no irresoluble factor in common), then f_2 must be divisible by g.*

For we can take Ω in the theorem to be the domain of all algebraic numbers.

Using Corollary 3 to Theorem I, we can deduce Theorem II from that particular case of it contained in the present corollary. For if f_1 and f_2 have no irreducible factor in common, then they have no irresoluble factor in common.

COROLLARY 2. *If f is an irreducible rational integral function of $x_1, x_2, \dots x_n$ belonging to the domain Ω, no power of f can have any irreducible factor distinct from f.*

Let g be any irreducible rational integral function of $x_1, x_2, \dots x_n$ belonging to the domain Ω which is a factor of f^m. Then by the theorem g must be a factor of f^{m-1} if $m>1$, therefore a factor of f^{m-2} if $m>2$, ... and finally a factor of f; i.e. g is not distinct from f.

COROLLARY 3. *Let f be a rational integral function of the n variables $x_1, x_2, \dots x_n$ which lies in the domain of rationality Ω and does not vanish identically; let $T_1, T_2, \dots T_s$ be all the distinct irreducible factors of f in Ω; and let $t_1, t_2, \dots t_r$ be all the distinct irresoluble factors of f. Also let $T_1^{\beta_1}, T_2^{\beta_2}, \dots T_s^{\beta_s}$ be the highest powers of $T_1, T_2, \dots T_s$ respectively which are factors of f; and let $t_1^{\alpha_1}, t_2^{\alpha_2}, \dots t_r^{\alpha_r}$ be the highest powers of $t_1, t_2, \dots t_r$ respectively which are factors of f. Then f differs by a non-vanishing constant factor only from each of the products*

$$T_1^{\beta_1} T_2^{\beta_2} \dots T_s^{\beta_s} \text{ and } t_1^{\alpha_1} t_2^{\alpha_2} \dots t_r^{\alpha_r}.$$

It will be sufficient to prove the first result, the second result being a particular case of the first.

Since $T_1^{\beta_1}, T_2^{\beta_2}, \dots T_s^{\beta_s}$ are factors of f and are the highest powers of $T_1, T_2, \dots T_s$ which are factors of f, we can write

$$f = T_1^{\beta_1} f_2,$$

where f_2 is a rational integral function lying in Ω which does not vanish identically and is not divisible by T_1. If i is any one of the integers $2, 3, \dots s$ we see from Corollary 2 that $T_1^{\beta_1}$ and $T_i^{\beta_i}$ have no irreducible factor in common. Therefore by the theorem $T_i^{\beta_i}$ is a factor of f_2, and it is necessarily the highest power of T_i which is a factor of f_2. Accordingly $T_2^{\beta_2}, T_3^{\beta_3}, \dots T_s^{\beta_s}$ are factors of f_2, and they are the highest powers of $T_2, T_3, \dots T_s$ which are factors of f_2.

If $s \not< 2$, it follows that we can write

$$f_2 = T_2^{\beta_2} f_3, \quad f = T_1^{\beta_1} T_2^{\beta_2} f_3,$$

where f_3 is a rational integral function lying in Ω which does not vanish identically and is divisible neither by T_1 nor by T_2. If i is any one of the integers 3, 4, ... s, then $T_2^{\beta_2}$ and $T_i^{\beta_i}$ have no irreducible factor in common. Therefore by the theorem $T_i^{\beta_i}$ is a factor of f_3, and it is necessarily the highest power of T_i which is a factor of f_3. Accordingly $T_3^{\beta_3}$, $T_4^{\beta_4}$, ... $T_s^{\beta_s}$ are factors of f_3, and they are the highest powers of T_3, T_4, ... T_s which are factors of f_3.

If $s \nless 3$, it follows that we can write

$$f_3 = T_3^{\beta_3} f_4, \quad f = T_1^{\beta_1} T_2^{\beta_2} T_3^{\beta_3} f_4,$$

where f_4 is a rational integral function lying in Ω which does not vanish identically and is not divisible by any one of the functions T_1, T_2, T_3; and we can show as before that $T_4^{\beta_4}$, $T_5^{\beta_5}$, ... $T_s^{\beta_s}$ are factors of f_4, and that they are the highest powers of T_4, T_5, ... T_s which are factors of f_4.

Proceeding in this way we see that

$$f = T_1^{\beta_1} T_2^{\beta_2} \ldots T_s^{\beta_s} f_{s+1},$$

where f_{s+1} is a rational integral function lying in Ω which does not vanish identically, and which is not divisible by any one of the functions T_1, T_2, ... T_s.

Since f has no irreducible factors distinct from T_1, T_2, ... T_s, therefore f_{s+1} (which lies in Ω) has no irreducible factors, i.e. f_{s+1} is a constant lying in Ω.

The theorem embodied in this corollary can be regarded as included in Corollary 4 to Theorem I.

Theorem III. *Let f_1, f_2 and g be rational integral functions of the n variables x_1, x_2, ... x_n lying in any domain of rationality Ω, no one of which vanishes identically; and let g be irreducible in Ω. Then if the product $f_1 f_2$ is divisible by g, at least one of the functions f_1 and f_2 must be divisible by g.*

From Theorem III we can deduce the corollaries which follow.

COROLLARY 1. *Let f_1, f_2, ... f_m and g be rational integral functions of x_1, x_2, ... x_n lying in any domain of rationality Ω, no one of which vanishes identically; and let g be irreducible in Ω. Then if the product $f_1 f_2 \ldots f_m$ is divisible by g, at least one of the functions f_1, f_2, ... f_m must be divisible by g.*

For if f_1 is not divisible by g, then by the theorem the product $f_2 f_3 \ldots f_m$ must be divisible by g. If in addition f_2 is not divisible by g, then the product $f_3 f_4 \ldots f_m$ must be divisible by g; and so on. Finally if no one of the functions f_1, f_2, ... f_{m-1} is divisible by g, then f_m must be divisible by g.

COROLLARY 2. *If f is a rational integral function of the n variables x_1, x_2, ... x_n which lies in any domain of rationality Ω and is not merely a constant, then f can be expressed in only one way as a product of distinct factors (unrepeated and repeated) which are irreducible in Ω.*

Let $$f = u_1 u_2 \ldots u_r = v_1 v_2 \ldots v_s,$$

where $s \nless r$, and where u_1, u_2, ... u_r, v_1, v_2, ... v_s are rational integral functions of x_1, x_2, ... x_n which lie in Ω and are irreducible in Ω, no one of them being merely a constant.

By Corollary 1 the function u_1 must be a factor of one at least of the functions $v_1, v_2, \dots v_s$, which we may suppose to be v_1. Then since u_1 and v_1 are both irreducible in Ω, we have

$$u_1 = c_1 v_1, \text{ and therefore } c_1 u_2 u_3 \dots u_r = v_2 v_3 \dots v_s,$$

where c_1 is a non-vanishing constant lying in Ω.

From the last equation we see that u_2 must be a factor of at least one of the functions $v_2, v_3, \dots v_s$, which we may suppose to be v_2; and we then have

$$u_2 = c_2 v_2, \text{ and therefore } c_1 c_2 u_3 u_4 \dots u_r = v_3 v_4 \dots v_s,$$

where c_2 is a non-vanishing constant lying in Ω.

Proceeding in this way we have finally

$$c_1 c_2 \dots c_{r-1} u_r = v_r v_{r+1} \dots v_s,$$

where $c_1, c_2, \dots c_{r-1}$ are non-vanishing constants lying in Ω.

Since u_r is irreducible in Ω, we must have

$$s = r, \quad u_r = c_r v_r,$$

where c_r is a non-vanishing constant lying in Ω.

Thus the irreducible factors occurring in the second expression for f are not distinct from those occurring in the first expression.

COROLLARY 3. *If f is a rational integral function of the n variables $x_1, x_2, \dots x_n$ which is not merely a constant, then f can be expressed in only one way as a product of distinct irresoluble factors (unrepeated and repeated).*

We deduce this result from Corollary 2 by taking Ω to be the domain of all algebraic numbers.

COROLLARY 4. *If f is a rational integral function of the n variables $x_1, x_2, \dots x_n$ which lies in any domain of rationality Ω and is not merely a constant, then every irresoluble factor of f is a factor of one and only one of the distinct irreducible factors of f.*

Let $f = u_1 u_2 \dots u_m$, where $u_1, u_2, \dots u_m$ are irreducible rational integral functions of $x_1, x_2, \dots x_n$ belonging to the domain Ω; and let g be an irresoluble rational integral function of $x_1, x_2, \dots x_n$. Then by Corollary 1 if g is a factor of f, it must be a factor of at least one of the functions $u_1, u_2, \dots u_m$. Moreover if g is a factor of two of these functions u_i and u_j, then by Corollary 2 to Theorem I each of these two functions is a factor of the other, i.e. they are not distinct from one another.

Theorems II and III can be regarded as equivalent to one another, for, as is shown in Exs. i and ii below, each of them can be deduced from the other.

Ex. i. *Deduction of Theorem III from Theorem II.*

Let f_1, f_2 and g be rational integral functions of $x_1, x_2, \dots x_n$ lying in Ω, no one of which vanishes identically; and let g be irreducible in Ω. Further let the product $f_1 f_2$ be divisible by g.

Then if g is not a factor of f_1, g and f_1 have no irreducible factor in common, and it follows from Theorem II that g must be a factor of f_2. Thus g must be a factor of at least one of the functions f_1 and f_2; and this is Theorem III.

Ex. ii. *Deduction of Theorem II from Theorem III.*

Let f_1, f_2 and g be rational integral functions of $x_1, x_2, \dots x_n$ lying in Ω, no one of which vanishes identically, and let g and f_1 be prime to one another in Ω. Further let the product $f_1 f_2$ be divisible by g.

Now let $$g = u_1 u_2 \dots u_r, \quad f_1 f_2 = gv = u_1 u_2 \dots u_r v,$$

where $u_1, u_2, \dots u_r$ are rational integral functions of $x_1, x_2, \dots x_n$ which lie in Ω and are irreducible in Ω, and where v is a rational integral function of $x_1, x_2, \dots x_n$ lying in Ω.

Since u_1 is a factor of $f_1 f_2$ and not a factor of f_1, therefore by Theorem III it is a factor of f_2, and we have

$$f_2 = u_1 g_1, \quad \text{and therefore} \quad f_1 g_1 = u_2 u_3 \dots u_r v,$$

where g_1 is a rational integral function of $x_1, x_2, \dots x_n$ lying in Ω.

Again since u_2 is a factor of $f_1 g_1$ and not a factor of f_1, we have

$$g_1 = u_2 g_2, \quad \text{and therefore} \quad f_1 g_2 = u_3 u_4 \dots u_r v,$$

where g_2 is a rational integral function of $x_1, x_2, \dots x_n$ lying in Ω.

Proceeding in this way we see that

$$f_2 = u_1 u_2 \dots u_r g_r = g g_r,$$

where g_r is a rational integral function of $x_1, x_2, \dots x_n$ lying in Ω.

Thus f_2 is divisible by g; and this is Theorem II.

The following lemmas are properties of functions of the n variables $x_1, x_2, \dots x_n$ which are immediate consequences of Theorems I—III. But in proving them we shall only assume Theorems I—III to be true for functions of the $n-1$ variables $x_2, x_3, \dots x_n$. From the way in which they are proved it follows that they can be used in the general proofs by induction of Theorems I—III, and on this account they are called lemmas.

Lemma A. *Let f be a rational integral function of the n variables $x_1, x_2, \dots x_n$ which lies in a domain of rationality Ω and does not vanish identically. Then f can always be expressed in the form*

$$f = g\phi,$$

where ϕ is a rational integral function of the $n-1$ variables $x_2, x_3, \dots x_n$ only lying in Ω, and where g is a rational integral function of $x_1, x_2, \dots x_n$ lying in Ω which is not divisible by any rational integral function of $x_2, x_3, \dots x_n$ only.

It is assumed that $n \not< 2$.

Let $$f = u_0 x_1^r + u_1 x_1^{r-1} + \dots + u_{r-1} x_1 + u_r,$$

where $u_0, u_1, \dots u_r$ are rational integral functions of $x_2, x_3, \dots x_n$ only lying in Ω; and let ϕ be the H.C.F. of $u_0, u_1, \dots u_r$. By Theorem I (for $n-1$ variables) we can choose ϕ so that it lies in Ω. We then have $f = g\phi$, where g is a rational integral function of $x_1, x_2, \dots x_n$ lying in Ω which by Ex. xxiii of § 185 is not divisible by any rational integral function of $x_2, x_3, \dots x_n$ only.

Lemma B. *Let f and g be rational integral functions of the n variables $x_1, x_2, \dots x_n$ lying in a domain of rationality Ω, neither of which vanishes identically; and let ϕ be a rational integral function of the $n-1$ variables $x_2, x_3, \dots x_n$ only which lies in Ω and is irreducible in Ω. Then if the product fg is divisible by ϕ, one at least of the functions f and g is divisible by ϕ.*

Let $f=u_0+u_1x_1+u_2x_1^2+\ldots,\quad g=v_0+v_1x_1+v_2x_1^2+\ldots,$

where $u_0, u_1, u_2, \ldots, v_0, v_1, v_2, \ldots$ are rational integral functions of $x_2, x_3, \ldots x_n$ only lying in Ω; and let the product fg be divisible by ϕ.

Now suppose that neither of the functions f and g is divisible by ϕ.

Then by Ex. xxiii of § 185 there is at least one of the functions $u_0, u_1, u_2, \ldots$ and at least one of the functions $v_0, v_1, v_2, \ldots$ which is not divisible by ϕ. Let u_i be the first of the functions $u_0, u_1, u_2, \ldots$ and v_j be the first of the functions $v_0, v_1, v_2, \ldots$ which is not divisible by ϕ. Then in the identical equation

$$(f-u_0-u_1x_1-\ldots-u_{i-1}x_1^{i-1})(g-v_0-v_1x_1-\ldots-v_{j-1}x_1^{j-1})$$
$$=(u_ix_1^i+u_{i+1}x_1^{i+1}+\ldots)(v_jx_1^j+v_{j+1}x_1^{j+1}+\ldots)$$

the expression on the left is divisible by ϕ; therefore the expression on the right is also divisible by ϕ; therefore by Ex. xxiii of § 185 the product u_iv_j is divisible by ϕ. By Theorem III (for $n-1$ variables) this is impossible, since neither u_i nor v_j is divisible by ϕ.

Hence it is impossible that neither of the functions f and g shall be divisible by ϕ.

Lemma C. *Let F and f be rational integral functions of the n variables $x_1, x_2, \ldots x_n$, neither of which vanishes identically; and let ϕ be a rational integral function of the $n-1$ variables $x_2, x_3, \ldots x_n$ only which does not vanish identically. Then if f is not divisible by any rational integral function of $x_2, x_3, \ldots x_n$ only (other than a constant) and is a factor of $F\phi$, it must be a factor of F.*

Let $F\phi=fg$, where g is a rational integral function of $x_1, x_2, \ldots x_n$. Then if ψ is any irresoluble factor of ϕ, not merely a constant, it must be a factor of fg, and therefore by Lemma B a factor of either f or g. It cannot be a factor of f, since it is a function of $x_2, x_3, \ldots x_n$ only. Accordingly it is a factor of g, and can be cancelled on both sides of the equation $F\phi=fg$. If the resulting equation is $F\phi_1=fg_1$, this can be treated in the same way; and after a succession of such cancellations the equation $F\phi=fg$ is reduced to $F=fg'$, where g' is a rational integral function of $x_1, x_2, \ldots x_n$. This shows that f must be a factor of F.

Ex. iii. *Let F, f_1, f_2 be rational integral functions of the n variables $x_1, x_2, \ldots x_n$, no one of which vanishes identically; and let ϕ_1 and ϕ_2 be rational integral functions of the $n-1$ variables $x_2, x_3, \ldots x_n$ only, neither of which vanishes identically. Then if F can be expressed in the form*

$$F=\frac{f_1}{\phi_1}\cdot\frac{f_2}{\phi_2},$$

where f_1 and ϕ_1 have no factor in common other than a constant, and f_2 and ϕ_2 have no factor in common other than a constant, f_1 must be divisible by ϕ_2, and f_2 must be divisible by ϕ_1.

This can be proved in the same way as Lemma C when we observe that f_1f_2 is divisible by $\phi_1\phi_2$. Writing $f_1=u_1\phi_2, f_2=u_2\phi_1$, we obtain

$$F=u_1u_2,$$

where u_1 and u_2 are rational integral functions of $x_1, x_2, \ldots x_n$ lying in the same domain as f_1, f_2, ϕ_1, ϕ_2.

Lemma D. *Let f_1 and f_2 be rational integral functions of the n variables $x_1, x_2, \ldots x_n$, neither of them vanishing identically, which lie in any domain of rationality Ω and are* *regular in* x_1. *Then* f_1 *and* f_2 *have a highest common factor* h *which has the following properties:*

(1) *It is a rational integral function of* $x_1, x_2, \ldots x_n$ *lying in* Ω.

(2) *It differs by a non-vanishing constant factor only from the product of all irreducible factors (unrepeated and repeated) common to* f_1 *and* f_2.

(3) *It differs by a non-vanishing constant factor only from the product of all irresoluble factors (unrepeated and repeated) common to* f_1 *and* f_2.

(4) *It satisfies an equation of the form*

$$P_1 f_1 + P_2 f_2 = h\phi, \quad \text{.......................................(A)}$$

where ϕ *is a rational integral function of* $x_2, x_3, \ldots x_n$ *only which lies in* Ω *and does not vanish identically, and where* P_1 *and* P_2 *are rational integral functions of* $x_1, x_2, \ldots x_n$ *lying in* Ω.

It is to be understood that in the products mentioned in (2) and (3) each irreducible or irresoluble factor occurs just as many times as it is repeated in both f_1 and f_2. In the particular case when f_1 or f_2 is merely a constant we interpret each of the products mentioned in (2) and (3) to be a non-vanishing constant or to be 1. Thus each of those products is determinate save for a non-vanishing constant factor.

We will suppose that the degree of f_2 in all the variables, i.e. in x_1, is not greater than that of f_1.

When f_1 is divisible by f_2, we have $f_1 = qf_2$, where q is a rational integral function of $x_1, x_2, \ldots x_n$ lying in Ω whose degree in x_1 is less than that of f_1. In this particular case Lemma D is clearly true when we take h to be f_2: and there exists an equation of the form (A) in which $P_1 = 0$, $P_2 = 1$ and $\phi = 1$.

In all other cases there exist rational integral equations in Ω of the forms

$$f_1 = q_2 f_2 + u_3 f_3, \quad v_2 f_2 = q_3 f_3 + u_4 f_4, \quad \ldots \quad v_{i-1} f_{i-1} = q_i f_i + u_{i+1} f_{i+1},$$
$$\ldots \quad v_{r-2} f_{r-2} = q_{r-1} f_{r-1} + u_r f_r, \quad v_{r-1} f_{r-1} = q_r f_r, \quad \text{......(B)}$$

where $f_1, f_2, f_3, \ldots f_r$ are rational integral functions of $x_1, x_2, \ldots x_n$ which lie in Ω and do not vanish identically, each function after the second having a lower degree in x_1 than the preceding function, and no one of these functions being divisible by any rational integral function of $x_2, x_3, \ldots x_n$ only other than a constant; where $q_2, q_3, \ldots q_r$ are rational integral functions in Ω of $x_1, x_2, \ldots x_n$ which do not vanish identically; and where $u_3, u_4, \ldots u_r$ and $v_2, v_3, \ldots v_{r-1}$ are rational integral functions of $x_2, x_3, \ldots x_n$ only which lie in Ω and do not vanish identically.

We can obtain such equations by dividing in succession f_1 by f_2, f_2 by f_3, ... f_{r-1} by f_r, arranging the dividend and divisor in descending powers of x_1, and performing the division as if x_1 were the only variable. If f_3 has a factor which is a rational integral function of $x_2, x_3, \ldots x_n$ only, then by Lemma A we can put $f_3 = f_3' w_3$, where w_3 is a rational integral function of $x_2, x_3, \ldots x_n$ only lying in Ω, and where f_3' is a rational integral function of $x_1, x_2, \ldots x_n$ lying in Ω which is not divisible by any rational integral function of $x_2, x_3, \ldots x_n$ only other than a constant; and we can then replace $u_3 f_3$ by $u_3' f_3'$, where $u_3' = u_3 w_3$. Similar reasoning applies to $f_4, f_5, \ldots f_r$. Thus $f_3, f_4, \ldots f_r$ can always be so determined that no one of the functions $f_1, f_2, \ldots f_r$ is divisible by any rational integral function of $x_2, x_3, \ldots x_n$ only other than a constant.

We will now show that all the conditions of Lemma D are satisfied when we take h to be f_r.

We first observe that every rational integral factor common to f_1 and f_2 must be a factor of $u_3 f_3$, and therefore by Lemma C a factor of f_3. Being a common factor of f_2 and

f_3, it must in the same way be a factor of f_4, and similarly a factor of $f_5, f_6, \dots f_r$. By similar reasoning f_r (and every rational integral factor of f_r) must be a factor of $f_{r-1}, f_{r-2}, \dots f_2, f_1$, and therefore a common factor of f_1 and f_2. Thus f_r is a common factor of f_1 and f_2 which is divisible by every common factor of f_1 and f_2; i.e. f_r is a highest common factor of f_1 and f_2, and moreover it lies in Ω.

To show that the condition (2) of the lemma is also satisfied when $h=f_r$, let $\psi_1, \psi_2, \dots \psi_s$ be the distinct irreducible factors common to f_1 and f_2, each irreducible factor occurring just as many times as it is repeated in both f_1 and f_2, and let $\psi=\psi_1\psi_2\dots\psi_s$. Then ψ_1, being a common factor of f_1 and f_2, is a factor of each one of the functions $f_1, f_2, \dots f_r$, and can be cancelled in all the equations (B). We can write

$$f_1=\psi_1 g_1, \quad f_2=\psi_1 g_2, \quad \dots \quad f_r=\psi_1 g_r,$$

and after cancellation of the factor ψ_1 we obtain a new set of equations of the same forms as the equations (B) in which $f_1, f_2, \dots f_r$ are replaced by $g_1, g_2, \dots g_r$. Since by Lemma B the function ψ_2 is a common factor of g_1 and g_2, these new equations show that ψ_2 is a factor of each one of the functions $g_1, g_2, \dots g_r$, and we can write

$$g_1=\psi_2 h_1, \quad g_2=\psi_2 h_2, \quad \dots \quad g_r=\psi_2 h_r;$$

so that $f_r=\psi_1 g_r=\psi_1\psi_2 h_r$, and is divisible by $\psi_1\psi_2$. Cancelling ψ_2 throughout in this second set of equations, we obtain a third set of equations of the same forms as the equations (B) in which $f_1, f_2, \dots f_r$ are replaced by $h_1, h_2, \dots h_r$. Since by Lemma B the function ψ_3 is a common factor of h_1 and h_2, these new equations show that ψ_3 is a factor of each one of the functions $h_1, h_2, \dots h_r$, and that therefore f_r is divisible by $\psi_1\psi_2\psi_3$. Proceeding in this way, we see that f_r is divisible by $\psi_1\psi_2\dots\psi_s$, i.e. by ψ.

Now every common factor of f_1 and f_2, when expressed as a product of irreducible factors, must be a product of a certain number of the functions $\psi_1, \psi_2, \dots \psi_s$ and a constant, and must therefore be a factor of ψ. Hence f_r, which is a common factor of f_1 and f_2, must be a factor of ψ.

Since ψ is a factor of f_r, and f_r is a factor of ψ, f_r can only differ from ψ by a constant factor. For if we write $f_r=\lambda\psi$, $\psi=\mu f_r$, where λ and μ are rational integral functions, we have $f_r=\lambda\mu f_r$, $\lambda\mu=1$, and therefore λ and μ are constants.

Thus the conditions (1) and (2) of Lemma D are both satisfied when $h=f_r$; and the condition (3) is also satisfied, as we see by taking Ω to be the domain of all algebraic numbers.

To show that the condition (4) of Lemma D is also satisfied when $h=f_r$, we observe that from the successive equations (B) we can deduce the successive equations

$$u_3u_4\dots u_i \cdot f_i=\lambda_i f_1+\mu_i f_2, \quad (i=3, 4, \dots r), \quad \dots\dots\dots\dots\dots\dots\dots\dots\text{(C)}$$

where $\lambda_3, \lambda_4, \dots \lambda_r$ and $\mu_2, \mu_3, \mu_4, \dots \mu_r$ are the rational integral functions of $x_1, x_2, \dots x_n$ lying in Ω which are defined by

$$\lambda_3=1, \quad \lambda_4=-q_3; \quad \lambda_{i+1}=u_i v_{i-1}\lambda_{i-1}-q_i\lambda_i, \quad (i+1=5, 6, \dots r); \quad \dots\dots\dots(1)$$

$$\mu_2=1, \quad \mu_3=-q_2; \quad \mu_{i+1}=u_i v_{i-1}\mu_{i-1}-q_i\mu_i, \quad (i+1=4, 5, \dots r); \quad \dots\dots\dots(2)$$

so that $$\mu_4=u_3v_2+q_2q_3.$$

The first two of the equations (B) show that (C) is true when $i=3$ or 4. Again if k is any integer such that $k \nless 3$, $k+2 \ngtr r$, and if (C) is true when $i=k$ and $i=k+1$, we see from (B) that

$$u_3u_4\dots u_{k+2}\cdot f_{k+2}=u_3u_4\dots u_{k+1}\cdot u_{k+2}f_{k+2}=u_3u_4\dots u_{k+1}\cdot(v_k f_k-q_{k+1}f_{k+1})$$
$$=u_{k+1}v_k(\lambda_k f_1+\mu_k f_2)-q_{k+1}(\lambda_{k+1}f_1+\mu_{k+1}f_2)=\lambda_{k+2}f_1+\mu_{k+2}f_2;$$

i.e. (C) is true when $i=k+2$. Thus (C) is true generally.

The last of the equations (C), obtained by giving to i the value r, is an equation of the form (A) in which $h=f_r$, $\phi=u_3u_4\ldots u_r$, $P_1=\lambda_r$, $P_2=\mu_r$. Thus all four of the conditions in Lemma D are satisfied when we take h to be f_r.

We have now completely proved Lemma D on the assumptions that Theorems I, II and III are true for functions of the $n-1$ variables $x_2, x_3, \ldots x_n$.

Ex. iv. *Degrees in x_1 of the functions $f_1, f_2, \ldots f_r$ and $q_2, q_3, \ldots q_r$.*

We will denote the degrees of $f_1, f_2, \ldots f_r$ in x_1 by $\rho_1, \rho_2, \ldots \rho_r$. These are a series of positive integers, each one after the second being less than the preceding, and the second being either less than or equal to the first. Unless f_1 and f_2 are both constants only the last number ρ_r can have the value 0.

The degrees in x_1 of the functions $q_2, q_3, \ldots q_r$ can be seen from the equations (B). Since the degree in x_1 of f_{i+1} is less than that of f_{i-1}, the equation $v_{i-1}f_{i-1}=q_if_i+u_{i+1}f_{i+1}$ shows that q_if_i has the same degree in x_1 as f_{i-1}.

Hence the degree of q_i in x_1 is $\rho_{i-1}-\rho_i$, and the degrees of $q_2, q_3, \ldots q_r$ in x_1 are positive integers, of which no one except the first can have the value 0.

Ex. v. *Degrees in x_1 of the functions $\lambda_3, \lambda_4, \ldots \lambda_r$ and $\mu_3, \mu_4, \ldots \mu_r$.*

The degrees of $\lambda_3, \lambda_4, \ldots \lambda_r$ in x_1 can be seen from the equations (1).

If $r>3$, it follows from the first two of these equations that the degree in x_1 of λ_4 exceeds that of λ_3, being equal to the degree of λ_3q_3.

If $i>3$ and $r>i$, it follows from the equation $\lambda_{i+1}=u_iv_{i-1}\lambda_{i-1}-q_i\lambda_i$ that if the degree in x_1 of λ_i exceeds that of λ_{i-1}, then the degree in x_1 of λ_{i+1} exceeds that of λ_i, being equal to the degree of λ_iq_i.

From these two results it follows that the degrees in x_1 of $\lambda_3, \lambda_4, \ldots \lambda_r$ are constantly increasing quantities, the degree of λ_{i+1} being always equal to the degree of λ_iq_i.

From the equations (2) we see in a similar way that the degrees in x_1 of $\mu_3, \mu_4, \ldots \mu_r$ are constantly increasing quantities, the degree of μ_{i+1} being always equal to the degree of μ_iq_i.

If now we denote the degree of λ_{i+1} in x_1 by σ_{i+1}, we have

$$\sigma_{i+1}-\sigma_i=\rho_{i-1}-\rho_i, \quad \sigma_i-\sigma_{i-1}=\rho_{i-2}-\rho_{i-1}, \ \ldots \ \sigma_4-\sigma_3=\rho_2-\rho_3, \quad \sigma_3=0,$$

and therefore $$\sigma_{i+1}=\rho_2-\rho_i.$$

And if we denote the degree of μ_{i+1} in x_1 by σ_{i+1}, we have

$$\sigma_{i+1}-\sigma_i=\rho_{i-1}-\rho_i, \quad \sigma_i-\sigma_{i-1}=\rho_{i-2}-\rho_{i-1}, \ \ldots \ \sigma_3-\sigma_2=\rho_1-\rho_2, \quad \sigma_2=0,$$

and therefore $$\sigma_{i+1}=\rho_1-\rho_i.$$

Thus the degrees of λ_{i+1} and μ_{i+1} in x_1 are respectively $\rho_2-\rho_i$ and $\rho_1-\rho_i$.

Ex. vi. *Degrees in x_1 of the functions P_1 and P_2 in Lemma D.*

If the degrees of f_1, f_2, h in x_1 are ρ_1, ρ_2, ρ, then the functions P_1 and P_2 can always be so chosen that their degrees in x_1 do not exceed $\rho_2-\rho$ and $\rho_1-\rho$ respectively; and when f_1 is not divisible by f_2, they can be so chosen that their degrees in x_1 are less than $\rho_2-\rho$ and $\rho_1-\rho$ respectively.

For we have $\rho=\rho_r$, and when we take P_1 and P_2 to be the functions λ_r and μ_r given by (1) and (2), the degrees of P_1 and P_2 in x_1 are $\rho_2-\rho_{r-1}$ and $\rho_1-\rho_{r-1}$. We always have

$$\rho_2-\rho_{r-1} \not< \rho_2-\rho_r, \quad \rho_1-\rho_{r-1} \not< \rho_1-\rho_r;$$

and if

$$\rho_2-\rho_{r-1}=\rho_2-\rho_r \quad \text{or} \quad \rho_1-\rho_{r-1}=\rho_1-\rho_r,$$

we must have $\rho_r=\rho_{r-1}$, which is only possible when $r=2$, i.e. when f_1 is divisible by f_2.

When f_1 is divisible by f_2, we have $r=2$, $h=f_2$, and we can put $P_1=0$, $P_2=1$, $\phi=1$.

Ex. vii. If f_2 is not divisible by f_1, so that $r>2$, and if f_i is any one of the functions $f_3, f_4, \ldots f_r$, we see from (C) that (when f_1 and f_2 are regular in x_1) there exists an identical equation of the form

$$Q_1 f_1 + Q_2 f_2 = f_i \phi_i,$$

where $\phi_i = u_3 u_4 \ldots u_i$ is a rational integral function of $x_2, x_3, \ldots x_n$ only lying in Ω which does not vanish identically, and where Q_1 and Q_2 are rational integral functions of $x_1, x_2, \ldots x_n$.

If the degrees of f_1, f_2, f_i in x_1 are ρ_1, ρ_2, ρ_i, then the degrees in x_1 of Q_1 and Q_2, when they are the functions λ_i and μ_i of the equation (C), are respectively less than $\rho_2-\rho_i$ and $\rho_1-\rho_i$.

Ex. viii. *When f_1 and f_2 are two rational integral functions in Ω of $x_1, x_2, \ldots x_n$, neither of them vanishing identically, which are both regular in x_1, and which have no irreducible* (*or no irresoluble*) *factor in common, the equation* (A) *assumes the form*

$$P_1 f_1 + P_2 f_2 = \phi, \quad \ldots\ldots\ldots\ldots\ldots\ldots\ldots\ldots\text{(A')}$$

where ϕ is a rational integral function of $x_2, x_3, \ldots x_n$ which does not vanish identically, and where P_1 and P_2 are rational integral functions of $x_1, x_2, \ldots x_n$.

The functions P_1 and P_2 can always be so chosen that their degrees in x_1 do not exceed the degrees of f_2 and f_1 in x_1; and when neither of the functions f_1 and f_2 is divisible by the other, i.e. when neither of them is a mere constant, the functions P_1 and P_2 can be so chosen that their degrees in x_1 are *less* respectively than the degrees of f_2 and f_1 in x_1.

Ex. ix. *Simplified forms of the equations* (B) *in Lemma D.*

The functions $u_3, u_4, \ldots u_r$ and $v_2, v_3, \ldots v_{r-1}$ can be so chosen that the two functions in each of the pairs $(v_2, u_4), (v_3, u_5), \ldots (v_{r-2}, u_r)$ have no irreducible factor in common, in which case by Corollary 3 to Theorem I (for $n-1$ variables) they have no irresoluble factor in common and are prime to one another. For if v_{i-1} and u_{i+1} have an irreducible factor ψ in common, then in the equation $v_{i-1} f_{i-1} = q_i f_i + u_{i+1} f_{i+1}$ the function ψ must be a factor of $q_i f_i$, and therefore by Lemma B it must be a factor of q_i. We can then cancel the factor ψ in v_{i-1}, u_{i+1} and q_i. By a succession of such cancellations we obtain a set of equations still of the form (B) in which the new additional conditions are satisfied.

Ex. x. From the equations (1) and (2) it follows that

$$\lambda_{i+1}\mu_i - \lambda_i\mu_{i+1} = (-1)^i u_3 u_4 \ldots u_i \,.\, v_2 v_3 \ldots v_{i-1}.$$

Ex. xi. Solving for f_1 and f_2 the equation (C) and the equation derived from it by replacing i by $i+1$, and making use of Ex. x, we obtain the identical equations

$$v_2 v_3 \ldots v_{i-1} \,.\, f_1 = (-1)^{i-1} (\mu_{i+1} f_i - \mu_i u_{i+1} f_{i+1}),$$

$$v_2 v_3 \ldots v_{i-1} \,.\, f_2 = (-1)^{i} \quad (\lambda_{i+1} f_i - \lambda_i u_{i+1} f_{i+1}).$$

We are now in a position to give the general proofs by induction of Theorems I—III. We assume that they are true generally for functions of $n-1$ variables, and show that they must then be true generally for functions of the n variables $x_1, x_2, \ldots x_n$.

Proof of Theorem I. Referring to sub-articles 1 and 2 of § 187, we see that Theorem I will be true for the functions f_1 and f_2 of $x_1, x_2, \ldots x_n$ if and only if it is true for the functions g_1 and g_2 of $y_1, y_2, \ldots y_n$ formed by regularising f_1 and f_2 with respect to all or any number of the variables. Hence Theorem I will be true generally if it is true in the particular case when both f_1 and f_2 are regular with respect to x_1. From Lemma D we see that Theorem I is true in this particular case. Therefore Theorem I is true generally.

Proof of Theorem II. Theorem II will be true if and only if it is true in the particular case when f_1, f_2 and g are all regular in x_1. In this particular case we see from Ex. xxiii of § 185 that no one of the functions f_1, f_2 and g is divisible by any rational integral function of $x_2, x_3, \ldots x_n$ only other than a constant. Further if g and f_1 have no irreducible factor in common, we see from Lemma D that the H.C.F. of g and f_1 is a constant, and that there exist identical equations of the forms

$$Pf_1 + Qg = \phi, \quad Pf_1f_2 + Qgf_2 = \phi f_2, \quad \ldots\ldots\ldots\ldots\ldots\ldots\text{(D)}$$

where ϕ is a rational integral function of $x_2, x_3, \ldots x_n$ only lying in Ω which does not vanish identically, and where P and Q are rational integral functions of $x_1, x_2, \ldots x_n$ lying in Ω.

The second of the equations (D) shows that if f_1f_2 is divisible by g, then ϕf_2 is divisible by g, and it follows from Lemma C that f_2 is divisible by g. Thus Theorem II is true in the particular case when f_1, f_2 and g are all regular in x_1, and is therefore true generally.

Proof of Theorem III. Theorem III follows from Theorem II as has been shown in Ex. i. Let g be an irreducible function belonging to the domain Ω in which f_1 and f_2 lie, and let f_1f_2 be divisible by g. Then if f_1 is not divisible by g, the functions f_1 and g have no irreducible factor in common, and therefore by Theorem II the function f_2 is divisible by g. Consequently one at least of the functions f_1 and f_2 is divisible by g.

Ex. xii. *Let $f_1, f_2, \ldots f_m$ be any m rational integral functions of the n variables $x_1, x_2, \ldots x_n$, no one of them vanishing identically, which lie in any domain of rationality Ω and are regular in x_1. Then they have a highest common factor h which lies in Ω and satisfies an equation of the form*

$$P_1f_1 + P_2f_2 + \ldots + P_mf_m = h\phi, \quad \ldots\ldots\ldots\ldots\ldots\ldots\ldots\ldots\text{(E)}$$

where ϕ is a rational integral function of $x_2, x_3, \ldots x_n$ only which lies in Ω and does not vanish identically, and where $P_1, P_2, \ldots P_m$ are rational integral functions of $x_1, x_2, \ldots x_n$ lying in Ω.

Assuming this to be true for any functions less than m in number, we can show that it is true for any m functions.

For then $f_1, f_2, \dots f_{m-1}$ have an H.C.F. k lying in Ω such that

$$q_1 f_1 + q_2 f_2 + \dots + q_{m-1} f_{m-1} = k\psi,$$

and k and f_m have an H.C.F. h lying in Ω such that

$$p_1 k + p_2 f_m = h\chi.$$

Here ψ and χ are rational integral functions of $x_2, x_3, \dots x_n$ only which lie in Ω and do not vanish identically; and $q_1, q_2, \dots q_{m-1}, p_1, p_2$ are rational integral functions of $x_1, x_2, \dots x_n$ lying in Ω.

By Ex. xxvi of § 185 the function h is an H.C.F. of $f_1, f_2, \dots f_m$; and when we eliminate k from the above two equations, we obtain an equation of the form (E) in which

$$P_1 = p_1 q_1, \quad P_2 = p_1 q_2, \ \dots \ P_{m-1} = p_1 q_{m-1}, \quad P_m = p_2\psi, \quad \phi = \psi\chi.$$

Ex. xiii. *If $f_1, f_2, \dots f_m$ are any m rational integral functions of the variables $x_1, x_2, \dots x_n$, no one of them vanishing identically, which lie in any domain of rationality Ω, there always exist rational integral identities in Ω of the form*

$$P_1 f_1 + P_2 f_2 + \dots + P_m f_m = 0. \quad \text{.................................(F)}$$

We obtain an identity of this form when $P_1 = f_2$, $P_2 = -f_1$, $P_3 = P_4 = \dots = P_m = 0$; and also (see Note 3 of § 130 a) when

$$\overline{\underline{P}}_m = [Q]_m^m \overline{\underline{f}}_m, \quad \text{......................................(3)}$$

where $[Q]_m^m$ is any skew-symmetric matrix of order m whose elements are rational integral functions in Ω of the variables $x_1, x_2, \dots x_n$.

It can be shown that for functions f of the most general form an identity of the form (F) is true when and only when the P's are given by some identical equation of the form (3).

§ 189. Miscellaneous properties of rational integral functions.

In the following examples Ω denotes any domain of rationality. The examples are deduced from the results obtained in the foregoing articles.

Ex. i. *Let f_1 and f_2 be two rational integral functions in Ω of the n variables $x_1, x_2, \dots x_n$. Then a necessary and sufficient condition that f_1 and f_2 shall have no finite common root is the existence of an identical equation of the form*

$$P_1 f_1 + P_2 f_2 = 1, \quad \text{..(A)}$$

where P_1 and P_2 are rational integral functions in Ω of the same variables.

There will be no loss of generality in supposing that f_1 and f_2 are both regular in x_1.

First suppose that there exists an identical equation of the form (A).

Then clearly f_1 and f_2 have no finite common root.

Next suppose that the functions f_1 and f_2 have no finite common root.

Then if one of them vanishes identically, the other must be a non-vanishing constant, and we clearly have a relation of the form (A). In the general case when neither f_1 nor f_2 vanishes identically, f_1 and f_2 have no common factor other than a non-vanishing constant, and by Lemma D of § 188 we have

$$P_1 f_1 + P_2 f_2 = \phi, \quad \text{......................................(A')}$$

where P_1, P_2, ϕ are rational integral functions in Ω; ϕ is a function of $x_2, x_3, \ldots x_n$ only which does not vanish identically; and the degrees in x_1 of P_1 and P_2 are less respectively than those of f_2 and f_1.

If ϕ is not merely a constant, we can ascribe such particular finite values to $x_2, x_3, \ldots x_n$ that $\phi=0$. The equation (A′) then becomes

$$P_1(x_1) f_1(x_1) = -P_2(x_1) f_2(x_1), \quad \text{.............................(A″)}$$

where $P_1(x_1)$, $P_2(x_1)$, $f_1(x_1)$, $f_2(x_1)$ are functions of x_1 only. Since the coefficients of the highest powers of x_1 in f_1 and f_2 are non-vanishing constants, the functions $f_1(x_1)$ and $f_2(x_1)$ have the same degrees in x_1 as f_1 and f_2. Consequently in (A″) the degree of $P_2(x_1)$ is less than the degree of $f_1(x_1)$. It follows that $f_1(x_1)$ has a factor in common with $f_2(x_1)$ which is not merely a constant. We can therefore ascribe such a finite value to x_1 that $f_1(x_1)$ and $f_2(x_1)$ both vanish; i.e. f_1 and f_2 have a finite common root, which is contrary to the hypothesis.

We conclude that ϕ in the equation (A′) must be a non-vanishing constant lying in Ω; and therefore (A′) can be put into the form (A).

Ex. ii. *If f_1 and f_2 are rational integral functions in Ω of the n variables $x_1, x_2, \ldots x_n$, there always exist identical equations of the form*

$$P_1 f_1 + P_2 f_2 = 0, \quad \text{...(B)}$$

where P_1 and P_2 are rational integral functions in Ω of the same variables, and do not both vanish identically.

This is obviously the case when both the functions f_1 and f_2 vanish identically; and when they do not both vanish identically, the identical equation $f_2 f_1 - f_1 f_2 = 0$ has the form (B).

When neither of the functions f_1, f_2 vanishes identically, then neither of the functions P_1, P_2 vanishes identically (except when they both vanish identically).

For if $P_2=0$, we have $P_1 f_1=0$, and therefore $P_1=0$.

If neither of the functions f_1, f_2 and neither of the functions P_1, P_2 vanishes identically, and if the functions f_1, f_2 have no irreducible factor in common and also the functions P_1, P_2 have no irreducible factor in common, then the equation (B) *must be*

$$f_2 f_1 - f_1 f_2 = 0. \quad \text{......................................(B′)}$$

For the equation $P_1 f_1 = -P_2 f_2$ shows that P_1 and f_2 have the same irreducible factors, and that P_2 and f_1 have the same irreducible factors; and these two sets of irreducible factors have no irreducible factor in common. Therefore by Corollary 2 to Theorem III in § 188 the functions P_1 and f_2 differ by a constant factor only, and the functions P_2 and f_1 differ by a constant factor only. It follows that we must have $P_1 = cf_2$, $P_2 = -cf_1$, where c is a non-vanishing constant lying in Ω.

Ex. iii. *Let f_1 and f_2 be rational integral functions of $x_1, x_2, \ldots x_n$, neither of which vanishes identically; let h be their* H.C.F.; *and let the degrees of f_1, f_2, h in all the variables (or in any one variable x_1) be ρ_1, ρ_2, ρ. Then in every identical relation of the form* (B) *in which P_1 and P_2 do not vanish identically, the degrees of P_1 and P_2 in all the variables (or in x_1) cannot be less than $\rho_2 - \rho$ and $\rho_1 - \rho$ respectively.*

Let H be the H.C.F. of P_1 and P_2. Then we can write

$$f_1 = hg_1, \quad f_2 = hg_2; \quad P_1 = HQ_1, \quad P_2 = HQ_2,$$

where g_1 and g_2 have no irreducible factor in common, and Q_1 and Q_2 have no irreducible factor in common; and we have

$$Q_1 g_1 + Q_2 g_2 = 0.$$

It follows from Ex. ii that $Q_1 = cg_2$, $Q_2 = -cg_1$, where c is a non-vanishing constant. Thus Q_1 and Q_2 have the same degrees $\rho_2 - \rho$ and $\rho_1 - \rho$ as g_2 and g_1; and the degrees of P_1 and P_2 cannot be less than the degrees of Q_1 and Q_2.

In particular if f_1 and f_2 have no irreducible (*or no irresoluble*) *factor in common, the degrees of P_1 and P_2 cannot be less than the degrees of f_2 and f_1 respectively.*

Ex. iv. *Let f_1 and f_2 be rational integral functions of $x_1, x_2, \dots x_n$ which have no irresoluble factor in common. Then if there exists an identical equation of the form* (B) *in which P_1 and P_2 do not both vanish identically, and if either P_1 has a lower degree than f_2 or P_2 has a lower degree than f_1 in all the variables or in any one of the variables, at least one of the functions f_1 and f_2 must vanish identically.*

This follows from Ex. iii.

Ex. v. *Let g be an irreducible rational integral function in Ω of the variables $x_1, x_2, \dots x_n$, and let f be any rational integral function in Ω of the same variables. Then g is a factor of f when and only when f vanishes for all finite roots of g.*

If g is a factor of f, then clearly f vanishes for all finite roots of g. We have therefore only to prove that if f vanishes for all finite roots of g, it must be divisible by g.

If f vanishes identically, this is obviously true. We may therefore suppose that f does not vanish identically; and there will be no loss of generality in assuming that f and g are both regular in x_1.

Then if f is not divisible by g, the H.C.F. of f and g is a non-vanishing constant, and we have

$$Pf + Qg = \phi,$$

where P, Q, ϕ are rational integral functions in Ω, and where ϕ is a function of $x_2, x_3, \dots x_n$ only which does not vanish identically. In this case we can (see Ex. xxi of § 185) choose the values of $x_2, x_3, \dots x_n$ so that ϕ does not vanish, and the coefficient of the highest power of x_1 in g does not vanish, and we can further choose the value of x_1 so that $g = 0$. We then have

$$\phi \neq 0, \quad g = 0, \quad Pf \neq 0, \quad f \neq 0.$$

Thus if f is not divisible by g, then f does not vanish for every finite root of g; and therefore if f vanishes for all finite roots of g, it must be divisible by g.

From Corollary 2 to Theorem I in § 188 we deduce the following result:

If g is irreducible in Ω but resoluble, and if f (*which lies in* Ω) *vanishes for all finite roots of any factor of g, then it must be divisible by g.*

Ex. vi. *Let f be a homogeneous rational integral function in Ω of the variables $x_1, x_2, \dots x_n$, and let g be a non-homogeneous irreducible rational integral function in Ω of the same variables. Then if f vanishes for all finite roots of g, it must vanish identically.*

For if f does not vanish identically, then by Ex. v it must have the non-homogeneous factor g, and by Ex. xvii of § 185 this is impossible.

Particular cases of this theorem are given in Ex. xviii of § 185 and Ex. x of § 187.

Ex. vii. *Let g be an irreducible rational integral function in Ω of the variables $x_1, x_2, \ldots x_n$; and let $g_1, g_2, \ldots g_m$ be any number of irreducible functions in Ω of the same variables, each of which is distinct from g (and does not vanish identically). Then there always exist finite roots of g for which no one of the functions $g_1, g_2, \ldots g_m$ vanishes.*

For if the product $g_1 g_2 \ldots g_m$ vanished for all finite roots of g, then by Ex. v the function g would be a factor of that product. Therefore by Corollary 1 to Theorem III of § 188, g would be a factor of one of the functions $g_1, g_2, \ldots g_m$, which is impossible; or by Corollary 2 to the same theorem, g would be one of the functions $g_1, g_2, \ldots g_m$, which is impossible.

Ex. viii. *Let F be a rational integral function in Ω of the variables $x_1, x_2, \ldots x_n$; let g be an irreducible function in Ω of the same variables; and let $g_1, g_2, \ldots g_m$ be any number of irreducible functions in Ω of the same variables, each of which is distinct from g. Then g is a factor of F when and only when F vanishes for all those finite roots of g for which no one of the functions $g_1, g_2, \ldots g_m$ vanishes.*

This theorem is obviously true when F vanishes identically. We will therefore suppose that F does not vanish identically.

Let the distinct irreducible factors of F be $f_1, f_2, \ldots f_r$.

If g is not a factor of F, then every one of the functions $f_1, f_2, \ldots f_r, g_1, g_2, \ldots g_m$ is distinct from g, and therefore by Ex. v there exist finite roots of g for which

$$f_1 f_2 \ldots f_r g_1 g_2 \ldots g_m \neq 0,$$

or for which $F \neq 0$ and $g_1 g_2 \ldots g_m \neq 0$. Consequently F does not vanish for all those finite roots of g for which $g_1 g_2 \ldots g_m \neq 0$.

And if g is a factor of F, then F vanishes for all finite roots of g, and therefore vanishes for all those finite roots of g for which $g_1 g_2 \ldots g_m \neq 0$.

Ex. ix. *Let g be an irreducible rational integral function in Ω of the variables $x_1, x_2, \ldots x_n$; and let $F_1, F_2, \ldots F_m$ be any rational integral functions in Ω of the same variables, no one of which vanishes identically, and no one of which is divisible by g. Then there always exist finite roots of g for which no one of the functions $F_1, F_2, \ldots F_m$ vanishes.*

This theorem is obviously true when $F_1, F_2, \ldots F_m$ are (non-vanishing) constants. When they are not all constants, let the various distinct irreducible factors of all these functions be $g_1, g_2, \ldots g_r$. By Ex. vii there exist finite roots of g for which no one of the functions $g_1, g_2, \ldots g_r$ vanishes; and for such roots no one of the functions $F_1, F_2, \ldots F_m$ vanishes; for each of these functions is the product of a non-vanishing constant and certain powers of $g_1, g_2, \ldots g_r$.

This theorem remains true when g is any irresoluble rational integral function, as we see by taking Ω to be the domain of all scalar numbers.

Ex. x. *If in Ex.* ix *F is any other rational integral function in Ω of the same variables, then g is a factor of F when and only when F vanishes for all those finite roots of g for which no one of the functions $F_1, F_2, \ldots F_m$ vanishes.*

Ex. xi. *An irreducible rational integral function of several variables cannot have a repeated irresoluble factor.*

Let Ω be any domain of rationality, let f be any rational integral function in Ω of the n variables $x_1, x_2, \ldots x_n$, and let f be expressed as a product of its distinct irresoluble factors $t_1, t_2, t_3, \ldots$ in the form

$$f = t_1^{\alpha} t_2^{\beta} t_3^{\gamma} \ldots,$$

where $\alpha, \beta, \gamma, \ldots$ are positive integers not less than 1. We may suppose that $x_1, x_2, \ldots x_n$ all actually occur in f, and we will write $f_i = \frac{\partial f}{\partial x_i}$ for the values 1, 2, ... n of i.

Since f_i is certainly divisible by $t_1^{\alpha-1}$ when t_1 contains x_i, and by t_1^α when t_1 does not contain x_i, we see that $f_1, f_2, \ldots f_n$ are all divisible by $t_1^{\alpha-1}$.

If t_1 contains x_i, then f_i is divisible by $t_1^{\alpha-1}$ but not by t_1^α; and since t_1 contains at least one of the variables $x_1, x_2, \ldots x_n$, we conclude that at least one of the functions $f_1, f_2, \ldots f_n$ is not divisible by t_1^α.

From these two results it follows that $t_1^{\alpha-1}$ is the highest power of t_1 which is a common factor of $f, f_1, f_2, \ldots f_n$; and applying the same argument to $t_2, t_3, \ldots$ we conclude that

$$h = t_1^{\alpha-1} t_2^{\beta-1} t_3^{\gamma-1} \ldots$$

is a highest common factor of $f, f_1, f_2, \ldots f_n$, and lies in Ω. Hence if

$$g = t_1 t_2 t_3 \ldots,$$

we have $f = gh$; and g and h are both rational integral functions in Ω.

Now if $\alpha > 1$, i.e. if t_1 is a repeated irresoluble factor of f, then both g and h are divisible by t_1; therefore neither of them is merely a constant, and f is reducible in Ω. We conclude that if f has any repeated irresoluble factor, it cannot be irreducible in Ω.

Ex. xii. Let ϕ and ψ be rational integral functions of several variables which do not vanish identically, and let

$$f = \frac{\phi}{\psi}.$$

Then if f is finite for all finite roots of ψ, it must be a rational integral function, i.e. ϕ must be divisible by ψ; and if f is finite and different from 0 for all finite roots of ϕ and ψ, it must be a finite non-zero constant.

This is evident when we express ϕ and ψ as products of their irresoluble factors.

§ 190. Extension of a domain by adjunction.

1. *Adjunction of any scalar number q.*

Let Ω be any domain of rationality, and let q be a scalar number which does not lie in Ω. Then if Ω' is the smallest possible domain of rationality which contains q as well as all the elements of Ω, we say that Ω' is formed from Ω by the adjunction of q, and we will denote it by (Ω, q). Clearly the elements of (Ω, q) consist of all numbers of the form

$$e = \frac{\phi(q)}{\psi(q)}, \quad \ldots\ldots\ldots\ldots(1)$$

where $\phi(x)$ and $\psi(x)$ are rational integral functions in Ω of the variable x, and $\psi(q) \neq 0$, or in other words:

The elements of (Ω, q) consist of all rational functions in Ω of the number q.

If $\alpha, \beta, \gamma, \ldots$ are several different numbers not lying in Ω, and if we denote by $(\Omega, \alpha, \beta, \gamma, \ldots)$ the extended domain formed from Ω by the successive adjunction of $\alpha, \beta, \gamma, \ldots$, then clearly:

The elements of $(\Omega, \alpha, \beta, \gamma, \ldots)$ consist of all rational functions in Ω of the numbers $\alpha, \beta, \gamma, \ldots$; and $(\Omega, \alpha, \beta, \gamma, \ldots)$ is independent of the order of arrangement of $\alpha, \beta, \gamma, \ldots$.

Ex. i. If q is a number lying in Ω, the domain (Ω, q) is identical with the domain Ω.

Ex. ii. If Ω and q both lie in a domain Ω', then (Ω, q) lies in Ω'.

Ex. iii. The unrestricted domain is formed from the domain of all real numbers by the adjunction of $\sqrt{-1}$.

2. *Adjunction of a number q immediately over Ω.*

A *number immediately over* Ω is one which is a root of a rational integral function (or equation) in Ω, but does not lie in Ω. Thus super-rational numbers, as defined in Note 3 of § 184, are numbers immediately over the domain Ω_1 of all rational numbers. When q is a number immediately over Ω, we can suppose it to be a root of an equation

$$f(x)=0, \qquad \text{...(2)}$$

where $f(x)$ is an irreducible rational integral function in Ω of some finite degree p. Since (see Corollary 2 to Theorem I of § 188) two irreducible functions in Ω of the single variable x which both have q as a root must be divisible by one another, the function $f(x)$ is uniquely determinate except for an arbitrary finite non-zero numerical factor, and q cannot satisfy any rational integral equation in Ω of degree lower than p.

Let e be any element of (Ω, q), and let it be given by (1). If $\psi(x)$ were not prime to $\phi(x)$, it would be divisible by $f(x)$, and we should have $\psi(q)=0$, which is not the case. Consequently the functions $f(x)$ and $\psi(x)$ are prime to one another, and therefore there exists an identity of the form

$$h(x).\psi(x)+k(x).f(x)=1;$$

where $h(x)$ and $k(x)$ are rational integral functions in Ω; and by putting $x=q$ we see that

$$\frac{1}{\psi(q)}=h(q), \quad e=h(q).\psi(q),$$

so that e is now a rational *integral* function of q in Ω.

Thus in this case the elements of (Ω, q) consist of all rational integral functions in Ω of the number q.

Accordingly the general formula for an element of (Ω, q) is now

$$e=\chi(q), \qquad \text{...(1')}$$

where $\chi(x)$ is any rational integral function in Ω of the variable x. When we divide $\chi(x)$ by $f(x)$, we obtain an identity of the form

$$\chi(x)=f(x).Q(x)+R(x),$$

where $Q(x)$ and $R(x)$ are rational integral functions in Ω, the degree of $R(x)$ being not greater than $p-1$; and when in this identity we put $x=q$, we see that $e=R(q)$. We conclude that:

The elements of (Ω, q) consist of all numbers of the form

$$e=a_0+a_1q+a_2q^2+\ldots+a_{p-1}q^{p-1}, \qquad \text{..............................(3)}$$

where $a_0, a_1, a_2, \ldots a_{p-1}$ are any numbers lying in Ω.

The number e given by (3) is 0 when and only when the coefficients $a_0, a_1, a_2, \ldots a_{p-1}$ all vanish; as otherwise q would satisfy a rational integral equation in Ω of degree lower than p. Consequently if

$$e'=b_0+b_1q+b_2q^2+\ldots+b_{p-1}q^{p-1}$$

is also an element of (Ω, q), we have $e'=e$ when and only when

$$b_0=a_0, \quad b_1=a_1, \quad b_2=a_2, \ \dots \ b_{p-1}=a_{p-1}.$$

If $e=\chi(q)$, as in (1′), is any element of (Ω, q), and if $q, q_1, q_2, \dots q_{p-1}$ are the p roots of the equation (2), then e is a root of the equation

$$\{x-\chi(q)\}\{x-\chi(q_1)\}\{x-\chi(q_2)\}\dots\{x-\chi(q_{p-1})\}=0.$$

Because the coefficients of the various powers of x on the left are all symmetric functions of the roots of (2), this is a rational integral equation in Ω of degree p; and we conclude that:

If q is a number immediately over Ω, then every element of the domain (Ω, q) is a number in or immediately over Ω; and if the irreducible function in Ω of which q is a root has degree p, then every element of (Ω, q) is a root of a rational integral function in Ω of degree p, which may of course be reducible.

Under these circumstances we call (Ω, q) a *domain of degree p* immediately over Ω.

Ex. iv. The domain of all scalar numbers has no number over it.

3. *Adjunction of several numbers immediately over Ω.*

If $\alpha, \beta, \gamma, \dots$ are several different numbers not lying in Ω which are roots of rational integral functions (or equations) in Ω, we see from sub-article 2 that the elements of $(\Omega, \alpha, \beta, \gamma, \dots)$ consist of all rational integral functions in Ω of the numbers $\alpha, \beta, \gamma, \dots$. We will now prove the following theorem:

Theorem. *If $\alpha, \beta, \gamma, \dots$ are r different numbers immediately over Ω, and if Ω' is the domain formed from Ω by the adjunction of $\alpha, \beta, \gamma, \dots$, then we can form Ω' from Ω by the adjunction of a single number q immediately over Ω, q being a root of an irreducible function in Ω of finite degree.*

Let the irreducible equations in Ω satisfied by $\alpha, \beta, \gamma, \dots$ respectively be

$$f_1(x)=0, \quad f_2(x)=0, \quad f_3(x)=0, \ \dots \ f_r(x)=0; \qquad \dots\dots(4)$$

where $f_1, f_2, f_3, \dots f_r$ are irreducible rational integral functions in Ω of finite degrees $p_1, p_2, p_3, \dots p_r$; and let

$$q=a\alpha+b\beta+c\gamma+\dots, \qquad \dots\dots(5)$$

where $a, b, c, \dots$ are arbitrary numbers lying in Ω. Also let

$$F(x)=(x-q)(x-q')(x-q'')\dots, \qquad \dots\dots(6)$$

where $q, q', q'', \dots$ are the $p_1p_2\dots p_r$ numbers obtained from q when α is either unaltered or replaced by another root of $f_1(x)$, β is either unaltered or replaced by another root of $f_2(x)$, and so on. Since $F(x)$ is a function of x in which the coefficients are symmetric in the roots of every one of the functions f, it is a rational integral function of x in Ω of degree $p_1p_2\dots p_r$; and because q satisfies the equation $F(x)=0$, it is a number immediately over Ω, and there must exist an *irreducible* function in Ω of finite degree which has q as a root.

Now by Ex. xiii of § 186 or Ex. xi of § 189 no one of the equations (4) has a repeated root. Therefore when we regard $q, q', q'', \dots$ as homogeneous linear functions of $a, b, c, \dots$, no two of them are identically equal, and no one of the differences $q-q', q-q'', q'-q'', \dots$ vanishes identically. Consequently we can assign such particular values to $a, b, c, \dots$ that $q, q', q'', \dots$ are all unequal. This having been done, let e be any number in Ω', i.e. any rational integral function in Ω of $\alpha, \beta, \gamma, \ldots$; let the substitutions which convert q into $q', q'', \ldots$ convert e into $e', e'', \ldots$; and let

$$G(x) = F(x) \cdot \left\{\frac{e}{x-q} + \frac{e'}{x-q'} + \frac{e''}{x-q''} + \ldots\right\} \ldots\ldots\ldots\ldots\ldots\ldots\ldots(7)$$

Since $G(x)$ is a rational integral function in Ω of x and the roots of the equations (4) which is symmetric in the roots of every one of those equations, therefore $G(x)$ as well as $F(x)$ is a rational integral function in Ω of the variable x. Putting $x=q$ in (7), we see that

$$e = \frac{G(q)}{F'(q)},$$

where $F'(x)$ is the first derivative of $F(x)$; and because $F'(q) \neq 0$, it follows that e (which is any number in Ω') is a rational function in Ω of the number q, i.e. it is a number lying in the domain (Ω, q). Conversely every number lying in (Ω, q) clearly lies in Ω'. Consequently Ω' is identical with the domain (Ω, q) formed by the adjunction of q to Ω, when the numbers $a, b, c, \ldots$ in (5) are suitably chosen.

A domain *which contains* Ω will be called a *domain immediately over* Ω when all its elements are numbers in or immediately over Ω, i.e. when every number in it is a root of a rational integral function in Ω. From the above theorem it follows that such a domain is one which can be formed from Ω by the adjunction of numbers immediately over Ω. A domain immediately over Ω_1 is an algebraic domain, and every algebraic domain can be formed by the adjunction of algebraic numbers to Ω_1.

Ex. v. All numbers in and immediately over Ω form a domain the elements of which are the roots of all rational integral functions in Ω. In particular all numbers in and immediately over Ω_1 form the domain of all algebraic numbers.

Ex. vi. Let $a, b, c, \ldots$ and $a', b', c', \ldots$ be two finite sets of algebraic numbers; let Ω be any domain of rationality in which $a, b, c, \ldots$ all lie; and let Ω' be the smallest domain of rationality which contains Ω and also all the numbers $a', b', c', \ldots$.

Then if Ω' is not the same as Ω, we can form it from Ω by the adjunction of a number q, not lying in Ω, which is a root of an irreducible function in Ω of finite degree.

For we can certainly form Ω' by the adjunction to Ω of a finite number of algebraic numbers which, being roots of irreducible functions in Ω_1, are necessarily roots of irreducible functions in Ω.

NOTE. *Domains and numbers over Ω.*

A *number over* Ω is usually understood to be a number immediately over Ω, and a *domain over* Ω to be a domain immediately over Ω.

If however Ω' is a domain immediately over Ω, Ω'' a domain immediately over Ω', Ω''' a domain immediately over Ω'', and so on, we might fitly call all such domains as $\Omega', \Omega'', \Omega''', \ldots$ domains over Ω, a *domain over* Ω being then one which can be formed from Ω by successive adjunctions each of which is the adjunction to a domain of a number immediately over it. When the term 'over' is used in this wider sense, all domains formed from Ω_1 by such successive adjunctions might be called *super-rational domains*, and all domains formed from the domain of all algebraic numbers by such successive adjunctions might be called *super-algebraic domains*; and *super-rational* and *super-algebraic numbers* would have corresponding wider meanings.

CHAPTER XXI

RESULTANTS AND ELIMINANTS OF RATIONAL INTEGRAL FUNCTIONS

[In §§ 191 and 192 precise definitions are given of the roots of a rational integral function of any number of variables; of infinite roots; of repeated roots; of general, special and particular functions; of the weights of the coefficients of a function; and of ordinary values of the variables or coefficients of a function. The next two articles deal with the resultants of rational integral functions; in § 193 we summarise in the form of two theorems the fundamental properties of a resultant of general functions; and in § 194 we consider the resultants of specialised functions. The remaining two articles deal with the common roots and the eliminants of rational integral functions; in § 195 we prove certain lemmas regarding the existence of common roots in which partial resultants and eliminants are used; and in § 196 we summarise in the form of two theorems the fundamental properties of a complete eliminant of general functions, from which the exact number of common roots is deduced. Some attention is devoted to the discriminant of a single function in § 194, and to the discriminant of a set of functions in § 196. The final proof of the theorems of §§ 193 and 196, based on the properties of symmetric functions, is given in § 200 at the end of the next chapter.]

§ 191. Roots of rational integral functions and equations.

The roots of any rational integral function f and the roots of the corresponding rational integral equation $f = 0$ will be defined to be identical. We shall usually speak of them as the roots of the function f.

1. *Roots of a homogeneous rational integral function.*

Let $f(x_1, x_2, \ldots x_n)$ be a homogeneous rational integral function of the n variables $x_1, x_2, \ldots x_n$ of non-zero degree r whose coefficients are *finite* numerical constants which are *not all zero.* We will define its roots with greater precision than before. It will be convenient to denote them by matrices.

The function f always has the *zero root*

$$[x_1 \, x_2 \ldots x_n] = 0. \qquad \text{(a)}$$

If $\alpha_1, \alpha_2, \ldots \alpha_n$ are finite numbers which are not all zero, the function f has a *finite non-zero* root given by $[x_1 \, x_2 \ldots x_n] = [\alpha_1 \, \alpha_2 \ldots \alpha_n]$ if and only if $f(\alpha_1, \alpha_2, \ldots \alpha_n) = 0$. When this condition is satisfied, it has finite non-zero roots given by $[x_1 \, x_2 \ldots x_n] = k\,[\alpha_1 \, \alpha_2 \ldots \alpha_n]$, where k is any finite non-zero number. We shall usually make no distinction between these roots and say that f has a finite non-zero root given by

$$[x_1 \, x_2 \ldots x_n] \equiv [\alpha_1 \, \alpha_2 \ldots \alpha_n] \qquad \text{(b)}$$

if and only if $f(\alpha_1, \alpha_2, \ldots \alpha_n) = 0$, the sign $\equiv$ indicating as usual that the matrices on the two sides of (b) differ only by a finite non-zero scalar factor.

If $[a_1 a_2 \dots a_n]$ are numbers which are not all finite, we can put

$$[a_1 a_2 \dots a_n] = \rho\,[\alpha_1 \alpha_2 \dots \alpha_n],$$

where ρ is an infinite scalar number and $\alpha_1, \alpha_2, \dots \alpha_n$ are all finite but not all zero; and we shall then say that f has an *infinite root* given by

$$[x_1 x_2 \dots x_n] = [a_1 a_2 \dots a_n] = \rho\,[\alpha_1 \alpha_2 \dots \alpha_n]$$

if and only if $f(\alpha_1, \alpha_2, \dots \alpha_n) = 0$. We shall however usually identify this root with that given by (b); and when this is done, the distinction between infinite roots and finite non-zero roots disappears, and all infinite roots can be left out of account.

Accordingly we define the roots of the function f or the equation $f = 0$ to be

(1) *the zero root given by* (a);

(2) *the non-zero roots given by congruences of the form* (b) *in which the α's are all finite but not all zero, and $f(\alpha_1, \alpha_2, \dots \alpha_n) = 0$.* ....(A)

A root given by a congruence of the form (b) is supposed to be known when the ratios $\alpha_1 : \alpha_2 : \dots : \alpha_n$ are known. Every finite root in which $x_n \neq 0$ may be supposed to be given by a congruence of the form

$$[x_1 x_2 \dots x_{n-1} x_n] \equiv [\alpha_1 \alpha_2 \dots \alpha_{n-1}\, 1],$$

where $\alpha_1, \alpha_2, \dots \alpha_{n-1}$ are finite numbers, every one of which may be 0, and $f(\alpha_1, \alpha_2, \dots \alpha_{n-1}, 1) = 0$.

NOTE 1. *Repeated roots of the homogeneous function f.*

Repeated roots are best regarded in the first instance from a geometrical standpoint. We can regard the roots of the function f or the equation $f=0$ as the 'points' of a 'surface f' defined by the equation $f=0$, a point A defined by the congruence $\underset{n}{\boxed{x}} \equiv \underset{n}{\boxed{a}}$ lying on the surface f if and only if the function f has a corresponding root a defined by the equivalent congruence $[x_1 x_2 \dots x_n] \equiv [a_1 a_2 \dots a_n]$. It is shown below that every straight line of homogeneous $(n-1)$-way space has exactly r points of intersection with the surface f of degree r. If i of these points coincide with A for *every* straight line drawn through A, we call a a root of f *repeated i times*; and if i is the greatest integer for which this is true, we say that a is a root of f *repeated exactly i times*, and call A an i-fold point of the surface f.

Let A and B be any two distinct points defined by the congruences $\underset{n}{\boxed{x}} \equiv \underset{n}{\boxed{a}}$, $\underset{n}{\boxed{x}} \equiv \underset{n}{\boxed{\beta}}$. Then the points of the straight line AB are given by the explicit congruence

$$\begin{bmatrix} x_1 \\ x_2 \\ \vdots \\ x_n \end{bmatrix} \equiv \begin{bmatrix} a_1 & \beta_1 \\ a_2 & \beta_2 \\ \dots & \dots \\ a_n & \beta_n \end{bmatrix} \begin{bmatrix} \lambda \\ \mu \end{bmatrix} \equiv \lambda \begin{bmatrix} a_1 \\ a_2 \\ \vdots \\ a_n \end{bmatrix} + \mu \begin{bmatrix} \beta_1 \\ \beta_2 \\ \vdots \\ \beta_n \end{bmatrix}, \qquad \text{.................(1)}$$

where λ and μ are arbitrary *finite* parameters which are not both zero. There is one and only one point on the straight line corresponding to each distinct non-zero value of $[\lambda\,\mu]$, the point A being given by $\mu=0$ or $[\lambda, \mu]\equiv[1, 0]$, and the point B by $\lambda=0$ or $[\lambda, \mu]=[0, 1]$. The points of intersection of the straight line with the surface f are given by the non-zero roots $[\lambda\ \mu]$ of the function

$$g(\lambda, \mu)=f(\lambda a_1+\mu\beta_1, \lambda a_2+\mu\beta_2, \dots \lambda a_n+\mu\beta_n) \quad\dots\dots(2)$$

$$=\lambda^r . f(a_1, a_2, \dots a_n)+\lambda^{r-1}\mu . \Delta f(a_1, a_2, \dots a_n)+\frac{1}{2!}\lambda^{r-2}\mu^2 . \Delta^2 f(a_1, a_2, \dots a_n)+\dots$$

$$+\lambda^{r-i}\mu^i . \Delta^i f(a_1, a_2, \dots a_n)+\dots+\mu^r . f(\beta_1, \beta_2, \dots \beta_n), \quad\dots(3)$$

where $$\Delta=\beta_1\frac{\partial}{\partial a_1}+\beta_2\frac{\partial}{\partial a_2}+\dots+\beta_n\frac{\partial}{\partial a_n}.$$

We can express $g(\lambda, \mu)$ as a product of linear factors in the form

$$g(\lambda, \mu)=(\lambda\mu_1-\mu\lambda_1)(\lambda\mu_2-\mu\lambda_2)\dots(\lambda\mu_r-\mu\lambda_r), \quad\dots\dots(4)$$

and identify the roots of g with the roots $[\lambda\ \mu]\equiv[\lambda_1\ \mu_1]$, $[\lambda_2\ \mu_2]$, ... $[\lambda_r\ \mu_r]$ of its linear factors; then $g(\lambda, \mu)$ has always exactly r non-zero roots, and therefore the straight line AB always intersects the surface f in exactly r points. There will be i of these points which coincide with A if i of the roots of g are given by $[\lambda\ \mu]\equiv[1\ 0]$, i.e. if i of the factors in (4) are equal to μ, i.e. if $g(\lambda, \mu)$ is divisible by μ^i, i.e. if the first $i-1$ terms of the expansion (3) vanish. This will be the case for every straight line drawn through A if and only if the first $i-1$ terms of the expansion (3) vanish identically when $\beta_1, \beta_2, \dots \beta_n$ are variables. If we understand a *repeated root* to be one which is repeated more than once, and make use of Euler's theorems regarding homogeneous functions, we conclude that:

(1) *The function f has a repeated root given by $[x_1\ x_2 \dots x_n]\equiv[a_1\ a_2 \dots a_n]$ when and only when the n first derivatives of f, viz. $\frac{\partial f}{\partial x_1}, \frac{\partial f}{\partial x_2}, \dots \frac{\partial f}{\partial x_n}$, all vanish for the values $a_1, a_2, \dots a_n$ of $x_1, x_2, \dots x_n$.*

(2) *It has a root $[x_1\ x_2 \dots x_n]\equiv[a_1\ a_2 \dots a_n]$ repeated i times when and only when all the $\binom{n+i-2}{i-1}$ derivatives of f of order $i-1$ vanish for the values $a_1, a_2, \dots a_n$ of $x_1, x_2, \dots x_n$; and this root is repeated exactly i times when and only when the derivatives of f of order i do not also all vanish for those values of $x_1, x_2, \dots x_n$.*

(3) *The function f of degree r cannot have a root repeated more than r times unless it vanishes identically.*

Strictly speaking 'root' must here mean 'non-zero root,' but this restriction can be removed by taking the first two of the above results as analytical definitions. It will be observed that, because f is homogeneous, the vanishing of all the derivatives of f of order $i-1$ is a necessary and sufficient condition for the vanishing of f and all its derivatives of orders 1, 2, ... $i-1$. The third result follows from the fact that the derivatives of f of order r are the coefficients of f multiplied by non-zero numerical constants.

In the equations (1), (2), (3), (4) we can replace $[\lambda\ \mu]$ by $[1\ \rho]$ where ρ is an arbitrary parameter which is *not necessarily finite*, $\rho=0$ giving the point A, and $\rho=\infty$ giving the point B. Then $[x_1\ x_2 \dots x_n]\equiv[a_1\ a_2 \dots a_n]$ is a root repeated i times if the expansion (3) in powers of ρ contains only terms divisible by ρ^i.

If A is a point for which $x_n=a_n\neq 0$, we can always find one and only one point on the straight line AB for which $x_n=0$, and take that for the point B. Hence in this case we can without loss of generality suppose that in (1) we have $a_n=1$, $\beta_n=0$.

The resultant of the n first derivatives $\frac{\partial f}{\partial x_1}, \frac{\partial f}{\partial x_2}, \dots \frac{\partial f}{\partial x_n}$ is called (see § 193. 3) the *discriminant* of f. Its vanishing is the necessary and sufficient condition that f shall have at least one repeated non-zero root.

It should be observed that i consecutive or coincident roots do not ordinarily form a root repeated i times.

Ex. i. The function f has the root $[x_1 x_2 \dots x_n] \equiv [a_1 a_2 \dots a_n]$ repeated i times, where $i \not> r$, when and only when the expansion of f in powers of $x_1 - a_1, x_2 - a_2, \dots x_n - a_n$ contains no term having total degree less than i in those quantities, and has therefore the form

$$f = X_i + X_{i+1} + \dots + X_r,$$

where X_s is a homogeneous rational integral function of $x_1 - a_1, x_2 - a_2, \dots x_n - a_n$ of degree s. This is the case when and only when $f(a_1 + \rho\xi_1, a_2 + \rho\xi_2, \dots a_n + \rho\xi_n)$ is divisible by ρ^i.

Ex. ii. The zero root $[x_1 x_2 \dots x_n] = 0$ of the function f is repeated exactly r times.

Ex. iii. If $i \not> r$, and if we express f in the form

$$f = U_r + x_n U_{r-1} + \dots + x_n^{r-s} U_s + \dots + x_n^{r-1} U_1 + x_n^r U_0, \quad \dots\dots\dots\dots\dots(5)$$

where U_s is a homogeneous rational integral function of $x_1, x_2, \dots x_{n-1}$ of degree s, then

$$[x_1 x_2 \dots x_{n-1} x_n] \equiv [0\, 0 \dots 0\, 1] \quad \dots\dots\dots\dots\dots\dots\dots\dots(6)$$

is a root of f repeated i times if and only if $U_0, U_1, \dots U_{i-1}$ all vanish identically, so that f has the form

$$f = U_r + x_n U_{r-1} + x_n^2 U_{r-2} + \dots + x_n^{r-i} U_i. \quad \dots\dots\dots\dots\dots(6')$$

The root (6) is repeated exactly i times when and only when f has the form (6′) and U_i does not vanish identically, i.e. when i is the lowest degree which a non-zero term of f can have in $x_1, x_2, \dots x_{n-1}$.

Ex. iv. If the coefficients of f are rational integral functions of an arbitrary parameter t, then in order that f may have a root of the form

$$[x_1 x_2 \dots x_{n-1} x_n] \equiv [ta_1, ta_2, \dots ta_{n-1}, 1] \quad \dots\dots\dots\dots\dots\dots(7)$$

repeated i times, where the a's are given finite constants and $i \not> r$, it is a necessary condition that f itself shall have the form

$$f = U_r + x_n U_{r-1} + \dots + x_n^{r-i} U_i + t \,.\, x_n^{r-i+1} U_{i-1} + \dots + t^{i-1} \,.\, x_n^{r-1} U_1 + t^i \,.\, x_n^r U_0. \quad \dots(7')$$

The coefficients of $U_r, U_{r-1}, \dots U_i$ in (7′) may be any given constants, and (7) will then be a root of (7′) repeated i times if the coefficients of $U_{i-1}, U_{i-2}, \dots U_0$ are suitably determined rational integral functions of t of degree $r - i$.

Ex. v. Let the coefficients of all the U's in (7′) be given finite numbers. Then if t is sufficiently small but not 0, (7) is a root of (7′) when and only when the finite quantities $a_1, a_2, \dots a_{n-1}$ satisfy the non-homogeneous equation of degree i,

$$U_i + U_{i-1} + \dots + U_0 = 0. \quad \dots\dots\dots\dots\dots\dots\dots\dots(8)$$

We conclude that when $n > 2$, the function (7′) has in general an infinite number of roots of the form (7) in which the a's are functions of t which are finite when $t = 0$; and that when $n = 2$ it has not more than i and in general exactly i such roots. For ordinary values of the coefficients of the functions U none of the a's vanish with t.

The same conclusion remains valid when the coefficients of the U's are rational integral functions of t, and none of the U's are divisible by t.

Ex. vi. If t is an arbitrary parameter which is finite and not 0, and if the coefficients of the U's are finite constants, the roots of the function

$$f = U_r + t \cdot x_n U_{r-1} + t^2 \cdot x_n^2 U_{r-2} + \ldots + t^r \cdot x_n^r U_0$$

in which $x_n \neq 0$ are given by

$$[x_1 x_2 \ldots x_{n-1} x_n] \equiv [ty_1, ty_2, \ldots ty_{n-1}, 1],$$

where $[y_1 y_2 \ldots y_{n-1}]$ is any finite root of the non-homogeneous equation

$$U_r + U_{r-1} + \ldots + U_0 = 0;$$

and the roots of f in which $x_n = 0$ are given by the homogeneous equation $U_r = 0$.

Ex. vii. The homogeneous function f given by (5) has a non-zero root in which $x_n = 0$, i.e. one of the form

$$[x_1 x_2 \ldots x_{n-1} x_n] \equiv [a_1 a_2 \ldots a_{n-1} 0], \quad \text{.............................(9)}$$

where the a's are finite, when and only when $U_r(a_1, a_2, \ldots a_{n-1}) = 0$; and if $i \not> r$, it has the root (9) repeated i times when and only when

all the derivatives of $U_r(a_1, a_2, \ldots a_{n-1})$ of order $i-1$ vanish,

all the derivatives of $U_{r-1}(a_1, a_2, \ldots a_{n-1})$ of order $i-2$ vanish,

.... ⋮

all the derivatives of $U_{r-i+2}(a_1, a_2, \ldots a_{n-1})$ of order 1 vanish,

$U_{r-i+1}(a_1, a_2, \ldots a_{n-1})$ vanishes.

Equivalent conditions are that $U_{r-i+k}(a_1, a_2, \ldots a_{n-1})$ and all its derivatives of orders $1, 2, \ldots k-1$ must vanish for the values $1, 2, \ldots i$ of k.

Ex. viii. Let

$$f = x_n^s \{U_r + x_n U_{r-1} + x_n^2 U_{r-2} + \ldots + x_n^r U_0\} \quad \text{.....................(10)}$$

be a homogeneous rational integral function of $x_1, x_2, \ldots x_n$ of degree $r+s$ which is divisible by x_n^s, the U's being homogeneous rational integral functions of $x_1, x_2, \ldots x_{n-1}$ whose degrees are indicated by their suffixes. Then if $i \not> r$, this function f has the root (9) repeated $s+i$ times when and only when the conditions of Ex. vii are satisfied. It always has this root repeated at least s times.

In this case the roots of f are the roots of its two factors; and the roots of the factor x_n^s are given by all sequences of the form (9), every one of them being repeated exactly s times.

2. *Roots of a non-homogeneous rational integral function.*

Let $f(x_1, x_2, \ldots x_{n-1})$ be a given non-homogeneous rational integral function of the $n-1$ variables $x_1, x_2, \ldots x_{n-1}$ of non-zero degree r in which the coefficients are finite numerical constants which are not all zero, and let the function obtained when we homogenise f without change of degree by the introduction of an additional variable x_n be

$$\begin{aligned} g &= g(x_1, x_2, \ldots x_{n-1}, x_n) \\ &= U_r + x_n U_{r-1} + \ldots + x_n^{r-1} U_1 + x_n^r U_0, \end{aligned} \quad \text{...........(11')}$$

where $U_s = U_s(x_1, x_2, \ldots x_{n-1})$ is a homogeneous rational integral function of degree s of the $n-1$ variables $x_1, x_2, \ldots x_{n-1}$, so that the given function is

$$f = f(x_1, x_2, \ldots x_{n-1}) = g(x_1, x_2, \ldots x_{n-1}, 1)$$
$$= U_r + U_{r-1} + \ldots + U_1 + U_0. \quad \text{.........................(11)}$$

If $\alpha_1, \alpha_2, \ldots \alpha_{n-1}$ are finite numbers, the function f given by (11) has a *finite root* given by

$$[x_1\, x_2 \ldots x_{n-1}] = [\alpha_1\, \alpha_2 \ldots \alpha_{n-1}] \quad \text{.................... (d)}$$

if and only if $f(\alpha_1, \alpha_2, \ldots \alpha_{n-1}) = g(\alpha_1, \alpha_2, \ldots \alpha_{n-1}, 1) = 0$; and this is the case if and only if the homogenised function g given by (11′) has a finite non-zero root in which $x_n \neq 0$ given by

$$[x_1\, x_2 \ldots x_{n-1}\, x_n] \equiv [\alpha_1\, \alpha_2 \ldots \alpha_{n-1}\, 1]. \quad \text{..................(d′)}$$

Thus in this case the function f has a root given by (d) if and only if finite numbers proportional to $\alpha_1, \alpha_2, \ldots \alpha_{n-1}, 1$ satisfy the equation $g = 0$.

If $\alpha_1, \alpha_2, \ldots \alpha_{n-1}$ are numbers which are not all finite, we can put

$$[a_1\, a_2 \ldots a_{n-1}] = \rho\,[\alpha_1\, \alpha_2 \ldots \alpha_{n-1}], \quad \text{.....................(12)}$$

where ρ is infinite and the α's are all finite but not all zero. Regarding the present case as a limit of the previous case, we will assert that f has an *infinite root* given by

$$[x_1\, x_2 \ldots x_{n-1}] = [a_1\, a_2 \ldots a_{n-1}] \quad \text{........................(13)}$$

if and only if finite numbers proportional to $a_1, a_2, \ldots a_{n-1}, 1$ satisfy the equation $g = 0$, i.e. if and only if g has a finite non-zero root in which $x_n \neq 0$ given by

$$[x_1\, x_2 \ldots x_{n-1}\, x_n] \equiv [\alpha_1 \alpha_2 \ldots \alpha_{n-1}\, 0]; \quad \text{.....................(e′)}$$

and this is so if and only if

$$U_r(\alpha_1, \alpha_2, \ldots \alpha_{n-1}) = 0. \quad \text{........................(14)}$$

NOTE 2. When $a_1, a_2, \ldots a_{n-1}$ are not all finite and are represented as in (12), we see from (11) that we cannot have $f(a_1, a_2, \ldots a_{n-1}) = 0$ unless the condition (14) is satisfied. Consequently (14) is a necessary condition that f shall have an infinite root given by (13). Our assertion makes this condition sufficient as well as necessary.

It follows that when f has the infinite root given by (13), it has infinite roots given by $[x_1\, x_2 \ldots x_{n-1}] = k\,[a_1\, a_2 \ldots a_{n-1}]$, where k is any finite non-zero number. We shall usually make no distinction between these roots, and say that f has an *infinite root* given by

$$[x_1\, x_2 \ldots x_{n-1}] \equiv \rho\,[\alpha_1\, \alpha_2 \ldots \alpha_{n-1}], \quad \text{or} \quad [x_1\, x_2 \ldots x_{n-1}] \equiv [\alpha_1\, \alpha_2 \ldots \alpha_{n-1}] \ldots\text{(e)}$$

(where the α's are all finite but not all zero), if and only if the condition (14) is satisfied, i.e. if and only if $[x_1\, x_2 \ldots x_{n-1}] \equiv [\alpha_1\, \alpha_2 \ldots \alpha_{n-1}]$ is a root of the homogeneous function formed by the terms of highest degree in f; this being so if and only if g has a finite non-zero root in which $x_n \neq 0$ given by (e'). In the first of the formulae (e) the sign $\equiv$ has its usual significance, ρ being infinite; in the second formula it is used in a special sense which is sufficiently indicated by calling the root 'infinite.'

Accordingly we define the roots of the non-homogeneous function f given by (11), *or the roots of the corresponding equation $f = 0$ to be:*

(1) *the finite roots given by equations of the form* (d);

(2) *the infinite roots given by congruences of the form* (e), *where the α's are all finite but not all zero, and $U_r(\alpha_1, \alpha_2, \dots \alpha_{n-1}) = 0$.* ....(B)

Every root, whether finite or infinite, is represented by a finite matrix.

NOTE 3. *Homogeneous and non-homogeneous variables.*

The terms 'homogeneous' and 'non-homogeneous' can be applied to the variables instead of the functions. In a non-homogeneous function f the variables are necessarily non-homogeneous; and the roots of f are defined as above. In a homogeneous function f we can regard the variables either as homogeneous or as non-homogeneous. In the former case the roots of f are defined as in sub-article 1, the zero root being excluded; in the latter case the roots of f are defined as in the present sub-article. When we speak of the zero root of a homogeneous function, we are regarding the variables as non-homogeneous; and when we speak of the common roots of a system of equations which are not all homogeneous, we must regard the variables as non-homogeneous.

In Analytical Geometry the equation $ax^2+2hxy+by^2=0$ represents a pair of points in a 1-way space when the variables are homogeneous, no meaning being attached to the zero root $[x\,y]=0$; and it represents a pair of straight lines in a 2-way space when the variables are non-homogeneous, the zero root corresponding to the origin. Again the equation $ax^2+by^2+cz^2+2fyz+2gzx+2hxy=0$ represents a pair of straight lines in a 2-way space when the variables are homogeneous, no meaning being attached to the zero root $[x\,y\,z]=0$; and it represents a quadric cone when the variables are non-homogeneous, the zero root corresponding to the origin.

3. *Correspondences between the roots of a non-homogeneous function f and the homogenised function g.*

As in (11) let $f(x_1, x_2, \dots x_{n-1})$ be a non-homogeneous rational integral function of degree r, and as in (11') let it be converted without change of degree into the homogeneous function $g(x_1, x_2, \dots x_{n-1}, x_n)$ by the introduction of the additional variable x_n, i.e. by the substitution $1 = x_n$. Then from sub-article 2 we see that:

There is a one-one correspondence between the finite roots of f and those (finite) non-zero roots of g in which $x_n \neq 0$, every two such corresponding roots having the forms

$$[x_1\,x_2 \dots x_{n-1}] = [\alpha_1\,\alpha_2 \dots \alpha_{n-1}], \quad [x_1\,x_2 \dots x_{n-1}\,x_n] \equiv [\alpha_1\,\alpha_2 \dots \alpha_{n-1}\,1];$$

also there is a one-one correspondence between the infinite roots of f and those (finite) non-zero roots of g in which $x_n = 0$, every two such corresponding roots having the forms

$$[x_1\, x_2 \ldots x_{n-1}] \equiv \rho\, [\alpha_1\, \alpha_2 \ldots \alpha_{n-1}], \quad [x_1\, x_2 \ldots x_{n-1}\, x_n] \equiv [\alpha_1\, \alpha_2 \ldots \alpha_n\, 0],$$

where ρ is infinite, and the α's are all finite but not all zero. ....(A′)

The function f has no other roots; the function g has only one other distinct (finite) root, viz. the zero root $[x_1\, x_2 \ldots x_n] = 0$, which is repeated r times.

More generally let $[l]_n^n$ be an undegenerate square matrix with constant elements, and let $f(x_1, x_2, \ldots x_{n-1})$ be converted without change of degree into the homogeneous function $g(y_1, y_2, \ldots y_{n-1}, y_n)$ by the linear substitution

$$\begin{bmatrix} x \\ 1 \end{bmatrix}_{n-1,1} = [l]_n^n \, [y]_n. \qquad \text{..........................(15)}$$

We can replace the substitution (15) by the successive substitutions $1 = x_n$, $[x]_n = [l]_n^n [y]_n$, and when we do this, we see that:

There is a one-one correspondence between the finite roots of f and those (finite) non-zero roots of g which satisfy the equation

$$l_{n1} y_1 + l_{n2} y_2 + \ldots + l_{nn} y_n = 1,$$

every two such corresponding roots having the forms

$$[x_1\, x_2 \ldots x_{n-1}] = [\alpha_1\, \alpha_2 \ldots \alpha_{n-1}], \quad [y_1\, y_2 \ldots y_n] \equiv [\beta_1\, \beta_2 \ldots \beta_n],$$

where

$$\begin{bmatrix} \alpha \\ 1 \end{bmatrix}_{n-1,1} = [l]_n^n \, [\beta]_n;$$

also there is a one-one correspondence between the infinite roots of f and those (finite) non-zero roots of g which satisfy the equation

$$l_{n1} y_1 + l_{n2} y_2 + \ldots + l_{nn} y_n = 0,$$

every two such corresponding roots having the forms

$$[x_1\, x_2 \ldots x_{n-1}] \equiv \rho\, [\alpha_1\, \alpha_2 \ldots \alpha_{n-1}], \quad [y_1\, y_2 \ldots y_n] \equiv [\beta_1\, \beta_2 \ldots \beta_n],$$

where ρ is infinite and

$$\begin{bmatrix} \alpha \\ 0 \end{bmatrix}_{n-1,1} = [l]_n^n \, [\beta]_n. \qquad \text{..........................(B′)}$$

The function f has no other roots; the function g has only one other distinct (finite) root, viz. the zero root $[y_1\, y_2 \ldots y_n] = 0$, which is repeated r times.

The first part of this theorem was proved directly in § 187.3, and the second part can be deduced from the first by an assertion similar to that made in sub-article 2, i.e. by regarding an infinite root of f as the limit of a finite root.

NOTE 4. *Repeated roots of the non-homogeneous function f.*

The non-homogeneous function f given by (11) has the finite root

$$[x_1\, x_2 \ldots x_{n-1}] = [a_1\, a_2 \ldots a_{n-1}]$$

repeated i times when and only when the homogenised function g given by (11′) has the corresponding root $[x_1\, x_2 \ldots x_{n-1}\, x_n] \equiv [a_1\, a_2 \ldots a_{n-1}\, 1]$ repeated i times; and it has the infinite root $[x_1\, x_2 \ldots x_{n-1}] \equiv \rho\, [a_1\, a_2 \ldots a_{n-1}]$ repeated i times when and only when g has the corresponding root $[x_1\, x_2 \ldots x_{n-1}\, x_n] \equiv [a_1\, a_2 \ldots a_{n-1}\, 0]$ repeated i times. The discriminant of f is the same as the discriminant of g, and its vanishing is the necessary and sufficient condition that f shall have at least one repeated root, finite or infinite.

NOTE 5. *Roots of a rational integral function which has infinite coefficients.*

In speaking of a rational integral function f of degree r it is usually assumed that the sequence $[a_1\, a_2\, a_3 \ldots]$ of its coefficients is a finite non-zero matrix. When this is not the case, we determine if possible numbers $b_1, b_2, b_3, \ldots$ proportional to $a_1, a_2, a_3, \ldots$ which are all finite and not all zero; then if g is the function *of degree r* formed from f when we replace the coefficients $a_1, a_2, a_3, \ldots$ by $b_1, b_2, b_3, \ldots$, we define the roots of f to be the roots of g.

If some or all the coefficients $a_1, a_2, a_3, \ldots$ of f are infinite, we can put

$$[a_1\, a_2\, a_3 \ldots] = \rho\, [b_1\, b_2\, b_3 \ldots],$$

where ρ is infinite and $b_1, b_2, b_3, \ldots$ are all finite and not all zero; and the roots of f are the roots of the function g formed in the manner just described. If ρ is very great but not infinite, then the roots of g are first approximations to the roots of f. Thus if f is the homogeneous function

$$f = x^3 + 2\lambda y^3 + \lambda^2 z^3 + \lambda^2 y z^2 + 2\lambda z x^2 + x y^2,$$

then when λ is infinite, the roots of f are the roots of the function

$$g = z^3 + y z^2,$$

which is the coefficient of the highest power of λ occurring in f. These roots (see Ex. vii) include roots in which $z = 0$ repeated twice. If λ is a very large number but not infinite, the roots of g are first approximations to the roots of f. Again if f is the non-homogeneous function

$$f = x^3 + 2\lambda y^3 + x y^2 + 2\lambda x^2 + \lambda^2 y + \lambda^2,$$

then when λ is infinite, the roots of f are the roots of the function of degree 3

$$g = y + 1,$$

which include infinite roots repeated twice.

NOTE 6. *Roots of a rational integral function which vanishes identically.*

When a rational integral function f vanishes identically, we must consider that every set of values of the variables is a root of f. The coefficients $a_1, a_2, a_3, \ldots$ of f all have the value 0 in this case, the ratios $a_1 : a_2 : a_3 : \ldots$ being indeterminate.

If however the coefficients $a_1, a_2, a_3, \ldots$ of a rational integral function f are all non-zero vanishing quantities, and if we can put

$$[a_1\, a_2\, a_3 \ldots] = \epsilon\, [b_1\, b_2\, b_3 \ldots],$$

where $b_1, b_2, b_3, \ldots$ are all finite and not all zero, then the roots of f must be defined as in Note 5.

The following examples on non-homogeneous functions correspond to and can be deduced from Exs. iii—viii.

Ex. ix. The non-homogeneous function f of $x_1, x_2, \ldots x_{n-1}$ given by

$$f = U_r + U_{r-1} + \ldots + U_1 + U_0, \quad \text{.........................(16)}$$

where U_s is a homogeneous rational integral function of degree s, has the zero root $[x_1\, x_2 \ldots x_{n-1}] = 0$ repeated i times, where $i \not> r$, when and only when $U_0, U_1, \ldots U_{i-1}$ all vanish identically, i.e. when and only when f contains no term whose degree in all the variables is less than i, or has the form

$$f = U_r + U_{r-1} + \ldots + U_i. \quad \text{.................................(17)}$$

Ex. x. If the coefficients of the non-homogeneous function f are rational integral functions of an arbitrary parameter t, then in order that f may have a root of the form

$$[x_1\, x_2 \ldots x_{n-1}] = t\, [a_1\, a_2 \ldots a_{n-1}] \quad \text{.............................(18)}$$

repeated i times, where the a's are finite constants, and $i \not> r$, it is a necessary condition that f itself shall have the form

$$f = U_r + U_{r-1} + \ldots + U_i + t \,.\, U_{i-1} + \ldots + t^{i-1} \,.\, U_1 + t^i \,.\, U_0. \quad \text{...........(18')}$$

The coefficients of $U_r, U_{r-1}, \ldots U_i$ in (18′) may be any given constants, and (18) will then be a root of (18′) repeated i times if the coefficients of $U_{i-1}, U_{i-2}, \ldots U_0$ are suitably determined rational integral functions of t of degree $r-i$.

When the coefficients of all the U's in (18′) are finite constants, and t is sufficiently small but not 0, the function (18′) has roots of the form (18) in which the a's are any finite quantities satisfying the equation

$$U_i + U_{i-1} + \ldots + U_0 = 0. \quad \text{.................................(19)}$$

Ex. xi. If $i \not> r$, the function f given by (16) has the infinite root

$$[x_1\, x_2 \ldots x_{n-1}] \equiv \rho\, [a_1\, a_2 \ldots a_{n-1}] \quad \text{.............................(20)}$$

when and only when $U_r(a_1, a_2, \ldots a_{n-1}) = 0$; and it has the infinite root (20) repeated i times, where $i \not> r$, when and only when the conditions of Ex. vii are satisfied.

Ex. xii. Let

$$f = 0 + 0 + \ldots + U_r + U_{r-1} + \ldots + U_0 \quad \text{.........................(21)}$$

be a rational integral function of $x_1, x_2, \ldots x_{n-1}$ of assigned degree $r+s$ in which the terms of degrees $r+1, r+2, \ldots r+s$ all vanish identically, U_s being homogeneous of degree s. Then if $i \not> r$, we see from Ex. viii that this function f has the infinite root (20) repeated $s+i$ times when and only when the conditions of Ex. vii are satisfied. It always has this root repeated at least s times.

In this case the roots of f are the roots of the function $U_r + U_{r-1} + \ldots + U_0$, and in addition every infinite root of the form (20) repeated exactly s times.

§ 192. General, special and particular rational integral functions.

1. *Definitions of general, special and particular functions.*

A rational integral function f of the n variables $x_1, x_2, \ldots x_n$ having degree r in all the variables is called a *general function of degree r* when every possible term of degree not greater than r occurs in it and the coefficients of all the $\binom{n+r}{r}$ terms are arbitrary parameters to which any particular values whatever can be ascribed. We *specialise* the function f when we give particular numerical values to some of the coefficients and leave the rest of them arbitrary, or when we replace some or all the coefficients by rational integral functions of a limited number of arbitrary parameters; and a function g derived from the general function f in this way is called a *special function of degree r.*

We *particularise* the function f when we give particular numerical values to all its coefficients, and a function g derived from f in this way is called a *particular function of degree r.* The actual degree of a special or particular function of degree r, i.e. the highest degree of a term having a non-zero coefficient, may be less than r.

A homogeneous rational integral function f of the $n+1$ variables $x_1, x_2, \ldots x_n, x_{n+1}$ is called a *general homogeneous function of degree r* when every possible term of degree r occurs in it and the coefficients of all the $\binom{n+r}{r}$ terms are arbitrary parameters to which any particular values whatever can be ascribed. By specialising or particularising the coefficients as before we obtain *special* and *particular homogeneous functions of degree r.* The actual degree of a special or particular homogeneous function of degree r which does not vanish identically is necessarily equal to r.

Ex. i. If $a, b, c, f, g, h, \lambda, \mu$ are arbitrary parameters, the functions

$$ax^2+2hxy+by^2+2gx+2fy+c,$$
$$a(x^2+y^2)+(a^2+b)\,x\,(y+1),$$
$$3x^2+6xy+7x+2$$

are respectively a general function and a special and particular function of degree 2 of the variables x and y; and the functions

$$ax^2+by^2+cz^2+2fyz+2gzx+2hxy,$$
$$\lambda(x^2+y^2+z^2)+(\lambda+2\mu)(yz+zx+xy),$$
$$3x^2+2yz$$

are respectively a general homogeneous function and a special and particular homogeneous function of degree 2 of the variables x, y, z.

2. *Weights of the coefficients of a rational integral function.*

When f is a *general homogeneous* rational integral function of degree r of the $n+1$ variables $x_1, x_2, \ldots x_n, x_{n+1}$, we will ascribe weights to the coefficients of the various terms with respect to $x_1, x_2, \ldots x_n, x_{n+1}$, the weight with respect to x_u of the coefficient of any term being the index of the power of x_u occurring in that term. These weights can be indicated by suffixes attached to the letters denoting the coefficients. Thus if f is expressed in the form

$$f=\Sigma a_{p_1 p_2 \ldots p_n p_{n+1}} x_1^{p_1} x_2^{p_2} \ldots x_n^{p_n} x_{n+1}^{p_{n+1}}, \quad (p_1+p_2+\ldots+p_n+p_{n+1}=r), \quad \ldots\ldots(1)$$

and if u is any one of the integers $1, 2, \ldots n, n+1$, then p_u is the weight of the coefficient $a_{p_1 p_2 \ldots p_n p_{n+1}}$ with respect to x_u. With this notation all the coefficients of f and the weights of every coefficient with respect to $x_1, x_2, \ldots x_n, x_{n+1}$ are shown without any ambiguity. The weight of a product of any number of the coefficients with respect to x_u is the sum of the weights of its factors with respect to x_u.

When f is a *general* rational integral function of degree r of the n variables $x_1, x_2, \ldots x_n$, we will ascribe weights to the coefficients of the various terms with respect to the variables $x_1, x_2, \ldots x_n$ and a variable x_{n+1} which will render f homogeneous of degree r when we apply to it the substitution $1=x_{n+1}$, these weights being defined as before. We will call x_{n+1} the 'absent variable' or the 'homogenising variable.' We generally express f in the form

$$f=\Sigma a_{p_1 p_2 \ldots p_n p_{n+1}} x_1^{p_1} x_2^{p_2} \ldots x_n^{p_n}, \quad (p_1+p_2+\ldots+p_n+p_{n+1}=r). \quad \ldots\ldots\ldots(2)$$

Then if u is any one of the integers $1, 2, \ldots n$, p_u is the weight of the coefficient $a_{p_1 p_2 \ldots p_n p_{n+1}}$ with respect to x_u, and p_{n+1} is the weight of that coefficient with respect to the absent variable x_{n+1}. We obtain (2) from (1) by the substitution $x_{n+1}=1$, and (1) from (2) by the homogenising substitution $1=x_{n+1}$.

We could of course ascribe any weights we please to the various coefficients of any rational integral function without using any special notation to indicate the weights, the weight of a product of any number of coefficients being always the sum of the weights attached to its factors. But it will usually be understood that the weights are those defined above.

3. *Ordinary values of variables and coefficients.*

By *ordinary values* of the variables $x_1, x_2, \ldots x_n$ we shall mean particular finite numerical values for which $g_1 \neq 0$, $g_2 \neq 0$, $g_3 \neq 0$, ..., where $g_1, g_2, g_3, \ldots$ are certain rational integral functions, finite in number, of the variables $x_1, x_2, \ldots x_n$ which do not vanish identically. We may speak of ordinary values of the variables even when the functions $g_1, g_2, g_3, \ldots$ are not specified if we know that it is possible to specify them in at least one way. *Ordinary values* of the arbitrary coefficients of general or special rational integral functions $f_1, f_2, \ldots f_m$ will be defined in the same way, $g_1, g_2, g_3, \ldots$ being now rational integral functions of those arbitrary coefficients. By *ordinary roots* of a rational integral function $f(x_1, x_2, \ldots x_n)$ with numerical coefficients we shall mean those finite roots $[x_1 x_2 \ldots x_n]$ for which $g_1 \neq 0$, $g_2 \neq 0$, $g_3 \neq 0$, ..., where $g_1, g_2, g_3, \ldots$ are rational integral functions of $x_1, x_2, \ldots x_n$ which do not vanish for all finite roots of f, the last condition being necessary in order that there may be ordinary roots.

A scalar number which is neither zero nor infinite will sometimes be called an *ordinary number* or an *ordinary scalar constant.*

Ex. ii. If a rational integral function of $x_1, x_2, \ldots x_n$ vanishes for all ordinary values of the variables, then it vanishes identically.

If two rational integral functions of $x_1, x_2, \ldots x_n$ are equal for all ordinary values of the variables, then they are identically equal.

If two rational integral functions of the coefficients of the general rational integral functions $f_1, f_2, \ldots f_m$ are equal for all ordinary values of the coefficients, then they are identically equal.

These theorems have been proved in Ex. xv of § 185.

Ex. iii. If a rational integral function F vanishes for all ordinary roots of an irresoluble function g, then it is divisible by g. Here ordinary roots are those for which $g_1 \neq 0$, $g_2 \neq 0$, $g_3 \neq 0$, ..., where $g_1, g_2, g_3, \ldots$ are not divisible by g.

If a rational integral function F lying in a domain of rationality Ω vanishes for all ordinary roots of an irreducible function g, then it is divisible by g. Here ordinary roots may be those for which $g_1 \neq 0, g_2 \neq 0, g_3 \neq 0, \ldots$, where $g_1, g_2, g_3, \ldots$ lie in Ω and are not divisible by g.

These theorems have been proved in Exs. viii and x of § 189.

Ex. iv. *If $f(z, x_1, x_2, \ldots x_n)$ is a rational integral function of $z, x_1, x_2, \ldots x_n$ which is divisible by z^p for all ordinary particular values of the x's, then it is divisible by z^p when $z, x_1, x_2, \ldots x_n$ are all arbitrary.*

Here ordinary values of the x's may be defined to be those for which $g_1(x_1, x_2, \ldots x_n) \neq 0$, $g_2(x_1, x_2, \ldots x_n) \neq 0, \ldots$, where the g's are given rational integral functions which do not vanish identically.

Let z^q be the highest power of z which is a factor of f when $z, x_1, x_2, \ldots x_n$ are all arbitrary, and let $q+s$ be the actual degree of f in z. Then there exists an identity in $z, x_1, x_2, \ldots x_n$ of the form

$$f = z^q \{z^s U_0 + z^{s-1} U_1 + \ldots + z U_{s-1} + U_s\},$$

where the U's are rational integral functions of the x's only, and where U_0 and U_s do not vanish identically.

Now let such ordinary particular values be given to the x's that $U_s \neq 0$. Then z^q is the highest power of z which is a factor of f. Since by hypothesis z^p is now a factor of f, we must have $q \nless p$; and this establishes the theorem.

Ex. v. *Let $x_1, x_2, \ldots$ and $y_1, y_2, \ldots$ be two sets of variables, the x's and y's; and let $f(x_1, x_2, \ldots)$ be an irresoluble function of the x's. Then if $F(x_1, x_2, \ldots y_1, y_2, \ldots)$ is a rational integral function of the x's and y's which is divisible by f for all ordinary particular values of the y's, it is divisible by f when the x's and y's are all arbitrary.*

We may suppose ordinary values of the y's to be those finite values for which

$$g_1(y_1, y_2, \ldots) \neq 0, \quad g_2(y_1, y_2, \ldots) \neq 0, \quad \ldots,$$

where $g_1, g_2, \ldots$ are rational integral functions of the y's which do not vanish identically. If F vanishes identically for all finite roots of f, it must clearly be divisible by f; for if we express it in the form

$$F(x_1, x_2, \ldots y_1, y_2, \ldots) = \Sigma U_{pq\ldots}(x_1, x_2, \ldots) \cdot y_1^p y_2^q \ldots,$$

each of the U's must in this case be divisible by f.

Now suppose that F is not divisible by f when the x's and y's are all arbitrary. Then we can determine a finite root $[x_1 x_2 \ldots] = [a_1 a_2 \ldots]$ of f for which F, regarded as a function of the y's, does not vanish identically; and when we substitute $a_1, a_2, \ldots$ for $x_1, x_2, \ldots$ we can determine such particular finite values of the y's that $F \neq 0, g_1 \neq 0, g_2 \neq 0, \ldots$. When we give these ordinary particular values to the y's, F does not vanish for all finite roots of f, and is not divisible by f. This is contrary to the hypothesis that F is divisible by f for all ordinary particular values of the y's.

It follows that F must be divisible by f when the x's and y's are all arbitrary.

Further if $F(x_1, x_2, \ldots y_1, y_2, \ldots)$ is divisible by f^p for all ordinary particular values of the y's, it is divisible by f^p when the x's and y's are all arbitrary.

For since F is divisible by f, we have $F = fF_1$, where F_1 is divisible by f^{p-1} for all ordinary particular values of the y's; if $p > 1$, F_1 is divisible by f, and we have $F_1 = fF_2$, $F = f^2 F_2$, where F_2 is divisible by f^{p-2} for all ordinary particular values of the y's; if $p > 2$, F_2 is divisible by f, and we have $F_2 = fF_3$, $F = f^3 F_3$; and so on.

The above theorems remain true when f is an irreducible function in Ω and F is any rational integral function of the x's and y's which lies in the same domain of rationality Ω.

Ex. vi. *Hilbert's Theorem generalised.* If $F(x_1, x_2, \ldots y_1, y_2, \ldots)$ is an irresoluble rational integral function of two sets of variables, the x's and the y's, then for all ordinary particular values of the y's it is an irresoluble function of the x's.

§ 193. Resultants of rational integral functions.

1. *Resultant of n general homogeneous functions of n variables.*

Let $f_i(x_1, x_2, \ldots x_{n-1}, x_n) = \Sigma a^{(i)}_{p_1 p_2 \ldots p_{n-1} p_n} x_1^{p_1} x_2^{p_2} \ldots x_{n-1}^{p_{n-1}} x_n^{p_n}$, ...(1)

$$(p_1 + p_2 + \ldots + p_{n-1} + p_n = r_i;\quad i = 1, 2, \ldots n),$$

be n general homogeneous rational integral functions of the n variables $x_1, x_2, \ldots x_{n-1}, x_n$ whose degrees in all the variables are $r_1, r_2, \ldots r_n$, these being all different from 0, and let

$$f_i(x_1, x_2, \ldots x_{n-1}, 1) = \Sigma a^{(i)}_{p_1 p_2 \ldots p_{n-1} p_n} x_1^{p_1} x_2^{p_2} \ldots x_{n-1}^{p_{n-1}}, \quad \text{.........(2)}$$

$$(p_1 + p_2 + \ldots + p_{n-1} + p_n = r_i;\quad i = 1, 2, \ldots n),$$

be the functions derived from them by the substitution $x_n = 1$; so that the functions (2) are n general rational integral functions of the $n-1$ variables $x_1, x_2, \ldots x_{n-1}$ of degrees $r_1, r_2, \ldots r_n$ which will be re-converted into the n functions (1) when we homogenise each of them without change of degree by the substitution $1 = x_n$. Also let

$$r = r_1 r_2 \ldots r_n;\quad s_i = \frac{r}{r_i},\quad (i = 1, 2, \ldots n), \quad \text{..............(3)}$$

so that r is the product of the degrees of all the n functions in either set, and s_i is the product of the degrees of all the functions except f_i in either set.

When the n general homogeneous functions (1) are particularised by assigning particular numerical values to their coefficients, they will not usually have a non-zero root in common; but the particular values can be so chosen that they do have at least one non-zero root in common. We will define a *resultant* R of these n functions (or of the corresponding equations $f_1 = 0$, $f_2 = 0$, ... $f_n = 0$) to be a rational integral function of their coefficients of the lowest possible total degree the vanishing of which is a necessary and sufficient condition that the n functions shall have *at least one non-zero root in common.* By this we mean that when the functions are particularised by assigning any particular values to their coefficients, the particularised functions will always have a non-zero root in common if and only if R vanishes for those particular values of the coefficients. The process of determining a resultant R is called the *elimination*

When any particular finite values are ascribed to u and v, the n functions F must have a common root; to each of their finite common roots there corresponds a finite non-zero common root of the n functions f; to each of their infinite common roots there corresponds a finite non-zero common root of the n functions f in which $[x_i x_j]=[0\,0]$; if all their common roots are infinite, then in all the finite non-zero common roots of the n functions f in which $x_i : x_j = u : v$ we have $[x_i x_j]=[0\,0]$.

Illustrations of these results are given in Exs. xvii—xx of § 196.

NOTE 5. *Information yielded by the x_i-resultant of the n general functions* (2).

We here anticipate the final proof of Theorem II of § 196. If $E(z;\ \kappa_1, \kappa_2, \dots \kappa_n)$ is the X-eliminant of the n general functions $f_1, f_2, \dots f_n$ of $x_1, x_2, \dots x_n$ as defined in that theorem, we obtain $R_i(t)$ from $E(z)$ by making the substitutions $z=-t$, $\kappa_i=1$, and putting all the other κ's equal to 0; in other words $R_i(t)$ is the X-eliminant when $X=x_i-t$.

If for any particular values of the coefficients of the n functions f we have

$$E(z)=\prod_{v=1}^{v=r-s}(\beta_{v1}\kappa_1+\beta_{v2}\kappa_2+\dots+\beta_{vn}\kappa_n)\,.\prod_{u=1}^{u=s}(z+a_{u1}\kappa_1+a_{u2}\kappa_2+\dots+a_{un}\kappa_n),$$

then for the same values of the coefficients we can write

$$R_i(t)=B\,.\prod_{u=1}^{u=s}(t-a_{ui}),\quad R(x_i)=B\,.\prod_{u=1}^{u=s}(x_i-a_{ui}),$$

where

$$B=\beta_{1i}\beta_{2i}\dots\beta_{r-s,\,i},$$

and where $R_i(t)$ is to be regarded as having degree r.

When $R_i(t)$ does not vanish identically, its r roots (finite and infinite) are the r values of x_i which occur in the r common roots of the n functions f. When $R_i(t)$ vanishes identically and $E(z)$ does not vanish identically, the n functions f have at least one infinite common root in which x_i is finite. When $E(z)$ vanishes identically, the n functions f have an infinite number of common roots; further they always have a common root in which x_i has any assigned value, but it may happen that every such common root is infinite.

§ 196. Eliminants of rational integral functions.

1. *Eliminant of n general homogeneous functions of n + 1 variables.*

Let

$$f_i(x_1,\ x_2,\ \dots\ x_n,\ x_{n+1})=\Sigma a^{(i)}_{p_1p_2\dots p_np_{n+1}}x_1^{p_1}x_2^{p_2}\dots x_n^{p_n}x_{n+1}^{p_{n+1}},\quad\dots(1)$$

and

$$f_i(x_1,\ x_2,\ \dots\ x_n,\ 1)\ =\Sigma a^{(i)}_{p_1p_2\dots p_np_{n+1}}x_1^{p_1}x_2^{p_2}\dots x_n^{p_n},\quad\dots\dots\dots(2)$$

$$(p_1+p_2+\dots+p_n+p_{n+1}=r_i;\ i=1,\ 2,\ \dots\ n),$$

be respectively n general homogeneous rational integral functions of the $n+1$ variables $x_1, x_2, \dots x_n, x_{n+1}$ and n general rational integral functions of the n variables $x_1, x_2, \dots x_n$ having in each case non-zero degrees $r_1, r_2, \dots r_n$ in all the variables; and let

$$r=r_1r_2\dots r_n;\quad s_i=\frac{r}{r_i},\quad (i=1,\ 2,\ \dots\ n).\quad\dots\dots\dots\dots\dots(3)$$

As in § 195 let the resultants of the $n+1$ sets of n general homogeneous functions of n variables obtained by putting $x_1=0$, $x_2=0$, ... $x_{n+1}=0$ in turn in the n functions (1) be denoted by

$$\rho_1,\ \rho_2,\ \ldots\ \rho_{n+1}, \qquad\qquad (4)$$

so that ρ_{n+1} is also the resultant of the n general homogeneous functions of n variables obtained when we retain only the terms of highest degree in each of the functions (2), and ρ_i for the values 1, 2, ... n of i is also the resultant of the n general functions obtained by putting $x_i=0$ in the n functions (2).

A *complete eliminant* of either set of n functions could be loosely defined to be a rational integral function of a single variable (or a homogeneous rational integral function of two variables) the determination of whose roots will yield all the common roots of the n functions, and a *partial eliminant* to be a similar function the determination of whose roots will yield all the simultaneous values which some of the variables can have in the common roots of the n functions. The eliminants precisely defined in Theorems I and II may be called *Poisson's eliminants.* It will be shown in § 199 that a knowledge of the complete *eliminant* of $n-1$ general homogeneous functions of n variables (or $n-1$ general functions of $n-1$ variables) enables us to *eliminate* all the variables from n general homogeneous functions of n variables (or n general functions of $n-1$ variables) and form their complete resultant or *eliminat.*

The fundamental properties of eliminants, by which we mean Poisson's eliminants, are summarised in the two theorems of this article. In proving these theorems for n functions we in the first place make the *provisional hypothesis* that they are true generally for functions not more than $n-1$ in number and that the two theorems of § 193 are true generally for functions not more than n in number. The general proof of both sets of theorems is completed by induction in § 200.

By the a's we shall mean the coefficients of the n functions (1) or (2), and by the κ's in Theorem I we shall mean the coefficients of the homogeneous linear function (1′). The weights of the coefficients with respect to $x_1, x_2, \ldots x_{n+1}$ will be defined as in § 192. 2. The weights of the a's are indicated by their suffixes. The weights of the κ's are not indicated in this way, but by the definition κ_i has weight 1 with respect to the variable x_i and weight 0 with respect to each of the other variables.

A 'common root' of the n functions (1) always means one whose elements are all finite.

Theorem I. *Let the n general homogeneous functions $f_1, f_2, \ldots f_n$ of the $n+1$ variables $x_1, x_2, \ldots x_{n+1}$ given by* (1) *have any non-zero degrees $r_1, r_2, \ldots r_n$;*

let $\kappa_1, \kappa_2, \ldots \kappa_{n+1}$ be $n+1$ auxiliary arbitrary parameters; and let X be the general homogeneous linear function

$$X = \kappa_1 x_1 + \kappa_2 x_2 + \ldots + \kappa_{n+1} x_{n+1}. \quad \ldots\ldots (1')$$

Then the n functions f have an X-eliminant $E(\kappa_1, \kappa_2, \ldots \kappa_{n+1}) = E$, uniquely determinate except for an arbitrary finite non-zero numerical factor, which can be defined by the following properties:

(*a*) *It is a rational integral function of the a's and κ's which is homogeneous of degree r in the κ's.*

(*b*) *When any particular finite values are ascribed to the a's which do not cause it to vanish identically, it becomes a function of the κ's only which can be expressed as a product of homogeneous linear factors in the form*

$$E(\kappa_1, \kappa_2, \ldots \kappa_{n+1}) = \prod_{u=1}^{u=r} (\alpha_{u1}\kappa_1 + \alpha_{u2}\kappa_2 + \ldots + \alpha_{u,n+1}\kappa_{n+1}). \quad \ldots (A)$$

(*c*) *Under the circumstances described in* (*b*), *i.e. when E does not vanish identically,*

$$[x_1 x_2 \ldots x_{n+1}] \equiv [\alpha_1 \alpha_2 \ldots \alpha_{n+1}] \quad \ldots\ldots (5)$$

is a non-zero common root of the n functions f if and only if

$$\alpha_1\kappa_1 + \alpha_2\kappa_2 + \ldots + \alpha_{n+1}\kappa_{n+1} \quad \ldots\ldots (6)$$

is one of the linear factors of E.

(*d*) *It has the lowest total degree which is possible for any rational integral function of the a's and κ's having the property* (*c*).

Moreover it is an irresoluble function of the a's and κ's when these are arbitrary; it is the resultant of the $n+1$ general homogeneous functions $f_1, f_2, \ldots f_n, X$; and it has the following further properties:

(*a'*) *Its own coefficients are rational numbers which we may suppose to be integers.*

(*b'*) *It is homogeneous in the coefficients of each one of the n functions f, its degree in the coefficients of any one function f_i being $s_i = \frac{r}{r_i}$, i.e. being the product of the degrees of the other $n-1$ functions.*

(*c'*) *It is isobaric of weight r with respect to each one of the variables $x_1, x_2, \ldots x_{n+1}$ in the coefficients of all the $n+1$ functions $f_1, f_2, \ldots f_n, X$; i.e. if j is any one of the integers $1, 2, \ldots n+1$, then in every term of E the sum of the weights of all the coefficients a in that term with respect to x_j (which is the sum of their suffixes p_j) and the index of κ_j in that term is equal to r.*

(d′) *The coefficients of* $\kappa_1^r, \kappa_2^r, \ldots \kappa_j^r, \ldots \kappa_{n+1}^r$ *in* E *are respectively*

$$\rho_1, \rho_2, \ldots \rho_j, \ldots \rho_{n+1},$$

provided only that the arbitrary numerical factors in all the functions ρ *are suitably chosen.*

(e′) *Whenever the* a*'s have such particular values that* $E(\kappa_1, \kappa_2, \ldots \kappa_{n+1})$ *vanishes identically, we can always determine a non-zero common root of the* n *functions* f *whose elements satisfy the equation*

$$k_1 x_1 + k_2 x_2 + \ldots + k_{n+1} x_{n+1} = 0 \quad \ldots\ldots\ldots\ldots\ldots\ldots\ldots(7)$$

in which $k_1, k_2, \ldots k_{n+1}$ *are any given finite numbers, i.e. we can determine a non-zero common root of the* $n+1$ *functions* $f_1, f_2, \ldots f_n$, X *when any particular finite values whatever are ascribed to the* κ*'s.*

(f′) *If* f_τ, *any one of the* n *functions* f, *is specialised into* g_τ, *where* g_τ *is a homogeneous function of degree* r_τ *formed by the product of two general homogeneous functions* ϕ_τ, ψ_τ *of* $x_1, x_2, \ldots x_{n+1}$ *of non-zero degrees* p, q, *so that*

$$g_\tau = \phi_\tau \psi_\tau, \quad p + q = r_\tau, \quad \ldots\ldots\ldots\ldots\ldots\ldots\ldots\ldots\ldots(8)$$

and if E, E', E'' *are respectively the* X*-eliminants of the three sets of* n *homogeneous functions*

$$f_1, f_2, \ldots g_\tau, \ldots f_n; \quad f_1, f_2, \ldots \phi_\tau, \ldots f_n; \quad f_1, f_2, \ldots \psi_\tau, \ldots f_n,$$

the first eliminant being unreduced, then the equation

$$E(\kappa_1, \kappa_2, \ldots \kappa_{n+1}) = E'(\kappa_1, \kappa_2, \ldots \kappa_{n+1}) \,.\, E''(\kappa_1, \kappa_2, \ldots \kappa_{n+1}) \quad \ldots\ldots(B)$$

is true, and is an identity in the κ*'s and the coefficients of the* $n+1$ *general homogeneous functions* $f_1, \ldots f_{\tau-1}, \phi_\tau, \psi_\tau, f_{\tau+1}, \ldots f_n$, *provided only that the arbitrary numerical factors in* E, E', E'' *are suitably chosen.*

Two important corollaries are added.

COROLLARY 1. *When we ascribe to the* a*'s any particular finite values which do not cause* $E(\kappa_1, \kappa_2, \ldots \kappa_{n+1})$ *to vanish identically, the* n *functions* f *have exactly* r *non-zero common roots corresponding one by one to the* r *linear factors of* E *in such a manner that* (5) *is a non-zero common root repeated exactly* k *times if and only if* (6) *is a linear factor of* E *repeated exactly* k *times.*

COROLLARY 2. *When the* a*'s have such particular finite values that* $E(\kappa_1, \kappa_2, \ldots \kappa_{n+1})$ *vanishes identically, the* n *functions* f *have an infinite number of non-zero common roots; and if* x_i *and* x_j *are any two of the* *variables, they have a non-zero common root in which the ratio $x_i : x_j$ has any assigned value, though it may happen that in every such common root we have* $[x_i\, x_j] = [0\, 0]$.

A knowledge of the X-eliminant E reduces the determination of the common roots of the n functions f, whenever the a's have such particular values that E does not vanish identically, to the solution of equations in a single variable; for we can determine those linear factors of E in which κ_i actually occurs by solving the equation $E = 0$, regarding κ_i as the only variable. If the highest power of κ_i which actually occurs in E is κ_i^p, where $p < r$, then the coefficient of κ_i^p is necessarily a factor of E, and we have $E = E'E''$, where E' and E'' are respectively the products of all those linear factors in which κ_i actually occurs and does not actually occur. Thus we can by inspection express E in the form

$$E = E_1 E_2 E_3 \ldots,$$

where $E_1, E_2, E_3, \ldots$ are regular in all the κ's which occur in them; and then all the linear factors of E, and therefore all the common roots of the n functions f, are found by solving the equations $E_1 = 0$, $E_2 = 0$, $E_3 = 0, \ldots,$ each of them being treated as an equation in a single variable.

If in E we put all the κ's equal to 0 except $\kappa_1, \kappa_2, \ldots \kappa_i$, we obtain a partial eliminant which serves to determine all the sets of simultaneous values which $x_1, x_2, \ldots x_i$ can have in the common roots of the n functions f.

In the notes which follow it is assumed that Theorem I just given is true generally for functions not more than $n - 1$ in number, and that Theorem I of § 193 is true generally for functions not more than n in number. From Note 6 it follows that in proving the above theorem for n functions on the same assumption in § 200 we shall only need to show that for every set of non-zero values of $r_1, r_2, \ldots r_n$ there exists a function $E(\kappa_1, \kappa_2, \ldots \kappa_{n+1})$ having the properties (a), (b), (c), (a'), (b'), (c'), (d'), (e') of the theorem.

From Note 4 it will be clear that the properties (c) and (d) alone are sufficient to define the function E.

Corollary 1, as shown in Note 7, is equivalent to a definition of 'repeated common roots.' Corollary 2 is proved in Note 8.

NOTE 1. *No rational integral function E of the a's and κ's having a lower degree than r in the κ's could have the property (c) of Theorem I.*

For when ordinary particular values are ascribed to the a's, we see from Lemma A of § 195 that the n functions f have at least r distinct non-zero common roots, and therefore E must have at least r distinct linear factors.

NOTE 2. *If for all non-zero values of $r_1, r_2, \ldots r_n$ there exists an 'X-eliminant' E which is defined to be a rational integral function of the a's and κ's having the properties (a), (b), (c), (d') of Theorem I, then it must also have the property (f') for all non-zero values of $r_1, r_2, \ldots r_n$.*

Let E, E', E'' be the 'X-eliminants' of the three sets of homogeneous functions

$$f_1, f_2, \ldots g_\tau, \ldots f_n;\ f_1, f_2, \ldots \phi_\tau, \ldots f_n;\ f_1, f_2, \ldots \psi_\tau, \ldots f_n, \ldots\ldots\ldots\ldots(9)$$

the first eliminant being unreduced, i.e. being derived from the 'X-eliminant' of $f_1, f_2, \ldots f_\tau, \ldots f_n$ by replacing the coefficients of f_τ by the coefficients of g_τ; let r, h, k be the products of the degrees of these three sets of functions, so that

$$h+k=r;$$

and let ρ_i, ρ_i', ρ_i'' be the resultants of the three sets of n homogeneous functions of n variables obtained from the three successive sets of functions (9) by putting $x_i=0$, the first resultant being unreduced, and i being any one of the integers $1, 2, \ldots n+1$. Using the property (d') of the theorem we will suppose that the arbitrary numerical constants in the functions ρ, ρ', ρ'' have been so chosen that for all values of i the coefficients of κ_i^r, κ_i^h, κ_i^k in E, E', E'' are ρ_i, ρ_i', ρ_i'', the numerical factors in the E's remaining arbitrary. By Theorem I of § 193 we then have

$$\rho_1=c_1\rho_1'\rho_2'',\quad \rho_2=c_2\rho_2'\rho_2'',\ \ldots\ \rho_{n+1}=c_{n+1}\rho'_{n+1}\rho''_{n+1}, \ldots\ldots\ldots\ldots\ldots\ldots(10)$$

where the c's are ordinary numerical constants, independent of the a's, which depend only on the values of the arbitrary numerical factors in E, E', E''.

From Ex. i of § 195 it will be clear that we can define ordinary values of the coefficients of the $n+1$ functions

$$f_1, \ldots f_{\tau-1}, \phi_\tau, \psi_\tau, f_{\tau+1}, \ldots f_n \ldots\ldots\ldots\ldots\ldots\ldots\ldots\ldots\ldots(11)$$

to be those for which the functions $f_1, f_2, \ldots \phi_\tau, \ldots f_n$ have exactly h distinct non-zero common roots in all of which every element is different from 0, the functions $f_1, f_2, \ldots \psi_\tau, \ldots f_n$ have exactly k distinct non-zero common roots in all of which every element is different from 0, and no non-zero root is common to both these sets of functions. For such ordinary values the functions ρ, ρ', ρ'' are all different from 0, and the n functions forming the first set in (9) have exactly r distinct non-zero common roots, these being the h distinct non-zero common roots of the second set of functions and the k distinct non-zero common roots of the third set of functions.

Let any such ordinary particular values be ascribed to the coefficients of the functions (11). Then from the properties (b), (c) of the theorem it follows that we can express E' and E'' as products of linear factors in the forms

$$E'=\prod_{u=1}^{u=h}(\lambda_{u1}\kappa_1+\lambda_{u2}\kappa_2+\ldots+\lambda_{u,n+1}\kappa_{n+1}),\quad E''=\prod_{v=1}^{v=k}(\mu_{v1}\kappa_1+\mu_{v2}\kappa_2+\ldots+\mu_{v,n+1}\kappa_{n+1}),$$

where the r linear factors of E' and E'', corresponding to the non-zero common roots of the second and third set of functions in (9), are distinct. Since these r factors correspond to the r distinct non-zero common roots of $f_1, f_2, \ldots g_\tau, \ldots f_n$, it follows from the properties (b), (c) that we must have

$$E=A\,.\prod_{u=1}^{u=h}(\lambda_{u1}\kappa_1+\lambda_{u2}\kappa_2+\ldots+\lambda_{u,n+1}\kappa_{n+1})\,.\prod_{v=1}^{v=k}(\mu_{v1}\kappa_1+\mu_{v2}\kappa_2+\ldots+\mu_{v,n+1}\kappa_{n+1}),$$

and therefore

$$E=A\,.\,E'E'', \ldots\ldots\ldots\ldots\ldots\ldots\ldots\ldots\ldots\ldots\ldots\ldots(12)$$

where A is, for the particular values ascribed to the a's, an ordinary numerical constant independent of the κ's, the equation (12) being an identity in the κ's.

Equating the coefficients of κ_i^r on both sides of (12) for all values of i, we see that

$$c_1=c_2=\ldots=c_{n+1}=A.$$

It follows that A has the same value for all values of the a's, and is an absolute constant which is independent of the a's and κ's, and depends only on the values of the arbitrary numerical factors in E, E', E''. Thus the equation (12) is an identity in the κ's which is true for all ordinary values of the a's, and it is therefore an identity in the a's and κ's when both sets of quantities are arbitrary. When the arbitrary numerical factors in E, E', E'' are suitably chosen, we have $A=1$; and then the equation

$$E=E'E'' \qquad\qquad (13)$$

is an identity in the a's and κ's.

NOTE 3. *If for all non-zero values of $r_1, r_2, \ldots r_n$ there exists a rational integral function $E(\kappa_1, \kappa_2, \ldots \kappa_{n+1})$ of the a's and κ's having the properties* (a), (b), (c), (b'), (d') *of Theorem I, then it must be irresoluble for all non-zero values of $r_1, r_2, \ldots r_n$.*

When we use Ex. ii of sub-article 2 the proof is the same as that of Note 3 to Theorem I in § 193; for by Note 2 the function E has the property (f') for all non-zero values of $r_1, r_2, \ldots r_n$.

NOTE 4. *If for any given non-zero values of $r_1, r_2, \ldots r_n$ there exists an irresoluble function E having the properties* (a), (b), (c) *of Theorem I, then:*

(1) *It must also have the property* (d).

(2) *Every rational integral function of the a's and κ's which has the properties* (c), (d) *differs from E only by an ordinary numerical factor.*

Let E' be any rational integral function of the a's and κ's which has the property (c). Then for all ordinary values of the a's all the linear factors of E are linear factors of E'. Therefore when the a's and κ's are regarded as variables, E' vanishes for all ordinary roots of the irresoluble function E, and must be divisible by E. Thus E' cannot have a lower total degree than E, i.e. E has the property (d).

Again if E', which cannot vanish identically, has the property (d), it cannot have a lower degree than E. Since it is divisible by E, it can only differ from E by an ordinary numerical factor.

NOTE 5. *If for any given non-zero values of $r_1, r_2, \ldots r_n$ there exists an irresoluble function $E(\kappa_1, \kappa_2, \ldots \kappa_{n+1})$ having the properties* (a), (b), (c), (e') *of Theorem I, then it must be a resultant of the $n+1$ functions $f_1, f_2, \ldots f_n, X$.*

Let any particular finite values be ascribed to the a's, the κ's remaining arbitrary.

First suppose that these particular values are such that E, regarded as a function of the κ's, does not vanish identically, and let E be expressed in the form (A). Then when any particular values are given to the κ's, it follows from the properties (b), (c) that E will vanish if and only if one of its linear factors, say $a_1\kappa_1+a_2\kappa_2+\ldots+a_{n+1}\kappa_{n+1}$, vanishes, i.e. if and only if the n functions f have a non-zero common root $[x_1 x_2 \ldots x_{n+1}]\equiv[a_1 a_2 \ldots a_{n+1}]$ which satisfies the equation $X=0$, i.e. if and only if $f_1, f_2, \ldots f_n, X$ have a non-zero common root.

Next suppose that the particular values given to the a's are such that E, regarded as a function of the κ's, vanishes identically. Then when any particular finite values whatever are given to the κ's, the function E vanishes, and from the property (e') it follows that the $n+1$ functions $f_1, f_2, \ldots f_n, X$ have a non-zero common root.

Thus for all particular finite values of the a's and κ's the $n+1$ functions $f_1, f_2, \ldots f_n, X$ have a non-zero common root when and only when E vanishes. Since E is irresoluble, it follows as in Note 4 to Theorem I in § 193 that E is a resultant of those $n+1$ functions.

NOTE 6. *If for all non-zero values of $r_1, r_2, \dots r_n$ there exists a function*

$$E(\kappa_1, \kappa_2, \dots \kappa_{n+1})$$

having the properties (a), (b), (c), (a′), (b′), (c′), (d′), (e′) of Theorem I, then for all non-zero values of $r_1, r_2, \dots r_n$, this function is irresoluble and has all the properties of the theorem; and every function having the properties (a), (b), (c) can only differ from it by an ordinary numerical factor.

It follows from Note 2 that E has the property (f'); from Note 3 that it is irresoluble; from Note 4 that it has the property (d), and that every function having the properties (a), (b), (c), (d) differs from it only by an ordinary numerical factor; and from Note 5 that it is a resultant of the $n+1$ functions $f_1, f_2, \dots f_n, X$. Thus it has all the properties mentioned in Theorem I.

NOTE 7. *Repeated common roots.*

When the a's have such particular values that the X-eliminant E does not vanish identically, Theorem I states that

$$[x_1 x_2 \dots x_{n+1}] \equiv [a_1 a_2 \dots a_{n+1}] \quad \dots\dots\dots\dots(5)$$

is a non-zero common root of the n functions f if and only if

$$a_1\kappa_1 + a_2\kappa_2 + \dots + a_{n+1}\kappa_{n+1} \quad \dots\dots\dots\dots(6)$$

is a linear factor of E. We *define* (5) to be a common root repeated exactly k times when (6) is a linear factor of E repeated exactly k times. Then, since E is always a product of r linear factors, Corollary 1 is necessarily true. Thus Corollary 1 can be interpreted to be a definition of 'repeated common roots' of the n functions f.

A 'repeated common root' of the n functions f, as thus defined, is necessarily a root of every individual function f, but it is not ordinarily a 'repeated root' of any one of them. On the other hand if a common root of the n functions f is a 'repeated root' of any one of them, it is necessarily a 'repeated common root' of all of them. In fact the following theorem is true whenever E does not vanish identically.

If (5) is a root of f_1 repeated exactly λ_1 times, a root of f_2 repeated exactly λ_2 times, … a root of f_n repeated exactly λ_n times, and if $\lambda = \lambda_1\lambda_2\dots\lambda_n$, then (5) is a common root of the n functions f repeated at least λ times, and ordinarily exactly λ times; i.e. (6) is a linear factor of E repeated at least λ times, and ordinarily exactly λ times.

This theorem is obviously true when any one of the integers $\lambda_1, \lambda_2, \dots \lambda_n$ is 0. We may therefore suppose them to be non-zero positive integers not greater than $r_1, r_2, \dots r_n$ respectively. From Ex. xiii of sub-article 2 we see that to prove the theorem, it is sufficient to show that if

$$[x_1 x_2 \dots x_n x_{n+1}] \equiv [0\,0 \dots 0\,1]$$

is a root of $f_1, f_2, \dots f_n$ repeated exactly $\lambda_1, \lambda_2, \dots \lambda_n$ times respectively, then E is divisible by κ^{λ}_{n+1}, but not ordinarily by any higher power of κ_{n+1}. Accordingly Exs. x and xi serve to establish the theorem for any non-zero root.

NOTE 8. *Cases in which E vanishes identically.*

Let the a's have such particular values that E, regarded as a function of the κ's, vanishes identically. Then from the property (c') we see that if we write down any homogeneous linear equation

$$k_1x_1 + k_2x_2 + \dots + k_{n+1}x_{n+1} = 0,$$

where the k's are given finite numbers, we can determine a non-zero common root of the n functions f which satisfies it; or in every plane there lies at least one non-zero common root. Since we can always construct a plane which contains no one of a given finite number of points, there must be an infinite number of non-zero common roots. Putting all the k's equal to 0 except k_i and k_j, we see that we can always determine a non-zero common root $[x_1 x_2 \ldots x_{n+1}]$ with finite elements such that

$$k_i x_i + k_j x_j = 0$$

whatever particular finite numbers k_i and k_j may be. Therefore, as stated in Corollary 2, we can determine one in which the ratio $x_i : x_j$ has any assigned value.

We shall not discuss these cases at present, but merely remark that all common roots are given by formulae of the form

$$[x_1 x_2 \ldots x_{n+1}] \equiv [\phi_1(t_1, t_2, \ldots), \quad \phi_2(t_1, t_2, \ldots), \ldots \phi_{n+1}(t_1, t_2, \ldots)],$$

where the t's are arbitrary parameters, and the ϕ's are definite functions, some of which may be 0's or absolute constants; there being p parameters t when the eliminants of every set of $n-p+1$ functions (treated as non-homogeneous functions of $n-p+1$ of the variables) which we can form from the n functions f all vanish identically, and there being some such formulae in which p parameters actually occur. If we take such a formula and substitute for the x's their values in terms of the t's in any rational integral equation $F(x_1, x_2, \ldots x_{n+1}) = 0$, it must be possible to determine the t's so that this equation is satisfied. We conclude that:

When E vanishes identically, it is always possible to determine non-zero common roots of the n functions f which are also roots of any other given rational integral function of the same variables.

In the special case when there are only two functions we have the theorem given in Ex. xxi.

NOTE 9. *Discriminant of a system of n homogeneous functions of $n+1$ variables.*

We may define the discriminant of the n general homogeneous functions f given by (1) to be a rational integral function of their coefficients a of the lowest possible total degree the vanishing of which is, for all particular finite values of the a's, a necessary and sufficient condition that the n functions f shall have a repeated non-zero common root. It is entirely distinct from the discriminant of a single function defined in § 194. 3 except in the special case when $n=1$. If we can find a rational integral function S of the a's which is either irresoluble or a product of unrepeated irresoluble factors when the a's are arbitrary, and which, for all ordinary particular values of the a's, vanishes when and only when the n functions f have a repeated non-zero common root, i.e. when and only when E has a repeated linear factor, then S is necessarily the discriminant; for it and the discriminant must have the same irresoluble factors. In this we may suppose ordinary values of the a's to be those for which E does not vanish identically, but they may be defined in any way we please.

Let D_i, for each of the values 1, 2, ... $n+1$ of i, be the discriminant of the *single function* $E(\kappa_1, \kappa_2, \ldots \kappa_{n+1})$ as defined in § 194. 3 when we treat it as a general function $E(\kappa_i)$ of the single variable κ_i, so that D_i is a rational integral function of the other n parameters κ in which the coefficients are rational integral functions of the a's. Whenever the a's have such particular values that $\rho_i \neq 0$, we can find all the linear factors of E by solving the equation $E(\kappa_i) = 0$ for κ_i, and for all such particular values of the a's the identical vanishing of D_i is a necessary and sufficient condition that E shall have a repeated linear factor. It can be shown that for all values of i

$$D_i = \sigma_i S,$$

where σ_i is homogeneous in $\kappa_1, \ldots \kappa_{i-1}, \kappa_{i+1}, \ldots \kappa_{n+1}$, and S is a function of the a's only which is the same for all the D's; and it can be shown moreover that $\sigma_1, \sigma_2, \ldots \sigma_{n+1}$ cannot all vanish identically unless E vanishes identically. The function S, if it has no repeated irresoluble factors, as is actually the case, must be the discriminant of the n functions f.

Another method of determining the discriminant can be derived from geometrical considerations. Let $J_i(x_1, x_2, \ldots x_{n+1})$ be the Jacobian of the n functions f with respect to the n variables $x_1, \ldots x_{i-1}, x_{i+1}, \ldots x_{n+1}$, so that in particular

$$J_{n+1}(x_1, x_2, \ldots x_{n+1}) = \frac{\partial (f_1, f_2, \ldots f_n)}{\partial (x_1, x_2, \ldots x_n)}.$$

If $[x_1 x_2 \ldots x_{n+1}] \equiv [a_1 a_2 \ldots a_{n+1}]$ is a common root a of the n functions f in which $x_i \neq 0$, and if we regard it as a point of intersection A of the corresponding n surfaces f, it follows from Note 1 of § 191 that we can draw a straight line through A which will meet all the surfaces a second time at the same point A if and only if $J_i(a_1, a_2, \ldots a_{n+1}) = 0$; and this is the condition that a shall be a repeated common root. Now suppose that the function E is given, and let

$$[x_1 x_2 \ldots x_{n+1}] \equiv [a_{u1} a_{u2} \ldots a_{u,n+1}], \quad (u = 1, 2, \ldots r),$$

be the r non-zero common roots of the n functions f when the absolute values of their elements are so chosen that the equation (A) of Theorem I is true. Then it can be shown, by evaluating the symmetric function of the common roots on the left, that for all ordinary values of the a's the equation

$$\prod_{u=1}^{u=r} J_i(a_{u1}, a_{u2}, \ldots a_{u,n+1}) = \rho_i S$$

is true, where ρ_i is the coefficient of κ_i^r in E, and S is a rational integral function of the a's, the same for all values of i. The vanishing of S is always a sufficient condition, and for ordinary values of the a's a necessary condition, that at least one of the non-zero common roots shall be repeated. Hence if, as is actually the case, S has no repeated irresoluble factors when the a's are arbitrary, it must be the discriminant of the n functions f.

The *unreduced* and *reduced discriminants* of n special homogeneous functions are defined in the same way as unreduced and reduced resultants in § 194.

2. *Eliminant of n general functions of n variables.*

Let $\quad f_i(x_1, x_2, \ldots x_n, x_{n+1}) = \Sigma a^{(i)}_{p_1 p_2 \ldots p_n p_{n+1}} x_1^{p_1} x_2^{p_2} \ldots x_n^{p_n} x_{n+1}^{p_{n+1}}, \ldots(1)$

and $\quad f_i(x_1, x_2, \ldots x_n, 1) \quad = \Sigma a^{(i)}_{p_1 p_2 \ldots p_n p_{n+1}} x_1^{p_1} x_2^{p_2} \ldots x_n^{p_n}, \ldots\ldots\ldots(2)$

$$(p_1 + p_2 + \ldots + p_n + p_{n+1} = r_i; \quad i = 1, 2, \ldots n),$$

be the functions defined in sub-article 1. By employing the substitution $x_{n+1} = 1$ and replacing the parameter κ_{n+1} by z in Theorem I we deduce from it the second theorem which follows.

Theorem II. *Let the n general functions $f_1, f_2, \dots f_n$ of the n variables $x_1, x_2, \dots x_n$ given by (2) have any non-zero degrees $r_1, r_2, \dots r_n$; let $\kappa_1, \kappa_2, \dots \kappa_n$, z be $n+1$ arbitrary finite parameters; and let X be the general linear function*

$$X = \kappa_1 x_1 + \kappa_2 x_2 + \dots + \kappa_n x_n + z. \quad \dots\dots\dots\dots\dots\dots\dots(2')$$

Then the n functions f have an X-eliminant $E(z; \kappa_1, \kappa_2, \dots \kappa_n) = E(z) = E$, uniquely determinate except for an arbitrary finite non-zero numerical factor, which can be defined by the following properties:

(*a*) *It is a rational integral function of the a's, the κ's and z which is homogeneous of degree r in z and the κ's, and therefore has the form*

$$E(z) = P_0 z^r + P_1 z^{r-1} + \dots + P_{r-s} z^s + \dots + P_{r-1} z + P_r, \quad \dots(15)$$

where P_s is a homogeneous rational integral function of the κ's of degree s whose coefficients are rational integral functions of the a's.

(*b*) *When any particular finite values are ascribed to the a's which do not cause E to vanish identically, it becomes a function of z whose coefficients are homogeneous functions of the κ's, and it can be expressed as a product of factors homogeneous and linear in z and the κ's in the form*

$$E(z) = P_{r-s}(\kappa_1, \kappa_2, \dots \kappa_n) \,.\, E_s(z)$$
$$= \prod_{v=1}^{v=r-s} (\beta_{v1}\kappa_1 + \beta_{v2}\kappa_2 + \dots + \beta_{vn}\kappa_n) \,.\, \prod_{u=1}^{u=s} (z + \alpha_{u1}\kappa_1 + \alpha_{u2}\kappa_2 + \dots + \alpha_{un}\kappa_n), \quad \dots\dots\dots(A')$$

where s is its actual degree in z; P_{r-s} being the product of all the linear factors which do not contain z, and $E_s(z)$ being the product of all the linear factors which do contain z; and consequently all the roots of $E(z)$, both finite and infinite, are homogeneous linear functions of the κ's.

In the ordinary case when $P_0 \neq 0$ we have $s = r$, and the equation (A′) is

$$E(z) = P_0 \,.\, \prod_{u=1}^{u=r} (z + \alpha_{u1}\kappa_1 + \alpha_{u2}\kappa_2 + \dots + \alpha_{un}\kappa_n). \quad \dots\dots\dots(A'')$$

(*c*) *Under the circumstances described in (b), i.e. when E does not vanish identically,*

$$[x_1 x_2 \dots x_n] = [\alpha_1 \alpha_2 \dots \alpha_n] \quad \dots\dots\dots\dots\dots\dots\dots\dots(5')$$

is a finite common root of the n functions f if and only if

$$z+\alpha_1\kappa_1+\alpha_2\kappa_2+\dots+\alpha_n\kappa_n \quad \textit{and} \quad z=-(\alpha_1\kappa_1+\alpha_2\kappa_2+\dots+\alpha_n\kappa_n) \qquad \dots\dots(6')$$

are respectively a linear factor and a finite root of $E(z)$*; and*

$$[x_1 x_2 \dots x_n] \equiv [\beta_1 \beta_2 \dots \beta_n], \qquad \dots\dots(5'')$$

where the β*'s are finite, is an infinite common root of the n functions f if and only if*

$$\beta_1\kappa_1+\beta_2\kappa_2+\dots+\beta_n\kappa_n \quad \textit{and} \quad z\equiv-(\beta_1\kappa_1+\beta_2\kappa_2+\dots+\beta_n\kappa_n) \qquad \dots\dots(6'')$$

are respectively a linear factor and an infinite root of $E(z)$.

(*d*) *The function E has the lowest total degree which is possible for any rational integral function of the* a*'s, the* κ*'s and* z *having the property* (*c*).

Moreover E is an irresoluble function of the a*'s, the* κ*'s and* z *when these are all arbitrary; it is the resultant of the* $n+1$ *general functions* $f_1, f_2, \dots f_n, X$*; it is the eliminant of Theorem I in which* κ_{n+1} *is replaced by* z*; and it has the following further properties:*

(*a'*) *Its own coefficients are rational numbers which we may suppose to be integers.*

(*b'*) *It is homogeneous in the coefficients of each of the n functions f, its degree in the coefficients of any one function* f_i *being* $s_i=\frac{r}{r_i}$.

(*c'*) *It is isobaric of weight r with respect to each one of the variables* $x_1, x_2, \dots x_n$ *and also with respect to the homogenising variable* x_{n+1} *in the coefficients of all the* $n+1$ *functions* $f_1, f_2, \dots f_n, X$. *Consequently in every term of E the sum of the suffixes* p_j *of the* a*'s in that term and the index of* κ_j *in that term is equal to r when* $j \neq n+1$*; also in every term of E the sum of the suffixes* p_{n+1} *of the* a*'s in that term and the index of* z *in that term is equal to r.*

(*d'*) *When the* ρ*'s are defined as in* (4), *the coefficients of* $\kappa_1^r, \kappa_2^r, \dots \kappa_n^r, z^r$ *in E are respectively* $\rho_1, \rho_2, \dots \rho_n, \rho_{n+1}$, *provided only that the arbitrary numerical factors in all the functions* ρ *are suitably chosen; in particular we have in* (15)

$$P_0=\rho_{n+1}. \qquad \dots\dots(16)$$

(e') *Whenever the a's have such particular values that* $E(z;\ \kappa_1, \kappa_2, \ldots \kappa_n)$ *vanishes identically, we can always determine a common root (finite or infinite) of the* $n+1$ *functions* $f_1, f_2, \ldots f_n, X$ *when any particular finite values whatever are ascribed to* z *and the* κ*'s.*

(f') *If* f_τ, *any one of the* n *functions* f, *is specialised into* g_τ, *where* g_τ *is a function of degree* r_τ *formed by the product of two general functions* ϕ_τ, ψ_τ *of* $x_1, x_2, \ldots x_n$ *of non-zero degrees* p, q, *so that*

$$g_\tau = \phi_\tau \psi_\tau, \quad p + q = r_\tau, \quad \ldots\ldots\ldots\ldots\ldots\ldots(8')$$

and if E, E', E'' *are respectively the* X*-eliminants of the three sets of* n *functions*

$$f_1, f_2, \ldots g_\tau, \ldots f_n;\ f_1, f_2, \ldots \phi_\tau, \ldots f_n;\ f_1, f_2, \ldots \psi_\tau, \ldots f_n, \quad \ldots\ldots(9')$$

the first of these eliminants being unreduced, then the equation

$$E(z;\ \kappa_1, \kappa_2, \ldots \kappa_n) = E'(z;\ \kappa_1, \kappa_2, \ldots \kappa_n) \,.\, E''(z;\ \kappa_1, \kappa_2, \ldots \kappa_n) \quad \ldots\ldots(B')$$

is true, and is an identity in z, *the* κ*'s and the coefficients of the* $n+1$ *general functions* $f_1, \ldots f_{\tau-1}, \phi_\tau, \psi_\tau, f_{\tau+1}, \ldots f_n$, *provided only that the arbitrary numerical factors in* E, E', E'' *are suitably chosen.*

We add two corollaries corresponding to those of Theorem I, Corollary 1 being equivalent to a definition of repeated common roots.

COROLLARY 1. *When the a's have any particular values which do not cause* $E(z;\ \kappa_1, \kappa_2, \ldots \kappa_n)$ *to vanish identically, the* n *functions* f *have exactly* r *common roots, finite and infinite; and if* E *is expressed in the form* (A′), *then* (5′) *is a finite common root repeated exactly* k *times if and only if the first expression in* (6′) *is a linear factor of* E *repeated exactly* k *times, and* (5″) *is an infinite common root repeated exactly* k *times if and only if the first expression in* (6″) *is a linear factor of* E *repeated exactly* k *times.*

COROLLARY 2. *When the a's have such particular values that*

$$E(z;\ \kappa_1, \kappa_2, \ldots \kappa_n)$$

vanishes identically, then the n *functions* f *have an infinite number of common roots; moreover if* x_i *and* x_j *are any two of the variables, they have a common root (finite or infinite) in which* x_i *has any assigned value, and also a common root (finite or infinite) in which the ratio* $x_i : x_j$ *has any assigned value.*

Theorem II can be proved directly by the methods used in the proof of Theorem I, but it is more convenient to deduce it from Theorem I.

A factor of $E(z)$ in (A′) which does not contain z is regarded as having the form $0 . z+\beta_1\kappa_1+\beta_2\kappa_2+\ldots+\beta_n\kappa_n$, so that $E(z)$ has an infinite root $z \equiv -(\beta_1\kappa_1+\beta_2\kappa_2+\ldots+\beta_n\kappa_n)$ corresponding to it. We can describe $E(z)$ as a function of z whose roots, when it does not vanish identically, are the values which the expression

$$z=-(x_1\kappa_1+x_2\kappa_2+\ldots+x_n\kappa_n)$$

can have when $[x_1 x_2 \ldots x_n]$ is a common root, finite or infinite, of the n functions f. By changing the sign of z in $E(z)$ we obtain the X-eliminant when

$$X=z-\kappa_1x_1-\kappa_2x_2-\ldots-\kappa_nx_n.$$

This is a function of z whose roots, when it does not vanish identically, are the values which the expression

$$z=x_1\kappa_1+x_2\kappa_2+\ldots+x_n\kappa_n$$

can have when $[x_1 x_2 \ldots x_n]$ is a common root, finite or infinite, of the n functions f.

Since in general we know all the common roots of the n functions f when the roots of $E(z)$ are known, $E(z)$ is a *complete eliminant.* If in $E(z)$ we put all the κ's equal to 0 except $\kappa_1, \kappa_2, \ldots \kappa_i$, we obtain a *partial eliminant* which serves to determine all the simultaneous values which $x_1, x_2, \ldots x_i$ can have in the common roots of the n functions f.

The properties (*c*) and (*d*) alone are sufficient to define the function $E(z)$.

The remarks of § 194 apply to eliminants as well as resultants, and we may therefore speak of *unreduced* and *reduced eliminants* of special functions without further defining them.

The *discriminant* of the n general functions (2) is the simplest function whose vanishing is, for all ordinary particular values of their coefficients, a necessary and sufficient condition that they shall have a repeated common root, ordinary values of the coefficients being those for which E does not vanish identically. It is the same as the discriminant of the corresponding general homogeneous functions (1) which is defined in Note 9.

Most of the following examples could be derived from the examples of §§ 193 and 194 by regarding the X-eliminant E as a resultant, so that

$$E=R(X, f_1, f_2, \ldots f_n).$$

In Exs. xiv and xv we have the simplest illustrations of Theorems I and II which are possible when there is more than one function and the degrees of the functions are not all equal to 1.

Ex. i. *Theorems I and II are obviously true in the special case when $n=1$.*

For a single general homogeneous function $f=f(x_1, x_2)$ of two variables of non-zero degree r we have

$$E=f(\kappa_2, -\kappa_1) \text{ when } X=\kappa_1 x_1+\kappa_2 x_2;$$

and for a single general function $f=f(x, 1)$ of a single variable x we have

$$E=f(z, -\kappa) \text{ when } X=\kappa x+z, \quad E=f(z, \kappa) \text{ when } X=z-\kappa x.$$

Ex. ii. *Special case in which the n functions f are all linear.*

If $f_1, f_2, \dots f_n$ are the n general homogeneous linear functions given by

$$\overline{\underline{f}}_n = [a]_n^{n+1} \overline{\underline{x}}_{n+1},$$

the X-eliminant of Theorem I is clearly the determinant

$$E=R(X, f_1, f_2, \dots f_n)=\begin{vmatrix} \kappa_1 & \kappa_2 & \dots & \kappa_{n+1} \\ a_{11} & a_{12} & \dots & a_{1,n+1} \\ a_{21} & a_{22} & \dots & a_{2,n+1} \\ \dots & \dots & \dots & \dots \\ a_{n1} & a_{n2} & \dots & a_{n,n+1} \end{vmatrix}=A_1\kappa_1+A_2\kappa_2+\dots+A_{n+1}\kappa_{n+1},$$

where $A_1, A_2, \dots A_{n+1}$ are the affected simple minor determinants of $[a]_n^{n+1}$.

This determinant has all the properties of the theorem, and every function which has the properties (a), (b), (c), (b') of Theorem I can only differ from it by an ordinary numerical factor.

The discriminant of these n functions f is 1, and the functions cannot have a repeated common root.

Ex. iii. *Special case in which $f_1, f_2, \dots f_n$ are respectively products of $r_1, r_2, \dots r_n$ general homogeneous linear factors.*

Let $\quad f_1=X_{11}X_{12}\dots X_{1r_1}, \quad f_2=X_{21}X_{22}\dots X_{2r_2}, \quad \dots \quad f_n=X_{n1}X_{n2}\dots X_{nr_n},$

where the X's are general homogeneous linear functions of $x_1, x_2, \dots x_{n+1}$, and let

$$X=\kappa_1 x_1+\kappa_2 x_2+\dots+\kappa_{n+1}x_{n+1}.$$

By selecting one factor from each of the n functions f we can form exactly r distinct sets of n of the linear factors X. Let the uth set of factors formed in this way be $(X_{1u_1}, X_{2u_2}, \dots X_{nu_n})$. Then if we use the same notation as in Ex. i of § 195, it follows from Ex. ii above that the X-eliminant of the uth set of n linear factors is the determinant

$$R(X, X_{1u_1}, X_{2u_2}, \dots X_{nu_n})=\begin{vmatrix} \kappa_1 & \kappa_2 & \dots & \kappa_{n+1} \\ a_{11} & a_{12} & \dots & a_{1,n+1} \\ a_{21} & a_{22} & \dots & a_{2,n+1} \\ \dots & \dots & \dots & \dots \\ a_{n1} & a_{n2} & \dots & a_{n,n+1} \end{vmatrix}=A_1\kappa_1+A_2\kappa_2+\dots+A_{n+1}\kappa_{n+1}; \quad \dots\dots\dots(17)$$

and by repeated applications of the formula (B) of Theorem I we see that the X-eliminant of the n functions f can be taken to be the product of all the r determinants of the type (17), so that

$$E=\Pi R(X, X_{1u_1}, X_{2u_2}, \dots X_{nu_n}), \quad \dots\dots\dots\dots\dots\dots\dots\dots\dots(18)$$

$$(u_1=1, 2, \dots r_1; \quad u_2=1, 2, \dots r_2; \quad \dots \quad u_n=1, 2, \dots r_n),$$

each of the r factors of the product (18) being expressible as in (17) as a homogeneous linear factor of $\kappa_1, \kappa_2, \ldots \kappa_{n+1}$.

When E does not vanish identically, the linear factors X of the uth set have a non-zero common root corresponding to the uth factor of E in (18) which we may call the uth non-zero common root of the n functions f and denote by

$$[x_1\, x_2 \ldots x_{n+1}] \equiv [a_{u1}\, a_{u2} \ldots a_{u,n+1}].$$

We can choose the absolute values of the a's so that

$$[a_{u1}\, a_{u2} \ldots a_{u,n+1}] = [A_1\, A_2 \ldots A_{n+1}], \quad \text{.........................(19)}$$

and the identity (18) then becomes

$$E = \prod_{u=1}^{u=r} (a_{u1}\kappa_1 + a_{u2}\kappa_2 + \ldots + a_{u,n+1}\kappa_{n+1}). \quad \text{.........................(18')}$$

Ex. iv. *Discriminant of the n functions f of Ex.* iii.

Let i and j be elements of the sequence $[1, 2, \ldots n+1]$; and let $u_i, v_i, w_i, \ldots$ be elements of the sequence $[1, 2, \ldots r_i]$. Then with the notation used in Ex. i of § 195 we can write

$$R(X_{iv_i}, X_{1u_1}, X_{2u_2}, \ldots X_{nu_n}) = \begin{vmatrix} b_{i1} & b_{i2} & \ldots & b_{i,n+1} \\ a_{11} & a_{12} & \ldots & a_{1,n+1} \\ a_{21} & a_{22} & \ldots & a_{2,n+1} \\ \ldots & \ldots & \ldots & \ldots \\ a_{n1} & a_{n2} & \ldots & a_{n,n+1} \end{vmatrix} = b_{i1}A_1 + b_{i2}A_2 + \ldots + b_{i,n+1}A_{n+1}; \quad \text{.........(20)}$$

and the determinant (20) vanishes identically if and only if $v_i = u_i$.

Now let $\quad X_{iv_i}, \; \dfrac{\partial f_i}{\partial x_j}, \;$ and the Jacobian $\; J_{n+1} = \dfrac{\partial(f_1, f_2, \ldots f_n)}{\partial(x_1, x_2, \ldots x_n)},$

when regarded as functions of $x_1, x_2, \ldots x_{n+1}$, be denoted respectively by

$$X_{iv_i}(x_1, x_2, \ldots x_{n+1}), \quad f_{ij}(x_1, x_2, \ldots x_{n+1}), \quad J_{n+1}(x_1, x_2, \ldots x_{n+1}).$$

Then when the a's are defined as in (19), it is easily seen from the properties of determinants that

$$X_{iv_i}(a_{u1}, a_{u2}, \ldots a_{u,n+1}) = R(X_{iv_i}, X_{1u_1}, X_{2u_2}, \ldots X_{nu_n}),$$
$$f_{ij}(a_{u1}, a_{u2}, \ldots a_{u,n+1}) = a_{ij} \,.\, P_{iu},$$
$$J_{n+1}(a_{u1}, a_{u2}, \ldots a_{u,n+1}) = a_{u,n+1} \,.\, Q_u,$$

where $\qquad P_{iu} = \Pi R(X_{iv_i}, X_{1u_1}, X_{2u_2}, \ldots X_{nu_n})$

(v_i receiving all the values $1, 2, \ldots r_i$ except u_i), and

$$Q_u = P_{1u}\, P_{2u} \ldots P_{nu}.$$

Here P_{iu} is, for a given value of u, the product of all the $r_i - 1$ determinants (20) which can be formed when $(X_{1u_1}, X_{2u_2}, \ldots X_{nu_n})$ is the uth set of n factors X selected one from each of the n functions f, and X_{iv_i} is any other factor of f_i different from the factor X_{iu_i} already selected; and if

$$p = (r_1 - 1) + (r_2 - 1) + \ldots + (r_n - 1),$$

i.e. if p is the total degree of $J_{n+1}(x_1, x_2, \ldots x_{n+1})$, then Q_u is, for a given value of u, the product of the p determinants (20) which can be formed when $(X_{1u_1}, X_{2u_2}, \ldots X_{nu_n})$ is the

uth set of n factors X selected one from each of the n functions f, and X_{iv_i} is any other factor of the functions f different from these. It follows that

$$\prod_{u=1}^{u=r} J_{n+1}(a_{u1}, a_{u2}, \dots a_{u,n+1}) = a_{1,n+1}\, a_{2,n+1} \dots a_{r,n+1} \,.\, Q_1 Q_2 \dots Q_r$$
$$= \rho_{n+1} \,.\, T^2, \quad \dots\dots(21)$$

where ρ_{n+1} is the coefficient of $\kappa^r{}_{n+1}$ in E, which is the resultant of the n functions formed by putting $x_{n+1}=0$ in $f_1, f_2, \dots f_n$; and where T is the product of the $\frac{1}{2}rp$ distinct determinants (20) which can be formed when $(X_{1u_1}, X_{2u_2}, \dots X_{nu_n})$ are any n factors X selected one from each of the n functions f, and X_{iv_i} is any other factor of the n functions f different from these.

Clearly $T=0$ is, for all ordinary values of the coefficients of the factors X, a necessary and sufficient condition that the n functions f shall have no repeated root; and from Note 9 we conclude that

$$S = T^2$$

is the unreduced discriminant of the n functions f.

Thus the unreduced discriminant of the n functions f of Ex. iii *is the square of the product of all the distinct determinants such as* (20) *which can be formed with the coefficients of* $n+1$ *linear factors of the functions f when two factors are selected from one of the functions and one factor from each of the other functions.*

Ex. v. *The special homogeneous functions* $f_1, f_2, \dots f_n$ *given by*

$$f_i = (x_i - c_{i1}x_{n+1})(x_i - c_{i2}x_{n+1}) \dots (x_i - c_{ir_i}x_{n+1}), \quad (i = 1, 2, \dots n).$$

This is a special case of Ex. iv which can be treated directly. If $r = r_1 r_2 \dots r_n \neq 0$ and $s_i = \frac{r}{r_i}$, there are exactly r non-zero common roots given by

$$[x_1\, x_2 \dots x_n\, x_{n+1}] \equiv [c_{1u_1}, c_{2u_2}, \dots c_{nu_n}, 1],$$
$$(u_1 = 1, 2, \dots r_1;\ u_2 = 1, 2, \dots r_2;\ \dots\ u_n = 1, 2, \dots r_n);$$

the X-eliminant E can be taken to be the product of r linear factors

$$E = \Pi\,(c_{1u_1}\kappa_1 + c_{2u_2}\kappa_2 + \dots + c_{nu_n}\kappa_n + \kappa_{n+1}),$$

the coefficient of κ_i^r in E being then ρ_i, where

$$\rho_{n+1} = 1;\quad \rho_i = (c_{i1}\, c_{i2} \dots c_{ir_i})^{s_i} \text{ when } i \neq n+1;$$

and if S is the unreduced discriminant, we have

$$S = T^2$$

where $$T = \Pi\,(c_{1u_1} - c_{1v_1})^{s_1} \,.\, \Pi\,(c_{2u_2} - c_{2v_2})^{s_2} \,.\, \dots \,.\, \Pi\,(c_{nu_n} - c_{nv_n})^{s_n},$$

the different values of $[u_i v_i]$ in the last product being the distinct corranged minors of order 2 of the sequence $[1, 2, \dots r_i]$. If $p = (r_1 - 1) + (r_2 - 1) + \dots + (r_n - 1)$, S is a product of rp factors.

In obtaining S directly we first observe that for the values $1, 2, \dots n$ of i and j we have $f_{ij} = 0$ when $j \neq i$, and

$$f_{ii}(c_{1u_1}, c_{2u_2}, \dots c_{nu_n}, 1) = \Pi\,(c_{iu_i} - c_{iv_i}), \quad (v_i = 1, 2, \dots u_{i-1}, u_{i+1}, \dots r_i);$$

therefore for any one given root we have

$$J_{n+1}(c_{1u_1}, c_{2u_2}, \dots c_{nu_n}, 1) = \Pi\,(c_{1u_1} - c_{1v_1}) \,.\, \Pi\,(c_{2u_2} - c_{2v_2}) \,.\, \dots \,.\, \Pi\,(c_{nu_n} - c_{nv_n}),$$

the total number of factors being p; and for all the r roots we have

$$\Pi J_{n+1}(c_{1u_1}, c_{2u_2}, \dots c_{nu_n}, 1) = S = T^2.$$

Ex. vi. *Special case when $f_i = Ax_j^{r_i}$ in Theorem I, A being an arbitrary parameter.*

Let the substitution $x_j=0$ convert X into X_j, and the general homogeneous functions $f_1, \dots f_{i-1}, f_{i+1}, \dots f_n$ into $\phi_1, \dots \phi_{i-1}, \phi_{i+1}, \dots \phi_n$. Then the unreduced X-eliminant of the n homogeneous functions

$$f_1, \dots f_{i-1}, Ax_j^{r_i}, f_{i+1}, \dots f_n$$

is
$$E = A^{s_i} E_{ij}^{r_i}, \quad \text{...(22)}$$

where E_{ij} is the X_j-eliminant of the $n-1$ general homogeneous functions $\phi_1, \dots \phi_{i-1}, \phi_{i+1}, \dots \phi_n$ of $x_1, \dots x_{j-1}, x_{j+1}, \dots x_n$.

This follows from Ex. iv of § 194. Or it is clear that E and E_{ij} must have the same linear factors for all ordinary values of the coefficients, and the result then follows from the dimensions of E and E_{ij}.

Ex. vii. *Special case when $f_i = x_j^p g_i$ in Theorem I, where g_i is a general homogeneous function of degree $r_i - p$.*

If E_{ij} is defined as in Ex. vi, the unreduced X-eliminant of the n homogeneous functions

$$f_1, \dots f_{i-1}, x_j^p g_i, f_{i+1}, \dots f_n$$

is
$$E = E' E_{ij}^p, \quad \text{...(23)}$$

where E' is the X-eliminant of $f_1, \dots f_{i-1}, g_i, f_{i+1}, \dots f_n$.

This follows from Ex. vi and formula (B) of Theorem I. When we regard the eliminants as resultants, we can replace (23) by

$$R(X, f_1, \dots f_i, \dots f_n) = R(X, f_1, \dots g_i, \dots f_n) \,.\, \{R(X_j, \phi_1, \dots \phi_{i-1}, \phi_{i+1}, \dots \phi_n)\}^p, \quad \text{......(24)}$$

where the ϕ's are defined as in Ex. vi.

Ex. viii. *Special case when the n functions f of Theorem I have the forms*

$$f_1 = x_1^{\lambda_1} . g_1(x_1, x_{n+1}), \quad f_2 = x_2^{\lambda_2} . g_2(x_2, x_{n+1}), \ \dots \ f_n = x_n^{\lambda_n} . g_n(x_n, x_{n+1}). \ \text{...(25)}$$

Here $\lambda_1, \lambda_2, \dots \lambda_n$ are positive integers not greater than $r_1, r_2, \dots r_n$ respectively, and g_i is a general homogeneous function of degree $r_i - \lambda_i$ of the two variables x_i, x_{n+1} only. In this case the unreduced X-eliminant can be evaluated by repeated applications of Ex. vii, or of formula (B) and Ex. vi. In order to put the result into a succinct form, we will express all the eliminants as resultants, and define $X_{\lambda\mu\nu\dots}$ to be the linear function derived from

$$X = \kappa_1 x_1 + \kappa_2 x_2 + \dots + \kappa_{n+1} x_{n+1}$$

by putting $x_\lambda = x_\mu = x_\nu = \dots = 0$. Then if

$$E = R(X, f_1, f_2, \dots f_n), \quad E' = R(X, g_1, g_2, \dots g_n),$$

it will be seen that:

The unreduced X-eliminant of the n special homogeneous functions (25) *is*

$$E = E' . \Pi\{R(X_1, g_2, g_3, \dots g_n)\}^{\lambda_1} . \Pi\{R(X_{12}, g_3, g_4, \dots g_n)\}^{\lambda_1\lambda_2} . \dots$$
$$\dots \Pi\{R(g_1, X_{23\dots n})\}^{\lambda_2\lambda_3\dots\lambda_n} . \{R(X_{123\dots n})\}^{\lambda_1\lambda_2\lambda_3\dots\lambda_n},$$

where $R(X_{123\dots n}) = \kappa_{n+1}$.

In each partial product only one typical factor is shown. For ordinary particular values of the coefficients of the n functions g, the functions $f_1, f_2, \dots f_n$ have the root

$$[x_1 x_2 \dots x_n x_{n+1}] \equiv [0 0 \dots 0 1]$$

repeated exactly $\lambda_1, \lambda_2, \dots \lambda_n$ times respectively, and E has the corresponding linear factor κ_{n+1} repeated exactly $\lambda_1 \lambda_2 \dots \lambda_n$ times; this root is therefore a 'common root' repeated exactly $\lambda_1 \lambda_2 \dots \lambda_n$ times.

Ex. ix. *Special case in which the n functions f of Theorem I have the forms given by*

$$f_i = U_{r_i}^{(i)} + x_{n+1} U_{r_i-1}^{(i)} + x_{n+1}^2 U_{r_i-2}^{(i)} + \dots + x_{n+1}^{r_i-\lambda_i} U_{\lambda_i}^{(i)}$$
$$+ t \,.\, x_{n+1}^{r_i-\lambda_i+1} U_{\lambda_i-1}^{(i)} + t^2 \,.\, x_{n+1}^{r_i-\lambda_i+2} U_{\lambda_i-2}^{(i)} + \dots + t^{\lambda_i} \,.\, x_{n+1}^{r_i} U_0^{(i)}, \quad \dots\dots(26)$$

where t is an arbitrary parameter; $\lambda_1, \lambda_2, \dots \lambda_n$ *are positive integers not greater than* $r_1, r_2, \dots r_n$ *respectively; and the U's are general homogeneous rational integral functions of* $x_1, x_2, \dots x_n$ *whose degrees are indicated by their suffixes.*

When any *ordinary* particular values are given to the coefficients of the U's, and t is sufficiently small, it follows from Ex. v of § 191 that these n functions f have exactly $\lambda_1 \lambda_2 \dots \lambda_n$ common roots of the form

$$[x_1 x_2 \dots x_n x_{n+1}] \equiv [ty_1, \; ty_2, \; \dots \; ty_n, \; 1], \quad \dots\dots\dots\dots\dots\dots\dots\dots(27)$$

where $y_1, y_2, \dots y_n$ are finite constants; the values of $y_1, y_2, \dots y_n$ being the common roots of the n non-homogeneous functions

$$g_i = U_{\lambda_i}^{(i)} + U_{\lambda_i-1}^{(i)} + \dots + U_0^{(i)}, \quad (i = 1, 2, \dots n),$$

which are $\lambda_1 \lambda_2 \dots \lambda_n$ in number and all finite; consequently (for these ordinary values of the coefficients) when t is arbitrary, these n functions f have exactly $\lambda_1 \lambda_2 \dots \lambda_n$ common roots of the form (27), where $y_1, y_2, \dots y_n$ are functions of t which are finite for all small values of t. We conclude that:

For all ordinary particular values of the coefficients of the U's the X-eliminant

$$E(\kappa_1, \kappa_2, \dots \kappa_{n+1})$$

of the n functions (26) *has exactly* $\lambda_1 \lambda_2 \dots \lambda_n$ *linear factors of the form*

$$(\kappa_1 y_1 + \kappa_2 y_2 + \dots + \kappa_n y_n)\, t + \kappa_{n+1}, \quad \dots\dots\dots\dots\dots\dots\dots\dots\dots(28)$$

where $y_1, y_2, \dots y_n$ *are functions of t which are finite for all small values of t.*

Ex. x. *Special case in which the n functions f of Theorem I have the forms given by*

$$f_i = U_{r_i}^{(i)} + x_{n+1} U_{r_i-1}^{(i)} + x_{n+1}^2 U_{r_i-2}^{(i)} + \dots + x_{n+1}^{r_i-\lambda_i} U_{\lambda_i}^{(i)}, \quad \dots\dots\dots\dots(29)$$

where $\lambda_1, \lambda_2, \dots \lambda_n$ *are positive integers not greater than* $r_1, r_2, \dots r_n$ *respectively, and the U's are general homogeneous rational integral functions of* $x_1, x_2, \dots x_n$ *whose degrees are indicated by their suffixes.*

We can obtain these functions from the functions (26) by putting $t=0$. Therefore from (28) we see that for all ordinary particular values of the coefficients of the U's their X-eliminant E has λ factors equal to κ_{n+1} and is divisible by κ_{n+1}^{λ}, where $\lambda = \lambda_1 \lambda_2 \dots \lambda_n$. Since E is a rational integral function of the κ's and the coefficients of the U's, it follows (see Ex. iv of § 192) that E is divisible by κ_{n+1}^{λ} when the κ's and the coefficients of the U's are arbitrary; and because it is not divisible by any higher power of κ_{n+1} when the U's are specialised as in Ex. viii, it is not divisible by any higher power of κ_{n+1} when the U's are general functions.

Thus the X-eliminant $E(\kappa_1, \kappa_2, \dots \kappa_{n+1})$ *of the n functions* (29) *is divisible by* κ_{n+1}^{λ}, *where* $\lambda = \lambda_1 \lambda_2 \dots \lambda_n$, *but not by any higher power of* κ_{n+1}.

Ex. xi. *Whenever, for particular values of their coefficients, the functions* $f_1, f_2, \dots f_n$ *of Theorem I have the root*

$$[x_1\, x_2 \dots x_n\, x_{n+1}] \equiv [0\, 0 \dots 0\, 1] \quad \dots\dots(30)$$

repeated $\lambda_1, \lambda_2, \dots \lambda_n$ *times respectively, their X-eliminant* $E(\kappa_1, \kappa_2, \dots \kappa_{n+1})$ *is divisible by* κ_{n+1}^{λ}, *where* $\lambda = \lambda_1 \lambda_2 \dots \lambda_n$, *but not ordinarily by any higher power of* κ_{n+1}; *consequently* (30) *is a common root repeated* λ *times, but ordinarily not more than* λ *times.*

This follows from the last example and Ex. iii of § 191; for $f_1, f_2, \dots f_n$ are in this case particularisations of the functions (29).

Ex. xii. *Effect of a linear transformation of the variables on the eliminant.*

Let the homogeneous linear transformation

$$\overline{\underline{x}}_{n+1} = [l]_{n+1}^{n+1}\, \overline{\underline{y}}_{n+1}, \quad \text{in which} \quad (l)_{n+1}^{n+1} = \Delta \quad \dots\dots(31)$$

convert the homogeneous functions $f_1, f_2, \dots f_n$ of $x_1, x_2, \dots x_{n+1}$ given by (1) into the homogeneous functions $g_1, g_2, \dots g_n$ of $y_1, y_2, \dots y_{n+1}$, and the homogeneous linear function X into Y, where

$$X = \lambda_1 x_1 + \lambda_2 x_2 + \dots + \lambda_{n+1} x_{n+1}, \quad Y = \mu_1 y_1 + \mu_2 y_2 + \dots + \mu_{n+1} y_{n+1}.$$

Then if E and E′ are respectively the X-eliminant of $f_1, f_2, \dots f_n$ *and the Y-eliminant of* $g_1, g_2, \dots g_n$, *we have*

$$E'(\mu_1, \mu_2, \dots \mu_{n+1}) = \Delta^r . E(\lambda_1, \lambda_2, \dots \lambda_{n+1}). \quad \dots\dots(\text{C})$$

We may suppose provisionally that $\Delta \neq 0$. Then if the transformation (31) converts $\overline{\underline{x}}_{n+1} = \overline{\underline{a}}_{n+1}$ into $\overline{\underline{y}}_{n+1} = \overline{\underline{\beta}}_{n+1}$, we have

$$\overline{\underline{a}}_{n+1} = [l]_{n+1}^{n+1}\, \overline{\underline{\beta}}_{n+1}, \quad \overline{\underline{\lambda}}_{n+1} = \overline{\underline{l}}_{n+1}^{n+1}\, \overline{\underline{\mu}}_{n+1}, \quad [a]_{n+1}\, \overline{\underline{\lambda}}_{n+1} = [\beta]_{n+1}\, \overline{\underline{\mu}}_{n+1};$$

and if $a_1\lambda_1 + a_2\lambda_2 + \dots + a_{n+1}\lambda_{n+1}$ is a linear factor of E, then $[x]_{n+1} = [a]_{n+1}$ is a common root of the n functions f, therefore $[y]_{n+1} = [\beta]_{n+1}$ is a common root of the n functions g, and therefore $\beta_1\mu_1 + \beta_2\mu_2 + \dots + \beta_{n+1}\mu_{n+1}$ is a linear factor of E'. If the a's and λ's have such particular values that E vanishes, then one of the linear factors of E must vanish, the corresponding linear factor of E' must vanish, and therefore E' must vanish. Thus E' vanishes whenever the irresoluble function E of the a's and λ's vanishes; and since E and E' have the same total degrees in the a's and λ's, it follows that we must have

$$E'(\mu_1, \mu_2, \dots \mu_{n+1}) = L . E(\lambda_1, \lambda_2, \dots \lambda_{n+1}),$$

where L is a function of the l's only. By considering the special case in which the functions f are all linear, we can show as in Ex. ix of § 193 that $L = \Delta^r$, and deduce that (C) is an identity in the a's and λ's.

When Theorem I has been completely proved, we can also treat the above theorem as a particular case of Ex. ix of § 193.

Ex. xiii. When particular values are given to the a's, let

$$[x_1\, x_2 \dots x_n x_{n+1}] = [a_1\, a_2 \dots a_n a_{n+1}] \quad \dots\dots(32)$$

be a non-zero common root of the n functions f in Ex. xii, the corresponding linear factor of E being

$$a_1\lambda_1 + a_2\lambda_2 + \dots + a_n\lambda_n + a_{n+1}\lambda_{n+1}, \quad \dots\dots(33)$$

and let a transformation (31) in which $\Delta \neq 0$ be so chosen that it converts (32) into

$$[y_1\, y_2 \ldots y_n\, y_{n+1}] = [0\, 0 \ldots 0\, 1]; \qquad \text{.........(32')}$$

so that (32′) is a common root of the n functions g, the corresponding linear factor of E' being μ_{n+1}. Then (32) will be a root of $f_1, f_2, \ldots f_n$ repeated $k_1, k_2, \ldots k_n$ times respectively if and only if (32′) is a root of $g_1, g_2, \ldots g_n$ repeated $k_1, k_2, \ldots k_n$ times respectively; and (33) will be a linear factor of E repeated k times if and only if μ_{n+1} is a linear factor of E' repeated k times.

Ex. xiv. *The eliminant of two homogeneous functions of degrees* 1 *and* 2 *of three variables.*

If
$$\phi = lx + my + nz, \quad \psi = ax^2 + by^2 + cz^2 + 2fyz + 2gzx + 2hxy$$
and
$$X = \lambda x + \mu y + \nu z,$$
then the X-eliminant of the two functions ϕ and ψ is

$$E(\lambda, \mu, \nu) = P\lambda^2 + Q\mu^2 + R\nu^2 + 2L\mu\nu + 2M\nu\lambda + 2N\lambda\mu,$$

where
$$P = bn^2 - 2fmn + cm^2, \qquad L = (-fl + gm + hn)\, l - amn,$$
$$Q = cl^2 - 2gnl + an^2, \qquad M = (fl - gm + hn)\, m - bnl,$$
$$R = am^2 - 2hlm + bl^2, \qquad N = (fl + gm - hn)\, n - clm.$$

We can obtain E in the way described in § 200; and if we write

$$S = (bc - f^2)\, l^2 + (ca - g^2)\, m^2 + (ab - h^2)\, n^2 + 2\,(gh - af)\, mn + 2\,(hf - bg)\, nl + 2\,(fg - ch)\, lm,$$

we have
$$QR - L^2 = l^2 S, \qquad RP - M^2 = m^2 S, \qquad PQ - N^2 = n^2 S,$$
$$MN - PL = mnS, \qquad NL - QM = nlS, \qquad LM - RN = lmS,$$

$$\begin{bmatrix} P & N & M \\ N & Q & L \\ M & L & R \end{bmatrix} \begin{bmatrix} l \\ m \\ n \end{bmatrix} = 0, \qquad \begin{vmatrix} P & N & M \\ N & Q & L \\ M & L & R \end{vmatrix} = PQR + 2LMN - PL^2 - QM^2 - RN^2 = 0,$$

these equations being identities in the coefficients of ϕ and ψ. When E does not vanish identically, it can be expressed in the form

$$E(\lambda, \mu, \nu) = (\lambda x_1 + \mu y_1 + \nu z_1)\,(\lambda x_2 + \mu y_2 + \nu z_2), \qquad \text{.........(34)}$$

and then $[x\, y\, z] \equiv [x_1\, y_1\, z_1]$, $[x_2\, y_2\, z_2]$ are the two non-zero common roots of ϕ and ψ.

From the partial eliminants we see that the possible values of $y : z$, $z : x$, $x : y$ in the common roots of ϕ and ψ are given respectively by

$$Ry^2 - 2Lyz + Qz^2 = 0, \quad Pz^2 - 2Mzx + Rx^2 = 0, \quad Qx^2 - 2Nxy + Py^2 = 0,$$

and it follows that the two non-zero common roots are given by

$$x : y : z = P : N + n\sqrt{-S} : M - m\sqrt{-S} = N - n\sqrt{-S} : Q : L + l\sqrt{-S}$$
$$= M + m\sqrt{-S} : L - l\sqrt{-S} : R$$
$$= P + M + N + (m - n)\sqrt{-S} : Q + N + L + (n - l)\sqrt{-S} : R + L + M + (l - m)\sqrt{-S},$$

one root corresponding to each choice of the radical $\sqrt{-S}$. We can also obtain these roots by solving the equation $E = 0$ for one of the parameters λ, μ, ν, and so determining the factors of E. One way of determining the absolute values of the elements of the common roots so that the equation (34) is true is

$$[x_1 y_1 z_1]=\frac{1}{\sqrt{R}}\{M+m\sqrt{-S},\ L-l\sqrt{-S},\ R\},\quad [x_2 y_2 z_2]=\frac{1}{\sqrt{R}}\{M-m\sqrt{-S},\ L+l\sqrt{-S},\ R\}.$$

If D_1, D_2, D_3 are the discriminants of E when we regard it as a function of λ, or μ, or ν only, we have

$$D_1=(m\nu-n\mu)^2 S,\quad D_2=(n\lambda-l\nu)^2 S,\quad D_3=(l\mu-m\lambda)^2 S.$$

The first factors on the right can only vanish identically when P, Q, R vanish respectively; and they cannot all vanish identically unless E vanishes identically. Consequently S must be the discriminant of the two functions ϕ and ψ, as is otherwise obvious.

The eliminant E is also the resultant of the three functions ϕ, ψ, X; and can also be written in the form

$$E=a(m\nu-n\mu)^2+b(n\lambda-l\nu)^2+c(l\mu-m\lambda)^2+2f(n\lambda-l\nu)(l\mu-m\lambda)$$
$$+2g(l\mu-m\lambda)(m\nu-n\mu)+2h(m\nu-n\mu)(n\lambda-l\nu).$$

The slant of scale 2 of the three functions ϕ, ψ, X (which will be defined in a later chapter) is a derangement of the matrix

$$[\omega]_6^7=\begin{bmatrix} a & l & 0 & 0 & \lambda & 0 & 0 \\ b & 0 & m & 0 & 0 & \mu & 0 \\ c & 0 & 0 & n & 0 & 0 & \nu \\ 2f & 0 & n & m & 0 & \nu & \mu \\ 2g & n & 0 & l & \nu & 0 & \lambda \\ 2h & m & l & 0 & \mu & \lambda & 0 \end{bmatrix},$$

and the matrix of the affected simple minor determinants of $[\omega]_6^7$ is

$$[0,\ \lambda E,\ \mu E,\ \nu E,\ -lE,\ -mE,\ -nE].$$

Thus, in accordance with a general theorem, the resultant of the three functions ϕ, ψ and X is the H. C. F. of the simple minor determinants of their slant. From the identity

$$[x^2,\ y^2,\ z^2,\ yz,\ zx,\ xy][\omega]_6^7=[\psi,\ x\phi,\ y\phi,\ z\phi,\ xX,\ yX,\ zX],$$

we see that the vanishing of E is a necessary condition that ϕ, ψ and X shall have a non-zero root in common.

Ex. XV. *The eliminant of two general functions of degrees* 1 *and* 2 *of two variables.*

If $$\phi=lx+my+n,\quad \psi=ax^2+2hxy+by^2+2gx+2fy+c$$
and $$X=z+\lambda x+\mu y,$$

the X-eliminant of the two functions ϕ and ψ is

$$E(z;\ \lambda,\ \mu)=E(z)=Rz^2+2(M\lambda+L\mu)z+(P\lambda^2+2N\lambda\mu+Q\mu^2)$$
$$=R(z+\lambda x_1+\mu y_1)(z+\lambda x_2+\mu y_2),$$

where

$$x_1=\frac{M+m\sqrt{-S}}{R},\quad y_1=\frac{L-l\sqrt{-S}}{R};\quad x_2=\frac{M-m\sqrt{-S}}{R},\quad y_2=\frac{L+l\sqrt{-S}}{R};$$

these being the common roots of ϕ and ψ. Putting $\mu=0$, $\lambda=-1$ and $\lambda=0$, $\mu=-1$ in $E(z)$, we see that the equations giving the two values of x and the two values of y are respectively

$$Rx^2-2Mx+P=0,\quad Ry^2-2Ly+Q=0.$$

Ex. xvi. If $\phi = lx+my+nz, \quad \psi = (x-y+z)(x+y+2z)+kz^2,$

we see from Ex. xiv that the (x, z)-resultant of ϕ and ψ vanishes identically without the identical vanishing of ϕ only in the following two cases:

$$(1)\quad m=l, \quad n=2l, \quad k=0; \quad \phi=l(x+y+2z), \quad \psi=(x-y+z)(x+y+2z); \quad E=0.$$

$$(2)\quad m=-l, \quad n=l, \quad k=0; \quad \phi=l(x-y+z), \quad \psi=(x-y+z)(x+y+2z); \quad E=0.$$

In both cases the complete eliminant E vanishes, and ϕ is a factor of ψ. There is no non-zero common root in which two variables are 0.

Ex. xvii. If $\phi = lx+my+nz, \quad \psi = 2xy+kz^2,$

we see from Ex. xiv that the (x, z)-resultant of ϕ and ψ vanishes identically without the identical vanishing of ϕ only in the following two cases:

$$(1)\quad m=0; \quad \phi=lx+nz, \quad \psi=2xy+kz^2; \quad E=\mu(-2n^2\lambda+kl^2\mu+2ln\nu).$$

$$(2)\quad l=0, \quad n=0, \quad k=0; \quad \phi=my, \quad \psi=2xy; \quad E=0.$$

In the first case E does not vanish identically, the non-zero common roots are

$$[x\,y\,z]\equiv[0\;1\;0], \quad [2n^2, \; -kl^2, \; -2ln],$$

and in one of the non-zero common roots (see Note 4 of § 195) we have $[x\,z]=[0\;0]$.

In the second case E vanishes identically, and ϕ is a factor of ψ; further in all the non-zero common roots in which $x:y=1:2$ we must have $[x\,y]=[0\;0]$.

Ex. xviii. If $\phi = x, \quad \psi = xy - z^2,$

then the (x, z)-resultant of ϕ and ψ vanishes identically whilst $E(\lambda, \mu, \nu) = -\mu^2$. The two non-zero common roots are

$$[x\,y\,z]\equiv[0\;1\;0], \quad [0\;1\;0],$$

and in both of them (see Note 4 of § 195) we have $[x\,z]=[0\;0]$.

Ex. xix. If $\phi = lx+my+n, \quad \psi = 2xy+k^2$

and ϕ does not vanish identically, the complete eliminant of ϕ and ψ can only vanish identically when $k=l=n=0$, and then ϕ is a factor of ψ.

Ex. xx. If $\phi = x, \quad \psi = xy-1,$

then the x-resultant of ϕ and ψ vanishes identically whilst $E(z; \lambda, \mu) = -\mu^2$. These two functions have only the two infinite common roots

$$[x\,y]\equiv[0\;1], \; [0\;1].$$

Ex. xxi. **Theorem.** *If $f_1(x, y, z)$ and $f_2(x, y, z)$ are any two particular homogeneous rational integral functions of the three variables x, y, z of non-zero degrees, they have an infinite number of common roots (i.e. their complete eliminant vanishes identically) when and only when they have a common factor of non-zero degree.*

First suppose that they have such a common factor. Then they clearly must have an infinite number of common roots.

Next suppose that they have an infinite number of common roots. Then we have to show that they have a common factor (of non-zero degree). In proving this we shall consider that f_1 and f_2 are both regular in each one of the variables, so that neither function has a finite non-zero root in which two of the elements are 0. This is allowable because we can always regularise the functions by an ordinary homogeneous linear transformation of the variables.

In this second case we shall now make the hypothesis that f_1 and f_2 have no common factor. Then by Ex. viii of § 188 there exists an identity of the form

$$P_1f_1+P_2f_2=\phi(y, z), \quad\text{.....................................(35)}$$

where P_1 and P_2 are homogeneous functions of x, y, z, and where ϕ is a homogeneous rational integral function of y and z only which does not vanish identically. Let β and γ be any two finite numbers not both zero so chosen that $\phi(\beta, \gamma)\neq 0$. Then by Corollary 2 to Theorem I the two functions f_1 and f_2 have a finite non-zero common root in which $y:z=\beta:\gamma$; and because they have no non-zero common root in which y and z are both 0, it follows that there exists a finite number α such that $[x\,y\,z]=[\alpha\,\beta\,\gamma]$ is a common root of f_1 and f_2. For these values of x, y, z the left-hand side of (35) vanishes whilst the right-hand side does not vanish. Therefore our hypothesis is impossible, and it follows that f_1 and f_2 have a common factor.

By the substitution $z=1$ we deduce that:

If $f_1(x, y)$ and $f_2(x, y)$ are any two particular rational integral functions of the two variables x and y of non-zero degrees, they have an infinite number of common roots (i.e. their complete eliminant vanishes identically) when and only when they have a common factor of non-zero degree.

Illustrations of these two theorems are given in Exs. xvi, xvii and xix.

Ex. xxii. *The eliminant of two homogeneous functions of degrees* 2 *and* 2 *of three variables.*

Let $\phi=ax^2+by^2+cz^2+fyz+gzx+hxy, \quad \Phi=Ax^2+By^2+Cz^2+Fyz+Gzx+Hxy,$

and $$X=\lambda x+\mu y+\nu z.$$

Then if $$M=\begin{bmatrix} a & b & c & f & g & h \\ A & B & C & F & G & H\end{bmatrix},$$

and if we use for the simple minor determinants of M the notations

$$P_{022}=bC-cB,\quad Q_{211}=gH-hG,\quad T_{211}=aF-fA,\quad L_{031}=bF-fB,\quad L_{013}=fC-cF,$$
$$P_{202}=cA-aC,\quad Q_{121}=hF-fH,\quad T_{121}=bG-gB,\quad L_{103}=cG-gC,\quad L_{301}=gA-aG,$$
$$P_{220}=aB-bA,\quad Q_{112}=fG-gF,\quad T_{112}=cH-hC,\quad L_{310}=aH-hA,\quad L_{130}=hB-bH,$$

the suffixes indicating the weights with respect to x, y, z, the X-eliminant of ϕ and Φ, which is also the resultant of ϕ, Φ and X, is the function

$$E(\lambda, \mu, \nu)=\Sigma\rho_{pqr}\lambda^p\mu^q\nu^r,\quad (p+q+r=4),$$

where there are three cyclically symmetric coefficients of each of the five types

$$\rho_{400}=\quad P^2_{022}-L_{031}L_{013},$$
$$\rho_{310}=\quad 2P_{022}T_{112}+(Q_{121}+T_{121})L_{013}-L_{031}L_{103},$$
$$\rho_{301}=-2P_{022}T_{121}+(Q_{112}-T_{112})L_{031}-L_{013}L_{130},$$
$$\rho_{220}=-2P_{022}P_{202}+T^2_{112}+(Q_{211}-T_{211})L_{013}+(Q_{121}+T_{121})L_{103},$$
$$\rho_{211}=-2P_{022}Q_{211}+2(P_{202}L_{031}+P_{220}L_{013})+(Q_{121}-T_{121})T_{112}$$
$$-(Q_{121}+T_{121})Q_{112}-L_{130}L_{103}.$$

Since there are six independent relations between the fifteen simple minor determinants of M, there are six independent relations between the fifteen coefficients ρ. These relations are the conditions that E, regarded as a homogeneous function of λ, μ, ν of degree 4 with arbitrary coefficients, shall be expressible as a product of linear factors.

The slant of scale 3 of the three functions X, ϕ, Φ is the matrix

$$\omega=[\omega]_{10}^{12}=\left[\begin{array}{c|cc|ccc|c|cc|c|cc}
\lambda & \cdot & \cdot & \cdot & \cdot & \cdot & a & \cdot & \cdot & A & \cdot & \cdot \\
\hline
\mu & \lambda & \cdot & \cdot & \cdot & \cdot & h & a & \cdot & H & A & \cdot \\
\nu & \cdot & \lambda & \cdot & \cdot & \cdot & g & \cdot & a & G & \cdot & A \\
\hline
\cdot & \mu & \cdot & \lambda & \cdot & \cdot & b & h & \cdot & B & H & \cdot \\
\cdot & \nu & \mu & \cdot & \lambda & \cdot & f & g & h & F & G & H \\
\cdot & \cdot & \nu & \cdot & \cdot & \lambda & c & \cdot & g & C & \cdot & G \\
\hline
\cdot & \cdot & \cdot & \mu & \cdot & \cdot & \cdot & b & \cdot & \cdot & B & \cdot \\
\cdot & \cdot & \cdot & \nu & \mu & \cdot & \cdot & f & b & \cdot & F & B \\
\cdot & \cdot & \cdot & \cdot & \nu & \mu & \cdot & c & f & \cdot & C & F \\
\cdot & \cdot & \cdot & \cdot & \cdot & \nu & \cdot & \cdot & c & \cdot & \cdot & C
\end{array}\right],$$

where each dot stands for 0. By putting $\mu=\nu=0$, it will be seen at once that this matrix is undegenerate of rank 10 when the coefficients of X, ϕ, Φ are arbitrary. The identity

$$[x^3,\ x^2y,\ x^2z,\ xy^2,\ xyz,\ xz^2,\ y^3,\ y^2z,\ yz^2,\ z^3]\,[\omega]_{10}^{12}$$
$$=[x^2X,\ xyX,\ xzX,\ y^2X,\ yzX,\ z^2X,\ x\phi,\ y\phi,\ z\phi,\ x\Phi,\ y\Phi,\ z\Phi]$$

shows that ω is degenerate whenever the three functions X, ϕ, Φ have a non-zero root in common, i.e. whenever E vanishes for particular values of their coefficients. Therefore E is a factor of every simple minor determinant of ω. Knowing this it is easily seen that it is the H.C.F. of the simple minor determinants of ω. In fact the normal to ω is the matrix

$$\omega'=[\omega']_2^{12}=\left[\begin{array}{c|cc|ccc|c|cc|c|cc}
a & h & g & b & f & c & -\lambda & -\mu & -\nu & \cdot & \cdot & \cdot \\
A & H & G & B & F & C & \cdot & \cdot & \cdot & -\lambda & -\mu & -\nu
\end{array}\right],$$

and therefore each of the 66 simple minor determinants of ω is the product of E and the anti-correspondent simple minor determinant of ω'. Thus we can determine E by evaluating any one of the 60 non-zero simple minor determinants of ω.

If $\phi=ax^2+by^2+cz^2+2fyz+2gzx+2hxy$, $\Phi=Ax^2+By^2+Cz^2+2Fyz+2Gzx+2Hxy$, the coefficients f, g, h, F, G, H being replaced by $2f$, $2g$, $2h$, $2F$, $2G$, $2H$, and if we put

$$T=(bC+cB-2fF)\,\lambda^2+(cA+aC-2gG)\,\mu^2+(aB+bA-2hH)\,\nu^2$$
$$-2\,(gH+hG-aF-fA)\,\mu\nu-2\,(hF+fH-bG-gB)\,\nu\lambda-2\,(fG+gF-cH-hC)\,\lambda\mu,$$

$$S=(bc-f^2)\,\lambda^2+(ca-g^2)\,\mu^2+(ab-h^2)\,\nu^2+2\,(gh-af)\,\mu\nu+2\,(hf-bg)\,\nu\lambda+2\,(fg-ch)\,\lambda\mu,$$

$$S'=(BC-F^2)\,\lambda^2+(CA-G^2)\,\mu^2+(AB-H^2)\,\nu^2$$
$$+2\,(GH-AF)\,\mu\nu+2\,(HF-BG)\,\nu\lambda+2\,(FG-CH)\,\lambda\mu,$$

we have
$$E(\lambda,\ \mu,\ \nu)=T^2-4SS'.$$

We can regard $E(\lambda, \mu, \nu)=0$ as the tangential equation of the 4-pointic locus of intersection of the two conics $\phi=0$, $\Phi=0$.

CHAPTER XXII

SYMMETRIC FUNCTIONS OF THE ELEMENTS OF SIMILAR SEQUENCES

[The first two articles deal with symmetric functions of the rn elements of r sequences each containing n arbitrary elements; in § 197 we show that all *monotypic* symmetric functions can be expressed as rational integral functions of those of order 1 (the σ's), and also as rational integral functions of those of degree 1 (the ϖ's, or the elementary symmetric functions); and in § 198 we consider the corresponding properties of *homogeneous monotypic* symmetric functions. The next article deals with the relations between the common roots and the coefficients of a set of rational integral functions having only a finite number of common roots; it is shown that every monotypic symmetric function of the common roots can be expressed in a unique way as a rational function of the coefficients. Finally in § 200 we show how resultants and eliminants can be determined, and complete the proof of the theorems of §§ 193 and 196. The relations between elementary symmetric functions are discussed in §§ 197 and 198, and the corresponding relations between the coefficients of an eliminant in § 199.]

§ 197. Symmetric functions of the elements of r similar sequences.

1. *Symmetric functions in general and monotypic symmetric functions.*

Let $$X = [x_1 x_2 \dots x_h \dots x_n] \quad \dots\dots\dots\dots(1)$$

be a sequence or one-rowed matrix of order n in which the elements are n scalar variables, and let

$$X_1 = [\alpha_{11}\alpha_{12} \dots \alpha_{1h} \dots \alpha_{1n}], \quad X_2 = [\alpha_{21}\alpha_{22} \dots \alpha_{2h} \dots \alpha_{2n}], \dots$$
$$\dots X_r = [\alpha_{r1}\alpha_{r2} \dots \alpha_{rh} \dots \alpha_{rn}], \quad \dots\dots\dots(2)$$

be r values of X in which all the rn elements are arbitrary. In the applications the r matrices (2) will usually be the common roots of a number of rational integral functions.

A *symmetric function* of the elements of the r matrices (2) will mean a rational integral function of those rn elements which remains unaltered when we give any permutation whatever to $X_1, X_2, \dots X_r$ without deranging the elements of any one of them. It is therefore an absolute invariant for all such permutations of $X_1, X_2, \dots X_r$. A *monotypic symmetric function* of the elements of the r matrices (2) will mean a symmetric function of the rn elements in which every term can be derived from any one term by some permutation of $X_1, X_2, \dots X_r$, and no term is repeated. Every symmetric function of the elements is a sum of a number of monotypic symmetric functions each multiplied by a numerical constant. Consequently the general properties of symmetric functions can be deduced from those of monotypic symmetric functions, and our attention will chiefly be directed to these.

A *monotypic symmetric function* S of the rn elements of the r sequences (2) is completely known when any one of its terms is known. Hence, when we disregard a numerical factor, every such function is representable without ambiguity in the form

$$S = \Sigma\{T_1 T_2 \dots T_r\}, \qquad \text{.............................(3)}$$

where

$$T_i = \alpha_{i1}^{\lambda_{i1}} \dots \alpha_{ih}^{\lambda_{ih}} \dots \alpha_{in}^{\lambda_{in}}, \quad T_j = \alpha_{j1}^{\lambda_{j1}} \dots \alpha_{jh}^{\lambda_{jh}} \dots \alpha_{jn}^{\lambda_{jn}}, \quad T_k = \alpha_{k1}^{\lambda_{k1}} \dots \alpha_{kh}^{\lambda_{kh}} \dots \alpha_{kn}^{\lambda_{kn}}, \quad (4)$$

the λ's being positive integers, some of which may be 0's. Here T_i is a product of elements selected from the matrix X_i, and $T_1 T_2 \dots T_r$ is any one term of S selected as a *typical term*.

Let $$S' = \Sigma'\{T_1 T_2 \dots T_r\} \qquad \text{..........................(3')}$$

be the sum of all the $r!$ terms which can be derived from the typical term $T_1 T_2 \dots T_r$ by giving all the possible $r!$ permutations to $X_1, X_2, \dots X_r$ (including the identical permutation which leaves their arrangement unaltered), the indices λ remaining unchanged. Clearly S has the same distinct terms as S', but it has no repeated term, whereas S' may have repeated terms. Thus another term of S' is $U_1 U_2 \dots U_r$, where

$$U_i = \alpha_{j1}^{\lambda_{i1}} \dots \alpha_{jh}^{\lambda_{ih}} \dots \alpha_{jn}^{\lambda_{in}}, \quad U_j = \alpha_{i1}^{\lambda_{j1}} \dots \alpha_{ih}^{\lambda_{jh}} \dots \alpha_{in}^{\lambda_{jn}}, \quad U_k = \alpha_{k1}^{\lambda_{k1}} \dots \alpha_{kh}^{\lambda_{kh}} \dots \alpha_{kn}^{\lambda_{kn}}, \quad (4')$$

but this is not another term of S if $[\lambda_{j1}\lambda_{j2} \dots \lambda_{jn}] = [\lambda_{i1}\lambda_{i2} \dots \lambda_{in}]$.

We can also define S' to be the sum of all the $r!$ terms which can be derived from the typical term $T_1 T_2 \dots T_r$ by giving all the possible $r!$ permutations to the r sequences

$$\lambda_1 = [\lambda_{11}\lambda_{12} \dots \lambda_{1h} \dots \lambda_{1n}], \quad \lambda_2 = [\lambda_{21}\lambda_{22} \dots \lambda_{2h} \dots \lambda_{2n}], \dots$$
$$\dots \lambda_r = [\lambda_{r1}\lambda_{r2} \dots \lambda_{rh} \dots \lambda_{rn}] \qquad \text{........(2')}$$

formed by the indices in $T_1, T_2, \dots T_r$, the α's remaining unaltered. Thus another term of S' is $T_1' T_2' \dots T_r'$, where

$$T_i' = \alpha_{i1}^{\lambda_{j1}} \dots \alpha_{ih}^{\lambda_{jh}} \dots \alpha_{in}^{\lambda_{jn}}, \quad T_j' = \alpha_{j1}^{\lambda_{i1}} \dots \alpha_{jh}^{\lambda_{ih}} \dots \alpha_{jn}^{\lambda_{in}}, \quad T_k' = \alpha_{k1}^{\lambda_{k1}} \dots \alpha_{kh}^{\lambda_{kh}} \dots \alpha_{kn}^{\lambda_{kn}}, \quad (4'')$$

but this is not another term of S if $\lambda_j = \lambda_i$. We obtain S from S' by discarding all repeated terms.

From the second way of developing S and S' we see what the total number of terms in S is. When the r sequences (2′) are all different, the terms of S' are all different, therefore S contains $r!$ terms, and we have $S' = S$; when the r sequences (2′) consist of τ distinct sequences repeated $r_1, r_2, \dots r_\tau$ times respectively, so that $r_1 + r_2 + \dots + r_\tau = r$, each distinct term of S' is repeated $r_1!\, r_2!\, \ldots r_\tau!$ times, therefore S contains $\dfrac{r!}{r_1!\, r_2!\, \ldots r_\tau!}$ terms, and we have $S' = r_1!\, r_2!\, \ldots r_\tau!\, S$; when the r sequences (2′) are all equal, all the terms of S' are equal, therefore S consists of one term only, and we have $S' = r!\, S$.

In representing S in the form given by (3) and (4) it is usual to omit all those elements α in (4) and in the development of S whose indices are 0's. In this case we can obtain all other terms of S from the typical term by making repeated interchanges of two elements of the sequence $[1, 2, \ldots r]$ in the *first* suffixes of the α's, the second suffixes and all the λ's remaining unchanged, and every repeated term thus obtained being discarded; and this is the plan usually followed. This shorter representation of S is however completely definite if and only if r, the number of the matrices (2), is given. When we develope S' and S from the typical term by permuting the sequences (2′) formed by the indices, we must retain the zero indices.

Ex. i. If $X = [x_1\, x_2\, x_3]$, then for the four sequences
$$X_1 = [a_{11}\, a_{12}\, a_{13}], \quad X_2 = [a_{21}\, a_{22}\, a_{23}], \quad X_3 = [a_{31}\, a_{32}\, a_{33}], \quad X_4 = [a_{41}\, a_{42}\, a_{43}],$$
we have
$$\Sigma\{a_{11}^2 a_{12} a_{21}\} = a_{11}^2 a_{12} a_{21} + a_{21}^2 a_{22} a_{31} + a_{31}^2 a_{32} a_{41} + a_{41}^2 a_{42} a_{11}$$
$$+ a_{11}^2 a_{12} a_{31} + a_{21}^2 a_{22} a_{41} + a_{31}^2 a_{32} a_{11} + a_{41}^2 a_{42} a_{21}$$
$$+ a_{11}^2 a_{12} a_{41} + a_{21}^2 a_{22} a_{11} + a_{31}^2 a_{32} a_{21} + a_{41}^2 a_{42} a_{31}.$$

The sequences formed by the indices in the typical term (and in every term) are
$$[2\,1\,0], \quad [1\,0\,0], \quad [0\,0\,0], \quad [0\,0\,0],$$
and we have
$$\Sigma'\{a_{11}^2 a_{12} a_{21}\} = 2\Sigma\{a_{11}^2 a_{12} a_{21}\}.$$

Ex. ii. When the number of variables, i.e. the number n, is small, it is simpler to use a single-suffix notation. Thus if in Ex. i we replace x_1, x_2, x_3 by x, y, z, so that $X = [x\,y\,z]$, then for the four sequences
$$X_1 = [\alpha_1\, \beta_1\, \gamma_1], \quad X_2 = [\alpha_2\, \beta_2\, \gamma_2], \quad X_3 = [\alpha_3\, \beta_3\, \gamma_3], \quad X_4 = [\alpha_4\, \beta_4\, \gamma_4],$$
we have
$$\Sigma\{\alpha_1^2 \beta_1 \alpha_2\} = \alpha_1^2 \beta_1 \alpha_2 + \alpha_2^2 \beta_2 \alpha_3 + \alpha_3^2 \beta_3 \alpha_4 + \alpha_4^2 \beta_4 \alpha_1$$
$$+ \alpha_1^2 \beta_1 \alpha_3 + \alpha_2^2 \beta_2 \alpha_4 + \alpha_3^2 \beta_3 \alpha_1 + \alpha_4^2 \beta_4 \alpha_2$$
$$+ \alpha_1^2 \beta_1 \alpha_4 + \alpha_2^2 \beta_2 \alpha_1 + \alpha_3^2 \beta_3 \alpha_2 + \alpha_4^2 \beta_4 \alpha_3,$$
the development from the typical term being obtained by permutations of the suffixes 1, 2, 3, 4.

If S is the monotypic symmetric function (3), then by the *weight of S with respect to the variable* x_h we mean the sum of the indices of those elements $\alpha_{1h}, \alpha_{2h}, \ldots \alpha_{rh}$ in any term of S which are values of x_h, i.e. the number $\lambda_{1h} + \lambda_{2h} + \ldots + \lambda_{rh}$ which is the same for every term of S. Thus if $u_1, u_2, \ldots u_n$ are the weights of S with respect to $x_1, x_2, \ldots x_n$, we have
$$u_h = \lambda_{1h} + \lambda_{2h} + \ldots + \lambda_{rh}, \quad (h = 1, 2, \ldots n) \ldots\ldots\ldots\ldots\ldots(5)$$

By the *degree of any term of S in the matrix or sequence X_i* we mean the sum of the indices of the elements $\alpha_{i1}, \alpha_{i2}, \ldots \alpha_{in}$ of X_i in that term, this number being in general different in different terms. The degrees of any term of S in the r separate matrices $X_1, X_2, \ldots X_r$ are the same r numbers for every term, except as regards their order of arrangement. The degrees of the typical term in the separate matrices $X_1, X_2, \ldots X_r$ are the numbers $v_1, v_2, \ldots v_r$ given by

$$v_k = \lambda_{k1} + \lambda_{k2} + \ldots + \lambda_{kn}, \quad (k = 1, 2, \ldots r). \quad \ldots\ldots\ldots\ldots\ldots(6)$$

The typical term is usually so chosen that the r numbers $v_1, v_2, \ldots v_r$ are arranged in descending order of magnitude, and may then be called a *standard typical term*.

By the *degree of S in a single matrix or sequence X_i* we mean the greatest degree which a term of S can have in X_i, this number being the same for every matrix X_i, and being the greatest of the sums (6). It will often be called simply the *degree of S*. When the typical term is chosen in the standard way, the degree of S in each and every separate matrix is the degree of the typical term in X_1, i.e. the number

$$s = \lambda_{11} + \lambda_{12} + \ldots + \lambda_{1n} = v_1. \quad \ldots\ldots\ldots\ldots\ldots\ldots\ldots\ldots(7)$$

By the *total weight* of S we mean the sum of its weights with respect to all the variables $x_1, x_2, \ldots x_n$, this number being the sum of all the indices in each term of S, i.e. the number w given by

$$w = u_1 + u_2 + \ldots + u_n = v_1 + v_2 + \ldots + v_r. \quad \ldots\ldots\ldots\ldots\ldots(8)$$

It is also the sum of the degrees of each term of S in the separate matrices $X_1, X_2, \ldots X_r$, and may be called the *total degree of each term of S* in all the r matrices.

The monotypic symmetric function S is *homogeneous* in the matrices $X_1, X_2, \ldots X_r$ when every one of its terms has the same degree in every one of those matrices. Thus S is homogeneous of degree s in the matrices if and only if

$$v_1 = v_2 = \ldots = v_r = s;$$

and this is the case when and only when its total weight is rs.

By the *order* of S we mean the number of matrices which contribute elements with non-zero indices to any one term of S, this number being the same for every term of S, and being the number of the integers $v_1, v_2, \ldots v_r$ which are different from 0. When the order of S is i, where of course $i \ngtr r$, and the typical term is chosen in the standard way, we have $\lambda_{kh} = 0$ when $k > i$, and $T_k = 1$ when $k > i$. Accordingly when elements with zero indices are omitted, the standard representation of a monotypic symmetric function S of order i is

$$S = \Sigma \{T_1 T_2 \ldots T_i\}. \quad \ldots\ldots\ldots\ldots\ldots\ldots\ldots\ldots\ldots(3'')$$

Here the typical term is so chosen that $X_1, X_2, \dots X_i$ are the matrices which contribute elements with non-zero indices to it; and we have

$$u_h = \lambda_{1h} + \lambda_{2h} + \dots + \lambda_{ih}, \quad (h = 1, 2, \dots n); \quad v_k = \lambda_{k1} + \lambda_{k2} + \dots + \lambda_{kn}, \quad (k = 1, 2, \dots i),$$

$$v_{i+1} = v_{i+2} = \dots = v_r = 0; \quad w = u_1 + u_2 + \dots + u_n = v_1 + v_2 + \dots + v_i.$$

The integers $v_1, v_2, \dots v_i$ are the *non-zero degrees* of each term of S in the separate matrices.

The abbreviated notation (3″) is definite if and only if the number of sequences, i.e. the number r, is given; but if $i \not> r$, the substitution $[\alpha_{r1}\alpha_{r2} \dots \alpha_{rn}] = 0$ converts the function $\Sigma\{T_1 T_2 \dots T_i\}$ of the r sequences $X_1, X_2, \dots X_r$ into the function $\Sigma\{T_1 T_2 \dots T_i\}$ of the $r-1$ sequences $X_1, X_2, \dots X_{r-1}$. If $i > r$, we must interpret (3′) to be 0.

Ex. iii. The monotypic symmetric functions S of Exs. i and ii have weights 3, 1, 0 with respect to the variables x_1, x_2, x_3 or x, y, z. In each case the function S has degree 3 in each of the matrices X_1, X_2, X_3, X_4; it has total weight 4; and it has order 2. The degrees of each term in the separate matrices X_1, X_2, X_3, X_4 are 3, 1, 0, 0; and the greatest of these numbers, i.e. the number 3, is the degree of S in each matrix.

2. *Monotypic symmetric functions of order* 1, *and of degree* 1.

If $p_1, p_2, \dots p_n$ are any n given positive integers which are not all 0, there is one and only one *monotypic symmetric function of order* 1 of the elements of the r matrices (2) whose weights with respect to the variables $x_1, x_2, \dots x_n$ are respectively $p_1, p_2, \dots p_n$. For this function we will use the notation $\sigma_{p_1 p_2 \dots p_n}$ in which the weights are indicated by the suffixes. It is the function of degree $p_1 + p_2 + \dots + p_n$ in each matrix given by

$$\sigma_{p_1 p_2 \dots p_n} = \Sigma\{T_1\} = \Sigma \alpha_{11}^{p_1} \alpha_{12}^{p_2} \dots \alpha_{1n}^{p_n}. \qquad \text{.....................(9)}$$

For the monotypic symmetric function of order 0 we will use the corresponding notation $\sigma_{00\dots0}$, which we must interpret to mean the integer r. By the σ's we mean the functions thus defined.

The *monotypic symmetric functions* of the elements of the r matrices (2) *of degree* 1 *in each matrix* are those which are expressible in the form (3′), where $T_1, T_2, \dots T_i$ each consist of one unrepeated element only, and $i \not> r$. If $q_1, q_2, \dots q_n$ are any n given positive integers which are not all 0, and which satisfy the condition

$$q_1 + q_2 + \dots + q_n \not> r,$$

there is one and only one such function whose weights with respect to the variables $x_1, x_2, \dots x_n$ are respectively $q_1, q_2, \dots q_n$. For this function we will use the notation $\varpi_{q_1 q_2 \dots q_n}$ in which the weights are indicated by the suffixes. If $i = q_1 + q_2 + \dots + q_n$ is its total weight, it is the monotypic symmetric function of order i given by

$$\varpi_{q_1 q_2 \dots q_n} = \Sigma\{T_1 T_2 \dots T_i\} = \Sigma \alpha_{1\mu_1} \alpha_{2\mu_2} \dots \alpha_{i\mu_i}, \qquad \text{...........(9′)}$$

where the first q_1 of the μ's are equal to 1, the next q_2 of the μ's are equal to 2, ... the last q_n of the μ's are equal to n; there being of course no μ equal to k when $q_k=0$. The typical term in (9′) contains i factors, one selected from each of the matrices $X_1, X_2, \dots X_i$, and of these i factors q_1 are values of x_1, q_2 values of $x_2, \dots q_n$ values of x_n. For the monotypic symmetric function of degree 0 in each matrix we will use the corresponding notation $\varpi_{00\dots 0}$, which we must interpret to mean the integer 1. When

$$q_1+q_2+\dots+q_n>r,$$

we must interpret $\varpi_{q_1 q_2 \dots q_n}$ to be 0. By the ϖ's we mean the functions thus defined. They are usually called the *elementary symmetric functions.*

In the notations for the σ's and ϖ's the number n of the variables is indicated (by the number of the suffixes), but not the number r of the sequences. They have definite meanings for every value of r, and they can be used when r is an arbitrary positive integer on the understanding that $\varpi_{q_1 q_2 \dots q_n}$ vanishes whenever $q_1+q_2+\dots+q_n>r$.

Ex. iv. For the four sequences X_1, X_2, X_3, X_4 of Ex. i we have

$$\sigma_{302}=\Sigma a_{11}{}^3 a_{13}{}^2=a_{11}{}^3 a_{13}{}^2+a_{21}{}^3 a_{23}{}^2+a_{31}{}^3 a_{33}{}^2+a_{41}{}^3 a_{43}{}^2,$$

$$\begin{aligned}\varpi_{201}=\Sigma a_{11}a_{21}a_{33}=a_{11}a_{21}a_{33}&+a_{21}a_{31}a_{43}+a_{31}a_{41}a_{13}+a_{41}a_{11}a_{23}\\&+a_{21}a_{31}a_{13}+a_{31}a_{41}a_{23}+a_{41}a_{11}a_{33}+a_{11}a_{21}a_{43}\\&+a_{31}a_{11}a_{23}+a_{41}a_{21}a_{33}+a_{11}a_{31}a_{43}+a_{21}a_{41}a_{13}.\end{aligned}$$

It will be observed that σ_{302}, ϖ_{201} for four values of $[x_1 x_2 x_3]$ are the same as ϖ_{32}, ϖ_{21} for four values of $[x_1 x_3]$, the variable with respect to which the weight is 0 being struck out.

Ex. v. For the four sequences X_1, X_2, X_3, X_4 of Ex. ii we have

$$\sigma_{302}=\Sigma a_1{}^3\gamma_1{}^2=a_1{}^3\gamma_1{}^2+a_2{}^3\gamma_2{}^2+a_3{}^3\gamma_3{}^2+a_4{}^3\gamma_4{}^2,$$

$$\begin{aligned}\varpi_{201}=\Sigma a_1 a_2\gamma_3=a_1a_2\gamma_3&+a_2a_3\gamma_4+a_3a_4\gamma_1+a_4a_1\gamma_2\\&+a_2a_3\gamma_1+a_3a_4\gamma_2+a_4a_1\gamma_3+a_1a_2\gamma_4\\&+a_3a_1\gamma_2+a_4a_2\gamma_3+a_1a_3\gamma_4+a_2a_4\gamma_1.\end{aligned}$$

NOTE 1. *Special case when* $n=1$.

In this case there is only a single variable. If we denote this by x, and its r values by $a_1, a_2, \dots a_r$, these being arbitrary parameters which may be the roots of an equation $f(x)=0$ of degree r, our r sequences are

$$X_1=[a_1],\quad X_2=[a_2],\ \dots\ X_r=[a_r],$$

and a general monotypic symmetric function of $a_1, a_2, \dots a_r$ has the form

$$S=\Sigma\,\{a_1{}^{\lambda_1}a_2{}^{\lambda_2}\dots a_r{}^{\lambda_r}\}.$$

The monotypic symmetric functions of order 1 and the elementary symmetric functions of degree 1 in each parameter have respectively the forms

$$\sigma_p=\Sigma a_1{}^p=a_1{}^p+a_2{}^p+\dots+a_r{}^p,$$

$$\varpi_q=\Sigma a_1 a_2\dots a_q,\quad (q=1, 2, \dots r),$$

whilst $\sigma_0=r$, $\varpi_0=1$, and $\varpi_q=0$ when $q>r$. In this case we have Newton's identical equations connecting the σ's and ϖ's, viz.

$$\sigma_1-\varpi_1=0,\quad \sigma_2-\sigma_1\varpi_1+2\varpi_2=0,\quad \sigma_3-\sigma_2\varpi_1+\sigma_1\varpi_2-3\varpi_3=0,\ \dots,$$

the general equation being

$$\sigma_p-\sigma_{p-1}\varpi_1+\sigma_{p-2}\varpi_2-\dots+(-1)^{p-1}.\sigma_1\varpi_{p-1}+(-1)^p.p\varpi_p=0, \quad\dots\dots\dots(10)$$

where $\varpi_i=0$ when $i>r$. The equations (10) enable us to express $\sigma_1, \sigma_2, \sigma_3, \dots$ in succession as rational integral functions of $\varpi_1, \varpi_2, \dots \varpi_r$, and also to express

$$\varpi_1, \varpi_2, \dots \varpi_r, \sigma_{r+1}, \sigma_{r+2}, \dots$$

in succession as rational integral functions of $\sigma_1, \sigma_2, \dots \sigma_r$. In fact we have

$$\sigma_1=\varpi_1,\quad \sigma_2=\varpi_1^2-2\varpi_2,\quad \sigma_3=\varpi_1^3-3\varpi_1\varpi_2+3\varpi_3,$$

$$\sigma_4=\varpi_1^4-4\varpi_1^2\varpi_2+4\varpi_1\varpi_3+2\varpi_2^2-4\varpi_4,$$

$$\sigma_5=\varpi_1^5-5\varpi_1^3\varpi_2+5(\varpi_1^2\varpi_3+\varpi_1\varpi_2^2)-5(\varpi_1\varpi_4+\varpi_2\varpi_3)+5\varpi_5,$$

$$\sigma_6=\varpi_1^6-6\varpi_1^4\varpi_2+(6\varpi_1^3\varpi_3+9\varpi_1^2\varpi_2^2)-(6\varpi_1^2\varpi_4+12\varpi_1\varpi_2\varpi_3+2\varpi_2^3)$$
$$+(6\varpi_1\varpi_5+6\varpi_2\varpi_4+3\varpi_3^2)-6\varpi_6,$$

$$\dots\dots\dots\dots\dots\dots\dots\dots\dots\dots,\quad\dots\dots(11)$$

and
$$\varpi_1=\sigma_1,\quad 2!\,\varpi_2=\sigma_1^2-\sigma_2,\quad 3!\,\varpi_3=\sigma_1^3-3\sigma_1\sigma_2+2\sigma_3,\ \dots,\quad\dots\dots\dots\dots(11')$$

the corresponding general formulae being

$$\sigma_p=\begin{vmatrix} \varpi_1 & 1 & 0 & \dots & 0 & 0 \\ 2\varpi_2 & \varpi_1 & 1 & \dots & 0 & 0 \\ 3\varpi_3 & \varpi_2 & \varpi_1 & \dots & 0 & 0 \\ \dots & \dots & \dots & \dots & \dots & \dots \\ (p-1)\varpi_{p-1} & \varpi_{p-2} & \varpi_{p-3} & \dots & \varpi_1 & 1 \\ p\varpi_p & \varpi_{p-1} & \varpi_{p-2} & \dots & \varpi_2 & \varpi_1 \end{vmatrix},\quad q!\,\varpi_q=\begin{vmatrix} \sigma_1 & 1 & 0 & \dots & 0 & 0 \\ \sigma_2 & \sigma_1 & 2 & \dots & 0 & 0 \\ \sigma_3 & \sigma_2 & \sigma_1 & \dots & 0 & 0 \\ \dots & \dots & \dots & \dots & \dots & \dots \\ \sigma_{q-1} & \sigma_{q-2} & \sigma_{q-3} & \dots & \sigma_1 & q-1 \\ \sigma_q & \sigma_{q-1} & \sigma_{q-2} & \dots & \sigma_2 & \sigma_1 \end{vmatrix}$$

$$\dots\dots(11'')$$

The equations (10), (11), (11′), (11″) are true for all positive integral values of r, provided that we interpret ϖ_i to be 0 when $i>r$; and they are identities in $a_1, a_2, \dots a_r$.

It is known from the theory of equations in a single variable that $\varpi_1, \varpi_2, \dots \varpi_r$ are independent functions to which any arbitrary values can be ascribed, and that every symmetric rational integral function f of $a_1, a_2, \dots a_r$ can be expressed in one and only one way as a rational integral function F of $\varpi_1, \varpi_2, \dots \varpi_r$, the coefficients of F being integers when the coefficients of f are integers.

NOTE 2. *Relations between the σ's and ϖ's for r sequences of n variables when r is arbitrary.*

Let $\kappa_1, \kappa_2, \dots \kappa_n$ be n auxiliary arbitrary parameters; and let the matrix X in (1) be replaced by the single quantity

$$t=\kappa_1x_1+\kappa_2x_2+\dots+\kappa_nx_n,$$

and the matrices $X_1, X_2, \dots X_r$ by the corresponding values $t_1, t_2, \dots t_r$ of t given by

$$t_i=\kappa_1a_{i1}+\kappa_2a_{i2}+\dots+\kappa_na_{in},\quad (i=1, 2, \dots r).$$

Then the monotypic symmetric functions of order 1 and the elementary symmetric functions of degree 1 of $t_1, t_2, \dots t_r$ are given respectively by

$$\sigma_p = \Sigma t_1^p = \sum_{i=1}^{i=r} (\kappa_1 a_{i1} + \kappa_2 a_{i2} + \dots + \kappa_n a_{in})^p$$

$$= \Sigma \frac{p!}{p_1!\, p_2! \dots p_n!} \sigma_{p_1 p_2 \dots p_n} \cdot \kappa_1^{p_1} \kappa_2^{p_2} \dots \kappa_n^{p_n}, \quad (p_1 + p_2 + \dots + p_n = p),$$

$$\varpi_q = \Sigma t_1 t_2 \dots t_q$$

$$= \Sigma \varpi_{q_1 q_2 \dots q_n} \cdot \kappa_1^{q_1} \kappa_2^{q_2} \dots \kappa_n^{q_n}, \quad (q_1 + q_2 + \dots + q_n = q; \; q \not> r),$$

where $\sigma_{p_1 p_2 \dots p_n}$ and $\varpi_{q_1 q_2 \dots q_n}$ are defined as in the text; and the second equation remains true when $q > r$ if we then interpret ϖ_q to be 0. These equations are identities in the κ's and the elements of the matrices (2). We could define the σ's and ϖ's for the matrices (2) in this way as being the coefficients in the expansions on the right.

When we substitute these values for $\sigma_p, \dots \varpi_q, \dots$ in the equations (11), regard the resulting equations as rational integral identities in the κ's, and equate the coefficients of corresponding terms on both sides, we obtain equations expressing the quantities $\sigma_{p_1 p_2 \dots p_n}$ as rational integral functions of the quantities $\varpi_{q_1 q_2 \dots q_n}$, these equations being identities in the a's. Since r does not occur in these equations, they are true when r is arbitrary, provided that we interpret a quantity $\varpi_{q_1 q_2 \dots q_n}$ for r sequences to be 0 when its total weight $q_1 + q_2 + \dots + q_n$ is greater than r. The general character of these equations is shown in Ex. vi. In the expression for $\sigma_{p_1 p_2 \dots p_n}$ the term of highest degree in the ϖ's is $\varpi_{100\dots0}^{p_1} \varpi_{010\dots0}^{p_2} \dots \varpi_{000\dots1}^{p_n}$; and if $p = p_1 + p_2 + \dots + p_n$, the term of lowest degree is $N \cdot \varpi_{p_1 p_2 \dots p_n}$ where N is the rational number given by $(p-1)!\, N = (-1)^{p-1} \cdot p_1!\, p_2! \dots p_n!$.

By substituting in the equations (11′) we can obtain equations expressing the quantities $\varpi_{q_1 q_2 \dots q_n}$ as rational integral functions of the quantities $\sigma_{p_1 p_2 \dots p_n}$.

In all these equations every term of the expression on the right must have the same weights with respect to $x_1, x_2, \dots x_n$ as the quantity on the left.

Ex. vi. By the process described in Note 2 we can obtain the following equations which are identities in the elements of the r sequences $X_1, X_2, \dots X_r$ for all values of r, provided that $\varpi_{q_1 q_2 \dots q_n} = 0$ when $q_1 + q_2 + \dots + q_n > r$. The equations are classified according to their total weights.

(1) $\sigma_1 = \varpi_1;$

(2) $\sigma_2 = \varpi_1^2 - 2\varpi_2,$

$\sigma_{11} = \varpi_{10}\varpi_{01} - \varpi_{11};$

(3) $\sigma_3 = \varpi_1^3 - 3\varpi_1\varpi_2 + 3\varpi_3,$

$\sigma_{21} = \varpi_{10}^2\varpi_{01} - (\varpi_{10}\varpi_{11} + \varpi_{01}\varpi_{20}) + \varpi_{21},$

$2\sigma_{111} = 2\varpi_{100}\varpi_{010}\varpi_{001} - (\varpi_{100}\varpi_{011} + \varpi_{010}\varpi_{101} + \varpi_{001}\varpi_{110}) + \varpi_{111};$

(4) $\sigma_4 = \varpi_1^4 - 4\varpi_1^2\varpi_2 + (4\varpi_1\varpi_3 + 2\varpi_2^2) - 4\varpi_4,$

$\sigma_{31} = \varpi_{10}^3\varpi_{01} - \varpi_{10}(\varpi_{10}\varpi_{11} + 2\varpi_{01}\varpi_{20}) + (\varpi_{10}\varpi_{21} + \varpi_{01}\varpi_{30} + \varpi_{11}\varpi_{20}) - \varpi_{31},$

$$3\sigma_{22} = 3\varpi_{10}^2\varpi_{01}^2 - 2(\varpi_{10}^2\varpi_{02} + 2\varpi_{10}\varpi_{01}\varpi_{11} + \varpi_{01}^2\varpi_{20}) + (2\varpi_{10}\varpi_{12} + 2\varpi_{01}\varpi_{21} + \varpi_{11}^2 + 2\varpi_{20}\varpi_{02}) - 2\varpi_{22},$$

$$3\sigma_{211} = 3\varpi_{100}^2\varpi_{010}\varpi_{001} - \{\varpi_{100}^2\varpi_{011} + 2\varpi_{100}(\varpi_{010}\varpi_{101} + \varpi_{001}\varpi_{110}) + 2\varpi_{010}\varpi_{001}\varpi_{200}\} + \{\varpi_{100}\varpi_{111} + (\varpi_{010}\varpi_{201} + \varpi_{001}\varpi_{210}) + \varpi_{011}\varpi_{200} + \varpi_{101}\varpi_{110}\} - \varpi_{211},$$

$$6\sigma_{1111} = 6\varpi_{1000}\varpi_{0100}\varpi_{0010}\varpi_{0001} - 2\Sigma_6\varpi_{1000}\varpi_{0100}\varpi_{0011} + \Sigma_4\varpi_{1000}\varpi_{0111} + \Sigma_3\varpi_{1100}\varpi_{0011} - \varpi_{1111}.$$

In the last equation Σ_6 means the symmetric sum of six terms of the type shown, and Σ_4 and Σ_3 have similar meanings.

The number of variables is indicated by the number of suffixes; but in any one of the equations we can supply any number of zero suffixes occupying the same positions throughout. Thus for three variables the complete set of equations of total weight 2 deduced from (2) is

$$(2')\qquad \sigma_{200}=\varpi_{100}^2-2\varpi_{200},\qquad \sigma_{020}=\varpi_{010}^2-2\varpi_{020},\qquad \sigma_{002}=\varpi_{001}^2-2\varpi_{002},$$

$$\sigma_{011}=\varpi_{010}\varpi_{001}-\varpi_{011},\quad \sigma_{101}=\varpi_{001}\varpi_{100}-\varpi_{101},\quad \sigma_{110}=\varpi_{100}\varpi_{010}-\varpi_{110}.$$

We can also interchange two suffixes occupying corresponding positions throughout any one of the equations. When there are only two sequences, the equation for σ_{31} becomes

$$\sigma_{31}=\varpi_{10}^3\varpi_{01}-\varpi_{10}(\varpi_{10}\varpi_{11}+2\varpi_{01}\varpi_{20})+\varpi_{11}\varpi_{20}.$$

NOTE 3. *Number of the functions ϖ and standard relations between them.*

For r sequences of n variables the total number of the elementary symmetric functions ϖ is

$$^nH_r=\binom{n+r}{r}=\binom{n+r}{n}, \qquad\qquad (a)$$

there being $\binom{n+i-1}{i}$ of total weight i, where $i \not> r$, and the total number being in accordance with the equation

$$^nH_r={}^{n-1}H_0+{}^{n-1}H_1+{}^{n-1}H_2+\ldots+{}^{n-1}H_r, \quad (\text{where } {}^{n-1}H_0=1).$$

In this enumeration we include $\varpi_{00\ldots0}$, but exclude the ϖ's of total weight greater than r, every one of which is 0. Since one of the ϖ's is the constant 1, and the rest are functions of rn arbitrary parameters, we may expect to have exactly

$$\binom{n+r}{r}-rn-1 \qquad\qquad (\beta)$$

independent relations between them, excluding the relation $\varpi_{00\ldots0}=1$. That this is actually the case can be seen from the equation

$$\prod_{i=1}^{i=r}(\kappa_1 a_{i1}+\kappa_2 a_{i2}+\ldots+\kappa_n a_{in}+1)=\Sigma\varpi_{q_1q_2\ldots q_n}\kappa_1^{q_1}\kappa_2^{q_2}\ldots\kappa_n^{q_n}, \qquad (12)$$

$$(q_1+q_2+\ldots+q_n \not> r),$$

which is an identity in the κ's and a's when $\varpi_{00\ldots0}=1$.

If we regard both sides of (12) as rational integral functions of the κ's and equate the coefficients of corresponding terms on both sides, we obtain $\binom{n+r}{r}-1$ equations, not counting the equation $\varpi_{00\ldots0}=1$, each having the form

$$\varpi_{q_1q_2\ldots q_n}=f(a), \quad (q_1+q_2+\ldots+q_n \not> r), \qquad (13)$$

where $f(a)$ is a rational integral function of the a's; and from these equations, which are in fact the definitions of the ϖ's, we have to eliminate the rn quantities a. To do this we will divide the equations (13) into three classes:

(i) the r equations in which $q_1+q_2+\ldots+q_{n-1}=0$;

(ii) the $r(n-1)$ equations in which $q_1+q_2+\ldots+q_{n-1}=1$;

(iii) the $\binom{n+r}{r}-rn-1$ equations in which $q_1+q_2+\ldots+q_{n-1}>1$;

and for the sake of brevity we will use for the ϖ's in which $q_1+q_2+\ldots+q_{n-1}=0$ the notation

$$\varpi_{00\ldots i}=\varpi_i.$$

The equations (i) express the fact that $a_{1n}, a_{2n}, \ldots a_{rn}$ are the roots $z_1, z_2, \ldots z_r$ of the equation of degree r

$$\phi(z)=z^r-\varpi_1 z^{r-1}+\varpi_2 z^{r-2}-\ldots+(-1)^r\varpi_r=0. \quad\ldots\ldots(14)$$

Those r of the equations (ii) in which $q_1=1$ can be expressed in the forms

$$\Sigma a_{i1}=\varpi_{10\ldots0}, \quad (i=1, 2, \ldots r),$$
$$\Sigma(\varpi_1-z_i)\,a_{i1}=\varpi_{10\ldots1},$$
$$\Sigma(\varpi_2-\varpi_1 z_i+z_i^2)\,a_{i1}=\varpi_{10\ldots2},$$
$$\Sigma(\varpi_3-\varpi_2 z_i+\varpi_1 z_i^2-z_i^3)\,a_{i1}=\varpi_{10\ldots3},$$
$$\ldots\ldots\ldots\ldots\ldots\ldots(15)$$

and they are r unconnected linear equations in $a_{11}, a_{21}, \ldots a_{r1}$ which can be used to express these quantities as rational functions of $z_1, z_2, \ldots z_r$ and those ϖ's in which

$$q_1+q_2+\ldots+q_{n-1}=0 \text{ or } 1.$$

In fact if h and k are positive integers, and if Π_k is the sum of the distinct products of total degree k which can be formed with $z_1, z_2, \ldots z_r$, it is known from the properties of symmetric functions of the roots of a function of one variable (see Burnside and Panton's *Theory of Equations*, Chapter XII) that

$$\Sigma\frac{z_i^h}{\phi'(z_i)}=0 \quad\text{when } h<r-1;$$

$$\Sigma\frac{z_i^{r-1}}{\phi'(z_i)}=1, \quad \Sigma\frac{z_i^{r-1+k}}{\phi'(z_i)}=\Pi_k;$$

$$\varpi_1-\Pi_1=0, \quad \varpi_2-\varpi_1\Pi_1+\Pi_2=0, \quad \varpi_3-\varpi_2\Pi_1+\varpi_1\Pi_2-\Pi_3=0, \quad \ldots;$$

and it is easily verified that the equations (15) have the unique solution

$$a_{i1}=\frac{\varpi_{10\ldots0}z_i^{r-1}-\varpi_{10\ldots1}z_i^{r-2}+\ldots+(-1)^{r-1}\varpi_{1,0,\ldots r-1}}{\phi'(z_i)}, \quad (i=1, 2, \ldots r). \quad\ldots(16)$$

Similar remarks are applicable to those r of the equations (ii) in which $q_j=1$, where j is any one of the integers $1, 2, \ldots n-1$; these being r unconnected linear equations in $a_{1j}, a_{2j}, \ldots a_{rj}$ which have a unique solution in which the value of a_{ij} is obtained from the value of a_{i1} given above by interchanging the first suffix 1 and the jth suffix 0 in each of the ϖ's in the numerator.

Thus the rn equations (i) and (ii) admit of unique solution for the rn quantities a_{ij} when arbitrary values are ascribed to the ϖ's in which $q_1+q_2+\ldots+q_{n-1}=0$ or 1; and we conclude that:

Those rn of the ϖ's, not counting $\varpi_{00\ldots0}$, in which $q_1+q_2+\ldots+q_{n-1}=0$ or 1 are independent functions to which any arbitrary values can be ascribed.

When we substitute for each quantity a_{ij} in which $j\neq n$ its value found by solving the equations (ii), and then multiply throughout by $\phi'(z_1)\,\phi'(z_2)\ldots\phi'(z_r)$, each of the equations (iii) assumes the form

$$\phi'(z_1)\,\phi'(z_2)\ldots\phi'(z_r)\,.\,\varpi_{q_1q_2\ldots q_n}=Q(z_1, z_2, \ldots z_r), \quad\ldots\ldots(17')$$

where Q is a symmetric function of $z_1, z_2, \ldots z_r$ in which the coefficients are rational integral functions of those ϖ's in which $q_1+q_2+\ldots+q_{n-1}=0$ or 1; and when we substitute for the symmetric functions of $z_1, z_2, \dots z_r$ occurring on both sides their values in terms of the coefficients of the function $\phi(z)$ given by (14), we can replace (17′) by

$$\Delta \,.\, \varpi_{q_1 q_2 \dots q_n} = S, \quad (q_1+q_2+\dots+q_{n-1} > 1), \qquad \text{(17)}$$

where S is a rational integral function of those ϖ's in which $q_1+q_2+\dots+q_{n-1}=0$ or 1, and where Δ is the discriminant of $\phi(z)$, and is therefore a rational integral function of those ϖ's in which $q_1+q_2+\dots+q_{n-1}=0$. The equations (17) are all independent, because the second factor on the left is not the same in any two of them; and we conclude that:

There are exactly $\binom{n+r}{r} - rn - 1$ *independent rational integral relations between the* ϖ*'s of r sequences of n variables, not counting the relation* $\varpi_{00\dots0}=1$; *and we can take these to be the relations* (17) *which express those* ϖ*'s in which* $q_1+q_2+\dots+q_{n-1}>1$ *as rational functions of the* ϖ*'s in which* $q_1+q_2+\dots+q_{n-1}=0$ *or* 1.

From the way in which the above results have been proved we see further that:

The relations between the ϖ*'s of r sequences of n variables are the necessary and sufficient conditions that the general function of degree r*

$$f(x_1, x_2, \dots x_n) = \Sigma \varpi_{q_1 q_2 \dots q_n} x_1^{q_1} x_2^{q_2} \dots x_n^{q_n}, \quad (q_1+q_2+\dots+q_n \not> r),$$

shall be expressible as a product of linear factors, when $\varpi_{00\dots0}=1$ *and the other coefficients are arbitrary.*

If we homogenise these relations by the substitution $1=\varpi_{00\dots0}$, they become the necessary and sufficient conditions that $f(x_1, x_2, \dots x_n)$ shall be expressible as a product of linear factors when all the coefficients are arbitrary and $\varpi_{00\dots0} \neq 0$.

Both sides of (17) are isobaric with respect to each one of the variables, the weights with respect to $x_1, x_2, \dots x_{n-1}, x_n$ being $q_1, q_2, \dots q_{n-1}, q_n + r(r-1)$; the left-hand side has total degree $2r-1$; and the expression S on the right is homogeneous of degree $q_1+q_2+\dots+q_{n-1}$ in those ϖ's for which $q_1+q_2+\dots+q_{n-1}=1$. Further the degree of S in the ϖ's for which $q_1+q_2+\dots+q_{n-1}=0$ is $r-1$ or r according as $q_n=0$ or $\neq 0$; therefore the total degree of S cannot exceed $2r-1$; and when we homogenise the relation (17) by the substitution $1=\varpi_{00\dots0}$, its degree is $2r-1$.

NOTE 4. *General character of the relations between the* ϖ*'s.*

By a relation between the ϖ's of r sequences of n variables we mean an equation of the form

$$F(\varpi_{q_1 q_2 \dots q_n}, \dots, \dots) = F(\varpi) = 0, \quad (q_1+q_2+\dots+q_n \not> r), \qquad \text{(18)}$$

where $F(\varpi)$ is a rational integral function of the ϖ's which does not vanish identically, the equation (18) being an identity in the a's. In every such relation we may and will suppose $F(\varpi)$ to be isobaric in the ϖ's with respect to each one of the variables. The standard relations (17) obtained in Note 3 are all of this form, but they are not necessarily the simplest set of independent relations. We can usually find relations of lower degrees and of smaller weights than the standard relations. The problem of finding the simplest possible set of independent relations has not been solved, though it has been shown that no relation can have total weight less than $r+2$.

Every relation between the ϖ's of r sequences of n variables (n being given) has reference to a particular value or particular values of r, which must be specified. There does not exist any relation which is true for all values of r, or for all values of r greater than a given value. On the other hand if s is any positive integer less than r, a relation between the ϖ's of r sequences of n variables becomes a relation between s sequences of n variables when in it we substitute 0 for every ϖ in which the sum of the suffixes (i.e. the total weight) is greater than s.

For if the ϖ's are defined as in the text, and if we use the letter Π with suffixes to denote the elementary symmetric functions of the elements of the $r+1$ sequences

$$X_1=[a_{11}a_{12}\dots a_{1n}],\ \dots\ X_r=[a_{r1}a_{r2}\dots a_{rn}],\quad X_{r+1}=[\xi_1\xi_2\dots\xi_n],\quad \dots\dots\dots(2'')$$

then each Π is a linear function of the ϖ's given by

$$\Pi_{q_1q_2\dots q_n}=\varpi_{q_1q_2\dots q_n}+\xi_1\,.\,\varpi_{q_1-1,q_2,\dots q_n}+\xi_2\,.\,\varpi_{q_1,q_2-1,\dots q_n}+\dots+\xi_n\,.\,\varpi_{q_1,q_2,\dots q_n-1},\quad (19)$$

the formula (19) being true for all integral values of the q's, provided that we interpret a Π or ϖ with a negative suffix to be 0, and of course consider a Π whose total weight is greater than $r+1$ to be 0 and a ϖ whose total weight is greater than r to be 0, the total weight being the sum of the suffixes. Thus the substitution $[\xi_1\xi_2\dots\xi_n]=0$ converts $\Pi_{q_1q_2\dots q_n}$ into $\varpi_{q_1q_2\dots q_n}$, and a relation between the Π's into a relation between the corresponding ϖ's; and this process can be repeated.

If $f(\Pi)=0$ is any relation between the Π's which have just been defined, and if we substitute for the Π's their values given by (19), we obtain a rational integral equation $g(\xi,\varpi)=0$ which is an identity in the ξ's and a's. If we arrange g as a rational integral function of the ξ's, the coefficient of every term is a rational integral function of the ϖ's which vanishes for all values of the a's; consequently when we equate to 0 the coefficient of any term which does not vanish identically, we must obtain one of the relations between the ϖ's. Thus from any given relation between the elementary symmetric functions Π of $r+1$ sequences of n variables we can deduce by means of the equations (19) a number of relations of lower weights between the elementary symmetric functions ϖ of r sequences of n variables.

It may happen that $f(\Pi)$ contains no Π of total weight $r+1$, and then $f(\varpi)=0$ is one of the relations between the ϖ's. We see from Ex. x that such cases are possible; for

$$\begin{vmatrix} 2\varpi_{200} & \varpi_{110} & \varpi_{101} & \varpi_{100} \\ \varpi_{110} & 2\varpi_{020} & \varpi_{011} & \varpi_{010} \\ \varpi_{101} & \varpi_{011} & 2\varpi_{002} & \varpi_{001} \\ 2\varpi_{100} & 2\varpi_{010} & 2\varpi_{001} & 3\varpi_{000} \end{vmatrix}=0$$

is a relation between the ϖ's of 3 sequences of 3 variables which does not involve any ϖ of total weight 3, and is therefore also a relation between the ϖ's of 2 sequences of 3 variables. Thus although there does not exist any relation between the ϖ's of r sequences of n variables which is true when r is arbitrary, there can exist relations which are true for two successive values of r.

Ex. vii. For 2 sequences of 2 variables we have the one relation

$$\frac{1}{2}\begin{vmatrix} 2\varpi_{20} & \varpi_{11} & \varpi_{10} \\ \varpi_{11} & 2\varpi_{02} & \varpi_{01} \\ \varpi_{10} & \varpi_{01} & 2\varpi_{00} \end{vmatrix}=(4\varpi_{20}\varpi_{02}-\varpi_{11}^2)-(\varpi_{10}^2\varpi_{02}-\varpi_{10}\varpi_{01}\varpi_{11}+\varpi_{01}^2\varpi_{20})=0,$$

or

$$\begin{vmatrix} 2\varpi_{02} & \varpi_{01} \\ \varpi_{01} & 2\varpi_{00} \end{vmatrix}.\,\varpi_{20}=\varpi_{10}^2\,.\,\varpi_{02}-\varpi_{10}\varpi_{11}\,.\,\varpi_{01}+\varpi_{11}^2\,.\,\varpi_{00}.$$

This relation (see Note 4) is not true for 3 sequences of 2 variables.

Ex. viii. For 2 sequences of 3 variables such as $[\alpha_1\beta_1\gamma_1]$, $[\alpha_2\beta_2\gamma_2]$, $[\alpha_3\beta_3\gamma_3]$ there are three independent relations which express the fact that the rank of the matrix

$$\begin{bmatrix} 2\varpi_{200} & \varpi_{110} & \varpi_{101} & \varpi_{100} \\ \varpi_{110} & 2\varpi_{020} & \varpi_{011} & \varpi_{010} \\ \varpi_{101} & \varpi_{011} & 2\varpi_{002} & \varpi_{001} \\ \varpi_{100} & \varpi_{010} & \varpi_{001} & 2\varpi_{000} \end{bmatrix} = \begin{bmatrix} \alpha_1 & \alpha_2 \\ \beta_1 & \beta_2 \\ \gamma_1 & \gamma_2 \\ 1 & 1 \end{bmatrix} \begin{bmatrix} \alpha_2 & \beta_2 & \gamma_2 & 1 \\ \alpha_1 & \beta_1 & \gamma_1 & 1 \end{bmatrix}$$

is not greater than 2. The three independent relations can be taken to be

$$\begin{vmatrix} 2\varpi_{200} & \varpi_{101} & \varpi_{100} \\ \varpi_{101} & 2\varpi_{002} & \varpi_{001} \\ \varpi_{100} & \varpi_{001} & 2\varpi_{000} \end{vmatrix} = 0, \quad \begin{vmatrix} 2\varpi_{020} & \varpi_{011} & \varpi_{010} \\ \varpi_{011} & 2\varpi_{002} & \varpi_{001} \\ \varpi_{010} & \varpi_{001} & 2\varpi_{000} \end{vmatrix} = 0, \quad \begin{vmatrix} \varpi_{110} & \varpi_{011} & \varpi_{010} \\ \varpi_{101} & 2\varpi_{002} & \varpi_{001} \\ \varpi_{100} & \varpi_{001} & 2\varpi_{000} \end{vmatrix} = 0,$$

the first two of these being deducible from the relation of Ex. vii by the insertion of zero suffixes. If we homogenise the relation of Ex. vii in the way described in § 198.1, we obtain the same equation as by equating to 0 the leading diagonal minor determinant of order 3 of the matrix given above.

For 2 sequences of n variables the relations between the ϖ's are those which express the fact that the symmetric matrix of order $n+1$ of the coefficients of the general quadratic function

$$f(x_1, x_2, \dots x_n) = \Sigma\varpi_{q_1 q_2 \dots q_n} x_1^{q_1} x_2^{q_2} \dots x_n^{q_n}, \quad (q_1+q_2+\dots+q_n \not> 2),$$

formed in the same way as the matrix given above shall have rank not greater than 2; for these are the conditions that the *quadratic* function f shall be expressible as a product of linear factors. Each relation has degree 3, being formed by equating a determinant of order 3 to 0.

Ex. ix. For 3 sequences of 2 variables such as $[\alpha_1\beta_1]$, $[\alpha_2\beta_2]$, $[\alpha_3\beta_3]$ there are exactly 3 independent relations between the ϖ's. When these are formed as in Note 3, they are the relations of weights (2, 6), (2, 7), (3, 6)

$$\begin{aligned} \Delta . \varpi_{20} = {} & \varpi_{10}^2 . (9\varpi_{03}^2 - 4\varpi_{01}\varpi_{02}\varpi_{03} + \varpi_{02}^3) + \varpi_{11}^2 . (\varpi_{02}^2 - 3\varpi_{01}\varpi_{03}) + \varpi_{12}^2 . (\varpi_{01}^2 - 3\varpi_{02}) \\ & + \varpi_{10}\varpi_{11} . (4\varpi_{01}^2\varpi_{03} - \varpi_{01}\varpi_{02}^2 - 3\varpi_{02}\varpi_{03}) + \varpi_{10}\varpi_{12} . (2\varpi_{02}^2 - 6\varpi_{01}\varpi_{03}) \\ & + \varpi_{11}\varpi_{12} . (9\varpi_{03} - \varpi_{01}\varpi_{02}), \end{aligned}$$

$$\begin{aligned} \Delta . \varpi_{21} = {} & \varpi_{10}^2 . (\varpi_{02}^2\varpi_{03} - 3\varpi_{01}\varpi_{03}^2) + \varpi_{11}^2 . (\varpi_{01}^2\varpi_{03} - 3\varpi_{02}\varpi_{03}) + \varpi_{12}^2 . (\varpi_{01}^3 - 4\varpi_{01}\varpi_{02} + 9\varpi_{03}) \\ & + \varpi_{10}\varpi_{11} . (9\varpi_{03}^2 - \varpi_{01}\varpi_{02}\varpi_{03}) + \varpi_{10}\varpi_{12} . (2\varpi_{01}^2\varpi_{03} - 6\varpi_{02}\varpi_{03}) \\ & + \varpi_{11}\varpi_{12} . (4\varpi_{02}^2 - 3\varpi_{01}\varpi_{03} - \varpi_{01}^2\varpi_{02}), \end{aligned}$$

$$\begin{aligned} \Delta . \varpi_{30} = {} & \varpi_{10}^3 . \varpi_{03}^2 - \varpi_{11}^3 . \varpi_{03} + \varpi_{12}^3 - \varpi_{10}^2\varpi_{11} . \varpi_{02}\varpi_{03} + \varpi_{10}^2\varpi_{12} . (\varpi_{02}^2 - 2\varpi_{01}\varpi_{03}) \\ & + \varpi_{11}^2\varpi_{10} . \varpi_{01}\varpi_{03} + \varpi_{11}^2\varpi_{12} . \varpi_{02} + \varpi_{12}^2\varpi_{10} . (\varpi_{01}^2 - 2\varpi_{02}) - \varpi_{12}^2\varpi_{11} . \varpi_{01} \\ & - \varpi_{10}\varpi_{11}\varpi_{12}(\varpi_{01}\varpi_{02} - 3\varpi_{03}), \end{aligned}$$

where

$$\Delta = \frac{1}{3} \begin{vmatrix} 3\varpi_{03} & 0 & \varpi_{02} & 0 \\ 2\varpi_{02} & 3\varpi_{03} & 2\varpi_{01} & \varpi_{02} \\ \varpi_{01} & 2\varpi_{02} & 3 & 2\varpi_{01} \\ 0 & \varpi_{01} & 0 & 3 \end{vmatrix}.$$

A simpler set of relations of weights (3, 2), (2, 3), (3, 3) obtained by the methods described in Note 6 is

$$\Delta' . \varpi_{10} = \varpi_{12}(3\varpi_{20} - \varpi_{10}^2) + \varpi_{21}(2\varpi_{10}\varpi_{01} - 3\varpi_{11}) + 3\varpi_{30}(3\varpi_{02} - \varpi_{01}^2),$$

$$\Delta' . \varpi_{01} = \varpi_{12}(2\varpi_{10}\varpi_{01} - 3\varpi_{11}) + \varpi_{21}(3\varpi_{02} - \varpi_{01}^2) + 3\varpi_{03}(3\varpi_{20} - \varpi_{10}^2),$$

$$\Delta' . \varpi_{11} = \varpi_{12}(3\varpi_{01}\varpi_{20} - \varpi_{10}\varpi_{11}) + \varpi_{21}(3\varpi_{10}\varpi_{02} - \varpi_{01}\varpi_{11}) - 3\varpi_{12}\varpi_{21} - 3\varpi_{30}\varpi_{01}\varpi_{02} - 3\varpi_{03}\varpi_{10}\varpi_{20} + 27\varpi_{30}\varpi_{03},$$

where
$$\Delta' = \frac{1}{2}\begin{vmatrix} 2\varpi_{20} & \varpi_{11} & \varpi_{10} \\ \varpi_{11} & 2\varpi_{02} & \varpi_{01} \\ \varpi_{10} & \varpi_{01} & 2 \end{vmatrix} = 4\varpi_{20}\varpi_{02} - \varpi_{11}^2 - \varpi_{01}^2\varpi_{20} - \varpi_{10}^2\varpi_{02} + \varpi_{10}\varpi_{01}\varpi_{11}.$$

Ex. x. Many relations between the ϖ's can be obtained by evaluating products of two matrices as in Ex. viii.

Thus for 3 sequences of 3 variables such as $[\alpha_1\beta_1\gamma_1]$, $[\alpha_2\beta_2\gamma_2]$, $[\alpha_3\beta_3\gamma_3]$ we have

$$\begin{bmatrix} \alpha_1 & \alpha_2 & \alpha_3 \\ \beta_1 & \beta_2 & \beta_3 \\ \gamma_1 & \gamma_2 & \gamma_3 \\ 1 & 1 & 1 \end{bmatrix} \begin{bmatrix} \alpha_2\alpha_3, \ldots, \ldots, & \beta_2\gamma_3 + \beta_3\gamma_2, \ldots, \ldots, & \alpha_2 + \alpha_3, \ldots, \ldots, & 1 \\ \alpha_3\alpha_1, \ldots, \ldots, & \beta_3\gamma_1 + \beta_1\gamma_3, \ldots, \ldots, & \alpha_3 + \alpha_1, \ldots, \ldots, & 1 \\ \alpha_1\alpha_2, \ldots, \ldots, & \beta_1\gamma_2 + \beta_2\gamma_1, \ldots, \ldots, & \alpha_1 + \alpha_2, \ldots, \ldots, & 1 \end{bmatrix} = [\omega]_4^{10},$$

where
$$[\omega]_4^{10} = \begin{bmatrix} 3\varpi_{300} & \varpi_{120} & \varpi_{102} & \varpi_{111} & 2\varpi_{201} & 2\varpi_{210} & 2\varpi_{200} & \varpi_{110} & \varpi_{101} & \varpi_{100} \\ \varpi_{210} & 3\varpi_{030} & \varpi_{012} & 2\varpi_{021} & \varpi_{111} & 2\varpi_{120} & \varpi_{110} & 2\varpi_{020} & \varpi_{011} & \varpi_{010} \\ \varpi_{201} & \varpi_{021} & 3\varpi_{003} & 2\varpi_{012} & 2\varpi_{102} & \varpi_{111} & \varpi_{101} & \varpi_{011} & 2\varpi_{002} & \varpi_{001} \\ \varpi_{200} & \varpi_{020} & \varpi_{002} & \varpi_{011} & \varpi_{101} & \varpi_{110} & 2\varpi_{100} & 2\varpi_{010} & 2\varpi_{001} & 3\varpi_{000} \end{bmatrix},$$

and therefore all the simple minor determinants of $[\omega]_4^{10}$ vanish. Only 7 of the relations thus obtained are independent, whereas there are 10 independent relations in all. We can however obtain 3 other relations by homogenising the 3 relations of Ex. ix in the way described in § 198. 2.

The elements of the 10 vertical rows of $[\omega]_4^{10}$ are the coefficients of the 10 derivatives of order 2 of the general cubic

$$f = \Sigma\varpi_{\lambda\mu\nu}x^\lambda y^\mu z^\nu, \quad (\lambda + \mu + \nu \not> 3)$$

with respect to x, y, z and the homogenising variable. Hence we could deduce the above relations from the fact that in this case (because $n \not< r$) f cannot be expressible as a product of linear factors unless its derivatives of order 2 have a common root (which may be any common root of the 3 linear factors). This argument is not applicable when (as in Ex. ix) we have $n < r$.

The construction of the matrix $[\omega]_4^{10}$ becomes more apparent when we consider the corresponding homogenised relations between the homogeneous ϖ's of 3 sequences of 4 variables.

Ex. xi. For 3 sequences of 4 variables the matrix $[\omega]_5^{15}$ whose 15 vertical rows are formed with the coefficients of the 15 derivatives of order 2 of the general function of degree 4

$$f = \Sigma\varpi_{pqrs}x^p y^q z^r w^s, \quad (p + q + r + s \not> 4),$$

with respect to x, y, z, w and the homogenising variable has degeneracy 2, and all its minor determinants of order 4 vanish.

This can be shown by expressing $[\omega]_5^{15}$ as a product; and it also follows from the fact that if f is expressible as a product of linear factors, its derivatives of order 3 must have two common roots (which may be any two common roots of the linear factors). Since the relations thus obtained are necessary but not sufficient conditions that f shall be expressible as a product of linear factors, they cannot include all the 22 independent relations between the ϖ's, and are equivalent to less than 22 independent relations.

In general whenever $n \nless r$, we can obtain some of the relations between the ϖ's of r sequences of n variables by using the fact that the matrix formed with the coefficients of the derivatives of order $r-1$ of the general function of degree r

$$f = \Sigma \varpi_{q_1 q_2 \ldots q_n} x_1^{q_1} x_2^{q_2} \ldots x_n^{q_n}, \quad (q_1+q_2+\ldots+q_n \ngtr r),$$

has degeneracy $n-r+1$, so that all its minor determinants of order r vanish; but we cannot obtain a complete set of independent relations in this way, except in the special case when $n=2$ as in Exs. vii and viii.

3. *Expression of any monotypic symmetric function as a function of the σ's of order* 1, *or as a function of the ϖ's of degree* 1.

The applications of symmetric functions which we shall make in the latter part of this chapter are based on the two lemmas which will now be given.

Using the notation $\lambda_u = [\lambda_{u1}\lambda_{u2} \ldots \lambda_{un}]$ in (3) and (4) let

$$T_i^{(\lambda_u)} = \alpha_{i1}^{\lambda_{u1}} \alpha_{i2}^{\lambda_{u2}} \ldots \alpha_{in}^{\lambda_{un}}, \text{ so that } T_i = \alpha_{i1}^{\lambda_{i1}} \alpha_{i2}^{\lambda_{i2}} \ldots \alpha_{in}^{\lambda_{in}} = T_i^{(\lambda_i)}.$$

Then if the elements of the sequences $\lambda_1, \lambda_2, \ldots \lambda_r$ formed by the indices in $T_1, T_2, \ldots T_r$ are regarded as arbitrary, and if we evaluate by direct multiplication the product

$$\Sigma\{T_1^{(\lambda_1)} . T_2^{(\lambda_2)} . T_3^{(\lambda_3)} . \ldots . T_i^{(\lambda_i)}\} . \Sigma\{T_i^{(\lambda_{i+1})}\},$$

we obtain for each of the values $1, 2, 3, \ldots r-1$ of i the equation

$$\begin{aligned}\Sigma\{T_1^{(\lambda_1)} . T_2^{(\lambda_2)} . T_3^{(\lambda_3)} . \ldots . T_{i+1}^{(\lambda_{i+1})}\} &= \Sigma\{T_1^{(\lambda_1)} . T_2^{(\lambda_2)} . T_3^{(\lambda_3)} . \ldots . T_i^{(\lambda_i)}\} . \Sigma\{T_{i+1}^{(\lambda_{i+1})}\} \\ &- \Sigma\{T_1^{(\lambda_1)} T_1^{(\lambda_{i+1})} . T_2^{(\lambda_2)} . T_3^{(\lambda_3)} . \ldots . T_i^{(\lambda_i)}\} - \Sigma\{T_1^{(\lambda_2)} T_1^{(\lambda_{i+1})} . T_2^{(\lambda_1)} . T_3^{(\lambda_3)} . \ldots . T_i^{(\lambda_i)}\} \\ &- \ldots - \Sigma\{T_1^{(\lambda_i)} T_1^{(\lambda_{i+1})} . T_2^{(\lambda_1)} . T_3^{(\lambda_2)} . \ldots . T_i^{(\lambda_{i-1})}\}, \quad \ldots\ldots (A)\end{aligned}$$

which is an identity in the elements of the r matrices $X_1, X_2, \ldots X_r$, and serves to express monotypic symmetric functions of order $i+1$ in terms of those of orders i and 1. From the $r-1$ equations (A) we can deduce in succession the equations $(a_2), (a_3), \ldots (a_r)$ of the set

$$N_1 . \Sigma\{T_1^{(\lambda_1)}\} = \Sigma\{T_1^{(\lambda_1)}\}, \quad \ldots\ldots (a_1)$$

$$N_2 . \Sigma\{T_1^{(\lambda_1)} . T_2^{(\lambda_2)}\} = \Sigma\{T_1^{(\lambda_1)}\} . \Sigma\{T_1^{(\lambda_2)}\} - \Sigma\{T_1^{(\lambda_1)} T_1^{(\lambda_2)}\}, \quad \ldots\ldots (a_2)$$

$$\begin{aligned}N_3 . \Sigma\{T_1^{(\lambda_1)} . T_2^{(\lambda_2)} . T_3^{(\lambda_3)}\} &= \Sigma\{T_1^{(\lambda_1)}\} . \Sigma\{T_1^{(\lambda_2)}\} . \Sigma\{T_1^{(\lambda_3)}\} \\ &- \Sigma\{T_1^{(\lambda_1)}\} . \Sigma\{T_1^{(\lambda_2)} T_1^{(\lambda_3)}\} - \Sigma\{T_1^{(\lambda_2)}\} . \Sigma\{T_1^{(\lambda_1)} T_1^{(\lambda_3)}\} - \Sigma\{T_1^{(\lambda_3)}\} . \Sigma\{T_1^{(\lambda_1)} T_1^{(\lambda_2)}\} \\ &+ 2\Sigma\{T_1^{(\lambda_1)} T_1^{(\lambda_2)} T_1^{(\lambda_3)}\}, \quad \ldots\ldots (a_3)\end{aligned}$$

$$\begin{aligned}N_4 . \Sigma\{T_1^{(\lambda_1)} . T_2^{(\lambda_2)} . T_3^{(\lambda_3)} . T_4^{(\lambda_4)}\} &= \Sigma\{T_1^{(\lambda_1)}\} . \Sigma\{T_1^{(\lambda_2)}\} . \Sigma\{T_1^{(\lambda_3)}\} . \Sigma\{T_1^{(\lambda_4)}\} \\ &- [\Sigma_6(\Sigma\{T_1^{(\lambda_1)}\} . \Sigma\{T_1^{(\lambda_2)}\} . \Sigma\{T_1^{(\lambda_3)} T_1^{(\lambda_4)}\})] \\ &+ [2\Sigma_4(\Sigma\{T_1^{(\lambda_1)}\} . \Sigma\{T_1^{(\lambda_2)} T_1^{(\lambda_3)} T_1^{(\lambda_4)}\}) + \Sigma_3(\Sigma\{T_1^{(\lambda_1)} T_1^{(\lambda_2)}\} . \Sigma\{T_1^{(\lambda_3)} T_1^{(\lambda_4)}\})] \\ &- 6\Sigma\{T_1^{(\lambda_1)} T_1^{(\lambda_2)} T_1^{(\lambda_3)} T_1^{(\lambda_4)}\}, \quad \ldots\ldots (a_4)\end{aligned}$$

....

where $N_1 = N_2 = N_3 = N_4 = \ldots = 1$. Here Σ_6 means the sum of the six terms which can be formed from the one given by permutations of the symbols $\lambda_1, \lambda_2, \lambda_3, \lambda_4$ when these are regarded as all different. The ith equation has the general form

$$N_i . \Sigma \{T_1^{(\lambda_1)} . T_2^{(\lambda_2)} \ldots\ldots T_i^{(\lambda_i)}\} = U_i - U_{i-1} + \ldots + (-1)^{i-k} U_k + \ldots + (-1)^{i-1} U_1, \quad \ldots\ldots(a_i)$$

where $N_i = 1$, U_k is a sum in which each term is a product of k monotypic symmetric functions of order 1, and in particular

$$U_i = \Sigma \{T_1^{(\lambda_1)}\} . \Sigma \{T_1^{(\lambda_2)}\} \ldots\ldots \Sigma \{T_1^{(\lambda_i)}\}, \quad \ldots\ldots\ldots\ldots\ldots(20)$$

$$U_1 = (i-1)! \; \Sigma \{T_1^{(\lambda_1)} T_1^{(\lambda_2)} \ldots T_1^{(\lambda_i)}\}. \quad \ldots\ldots\ldots\ldots\ldots(21)$$

These equations are identities in the rn elements of the r sequences $X_1, X_2, \ldots X_r$ when the elements of $\lambda_1, \lambda_2, \ldots \lambda_r$ are arbitrary. In order that they may remain true when any particular values whatever are assigned to the indices forming the sequences $\lambda_1, \lambda_2, \ldots \lambda_r$, we must supply numerical factors $N_1, N_2, \ldots N_r$ on the left, this being necessary because in such cases the Σ's are defined to contain no repeated terms. If in the ith equation the i sequences consist of τ distinct sequences repeated $r_1, r_2, \ldots r_\tau$ times respectively, the numerical factor N_i is given by

$$N_i = r_1! \; r_2! \ldots r_\tau! \quad \ldots\ldots\ldots\ldots\ldots\ldots\ldots\ldots(22)$$

Accordingly it is equal to 1 only when $\lambda_1, \lambda_2, \ldots \lambda_i$ are all different, and it is equal to $i!$ when they are all equal. In the first equation we always have $N_1 = 1$. No additional numerical factors are required on the right because each sum is of order 1. When these numerical factors are supplied on the left, the equations serve to express all monotypic symmetric functions in terms of those of order 1, i.e. in terms of the σ's.

Ex. xii. If T_1', T_1'', T_1''' are the products into which the substitutions of X_1 for X_2, X_3, X_4 convert T_2, T_3, T_4 respectively, we can express the fourth equation (a_4) in the form

$$N_4 . \Sigma \{T_1 T_2 T_3 T_4\} = U_4 - U_3 + U_2 - U_1,$$

where

$$U_4 = \Sigma T_1 . \Sigma T_1' . \Sigma T_1'' . \Sigma T_1''',$$

$$U_3 = \Sigma T_1 . \Sigma T_1' . \Sigma T_1'' T_1''' + \Sigma T_1 . \Sigma T_1'' . \Sigma T_1''' T_1' + \Sigma T_1 . \Sigma T_1''' . \Sigma T_1' T_1''$$
$$+ \Sigma T_1'' . \Sigma T_1''' . \Sigma T_1 T_1' + \Sigma T_1''' . \Sigma T_1' . \Sigma T_1 T_1'' + \Sigma T_1' . \Sigma T_1'' . \Sigma T_1 T_1''',$$

$$U_2 = 2 \{\Sigma T_1 . \Sigma T_1' T_1'' T_1''' + \Sigma T_1' . \Sigma T_1 T_1'' T_1''' + \Sigma T_1'' . \Sigma T_1 T_1' T_1''' + \Sigma T_1''' . \Sigma T_1 T_1' T_1''\}$$
$$+ \{\Sigma T_1 T_1' . \Sigma T_1'' T_1''' + \Sigma T_1 T_1'' . \Sigma T_1''' T_1' + \Sigma T_1 T_1''' . \Sigma T_1' T_1''\},$$

$$U_1 = 6 \Sigma T_1 T_1' T_1'' T_1'''.$$

The possible values of N_4 are 1, 2, 4, 6 and 24.

NOTE 5. *Extension of the validity of the foregoing equations.*

By putting some of the sequences $X_1, X_2, \ldots X_r$ equal to 0 we see that the equation (a_i) remains true when $r < i$, provided that we interpret $\Sigma \{T_1 T_2 \ldots T_i\}$ to be 0 when the number of those sequences is less than i. With this understanding the equations $(a_1), (a_2), (a_3), \ldots$

are true when r is an arbitrary positive integer. In the same way we see that we can use the equation (A) when there are only i sequences, provided that we interpret the sum on the left to be 0. Thus for i sequences $X_1, X_2, \dots X_i$ we have

$$\Sigma\{T_1^{(\lambda_1)}.T_2^{(\lambda_2)}.T_3^{(\lambda_3)}\dots\dots T_i^{(\lambda_i)}\}.\Sigma\{T_1^{(\lambda_{i+1})}\}$$
$$=\Sigma\{T_1^{(\lambda_1)}T_1^{(\lambda_{i+1})}.T_2^{(\lambda_2)}.T_3^{(\lambda_3)}\dots\dots T_i^{(\lambda_i)}\}+\Sigma\{T_1^{(\lambda_2)}T_1^{(\lambda_{i+1})}.T_2^{(\lambda_1)}.T_3^{(\lambda_3)}\dots\dots T_i^{(\lambda_i)}\}$$
$$+\dots+\Sigma\{T_1^{(\lambda_i)}T_1^{(\lambda_{i+1})}.T_2^{(\lambda_1)}.T_3^{(\lambda_2)}\dots\dots T_i^{(\lambda_{i-1})}\}, \quad\dots\dots\text{(B)}$$

when the elements of $\lambda_1, \lambda_2, \dots \lambda_{i+1}$ are arbitrary. When these elements are given integers, we must supply the appropriate numerical factor to each sum as in the sums on the left in formulae (a_1), (a_2), (a_3), ...

NOTE 6. *Alternative methods of determining the relations between the ϖ's of i sequences of n variables.*

We can determine the $\binom{n+i}{i}-in-1$ independent relations between the ϖ's of i sequences of n variables by applying the identity (a_{i+1}) to i sequences, the expression on the left being then 0. The relations of various weights are obtained by assigning various sets of values to the elements of the sequences $\lambda_1, \lambda_2, \dots \lambda_{i+1}$, these being so chosen that the total weight is not less than $r+2$ and the weight in each individual variable is not greater than r. We can also obtain them by applying formula (B) to i sequences. Illustrations are given in Exs. xiii, xiv and xv.

Now let S be any monotypic symmetric function whatever of the elements of the r sequences of n variables given in (2); let its order be i, where $i \not> r$; let its weights with respect to the n variables $x_1, x_2, \dots x_n$ be $u_1, u_2, \dots u_n$; let its total weight be $w = u_1 + u_2 + \dots + u_n$; and let it be expressed in the form

$$S = S_{u_1 u_2 \dots u_n} = \Sigma\{T_1 T_2 \dots T_i\}, \quad\dots\dots\dots\dots\dots\dots\dots(23)$$

where the T's are given by (4). Then in obtaining the equation (a_i), in which every sum on the right is one of the functions σ, we have proved the following theorem which we call Lemma A.

Lemma A. *The monotypic symmetric function S given by (23) can be expressed in the form*

$$S = F(\sigma_{p_1 p_2 \dots p_n}, \dots, \dots) = F(\sigma), \quad\dots\dots\dots\dots\dots\dots\dots\text{(a)}$$

where $F(\sigma)$ is a rational integral function of those functions σ of order 1 which have total weight not greater than w, and has the following properties:

(1) *The coefficients of $F(\sigma)$ are rational numbers.*

(2) *The function $F(\sigma)$ is isobaric in the σ's with respect to each one of the variables $x_1, x_2, \dots x_n$, the weight of every one of its terms with respect to x_h (i.e. the value of Σp_h in every term of F) being equal to the weight u_h of S with respect to x_h.*

(3) *The degree of $F(\sigma)$ in the σ's is equal to the order i of S.*

(4) *In every quantity $\sigma_{p_1 p_2 \dots p_n}$ occurring in $F(\sigma)$, p_h is a sum of certain of the integers $\lambda_{1h}, \lambda_{2h}, \dots \lambda_{ih}$.*

(5) *If N_i is the number defined in (22), the term of highest degree in $F(\sigma)$ is*

$$\frac{1}{N_i} \cdot \sigma_{\lambda_{11}\lambda_{12}\dots\lambda_{1n}} \sigma_{\lambda_{21}\lambda_{22}\dots\lambda_{2n}} \dots \sigma_{\lambda_{i1}\lambda_{i2}\dots\lambda_{in}}, \quad \dots\dots\dots\dots\dots(20')$$

and the term of lowest degree in $F(\sigma)$ is

$$(-1)^{i-1} \cdot \frac{(i-1)!}{N_i} \cdot \sigma_{u_1 u_2 \dots u_n}. \quad \dots\dots\dots\dots\dots(21')$$

By substituting for every function σ occurring on the right in the equation (a) of Lemma A its value in terms of the ϖ's as determined in Note 2 we obtain the second theorem which we call Lemma B.

Lemma B. *The monotypic symmetric function S given by (23) can be expressed in the form*

$$S = G(\varpi_{q_1 q_2 \dots q_n}, \dots, \dots) = G(\varpi), \quad \dots\dots\dots\dots\dots(b)$$

where $G(\varpi)$ is a rational integral function of those functions ϖ of degree 1 which have total weight not greater than w, and has the following properties:

(1) *The coefficients of $G(\varpi)$ are rational numbers.*

(2) *The function $G(\varpi)$ is isobaric in the ϖ's with respect to each one of the variables $x_1, x_2, \dots x_n$, the weight of every one of its terms with respect to x_h (i.e. the value of Σq_h in every term of G) being equal to the weight u_h of S with respect to x_h.*

(3) *The degree of $G(\varpi)$ in the ϖ's cannot be greater than the total weight w of S, and cannot be less than the degree of S in each of the matrices $X_1, X_2, \dots X_r$.*

(4) *The terms of the highest possible degree w in $G(\varpi)$ all differ by numerical factors only from the product $\varpi_{100\dots0}^{u_1} \varpi_{010\dots0}^{u_2} \dots \varpi_{000\dots1}^{u_n}$, but such terms will usually cancel except when S has order 1.*

(5) *When r is arbitrary, the term of lowest degree in $G(\varpi)$ is $\varpi_{u_1 u_2 \dots u_n}$ multiplied by a non-zero numerical factor, but this term vanishes when r is given and is less than w.*

(6) *When r is given and is less than w, the equation* (b) *expresses S in terms of the ϖ's which have total weight not greater than r; for the other ϖ's all vanish.*

The second property in (3) follows from the fact that every ϖ has degree 1 in each matrix; hence if the degree of $G(\varpi)$ is μ, no term of S can have degree greater than μ in any matrix. The second property in (4) follows from the fact that no other product of the ϖ's occurring in G can have degree w in each matrix; hence if such terms do not cancel, S must have degree w in each matrix, and this is only possible when its order is 1.

NOTE 7. *Insertion and removal of zero suffixes.*

In the equations (a) and (b) we can insert or remove k zero suffixes occupying corresponding positions in all the σ's or ϖ's. We then obtain equations true for $n+k$ or $n-k$ variables, the k variables which have been added or removed being such as do not actually occur in S. Consequently in writing down these equations we may always suppose the number of variables to be equal to i, the order of S.

NOTE 8. *Cases in which the function $G(\varpi)$ of Lemma B is unique.*

From Note 4 we see that when r is arbitrary, the function $G(\varpi)$ in (b) is uniquely determinate. For particular values of r, when ϖ's of total weight greater than r are replaced by 0's, the function $G(\varpi)$ is unique if and only if there is no relation between the ϖ's of weights $u_1, u_2, \ldots u_n$; in particular it is unique whenever its total weight is less than $r+2$, i.e. when $r \not< w-1$.

Ex. xiii. When there is only one variable $[x]$ and r is arbitrary, we have for the r sequences $[a_1]$, $[a_2]$, $[a_3]$, ... in accordance with Lemma A:

$$\Sigma a_1^2 a_2 = \sigma_2\sigma_1 - \sigma_3, \quad 2\Sigma a_1^2 a_2^2 = \sigma_2^2 - \sigma_4, \quad 2\Sigma a_1^2 a_2 a_3 = \sigma_2\sigma_1^2 - \sigma_2^2 - 2\sigma_1\sigma_3 + 2\sigma_4;$$

and therefore in accordance with Lemma B:

$$\Sigma a_1^2 a_2 = \varpi_2\varpi_1 - 3\varpi_3, \quad \Sigma a_1^2 a_2^2 = \varpi_2^2 - 2\varpi_1\varpi_3 + 2\varpi_4, \quad \Sigma a_1^2 a_2 a_3 = \varpi_1\varpi_3 - 4\varpi_4.$$

By inserting zero suffixes we see from the last result that for r sequences of three variables we have

$$\Sigma a_1^2 a_2 a_3 = \varpi_{100}\varpi_{300} - 4\varpi_{400}, \quad \Sigma \beta_1^2\beta_2\beta_3 = \varpi_{010}\varpi_{030} - 4\varpi_{040}.$$

Ex. xiv. When there are two variables $[x, y]$ and r is arbitrary, we have for the r sequences $[a_1\beta_1]$, $[a_2\beta_2]$, $[a_3\beta_3]$, ... in accordance with Lemma A:

$$\Sigma a_1^2\beta_2 = \sigma_{20}\sigma_{01} - \sigma_{21}, \quad \Sigma a_1^2\beta_1\beta_2 = \sigma_{21}\sigma_{01} - \sigma_{22}, \quad \Sigma a_1^2\beta_2^2 = \sigma_{20}\sigma_{02} - \sigma_{22},$$

$$2\Sigma a_1^2\beta_2\beta_3 = \sigma_{20}\sigma_{01}^2 - \sigma_{20}\sigma_{02} - 2\sigma_{01}\sigma_{21} + 2\sigma_{22},$$

$$\Sigma a_1\beta_1 a_2\beta_3 = \sigma_{11}\sigma_{10}\sigma_{01} - \sigma_{11}^2 - \sigma_{10}\sigma_{12} - \sigma_{01}\sigma_{21} + 2\sigma_{22},$$

and therefore in accordance with Lemma B:

(1) $\Sigma a_1^2\beta_2 = (\varpi_{10}\varpi_{11} - \varpi_{01}\varpi_{20}) + \varpi_{21}$,

(2) $3\Sigma a_1^2\beta_1\beta_2 = (\varpi_{10}\varpi_{01}\varpi_{11} + 2\varpi_{10}^2\varpi_{02} - \varpi_{01}^2\varpi_{20}) + (\varpi_{01}\varpi_{21} - 2\varpi_{10}\varpi_{12} - 2\varpi_{20}\varpi_{02} - \varpi_{11}^2) + 2\varpi_{22}$,

(3) $3\Sigma a_1^2\beta_2^2 = 4(\varpi_{10}\varpi_{01}\varpi_{11} - \varpi_{10}^2\varpi_{02} - \varpi_{01}^2\varpi_{20}) + (10\varpi_{20}\varpi_{02} - 2\varpi_{10}\varpi_{12} - 2\varpi_{01}\varpi_{21} - \varpi_{11}^2) + 2\varpi_{22}$,

(4) $3\Sigma a_1^2\beta_2\beta_3 = (\varpi_{10}^2\varpi_{02} - \varpi_{10}\varpi_{01}\varpi_{11} + \varpi_{01}^2\varpi_{20}) + (2\varpi_{10}\varpi_{12} - \varpi_{01}\varpi_{21} - 4\varpi_{02}\varpi_{20} + \varpi_{11}^2) - 2\varpi_{22}$,

(5) $3\Sigma a_1\beta_1 a_2\beta_3 = (\varpi_{10}\varpi_{01}\varpi_{11} - \varpi_{10}^2\varpi_{02} - \varpi_{01}^2\varpi_{20}) + (4\varpi_{20}\varpi_{02} - \varpi_{11}^2 + \varpi_{01}\varpi_{21} + \varpi_{10}\varpi_{12}) - 4\varpi_{22}$.

When $r=2$, and we neglect ϖ's of weights 4 and 3, the equations (2) and (3) cease to be unique. In this case we have by Ex. vii the relation

$$\varpi_{10}^2\varpi_{02} - \varpi_{10}\varpi_{01}\varpi_{11} + \varpi_{01}^2\varpi_{20} = 4\varpi_{20}\varpi_{02} - \varpi_{11}^2,$$

and the equations (2) and (3) can be replaced by

$$(2') \ \Sigma a_1^2\beta_1\beta_2 = \varpi_{02}(\varpi_{10}^2 - 2\varpi_{02}), \qquad (3') \ \Sigma a_1^2\beta_2^2 = \varpi_{11}^2 - 2\varpi_{20}\varpi_{02}.$$

Again when $r=2$, the left-hand sides of the equations (4) and (5) vanish (or we may consider that $a_3=\beta_3=0$), and each of these equations becomes the relation of Ex. vii given above, which we could have obtained in this way.

Ex. xv. When there are three variables $[x, y, z]$ and r is arbitrary, we have for the r sequences $[a_1\beta_1\gamma_1]$, $[a_2\beta_2\gamma_2]$, $[a_3\beta_3\gamma_3]$, ... in accordance with Lemma A:

$$\Sigma a_1^2\beta_2\gamma_3=\sigma_{200}\,\sigma_{010}\,\sigma_{001}-\sigma_{200}\,\sigma_{011}-\sigma_{201}\,\sigma_{010}-\sigma_{210}\,\sigma_{001}+2\sigma_{211},$$

and therefore in accordance with Lemma B:

$$3\Sigma a_1^2\beta_2\gamma_3=\{\varpi^2_{100}\,\varpi_{011}-\varpi_{100}\,(\varpi_{010}\,\varpi_{101}-\varpi_{001}\,\varpi_{110})+2\varpi_{010}\,\varpi_{001}\,\varpi_{200}\}$$
$$+\{2\varpi_{101}\,\varpi_{110}+2\varpi_{100}\,\varpi_{111}-4\varpi_{011}\,\varpi_{200}-\varpi_{010}\,\varpi_{201}-\varpi_{001}\,\varpi_{210}\}-2\varpi_{211}.$$

When $r=2$, the last equation becomes one of the relations of Ex. viii, viz.

$$\varpi^2_{100}\,\varpi_{011}-\varpi_{100}\,(\varpi_{010}\,\varpi_{101}-\varpi_{001}\,\varpi_{110})+2\varpi_{010}\,\varpi_{001}\,\varpi_{200}=4\varpi_{011}\,\varpi_{200}-2\varpi_{101}\,\varpi_{110}.$$

Thus this relation of weights 2, 1, 1 between the ϖ's of 2 sequences of 3 variables can be obtained by applying the identity (a_3) to 2 sequences of 3 variables, putting $\lambda_1=[2\,0\,0]$, $\lambda_2=[0\,1\,0]$, $\lambda_3=[0\,0\,1]$; and it can also be obtained from the identity (B) in the same way when $i=2$.

Ex. xvi. *Relations between the ϖ's of 2 sequences of n variables.*

When we have 2 sequences of n variables, the left-hand side of (a_3) vanishes, and we therefore have relations between the ϖ's of the form

$$F(\varpi_{q_1q_2\dots q_n}, \dots, \dots)=F(\varpi)=0, \quad (q_1+q_2+\dots+q_n \not> 2),$$

where

$$F(\varpi)=\Sigma\{T_1^{(\lambda_1)}\}.\Sigma\{T_1^{(\lambda_2)}\}.\Sigma\{T_1^{(\lambda_3)}\}$$
$$-\Sigma\{T_1^{(\lambda_1)}\}.\Sigma\{T_1^{(\lambda_2)}\,T_1^{(\lambda_3)}\}-\Sigma\{T_1^{(\lambda_2)}\}.\Sigma\{T_1^{(\lambda_1)}\,T_1^{(\lambda_3)}\}-\Sigma\{T_1^{(\lambda_3)}\}.\Sigma\{T_1^{(\lambda_1)}\,T_1^{(\lambda_2)}\}$$
$$+2\Sigma\{T_1^{(\lambda_1)}\,T_1^{(\lambda_2)}\,T_1^{(\lambda_3)}\},$$

and the sequences λ_1, λ_2, λ_3 are so chosen that the total weight of $F(\varpi)$ is not less than 4, and its weight in each individual variable is not greater than 2. For other values of λ_1, λ_2, λ_3 the function $F(\varpi)$ vanishes identically.

Thus if $n=2$, and we put $\lambda_1=[1\,0]$, $\lambda_2=[0\,1]$, $\lambda_3=[1\,1]$, we obtain the relation of weights 2, 2 given in Ex. vii. Again if $n=3$, and we put $\lambda_1=[1\,0\,0]$, $\lambda_2=[0\,1\,0]$, $\lambda_3=[0\,0\,2]$, or $\lambda_1=[0\,0\,1]$, $\lambda_2=[0\,0\,1]$, $\lambda_3=[1\,1\,0]$, we obtain the relation of weights 1, 1, 2 given in Ex. viii.

We can obtain the same results from the equation

$$\Sigma\{T_1^{(\lambda_1)}.\,T_2^{(\lambda_2)}\}.\Sigma\{T_1^{(\lambda_3)}\}=\Sigma\{T_1^{(\lambda_1)}\,T_1^{(\lambda_3)}.\,T_2^{(\lambda_2)}\}+\Sigma\{T_1^{(\lambda_2)}\,T_1^{(\lambda_3)}.\,T_2^{(\lambda_1)}\},$$

which is the equation (B) when $i=2$, and is true for 2 sequences of n variables.

NOTE 9. *Other methods of expressing the monotypic symmetric function S of Lemma B in the form $G(\varpi)$.*

The method used in the proof of Lemma B, besides being very laborious in application, does not usually lead to the simplest possible rational integral function $G(\varpi)$, except in the special case when $G(\varpi)$ is unique. It is often simpler to use other methods such as:

(1) to assume an equation of the form $S=G(\varpi)$ in which the coefficients of $G(\varpi)$ are constants whose values are to be determined by ascribing various sets of particular values to the elements of the sequences;

(2) to make a classification of the products which can be formed with the ϖ's, each product being expressed as a rational integral function of the elements of the sequences.

Ex. xvii. In the case of the three sequences $[\alpha_1\beta_1\gamma_1]$, $[\alpha_2\beta_2\gamma_2]$, $[\alpha_3\beta_3\gamma_3]$ we have equations of the following types for homogeneous monotypic symmetric functions of degree 2 :

$$\Sigma\alpha_1^2\alpha_2^2\beta_3^2 = \varpi^2_{210} - 2\varpi_{300}\varpi_{120}, \qquad \Sigma\alpha_1^2\alpha_2^2\gamma_3^2 = \varpi^2_{201} - 2\varpi_{300}\varpi_{102},$$

$$\Sigma\alpha_1^2\alpha_2^2\beta_3\gamma_3 = \varpi_{210}\varpi_{201} - \varpi_{300}\varpi_{111}, \qquad \Sigma\alpha_1^2\beta_2^2\alpha_3\beta_3 = \varpi_{210}\varpi_{120} - 3\varpi_{300}\varpi_{030},$$

$$\Sigma\alpha_1^2\beta_2^2\alpha_3\gamma_3 = \varpi_{210}\varpi_{111} - \varpi_{201}\varpi_{120} - \varpi_{300}\varpi_{021}, \quad \Sigma\alpha_1^2\alpha_2\gamma_2\beta_3\gamma_3 = \varpi_{210}\varpi_{102} - \varpi_{300}\varpi_{012},$$

$$\Sigma\alpha_1^2\beta_2\gamma_2\beta_3\gamma_3 + \Sigma\alpha_1\beta_1\alpha_2\gamma_2\beta_3\gamma_3 = \varpi_{120}\varpi_{102}.$$

These results are easily obtained by the second method of Note 9. The general method is only required for the evaluation of the one function $\Sigma\alpha_1\beta_1\alpha_2\gamma_2\beta_3\gamma_3$.

§ 198. Homogeneous symmetric functions of the elements of *r* similar sequences.

1. *Homogenisation of a monotypic symmetric function.*

Let
$$X = [x_1 x_2 \dots x_n], \quad Y = [y_1 y_2 \dots y_n y_{n+1}],$$

where the x's and y's are scalar variables;

let $\quad X_k = [\alpha_{k1}\alpha_{k2}\dots\alpha_{kn}], \quad Y_k = [\beta_{k1}\beta_{k2}\dots\beta_{kn}\beta_{k,n+1}], \quad (k = 1, 2, \dots r),$

be respectively r values of X and r values of Y in which all the elements are arbitrary;

and let
$$S' = S'_{u_1 u_2 \dots u_n} = \Sigma\{T_1' T_2' \dots T_r'\}, \quad \dots\dots\dots\dots\dots\dots(\text{A}')$$

where
$$T_k' = \alpha_{k1}^{\lambda_{k1}} \alpha_{k2}^{\lambda_{k2}} \dots \alpha_{kn}^{\lambda_{kn}}, \quad (k = 1, 2, \dots r), \dots\dots\dots\dots\dots\dots(\text{a}')$$

be any monotypic symmetric function of the elements of $X_1, X_2, \dots X_r$ which has degree s in each matrix and weights $u_1, u_2, \dots u_n$ with respect to $x_1, x_2, \dots x_n$, so that

$$\lambda_{1h} + \lambda_{2h} + \dots + \lambda_{rh} = u_h, \quad (h = 1, 2, \dots n).$$

Also, as is clearly possible, let $\lambda_{1,n+1}, \lambda_{2,n+1}, \dots \lambda_{r,n+1}$ be positive integers, one at least of which is 0, so chosen that

$$\lambda_{k1} + \lambda_{k2} + \dots + \lambda_{kn} + \lambda_{k,n+1} = s, \quad (k = 1, 2, \dots r);$$

and let u_{n+1} be the positive integer determined by the equation

$$\lambda_{1,n+1} + \lambda_{2,n+1} + \dots + \lambda_{r,n+1} = u_{n+1},$$

so that
$$u_1 + u_2 + \dots + u_n + u_{n+1} = rs.$$

Then the equation

$$S = S_{u_1 u_2 \dots u_n u_{n+1}} = \Sigma\{T_1 T_2 \dots T_r\}, \quad \dots\dots\dots\dots\dots\dots(\text{A})$$

where
$$T_k = \beta_{k1}^{\lambda_{k1}} \beta_{k2}^{\lambda_{k2}} \dots \beta_{kn}^{\lambda_{kn}} \beta_{k,n+1}^{\lambda_{k,n+1}}, \quad (k = 1, 2, \dots r), \quad \dots\dots\dots(\text{a})$$

defines a uniquely determinate monotypic symmetric function S of the elements of the r matrices or sequences $Y_1, Y_2, \dots Y_r$ which is *homogeneous* of degree s in each of these matrices according to the definition of homogeneity given in § 197, and which has weights $u_1, u_2, \ldots u_n, u_{n+1}$ with respect to $y_1, y_2, \ldots y_n, y_{n+1}$; and because one at least of the integers $\lambda_{1,n+1}, \lambda_{2,n+1}, \ldots \lambda_{r,n+1}$ is 0, S is not divisible by the product

$$\varpi_{n+1} = \beta_{1,n+1}\,\beta_{2,n+1}\ldots\beta_{r,n+1}. \quad\ldots\ldots\ldots\ldots(1)$$

We call S the homogenised monotypic symmetric function of degree s (in each of the Y's) derived from S' by the substitution

$$[x_1\, x_2 \ldots x_n\, 1] = [y_1\, y_2 \ldots y_n\, y_{n+1}].$$

To obtain it from S' we introduce into each of the products $T_1', T_2', \ldots T_r'$ expressed in terms of the new variables an additional factor which will raise its total degree to s, the new factor being a power of the value of y_{n+1} appropriate to that product. Conversely S, which may be any monotypic symmetric function of the elements of the r matrices $Y_1, Y_2, \ldots Y_r$ which is homogeneous of degree s in each matrix, has weights $u_1, u_2, \ldots u_n, u_{n+1}$, and is not divisible by ϖ_{n+1}, is re-converted into S' by the substitution $[y_1 y_2 \ldots y_n y_{n+1}] = [x_1 x_2 \ldots x_n\, 1]$, i.e. by the substitutions $\beta_{ki} = \alpha_{ki}$ (when $k \neq n+1$), $\beta_{k,n+1} = 1$. Thus these substitutions establish a one-one correspondence between all monotypic symmetric functions of the elements of the matrices $X_1, X_2, \ldots X_r$, and all those monotypic symmetric functions of the elements of the matrices $Y_1, Y_2, \ldots Y_r$ which are homogeneous in each matrix and are not divisible by ϖ_{n+1}, corresponding symmetric functions having equal degrees in their matrices.

More generally if we choose the new indices and u_{n+1} so that

$$\lambda_{k1} + \lambda_{k2} + \ldots + \lambda_{k,n+1} = s + t, \quad \lambda_{1,n+1} + \lambda_{2,n+1} + \ldots + \lambda_{r,n+1} = u_{n+1},$$

then S is homogeneous of degree $s + t$ in each of the matrices $Y_1, Y_2, \ldots Y_r$, it is divisible by ϖ^t_{n+1} but not by any higher power of ϖ_{n+1}, and it is convertible into S' by the same substitutions as before. We call it the homogenised symmetric function of degree $s + t$ in each matrix which corresponds to S'.

We can of course write

$$[y_1 y_2 \ldots y_{n+1}] = [x_1 x_2 \ldots x_{n+1}], \quad [\beta_{k1}\beta_{k2} \ldots \beta_{k,n+1}] = [\alpha_{k1}\alpha_{k2} \ldots \alpha_{k,n+1}].$$

Then to obtain S from S' we homogenise S' by the introduction of a new variable x_{n+1}, or by the substitution $1 = x_{n+1}$.

We shall usually suppose homogenisation to be effected without change of degree.

Ex. i. To the monotypic symmetric functions of degree 3

$$S_1' = \Sigma a_{11}^2 a_{12} a_{21}, \quad S_2' = \Sigma a_{11}^2 a_{22} a_{23} a_{31}^3$$

of the elements of the three sequences $[a_{11} a_{12} a_{13}]$, $[a_{21} a_{22} a_{23}]$, $[a_{31} a_{32} a_{33}]$ there correspond the homogeneous monotypic symmetric functions of degree 3

$$S_1 = \Sigma a_{11}^2 a_{12} a_{21} a_{24}^2 a_{34}^3, \quad S_2 = \Sigma a_{11}^2 a_{14} a_{22} a_{23} a_{24} a_{31}^3$$

of the elements of the three sequences $[a_{11} a_{12} a_{13} a_{14}]$, $[a_{21} a_{22} a_{23} a_{24}]$, $[a_{31} a_{32} a_{33} a_{34}]$.

Ex. ii. To the monotypic symmetric functions of degree 3

$$S_1' = \Sigma a_1^2 \beta_1 a_2, \quad S_2' = \Sigma a_1^2 \beta_2 \gamma_2 a_3^3$$

of the elements of the three sequences $[a_1\beta_1\gamma_1]$, $[a_2\beta_2\gamma_2]$, $[a_3\beta_3\gamma_3]$ there correspond the homogeneous monotypic symmetric functions of degree 3

$$S_1 = \Sigma a_1^2 \beta_1 a_2 \delta_2^2 \delta_3^3, \quad S_2 = \Sigma a_1^2 \delta_1 \beta_2 \gamma_2 \delta_2 a_3^3$$

of the elements of the three sequences $[a_1\beta_1\gamma_1\delta_1]$, $[a_2\beta_2\gamma_2\delta_2]$, $[a_3\beta_3\gamma_3\delta_3]$.

Ex. iii. *Homogenisation of the elementary symmetric functions.*

Let the elementary symmetric functions of the elements of the r sequences $X_1, X_2, \dots X_r$ and those of the elements of the r sequences $Y_1, Y_2, \dots Y_r$ be denoted respectively by

$$\varpi'_{q_1 q_2 \dots q_n}, \ (q_1 + q_2 + \dots + q_n \not> r); \quad \varpi_{q_1 q_2 \dots q_{n+1}}, \ (q_1 + q_2 + \dots + q_{n+1} \not> r).$$

Then when $q_1, q_2, \dots q_n$ are given and are not all 0, the symmetric function of degree 1 in the Y's which we obtain by homogenising $\varpi'_{q_1 q_2 \dots q_n}$ is the function $\varpi_{q_1 q_2 \dots q_{n+1}}$ in which q_{n+1} has the value given by $q_1 + q_2 + \dots + q_n + q_{n+1} = r$. Thus there is a one-one correspondence between all the elementary symmetric functions of the elements of the X's (with the exception of $\varpi'_{00\dots0}$) and all those elementary symmetric functions of the elements of the Y's (with the exception of $\varpi_{00\dots0r}$) which have total weight r, two corresponding functions having the same degree 1 in each of their matrices. We may consider further that to the function $\varpi'_{00\dots0} = 1$ of degree 0 there corresponds the homogeneous function $\varpi_{00\dots0r} = \varpi_{n+1}$ of degree 1.

It will be observed that an elementary symmetric function of degree 1 of the elements of r sequences is homogeneous when and only when its total weight is equal to r, so that it has order r.

2. *Relations between the homogeneous ϖ's of r sequences of $n+1$ variables.*

By the *homogeneous* ϖ's or the *homogeneous* elementary symmetric functions of the elements of r similar sequences we shall mean (see Ex. iii) those of the ϖ's which have total weight r, i.e. those in which the sums of the suffixes are equal to r. All independent rational integral relations between these homogeneous ϖ's may be supposed to be homogeneous in them.

Let $F(\varpi')$ be any given rational integral function of degree μ of the elementary symmetric functions of the r sequences $X_1, X_2, \dots X_r$ defined in sub-article 1, and let $G(\varpi)$ be the function which we obtain when we first make $F(\varpi')$ homogeneous of degree μ in the functions ϖ' by the substitution $1 = \varpi'_{00\dots0}$, and then substitute for every quantity $\varpi'_{q_1 q_2 \dots q_n}$ the corresponding quantity $\varpi_{q_1 q_2 \dots q_n q_{n+1}}$ in which q_{n+1} has the value given by

$$q_1 + q_2 + \dots + q_n + q_{n+1} = r.$$

Then $G(\varpi)$ is a homogeneous rational integral function of those elementary symmetric functions of the elements of the r sequences $Y_1, Y_2, \dots Y_r$ which have total weight r, and we will call it the function of the homogeneous ϖ's of the r sequences $Y_1, Y_2, \dots Y_r$ obtained by the homogenisation of the function $F(\varpi')$ without change of degree. There is a one-one correspondence between all such functions as $F(\varpi')$ and all such homogeneous functions as $G(\varpi)$.

If $F(\varpi')$ when expressed as a rational integral function of the α's has the form

$$F(\varpi') = \phi(\alpha_{11}, \alpha_{12}, \ldots \alpha_{1n};\ \ldots;\ \ldots;\ \alpha_{r1}, \alpha_{r2}, \ldots \alpha_{rn}), \qquad \text{(B')}$$

then $G(\varpi)$ is the homogeneous rational integral function of the β's given by

$$G(\varpi) = \varpi^{\mu}{}_{n+1} \cdot \phi\left(\frac{\beta_{11}}{\beta_{1,n+1}}, \frac{\beta_{12}}{\beta_{1,n+1}}, \ldots \frac{\beta_{1n}}{\beta_{1,n+1}};\ \ldots;\ \ldots;\ \frac{\beta_{r1}}{\beta_{r,n+1}}, \frac{\beta_{r2}}{\beta_{r,n+1}}, \ldots \frac{\beta_{rn}}{\beta_{r,n+1}}\right); \ldots \text{(B)}$$

and it follows that the equation $G(\varpi) = 0$ is an identity in the β's when and only when the equation $F(\varpi') = 0$ is an identity in the α's. Consequently the homogeneous relations between the elementary symmetric functions ϖ of total weight r of the r sequences $Y_1, Y_2, \ldots Y_r$ bear a one-one correspondence to the relations between the elementary symmetric functions ϖ' of the r sequences $X_1, X_2, \ldots X_r$, and are derivable from the latter relations by the process of homogenisation just described. From Note 3 of § 197 we now conclude that:

There are exactly $\binom{n+r}{r} - rn - 1$ *independent relations between the* $\binom{n+r}{r}$ *homogeneous* ϖ*'s (of total weight* r*) of* r *sequences of* $n+1$ *variables; and these relations are the necessary and sufficient conditions that the general homogeneous rational integral function of degree* r

$$f(y_1, y_2, \ldots y_{n+1}) = \Sigma \varpi_{q_1 q_2 \ldots q_{n+1}} y_1^{q_1} y_2^{q_2} \ldots y_{n+1}^{q_{n+1}}, \quad (q_1 + q_2 + \ldots + q_{n+1} = r),$$

shall be expressible as a product of linear factors.

The independent relations when determined as in Note 3 of § 197 all have degree $2r-1$ in the homogeneous ϖ's. If i and j are any two different integers selected from $1, 2, \ldots n+1$, we can take the $rn+1$ independent homogeneous ϖ's to be

(1) those $r+1$ in which $q_i + q_j = r$, (2) those $r(n-1)$ in which $q_i + q_j = r-1$;

and the independent relations are then those which express each of the other homogeneous ϖ's as a rational function of these $rn+1$ independent homogeneous ϖ's.

Another aspect of these relations is shown in Note 1 of § 199.

Ex. iv. If $F(\varpi')$ is isobaric of weights $u_1, u_2, \ldots u_n$ with respect to $x_1, x_2, \ldots x_n$, then $G(\varpi)$ is isobaric of weights $u_1, u_2, \ldots u_n$ with respect to $y_1, y_2, \ldots y_n$, and has total weight $r\mu$.

Ex. v. From the relation of Ex. vii of § 197

$$(4\varpi_{20} \cdot \varpi_{02} - \varpi_{11}^2) - (\varpi_{10}^2 \varpi_{02} - \varpi_{10}\varpi_{01}\varpi_{11} + \varpi_{01}^2 \varpi_{20}) = 0$$

between the ϖ's of the two sequences $[x_1 y_1]$, $[x_2 y_2]$ we deduce by homogenisation the relation

$$4\varpi_{200}\varpi_{020}\varpi_{002} + \varpi_{011}\varpi_{101}\varpi_{110} - \varpi_{200}\varpi_{011}^2 - \varpi_{020}\varpi_{101}^2 - \varpi_{002}\varpi_{110}^2 = 0$$

between the homogeneous ϖ's of the two sequences $[\alpha_1\beta_1\gamma_1]$, $[\alpha_2\beta_2\gamma_2]$.

Ex. vi. From the fourth of the relations given in Ex. ix of § 197 we deduce by homogenisation the relation

$$\frac{1}{2}\begin{vmatrix} 2\varpi_{201} & \varpi_{111} & \varpi_{102} \\ \varpi_{111} & 2\varpi_{021} & \varpi_{012} \\ \varpi_{102} & \varpi_{012} & 2\varpi_{003} \end{vmatrix} \varpi_{102}$$

$$=(2\varpi_{210}\varpi_{012}\varpi_{102}-3\varpi_{300}\varpi_{012}{}^2-\varpi_{120}\varpi_{102}{}^2)\,.\,\varpi_{003}+3\,(3\varpi_{300}\varpi_{021}+\varpi_{201}\varpi_{120}-\varpi_{210}\varpi_{111})\,.\,\varpi_{003}{}^2$$

between the homogeneous ϖ's of the three sequences $[\alpha_1\beta_1\gamma_1]$, $[\alpha_2\beta_2\gamma_2]$, $[\alpha_3\beta_3\gamma_3]$.

3. *Expression of any homogeneous monotypic symmetric function of the elements of r sequences as a rational function of the homogeneous ϖ's.*

First let S, as given by (A), be any monotypic symmetric function of the elements of the r sequences $Y_1, Y_2, \ldots Y_r$ which is homogeneous of degree s in each of those sequences and is *not divisible by* ϖ_{n+1}; and let S', as given by (A′), be the monotypic symmetric function of degree s of the elements of the r sequences $X_1, X_2, \ldots X_r$ from which S is derived by homogenisation.

By Lemma B of § 197 there exists an equation of the form

$$S' = F(\varpi'_{q_1 q_2 \ldots q_n}, \ldots, \ldots) = F(\varpi'), \quad (q_1+q_2+\ldots+q_n \not> r), \ \ldots(\mathrm{C}')$$

where $F(\varpi')$ is a rational integral function of degree μ of the quantities ϖ', μ being some integer which is not less than s, the equation (C′) being an identity in the elements of $X_1, X_2, \ldots X_r$. If in (C′) we make the substitutions

$$\alpha_{ki} = \frac{\beta_{ki}}{\beta_{k,\,n+1}}, \quad (k=1, 2, \ldots r;\ i=1, 2, \ldots n),$$

and then multiply both sides by $\{\beta_{1,n+1}\beta_{2,n+1}\ldots\beta_{n,n+1}\}^{\mu} = \varpi^{\mu}{}_{n+1}$, we obtain the equation

$$\varpi^{\mu-s}_{n+1}\,.\,S = G(\varpi_{q_1 q_2 \ldots q_{n+1}}, \ldots, \ldots) = G(\varpi), \quad (q_1+q_2+\ldots+q_{n+1}=r) \ \ldots(\mathrm{c})$$

or

$$\varpi^{\mu}{}_{n+1}\,S = \varpi^{s}{}_{n+1}\,G(\varpi), \ \ldots\ldots\ldots\ldots\ldots\ldots\ldots\ldots(\mathrm{C})$$

where $G(\varpi)$ is a *homogeneous* rational integral function of degree μ of those of the quantities ϖ which have total weight r, i.e. of the homogeneous ϖ's; in fact $G(\varpi)$ is the function obtained when we homogenise $F(\varpi')$ in the way described in sub-article 2.

When we regard $G(\varpi)$ as a rational integral function of the homogeneous ϖ's, it is not divisible by the quantity $\varpi_{00\ldots0r} = \varpi_{n+1}$, though it may be possible to separate off this factor, or even the factor $\varpi^{\mu-s}_{n+1}$ when we transform $G(\varpi)$ by means of the relations between the homogeneous ϖ's. But since the equation (c) is an identity in the β's, $G(\varpi)$ must certainly be divisible by $\varpi^{\mu-s}_{n+1}$ when it is regarded as a function of the β's. The quotient, being the monotypic symmetric function S of the β's, can be expressed as a rational integral function of the ϖ's whose total weights do not exceed r, though not necessarily of those whose total weights are equal to r.

Next let S be homogeneous of degree s in each of the matrices $Y_1, Y_2, \dots Y_r$ and be *divisible by* ϖ^t_{n+1} but not by any higher power of ϖ_{n+1}. Then it is derivable by homogenisation from a function S' of degree $\sigma = s - t$, and if S' is expressed in the form (C′), where $F(\varpi')$ has degree μ in the quantities ϖ', μ being not greater than σ, we have

$$S = \varpi^t_{n+1} S_1, \quad \varpi^{\mu-\sigma}_{n+1} S_1 = G(\varpi), \quad \dots\dots\dots\dots(\mathrm{c}_1)$$

where $G(\varpi)$ is the function of degree μ in the ϖ's formed by the homogenisation of $F(\varpi')$; and the equation (C) still holds good.

In the former case $G(\varpi)$, regarded as a function of the β's, was divisible by $\varpi^{\mu-s}_{n+1}$, and therefore $\varpi^s_{n+1} G(\varpi)$ was divisible by ϖ^{μ}_{n+1}; in this latter case $G(\varpi)$, regarded as a function of the β's, is divisible by $\varpi^{\mu-\sigma}_{n+1}$, and therefore $\varpi^s_{n+1} G(\varpi)$ is divisible by $\varpi^{\mu+t}_{n+1}$.

Ex. vii. If we suppress the dashes as being superfluous, we have the following pairs of corresponding results for the two sets of 3 sequences

$$[x_1 y_1],\ [x_2 y_2],\ [x_3 y_3] \text{ and } [a_1\beta_1\gamma_1],\ [a_2\beta_2\gamma_2],\ [a_3\beta_3\gamma_3]:$$

(1) $\Sigma x_1^2 y_2 = \varpi_{11}\varpi_{10} - \varpi_{20}\varpi_{01} - \varpi_{21}$; $\Sigma a_1^2\beta_2\gamma_2\gamma_3^2 = \varpi_{111}\varpi_{102} - \varpi_{201}\varpi_{012} - \varpi_{210}\varpi_{003}$.

(2) $\Sigma x_1^2 x_2^2 y_3 = \varpi_{21}\varpi_{20} - \varpi_{30}\varpi_{11}$; $\Sigma a_1^2 a_2^2\beta_3\gamma_3 = \varpi_{210}\varpi_{201} - \varpi_{300}\varpi_{111}$.

(3) $3\Sigma x_1^2 y_2 y_3 = (\varpi_{10}^2\varpi_{02} - \varpi_{10}\varpi_{01}\varpi_{11} + \varpi_{01}^2\varpi_{20}) + (2\varpi_{10}\varpi_{12} - \varpi_{01}\varpi_{21} - 4\varpi_{02}\varpi_{20} + \varpi_{11}^2)$;

$$3\varpi_{003}\,.\,\Sigma a_1^2\beta_2\gamma_2\beta_3\gamma_3 = (\varpi^2_{102}\varpi_{021} - \varpi_{102}\varpi_{012}\varpi_{111} + \varpi^2_{012}\varpi_{201})$$
$$+ (2\varpi_{102}\varpi_{120} - \varpi_{012}\varpi_{210} - 4\varpi_{021}\varpi_{201} + \varpi^2_{111})\,.\,\varpi_{003}.$$

In the last result the first bracket on the right can be expressed in the form

$$[4\varpi_{021}\varpi_{201} - \varpi^2_{111} + (\varpi_{102}\varpi_{120} + \varpi_{012}\varpi_{210}) - 3\Sigma a_1\beta_1 a_2\gamma_2\beta_3\gamma_3]\,.\,\varpi_{003},$$

and is then divisible by ϖ_{003}. We can express $\Sigma a_1\beta_1 a_2\gamma_2\beta_3\gamma_3$ as a rational integral function of the ϖ's (though not of the homogeneous ϖ's) by using the formula

$$45\Sigma a_1\beta_1 a_2\gamma_2\beta_3\gamma_3 = 45\sigma_{011}\sigma_{101}\sigma_{110} - 45(\sigma_{011}\sigma_{211} + \sigma_{101}\sigma_{121} + \sigma_{110}\sigma_{112}) + 90\sigma_{222}.$$

§ 199. Monotypic symmetric functions of common roots.

In this article we shall assume that the theorems of §§ 193 and 196 are true generally for functions not more than n in number. Accordingly Theorems I and II of this article will be true when n is replaced by $n-1$ if the theorems of §§ 193 and 196 are true generally for functions not more than $n-1$ in number; and we shall know that they are true for all positive integral values of n when we have completed the proofs of the theorems of §§ 193 and 196 in § 200.

1. *Homogeneous symmetric functions of the common roots of n general homogeneous functions of n + 1 variables.*

Let $f_1, f_2, \dots f_n$ be n general homogeneous rational integral functions of the $n+1$ variables $x_1, x_2, \dots x_{n+1}$ of non-zero degrees $r_1, r_2, \dots r_n$ given by

$$f_\tau = \Sigma a^{(\tau)}_{p_1 p_2 \dots p_{n+1}} x_1^{p_1} x_2^{p_2} \dots x_{n+1}^{p_{n+1}}, \quad (p_1 + p_2 + \dots + p_{n+1} = r_\tau;\ \tau = 1, 2, \dots n); \quad \dots\dots\dots(1)$$

let their coefficients be called the a's; let $r_1 r_2 \ldots r_n = r$; and let a given eliminant of these n functions, as defined in Theorem I of § 196, be

$$E(\kappa_1, \kappa_2, \ldots \kappa_{n+1}) = \Sigma \rho_{q_1 q_2 \ldots q_{n+1}} \kappa_1^{q_1} \kappa_2^{q_2} \ldots \kappa_{n+1}^{q_{n+1}}, \quad (q_1 + q_2 + \ldots + q_{n+1} = r); \quad \ldots\ldots\ldots\ldots(2)$$

so that $\rho_{q_1 q_2 \ldots q_{n+1}}$ is a homogeneous rational integral function of the a's which is homogeneous of degree $\frac{r}{r_\tau}$ in the coefficients of f_τ, and isobaric of weight $r - q_h$ with respect to the variable x_h. For the coefficients of $\kappa_1^r, \kappa_2^r, \ldots \kappa_{n+1}^r$ we will use the alternative notations $\rho_1, \rho_2, \ldots \rho_{n+1}$. We know that ρ_i is the resultant of the n homogeneous functions derived from $f_1, f_2, \ldots f_n$ by rejecting all terms which contain x_i, and that it is a homogeneous rational integral function of the a's which is homogeneous of degree $\frac{r}{r_\tau}$ in the coefficients of f_τ, isobaric of weight r with respect to the variable x_h when $h \neq i$, and isobaric of weight 0 with respect to x_i.

Whenever the a's have any particular values which do not cause E to vanish identically, the n functions f have exactly r non-zero common roots. We will denote these by

$$[x_1 x_2 \ldots x_{n+1}] \equiv [\alpha_{11} \alpha_{12} \ldots \alpha_{1, n+1}], \; [\alpha_{21} \alpha_{22} \ldots \alpha_{2, n+1}], \; \ldots \; [\alpha_{r1} \alpha_{r2} \ldots \alpha_{r, n+1}], \; \ldots(3)$$

and we will suppose that the absolute values of the α's are always so chosen that the equation

$$\prod_{k=1}^{k=r} (\alpha_{k1} \kappa_1 + \alpha_{k2} \kappa_2 + \ldots + \alpha_{k, n+1} \kappa_{n+1}) = \Sigma \rho_{q_1 q_2 \ldots q_{n+1}} \kappa_1^{q_1} \kappa_2^{q_2} \ldots \kappa_{n+1}^{q_{n+1}} \ldots (4)$$

is true and is an identity in the κ's. This is always possible because the coefficients of the various terms in the expansion of the product on the left always bear to one another the same ratios as the coefficients of the corresponding terms on the right.

By a *symmetric function of the r common roots* of the n functions f we shall mean a symmetric function of the $r(n+1)$ elements of the r one-rowed matrices (3); and by the kth common root we shall mean the kth of the matrices (3). As in § 197 the elementary symmetric function of these r common roots which has weights $q_1, q_2, \ldots q_{n+1}$ with respect to $x_1, x_2, \ldots x_{n+1}$ will be denoted by $\varpi_{q_1 q_2 \ldots q_{n+1}}$, where $q_1 + q_2 + \ldots + q_{n+1} \not> r$. When we expand the product on the left in (4) as a rational integral function of the κ's, the coefficients of the various terms are those of the ϖ's in which

$$q_1 + q_2 + \ldots + q_{n+1} = r,$$

i.e. they are those monotypic symmetric functions of the r common roots which are homogeneous of degree 1 in every one of the common roots.

When we equate coefficients of corresponding terms on both sides of (4), we see that for all particular values of the a's which do not cause E to vanish identically we have

$$\varpi_{q_1 q_2 \dots q_{n+1}} = \rho_{q_1 q_2 \dots q_{n+1}}, \quad (q_1 + q_2 + \dots + q_{n+1} = r), \quad \dots\dots\dots(5)$$

where the ϖ on the left is a function of the *common roots*, and the ρ on the right is a function of the *coefficients* of the n functions f. For the coefficients of $\kappa_1{}^r, \kappa_2{}^r, \dots \kappa^r{}_{n+1}$ in the product on the left in (4) we will use the alternative notations $\varpi_1, \varpi_2, \dots \varpi_{n+1}$, so that

$$\varpi_i = \alpha_{1i}\, \alpha_{2i} \dots \alpha_{ri}, \quad (i = 1, 2, \dots n+1);$$

and then particular cases of (5) are

$$\varpi_1 = \rho_1, \quad \varpi_2 = \rho_2, \dots \varpi_i = \rho_i, \dots \varpi_{n+1} = \rho_{n+1}. \quad \dots\dots\dots\dots(5')$$

Our object is to investigate the properties of those monotypic symmetric functions of the r common roots which, like the ϖ's in (5), are homogeneous (of the same degree) in every common root.

Let $$S = S_{u_1 u_2 \dots u_{n+1}} = \Sigma \{T_1 T_2 \dots T_k \dots T_r\}, \dots\dots\dots\dots\dots\dots(A)$$

where $$T_k = \alpha_{k1}^{\lambda_{k1}}\, \alpha_{k2}^{\lambda_{k2}} \dots \alpha_{k,\, n+1}^{\lambda_{k,\, n+1}}, \quad \dots\dots\dots\dots\dots\dots\dots\dots\dots(a)$$

be any *monotypic* symmetric function of the r common roots (3) which is *homogeneous of degree s* in each root, i.e. in the elements of each one of the matrices (3), and has *weights* $u_1, u_2, \dots u_{n+1}$ with respect to the variables $x_1, x_2, \dots x_{n+1}$, so that its total weight is rs, and

$$\lambda_{k1} + \lambda_{k2} + \dots + \lambda_{k,\, n+1} = s, \quad (k = 1, 2, \dots r),$$

$$\lambda_{1h} + \lambda_{2h} + \dots + \lambda_{rh} = u_h, \quad (h = 1, 2, \dots n+1),$$

$$u_1 + u_2 + \dots + u_{n+1} = rs;$$

and let *ordinary values* of the a's be those for which every element of every common root is a finite number different from 0.

If x_i is any one of the variables $x_1, x_2, \dots x_{n+1}$, we can write for all ordinary particular values of the a's and for the values $1, 2, \dots n+1$ of k

$$[\alpha_{k1} \dots \alpha_{k,\, i-1}\, \alpha_{k,\, i+1} \dots \alpha_{k,\, n+1}] = \alpha_{ki}\, [x_{k1} \dots x_{k,\, i-1}\, 1\, x_{k,\, i+1} \dots x_{k,\, n+1}],$$

and we then have

$$T_k = \alpha_{ki}^{s}\, T_k', \text{ where } T_k' = x_{k1}^{\lambda_{k1}} \dots x_{k,\, i-1}^{\lambda_{k,\, i-1}}\, x_{k,\, i+1}^{\lambda_{k,\, i+1}} \dots x_{k,\, n+1}^{\lambda_{k,\, n+1}}, \quad \dots\dots(a')$$

and $$S = \varpi_i{}^s\, S_i', \quad \dots\dots\dots\dots\dots\dots\dots\dots\dots\dots(6)$$

where $$S_i' = \Sigma \{T_1' T_2' \dots T_k' \dots T_r'\}. \quad \dots\dots\dots\dots\dots\dots(A')$$

Here S_i' is a monotypic symmetric function of the elements of the r matrices

$$[x_1 \dots x_{i-1}\, x_{i+1} \dots x_{n+1}] = [x_{k1} \dots x_{k,\, i-1}\, x_{k,\, i+1} \dots x_{k,\, n+1}], \quad (k = 1, 2, \dots r), \dots(3')$$

which has weights $u_1, \dots u_{i-1}, u_{i+1}, \dots u_{n+1}$ with respect to the variables $x_1, \dots x_{i-1}, x_{i+1}, \dots x_{n+1}$, and total weight $rs - u_i$. The degree of S_i' in each of the matrices (3′) is s when ϖ_i is not a factor of S, and $s-t$ when ϖ_i is a factor of S repeated exactly t times, but it is not in general homogeneous in those matrices. We obtain S when we homogenise S' by the substitution $1 = x_i$. If we denote each of the elementary symmetric functions of the rn elements of the r matrices (3′) by a letter ϖ' with suffixes indicating its weights with respect to $x_1, \dots x_{i-1}, x_{i+1}, \dots x_{n+1}$, then as particular cases of (6) we have

$$\varpi_{q_1 q_2 \dots q_{n+1}} = \varpi_i \,.\, \varpi'_{q_1 \dots q_{i-1} q_{i+1} \dots q_{n+1}}, \quad (q_1 + q_2 + \dots + q_{n+1} = r), \quad \dots(6')$$

where the function ϖ' has total weight $r - q_i$.

By Lemma B of § 197 we can express S_i' in the form

$$S_i' = F_i(\varpi'_{q_1 \dots q_{i-1} q_{i+1} \dots q_{n+1}}, \dots, \dots) = F_i(\varpi') \quad \dots\dots\dots\dots(\mathrm{B}')$$

as a rational integral function of the quantities ϖ', in each of which we have $q_1 + \dots + q_{i-1} + q_{i+1} + \dots + q_{n+1} \not> r$. Every term of F_i is a product of certain of the quantities ϖ' in which

$$\Sigma q_h = u_h, \quad (h = 1, \dots i-1, i+1, \dots n+1),$$

i.e. F_i has weights $u_1, \dots u_{i-1}, u_{i+1}, \dots u_{n+1}$ with respect to $x_1, \dots x_{i-1}, x_{i+1}, \dots x_{n+1}$, and therefore total weight $rs - u_i$. We will denote the degree of F_i in the quantities ϖ' by μ_i. We do not know μ_i, but if ϖ_i is a factor of S repeated t times exactly, so that $s-t$ is the degree of S_i' in each of the matrices (3′), then by Lemma B of § 197 we must have $\mu_i \not< s-t$; and if ϖ_i is not a factor of S, then $\mu_i \not< s$.

From (5), (5′), (6), (6′) we see that for all ordinary values of the a's

$$\varpi'_{q_1 \dots q_{i-1} q_{i+1} \dots q_{n+1}} = \frac{\varpi_{q_1 q_2 \dots q_{n+1}}}{\varpi_i} = \frac{\rho_{q_1 q_2 \dots q_{n+1}}}{\rho_i}, \quad (q_1 + q_2 + \dots + q_{n+1} = r);$$

and when we substitute these values for the quantities ϖ' in (B′), we obtain

$$\varpi_i^{\mu_i} S_i' = G_i(\varpi_{q_1 q_2 \dots q_{n+1}}, \dots, \dots) = G_i(\varpi), \quad \varpi_i^{\mu_i} S = \varpi_i^{s}\, G_i(\varpi), \dots (\mathrm{B})$$

$$\rho_i^{\mu_i} S_i' = G_i(\rho_{q_1 q_2 \dots q_{n+1}}, \dots, \dots) = G_i(\rho), \quad \rho_i^{\mu_i} S = \rho_i^{s}\, G_i(\rho), \dots (\mathrm{C})$$

where $G_i(\varpi)$ is a homogeneous rational integral function of degree μ_i of those ϖ's in which $q_1 + q_2 + \dots + q_{n+1} = r$, and $G_i(\rho)$ is the same function of the corresponding ρ's. The terms of $G_i(\varpi)$ correspond one by one to the terms of $F_i(\varpi')$. We obtain $G_i(\varpi)$ from $F_i(\varpi')$ by replacing each of the quantities ϖ' by the corresponding quantity ϖ, and supplying in each term an additional factor of the form ϖ_i^{ν} which will raise its total degree in the ϖ's to μ_i. Thus in every term of $G_i(\varpi)$, which is a product of certain of the

ϖ's, and therefore also in every term of $G_i(\rho)$, which is a product of corresponding ρ's, we have

$$\left.\begin{aligned}&\Sigma q_h = u_h, \quad (h = 1, \ldots i-1, i+1, \ldots n+1),\\ &\Sigma(q_1 + \ldots + q_{i-1} + q_{i+1} + \ldots + q_{n+1}) = rs - u_i,\\ &\Sigma q_i = r\mu_i - (rs - u_i) = u_i - r(s - \mu_i).\end{aligned}\right\} \quad \ldots\ldots\ldots\ldots(7)$$

Let the ρ's and each term of $G_i(\rho)$ be expressed as rational integral functions of the a's. Then if f_τ is any one of the n functions f, the ρ's are all homogeneous of degree $\frac{r}{r_\tau}$ in the coefficients of f_τ; therefore every term of $G_i(\rho)$ is homogeneous of degree $\mu_i \frac{r}{r_\tau}$ in the coefficients of f_τ. Again if x_h is any one of the variables $x_1, x_2, \ldots x_{n+1}$, then $\rho_{q_1 q_2 \ldots q_{n+1}}$ is isobaric of weight $r - q_h$ with respect to x_h in all the a's; therefore from (7) we see that the weight of every term of $G_i(\rho)$ with respect to x_h in all the a's is

$$\Sigma(r - q_h) = r\mu_i - \Sigma q_h = r\mu_i - u_h \quad \text{when } h \neq i,$$

$$\Sigma(r - q_i) = r\mu_i - \Sigma q_i = rs - u_i \quad \text{when } h = i.$$

Thus $G_i(\rho)$, regarded as a rational integral function of the a's, is homogeneous of degree $\mu_i \frac{r}{r_\tau}$ in the coefficients of f_τ, isobaric of weight $r\mu_i - u_h$ in all the a's with respect to x_h when $h \neq i$, and isobaric of weight $rs - u_i$ in all the a's with respect to x_i.

If x_j is any other of the variables $x_1, x_2, \ldots x_{n+1}$ different from x_i, we can obtain in the same way equations of the forms (B) and (C) in which i is everywhere replaced by j, and from these two sets of equations we deduce the equations

$$\rho_i^s \rho_j^{\mu_j} . G_i(\rho) = \rho_j^s \rho_i^{\mu_i} . G_j(\rho),$$

in which μ_j like μ_i is a positive integer; and this equation, being true for all ordinary values of the a's, is an identity in the a's. Since ρ_i and ρ_j are irresoluble functions of the a's, it follows that $G_i(\rho)$ must be divisible by $\rho_i^{\mu_i}$. Therefore there must exist a rational integral function H of the a's such that

$$\rho_i^s . G_i(\rho_{q_1 q_2 \ldots q_{n+1}}, \ldots, \ldots) = \rho_i^{\mu_i} . H(a_{p_1 p_2 \ldots p_{n+1}}, \ldots, \ldots), \quad \ldots\ldots(D)$$

this equation being an identity in the a's, and being analogous to the equation (C) of § 198.

The function H, like the other factors on both sides of (D), must be homogeneous in the coefficients of each one of the n functions f, and isobaric in the coefficients of all those functions with respect to each one of the variables $x_1, x_2, \dots x_{n+1}$. If its degree in the coefficients of f_τ is s_τ, we see by equating the degrees of both sides of (D) that

$$s\frac{r}{r_\tau} + \mu_i \frac{r}{r_\tau} = \mu_i \frac{r}{r_\tau} + s_\tau, \text{ or } s_\tau = \frac{rs}{r_\tau};$$

and if v_h is its weight in all the a's with respect to x_h, then by equating the weights on both sides of (D) we see that

$$sr + (\mu_i r - u_h) = \mu_i r + v_h, \text{ or } v_h = rs - u_h \text{ when } h \neq i;$$

$$0 + (rs - u_i) = 0 + v_i, \text{ or } v_i = rs - u_i \text{ when } h = i.$$

The equation $S = H$ which follows from (C) and (D), being true for all ordinary particular values of the a's, must be true for all values of the a's which do not cause E to vanish identically, because for every such set of values of the a's each side of the equation has a unique finite value.

We have now proved the following theorem.

Theorem I. *If S is any monotypic symmetric function of the r common roots of the n general homogeneous functions f given by* (1) *which is homogeneous of degree s in each root and has weights $u_1, u_2, \dots u_{n+1}$ with respect to the variables $x_1, x_2, \dots x_{n+1}$, then there exists a rational integral function H of the coefficients of $f_1, f_2, \dots f_n$ which has the following properties:*

(1) *It is homogeneous in the coefficients of each one of the n functions f, its degree in the coefficients of f_τ being $\frac{rs}{r_\tau}$.*

(2) *It is isobaric in the coefficients of all the n functions f with respect to each one of the variables $x_1, x_2, \dots x_{n+1}$, its weight with respect to x_h being $rs - u_h$ and its total weight being nrs.*

(3) *For all particular values of the a's which do not cause the eliminant E to vanish identically we have*

$$S = H,$$

provided only that the absolute values of the elements of the common roots are always so chosen that the equation (4) *is true.*

We may call S and H *equivalent functions* of the *common roots* and the *coefficients* of the n functions f. It is very often possible to express H as a homogeneous rational integral function of degree μ of the ρ's, i.e. of the coefficients of the eliminant E, but we cannot assert that this is always possible.

Ex. i. The function H is uniquely determinate when the arbitrary numerical factor in E is given; for if there were another function H' with the same properties, the equation $H' = H$ would be true for all ordinary values of the a's, and would be an identity in the a's.

Ex. ii. *If S is the elementary symmetric function $\varpi_{q_1 q_2 \cdots q_{n+1}}$ in which*

$$q_1 + q_2 + \ldots + q_{n+1} = r,$$

then $H = \rho_{q_1 q_2 \cdots q_{n+1}}$; in particular if S is the function ϖ_i, then $H = \rho_i$.

Ex. iii. If the a's have such particular values that all the coefficients $\rho_{q_1 q_2 \cdots q_{n+1}}$ of the eliminant E vanish except ρ_i, we see from (D) that H is expressible as a function of ρ_i only, and that in this case we must have $H = A\rho_i^s$, and therefore $S = A\varpi_i^s$, where A is a constant. That such values of the a's are possible will be clear from Ex. xviii of § 196. We conclude that:

The function H vanishes whenever E vanishes identically.

Ex. iv. If S can be expressed as a product of two monotypic symmetric functions in the form $S = S_1 S_2$, we must have $H = H_1 H_2$, where H_1 and H_2 are the functions of the coefficients equivalent to S_1 and S_2 respectively.

In this case however either S_1 or S_2 must have the form $\varpi_1^{k_1} \varpi_2^{k_2} \varpi_3^{k_3} \ldots$; for if the product $S_1 S_2$ of two monotypic symmetric functions of the elements of the r sequences (3) is itself a monotypic symmetric function of those elements whose weight with respect to x_i is not 0, then every term either of S_1 or of S_2 must contain all the x_i-elements $a_{1i}, a_{2i}, a_{3i}, \ldots$, and must therefore be divisible by ϖ_i. If this were not so, the product $S_1 S_2$ would contain terms of different types, some involving p and some involving q of the x_i-elements, where $p \neq q$.

It appears probable that a monotypic symmetric function of the elements of the r sequences (3) which is not divisible by any one of the products $\varpi_1, \varpi_2, \varpi_3, \ldots$ is irresoluble. This is obviously true of the elementary symmetric functions.

Ex. v. If $\varpi_i^t = (a_{1i} a_{2i} \ldots a_{ri})^t$ is a factor of every term of S, then $S \div \varpi_i^t$ is a monotypic symmetric function of the r common roots (3) which is homogeneous of degree $\sigma = s - t$ in every common root, has weights $u_1, u_2, \ldots u_i' = u_i - rt, \ldots u_{n+1}$ with respect to $x_1, x_2, \ldots x_i, \ldots x_{n+1}$, and has total weight $r\sigma$. Applying Theorem I to this function, we see that in this case

$$H = \rho_i^t K, \qquad \text{(E)}$$

where K is a rational integral function of the a's which is homogeneous of degree $\frac{r\sigma}{r_\tau}$ in the coefficients of f_τ, is isobaric of weight $r\sigma - u_h$ in all the a's with respect to x_h when $h \neq i$, is isobaric of weight $r\sigma - u_i' = rs - u_i$ in all the a's with respect to x_i, and has total weight $r\sigma$.

If moreover ϖ_i^t is the highest power of ϖ_i which is a factor of S, then K cannot be divisible by the irresoluble factor ρ_i for general functions f, i.e. ρ_i^t is the highest power of ρ_i which is a factor of H, and if $\sigma = s - t$, we have $\mu_i \not< \sigma$ and

$$G_i(\rho) = \rho_i^{\mu_i - \sigma} K = \rho_i^{\mu_i - s} H, \qquad \text{(8)}$$

the equation (8) being an identity in the a's, but not necessarily an identity in the ρ's.

In the particular case when ϖ_i is not a factor of every term of S, then H is not divisible by ρ_i, and we have $\mu_i \not< s$ and

$$G_i(\rho) = \rho_i^{\mu_i - s} H. \qquad \text{(8')}$$

Ex. vi. From (C) and (D) we see that the equation

$$\rho_i^s S_i' = H \qquad \text{(9)}$$

is true for all particular values of the a's which do not cause E to vanish identically; and when ρ_i vanishes for such values of the a's, the equation means that S_i' is infinite.

If S is not divisible by ϖ_i, S_i' may be any monotypic symmetric function of the elements of the r matrices (3′) which has degree s (but is not necessarily homogeneous) in each of those matrices, has weights $u_1, \ldots u_{i-1}, u_{i+1}, \ldots u_{n+1}$ with respect to

$$x_1, \ldots x_{i-1}, x_{i+1}, \ldots x_{n+1},$$

and has total weight $rs-u_i$. In this case H is not divisible by the irresoluble factor ρ_i.

If S is divisible by ϖ_i^t, but not by any higher power of ϖ_i, then H is divisible by ρ_i^t, and we can replace (9) by

$$\rho_i^{s-t} S_i' = K, \quad \text{...(9′)}$$

where K has the properties described in Ex. v, and is not divisible by ρ_i. In this case S_i' may be any monotypic symmetric function of the elements of the r matrices (3′) which has degree $\sigma = s-t$ in each of those matrices, has weights $u_1, \ldots u_{i-1}, u_{i+1}, \ldots u_{n+1}$ with respect to $x_1, \ldots x_{i-1}, x_{i+1}, \ldots x_{n+1}$, and has total weight $r\sigma - u_i' = rs - u_i$.

NOTE 1. *Relations between the coefficients of the eliminant of n homogeneous functions of $n+1$ variables.*

If
$$F(\varpi_{q_1 q_2 \ldots q_{n+1}}, \ldots, \ldots) = F(\varpi) = 0, \quad (q_1+q_2+\ldots+q_{n+1}=r) \quad \text{............(10)}$$

is one of the rational integral relations between the homogeneous elementary symmetric functions of any r sequences of $n+1$ variables, then

$$F(\rho_{q_1 q_2 \ldots q_{n+1}}, \ldots, \ldots) = F(\rho) = 0, \quad (q_1+q_2+\ldots+q_{n+1}=r) \quad \text{............(10′)}$$

is a rational integral relation between the coefficients of the eliminant (2) of the n homogeneous functions (1) which is an identity in the coefficients a of those functions; for from (5) we see that the equation (10′) is true for all ordinary values of the a's. The same result follows from § 198.2; for (10′) is one of the conditions that $E(\kappa_1, \kappa_2, \ldots \kappa_{n+1})$ shall be expressible as a product of linear factors; it is therefore necessarily true for all ordinary values of the a's, and must be an identity in the a's. There may be other relations between the ρ's which are identities in the a's; and there always are other relations between the ρ's and a's which are identities in the a's.

NOTE 2. *Derivation of the function $H(a)$ from the function $G_i(\varpi)$.*

When we have obtained by homogenisation the equation

$$\varpi_i^{\mu_i} S = \varpi_i^s G_i(\varpi),$$

in which we may take i to be $n+1$, the function H is given by the equation

$$\rho_i^{\mu_i} H = \rho_i^s G_i(\rho)$$

which is an identity in the a's but not necessarily an identity in the ρ's.

If ϖ_i is not a factor of S, we have $\mu_i \nless s$, and the equation (8′) shows that $G_i(\rho)$, regarded as a function of the a's, is divisible by $\rho_i^{\mu_i - s}$; and after separating off this factor from it we obtain H. It may be possible to separate off this factor when we transform $G_i(\rho)$ by means of the relations between the ρ's; and in such cases H is expressible as a rational integral function of the ρ's. It must always be possible to separate off this factor when we transform $G_i(\rho)$ by means of the relations between the ρ's and a's.

If S is divisible by ϖ_i^t but not by any higher power of ϖ_i, we have $\mu_i \nless s-t$. In this case the equation (8) shows that $G_i(\rho)$, regarded as a function of the a's, is divisible by $\rho_i^{\mu_i - s + t}$, and when we separate off from it the factor $\rho_i^{\mu_i - s}$, we obtain the function

$$H = \rho_i^t K.$$

NOTE 3. *Derivation of the function $H(a)$ from the function $F_i(\varpi')$ in* (B′).

Both when S is not and when it is divisible by ϖ_i, H is the rational integral function of the a's given by

$$H=\rho_i^s \cdot F_i\left(\frac{\rho_{q_1 q_2 \dots q_{n+1}}}{\rho_i}, \dots, \dots\right)=\frac{G_i(\rho)}{\rho_i^{\mu_i - s}}, \quad \dots\dots\dots\dots\dots\dots\dots\dots \text{(F)}$$

where $q_1+q_2+\dots+q_{n+1}=r$, and s is the degree of S.

Ex. vii. *Homogeneous symmetric functions of the two non-zero common roots* $[a_1\beta_1\gamma_1]$, $[a_2\beta_2\gamma_2]$ *of the two homogeneous functions* ϕ *and* ψ *of Ex.* xiv *in* § 196.

(1) $S=\Sigma a_1^2\beta_2\gamma_2$. For the two sequences $[x_1y_1]$, $[x_2y_2]$ we have

$$S'=\Sigma x_1^2 y_2=\varpi_{10}\varpi_{11}-\varpi_{01}\varpi_{20};$$

and by homogenisation we deduce for the two common roots the equation

$$\begin{aligned} S=\Sigma a_1^2\beta_2\gamma_2&=\varpi_{101}\varpi_{110}-\varpi_{011}\varpi_{200}\\ &=\rho_{101}\rho_{110}-\rho_{011}\rho_{200}=4MN-PL\\ &=H. \end{aligned}$$

(2) $S=\Sigma a_1^2\beta_2^2$. For the two sequences $[x_1y_1]$, $[x_2y_2]$ we have (see Ex. xiv of § 197)

$$3S'=3\Sigma x_1^2y_2^2=4\,(\varpi_{10}\varpi_{01}\varpi_{11}-\varpi_{10}^2\varpi_{02}-\varpi_{01}^2\varpi_{20})+(10\varpi_{20}\varpi_{02}-\varpi_{11}^2), \quad \dots\dots\dots(11)$$

and by homogenisation we deduce for the two common roots the equation

$$3\varpi_{002}\cdot\Sigma a_1^2\beta_2^2=4\,(\varpi_{011}\varpi_{101}\varpi_{110}-\varpi_{200}\varpi_{011}^2-\varpi_{020}\varpi_{101}^2)+(10\varpi_{200}\varpi_{020}-\varpi_{110}^2)\,\varpi_{002}.$$

Therefore the value of S in terms of the coefficients of ϕ and ψ is given by

$$3\rho_{002}\cdot\Sigma a_1^2\beta_2^2=4\,(\rho_{011}\rho_{101}\rho_{110}-\rho_{200}\rho_{011}^2-\rho_{020}\rho_{101}^2)+(10\rho_{200}\rho_{020}-\rho_{110}^2)\,\rho_{002},$$

or

$$3R\,.\,\Sigma a_1^2\beta_2^2=16\,(2LMN-PL^2-QM^2)+(10PQ-4N^2)\,R.$$

Using the relation $PQR+2LMN-PL^2-QM^2-RN^2=0$, it follows that

$$\begin{aligned} S=\Sigma a_1^2\beta_2^2&=4N^2-2PQ\\ &=H. \end{aligned}$$

We could use the relation of Ex. vii of § 197 to replace (11) by

$$S'=\Sigma x_1^2y_2^2=\varpi_{11}^2-2\varpi_{20}\varpi_{02},$$

and we should then have

$$\begin{aligned} S=\Sigma a_1^2\beta_2^2&=\varpi_{110}^2-2\varpi_{200}\varpi_{020}\\ &=\rho_{110}^2-2\rho_{200}\rho_{020}=4N^2-2PQ\\ &=H. \end{aligned}$$

It is to be understood that the absolute values of the elements of the common roots are chosen in accordance with the equation (4). Both the above functions S are expressible as rational integral functions of the coefficients of the eliminant.

2. *Symmetric functions of the common roots of n general functions of n variables.*

Let $f_1, f_2, \dots f_n$ be n general rational integral functions of the n variables $x_1, x_2, \dots x_n$ of non-zero degrees $r_1, r_2, \dots r_n$ given by

$$f_\tau=\Sigma a_{p_1p_2\dots p_np_{n+1}}x_1^{p_1}x_2^{p_2}\dots x_n^{p_n}, \quad (p_1+p_2+\dots+p_n+p_{n+1}=r_\tau;\ \tau=1, 2, \dots n); \quad \dots\dots(12)$$

let their coefficients be called the a's; and let $r_1 r_2 \dots r_n = r$. We may regard these functions as derived from the homogeneous functions (1) by the substitution $x_{n+1} = 1$, and define $\rho_1, \rho_2, \dots \rho_{n+1}$ as before, so that ρ_{n+1} is the resultant of the n functions formed when we retain only the terms of the highest degree in each of the n functions (12). We may use the same notation for the eliminant as before if we replace κ_{n+1} by z. When the n functions f have no infinite common roots, they have exactly r common roots which we will denote by

$$[x_1 x_2 \dots x_n] = [x_{11} x_{12} \dots x_{1n}],\ [x_{21} x_{22} \dots x_{2n}],\ \dots\ [x_{r1} x_{r2} \dots x_{rn}]. \quad \dots\dots(13)$$

By a monotypic symmetric function of these r common roots we mean a monotypic symmetric function of the elements of the r matrices (13).

Putting $i = n+1$ in the first part of Ex. vi we obtain the following second theorem:

Theorem II. *If S is any monotypic symmetric function of the r common roots of the n general functions f given by* (12) *which has degree s in each root, weights $u_1, u_2, \dots u_n$ in those roots with respect to $x_1, x_2, \dots x_n$, and total weight $w = u_1 + u_2 + \dots + u_n = rs - u_{n+1}$, then there exists a rational integral function H of the a's which has the following properties:*

(1) *It is homogeneous in the coefficients of each one of the n functions f, its degree in the coefficients of f_τ being $\dfrac{rs}{r_\tau}$.*

(2) *It is isobaric in the coefficients of all the n functions f with respect to each one of the variables $x_1, x_2, \dots x_n$, and also with respect to the homogenising variable x_{n+1}, its weight with respect to x_h being $rs - u_h$, and its total weight with respect to $x_1, x_2, \dots x_{n+1}$ being nrs.*

(3) *The equation* $$\rho^s_{n+1} S = H$$ *is true whenever $\rho_{n+1} \neq 0$, i.e. whenever the a's have such particular values that the n functions f have no infinite common roots.*

We first express S as a rational integral function of the ϖ's of the r common roots (13) in the form

$$S = F(\varpi_{q_1 q_2 \dots q_n}, \dots, \dots) = F(\varpi), \quad (q_1 + q_2 + \dots + q_n \ngtr r). \quad \dots(14)$$

Then if μ is the degree of $F(\varpi)$, it follows from (D) that H is the rational integral function of the a's given by

$$\rho^{\mu-s}_{n+1} H = G(\rho_{q_1 q_2 \dots q_{n+1}}, \dots, \dots) = G(\rho), \quad (q_1 + q_2 + \dots + q_n + q_{n+1} = r),$$

where $G(\rho)$ is the function which we derive from $F(\varpi)$ when we make the substitutions

$$\varpi_{q_1 q_2 \dots q_n} = \frac{\rho_{q_1 q_2 \dots q_n q_{n+1}}}{\rho_{n+1}}, \quad (q_1 + q_2 + \dots + q_n + q_{n+1} = r),$$

and multiply throughout by $\rho^{\mu}{}_{n+1}$. In other words H is the rational integral function of the a's given by

$$H = \rho^{s}{}_{n+1} \, . \, F\left(\frac{\rho_{q_1 q_2 \dots q_n q_{n+1}}}{\rho_{n+1}}, \; \dots, \; \dots\right), \quad \dots\dots\dots\dots\dots(\mathrm{F}')$$

where $q_1 + q_2 + \dots + q_n + q_{n+1} = r$.

Ex. viii. *Symmetric functions of the two common roots* $[x_1 y_1]$, $[x_2 y_2]$ *of the two functions* ϕ *and* ψ *of Ex.* xv *in* § 196.

(1) If $S = \Sigma x_1^2 y_2 = \varpi_{10}\varpi_{11} - \varpi_{01}\varpi_{20}$, we have $s = 2$, $\mu = 2$, and

$$\begin{aligned} \rho^2{}_{002} S &= \rho^2{}_{002} \, . \, \frac{\rho_{101}\rho_{110} - \rho_{011}\rho_{200}}{\rho^2{}_{002}} = \rho_{101}\rho_{110} - \rho_{011}\rho_{200} \\ &= 4MN - PL \\ &= H. \end{aligned}$$

(2) If $S = \Sigma x_1^2 y_2^2 = \frac{1}{3}\left[4\left(\varpi_{10}\varpi_{01}\varpi_{11} - \varpi_{10}^2\varpi_{02} - \varpi_{01}^2\varpi_{20}\right) + \left(10\varpi_{20}\varpi_{02} - \varpi_{11}^2\right)\right]$, then $s = 2$, $\mu = 3$, and we have

$$\begin{aligned} \rho^2{}_{002} S &= \rho^2{}_{002} \, . \, \frac{4\left(\rho_{101}\rho_{011}\rho_{110} - \rho^2{}_{101}\rho_{020} - \rho^2{}_{011}\rho_{200}\right) + \left(10\rho_{200}\rho_{002} - \rho^2{}_{110}\right)}{3\rho^3{}_{002}} \\ &= 4N^2 - 2PQ = H. \end{aligned}$$

§ 200. Determination of resultants and eliminants.

1. *Determination of the resultant of n general homogeneous rational integral functions of n variables.*

By Ex. i of § 193 we may suppose that $n > 1$. We will make the hypothesis that the theorems of §§ 193 and 196 are true generally for functions less than n in number, and suppose that it is required to find the resultant of n general homogeneous rational integral functions $f_1, f_2, \dots f_n$ of the n variables $x_1, x_2, \dots x_n$ which have non-zero degrees $r_1, r_2, \dots r_n$ in all the variables and are given by

$$f_i = f_i(x_1, x_2, \dots x_n) = \Sigma a^{(i)}_{p_1 p_2 \dots p_n} x_1^{p_1} x_2^{p_2} \dots x_n^{p_n}, \quad \dots\dots\dots(1)$$
$$(p_1 + p_2 + \dots + p_n = r_i; \; i = 1, 2, \dots n).$$

Let $$r = r_1 r_2 \dots r_n; \quad s_i = \frac{r}{r_i}, \quad (i = 1, 2, \dots n);$$

let $\kappa_1, \kappa_2, \dots \kappa_n$ be n auxiliary arbitrary parameters; and let X be the homogeneous linear function

$$X = \kappa_1 x_1 + \kappa_2 x_2 + \dots + \kappa_n x_n. \dots\dots\dots\dots\dots\dots\dots(2)$$

Selecting any one f_i of the n functions f, let

$$E_i(\kappa_1, \kappa_2, \dots \kappa_n) = \Sigma \rho_{q_1 q_2 \dots q_n} \kappa_1^{q_1} \kappa_2^{q_2} \dots \kappa_n^{q_n}, \quad (q_1 + q_2 + \dots + q_n = s_i),$$

be a given X-eliminant of the remaining $n - 1$ functions

$$f_1, \dots f_{i-1}, f_{i+1}, \dots f_n. \dots\dots\dots\dots\dots\dots\dots(3)$$

Let ordinary values of the coefficients of the $n-1$ functions (3) be finite values which do not cause E_i, regarded as a function of the κ's, to vanish identically; and when any ordinary particular values are given to their coefficients, let

$$[x_1 x_2 \ldots x_n] \equiv [\alpha_{u1} \alpha_{u2} \ldots \alpha_{un}], \quad (u = 1, 2, \ldots s_i), \quad \ldots\ldots\ldots\ldots(4)$$

be the s_i non-zero common roots of the $n-1$ functions (3) when the absolute values of the α's are so chosen that the equation

$$\prod_{u=1}^{u=s_i} (\kappa_1 \alpha_{u1} + \kappa_2 \alpha_{u2} + \ldots + \kappa_n \alpha_{un}) = E_i(\kappa_1, \kappa_2, \ldots \kappa_n) \quad \ldots\ldots\ldots(5)$$

is true, and is an identity in the κ's.

Then by ordinary multiplication we obtain an identity in the coefficients of f_i of the form

$$\prod_{u=1}^{u=s_i} f_i(\alpha_{u1}, \alpha_{u2}, \ldots \alpha_{un}) = \Sigma A_{\lambda_1 \lambda_2 \ldots \lambda_n} S_{\lambda_1 \lambda_2 \ldots \lambda_n}, \quad \ldots\ldots\ldots\ldots(6)$$

where $A_{\lambda_1 \lambda_2 \ldots \lambda_n}$ is a product of s_i of the coefficients of f_i (some or all of which may be repeated) which has weights $\lambda_1, \lambda_2, \ldots \lambda_n$ in those coefficients with respect to $x_1, x_2, \ldots x_n$, so that in every term of (6)

$$\lambda_1 + \lambda_2 + \ldots + \lambda_n = s_i r_i = r;$$

and where $S_{\lambda_1 \lambda_2 \ldots \lambda_n}$ is a monotypic symmetric function of the s_i common roots (4) which

(1) is homogeneous of degree r_i in each root,

(2) has weights $\lambda_1, \lambda_2, \ldots \lambda_n$ with respect to $x_1, x_2, \ldots x_n$.

By Theorem I of § 199 we have for all ordinary values of the coefficients of the $n-1$ functions (3)

$$S_{\lambda_1 \lambda_2 \ldots \lambda_n} = H_{\mu_1 \mu_2 \ldots \mu_n}, \quad (\mu_j = s_i r_i - \lambda_j = r - \lambda_j;\ j = 1, 2, \ldots n), \ \ldots(7)$$

where $H_{\mu_1 \mu_2 \ldots \mu_n}$ is a uniquely determinate rational integral function of the coefficients of the $n-1$ functions (3) which is

(1) homogeneous of degree $\dfrac{s_i r_i}{r_\tau} = \dfrac{r}{r_\tau}$ in the coefficients of f_τ,

$$(\tau = 1, \ldots i-1, i+1, \ldots n),$$

(2) isobaric of weight μ_j with respect to x_j in the coefficients of all the $n-1$ functions (3),

$$(j = 1, 2, \ldots n).$$

Now substituting the functions H for the functions S on the right in (6), let

$$\Sigma A_{\lambda_1 \lambda_2 \ldots \lambda_n} H_{\mu_1 \mu_2 \ldots \mu_n} = R(f_1, f_2, \ldots f_n). \quad \ldots\ldots\ldots\ldots\ldots(8)$$

Then R is a uniquely determinate rational integral function of the coefficients of $f_1, f_2, \dots f_n$ which is

(1) homogeneous of degree s_τ in the coefficients of f_τ, $(\tau = 1, 2, \dots n)$,

(2) isobaric of weight $\lambda_j + \mu_j = r$ with respect to x_j in the coefficients of all the n functions $f_1, f_2, \dots f_n$, $(j = 1, 2, \dots n)$;

and for all ordinary particular values of the coefficients of the $n-1$ functions (3) the equation

$$R(f_1, f_2, \dots f_n) = \prod_{u=1}^{u=s_i} f_i(\alpha_{u1}\alpha_{u2}\dots\alpha_{un}) \quad \dots\dots\dots\dots\dots(9)$$

is true, and is an identity in the coefficients of f_i.

The equation (9) shows that whenever the coefficients of the n functions (1) have such particular values that E_i does not vanish identically, the vanishing of R is a necessary and sufficient condition that at least one of the non-zero common roots of the $n-1$ functions (3) shall be a root of f_i, i.e. a necessary and sufficient condition that the n functions (1) shall have at least one non-zero root in common. When we give such particular values to the coefficients of the functions (1) that E_i vanishes identically, it follows from Ex. iii of § 199 that all the functions H vanish, and therefore R vanishes. In this case (see Note 8 of § 196) the $n-1$ functions (3) have common roots in which the ratios $x_1 : x_2 : \dots : x_n$ are functions of one or more arbitrary parameters $t_1, t_2, \dots$, which can be so chosen that the equation $f_i = 0$ is satisfied; consequently the n functions f have non-zero roots in common. Thus in all cases the vanishing of R is a necessary and sufficient condition that the n functions f shall have a non-zero root in common.

We have now shown that for each of the values $1, 2, \dots n$ of i and for every set of non-zero values of $r_1, r_2, \dots r_n$ there exists a function $R(f_1, f_2, \dots f_n)$ which has the properties (a), (a'), (b'), (c'), (e') of Theorem I of § 193; and by Note 5 of § 193 it follows that that theorem is true generally for n functions, and that the function $R(f_1, f_2, \dots f_n)$ determined above is the resultant of the n general homogeneous functions f given by (1). It follows that Theorem II of § 193 is true generally for n functions, and that $R(f_1, f_2, \dots f_n)$ is also the resultant of the n general functions derived from the n functions (1) by the substitution $x_n = 1$.

NOTE 1. *On the property (d') of Theorem I of* § 193.

We can modify the general argument so as to make Note 2 of § 193 superfluous.

Let A_{ij} and ρ_{ij} be respectively the coefficients of $x_j^{r_i}$ in f_i and of $\kappa_j^{s_i}$ in E_i. Then the series on the right in (6) contains one and only one term in which $A_{\lambda_1\lambda_2\dots\lambda_n} = A_{ij}^{s_i}$, the corresponding value of $S_{\lambda_1\lambda_2\dots\lambda_n}$ being $(a_{1j}a_{2j}\dots a_{nj})^{r_i}$. From (5) we see that $a_{1j}a_{2j}\dots a_{nj} = \rho_{ij}$; therefore the corresponding value of $H_{\mu_1\mu_2\dots\mu_n}$ in (8) is $\rho_{ij}^{r_i}$. Accordingly when $R(f_1, f_2, \dots f_n)$ is defined as in the text, the coefficient of $A_{ij}^{s_i}$ in R is $\rho_{ij}^{r_i}$, this being true for all the values $1, 2, \dots n$ of j.

We conclude that for all non-zero values of $r_1, r_2, \ldots r_n$ there exist functions $R_1(f_1, f_2, \ldots f_n)$, $R_2(f_1, f_2, \ldots f_n)$, ... $R_n(f_1, f_2, \ldots f_n)$ which have the properties (d') and (e') of the theorem for the values 1, 2, ... n of i respectively, and which also have the properties (a), (a'), (b'), (c') of the theorem; and from this it follows as in Note 5 of § 193 that the theorem is true generally for n functions, and that any one of the n functions R can be taken to be the resultant.

NOTE 2. *Determination of the resultant of n general rational integral functions of $n-1$ variables.*

Let $$f_i = \Sigma a^{(i)}_{p_1 p_2 \ldots p_n} x_1^{p_1} x_2^{p_2} \ldots x_{n-1}^{p_{n-1}}, \quad (i=1, 2, \ldots n), \qquad \ldots\ldots(1')$$

be the n general functions derived from the functions (1) by the substitution $x_n = 1$. To find their resultant it is best to replace them by the corresponding homogeneous functions (1). We can however determine their resultant directly by using Theorem II of § 199, and this will of course also be the resultant of the n functions (1).

Let ρ_{in} be the resultant of the $n-1$ functions formed by retaining only the terms of highest degree in each of the $n-1$ functions

$$f_1, \ldots f_{i-1}, f_{i+1}, \ldots f_n, \qquad \ldots\ldots(3')$$

and when $\rho_{in} \neq 0$, let the s_i common roots of the $n-1$ functions (3′) be

$$[x_1 x_2 \ldots x_{n-1}] = [x_{u1} x_{u2} \ldots x_{u,n-1}], \quad (u=1, 2, \ldots s_i). \qquad \ldots\ldots(4')$$

Then by ordinary multiplication we obtain an identity in the coefficients of f_i of the form

$$\prod_{u=1}^{u=s_i} f_i(x_{u1}, x_{u2}, \ldots x_{u,n-1}, 1) = \Sigma A_{\lambda_1 \lambda_2 \ldots \lambda_n} S_{\lambda_1 \lambda_2 \ldots \lambda_{n-1}}, \qquad \ldots\ldots(6')$$
$$(\lambda_1 + \lambda_2 + \ldots + \lambda_n = r),$$

where $A_{\lambda_1 \lambda_2 \ldots \lambda_n}$ has the same properties as before when we regard x_n as a homogenising variable, and where $S_{\lambda_1 \lambda_2 \ldots \lambda_{n-1}}$ is a monotypic symmetric function of the s_i common roots (4′) which

(1) has degree not greater than r_i in each root,

(2) has weights $\lambda_1, \lambda_2, \ldots \lambda_{n-1}$ in those roots with respect to the variables $x_1, x_2, \ldots x_{n-1}$.

Let the degree of $S_{\lambda_1 \lambda_2 \ldots \lambda_{n-1}}$ in each root be s, and let its total weight be

$$w = \lambda_1 + \lambda_2 + \ldots + \lambda_{n-1} = ss_i - \lambda_n' = \frac{rs}{r_i} - \lambda_n' = r - \lambda_n.$$

Then from Theorem II of § 199 we see that when $\rho_{in} \neq 0$, we can write

$$\rho_{in}{}^s S_{\lambda_1 \lambda_2 \ldots \lambda_{n-1}} = K, \qquad \ldots\ldots(10)$$

where K is a rational integral function of the coefficients of the $n-1$ functions (3′) which is not divisible by ρ_{in} and

(1) is homogeneous of degree $\dfrac{s_i s}{r_\tau} = \dfrac{rs}{r_i r_\tau}$ in the coefficients of f_τ, $(\tau = 1, \ldots i-1, i+1, \ldots n)$,

(2) is isobaric of weight $s_i s - \lambda_j = \dfrac{rs}{r_i} - \lambda_j$ with respect to x_j when $j \neq n$, and isobaric of weight $s_i s - \lambda_n' = \dfrac{rs}{r_i} - \lambda_n'$ with respect to x_n, its total weight being

$$(n-1)\, ss_i = (n-1)\frac{rs}{r_i}.$$

Now ρ_{in} is a function of the coefficients of the $n-1$ functions (3′) which is homogeneous of degree $\frac{r}{r_i r_\tau}$ in the coefficients of f_τ, $(\tau=1, \ldots i-1, i+1, \ldots n)$, isobaric of weight $\frac{r}{r_i}$ with respect to x_j when $j \neq n$, and isobaric of weight 0 with respect to x_n. Hence when we multiply both sides of (10) by $\rho_{in}^{r_i - s}$, we obtain

$$\rho_{in}^{r_i} S_{\lambda_1 \lambda_2 \ldots \lambda_{n-1}} = \rho_i^{r_i - s} K = H_{\mu_1 \mu_2 \ldots \mu_n}, \quad \ldots\ldots(10')$$

$$(\mu_j = r - \lambda_j;\ j=1, 2, \ldots n),$$

where $H_{\mu_1 \mu_2 \ldots \mu_n}$ has the same properties as before; and when we multiply both sides of (6′) by $\rho_{in}^{r_i}$, and define $R(f_1, f_2, \ldots f_n)$ by (8) as before, we see that the equation

$$R(f_1, f_2, \ldots f_n) = \rho_{in}^{r_i} \prod_{u=1}^{u=s_i} f_i(x_{u1}, x_{u2}, \ldots x_{u,n-1}, 1)$$

is true whenever $\rho_{in} \neq 0$, and that $R(f_1, f_2, \ldots f_n)$ is the resultant of the n general functions f as defined in Theorem II of § 193.

2. *Determination of the eliminant of n general homogeneous functions of n + 1 variables.*

By Ex. i of § 196 we may suppose that $n > 1$. We will make the hypothesis that the theorems of § 196 regarding eliminants and the theorems of § 193 regarding resultants are true generally for functions less than n in number and for functions not more than n in number respectively, and suppose that it is required to find the X-eliminant of n general homogeneous functions $f_1, f_2, \ldots f_n$ of the $n+1$ variables $x_1, x_2, \ldots x_{n+1}$ having non-zero degrees $r_1, r_2, \ldots r_n$ in all the variables and given by

$$f_i = f_i(x_1, x_2, \ldots x_n, x_{n+1}) = \Sigma a^{(i)}_{p_1 p_2 \ldots p_n p_{n+1}} x_1^{p_1} x_2^{p_2} \ldots x_n^{p_n} x_{n+1}^{p_{n+1}}, \quad (11)$$

$$(p_1 + p_2 + \ldots + p_n + p_{n+1} = r_i;\ i = 1, 2, \ldots n),$$

when
$$X = \kappa_1 x_1 + \kappa_2 x_2 + \ldots + \kappa_n x_n + \kappa_{n+1} x_{n+1}, \quad \ldots\ldots(12)$$

the κ's being arbitrary parameters to which any values can be ascribed. We shall write

$$\kappa = \kappa_{n+1};\quad r = r_1 r_2 \ldots r_n;\quad s_i = \frac{r}{r_i},\ (i = 1, 2, \ldots n),$$

and denote the resultants of the $n+1$ successive sets of n homogeneous functions of n variables formed by putting $x_1 = 0, x_2 = 0, \ldots x_n = 0, x_{n+1} = 0$ in turn in the n functions (11) by

$$\rho_1, \rho_2, \ldots \rho_n, \rho_{n+1}. \quad \ldots\ldots(13)$$

If we suppose that $\kappa_{n+1} \neq 0$, we can solve the equation

$$\kappa_1 x_1 + \kappa_2 x_2 + \ldots + \kappa_n x_n + \kappa_{n+1} x_{n+1} = 0 \quad \ldots\ldots(14)$$

for x_{n+1}, make the substitution

$$x_{n+1} = -\frac{\kappa_1 x_1 + \kappa_2 x_2 + \ldots + \kappa_n x_n}{\kappa_{n+1}} = -\frac{\kappa_1 x_1 + \kappa_2 x_2 + \ldots + \kappa_n x_n}{\kappa} \qquad (15)$$

in the n functions f, render each of the resulting functions integral in the κ's by multiplying it by a suitable power of κ, and so obtain the n homogeneous functions $F_1, F_2, \ldots F_n$ of $x_1, x_2, \ldots x_n$ given by

$$F_i = \kappa^{r_i} f_i\left(x_1, x_2, \ldots x_n, -\frac{\kappa_1 x_1 + \kappa_2 x_2 + \ldots + \kappa_n x_n}{\kappa}\right)$$

$$= \Sigma(-1)^{p_{n+1}} a^{(i)}_{p_1 p_2 \ldots p_n p_{n+1}} x_1^{p_1} x_2^{p_2} \ldots x_n^{p_n} (\kappa_1 x_1 + \kappa_2 x_2 + \ldots + \kappa_n x_n)^{p_{n+1}} \kappa^{r_i - p_{n+1}}.$$

$$\ldots\ldots\ldots(16)$$

When we express these in the form of general homogeneous functions, we have

$$F_i = F_i(x_1, x_2, \ldots x_n) = \Sigma A^{(i)}_{q_1 q_2 \ldots q_n} x_1^{q_1} x_2^{q_2} \ldots x_n^{q_n}, \ \ldots\ldots\ldots(11')$$

$$(q_1 + q_2 + \ldots + q_n = r_i; \quad i = 1, 2, \ldots n),$$

the coefficient $A^{(i)}_{q_1 q_2 \ldots q_n}$ being a homogeneous rational integral function of the κ's of degree r_i given by

$$A^{(i)}_{q_1 q_2 \ldots q_n} = \Sigma(-1)^{p_{n+1}} N a^{(i)}_{p_1 p_2 \ldots p_n p_{n+1}} \kappa_1^{\lambda_1} \kappa_2^{\lambda_2} \ldots \kappa_n^{\lambda_n} \kappa^{\lambda_{n+1}}, \quad \ldots(17)$$

$$\left(N = \frac{p_{n+1}!}{\lambda_1! \lambda_2! \ldots \lambda_n!} = \frac{(\lambda_1 + \lambda_2 + \ldots + \lambda_n)!}{\lambda_1! \lambda_2! \ldots \lambda_n!}\right),$$

where the p's have all positive integral values consistent with the conditions

$$p_1 + p_2 + \ldots + p_n + p_{n+1} = r_i, \quad p_1 \not> q_1, \quad p_2 \not> q_2, \quad \ldots p_n \not> q_n, \ \ldots(18)$$

and $$\lambda_1 = q_1 - p_1, \quad \lambda_2 = q_2 - p_2, \quad \ldots \lambda_n = q_n - p_n, \quad \lambda_{n+1} = r_i - p_{n+1}, \quad (19)$$

or where the λ's have all positive integral values consistent with the conditions

$$\lambda_1 + \lambda_2 + \ldots + \lambda_n + \lambda_{n+1} = r_i, \quad \lambda_1 \not> q_1, \quad \lambda_2 \not> q_2, \quad \ldots \lambda_n \not> q_n, \ \ldots(18')$$

and $$p_1 = q_1 - \lambda_1, \quad p_2 = q_2 - \lambda_2, \quad \ldots p_n = q_n - \lambda_n, \quad p_{n+1} = r_i - \lambda_{n+1}. \quad (19')$$

Since f_i has one and only one coefficient a corresponding to each possible set of values of the p's, it follows that $a^{(i)}_{p_1 p_2 \ldots p_n p_{n+1}}$ will always have the same meaning in (17) as in (11), whatever the particular values of $p_1, p_2, \ldots p_{n+1}$ may be.

The unreduced resultant $R(F_1, F_2, \ldots F_n)$ of the n special homogeneous functions $F_1, F_2, \ldots F_n$ of the n variables $x_1, x_2, \ldots x_n$ given by (11′) will be denoted by R_{n+1}.

More generally if x_i is any one of the variables $x_1, x_2, \ldots x_{n+1}$, and if we solve the equation (14) for x_i, substitute the value thus obtained for x_i in the functions $f_1, f_2, \ldots f_n$, and then render the resulting functions integral in the κ's by multiplying them by $\kappa_i^{r_1}, \kappa_i^{r_2}, \ldots \kappa_i^{r_n}$ respectively, the resultant of the n special homogeneous functions of the n variables $x_1, \ldots x_{i-1}, x_{i+1}, \ldots x_{n+1}$ thus obtained will be denoted by R_i.

We shall investigate the properties of the functions $R_1, R_2, \ldots R_{n+1}$, usually deducing the properties of R_i from the properties of R_{n+1}. We shall show finally that for each of the values $1, 2, \ldots n+1$ of i the function R_i can be expressed in the form

$$R_i = \kappa_i^{(n-1)r} E(\kappa_1, \kappa_2, \ldots \kappa_{n+1}),$$

where E is a homogeneous rational integral function of the κ's of degree r in which the coefficients are rational integral functions of the coefficients a of $f_1, f_2, \ldots f_n$; and that E has all the properties enumerated in Theorem I of § 196, and is the required X-eliminant. For convenience the successive steps by which the final conclusion is reached will be presented in the form of examples on the properties of $R_1, R_2, \ldots R_{n+1}$.

We shall mean by the a's the coefficients of the n functions f in (11), and by the A's the coefficients of the n functions F in (11′). The unreduced resultant $R(F_1, F_2, \ldots F_n)$ can be expressed as a rational integral function of the A's having the properties described in Theorem I of § 193; but it is to be regarded as a rational integral function of the a's and κ's derived from that expression by the substitutions (17). When it is expressed as a function of the A's, it is a sum of terms each of which is a product of $s_1 s_2 \ldots s_n$ of the A's, s_i of them being coefficients of F_i. When we make the substitutions (17) for all the A's, each of the original terms becomes a product of expressions of the form (17), and when we multiply out each of these products, we obtain an expression for R_{n+1} as a sum ΣT of 'uncollected terms' all having the form

$$T = T_1 T_2 T_3 \ldots, \qquad \text{.............................(20)}$$

where each of the factors on the right is a term of such a sum as that on the right in (17). For a given term T and a given value of h we will denote the sum of the suffixes p_h of the a's occurring as factors in $T_1, T_2, T_3, \ldots$ by Σp_h, and the sum of the indices λ_h of the powers of κ_h occurring as factors in $T_1, T_2, T_3, \ldots$ by $\Sigma\lambda_h$. It will appear from Ex. xii that when we proceed to collect the terms of R_{n+1}, only those which are divisible by $\kappa^{(n-1)r}$ will remain with non-zero coefficients.

The weights of the A's with respect to $x_1, x_2, \ldots x_n$, and the weights of the a's with respect to $x_1, x_2, \ldots x_n, x_{n+1}$ are indicated by their suffixes. The weights of the κ's are not indicated in this way, but by definition (see § 192. 2) the parameter κ_j has weight j with respect to x_j, and weight 0 with respect to each of the other variables; in particular κ or κ_{n+1} has weight 1 with respect to x_{n+1}, and weight 0 with respect to each one of the other variables.

Coefficients and dimensions of R_{n+1}.

Ex. i. *When R_{n+1} is expressed as a rational integral function of the a's and κ's, its own coefficients are rational numbers which we may suppose to be integers.*

For when R_{n+1} is expressed as a rational integral function of the A's, we see from Theorem I of § 193 for n functions that its own coefficients are rational numbers which we may suppose to be integers; and from (17) we see that each of the A's is a rational integral function of the a's and κ's in which the coefficients are integers.

Ex. ii. *When R_{n+1} is expressed as a function of the a's and κ's, it is*

(1) *homogeneous of degree nr in the κ's;*

(2) *homogeneous of degree s_i in the coefficients of f_i, where f_i is any one of the n functions f in* (11).

When R_{n+1} is expressed as a function of the A's, then by Theorem I of § 193 for n functions it is homogeneous of degree s_i in the coefficients of F_i; and it follows from (17) that the above two results are true.

Ex. iii. *When R_{n+1} is expressed as a function of the a's and κ's, it is*

(1) *isobaric of weight r with respect to each one of the n variables $x_1, x_2, \dots x_n$ in the coefficients of all the $n+1$ functions $f_1, f_2, \dots f_n, X$;*

(2) *isobaric of weight nr with respect to the variable x_{n+1} in the coefficients of all the $n+1$ functions $f_1, f_2, \dots f_n, X$.*

Let x_h be any one of the variables $x_1, x_2, \dots x_n$. When R_{n+1} is expressed as a function of the A's, we see from Theorem I of § 193 for n functions that in every term (which is a product of certain of the A's) we have $\Sigma q_h = r$. Now in every term of (17) we have $q_h = p_h + \lambda_h$. Hence when R_{n+1} is expressed as a function of the a's and κ's by means of the equations (17), the terms being uncollected, then in every term (which is a product of certain of the a's and certain powers of the κ's) we have $\Sigma p_h + \Sigma\lambda_h = r$; i.e. every term has weight r with respect to x_h in the a's and κ's.

To prove the second result, we observe that by Ex. ii we can express R_{n+1} in the form

$$R_{n+1} = \Sigma P_1 P_2 \dots P_n,$$

where P_i is a product of s_i of the coefficients A of F_i. Now in every term of the sum (17) we have $p_{n+1} + \lambda_{n+1} = r_i$. Therefore every coefficient A of F_i is isobaric of weight r_i in the a's and κ's with respect to x_{n+1}. Therefore with respect to x_{n+1} the factor P_i is isobaric of weight $s_i r_i = r$ in the a's and κ's, and the product $P_1 P_2 \dots P_n$ is isobaric of weight nr in the a's and κ's.

Terms of R_i of the highest degrees in $\kappa_1, \kappa_2, \dots \kappa_{n+1}$ respectively.

Ex. iv. *If h is any one of the integers 1, 2, ... n, the actual degree of R_{n+1} in κ_h is r; and if S_h is the sum of all the terms of R_{n+1} which contain $\kappa_h{}^r$, then*

$$S_h = \kappa^{(n-1)r} \,.\, \rho_h \kappa_h{}^r.$$

When R_{n+1} is expressed as a sum of uncollected terms of the form (20), we see from Ex. iii that in every term T we have

$$\Sigma p_h + \Sigma\lambda_h = r,$$

$\Sigma\lambda_h$ being the total index of κ_h in T. It follows in the first place that $\Sigma\lambda_h \ngtr r$, and therefore no term which actually occurs can have a higher degree than r in κ_h. It follows in the second place that we can have $\Sigma\lambda_h = r$ in a term T when and only when $\Sigma p_h = 0$, i.e. when and only when the suffix p_h of *every* a occurring as a factor in T is 0; consequently S_h is what $R(F_1, F_2, \ldots F_n)$ becomes when in it we substitute 0 for every coefficient $a_{p_1 p_2 \ldots p_{n+1}}$ in which $p_h \neq 0$.

Let these substitutions convert $f_1, f_2, \ldots f_n$ into $g_1, g_2, \ldots g_n$, and $F_1, F_2, \ldots F_n$ into $G_1, G_2, \ldots G_n$, so that

$$S_h = R(G_1, G_2, \ldots G_n).$$

Then $g_1, g_2, \ldots g_n$ are the n general homogeneous functions of n variables derived from $f_1, f_2, \ldots f_n$ by putting $x_h = 0$, and have a resultant $R(g_1, g_2, \ldots g_n)$ which differs from ρ_h only by an ordinary numerical factor; and $G_1, G_2, \ldots G_n$ are derived from $g_1, g_2, \ldots g_n$ in the same way as $F_1, F_2, \ldots F_n$ are derived from $f_1, f_2, \ldots f_n$. If we denote by $b_{p_1 \ldots p_{h-1} p_{h+1} \ldots p_{n+1}}$ the coefficient $a_{p_1 p_2 \ldots p_{n+1}}$ in which $p_h = 0$, we have

$$g_\tau = \Sigma b^{(\tau)}_{p_1 \ldots p_{h-1} p_{h+1} \ldots p_{n+1}} x_1^{p_1} \ldots x_{h-1}^{p_{h-1}} x_{h+1}^{p_{h+1}} \ldots x_{n+1}^{p_{n+1}},$$

$$G_\tau = \Sigma(-1)^{p_{n+1}} b^{(\tau)}_{p_1 \ldots p_{h-1} p_{h+1} \ldots p_{n+1}} x_1^{p_1} \ldots x_{h-1}^{p_{h-1}} x_{h+1}^{p_{h+1}} \ldots x_n^{p_n} \xi_{n+1}^{p_{n+1}} \kappa^{r_\tau - p_{n+1}},$$

$$(p_1 + \ldots + p_{h-1} + p_{h+1} + \ldots + p_{n+1} = r_\tau),$$

where
$$\xi_{n+1} = \kappa_1 x_1 + \kappa_2 x_2 + \ldots + \kappa_n x_n.$$

Since $G_1, G_2, \ldots G_n$ can evidently be derived from $g_1, g_2, \ldots g_n$ by the homogeneous linear substitution

$$x_i = \kappa x_i, \quad (i = 1, \ldots h-1, h+1, \ldots n); \quad x_{n+1} = -(\kappa_1 x_1 + \kappa_2 x_2 + \ldots + \kappa_n x_n)$$

whose determinant is

$$\Delta = (-1)^{n+1-h} \kappa^{n-1} \kappa_h,$$

it follows from Ex. ix of § 193 that when the arbitrary constant in ρ_h is suitably determined, we have

$$S_h = R(G_1, G_2, \ldots G_n) = \Delta^r R(g_1, g_2, \ldots g_n) = \kappa^{(n-1)r} \cdot \rho_h \kappa_h^r.$$

Ex. v. *The actual degree of R_{n+1} in κ_{n+1} is nr; and if S_{n+1} is the sum of all the terms of R_{n+1} which contain κ_{n+1}^{nr}, then*

$$S_{n+1} = \kappa^{(n-1)r} \cdot \rho_{n+1} \kappa_{n+1}^r.$$

In this case S_{n+1} is what $R(F_1, F_2, \ldots F_n)$ becomes when in it we substitute 0 for every coefficient $a_{p_1 p_2 \ldots p_{n+1}}$ of the functions $f_1, f_2, \ldots f_n$ in which $p_{n+1} \neq 0$. If we write $b_{p_1 p_2 \ldots p_n}$ for the coefficient $a_{p_1 p_2 \ldots p_{n+1}}$ in which $p_{n+1} = 0$, these substitutions convert f_τ and F_τ into the functions

$$g_\tau = \Sigma b^{(\tau)}_{p_1 p_2 \ldots p_n} x_1^{p_1} x_2^{p_2} \ldots x_n^{p_n}, \quad (p_1 + p_2 + \ldots + p_n = r_\tau),$$

$$G_\tau = \Sigma b^{(\tau)}_{p_1 p_2 \ldots p_n} x_1^{p_1} x_2^{p_2} \ldots x_n^{p_n} \kappa^{r_\tau}.$$

Since the substitutions $\kappa x_1, \kappa x_2, \ldots \kappa x_n$ for $x_1, x_2, \ldots x_n$ convert $g_1, g_2, \ldots g_n$ into $G_1, G_2, \ldots G_n$, we have

$$S_{n+1} = R(G_1, G_2, \ldots G_n) = \kappa^{nr} R(g_1, g_2, \ldots g_n) = \kappa^{(n-1)r} \cdot \rho_{n+1} \kappa_{n+1}^r$$

when the arbitrary constant in ρ_{n+1} is suitably chosen.

Ex. vi. *When R_i is arranged as a rational integral function of the κ's, the terms of highest degrees in $\kappa_1, \kappa_2, \ldots \kappa_{n+1}$ respectively which actually occur in it are*

$$\kappa_i^{(n-1)r}\,.\,\rho_1\kappa_1^r, \quad \kappa_i^{(n-1)r}\,.\,\rho_2\kappa_2^r, \ldots \kappa_i^{(n-1)r}\,.\,\rho_{n+1}\kappa_{n+1}^r,$$

provided that the arbitrary numerical factors in $\rho_1, \rho_2, \ldots \rho_{n+1}$ are suitably chosen.

This follows from Exs. iv and v, and is true for each of the values 1, 2, ... $n+1$ of i. The actual degree of R_i in κ_i is nr, and its actual degree in each of the other κ's is r.

When we replace R_{n+1} by R_i in Ex. iv, we shall have

$$\Delta=(-1)^{i-h}\kappa_i^{n-1}\kappa_h.$$

From this it is clear that in the above theorem $\rho_1, \rho_2, \ldots \rho_{n+1}$ will be the same for all values of i, provided that the arbitrary numerical factors in $R_1, R_2, \ldots R_{n+1}$ are suitably chosen.

Linear factors of R_i for any particular values whatever of the a's.

Ex. vii. *When $n>1$, R_i is divisible by κ_i even if the a's are all arbitrary.*

We will prove this for the case in which $i=n+1$, writing $\kappa_{n+1}=\kappa$.

When we put $\kappa=0$, we have in (16)

$$F_i=c_i(\kappa_1x_1+\kappa_2x_2+\ldots+\kappa_nx_n)^r,$$

where c_i is the coefficient of $x_{n+1}^{r_i}$ in f_i, and is a numerical constant; therefore every root $[x_1x_2\ldots x_n]$ of the homogeneous linear function

$$H=\kappa_1x_1+\kappa_2x_2+\ldots+\kappa_nx_n$$

is a common root of the n functions F for all particular finite values of $\kappa_1, \kappa_2, \ldots \kappa_n$ and the a's. If $n>1$, the function H certainly has a non-zero root for all ordinary values of $\kappa_1, \kappa_2, \ldots \kappa_n$. Therefore R_{n+1} vanishes when $\kappa=0$ for all finite values of the a's and all ordinary particular values of $\kappa_1, \kappa_2, \ldots \kappa_n$; and it follows that κ is a factor of R_{n+1} when the a's and κ's are all arbitrary.

In the special case when $n=1$, there is only one function $f=f(x_1, x_2)$ of degree r, the corresponding function F being $F=x_1^r.f(\kappa_2, -\kappa_1)$; and we have

$$R_1=R_2=f(\kappa_2, -\kappa_1).$$

In this case R_1 is divisible by κ_1 when and only when the coefficient of x_1^r in f vanishes, i.e. when f has the root $[x_1x_2]\equiv[1\,0]$, or R_1 has the root $[\kappa_1\kappa_2]\equiv[0\,1]$; and R_2 is divisible by κ_2 when and only when the coefficient of x_2^r in f vanishes, i.e. when and only when f has the root $[x_1x_2]\equiv[0\,1]$, or R_2 has the root $[\kappa_1\kappa_2]\equiv[1\,0]$.

Ex. viii. *Let any particular finite values whatever be ascribed to the a's, so that the R's are rational integral functions of the κ's only. Then if*

$$[x_1x_2\ldots x_{n+1}]\equiv[a_1a_2\ldots a_{n+1}] \quad \text{.................................(a)}$$

is a non-zero common root of the n functions f, each of the functions $R_1, R_2, \ldots R_{n+1}$ has the linear factor

$$L=a_1\kappa_1+a_2\kappa_2+\ldots+a_{n+1}\kappa_{n+1}. \quad \text{.............................(b)}$$

We assume that the a's are all finite, and show that L is a factor of R_{n+i}.

First suppose that $a_1, a_2, \dots a_n$ all vanish, so that $a_{n+1} \neq 0$. Then if $n > 1$, it follows from Ex. vii that κ_{n+1} is a factor of R_{n+1}, i.e. L is a factor of R_{n+1}. If $n=1$, we have by supposition $a_1=0$, $a_2 \neq 0$, and it follows again from Ex. vii that κ_2 is a factor of R_2, i.e. L is a factor of R_{n+1}.

Next consider the general case in which $a_1, a_2, \dots a_n$ do not all vanish, n being either greater than 1 or equal to 1. Let any particular finite values be ascribed to $\kappa_1, \kappa_2, \dots \kappa_{n+1}$ which make the expression L vanish. Then from (11) and (16) we see that

$$F_i(a_1, a_2, \dots a_n) = \kappa^{r_i} . f_i(a_1, a_2, \dots a_{n+1}) = 0;$$

therefore $[x_1 x_2 \dots x_n] \equiv [a_1 a_2 \dots a_n]$ is a non-zero common root of the n functions F, and R_{n+1} vanishes. Thus when we regard the κ's as variables, R_{n+1} vanishes for all roots of the irresoluble function L, and is therefore divisible by L.

Since R_i, which is homogeneous of degree nr in the κ's, cannot have more than nr linear factors, it follows that:

If R_i does not vanish identically, the n functions f cannot have more than nr distinct non-zero common roots.

Ex. ix. *Let any particular finite values whatever be ascribed to the a's, and let*

$$[\kappa_1 \kappa_2 \dots \kappa_{n+1}] \equiv [k_1 k_2 \dots k_{n+1}] \quad \text{.................................(c)}$$

be any non-zero root of R_i, when R_i is regarded as a function of the κ's. Then:

(1) *If $k_i \neq 0$, the n functions f have at least one non-zero common root*

$$[x_1 x_2 \dots x_{n+1}] \equiv [a_1 a_2 \dots a_{n+1}] \quad \text{... ....(a)}$$

in which $a_1, \dots a_{i-1}, a_{i+1}, \dots a_{n+1}$ are all finite and not all 0, *and satisfy the equation*

$$k_1 a_1 + k_2 a_2 + \dots + k_{n+1} a_{n+1} = 0; \quad \text{.................................(d)}$$

and (c) *is a root of the corresponding linear factor $L = a_1 \kappa_1 + a_2 \kappa_2 + \dots + a_{n+1} \kappa_{n+1}$ of R_i, which is in this case distinct from κ_i.*

(2) *If $k_i = 0$, R_i has the linear factor κ_i of which* (c) *is a root.*

We will prove the first result for the case in which $i = n+1$, so that $k_{n+1} \neq 0$.

When the κ's have the particular values (c), the n functions F given by (16), whose resultant vanishes, certainly have a non-zero common root. If $[x_1 x_2 \dots x_n] \equiv [a_1 a_2 \dots a_n]$ is any such common root, we can choose a_{n+1} so that the equation (d) is satisfied, and we see from (16) that (a) is then a common root of the n functions f in which $a_1, a_2, \dots a_n$ are not all 0. By Ex. viii the linear function L given by (b) is then a factor of R_{n+1}, and from (d) it follows that (c) is a root of L.

The second result follows from Ex. vii both when $n > 1$ and when $n=1$.

We conclude that when any particular finite values are ascribed to the a's:

Every root of R_i is a root of a linear factor of R_i; a root of R_i in which $\kappa_i \neq 0$ being a root of a linear factor distinct from κ_i, and a root of R_i in which $\kappa_i = 0$ being a root of the linear factor κ_i.

These results remain true when R_i, regarded as a function of the κ's, vanishes identically, and in that case (c) is a root of R_i for all values of $k_1, k_2, \dots k_{n+1}$. It follows that:

When the a's have such particular values that R_i vanishes identically, the n functions f have an infinite number of non-zero common roots to each of which there corresponds a linear factor of R_i as in Ex. viii.

Factors of R_i for ordinary particular values of the a's.

Ex. x. *If we ascribe to the a's any finite particular values which do not cause R_i, regarded as a function of the κ's, to vanish identically, then R_i can be expressed as a product of homogeneous linear factors; and if*

$$L = a_1\kappa_1 + a_2\kappa_2 + \dots + a_{n+1}\kappa_{n+1} \text{ ....(b)}$$

is a linear factor of R_i distinct from κ_i, so that $a_1, \dots a_{i-1}, a_{i+1} \dots a_{n+1}$ are not all 0, then

$$[x_1 x_2 \dots x_{n+1}] \equiv [a_1 a_2 \dots a_{n+1}] \text{ ....(a)}$$

is a non-zero common root of the n functions f.

First let $L_1, L_2, \dots L_p$, where $p \not> nr$, be all the distinct linear factors of R_i, and let $\phi(\kappa_1, \kappa_2, \dots \kappa_{n+1})$ be any irresoluble factor of R_i. Then by Ex. ix every root of ϕ is a root of one of the linear factors $L_1, L_2, \dots L_p$. Therefore the product $L_1 L_2 \dots L_p$ vanishes for all roots of ϕ, and must be divisible by ϕ. Consequently ϕ itself must be one of the linear factors $L_1, L_2, \dots L_p$.

Next let the n functions f have exactly q distinct non-zero common roots of the form (a) in which $a_1, \dots a_{i-1}, a_{i+1}, \dots a_{n+1}$ are not all 0, where $q \not> nr$ by Ex. viii; let $L_1, L_2, \dots L_q$ be the q distinct linear factors of R_i, all distinct from κ_i, which correspond to them by Ex. viii; let L be any linear factor of R distinct from κ_i; and let ordinary roots of L be those in which $\kappa_i \neq 0$. Then by Ex. ix every ordinary root of L is a root of one of the linear factors $L_1, L_2, \dots L_q$. Therefore the product $L_1 L_2 \dots L_q$ vanishes for all ordinary roots of the irresoluble function L, and must be divisible by L. Consequently L itself must be one of the linear factors $L_1, L_2, \dots L_q$, i.e. L must have the form (b), where (a) is a non-zero common root of the n functions f.

If κ_i is one of the linear factors of R_i, there may not be a corresponding root of the n functions f of the form (a); but if the n functions f have a non-zero common root in which every element vanishes except x_i, then by Ex. viii the function R_i must have the corresponding linear factor κ_i.

Ex. xi. *When any ordinary particular values are ascribed to the a's, R_i is a function of the κ's which is divisible by $\kappa_i^{(n-1)r}$, and the n functions f have exactly r distinct non-zero common roots to which the other r linear factors of R_i correspond.*

We will consider *ordinary values* of the a's to be those for which $\rho_1 \neq 0, \rho_2 \neq 0, \dots \rho_{n+1} \neq 0$, and at the same time one of the discriminants σ of Lemma A of § 195 does not vanish.

Let any such ordinary particular values be ascribed to the a's. Then by Lemma A of § 195 the n functions f have at least r distinct non-zero common roots

$$[x_1 x_2 \dots x_{n+1}] \equiv [a_{u1} a_{u2} \dots a_{u,n+1}], \quad (u = 1, 2, \dots r), \text{ ....(21)}$$

in which the a's are all different from 0, but no non-zero common root in which any one of the elements has the value 0; and it follows from Ex. viii that R_i can be expressed in the form

$$R_i = \phi(\kappa_1, \kappa_2, \dots \kappa_{n+1}) \cdot \prod_{u=1}^{u=r} (a_{u1}\kappa_1 + a_{u2}\kappa_2 + \dots + a_{u,n+1}\kappa_{n+1}), \text{ ....(22)}$$

where ϕ is a homogeneous rational integral function of the κ's of degree $(n-1)r$, which by Ex. x is a product of linear factors. If j is any one of the integers $1, \dots i-1, i+1, \dots n+1$, and if we equate the terms of highest degree in κ_j which actually occur on both sides of (22), we see from Ex. vi that κ_j cannot actually occur in ϕ; and if we equate the terms of highest degree in κ_i which actually occur on both sides, we see that the term $\kappa_i^{(n-1)r}$ must occur in ϕ with a non-zero coefficient. It follows that ϕ differs by an ordinary numerical factor only from $\kappa_i^{(n-1)r}$.

If the n functions f had another non-zero common root

$$[x_1 x_2 \dots x_{n+1}] \equiv [a_1 a_2 \dots a_{n+1}]$$

distinct from the r common roots (21), then by Ex. viii

$$a_1 \kappa_1 + a_2 \kappa_2 + \dots + a_{n+1} \kappa_{n+1}$$

would be a linear factor of ϕ, and therefore of κ_i; but this is impossible because $a_1, a_2, \dots a_{n+1}$ are all different from 0.

Derivation of the eliminant E from the resultants $R_1, R_2, \dots R_{n+1}$.

Ex. xii. *The resultant R_i, regarded as a function of the a's and κ's, is divisible by $\kappa_i^{(n-1)r}$.*

It has been shown in Ex. xi that $\kappa_i^{(n-1)r}$ is a factor of R_i for all ordinary values of the a's and the parameters $\kappa_1, \dots \kappa_{i-1}, \kappa_{i+1}, \dots \kappa_{n+1}$; and it follows (see Ex. iv of § 192) that it is a factor of R_i when all the a's and κ's are arbitrary.

Ex. xiii. *When the arbitrary numerical factors in them have been suitably determined, we can express all the resultants $R_1, R_2, \dots R_{n+1}$ in the forms*

$$R_i = \kappa_i^{(n-1)r} . E(\kappa_1, \kappa_2, \dots \kappa_{n+1}), \quad (i = 1, 2, \dots n+1), \quad \dots\dots\dots\dots\dots\dots (23)$$

where E is the same for all the R's, and is a rational integral function of the a's and κ's which is homogeneous of degree r in the κ's; and the coefficients of $\kappa_1^r, \kappa_2^r, \dots \kappa^r_{n+1}$ in E are the resultants $\rho_1, \rho_2, \dots \rho_{n+1}$ defined in (13) *when the arbitrary numerical factors in them have been suitably determined.*

If i and j are any two of the integers $1, 2, \dots n+1$, it follows from Ex. xii that when the arbitrary numerical factors in R_i and R_j are fixed, we have

$$R_i = \kappa_i^{(n-1)r} . E_i(\kappa_1, \kappa_2, \dots \kappa_{n+1}), \quad R_j = \kappa_j^{(n-1)r} . E_j(\kappa_1, \kappa_2, \dots \kappa_{n+1}),$$

where E_i and E_j are rational integral functions of the a's and κ's which are both homogeneous of degree r in the κ's. Let the coefficients of $\kappa_1^r, \kappa_2^r, \dots \kappa^r_{n+1}$ in E_i and E_j respectively be

$$\rho_1, \rho_2, \dots \rho_{n+1} \quad \text{and} \quad \rho_1', \rho_2', \dots \rho'_{n+1},$$

and let

$$\frac{\rho_1'}{\rho_1} = c_1, \quad \frac{\rho_2'}{\rho_2} = c_2, \dots \frac{\rho'_{n+1}}{\rho_{n+1}} = c_{n+1}.$$

Then from Ex. vi it follows that $c_1, c_2, \dots c_{n+1}$ are absolute constants, the same for all particular values of the a's and κ's (and also that they are all equal).

Now let any ordinary particular values be ascribed to the a's, ordinary values being defined as in Ex. xi. Then the n functions f have exactly r non-zero common roots

$$[x_1 x_2 \dots x_{n+1}] \equiv [a_{u1} a_{u2} \dots a_{u,n+1}], \quad (u = 1, 2, \dots r),$$

in which we may suppose each of the a's to have a definite finite non-zero value; and from Ex. xi we see that for all values of the κ's which are such that $\kappa_i \neq 0$, $\kappa_j \neq 0$ we have

$$E_i = A \prod_{u=1}^{u=r} (a_{u1}\kappa_1 + a_{u2}\kappa_2 + \dots + a_{u,n+1}\kappa_{n+1}), \quad E_j = B \prod_{u=1}^{u=r} (a_{u1}\kappa_1 + a_{u2}\kappa_2 + \dots + a_{u,n+1}\kappa_{n+1}),$$

where A and B are, for the particular values given to the a's, ordinary numerical constants independent of the κ's. It follows that we have

$$E_j(\kappa_1, \kappa_2, \dots \kappa_{n+1}) = cE_i(\kappa_1, \kappa_2, \dots \kappa_{n+1}), \quad \dots\dots\dots\dots(24)$$

where c is an ordinary numerical constant independent of the κ's; and this equation, being true for all ordinary values of the κ's, is an identity in the κ's. When we equate the coefficients of $\kappa_1{}^r, \kappa_2{}^r, \dots \kappa^r{}_{n+1}$ on both sides of (24), we obtain

$$c_1 = c_2 = \dots = c_{n+1} = c\,;$$

and this shows that c (like $c_1, c_2, \dots$) is an absolute constant independent both of the κ's and of the particular values given to the a's. Consequently the equation (24) is an identity in the κ's which is true for all ordinary values of the a's, and is therefore an identity in the a's and κ's when both sets of quantities are arbitrary.

When the arbitrary numerical factor in R_j is suitably chosen, we shall have $c=1$, and we can then write

$$E_i(\kappa_1, \kappa_2, \dots \kappa_{n+1}) = E_j(\kappa_1, \kappa_2, \dots \kappa_{n+1}) = E(\kappa_1, \kappa_2, \dots \kappa_{n+1}).$$

Giving to j the values $1, \dots i-1, i+1, \dots n+1$ in turn, we obtain all the equations (23).

The function $E(\kappa_1, \kappa_2, \dots \kappa_{n+1})$ of Ex. xiii has the properties (a) and (d') of Theorem I of § 196. It follows from Exs. viii and x that it has the properties (b) and (c); from Exs. i, ii and iii that it has the properties (a'), (b') and (c'); and from Ex. ix that it has the property (e'). Thus it has all the properties (a), (b), (c), (a'), (b'), (c'), (d'), (e') of Theorem I of § 196. Hence by Note 6 of § 196 the theorems of § 196 are true generally for n functions, and $E(\kappa_1, \kappa_2, \dots \kappa_{n+1})$ is the required X-eliminant of the n functions f given by (11).

Note. *Determination of the X-eliminant of the n general functions $f_1, f_2, \dots f_n$ given by*

$$f_i = f_i(x_1, x_2, \dots x_n, 1), \quad (i=1, 2, \dots n), \quad \dots\dots\dots\dots(11')$$

when
$$X = z + \kappa_1 x_1 + \kappa_2 x_2 + \dots + \kappa_n x_n.$$

We can suppose the n functions (11′) to be derived from the n functions (11) by the substitution $x_{n+1}=1$. Then the required X-eliminant is obtained by putting $\kappa_{n+1}=z$ in the eliminant $E(\kappa_1, \kappa_2, \dots \kappa_{n+1})$ of the homogenised functions (11).

To find the new X-eliminant directly, we may homogenise the n functions (11′) by the substitution

$$1 = -\frac{\kappa_1 x_1 + \kappa_2 x_2 + \dots + \kappa_n x_n}{z},$$

and so form the n homogeneous functions $F_1, F_2, \dots F_n$ of $x_1, x_2, \dots x_n$ given by

$$F_i = z^{r_i} f_i\left(x_1, x_2, \dots x_n, -\frac{\kappa_1 x_1 + \kappa_2 x_2 + \dots + \kappa_n x_n}{z}\right).$$

Then the unreduced resultant of $F_1, F_2, \dots F_n$ can be expressed in the form

$$R(F_1, F_2, \dots F_n) = z^{(n-1)r} \,.\, E(z\,;\, \kappa_1, \kappa_2, \dots \kappa_n)\,;$$

and $E(z\,;\, \kappa_1, \kappa_2, \dots \kappa_n)$ is the required X-eliminant of the n general functions (11′).

3. *Final proof of the theorems of* §§ 193 *and* 196.

In sub-article 1 it has been shown that if the theorems of §§ 193 and 196 are true generally for functions less than n in number, then the theorems of § 193 are true generally for n functions. In sub-article 2 it has been shown that if the theorems of § 196 are true generally for functions less than n in number, and the theorems of § 193 are true generally for functions not more than n in number, then the theorems of § 196 are true generally for n functions.

Thus if the theorems of §§ 193 and 196 are true generally for functions less than n in number, then both sets of theorems are true generally for n functions. Since by Ex. i of § 193 and Ex. i of § 196 both sets of theorems are true generally when $n = 1$, it follows that both sets of theorems are true generally when n is any non-zero positive integer; and the proof of the theorems is now complete.

CHAPTER XXIII

THE POTENT DIVISORS OF A RATIONAL INTEGRAL FUNCTIONAL MATRIX

[In § 201 we define rational integral functional matrices and consider some of their preliminary properties. In §§ 202—4 we define the irreducible and irresoluble divisors of a rational integral functional matrix and prove their fundamental properties; we also consider the special properties of those minor determinants of such a matrix which are regular with respect to a given irreducible or irresoluble divisor. In §§ 205—6 we define the maximum and potent divisors and the maximum and potent factors of a rational integral functional matrix; and we consider some properties of the potent factors and potent divisors of a product of two or more such matrices. In § 207 we show that the potent divisors of a compartite matrix are the potent divisors of its several parts; and in § 208 we determine the irreducible divisors of all orders of a complete matrix of minor determinants. In §§ 209—10 we effect the reduction of any rational integral functional matrix to one whose successive leading minors are regular by equigradent transformations, the reduction being symmetric when the given matrix is symmetric. The remaining two articles are concerned with homogeneous linear transformations of the variables in a rational integral functional matrix, and with the methods of homogenising such a matrix.]

§ 201. **Rational integral functional matrices.**

A matrix $A = [a]_m^n$ whose elements are rational integral functions of several scalar variables $x, y, z, \ldots$ will be called a *rational integral functional matrix* or *a matrix which is a rational integral function of the scalar variables* $x, y, z, \ldots$. Every such matrix can be expressed as the sum of a number of terms in the form

$$A = [a]_m^n = \Sigma A_{ijk\ldots} x^i y^j z^k \ldots, \quad \ldots\ldots\ldots\ldots\ldots\ldots\ldots(1)$$

where $i, j, k, \ldots$ are positive integers each of which may be 0; where the values of $i, j, k, \ldots$ are not all the same in any two terms of the sum; and where $A_{ijk\ldots}$ is a matrix of the form $[a]_m^n$ whose elements are constants. The matrix $A_{ijk\ldots}$ is then called the *coefficient* of $x^i y^j z^k \ldots$ in the matrix A. Each functional element of A which does not vanish identically has a certain degree in all the variables $x, y, z, \ldots$, and the highest of such degrees will be called the degree of the matrix A in all the variables. Similarly each functional element of A which does not vanish identically has a certain degree in any one variable x, and the highest of such degrees will be called the degree of the matrix A in the one variable x. In other words the *degree of A in all the variables $x, y, z, \ldots$* is the greatest value of the sum $i+j+k+\ldots$ in those terms of A whose coefficients do not vanish identically, and the *degree of A in the one variable x* is the greatest value of i in those terms of A whose coefficients do not vanish identically.

The matrix A is *homogeneous* when its elements are homogeneous rational integral functions of the variables all having the same degree in all the variables, i.e. when the sum $i+j+k+\ldots$ has the same value in every term of A.

If the functional elements of A, or the elements of all such matrices as $A_{ijk\ldots}$, all lie in a domain of rationality Ω, then the matrix A will be said to *lie in* or to *belong to* the domain Ω.

As in § 74 the *rank* of the matrix A is the greatest order of a derived determinant which does not vanish identically.

Ex. i. *If the rational integral functional matrix $\phi=[a]_m^n$ has rank r, and if $s \nless 1$ and $s \ngtr r$, then every complete matrix $\Phi=[A]_\mu^\nu$ of the minor determinants of ϕ of order s has rank $\binom{r}{s}$.*

Since ϕ cannot have rank greater than r for any set of particular values of the variables, it follows from § 73 that Φ cannot have rank greater than $\binom{r}{s}$ for any set of particular values of the variables. Therefore all minor determinants of the functional matrix Φ of order greater than $\binom{r}{s}$ vanish identically.

Again there exist sets of values of the variables for which ϕ, regarded as a matrix whose elements are constants, has rank r. Therefore by § 73 there exist sets of values of the variables for which Φ, regarded as a matrix whose elements are constants, has rank $\binom{r}{s}$. Therefore the functional matrix Φ has minor determinants of order $\binom{r}{s}$ which do not vanish identically.

Ex. ii. *If $[a]_p^q$ and $[b]_m^n$ are rational integral functional matrices, and if there exists an identical relation of the form*

$$[b]_m^n=[h]_m^p\,[a]_p^q\,[k]_q^n,$$

where $[h]_m^p$ and $[k]_q^n$ are undegenerate rational integral functional matrices of ranks p and q, then the functional matrices $[a]_p^q$ and $[b]_m^n$ have the same rank.

This has been proved in § 131.

A matrix $[a]_m^n$ whose elements are all functions of a single scalar variable x will often be called an *x-matrix*. In particular a matrix $[a]_m^n$ whose elements are all rational integral functions of a single scalar variable x

will often be called a *rational integral x-matrix*. Every rational integral x-matrix can be expressed in the form

$$[a]_m^n = x^r[\alpha]_m^n + x^{r-1}[\beta]_m^n + \dots + [\kappa]_m^n, \quad \dots\dots\dots\dots\dots(2)$$

where $[\alpha]_m^n, [\beta]_m^n, \dots [\kappa]_m^n$ are matrices whose elements are constants, and are the coefficients of $x^r, x^{r-1}, \dots x^0$ in $[a]_m^n$. The matrix $[\kappa]_m^n$ is then the constant term in $[a]_m^n$.

The following examples contain some properties of rational integral x-matrices.

Ex. iii. *Let $[a]_m^r$ and $[b]_r^n$ be rational integral x-matrices, neither of which vanishes identically, and let their degrees in x be p and q. Then if the coefficient of the highest power of x in either matrix is undegenerate and has rank r, the product matrix $[a]_m^r[b]_r^n$ is a rational integral x-matrix of degree $p+q$.*

Let $\quad [a]_m^r = x^p[a]_m^r + x^{p-1}[a']_m^r + \dots, \quad [b]_r^n = x^q[\beta]_r^n + x^{q-1}[\beta']_r^n + \dots,$

where $[a]_m^r, [a']_m^r, \dots [\beta]_r^n, [\beta']_r^n, \dots$ are matrices with constant elements, and where neither $[a]_m^r$ nor $[\beta]_r^n$ has rank 0. Then the coefficient of x^{p+q} in $[a]_m^r[b]_r^n$ is $[a]_m^r[\beta]_r^n$.

If in the product $[a]_m^r[\beta]_r^n$ either factor matrix has rank r, then by § 131 the product matrix has the same rank as the other factor matrix, and therefore does not vanish.

More generally let $[a]_m^r$ and $[b]_r^n$ be rational integral functional matrices, neither of which vanishes identically, and let their degrees in all the variables be p and q. Then if the terms of highest degree in either matrix form an undegenerate matrix of rank r, the product matrix $[a]_m^r[b]_r^n$ has degree $p+q$ in all the variables.

We can write $\quad [a]_m^r = [a]_m^r + [a']_m^r + \dots, \quad [b]_r^n = [\beta]_r^n + [\beta']_r^n + \dots,$

where $[a]_m^r, [a']_m^r, \dots$ are homogeneous matrices of degrees $p, p-1, \dots$, and $[\beta]_r^n, [\beta']_r^n, \dots$ are homogeneous matrices of degrees $q, q-1, \dots$. Then the terms of degree $p+q$ in the product $[a]_m^r[b]_r^n$ form the homogeneous matrix $[a]_m^r[\beta]_r^n$ which does not vanish identically.

Ex. iv. *Let $[a]_m^r$ and $[b]_r^n$ be rational integral x-matrices, and let the degree of the product $[a]_m^r[b]_r^n$ in x be less than the degree in x of either one of the factors. Then if the coefficient of the highest power of x in one of the factors is an undegenerate matrix of rank r, the other factor matrix must vanish identically.*

This follows immediately from the first result in Ex. iii.

Ex. v. *Let $[a]_m^m$ and $[b]_m^m$ be square rational integral x-matrices of the same order m. Then if the coefficient of the highest power of x in $[b]_m^m$ is undegenerate, we can in one and only one way express $[a]_m^m$ in each of the forms*

$$[a]_m^m = [q]_m^m[b]_m^m + [r]_m^m, \quad [a]_m^m = [b]_m^m[q']_m^m + [r']_m^m,$$

where $[q]_m^m$, $[q']_m^m$, $[r]_m^m$, $[r']_m^m$ *are rational integral x-matrices, and the last two have lower degrees in x than* $[b]_m^m$.

It will be sufficient to prove the first result, the proof of the second result being similar.

Let $$[a]_m^m = x^u [\alpha]_m^m + x^{u-1} [\alpha']_m^m + \ldots, \quad [b]_m^m = x^v [\beta]_m^m + x^{v-1} [\beta']_m^m + \ldots,$$

where $[\alpha]_m^m$ does not vanish, $[\beta]_m^m$ is undegenerate, and u and v are positive integers, being the degrees in x of $[a]_m^m$ and $[b]_m^m$.

If $u < v$, we can express $[a]_m^m$ in the form specified by putting $[q]_m^m = 0$, $[r]_m^m = [a]_m^m$.

If $u \nless v$, we can solve the equation $[h]_m^m [\beta]_m^m = [\alpha]_m^m$ for $[h]_m^m$, and have

$$[a]_m^m - x^{u-v} [h]_m^m [b]_m^m = x^{u-1} [\gamma]_m^m + x^{u-2} [\gamma']_m^m + \ldots = [a']_m^m,$$

where the elements of $[\gamma]_m^m$, $[\gamma']_m^m$, ... are constants.

If $u - 1 \nless v$, we can then solve the equation $[h']_m^m [\beta]_m^m = [\gamma]_m^m$ for $[h']_m^m$, and have

$$[a']_m^m - x^{u-v-1} [h']_m^m [b]_m^m = x^{u-2} [\delta]_m^m + x^{u-3} [\delta']_m^m + \ldots = [a'']_m^m,$$

where the elements of $[\delta]_m^m$, $[\delta']_m^m$, ... are constants.

Proceeding in this way we have finally

$$[a]_m^m = \{x^{u-v} [h]_m^m + x^{u-v-1} [h']_m^m + \ldots\} [b]_m^m + [r]_m^m,$$

or $$[a]_m^m = [q]_m^m [b]_m^m + [r]_m^m, \quad \ldots\ldots\ldots\ldots\ldots\ldots\ldots\ldots(3)$$

where $[r]_m^m$ is a rational integral x-matrix whose degree in x is less than v.

Thus there is always one way of expressing $[a]_m^m$ in the form (3).

There cannot be any second way of doing this; for if we also had

$$[a]_m^m = [q']_m^m [b]_m^m + [r']_m^m, \quad \ldots\ldots\ldots\ldots\ldots\ldots\ldots\ldots(4)$$

where $[q']_m^m$ and $[r']_m^m$ are rational integral x-matrices, the degree of $[r']_m^m$ in x being less than v, we should have

$$\{[q]_m^m - [q']_m^m\} [b]_m^m = [r']_m^m - [r]_m^m. \quad \ldots\ldots\ldots\ldots\ldots\ldots(5)$$

Then since the degree in x of the matrix on the right in (5) is less than the degree in x of $[b]_m^m$, it follows from Ex. iv that

$$[q]_m^m - [q']_m^m = 0, \quad \text{and therefore} \quad [r']_m^m - [r]_m^m = 0,$$

i.e. we have $$[q']_m^m = [q]_m^m, \quad [r']_m^m = [r]_m^m$$

Ex. vi. *If in the theorem of Ex.* v *the matrices* $[a]_m^m$ *and* $[b]_m^m$ *both lie in a domain of rationality* Ω, *then the matrices* $[q]_m^m$, $[r]_m^m$, $[q']_m^m$, $[r']_m^m$ *also lie in* Ω.

For in the proof of the theorem the matrices $[h]_m^m$, $[\gamma]_m^m$, $[\gamma']_m^m$, ..., $[h']_m^m$, $[\delta]_m^m$, $[\delta']_m^m$, ... necessarily lie in Ω.

Ex. vii. If $[a]_r^m$ and $[b]_r^m$ are rational integral x-matrices, and if the coefficient of the highest power of x in $[b]_r^m$ has rank r (so that $r \not> m$), we can express $[a]_r^m$ in the form

$$[a]_r^m = [b]_r^m [p]_m^m + [q]_r^m,$$

where $[p]_m^m$ and $[q]_r^m$ are rational integral x-matrices, and where $[q]_r^m$ has a lower degree in x than $[b]_r^m$. If $r < m$, this can be done in many ways. If $[a]_r^m$ and $[b]_r^m$ lie in any domain of rationality Ω, we can so choose $[p]_m^m$ and $[q]_r^m$ that they also lie in Ω.

§ 202. The irreducible, irresoluble and linear divisors of a rational integral functional matrix.

Let $A = [a]_m^n$ be a rational integral functional matrix of rank r lying in any domain of rationality Ω, its elements being rational integral functions in Ω of certain scalar variables $x, y, z, \ldots$; let t be a rational integral function in Ω of the same variables which is irreducible in Ω; and let i be any one of the integers $1, 2, \ldots r$. Then if the irreducible function t is a common factor of all minor determinants of A of order i, it will be called an *irreducible divisor* of A of order i. Since every minor determinant of A of order $i+s$ can be expanded in terms of minor determinants of order i, every irreducible divisor of A of order i is also an irreducible divisor of A of each of the orders $i+1, i+2, \ldots r$; and therefore t is an irreducible divisor of A (of some order) when and only when it is an irreducible divisor of order r. Accordingly the *irreducible divisors* of A are the irreducible factors of the H.C.F. of all minor determinants of A of order r; and the irreducible divisors of A of order i are the irreducible factors of the H.C.F. of all minor determinants of A of order i.

In particular, (taking Ω to be the domain of all scalar numbers), if t is an irresoluble rational integral function of the variables $x, y, z, \ldots$, and if the irresoluble function t is a common factor of all minor determinants of A of order i, it will be called an *irresoluble divisor* of A of order i. Every irresoluble divisor of A of order i is also an irresoluble divisor of A of each of the orders $i+1, i+2, \ldots r$; and therefore t is an irresoluble divisor of A (of some order) when and only when it is an irresoluble divisor of order r. Accordingly the *irresoluble divisors* of A are the irresoluble factors of the H.C.F. of all minor determinants of A of order r; and the irresoluble divisors of A of order i are the irresoluble factors of the H.C.F. of all minor determinants of A of order i.

An irresoluble divisor of A which is linear in the variables $x, y, z, \ldots$ will be called a *linear divisor* of A. When A is rational and integral in a single variable or rational integral and homogeneous in two variables, the irresoluble divisors of A are identical with its linear divisors.

If t is any irreducible or irresoluble or linear divisor of the matrix A, and if t^{d_i} is the highest power of t which is a common factor of all minor determinants of A of order i, i being any one of the integers $1, 2, \ldots r$, we will call d_i the *maximum index of order* i of the divisor t for the matrix A.

NOTE. When the matrix A has rank 0, we shall consider that A has no irreducible or irresoluble divisors.

For we must in this case (see Ex. xxvii of § 185) regard 0 as the H.C.F. of the minor determinants of order i, and 0 is not regarded as an irreducible function. In general formulae it must be possible to replace every irreducible and irresoluble divisor by 0 in the limiting case when A has rank 0.

When the matrix A has rank r, we shall consider for similar reasons that it has no irreducible or irresoluble divisors of order greater than r.

Ex. i. Let A be a rational integral functional matrix in Ω of rank r, where $r \not< 1$; let t be an irreducible function in Ω; and let $g_1, g_2, \ldots g_p$ be a number of irreducible functions in Ω which are all distinct from t. Then t is an irreducible divisor of A

(1) when and only when A has rank less than r for every finite root of t;

(2) when and only when A has rank less than r for every finite root of t which is not a root of any one of the functions $g_1, g_2, \ldots g_p$.

This follows from Exs. v and viii of § 189.

If t is not an irreducible divisor of A, then there exist finite roots of t for which A has rank r; and this remains true when $r=0$.

Ex. ii. Let $A, B, C, \ldots$ be rational integral functional matrices in Ω whose ranks are $\alpha, \beta, \gamma, \ldots$; let t be any irreducible rational integral function in Ω; and let $\alpha', \beta', \gamma', \ldots$ be positive integers which are not greater than $\alpha, \beta, \gamma, \ldots$ respectively. Then if t is not a common factor of all minor determinants of A of order α', nor a common factor of all minor determinants of B of order β', nor a common factor of all minor determinants of C of order $\gamma', \ldots$, there must exist finite roots of t for which $A, B, C, \ldots$ have ranks not less than $\alpha', \beta', \gamma', \ldots$.

We will now proceed to establish the following important theorem:

Theorem I. *Let $A = [a]_m^n$ be any matrix of rank r, r being not less than 1, whose elements are rational integral functions in the domain of rationality Ω of the scalar variables $x, y, z, \ldots$; let t be any irreducible divisor in Ω (or any irresoluble divisor) of the matrix A; let $t^{d_1}, t^{d_2}, \ldots t^{d_i}, \ldots t^{d_r}$ be the highest powers of t which are common factors of all minor determinants of A*

of orders 1, 2, ... i, ... r *respectively; and let* $d_0 = 0$. *Then for the values* 1, 2, ... $r-1$ *of* i *we have*

$$d_{i+1} - d_i \nless d_i - d_{i-1}. \qquad \text{.........................(A)}$$

The index d_i, where $i \nless 1$, has the value 0 when and only when t is not a common factor of all minor determinants of order i; and when $d_i = 0$, all the preceding indices $d_1, d_2, \dots d_{i-1}$ clearly have the value 0. If $d_i \neq 0$, then t is a common factor of all minor determinants of order i, and therefore a common factor of all minor determinants of each of the orders $i, i+1, \dots r$; and in this case the inequalities (A) show that the indices $d_i, d_{i+1}, \dots d_r$ form a series of constantly increasing integers.

Since $d_0 = 0$, the first of the inequalities (A) is

$$d_2 - d_1 \nless d_1, \quad \text{or} \quad d_2 \nless 2d_1.$$

In proving Theorem I we shall make use of the lemmas given below.

LEMMA A. *If* $[a]_m^n = [a]_m^n + [\beta]_m^n$, *then every minor determinant of order* k *of the matrix* $[a]_m^n$ *can be expressed as a sum of determinants of the* $k+1$ *types*

$$(a)_k^k, \quad (a, \beta)_k^{k-1,1}, \ \dots \ (a, \beta)_k^{k-s,s}, \ \dots \ (\beta)_k^k, \qquad \text{.......................(1)}$$

where s *receives the values* 0, 1, 2, ... k.

Here for the sake of brevity every determinant of the form $(a_{uv}, \beta_{uw})_k^{k-s,s}$, where $[u]_k$ is a minor of the sequence $[1\,2 \dots m]$, and where $[v]_{k-s}$ and $[w]_s$ are minors of the sequence $[1\,2 \dots n]$ having no element in common, is called a determinant of the type $(a, \beta)_k^{k-s,s}$. The truth of this lemma is seen when we observe that the minor determinant

$$(a_{pq})_k^k = \begin{vmatrix} a_{p_1q_1} + \beta_{p_1q_1}, & a_{p_1q_2} + \beta_{p_1q_2}, & \dots & a_{p_1q_k} + \beta_{p_1q_k} \\ a_{p_2q_1} + \beta_{p_2q_1}, & a_{p_2q_2} + \beta_{p_2q_2}, & \dots & a_{p_2q_k} + \beta_{p_2q_k} \\ \dots & \dots & \dots & \dots \\ a_{p_kq_1} + \beta_{p_kq_1}, & a_{p_kq_2} + \beta_{p_kq_2}, & \dots & a_{p_kq_k} + \beta_{p_kq_k} \end{vmatrix}$$

can be expressed as a sum of determinants of order k each vertical row of which contains a's only or β's only.

If $[a]_m^n$ has rank ρ and $[\beta]_m^n$ has rank σ, then all those determinants of the various types (1) vanish for which $k - s > \rho$ or $s > \sigma$. This is seen by expanding each determinant of the types (1) in terms of the simple minor determinants of the vertical rows which contain β's only.

LEMMA B. *Let* $\Delta_i = (a_{pq})_i^i$ *be any minor determinant of order* i *of the matrix* A *in Theorem I which does not vanish identically, so that* $i \nless 1$ *and* $i \ngtr r$; *let* u *and* v *be any elements of the respective sequences* $[1\,2 \dots m]$ *and* $[1\,2 \dots n]$; *and let*

$$P_{uv} = \begin{pmatrix} q_1\, q_2 \dots q_i\, v \\ a \\ p_1\, p_2 \dots p_i\, u \end{pmatrix} = \begin{vmatrix} a_{p_1q_1} \dots a_{p_1q_i} & a_{p_1v} \\ \dots & \dots \\ a_{p_iq_1} \dots a_{p_iq_i} & a_{p_iv} \\ a_{uq_1} \dots a_{uq_i} & a_{uv} \end{vmatrix}, \quad P'_{uv} = \begin{vmatrix} a_{p_1q_1} \dots a_{p_1q_i} & a_{p_1v} \\ \dots & \dots \\ a_{p_iq_1} \dots a_{p_iq_i} & a_{p_iv} \\ a_{uq_1} \dots a_{uq_i} & 0 \end{vmatrix}.$$

Then $$\Delta_i [a]_m^n = [P]_m^n - [P']_m^n, \quad \text{.......................................(2)}$$

this equation being an identity in the elements of A, and therefore an identity in the variables $x, y, z, \ldots$. Also the functional matrices $[P]_m^n$ and $[P']_m^n$ lie in the domain Ω and have respectively ranks $r-i$ and i.

That the equation (2) is an identity in the elements of A has been shown in Theorem I of § 116. This can also be seen immediately by equating corresponding elements on both sides. If $[A_{pq}]_i^i$ is the reciprocal matrix of $[a_{pq}]_i^i$, it has also been shown in Theorem I of § 116 that the equation

$$[P']_m^n = -[a_{1q}]_m^i \boxed{A_{pq}}_i^i [a_{p1}]_i^n \quad \text{..................................(3)}$$

is an identity in the elements of A, and therefore an identity in the variables $x, y, z, \ldots$; and it follows that the matrix $[P']_m^n$ has rank i. That the matrix $[P]_m^n$ has rank $r-i$ has been shown in Ex. viii of § 116.

LEMMA C. *With the notation used in Theorem I and Lemma B we have the following results:*

(1) *If $k \nless i$ and $k \ngtr r$, every minor determinant $(a_{uv})_k^k$ of $[a]_m^n$ of order k satisfies an identity of the form*

$$\Delta_i^k (a_{uv})_k^k = \Sigma (P, P')_k^{k-i,i} + \Sigma (P, P')_k^{k-i+1,i-1} + \ldots + \Sigma (P, P')_k^{k-i+s,i-s} + \ldots, \quad \ldots(4)$$

where the final value of s is the smaller of the two numbers i and $r-k$.

(2) *If $k \ngtr i$, every minor determinant $(P'_{uv})_k^k$ of $[P']_m^n$ of order k satisfies an identity of the form*

$$(P'_{uv})_k^k = (-1)^k \Delta_i^k (a_{uv})_k^k + \Delta_i^{k-1} \Sigma (a, P)_k^{k-1,1} + \ldots + \Delta_i^{k-s} \Sigma (a, P)_k^{k-s,s} + \ldots, \quad \ldots(5)$$

where the final value of s is the smaller of the two numbers k and $r-i$.

(3) *If $k \ngtr r-i$, every minor determinant $(P_{uv})_k^k$ of $[P]_m^n$ of order k satisfies an identity of the form*

$$(P_{uv})_k^k = \Delta_i^k (a_{uv})_k^k + \Delta_i^{k-1} \Sigma (a, P')_k^{k-1,1} + \ldots + \Delta_i^{k-s} \Sigma (a, P')_k^{k-s,s} + \ldots, \quad \ldots\ldots(6)$$

where the final value of s is the smaller of the two numbers k and i.

We deduce these results from Lemma A by using the identities for $[a_{uv}]_k^k$, $[P'_{uv}]_k^k$, $[P_{uv}]_k^k$ given by

$$\Delta_i [a_{uv}]_k^k = [P_{uv}]_k^k - [P'_{uv}]_k^k. \quad \text{..................................(2')}$$

Each sum in formulae (4), (5) and (6) is a sum of a number of determinants of the type shown, as defined in Lemma A.

We may observe that $(a_{uv})_k^k$ vanishes identically when $k > r$, $(P'_{uv})_k^k$ vanishes identically when $k > i$, and $(P_{uv})_k^k$ vanishes identically when $k > r-i$. When $k < i$, the first term on the right in (4) is $(-1)^k (P'_{uv})_k^k$ corresponding to the value $i-k$ of s.

LEMMA D. *If in Theorem I the first i of the inequalities* (A) *are true, then:*

(a) *Every element of* $[P]^n_m$ *contains the factor* $t^{d_{i+1}}$.

(b) *Every element of* $[P']^n_m$ *contains the factor* $t^{d_i+d_1}$.

(c) *Every minor determinant of* $[P']^n_m$ *of order i contains the factor* $t^{(i+1)d_i}$.

(d) *Every minor determinant of* $[P']^n_m$ *of order k, where $k \not> i$, contains the factor* $t^{kd_i+d_k}$.

The property (a) follows from the fact that every non-vanishing element of $[P]^n_m$ is a minor determinant of $[a]^n_m$ of order $i+1$.

The property (b), which is a particular case of the property (d), follows from the identity $P'_{uv}=P_{uv}-\Delta_i a_{uv}$ and the inequality $d_{i+1}-d_i \not< d_1$.

The property (c), which is also a particular case of the property (d), follows from the identity

$$(P'_{uv})^i_i=(-1)^i \det [a_{uq}]^i_i \boxed{A_{pq}}^i_i [a_{pv}]^i_i=(-1)^i (a_{uq})^i_i (A_{pq})^i_i (a_{pv})^i_i$$

$$=(-1)^i \Delta_i^{i-1} (a_{uq})^i_i (a_{pv})^i_i$$

derived from (3).

To prove the property (d) we use the identical equation (5) of Lemma C. Expanding every determinant in each sum in terms of the simple minor determinants of the vertical rows containing the P's, it follows from the property (a) that the equation has the form

$$(P'_{uv})^k_k=t^{\lambda_0}U_0+t^{\lambda_1}U_1+\ldots+t^{\lambda_s}U_s+\ldots, \quad \text{.........................(7)}$$

where $U_0, U_1, \ldots U_s, \ldots$ are rational integral functions in Ω of the variables $x, y, z, \ldots$, and where

$$\lambda_0=kd_i+d_k, \quad \lambda_1=d_{i+1}+(k-1)\,d_i+d_{k-1}, \quad \ldots \quad \lambda_s=sd_{i+1}+(k-s)\,d_i+d_{k-s}, \quad \ldots.$$

The first i of the inequalities (A) show that

$$\lambda_s \not< \lambda_{s-1} \not< \lambda_{s-2} \not< \ldots \not< \lambda_0.$$

Hence from (7) it follows that the minor determinant $(P'_{uv})^k_k$ contains the factor t^{λ_0}, i.e. the factor $t^{kd_i+d_k}$.

We are now in a position to prove Theorem I. We shall consider that t is an irreducible divisor in Ω; for if the theorem is true for irreducible divisors, it follows by taking Ω to be the domain of all scalar numbers that it is true for irresoluble divisors.

The first of the inequalities (A) is clearly true. For if t^{d_1} is a factor of every element of $[a]^n_m$, we have

$$[a]^n_m=t^{d_1}[b]^n_m,$$

where $[b]^n_m$ is a matrix whose elements are rational integral functions in Ω

of the variables $x, y, z, \ldots$; and if $\Delta_2 = (a_{uv})_2^2$ is any minor determinant of $[a]_m^n$ of order 2, we have

$$[a_{uv}]_2^2 = t^{d_1}[b_{uv}]_2^2, \quad \Delta_2 = t^{2d_1}(b_{uv})_2^2.$$

Thus every minor determinant of $[a]_m^n$ of order 2 is divisible by t^{2d_1}; and if t^{d_2} is the highest power of t which is a factor of all minor determinants of $[a]_m^n$ of order 2, we must have

$$d_2 \not< 2d_1, \quad \text{i.e.} \quad d_2 - d_1 \not< d_1 - d_0.$$

Hence we can prove Theorem I by induction. We will make the hypothesis that the first i of the inequalities (A) are true, and show that, if $i+2 \not> r$, the $(i+1)$th inequality, viz.

$$d_{i+2} - d_{i+1} \not< d_{i+1} - d_i$$

must then also be true.

Let $\Delta_i = (a_{pq})_i^i$ be any one of the non-vanishing minor determinants of order i of the matrix A in which the highest power of t occurring as a factor is t^{d_i}. Then if Δ_{i+2} is any minor determinant of A of order $i+2$, we have by Lemma C an identity of the form

$$\Delta_i^{i+2}\Delta_{i+2} = \Sigma(P, P')_{i+2}^{2,\, i} + \Sigma(P, P')_{i+2}^{3,\, i-1} + \ldots + \Sigma(P, P')_{i+2}^{2+s,\, i-s} + \ldots \quad \ldots(4')$$

where the final value of s is the smaller of the two numbers i and $r-i-2$.

Expanding each determinant on the right in terms of the simple minor determinants of the vertical rows containing the P's, and using the properties (a), (b), (c), (d) of Lemma D, we see that the equation (4′) has the form

$$t^{\lambda}\Delta_{i+2}V = t^{\lambda_0}V_0 + t^{\lambda_1}V_1 + \ldots + t^{\lambda_s}V_s + \ldots, \quad \ldots\ldots\ldots\ldots(8)$$

where $V, V_0, V_1, \ldots V_s, \ldots$ are rational integral functions in Ω of the variables $x, y, z, \ldots$, and where

$$\lambda = (i+2)d_i, \quad \lambda_0 = 2d_{i+1} + (i+1)d_i; \quad \lambda_1 = 3d_{i+1} + (i-1)d_i + d_{i-1},$$
$$\ldots\ \lambda_s = (s+2)d_{i+1} + (i-s)d_i + d_{i-s}, \ldots. \quad \ldots\ldots\ldots\ldots(9)$$

Since Δ_i is not divisible by a higher power of t than t^{d_i}, the function V is not divisible by t.

The first i of the inequalities (A), which are true by the hypothesis, show that

$$\lambda_s \not< \lambda_{s-1} \not< \ldots \not< \lambda_1 \not< \lambda_0 \not< \lambda,$$

and when we divide both sides of the equation (9) by t^{λ}, we obtain the identity

$$\Delta_{i+2}V = t^{\mu_0}V_0 + t^{\mu_1}V_1 + \ldots + t^{\mu_s}V_s + \ldots, \quad \ldots\ldots\ldots\ldots(10)$$

where

$$\mu_0 = 2d_{i+1} - d_i, \quad \mu_1 = 3(d_{i+1} - d_i) + d_{i-1}, \ldots \mu_s = (s+2)(d_{i+1} - d_i) + d_{i-s}, \ldots,$$

and
$$\mu_s \not< \mu_{s-1} \not< \mu_{s-2} \not< \ldots \not< \mu_1 \not< \mu_0.$$

Since the right-hand side of (10) is divisible by t^{μ_0}, therefore $\Delta_{i+2}V$ is divisible by t^{μ_0}; and since V is not divisible by t, it follows from Corollary 3 to Theorem III in § 188 that Δ_{i+2} is divisible by t^{μ_0}.

Thus every minor determinant of the matrix A of order $i+2$ is divisible by t^{μ_0}, i.e. by $t^{2d_{i+1}-d_i}$; and if $t^{d_{i+2}}$ is the highest power of t which is a common factor of all minor determinants of A of order $i+2$, we must have

$$d_{i+2} \not< 2d_{i+1} - d_i, \quad \text{or} \quad d_{i+2} - d_{i+1} \not< d_{i+1} - d_i.$$

This completes the proof by induction of Theorem I.

NOTE. *Alternative proof of Theorem I for the special case when the elements of the matrix A are rational integral functions of a single variable x.*

Another proof for this special case is given in Note 2 of § 219.

Ex. iii. From the inequalities (A) we see that if u, v, w are positive integers such that $u \not< v$, $u+w \not> r$, we have

$$d_{u+w} - d_u \not< d_{v+w} - d_v. \quad \ldots\ldots\ldots\ldots\ldots\ldots\text{(B)}$$

Ex. iv. From the inequalities (A) we see that if $i \not< 0$, $i+k \not> r$, then

$$d_{i+k} - d_i \not< k(d_{i+1} - d_i). \quad \ldots\ldots\ldots\ldots\ldots\ldots\text{(C)}$$

Ex. v. If t first occurs as a common factor in the minor determinants of A of order i, and if the highest power of t which is a common factor of all minor determinants of order i is t^α, then the highest powers of t which are common factors of all minor determinants of orders i, $i+1$, $i+2$, $i+3$, ... respectively, so long as no one of these orders exceeds r, are

$$t^\alpha, \quad t^{2\alpha+\beta}, \quad t^{3\alpha+2\beta+\gamma}, \quad t^{4\alpha+3\beta+2\gamma+\delta}, \ldots,$$

where β, γ, δ, ... are positive integers, each one of which may be 0.

Ex. vi. *Special proof of the last of the inequalities* (A) *for an undegenerate square matrix* $[a]_r^r$.

Let t be any irreducible (or irresoluble) divisor of $[a]_r^r$, and let Δ_{r-2} be any minor determinant of $[a]_r^r$ of order $r-2$ which is not divisible by a higher power of t than $t^{d_{r-2}}$. Without loss of generality we may assume that $\Delta_{r-2} = (a)_{r-2}^{r-2}$. Then in the identity

$$(a)_{r-2}^{r-2}(a)_r^r = \Delta_{r-2}(a)_r^r = \begin{vmatrix} \begin{pmatrix} 1, 2, \ldots r-2, r-1 \\ a \\ 1, 2, \ldots r-2, r-1 \end{pmatrix} & \begin{pmatrix} 1, 2, \ldots r-2, r \\ a \\ 1, 2, \ldots r-2, r-1 \end{pmatrix} \\ \begin{pmatrix} 1, 2, \ldots r-2, r-1 \\ a \\ 1, 2, \ldots r-2, r \end{pmatrix} & \begin{pmatrix} 1, 2, \ldots r-2, r \\ a \\ 1, 2, \ldots r-2, r \end{pmatrix} \end{vmatrix}$$

given by § 110, the highest power of t which is a factor of the left-hand side is $t^{d_r+d_{r-2}}$; and $t^{2d_{r-1}}$ is certainly a factor of the right-hand side. We have therefore

$$d_r + d_{r-2} \not< 2d_{r-1}, \quad \text{or} \quad d_r - d_{r-1} \not< d_{r-1} - d_{r-2}.$$

Ex. vii. *If* $[A]_\mu^\nu$ *is a complete matrix of the minor determinants of order i of the matrix* $A=[a]_m^n$ *of Theorem I whose rank is r, and if* $i+1 \not> r$, *then every minor determinant of* $[A]_\mu^\nu$ *of order 2 is divisible by* $t^{d_{i-1}+d_{i+1}}$.

We observe that if $i+1>r$, then every minor determinant of $[A]_\mu^\nu$ of order 2 vanishes identically, but if $i+1 \not> r$, then $[A]_\mu^\nu$ has minor determinants of order 2 which do not vanish identically.

We prove the above theorem by showing that if $[p]_i$, $[u]_i$ are minors of the sequence $[1\,2\,...\,m]$ and $[q]_i$, $[v]_i$ are minors of the sequence $[1\,2\,...\,n]$, and if

$$A_{pq}=(a_{pq})_i^i,\quad A_{uv}=(a_{uv})_i^i,\quad A_{pv}=(a_{pv})_i^i,\quad A_{uq}=(a_{uq})_i^i,$$

then $t^{d_{i-1}+d_{i+1}}$ is a factor of the determinant

$$D=\begin{vmatrix} A_{pq} & A_{pv} \\ A_{uq} & A_{uv} \end{vmatrix}=A_{pq}A_{uv}-A_{uq}A_{pv}.$$

Let $\Delta=A_{pq}=(a_{pq})_i^i$, and let $[P]_m^n$ and $[P']_m^n$ be defined as in Lemma B, so that $P_{u_h v_k}=\begin{pmatrix} q_1\, q_2 \ldots q_i\, v_k \\ a \\ p_1\, p_2 \ldots p_i\, u_h \end{pmatrix}$, and $P'_{u_h v_k}$ is obtained from $P_{u_h v_k}$ by substituting 0 for $a_{u_h v_k}$.

Then
$$\Delta\,[a_{uv}]_i^i=[P_{uv}]_i^i-[P'_{uv}]_i^i.$$

By § 110 we have the identities

$$(P_{uv})_i^i = \Delta^{i-1}\begin{pmatrix} a_{pq}, & a_{pv} \\ a_{uq}, & a_{uv} \end{pmatrix}_{i,i}^{i,i},\quad (P'_{uv})_i^i = \Delta^{i-1}\begin{pmatrix} a_{pq}, & a_{pv} \\ a_{uq}, & 0 \end{pmatrix}_{i,i}^{i,i},$$

the second being derived from the first by the substitutions $[a_{uv}]_i^i=0$.

Thus
$$\det\{[P_{uv}]_i^i - \Delta\,[a_{uv}]_i^i\} = \Delta^{i-1}\begin{pmatrix} a_{pq}, & a_{pv} \\ a_{uq}, & 0 \end{pmatrix}_{i,i}^{i,i},$$

or
$$\det\{\Delta\,[a_{uv}]_i^i - [P_{uv}]_i^i\} = \Delta^{i-1}A_{pv}A_{uq}. \quad\ldots\ldots(11)$$

Expanding the left-hand side of (11) in ascending powers of Δ, we obtain a result of the form

$$\Delta^i A_{uv}-\Delta^{i-1}A_{pv}A_{uq}=\Delta^{i-1}D$$
$$=\Delta^{i-1}\Sigma\,(a,\,P)_i^{i-1,1}+\Delta^{i-2}\Sigma\,(a,\,P)_i^{i-2,2}+\ldots+\Delta^{i-s}\Sigma\,(a,\,P)_i^{i-s,s}+\ldots.\quad\ldots(12)$$

Now every determinant of the type $(a,\,P)_i^{i-s,s}$ in (12) is a sum of products of a minor determinant of $[a_{uv}]_i^i$ of order $i-s$ and a minor determinant of $[P_{uv}]_i^i$ of order s, and every minor determinant of $[P_{uv}]_i^i$ of order s is by § 110 the product of Δ^{s-1} and a minor determinant of $\begin{bmatrix} a_{pq}, & a_{pv} \\ a_{uq}, & a_{uv} \end{bmatrix}_{i,i}^{i,i}$ of order $i+s$ containing Δ.

Consequently in (12) we can write

$$\Delta^{i-s}\Sigma\,(a,\,P)_i^{i-s,s}=\Delta^{i-1}\Sigma\,\delta_{i-s}\delta_{i+s},$$

where δ_{i-s} and δ_{i+s} are minor determinants of $[a]_m^n$ of orders $i-s$ and $i+s$; and we see that Δ^{i-1} is a factor of both sides.

Since (12) is an identity in the elements of $[a]_m^n$, and Δ is a function of those elements which does not vanish identically, we can cancel the factor Δ^{i-1} common to both sides, and we then obtain a result of the form

$$D=t^{\tau_1}U_1+t^{\tau_2}U_2+\ldots+t^{\tau_s}U_s+\ldots, \quad\ldots\ldots(13)$$

where U_1, U_2, ... U_s, ... are rational integral functions in Ω of the variables x, y, z, ..., and where

$$\tau_s=d_{i-s}+d_{i+s}.$$

In (12) and (13) we have $s \not> i$, $s \not> r-i$, i.e. $i-s \not< 0$, $i+s \not> r$, and the inequalities (A) show that

$$\tau_1 \not> \tau_2 \not> \tau_3 \not> \ldots \not> \tau_s \not> \ldots.$$

Therefore D is divisible by t^{τ_1}, i.e. by $t^{d_{i-1}+d_{i+1}}$.

Ex. viii. *If the irreducible (or irresoluble) function t is not a factor of the minor determinant $\Delta_i=(a_{pq})_i^i$, and if t^a is a factor of all those minor determinants $[a]_m^n$ of order $i+1$ which contain Δ_i, then t^a is a factor of all minor determinants of $[a]_m^n$ of order $i+1$, and t^{ka} is a factor of all minor determinants of $[a]_m^n$ of order $i+k$.*

Let Δ_{i+k} be any minor determinant of $[a]_m^n$ of order $i+k$, where $k \not< 1$ and $i+k \not> r$. Then by Lemma C we have an identity of the form

$$\Delta_i^{i+k}\Delta_{i+k}=\Sigma\,(P, P')_{i+k}^{k,\,i}+\Sigma\,(P, P')_{i+k}^{k+1,\,i-1}+\ldots+\Sigma\,(P, P')_{i+k}^{k+s,\,i-s}+\ldots, \quad\ldots(14)$$

the final value of s being the smaller of the two numbers i and $r-i-k$. Since each of the P's is divisible by t^a, this identity has the form

$$\Delta_{i+k}U=t^{ka}U_0+t^{(k+1)a}U_1+\ldots+t^{(k+s)a}U_s+\ldots,$$

where U, U_0, U_1, ... U_s, ... are rational integral functions in Ω, and where U is not divisible by t. It follows from Corollary 2 to Theorem III in § 188 that Δ_{i+k} is divisible by t^{ka}.

The theorem of this example corresponds to Theorem I of § 71.

Ex. ix. *If the irreducible (or irresoluble) function t is not a factor of the minor determinant $\Delta_i=(a_{pq})_i^i$, and if t^a is a factor of all those minor determinants of $[a]_m^n$ of order $i+k$ which contain Δ_i, then t^a is a factor of all minor determinants of $[a]_m^n$ of order $i+k$.*

We again use the identity (14).

If $(P_{uv})_k^k$ is any minor determinant of $[P]_m^n$ of order k, we have by § 110

$$(P_{uv})_k^k=\Delta_i^{k-1}\begin{pmatrix} q_1 \ldots q_i\, v_1 \ldots v_k \\ a \\ p_1 \ldots p_i\, u_1 \ldots u_k\end{pmatrix}=\Delta_i^{k-1}\begin{pmatrix} a_{pq}, & a_{pv} \\ a_{uq}, & a_{uv}\end{pmatrix}_{i,k}^{i,k}.$$

Since the second factor on the right is divisible by t^a, we see that every minor determinant of $[P]_m^n$ of order k is divisible by t^a. Moreover every minor determinant of $[P]_m^n$ of order greater than k can be expanded in terms of minor determinants of order k, and is therefore divisible by t^a. Hence from (14) it follows that Δ_{i+k} is divisible by t^a.

The theorem of this example corresponds to Theorems IIa and IIb of § 71.

Ex. x. *As in the text let* $[a]_m^n$ *have rank* r, *where* $r \nless 1$; *let* $\Delta_i = (a_{pq})_i^i$ *be a minor determinant of* $[a]_m^n$ *of order* i *which does not vanish identically, and let* t *be an irreducible (or irresoluble) divisor of* $[a]_m^n$. *Then if* $[P]_m^n$ *and* $[P']_m^n$ *are defined as in Lemma B, we have the following results in which* κ *is a positive integer:*

(1) *If* $k = 2\kappa \ngtr r - i$, *every minor determinant* $(P_{uv})_k^k$ *of* $[P]_m^n$ *of order* k *is divisible by* $t^{kd_i + 2d_\kappa}$.

(2) *If* $k = 2\kappa + 1 \ngtr r - i$, *every minor determinant* $(P_{uv})_k^k$ *of* $[P]_m^n$ *of order* k *is divisible by* $t^{kd_i + d_\kappa + d_{\kappa+1}}$.

By Lemma C there exists an identity of the form

$$(P_{uv})_k^k = \Delta_i^k (a_{uv})_k^k + \dots + \Delta_i^{k-s} \Sigma (a, P')_k^{k-s, s} + \dots + (P'_{uv})_k^k.$$

Using the property (d) of Lemma D, we can write this in the form

$$(P_{uv})_k^k = t^{\lambda_0} U_0 + t^{\lambda_1} U_1 + \dots + t^{\lambda_s} U_s + \dots + t^{\lambda_k} U_k, \quad \text{.....................(15)}$$

where $U_0, U_1, \dots U_s, \dots U_k$ are rational integral functions in Ω of the variables $x, y, z, \dots$, the functions $U_{i+1}, U_{i+2}, \dots U_k$ vanishing identically when $k > i$; and where

$$\lambda_0 = kd_i + d_k, \ \dots \ \lambda_s = kd_i + d_{k-s} + d_s, \ \dots \ \lambda_k = kd_i + d_k.$$

We observe that $$\lambda_{k-s} = \lambda_s.$$

Now $\lambda_{s-1} \nless \lambda_s$ when $d_{k-s+1} - d_{k-s} \nless d_s - d_{s-1}$. Therefore $\lambda_{s-1} \nless \lambda_s$ when $k - s + 1 \nless s$, or $2s \ngtr k + 1$. Thus when $k = 2\kappa$, the least of the numbers $\lambda_0, \lambda_1, \dots \lambda_k$ is λ_κ; and when $k = 2\kappa + 1$, the least of these numbers are the two equal numbers λ_κ and $\lambda_{\kappa+1}$. Hence the identity (15) shows that the results (1) and (2) given above are true.

In the following second theorem, which will be used in the next article, Ω again denotes any domain of rationality.

Theorem II. *Let* $A = [a]_m^n$ *be any matrix of rank* r, *where* $r \nless 1$, *whose elements are rational integral functions in* Ω *of the scalar variables* $x, y, z, \dots$; *let* $\Delta_i = (a_{pq})_i^i$ *and* $D_{i+1} = (a_{uv})_{i+1}^{i+1}$ *be any two minor determinants of* A *of orders* i *and* $i+1$ *which do not vanish identically, so that* $i + 1 \ngtr r$; *let* t *be any irreducible (or irresoluble) divisor of* A; *let* $d_1', d_2', \dots d'_{r+1}$ *be the indices of the highest powers of* t *which are common factors of all minor determinants of* D_{i+1} *of orders* $1, 2, \dots i+1$ *respectively, so that in particular* d'_{i+1} *is the index of the highest power of* t *which is a factor of* D_{i+1}; *and let* $d_0' = 0$.

Further let δ_i *be the index of the highest power of* t *which is a factor of* Δ_i, *and let* δ_{i+1} *be the index of the highest power of* t *which is a common factor of those minor determinants of* A *of order* $i+1$ *which contain* Δ_i.

Then $$\delta_{i+1} - \delta_i \nless 0, \quad \delta_{i+1} - \delta_i \ngtr d'_{i+1} - d_i'. \quad \text{..................(D)}$$

As before we can confine ourselves to the case in which t is an irreducible divisor of A in Ω.

The first of the inequalities (D) is obviously true.

To prove the second inequality let $P_{uv} = \begin{pmatrix} q_1 q_2 \dots q_i v \\ a \\ p_1 p_2 \dots p_i u \end{pmatrix}$, so that δ_{i+1} is the index of the highest power of t which is a common factor of all the elements of $[P]_m^n$.

Then if $[A_{pq}]_i^i$ is the reciprocal of $[a_{pq}]_i^i$, we have by § 116 the equations

$$\Delta_i [a]_m^n - [P]_m^n = [a_{1q}]_m^i \overline{\underline{A_{pq}}}_i^i [a_{p1}]_i^n,$$

$$\Delta_i [a_{uv}]_{i+1}^{i+1} - [P_{uv}]_{i+1}^{i+1} = [a_{uq}]_{i+1}^i \overline{\underline{A_{pq}}}_i^i [a_{pv}]_i^{i+1},$$

which are identities in $x, y, z, \dots$; and from the second equation we deduce the identical equation

$$\det \{\Delta_i [a_{uv}]_{i+1}^{i+1} - [P_{uv}]_{i+1}^{i+1}\} = 0, \qquad \dots\dots\dots\dots\dots\dots(16)$$

which is Kronecker's identity given in Ex. xvii of § 116.

Let δ'_{i+1} be the index of the highest power of t which is a common factor of all the elements of $[P_{uv}]_{i+1}^{i+1}$. Then $\delta_{i+1} \not> \delta'_{i+1}$, and therefore the second of the inequalities (D) will have been proved if we can show that

$$\delta'_{i+1} - \delta_i \not> d'_{i+1} - d'_i. \qquad \dots\dots\dots\dots\dots\dots\dots\dots(D')$$

Now putting $[a_{uv}]_{i+1}^{i+1} = [b]_{i+1}^{i+1}$, $[P_{uv}]_{i+1}^{i+1} = [Q]_{i+1}^{i+1}$; writing (16) in the form

$$\det \{\Delta_i [b]_{i+1}^{i+1} - [Q]_{i+1}^{i+1}\} = 0;$$

and expanding the determinant on the left in powers of Δ, we obtain an identity of the form

$$\Delta_i^{i+1} D_{i+1} = \Delta_i^i \Sigma (b, Q)_{i+1}^{i,1} + \dots + \Delta_i^{i+1-s} \Sigma (b, Q)_{i+1}^{i+1-s, s} + \dots + (-1)^i (Q)_{i+1}^{i+1}.$$

When we further expand each determinant of the type $(b, Q)_{i+1}^{i+1-s, s}$ in terms of the simple minor determinants of the vertical rows containing the b's, we see that the last identity has the form

$$t^{\tau_0} U_0 = t^{\tau_1} U_1 + t^{\tau_2} U_2 + \dots + t^{\tau_s} U_s + \dots + t^{\tau_{i+1}} U_{i+1}, \quad \dots\dots\dots(17)$$

where $U_0, U_1, \dots U_s, \dots$ are rational integral functions in Ω of the variables $x, y, z, \dots$, the function U_0 being not divisible by t; where $U_{r-i+1}, U_{r-i+2}, \dots U_{i+1}$ vanish identically when $i + 1 > r - i$; and where

$$\tau_0 = (i+1)\delta_i + d'_{i+1}, \quad \tau_1 = \delta'_{i+1} + i\delta_i + d'_i, \dots$$
$$\tau_s = s\delta'_{i+1} + (i+1-s)\delta_i + d'_{i+1-s}, \dots .$$

With this notation we have

$$\tau_{s+1} - \tau_s = (\delta'_{i+1} - \delta_i) - (d'_{i-s+1} - d'_{i-s}), \quad (s = 0, 1, 2, \dots i).$$

But by Theorem I $\quad d'_{i+1} - d_i' \nless d'_{i-s+1} - d'_{i-s}.$

Therefore $\quad \tau_{s+1} - \tau_s \nless (\delta'_{i+1} - \delta_i) - (d'_{i+1} - d_i'). \quad \dots\dots\dots\dots\dots(18)$

If we had $\delta'_{i+1} - \delta_i > d'_{i+1} - d_i'$, it would follow from (18) that $\tau_1, \tau_2, \dots \tau_s$ are all greater than τ_0, and then from (17) that U_0 is divisible by t, which is not the case.

Accordingly we must have $\delta'_{i+1} - \delta_i \ngtr d'_{i+1} - d_i'$; and this establishes Theorem II.

§ 203. Regular minors of a rational integral functional matrix.

Let $A = [a]_m^n$ be a matrix of rank r whose elements are rational integral functions of the scalar variables $x, y, z, \dots$, all lying in any domain of rationality Ω; let t be any one of the irreducible (or irresoluble) divisors of A; and let $d_1, d_2, \dots d_i, \dots d_r$ be the indices of the highest powers of t which are common factors of all minor determinants of A of orders $1, 2, \dots i, \dots r$ respectively. Then any minor determinant $\Delta_i = (a_{pq})_i^i$ of A of order i, where $i \ngtr r$, is said to be *regular* with respect to the divisor t when it does not vanish identically and is not divisible by a higher power of t than t^{d_i}. The square minor matrix $[a_{pq}]_i^i$ will then also be said to be regular with respect to the irreducible (or irresoluble) divisor t. Further any horizontal or vertical minor of A of reduced order i, where $i \ngtr r$, will be said to be *regular* with respect to t when one of its simple minor determinants (of order i) is regular with respect to t.

If t is an irreducible (or irresoluble) function which is not a divisor of A, any minor determinant of A of order i which does not vanish identically and is not divisible by t will be said to be *regular* with respect to t. In fact we could regard t in this case as an irreducible (or irresoluble) divisor of A for which $d_1 = d_2 = \dots = d_r = 0$. In the particular case when all the elements of A are constants the term *regular* will have the meaning ascribed to it in § 117 *a*.

We have used the term 'regular' in an entirely different sense in § 187. 2. The sense in which it is used will always be clear from the context.

With respect to the regular minors of A we can prove the following theorem which corresponds to Ex. i of § 116:

Theorem. *If* $i + 1 \ngtr r$ *and* $i \nless 1$, *every minor determinant of order* $i + 1$ *of the rational integral functional matrix* A *which is regular with respect* *to the irreducible (or irresoluble) divisor t contains at least one minor determinant of order i which is regular with respect to t; also every minor determinant of order i of A which is regular with respect to t is contained in at least one minor determinant of order $i+1$ of A which is regular with respect to t.*

We shall prove the theorem for the case in which t is an irreducible divisor in Ω. The truth of the theorem for the case in which t is an irresoluble divisor then follows when we take Ω to be the domain of all algebraic numbers.

In the first place let $\Delta_i = (a_{pq})_i^i$ and $D_{i+1} = (a_{uv})_{i+1}^{i+1}$ be any two minor determinants of A of orders i and $i+1$ which do not vanish identically.

Let d'_{i+1} be the index of the highest power of t which is a factor of D_{i+1}, and let d_i' be the index of the highest power of t which is a common factor of all those minor determinants of A of order i which are contained in D_{i+1}.

Also let δ_i be the index of the highest power of t which is a factor of Δ_i, and let δ_{i+1} be the index of the highest power of t which is a common factor of all those minor determinants of A of order $i+1$ which contain Δ_i.

Then by Theorem II of § 202 we have

$$\delta_{i+1} - \delta_i \not> d'_{i+1} - d_i'. \quad \text{.........................(1)}$$

Now let both the determinants Δ_i and D_{i+1} be regular with respect to t so that

$$\delta_i = d_i, \quad d'_{i+1} = d_{i+1}. \quad \text{.........................(2)}$$

Then the inequality (1) becomes

$$d_i + d_{i+1} \not< d_i' + \delta_{i+1}. \quad \text{.........................(3)}$$

From (3) and the obvious inequalities

$$\delta_{i+1} \not< d_{i+1}, \quad d_i' \not< d_i, \quad \text{.........................(4)}$$

we deduce in turn that $$d_i \not< d_i', \quad d_{i+1} \not< \delta_{i+1}; \quad \text{.........................(5)}$$

and from (4) and (5) it follows that

$$d_i' = d_i, \quad \delta_{i+1} = d_{i+1}. \quad \text{.........................(6)}$$

The first of the equations (6) shows that if D_{i+1} is regular with respect to t, then it contains a minor determinant of order i which is regular with respect to t; and the second of the equations (6) shows that if Δ_i is regular with respect to t, then it is contained in a minor determinant of A of order $i+1$ which is regular with respect to t.

Ex. i. If the minor determinants $\Delta_i = (a_{pq})_i^i$ and $D_{i+1} = (a_{uv})_{i+1}^{i+1}$ are regular with respect to the divisor t, then one at least of the minor determinants of A of order $i+1$ which contains Δ_i and is regular with respect to t is a minor determinant of the matrix

$$M = \begin{bmatrix} a_{pq}, & a_{pv} \\ a_{uq}, & a_{uv} \end{bmatrix}_{i,i+1}^{i,i+1}.$$

We see this by applying the theorem of the text to the matrix M. We can also see it by using the inequality $\delta'_{i+1} - \delta_i \not> d'_{i+1} - d_i'$, which is the inequality (D′) of § 202, in place of the inequality (1) in the above proof; in which case we obtain

$$d_i' = d_i, \quad \delta'_{i+1} = d_{i+1}.$$

It follows that $$\delta'_{i+1} = \delta_{i+1} = d_{i+1}. \quad \text{.......................................(7)}$$

Ex. ii. *If the rational integral functional matrix A is convertible into the matrix B by an equigradent transformation, then B is also a rational integral functional matrix, and:*

(1) *The matrices A and B have the same rank r.*

(2) *The matrices A and B have the same irresoluble divisors.*

(3) *If t is any irresoluble divisor of A and B, and if* $i \not> r$, *the maximum indices* $d_1, d_2, \ldots d_i, \ldots d_r$ *of t of orders* 1, 2, ... i, ... r *are the same for both the matrices A and B.*

Let $A = [a]_p^q$, $B = [b]_m^n$, and let

$$[b]_m^n = [h]_m^p [a]_p^q [k]_q^n, \quad [a]_p^q = \overline{\underline{H}}_p^m [b]_m^n \overline{\underline{K}}_n^q, \quad \text{.......................(8)}$$

where $[h]_m^p$, $[k]_q^n$ are undegenerate matrices of ranks p and q with constant elements, and $\overline{\underline{H}}_p^m$, $\overline{\underline{K}}_n^q$ are undegenerate matrices of ranks p and q with constant elements inverse to them. When we equate correspondingly formed complete matrices of the minor determinants of order i on both sides of the equations (8), we see that every minor determinant of B (or A) of order i is a homogeneous linear function of the minor determinants of A (or B) of order i.

Hence if the minor determinants of A (or B) all vanish identically, then the minor determinants of B (or A) all vanish identically. Therefore A and B must have the same rank, which we denote by r.

Again if $i \not> r$, the common factors of all minor determinants of A of order i are identical with the common factors of all minor determinants of B of order i; and this proves the properties (2) and (3) of the theorem.

Let the equigradent transformation converting A into B be a transformation in Ω, Ω being any domain of rationality; and let one of the matrices A and B lie in Ω. Then both these matrices lie in Ω, and:

(4) *The matrices A and B have the same irreducible divisors in Ω.*

(5) *If t is any irreducible divisor of A and B, and if* $i \not> r$, *the maximum indices* $d_1, d_2, \ldots d_i, \ldots d_r$ *of t of orders* 1, 2, ... i, ... r *are the same for both the matrices A and B.*

In this case the matrices $[h]_m^p$, $[k]_q^n$ lie in Ω, and we can choose the inverse matrices $\underline{\overline{H}}_p^m$, $\underline{\overline{K}}_n^q$ so that they also lie in Ω. Therefore if one of the matrices A and B lies in Ω, the other also lies in Ω. The same reasoning as before establishes the properties (4) and (5).

Ex. iii. *If $A=[a]_m^n$ is a rational integral functional matrix in Ω of rank r, where $r \not< 1$, and if t is any one of its irreducible (or irresoluble) divisors, we can always determine a series of r minor determinants*

$$\Delta_1=a_{p_1q_1},\quad \Delta_2=(a_{pq})_2^2,\quad \dots\quad \Delta_i=(a_{pq})_i^i,\quad \dots\quad \Delta_r=(a_{pq})_r^r$$

which are all regular with respect to t, and each one of which after the first is contained in the preceding determinant.

This follows from the theorem of the text. We can clearly choose Δ_1 to be any element of A which is regular with respect to t; or as an alternative we can choose Δ_r to be any minor determinant of A of order r which is regular with respect to t.

Ex. iv. *The matrix A of Ex.* ii *can be converted by derangements of its horizontal and vertical rows into a similar matrix $[b]_m^n$ in which the series of r leading minor determinants*

$$\Delta_1'=b_{11},\quad \Delta_2'=(b)_2^2,\quad \dots\quad \Delta_i'=(b)_i^i,\quad \dots\quad \Delta_r'=(b)_r^r$$

are all regular with respect to t.

We deduce this result from Ex. iii when we form B by bringing the p_1th, p_2th, ... p_rth horizontal rows of A into the 1st, 2nd, ... rth positions, and then bringing the q_1th, q_2th, ... q_rth vertical rows into the 1st, 2nd, ... rth positions.

By Ex. ii the matrices A and B have the same rank, lie in the same domain of rationality, and have the same irreducible (or irresoluble) divisors. Moreover the highest power of t which is a common factor of all minor determinants of A of order i is the same as the highest power of t which is a common factor of all minor determinants of B of order i. Since $\Delta_i'=\pm\Delta_i$, and Δ_i is a regular minor determinant of A, therefore Δ_i' is a regular minor determinant of B.

Ex. v. Let the matrix $A=[a]_m^n$ have rank r, where $r \not< 1$; let t be any one of the irreducible (or irresoluble) divisors of A; and let $d_1, d_2, \dots d_r$ be the maximum indices of t of orders 1, 2, ... r for the matrix A.

Then if $A'=[a_{pq}]_r^r$ is a square minor of A of order r which is regular with respect to t, the maximum indices of t of orders 1, 2, ... r for the matrix A' are $d_1, d_2, \dots d_r$, i.e. they are the same as for the matrix A.

More generally if $s \not> r$, and if $A'=[a_{pq}]_s^s$ is a square minor of A of order s which is regular with respect to t, then the maximum indices of t of orders 1, 2, ... s for the matrix A' are $d_1, d_2, \dots d_s$.

These results follow immediately from the theorem of the text.

Ex. vi. *If $\Delta_i=(a_{pq})_i^i$ is a minor determinant of A of order i which is regular with respect to t, and if $i+s \not> r$, the highest power of t which is a common factor of all minor determinants of A of order $i+s$ is the same as the highest power of t which is a common factor of all those minor determinants of A of order $i+s$ which contain Δ_i.*

For by the theorem there exists a minor determinant Δ_{i+s} of A of order $i+s$ which contains Δ_i and is regular with respect to t. Then all minor determinants of A of order $i+s$ containing Δ_i are divisible by $t^{d_{i+s}}$, and one of them, viz. Δ_{i+s}, is certainly not divisible by a higher power of t.

Ex. vii. *If $H=[a_{p1}]_i^n$ is a horizontal minor (or if $K=[a_{1q}]_m^i$ is a vertical minor) of A of reduced order i which is regular with respect to t, and if $i+s \ngtr r$, then the highest power of t which is a common factor of all minor determinants of A of order $i+s$ is the same as the highest power of t which is a common factor of all those minor determinants of A of order $i+s$ which involve all the i horizontal rows of H (or all the i vertical rows of K).*

This follows from Ex. vi.

§ 204. Regular minors of a rational integral functional matrix which is symmetric.

In the following examples $A=[a]_m^m$ will denote a symmetric matrix whose elements are rational integral functions in Ω of the variables x, $y, z, \dots$, Ω being any domain of rationality. The rank of A will, unless it is otherwise defined, be denoted by r. It will be supposed that t is any irreducible divisor of A in Ω. In the special case when Ω is taken to be the domain of all scalar numbers t will be any irresoluble divisor of A. The indices of the highest powers of t which are common factors of all minor determinants of A of orders $1, 2, \dots r$ respectively will be denoted by $d_1, d_2, \dots d_r$; and we shall use the convention that $d_0=0$. When any minor determinant of A is said to be regular, it will be understood that it is regular with respect to the irreducible divisor t.

These examples correspond to the theorems given in § 126.

Ex. i. *Every symmetric matrix $A=[a]_m^m$ of rank 1 has a regular diagonal element.*

If a_{ij} is any non-diagonal element, the equation $\begin{vmatrix} a_{ii} & a_{ij} \\ a_{ji} & a_{jj} \end{vmatrix}=0$ is an identity in the variables $x, y, z, \dots$ and leads to the identical equation

$$a_{ii}a_{jj}=a_{ij}^2. \quad\dots\dots(1)$$

If a_{ij} is regular, the right-hand side of (1) does not vanish identically, and the highest power of t occurring as a factor in it is t^{2d_1}. It follows that neither a_{ii} nor a_{jj} vanishes identically; and since each of these elements must contain the factor t^{d_1}, this must be the highest power of t occurring as a factor in each of them, i.e. each of them is regular. Thus if there is a regular non-diagonal element a_{ij}, there are at least two regular diagonal elements a_{ii}, a_{jj}. And if there is no regular non-diagonal element, there must necessarily be at least one regular diagonal element.

Ex. ii. *Every symmetric matrix $A=[a]_m^m$ of rank r, where $r \nless 1$, has at least one regular diagonal minor determinant of order r.*

Let $[A]_\mu^\mu$ be a complete matrix of the minor determinants of A of order r with a common scheme of formation for horizontal and vertical rows. Then $[A]_\mu^\mu$ is a symmetric matrix of rank 1, and by Ex. i has a regular diagonal element which is a regular diagonal minor determinant of A of order r.

As an immediate consequence we see that:

The H.C.F. of all minor determinants of order r of a symmetric matrix of rank r is the same as the H.C.F. of all its diagonal minor determinants of order r.

Ex. iii. *Let* $\Delta_{s-1}=\begin{pmatrix} p_1\, p_2 \dots p_{s-1} \\ a \\ p_1\, p_2 \dots p_{s-1}\end{pmatrix}$ *and* $\Delta_{s+1}=\begin{pmatrix} p_1\, p_2 \dots p_{s-1}\, uv \\ a \\ p_1\, p_2 \dots p_{s-1}\, uv\end{pmatrix}$ *be two diagonal minor determinants of the symmetric matrix* $A=[a]_m^m$ *of orders* $s-1$ *and* $s+1$ (*where* $s-1 \nless 1, s+1 \ngtr m$) *such that the second contains the first as a minor, and let*

$$A_{uu}=\begin{pmatrix} p_1 \dots p_{s-1}\, u \\ a \\ p_1 \dots p_{s-1}\, u\end{pmatrix},\quad A_{vv}=\begin{pmatrix} p_1 \dots p_{s-1}\, v \\ a \\ p_1 \dots p_{s-1}\, v\end{pmatrix},\quad A_{uv}=\begin{pmatrix} p_1 \dots p_{s-1}\, v \\ a \\ p_1 \dots p_{s-1}\, u\end{pmatrix},\quad A_{vu}=\begin{pmatrix} p_1 \dots p_{s-1}\, u \\ a \\ p_1 \dots p_{s-1}\, v\end{pmatrix},$$

so that A_{uu}, A_{vv} *are the two diagonal minor determinants, and* A_{uv}, A_{vu} *are the two non-diagonal minor determinants of* A *of order* s *which are contained in* Δ_{s+1} *and contain* Δ_{s-1}.

Then if A_{uv} *is regular and either* A_{uu} *or* A_{vv} *is non-regular with respect to the irreducible (or irresoluble) divisor* t *of* A, *both* Δ_{s-1} *and* Δ_{s+1} *are regular with respect to* t.

By § 110 we have the identical equation

$$\Delta_{s-1}\Delta_{s+1}=\begin{vmatrix} A_{uu} & A_{uv} \\ A_{vu} & A_{vv}\end{vmatrix}=A_{uu}A_{vv}-A^2_{uv}. \qquad \text{.........................(2)}$$

If A_{uv} is regular, and A_{uu} and A_{vv} are not both regular with respect to t, the identity (2) can be written in the form

$$\Delta_{s-1}\Delta_{s+1}=t^{2d_s}(tU+V),$$

where U and V are rational integral functions in Ω of $x, y, z, \dots$, and where V is not divisible by t. Therefore the highest power of t occurring as a factor in the product $\Delta_{s-1}\Delta_{s+1}$ is t^{2d_s}. Since Δ_{s-1} certainly contains the factor $t^{d_{s-1}}$ and Δ_{s+1} certainly contains the factor $t^{d_{s+1}}$, and since by Theorem I of § 202 we have $d_{s-1}+d_{s+1} \nless 2d_s$, it follows that $t^{d_{s-1}}$ and $t^{d_{s+1}}$ are the highest powers of t occurring as factors in Δ_{s-1} and Δ_{s+1} respectively, i.e. Δ_{s-1} and Δ_{s+1} are both regular with respect to t.

It also follows that $d_{s-1}+d_{s+1}=2d_s$, or $d_{s+1}-d_s=d_s-d_{s-1}$.

Ex. iv. *If a diagonal minor determinant of order* $s+1$ *of the symmetric matrix* A *which is regular with respect to the irreducible (or irresoluble) divisor* t *of* A *does not contain any diagonal minor determinant of order* s *which is regular with respect to* t, *then it certainly contains a diagonal minor determinant of order* $s-1$ *which is regular with respect to* t.

It is assumed that $r \nless 3$, $s-1 \nless 1$, $s+1 \ngtr r$.

Let Δ_{s+1} be a regular diagonal minor determinant of A of order $s+1$ which does not contain any regular diagonal minor determinant of order s. By the theorem of § 203 it contains a regular non-diagonal minor determinant Δ_s of order s, and we can by like derangements of the horizontal and vertical rows of A (which leave its symmetry unimpaired) express Δ_{s+1} and Δ_s in the forms given to Δ_{s+1} and A_{uv} in Ex. iii. Then by that example Δ_{s-1} is regular; and it is a diagonal minor determinant of A contained in Δ_{s+1}.

Ex. v. *If a diagonal minor determinant of order $s-1$ of the symmetric matrix A which is regular with respect to the irreducible (or irresoluble) divisor t of A is not contained in any diagonal minor determinant of A of order s which is regular with respect to t, then it is certainly contained in a diagonal minor determinant of A of order $s+1$ which is regular with respect to t.*

It is assumed that $r \nless 3$, $s-1 \nless 1$, $s+1 \ngtr r$.

Let Δ_{s-1} be a regular diagonal minor determinant of A of order $s-1$ which is not contained in any regular diagonal minor determinant of A of order s. By the theorem of § 203 it is contained in a regular non-diagonal minor determinant Δ_s of A of order s, and we can by like derangements of the horizontal and vertical rows of A (which leave its symmetry unimpaired) express Δ_{s-1} and Δ_s in the forms given to Δ_{s-1} and A_{uv} in Ex. iii. Then by that example Δ_{s+1} is regular; and it is a diagonal minor determinant of A which contains Δ_{s-1}.

Ex. vi. *If the symmetric matrix $A=[a]_m^m$ has rank r, then every diagonal minor determinant of A of order $r-1$ which is regular with respect to the irreducible (or irresoluble) divisor t of A is contained in a diagonal minor determinant of A of order r which is regular with respect to t.*

It is assumed that $r \nless 2$.

Let $\Delta_{r-1}=(a_{pp})_{r-1}^{r-1}$ be a regular diagonal minor determinant of A of order $r-1$. Then if the uth horizontal row and the vth vertical row of A are rows which do not occur in Δ_{r-1}, and if

$$A_{uu}=\begin{pmatrix} p_1 \dots p_{r-1}\, u \\ a \\ p_1 \dots p_{r-1}\, u \end{pmatrix}, \quad A_{vv}=\begin{pmatrix} p_1 \dots p_{r-1}\, v \\ a \\ p_1 \dots p_{r-1}\, v \end{pmatrix}, \quad A_{uv}=\begin{pmatrix} p_1 \dots p_{r-1}\, v \\ a \\ p_1 \dots p_{r-1}\, u \end{pmatrix}, \quad A_{vu}=\begin{pmatrix} p_1 \dots p_{r-1}\, u \\ a \\ p_1 \dots p_{r-1}\, v \end{pmatrix},$$

the equation $\begin{vmatrix} A_{uu} & A_{uv} \\ A_{vu} & A_{vv} \end{vmatrix}=0$ is by § 110 an identity in the variables $x, y, z, \dots$, and leads to the identical equation

$$A_{uu}A_{vv}=A^2_{uv}. \qquad \text{(3)}$$

If there is a regular non-diagonal minor determinant of A of order r which contains Δ_{r-1} as a minor, we may select it for A_{uv}. Then A_{uv} does not vanish identically, and the highest power of t occurring as a factor on the right in (3) is t^{2d_r}. It follows from (3) that neither A_{uu} nor A_{vv} vanishes identically, and since each of these determinants must contain the factor t^{d_r}, this must be the highest power of t occurring as a factor in each of them. Therefore A_{uu} and A_{vv} are regular diagonal minor determinants of A of order r containing Δ_{r-1}. And if there is no regular non-diagonal minor determinant of A of order r which contains Δ_{r-1} as a minor, then by the theorem of § 203 there is certainly at least one diagonal minor determinant of A of order r which contains Δ_{r-1}.

The above theorem is that which replaces the theorem of Ex. v when $s=r-1$.

Ex. vii. *If no one of the diagonal elements $a_{11}, a_{22}, \dots a_{mm}$ of the symmetric matrix $A=[a]_m^m$ is regular with respect to the irreducible (or irresoluble) divisor t of A, and if a_{ij} is a non-diagonal element which is regular with respect to t, then*

$$\Delta_2=\begin{vmatrix} a_{ii} & a_{ij} \\ a_{ji} & a_{jj} \end{vmatrix}=a_{ii}a_{jj}-a_{ij}^2 \qquad \text{(4)}$$

is a diagonal minor determinant of A of order 2 which is regular with respect to t.

It is assumed that $r \not< 2$.

By § 203 there must exist some non-diagonal element a_{ij} which is regular with respect to t. Since t^{2d_1} is the highest power of t which is a factor of a_{ij}, and since the product $a_{ii}a_{jj}$ is divisible by a higher power of t than this, we see from (4) that

$$\Delta_2 = t^{2d_1}(tU + V), \qquad \text{...(5)}$$

where U and V are rational integral functions in Ω of $x, y, z, \ldots$, and where V does not vanish identically and is not divisible by t.

Let d_2' be the index of the highest power of t which is a factor of Δ_2. Then from (5) we see that

$$d_2' = 2d_1.$$

Now $d_2 - d_1 \not< d_1 - d_0$, i.e. $d_2 \not< 2d_1$; and therefore $d_2 \not< d_2'$. Since $d_2' \not< d_2$, it follows that $d_2' = d_2 = 2d_1$; and this shows that Δ_2 is regular with respect to t.

The above theorem is the special form assumed by the theorem of Ex. v when $s = 1$.

Ex. viii. *If $A = [a]_m^m$ is a symmetric matrix of rank r, where $r \not< 1$, whose elements are rational integral functions in Ω of the variables $x, y, z, \ldots$, Ω being any domain of rationality, we can always determine a series of diagonal minor determinants*

$$\Delta_0 = 1, \quad \Delta_1 = a_{p_1p_1}, \quad \Delta_2 = (a_{pp})_2^2, \quad \ldots \quad \Delta_s = (a_{pp})_s^s, \quad \ldots \quad \Delta_r = (a_{pp})_r^r$$

of orders 0, 1, 2, ... s, ... r *such that:*

(1) *Each determinant after the first contains the preceding determinant.*

(2) *The first and last determinants Δ_0 and Δ_r are regular with respect to the given irreducible (or irresoluble) divisor t of A.*

(3) *No two consecutive determinants are both non-regular with respect to t.*

(4) *If Δ_s is non-regular with respect to t, then all diagonal minor determinants of A of order s are non-regular with respect to t, but those two non-diagonal minor determinants of A of order s which contain Δ_{s-1} and are contained in Δ_{s+1}, viz.*

$$D_s = \begin{pmatrix} p_1\, p_2 \ldots p_{s-1}\, p_{s+1} \\ a \\ p_1\, p_2 \ldots p_{s-1}\, p_s \end{pmatrix}, \quad D_s' = \begin{pmatrix} p_1\, p_2 \ldots p_{s-1}\, p_s \\ a \\ p_1\, p_2 \ldots p_{s-1}\, p_{s+1} \end{pmatrix},$$

are regular with respect to t.

Taking Δ_r to be any regular diagonal minor determinant of A of order r, we can determine $\Delta_{r-1}, \Delta_{r-2}, \ldots \Delta_1$ in turn. When Δ_{s+1} has been determined and is regular, we choose Δ_s to be regular if this is possible. If this is not possible, then if $s+1 \not< 3$, we determine Δ_s and Δ_{s-1} in the manner described in Ex. iv; and if $s = 1$, we choose Δ_s to be one of the two diagonal elements of Δ_{s+1}.

Or starting with Δ_1, we take Δ_1 to be any regular diagonal element of A, if this is possible; and when this is not possible, we choose Δ_1 and Δ_2 in the manner described in Ex. vii. We can then proceed to determine the remaining determinants in succession. When Δ_{s-1} has been determined and is regular, we choose Δ_s to be regular if this is possible. If this is not possible, then if $s+1 \not> r$, we determine Δ_s and Δ_{s+1} in the manner described in Ex. v; and if $s = r-1$, we determine Δ_s in the manner described in Ex. vi.

Thus in the above series of determinants Δ_r can be any regular diagonal minor determinant of A of order r which we choose to select. Or as an alternative Δ_1 can be any selected regular diagonal element of A, when one exists; and when A has no regular diagonal element, Δ_1 can be any diagonal element of A which lies in the same horizontal or vertical row as a regular non-diagonal element.

Ex. ix. *Let* $A=[a]_m^m$ *be any symmetric rational integral functional matrix in* Ω *of rank* r, *where* $r \nless 1$, *and let* t *be any one of its irreducible (or irresoluble) divisors. Then by like derangements of its horizontal and vertical rows we can convert* A *into a symmetric matrix* $B=[b]_m^m$ *whose leading diagonal minor determinants*

$$\Delta_0'=1,\quad \Delta_1'=b_{11},\quad \Delta_2'=(b)_2^2,\ \dots\ \Delta_s'=(b)_s^s,\ \dots\ \Delta_r'=(b)_r^r$$

have the following properties:

(1) *The first and last determinants* Δ_0' *and* Δ_r' *are regular with respect to* t.

(2) *No two consecutive determinants are both non-regular with respect to* t.

(3) *If* Δ_s' *is non-regular with respect to* t, *then all diagonal minor determinants of* B *of order* s, *including the two diagonal minor determinants*

$$B_{11}=\Delta_s'=\begin{pmatrix}1,2,\dots s-1,s\\ b\\ 1,2,\dots s-1,s\end{pmatrix},\quad B_{22}=\begin{pmatrix}1,2,\dots s-1,s+1\\ b\\ 1,2,\dots s-1,s+1\end{pmatrix},$$

are non-regular with respect to t, *but the two non-diagonal minor determinants of order* s

$$B_{12}=\begin{pmatrix}1,2,\dots s-1,s+1\\ b\\ 1,2,\dots s-1,s\end{pmatrix},\quad B_{21}=\begin{pmatrix}1,2,\dots s-1,s\\ b\\ 1,2,\dots s-1,s+1\end{pmatrix}$$

are regular with respect to t.

Let A be the matrix of Ex. viii. Then B will be a matrix having the properties stated above when we form it by bringing the p_1th, p_2th, ... p_rth horizontal and vertical rows of A into the 1st, 2nd, ... rth positions.

The matrix A is converted into B by a symmetric equigradent transformation or a symmetric derangement of the form

$$\overline{\underline{h}}_m^m\,[a]_m^m\,[h]_m^m=[b]_m^m,$$

where $[h]_m^m$ is a derangement of the unit matrix $[1]_m^m$.

Ex. x. In Ex. ix suppose that Δ_s' is non-regular with respect to t; let the symmetric matrix $B=[b]_m^m$ be converted into the symmetric matrix $C=[c]_m^m$ by adding in succession the $(s+1)$th horizontal row to the sth horizontal row and the $(s+1)$th vertical row to the sth vertical row; and let

$$\Delta_0''=1,\quad \Delta_1''=c_{11},\quad \Delta_2''=(c)_2^2,\ \dots\ \Delta_s''=(c)_s^s,\ \dots\ \Delta_r''=(c)_r^r.$$

Then by Ex. ii of § 203 the matrices B and C have the same rank, lie in the same domain of rationality, and have the same irreducible (and irresoluble) divisors. Also the maximum indices $d_1, d_2, \dots d_r$ of orders 1, 2, ... r of the irreducible (or irresoluble) divisor t are the same for the matrix C as for the matrix B.

As shown in the proof of Theorem I c of § 147 we have

$$\Delta_i''=\Delta_i' \text{ when } i\neq s; \qquad\qquad (6)$$

and

$$\Delta_s''=B_{11}+2B_{12}+B_{22}. \qquad\qquad (7)$$

Since t^{d_s} is the highest power of t which is a factor of B_{12}, and both B_{11} and B_{22} are divisible by a higher power of t, the identity (7) has the form

$$\Delta_s''=t^{d_s}(tU+V), \qquad\qquad (8)$$

where U and V are rational integral functions in Ω, and V is not divisible by t. The identity (8) shows that t^{d_s} is the highest power of t which is a factor of Δ_s''. Thus the non-regular determinant Δ_s' has been replaced by the regular determinant Δ_s''. From (6) we see further that when $i \neq s$, Δ_i'' is regular or non-regular according as Δ_i' is regular or non-regular.

By a succession of such symmetric equigradent transformations we can convert B into a matrix $C=[c]_m^m$ *in which the r leading diagonal minor determinants* $c_{11}, (c)_2^2, \dots (c)_r^r$ *are all regular with respect to t.*

Ex. xi. *Let* $A=[a]_m^m$ *be a symmetric rational integral functional matrix of rank r lying in a domain of rationality* Ω, *where* $r \nless 1$, *and let t be any one of its irreducible* (*or irresoluble*) *divisors. Then we can convert A into a similar symmetric rational integral functional matrix* $C=[c]_m^m$ *of rank r lying in* Ω, *whose first r leading diagonal minor determinants*

$$\Delta_1=c_{11}, \quad \Delta_2=(c)_2^2, \quad \dots \quad \Delta_s=(c)_s^s, \quad \dots \quad \Delta_r=(c)_r^r$$

are all regular with respect to the divisor t, by a symmetric unitary equigradent transformation of the form

$$\overline{\underline{h}}_m^m [a]_m^m [h]_m^m = [c]_m^m, \quad \dots\dots\dots\dots(9)$$

where $(h)_m^m=1$, *and where every element of* $[h]_m^m$ *is either* 0 *or* 1.

To do this we apply to A in succession the transformations indicated in Exs. ix and x. The form of the matrix $[h]_m^m$ in the resultant transformation (9) has been considered in the proof of Theorem Ic of § 147.

By Ex. ii of § 203 the matrix C has the same irreducible and irresoluble divisors as the matrix A, and the maximum indices of orders 1, 2, ... r of every irreducible or irresoluble divisor are the same for the matrix C as for the matrix A.

§ 205. The potent divisors and potent factors of a rational integral functional matrix.

1. *Maximum and potent indices. Maximum and potent divisors.*

Let $A=[a]_m^n$ be a rational integral functional matrix of rank r lying in any domain of rationality Ω, its elements being rational integral functions in Ω of certain scalar variables $x, y, z, \dots$; and let t be any one of the irreducible (or irresoluble) divisors of A as defined in § 202. Let $d_1, d_2, \dots d_r$ be the indices of the highest powers of t which are common factors of all minor determinants of A of orders 1, 2, ... r respectively; and let the convention $d_0=0$ be adopted. Then if 0 is regarded as a positive integer, $d_0, d_1, d_2, \dots d_r$ are positive integers satisfying the inequalities

$$d_r \nless d_{r-1} \nless d_{r-2} \nless \dots \nless d_2 \nless d_1 \nless d_0. \quad \dots\dots\dots\dots(1)$$

They will be called the *maximum indices* of orders 0, 1, 2, ... r of the irreducible (or irresoluble) divisor t for the matrix A.

Let $\quad e_1=d_1-d_0=d_1, \quad e_2=d_2-d_1, \quad e_3=d_3-d_2, \dots e_r=d_r-d_{r-1}.$

Then if 0 is included amongst positive integers, we know by Theorem I of § 202 that $e_1, e_2, e_3, \dots e_r$ are positive integers satisfying the inequalities

$$e_r \nless e_{r-1} \nless e_{r-2} \nless \dots \nless e_3 \nless e_2 \nless e_1. \quad \dots\dots\dots\dots(2)$$

They will be called the *potent indices* of orders 1, 2, 3, ... r of the irreducible (or irresoluble) divisor t for the matrix A.

From the inequalities (2) we see that if $d_i \neq 0$, then $d_i, d_{i+1}, \ldots d_r$ are a series of constantly increasing positive integers.

If $d_i \neq 0$, we will call t^{d_i} the *maximum divisor* of order i for the matrix A corresponding to t; if $d_i = 0$, we shall consider that A has no maximum divisor of order i corresponding to t.

If $e_i \neq 0$, we will call t^{e_i} the *potent divisor* of order i for the matrix A corresponding to t; if $e_i = 0$, we shall consider that A has no potent divisor of order i corresponding to t.

If $i+1 \not> r$, we see from the inequalities (1) that:

Any maximum divisor t^{d_i} of order i is a factor of the corresponding maximum divisor $t^{d_{i+1}}$ of order $i+1$. ....(A)

Similarly if $i+1 \not> r$, we see from the inequalities (2) that:

Any potent divisor t^{e_i} of order i is a factor of the corresponding potent divisor $t^{e_{i+1}}$ of order $i+1$. ....(B)

A potent divisor t^e corresponding to any irreducible (or irresoluble) divisor t of the matrix A will be said to be *repeated* s times when it is a potent divisor of s different orders; and it will be said to be *unrepeated* when it is a potent divisor of only one order. If $e_1, e_2, \ldots e_r$ are the potent indices of t of orders $1, 2, \ldots r$, there is a potent divisor repeated s times corresponding to the irreducible (or irresoluble) divisor t when and only when s consecutive integers in the series $e_1, e_2, \ldots e_r$ are equal and not zero.

NOTE. When we speak of the potent divisors of a matrix A without mentioning the domain in which they lie, we shall as a rule understand that they are the potent divisors corresponding to the various irresoluble divisors of A; and this will be so whether the domain in which A lies is or is not specified.

When A lies in a specified domain of rationality Ω, the potent divisors of A may in general be either those corresponding to its various irreducible divisors in Ω or those corresponding to its various irresoluble divisors. When the first interpretation is to be given to them, we shall as a rule call them potent divisors in Ω.

2. *Maximum and potent factors.*

Let $D_1, D_2, \ldots D_r$ be the highest common factors of all minor determinants of A of orders $1, 2, \ldots r$ respectively. They are rational integral functions of the variables $x, y, z, \ldots$, each of which is determinate save for a constant factor to which any finite non-zero value can be ascribed. The constant factors can be so chosen that all the functions $D_1, D_2, \ldots D_r$ lie in the domain Ω, and we shall suppose that this has been done. We will also introduce the convention that $D_0 = 1$. The functions $D_0, D_1, D_2, \ldots D_r$ will be called the *maximum factors* of the matrix A of orders $0, 1, 2, \ldots r$. The irreducible and irresoluble divisors of the matrix A are the irreducible and irresoluble factors of the function D_r; and the irreducible and irresoluble divisors of A of order i are the irreducible and irresoluble factors of the function D_i.

If t is any irreducible (or irresoluble) divisor of A whose maximum indices of orders 0, 1, 2, ... r are $d_0, d_1, d_2, \ldots d_r$, then $t^{d_0}, t^{d_1}, t^{d_2}, \ldots t^{d_r}$ are the highest powers of t which are factors of $D_0, D_1, D_2, \ldots D_r$ respectively.

If $t, t', t'', \ldots$ are all the irreducible (or all the irresoluble) divisors of the matrix A, and if the maximum indices of $t, t', t'', \ldots$ of orders 1, 2, ... r are $d_1, d_2, \ldots d_r$; $d_1', d_2', \ldots d_r'$; $d_1'', d_2'', \ldots d_r''$; ... respectively, and if $i+1 \not> r$, we can express D_i and D_{i+1} in the forms

$$D_i = a_i t^{d_i} t'^{d_i'} t''^{d_i''} \ldots, \quad D_{i+1} = a_{i+1} t^{d_{i+1}} t'^{d'_{i+1}} t''^{d''_{i+1}} \ldots, \quad \text{.........(3)}$$

where a_i and a_{i+1} are non-zero constants lying in Ω; and from the inequalities (1) we see (as is otherwise evident) that D_{i+1} is divisible by D_i for the values 0, 1, 2, ... $r-1$ of i.

Thus each of the maximum factors $D_0, D_1, D_2, \ldots D_r$ after the first is divisible by the preceding. ....(C)

Hence we can write

$$D_1 = D_0 E_1, \quad D_2 = D_1 E_2, \ldots D_{i+1} = D_i E_{i+1}, \ldots D_r = D_{r-1} E_r, \quad \ldots(4)$$

where $E_1, E_2, \ldots E_r$ are rational integral functions of $x, y, z, \ldots$ lying in the domain Ω; and we then have

$$D_1 = E_1, \quad D_2 = E_1 E_2, \ \ldots \ D_i = E_1 E_2 \ldots E_i, \ \ldots \ D_r = E_1 E_2 \ldots E_i \ldots E_r. \quad (5)$$

The functions $E_1, E_2, \ldots E_r$ determined in this manner will be called the *potent factors* of the matrix A of orders 1, 2, ... r. Each of them is determinate save for a non-zero constant factor lying in Ω to which any value may be ascribed.

If $t, t', t'', \ldots$ are all the irreducible (or all the irresoluble) divisors of the matrix A, and if the potent indices of $t, t', t'', \ldots$ of orders 1, 2, ... r are $e_1, e_2, \ldots e_r$; $e_1', e_2', \ldots e_r'$; $e_1'', e_2'', \ldots e_r''$; ... respectively, and if $i+1 \not> r$, we see from (3) that

$$E_i = b_i t^{e_i} t'^{e_i'} t''^{e_i''} \ldots, \quad E_{i+1} = b_{i+1} t^{e_{i+1}} t'^{e'_{i+1}} t''^{e''_{i+1}} \ldots, \quad \text{.........(6)}$$

where b_i and b_{i+1} are non-zero constants lying in Ω; and from the inequalities (2) it follows that E_{i+1} is divisible by E_i for the values 1, 2, ... $r-1$ of i.

Thus each of the potent factors $E_1, E_2, E_3, \ldots E_r$ after the first is divisible by the preceding. ....(D)

The irreducible and irresoluble divisors of the matrix A are the irreducible and irresoluble factors of the function E_r; and the irreducible and irresoluble divisors of A of order i are the irreducible and irresoluble factors of the function E_i.

If t is any one irreducible (or irresoluble) divisor of A, and if $e_1, e_2, \dots e_i, \dots e_r$ are the potent indices of t of orders $1, 2, \dots i, \dots r$, then t^{e_i} is the highest power of t which is a factor of E_i.

Ex. i. The function D_i is the product of all the maximum divisors of A of order i; and this is true both when the maximum divisors are those corresponding to the various irreducible divisors of A, and when they are those corresponding to the various irresoluble divisors of A.

Ex. ii. The function E_i is the product of all the potent divisors of A of order i; and this is true both when the potent divisors are those corresponding to the various irreducible divisors of A, and when they are those corresponding to the various irresoluble divisors of A.

Ex. iii. Every maximum divisor of A of order i is a factor of D_i but not a factor of D_{i-1}.

Every potent divisor of A which is a factor of E_i but not a factor of E_{i-1} is a potent divisor of order i.

We have now defined four sets of quantities, viz.

(1) The maximum factors of A of all orders.

(2) The potent factors of A of all orders.

(3) The maximum divisors of A of all orders.

(4) The potent divisors of A of all orders.

All these quantities are known when all the quantities composing any one set are known. It is generally most convenient to regard the potent divisors as fundamental. When these are known, the maximum divisors are known from the equations

$$t^{d_i} = t^{e_1} t^{e_2} \dots t^{e_i}, \quad d_i = e_1 + e_2 + \dots + e_i; \quad \dots\dots\dots\dots\dots(7)$$

and the potent factors and maximum factors are known from the equations (6) and (3).

Ex. iv. The matrix

$$[a]_4^4 = \begin{bmatrix} 1, & 0\ , & 0 & , & 0 \\ 0, & x+1, & 0 & , & 0 \\ 0, & 0\ , & (x+2)(x^3+1), & & 0 \\ 0, & 0\ , & 0 & , & (x^2-4)(x^3+1)(x+1) \end{bmatrix}$$

has rank 4 and lies in the domain Ω of all rational numbers. Expressed in terms of irreducible factors belonging to this domain, we have

$$D_1 = 1, \quad E_1 = 1,$$
$$D_2 = (x+1), \quad E_2 = x+1,$$
$$D_3 = (x+1)^2 (x+2)(x^2-x+1), \quad E_3 = (x+1)(x+2)(x^2-x+1),$$
$$D_4 = (x+1)^4 (x+2)^2 (x^2-x+1)^2 (x-2), \quad E_4 = (x+1)^2 (x+2)(x^2-x+1)(x-2).$$

The potent divisors corresponding to these irreducible factors are shown in the following table.

Irreducible divisors in Ω.	*Maximum indices.*	*Potent indices.*	*Maximum divisors in Ω.*	*Potent divisors in Ω.*
$T_1=x+1$	0, 1, 2, 4	0, 1, 1, 2	—, T_1, T_1^2, T_1^4	—, T_1, T_1, T_1^2
$T_2=x+2$	0, 0, 1, 2	0, 0, 1, 1	—, —, T_2, T_2^2	—, —, T_2, T_2
$T_3=x^2-x+2$	0, 0, 1, 2	0, 0, 1, 1	—, —, T_3, T_3^2	—, —, T_3, T_3
$T_4=x-2$	0, 0, 0, 1	0, 0, 0, 1	—, —, —, T_4	—, —, —, T_4

There are eight potent divisors in Ω, viz.

T_1^2, T_2, T_3, T_4 of order 4;
T_1, T_2, T_3 of order 3;
T_1 of order 2.

The potent divisors T_1, T_2, T_3 are repeated; and the potent divisors T_1^2 and T_4 are unrepeated.

Expressed in terms of irresoluble factors we have

$D_1=1$, $E_1=1$,
$D_2=x+1$, $E_2=x+1$,
$D_3=(x+1)^2(x+2)(x+\omega_1)(x+\omega_2)$, $E_3=(x+1)(x+2)(x+\omega_1)(x+\omega_2)$,
$D_4=(x+1)^4(x+2)^2(x+\omega_1)^2(x+\omega_2)^2(x-2)$, $E_4=(x+1)^2(x+2)(x+\omega_1)(x+\omega_2)(x-2)$,

where ω_1 and ω_2 are the two imaginary cube roots of 1.

The potent divisors corresponding to these irresoluble factors are shown in the following table.

Irresoluble divisors.	*Maximum indices.*	*Potent indices.*	*Maximum divisors.*	*Potent divisors.*
$t_1=x+1$	0, 1, 2, 4	0, 1, 1, 2	—, t_1, t_1^2, t_1^4	—, t_1, t_1, t_1^2
$t_2=x+2$	0, 0, 1, 2	0, 0, 1, 1	—, —, t_2, t_2^2	—, —, t_2, t_2
$t_3=x+\omega_1$	0, 0, 1, 2	0, 0, 1, 1	—, —, t_3, t_3^2	—, —, t_3, t_3
$t_4=x+\omega_2$	0, 0, 1, 2	0, 0, 1, 1	—, —, t_4, t_4^2	—, —, t_4, t_4
$t_5=x-2$	0, 0, 0, 1	0, 0, 0, 1	—, —, —, t_5	—, —, —, t_5

There are ten potent divisors, viz.

$$t_1^2,\ t_2,\ t_3,\ t_4,\ t_5 \quad \text{of order } 4;$$
$$t_1,\ t_2,\ t_3,\ t_4 \quad \text{of order } 3;$$
$$t_1 \quad \text{of order } 2.$$

The potent divisors t_1, t_2, t_3, t_4 are repeated; and the potent divisors t_1^2 and t_5 are unrepeated.

In this example the elements of $[a]_4^4$ are functions of the single variable x. Therefore every irreducible divisor in Ω is either linear or quadratic, and all the irresoluble divisors are linear.

Ex. v. Let
$$A=[a]_m^m=X[1]_m^m=\begin{bmatrix} X & 0 & \dots & 0 \\ 0 & X & \dots & 0 \\ \dots & \dots & \dots & \dots \\ 0 & 0 & \dots & X \end{bmatrix},$$

where X is a rational integral function of the variables $x_1, x_2, \dots x_n$ which is irresoluble (or is irreducible in Ω).

Then A has rank m; it has only one irresoluble (or irreducible) divisor X; and it has m potent divisors corresponding to the irresoluble (or irreducible) divisor X, viz.

$$X,\ X,\ \dots\ X,\ X \quad \textit{of orders} \quad 1,\ 2,\ \dots\ m-1,\ m.$$

For in this case we have

$$D_1=X,\quad D_2=X^2,\ \dots\ D_m=X^m;\quad E_1=X,\quad E_2=X,\ \dots\ E_m=X.$$

The potent divisor X is repeated m times.

Ex. vi. Let
$$A=[a]_m^{m+1}=\begin{bmatrix} X & Y & 0 & \dots & 0 & 0 \\ 0 & X & Y & \dots & 0 & 0 \\ \dots & \dots & \dots & \dots & \dots & \dots \\ 0 & 0 & 0 & \dots & X & Y \end{bmatrix},$$

where X and Y are rational integral functions of the variables $x_1, x_2, \dots x_n$, and are moreover distinct irresoluble functions (or distinct irreducible functions in Ω).

Then A has rank m; it has no irresoluble (and no irreducible) divisor; and it has therefore no potent divisor.

For the matrix of the affected simple minor determinants of A is

$$[k_1 k_2 k_3 \dots k_{m+1}]=(-1)^m[Y^m,\ -XY^{m-1},\ X^2Y^{m-2},\ \dots\ (-1)^m X^m];$$

and the H.C.F. of the minor determinants of A of order m is 1. Accordingly we have

$$D_1=D_2=\dots=D_m=1;\quad E_1=E_2=\dots=E_m=1.$$

There is only one distinct connection between the vertical rows of A, viz. the connection

$$[a]_m^{m+1}\ \overline{k}\rfloor_{m+1}=0.$$

When the irresoluble (or irreducible) functions X and Y are not distinct, we can write $Y=\kappa X$, where κ is a non-zero constant, and we have

$$D_1=X,\quad D_2=X^2,\ \dots\ D_m=X^m;\quad E_1=X,\quad E_2=X,\ \dots\ E_m=X.$$

Thus in this special case there is just one irresoluble (or irreducible) divisor X; and there are m potent divisors corresponding to this irresoluble (or irreducible) divisor X, viz.

$$X, \; X, \; \dots \; X, \; X \quad \text{of orders } 1, \; 2, \; \dots \; m-1, \; m.$$

Ex. vii. Let

$$A=[a]_{m+1}^{m}=\begin{bmatrix} Y & 0 & \dots & 0 \\ X & Y & \dots & 0 \\ 0 & X & \dots & 0 \\ \dots & \dots & \dots & \dots \\ 0 & 0 & \dots & Y \\ 0 & 0 & \dots & X \end{bmatrix},$$

where X and Y are defined as in Ex. vi.

Then A has rank m; it has no irresoluble (and no irreducible) divisor; and it has therefore no potent divisor.

For the matrix of the affected simple minor determinants of A is

$$[h_1 h_2 h_3 \dots h_{m+1}]=(-1)^m [X^m, \; -X^{m-1}Y, \; X^{m-2}Y^2, \; \dots \; (-1)^m Y^m],$$

the signs on the right being alternately $+$ and $-$, and we have

$$D_1=D_2=\dots=D_m=1; \quad E_1=E_2=\dots=E_m=1.$$

There is only one distinct connection between the horizontal rows of A, viz. the connection

$$[h_1 h_2 h_3 \dots h_{m+1}]\,[a]_{m+1}^{m}=0.$$

When the irresoluble (or irreducible) functions X and Y are not distinct, we have

$$D_1=X, \quad D_2=X^2, \; \dots \; D_m=X^m; \quad E_1=X, \quad E_2=X, \; \dots \; E_m=X.$$

In this special case there is just one irresoluble (or irreducible) divisor X; and there are m potent divisors corresponding to this divisor X, viz. the potent divisor X repeated m times.

Ex. viii. Let

$$A=[a]_m^m=\begin{bmatrix} X & Y & 0 & \dots & 0 & 0 \\ 0 & X & Y & \dots & 0 & 0 \\ \dots & \dots & \dots & \dots & \dots & \dots \\ 0 & 0 & 0 & \dots & X & Y \\ 0 & 0 & 0 & \dots & 0 & X \end{bmatrix},$$

where X and Y are rational integral functions of the variables $x_1, x_2, \dots x_n$, and are moreover distinct irresoluble functions (or distinct irreducible functions in Ω).

Then the matrix A has rank m; it has only one irresoluble (or irreducible) divisor X; and it has only one potent divisor corresponding to this irresoluble (or irreducible) divisor X, viz. the potent divisor X^m of order m.

For we have $(a)_m^m=X^m$; and since by Ex. vi the simple minor determinants of the minor matrix $[a]_{m-1}^{m}$ differ in signs only from $X^{m-1}, X^{m-2}Y, X^{m-3}Y^2, \dots Y^{m-1}$, the H.C.F. of the minor determinants of A of order $m-1$ is 1. Thus in this case we have

$$D_1=1, \quad D_2=1, \; \dots \; D_{m-1}=1, \quad D_m=X^m; \quad E_1=1, \quad E_2=1, \; \dots \; E_{m-1}=1, \quad E_m=X^m.$$

When the irresoluble (or irreducible) functions X and Y are not distinct, so that $Y=\kappa X$, where κ is a non-zero constant, we have

$$D_1=X, \quad D_2=X^2, \; \dots \; D_m=X^m; \quad E_1=X, \quad E_2=X, \; \dots \; E_m=X.$$

Thus in this special case there is just one irresoluble (or irreducible) divisor X whose potent indices are 1, 1, ... 1, 1; and there are m potent divisors corresponding to this divisor X, viz. the potent divisors

$$X,\ X,\ \dots\ X,\ X \text{ of orders } 1,\ 2,\ \dots\ m-1,\ m.$$

The potent divisor X is repeated m times.

Ex. ix. *Potent divisors of an undegenerate quasi-scalar matrix.*

Let
$$\phi = {}^1[a]_r = \begin{bmatrix} a_1 & 0 & \dots & 0 \\ 0 & a_2 & \dots & 0 \\ \dots & \dots & \dots & \dots \\ 0 & 0 & \dots & a_r \end{bmatrix}$$

be an undegenerate quasi-scalar matrix of rank r whose diagonal elements $a_1, a_2, \dots a_r$ are rational integral functions in the domain of rationality Ω of certain variables $x, y, z, \dots$, no one of them vanishing identically. Then we have the following theorem:

Theorem. *The irreducible and irresoluble divisors of ϕ are the irreducible and irresoluble factors of the functions $a_1, a_2, \dots a_r$.*

If t is any irreducible (or irresoluble) divisor of ϕ, and if l_i is the index of the highest power of t which is a factor of a_i, then $l_1, l_2, \dots l_r$ when arranged in ascending order of magnitude are the potent indices of t of orders 1, 2, ... r for the matrix ϕ.

Let $D_1, D_2, \dots D_r$ be the maximum factors of ϕ of orders 1, 2, ... r, so that D_i is the H.C.F. of all the functions $\Delta_i = a_{u_1} a_{u_2} \dots a_{u_i}$, where $[u_1 u_2 \dots u_i]$ is any minor of order i of the sequence $[1\, 2 \dots r]$.

To prove the first part of the theorem it is sufficient to observe that

$$D_r = a_1 a_2 \dots a_r.$$

To prove the second part of the theorem, let $e_1, e_2, \dots e_r$ be the integers $l_1, l_2, \dots l_r$ so arranged that

$$e_r \nless e_{r-1} \nless e_{r-2} \nless \dots \nless e_2 \nless e_1.$$

Then if $d_1, d_2, \dots d_r$ are the highest powers of t which are factors of $D_1, D_2, \dots D_r$ respectively, we have

$$d_1 = e_1, \quad d_2 = e_1 + e_2, \quad \dots \quad d_i = e_1 + e_2 + \dots + e_i, \quad \dots \quad d_r = e_1 + e_2 + \dots + e_r.$$

Thus if $i \ngtr r$, we have $e_i = d_i - d_{i-1}$, i.e. e_i is the potent index of t of order i.

Ex. x. *Potent divisors of any quasi-scalar matrix.*

Any quasi-scalar matrix ψ of rank r, where $r \nless 1$, can be reduced by striking out rows of 0's to an undegenerate quasi-scalar matrix ϕ of the form considered in Ex. ix; and ψ has then the same maximum and potent factors and the same maximum and potent divisors as ϕ.

Accordingly the irreducible and irresoluble divisors of ψ are the irreducible and irresoluble factors of those of its diagonal elements which do not vanish identically.

Also if t is any irreducible (or irresoluble) divisor of ψ, and if the indices of the highest powers of t which are factors of the r non-vanishing diagonal elements of ψ are $e_1, e_2, \dots e_r$ when arranged in ascending order of magnitude, then $e_1, e_2, \dots e_r$ are the potent indices of t of orders 1, 2, ... r for the matrix ϕ.

If e_s is the first of the integers $e_1, e_2, \dots e_r$ which is not 0, then the potent divisors of ϕ corresponding to the irreducible (or irresoluble) divisor t are

$$t^{e_s},\ t^{e_{s+1}},\ \dots\ t^{e_r} \text{ of orders } s,\ s+1,\ \dots\ r.$$

A quasi-scalar matrix which vanishes identically has no potent divisors.

Ex. xi. If $a_1, a_2, \dots a_m$ are rational integral functions of $x, y, z, \dots$, and if $b_1, b_2, \dots b_m$ are the same functions arranged in a different order, then the two quasi-scalar matrices ${}^1[a]_m$ and ${}^1[b]_m$ have the same potent divisors of all orders, i.e. they are equipotent.

NOTE. If a rational integral functional matrix A has rank r, and if we speak of a maximum or potent factor of A of any order greater than r, it must (see Ex. xxvii of § 185) be considered to have the value 0.

We shall however consider that a matrix of rank r has no maximum or potent factor, and no maximum or potent divisor of order greater than r. We shall also consider that a matrix of rank 0 has no maximum or potent factor, and no maximum or potent divisor.

§ 206. Potent factors and potent divisors of a product of two or more matrices.

Theorem I. *If* $[c]_m^n = [a]_m^r [b]_r^n$, *or* $C = AB$ ....(A)

is a standard product of two matrices $A = [a]_m^r$ *and* $B = [b]_r^n$ *whose elements are rational integral functions in the domain* Ω *of the scalar variables* $x, y, z, \dots$, *then:*

(1) *The maximum factor of any order s of C is divisible by the product of the maximum factors of order s of A and B.*

(2) *The potent factor of any order s of C is divisible by the potent factors of order s of A and B.*

By Theorem III of § 71, which is clearly applicable to rational integral functional matrices, the rank (i.e. the number of maximum or potent factors) of C cannot exceed the rank (i.e. the number of maximum or potent factors) of either A or B.

Let s be any positive integer which does not exceed the rank of C; and let A_s, B_s, C_s be the highest common factors of all minor determinants of order s of A, B, C. By equating correspondingly formed complete matrices of the minor determinants of order s on both sides of (A), we see that all minor determinants of C of order s are divisible by the product $A_s B_s$. Therefore C_s is divisible by the product $A_s B_s$; and this proves the first part of the theorem.

Now let t be any irresoluble divisor of A and B (or any irreducible divisor in Ω of A and B), and let the highest powers of t occurring as factors in A_s, B_s, C_s be respectively $\alpha_s, \beta_s, \gamma_s$. The second part of the theorem will be proved if we can show that

$$\gamma_s - \gamma_{s-1} \nless \alpha_s - \alpha_{s-1}, \text{ and } \gamma_s - \gamma_{s-1} \nless \beta_s - \beta_{s-1}. \qquad \text{............(1)}$$

Let $(b_{pq})_{s-1}^{s-1}$ and $(c_{uv})_s^s$ be minor determinants of B and C of orders $s-1$ and s which are regular with respect to t; let

$$M = \begin{bmatrix} b_{pq}, & b_{pv} \\ c_{uq}, & c_{uv} \end{bmatrix}_{s-1,\, s}^{s-1,\, s};$$

and let β_s' be the index of the highest power of t which is a common factor of all those minor determinants of M of order s which contain $(b_{pq})_{s-1}^{s-1}$. Then by Theorem II of § 202 we have

$$\beta_s' - \beta_{s-1} \not> \gamma_s - \gamma_{s-1}. \qquad\qquad (2)$$

But from (A) we see that

$$[c_{u_iq_1}\, c_{u_iq_2} \ldots c_{u_iq_{s-1}}\, c_{u_iv_j}] = a_{u_i1}[b_{1q_1}\, b_{1q_2} \ldots b_{1q_{s-1}}\, b_{1v_j}]$$
$$+ a_{u_i2}[b_{2q_1} b_{2q_2} \ldots b_{2q_{s-1}} b_{2v_j}] + \ldots + a_{u_ir}[b_{rq_1} b_{rq_2} \ldots b_{rq_{s-1}} b_{rv_j}].$$

Therefore

$$\begin{vmatrix} b_{p_1q_1} & \ldots & b_{p_1q_{s-1}} & b_{p_1v_j} \\ \ldots & \ldots & \ldots & \ldots \\ b_{p_{s-1}q_1} & \ldots & b_{p_{s-1}q_{s-1}} & b_{p_{s-1}v_j} \\ c_{u_iq_1} & \ldots & c_{u_iq_{s-1}} & c_{u_iv_j} \end{vmatrix} = a_{u_i1}\Delta_1 + a_{u_i2}\Delta_2 + \ldots + a_{u_ir}\Delta_r,$$

where $\Delta_1, \Delta_2, \ldots \Delta_r$ are minor determinants of order s of $[b]_r^n$, and are therefore divisible by t^{β_s}. Consequently every minor determinant of M of order s which contains $(b_{pq})_{s-1}^{s-1}$ is divisible by t^{β_s}, and we have

$$\beta_s' \not< \beta_s. \qquad\qquad (3)$$

From (2) and (3) it follows that

$$\beta_s - \beta_{s-1} \not> \gamma_s - \gamma_{s-1}.$$

Thus the second of the inequalities (1) is true; and in the same way we can show that the first of those inequalities is true.

This proves the second part of the theorem.

It should be observed that the maximum factor of C of order s is in general *not equal to the product* of the maximum factors of A and B of order s. Also the potent factor of C of order s is in general *not divisible by the product* of the potent factors of A and B of order s. These facts are illustrated in Ex. i below.

The converse of Theorem I for rational integral x-matrices is proved in § 221.

Ex. i. Let

$$A = \begin{bmatrix} a, & 0, & 0 \\ 0, & a^2b, & 0 \\ 0, & 0, & a^3bc \end{bmatrix}, \quad B = \begin{bmatrix} ba\beta, & 0, & 0 \\ 0, & ba\beta, & 0 \\ 0, & 0, & a \end{bmatrix}, \quad C = \begin{bmatrix} aba\beta, & 0, & 0 \\ 0, & a^2b^2a\beta, & 0 \\ 0, & 0, & a^3bca \end{bmatrix},$$

so that $C = AB$, where a, b, c, a, β are distinct irresoluble functions. Then the maximum factors of orders 1, 2, 3 of A, B, C are respectively

$$a,\ a^3b,\ a^6b^2c;\ \ a,\ ba^2\beta,\ b^2a^3\beta^2;\ \ aba,\ a^3b^2a^2\beta,\ a^6b^4ca^3\beta^2;$$

and the potent factors of orders 1, 2, 3 of A, B, C are respectively

$$a,\ a^2b,\ a^3bc;\ \ a,\ ba\beta,\ ba\beta:\ \ aba\ \ a^2ba\beta,\ a^3b^2ca\beta.$$

Here the potent factor $ab\alpha$ of C of order 1 is divisible by a and α, but is not equal to the product $a \times \alpha$; and the potent factor $a^2 b\alpha\beta$ of C of order 2 is divisible by $a^2 b$ and $b\alpha\beta$, but is not divisible by the product $a^2 b \times b\alpha\beta$.

Ex. ii. *If either one of the matrices A and B in Theorem I has rank r, then every irresoluble (or irreducible) divisor of C must be an irresoluble (or irreducible) divisor of at least one of the matrices A and B.*

It will be sufficient to consider the case in which A has rank r. In this case by § 131 the matrices B and C have the same rank which we will denote by s.

Let t be a rational integral function of the variables $x, y, z, \ldots$ which is irresoluble (or irreducible in Ω) and is not a divisor of either of the matrices A and B. By Ex. i of § 202 there exist finite roots of t for which A and B have ranks r and s respectively. Therefore by § 131 there exist finite roots of t for which C has rank s. Therefore by Ex. i of § 202 the function t cannot be an irresoluble (or irreducible) divisor of C.

Ex. iii. *If the matrix A in Theorem I has rank r, and if t is an irresoluble (or irreducible) divisor of B which is not also a divisor of A, then the matrices B and C have the same potent divisors of all orders corresponding to the divisor t.*

Let the common rank of B and C be s, and let d_i be the index of the highest power of t which is a common factor of all minor determinants of B of order i, i being any one of the integers $1, 2, \ldots s$.

Equating correspondingly formed complete matrices of the minor determinants of order i on both sides of the equation (A), we obtain an identical equation in $x, y, z, \ldots$ of the form

$$[C]_\mu^\nu = [A]_\mu^\rho [B]_\rho^\nu, \quad \text{where} \quad \mu = \binom{m}{i}, \quad \nu = \binom{n}{i}, \quad \rho = \binom{r}{i}, \quad \ldots\ldots\ldots\ldots(4)$$

and where $[B]_\rho^\nu$ and $[C]_\mu^\nu$ have rank $\binom{s}{i}$, and $[A]_\mu^\rho$ has rank ρ.

Since t^{d_i} is a factor of every element of $[B]_\rho^\nu$, it is a factor of every element of $[C]_\mu^\nu$, and we can write

$$[C]_\mu^\nu = t^{d_i} [C']_\mu^\nu, \quad [B]_\rho^\nu = t^{d_i} [B']_\rho^\nu,$$

and replace the equation (4) by

$$[C']_\mu^\nu = [A]_\mu^\rho [B']_\rho^\nu, \quad \ldots\ldots\ldots\ldots\ldots\ldots\ldots\ldots(5)$$

where t is not a common factor of all elements of $[B']_\rho^\nu$.

We will now suppose that t is an *irreducible* divisor of B in Ω, so that $[B']_\rho^\nu$ and $[C']_\mu^\nu$ are rational integral functional matrices in Ω.

Since t is not a divisor of A, there are finite roots of t for which A has rank r; therefore by § 73 there are finite roots of t for which $[A]_\mu^\rho$ has rank ρ; therefore t is not one of the irreducible divisors of $[A]_\mu^\rho$. Hence $[A]_\mu^\rho$ has at least one minor determinant of order ρ which does not vanish identically and is not divisible by t.

And because t is not a common factor of all elements of $[B']_\rho^\nu$, therefore $[B']_\rho^\nu$ has at least one element which does not vanish identically and is not divisible by t.

It follows by Ex. ix of § 189 that there exist finite roots of t for which $[A]_\mu^\rho$ has a non-vanishing minor determinant of order ρ and $[B']_\rho^\nu$ has a non-vanishing element, i.e. for which $[A]_{\mu}^{\rho}$ has rank ρ and $[B']_{\rho}^{\nu}$ has rank not less than 1; and for such roots of t we see from (5) that $[C']_{\mu}^{\nu}$ has rank not less than 1. Thus there exist finite roots of t for which $[C']_{\mu}^{\nu}$ does not vanish; and therefore t is not a factor of the elements of $[C']_{\mu}^{\nu}$. Consequently t^{d_i} is the highest power of t which is a common factor of all the elements of $[C]_{\mu}^{\nu}$, i.e. d_i is also the index of the highest power of t which is a common factor of all the minor determinants of C of order i.

Thus the maximum indices of t of orders 1, 2, ... s are the same for the matrix C as for the matrix B, and therefore the potent indices of t of orders 1, 2, ... s are the same for the matrix C as for the matrix B; i.e. the matrices B and C have the same potent divisors of all orders corresponding to the irreducible divisor t.

When the divisor t is *irresoluble*, we obtain the same result by taking Ω to be the domain of all scalar numbers.

Ex. iv. *If the matrix B in Theorem I has rank r, and if t is an irresoluble (or irreducible) divisor of A which is not also a divisor of B, then the matrices A and C have the same potent divisors of all orders corresponding to the divisor t.*

The proof of Ex. iv is similar to that of Ex. iii.

Ex. v. *If either one of the factor matrices A and B in Theorem I has rank r and moreover has no potent divisors (i.e. no irresoluble or no irreducible divisors), then the product matrix C has the same rank and the same potent divisors of all orders as the other factor matrix.*

This follows from Exs. iii and iv.

Ex. vi. *If $M = ABC \dots K$ is a standard product of any number of rational integral functional matrices, then:*

(1) *The maximum factor of any order s of the product matrix M is divisible by the product of the maximum factors of order s of the factor matrices $A, B, C, \dots K$.*

(2) *The potent factor of any order s of the product matrix M is divisible by the potent factor of order s of each of the factor matrices $A, B, C, \dots K$.*

This follows by repeated applications of Theorem I.

Theorem II. *If there exists an identical equation of the form*

$$[h]_m^p \, [a]_p^q \, [k]_q^n = [b]_m^n, \quad \text{or} \quad HAK = B, \qquad \dots\dots\dots\dots\dots \text{(B)}$$

where $A = [a]_p^q$, $B = [b]_m^n$, $H = [h]_m^p$, $K = [k]_q^n$ are matrices whose elements are rational integral functions of certain scalar variables $x, y, z, \dots$, and where the matrices H and K have ranks p and q equal to their respective passivities, then:

(1) *The matrices A and B have the same rank.*

(2) *The irresoluble (or irreducible) divisors of B are the irresoluble (or irreducible) divisors of the matrices H, A and K.*

(3) *If t is an irresoluble (or irreducible) divisor of B which is not a divisor of either H or K, then the matrices A and B have the same potent divisors of all orders corresponding to the irresoluble (or irreducible) divisor t.*

By an irreducible divisor is meant one which lies in and is irreducible in any domain of rationality Ω in which all four of the matrices A, B, H, K lie, i.e. in which the three matrices H, A, K lie.

The first part of the theorem has been proved in § 131.

The second part of the theorem follows from Ex. ii. For by Ex. ii the irresoluble (or irreducible) divisors of B are those of H and AK, and the irresoluble (or irreducible) divisors of AK are those of A and K.

The third part of the theorem follows from Exs. iii and iv, and can also be proved directly in the same way as Ex. iii. By Ex. iii the matrices B and AK have the same potent divisors of all orders corresponding to t, and by Ex. iv the matrices AK and A have the same potent divisors of all orders corresponding to t. Therefore the matrices B and A have the same potent divisors of all orders corresponding to the irresoluble (or irreducible) divisor t.

Ex. vii. *If in the equation* (B) *the matrices H and K have ranks p and q, and if each of them has no potent divisors, then the matrix B has the same rank and the same potent divisors of all orders as the matrix A.*

This follows from Theorem II as a particular case.

This case occurs when and only when each of the matrices H and K has no irresoluble (or no irreducible) divisors. The matrix B then has the same irresoluble (and the same irreducible) divisors as the matrix A; and it also has the same maximum divisors, the same maximum factors and the same potent factors of all orders as the matrix A. It is said to be *equipotent* with A.

§ 207. The irreducible divisors and potent divisors of a compartite matrix.

Theorem I. *The irreducible (or irresoluble) divisors of a compartite matrix are the irreducible (or irresoluble) divisors of its various parts.*

It will be sufficient to prove this theorem for the irreducible divisors; and there will be no loss of generality in supposing the compartite matrix to be in standard form.

Let
$$\phi = \left[\begin{matrix} a, & 0, & \dots & 0 \\ 0, & b, & \dots & 0 \\ \dots & \dots & \dots & \dots \\ 0, & 0, & \dots & c \end{matrix}\right]_{m,\,p,\,\dots\,u}^{n,\,q,\,\dots\,v}$$

be a compartite matrix in standard form whose elements are rational integral functions in Ω of certain scalar variables $x, y, z, \dots$, Ω being any domain of rationality; and let the ranks of its parts $[a]_m^n, [b]_p^q, \dots [c]_u^v$ be $\alpha, \beta, \dots \gamma$, so that by § 100 the rank of ϕ is

$$\rho = \alpha + \beta + \dots + \gamma.$$

The minor determinants of ϕ of order ρ which do not vanish identically are those of the forms

$$\Delta_\rho = A_\alpha B_\beta \dots C_\gamma,$$

where A_α, B_β, ... C_γ are minor determinants of $[a]_m^n$, $[b]_p^q$, ... $[c]_u^v$ of orders α, β, ... γ which do not vanish identically.

Let t be any irreducible function in Ω.

Then if t is an irreducible divisor of the part $[a]_m^n$, it is a factor of all such determinants as A_α, and therefore a factor of all such determinants as Δ_ρ. Thus if t is an irreducible divisor of any one of the parts of ϕ, it is an irreducible divisor of ϕ.

On the other hand if t is not an irreducible divisor of any one of the parts $[a]_m^n$, $[b]_p^q$, ... $[c]_u^v$ of ϕ, we can determine A_α, B_β, ... C_γ so that no one of them is divisible by t; and then Δ_ρ is not divisible by t, and t is not an irreducible divisor of ϕ.

Consequently t is an irreducible divisor of ϕ when and only when it is an irreducible divisor of at least one of the parts of ϕ.

Theorem II a. *The potent divisors of the compartite matrix*

$$\phi = \begin{bmatrix} a, & 0 \\ 0, & b \end{bmatrix}_{m,\,p}^{n,\,q}$$

are the potent divisors of the two parts $A = [a]_m^n$ *and* $B = [b]_p^q$.

We assume that ϕ is a matrix whose elements are rational integral functions in Ω of the variables $x, y, z, \ldots, \Omega$ being any domain of rationality, and we will show that this theorem is true for the potent divisors of ϕ corresponding to its various irreducible divisors in Ω. It will then follow as a particular case that the theorem is true for the potent divisors of ϕ corresponding to its various irresoluble divisors.

Let the parts A and B have ranks α and β, so that ϕ has rank ρ, where $\rho = \alpha + \beta$; and let t be any one of the irreducible divisors in Ω of ϕ, so that t by Theorem I is any rational integral function in Ω which is an irreducible divisor in Ω of at least one of the parts A and B.

Let $d_1', d_2', \ldots d_\alpha'$ and $e_1', e_2', \ldots e_\alpha'$ be the maximum and potent indices of t of orders $1, 2, \ldots \alpha$ for the matrix A. When t is an irreducible divisor of B only, these quantities all have the value 0.

Let $d_1'', d_2'', \ldots d_\beta''$ and $e_1'', e_2'', \ldots e_\beta''$ be the maximum and potent indices of t of orders $1, 2, \ldots \beta$ for the matrix B. When t is an irreducible divisor of A only, these quantities all have the value 0.

Let $d_1, d_2, \ldots d_\rho$ be the maximum indices of t of orders $1, 2, \ldots \rho$ for the matrix ϕ, and let $d_0 = 0$.

Let $e_1, e_2, \dots e_\rho$ be the integers $e_1', e_2', \dots e_\alpha', e_1'', e_2'', \dots e_\beta''$ arranged in ascending order of magnitude, i.e. so arranged that

$$e_1 \not> e_2 \not> e_3 \not> \dots \not> e_{\rho-2} \not> e_{\rho-1} \not> e_\rho.$$

We will prove the theorem by showing that $e_1, e_2, \dots e_\rho$ are the potent indices of t of orders $1, 2, \dots \rho$ for the matrix ϕ; i.e. by showing that if $s \not< 1$ and $s \not> \rho$, then $e_s = d_s - d_{s-1}$. The special case in which $s = 1$ is considered in Ex. i below.

Ex. i. *The potent index of t of order 1 for the matrix ϕ is e_1.*

The potent index of t of order 1 for the matrix ϕ is the index of the highest power of t which is a common factor of all the elements of ϕ, i.e. a common factor of all elements of A and B; and this is the smaller of the two integers d_1' and d_1'', i.e. the smaller of the two integers e_1' and e_1'', i.e. the integer e_1. We have therefore

$$e_1 = d_1 = d_1 - d_0.$$

If we consider that $A_0 = 1$ and $B_0 = 1$, and if s is any one of the integers $1, 2, \dots \rho$, the minor determinants of ϕ of order s which do not vanish identically are of the types

$$A_h B_{s-h}, \dots A_{u+1} B_{v-1},\ A_u B_v,\ A_{u-1} B_{v+1}, \dots A_{s-k} B_k, \quad \text{.........(a)}$$

where $u + v = s$; $A_h, \dots A_{u+1}, A_u, A_{u-1}, \dots A_{s-k}$ are minor determinants of A of orders $h, \dots u+1, u, u-1, \dots s-k$ which do not vanish identically; and $B_{s-h}, \dots B_{v-1}, B_v, B_{v+1}, \dots B_k$ are minor determinants of B of orders $s-h, \dots v-1, v, v+1, \dots k$ which do not vanish identically. We obtain all terms of the series (a) from the term $A_u B_v$ by giving to u and v all integral values consistent with the conditions

$$u + v = s, \quad u \not< 0, \quad u \not> \alpha, \quad v \not< 0, \quad v \not> \beta.$$

The possible values of u and v are given separately by

$$u \not< 0, \quad u \not< s - \beta, \quad u \not> \alpha, \quad u \not> s;$$

$$v \not< 0, \quad v \not< s - \alpha, \quad v \not> \beta, \quad v \not> s.$$

The integer h in the series (a) is the greatest possible value of u, and is the smaller of the two integers s and α; the integer k is the greatest possible value of v, and is the smaller of the two integers s and β; and it is assumed that $s \not< 2$, and $s \not> \rho$, so that $\rho \not< 2$.

The index of the highest power of t which is a common factor of minor determinants of ϕ of the type $A_u B_v$ in (a), where u and v are given integers, is $d_u' + d_v''$. Consequently d_s is the smallest of the integers

$$d_h' + d''_{s-h}, \dots d'_{u+1} + d''_{v-1},\ d_u' + d_v'',\ d'_{u-1} + d''_{v+1}, \dots d'_{s-k} + d_k''. \quad \text{...(a')}$$

Now since $e'_{u+2} \not< e'_{u+1}$, $e_v'' \not< e''_{v-1}$ and $e''_{v+2} \not< e''_{v+1}$, $e_u' \not< e'_{u-1}$, we see that

if $e'_{u+1} > e_v''$, then $e'_{u+2} > e''_{v-1}$; and if $e''_{v+1} > e_u'$, then $e''_{v+2} > e'_{u-1}$.

Therefore if $d_u' + d_v'' < d'_{u+1} + d''_{v-1}$, then $d'_{u+1} + d''_{v-1} < d'_{u+2} + d''_{v-2}$;

and if $d_u' + d_v'' < d'_{u-1} + d''_{v+1}$, then $d'_{u-1} + d''_{v+1} < d'_{u-2} + d''_{v+2}$.

Hence those terms of the series (a′) *which have the smallest value* d_s *are consecutive terms; the terms preceding them constantly diminish from the first; and the terms following them constantly increase up to the last.* ......(A)

We observe further that if $d_u' + d_v'' = d'_{u+1} + d''_{v-1}$, *then* $e'_{u+1} = e_v''$. ...(B)

Now let $d_u' + d_v''$ be any one of the terms in the series (a′) which has the smallest value d_s. Then in general we have

$$d_u' + d_v'' \not> d'_{u+1} + d''_{v-1} \text{ and } d_u' + d_v'' \not> d'_{u-1} + d''_{v+1},$$

i.e. $e'_{u+1} \not< e_v''$, and $e''_{v+1} \not< e_u'$. ......(1)

Since also $e'_{u+1} \not< e_u'$, and $e''_{v+1} \not< e_v''$, ...(2)

we see that neither e'_{u+1} nor e''_{v+1} is less than any one of the numbers $e_1', e_2', \ldots e_u', e_1'', e_2'', \ldots e_v''$.

Thus if $d_u' + d_v''$ *is any one of the terms in the series* (a′) *which has the smallest value* d_s, *then* $e_1', e_2', \ldots e_u', e_1'', e_2'', \ldots e_v''$ *are the integers* $e_1, e_2, \ldots e_s$ *arranged in some order.* ......(C)

In proving this result we have implicitly assumed that $d_u' + d_v''$ is neither the first nor the last term in the series (a′); but (as is shown in Exs. ii—iv) the result is true in all cases.

Ex. ii. Suppose that there are at least two terms in the series (a′), and that the first term has the smallest value d_s.

If $h=s$, the first two terms of (a′) are d_s', $d'_{s-1}+d_1''$; and we have $d_s' \not> d'_{s-1}+d_1''$, or $e_1'' \not< e_s'$. Thus e_1'' is not less than any one of the numbers $e_1', e_2', \ldots e_s'$; and therefore these are the numbers $e_1, e_2, \ldots e_s$. In this special case of (C) we have $u=s$, $v=0$.

If $h=a$, the first two terms of (a′) are $d_a'+d''_{s-a}$, $d'_{a-1}+d''_{s-a+1}$; and we have

$$d_a'+d''_{s-a} \not> d'_{a-1}+d''_{s-a+1}, \text{ or } e''_{s-a+1} \not< e_a'.$$

Thus e''_{s-a+1} is not less than any one of the numbers $e_1', e_2', \ldots e_a', e_1'', e_2'', \ldots e''_{s-a}$; and therefore these are the numbers $e_1, e_2, \ldots e_s$ arranged in some order. In this special case of (C) we have $u=a$, $v=s-a$.

In both cases the result (C) is true.

Ex. iii. Suppose that there are at least two terms in the series (a′), and that the last term has the smallest value d_s.

The proof that the result (C) is true in these special cases is similar to the proof of Ex. ii.

Ex. iv. Suppose that there is only one term in the series (a′).

If $h=s$, we have $k=0$, $\beta=0$, and the series (a′) consists of the single term d_s'. Further $e_1, e_2, \ldots e_s$ are the numbers $e_1', e_2', \ldots e_s'$; and the result (C) is true.

If $h=a$, we have $k=s-a=\beta$, $s=a+\beta$, and the series (a′) consists of the single term $d_a'+d_\beta''$. Further $e_1, e_2, \dots e_s$ are the numbers $e_1', e_2', \dots e_a', e_1'', e_2'', \dots e_\beta''$ arranged in some order; and the result (C) is true.

From (C) we see that if $e_u' \not< e_v''$, then $e_s = e_u'$; and if $e_v'' \not< e_u'$, then $e_s = e_v''$. Our object is to show that in both cases we have $e_s = d_s - d_{s-1}$ when $s \not< 2$, this having already been shown in Ex. i to be true when $s = 1$. To do this we must consider the value of d_{s-1}; and this can be obtained from the series which replace (a) and (a′) when s is replaced by $s-1$.

We may suppose without loss of generality that $\alpha \not< \beta$. Then there are three possible cases.

CASE I. $s \not< 2$, $s \not> \beta$.

The non-vanishing minor determinants of ϕ of order s are of the types

$$A_{h-1}B_{s-h}, \dots A_u B_{v-1},\ A_{u-1}B_v,\ A_{u-2}B_{v+1}, \dots A_{s-k}B_{k-1}\,; \quad \dots\dots(\mathrm{b}_1)$$

and d_{s-1} is the smallest of the integers

$$d'_{h-1}+d''_{s-h}, \dots d_u'+d''_{v-1},\ d'_{u-1}+d_v'',\ d'_{u-2}+d''_{v+1}, \dots d'_{s-k}+d''_{k-1}. \quad \dots(\mathrm{b}_1')$$

In this case we have $h=s$ and $k=s$ both in (a′) and in (b₁′); and the number of terms in (b₁′) is less by one than the number of terms in (a′).

CASE II. $s \not< 2$, $s > \beta$, $s \not> \alpha$.

The non-vanishing minor determinants of ϕ of order s are of the types

$$A_{h-1}B_{s-h}, \dots A_u B_{v-1},\ A_{u-1}B_v,\ A_{u-2}B_{v+1}, \dots A_{s-k-1}B_k\,; \quad \dots\dots(\mathrm{b}_2)$$

and d_{s-1} is the smallest of the integers

$$d'_{h-1}+d''_{s-h}, \dots d_u'+d''_{v-1},\ d'_{u-1}+d_v'',\ d'_{u-2}+d''_{v+1}, \dots d'_{s-k-1}+d_k''. \quad \dots(\mathrm{b}_2')$$

In this case we have $h=s$ and $k=\beta$ both in (a′) and in (b₂′); and the number of terms in (b₂′) is equal to the number of terms in (a′).

CASE III. $s \not< 2$, $s > \alpha$, $s \not> \rho$.

The non-vanishing minor determinants of ϕ of order s are of the types

$$A_h B_{s-h-1}, \dots A_{u+1}B_{v-2},\ A_u B_{v-1},\ A_{u-1}B_v, \dots A_{s-k-1}B_k\,; \quad \dots\dots(\mathrm{b}_3)$$

and d_{s-1} is the smallest of the integers

$$d_h'+d''_{s-h-1}, \dots d'_{u+1}+d''_{v-2},\ d_u'+d''_{v-1},\ d'_{u-1}+d_v'', \dots d'_{s-k-1}+d_k''. \quad \dots(\mathrm{b}_3')$$

In this case we have $h=\alpha$ and $k=\beta$ both in (a′) and in (b₃′); and the number of terms in (b₃′) is greater by one than the number of terms in (a′).

In each of the series (b₁′), (b₂′), (b₃′) those terms which have the smallest value d_{s-1} are consecutive; the preceding terms constantly diminish from the first; and the following terms constantly increase up to the last.

As before let $d_u'+d_v''$ be one of the terms in the series (a′) which has the smallest value d_s, where now $s \not< 2$, $s \not> \rho$.

First suppose that $e_u' \nless e_v''$, *or* $d'_{u-1} + d_v'' \ngtr d_u' + d''_{v-1}$, *so that* $e_s = e_u'$.

Then if $d'_{u-1} + d_v''$ is the last term in (b_1'), (b_2') or (b_3'), the above inequality shows that it must have the smallest value d_{s-1}; and if $d'_{u-1} + d_v''$ occurs in (b_1'), (b_2') or (b_3'), and is not the last term, the above inequality and the inequality $e''_{v+1} \nless e'_{u-1}$, or $d'_{u-1} + d_v'' \ngtr d'_{u-2} + d''_{v+1}$, which follows from the second inequality in (1), show that $d'_{u-1} + d_v''$ has the value d_{s-1}.

Thus if the term $d'_{u-1} + d_v''$ occurs in (b_1'), (b_2') or (b_3'), we have

$$d_s = d_u' + d_v'',\ d_{s-1} = d'_{u-1} + d_v'';\ \text{and } d_s - d_{s-1} = d_u' - d'_{u-1} = e_u' = e_s.$$

Now the term $d'_{u-1} + d_v''$ always does occur in the series (b_2') and (b_3'); it always does occur in (b_1') except when $d_u' + d_v''$ is the last term in (a'); and it is shown in Ex. v that the equation $d_s - d_{s-1} = e_s$ is still true in the last exceptional case.

Consequently we always have $e_s = d_s - d_{s-1}$.

Next suppose that $e_v'' \nless e_u'$, *or* $d_u' + d''_{v-1} \ngtr d'_{u-1} + d_v''$, *so that* $e_s = e_v''$.

Then if $d_u' + d''_{v-1}$ is the first term in (b_1'), (b_2') or (b_3'), the above inequality shows that it must have the smallest value d_{s-1}; and if $d_u' + d''_{v-1}$ occurs in (b_1'), (b_2') or (b_3'), and is not the first term, the above inequality and the inequality $e'_{u+1} \nless e''_{v-1}$, or $d_u' + d''_{v-1} \ngtr d'_{u+1} + d''_{v-2}$, which follows from the first inequality in (1), show that $d_u' + d''_{v-1}$ has the smallest value d_{s-1}.

Thus if the term $d_u' + d''_{v-1}$ occurs in (b_1'), (b_2') or (b_3'), we have

$$d_s = d_u' + d_v'',\ d_{s-1} = d_u' + d''_{v-1};\ \text{and } d_s - d_{s-1} = d_v'' - d''_{v-1} = e_v'' = e_s.$$

Now the term $d_u' + d''_{v-1}$ always does occur in the series (b_3'); it always does occur in the series (b_1') and (b_2') except when $d_u' + d_v''$ is the first term in (a'); and it is shown in Ex. vi that the equation $d_s - d_{s-1} = e_s$ is still true in the last two exceptional cases.

Consequently we always have $e_s = d_s - d_{s-1}$.

We have now shown that the equation $e_s = d_s - d_{s-1}$ is always true when $s \nless 2$ and $s \ngtr \rho$; and in Ex. i we have shown that it is true when $s = 1$ if $\rho \nless 1$. Therefore $e_1, e_2, \dots e_\rho$ are the potent indices of t of orders $1, 2, \dots \rho$ for the matrix ϕ; and this establishes Theorem II *a*.

Ex. v. In Case I suppose that $d_u' + d_v''$ has the value d_s, and is the last term in (a'). Since the last term in (a') is d_s'', we now have $u = 0$, $v = s$, and $d_s'' \ngtr d_1' + d''_{s-1}$, or $e_1' \nless e_s''$. Therefore $e_1, e_2, \dots e_s$ are the integers $e_1'', e_2'', \dots e_s''$.

The last term in (b_1') is d''_{s-1}, and the inequality $e_1' \nless e''_{s-1}$, or $d''_{s-1} \ngtr d_1' + d''_{s-2}$, which follows from the inequality $e_1' \nless e_s''$, shows that the last term d''_{s-1} in (b_1') has the smallest value d_{s-1}.

Therefore $d_s = d_s''$, $d_{s-1} = d''_{s-1}$; and $d_s - d_{s-1} = d_s'' - d''_{s-1} = e_s'' = e_s$.

Ex. vi. In Case I or II suppose that $d_u'+d_v''$ has the value d_s, and is the first term in (a'). Since the first term in (a') is d_s', we now have $u=s$, $v=0$, and $d_s' \not> d'_{s-1}+d_1''$, or $e_1'' \not< e_s'$. Therefore $e_1, e_2, \ldots e_s$ are the integers $e_1', e_2', \ldots e_s'$.

The first term in (b_1') or (b_2') is d'_{s-1}, and the inequality $e_1'' \not< e'_{s-1}$, or $d'_{s-1} \not> d'_{s-2}+d_1''$, which follows from the inequality $e_1'' \not< e_s'$, shows that the first term d'_{s-1} in (b_1') or (b_2') has the smallest value d_{s-1}.

Therefore $d_s=d_s'$, $d_{s-1}=d'_{s-1}$; and $d_s-d_{s-1}=d_s'-d'_{s-1}=e_s'=e_s$.

Theorem II b. *The potent divisors of any compartite matrix are the potent divisors of its several parts.*

We will suppose that

$$\phi = \begin{bmatrix} a, & 0, & 0, & \ldots 0 \\ 0, & b, & 0, & \ldots 0 \\ 0, & 0, & c, & \ldots 0 \\ \multicolumn{4}{c}{\ldots\ldots\ldots\ldots} \\ 0, & 0, & 0, & \ldots d \end{bmatrix}_{m,\, p,\, u,\, \ldots h}^{n,\, q,\, v,\, \ldots k}$$

is a compartite matrix in standard form whose elements are rational integral functions in Ω of the variables $x, y, z, \ldots$, Ω being any domain of rationality, and that t is any irreducible divisor in Ω of the matrix ϕ, so that t by Theorem I is any irreducible divisor in Ω of at least one of the parts of ϕ. In the particular case when we take Ω to be the domain of all scalar numbers t is any irresoluble divisor of ϕ.

Let the ranks of the parts $A=[a]_m^n$, $B=[b]_p^q$, $C=[c]_u^v, \ldots D=[d]_h^k$ be $\alpha, \beta, \gamma, \ldots \delta$, so that the rank of ϕ is ρ, where $\rho=\alpha+\beta+\gamma+\ldots+\delta$.

By Theorem II a the ρ potent indices of t for the matrix ϕ are the α potent indices of t for the matrix A and the $\beta+\gamma+\ldots+\delta$ potent indices of t for the matrix

$$\phi' = \begin{bmatrix} b, & 0, & \ldots 0 \\ 0, & c, & \ldots 0 \\ \multicolumn{3}{c}{\ldots\ldots\ldots} \\ 0, & 0, & \ldots d \end{bmatrix}_{p,\, u,\, \ldots h}^{q,\, v,\, \ldots k}.$$

By the same theorem the $\beta+\gamma+\ldots+\delta$ potent indices of t for the matrix ϕ' are the β potent indices of t for the matrix B and the $\gamma+\ldots+\delta$ potent indices of t for the matrix

$$\phi'' = \begin{bmatrix} c, & \ldots 0 \\ \multicolumn{2}{c}{\ldots\ldots} \\ 0, & \ldots d \end{bmatrix}_{u,\, \ldots h}^{v,\, \ldots k}.$$

Proceeding in this way we see that the ρ potent indices of t for the matrix ϕ are the α potent indices of t for the matrix A, the β potent indices of t for the matrix B, the γ potent indices of t for the matrix $C, \ldots$ and the δ potent indices of t for the matrix D; and if $e_1, e_2, \ldots e_\rho$ are the potent indices of t for the several parts of ϕ arranged in ascending order of magnitude, then these are the potent indices of t of orders 1, 2, ... ρ for the matrix ϕ.

If e_i is the first of the integers $e_1, e_2, \ldots e_\rho$ which is not 0, then the potent divisors of ϕ corresponding to the irreducible (or irresoluble) divisor t are

$$t^{e_i},\ t^{e_{i+1}}, \ldots t^{e_\rho} \quad \text{of orders } i, i+1, \ldots \rho;$$

and these are the potent divisors corresponding to t of the various parts of ϕ so arranged that their indices are in ascending order of magnitude.

NOTE. Theorem IIb is true both when the compartite matrix ϕ is and when it is not in standard form. There is clearly no loss of generality in supposing it to be in standard form.

Ex. vii. *Potent divisors of the undegenerate square matrix of order m,*

$$\phi = \lambda\,[1]_m^m + \mu \begin{bmatrix} 0, & 1 \\ 0, & 0 \end{bmatrix}_{m-p,\ p}^{p,\ m-p} = \begin{bmatrix} \lambda & 0 & 0 & \ldots & \mu & 0 & 0 & \ldots & 0 & 0 \\ 0 & \lambda & 0 & \ldots & 0 & \mu & 0 & \ldots & 0 & 0 \\ 0 & 0 & \lambda & \ldots & 0 & 0 & \mu & \ldots & 0 & 0 \\ \ldots & & & & & & & & & \\ 0 & 0 & 0 & \ldots & \lambda & 0 & 0 & \ldots & \mu & 0 \\ 0 & 0 & 0 & \ldots & 0 & \lambda & 0 & \ldots & 0 & \mu \\ 0 & 0 & 0 & \ldots & 0 & 0 & \lambda & \ldots & 0 & 0 \\ \ldots & & & & & & & & & \\ 0 & 0 & 0 & \ldots & 0 & 0 & 0 & \ldots & \lambda & 0 \\ 0 & 0 & 0 & \ldots & 0 & 0 & 0 & \ldots & 0 & \lambda \end{bmatrix}$$

in which all elements are 0 *except those lying in the leading diagonal and those which lie in any one row parallel to the leading diagonal and separated from the leading diagonal by* $p-1$ *rows of* 0's.

We regard λ and μ as distinct irreducible rational integral functions in a domain of rationality Ω of certain scalar variables $x, y, z, \ldots$. They may however be independent variables, in which case Ω is the domain of all rational numbers. The maximum and potent factors of orders 1, 2, ... m of ϕ will be denoted by $D_1, D_2, \ldots D_m$ and $E_1, E_2, \ldots E_m$ respectively. The potent divisors will be those corresponding to the irreducible divisors of ϕ in Ω (of which there is only one, viz. λ).

We consider that p and m are positive integers such that $p \nless 1$, $p \ngtr m$. We exclude the case $p=0$ in which ϕ is the scalar matrix $(\lambda+\mu).[1]_m^m$. We can then write

$$m = np + q,$$

where p, q, n are positive integers such that

$$p \nless 1,\quad n \nless 1,\quad q \nless 0,\quad q \ngtr p-1;$$

and (the constituent matrices being scalar) we can express ϕ in the form

$$\phi = \begin{bmatrix} \lambda, & \mu, & 0, & \dots & 0, & 0, & 0 \\ 0, & \lambda, & \mu, & \dots & 0, & 0, & 0 \\ 0, & 0, & \lambda, & \dots & 0, & 0, & 0 \\ \dots & \dots & \dots & \dots & \dots & \dots & \dots \\ 0, & 0, & 0, & \dots & \lambda, & \mu, & 0 \\ 0, & 0, & 0, & \dots & 0, & \lambda, & \mu.\epsilon \\ 0, & 0, & 0, & \dots & 0, & 0, & \lambda.1 \end{bmatrix}_{p, p, p, \dots p, p, q}^{p, p, p, \dots p, p, q}, \quad \dots\dots(3)$$

where

$$[\epsilon]_p^q = \begin{bmatrix} 1 \\ 0 \end{bmatrix}_{q, p-q}^{q}. \quad \dots\dots(4)$$

Let i be any one of the integers $1, 2, \dots p-q$. Then if we regard the $(q+i)$th vertical row in each of the first n sets of vertical rows in (3) and the $(q+i)$th horizontal row in each of the first n sets of horizontal rows in (3), we see that the elements common to these n vertical and horizontal rows of ϕ form the square matrix of order n

$$[\omega]_n^n = \begin{bmatrix} \lambda & \mu & 0 & \dots & 0 & 0 \\ 0 & \lambda & \mu & \dots & 0 & 0 \\ \dots & \dots & \dots & \dots & \dots & \dots \\ 0 & 0 & 0 & \dots & \lambda & \mu \\ 0 & 0 & 0 & \dots & 0 & \lambda \end{bmatrix},$$

and that all other elements in these rows of ϕ are 0's.

Again let i be any one of the integers $1, 2, \dots q$. Then if we regard the ith vertical row in each of the $n+1$ sets of vertical rows in (3) and the ith horizontal row in each of the $n+1$ sets of horizontal rows in (3), we see that the elements common to these $n+1$ sets of vertical and horizontal rows of ϕ form the square matrix of order $n+1$

$$[\omega]_{n+1}^{n+1} = \begin{bmatrix} \lambda & \mu & 0 & \dots & 0 & 0 \\ 0 & \lambda & \mu & \dots & 0 & 0 \\ \dots & \dots & \dots & \dots & \dots & \dots \\ 0 & 0 & 0 & \dots & \lambda & \mu \\ 0 & 0 & 0 & \dots & 0 & \lambda \end{bmatrix},$$

and that all other elements in these rows of ϕ are 0's.

Consequently ϕ is a compartite matrix having exactly p parts of which $p-q$ are equal to $[\omega]_n^n$ and q are equal to $[\omega]_{n+1}^{n+1}$; and we can convert ϕ by a symmetric derangement of its horizontal and vertical rows into a compartite matrix of standard form whose parts are those just described. By Ex. viii of § 205 each of the parts $[\omega]_n^n$ has only the one potent divisor λ^n, and each of the parts $[\omega]_{n+1}^{n+1}$ has only the one potent divisor λ^{n+1}. Hence from Theorem II *b* we conclude that:

The matrix ϕ has only the one irreducible divisor λ, and it has exactly p potent divisors of which $p-q$ are equal to λ^n and q are equal to λ^{n+1}.

It follows that the potent factors and the maximum factors of ϕ are

$$E_1 = E_2 = \dots = E_{m-p} = 1;\ E_{m-p+1} = E_{m-p+2} = \dots = E_{m-q} = \lambda^n;$$
$$E_{m-q+1} = E_{m-q+2} = \dots = E_m = \lambda^{n+1}; \quad \dots\dots(5)$$

$$D_1 = D_2 = \dots = D_{m-p} = 1;\ D_{m-p+1} = \lambda^n,\ D_{m-p+2} = \lambda^{2n},\ \dots\ D_{m-q} = \lambda^{(p-q)n};$$
$$D_{m-q+1} = \lambda^{(p-q+1)n+1},\ D_{m-q+2} = \lambda^{(p-q+2)n+2},\ \dots\ D_m = \lambda^{pn+q} = \lambda^m. \quad \dots(6)$$

In the particular case when $q=0$, so that $m=np$, the matrix ϕ has exactly p potent divisors each of which is equal to λ^n. We obtain the matrix of Ex. viii in § 205 when $q=0$, $p=1$, $m=n$.

Ex. viii. *Conjugate reciprocal of the matrix* ϕ *of Ex.* vii.

It can be at once verified by actual multiplication that the conjugate reciprocal of ϕ is the square matrix

$$\Phi=\left[\begin{matrix} \lambda^{m-1}, & -\lambda^{m-2}\mu, & \lambda^{m-3}\mu^2, & \ldots & (-1)^{n-1}\lambda^{m-n}\mu^{n-1}, & (-1)^n\lambda^{m-n-1}\mu^n\,.\,\epsilon \\ 0, & \lambda^{m-1}, & -\lambda^{m-2}\mu, & \ldots & (-1)^{n-2}\lambda^{m-n+1}\mu^{n-2}, & (-1)^{n-1}\lambda^{m-n}\mu^{n-1}\,.\,\epsilon \\ 0, & 0, & \lambda^{m-1}, & \ldots & (-1)^{n-3}\lambda^{m-n+2}\mu^{n-3}, & (-1)^{n-2}\lambda^{m-n+1}\mu^{n-2}\,.\,\epsilon \\ \cdots & \cdots & \cdots & \cdots & \cdots & \cdots \\ 0, & 0, & 0, & \ldots & \lambda^{m-1}, & -\lambda^{m-2}\mu\,.\,\epsilon \\ 0, & 0, & 0, & \ldots & 0, & \lambda^{m-1}\,.\,1 \end{matrix}\right]^{p,\,p,\,p,\,\ldots\,p,\,q}_{p,\,p,\,p,\,\ldots\,p,\,q},$$

where as before $$[\epsilon]_p^q=\begin{bmatrix}1\\0\end{bmatrix}^q_{q,\,p-q}$$

Since the elements of Φ are the minor determinants of ϕ of order $n-1$, it follows that

$$D_{m-1}=\lambda^{m-n-1} \text{ when } q\neq 0, \qquad D_{m-1}=\lambda^{m-n} \text{ when } q=0;$$

and because $D_m=\lambda^m$ in all cases, it follows that

$$E_m=\lambda^{n+1} \text{ when } q\neq 0, \qquad E_m=\lambda^n \text{ when } q=0.$$

Because ϕ has only the one irreducible divisor λ, E_m is its one potent divisor of order m. These results agree with Ex. vii.

Ex. ix. *Minor determinants of all orders of the matrix* ϕ *of Ex.* vii.

The expansion of every non-vanishing minor determinant of ϕ is (when we disregard a possible negative sign) the product of a number of the elements λ and μ so chosen that no two of the elements lie in the same horizontal or vertical row of ϕ. Or again a general formula for any non-vanishing minor determinant Δ of ϕ is

$$\Delta=P_0Q_0P_1Q_1P_2Q_2\ldots P_{n-1}Q_{n-1}P_n, \quad \ldots\ldots\ldots\ldots(7)$$

where P_0, Q_0, P_1, Q_1, P_2, Q_2, ... P_{n-1}, Q_{n-1}, P_n are minor determinants of the $2n+1$ successive constituent matrices

$$\lambda[1]_q^q,\ \mu[\epsilon]_p^q,\ \lambda[1]_p^p,\ \mu[1]_p^p,\ \lambda[1]_p^p,\ \mu[1]_p^p,\ \ldots\ \lambda[1]_p^p,\ \mu[1]_p^p,\ \lambda[1]_p^p$$

of ϕ shown in (3), where we start at the bottom and pass upwards and towards the left, these determinants being so chosen that no two of them contain a horizontal or vertical row in common, and a determinant of order 0 being considered to have the value 1.

By solving the inequalities to be satisfied by the indices of λ and μ in the factors on the right in (7) it can be shown that:

The values of the non-vanishing minor determinants of ϕ *are given by the formula*

$$\Delta=\lambda^u\mu^v, \quad \ldots\ldots\ldots\ldots(8)$$

where u *and* v *are integers whose possible values are those consistent with the conditions*

$$u \nless 0, \quad v \nless 0, \quad u+v \ngtr np+q,$$

$$nv+(n-1)\,u \ngtr n\{(n-1)\,p+q\}, \quad (n+1)\,v+nu \ngtr n\{np+q\}.$$

Writing $u+v=i$, and eliminating v, it follows that:

The values of the non-vanishing minor determinants of ϕ of order i are given by the formula

$$\Delta_i = \lambda^u \mu^{i-u}, \qquad (9)$$

where u and i are integers whose possible values are those consistent with the conditions

$$u \not< 0, \quad u \not< (n+1)\,i - n\{np+q\}, \quad u \not> i, \quad u \not> ni - n\{(n-1)\,p+q\},$$

which include the necessary conditions $i \not< 0$, $i \not> np+q$.

Since $$ni - n\{(n-1)\,p+q\} = n\{i-m+p\},$$

and

$$(n+1)\,i - n\{np+q\} = (n+1)\,i - nm = n\,(i-m+p) + (i-m+q) = n\,(p-q) + (n+1)\,(i-m+q),$$

the possible values of u in formula (9) are dependent on the value of i in the manner shown in the following table:

	Value of i		*Possible values of u*	
Case I.	$i \not< 0,$	$i \not> m-p$	$u \not< 0,$	$u \not> i$
Case II.	$i \not< m-p,$	$i \not> m-q$	$u \not< n\,(i-m+p),$	$u \not> i$
Case III.	$i \not< m-q,$	$i \not> m$	$u \not< (n+1)\,i - nm,$	$u \not> i$

The values of $D_1, D_2, \ldots D_n$ can be constructed from this table, and we thus obtain independent proofs of the results of Ex. vii. The corresponding table for the matrix $[\omega]_n^n$ of Ex. vii is obtained by putting $q=0$, $p=1$, $m=n$.

§ 208. Irreducible divisors and maximum factors of a complete matrix of minor determinants.

In the examples which follow ϕ is a rational integral functional matrix of rank r lying in a domain of rationality Ω; Φ is a complete matrix of the minor determinants of ϕ of order s, where s is a given positive integer not greater than r; and we write

$$\phi = [a]_m^n, \quad \mu = \binom{m}{s}, \quad \nu = \binom{n}{s}, \quad \rho = \binom{r}{s}, \quad \Phi = [A]_\mu^\nu.$$

When the variables are arbitrary, the matrix Φ has rank ρ, the maximum and potent factors of ϕ of any order i will be denoted by D_i and E_i, and the maximum and potent factors of Φ of any order i will be denoted by D_i' and E_i'. When for particular values of the variables the matrix ϕ has rank

$$0,\ 1,\ \ldots\ s-1,\ s,\ \ldots\ i-1,\ i,\ \ldots\ r-2,\ r-1,\ r, \qquad (1)$$

then by § 73 the matrix Φ has the corresponding rank

$$0,\ 0,\ \ldots\ 0,\ 1,\ \ldots\ \binom{i-1}{s},\ \binom{i}{s},\ \ldots\ \binom{r-2}{s},\ \binom{r-1}{s},\ \binom{r}{s}. \qquad (1')$$

By the irreducible divisors of ϕ and Φ we mean those which lie in Ω and are irreducible in Ω.

Ex. i. *The matrix* Φ *has the same irreducible divisors as* ϕ.

Let t and T be irreducible divisors of ϕ and Φ respectively in their common domain of rationality Ω. Then for all finite roots of t the matrix ϕ has rank less than r, and therefore Φ has rank less than ρ; consequently t is an irreducible divisor of Φ. Again for all roots of T the matrix Φ has rank less than ρ, and therefore ϕ has rank less than r; consequently T is an irreducible divisor of ϕ.

Thus every irreducible divisor of ϕ is also an irreducible divisor of Φ, and conversely every irreducible divisor of Φ is also an irreducible divisor of ϕ.

Ex. ii. *If i is any integer not greater than r and not less than s, the irreducible divisors of Φ of each of the orders*

$$\binom{i-1}{s}+1,\quad \binom{i-1}{s}+2, \ldots \binom{i}{s}-1,\quad \binom{i}{s} \qquad \text{.......................(2)}$$

are identical with the irreducible divisors of ϕ of order i, i.e. with the irreducible factors of D_i and E_i.

Let j be any one of the integers (2); let t be any irreducible divisor of ϕ of order i; and let T be any irreducible divisor of Φ of order j.

For all roots of t the matrix ϕ has rank less than i, therefore Φ has rank less than $\binom{i}{s}$, i.e. rank not greater than $\binom{i-1}{s}$, and therefore all minor determinants of Φ of order j vanish; consequently t is an irreducible divisor of Φ of order j. Again for all roots of T all minor determinants of Φ of order j vanish, therefore Φ has rank less than j, therefore the rank of Φ cannot exceed $\binom{i-1}{s}$, therefore the rank of ϕ cannot exceed $i-1$, therefore all minor determinants of ϕ of order i vanish; consequently T is an irreducible divisor of ϕ of order i.

Thus every irreducible divisor of ϕ of order i is also an irreducible divisor of Φ of order j, and conversely every irreducible divisor of Φ of order j is also an irreducible divisor of ϕ of order i.

Ex. iii. *If $i \nless 0$, $i \ngtr r$, and if $u=\binom{i}{s}$, $v=\binom{i-1}{s-1}$, then D_u' is a factor of D_i^v.*

Let $\delta=(b)_i^i$ be any minor determinant of ϕ of order i which does not vanish identically; let $\psi=[b]_i^i$, so that ψ is an undegenerate square minor of ϕ of order i; and let $\Psi=[B]_u^u$ be a complete matrix of the minor determinants of ψ of order s. Then $(B)_u^u$ is a minor determinant of Φ of order u, and by § 120 we have

$$(B)_u^u = \pm\delta^v.$$

Therefore D_u', which must be a factor of $(B)_u^u$, is a factor of δ^v; and this is true for all such determinants as δ. Consequently D_u' must be a factor of D_i^v.

When D_i has no repeated irreducible factors (in which case $D_{i-1}=1$), the maximum factor of Φ of order $\binom{i-1}{s}$ is 1, and the maximum factors of Φ of the $\binom{i-1}{s-1}$ orders

$$\binom{i-1}{s}+1,\quad \binom{i-1}{s}+2, \ldots \binom{i-1}{s}+k, \ldots \binom{i}{s} \qquad \text{.....................(3)}$$

are divisible respectively by

$$D_i,\quad D_i^2, \ldots D_i^k, \ldots D_i^v. \qquad \text{....................................(4)}$$

Consequently in this case D_u' is also divisible by D_i^v, and we can write

$$D_u' = D_i^v.$$

It follows that in this case we have in (4) the maximum factors of Φ of the respective orders (3).

Ex. iv. *Maximum factors of the reciprocal of an undegenerate square matrix.*

In the special case when $\phi = [a]_m^m$ is an undegenerate square matrix and $\Phi = [A]_m^m$ is the reciprocal of ϕ, we have $D_m = (a)_m^m$, and it follows at once from § 124. 4 that for the values 1, 2, ... m of i

$$D_i' = D_m^{i-1} D_{m-i},$$

and in particular

$$D_1' = D_m^0 D_{m-1} = D_{m-1}, \quad D_m' = D_m^{m-1} D_0 = D_m^{m-1}.$$

The possible ranks of Φ for particular values of the variables are 0, 1, m. The irreducible divisors of Φ of order 1 are identical with the irreducible divisors of ϕ of order $m-1$, and the irreducible divisors of Φ of any one of the orders 2, 3, ... m are identical with the irreducible divisors of ϕ of order m.

§ 209. Reduction of a rational integral functional matrix to one whose successive leading minors are all regular.

Throughout the present article $A = [a]_m^n$ is a given matrix of rank r whose elements are rational integral functions in Ω of certain scalar variables $x, y, z, \ldots$, Ω being any domain of rationality. The letters $t, t', t'', \ldots, \tau, \tau', \tau'', \ldots$ will denote any given distinct irreducible functions in Ω, which may be irreducible divisors of A; and the indices of the highest powers of these functions which are common factors of all simple minor determinants of A of order i, where $i \not> r$, will be denoted by $d_i, d_i', d_i'', \ldots \delta_i, \delta_i', \delta_i'', \ldots$. If t is not an irreducible divisor of A, then $d_i = 0$; and if τ is not an irreducible divisor of A, then $\delta_i = 0$. The domain of all rational numbers will be denoted by Ω_1. The maximum factors of A of orders 1, 2, ... r will be denoted by $D_1, D_2, \ldots D_r$. The chief results to be established are contained in the three theorems which follow.

Theorem I. *We can always convert the matrix $A = [a]_m^n$ into a similar equipotent matrix $B = [b]_m^n$ whose first r leading horizontal minors $[b]_1^n, [b]_2^n, \ldots [b]_r^n$ are all regular with respect to every one of the given distinct irreducible functions $t, t', t'', \ldots$ by a unitary equigradent transformation in Ω_1 of the form*

$$\begin{bmatrix} 1 & h_{12} & h_{13} & \ldots & h_{1m} \\ 0 & 1 & h_{23} & \ldots & h_{2m} \\ 0 & 0 & 1 & \ldots & h_{3m} \\ \cdots & \cdots & \cdots & \cdots & \cdots \\ 0 & 0 & 0 & \ldots & 1 \end{bmatrix} [a]_m^n = [b]_m^n \quad \ldots\ldots\ldots\ldots\ldots\ldots\ldots\ldots\text{(A)}$$

in which every element h of the prefactor on the left can be a positive integer.

Theorem II. *We can always convert the matrix $A = [a]_m^n$ into a similar equipotent matrix $B = [b]_m^n$ whose first r leading vertical minors $[b]_m^1, [b]_m^2, \ldots [b]_m^r$ are all regular* *with respect to every one of the given distinct irreducible functions $t, t', t'', \ldots$ by a unitary equigradent transformation in Ω_1 of the form*

$$[a]_m^n \begin{bmatrix} 1 & 0 & 0 & \ldots & 0 \\ k_{21} & 1 & 0 & \ldots & 0 \\ k_{31} & k_{32} & 1 & \ldots & 0 \\ \ldots & \ldots & \ldots & \ldots & \ldots \\ k_{n1} & k_{n2} & k_{n3} & \ldots & 1 \end{bmatrix} = [b]_m^n \qquad \text{.........................(B)}$$

in which every element k of the post-factor on the left can be a positive integer.

Theorem III. *We can always convert the matrix $A=[a]_m^n$ into a similar equipotent matrix $C=[c]_m^n$ whose first r leading diagonal minor determinants $c_{11}, (c)_2^2, (c)_3^3, \ldots (c)_r^r$ are all regular with respect to every one of the given distinct irreducible functions $t, t', t'', \ldots$ by a unitary equigradent transformation in Ω_1 of the form*

$$\begin{bmatrix} 1 & h_{12} & h_{13} & \ldots & h_{1m} \\ 0 & 1 & h_{23} & \ldots & h_{2m} \\ 0 & 0 & 1 & \ldots & h_{3m} \\ \ldots & \ldots & \ldots & \ldots & \ldots \\ 0 & 0 & 0 & \ldots & 1 \end{bmatrix} [a]_m^n \begin{bmatrix} 1 & 0 & 0 & \ldots & 0 \\ k_{21} & 1 & 0 & \ldots & 0 \\ k_{31} & k_{32} & 1 & \ldots & 0 \\ \ldots & \ldots & \ldots & \ldots & \ldots \\ k_{n1} & k_{n2} & k_{n3} & \ldots & 1 \end{bmatrix} = [c]_m^n \qquad \text{...............(C)}$$

in which all the elements h and k of the prefactor and post-factor on the left can be positive integers.

NOTE 1. *The matrices B in Theorems I and II and the matrix C in Theorem III lie in Ω and are equipotent with A.*

This follows from Ex. vii of § 206. The properties of equipotent matrices are enumerated in § 213. The matrix B (or C) has the same irreducible divisors as A, and these irreducible divisors have the same maximum indices in B (or C) as in A.

NOTE 2. *Special case in which $t, t', t'', \ldots$ are all the irreducible divisors of A.*

If $r=m$, then in this case the H. C. F. of all the simple minor determinants of $[b]_i^n$ in Theorem I or $[c]_i^n$ in Theorem III is D_i, the common maximum factor of order i of A and B or of A and C; but this is not necessarily so when $r < m$, for then the simple minor determinants of $[b]_i^n$ or $[c]_i^n$ may have a common factor which is not a factor of D_i.

Also if $r=n$, then in this case the H.C.F. of all the simple minor determinants of $[b]_m^i$ in Theorem II or $[c]_m^i$ in Theorem III is D_i.

NOTE 3. *Special case in which the elements of A are constants.*

Theorems I—III are still true, but in the enunciations we must replace 'regular with respect to $t, t', t'', \ldots$' by 'regular' simply. The proofs of the special theorems for this case are simplifications of the proofs of the general theorems.

We will prove Theorems I—III with the help of the examples and lemmas which will now be given.

Ex. i. *If P, Q and T are given rational integral functions of x, y, z, ..., and if Q does not vanish identically and is not divisible by T, there cannot be more than one constant λ which is such that $P+\lambda Q$ is divisible by T.*

For if $P+\lambda Q$ were divisible by T for two different values λ_1 and λ_2 of the constant λ, it would follow that $(\lambda_1-\lambda_2)\, Q$ is divisible by T, and this is impossible.

Ex. ii. *Let the first horizontal row of the matrix $A=[a]_m^n$ be regular with respect to $t, t', t'', \ldots$, but not regular with respect to τ; and let the pth horizontal row of A be regular with respect to τ. Then by adding to the first horizontal row the pth horizontal row multiplied by a non-zero constant λ in Ω_1 (which can always be a positive integer) we can convert A into a similar equipotent matrix $B=[b]_m^n$ whose first horizontal row is regular with respect to $t, t', t'', \ldots$ and τ.*

The simple special case in which τ is the only irreducible function to be considered (no mention being made of $t, t', t'', \ldots$) is included in the general case.

Let the matrix $B=[b]_m^n$ be formed in the manner described, λ being any constant; and let the element a_{pq} of the pth horizontal row of A be regular with respect to τ. Then we have

$$b_{1q}=a_{1q}+\lambda a_{pq}=\tau^{\delta_1}\,(P\tau+\lambda Q),$$

where Q is not divisible by τ. Since $P\tau+\lambda Q$ is only divisible by τ when $\lambda=0$, the element b_{1q} of the first horizontal row of B is regular with respect to τ for all non-zero values of λ. Again let the elements $a_{1u}, a_{1v}, a_{1w}, \ldots$ (not necessarily all different) of the first horizontal row of A be regular with respect to $t, t', t'', \ldots$ respectively. Then we have (with new meanings for P and Q)

$$b_{1u}=a_{1u}+\lambda a_{pu}=t^{d_1}\,(P+\lambda Q),$$

where P is not divisible by t. If Q is divisible by t, then $P+\lambda Q$ is not divisible by t; and if Q is not divisible by t, then by Ex. i there cannot be more than one value of λ for which $P+\lambda Q$ is divisible by t. Therefore b_{1u} is regular with respect to t for all non-zero values of λ with the possible exception of one; similarly b_{1v} is regular with respect to t' for all non-zero values of λ with the possible exception of one; and so on.

It follows that $[b]_1^n$ is regular with respect to $t, t', t'', \ldots$ and τ for all non-zero values of λ with the possible exception of a finite number of values, one corresponding to each of the irreducible functions $t, t', t'', \ldots$; and we can always choose λ to be a positive integer such that $[b]_1^n$ is regular with respect to $t, t', t'', \ldots$ and τ. Except in very special cases (as when $P+Q$ is divisible by t in the expression for b_{1u}) we can take λ to be 1.

Ex. iii. *Let the first $i-1$ leading horizontal minors of the matrix $A=[a]_m^n$ be all regular with respect to $t, t', t'', \ldots$ and τ, and let the ith leading horizontal minor $[a]_i^n$ be regular with respect to $t, t', t'', \ldots$, but not regular with respect to τ; i being not greater than r; also let the horizontal minor of A of reduced order i formed with the first $i-1$ and the pth horizontal rows be regular with respect to τ. Then by adding to the ith horizontal row the pth horizontal row multiplied by a non-zero constant λ in Ω_1 (which can always be a positive integer) we can convert A into a similar equipotent matrix $B=[b]_m^n$ in which $[b]_{i-1}^n=[a]_{i-1}^n$ and at the same time the ith leading horizontal minor $[b]_i^n$ is regular with respect to $t, t', t'', \ldots$ and τ.*

The simple special case in which τ is the only irreducible function to be considered (no mention being made of $t, t', t'', \ldots$) is included in the general case.

By Ex. vii of § 203 there must exist some horizontal minor of A of order i containing $[a]_{i-1}^{n}$ which is regular with respect to τ, and we suppose such a minor to be that formed with $[a]_{i-1}^{n}$ and the pth horizontal row, where of necessity $p > i$. Let the matrix $B = [b]_m^n$ be formed in the manner described, λ being any constant, and let the minor determinant

$$\Delta = \begin{pmatrix} q_1, q_2, \ldots q_i \\ a \\ 1, 2, \ldots i-1, p \end{pmatrix}$$

of the aforesaid horizontal minor of A be regular with respect to τ. Then the simple minor determinant of $[b]_i^n$ formed with the same vertical rows as Δ is

$$\Delta' = \begin{pmatrix} q_1 q_2 \ldots q_i \\ b \\ 1, 2, \ldots i-1, i \end{pmatrix} = \begin{pmatrix} q_1 q_2 \ldots q_i \\ a \\ 1, 2, \ldots i-1, i \end{pmatrix} + \lambda \begin{pmatrix} q_1 q_2 \ldots q_i \\ a \\ 1, 2, \ldots i-1, p \end{pmatrix} = \tau^{\delta_i}(P\tau + \lambda Q),$$

where Q is not divisible by τ; and Δ' is regular with respect to τ for all non-zero values of λ. Again let the simple minor determinants

$$\Delta_1 = \begin{pmatrix} u_1 u_2 \ldots u_i \\ a \\ 1, 2, \ldots i-1, i \end{pmatrix}, \quad \Delta_2 = \begin{pmatrix} v_1 v_2 \ldots v_i \\ a \\ 1, 2, \ldots i-1, i \end{pmatrix}, \quad \Delta_3 = \begin{pmatrix} w_1 w_2 \ldots w_i \\ a \\ 1, 2, \ldots i-1, i \end{pmatrix}, \ldots$$

of $[a]_i^n$, which are not necessarily all different, be regular with respect to $t, t', t'', \ldots$ respectively, and let $\Delta_1', \Delta_2', \Delta_3', \ldots$ be the correspondingly formed simple minor determinants of $[b]_i^n$. Then we have (with different meanings for P and Q)

$$\Delta_1' = \begin{pmatrix} u_1 u_2 \ldots u_i \\ b \\ 1, 2, \ldots i-1, i \end{pmatrix} = \begin{pmatrix} u_1 u_2 \ldots u_i \\ a \\ 1, 2, \ldots i-1, i \end{pmatrix} + \lambda \begin{pmatrix} u_1 u_2 \ldots u_i \\ a \\ 1, 2, \ldots i-1, p \end{pmatrix} = t^{d_i}(P + \lambda Q),$$

where P is not divisible by t. If Q is divisible by t, then Δ_1' is regular with respect to t for all values of λ; and if Q is not divisible by t, then by Ex. i the determinant Δ_1' is regular with respect to t for all values of λ with the possible exception of one. Thus the simple minor determinant Δ_1' of $[b]_i^n$ is regular with respect to t for all non-zero values of λ with the possible exception of one; similarly Δ_2' is regular with respect to t' for all non-zero values of λ with the possible exception of one; and so on.

It follows that $[b]_i^n$ is regular with respect to $t, t', t'', \ldots$ and τ for all non-zero values of λ with the possible exception of a finite number of values, one corresponding to each of the irreducible functions $t, t', t'', \ldots$; and therefore we can always choose λ to be a positive integer such that $[b]_i^n$ is regular with respect to $t, t', t'', \ldots$ and τ. Except in very special cases (as when $P+Q$ is divisible by t in the expression for Δ_1') we can take λ to be 1.

Ex. iv. *Let the first $i-1$ leading diagonal minor determinants $a_{11}, (a)_2^2, \ldots (a)_{i-1}^{i-1}$ of the matrix $A = [a]_m^n$ be regular with respect to $t, t', t'', \ldots$ and τ; and let the ith leading diagonal minor determinant $(a)_i^i$ be regular with respect to $t, t', t'', \ldots$, but not regular with respect to τ; i being not greater than r. Then if the uth and vth horizontal and vertical rows of A are suitably chosen, we can convert A into a similar equipotent matrix $C = [c]_m^n$ in which $[c]_{i-1}^{i-1} = [a]_{i-1}^{i-1}$ and the ith leading diagonal minor determinant $(c)_i^i$ is also regular* *with respect to t, t', t'', ... and τ by first adding to the ith horizontal row of A the uth horizontal row multiplied by a constant λ in Ω_1, and then adding to the ith vertical row the vth vertical row multiplied by a constant μ in Ω_1. The constants λ and μ can always be positive integers.*

The simple special case in which τ is the only irreducible function to be considered (no mention being made of t, t', t'', ...) is included in the general case.

In this transformation only the ith horizontal and vertical rows of A are changed. We will write

$$A_{uv}=\begin{pmatrix}1,2,\ldots i-1, v\\ a\\ 1,2,\ldots i-1, u\end{pmatrix},\quad B_{uv}=\begin{pmatrix}1,2,\ldots i-1, v\\ b\\ 1,2,\ldots i-1, u\end{pmatrix},\quad C_{uv}=\begin{pmatrix}1,2,\ldots i-1, v\\ c\\ 1,2,\ldots i-1, u\end{pmatrix},$$

so that

$$(c)_i^i=C_{ii}=A_{ii}+\lambda A_{ui}+\mu A_{iv}+\lambda\mu A_{uv}.$$

The uth and vth horizontal and vertical rows of A must always be so chosen that A_{uv} is one of those minor determinants of A of order i containing $(a)_{i-1}^{i-1}$ which are regular with respect to τ. Consequently we shall always have $u \nless i$, $v \nless i$, and one at least of the two integers u and v must be greater than i.

Case I. When A has a minor determinant A_{ui} of order i containing $(a)_{i-1}^{i-1}$ and lying in the same vertical minor $[a]_m^i$ as $(a)_i^i$ which is regular with respect to τ.

In this case we have $v=i$, $u>i$, and we form C by adding to the ith horizontal row of A the uth horizontal row multiplied by a constant λ; so that $\mu=0$ and

$$(c)_i^i=C_{ii}=A_{ii}+\lambda A_{ui}.$$

Then by the reasoning applied to b_{1q} in Ex. ii the determinant $(c)_i^i$ is regular with respect to t, t', t'', ... and τ for all non-zero values of λ with the possible exception of a finite number of values, one corresponding to each of the irreducible functions t, t', t'', ...

Case II. When A has a minor determinant A_{iv} of order i containing $(a)_{i-1}^{i-1}$ and lying in the same horizontal minor $[a]_i^n$ as $(a)_i^i$ which is regular with respect to τ.

In this case we have $u=i$, $v>i$, and we form C by adding to the ith vertical row of A the vth vertical row multiplied by a constant μ; so that $\lambda=0$ and

$$(c)_i^i=C_{ii}=A_{ii}+\mu A_{iv}.$$

Then $(c)_i^i$ is regular with respect to t, t', t'', ... and τ for all non-zero values of μ with the possible exception of a finite number of values, one corresponding to each of the irreducible functions t, t', t'', ...

Case III. When none of the minor determinants of A of order i containing $(a)_{i-1}^{i-1}$ and lying in the same horizontal or the same vertical minor as $(a)_i^i$ are regular with respect to τ.

In this case we select any minor determinant A_{uv} of A of order i containing $(a)_{i-1}^{i-1}$ which is regular with respect to τ; and we have $u>i$ and $v>i$. We first add to the ith horizontal row of A the uth horizontal row multiplied by a constant λ, thus converting A into a similar matrix $B=[b]_m^n$ in which $[b]_{i-1}^n=[a]_{i-1}^n$ and

$$(b)^i=B_{ii}=A_{ii}+\lambda A_{ui},\quad B_{iv}=A_{iv}+\lambda A_{uv}.$$

Choosing λ so that $(b)_i^i$ is regular with respect to $t, t', t'', \ldots$, and B_{iv} is regular with respect to τ, we see that λ can have any non-zero value with the possible exception of a finite number of values, one corresponding to each of the irreducible functions $t, t', t'', \ldots$. The matrix B now falls under Case II. Accordingly we next add to the ith vertical row of B the vth vertical row multiplied by a constant μ, thus converting B into a similar matrix $C=[c]_m^n$ in which

$$[c]_m^{i-1}=[b]_m^{i-1}, \quad [c]_{i-1}^{i-1}=[b]_{i-1}^{i-1}=[a]_{i-1}^{i-1},$$

$$(c)_i^i = B_{ii}+\mu B_{iv}= A_{ii}+\lambda A_{ui}+\mu A_{vi}+\lambda\mu A_{uv}.$$

Choosing μ so that $(c)_i^i$ is regular with respect to $t, t', t'', \ldots$ and τ, we see as in Case II that μ can have any non-zero value with the possible exception of a finite number of values, one corresponding to each of the irreducible functions $t, t', t'', \ldots$.

Lemma 1a. *We can always convert the matrix $A=[a]_m^n$ into a similar equipotent matrix $B=[b]_m^n$ whose first horizontal row is regular with respect to any number of given irreducible functions $t, t', t'', \ldots$ by an equigradent transformation in Ω_1 of the form*

$$\begin{bmatrix}1, & h\\ 0, & 1\end{bmatrix}_{1,m-1}^{1,m-1}[a]_m^n=[b]_m^n \quad \ldots\ldots\ldots\ldots(a)$$

in which every element of the constituent matrix $[h]_1^{m-1}$ can be a positive integer (which may be 0). Only the elements of the first horizontal row of A are changed.

We obtain the transformation (a) by successive applications of Ex. ii, making $[b]_1^n$ first regular with respect to t; then regular with respect to t and t'; then regular with respect to $t, t', t'', \ldots$; and so on.

The functions $t, t', t'', \ldots$ may be all or any number of the irreducible divisors of A. In the particular case when they are all the irreducible divisors of A we have

$$[1,\ h]_1^{1,n-1}[a]_m^n=[b]_1^n=D_1\,.\,[\beta]_1^n,$$

where $[\beta]_1^n$ is a non-zero matrix which is impotent (see § 213·2) with respect to all the irreducible divisors of A or B.

Lemma 1b. *If $i \not> r$, and if the first $i-1$ leading horizontal minors $[a]_1^n, [a]_2^n, \ldots [a]_{i-1}^n$ of the matrix $A=[a]_m^n$ are all regular with respect to every one of the given irreducible functions $t, t', t'', \ldots$, we can convert A into a similar equipotent matrix $B=[b]_m^n$ in which $[b]_{i-1}^n=[a]_{i-1}^n$ and at the same time the ith leading horizontal minor $[b]_i^n$ is also regular with respect to every one of the functions $t, t', t'', \ldots$ by an equigradent transformation in Ω_1 of the form*

$$\begin{bmatrix}1, & 0, & 0\\ 0, & 1, & h\\ 0, & 0, & 1\end{bmatrix}_{i-1,1,m-i}^{i-1,1,m-i}[a]_m^n=[b]_m^n \quad \ldots\ldots\ldots\ldots(a')$$

in which every element of the constituent matrix $[h]_1^{m-1}$ can be a positive integer. Only the elements of the ith horizontal row of A are changed.

We obtain the transformation (a′) by successive applications of Ex. iii, making $[b]_i^n$ first regular with respect to t; then regular with respect to t and t'; then regular with respect to t, t', t''; and so on.

Proof of Theorem I. By applying to the matrix $A=[a]_m^n$ in succession first Lemma 1a, and then Lemma 1b for the values 2, 3, ... r of i, we obtain a resultant transformation of the form (A) converting A into a similar matrix $B=[b]_m^n$ having the properties mentioned in Theorem I.

Lemma 2a. *We can always convert the matrix $A=[a]_m^n$ into a similar equipotent matrix $B=[b]_m^n$ whose first vertical row is regular with respect to any number of given irreducible functions $t, t', t'', \ldots$ by an equigradent transformation in Ω_1 of the form*

$$[a]_m^n \begin{bmatrix} 1, & 0 \\ k, & 1 \end{bmatrix}_{1,\, n-1}^{1,\, n-1} = [b]_m^n \quad \text{.................................(b)}$$

in which every element of the constituent matrix $[k]_{n-1}^1$ can be a positive integer (which may be 0). Only the elements of the first vertical row of A are changed.

The proof is similar to that of Lemma 1a.

Lemma 2b. *If $i \ngtr r$, and if the first $i-1$ leading vertical minors $[a]_m^1, [a]_m^2, \ldots [a]_m^{i-1}$ of the matrix $A=[a]_m^n$ are all regular with respect to every one of the given irreducible functions $t, t', t'', \ldots$, we can convert A into a similar equipotent matrix $B=[b]_m^n$ in which $[b]_m^{i-1}=[a]_m^{i-1}$ and at the same time the ith leading vertical minor $[b]_m^i$ is also regular with respect to every one of the functions $t, t', t'', \ldots$ by an equigradent transformation in Ω_1 of the form*

$$[a]_m^n \begin{bmatrix} 1, & 0, & 0 \\ 0, & 1, & 0 \\ 0, & k, & 1 \end{bmatrix}_{i-1,\, 1,\, n-i}^{i-1,\, 1,\, n-i} = [b]_m^n \quad \text{.............................(b′)}$$

in which every element of the constituent matrix $[k]_{n-i}^1$ can be a positive integer. Only the ith vertical row of A is changed.

The proof is similar to that of Lemma 1b.

Proof of Theorem II. By applying to the matrix $A=[a]_m^n$ in succession first Lemma 2a, and then Lemma 2b for the values 2, 3, ... r of i, we obtain a resultant transformation of the form (B) converting A into a similar matrix $B=[b]_m^n$ having the properties mentioned in Theorem II.

Lemma 3a. *We can always convert the matrix $A=[a]_m^n$ into a similar equipotent matrix $C=[c]_m^n$ whose leading element c_{11} is regular with respect to any number of given irreducible functions $t, t', t'', \ldots$ by an equigradent transformation in Ω_1 of the form*

$$\begin{bmatrix} 1, & h \\ 0, & 1 \end{bmatrix}_{1,\, m-1}^{1,\, m-1} \cdot [a]_m^n \begin{bmatrix} 1, & 0 \\ k, & 1 \end{bmatrix}_{1,\, n-1}^{1,\, n-1} = [c]_m^n \quad \text{..........................(c)}$$

in which every element of the constituent matrices $[h]_1^{m-1}$ and $[k]_{n-1}^1$ can be a positive integer (which may be 0). The only elements of A which are changed are those which lie in the first horizontal and vertical rows.

We can obtain the transformation (c) by successive applications of Ex. iv for the special case when $i=1$, first making c_{11} regular with respect to t; then making it regular with respect to t and t'; then making it regular with respect to t, t', t''; and so on.

Or we can first apply Lemma 1*a* to convert $[a]_m^n$ into a matrix $[b]_m^n$ whose first horizontal row is regular with respect to $t, t', t'', \dots$; and then apply Lemma 2*a* to convert $[b]_1^n$ into a matrix $[c]_1^n$ whose first vertical row, i.e. the element c_{11}, is regular with respect to $t, t', t'', \dots$. If the two transformations thus obtained are

$$\begin{bmatrix} 1, & h \\ 0, & 1 \end{bmatrix}_{1,\,m-1}^{1,\,m-1} [a]_m^n = [b]_m^n, \quad [b]_1^n \begin{bmatrix} 1, & 0 \\ k, & 1 \end{bmatrix}_{1,\,n-1}^{1,\,n-1} = [c]_1^n,$$

then the matrix $[c]_m^n$ defined by (c) has the required properties.

Lemma 3b. *If $i \not> r$, and if the first $i-1$ leading diagonal minor determinants $a_{11}, (a)_2^2, \dots (a)_{i-1}^{i-1}$ of the matrix $A=[a]_m^n$ are all regular with respect to every one of the given irreducible functions $t, t', t'', \dots$, we can convert A into a similar equipotent matrix $C=[c]_m^n$ in which $[c]_{i-1}^{i-1}=[a]_{i-1}^{i-1}$ and at the same time the ith leading diagonal minor determinant $(c)_i^i$ is also regular with respect to every one of the functions $t, t', t'', \dots$ by an equigradent transformation in Ω_1 of the form*

$$\begin{bmatrix} 1, & 0, & 0 \\ 0, & 1, & h \\ 0, & 0, & 1 \end{bmatrix}_{i-1,\,1,\,m-i}^{i-1,\,1,\,m-i} [a]_m^n \begin{bmatrix} 1, & 0, & 0 \\ 0, & 1, & 0 \\ 0, & k, & 1 \end{bmatrix}_{i-1,\,1,\,n-i}^{i-1,\,1,\,n-i} = [c]_m^n \quad \dots\dots\dots\dots\dots\dots (c')$$

in which every element of the constituent matrices $[h]_1^{m-i}$ and $[k]_{n-i}^1$ can be a positive integer. The only elements of A which are changed are those which lie in the ith horizontal and vertical rows.

We can obtain the transformation (c′) by successive applications of Ex. iv.

Or, because the first $i-1$ leading horizontal minors of A are regular with respect to $t, t', t'', \dots$, we can apply Lemma 1*b* to convert $[a]_m^n$ into a matrix $[b]_m^n$ in which $[b]_{i-1}^n=[a]_{i-1}^n$ and $[b]_i^n$ is regular with respect to $t, t', t'', \dots$. Then, because the first $i-1$ leading vertical minors of $[b]_i^n$ are regular with respect to $t, t', t'', \dots$, we can apply Lemma 2*b* to convert $[b]_i^n$ into a matrix $[c]_i^n$ in which $[c]_i^{i-1}=[b]_i^{i-1}$ and $[c]_i^i$ is regular with respect to $t, t', t'', \dots$. By applying this second transformation to $[b]_m^n$ instead of $[b]_i^n$ we obtain (c′).

Proof of Theorem III. By applying to the matrix $A=[a]_m^n$ in succession first Lemma 3*a*, and then Lemma 3*b* for the values $2, 3, \dots r$ of i, we obtain a resultant transformation of the form (C) converting A into a similar matrix $C=[c]_m^n$ having the properties mentioned in Theorem III.

Ex. v. We will apply the foregoing theorems to the matrix of rank 3

$$A=[a]_3^4=(x+1)\begin{bmatrix} 2x^3-x^2+1, & 2x^2-x-1, & 2x^3+x^2-x, & 2x^3-5x^2+2x+3 \\ x^2-x+2, & 3x^2+2x-5, & x^2+1, & x^2-3x+4 \\ 2x^2-2x+1, & x-1, & 2x^2-1, & 2x^2-6x+5 \end{bmatrix}$$

whose maximum factors of orders 1, 2, 3 are

$$D_1=x+1,\quad D_2=(x+1)^2(x-1),\quad D_3=(x+1)^3(x-1)^2,$$

and whose irreducible divisors in Ω_1 are

$$t_1=x-1,\quad t_2=x+1.$$

The leading element a_{11} of A is regular with respect to both t_1 and t_2; the leading diagonal minor determinant of order 2, viz.

$$\Delta=\begin{pmatrix}12\\ a\\ 12\end{pmatrix}=(x+1)^3(x-1)(6x^3-x^2-3x+3),$$

is regular with respect to t_1 but not with respect to t_2; and the minor determinant of order 2

$$\Delta'=\begin{pmatrix}12\\ a\\ 23\end{pmatrix}=-3(x+1)^2(x-1)(2x^3+x^2-2x+1)$$

lying in the same vertical minor of A as Δ is regular with respect to t_2. Accordingly we can convert A into a similar equipotent matrix $B=[b]_3^4$ whose leading diagonal minor determinants of orders 1 and 2 are regular with respect to both t_1 and t_2 by adding to the second horizontal row of A the third horizontal row multiplied by a constant λ. We then have

$$(b)_2^2=\begin{pmatrix}12\\ a\\ 12\end{pmatrix}+\lambda\begin{pmatrix}12\\ a\\ 13\end{pmatrix}=(x+1)^2(x-1)\{(x+1)(6x^3-x^2-3x+3)-\lambda x^2(2x-1)\}.$$

The minor determinant $(b)_2^2$ is regular with respect to t_2 except when $\lambda=0$, and regular with respect to t_1 except when $\lambda=10$. Therefore if we give to the constant λ any value except 0 and 10, then $(b)_2^2$ as well as b_{11} will be regular with respect to both the divisors t_1 and t_2. Giving to λ the value 1, we obtain the unitary equigradent transformation

$$\begin{bmatrix}1&0&0\\0&1&1\\0&0&1\end{bmatrix}[a]_3^4=(x+1)\begin{bmatrix}2x^3-x^2+1, & 2x^2-x-1, & 2x^3+x^2-x, & 2x^3-5x^2+2x+3\\ 3x^2-3x+3, & 3x^2+3x-6, & 3x^2\quad, & 3x^2-9x+9\\ 2x^2-2x+1, & x-1\quad, & 2x^2-1\quad, & 2x^2-6x+15\end{bmatrix}$$

$$=[b]_3^4,$$

converting A into a similar matrix $B=[b]_3^4$ whose leading diagonal minor determinants of orders 1 and 2 are regular with respect to both the divisors t_1 and t_2.

In this case the leading diagonal minor determinants of B of orders 1, 2 and 3 are all regular with respect to both the divisors t_1 and t_2; and no further transformation is required to secure this.

§ 210. Reduction of a symmetric matrix to one whose successive leading diagonal minor determinants are all regular.

Throughout the present article $A=[a]_m^m$ is a given symmetric matrix of rank r whose elements are rational integral functions in Ω of certain scalar variables, Ω being any domain of rationality. The letters $t, t', t'', \ldots, \tau, \tau', \tau'', \ldots$ will denote any given distinct irreducible functions in Ω, which may be irreducible divisors of A; and the indices of the highest powers of these functions which are common factors of all minor determinants of A of order i, where $i \not> r$, will be denoted by $d_i, d_i', d_i'', \ldots, \delta_i, \delta_i', \delta_i'', \ldots$. The domain of all rational numbers will be denoted by Ω_1. The chief result to be established is that given by the following theorem.

Theorem. *We can always convert the symmetric matrix* $A=[a]_m^m$ *into a similar equipotent symmetric matrix* $B=[b]_m^m$ *whose first r leading diagonal minor determinants* b_{11}, $(b)_2^2, (b)_3^3, \ldots (b)_r^r$ *are all regular with respect to every one of the distinct irreducible functions* $t, t', t'', \ldots$ *by a symmetric unitary equigradent transformation in* Ω_1 *of the form*

$$\begin{bmatrix} 1 & h_{12} & h_{13} & \ldots & h_{1m} \\ 0 & 1 & h_{23} & \ldots & h_{2m} \\ 0 & 0 & 1 & \ldots & h_{3m} \\ \ldots & \ldots & \ldots & \ldots & \ldots \\ 0 & 0 & 0 & \ldots & 1 \end{bmatrix} [a]_m^m \begin{bmatrix} 1 & 0 & 0 & \ldots & 0 \\ h_{12} & 1 & 0 & \ldots & 0 \\ h_{13} & h_{23} & 1 & \ldots & 0 \\ \ldots & \ldots & \ldots & \ldots & \ldots \\ h_{1m} & h_{2m} & h_{3m} & \ldots & 1 \end{bmatrix} = [b]_m^m \quad \ldots\ldots\ldots\ldots(\text{A})$$

in which every element h of the prefactor and post-factor on the left can be a positive integer (which may be 0).

The matrix B given by (A) always lies in Ω and (see Ex. vii of § 206) is equipotent with A. Hence it has the same irreducible divisors as A, and those irreducible divisors have the same maximum indices in B as in A. The theorem remains true in the special case when all the elements of A are constants, but in the enunciation we then replace 'regular with respect to $t, t', t'', \ldots$' by 'regular' simply.

The theorem will be proved with the help of the examples and lemmas which follow.

Ex. i. *If* P, Q, R *and* T *are given rational integral functions of* $x, y, z, \ldots$, *and if* R *does not vanish identically and is not divisible by* T, *there cannot be more than two different constants* λ *which are such that* $P+\lambda Q+\lambda^2 R$ *is divisible by* T.

For if the above expression were divisible by T for three different values $\lambda_1, \lambda_2, \lambda_3$ of the constant λ, it would follow that

$$Q+(\lambda_2+\lambda_3)R,\quad Q+(\lambda_3+\lambda_1)R,\quad Q+(\lambda_1+\lambda_2)R \quad \text{and} \quad (\lambda_2-\lambda_3)R,\quad (\lambda_3-\lambda_1)R,\quad (\lambda_1-\lambda_2)R$$

are all divisible by T; and this is impossible.

Ex. ii. *If the leading element* a_{11} *of the symmetric matrix* $A=[a]_m^m$ *is regular with respect to* $t, t', t'', \ldots$, *but is not regular with respect to* τ, *we can convert* A *by a symmetric equigradent transformation of the form* (A) *into a similar symmetric matrix* $B=[b]_m^m$ *whose leading element* b_{11} *is regular with respect to* $t, t', t'', \ldots$ *and* τ.

The simple special case in which τ is the only irreducible function to be considered (no mention being made of $t, t', t'', \ldots$) is included in the general case.

Case I. When A has a diagonal element a_{uu} *which is regular with respect to* τ.

If we add to the first horizontal row of A the uth horizontal row multiplied by a constant λ, and afterwards add to the first vertical row the uth vertical row multiplied by λ, we convert A into a similar equipotent symmetric matrix $B=[b]_m^m$ in which

$$b_{11}=a_{11}+2\lambda a_{u1}+\lambda^2 a_{uu}=\tau^{\delta_1}\{\tau a_{11}+2\lambda a_{u1}+\lambda^2 a_{uu}\},$$

where a_{uu} is not divisible by τ. Then b_{11} is regular with respect to τ except when $\lambda=0$, and when $2a_{u1}+\lambda a_{uu}$ is divisible by τ. Since by Ex. i of § 209 there cannot be more than one value of λ for which $2a_{u1}+\lambda a_{uu}$ is divisible by τ, we see that b_{11} is regular with respect to τ for all non-zero values of λ with the possible exception of one. Again we can write (with a change in the meanings of the a's)

$$b_{11}=t^{d_1}\{a_{11}+2\lambda a_{u1}+\lambda^2 a_{uu}\},$$

where now a_{11} is not divisible by t. If a_{uu} is not divisible by t, then by Ex. i above b_{11} is regular with respect to t for all values of λ with the possible exception of two; if a_{uu} is divisible by t and a_{u1} is not divisible by t, then by Ex. i of § 209 there cannot be more than one value of λ for which $a_{11}+2\lambda a_{u1}$ is divisible by t, and therefore b_{11} is regular with respect to t for all values of λ with the possible exception of one; and if a_{uu} and a_{u1} are both divisible by t, then b_{11} is regular with respect to t for all values of λ. Thus b_{11} is regular with respect to t for all values of λ with the possible exception of two; similarly b_{11} is regular with respect to t' for all values of λ with the possible exception of two; and so on.

It follows that in Case I the leading element b_{11} of B is regular with respect to $t, t', t'', \ldots$ and τ for all non-zero values of λ with the possible exception of a finite number of values, one corresponding to τ and two corresponding to each one of the functions $t, t', t'', \ldots$. Usually all non-zero values of λ in Ω_1 are admissible, and we can choose λ to be 1. In all cases we can choose λ to be some non-zero positive integer.

Case II. When A has no diagonal element which is regular with respect to τ.

Let a_{uv}, where $u>v$ and $u>1$, be a non-diagonal element of A which is regular with respect to τ. Then if we add to the vth horizontal row of A the uth horizontal row multiplied by a constant λ, and afterwards add to the vth vertical row the uth vertical row multiplied by λ, we convert A into a similar symmetric matrix $A'=[a']_m^m$ in which

$$a'_{vv}=a_{vv}+2\lambda a_{uv}+\lambda^2 a_{uu}=\tau^{\delta_1}\{\tau a_{vv}+2\lambda a_{uv}+\tau\lambda^2 a_{uu}\},$$

where a_{uv} is not divisible by τ; and a'_{vv} is necessarily regular with respect to τ for all non-zero values of λ.

If $v\neq 1$, i.e. if a_{uv} does not lie in the first vertical row of A, we have $a'_{11}=a_{11}$, and we can treat the matrix A' just as we treated the matrix A in Case I, thus converting it into a symmetric matrix $B=[b]_m^m$ in which b_{11} is regular with respect to $t, t', t'', \ldots$ and τ. The resultant transformation converting A into B has the form (A).

If $v=1$, i.e. if a_{uv} lies in the first vertical row of A, we take the matrix A' obtained above to be the matrix $B=[b]_m^m$, so that

$$b_{11}=a_{11}+2\lambda a_{u1}+\lambda^2 a_{uu}=\tau^{\delta_1}\{\tau a_{11}+2\lambda a_{u1}+\tau\lambda^2 a_{uu}\},$$

where a_{u1} is not divisible by τ. Then b_{11} is regular with respect to τ for all non-zero values of λ, and just as in Case I we can choose λ so that b_{11} is also regular with respect to $t, t', t'', \ldots$.

Ex. iii. *Let the first $i-1$ leading diagonal minor determinants of A be all regular with respect to the given irreducible functions $t, t', t'', \ldots$ and τ, and let the ith leading diagonal minor determinant $(a)_i^i$ be regular with respect to $t, t', t'', \ldots$, but not regular with respect to τ; i being not greater than r. Then we can convert A into a similar equipotent symmetric matrix $B=[b]_m^m$ in which $[b]_{i-1}^{i-1}=[a]_{i-1}^{i-1}$ and at the same time the ith leading diagonal minor determinant $(b)_i^i$ is also regular with respect to $t, t', t'', \ldots$ and τ by a symmetric equigradent transformation in Ω_1 of the form*

$$\begin{bmatrix}1, & 0\\ 0, & \omega\end{bmatrix}_{i-1,m-i+1}^{i-1,m-i+1}[a]_m^m\,\overline{\underline{\begin{matrix}1, & 0\\ 0, & \omega\end{matrix}}}_{i-1,m-i+1}^{i-1,m-i+1}=[b]_m^m,$$

where $[\omega]_{m-i+1}^{m-i+1}$ is a matrix having the same form as the prefactor on the left in (A).

The simple special case in which τ is the only irreducible function to be considered (no mention being made of $t, t', t'', \ldots$) is included in the general case.

For the minor determinants of A of order i containing $(a)_{i-1}^{i-1}$ we will use the notation

$$A_{uv} = \begin{pmatrix} 1, 2, \ldots i-1, v \\ a \\ 1, 2, \ldots i-1, u \end{pmatrix}, \quad \text{where} \quad u \nless i, \; v \nless i.$$

Case I. When A has a diagonal minor determinant A_{uu} of order i containing $(a)_{i-1}^{i-1}$ which is regular with respect to τ, where of course $u > i$.

If we add to the ith horizontal row of A the uth horizontal row multiplied by a constant λ, and afterwards add to the ith vertical row the uth vertical row multiplied by λ, we convert A into a similar symmetric matrix $B = [b]_m^m$ in which $[b]_{i-1}^{i-1} = [a]_{i-1}^{i-1}$ and

$$(b)_i^i = A_{ii} + 2\lambda A_{ui} + \lambda^2 A_{uu} = \tau^{\delta_i}\{\tau a_{ii} + 2\lambda a_{ui} + \lambda^2 a_{uu}\},$$

where a_{uu} is not divisible by τ. By the same reasoning as that used for the element b_{11} in Case I of Ex. ii we see that $(b)_i^i$ is regular with respect to τ for all non-zero values of λ with the possible exception of one. Again we can write (with a change in the a's)

$$(b)_i^i = t^{d_i}\{a_{ii} + 2\lambda a_{ui} + \lambda^2 a_{uu}\},$$

where a_{ii} is not divisible by t, and by the same reasoning as that used for the element b_{11} in Case I of Ex. ii we see that $(b)_i^i$ is regular with respect to $t, t', t'', \ldots$ and τ for all non-zero values of λ with the possible exception of a finite number of values. In general all non-zero values of λ in Ω_1 are admissible, and we can take λ to be 1. In all cases we can choose λ to be some positive integer.

Case II. When A has no diagonal minor determinant of order i containing $(a)_{i-1}^{i-1}$ which is regular with respect to τ.

Let A_{uv}, where $u > v$, $v \nless i$, $u > i$, be a non-diagonal minor determinant of A of order i containing $(a)_{i-1}^{i-1}$ which is regular with respect to τ. Then if we add to the vth horizontal row of A the uth horizontal row multiplied by a constant λ, and afterwards add to the vth vertical row the uth vertical row multiplied by λ, we convert A into a similar symmetric matrix $A' = [a']_m^m$ in which $[a']_{i-1}^{i-1} = [a]_{i-1}^{i-1}$ and

$$A'_{vv} = A_{vv} + 2\lambda A_{uv} + \lambda^2 A_{uu} = \tau^{\delta_i}\{\tau a_{vv} + 2\lambda a_{uv} + \tau\lambda^2 a_{uu}\},$$

where a_{uv} is not divisible by τ; and A'_{vv} is a diagonal minor determinant of A' of order i containing $(a')_{i-1}^{i-1}$ which is necessarily regular with respect to τ for all non-zero values of λ.

If $v \neq i$, i.e. if A_{uv} does not lie in the same vertical minor of A as $(a)_i^i$, we have $[a']_i^i = [a]_i^i$, and we can treat the matrix A' just as we treated the matrix A in Case I, thus converting it into a symmetric matrix $B = [b]_m^m$ in which $[b]_{i-1}^{i-1} = [a]_{i-1}^{i-1}$, and $(b)_i^i$ is regular with respect to $t, t', t'', \ldots$ and τ. The resultant transformation converting A into B has then the form given in the enunciation.

If $v = i$, i.e. if A_{uv} lies in the same vertical minor of A as $(a)_i^i$, we can take the matrix A' obtained above to be the matrix $B = [b]_m^m$, so that

$$(b)_i^i = A_{ii} + 2\lambda A_{ui} + \lambda^2 A_{uu} = \tau^{\delta_i}\{\tau a_{ii} + 2\lambda a_{ui} + \tau\lambda^2 a_{uu}\},$$

where a_{ui} is not divisible by τ. Then $(b)^i_i$ is regular with respect to τ for all non-zero values of λ, and just as in Case I we can choose λ so that $(b)^i_i$ is also regular with respect to $t, t', t'', \ldots$.

Lemma 1. *We can always convert the symmetric matrix $A=[a]^m_m$ into a similar equipotent symmetric matrix $B=[b]^m_m$ whose leading element b_{11} is regular with respect to any number of given irreducible functions $t, t', t'', \ldots$ by a symmetric equigradent transformation in Ω_1 of the form* (A) *in which every element h of the prefactor and post-factor on the left can be a positive integer.*

We obtain the required transformation by successive applications of Ex. ii, making b_{11} first regular with respect to t; then regular with respect to t and t'; then regular with respect to t, t', t''; and so on.

Lemma 2. *If $i \not> r$, and if the first $i-1$ leading diagonal minor determinants a_{11}, $(a)^2_2, \ldots (a)^{i-1}_{i-1}$ of the symmetric matrix $A=[a]^m_m$ are all regular with respect to every one of the given irreducible functions $t, t', t'', \ldots$, then we can convert A into a similar equipotent symmetric matrix $B=[b]^m_m$ in which $[b]^{i-1}_{i-1}=[a]^{i-1}_{i-1}$ and at the same time the ith leading diagonal minor determinant $(b)^i_i$ is also regular with respect to every one of the functions $t, t', t'', \ldots$ by a symmetric equigradent transformation in Ω_1 of the form*

$$\begin{bmatrix}1, & 0\\ 0, & \omega\end{bmatrix}^{i-1,\, m-i+1}_{i-1,\, m-i+1} [a]^m_m \begin{bmatrix}1, & 0\\ 0, & \omega\end{bmatrix}^{i-1,\, m-i+1}_{i-1,\, m-i+1} = [b]^m_m, \quad \text{......................(B)}$$

where $[\omega]^{m-i+1}_{m-i+1}$ is a matrix having the same general form as the prefactor on the left in (A), *in which every one of the elements h can be a positive integer.*

We obtain the required transformation by successive applications of Ex. iii, making $(b)^i_i$ first regular with respect to t; then regular with respect to t and t'; then regular with respect to t, t', t''; and so on.

PROOF OF THE THEOREM. By applying to the symmetric matrix $A=[a]^m_m$ in succession first Lemma 1, and then Lemma 2 for the values $2, 3, \ldots r$ of i, we obtain a resultant symmetric equigradent transformation of the form (A) converting A into a similar symmetric matrix $B=[b]^m_m$ having the properties mentioned in the theorem given at the commencement of this article.

§ 211. Homogeneous linear transformations of the variables in a rational integral functional matrix.

If $[l]^k_k$ and $\overline{\underline{L}}^k_k$ are two mutually inverse undegenerate square matrices with constant elements lying in any domain of rationality Ω, it has been shown in § 187 that the two mutually inverse transformations

$$\overline{\underline{x}}_k = [l]^k_k \, \overline{\underline{y}}_k, \quad \overline{\underline{y}}_k = \overline{\underline{L}}^k_k \, \overline{\underline{x}}_k \quad \text{..............................(A)}$$

establish a one-one correspondence between all rational integral functions of the k variables $x_1, x_2, \ldots x_k$ which lie in Ω and all rational integral functions of the k variables $y_1, y_2, \ldots y_k$ which lie in Ω, two such functions of the x's and y's corresponding when and only when they are convertible into one another by the substitutions (A).

If $X_1, X_2, \dots X_k$ and $Y_1, Y_2, \dots Y_k$ are the homogeneous linear functions of $x_1, x_2, \dots x_k$ and of $y_1, y_2, \dots y_k$ defined by the equations

$$\overline{\underline{X}}_k = \overline{\underline{L}}^k_k\, \overline{\underline{x}}_k, \quad \overline{\underline{Y}}_k = [l]^k_k\, \overline{\underline{y}}_k,$$

or
$$X_i = L_{1i}x_1 + L_{2i}x_2 + \dots + L_{ki}x_k, \quad Y_i = l_{i1}y_1 + l_{i2}y_2 + \dots + l_{ik}y_k,$$

two corresponding functions U and V of the x's and y's have the forms

$$U = \Sigma a\, x_1^{p_1} x_2^{p_2} \dots x_k^{p_k} = \Sigma b X_1^{q_1} X_2^{q_2} \dots X_k^{q_k},$$

$$V = \Sigma a\, Y_1^{p_1} Y_2^{p_2} \dots Y_k^{p_k} = \Sigma b\, y_1^{q_1}\, y_2^{q_2} \dots y_k^{q_k};$$

the first of the transformations (A) converts U into V; the second of the transformations (A) converts V into U; each of the functions U and V is uniquely determinate when the other is given; each of them vanishes identically when and only when the other vanishes identically; and when U and V do not vanish identically, the degree of V in all the y's is the same as the degree of U in all the x's. Further V is irreducible in Ω when and only when U is irreducible in Ω.

Again when U and V do not vanish identically and are expressed as products of factors irreducible in Ω, we have

$$U = cu_1^{r_1} u_2^{r_2} u_3^{r_3} \dots, \quad V = cv_1^{r_1} v_2^{r_2} v_3^{r_3} \dots,$$

where $u_1, u_2, u_3, \dots$ are distinct irreducible functions in Ω of $x_1, x_2, \dots x_k$; $v_1, v_2, v_3, \dots$ are the corresponding distinct irreducible functions in Ω of $y_1, y_2, \dots y_k$; and c is a non-vanishing constant lying in Ω.

It follows that the two mutually inverse transformations (A) *establish a one-one correspondence between all matrices ϕ whose elements are rational integral functions in Ω of the variables $x_1, x_2, \dots x_k$ and all matrices ψ whose elements are rational integral functions in Ω of the variables $y_1, y_2, \dots y_k$, two corresponding matrices ϕ and ψ being convertible into one another by the substitutions* (A).

Any two corresponding matrices ϕ and ψ have the forms

$$\phi = [a]^n_m = \Sigma [a]^n_m\, x_1^{p_1}\, x_2^{p_2} \dots x_k^{p_k} = \Sigma [\beta]^n_m\, X_1^{q_1}\, X_2^{q_2} \dots X_k^{q_k},$$

$$\psi = [b]^n_m = \Sigma [a]^n_m\, Y_1^{p_1}\, Y_2^{p_2} \dots Y_k^{p_k} = \Sigma [\beta]^n_m\, y_1^{q_1}\, y_2^{q_2} \dots y_k^{q_k},$$

where $[a]^n_m$ and $[\beta]^n_m$ are matrices whose elements are constants in Ω, the coefficients $[\beta]^n_m$ being uniquely determinate when the coefficients $[a]^n_m$ are known, and the coefficients $[a]^n_m$ being uniquely determinate when the coefficients $[\beta]^n_m$ are known. The first of the transformations (A) converts ϕ into ψ; the second of the transformations (A) converts ψ into ϕ; each of the matrices ϕ and ψ vanishes identically when and only when the other vanishes identically; and when ϕ and ψ do not vanish identically, their degrees in all the variables are the same, and each of them is homogeneous when and only when the other is homogeneous.

Since correspondingly formed minor determinants of ϕ and ψ of any given order i are corresponding functions of the x's and y's in the sense just described, we see that any two corresponding matrices ϕ and ψ have the following properties:

(1) The matrices ϕ and ψ are similar and have equal ranks.

(2) The maximum factors D_i and D_i' of ϕ and ψ of any given order i are (when their arbitrary non-vanishing constant factors are suitably chosen) corresponding functions in Ω convertible into one another by the transformations (A).

(3) The potent factors E_i and E_i' of ϕ and ψ of any given order i are (when their arbitrary non-vanishing constant factors are suitably chosen) corresponding functions in Ω convertible into one another by the transformations (A).

(4) There is a one-one correspondence between the irreducible divisors in Ω of ϕ and ψ, any two corresponding irreducible divisors u and v of ϕ and ψ being convertible each into the other by the transformations (A). Also the maximum and potent indices of orders $1, 2, 3, \dots i, \dots$ of v for the matrix ψ are the same as the maximum and potent indices of orders $1, 2, 3, \dots i, \dots$ of u for the matrix ϕ.

(5) There is a one-one correspondence between the maximum and potent divisors of ϕ of any order i and the maximum and potent divisors of ψ of the same order i, any two corresponding maximum (or potent) divisors of ϕ and ψ being convertible each into the other by the substitutions (A).

Ex. i. If $q_1, q_2, \dots q_k$ and $Q_1, Q_2, \dots Q_k$ are two sets of constants in Ω, neither set consisting of 0's only, which satisfy the equations

$$\overline{Q}_k = \overline{l}^{\,k}_{k}\,\overline{q}_k, \quad \overline{q}_k = [L]^k_k\,\overline{Q}_k, \quad \text{.........................(B)}$$

and if $$u = q_1x_1 + q_2x_2 + \dots + q_kx_k = Q_1X_1 + Q_2X_2 + \dots + Q_kX_k,$$

and $$v = Q_1y_1 + Q_2y_2 + \dots + Q_ky_k = q_1Y_1 + q_2Y_2 + \dots + q_kY_k,$$

then v is a linear (or irreducible) divisor of ψ when and only when u is a linear (or irreducible) divisor of ϕ.

For u and v are corresponding functions of the x's and y's.

If $q_1, q_2, \dots q_k$ and $Q_1, Q_2, \dots Q_k$ are two given sets of constants in Ω, neither set consisting of 0's only, then by § 132 we can determine two mutually inverse undegenerate square matrices $[l]^k_k$ and $\overline{L}^{\,k}_{k}$ in Ω so that the equations (B) are satisfied; and if $q_1, q_2, \dots q_k$ are so chosen that u is not a linear divisor of the given matrix $\phi = [a]^n_m$, which is possible whenever ϕ has rank greater than 0, then the given expression v is not a linear divisor of the corresponding matrix $\psi = [b]^n_m$ into which ϕ is converted by the first of the transformations (A).

Thus if $\phi = [a]^n_m$ is a given rational integral functional matrix whose rank is greater than 0, we can convert it by a homogeneous linear transformation in Ω of the variables into a matrix $\psi = [b]^n_m$ of which the given expression v is not a linear divisor.

In particular y_i is a linear divisor of ψ when and only when

$$X_i = L_{1i}x_1 + L_{2i}x_2 + \dots + L_{ki}x_k$$

is a linear divisor of ϕ; and when ϕ is given and has rank greater than 0, it is obvious that we can form an undegenerate matrix $\overline{L}^{\,k}_{k}$ in Ω such that X_i is not a linear divisor of ϕ.

Thus if $\phi = [a]^n_m$ is a given rational integral functional matrix whose rank is greater than 0, we can convert it by a homogeneous linear transformation in Ω of the variables into a matrix $\psi = [b]^n_m$ of which y_i is not a linear divisor.

Ex. ii. *Transformations of a matrix which is homogeneous in two variables x and y.*

Let $\begin{bmatrix} p & q \\ u & v \end{bmatrix}$ and $\begin{bmatrix} P & U \\ Q & V \end{bmatrix}$ be two mutually inverse undegenerate matrices with constant elements, and let

$$X = px + qy, \quad Y = ux + vy, \quad X' = Px' + Uy', \quad Y' = Qx' + Vy'.$$

Then any two homogeneous matrices $\phi=[a]_m^n$, $\phi'=[a']_m^n$ whose elements are rational integral functions of x, y and of x', y' of degree r, and which correspond for the transformations

$$\begin{bmatrix} x \\ y \end{bmatrix}=\begin{bmatrix} P & U \\ Q & V \end{bmatrix}\begin{bmatrix} x' \\ y' \end{bmatrix}, \quad \begin{bmatrix} x' \\ y' \end{bmatrix}=\begin{bmatrix} p & q \\ u & v \end{bmatrix}\begin{bmatrix} x \\ y \end{bmatrix} \quad \text{.........................(C)}$$

have the forms

$$\phi=[a]_m^n= x^r[a]_m^n+ x^{r-1}y[\beta]_m^n+\dots+y^r[\kappa]_m^n$$

$$= X^r[a']_m^n+X^{r-1}Y[\beta']_m^n+\dots+Y^r[\kappa']_m^n, \quad \text{..................(1)}$$

$$\phi'=[a']_m^n= x'^r[a']_m^n+ x'^{r-1}y'[\beta']_m^n+\dots+y'^r[\kappa']_m^n$$

$$=X'^r[a]_m^n+X'^{r-1}Y'[\beta]_m^n+\dots+Y'^r[\kappa]_m^n. \quad \text{..................(2)}$$

The first of the transformations (C) converts ϕ into ϕ'; and the second of the transformations (C) converts ϕ' into ϕ. The identity (2) enables us to express the coefficients $[a']_m^n$, $[\beta']_m^n$, ... in terms of the coefficients $[a]_m^n$, $[\beta]_m^n$, ...; and the identity (1) enables us to express the coefficients $[a]_m^n$, $[\beta]_m^n$, ... in terms of the coefficients $[a']_m^n$, $[\beta']_m^n$, ...

Let (λ, μ) and (λ', μ') be two pairs of constants such that

$$\begin{bmatrix} \lambda' \\ \mu' \end{bmatrix}=\begin{bmatrix} P & Q \\ U & V \end{bmatrix}\begin{bmatrix} \lambda \\ \mu \end{bmatrix}, \quad \begin{bmatrix} \lambda \\ \mu \end{bmatrix}=\begin{bmatrix} p & u \\ q & v \end{bmatrix}\begin{bmatrix} \lambda' \\ \mu' \end{bmatrix}; \quad \text{.....................(C')}$$

and for the values 1, 2, 3, ... of i let (λ_i, μ_i) and (λ_i', μ_i') be two pairs of constants such that

$$\begin{bmatrix} \lambda_i' \\ \mu_i' \end{bmatrix}=\begin{bmatrix} P & Q \\ U & V \end{bmatrix}\begin{bmatrix} \lambda_i \\ \mu_i \end{bmatrix}, \quad \begin{bmatrix} \lambda_i \\ \mu_i \end{bmatrix}=\begin{bmatrix} p & u \\ q & v \end{bmatrix}\begin{bmatrix} \lambda_i' \\ \mu_i' \end{bmatrix}.$$

Then $(\lambda_1'x'+\mu_1'y')$, $(\lambda_2'x'+\mu_2'y')$, ... $(\lambda_s'x'+\mu_s'y')$ are the distinct linear (or irresoluble) divisors of ϕ' when and only when $(\lambda_1x+\mu_1y)$, $(\lambda_2x+\mu_2y)$, ... $(\lambda_sx+\mu_sy)$ are the distinct linear (or irresoluble) divisors of ϕ. Also $(\lambda'x'+\mu'y')^\tau$ is a potent divisor of ϕ' of order i when and only when $(\lambda x+\mu y)^\tau$ is a potent divisor of ϕ of order i. In particular y' is a linear divisor of ϕ' when and only when $Y=ux+vy$ is a linear divisor of ϕ.

If the matrix ϕ is given and has rank greater than 0, we can choose the constants p, q, u, v so that $pv-qu\neq0$, and $ux+vy$ is not a linear divisor of ϕ; then P, Q, U, V are constants such that $PV-QU\neq0$, and y' is not a linear divisor of ϕ'. We can also choose the constants p, q, u, v so that $pv-qu=1$, and $ux+vy$ is not a linear divisor of ϕ; then y' is not a linear divisor of ϕ', and we have

$$\begin{bmatrix} P & U \\ Q & V \end{bmatrix}=\begin{bmatrix} v, & -q \\ -u, & p \end{bmatrix}. \quad \text{...................................(3)}$$

Ex. iii. *Transformations of a matrix which is homogeneous and linear in two variables x and y.*

Using the same transformations (C) as in Ex. ii, we have

$$\phi=[a]_m^n=x[a]_m^n+y[\beta]_m^n=X[a']_m^n+Y[\beta']_m^n,$$

$$\phi'=[a']_m^n=x'[a']_m^n+y'[\beta']_m^n=X'[a]_m^n+Y'[\beta]_m^n.$$

In this case each of the two pairs of coefficients $[a]_m^n$, $[\beta]_m^n$ and $[a']_m^n$, $[\beta']_m^n$ is given in terms of the other pair by the equations

$$[a']_m^n = P[a]_m^n + Q[\beta]_m^n, \quad [\beta']_m^n = U[a]_m^n + V[\beta]_m^n,$$

$$[a]_m^n = p[a']_m^n + u[\beta']_m^n, \quad [\beta]_m^n = q[a']_m^n + v[\beta']_m^n.$$

When $pv - qu = 1$, so that the equation (3) is true, these equations are

$$[a']_m^n = v[a]_m^n - u[\beta]_m^n, \quad [\beta']_m^n = -q[a]_m^n + p[\beta]_m^n,$$

$$[a]_m^n = p[a']_m^n + u[\beta']_m^n, \quad [\beta]_m^n = \;\; q[a']_m^n + v[\beta']_m^n.$$

§ 212. Homogenisation of a rational integral functional matrix.

1. *Homogenisation by a linear transformation of the variables.*

Let $[l]_{k+1}^{k+1}$ and $\underline{\overline{L}}_{k+1}^{k+1}$ be two mutually inverse undegenerate square matrices whose elements are constants lying in any domain of rationality Ω; and let $X_1, X_2, \ldots X_{k+1}$ and $Y_1, Y_2, \ldots Y_{k+1}$ be respectively the linear functions of $x_1, x_2, \ldots x_k$ and the homogeneous linear functions of $y_1, y_2, \ldots y_{k+1}$ defined by the equations

$$\underline{\overline{X}}_{k+1} = \underline{\overline{L}}_{k+1}^{k+1} \underline{\overline{\begin{matrix} x \\ 1 \end{matrix}}}_{k,\,1}, \quad \underline{\overline{Y}}_{k+1} = [l]_{k+1}^{k+1} \underline{\overline{y}}_{k+1},$$

or $$X_i = L_{1i}x_1 + L_{2i}x_2 + \ldots + L_{ki}x_k + L_{k+1,\,i}, \quad Y_i = l_{i1}y_1 + l_{i2}y_2 + \ldots + l_{i,\,k+1}y_{k+1}.$$

If r and s are any two given positive integers such that $r \not< s$, it has been shown in Note 3 of § 187.3 that the two mutually inverse transformations

$$\underline{\overline{\begin{matrix} x \\ 1 \end{matrix}}}_{k,\,1} = [l]_{k+1}^{k+1} \underline{\overline{y}}_{k+1}, \quad \underline{\overline{y}}_{k+1} = \underline{\overline{L}}_{k+1}^{k+1} \underline{\overline{\begin{matrix} x \\ 1 \end{matrix}}}_{k,\,1} \quad \ldots\ldots\ldots\ldots\ldots\ldots\ldots\text{(A)}$$

establish a one-one correspondence between

(1) all rational integral functions in Ω of the k variables $x_1, x_2, \ldots x_k$ having degree s in all the variables;

(2) all those homogeneous rational integral functions in Ω of the $k+1$ variables $y_1, y_2, \ldots y_{k+1}$ having degree r in all the variables which are divisible by Y_{k+1}^{r-s} and are not divisible by any higher power of Y_{k+1};

two such functions corresponding when and only when they are convertible into one another by the transformations (A)

Any two corresponding functions U and V' of degrees s and r of the x's and y's have the forms

$$U = \quad \Sigma a\, x_1^{p_1} x_2^{p_2} \ldots x_k^{p_k} \quad = \Sigma b X_1^{q_1} X_2^{q_2} \ldots X_{k+1}^{q_{k+1}},$$

$$V' = Y_{k+1}^{r-s} \Sigma a\, Y_1^{p_1} Y_2^{p_2} \ldots Y_k^{p_k} Y_{k+1}^{p_{k+1}} = \Sigma b\; y_1^{q_1} y_2^{q_2} \ldots y_{k+1}^{q_{k+1}},$$

where $$p_1 + p_2 + \ldots + p_k + p_{k+1} = s, \quad q_1 + q_2 + \ldots + q_{k+1} = r;$$

and each of them vanishes identically when and only when the other vanishes identically. We call V' the homogenised function of degree r corresponding to U.

If $V=\Sigma a Y_1^{p_1} Y_2^{p_2} \dots Y_{k+1}^{p_{k+1}}=\Sigma c y_1^{s_1} y_2^{s_2} \dots y_{k+1}^{s_{k+1}}$, where $s_1+s_2+\dots+s_{k+1}=s$,

we have $$V'=Y_{k+1}^{r-s} V,$$

V being the homogenised function of degree s corresponding to U, or the function obtained by homogenising U without raising its degree, which is not divisible by the function Y_{k+1}.

When U and V' do not vanish identically, and when we express them as products of irreducible factors in Ω, we have

$$U=cu_1^{\tau_1}u_2^{\tau_2}u_3^{\tau_3}\dots, \quad V'=cY_{k+1}^{r-s}v_1^{\tau_1}v_2^{\tau_2}v_3^{\tau_3}\dots,$$

where $u_1, u_2, u_3, \dots$ are distinct irreducible functions in Ω of $x_1, x_2, \dots x_k$; $v_1, v_2, v_3, \dots$ are homogeneous irreducible functions in Ω of $y_1, y_2, \dots y_{k+1}$ which are distinct from one another and distinct from the irreducible function Y_{k+1}; and c is a non-vanishing constant in Ω. The functions $v_1, v_2, v_3, \dots$ are the homogenised functions corresponding to and having the same degrees as $u_1, u_2, u_3, \dots$.

In particular U is a non-vanishing constant c when and only when

$$U=c, \quad V'=cY_{k+1}^{r-s}.$$

If r is any given positive integer, it follows that:

The two mutually inverse transformations (A) *establish a one-one correspondence between*

(1) *all matrices ϕ which are rational integral functions in Ω of degree r of the k variables $x_1, x_2, \dots x_k$;*

(2) *all those matrices ψ which are homogeneous rational integral functions in Ω of the same degree r of the $k+1$ variables $y_1, y_2, \dots y_{k+1}$, and whose elements are not all divisible by the homogeneous linear function Y_{k+1};*

two such matrices ϕ and ψ corresponding when and only when they are convertible into one another by the transformations (A).

Any two corresponding matrices ϕ and ψ of degree r have the forms

$$\phi=[a]_m^n=\Sigma\,[a]_m^n\, x_1^{p_1} x_2^{p_2}\dots x_k^{p_k} \qquad =\Sigma\,[\beta]_m^n\, X_1^{q_1} X_2^{q_2}\dots X_{k+1}^{q_{k+1}},$$

$$\psi=[b]_m^n=\Sigma\,[a]_m^n\, Y_1^{p_1} Y_2^{p_2}\dots Y_k^{p_k} Y_{k+1}^{p_{k+1}}=\Sigma\,[\beta]_m^n\, y_1^{q_1} y_2^{q_2}\dots y_{k+1}^{q_{k+1}},$$

where $$p_1+p_2+\dots+p_k+p_{k+1}=r, \quad q_1+q_2+\dots+q_{k+1}=r;$$

and each of them vanishes identically when and only when the other vanishes identically. The first of the transformations (A) converts ϕ into ψ, and the second of the transformations (A) converts ψ into ϕ. When ϕ and ψ do not vanish identically, there are terms in the first expression for ϕ with non-vanishing coefficients in which $p_1+p_2+\dots+p_k=r$; and in the first expression for ψ there are corresponding terms in which $p_{k+1}=0$, and these are not divisible by Y_{k+1}. In this case there cannot be two different homogeneous matrices which are derived from ψ by the second of the substitutions (A); consequently each of the matrices ϕ and ψ is uniquely determinate when the other matrix is given. The coefficients $[\beta]_m^n$ are uniquely determinate when the coefficients $[a]_m^n$ are known; and the coefficients $[a]_m^n$ are uniquely determinate when the coefficients $[\beta]_m^n$ are known. We call ψ the homogenised matrix which corresponds to ϕ and has the same degree as ϕ.

Any two corresponding elements a_{ij} and b_{ij} of ϕ and ψ have the forms

$$a_{ij}=\Sigma a_{ij} x_1^{p_1} x_2^{p_2} \dots x_k^{p_k}, \quad b_{ij}=\Sigma a_{ij} Y_1^{p_1} Y_2^{p_2} \dots Y_k^{p_k} Y_{k+1}^{p_{k+1}};$$

and b_{ij} is the homogenised function of degree r corresponding to a_{ij}. If a_{ij} has degree r, b_{ij} is not divisible by Y_{k+1}; and if a_{ij} has degree s, where $s<r$, then b_{ij} is divisible by Y_{k+1}^{r-s} and is not divisible by any higher power of Y_{k+1}. One of two corresponding elements a_{ij} and b_{ij} vanishes identically when and only when the other vanishes identically.

Ex. i. Two corresponding matrices ϕ and ψ have equal values for corresponding sets of finite values of the x's and y's satisfying the equations (A).

Ex. ii. One of two corresponding matrices ϕ and ψ is a matrix $[c]_m^n$ with constant elements when and only when

$$\phi=[c]_m^n, \quad \psi=[c]_m^n.$$

Let F be any homogeneous rational integral function in Ω of the elements of ϕ having degree i in those elements, and let G' be the same function of the elements of ψ. Then G' is a homogeneous rational integral function in Ω of the variables $y_1, y_2, \dots y_{k+1}$ having degree ir in all the variables. Since the second of the transformations (A) converts G' into F, and there are not two different homogeneous rational integral functions of $y_1, y_2, \dots y_{k+1}$ of the same degree which are converted into F by that transformation, G' must be the homogenised function of degree ir corresponding to F. Hence G' vanishes identically when and only when F vanishes identically; and when F and G' do not vanish identically and F has degree s in the variables $x_1, x_2, \dots x_k$, we can write

$$F=\Sigma a x_1^{p_1} x_2^{p_2} \dots x_k^{p_k}, \quad G'=Y_{k+1}^{ir-s} G=Y_{k+1}^{ir-s} \Sigma a Y_1^{p_1} Y_2^{p_2} \dots Y_k^{p_k} Y_{k+1}^{p_{k+1}},$$

where $\quad p_1+p_2+\dots+p_k \ngtr s, \quad p_1+p_2+\dots+p_k+p_{k+1}=s;$

and where there are terms in which $p_1+p_2+\dots+p_k=s$, $p_{k+1}=0$.

The function $G=\Sigma a Y_1^{p_1} Y_2^{p_2} \dots Y_k^{p_k} Y_{k+1}^{p_{k+1}}$ is then the homogenised function of $y_1, y_2, \dots y_{k+1}$ which corresponds to F and has the same degree as F, and it is not divisible by the linear function Y_{k+1}. When F and G' do not vanish identically and are expressed as products of factors irreducible in Ω, we have

$$F=cu_1^{\tau_1} u_2^{\tau_2} u_3^{\tau_3} \dots, \quad G'=cY_{k+1}^{ir-s} v_1^{\tau_1} v_2^{\tau_2} v_3^{\tau_3} \dots, \quad \dots\dots\dots\dots\dots\dots (\text{A}')$$

where c is a non-vanishing constant in Ω; $u_1, u_2, u_3, \dots$ are distinct irreducible functions in Ω of $x_1, x_2, \dots x_k$; $v_1, v_2, v_3, \dots$ are homogeneous irreducible functions in Ω of $y_1, y_2, \dots y_{k+1}$ distinct from one another and distinct from the irreducible function Y_{k+1}; and $v_1, v_2, v_3, \dots$ are the homogenised functions which correspond to and have the same degrees as $u_1, u_2, u_3, \dots$.

Taking F and G' to be correspondingly formed minor determinants of ϕ and ψ of order i, we draw the following conclusions:

(1) Two corresponding matrices ϕ and ψ have the same rank, which will be denoted by ρ.

(2) There is a one-one correspondence between the irreducible divisors of ϕ and those (homogeneous) irreducible divisors of ψ which are distinct from Y_{k+1}, any two corresponding irreducible divisors of ϕ and ψ being convertible into one another by the transformations (A).

(3) If u and v are a pair of corresponding irreducible divisors of ϕ and ψ (v being distinct from Y_{k+1}), the maximum and potent indices of v of orders $1, 2, \dots \rho$ for the matrix ψ are respectively the same as the maximum and potent indices of u of orders $1, 2, \dots \rho$ for the matrix ϕ.

When the homogeneous linear function Y_{k+1} is not an irreducible divisor of ψ, there is a one-one correspondence between all the irreducible divisors of ϕ and all the irreducible divisors of ψ; also there is a one-one correspondence between the maximum and potent divisors of orders $1, 2, \dots \rho$ and the maximum and potent factors of orders $1, 2, \dots \rho$ of ϕ and ψ; any two corresponding factors or divisors of the same order having the same degree, and being convertible each into the other by the substitutions (A).

NOTE. *Homogenisation with change of degree.*

If r and s are any two positive integers such that $r \nless s$, we can show in the same way that the two mutually inverse transformations (A) establish a one-one correspondence between

(1) all matrices ϕ which are rational integral functions in Ω of degree s of the k variables $x_1, x_2, \dots x_k$;

(2) all those matrices ψ' which are homogeneous rational integral functions in Ω of degree r of the $k+1$ variables $y_1, y_2, \dots y_{k+1}$, and whose elements are all divisible by Y_{k+1}^{r-s} and not all divisible by any higher power of Y_{k+1};

two such matrices ϕ and ψ' corresponding when and only when they are convertible into one another by the transformations (A).

2. *Homogenisation by the introduction of a new variable.*

In the particular case when $[l]_{k+1}^{k+1}=[1]_{k+1}^{k+1}$, the transformations (A) become

$$x_1=y_1,\ x_2=y_2,\ \dots\ x_k=y_k,\ 1=y_{k+1};\quad y_1=x_1,\ y_2=x_2,\ \dots\ y_k=x_k,\ y_{k+1}=1;\ \dots\text{(B)}$$

and we have

$$[X_1 X_2 \dots X_k X_{k+1}]=[x_1 x_2 \dots x_k 1],\quad [Y_1 Y_2 \dots Y_k Y_{k+1}]=[y_1 y_2 \dots y_k y_{k+1}].$$

In this case the equations (A′) are replaced by

$$F=cu_1^{\tau_1}u_2^{\tau_2}u_3^{\tau_3}\dots,\quad G'=y_{k+1}^{ir-s}v_1^{\tau_1}v_2^{\tau_2}v_3^{\tau_3}\dots,\quad \dots\dots\dots\text{(B′)}$$

and we have the same results as before, Y_{k+1} being now y_{k+1}.

We can replace $y_1, y_2, \dots y_k, y_{k+1}$ by $x_1, x_2, \dots x_k, x_{k+1}$, regarding x_{k+1} as a new variable by means of which the homogenisation is effected. Then instead of (A) we have the transformations

$$1=x_{k+1};\quad x_{k+1}=1;\ \dots\dots\dots\text{(C)}$$

two corresponding matrices ϕ and ψ of degree r have the forms

$$\phi=[a]_m^n=\Sigma\,[a]_m^n x_1^{p_1}x_2^{p_2}\dots x_k^{p_k},\quad \psi=[b]_m^n=\Sigma\,[a]_m^n x_1^{p_1}x_2^{p_2}\dots x_k^{p_k}x_{k+1}^{p_{k+1}},$$

where $p_1+p_2+\dots+p_k+p_{k+1}=r$; and the equations (A′) are replaced by

$$F=cu_1^{\tau_1}u_2^{\tau_2}u_3^{\tau_3}\dots,\quad G'=x_{k+1}^{ir-s}v_1^{\tau_1}v_2^{\tau_2}v_3^{\tau_3}\dots\quad \dots\dots\dots\text{(C′)}$$

We have the same results as before, Y_{k+1} being replaced by x_{k+1}.

CHAPTER XXIV

EQUIPOTENT TRANSFORMATIONS OF RATIONAL INTEGRAL FUNCTIONAL MATRICES

[The first three articles of this chapter (with Note 4 of § 219) deal with rational integral functional matrices in general; the remaining six articles deal with rational integral x-matrices. In §§ 213 and 214 we define equipotent matrices and impotent matrices, and describe the chief features of equipotent transformations. In §§ 215 and 216 we investigate certain properties of one-rowed matrices connected with the long rows of an undegenerate matrix; in § 217 we use these properties in constructing certain uniquely determinate equipotent transformations of a given rational integral x-matrix; and in § 218 we show that every such matrix can be derived from an undegenerate square matrix by equipotent transformations. Methods of reducing a given rational integral x-matrix whose potent factors are known to its standard form by equipotent transformations are obtained in §§ 216—8. The usual method of effecting the reduction, which is applicable when the potent factors are not known, is described in § 219. In § 220 we describe transformations which are necessary and sufficient conditions that two given rational integral x-matrices shall be equipotent; and in § 221 we find the circumstances under which it is possible to convert one given rational integral x-matrix into another by a rational integral transformation.]

§ 213. Equipotent and impotent rational integral functional matrices.

1. *Equipotent matrices.*

Two matrices $A=[a]_p^q$, $B=[b]_m^n$ whose elements are rational integral functions of certain scalar variables $x, y, z, \ldots$ will be said to be *equipotent* with one another when and only when any one of the following sets of equivalent conditions is satisfied:

(a) *They have the same rank and the same potent divisors.*

(b) *They have the same rank and the same potent factors.*

(c) *They have the same rank and the same maximum divisors.*

(d) *They have the same rank and the same maximum factors.*

(e) *They have the same potent factors of all orders.*

(f) *They have the same maximum factors of all orders.*

It will generally be understood that the potent and maximum divisors are those corresponding to the various irresoluble divisors of A and B; they may however be those corresponding to the various irreducible divisors in Ω

of A and B, where Ω is any domain of rationality in which both A and B lie. It is also to be understood that a matrix of rank r has no potent or maximum divisor of any order greater than r and no potent or maximum factor of any order greater than r, and that a matrix of rank 0 has no potent or maximum divisor and no potent or maximum factor. Further in (a), (b), (c), (d) it is to be understood that a repeated potent or maximum divisor and a repeated potent or maximum factor is repeated the same number of times in each of the two matrices.

If A and B have the same potent (or maximum) factors of all orders, they have the same number of potent (or maximum) factors, i.e. they have the same rank. Consequently the conditions (e) and (f) are equivalent to the conditions (b) and (d) respectively, and we only have to prove the mutual equivalence of the conditions (a), (b), (c), (d).

When A and B have the same rank r and the same potent divisors, they have the same potent divisors of all orders. For if

$$t^{e_s},\ t^{e_{s+1}},\ \dots\ t^{e_i},\ \dots\ t^{e_{r-1}},\ t^{e_r}$$

are their common potent divisors corresponding to the irresoluble (or irreducible) divisor t and arranged in such an order that

$$e_s \not> e_{s+1} \not> \dots \not> e_i \not> \dots \not> e_{r-1} \not> e_r,$$

where $e_s > 0$, then t^{e_i} is necessarily a potent divisor of order i for both matrices. Consequently t has the same potent and maximum indices of orders $1, 2, \dots r$ for both the matrices A and B; and A and B have the same maximum divisors, the same potent factors and the same maximum factors of all orders. Hence when the conditions (a) are satisfied, the conditions (a), (b), (c), (d) are all satisfied.

Again when A and B have the same rank and the same potent factors, they have the same potent factors of all orders. For if

$$E_1,\ E_2,\ \dots\ E_i,\ \dots\ E_{r-1},\ E_r$$

are their common potent factors arranged in such an order that each of them after the first is divisible by the preceding, then E_i is necessarily the potent factor of order i of both matrices, and the matrices A and B have the same maximum factors, and the same maximum divisors and the same potent divisors of all orders. Hence when the conditions (b) are satisfied, the conditions (a), (b), (c), (d) are all satisfied.

Reasoning in a similar way from the conditions (c) and (d), we see that if any one of the four sets of conditions (a), (b), (c), (d) is satisfied, then they are all satisfied. Consequently the conditions (a), (b), (c), (d) are mutually equivalent. We shall usually regard (a) as the fundamental definition of equipotence.

Ex. i. *The two matrices A and B may have the same potent divisors when their ranks are not the same. In such a case they are not equipotent.*

If A and B have the same potent divisors, and if the ranks of A and B are r and s respectively, then the potent divisors of A of order $r-i$ are the same as the potent divisors of B of order $s-i$ for all permissible values of i; and if $r \nless s$ the matrix A cannot have a potent divisor of any one of the orders $1, 2, \ldots r-s$. Hence when r and s are different the two matrices have not the same potent divisors of all orders except in the special case when each matrix has no potent divisors.

Ex. ii. *If A and B are both impotent, their ranks are not necessarily the same, and they are therefore not necessarily equipotent.*

Ex. iii. *If A and B are not both impotent, and if they have the same potent (or maximum) divisors of all orders, then they must have equal ranks and must be equipotent.*

We do not include 1 amongst the irresoluble (or irreducible) divisors of a matrix. If we did include it, then A and B would be equipotent when and only when they have the same potent (or maximum) divisors of all orders.

Ex. iv. Since a matrix of rank 0 has no potent divisors, all matrices of rank 0 can be regarded as being equipotent with one another.

Ex. v. *Any two matrices which differ only by rows of 0's (or any two conventionally equal matrices) are equipotent with one another.*

For they have the same ranks and the same maximum factors.

Ex. vi. All derangements of a rational integral functional matrix A are equipotent with A.

Ex. vii. *If the matrix A can be converted into the matrix B by an equigradent transformation, then A and B are equipotent.*

This has been proved in Ex. ii of § 203, and is included in Theorem II of § 214.

NOTE. *Correction to* § 74. The definition of equipotence given above should be substituted for that given in § 74. When that is done, all the four results mentioned in Ex. iv of § 74 are necessarily true, the last result following from Ex. v of § 214; but the last result is not true universally when the definition of § 74 is adhered to. It will be shown in § 214 that if two matrices are equipotent according to the definition of § 74, then they are equipotent according to the definition of this article; but by Note 4 of § 219 the converse is not true generally when there are more than two variables.

2. *Impotent matrices.*

A rational integral functional matrix $A=[a]_m^n$ of any rank r will be said to be *impotent* when it has no potent divisors. If $r=0$, the matrix A is necessarily impotent. If $r>0$, it is impotent when and only when its potent factors $E_1, E_2, \ldots E_r$ are all non-vanishing constants, i.e. when and only when its maximum factors $D_1, D_2, \ldots D_r$ are all non-vanishing constants; and this is the case when and only when D_r, the H.C.F. of its minor determinants of order r, is a non-vanishing constant which we can take to be 1; or when and only when E_r is a non-vanishing constant which we can take to be 1. A matrix whose elements are chosen at random will usually be impotent when it is not square and not impotent when it is square.

We shall also speak of a matrix as being *impotent* with respect to a particular irresoluble (or irreducible) function t when t is not one of its irresoluble (or irreducible) divisors, so that no one of its potent divisors is a power of t; and we shall sometimes speak of a matrix as being *impotent* with respect to a particular variable x when all its irresoluble (or irreducible) divisors are independent of x.

Ex. viii. A rational integral functional matrix A of rank r is impotent when and only when every irresoluble (or irreducible) function of the variables which occur in it has at least one finite root for which A has rank r. Also it is impotent with respect to a particular irresoluble (or irreducible) function t of those variables when and only when t has at least one finite root for which A has rank r.

These results have been proved in Ex. i of § 202.

Ex. ix. *If the rational integral functional matrix* $\phi=[a]_m^n$ *has rank* r *and is impotent, and if* $s \not> r$ *and* $s \not< 1$, *then every complete matrix* $\Phi=[A]_\mu^\nu$ *of the minor determinants of* ϕ *of order* s *is impotent.*

For when ϕ is impotent, it has rank r for at least one finite root of every irresoluble function f of the variables which occur in it; therefore by § 73 the matrix Φ has rank $\binom{r}{s}$ for at least one finite root of f; therefore by Ex. i of § 201 and Ex. viii above the matrix Φ is impotent. The result also follows from § 208, and we see further that:

If ϕ *is impotent with respect to a particular irresoluble (or irreducible) function* t, *then* Φ *is also impotent with respect to* t.

Ex. x. *If the rational integral functional matrix* $A=[a]_m^n$ *is undegenerate and impotent, then every long minor of* A *is undegenerate and impotent.*

Let A' be any long minor of A formed with r of the long rows of A. Then A' has rank r and is *undegenerate*, and every simple minor determinant of A can be expanded in terms of the minor determinants of A' of order r. Hence if f is any common factor of the minor determinants of A' of order r, then f is also a common factor of all the simple minor determinants of A, and must therefore be a non-vanishing constant, i.e. the matrix A' is *impotent.*

Ex. xi. *If the rational integral functional matrices* $[a]_m^r$ *and* $[b]_r^n$ *are both impotent and both have rank* r, *then the product*

$$[c]_m^n=[a]_m^r\,[b]_r^n \quad\text{...(1)}$$

is also impotent and has rank r.

By § 131 the matrix $[c]_m^n$ has rank r. When we equate correspondingly formed complete matrices of the minor determinants of order r on both sides of (1), we obtain an equation of the form

$$[C]_\mu^\nu=[A]_\mu^1\,[B]_1^\nu, \quad \text{where} \quad \mu=\binom{m}{r}, \quad \nu=\binom{n}{r},$$

and all three matrices have rank 1. If the elements of $[C]^{\nu}_{\mu}$ had a common irresoluble factor g, then g would be a common factor of all elements of the matrix

$$A_{i1}\,[B_{11}\,B_{12}\ldots B_{1\nu}],$$

and since $B_{11}, B_{12}, \ldots B_{1\nu}$ have no irresoluble factor in common, g would be a factor of A_{i1}. In the same way we see that g would be a common factor of $A_{11}, A_{21}, \ldots A_{\mu 1}$; and this is impossible, since these functions have no common factor other than a non-vanishing constant. Thus the elements of $[C]^{\nu}_{\mu}$, which are the minor determinants of $[c]^{n}_{m}$ of order r, have no common factor other than a constant, i.e. the matrix $[c]^{n}_{m}$ is impotent.

The above theorem can be deduced from Ex. ii of § 206, and is included as a special case in Theorem I of § 214.

Ex. xii. *Properties of a square matrix which is undegenerate and impotent.*

Let $A=[a]^{m}_{m}$ be a square matrix whose elements are rational integral functions of the scalar variables $x, y, z, \ldots$. Then when A is undegenerate and impotent, it has the following properties:

(1) *Its determinant* $\Delta=(a)^{m}_{m}$ *is a non-zero constant.*

For this is the necessary and sufficient condition that A shall be undegenerate and impotent.

(2) *Its reciprocal, conjugate reciprocal and inverse are also square rational integral functional matrices, and are also undegenerate and impotent.*

If $\Delta=(a)^{m}_{m}$, and if $[A]^{m}_{m}$ is the reciprocal of A, we have

$$[a]^{m}_{m}\,\overline{\underline{A}}{}^{m}_{m}=[1]^{m}_{m},\quad (A)^{m}_{m}=\Delta^{m-1};$$

and since Δ is a non-zero constant, it follows that $(A)^{m}_{m}$ is a non-zero constant, i.e. $[A]^{m}_{m}$ is undegenerate and impotent. Again the inverse of A is the rational integral functional matrix $\frac{1}{\Delta}\overline{\underline{A}}{}^{m}_{m}$ whose determinant is the non-zero constant $\frac{1}{\Delta}$; therefore it also is undegenerate and impotent.

(3) *Every complete matrix* $[A]^{\mu}_{\mu}$ *of the minor determinants of* A *of any order* s *is undegenerate and impotent.*

This has been proved in Ex. ix. It also follows from the theorem of § 120, which shows that $(A)^{\mu}_{\mu}$ is a non-zero constant.

(4) *Every horizontal and vertical minor of* A *is an undegenerate impotent matrix and has an inverse matrix which is undegenerate and impotent.*

Let $[a_{p1}]^{m}_{r}$ be any horizontal minor of A of reduced order r. Then if $\overline{\underline{A}}{}^{m}_{m}$ is the inverse matrix of A, we have

$$[a]^{m}_{m}\,\overline{\underline{A}}{}^{m}_{m}=[1]^{m}_{m},\quad [a_{p1}]^{m}_{r}\,\overline{\underline{A_{p1}}}{}^{r}_{m}=[1]^{r}_{r},\quad \det[a_{p1}]^{m}_{r}\,\overline{\underline{A_{p1}}}{}^{r}_{m}=1.$$

The last equation shows that every common factor of the simple minor determinants of either of the undegenerate rational integral functional matrices $[a_{p1}]^{m}_{r}$, $\overline{\underline{A_{p1}}}{}^{r}_{m}$ must be a factor of 1, and must therefore be a non-vanishing constant. Consequently $[a_{p1}]_r^m$ is undegenerate and impotent and has an inverse matrix $\overline{\underline{A_{p1}}}{}_m^r$ which is also undegenerate and impotent.

(5) *If A is homogeneous in the variables, it can only be impotent when its degree in all the variables is 0, i.e. when all its elements are constants.*

This follows from the first property.

Ex. xiii. *If the square matrix $[h]_m^m$ is undegenerate and impotent, then every square matrix $[a]_m^m$ of order m which is rational and integral in the variables can be expressed uniquely in the forms*

$$[a]_m^m=[u]_m^m[h]_m^m,\quad [a]_m^m=[h]_m^m[v]_m^m, \qquad\text{.............................(2)}$$

where $[u]_m^m$ and $[v]_m^m$ are rational and integral in the variables.

For the inverse of $[h]_m^m$ is a uniquely determinate matrix $\overline{\underline{H}}{}_m^m$ which is rational and integral in the variables, and is also undegenerate and impotent; and the respective equations (2) are satisfied identically when and only when

$$[u]_m^m=[a]_m^m\overline{\underline{H}}{}_m^m,\quad [v]_m^m=\overline{\underline{H}}{}_m^m[a]_m^m.$$

More generally if $[h]_m^m$ and $[k]_n^n$ are given square matrices which are undegenerate and impotent, every matrix $[a]_m^n$ of orders m and n which is rational and integral in the variables can be expressed uniquely in the form

$$[a]_m^n=[h]_m^m[b]_m^n[k]_n^n, \qquad\text{....................................(3)}$$

where $[b]_m^n$ is rational and integral in the variables.

For if $\overline{\underline{H}}{}_m^m$ and $\overline{\underline{K}}{}_n^n$ are the inverses of $[h]_m^m$ and $[k]_n^n$, the equation (3) is satisfied identically when and only when

$$[b]_m^n=\overline{\underline{H}}{}_m^m[a]_m^n\overline{\underline{K}}{}_n^n.$$

§ 214. Equipotent transformations of a rational integral functional matrix.

From the theorems of § 206 we deduce the two theorems which follow, the first of which is included in the second.

Theorem I. *If $[a]_m^r$, $[b]_r^n$, $[c]_m^n$ are matrices whose elements are rational integral functions of the scalar variables x, y, z, ..., and if the equation*

$$[a]_m^r[b]_r^n=[c]_m^n \qquad\text{.............................(A)}$$

is an identity in the variables x, y, z, ..., then:

(1) *If $[a]_m^r$ has rank r and is impotent, $[c]_m^n$ is equipotent with $[b]_r^n$.*

(2) *If $[b]_r^n$ has rank r and is impotent, $[c]_m^n$ is equipotent with $[a]_m^r$.*

Suppose that $[a]^r_m$ has rank r and is impotent, and let $[b]^n_r$ have rank s. Then by § 131 the matrix $[c]^n_m$ also has rank s; and by Exs. ii and iii of § 206 the matrices $[c]^n_m$ and $[b]^n_r$ have the same irresoluble divisors, and the same potent divisors of orders 1, 2, ... s, i.e. they are equipotent.

Theorem II. *If $[a]^q_p$, $[b]^n_m$, $[h]^p_m$, $[k]^n_q$ are matrices whose elements are rational integral functions of the scalar variables $x, y, z, \ldots$, the matrices $[h]^p_m$ and $[k]^n_q$ being impotent and having ranks p and q respectively, and if the equation*

$$[h]^p_m [a]^q_p [k]^n_q = [b]^n_m \quad \text{.........................(B)}$$

is an identity in the variables $x, y, z, \ldots$, then the matrices $[a]^q_p$ and $[b]^n_m$ are equipotent with one another.

Let the matrix $[a]^q_p$ have rank r. Then by § 131 the matrix $[b]^n_m$ also has rank r, and by Ex. vii of § 206 the matrices $[a]^q_p$ and $[b]^n_m$ have the same irresoluble divisors, and the same potent divisors of orders 1, 2, ... r, i.e. they are equipotent.

On account of these properties a relation of the form (B) in which $[h]^p_m$ and $[k]^n_q$ are undegenerate impotent matrices having ranks p and q equal to their respective passivities will be called an *equipotent transformation* converting the rational integral functional matrix $[a]^q_p$ into the rational integral functional matrix $[b]^n_m$. Further (B) will be called an equipotent transformation in Ω when Ω is any domain of rationality in which both the matrices $[h]^p_m$ and $[k]^n_q$ lie. Every equigradent transformation is also an equipotent transformation.

NOTE 1. *Equipotent transformations of a matrix into a similar matrix.*

An equipotent transformation converting $[a]^n_m$ into a similar matrix $[b]^n_m$ has the form

$$[h]^m_m [a]^n_m [k]^n_n = [b]^n_m, \quad \text{.................................(C)}$$

where $[h]^m_m$ and $[k]^n_n$ are square rational integral functional matrices which are undegenerate and impotent, i.e. where $(h)^m_m$ and $(k)^n_n$ are non-zero constants.

If $\overline{\underline{H}}{}^m_m$ and $\overline{\underline{K}}{}^n_n$ are the inverse matrices of $[h]^m_m$ and $[k]^n_n$, the determinants $(H)^m_m$ and $(K)^n_n$ are also non-zero constants (see Ex. xii of § 213), and $\overline{\underline{H}}{}^m_m$ and $\overline{\underline{K}}{}^n_n$ are square rational integral functional matrices which are undegenerate and impotent. By prefixing and postfixing these matrices on both sides of (C) we deduce the identical equation

$$\overline{\underline{H}}{}^m_m [b]^n_m \overline{\underline{K}}{}^n_n = [a]^n_m, \quad \text{.................................(C')}$$

which is an equipotent transformation converting $[b]_m^n$ into $[a]_m^n$; and each of the equations (C) and (C′) is true when and only when the other is true. We call (C′) the transformation *inverse* to (C).

Without using Theorem II we can show in a simpler way that two similar rational integral functional matrices $A=[a]_m^n$ and $B=[b]_m^n$ satisfying an equation of the form (C) are equipotent with one another. From (C) we see that the rank of B cannot exceed the rank of A, and from (C′) we see that the rank of A cannot exceed the rank of B. Therefore A and B have the same rank. Let their common rank be r, let s be any positive integer which is not greater than r, and let α_s and β_s be the highest common factors of the minor determinants of order s of A and B respectively. When we equate correspondingly formed complete matrices of the minor determinants of order s on both sides of (C), we see that α_s is a common factor of all minor determinants of B of order s. Therefore α_s is a factor of β_s. Treating the equation (C′) in the same way, we see that β_s is a factor of α_s. It follows that α_s and β_s can only differ by a non-zero constant factor. Thus the matrices A and B have the same rank and the same maximum factors, and they are therefore equipotent with one another.

NOTE 2. *Composition of equipotent transformations.*

Any number of equipotent transformations applied in succession to a rational integral functional matrix are together equivalent to a single resultant equipotent transformation applied to that matrix. For if

$$[h']_r^u [a]_u^v [k']_v^s = [a']_r^s, \quad [h'']_p^r [a']_r^s [k'']_s^q = [a'']_p^q, \quad [h''']_m^p [a'']_p^q [k''']_q^n = [a''']_m^n$$

are equipotent transformations converting

$$[a]_u^v \text{ into } [a']_r^s, \quad [a']_r^s \text{ into } [a'']_p^q, \quad [a'']_p^q \text{ into } [a''']_m^n$$

respectively, we have

$$[h]_m^u [a]_u^v [k]_v^n = [a''']_m^n, \quad \text{......................................}(1)$$

where

$$[h]_m^u = [h''']_m^p [h'']_p^r [h']_r^u, \quad [k]_v^n = [k']_v^s [k'']_s^q [k''']_q^n.$$

By Theorem I the matrix $[h]_m^u$ is equipotent with $[h'']_p^r [h']_r^u$, and this latter matrix is equipotent with $[h']_r^u$. Thus $[h]_m^u$, being equipotent with $[h']_r^u$, is an undegenerate impotent matrix of rank u. Similarly $[k]_v^n$ is an undegenerate impotent matrix of rank v. Consequently (1) is an equipotent transformation converting $[a]_u^v$ into $[a''']_m^n$; and it is the resultant of the three successive equipotent transformations given above.

If the component transformations all lie in any domain of rationality Ω, then the resultant transformation also lies in Ω. Also if the component transformations are all symmetric, then the resultant transformation is symmetric.

NOTE 3. *Elementary equipotent transformations.*

The following will be called *elementary* equipotent transformations of a rational integral functional matrix.

(*a*) The insertion of an additional horizontal or vertical row of 0's in any position.

(*b*) The interchange of any two parallel rows (horizontal or vertical).

(*c*) The addition to any row of any parallel row multiplied by a rational integral function of the variables.

(*d*) The multiplication of any horizontal or vertical row by a non-zero scalar constant.

All transformations of the types (*b*), (*c*), (*d*) have the same form as (C) in Note 1, and convert any matrix into a similar matrix.

A transformation of the type (*c*) has one of the forms

$$[h]_m^m [a]_m^n = [b]_m^n, \quad [a]_m^n [k]_n^n = [b]_m^n,$$

where $[h]_m^m$ is formed from $[1]_m^m$ by replacing one of its zero elements by a rational integral function σ of the variables, and where $[k]_n^n$ is formed from $[1]_n^n$ by replacing one of its zero elements by a rational integral function τ of the variables. The inverse transformations have the same forms, and to obtain them we replace σ by $-\sigma$ and τ by $-\tau$ respectively.

Transformations of the types (*b*) and (*d*) have been described in Note 3 of § 141. The inverse of a transformation of any one of the types (*b*), (*c*), (*d*) is a transformation of the same type.

NOTE 4. *Unitary equipotent transformations and quasi-scalar equipotent transformations.*

By a *unitary equipotent transformation* will be meant one which is a resultant of a number of successive transformations of the type (*c*) of Note 3. The inverse of such a transformation and the resultant of any number of such transformations are again unitary equipotent transformations. Transformations of the type (*c*) are *elementary* unitary equipotent transformations. Whenever

$$[h]_m^m [a]_m^n [k]_n^n = [b]_m^n$$

is a unitary equipotent transformation, we must have $(h)_m^m = 1$ and $(k)_n^n = 1$.

A transformation of $[a]_m^n$ is the resultant of a number of successive transformations of the type (*d*) of Note 3 when and only when it has the form

$$^1[h]_m [a]_m^n \, ^1[k]_n = [b]_m^n,$$

where $^1[h]_m$ and $^1[k]_n$ are quasi-scalar matrices whose diagonal elements $h_1, h_2, \dots h_m$ and $k_1, k_2, \dots k_n$ are non-zero constants. Such a transformation will be called a *quasi-scalar equipotent* (or *equigradent*) *transformation*. The inverse transformation is obtained by replacing $h_1, h_2, \dots h_m$ and $k_1, k_2, \dots k_n$ by their reciprocals, and it is also a quasi-scalar equipotent (or equigradent) transformation. The resultant of any number of such transformations is also of the same form.

NOTE 5. *Equipotent transformations of a compartite matrix.*

Theorem. *If ϕ and ψ are two compartite matrices with the same number of corresponding parts, and if each part of ϕ can be converted into the corresponding part of ψ by an equipotent transformation, then ϕ can be converted into ψ by an equipotent transformation.*

There will clearly be no loss of generality in supposing that ϕ and ψ are compartite matrices of standard forms, and that the first, second, ... last parts of ϕ correspond respectively to the first, second, ... last parts of ψ.

Let $$\phi = \begin{bmatrix} a, & 0, & \dots & 0 \\ 0, & b, & \dots & 0 \\ \dots & \dots & \dots & \dots \\ 0, & 0, & \dots & c \end{bmatrix}_{l,\, m,\, \dots\, n}^{p,\, q,\, \dots\, r}, \quad \psi = \begin{bmatrix} \alpha, & 0, & \dots & 0 \\ 0, & \beta, & \dots & 0 \\ \dots & \dots & \dots & \dots \\ 0, & 0, & \dots & \gamma \end{bmatrix}_{L,\, M,\, \dots\, N}^{P,\, Q,\, \dots\, R}$$

be two compartite matrices of standard forms having the same number of parts. Also let

$$H = \begin{bmatrix} h, & 0, & \dots & 0 \\ 0, & h', & \dots & 0 \\ \dots & \dots & \dots & \dots \\ 0, & 0, & \dots & h'' \end{bmatrix}_{L,\, M,\, \dots\, N}^{l,\, m,\, \dots\, n}, \qquad K = \begin{bmatrix} k, & 0, & \dots & 0 \\ 0, & k', & \dots & 0 \\ \dots & \dots & \dots & \dots \\ 0, & 0, & \dots & k'' \end{bmatrix}_{p,\, q,\, \dots\, r}^{P,\, Q,\, \dots\, R}$$

be compartite matrices of standard forms, H having the same number of parts as ϕ, and K having the same number of parts as ψ. Then we have

$$H\phi K = \psi \quad \dots\dots(2)$$

when and only when

$$[h]_L^l [a]_l^p [k]_p^P = [\alpha]_L^P, \quad [h']_M^m [b]_m^q [k']_q^Q = [\beta]_M^Q, \quad \dots \quad [h'']_N^n [c]_n^r [k'']_r^R = [\gamma]_N^R. \quad \dots\dots(3)$$

Now let the elements of ϕ, ψ, H, K be rational integral functions of the scalar variables $x, y, z, \dots$. Then (2) is an equipotent transformation converting ϕ into ψ when and only when H has rank $l+m+\dots+n$ and is impotent and also K has rank $p+q+\dots+r$ and is impotent. Referring to the theorem of § 100 and to Theorem II b of § 207, we see that this is the case when and only when

$[h]_L^l$, $[h']_M^m$, ... $[h'']_N^n$ have ranks $l, m, \dots n$ and are impotent,

$[k]_p^P$, $[k']_q^Q$, ... $[k'']_r^R$ have ranks $p, q, \dots r$ and are impotent,

i.e. when and only when the equations (3) are equipotent transformations converting the first, second, ... last parts of ϕ into the first, second, ... last parts of ψ.

Thus (2) *is an equipotent transformation when and only when all the equations* (3) *are equipotent transformations.*

It will be clear also that (2) is a derangement, a quasi-scalar equipotent transformation, or a unitary equipotent transformation when and only when the equations (3) are all derangements, or all quasi-scalar equipotent transformations, or all unitary equipotent transformations.

Ex. i. If $[e]_2^2 = \begin{bmatrix} 1, & 0 \\ 0, & (x+1)(x+2) \end{bmatrix}$,

$$[b]_2^2 = \begin{bmatrix} x^4+7x^3+15x^2+12x+4, & x^4+5x^3+9x^2+6x \\ x^4+8x^3+22x^2+27x+14, & x^4+6x^3+14x^2+15x+4 \end{bmatrix},$$

$$[a]_3^3 = \begin{bmatrix} 8x^2+28x+24, & 8x^2+20x+8, & 10x^2+32x+24 \\ 8x^2+32x+32, & 8x^2+24x+8, & 10x^2+36x+30 \\ 6x^2+12x, & 6x^2+4x+4, & 7x^2+13x+3 \end{bmatrix},$$

each of the following equations is an equipotent transformation of a matrix whose elements are rational integral functions of the single variable x:

(1) $$\begin{bmatrix} x+2, & 1 \\ x+3, & 1 \end{bmatrix} [e]_2^2 \begin{bmatrix} x+2, & x \\ 1, & 1 \end{bmatrix} = \begin{bmatrix} 2x^2+7x+6, & 2x^2+5x+2 \\ 2x^2+8x+8, & 2x^2+6x+2 \end{bmatrix};$$

(2) $$\begin{bmatrix} x+2, & x \\ x+3, & x+1 \end{bmatrix} [e]_2^2 \begin{bmatrix} x+2, & x \\ 1, & 1 \end{bmatrix} = \begin{bmatrix} x^3+4x^2+6x+4, & x^3+4x^2+4x \\ x^3+5x^2+10x+8, & x^3+5x^2+8x+2 \end{bmatrix};$$

(3) $$\begin{bmatrix} x+2, & x \\ x+3, & x+1 \end{bmatrix} [e]_2^2 \begin{bmatrix} x+2, & x \\ x+4, & x+2 \end{bmatrix} = [b]_2^2;$$

$$(4)\quad \begin{bmatrix} 2x+4, & 2 \\ 2x+6, & 2 \\ 2x-1, & 1 \end{bmatrix} [e]_2^2 \begin{bmatrix} 2x+4, & 2x, & 2x+3 \\ 2\ , & 2\ , & 3 \end{bmatrix} = [a]_3^3;$$

$$(5)\quad \begin{bmatrix} 2x+4, & 2 \\ 2x+6, & 2 \end{bmatrix} \begin{bmatrix} 1, & 0 & , 0 \\ 0, & (x+1)(x+2), & 0 \end{bmatrix} \begin{bmatrix} 2x+4, & 2x, & 2x+3 \\ 2\ , & 2\ , & 3 \\ 1\ , & 1\ , & 1 \end{bmatrix} = [a]_2^3;$$

$$(6)\quad \begin{bmatrix} x^2+2x-1, & -x^2-\ x+2 \\ x^2+3x\ , & -x^2-2x+2 \\ x^2-2\ , & -x^2+\ x+1 \end{bmatrix} [b]_2^2 \begin{bmatrix} x^2+3x+4, & x^2+x\ , & x^2+2x+3 \\ -x^2-5x-6, & -x^2-3x+2, & -x^2-4x-3 \end{bmatrix} = 2[a]_3^3.$$

The matrices on the right and the middle factor matrices on the left in these equations are all equipotent.

Ex. ii. When there are two variables x and y, the equation

$$\begin{bmatrix} x+1, & y+1 \\ 0\ , & x \\ y\ , & 0 \end{bmatrix} \begin{bmatrix} 3x+4y+1, & 2x+y+1 \\ x+2y+1, & x+y+2 \end{bmatrix} \begin{bmatrix} x, & 0, & 1 \\ 0, & y, & 1 \end{bmatrix}$$

$$= \begin{bmatrix} x(3x^2+5xy+2y^2+5x+7y+2), & y(2x^2+2xy+y^2+4x+4y+3), & 5x^2+7xy+3y^2+9x+11y+5 \\ x^2(x+2y+1)\ , & xy(x+y+2)\ , & x(2x+3y+3) \\ xy(3x+4y+1)\ , & y^2(2x+y+1)\ , & y(5x+5y+2) \end{bmatrix}$$

is an equipotent transformation of the middle factor matrix on the left.

The matrix on the right and the middle factor matrix on the left are equipotent, their common rank being 2, and their common maximum and potent factors of orders 1 and 2 being given by

$$D_1=E_1=1,\quad D_2=E_2=x^2+2xy+2y^2+4x+6y+1.$$

Ex. iii. *If we can convert the matrix*

$$\phi=[b]_m^n \quad \textit{into} \quad \phi'=[\beta]_m^n$$

by equipotent transformations, then we can also convert the matrix

$$\psi=\begin{bmatrix} a, & 0 \\ 0, & b \end{bmatrix}_{r,\,m}^{s,\,n} \quad \textit{into} \quad \psi'=\begin{bmatrix} a, & 0 \\ 0, & \beta \end{bmatrix}_{r,\,m}^{s,\,n}$$

by equipotent transformations. Moreover if the first conversion can be effected by elementary equipotent transformations applied to the m horizontal and the n vertical rows of ϕ, then the second conversion can be effected by the same elementary equipotent transformations applied to the last m horizontal rows and the last n vertical rows of ψ.

This follows from Note 5. For each of the equations

$$[h]_m^m[b]_m^n[k]_n^n=[\beta]_m^n,\quad \begin{bmatrix} 1, & 0 \\ 0, & h \end{bmatrix}_{r,\,m}^{r,\,m} \begin{bmatrix} a, & 0 \\ 0, & b \end{bmatrix}_{r,\,m}^{s,\,n} \begin{bmatrix} 1, & 0 \\ 0, & k \end{bmatrix}_{s,\,n}^{s,\,n} = \begin{bmatrix} a, & 0 \\ 0, & \beta \end{bmatrix}_{r,\,m}^{s,\,n}$$

is a necessary consequence of the other, and each of them is an equipotent transformation when and only when the other is an equipotent transformation.

Ex. iv. *If the matrix A can be converted into the matrix B and the matrix B into the matrix C by equipotent transformations in Ω, then A can be converted into C by an equipotent transformation in Ω.*

This follows from Note 2.

Ex. v. *If* $\phi=[a]_m^n$ *is a given rational integral functional matrix of rank r, we can always determine an undegenerate square rational integral functional matrix* $[b]_r^r$ *of order and rank r which is equipotent with* ϕ.

For if $E_1, E_2, \ldots E_r$ are the potent factors of ϕ of orders 1, 2, ... r, these conditions are satisfied when

$$[b]_r^r = {}^1[E]_r.$$

Ex. vi. *If there exists a rational integral identity of the form*

$$[h]_m^r [b]_r^r [k]_r^n = [a]_m^n \quad \text{.....................................(4)}$$

in which the elements of all matrices are rational integral functions of the variables x, y, z, ..., and the square matrix $[b]_r^r$ *is undegenerate, then* $[a]_m^n$ *is equipotent with* $[b]_r^r$ *if and only if* $[h]_m^r$ *and* $[k]_r^n$ *are undegenerate and impotent matrices of rank r.*

Clearly $[a]_m^n$ has rank r when and only when $[h]_m^r$ and $[k]_r^n$ both have rank r. This condition being satisfied, let D_r, H_r, K_r be the potent factors of order r of $[a]_m^n, [h]_m^r, [k]_r^n$. Then from (4) we see that D_r must be divisible by $H_r K_r (b)_r^r$. Hence if $[a]_m^n$ is equipotent with $[b]_r^r$, so that $D_r=(b)_r^r$, then both H_r and K_r must be non-zero constants, i.e. $[h]_m^r$ and $[k]_r^n$ must be impotent. Conversely if $[h]_m^r$ and $[k]_r^n$ are both impotent and both have rank r, then by Theorem II the matrix $[a]_m^n$ is equipotent with $[b]_r^r$.

§ 215. Properties of one-rowed matrices connected with the long rows of an undegenerate rational integral functional matrix.

In the present article all functions and all matrices are rational and integral in certain scalar variables $x, y, z, \ldots$; $A=[a]_r^n$ is an undegenerate matrix of rank r lying in a given domain of rationality Ω; $D_1, D_2, \ldots D_r$ and $E_1, E_2, \ldots E_r$ are the maximum and potent factors of A of orders 1, 2, ... r so chosen as to lie in Ω; and irreducible functions are those which lie in Ω and are irreducible in Ω. Consistently with the general definition of a primitive matrix of any orders given in a later chapter we may here define a *primitive one-rowed matrix* to be one which is undegenerate (or not 0) and impotent; thus it is a one-rowed matrix whose elements have no common factor other than a non-zero constant.

Lemma. *If* $A=[a]_r^n$ *is an undegenerate matrix of rank r which lies in Ω and is impotent with respect to the irreducible function t, and if* $r<n$, *then by adding final horizontal rows to A we can form in succession undegenerate matrices* $[a]_{r+1}^n, [a]_{r+2}^n, \ldots [a]_n^n$ *of ranks* $r+1, r+2, \ldots n$ *which lie in Ω and are impotent with respect to t, the last of them being an undegenerate square matrix of order n.*

We may suppose without loss of generality that the leading simple minor determinant $\Delta_r = (a)_r^r$ of A does not vanish identically and is not divisible by t. We can then certainly form a matrix

$$A' = \begin{bmatrix} a \\ b \end{bmatrix}_{r,1}^{n}$$

lying in Ω in which the leading simple minor determinant

$$\Delta_{r+1} = \begin{pmatrix} a \\ b \end{pmatrix}_{r,1}^{r+1}$$

does not vanish identically. If Δ_{r+1} is not divisible by t, we can take $[a]_{r+1}^n$ to be A'. If Δ_{r+1} is divisible by t, and if in A' we replace the element $b_{1,r+1}$ by $b_{1,r+1}+1$, we form a matrix $[a]_{r+1}^n$ in which

$$(a)_{r+1}^{r+1} = \Delta_{r+1} + \Delta_r = t \, . \, \Delta'_{r+1} + \Delta_r,$$

so that $(a)_{r+1}^{r+1}$ does not vanish identically and is not divisible by t. Thus we can in all cases determine an undegenerate matrix $[a]_{r+1}^n$ of rank $r+1$ which lies in Ω and is impotent with respect to t; and from this fact the lemma follows by induction.

Ex. i. If $[a]_r^n = [a, a]_r^{r, n-r}$, where $(a)_r^r$ does not vanish identically and is not divisible by t, the conditions of the lemma are clearly satisfied when

$$[a]_n^n = \begin{bmatrix} a, & a \\ 0, & 1 \end{bmatrix}_{r, n-r}^{r, n-r}.$$

Theorem I. *If $A = [a]_r^n$ is an undegenerate matrix of rank r whose elements are rational integral functions of $x, y, z, \ldots$, then in every rational integral identity in $x, y, z, \ldots$ of the form*

$$[\phi]_1^r [a]_r^n = \omega \, [u]_1^n \quad \ldots\ldots\ldots\ldots\ldots\ldots\ldots\ldots\ldots (A)$$

in which $[\phi]_1^r$ is a non-zero impotent one-rowed matrix (i.e. is primitive), the function ω must be a factor of E_r, the potent factor of A of order r.

Let t be any one of the irresoluble factors of ω, and let t^s be the highest power of t which is a factor of ω; also let $[\phi]_r^r$ be an undegenerate square matrix, constructed as in the lemma, which is impotent with respect to t, and let

$$[\phi]_r^r [a]_r^n = \begin{bmatrix} \omega \, . \, u \\ b \end{bmatrix}_{1, r-1}^{n} = [c]_r^n. \quad \ldots\ldots\ldots\ldots\ldots\ldots (1)$$

Since all simple minor determinants of $[b]_{r-1}^n$, and all minor determinants of $[c]_r^n$ of order $r-1$, are divisible by D_{r-1}, we obtain by equating the maximum factors of order r on both sides of (1) an equation of the form

$$(\phi)_r^r D_r = \omega D_{r-1} \, . \, F(x, y, z, \ldots),$$

or

$$(\phi)_r^r E_r = \omega \, . \, F(x, y, z, \ldots); \quad \ldots\ldots\ldots\ldots\ldots\ldots (2)$$

and because ω is divisible by t^s, it follows from (2) that t^s must be a factor of E_r. Hence by expressing ω in the form

$$\omega = t_1^{s_1} t_2^{s_2} t_3^{s_3} \ldots,$$

where $t_1, t_2, t_3, \ldots$ are distinct irresoluble functions, we see that ω must be a factor of E_r.

Ex. ii. *If* $[\phi]_1^r$ *and* $[a]_r^n$ *are undegenerate impotent matrices of ranks* 1 *and* r, *then the product*

$$[\phi]_1^r [a]_r^n = [b_1 b_2 \ldots b_n]$$

is undegenerate and impotent, i.e. it is primitive.

For every common factor ω of the elements of $[b]_n$ must be a factor of E_r, i.e. a factor of 1.

Ex. iii. *If* $[\phi]_i^r$ *and* $[a]_r^n$ *are undegenerate impotent matrices of ranks* i *and* r, *then the product*

$$[\phi]_i^r [a]_r^n = [b]_i^n$$

is an undegenerate and impotent matrix of rank i.

That the rational integral functional matrix $[b]_i^n$ has rank i follows from § 131.

Let t be any irresoluble function of the variables $x, y, z, \ldots$; and let p and q be simple minor determinants of $[\phi]_i^r$ and $[a]_r^n$ which are not divisible by t. Then there are finite roots of t for which $pq \neq 0$, i.e. for which $[\phi]_i^r$ and $[a]_r^n$ have ranks i and r and therefore $[b]_i^n$ has rank i; and it follows that t is not an irresoluble divisor of $[b]_i^n$. Thus $[b]_i^n$ has no irresoluble divisor and is impotent.

Further if $[\phi]_i^r$ *and* $[a]_r^n$ *have ranks* i *and* r *and are impotent with respect to any irresoluble (or irreducible) function* t, *then* $[b]_i^n$ *has rank* i *and is impotent with respect to* t.

Ex. iv. *If* $A = [a]_r^n$ *is an undegenerate matrix of rank* r *whose elements are constants, and if* $[u]_1^n$ *is any one-rowed matrix connected with the horizontal rows of* A *whose elements are rational integral functions of the variables* $x, y, z, \ldots$, *then in every term* $[h]_1^n . x^\alpha y^\beta z^\gamma \ldots$ *of* $[u]_1^n$ *the coefficient* $[h]_1^n$ *is a one-rowed matrix connected with the horizontal rows of* A.

First suppose that $[u]_1^n$ is primitive. Then since in this case $E_r = 1$, we see from Theorem I that there exists a rational integral identity in $x, y, z, \ldots$ of the form

$$[u]_1^n = [\phi]_1^r [a]_r^n,$$

and by equating the coefficients of $x^\alpha y^\beta z^\gamma \ldots$ on both sides we see that there exists a relation of the form

$$[h]_1^n = [\lambda]_1^r [a]_r^n,$$

where $[\lambda]_1^r$ is a matrix with constant elements.

Next suppose that $[u]_1^n$ does not vanish identically but is not primitive. Then we can write

$$[u]_1^n = Q . [v]_1^n,$$

where Q is the H.C.F. of the elements of $[u]_1^n$, and where $[v]_1^n$ is primitive. The coefficient of any term of $[u]_1^n$ is a sum of the coefficients of certain terms of $[v]_1^n$ each multiplied by a scalar constant. Since the coefficients in $[v]_1^n$ are all connected with the horizontal rows of A, it follows that the coefficients in $[u]_1^n$ are all connected with the horizontal rows of A.

Lastly when $[u]_1^n$ vanishes identically, the theorem is obviously true.

Ex. v. *If* $A=[a]_m^n$ *is any matrix whose elements are constants, and if* $[u]_1^n$ *is any one-rowed matrix connected with the horizontal rows of A whose elements are rational integral functions of the variables* $x, y, z, \ldots$, *then in every term* $[h]_1^n \,.\, x^\alpha y^\beta z^\gamma \ldots$ *of* $[u]_1^n$ *the coefficient* $[h]_1^n$ *is a one-rowed matrix connected with the horizontal rows of A.*

This follows from Ex. iv, but can be proved more simply as follows.

Let r be the rank of A. Then every minor determinant of order $r+1$ of the matrix $\begin{bmatrix} a \\ u \end{bmatrix}_{m,1}^n$ vanishes identically. Therefore every minor determinant of order $r+1$ of the matrix $\begin{bmatrix} a \\ h \end{bmatrix}_{m,1}^n$ must vanish, and therefore this last matrix has the same rank r as $[a]_m^n$.

Theorem II. *If* $A=[a]_r^n$ *is an undegenerate matrix of rank r whose elements are rational integral functions in* Ω *of the scalar variables* $x, y, z, \ldots$, *we can always determine a rational integral equation in* Ω *of the form*

$$[\phi_1 \phi_2 \ldots \phi_r]\,[a]_r^n = E_r\,[u_1 u_2 \ldots u_n], \quad \ldots\ldots\ldots\ldots\ldots\ldots\text{(B)}$$

where E_r *is the potent factor of A of order r, and where* $[\phi]_r$ *and* $[u]_n$ *are undegenerate and impotent (i.e. primitive) one-rowed matrices, the equation being an identity in* $x, y, z, \ldots$.

Let $\nu = \binom{n}{r-1}$, and let $[A]_r^\nu$ be a complete matrix of the affected minor determinants of A of order $r-1$ in which the elements of the ith horizontal row are the simple minor determinants of the minor of A formed by striking out its ith horizontal row, this being true for all the values $1, 2, \ldots r$ of i. Then

$$\overline{A}_\nu^{\,r}\,[a]_r^n = [B]_\nu^n, \quad \ldots\ldots\ldots\ldots\ldots\ldots\ldots\ldots\text{(3)}$$

where the non-zero elements of $[B]_\nu^n$ are the minor determinants of A of order r, all of which occur. In fact when we disregard signs, the matrix $[B]_\nu^n$ contains exactly r elements equal to each minor determinant of A of order r, these r elements occurring in r different horizontal rows; and the remaining $(r-1)\nu$ elements of $[B]_\nu^n$ are 0's, there being $r-1$ zero elements in each horizontal row and $\binom{n-1}{r-2}$ zero elements in each vertical row.

Since D_{r-1} is the H.C.F. of all the elements of $\overline{\underline{A}}{}^{r}_{\nu}$, it follows from Lemma 1 *a* of § 209 that we can determine positive integers $h_2, h_3, \ldots h_\nu$ such that

$$[1, h_2, h_3, \ldots h_\nu]\,\overline{\underline{A}}{}^{r}_{\nu} = D_{r-1}\,[\alpha_1\,\alpha_2 \ldots \alpha_r], \quad \ldots\ldots\ldots\ldots\ldots(4)$$

where $[\alpha]_r$ does not vanish identically and is impotent with respect to every irreducible divisor of A. When we prefix the matrix $[1, h_2, h_3, \ldots h_\nu]$ on both sides of (3) and observe that every element of $[B]^n_\nu$ is divisible by D_r, we obtain an equation of the form

$$D_{r-1}\,[\alpha_1\,\alpha_2 \ldots \alpha_r]\,[a]^n_r = D_r\,[\beta_1\,\beta_2 \ldots \beta_n],$$

or

$$[\alpha_1\,\alpha_2 \ldots \alpha_r]\,[a]^n_r = E_r\,[\beta_1\,\beta_2 \ldots \beta_n], \ldots\ldots\ldots\ldots\ldots\ldots(5)$$

where $[\beta]_n$ does not vanish identically and $[\alpha]_r$ is the same matrix as before. Replacing the equation (5) by

$$Q\,[\phi_1\,\phi_2 \ldots \phi_r]\,[a]^n_r = E_r\,[\beta_1\,\beta_2 \ldots \beta_n], \quad \ldots\ldots\ldots\ldots\ldots\ldots(6)$$

where Q is the H.C.F. of all the elements of $[\alpha]_r$, and where $[\phi]_r$ is impotent, we see that Q must be a common factor of all the elements of $[\beta]_n$, because it has no irreducible factor in common with E_r; and when we cancel Q on both sides of (6), we obtain an equation of the form (B) in which $[\phi]_r$ is impotent. By Theorem I the matrix $[u]_n$ must also be impotent.

Ex. vi. We can put $\overline{\underline{A}}{}^{r}_{\nu} = D_{r-1}\,\overline{\underline{A'}}{}^{r}_{\nu}$, $[B]^n_\nu = D_r\,[B']^n_\nu$, where the elements of $\overline{\underline{A'}}{}^{r}_{\nu}$, and also the elements of $[B']^n_\nu$, have no common factor other than a non-zero constant, and replace (3) by

$$\overline{\underline{A'}}{}^{r}_{\nu}\,[a]^n_r = E_r\,[B']^n_\nu. \ldots\ldots\ldots\ldots\ldots\ldots\ldots\ldots\ldots\ldots(3')$$

We can then determine positive integers $h_2, h_3, \ldots h_\nu$ such that

$$[1, h_2, h_3, \ldots h_\nu]\,\overline{\underline{A'}}{}^{r}_{\nu} = [a_1\,a_2 \ldots a_r], \quad \ldots\ldots\ldots\ldots\ldots\ldots\ldots(4')$$

where $[a]_r$ does not vanish identically and is impotent with respect to all the irreducible divisors of A, and obtain the equation (B) by prefixing $[1, h_2, h_3, \ldots h_\nu]$ on both sides of (3′).

Ex. vii. By § 208 the matrix $\overline{\underline{A}}{}^{r}_{\nu}$ in Theorem II has the same irreducible divisors as A, and therefore every irreducible divisor of $\overline{\underline{A'}}{}^{r}_{\nu}$ in Ex. vi is an irreducible divisor of A.

Ex. viii. For the matrix $A=[a\,b\,c\,d\,e]_{123}$ the equation (3) becomes

$$\begin{bmatrix} (ab)_{23} & (ab)_{31} & (ab)_{12} \\ (ac)_{23} & (ac)_{31} & (ac)_{12} \\ (ad)_{23} & (ad)_{31} & (ad)_{12} \\ (ae)_{23} & (ae)_{31} & (ae)_{12} \\ (bc)_{23} & (bc)_{31} & (bc)_{12} \\ (bd)_{23} & (bd)_{31} & (bd)_{12} \\ (be)_{23} & (be)_{31} & (be)_{12} \\ (cd)_{23} & (cd)_{31} & (cd)_{12} \\ (ce)_{23} & (ce)_{31} & (ce)_{12} \\ (de)_{23} & (de)_{31} & (de)_{12} \end{bmatrix} \begin{bmatrix} a_1 & b_1 & c_1 & d_1 & e_1 \\ a_2 & b_2 & c_2 & d_2 & e_2 \\ a_3 & b_3 & c_3 & d_3 & e_3 \end{bmatrix} = \begin{bmatrix} 0 & 0 & (abc) & (abd) & (abe) \\ 0 & (acb) & 0 & (acd) & (ace) \\ 0 & (adb) & (adc) & 0 & (ade) \\ 0 & (aeb) & (aec) & (aed) & 0 \\ (bca) & 0 & 0 & (bcd) & (bce) \\ (bda) & 0 & (bdc) & 0 & (bde) \\ (bea) & 0 & (bec) & (bed) & 0 \\ (cda) & (cdb) & 0 & 0 & (cde) \\ (cea) & (ceb) & 0 & (ced) & 0 \\ (dea) & (deb) & (dec) & 0 & 0 \end{bmatrix}$$

when we change the signs of all elements in the 2nd, 4th, 6th and 9th horizontal rows.

Ex. ix. In the special case of an undegenerate square matrix $A=[a]_r^r$ we can take D_r to be $(a)_r^r$. Then if $[A]_r^r$ is the reciprocal of A, and if we put

$$[A]_r^r = D_{r-1}\,[a]_r^r,$$

we see by postfixing $\underline{\overline{A}}{}_r^r$ on both sides that we have an identity of the form

$$[\phi_1\,\phi_2 \ldots \phi_r]\,[a]_r^r = E_r\,[u_1\,u_2 \ldots u_r]$$

when and only when

$$[\phi_1\,\phi_2 \ldots \phi_r] = [u_1\,u_2 \ldots u_r]\,\underline{\overline{a}}{}_r^r,$$

where $u_1, u_2, \ldots u_r$ are arbitrary. We can choose $u_1, u_2, \ldots u_r$ to be positive integers such that $[\phi_1\,\phi_2 \ldots \phi_r]$ is impotent.

§ 216. Properties of one-rowed matrices connected with the long rows of an undegenerate rational integral x-matrix.

Throughout the present article $A=[a]_r^n$ is a given undegenerate rational integral x-matrix of rank r lying in a given domain of rationality Ω. The maximum and potent factors of A of orders $1, 2, \ldots r$ are $D_1, D_2, \ldots D_r$ and $E_1, E_2, \ldots E_r$; and Δ_{r-1} is the H.C.F. of the simple minor determinants of the leading horizontal minor $[a]_{r-1}^n$, all these functions being so chosen as to lie in Ω.

NOTE 1. It is to be observed however that if we put $D_r=0$, Theorems I and III remain true when $[a]_r^n$ is degenerate, provided that $[a]_{r-1}^n$ is undegenerate of rank $r-1$, and that $f(x)$ does not vanish identically. The condition that $f(x)$ is to be a factor of D_r then imposes no restrictions on $f(x)$.

Theorem I. *If $f(x)$ is a rational integral function of x of degree p which is a factor of D_r and has no linear factor in common with Δ_{r-1}, then there exist uniquely determinate rational integral identities in x of the forms*

$$[\phi,\,1]_1^{r-1,1}\,[a]_r^n = f(x)\,.\,[b]_1^n, \quad \ldots\ldots\ldots\ldots\ldots\ldots\ldots\ldots\ldots\ldots\text{(A)}$$

$$\begin{bmatrix} 1, & 0 \\ \phi, & 1 \end{bmatrix}_{r-1,1}^{r-1,1} [a]_r^n = \begin{bmatrix} 1, & 0 \\ 0, & f(x) \end{bmatrix}_{r-1,1}^{r-1,1} \begin{bmatrix} a \\ b \end{bmatrix}_{r-1,1}^n = \begin{bmatrix} a \\ f(x)\,.\,b \end{bmatrix}_{r-1,1}^n, \ldots\text{(A}'\text{)}$$

where $[\phi]_1^{r-1}$ is a rational integral x-matrix whose degree in x is less than p.

We assume (see Note 4) that p is not less than 1, and that

$$f(x) = (x - \alpha)(x - \beta)(x - \gamma)\dots,$$

where the linear factors are not necessarily all different. If there exists an identity in x of the form (A) and also a second identity

$$[\phi', 1]_1^{r-1,1} [a]_r^n = f(x) . [b']_1^n$$

of the same form as (A) in which $[\phi']_1^{r-1}$ also has degree less than p,

and if $$[\psi]_1^{r-1} = [\phi']_1^{r-1} - [\phi]_1^{r-1}, \quad [c]_1^n = [b']_1^n - [b]_1^n,$$

then $$[\psi]_1^{r-1} [a]_{r-1}^n = f(x) . [c]_1^n. \quad \dots\dots\dots\dots\dots\dots\text{(a)}$$

Since $[a]_{r-1}^n$ is undegenerate when $x = \alpha, \beta, \gamma, \dots$, it follows in succession from the last equation (cancelling $x - \alpha$, then $x - \beta$, then $x - \gamma, \dots$ on both sides) that $x - \alpha$, $(x - \alpha)(x - \beta)$, $(x - \alpha)(x - \beta)(x - \gamma)$, $\dots f(x)$ are factors of $[\psi]_1^{r-1}$, i.e. $[\psi]_1^{r-1}$ is divisible by $f(x)$; and because $[\psi]_1^{r-1}$ has a lower degree in x than $f(x)$, this is only possible when $[\psi]_1^{r-1} = 0$. Consequently $[\phi']_1^{r-1}$ is identical with $[\phi]_1^{r-1}$. We conclude that there cannot be two different identities such as (A), and that if there is any such identity, it must be uniquely determinate. This conclusion would remain valid under the circumstances described in Note 1. It remains to show that there always does exist an identity of the form (A) in which $[\phi]_1^{r-1}$ has degree less than p. In proving this we shall consider first a special case, and then the general case.

Case I. When $f(x) = x - a$.

Here $x - a$ is any linear factor of D_r which is not a factor of Δ_{r-1}.

Let $[a]_r^n$ become $[k]_r^n$ when $x = a$. Then $[k]_r^n$ is a matrix of rank $r - 1$ with constant elements in which the minor $[k]_{r-1}^n$ is undegenerate of rank $r - 1$, and therefore there exists one and only one matrix $[\lambda]_1^{r-1}$ with constant elements such that

$$[\lambda, 1]_1^{r-1,1} [k]_r^n = 0,$$

i.e. such that $[\lambda, 1]_1^{r-1,1} [a]_r^n$ is divisible by $x - a$. Accordingly there exist rational integral identities of the forms

$$[\lambda, 1]_1^{r-1,1} [a]_r^n = (x - a) . [u]_1^n, \quad \dots\dots\dots\dots\dots\dots\text{(1)}$$

$$\begin{bmatrix} 1, & 0 \\ \lambda, & 1 \end{bmatrix}_{r-1,1}^{r-1,1} [a]_r^n = \begin{bmatrix} 1, & 0 \\ 0, & x - a \end{bmatrix}_{r-1,1}^{r-1,1} \begin{bmatrix} a \\ u \end{bmatrix}_{r-1,1}^n, \quad \dots\dots\dots\dots\text{(1')}$$

where $[\lambda]_1^{r-1}$ is a matrix with constant elements, and in these the matrix $[\lambda]_1^{r-1}$ must have the unique value found above. Thus the theorem is true in this case; and it would clearly also be true under the circumstances described in Note 1, when the only restriction on $x - a$ is that it is not a factor of Δ_{r-1}.

Case II. When $f(x)=(x-a)(x-\beta)(x-\gamma)\ldots$.

For the sake of brevity we will write

$$f_1(x)=x-a,\quad f_2(x)=(x-a)(x-\beta),\quad f_3(x)=(x-a)(x-\beta)(x-\gamma),\ \ldots.$$

We first determine the identities (1) and (1′), and observe that (1′) is a unitary equipotent transformation of the matrix A.

Let Δ_r' be the H.C.F. of the simple minor determinants of the postfactor on the right in (1′). Then by equating correspondingly formed complete matrices of the simple minor determinants of both sides in (1′) we see that

$$D_r=(x-a)\,\Delta_r'.$$

Since $(x-a)(x-\beta)$ is a factor of D_r, it follows that $x-\beta$ is a factor of Δ_r' which is not a factor of Δ_{r-1}, and therefore by Case I there exists an identity of the form

$$[\lambda',\ 1]_1^{r-1,1}\begin{bmatrix} a \\ u \end{bmatrix}_{r-1,1}^{n}=(x-\beta)\,.\,[v]_1^n,\ \ldots\ldots\ldots\ldots(1'')$$

where the elements of $[\lambda']_1^{r-1}$ are constants. Multiplying (1″) by $x-a$ and using (1), we deduce the identities

$$[\mu,\ 1]_1^{r-1,1}[a]_r^n=(x-a)(x-\beta)\,.\,[v]_1^n,\ \ldots\ldots\ldots\ldots(2)$$

$$\begin{bmatrix} 1, & 0 \\ \mu, & 1 \end{bmatrix}_{r-1,1}^{r-1,1}[a]_r^n=\begin{bmatrix} 1, & 0 \\ 0, & f_2(x) \end{bmatrix}_{r-1,1}^{r-1,1}\begin{bmatrix} a \\ v \end{bmatrix}_{r-1,1}^{n},\ \ldots\ldots\ldots\ldots(2')$$

where $$[\mu]_1^{r-1}=(x-a)\,[\lambda']_1^{r-1}+[\lambda]_1^{r-1};$$

and (2′) is a unitary equipotent transformation of the matrix A. Under the circumstances described in Note 1 the postfactor on the right in (1′) would be degenerate, and by Case I we should still obtain an identity of the form (1″) from which identities of the forms (2) and (2′) could be deduced.

Let Δ_r'' be the H.C.F. of the simple minor determinants of the postfactor on the right in (2′). Then by equating correspondingly formed complete matrices of the simple minor determinants of both sides in (2′) we see that

$$D_r=(x-a)(x-\beta)\,\Delta_r''.$$

Since $(x-a)(x-\beta)(x-\gamma)$ is a factor of D_r, it follows that $x-\gamma$ is a factor of Δ_r'' which is not a factor of Δ_{r-1}, and therefore by Case I there exists an identity of the form

$$[\lambda'',\ 1]_1^{r-1,1}\begin{bmatrix} a \\ v \end{bmatrix}_{r-1,1}^{n}=(x-\gamma)\,.\,[w]_1^n,\ \ldots\ldots\ldots\ldots(2'')$$

where the elements of $[\lambda'']_1^{r-1}$ are constants. Multiplying (2″) by $(x-a)(x-\beta)$ and using (1″) and (1), we deduce the identities

$$[\nu,\ 1]_1^{r-1,1}[a]_r^n=(x-a)(x-\beta)(x-\gamma)\,.\,[w]_1^n,\ \ldots\ldots\ldots\ldots(3)$$

$$\begin{bmatrix} 1, & 0 \\ \nu, & 1 \end{bmatrix}_{r-1,1}^{r-1,1}[a]_r^n=\begin{bmatrix} 1, & 0 \\ 0, & f_3(x) \end{bmatrix}_{r-1,1}^{r-1,1}\begin{bmatrix} a \\ w \end{bmatrix}_{r-1,1}^{n},\ \ldots\ldots\ldots\ldots(3')$$

where $$[\nu]_1^{r-1}=(x-a)(x-\beta)\,[\lambda'']_1^{r-1}+[\mu]_1^{r-1}$$
$$=(x-a)(x-\beta)\,[\lambda'']_1^{r-1}+(x-a)\,[\lambda']_1^{r-1}+[\lambda]_1^{r-1},$$

being a matrix whose degree in x is less than 3; and (3′) is a unitary equigradent transformation of the matrix A. Under the circumstances described in Note 1 we should have $\Delta_r''=0$, and we could still obtain in succession the identities (2″), (3), (3′).

Continuing in this way we finally obtain identities of the forms (A) and (A′) in which $[\phi]_1^{r-1}$ has degree less than p; and this establishes the theorem.

The equation (A′) is a unitary equipotent transformation of the matrix A. Consequently the matrix $\begin{bmatrix} a \\ b \end{bmatrix}_{r-1,1}^{n}$ has the same rank r as $[a]_r^n$, and if D_r' is the H.C.F. of its simple minor determinants, we have

$$D_r=f(x)\,D_r'. \qquad\qquad (4)$$

Theorem I is equivalent to the statement that there exists a uniquely determinate rational integral x-matrix $[\phi]_1^{r-1}$ which has a lower degree in x than $f(x)$, and is such that

$$[\phi]_1^{r-1}[a]_r^n \text{ is divisible by } f(x). \qquad\qquad (A'')$$

NOTE 2. *Actual determination of the identity* (A).

If we assume for each element of $[\phi]_1^{r-1}$ an expression of the form $c_0x^{p-1}+c_1x^{p-2}+\dots+c_p$, where the coefficients c are to be determined, then divide each of the elements of $[\phi,1]_1^{r-1,1}[a]_r^n$ by $f(x)$ and equate the remainders to 0, i.e. equate the coefficients of x^{p-1} $x^{p-2}, \dots x^0$ in each remainder to 0, we obtain np linear equations which have to be satisfied by the $(r-1)p$ unknown coefficients. By the theorem this system of equations has a unique solution. Hence by solving the equations, or any $(r-1)p$ of them which are unconnected, we can find the matrix $[\phi]_1^{r-1}$. By using this method we avoid the necessity for finding the linear factors of $f(x)$.

NOTE 3. *Domain of rationality of the matrix* $[\phi]_1^{r-1}$ *in* (A) *and* (A′).

From Note 2 it will be clear that if $f(x)$ lies in the same domain of rationality Ω as $[a]_r^n$, i.e. if Ω is any domain of rationality in which both $[a]_r^n$ and $f(x)$ lie, then $[\phi]_1^{r-1}$ lies in Ω. This can also be proved in the following way. From the mode of its formation it is clear that $[\phi]_1^{r-1}$ lies in the domain Ω' formed by the adjunction of $\alpha, \beta, \gamma, \dots$ to Ω. Since $\alpha, \beta, \gamma, \dots$ all lie immediately over Ω, the domain Ω' can be formed from Ω by the adjunction of a certain single element q of Ω' which is a root of an irreducible equation $g(x)=0$ in Ω having a certain finite degree s, and we can write

$$[\phi]_1^{r-1}=[\psi]_1^{r-1}+q[\psi']_1^{r-1}+q^2[\psi'']_1^{r-1}+\dots,$$

$$[b]_1^n\;=[c]_1^n\;+q[c']_1^n\;+q^2[c'']_1^n\;+\dots,$$

where there are s terms on the right in each equation, and where all the matrix coefficients lie in Ω. Substituting these values in (A), we see that

$$[\psi]_1^{r-1,1}[a]_r^n=f(x)\,.\,[c]_1^n, \qquad\qquad (5)$$

$$[\psi']_1^{r-1}[a]_{r-1}^n=f(x)\,.\,[c']_1^n,\quad [\psi'']_1^{r-1}[a]_{r-1}^n=f(x)\,.\,[c'']_1^n,\ \dots,\ \dots \qquad (6)$$

As in the case of the equation (a) the $s-1$ equations (6) show that $[\psi']_1^{r-1}$, $[\psi'']_1^{r-1}, \dots$ all vanish, i.e. $[\phi]_1^{r-1}$ must lie in Ω, the equation (5) being the same as (A).

NOTE 4. *Special case when $p=0$.*

Theorem I remains true when $f(x)$ is a non-zero constant, provided that we interpret $[\phi]_1^{r-1}$ to be 0 when its degree in x is less than 0, this being in accordance with the usual convention that a rational integral functional matrix whose degree is less than 0 is one which vanishes identically.

NOTE 5. *Degree in x of the matrix $[b]_1^n$ in (A) and (A').*

The degree in x of $[b]_1^n$ is less than the degree of $[a]_r^n$. For if this were not so, the degree in x of the right-hand side of (A) would be not less than the sum of the degrees of $[a]_r^n$ and $f(x)$, and therefore greater than the sum of the degrees of $[a]_r^n$ and $[\phi, 1]_1^{r-1,1}$, and greater than the degree of the left-hand side; which is impossible.

Corollary to Theorem I. *If $[a]_{r-1}^n$ is impotent and $[a]_r^n$ is undegenerate, then there exists a uniquely determinate rational integral identity in Ω (the domain of rationality of A) of the form*

$$[\phi]_1^{r-1,1}[a]_r^n = E_r . [b]_1^n, \quad \text{......................(B)}$$

where the degree in x of $[\phi]_1^{r-1}$ is less than the degree of E_r. In (B) the degree in x of $[b]_1^n$ is less than the degree of $[a]_r^n$, and $\begin{bmatrix} a \\ b \end{bmatrix}_{r-1,1}^n$ is an undegenerate and impotent matrix of rank r, so that in particular $[b]_1^n$ is impotent.

It is assumed that E_r, which is now the same as D_r, is given and lies in Ω. This corollary is that particular case of Theorem I in which $\Delta_{r-1}=1$ and $f(x)$ is taken to be D_r. That $\begin{bmatrix} a \\ b \end{bmatrix}_{r-1,1}^n$ is impotent follows from the equation (4) which shows that $D_r'=1$.

Ex. i. *Theorem I cannot be extended to matrices whose elements are functions of more than one variable.*

Consider for example the matrix

$$A=[a]_2^2=\begin{bmatrix} 3x+4y+1, & 2x+y+1 \\ x+2y+1, & x+y+2 \end{bmatrix}$$

which lies in the domain Ω_1 of all rational numbers, and for which

$$D_1=E_1=1, \quad D_2=E_2=x^2+2xy+2y^2+4x+6y+1=t,$$

where t is irreducible in Ω_1. If x is arbitrary, the matrix

$$\phi=[y(ax+b)+(cx+d),\ 1][a]_2^2$$

is divisible by t when $y=0$ and when $y=2$ if and only if $a=\frac{5}{9}$, $b=\frac{19}{9}$, $c=-1$, $d=-4$. But when x and y are both variable, ϕ is not divisible by t when a, b, c, d have these values.

Ex. ii. *If $[a]_{r-1}^n$ is any undegenerate and impotent rational integral x-matrix in Ω of rank $r-1$, $r-1$ being less than n, then by adding to it a final horizontal row we can form an undegenerate and impotent rational integral x-matrix $\begin{bmatrix} a \\ b \end{bmatrix}_{r-1,1}^n$ of rank r lying in Ω.*

By adding a final horizontal row to $[a]_{r-1}^n$ we can clearly form an undegenerate rational integral x-matrix $[a]_r^n$ of rank r lying in Ω. If $[a]_r^n$ is not impotent, then there exists an identity in Ω of the form (B), where E_r is the maximum (or potent) factor of $[a]_r^n$ of order r, and the Corollary to Theorem I shows that the above result is true.

Ex. iii. *If $[h]_r^n$ is any undegenerate and impotent rational integral x-matrix of rank r lying in Ω, and if $r<n$, then by adding final horizontal rows to $[h]_r^n$ we can form undegenerate and impotent rational integral x-matrices $[h]_{r+1}^n$, $[h]_{r+2}^n$, ... $[h]_n^n$ of ranks $r+1$, $r+2$, ... n lying in Ω; in particular we can form an undegenerate square matrix $[h]_n^n$, rational and integral in x, which lies in Ω and is impotent.*

This follows from Ex. ii. We can clearly enunciate a similar theorem regarding the completion of an undegenerate and impotent x-matrix $[h]_m^r$ of rank r, where $r<m$, by the addition of final vertical rows.

The above theorem does not remain generally true when there are more than two variables; for if there are two variables x and y, and if X and Y are rational integral functions of x and y, the first of the two matrices

$$[x,\ y],\quad \begin{bmatrix} x, & y \\ X, & Y \end{bmatrix}$$

is impotent, but the second cannot be both undegenerate and impotent.

Ex. iv. *Every undegenerate and impotent rational integral x-matrix of rank r lying in Ω has an inverse which is also an undegenerate and impotent rational integral x-matrix of rank r lying in Ω.*

Let $[h]_r^n$ be an undegenerate and impotent rational integral x-matrix of rank r lying in Ω. Then if we form as in Ex. iii a square rational integral x-matrix

$$\begin{bmatrix} h \\ u \end{bmatrix}_{r,\,n-r}^{n}$$

in Ω which is undegenerate and impotent, its inverse $\overline{\underline{H,\ U}}_{n}^{r,\,n-r}$ is also a square rational integral x-matrix in Ω which is undegenerate and impotent; and we have

$$\begin{bmatrix} h \\ u \end{bmatrix}_{r,\,n-r}^{n} \overline{\underline{H,\ U}}_{n}^{r,\,n-r} = [1]_n^n, \quad [h]_r^n \overline{\underline{H}}_n^r = [1]_r^r.$$

Thu $\overline{\underline{H}}_n^r$ is a matrix inverse to $[h]_r^n$ which has the required properties.

Theorem II. *If $A=[a]_r^n$ is an undegenerate rational integral x-matrix of rank r, then in every rational integral identity in x of the form*

$$[\phi]_1^r [a]_r^n = f(x).[u]_1^n \quad \text{.........................(C)}$$

in which $[\phi]_1^r$ is a primitive (i.e. undegenerate and impotent) one-rowed matrix, the function $f(x)$ must be a factor of E_r, the potent factor of A of order r.

Let $[\phi]_r^r$ be a square rational integral x-matrix, formed by adding final horizontal rows to $[\phi]_1^r$ as in Ex. iii, which is undegenerate and impotent, and let

$$[\phi]_r^r\,[a]_r^n = \begin{bmatrix} f(x)\,.\,u \\ b \end{bmatrix}_{1,\,r-1}^n = [c]_r^n. \quad \text{..................(C}')$$

Then $[c]_r^n$ is a matrix of rank r having the same maximum and potent factors as $[a]_r^n$; and since all minor determinants of $[b]_{r-1}^n$ of order $r-1$ are divisible by D_{r-1}, we see by equating the maximum factors of order r on both sides of (C′) that

$$D_r = D_{r-1} f(x)\, g(x), \quad E_r = f(x)\, g(x),$$

where $g(x)$ is a rational integral function of x; i.e. $f(x)$ is a factor of E_r. If $[u]_1^n$ is impotent, then clearly E_1 must be a factor of $f(x)$.

This theorem is a particular case of Theorem I of § 215.

Ex. v. If $[a]_r^n$ is impotent as well as $[\phi]_1^r$, then $[\phi]_1^r\,[a]_r^n$ is impotent.

More generally if $[\phi]_i^r$ and $[a]_r^n$ are undegenerate and impotent x-matrices of ranks i and r, then $[\phi]_i^r\,[a]_r^n$ is an undegenerate and impotent x-matrix of rank i.

Theorem III. *If $A = [a]_r^n$ is an undegenerate rational integral x-matrix in Ω of rank r whose leading horizontal minor $[a]_{r-1}^n$ is regular with respect to all the irreducible divisors of A, and if $f(x)$ is any rational integral function of x lying in Ω which is a factor of E_r, then there exists a rational integral identity in Ω of the form*

$$[\phi,\ 1]_1^{r-1,1}\,[a]_r^n = f(x)\,.\,[b]_1^n, \quad \text{....................(D)}$$

where $[\phi]_1^{r-1}$ is a rational integral x-matrix having a lower degree in x than $f(x)$.

As usual D_i and E_i are the maximum and potent factors of A of order i. In the present case $D_1, D_2, \ldots D_{r-1}$ and $E_1, E_2, \ldots E_{r-1}$ are also the maximum and potent factors of $[a]_{r-1}^n$ of orders $1, 2, \ldots r-1$; for every irreducible divisor of $[a]_{r-1}^n$ is necessarily an irreducible divisor of A; in particular D_{r-1} is the H.C.F. of the simple minor determinants of $[a]_{r-1}^n$.

Let $[A]_r^\nu$, where $\nu = \binom{n}{r-1}$, be a complete matrix of the affected minor determinants of A of order $r-1$ in which the 1st, 2nd, … rth horizontal rows are composed of the simple minor determinants of the minor matrices of A

formed by striking out its 1st, 2nd, ... rth horizontal rows, so that as in Theorem II of § 215

$$\underline{\overline{A}}{}^{r}_{\nu}\,[a]^{n}_{r} = [B]^{n}_{\nu} = D_r\,[\beta]^{n}_{\nu}, \quad \text{.......................(7)}$$

where the non-zero elements of $[B]^{n}_{\nu}$ are the minor determinants of A of order r, all of which are divisible by D_r. Writing

$$\underline{\overline{A}}{}^{r}_{\nu} = D_{r-1}\,\underline{\overline{\alpha}}{}^{r}_{\nu}, \quad D_r = D_{r-1}E_r,$$

we can replace the equation (7) by

$$\underline{\overline{\alpha}}{}^{r}_{\nu}\,[a]^{n}_{r} = E_r\,[\beta]^{n}_{\nu}. \quad \text{..........................(8)}$$

Because the highest common factors of all elements in the last vertical rows of $\underline{\overline{A}}{}^{r}_{\nu}$ and $\underline{\overline{\alpha}}{}^{r}_{\nu}$ are respectively D_{r-1} and 1, therefore (see Ex. xii of § 188) there exist rational integral functions $\psi_1, \psi_2, \ldots \psi_\nu$ in Ω such that

$$\psi_1\alpha_{r1} + \psi_2\alpha_{r2} + \ldots + \psi_\nu\alpha_{r\nu} = 1\,;$$

and by prefixing $[\psi]_\nu$ on both sides of (8) we obtain an identity in Ω of the form

$$[\phi',\ 1]^{r-1,1}_{1}\,[a]^{n}_{r} = E_r\,.\,[\beta']^{n}_{1} = f(x)\,.\,[b']^{n}_{1}. \quad \text{..............(9)}$$

Let $[\phi]^{r-1}_{1}$ be the remainder obtained when $[\phi']^{r-1}_{1}$ is divided by $f(x)$. Then from (9) we deduce an identity of the form (D) in which the degree $[\phi]^{r-1}_{1}$ is less than the degree of $f(x)$. In fact if

$$[\phi']^{r-1}_{1} = f(x)\,.\,[\psi']^{r-1}_{1} + [\phi]^{r-1}_{1},$$

the identity (D) is true when

$$[b]^{n}_{1} = [b']^{n}_{1} - [\psi']^{r-1}_{1}\,[a]^{n}_{r-1}.$$

Theorem III clearly remains true under the circumstances described in Note 1, E_r and $[b']^{n}_{1}$ being then 0, and $f(x)$ being any rational integral function of x which does not vanish identically.

COROLLARY 1. *If the leading horizontal minor* $[a]^{n}_{r-1}$ *of the x-matrix* $A=[a]^{n}_{r}$ *is undegenerate and impotent, and if* $f(x)$ *lies in* Ω *and is a factor of* D_r, *then there exists a rational integral identity in* Ω *of the form* (D) *in which the degree of* $[\phi]^{r-1}_{1}$ *in x is less than the degree of* $f(x)$.

For in this case $[a]^{n}_{r-1}$ is necessarily regular with respect to all the irreducible divisors of A, and D_r is the same as E_r. This result also follows from the Corollary to Theorem I; and it remains true when A is degenerate and $f(x)$ is any non-vanishing function lying in Ω. In this case the identity (D) is uniquely determinate.

COROLLARY 2. *When the matrix $A=[a]_r^n$ has rank r and $[a]_{r-1}^n$ is regular with respect to all the irreducible divisors of A, there exists a rational integral identity in Ω of the form*

$$[\phi,\ 1]_1^{r-1,1}[a]_r^n=E_r.[b]_1^n, \quad\text{.................................(D')}$$

where $[\phi]_1^{r-1}$ has a lower degree in x than E_r, and where $[b]_1^n$ is primitive (i.e. undegenerate and impotent).

This is the particular case of Theorem III in which $f(x)$ is E_r. It follows from Theorem II that $[b]_1^n$ must be primitive. When E_r is a non-zero constant, we have $[\phi]_1^{r-1}=0$.

COROLLARY 3. *When the matrix $A=[a]_r^n$ has rank r and the successive leading horizontal minors of A are all regular with respect to every one of the irreducible divisors of A, then for each of the values 1, 2, ... r of i there exists a rational integral identity in Ω of the form*

$$[\phi,\ 1]_1^{i-1,1}[a]_i^n=E_i.[b]_1^n, \quad\text{.................................(D'')}$$

where $[\phi]_1^{i-1}$ has a lower degree in x than E_i, and where $[b]_1^n$ is primitive (i.e. undegenerate and impotent).

For in Corollary 2 we can replace r by i.

COROLLARY 4. *When the x-matrix $A=[a]_r^n$ has rank r and the successive leading horizontal minors of A are all regular with respect to every one of the irreducible divisors of A, there exists a rational integral identity in Ω of the form*

$$\begin{bmatrix} 1 & 0 & 0 & \dots & 0 \\ \phi_{21} & 1 & 0 & \dots & 0 \\ \phi_{31} & \phi_{32} & 1 & \dots & 0 \\ \dots & \dots & \dots & \dots & \dots \\ \phi_{r1} & \phi_{r2} & \phi_{r3} & \dots & 1 \end{bmatrix}[a]_r^n = \begin{bmatrix} E_1 & 0 & 0 & \dots & 0 \\ 0 & E_2 & 0 & \dots & 0 \\ 0 & 0 & E_3 & \dots & 0 \\ \dots & \dots & \dots & \dots & \dots \\ 0 & 0 & 0 & \dots & E_r \end{bmatrix}[b]_r^n=[c]_r^n, \quad\text{............(E)}$$

where the ϕ's in the ith horizontal row of the prefactor on the right have a lower degree in x than E_i, and where $[b]_r^n$ is an undegenerate and impotent matrix of rank r.

If we construct r equations of the form (D''), one corresponding to each one of the values 1, 2, ... r of i, these are together equivalent to a single equation of the form (E); and when (E) is any equation constructed in this way, we see by equating correspondingly formed complete matrices of the minor determinants of order r on both sides that $[b]_r^n$ must be an undegenerate and impotent matrix of rank r.

The equation (E) is a unitary equipotent transformation converting $[a]_r^n$ into a similar equipotent matrix $[c]_r^n$ in which the 1st, 2nd, ... rth horizontal rows are divisible by E_1, E_2, ... E_r respectively.

COROLLARY 5. *If $A=[a]_r^n$ is any undegenerate rational integral x-matrix of rank r lying in Ω, and if*

$$[E]_r^r={}^1[E]_r=\begin{bmatrix} E_1 & 0 & \dots & 0 \\ 0 & E_2 & \dots & 0 \\ \dots & \dots & \dots & \dots \\ 0 & 0 & \dots & E_r \end{bmatrix},$$

then there exist unitary equipotent transformations in Ω *of the form*

$$[h]_r^r [a]_r^n = [E]_r^r [b]_r^n = [c]_r^n, \quad \text{..............................(E')}$$

where $[h]_r^r$ *and* $[b]_r^n$ *are undegenerate and impotent rational integral x-matrices of rank* r.

We first apply Theorem I of § 209 to transform A by a unitary equigradent transformation in Ω into a similar matrix $A' = [a']_r^n$ whose successive leading horizontal minors are all regular with respect to all the irreducible divisors of A, and then apply Corollary 4 to the matrix A'. In this way we obtain a resultant transformation of the form (E′) such that in the ith horizontal row of $[h]_r^r$ every non-diagonal element has a lower degree in x than E_i whilst the diagonal element differs from 1 by terms having a lower degree in x than E_i. Since $(h)_r^r = 1$ and $(E)_r^r = D_r$, we see by equating the maximum factors of order r of the two sides of (E′) that $[b]_r^n$ is impotent.

The unitary equipotent transformation (E′) converts $[a]_r^n$ into a similar equipotent matrix $[c]_r^n$ in which the 1st, 2nd, ... rth horizontal rows are divisible by $E_1, E_2, \dots E_r$ respectively.

Theorem IV. *If* $A = [a]_r^r$ *is an undegenerate square matrix whose elements are rational integral functions in* Ω *of the single variable* x *and whose potent factors of orders* 1, 2, ... r *are* $E_1, E_2, \dots E_r$, *and if as in Corollary* 5 *above we write* $[E]_r^r = {}^1[E]_r$, *then there always exist mutually inverse equipotent transformations in* Ω *of the forms*

$$[h]_r^r [a]_r^r [k]_r^r = [E]_r^r, \quad \overline{\underline{H}}{}_r^r \, [E]_r^r \, \overline{\underline{K}}{}_r^r = [a]_r^r. \quad \text{............(F)}$$

Thus the matrix A can be converted into or derived from the equipotent quasi-scalar matrix ${}^1[E]_r$, which we call the *standard reduced form* of A, by equipotent transformations in Ω.

By Corollary 5 to Theorem III there exist unitary equipotent transformations of A in Ω of the form

$$[h]_r^r [a]_r^r = [E]_r^r \, \overline{\underline{K}}{}_r^r, \quad \text{..........................(10)}$$

where $[h]_r^r$ and $\overline{\underline{K}}{}_r^r$ are undegenerate and impotent square matrices. If we denote the inverses of these two matrices by $\overline{\underline{H}}{}_r^r$ and $[k]_r^r$ respectively, we can deduce the first of the equations (F) from (10) by postfixing $[k]_r^r$ on both sides, and the second of the equations (F) from (10) by prefixing $\overline{\underline{H}}{}_r^r$ on both sides.

When the transformations (F) are obtained in this way by the use of Theorem III, the degrees of the elements of $[h]_r^r$ are subject to the limits described in Corollary 5 above. In the next article we shall obtain more special transformations of the same character by the use of Theorem I.

§ 217. Some special unitary equipotent transformations of rational integral *x*-matrices.

Throughout this article $A = [a]_m^n$ is a given rational integral x-matrix of rank r lying in the domain of rationality Ω; $t, t', t'', \dots$ are any number of the distinct irreducible divisors of A; and D_s and E_s are respectively the products of the highest powers of $t, t', t'', \dots$ by which the maximum and potent factors of A of order s are divisible, s having any one of the values $1, 2, \dots r$. If $d_s, d_s', d_s'', \dots$ and $e_s, e_s', e_s'', \dots$ are the maximum and potent indices of $t, t', t'', \dots$ of order s in A, then

$$D_s = t^{d_s} t'^{d_s'} t''^{d_s''} \dots, \quad E_s = t^{e_s} t'^{e_s'} t''^{e_s''} \dots, \quad D_s = E_1 E_2 \dots E_s.$$

In the particular case when $t, t', t'', \dots$ are *all* the distinct irreducible divisors of A, D_s and E_s are the maximum and potent factors of A of order s. For the values $1, 2, \dots r$ of i we will write

$$E_{i1} = \frac{E_i}{E_1}, \quad E_{i2} = \frac{E_i}{E_2}, \quad \dots \quad E_{i,i-1} = \frac{E_i}{E_{i-1}}, \quad E_{ii} = \frac{E_i}{E_i} = 1,$$

these being all rational integral functions of x; and whenever $i > j$ and E_i has the same degree in x as E_j, we may suppose that $E_i = E_j$, so that $E_{ij} = 1$.

It will be convenient to assign a definite meaning to E_{ij} whenever i and j are positive integers not less than 1 and $i \nless j$. Accordingly for all such values of i and j we will use the definitions $E_{ii} = 1$,

$$E_{ij} = \frac{E_i}{E_j} \text{ when } i > j \text{ and } i \ngtr r, \quad E_{ij} = 0 \text{ when } i > j \text{ and } i > r. \quad \dots(1)$$

Then E_{ij} is a rational integral function of x whenever $i \nless j$, but it vanishes identically when $i > r$ except when $j = i$. These definitions are allowable because it is possible to regard D_i and E_i as being 0 whenever $i > r$. In the lemmas and theorems which follow we could regard E_{ij} as having degree not greater than 0 whenever $i < j$, and E_0 as being 1.

For the letters u, v, E we shall use the definitions

$$[u]_i^i = \begin{bmatrix} 1 & 0 & 0 & \dots & 0 \\ u_{21} & 1 & 0 & \dots & 0 \\ u_{31} & u_{32} & 1 & \dots & 0 \\ \dots & \dots & \dots & \dots & \dots \\ u_{i1} & u_{i2} & u_{i3} & \dots & 1 \end{bmatrix}, \quad [v]_i^i = \begin{bmatrix} 1 & v_{12} & v_{13} & \dots & v_{1i} \\ 0 & 1 & v_{23} & \dots & v_{2i} \\ 0 & 0 & 1 & \dots & v_{3i} \\ \dots & \dots & \dots & \dots & \dots \\ 0 & 0 & 0 & \dots & 1 \end{bmatrix}, \quad \dots\dots(2)$$

and

$$[E]_i^i = {}^1[E]_i = \begin{bmatrix} E_1 & 0 & \dots & 0 \\ 0 & E_2 & \dots & 0 \\ \dots & \dots & \dots & \dots \\ 0 & 0 & \dots & E_i \end{bmatrix}, \quad \dots\dots\dots\dots\dots\dots\dots(3)$$

the notation (2) to hold good for all positive integral values of i, and the notation (3) for the values 1, 2, ... r of i. If ϕ is any letter other than u, v and E, the notation $[\phi]_p^q$ will have its usual signification; in particular $[E_i]_p^p$ will mean the scalar matrix $E_i[1]_p^p$.

LEMMA 1. *If $A=[a]_m^n$ is a rational integral x-matrix of rank r lying in Ω whose leading horizontal minors of orders* 1, 2, ... *r are all regular with respect to every one of the given irreducible divisors t, t′, t″, ..., and if for any one of the values* 1, 2, ... *r of i there exists a rational integral identity in Ω of the form*

$$[u]_m^m[a]_m^n = \begin{bmatrix} E, & 0 \\ 0, & E_i \end{bmatrix}_{i,m-i}^{i,m-i} \begin{bmatrix} a \\ \beta \end{bmatrix}_{i,m-i}^n = [b]_m^n, \quad \text{.....................(a)}$$

where $[u]_m^m$ is a matrix defined as in (2), *then:*

(1) *The matrix $[a]_i^n$ is impotent with respect to t, t′, t″, ...*

(2) *If s is any one of the numbers i+1, i+2, ... r, and if*

$$A_i = \begin{bmatrix} a \\ \beta \end{bmatrix}_{i,m-i}^n,$$

the product of the highest powers of t, t′, t″, ... which are common factors of all those minor determinants of A_i of order s which involve all the first i horizontal rows of A_i is the function Δ_s given by

$$\Delta_s = E_{i+1,i}E_{i+2,i}\dots E_{si};$$

and Δ_s is also the product of the highest powers of t, t′, t″, ... which are common factors of all simple minor determinants of the leading horizontal minor of A_i of order s.

(3) *In particular all those minor determinants of A_i of order i+1 which involve all the first i horizontal rows of A_i are divisible by $E_{i+1,i}$.*

The equation (a) is a unitary equipotent transformation converting $[a]_m^n$ into $[b]_m^n$; and by striking out corresponding final horizontal rows on both sides of (a) we obtain unitary equipotent transformations converting the successive leading horizontal minors of $[a]_m^n$ into the corresponding leading horizontal minors of $[b]_m^n$. Moreover we can obtain $[b]_m^n$ from A_i by multiplying the sth horizontal row of A_i by E_s when $s \not> i$, and by E_i when $s \not< i$. Consequently $[b]_m^n$ is a matrix equipotent with $[a]_m^n$ whose successive leading horizontal minors of orders 1, 2, ... r are also all regular with respect to t, t', t'', ..., and which has the additional property that its sth horizontal row is divisible by E_s when $s \not> i$, and by E_i when $s \not< i$.

First let Δ_i be the product of the highest powers of t, t', t'', ... which are common factors of all simple minor determinants of $[a]_i^n$. Then by equating the maximum factors of order i of $[a]_i^n$ and $[b]_i^n$ we see that

$$D_i = E_1E_2\dots E_i\Delta_i = D_i\Delta_i, \quad \text{or} \quad \Delta_i = 1,$$

i.e. $[a]_i^n$ is impotent with respect to t, t', t'', ...

Next let s be any one of the integers $i+1, i+2, \ldots r$, and let Δ_s be the product of the highest powers of $t, t', t'', \ldots$ which are common factors of all simple minor determinants of the leading horizontal minor of A_i of order s. Then by equating the maximum factors of order s of $[a]_s^n$ and $[b]_s^n$ we see that

$$D_s = E_1 E_2 \ldots E_i E_i^{s-i} \Delta_s, \quad \text{or} \quad \Delta_s = E_{i+1,i} E_{i+2,i} \ldots E_{si}.$$

Moreover if Δ is any non-vanishing minor determinant of A_i of order s which involves all the first i horizontal rows of A_i, then because the corresponding minor determinant of $[b]_m^n$ is divisible by D_s, we see that $E_1 E_2 \ldots E_i E_i^{s-i}\Delta$ is divisible by D_s, i.e. Δ is divisible by $E_{i+1,i} E_{i+2,i} \ldots E_{si}$; and this completes the proof of the lemma.

Ex. i. *Generalisation of Lemma* 1. Lemma 1 clearly remains true when the elements of A and $[u]_m^m$ are rational integral functions of any number of scalar variables.

Lemma 2. *If* $A=[a]_m^n$ *is a rational integral x-matrix of rank r lying in* Ω *whose leading horizontal minors of orders* 1, 2, ... *r are all regular with respect to every one of the given irreducible divisors* $t, t', t'', \ldots$, *and if for any one of the values* 1, 2, ... $r-1$ *of i there exists a rational integral identity in* Ω *having the general form*

$$[u]_m^m [a]_m^n = \begin{bmatrix} E, & 0 \\ 0, & E_i \end{bmatrix}_{i,m-i}^{i,m-i} \begin{bmatrix} a \\ \beta \end{bmatrix}_{i,m-i}^{n} = [b]_m^n, \quad \ldots\ldots\ldots\ldots\ldots\ldots (\mathrm{a}_i)$$

where $[u]_m^m$ *is a matrix defined as in* (2) *whose non-diagonal elements* u_{sj} (*in all of which* $s>j$) *satisfy the conditions that:*

(1) u_{sj} *has a lower degree in x than* E_{sj} *when* $s \not> i$,

(2) u_{sj} *has a lower degree in x than* E_{ij} *when* $s>i$ *and* $j<i$,

(3) $u_{sj}=0$ *when* $s>i$ *and* $j \not< i$,

then there also exists a rational integral identity in Ω *having the general form*

$$[u]_m^m [a]_m^n = \begin{bmatrix} E, & 0 \\ 0, & E_{i+1} \end{bmatrix}_{i+1,m-i-1}^{i+1,m-i-1} \begin{bmatrix} a \\ \beta \end{bmatrix}_{i+1,m-i-1}^{n} = [b]_m^n, \quad \ldots\ldots\ldots (\mathrm{a}_{i+1})$$

where $[u]_m^m$ *is a matrix defined as in* (2) *whose non-diagonal elements* u_{sj} (*in all of which* $s>j$) *satisfy the conditions that:*

(1′) u_{sj} *has a lower degree in x than* E_{sj} *when* $s \not> i+1$,

(2′) u_{sj} *has a lower degree in x than* $E_{i+1,j}$ *when* $s>i+1$ *and* $j<i+1$,

(3′) $u_{sj}=0$ *when* $s>i+1$ *and* $j \not< i+1$,

and where the leading horizontal minors $[u]_i^m, [a]_i^n, [b]_i^n$ *are the same in* (a_{i+1}) *as in* (a_i).

As usual a rational integral function of x whose degree in x is less than 0 must be interpreted to be 0, and a rational integral x-matrix whose degree in x is less than 0 must be interpreted to be one every element of which is 0. Thus if $E_s = E_j$, then in (a_i) and (a_{i+1}) we must have $u_{sj}=0$.

We will express (a_i) in the form

$$\begin{bmatrix} u, & 0 \\ \phi, & 1 \end{bmatrix}_{i,m-i}^{i,m-i} [a]_m^n = \begin{bmatrix} E, & 0 \\ 0, & E_i \end{bmatrix}_{i,m-i}^{i,m-i} \begin{bmatrix} a \\ \beta \end{bmatrix}_{i,m-i}^{n} = [b]_m^n, \quad \ldots\ldots\ldots\ldots\ldots (\mathrm{b}_i)$$

where $[u]_i^i$ is a matrix defined as in (2) whose non-diagonal elements u_{sj} (in all of which $s>j$ and $j<i$) satisfy the condition that

$$u_{sj} \text{ has a lower degree in } x \text{ than } E_{ij},$$

and where $[\phi]_{m-i}^i$ is a matrix whose elements ϕ_{sj} (in all of which $j \not> i$) satisfy the condition that

$$\phi_{sj} \text{ has a lower degree in } x \text{ than } E_{ij},$$

so that in particular

$$\phi_{si}=0 \text{ for all permissible values of } s;$$

and we will express (a_{i+1}) in the form

$$\begin{bmatrix} u, & 0 \\ \phi', & 1 \end{bmatrix}_{i,\, m-i}^{i,\, m-i} [a]_m^n = \begin{bmatrix} E, & 0 \\ 0, & E_{i+1} \end{bmatrix}_{i,\, m-i}^{i,\, m-i} \begin{bmatrix} a \\ \beta' \end{bmatrix}_{i,\, m-i}^{n}, \quad \text{..............}(b_{i+1})$$

where $[u]_i^i$ and $[a]_i^n$ are the same matrices as in (b_i), and where $[\phi']_{m-i}^i$ is a matrix whose elements ϕ'_{sj} (in all of which $j \not> i$) satisfy the condition that

$$\phi'_{sj} \text{ has a lower degree in } x \text{ than } E_{i+1,j}.$$

To prove the lemma we will assume that there exists a rational integral identity in Ω of the form (b_i), and show that there must then also exist a rational integral identity in Ω of the form (b_{i+1}). For the postfactor on the left in (b_i) we will use the notation

$$A_i = \begin{bmatrix} a \\ \beta \end{bmatrix}_{i,\, m-i}^{n}.$$

By Lemma 1 all those minor determinants of A_i of order $i+1$ which involve all the first i horizontal rows of A_i are divisible by $E_{i+1,i}$, whilst the horizontal minor $[a]_i^n$ is impotent with respect to $t, t', t'', \ldots$. Hence by applying Theorem I of § 216 in turn to each of the $m-i$ horizontal minors of A_i of order $i+1$ which contain $[a]_i^n$ we see that there exists a *uniquely determinate* rational integral identity in Ω of the form

$$[\lambda,\ 1]_{m-i}^{i,m-i} \begin{bmatrix} a \\ \beta \end{bmatrix}_{i,\, m-i}^{n} = [\lambda]_{m-i}^i [a]_i^n + [\beta]_{m-i}^n = E_{i+1,i}\, [\beta']_{m-i}^n, \quad \text{............}(4)$$

where $[\lambda]_{m-i}^i$ has a lower degree in x than $E_{i+1,i}$ (and vanishes when $E_{i+1}=E_i$), and where $[a]_i^n$ and $[\beta]_{m-i}^n$ are the same as in (b_i).

Multiplying both sides of (4) by E_i, we obtain the identity

$$[\lambda]_{m-i}^i \begin{bmatrix} E_{i1} & 0 & \ldots & 0 \\ 0 & E_{i2} & \ldots & 0 \\ \cdots & \cdots & \cdots & \cdots \\ 0 & 0 & \ldots & E_{ii} \end{bmatrix} \begin{bmatrix} E_1 & 0 & \ldots & 0 \\ 0 & E_2 & \ldots & 0 \\ \cdots & \cdots & \cdots & \cdots \\ 0 & 0 & \ldots & E_i \end{bmatrix} [a]_i^n + E_i [\beta]_{m-i}^n = E_{i+1} [\beta']_{m-i}^n,$$

or

$$[\lambda']_{m-i}^i [b]_i^n + E_i [\beta]_{m-i}^n = E_{i+1} [\beta']_{m-i}^n,$$

or

$$[\lambda',\ 1]_{m-i}^{i,\, m-i} [b]_m^n = E_{i+1} [\beta']_{m-i}^n, \quad \text{..........................}(4')$$

where

$$[\lambda']_{m-i}^i = [\lambda]_{m-i}^i \begin{bmatrix} E_{i1} & 0 & \ldots & 0 & 0 \\ 0 & E_{i2} & \ldots & 0 & 0 \\ \cdots & \cdots & \cdots & \cdots & \cdots \\ 0 & 0 & \ldots & E_{i,i-1} & 0 \\ 0 & 0 & \ldots & 0 & 1 \end{bmatrix},$$

or

$$\lambda'_{sj} = E_{ij}\lambda_{sj}, \quad (j=1,\ 2,\ \ldots\ i); \quad \text{.................................}(5)$$

and when we associate with (4′) the equation

$$[b]_i^n = [E]_i^i [a]_i^n, \quad \text{.......................................(4'')}$$

we obtain the identity

$$\begin{bmatrix} 1, & 0 \\ \lambda', & 1 \end{bmatrix}_{i,\, m-i}^{i,\, m-i} [b]_m^n = \begin{bmatrix} E, & 0 \\ 0, & E_{i+1} \end{bmatrix}_{i,\, m-i}^{i,\, m-i} \begin{bmatrix} a \\ \beta' \end{bmatrix}_{i,\, m-i}^{n}, \quad \text{.....................(6)}$$

or

$$\begin{bmatrix} 1, & 0 \\ \lambda', & 1 \end{bmatrix}_{i,\, m-i}^{i,\, m-i} \begin{bmatrix} u, & 0 \\ \phi, & 1 \end{bmatrix}_{i,\, m-i}^{i,\, m-i} [a]_m^n = \begin{bmatrix} E, & 0 \\ 0, & E_{i+1} \end{bmatrix}_{i,\, m-i}^{i,\, m-i} \begin{bmatrix} a \\ \beta' \end{bmatrix}_{i,\, m-i}^{n}. \quad \text{............(6')}$$

Now (6′) is the identity (b_{i+1}) in which

$$[\phi']_{m-i}^i = [\lambda']_{m-i}^i [u]_i^i + [\phi]_{m-i}^i, \quad \text{..............................(7)}$$

or

$$\phi'_{sj} = \lambda'_{sj} + \lambda'_{s,j+1} u_{j+1,j} + \lambda'_{s,j+2} u_{j+2,j} + \ldots + \lambda'_{si} u_{ij} + \phi_{sj}; \quad \text{..............(7')}$$

and because λ'_{sj} has a lower degree in x than $E_{ij} E_{i+1,i}$, i.e. than $E_{i+1,j}$, we see that in (7)

$$\phi'_{sj} = \phi_{sj} + \text{terms having a lower degree in } x \text{ than } E_{i+1,j},$$

and therefore ϕ'_{sj} has a lower degree in x than $E_{i+1,j}$. Thus from the given identity (b_i) we have deduced an identity (b_{i+1}) in which the prescribed conditions are satisfied.

It will be seen that (a_{i+1}) *is obtained by multiplying both sides of* (a_i) *by the prefactor*

$$\begin{bmatrix} 1, & 0 \\ \lambda', & 1 \end{bmatrix}_{i,\, m-i}^{i,\, m-i}.$$

Ex. ii. When $i=1$, we have

$$[\phi]_{m-1}^1 = 0, \quad [\lambda']_{m-1}^1 = [\lambda]_{m-1}^1, \quad [\phi']_{m-1}^1 = [\lambda]_{m-1}^1.$$

Ex. iii. In the corresponding lemma for vertical minors we must replace (a_i) by

$$[a]_m^n [v]_n^n = [a, \beta]_m^{i,\, n-i} \begin{bmatrix} E, & 0 \\ 0, & E_i \end{bmatrix}_{i,\, n-i}^{i,\, n-i} = [b]_m^n,$$

where v_{js} has a lower degree than E_{sj} when $s \not> i$, and a lower degree than E_{ij} when $s \not> i$ (being 0 when $j \not< i$); and we obtain the identity corresponding to (a_{i+1}) by multiplying both sides by the postfactor

$$\begin{bmatrix} 1, & \lambda' \\ 0, & 1 \end{bmatrix}_{i,\, n-i}^{i,\, n-i}, \quad \text{where } \lambda'_{js} = E_{ij} \lambda_{js},$$

and where $[\lambda]_i^{n-i}$ is the uniquely determinate matrix whose elements have a lower degree than $E_{i+1,i}$, and which is such that

$$[a, \beta]_m^{i,\, n-i} \begin{bmatrix} \lambda \\ 1 \end{bmatrix}_{i,\, n-i}^{n-i} \text{ is divisible by } E_{i+1,i}.$$

Lemma 3. *If* $A = [a]_m^n$ *is the matrix of Lemmas* 1 *and* 2, *and if* i *is any one of the integers* 1, 2, ... r, *then there cannot exist two different identities of the form* (a_i) *in which the conditions* (1), (2), (3) *of Lemma* 2 *are satisfied.*

It is to be understood that a definite choice has been made of the functions $E_1, E_2, \ldots E_r$, or that fixed values have been given to the arbitrary numerical factors occurring in them.

Suppose that (a_i) is an actual identity satisfying the prescribed conditions, and let (a_i') be any other identity of the same form satisfying the same conditions; let (a_i) be expressed in the form (b_i), and when (a_i') is expressed in a similar form (b_i'), let (b_i') be derivable from (b_i) by replacing u, ϕ, a, β by u', ϕ', a', β'; further let

$$[U]_i^i=[u]_i^i-[u']_i^i,\quad [\Phi]_{m-i}^i=[\phi]_{m-i}^i-[\phi']_{m-i}^i,$$

$$[\xi]_i^n=[a]_i^n-[a']_i^n,\qquad [\eta]_i^n=[\beta]_i^n-[\beta']_i^n,$$

so that

$$U_{sj}=0 \text{ when } s \not> j,\quad U_{sj} \text{ has a lower degree than } E_{sj} \text{ when } s>j,$$

$$\Phi_{sj}=0 \text{ when } j=i,\quad \Phi_{sj} \text{ has a lower degree than } E_{ij} \text{ when } j<i.$$

Then by subtracting (a_i') from (a_i) or (b_i') from (b_i) we obtain an equation which is equivalent to the two identities

$$[U]_i^i[a]_i^n=[E]_i^i[\xi]_i^n,\quad [\Phi]_{m-i}^i[a]_i^n=E_i[\eta]_{m-i}^n. \quad\text{.....................}(c_i)$$

First suppose that the elements of the sth horizontal row of $[U]_i^i$ are not all 0's (where $s>1$ and $s \not> i$), and let U_{sk} be the last of those elements which is not 0 (where $k<s$), so that

$$[U_{s1}\,U_{s2}\ldots U_{sk}]\,[a]_k^n=E_s\,[\xi_{s1}\,\xi_{s2}\ldots\xi_{sn}].$$

Then if Q is the H.C.F. of U_{s1}, U_{s2}, ... U_{sk}, and if we write

$$[U_{s1}\,U_{s2}\ldots U_{sk}]=Q\,[l_{s1}\,l_{s2}\ldots l_{sk}], \quad\text{................................}(8)$$

$$[l_{s1}\,l_{s2}\ldots l_{sk}]\,[a]_k^n=\omega\,[\lambda_{s1}\,\lambda_{s2}\ldots\lambda_{sk}], \quad\text{............................}(9)$$

where the one-rowed matrices on the right are primitive, we have

$$E_s\,[\xi_{s1}\,\xi_{s2}\ldots\xi_{sn}]=Q\omega\,[\lambda_{s1}\,\lambda_{s2}\ldots\lambda_{sk}]. \quad\text{..........................}(10)$$

If further Q' and ω' are the products of the highest powers of t, t', t'', ... which are factors of Q and ω respectively, we see from (10) that $Q'\omega'$ is divisible by E_s, i.e. by $E_{sk}E_k$; moreover by Theorem I of § 215 or Theorem II of § 216 it follows from (9) that E_k is divisible by ω'; therefore Q' must be divisible by E_{sk}. But this is impossible because Q and Q', which do not vanish identically and are factors of U_{sk}, have each a lower degree in x than E_{sk}.

We conclude that every element of $[U]_i^i$ must be 0.

Next suppose that the elements of the sth horizontal row of $[\Phi]_{m-i}^i$ are not all 0's, and let Φ_{sk} be the last of those elements which is not 0 (where $k<i$), so that

$$[\Phi_{s1}\,\Phi_{s2}\ldots\Phi_{sk}]\,[a]_k^n=E_i\,[\eta_{s1}\,\eta_{s2}\ldots\eta_{sn}].$$

Then if Q is the H.C.F. of Φ_{s1}, Φ_{s2}, ... Φ_{sk}, and if we write

$$[\Phi_{s1}\,\Phi_{s2}\ldots\Phi_{sk}]=Q\,[l_{s1}\,l_{s2}\ldots l_{sk}], \quad\text{................................}(8')$$

$$[l_{s1}\,l_{s2}\ldots l_{sk}]\,[a]_k^n=\omega\,[\lambda_{s1}\,\lambda_{s2}\ldots\lambda_{sk}], \quad\text{............................}(9')$$

where the one-rowed matrices on the right are primitive, we have

$$E_i\,[\eta_{s1}\,\eta_{s2}\ldots\eta_{sn}]=Q\omega\,[\lambda_{s1}\,\lambda_{s2}\ldots\lambda_{sk}]. \quad\text{..........................}(10')$$

If further Q' and ω' are the products of the highest powers of t, t', t'', ... which are factors of Q and ω respectively, we see from (10′) that $Q'\omega'$ is divisible by E_i, i.e. by $E_{ik}E_k$;

moreover by Theorem I of § 215 or Theorem II of § 216 it follows from (9') that E_k is divisible by ω'; therefore Q' must be divisible by E_{ik}. But this is impossible because Q', which is a factor of Φ_{sk}, has a lower degree in x than E_{ik}.

We conclude that every element of $[\Phi]_{m-i}^{i}$ must be 0.

It has now been shown that we must have $[u']_i^i=[u]_i^i$, $[\phi']_{m-i}^i=[\phi]_{m-i}^i$, i.e. the identity ($b_i'$) or ($a_i'$) must be the same as the identity (b_i) or (a_i); and this proves the lemma.

It follows that when the identity (a_i) is given, and $i \ngtr r-1$, there exists *one and only one* identity of the form (a_{i+1}), viz. that obtained in Lemma 2.

Theorem I a. *If $A=[a]_m^n$ is a rational integral x-matrix of rank r lying in Ω whose leading horizontal minors of orders 1, 2, ... r are all regular with respect to every one of the given irreducible divisors $t, t', t'', \ldots$, then there exists one and only one unitary equipotent transformation of A in Ω of the form*

$$[u]_m^m [a]_m^n = \begin{bmatrix} E, & 0 \\ 0, & E_r \end{bmatrix}_{r,m-r}^{r,m-r} [\alpha]_m^n = [b]_m^n, \quad \text{...............(A)}$$

where $[u]_m^m$ is a matrix defined as in (2) whose non-diagonal elements u_{sj} (in all of which $s>j$) satisfy the conditions that:

(1) *u_{sj} has a lower degree in x than E_{sj} when $s \ngtr r$;*

(2) *u_{sj} has a lower degree in x than E_{rj} when $s \nless r$, and in particular $u_{sj}=0$ when $j \nless r$.*

The leading horizontal minors of orders 1, 2, ... r of $[b]_m^n$ are all regular with respect to $t, t', t'', \ldots$; and the sth horizontal row of $[b]_m^n$ is divisible by E_s when $s \ngtr r$, and by E_r when $s \nless r$.

The functions $E_1, E_2, \ldots E_r$ are the potent factors of A of orders 1, 2, ... r when and only when $t, t', t'', \ldots$ are all the distinct irreducible divisors of A.

The theorem is proved by repeated applications of Lemmas 1, 2 and 3. Since E_1 is a factor of every element of A, we can write

$$[a]_m^n = E_1 [\alpha]_m^n,$$

or

$$\begin{bmatrix} 1, & 0 \\ 0, & 1 \end{bmatrix}_{1,m-1}^{1,m-1} [a]_m^n = \begin{bmatrix} E, & 0 \\ 0, & E_1 \end{bmatrix}_{1,m-1}^{1,m-1} [\alpha]_m^n \text{..................}(a_1)$$

This is the transformation (a_i) of Lemma 2 for the case in which $i=1$, and it is necessarily unique. By Lemma 1 the matrix $[\alpha]_1^n$ is impotent with respect to $t, t', t'', \ldots$, and all those minor determinants of order 2 of the matrix $[\alpha]_m^n$ which involve the first horizontal row of $[\alpha]_m^n$ are divisible by E_{21}.

Therefore by Lemma 2 we can derive from (a_1) a transformation (a_2) which is the transformation (a_i) of Lemma 2 for the case in which $i=2$, and by Lemma 3 this transformation is unique. When we obtain in this way in succession transformations (a_1), (a_2), ... (a_r) all of the same type as (a_i) in Lemma 2, the last of them, for which $i=r$, is the transformation (A), and by Lemma 3 there is only one such transformation.

In proving Theorem I *a* we have at the same time proved the more general theorem that:

For each of the values 1, 2, ... *r of i there exists one and only one transformation* (a_i) *defined as in Lemma* 2; *and when* $i \not> r-1$, *the transformation* (a_{i+1}) *must always be derivable from* (a_i) *in the way described in Lemma* 2.

Ex. iv. If in Theorem I *a* the leading diagonal minor determinants of A of orders 1, 2, ... r are all regular with respect to $t, t', t'', \ldots$, then the leading diagonal minor determinants of $[b]_m^n$ of orders 1, 2, ... r are also all regular with respect to $t, t', t'', \ldots$.

This follows from the equations $[u]_i^i [a]_i^i = [b]_i^i$, $(a)_i^i = (b)_i^i$.

Ex. v. If $A=[a]_m^n$ is any rational integral x-matrix of rank r lying in Ω, and if $t, t', t'', \ldots$ are any number of its distinct irresoluble divisors, we can convert it into an equipotent matrix $[b]_m^n$ having the properties described in Theorem I *a* by applying to it in succession an equigradent transformation of the form described in Theorem I of § 209 and a unitary equipotent transformation in Ω of the form described in Theorem I *a* above. We thus obtain a resultant unitary equipotent transformation in Ω of the form

$$[p]_m^m [a]_m^n = \begin{bmatrix} E, & 0 \\ 0, & E_r \end{bmatrix}_{r,\, m-r}^{r,\, m-r} [a]_m^n = [b]_m^n, \quad \ldots\ldots\ldots\ldots\text{(A')}$$

where

$$[p]_m^m = \begin{bmatrix} 1 & 0 & 0 & \ldots & 0 \\ u_{21} & 1 & 0 & \ldots & 0 \\ u_{31} & u_{32} & 1 & \ldots & 0 \\ \ldots & \ldots & \ldots & \ldots & \ldots \\ u_{m1} & u_{m2} & u_{m3} & \ldots & 1 \end{bmatrix} \begin{bmatrix} 1 & h_{12} & h_{13} & \ldots & h_{1m} \\ 0 & 1 & h_{23} & \ldots & h_{2m} \\ 0 & 0 & 1 & \ldots & h_{3m} \\ \ldots & \ldots & \ldots & \ldots & \ldots \\ 0 & 0 & 0 & \ldots & 1 \end{bmatrix},$$

each element h_{js} being either 0 or a positive integer, and each element u_{sj} satisfying the conditions of Theorem I *a*.

When $t, t', t'', \ldots$ are all the irreducible divisors of A, the function E_i is the potent factor of A of order i.

Theorem I b. *If* $A=[a]_m^n$ *is a rational integral x-matrix of rank r lying in* Ω *whose leading vertical minors of orders* 1, 2, ... *r are all regular with respect to every one of the given irreducible divisors* $t, t', t'', \ldots$, *then there exists one and only one unitary equipotent transformation of A in* Ω *of the form*

$$[a]_m^n [v]_n^n = [\alpha]_m^n \begin{bmatrix} E, & 0 \\ 0, & E_r \end{bmatrix}_{r,\, n-r}^{r,\, n-r} = [b]_m^n, \quad \ldots\ldots\ldots\ldots\text{(B)}$$

where $[v]_n^n$ *is a matrix defined as in* (2) *whose non-diagonal elements* v_{js} (*in all of which* $s > j$) *satisfy the conditions that:*

(1) v_{js} *has a lower degree in* x *than* E_{sj} *when* $s \not> r$;

(2) v_{js} *has a lower degree in* x *than* E_{rj} *when* $s \not< r$, *and in particular* $v_{js} = 0$ *when* $j \not< r$.

The leading vertical minors of orders 1, 2, ... r *of* $[b]_m^n$ *are all regular with respect to* $t, t', t'', \ldots$; *and the* sth *vertical row of* $[b]_m^n$ *is divisible by* E_s *when* $s \not> r$, *and by* E_r *when* $s \not< r$.

We can obtain this theorem by equating the conjugates of both sides in (A), or we can give an independent proof similar to the proof of Theorem Ia.

Ex. vi. If in Theorem Ib the leading diagonal minor determinants of A of orders 1, 2, ... r are all regular with respect to $t, t', t'', \ldots$, then the leading diagonal minor determinants of $[b]_m^n$ of orders 1, 2, ... r are also all regular with respect to $t, t', t'', \ldots$.

Ex. vii. If $A = [a]_m^n$ is any rational integral x-matrix of rank r lying in Ω, and if $t, t', t'', \ldots$ are any number of its distinct irreducible divisors, we can convert it into an equipotent matrix $[b]_m^n$ having the properties described in Theorem Ib by applying to it in succession an equigradent transformation of the form described in Theorem II of § 209 and a unitary equipotent transformation of the form described in Theorem Ib above. We thus obtain a resultant unitary equipotent transformation in Ω of the form

$$[a]_m^n [q]_n^n = [a]_m^n \begin{bmatrix} E, & 0 \\ 0, & E_r \end{bmatrix}_{r,\, n-r}^{r,\, n-r} = [b]_m^n, \quad \text{.......................(B')}$$

where

$$[q]_n^n = \begin{bmatrix} 1 & 0 & 0 & \ldots & 0 \\ k_{21} & 1 & 0 & \ldots & 0 \\ k_{31} & k_{32} & 1 & \ldots & 0 \\ \cdots & \cdots & \cdots & \cdots & \cdots \\ k_{n1} & k_{n2} & k_{n3} & \ldots & 1 \end{bmatrix} \begin{bmatrix} 1 & v_{12} & v_{13} & \ldots & v_{1n} \\ 0 & 1 & v_{23} & \ldots & v_{2n} \\ 0 & 0 & 1 & \ldots & v_{3n} \\ \cdots & \cdots & \cdots & \cdots & \cdots \\ 0 & 0 & 0 & \ldots & 1 \end{bmatrix},$$

each element k_{sj} being either 0 or a positive integer, and each element v_{js} satisfying the conditions of Theorem Ib.

When $t, t', t'', \ldots$ are all the irreducible divisors of A, the function E_i is the potent factor of A of order i.

Theorem I c. *If* $A = [a]_m^n$ *is a rational integral* x*-matrix of rank* r *lying in* Ω *whose leading diagonal minor determinants of orders* 1, 2, ... r *are all regular with respect to every one of the given irreducible divisors* $t, t', t'', \ldots$, *then there exists one and only one unitary equipotent transformation of* A *in* Ω *of the form*

$$[u]_m^m [a]_m^n [v]_n^n = [c]_m^n, \quad \text{..........................(C)}$$

where $[c]_m^n$ *is a rational integral x-matrix in which*

(1) *the sth horizontal row is divisible by* E_s *when* $s \not> r$ *and by* E_r *when* $s \not< r$,

(2) *the sth vertical row is divisible by* E_s *when* $s \not> r$ *and by* E_r *when* $s \not< r$,

and where $[u]_m^m$ *and* $[v]_n^n$ *are matrices defined as in* (2) *whose elements are subject respectively to the conditions specified in Theorem I a and Theorem I b.*

The leading diagonal minor determinants of $[c]_m^n$ *of orders* 1, 2, ... *r are all regular with respect to* $t, t', t'', \ldots$.

In this case we can determine identities of both the forms (A) and (B), and when $[u]_m^m$ and $[v]_n^n$ are the matrices thus found, we clearly have an identity of the form (C). By Theorems I *a* and I *b* no other values of $[u]_m^m$ and $[v]_n^n$ are possible in a relation of the form (C). It can be seen from the particular case of Ex. xi that in general the non-diagonal elements of $[c]_m^n$ do not vanish.

When A is a symmetric matrix $[a]_m^m$, so that $n = m$, the matrix $[v]_m^m$ must be the conjugate of $[u]_m^m$, and the transformation (C) is symmetric.

Ex. viii. If $A=[a]_m^n$ is any rational integral x-matrix of rank r lying in Ω, and if $t, t', t'', \ldots$ are any number of its distinct irreducible divisors, we can convert it into an equipotent matrix $[c]_m^n$ having the properties described in Theorem I *c* by applying to it in succession an equigradent transformation of the form described in Theorem III of § 209 and a unitary equipotent transformation of the form described in Theorem I *c* above. We thus obtain a resultant unitary equipotent transformation in Ω of the form

$$[p]_m^m [a]_m^n [q]_n^n = [c]_m^n, \quad \text{.................................(C}')$$

where $[c]_m^n$ has the properties described in Theorem I *c*, and where $[p]_m^m$ and $[q]_n^n$ have respectively the forms and properties described in Exs. v and vii.

When A is the symmetric matrix $[a]_m^m$, so that $n=m$, then by § 210 the component equigradent transformations can be symmetric, and then (C′) is also a symmetric transformation.

Ex. ix. *The transformation* (A) *for the matrix* $A=[a]_3^3$ *of rank* 3 *given by*

$$[a]_3^3 = \begin{bmatrix} x+1, & 0, & x^3 \\ 0, & x+1, & 0 \\ 0, & 0, & x^3 \end{bmatrix}. \quad \text{..............................(11)}$$

We will take Ω to be the domain of all rational numbers, and $t, t', t'', \ldots$ to be all the irreducible divisors of A. The potent factors of A of orders 1, 2, 3 are

$$E_1=1, \quad E_2=x+1, \quad E_3=x^3(x+1);$$

there are only two irreducible divisors, viz. x and $x+1$; and the leading horizontal minors of A of orders 1, 2, 3 are all regular with respect to both these divisors. There will be three successive transformations (a_1), (a_2), (a_3) corresponding to the transformation (a_i) of Lemma 2, the last of these being the transformation (A).

We may regard (11) as the transformation (a_1) when we replace $[a]_3^3$ by $[1]_3^3[a]_3^3$.

To deduce (a_2) we first find the uniquely determinate matrix $[\lambda]_2^1$ whose elements have a lower degree in x than the function $E_{21}=x+1$, i.e. are constants, and which is such that

$$[\lambda,\ 1]_2^{1,2}[a]_3^3 \text{ is divisible by } E_{21};$$

and (using Note 2 of § 216) we easily see that $\lambda_{11}=0$, $\lambda_{21}=-1$. We next determine the matrix $[\lambda']_2^1$ given by $\lambda'_{sj}=E_{ij}\lambda_{sj}$ when $i=1$, and find that $[\lambda']_2^1=[\lambda]_2^1$. Finally by prefixing the matrix

$$\begin{bmatrix}1, & 0\\ \lambda', & 1\end{bmatrix}_{1,2}^{1,2}$$

on both sides of (11) we find the transformation (a_2) to be

$$\begin{bmatrix}1, & 0, & 0\\ 0, & 1, & 0\\ -1, & 0, & 1\end{bmatrix}[a]_3^3=\begin{bmatrix}1, & 0, & 0\\ 0, & x+1, & 0\\ 0, & 0, & x+1\end{bmatrix}[b]_3^3,$$

where

$$[b]_3^3=\begin{bmatrix}x+1, & 0, & x^3\\ 0, & 1, & 0\\ -1, & 0, & 0\end{bmatrix}. \qquad \text{.................................(11')}$$

To deduce (a_3) we first find the uniquely determinate matrix $[\lambda]_1^2$ whose elements have a lower degree in x than the function $E_{32}=x^3$, and which is such that

$$[\lambda,\ 1]_1^{2,1}[b]_3^3 \text{ is divisible by } E_{32};$$

and (proceeding as in Note 2 of § 216) we find that $\lambda_{11}=x^2-x+1$, $\lambda_{12}=0$. We next determine the matrix $[\lambda']_1^2$ given by $\lambda'_{sj}=E_{ij}\lambda_{sj}$ when $i=2$, and find that $\lambda'_{11}=E_{21}\lambda_{11}=x^3+1$, $\lambda'_{12}=0$. Finally by prefixing the matrix

$$\begin{bmatrix}1, & 0\\ \lambda', & 1\end{bmatrix}_{2,1}^{2,1}$$

on both sides of (11') we find the transformation (a_3) or (A) to be

$$\begin{bmatrix}1, & 0, & 0\\ 0, & 1, & 0\\ x^3, & 0, & 1\end{bmatrix}\begin{bmatrix}x+1, & 0, & x^3\\ 0, & x+1, & 0\\ 0, & 0, & x^3\end{bmatrix}=\begin{bmatrix}1, & 0, & 0\\ 0, & x+1, & 0\\ 0, & 0, & x^3(x+1)\end{bmatrix}\begin{bmatrix}x+1, & 0, & x^3\\ 0, & 1, & 0\\ 1, & 0, & x^2-x+1\end{bmatrix}.$$

......(11'')

Ex. x. *A transformation of the form* (A') *for the matrix of rank* 3,

$$A=[a]_3^3=\begin{bmatrix}x+1, & 0, & 0\\ 0, & x+1, & 0\\ 0, & 0, & x^3\end{bmatrix}.$$

The potent factors of A are

$$E_1=1, \quad E_2=x+1, \quad E_3=x^3(x+1).$$

Applying Theorem I of § 209 we obtain the equigradent transformation

$$\begin{bmatrix} 1 & 0 & 1 \\ 0 & 1 & 0 \\ 0 & 0 & 1 \end{bmatrix} [a]_3^3 = \begin{bmatrix} x+1, & 0, & x^3 \\ 0, & x+1, & 0 \\ 0, & 0, & x^3 \end{bmatrix} = [b]_3^3,$$

and when this is followed by the transformation (11″) of Ex. ix, we obtain the resultant equipotent transformation

$$\begin{bmatrix} 1, & 0, & 1 \\ 0, & 1, & 0 \\ x^3, & 0, & x^3+1 \end{bmatrix} [a]_3^3 = \begin{bmatrix} 1, & 0, & 0 \\ 0, & x+1, & 0 \\ 0, & 0, & x^3(x+1) \end{bmatrix} \begin{bmatrix} x+1, & 0, & x^3 \\ 0, & 1, & 0 \\ 1, & 0, & x^2-x+1 \end{bmatrix} \quad \text{....(12)}$$

Ex. xi. *The transformations* (A), (B), (C) *for the matrix* $A=[a]_3^4$ *of rank* 3 *given by*

$$[a]_3^4=(x+1)\,[b]_3^4, \quad \text{.......................................(13)}$$

where

$$[b]_3^4 = \begin{bmatrix} 2x^3-x^2+1, & 2x^2-x-1, & 2x^3+x^2-x, & 2x^3-5x^2+2x+3 \\ 3x^2-3x+3, & 3x^2+3x-6, & 3x^2, & 3x^2-9x+9 \\ 2x^2-2x+1, & x-1, & 2x^2-1, & 2x^2-6x+5 \end{bmatrix}.$$

We will take Ω to be the domain of all rational numbers, and $t, t', t'', \ldots$ to be all the irreducible divisors of A. The matrix A (see Ex. v of § 209) has only two irreducible divisors $x+1$ and $x-1$, its leading diagonal minor determinants of orders 1, 2, 3 are all regular with respect to both these divisors, and its potent factors of orders 1, 2, 3 are

$$E_1=x+1, \quad E_2=x^2-1, \quad E_3=x^2-1.$$

There will be three successive transformations (a_1), (a_2), (a_3) of the same type as (a_i) in Lemma 2, the last of these being the transformation (A). We may regard (13) as the transformation (a_1) when we replace $[a]_3^4$ by $[1]_3^3[a]_3^4$. The matrix $[\lambda]_2^1$ on which the determination of (a_2) by Lemma 2 depends has a lower degree in x than the function $E_{21}=x-1$, and is such that

$$[\lambda, 1]_2^{1,2}\,[b]_3^4 \text{ is divisible by } E_{21}.$$

We have $\lambda_{11}=-\frac{3}{2}$, $\lambda_{21}=-\frac{1}{2}$, $[\lambda']_2^1=[\lambda]_2^1$, and when we prefix

$$\begin{bmatrix} 1, & 0 \\ \lambda', & 1 \end{bmatrix}_{1,2}^{1,2}$$

on both sides of (13), we obtain as the transformation (a_2)

$$\begin{bmatrix} 1, & 0, & 0 \\ -\frac{3}{2}, & 1, & 0 \\ -\frac{1}{2}, & 0, & 1 \end{bmatrix} [a]_3^4 = \begin{bmatrix} x+1, & 0, & 0 \\ 0, & x^2-1, & 0 \\ 0, & 0, & x^2-1 \end{bmatrix} [c]_3^4, \quad \text{...............(14)}$$

where

$$[c]_3^4 = \begin{bmatrix} 2x^3-x^2+1, & 2x^2-x-1, & 2x^3+x^2-x, & 2x^3-5x^2+2x+3 \\ -\frac{3}{2}(2x^2-x+1), & \frac{9}{2}, & -\frac{3}{2}x(2x+1), & -\frac{3}{2}(2x^2-5x+3) \\ -\frac{1}{2}(2x^2-3x+1), & -\frac{1}{2}(2x-1), & -\frac{1}{2}(2x^2-x-2), & -\frac{1}{2}(2x^2-7x+7) \end{bmatrix}.$$

In the present case this is also the transformation (a_3) or (A).

There will be three successive transformations (b_1), (b_2), (b_3) for the vertical rows, the last of these being the transformation (B). We may regard (13) as the transformation (b_1) when we replace $[a]_3^4$ by $[a]_3^4 [1]_4^4$. The matrix $[\lambda]_1^3$ on which the determination of (b_2) depends has a lower degree than the function $E_{21}=x-1$, and is such that

$$[b]_3^4 \begin{bmatrix} \lambda \\ 1 \end{bmatrix}_{1,3}^{3} \text{ is divisible by } E_{21}.$$

We have $\lambda_{11}=0$, $\lambda_{12}=-1$, $\lambda_{13}=-1$, $[\lambda']_1^3=[\lambda]_1^3$, and when we postfix

$$\begin{bmatrix} 1, & \lambda' \\ 0, & 1 \end{bmatrix}_{1,3}^{1,3}$$

on both sides of (13), we obtain as the transformation (b_2)

$$[a]_3^4 \begin{bmatrix} 1, & 0, & -1, & -1 \\ 0, & 1, & 0, & 0 \\ 0, & 0, & 1, & 0 \\ 0, & 0, & 0, & 1 \end{bmatrix} = [c]_3^4 \begin{bmatrix} x+1, & 0, & 0, & 0 \\ 0, & x^2-1, & 0, & 0 \\ 0, & 0, & x^2-1, & 0 \\ 0, & 0, & 0, & x^2-1 \end{bmatrix}, \quad \text{.........(15)}$$

where

$$[c]_3^4 = \begin{bmatrix} 2x^3-x^2+1, & 2x+1, & 2x+1, & -2(2x+1) \\ 3x^2-3x+3, & 3(x+2), & 3, & -6 \\ 2x^2-2x+1, & 1, & 2, & -4 \end{bmatrix}.$$

In the present case this is also the transformation (b_3) or (B).

From (13) and (14) we see that the transformation (C) is

$$\begin{bmatrix} 1, & 0, & 0 \\ -\tfrac{3}{2}, & 1, & 0 \\ -\tfrac{1}{2}, & 0, & 1 \end{bmatrix} [a]_3^4 \begin{bmatrix} 1, & 0, & -1, & -1 \\ 0, & 1, & 0, & 0 \\ 0, & 0, & 1, & 0 \\ 0, & 0, & 0, & 1 \end{bmatrix} = [e]_3^4, \quad \text{..................(16)}$$

where the elements of the matrix on the right are easily found. From (16) we can deduce a transformation of the form (C′) for the matrix of Ex. v in § 209.

Theorem II. *If $A=[a]_r^r$ is an undegenerate square rational integral x-matrix of rank r lying in Ω whose potent factors of orders 1, 2, ... r are $E_1, E_2, \ldots E_r$, and if*

$$[E]_r^r = {}^1[E]_r = \begin{bmatrix} E_1 & 0 & \ldots & 0 \\ 0 & E_2 & \ldots & 0 \\ \ldots & \ldots & \ldots & \ldots \\ 0 & 0 & \ldots & E_r \end{bmatrix},$$

we can always determine two mutually inverse equipotent transformations in Ω of the forms

$$[h]_r^r [a]_r^r [k]_r^r = [E]_r^r, \quad \overline{\underline{H}}_r^r [E]_r^r \overline{\underline{K}}_r^r = [a]_r^r. \quad \text{...........(D)}$$

Thus the undegenerate square matrix A can be converted into or derived from the *standard form* $[E]_r^r$ by equipotent transformations in Ω.

Taking $t, t', t'', \ldots$ to be all the irreducible divisors of A, and applying to A in succession an equigradent transformation of the form described in Theorem I of § 209 and an equipotent transformation of the form described in Theorem Ia of the present article, we obtain as in Ex. v a resultant equipotent transformation of A in Ω of the form

$$[h]_r^r\,[a]_r^r = [E]_r^r\,\overline{\underline{K}}_r^r, \quad\ldots\ldots\ldots\ldots\ldots\ldots\ldots\ldots\text{(d)}$$

where $[h]_r^r$ is undegenerate and impotent, in fact where $(h)_r^r = 1$. Equating the determinants of both sides in (d), we see that $\overline{\underline{K}}_r^r$ is also undegenerate and impotent. Let $\overline{\underline{H}}_r^r$, $[k]_r^r$ be the inverses of $[h]_r^r$, $\overline{\underline{K}}_r^r$. Then by postfixing $[k]_r^r$ or prefixing $\overline{\underline{H}}_r^r$ on both sides of (d) we obtain the transformations (D). When the transformations (D) are determined in this way, then in the ith horizontal row of $[h]_r^r$ each non-diagonal element has a lower degree in x than E_{i1}, and each diagonal element differs from 1 by terms having a lower degree in x than E_{i1}.

If we apply Theorem II of § 209 and Theorem Ib of the present article, we obtain transformations (D) in which the degrees of the elements of $[k]_r^r$ are subject to the same limits.

Ex. xii. For the undegenerate square matrix A of Ex. x we have found the transformation (12) which is one of the form (d). When we postfix on both sides of (12) the inverse of the postfactor on the right we obtain the equipotent transformation

$$\begin{bmatrix} 1, & 0, & 1 \\ 0, & 1, & 0 \\ x^3, & 0, & x^3+1 \end{bmatrix}\begin{bmatrix} x+1, & 0, & 0 \\ 0, & x+1, & 0 \\ 0, & 0, & x^3 \end{bmatrix}\begin{bmatrix} x^2-x+1, & 0, & -x^3 \\ 0, & 1, & 0 \\ -1, & 0, & x+1 \end{bmatrix} = \begin{bmatrix} 1, & 0, & 0 \\ 0, & x+1, & 0 \\ 0, & 0, & x^3(x+1) \end{bmatrix}$$

corresponding to the first of the transformations (D).

§ 218. Derivation of any given rational integral x-matrix from an undegenerate square matrix by equipotent transformations.

In the following theorem $\phi = [a]_m^n$ is a given rational integral x-matrix of rank r lying in any domain of rationality Ω; and $D_1, D_2, \ldots D_r$ and $E_1, E_2, \ldots E_r$ are the maximum and potent factors of ϕ of orders $1, 2, \ldots r$ chosen in any manner so as to lie in Ω. In proving the theorem we shall make use of some of the results obtained in the chapter dealing with primitive degrees and minimum degrees of connection.

Theorem. *We can always convert the x-matrix $\phi = [a]_m^n$ by an equipotent transformation in Ω into a similar matrix ψ of the form*

$$\psi = \begin{bmatrix} b, & 0 \\ 0, & 0 \end{bmatrix}_{r,\, m-r}^{r,\, n-r}; \quad\ldots\ldots\ldots\ldots\ldots\ldots\ldots\ldots\ldots\text{(1)}$$

and $[b]_r^r$ is then an undegenerate square matrix in Ω equipotent with ϕ from which ϕ can be derived by an equipotent transformation in Ω.

Let $$[\lambda]^m_{m-r}\,[a]^n_m=0,\quad [a]^n_m\,[\mu]^{n-r}_n=0 \quad\text{.............................(2)}$$

be complete sets of unconnected connections between the horizontal and vertical rows of ϕ so chosen that $[\lambda]^m_{m-r}$ and $[\mu]^{n-r}_n$ are undegenerate and impotent rational integral x-matrices in Ω of ranks $m-r$ and $n-r$. This is always possible, for it is the case when the equations (2) are complete primitive sets of connections in Ω of minimum degrees between the horizontal and vertical rows of ϕ. Using Ex. ii of § 216, let

$$[p]^m_m=\begin{bmatrix} p \\ \lambda \end{bmatrix}^m_{r,\,m-r},\quad [q]^n_n=[q,\,\mu]^{r,\,n-r}_n$$

be undegenerate and impotent square matrices in Ω, rational and integral in x, formed by adding initial horizontal rows to $[\lambda]^m_{m-r}$ and initial vertical rows to $[\mu]^{n-r}_n$. Then if

$$[b]^r_r=[p]^m_r\,[a]^n_m\,[q]^r_n, \quad\text{.......................................(3)}$$

we have $$[p]^m_m\,[a]^n_m\,[q]^n_n=\begin{bmatrix} b, & 0 \\ 0, & 0 \end{bmatrix}^{r,\,n-r}_{r,\,m-r}; \quad\text{..............................(4)}$$

and (4) is an equipotent transformation in Ω converting ϕ into a matrix of the form ψ. If further $\overline{\underline{P}}{}^m_m$, $\overline{\underline{Q}}{}^n_n$ are the inverses of $[p]^m_m$, $[q]^n_n$, we deduce from (4) the equations

$$\overline{\underline{P}}{}^m_m\begin{bmatrix} b, & 0 \\ 0, & 0 \end{bmatrix}^{r,\,n-r}_{r,\,m-r}\overline{\underline{Q}}{}^n_n=[a]^n_m,\quad \overline{\underline{P}}{}^r_m\,[b]^r_r\,\overline{\underline{Q}}{}^n_r=[a]^n_m; \quad\text{...............(5)}$$

and the second of the equations (5) shows that the matrix $[b]^r_r$ found in (4) is an undegenerate square matrix in Ω equipotent with ϕ from which ϕ can be derived by an equipotent transformation in Ω. Moreover $[b]^r_r$ must have these properties in every equipotent transformation in Ω of the form (4).

From the above we see that there always exist pairs of mutually inverse equipotent transformations in Ω of the forms

$$[h]^m_m\begin{bmatrix} b, & 0 \\ 0, & 0 \end{bmatrix}^{r,\,n-r}_{r,\,m-r}[k]^n_n=[a]^n_m, \quad\text{..............................(A)}$$

$$\overline{\underline{H}}{}^m_m\,[a]^n_m\,\overline{\underline{K}}{}^n_n=\begin{bmatrix} b, & 0 \\ 0, & 0 \end{bmatrix}^{r,\,n-r}_{r,\,m-r}, \quad\text{..................................(B)}$$

in which $[h]^m_m$, $\overline{\underline{H}}{}^m_m$ and $[k]^n_n$, $\overline{\underline{K}}{}^n_n$ are two pairs of mutually inverse undegenerate and impotent x-matrices in Ω, so that

$$\overline{\underline{H}}{}^m_m\,[h]^m_m=[h]^m_m\,\overline{\underline{H}}{}^m_m=[1]^m_m,\quad [k]^n_n\,\overline{\underline{K}}{}^n_n=\overline{\underline{K}}{}^n_n\,[k]^n_n=[1]^n_n; \quad\text{...........(C)}$$

and from these there always follow the two transformations

$$[h]^r_m\,[b]^r_r\,[k]^n_r=[a]^n_m, \quad\text{....................................(A')}$$

$$\overline{\underline{H}}{}^m_r\,[a]^n_m\,\overline{\underline{K}}{}^r_n=[b]^r_r, \quad\text{....................................(B')}$$

in which $[h]^r_m$, $\overline{\underline{H}}{}^m_r$ and $[k]^n_r$, $\overline{\underline{K}}{}^r_n$ are two pairs of mutually inverse undegenerate and impotent x-matrices in Ω of rank r, so that

$$\overline{\underline{H}}{}^m_r\,[h]^r_m=[1]^r_r,\quad [k]^n_r\,\overline{\underline{K}}{}^r_n=[1]^r_r. \quad\text{..........................(C')}$$

The matrix $[b]^r_r$ in all such sets of transformations is an undegenerate square matrix in Ω equipotent with ϕ from which ϕ can be derived by the equipotent transformation (A′). The transformation (B′) is not equipotent (except when $m=n=r$), but it can always be derived from (A′) by applying the prefactor $\overline{\underline{H}}{}^m_r$ and the postfactor $\overline{\underline{K}}{}^r_n$ to both sides, and it is the transformation inverse to (A′). We can regard (A′) as the most fundamental of the four transformations (A), (B), (A′), (B′) because when (A′) is given, we can construct undegenerate and impotent square matrices $[h]^m_m$ and $[k]^n_n$ by adding final vertical and horizontal rows to $[h]^r_m$ and $[k]^n_r$, and we then have a transformation of the form (A) from which (B) and (B′) can be deduced.

We will now consider some properties of the transforming factors in such transformations as (A), (B), (A′), (B′).

First let $$[h]^m_m = [h, u]^{r,\, m-r}_m, \quad [k]^n_n = \begin{bmatrix} k \\ v \end{bmatrix}^n_{r,\, n-r} \quad \text{..........................(a)}$$

Then if μ and ν are positive integers such that $\mu \not> m-r$ and $\nu \not> n-r$, we see from (A′) that in (A) we have

$$[h, u]^{r,\mu}_m \begin{bmatrix} b, & 0 \\ 0, & 1 \end{bmatrix}^{r,\mu}_{r,\mu} \begin{bmatrix} k, & 0 \\ 0, & 1 \end{bmatrix}^{n,\mu}_{r,\mu} = [a, u]^{n,\mu}_m, \quad \text{....................(6)}$$

$$\begin{bmatrix} h, & 0 \\ 0, & 1 \end{bmatrix}^{r,\nu}_{m,\nu} \begin{bmatrix} b, & 0 \\ 0, & 1 \end{bmatrix}^{r,\nu}_{r,\nu} \begin{bmatrix} k \\ v \end{bmatrix}^n_{r,\nu} = \begin{bmatrix} a \\ v \end{bmatrix}^n_{m,\nu} \quad \text{........................(6′)}$$

Because these are equipotent transformations in Ω, and because the potent divisors of the middle factor matrices on the left are respectively those of $[b]^r_r$, $[1]^\mu_\mu$ and those of $[b]^r_r$, $[1]^\nu_\nu$, we see that the transforming matrices in (A) have the following properties:

If $\mu \not> m-r$, *then* $[a, u]^{n,\mu}_m$ *is a matrix in* Ω *of rank* $r+\mu$ *which has the same potent divisors as* $[a]^n_m$, *i.e. its potent factors of orders* $1, 2, \ldots \mu, \mu+1, \mu+2, \ldots \mu+r$ *are respectively* $1, 1, \ldots 1, E_1, E_2, \ldots E_r$. *In particular* $[a, u]^{n,\, m-r}_m$ *is an undegenerate matrix in* Ω *of rank* m *which has the same potent divisors as* $[a]^n_m$ ....(a)

If $\nu \not> n-r$, *then* $\begin{bmatrix} a \\ v \end{bmatrix}^n_{m,\nu}$ *is a matrix in* Ω *of rank* $r+\nu$ *which has the same potent divisors as* $[a]^n_m$, *i.e. its potent factors of orders* $1, 2, \ldots \nu, \nu+1, \nu+2, \ldots \nu+r$ *are respectively* $1, 1, \ldots 1, E_1, E_2, \ldots E_r$. *In particular* $\begin{bmatrix} a \\ v \end{bmatrix}^n_{m,\, n-r}$ *is an undegenerate matrix in* Ω *of rank* n *which has the same potent divisors as* $[a]^n_m$. ....(a′)

Suppose that (A′) alone is a given equipotent transformation in Ω, and that $[u]^\mu_m$, $[v]^n_\nu$ are rational integral x-matrices in Ω whose elements are to be determined, where $\mu \not> m-r$ and $\nu \not> n-r$. Then whenever the matrices

$$[h, u]^{r,\mu}_m, \quad \begin{bmatrix} k \\ v \end{bmatrix}^n_{r,\nu} \quad \text{....................................(7)}$$

are undegenerate and impotent, the matrices

$$[a, u]^{n,\mu}_m, \quad \begin{bmatrix} a \\ v \end{bmatrix}^n_{m,\nu} \quad \text{....................................(8)}$$

have the properties described in (a) and (a′); conversely whenever the matrices (8) have the properties described in (a) and (a′), the matrices (7) are undegenerate and impotent, as we see by equating correspondingly formed complete matrices of the minor determinants of orders $r+\mu$ and $r+\nu$ on both sides of (6) and (6′).

Next let
$$\overline{\underline{H}}{}^{m}_{m} = \overline{\underline{\begin{matrix} H \\ U \end{matrix}}}{}^{m}_{r,\, m-r}, \quad \overline{\underline{K}}{}^{n}_{n} = \overline{\underline{K,\ V}}{}^{r,\, n-r}_{n} \quad \text{..........................}(\beta)$$

Then (B) is equivalent to (B′) together with the equations

$$\overline{\underline{U}}{}^{m}_{m-r}\,[a]^{n}_{m} = 0, \quad [a]^{n}_{m}\,\overline{\underline{V}}{}^{n-r}_{n} = 0. \quad \text{..........................}(9)$$

Thus the transforming matrices in (B) *are always such that the equations* (9) *represent complete sets of unconnected connections between the horizontal and vertical rows of* $[a]^{n}_{m}$.
......(b)

The property (b) was used in obtaining the transformation (4), i.e. in proving the existence of transformations of the forms (A) and (B).

Ex. i. *When* ϕ *is a symmetric (or skew-symmetric) matrix* $[a]^{m}_{m}$, *so that* $n=m$, *there exist symmetric equipotent transformations in* Ω *of the forms* (A) *and* (B); *and in these* $[b]^{r}_{r}$ *is a symmetric (or skew-symmetric) matrix.*

For in these cases the matrices $[\lambda]^{m}_{m-r}$, $[\mu]^{m-r}_{m}$ in (2) can be chosen to be mutually conjugate.

Ex. ii. *There exist equipotent transformations in* Ω *of the forms* (A) *and* (B) *when* $[b]^{r}_{r}$ *is any given undegenerate square matrix in* Ω, *rational and integral in* x, *which is equipotent with* ϕ, *and in particular when* $[b]^{r}_{r} = {}^{1}[E]_{r}$.

This follows from Theorem IV of § 216 or Theorem II of § 217, which show that every such matrix $[b]^{r}_{r}$ can be converted into or derived from ${}^{1}[E]_{r}$ by equipotent transformations in Ω. For there certainly exists an equipotent transformation in Ω of the form (A′) in which $[b]^{r}_{r}$ is some undegenerate square matrix in Ω equipotent with ϕ, therefore there exists such a transformation when $[b]^{r}_{r}$ is replaced by ${}^{1}[E]_{r}$, therefore there exists such a transformation when ${}^{1}[E]_{r}$ is replaced by any given undegenerate square matrix $[b]^{r}_{r}$ which lies in Ω and is equipotent with ϕ; and from (A′) we can deduce (A) and (B).

If however ϕ is symmetric, so that $n=m$, we cannot conclude that there exist symmetric equipotent transformations in Ω of the forms (A) and (B) when $[b]^{r}_{r}$ is *any* given symmetric matrix in Ω of order r which is equipotent with ϕ.

Ex. iii. *If* $[a]^{n}_{m}$ *is a rational integral x-matrix in* Ω *of rank* r, *and if* $r < m$, *then by adding final vertical rows to* $[a]^{n}_{m}$ *we can construct rational integral x-matrices*

$$[a, u]^{n,1}_{m}, \quad [a, u]^{n,2}_{m}, \quad \dots \quad [a, u]^{n,\, m-r}_{m}$$

in Ω *of ranks* $r+1, r+2, \dots m$ *which have the same potent divisors as* $[a]^{n}_{m}$. *The matrices of the added rows are necessarily undegenerate and impotent.*

We deduce this theorem from the remarks which follow the results (a) and (a′). The last part of the theorem can be proved independently by observing that if $\mu \ngtr m-r$, and if $\psi=[a, u]_m^{n,\mu}$ has rank $r+\mu$, the only non-vanishing minor determinants of ψ of order $r+\mu$ are those which involve all the vertical rows of $[u]_m^{\mu}$. If then the potent factor of ψ of order $r+\mu$ is D_r, the matrix $[u]_m^{\mu}$ must be impotent as well as undegenerate.

Similarly if $r < n$, then by adding final horizontal rows to $[a]_m^n$ we can construct rational integral x-matrices

$$\begin{bmatrix} a \\ v \end{bmatrix}_{m,1}^{n}, \begin{bmatrix} a \\ v \end{bmatrix}_{m,2}^{n}, \dots \begin{bmatrix} a \\ v \end{bmatrix}_{m,n-r}^{n}$$

in Ω of ranks $r+1, r+2, \dots n$ which have the same potent divisors as $[a]_m^n$. The matrices of the added rows are necessarily undegenerate and impotent.

§ 219. Reduction of any given rational integral x-matrix to a standard form by equipotent transformations.

The usual method of reducing a given rational x-matrix ϕ to a standard form by equipotent transformations is that indicated in the proof of Theorem I which follows. The reduction has been effected in other ways in the preceding articles—for an undegenerate square matrix in Theorem IV of § 216 and Theorem II of § 217, and for any x-matrix in Ex. ii of § 218—but it has hitherto always been supposed that the potent factors of ϕ are known. The present method is independent of the results of §§ 215—218, and can be used when the potent factors of ϕ have not been previously determined; in fact it is in many cases the easiest way of determining them.

Theorem I. *If $\phi=[a]_m^n$ is a matrix of rank r whose elements are rational integral functions in Ω of a single variable x (r being not less than 1, and Ω being any domain of rationality), then ϕ can be converted by derangements and elementary unitary equipotent transformations in Ω into a similar quasi-scalar matrix Φ of the form*

$$\Phi=[E]_m^n=\begin{bmatrix} E_1 & 0 & \dots & 0 & 0 & \dots & 0 \\ 0 & E_2 & \dots & 0 & 0 & \dots & 0 \\ \dots & \dots & \dots & \dots & \dots & \dots & \dots \\ 0 & 0 & \dots & E_r & 0 & \dots & 0 \\ 0 & 0 & \dots & 0 & 0 & \dots & 0 \\ \dots & \dots & \dots & \dots & \dots & \dots & \dots \\ 0 & 0 & \dots & 0 & 0 & \dots & 0 \end{bmatrix}, \qquad \dots\dots\dots\dots\text{(A)}$$

where $E_1, E_2, \dots E_r$ are rational integral functions of x lying in the domain Ω which do not vanish identically and are such that

$$E_{i+1} \text{ is divisible by } E_i, \quad (i=1, 2, \dots r-1).$$

The functions $E_1, E_2, \dots E_r$ are then the potent factors of ϕ of orders 1, 2, ... r.

We shall use the notations

$$\Phi = \begin{bmatrix} E, & 0 \\ 0, & 0 \end{bmatrix}^{r, n-r}_{r, m-r}, \quad [E]^r_r = {}^1[E]_r = \begin{bmatrix} E_1 & 0 & \dots & 0 \\ 0 & E_2 & \dots & 0 \\ \dots & \dots & \dots & \dots \\ 0 & 0 & \dots & E_r \end{bmatrix},$$

and call Φ the *reduced standard form* of $[a]^n_m$ for equipotent transformation. Lemmas 1—4 will furnish the successive steps by which ϕ is reduced to this form.

Lemma 1. *If the x-matrix $\phi=[a]^n_m$ has a non-zero element a_{ij} which is not a factor of all other elements, then we can convert ϕ by elementary unitary equipotent transformations in Ω into a similar matrix $\psi=[b]^n_m$ lying in Ω and containing at least one non-zero element which has a lower degree in x than a_{ij}.*

The element a_{ij} clearly cannot be a constant; and ϕ must contain at least one *non-zero* element which is not divisible by a_{ij}. The lemma is obviously true when ϕ contains any non-zero element which has a lower degree in x than a_{ij}. We will consider three cases which include all possible cases.

Case I. *Suppose that there is a non-zero element a_{iq} in the same horizontal row of ϕ as a_{ij} which is not divisible by a_{ij}.*

In this case there exists a uniquely determinate equation of the form

$$a_{iq} = Qa_{ij} + R,$$

where Q and R are rational integral functions of x lying in Ω, and where R does not vanish identically and has a lower degree in x than a_{ij}. If the degree of a_{iq} is less than that of a_{ij}, we have $Q=0$, $R=a_{iq}$. If the degree of a_{iq} is not less than that of a_{ij}, then Q and R are the quotient and remainder in the division of a_{iq} by a_{ij}. By adding to the qth vertical row of ϕ the jth vertical row multiplied by $-Q$, we form an equipotent matrix $[b]^n_m$ lying in Ω in which $b_{iq}=R$; and b_{iq} has then a lower degree in x than a_{ij}.

Case II. *Suppose that there is a non-zero element a_{pj} in the same vertical row of ϕ as a_{ij} which is not divisible by a_{ij}.*

In this case there exists a uniquely determinate equation of the form

$$a_{pj} = Qa_{ij} + R,$$

where Q and R are rational integral functions of x lying in Ω, and where R does not vanish identically and has a lower degree in x than a_{ij}. By adding to the pth horizontal row of ϕ the ith horizontal row multiplied by $-Q$, we form an equipotent matrix $[b]^n_m$ lying in Ω in which $b_{pj}=R$; and b_{pj} has then a lower degree in x than a_{ij}.

Case III. *Suppose that all elements of ϕ in the same horizontal row as a_{ij} and all elements of ϕ in the same vertical row as a_{ij} are divisible by a_{ij}, but that there is some other element a_{pq} of ϕ which does not vanish identically and is not divisible by a_{ij}.*

In this case we can write

$$a_{pj} = Qa_{ij},$$

where Q is a rational integral function of x lying in Ω, which may vanish identically. If we first add $-Q$ times the ith horizontal row of ϕ to the pth horizontal row, and then in the matrix thus formed add the pth horizontal row to the ith horizontal row, we form an equipotent matrix $[c]_m^n$ lying in Ω in which

$$\begin{bmatrix} c_{i1} & c_{i2} & \dots & c_{ij} & \dots & c_{iq} & \dots & c_{in} \\ c_{p1} & c_{p2} & \dots & c_{pj} & \dots & c_{pq} & \dots & c_{pn} \end{bmatrix} = \begin{bmatrix} 1-Q, & 1 \\ -Q, & 1 \end{bmatrix} \begin{bmatrix} a_{i1} & a_{i2} & \dots & a_{ij} & \dots & a_{iq} & \dots & a_{in} \\ a_{p1} & a_{p2} & \dots & a_{pj} & \dots & a_{pq} & \dots & a_{pn} \end{bmatrix},$$

whilst every other element of $[c]_m^n$ is the same as the corresponding element of $[a]_m^n$; and we see that

$$c_{ij} = a_{ij}, \quad c_{iq} = (1-Q)\, a_{iq} + a_{pq}.$$

Since a_{iq} is divisible by a_{ij}, and a_{pq} is not divisible by a_{ij}, therefore c_{iq} is not divisible by a_{ij}, i.e. not divisible by c_{ij}. Thus c_{ij} is a non-zero element of $[c]_m^n$ which is not a factor of c_{iq}. Therefore by Case I we can convert $[c]_m^n$ by elementary unitary equipotent transformations in Ω into a matrix $[b]_m^n$ lying in Ω in which the element b_{iq} does not vanish identically and has a lower degree in x than c_{ij}, i.e. than a_{ij}; and $[a]_m^n$ can be converted into $[b]_m^n$ by elementary unitary equipotent transformations in Ω.

Lemma 2. *If the elements of ϕ have no factor in common other than a non-zero constant, then we can convert ϕ by elementary unitary equipotent transformations in Ω into a similar matrix $\psi = [b]_m^n$ lying in Ω one of whose elements is a non-zero constant.*

If no element of ϕ is a non-zero constant, let s, where $s > 0$, be the lowest degree in x of the various non-zero elements of ϕ; and let a_{ij} be one of the non-zero elements of ϕ having this lowest degree s. Then a_{ij} is not a factor of all other elements, and therefore by Lemma 1 we can convert ϕ by elementary unitary equipotent transformations in Ω into a similar matrix ϕ' one of whose non-zero elements has a lower degree than s in x.

The matrix ϕ', being equipotent with ϕ, has the same character as ϕ. If no one of its elements is a non-zero constant, let s', where $s' < s$ and $s' > 0$, be the lowest degree in x of its various non-zero elements. Then by Lemma 1 we can convert ϕ' by elementary unitary equipotent transformations in Ω into a similar matrix ϕ'' one of whose non-zero elements has a lower degree than s' in x.

If no element of ϕ'' is a non-zero constant, we can treat ϕ'' in the same way.

A succession of such elementary unitary equipotent transformations in Ω will ultimately convert ϕ into a similar matrix ψ one of whose elements is a non-zero constant.

Lemma 3. *If $r \nless 1$, and if the elements of the x-matrix $\phi = [a]_m^n$ have no factor in common other than a non-zero constant, then we can convert ϕ by elementary unitary equipotent transformations in Ω and derangements into a similar matrix ψ of the form*

$$\psi = \begin{bmatrix} k, & 0 \\ 0, & u \end{bmatrix}_{1,\, m-1}^{1,\, n-1}, \quad \dots\dots\dots\dots\text{(a)}$$

where k is a non-zero constant in Ω, and where $[u]_{m-1}^{n-1}$ is a rational integral x-matrix of rank $r-1$ lying in Ω.

By Lemma 2 we can convert ϕ by elementary unitary equipotent transformations in Ω into a similar matrix one of whose elements is a non-zero constant k, and we can bring that element to the leading position by derangements of horizontal and vertical rows. If we denote the matrix last formed by $[k]_m^n$, so that $k_{11}=k$; add to the 2nd, 3rd, ... mth horizontal rows the first horizontal row multiplied by

$$-\frac{k_{21}}{k},\ -\frac{k_{31}}{k},\ \ldots\ -\frac{k_{m1}}{k}$$

respectively; and then add to the 2nd, 3rd, ... nth vertical rows the first vertical row multiplied by

$$-\frac{k_{12}}{k},\ -\frac{k_{13}}{k},\ \ldots\ -\frac{k_{1n}}{k}$$

respectively; we shall convert $[k]_m^n$ by elementary unitary equipotent transformations into the form shown in (a). Since the matrix ψ thus obtained is equipotent with ϕ, it has rank r; and it follows from the theorem of § 200 that the part $[u]_{m-1}^{n-1}$ has rank $r-1$.

COROLLARY. *Under the same circumstances we can convert ϕ by elementary equipotent transformations in Ω into a similar matrix ψ' of the form*

$$\psi'=\begin{bmatrix}1, & 0\\ 0, & v\end{bmatrix}_{1,\,m-1}^{1,\,n-1}, \qquad\text{.....................................(a')}$$

where $[v]_{m-1}^{n-1}$ is a rational integral x-matrix of rank $r-1$ lying in Ω.

For if we divide every element of ψ by the non-zero constant k, we convert it into ψ', where $[v]_{m-1}^{n-1}=\frac{1}{k}[u]_{m-1}^{n-1}$.

Lemma 4. *If $r \nless 1$, then we can always convert ϕ by elementary unitary equipotent transformations in Ω and derangements into a similar matrix ψ of the form*

$$\psi=\begin{bmatrix}E_1, & 0\\ 0, & E_1 b\end{bmatrix}_{1,\,m-1}^{1,\,n-1}, \qquad\text{.....................................(b)}$$

where E_1 is a highest common factor in Ω of all the elements of ϕ, and $[b]_{m-1}^{n-1}$ is a rational integral x-matrix of rank $r-1$ lying in Ω.

If η is any given H.C.F. of all the elements of ϕ which lies in Ω, we can write $\phi=\eta\,[a]_m^n$, and reduce $[a]_m^n$ to the form (a) of Lemma 3. The same transformations reduce $\eta\,[a]_m^n$ to the form (b), where $E_1=k\eta$ and $[b]_{m-1}^{n-1}=\frac{E_1}{k\eta}[b]_{m-1}^{n-1}=\frac{1}{k}[u]_{m-1}^{n-1}$. Then E_1 is a highest common factor in Ω of all the elements of ϕ, and $[b]_{m-1}^{n-1}$ lies in Ω and has the same rank as $[u]_{m-1}^{n-1}$, i.e. has rank $r-1$.

If $f(x)$ is any scalar function, the transformations which convert ϕ into ψ also convert $f(x)\,.\,\phi$ into $f(x)\,.\,\psi$.

COROLLARY. *Under the same circumstances we can always convert ϕ by elementary equipotent transformations in Ω into a matrix of the form* (b) *in which E_1 is any given* H.C.F. *in Ω of all the elements of ϕ.*

For we merely have to multiply all elements of ψ by a certain non-zero scalar constant lying in Ω.

We can now prove Theorem I; and in the proof it will be understood that every equipotent transformation employed is either an elementary unitary equipotent transformation in Ω or a derangement.

If $r \nless 1$, then by Lemma 4 we can convert ϕ by equipotent transformations into the form

$$\phi_1 = \begin{bmatrix} E_1, & 0 \\ 0, & E_1 b \end{bmatrix}_{1, m-1}^{1, n-1},$$

where E_1 is a highest common factor in Ω of all the elements of ϕ, and where $[b]_{m-1}^{n-1}$ lies in Ω and has rank $r-1$.

If $r \nless 2$, then by Lemma 4 and by Ex. iii of § 214 we can convert $[b]_{m-1}^{n-1}$ and ϕ_1 by equipotent transformations into the respective forms

$$\psi_2 = \begin{bmatrix} \eta_2, & 0 \\ 0, & \eta_2 c \end{bmatrix}_{1, m-2}^{1, n-2} \quad \text{and} \quad \phi_2 = \begin{bmatrix} E_1, & 0, & 0 \\ 0, & E_2, & 0 \\ 0, & 0, & E_2 c \end{bmatrix}_{1, 1, m-2}^{1, 1, n-2},$$

where η_2 is a highest common factor in Ω of all the elements of $[b]_{m-1}^{n-1}$, where $[c]_{m-2}^{n-2}$ lies in Ω and has rank $r-2$, and where $E_2 = E_1\eta_2$. Therefore by Ex. iv of § 214 we can convert ϕ into ϕ_2 by equipotent transformations.

If $r \nless 3$, then by Lemma 4 and by Ex. iii of § 214 we can convert $[c]_{m-2}^{n-2}$ and ϕ_2 by equipotent transformations into the respective forms

$$\psi_3 = \begin{bmatrix} \eta_3, & 0 \\ 0, & \eta_3 d \end{bmatrix}_{1, m-3}^{1, n-3} \quad \text{and} \quad \phi_3 = \begin{bmatrix} E_1, & 0, & 0, & 0 \\ 0, & E_2, & 0, & 0 \\ 0, & 0, & E_3, & 0 \\ 0, & 0, & 0, & E_3 d \end{bmatrix}_{1, 1, 1, m-3}^{1, 1, 1, n-3},$$

where η_3 is a highest common factor in Ω of all the elements of $[c]_{m-2}^{n-2}$, where $[d]_{m-3}^{n-3}$ lies in Ω and has rank $r-3$, and where $E_3 = E_2\eta_3$. Therefore by Ex. iv of § 214 we can convert ϕ into ϕ_3 by equipotent transformations.

Proceeding in this way we see finally at the rth stage that we convert ϕ by a succession of equipotent transformations in Ω (each of which is either an elementary unitary transformation or a derangement) into the matrix $\phi_r = \Phi$ of Theorem I; and because $\eta_2, \eta_3, \ldots \eta_r$ are rational integral functions of x lying in Ω which do not vanish identically, and

$$E_2 = E_1\eta_2, \quad E_3 = E_2\eta_3, \quad \ldots \quad E_r = E_{r-1}\eta_r,$$

therefore each of the functions $E_1, E_2, \ldots E_r$ after the first is divisible by the preceding. Moreover since ϕ is equipotent with Φ, and $E_1, E_2, \ldots E_r$ are clearly the potent factors of Φ of orders $1, 2, \ldots r$, they are also the potent factors of ϕ of orders $1, 2, \ldots r$.

By applying this reduction to any given rational integral x-matrix, we can determine both its rank and its maximum and potent factors.

Corollary to Theorem I. *We can always convert ϕ by elementary equipotent transformations in Ω into a matrix of the same form as Φ in which $E_1, E_2, \ldots E_r$ are the potent factors of ϕ of orders 1, 2, ... r chosen in any given manner so as to lie in Ω.*

Since all potent factors of ϕ of any given order i differ from one another only by non-zero constant scalar factors, it is only necessary to multiply the 1st, 2nd, ... rth horizontal (or vertical) rows of the matrix Φ obtained in Theorem I by certain non-zero scalar constants lying in Ω, i.e. to apply additional elementary quasi-scalar transformations in Ω.

Ex. i. *Application of the method of reduction just described to the matrix*

$$A=[a]_3^3=\begin{bmatrix} 2x^4+x^3-x^2+x+1, & 2x^3+x^2+2x-1, & 2x^3+x^2-2x-1 \\ 2x^3-x+1 & , \quad 2x^2-2 & , \quad 2x^2-2 \\ 2x^3-x+1 & , \quad x^2-1 & , \quad 2x^2-2 \end{bmatrix}.$$

Observing that the H.C.F. of all the elements of A is $x-1$, we have

$$[a]_3^3=(x+1)\,[a]_3^3, \text{ where } [a]_3^3=\begin{bmatrix} 2x^3-x^2+1, & 2x^2-x-1, & 2x^2-x-1 \\ 2x^2-2x+1, & 2x-2 & , \quad 2x-2 \\ 2x^2-2x+1, & x-1 & , \quad 2x-2 \end{bmatrix}.$$

Then using the equation $2x^2-2x+1=2x\,(x-1)+1$, or $a_{31}=2x\,a_{32}+1$, to reduce $[a]_3^3$ to a matrix containing a non-zero element of lower degree in x than the element $a_{32}=x-1$, we have

$$[a]_3^3\begin{bmatrix} 1 & , 0, & 0 \\ -2x, & 1, & 0 \\ 0 & , 0, & 1 \end{bmatrix}=\begin{bmatrix} -2x^3+x^2+2x+1, & 2x^2-x-1, & 2x^2-x-1 \\ -2x^2+2x+1 & , \quad 2x-2 & , \quad 2x-2 \\ 1 & , \quad x-1 & , \quad 2x-2 \end{bmatrix}=[a']_3^3,$$

$$\begin{bmatrix} 0, & 0, & 1 \\ 1, & 0, & 0 \\ 0, & 1, & 0 \end{bmatrix}[a']_3^3=\begin{bmatrix} 1 & , \quad x-1 & , \quad 2x-2 \\ -2x^3+x^2+2x+1, & 2x^2-x-1, & 2x^2-x-1 \\ -2x^2+2x+1 & , \quad 2x-2 & , \quad 2x-2 \end{bmatrix}=[a'']_3^3,$$

$$\begin{bmatrix} 1 & , 0, & 0 \\ 2x^3-x^2-2x-1, & 1, & 0 \\ 2x^2-2x-1 & , 0, & 1 \end{bmatrix}[a'']_3^3\begin{bmatrix} 1, & -x+1, & -2x+2 \\ 0, & 1 & , \quad 0 \\ 0, & 0 & , \quad 1 \end{bmatrix}$$

$$=\begin{bmatrix} 1, & 0 & , \quad 0 \\ 0, & 2x^4-3x^3+x^2 & , \quad 4x^4-6x^3+x+1 \\ 0, & 2x^3-4x^2+3x-1, & 4x^3-8x^2+4x \end{bmatrix},$$

and therefore

$$\begin{bmatrix} 0, & 0, & 1 \\ 1, & 0, & 2x^3-x^2-2x-1 \\ 0, & 1, & 2x^2-2x-1 \end{bmatrix} [a]_3^3 \begin{bmatrix} 1, & -x+1, & -2x+2 \\ -2x, & 2x^2-2x+1, & 4x^2-4x \\ 0, & 0, & 1 \end{bmatrix} = \begin{bmatrix} x+1, & 0 \\ 0, & (x+1)\,b \end{bmatrix}_{1,2}^{1,2}, \qquad \text{......(1)}$$

where $$[b]_2^2 = \begin{bmatrix} 2x^4-3x^3+x^2, & 4x^4-6x^3+x+1 \\ 2x^3-4x^2+3x-1, & 4x^3-8x^2+4x \end{bmatrix} = B;$$

and this completes the first stage of the reduction.

Next observing that the H.C.F. of all the elements of B is $x-1$, we have

$$[b]_2^2 = (x-1)\,[\beta]_2^2, \quad \text{where } [\beta]_2^2 = \begin{bmatrix} 2x^3-x^2, & 4x^3-2x^2-2x-1 \\ 2x^2-2x+1, & 4x^2-4x \end{bmatrix}.$$

Then using the equation $4x^2-4x=2(2x^2-2x+1)-2$, or $2\beta_{21}-\beta_{22}=2$, to reduce $[\beta]_2^2$ to a matrix containing a non-zero element of lower degree in x than the element $\beta_{21}=2x^2-2x+1$, we have

$$[\beta]_2^2 \begin{bmatrix} 1, & 2 \\ 0, & -1 \end{bmatrix} = \begin{bmatrix} 2x^3-x^2, & 2x+1 \\ 2x^2-2x+1, & 2 \end{bmatrix} = [\beta']_2^2,$$

$$\begin{bmatrix} 0, & 1 \\ 1, & 0 \end{bmatrix} [\beta']_2^2 \begin{bmatrix} 0, & 1 \\ 1, & 0 \end{bmatrix} = \begin{bmatrix} 2, & 2x^2-x+1 \\ 2x+1, & 2x^3-x^2 \end{bmatrix} = [\beta'']_2^2,$$

$$\begin{bmatrix} 1, & 0 \\ -2x-1, & 2 \end{bmatrix} [\beta'']_2^2 \begin{bmatrix} 1, & -2x^2+2x-1 \\ 0, & 2 \end{bmatrix} = \begin{bmatrix} 2, & 0 \\ 0, & 2 \end{bmatrix},$$

and therefore

$$\begin{bmatrix} 0, & 1 \\ 2, & -2x-1 \end{bmatrix} [b]_2^2 \begin{bmatrix} 2, & 4x^2-4x \\ -1, & -2x^2+2x-1 \end{bmatrix} = \begin{bmatrix} 2(x-1), & 0 \\ 0, & 2(x-1) \end{bmatrix},$$

$$\begin{bmatrix} 1, & 0, & 0 \\ 0, & 0, & 1 \\ 0, & 2, & -2x-1 \end{bmatrix} \begin{bmatrix} x+1, & 0 \\ 0, & (x+1)\,b \end{bmatrix}_{1,2}^{1,2} \begin{bmatrix} 1, & 0, & 0 \\ 0, & 2, & 4x^2-4x \\ 0, & -1, & -2x^2+2x-1 \end{bmatrix}$$

$$= \begin{bmatrix} x+1, & 0, & 0 \\ 0, & 2(x^2-1), & 0 \\ 0, & 0, & 2(x^2-1) \end{bmatrix};$$

and from the last equation and (1) it follows that

$$\begin{bmatrix} 0, & 0, & 1 \\ 0, & 1, & 2x^2-2x-1 \\ 2, & -2x-1, & -1 \end{bmatrix} [a]_3^3 \begin{bmatrix} 1, & 0, & 2x-2 \\ -2x, & 2, & 0 \\ 0, & -1, & -2x^2+2x-1 \end{bmatrix}$$

$$= \begin{bmatrix} x+1, & 0, & 0 \\ 0, & 2(x^2-1), & 0 \\ 0, & 0, & 2(x^2-1) \end{bmatrix}. \qquad \text{......(2)}$$

This completes the second stage in the process of reduction; and in the present case no further stage is required.

From (2) we see that A has rank 3, and that the potent factors E_1, E_2, E_3 and the maximum factors D_1, D_2, D_3 of A are

$$E_1=x+1, \quad E_2=x^2-1, \quad E_3=x^2-1,$$
$$D_1=x+1, \quad D_2=(x+1)(x^2-1), \quad D_3=(x+1)(x^2-1)^2.$$

Theorem II. *If $\phi=[a]_m^n$ is the matrix of rank r described in Theorem I, and if $E_1, E_2, \ldots E_r$ are the potent factors of ϕ of orders $1, 2, \ldots r$ chosen in any manner so as to lie in Ω, there always exist mutually inverse equipotent transformations in Ω of the forms*

$$[h]_m^m \begin{bmatrix} E, & 0 \\ 0, & 0 \end{bmatrix}_{r,m-r}^{r,n-r} [k]_n^n=[a]_m^n, \quad \overline{\underline{H}}_m^m [a]_m^n \overline{\underline{K}}_n^n = \begin{bmatrix} E, & 0 \\ 0, & 0 \end{bmatrix}_{r,m-r}^{r,n-r}, \ldots\text{(B)}$$

and therefore also mutually inverse transformations in Ω of the forms

$$[h]_m^r [E]_r^r [k]_r^n=[a]_m^n, \quad \overline{\underline{H}}_r^m [a]_m^n \overline{\underline{K}}_n^r=[E]_r^r, \ \ldots\ldots\ldots\text{(B}')$$

the first of the transformations (B′) *being equipotent.*

This theorem is an immediate consequence of the Corollary to Theorem I. The transforming matrix factors in (B) and (B′) have the same properties as those in the corresponding transformations (A), (B), (A′), (B′) of § 218; in particular the other three transformations of Theorem II can be derived from the first of the transformations (B′).

Ex. ii. *If $[a]_m^n$ is a rational integral x-matrix in Ω, it can be expressed in the form*

$$[a]_m^n=[h]_m^r [k]_r^n, \ \ldots\ldots\ldots\ldots\ldots\ldots\ldots\ldots\text{(C)}$$

where $[h]_m^r$ and $[k]_r^n$ are undegenerate and impotent rational integral x-matrices in Ω of rank r, when and only when it is an impotent matrix of rank r.

For when $[a]_m^n$ is impotent and has rank r, we can put $E_1=E_2=\ldots=E_r=1$, and the first of the equations (B′) then becomes (C). Conversely when there exists an identity of the form (C), $[a]_m^n$ must be equipotent with $[1]_r^r$, i.e. it must be an impotent matrix of rank r.

Ex. iii. *If $[a]_m^n$ is a rational integral x-matrix in Ω of rank r, it can always be expressed in the form*

$$[a]_m^n=[p]_m^r [q]_r^n, \ \ldots\ldots\ldots\ldots\ldots\ldots\ldots\ldots\text{(D)}$$

where $[p]_m^r$ and $[q]_r^n$ are undegenerate rational integral x-matrices in Ω of rank r.

Evidently either one of the factor matrices on the right can be impotent.

NOTE 1. *When ϕ is symmetric, the transformations of Theorems I and II are not in general symmetric.*

Accordingly it has not been proved in this article that a symmetric x-matrix can be reduced to its standard form by symmetric equipotent transformations.

NOTE 2. *Alternative proof of Theorem I of § 202 for matrices whose elements are rational integral functions of a single variable x.*

The reduction carried out in Theorem I affords independent proofs of Theorem I of § 202 and the equivalent results (B) and (D) of § 205 for any rational integral x-matrix. For since the matrix ϕ of Theorem I is equipotent with Φ, we can take the functions $E_1, E_2, \dots E_r$ obtained in the theorem to be the potent factors of ϕ of orders $1, 2, \dots r$; and the fact that E_{i+1} is divisible by E_i is equivalent to the results cited.

Since the H.C.F. of the minor determinants of Φ of order i is $E_1 E_2 \dots E_i$, we can take the maximum factors of ϕ of orders $1, 2, \dots r$ to be the functions $D_1, D_2, \dots D_r$ given by

$$D_1=E_1, \quad D_2=D_1E_2, \quad \dots \quad D_{i+1}=D_iE_{i+1}, \quad \dots \quad D_r=D_{r-1}E_r,$$

the E's being the functions obtained in Theorem I. If t is any irreducible (or irresoluble) divisor of ϕ, and if $d_1, d_2, \dots d_r$ are the indices of the highest powers of t which are factors of $D_1, D_2, \dots D_r$, the convention $d_0=1$ being adopted as usual, then

$$d_1-d_0, \quad d_2-d_1, \quad \dots \quad d_i-d_{i-1}, \quad d_{i+1}-d_i, \quad \dots \quad d_r-d_{r-1}$$

are the indices of the highest powers of t which are factors of $E_1, E_2, \dots E_i, E_{i+1}, \dots E_r$; and because the function E_{i+1} is divisible by E_i, it follows that

$$d_{i+1}-d_i \nless d_i-d_{i-1},$$

this being Theorem I of § 202.

NOTE 3. *Another method of reducing a rational integral x-matrix ϕ to its standard form by equipotent transformations.*

Yet another method is indicated in Exs. iv—vi which follow. We can first reduce ϕ to a quasi-scalar matrix, and then reduce the quasi-scalar matrix to its standard form.

Ex. iv. *Conversion of the rational integral x-matrix $\phi=[a]_m^n$ of Theorem I into a similar quasi-scalar matrix by equipotent transformations in Ω.*

If it is desired simply to convert ϕ into some quasi-scalar matrix whose diagonal elements are unspecified, we may proceed as follows.

Let a_{ij} be one of the non-zero elements of ϕ of lowest degree. If a_{ij} is not a common factor of all elements in the jth vertical row of ϕ, the highest common factor H of all those elements has a lower degree in x than a_{ij}, and there exists a rational integral identity in Ω of the form

$$u_{i1}a_{1j}+u_{i2}a_{2j}+\dots+u_{im}a_{mj}=H,$$

in which $[u_{i1}u_{i2}\dots u_{im}]$ is a primitive one-rowed matrix from which (see Ex. iii of § 216) we can form an undegenerate and impotent square matrix $[u]_m^m$ in Ω by the addition of horizontal rows. Then

$$[u]_m^m [a]_m^n=[b]_m^n$$

is an equipotent transformation in Ω converting ϕ into the matrix $[b]_m^n$ in which $b_{ij}=H$; and we can treat $[b]_m^n$ in the same way. We can use a similar process when a_{ij} is not a common factor of all elements in the ith horizontal row of ϕ. A succession of such transformations and a final derangement will convert ϕ into a similar matrix in which the leading element is not 0 and is a factor of all elements in the leading horizontal and vertical rows. Further unitary equipotent transformations in Ω will then convert ϕ into the form

$$\phi_1=\begin{bmatrix} a, & 0 \\ 0, & b \end{bmatrix}_{1,\, m-1}^{1,\, n-1},$$

where the leading element a_{11} is not 0. We can then apply the same processes to $[b]_{m-1}^{n-1}$, and continuing in this way we shall finally convert ϕ into a quasi-scalar matrix of the form

$$\psi=\begin{bmatrix} a, & 0 \\ 0, & 0 \end{bmatrix}_{r,\, m-r}^{r,\, n-r},$$

where $[a]_r^r$ is an undegenerate quasi-scalar matrix.

Ex. v. *Reduction of an undegenerate quasi-scalar x-matrix of rank r to the standard form* ${}^1[E]_r$ *by equipotent transformations.*

Let $\phi={}^1[a]_r$ be an undegenerate quasi-scalar matrix of rank r whose diagonal elements $a_1, a_2, \ldots a_r$ are rational integral functions in Ω of the single variable x; let t be any one of the irreducible divisors of ϕ; and let the potent indices of t of orders 1, 2, ... r be $e_1, e_2, \ldots e_r$. The leading diagonal minor determinants of ϕ are all regular with respect to t when and only when the highest powers of t which are factors of $a_1, a_2, \ldots a_r$ are $t^{e_1}, t^{e_2}, \ldots t^{e_r}$, and in such a case we shall say that ϕ is *completely regular* with respect to t. When ϕ is completely regular with respect to all its irreducible divisors, it has the standard form ${}^1[E]_r$, where $E_1, E_2, \ldots E_r$ are its potent factors of orders 1, 2, ... r.

We can clearly render ϕ completely regular with respect to any one of its irreducible divisors by a symmetric derangement. Accordingly we will make the hypothesis that ϕ is completely regular with respect to a number of irreducible divisors $\tau, \tau', \tau'', \ldots$, but not completely regular with respect to t, and show that equipotent transformations in Ω can then be determined which will render ϕ completely regular with respect to t as well as with respect to $\tau, \tau', \tau'', \ldots$.

Let $T_1, T_2, \ldots T_r$ be the products of the highest powers of $\tau, \tau', \tau'', \ldots$ which are factors of $E_1, E_2, \ldots E_r$ respectively, so that by the hypothesis we can write

$$a_1=T_1 t^{\eta_1} U_1, \quad a_2=T_2 t^{\eta_2} U_2, \quad \ldots \quad a_r=T_r t^{\eta_r} U_r,$$

where $[\eta_1 \eta_2 \ldots \eta_r]$ is some derangement of $[e_1 e_2 \ldots e_r]$, and where no one of the functions U is divisible by $\tau, \tau', \tau'', \ldots$ or t. Also let i and j be two of the integers 1, 2, ... r so chosen that $i<j$ but $\eta_i>\eta_j$; let H be the H.C.F. of U_i and U_j lying in Ω; and let

$$\eta_i-\eta_j=\epsilon, \quad T_j=ST_i, \quad U_i=Hu, \quad U_j=Hu', \quad Su'=v, \quad K=T_i t^{\eta_j} H;$$

so that the part of ϕ formed by its ith and jth horizontal and vertical rows is

$$\Phi_{ij}=\begin{bmatrix} a_i & 0 \\ 0 & a_j \end{bmatrix}=\begin{bmatrix} T_i t^{\eta_i} U_i, & 0 \\ 0 \ , & T_j t^{\eta_j} U_j \end{bmatrix}=K \cdot \phi_{ij},$$

where

$$\phi_{ij}=\begin{bmatrix} t^{\epsilon} u, & 0 \\ 0 \ , & v \end{bmatrix}.$$

Then because K is the H.C.F. of a_i and a_j, or because $t^{\epsilon}u$ and v are prime to one another, there exist rational integral identities in Ω of the forms

$$h_{11} a_i+h_{12} a_j=K, \quad h_{11} t^{\epsilon} u+h_{12} v=1,$$

where $[h_{11} h_{12}]$ is a primitive one-rowed matrix from which we can form the undegenerate and impotent square matrix

$$[h]_2^2=\begin{bmatrix} h_{11}, & h_{12} \\ -v, & t^{\epsilon} u \end{bmatrix}.$$

If we transform ϕ_{ij} by first adding the 2nd vertical row to the 1st vertical row, then prefixing $[h]_2^2$, and then adding to the 2nd vertical row the 1st vertical row multiplied by $-h_{12}v$, we shall obtain a resultant equipotent transformation in Ω which converts ϕ_{ij} and Φ_{ij} respectively into

$$\psi_{ij}=\begin{bmatrix}1, & 0\\ 0, & t^{\epsilon}uv\end{bmatrix} \text{ and } \Psi_{ij}=\begin{bmatrix}T_i t^{\eta_j} V_i, & 0\\ 0\;\;, & T_j t^{\eta_i} V_j\end{bmatrix},$$

where $V_i=H$, and $V_j=Huu'$; the form of Ψ_{ij} being that of Φ_{ij} when the indices η_i and η_j of t are interchanged. By a succession of such transformations we can convert ϕ into a similar quasi-scalar matrix ψ which is completely regular with respect to t as well as with respect to $\tau, \tau', \tau'', \ldots$.

Repetitions of this process will make ϕ completely regular with respect to all its irreducible divisors, i.e. will reduce ϕ to the standard form.

Ex. vi. *If ϕ is a quasi-scalar x-matrix in Ω whose elements are expressed as products of potent divisors, the interchange of two potent divisors which are powers of the same irreducible divisor t can be effected by an equipotent transformation in Ω.*

If in Ex. v we determine a rational integral identity in Ω of the form

$$k_{11}u+k_{12}t^{\epsilon}v=1$$

and use the undegenerate and impotent square matrix

$$[k]_2^2=\begin{bmatrix}k_{11}, & k_{12}\\ t^{\epsilon}v, & -u\end{bmatrix}$$

we can by the methods of Ex. v also convert

$$\phi'_{ij}=\begin{bmatrix}u, & 0\\ 0, & t^{\epsilon}v\end{bmatrix} \text{ into } \psi_{ij}=\begin{bmatrix}1, & 0\\ 0, & t^{\epsilon}uv\end{bmatrix}.$$

It follows that we can determine equipotent transformations in Ω which will convert ϕ_{ij} into ϕ'_{ij}, and

$$\Phi_{ij}=\begin{bmatrix}T_i t^{\eta_i} U_i, & 0\\ 0\;\;, & T_j t^{\eta_j} U_j\end{bmatrix} \text{ into } \Psi'_{ij}=\begin{bmatrix}T_i t^{\eta_j} U_i, & 0\\ 0\;\;, & T_j t^{\eta_i} U_j\end{bmatrix};$$

and this proves the theorem.

By a number of such interchanges and by symmetric derangements we can reduce ϕ to its standard form. An illustration is furnished by Ex. xii of § 217.

NOTE 4. *The reductions of Theorems I and II cannot in general be effected when the elements of the matrix $\phi=[a]_m^n$ are rational integral functions of two or more variables.*

Let $\phi=[a]_2^2$ be an undegenerate square matrix of order 2 whose elements are rational integral functions in Ω of two or more variables and have no common factor other than a non-zero constant, let $\overline{A}_2^2$ be the conjugate reciprocal of ϕ, and let $(a)_2^2=\Delta$, so that

$$D_1=E_1=1, \quad D_2=E_2=(a)_2^2=\Delta.$$

Then we can reduce ϕ by an equipotent transformation to the standard form

$$^1[E]_2=\begin{bmatrix}1, & 0\\ 0, & \Delta\end{bmatrix} \quad\ldots\ldots\ldots\ldots\ldots\ldots\ldots\ldots(3)$$

if and only if we can determine a rational integral identity in the variables of the form

$$\begin{bmatrix}a, & b\\ l, & m\end{bmatrix}[a]_2^2=\begin{bmatrix}1, & 0\\ 0, & \Delta\end{bmatrix}\begin{bmatrix}\alpha, & \beta\\ \lambda, & \mu\end{bmatrix}, \quad\ldots\ldots\ldots\ldots\ldots\ldots(4)$$

where $am-bl=1$; for this is clearly a necessary condition, and when it is satisfied, we have $\alpha\mu-\beta\lambda=1$, and the prefactor on the left and the postfactor on the right are undegenerate and impotent square matrices. This is possible when and only when we can determine rational integral functions a, b, l, m, λ, μ of the variables such that:

$$\text{(i)}\quad [l, m]\,[a]_2^2=\Delta\,[\lambda, \mu],$$

$$\text{(ii)}\quad am-bl\quad =1.$$

The condition (i) is satisfied when and only when

$$[l, m]=[\lambda, \mu]\,\overline{\underline{A}}_2^2=[\lambda, \mu]\begin{bmatrix} a_{22}, & -a_{12} \\ -a_{21}, & a_{11}\end{bmatrix}, \quad\ldots\ldots\ldots\ldots(5)$$

where λ and μ are arbitrary rational integral functions of the variables; and the condition (ii) is then also satisfied if and only if the functions a and b are such that

$$a\,\{\mu a_{11}-\lambda a_{12}\}+b\,\{\mu a_{21}-\lambda a_{22}\}=1. \quad\ldots\ldots\ldots\ldots(6)$$

Accordingly we can reduce ϕ to the standard form (3) by an equipotent transformation when and only when there exists an identity of the form (6) in which λ, μ, a, b are rational integral functions of the variables.

Now when the elements of ϕ contain no constant terms, there cannot be any identity of the form (6); for the left-hand side vanishes when zero values are ascribed to all the variables. Thus there certainly exist undegenerate square matrices $[a]_2^2$ which cannot be reduced to the standard form ${}^1[E]_2$ by an equipotent transformation; and clearly a similar result is true for undegenerate square matrices of any order when there are at least two variables.

Ex. vii. For the matrix

$$\phi=[a]_2^2=\begin{bmatrix} 3x+4y+1, & 2x+y+1 \\ x+2y+1, & x+y+2\end{bmatrix},$$

which lies in the domain Ω_1 of all rational numbers, we have

$$D_1=E_1=1,\quad D_2=E_2=x^2+2xy+y^2+4x+6y+1=\Delta,$$

and the equations (5) and (6) are

$$[l, m]=[\lambda, \mu]\,\overline{\underline{A}}_2^2=[\lambda, \mu]\begin{bmatrix} x+y+2 & -(2x+y+1) \\ -(x+2y+1), & 3x+4y+1\end{bmatrix}, \quad\ldots\ldots\ldots(5')$$

$$a\,\{\mu\,(3x+4y+1)-\lambda\,(2x+y+1)\}+b\,\{\mu\,(x+2y+1)-\lambda\,(x+y+2)\}=1. \quad\ldots\ldots(6')$$

We can convert ϕ into the standard form (3) by an equipotent transformation when and only when there exist rational integral functions a, b, λ, μ satisfying the identity (6′).

For given functions λ and μ (see Ex. i of § 189) there exists an identity of the form (6′) when and only when the two expressions multiplying a and b in (6′) have no finite root in common. If λ and μ are constants, this is the case when and only when

$$\lambda^2-2\lambda\mu+2\mu^2=0. \quad\ldots\ldots\ldots\ldots(7)$$

If a, b, λ, μ are all constants, the condition (6′) is satisfied when and only when

$$\lambda^2-2\lambda\mu+2\mu^2=0,\quad 2\lambda^2 a=\mu,\quad 2\lambda^2 b=-(\lambda+\mu);$$

in particular it is satisfied when

$$\lambda=1,\quad \mu=\tfrac{1}{2}(1\pm\sqrt{-1}),\quad a=\tfrac{1}{2}\mu,\quad b=\tfrac{1}{2}(\mu+1).$$

Thus when λ, μ, a, b have the values last given, and l and m are the linear functions given by (5′), we have an identity of the form (4) in which $am-bl=1$, $a\mu-\beta\lambda=1$, from which we can deduce an equipotent transformation converting ϕ into the standard form (3); but this is not a transformation in Ω_1.

Again if λ and μ are constants, and a and b are the linear functions

$$a=px+qy+r, \quad b=ux+vy+w,$$

there exists an identity of the form (6′) when and only when

$$\begin{bmatrix} 2\lambda-3\mu, & 0, & 0, & \lambda-\mu, & 0, & 0 \\ 0, & \lambda-4\mu, & 0, & 0, & \lambda-2\mu, & 0 \\ \lambda-4\mu, & 2\lambda-3\mu, & 0, & \lambda-2\mu, & \lambda-\mu, & 0 \\ \lambda-\mu, & 0, & 2\lambda-3\mu, & 2\lambda-\mu, & 0, & \lambda-\mu \\ 0, & \lambda-\mu, & \lambda-4\mu, & 0, & 2\lambda-\mu, & \lambda-2\mu \end{bmatrix} \begin{bmatrix} p \\ q \\ r \\ u \\ v \\ w \end{bmatrix} = 0, \quad \text{.........(8)}$$

and
$$(\lambda-\mu)\,r+(2\lambda-\mu)\,w+1=0. \quad \text{.........(9)}$$

The maximum factors of orders 1, 2, 3, 4, 5 of the prefactor on the left in (8) are

$$1,\ 1,\ 1,\ 1,\ (\lambda^2-2\lambda\mu+2\mu^2)^2;$$

and the matrix of its affected simple minor determinants is

$$(\lambda^2-2\lambda\mu+2\mu^2)^2 . [\mu-\lambda,\ 2\mu-\lambda,\ \mu-2\lambda,\ 2\lambda-3\mu,\ \lambda-4\mu,\ \lambda-\mu]. \quad \text{.........(10)}$$

If $\lambda^2-2\lambda\mu+2\mu^2 \neq 0$, there is only one distinct non-zero solution of (8) given by the matrix (10), and this solution is inconsistent with (9). We are therefore again led to the necessary condition (7), and when this is satisfied, we can determine equipotent transformations, not in Ω_1, which convert ϕ into the standard form (3).

Ex. viii. *Modification of Theorem II when there are several variables* $x, y_1, y_2, y_3, \ldots$.

In this case we can replace the equations (B) by

$$[h]_m^m \begin{bmatrix} E, & 0 \\ 0, & 0 \end{bmatrix}_{r,\,m-r}^{r,\,n-r} [k]_n^n = Y[a]_m^n, \quad [h']_m^m [a]_m^n [k']_n^n = Y' \begin{bmatrix} E, & 0 \\ 0, & 0 \end{bmatrix}_{r,\,m-r}^{r,\,n-r}, \quad \text{......(11)}$$

where
$$[h]_m^m [h']_m^m = Y_1 [1]_m^m, \quad [k']_n^n [k]_n^n = Y_2 [1]_n^n, \quad \text{.........(12)}$$

the prefactors and postfactors on the left in (11) being undegenerate square matrices whose elements are rational integral functions of all the variables, and whose determinants are independent of x, and Y, Y', Y_1, Y_2 being non-zero rational integral functions of $y_1, y_2, y_3, \ldots$ only.

There still exist equations of these forms when E_i means the product of all those potent divisors of $[a]_m^n$ of order i which actually contain x, E_i being 1 or any assigned non-zero constant in Ω when none of the potent divisors of order i contain x.

§ 220. Necessary and sufficient conditions for the equipotence of two rational integral x-matrices.

Theorem. *Any two rational integral x-matrices* $A=[a]_m^n$ *and* $B=[b]_\mu^\nu$ *are equipotent with one another when and only when there exists a relation of the form*

$$[h]_u^m [a]_m^n [k]_n^v = [p]_u^\mu [b]_\mu^\nu [q]_\nu^v, \quad \text{.........(A)}$$

where the prefactors and postfactors are undegenerate and impotent x-matrices whose ranks are equal to their passivities.

This theorem remains true when we restrict u to be the larger of the two integers m and μ, *and v to be the larger of the two integers n and* ν.

The theorem simply states that A and B are equipotent when and only when there exists a rational integral x-matrix $C=[c]_u^v$ such that both A and B can be converted into C by equipotent transformations.

First suppose that there exists a relation of the form (A), and let $C=[c]_u^v$ be the common value of the two product matrices. Then by Theorem II of § 214 the matrices A and B are both equipotent with C, and are therefore equipotent with one another.

Next suppose that A and B are equipotent with one another; let their common rank be r; let their common potent factors be $E_1, E_2, \ldots E_r$; and let

$$[E]_r^r = \begin{bmatrix} E_1 & 0 & \ldots & 0 \\ 0 & E_2 & \ldots & 0 \\ \ldots & \ldots & \ldots & \ldots \\ 0 & 0 & \ldots & E_r \end{bmatrix}, \quad [E]_u^v = \begin{bmatrix} E, & 0 \\ 0, & 0 \end{bmatrix}_{r,\, u-r}^{r,\, v-r},$$

the second notation being employed only when $u \nless r$ and $v \nless r$. Then by the second of the formulae (B) in § 219 or by Ex. ii of § 218 we have

$$[E]_m^n = [h]_m^m [a]_m^n [k]_n^n, \quad [E]_\mu^\nu = [p]_\mu^\mu [b]_\mu^\nu [q]_\nu^\nu,$$

where the prefactors and postfactors on the right are undegenerate and impotent. If $u \nless m$, $u \nless \mu$ and $v \nless n$, $v \nless \nu$, we deduce that

$$\begin{bmatrix} h \\ 0 \end{bmatrix}_{m,\, u-m}^{m} [a]_m^n [k, 0]_n^{n,\, v-n} = \begin{bmatrix} p \\ 0 \end{bmatrix}_{\mu,\, u-\mu}^{\mu} [b]_\mu^\nu [q, 0]_\nu^{\nu,\, v-\nu} = [E]_u^v;$$

and this is a relation of the form (A).

In the above theorem we can introduce the additional restriction that the prefactors and postfactors on the two sides of (A) lie in any domain of rationality in which A and B both lie.

COROLLARY 1. *Two similar rational integral x-matrices* $[a]_m^n$ *and* $[b]_m^n$ *both lying in the domain* Ω *are equipotent with one another when and only when there exists an equipotent transformation in* Ω *of the form*

$$[h]_m^m [a]_m^n [k]_n^n = [b]_m^n,$$

$[h]_m^m$ *and* $[k]_n^n$ *being undegenerate and impotent rational integral x-matrices in* Ω.

COROLLARY 2. *If* $[b]_r^r$ *is an undegenerate square rational integral x-matrix lying in the domain* Ω, $[a]_m^n$ *will be a rational integral x-matrix in* Ω *equipotent with it when and only when there exists an equipotent transformation in* Ω *of the form*

$$[h]_m^r [b]_r^r [k]_r^n = [a]_m^n,$$

$[h]_m^r$ *and* $[k]_r^n$ *being undegenerate and impotent rational integral x-matrices of rank r lying in* Ω.

§ 221. Rational integral transformations of a rational integral x-matrix.

The circumstances under which one given rational integral x-matrix can be converted into another by a rational integral transformation of minimum passivities are given in the following theorem:

Theorem. *If $A=[a]_m^n$ and $B=[b]_\mu^\nu$ are two matrices whose elements are rational integral functions of the single variable x, there exists a relation of the form*

$$[h]_\mu^m [a]_m^n [k]_n^\nu = [b]_\mu^\nu \quad \text{.................................(A)}$$

in which the elements of $[h]_\mu^m$ and $[k]_n^\nu$ are also rational integral functions of x when and only when the following two conditions are satisfied:

(1) *The rank (or number of potent factors) of B does not exceed the rank (or number of potent factors) of A.*

(2) *The potent factor of each order i of B is divisible by the potent factor of the same order i of A.*

First suppose that there exists a relation of the form (A). Then by § 133 and Ex. vi of § 206 the conditions (1) and (2) are satisfied.

Next suppose that the conditions (1) and (2) are satisfied. Let the ranks of A and B be r and s, where $s \not> r$; let the potent factors of A and B be respectively $E_1, E_2, \ldots E_r$ and $F_1, F_2, \ldots F_s$; and let

$$F_i = \eta_i E_i, \quad (i=1, 2, \ldots s),$$

where η_i is a rational integral function of x which does not vanish identically. Also let

$$[E]_r^r = \begin{bmatrix} E_1 & 0 & \ldots & 0 \\ 0 & E_2 & \ldots & 0 \\ \ldots & \ldots & \ldots & \ldots \\ 0 & 0 & \ldots & E_r \end{bmatrix}, \quad [F]_s^s = \begin{bmatrix} F_1 & 0 & \ldots & 0 \\ 0 & F_2 & \ldots & 0 \\ \ldots & \ldots & \ldots & \ldots \\ 0 & 0 & \ldots & F_s \end{bmatrix}, \quad [\eta]_s^s = \begin{bmatrix} \eta_1 & 0 & \ldots & 0 \\ 0 & \eta_2 & \ldots & 0 \\ \ldots & \ldots & \ldots & \ldots \\ 0 & 0 & \ldots & \eta_s \end{bmatrix},$$

$$[E]_m^n = \begin{bmatrix} E, & 0 \\ 0, & 0 \end{bmatrix}_{r,\, m-r}^{r,\, n-r}, \quad [F]_\mu^\nu = \begin{bmatrix} F, & 0 \\ 0, & 0 \end{bmatrix}_{s,\, \mu-s}^{s,\, \nu-s}.$$

Then by § 219 we have

$$[E]_m^n = [h']_m^m [a]_m^n [k']_n^n, \quad [b]_\mu^\nu = [h'']_\mu^s [F]_s^s [k'']_s^\nu, \quad \text{.................(1)}$$

where the prefactors and postfactors on the right are undegenerate and impotent rational integral x-matrices whose ranks are equal to their passivities. From the first of the equations (1) we deduce that

$$[E]_s^s = [h']_s^m [a]_m^n [k']_n^s, \quad \text{...................................(2)}$$

where the prefactor and postfactor on the right both have rank s. We also have

$$[F]_s^s = [\eta]_s^s [E]_s^s. \quad \text{......................................(3)}$$

When we substitute for $[F]_s^s$ and $[E]_s^s$ their values given by (3) and (2), it follows from the second of the equations (1) that there exists a relation of the form (A) in which

$$[h]_\mu^m = [h'']_\mu^s [\eta]_s^s [h']_s^m, \quad [k]_n^\nu = [k']_n^s [k'']_s^\nu, \quad \text{........................(4)}$$

both these matrices having rank s.

CHAPTER XXV

RATIONAL INTEGRAL FUNCTIONS OF A SQUARE MATRIX

[In § 222 we define the latent roots and also the characteristic matrix and the characteristic determinant of a given square matrix. In § 223 we define the rational integral function $f(\phi)$ of a square matrix ϕ which corresponds to a given rational integral function $f(x)$ of a scalar variable x; we show that the product of two such rational integral functions of ϕ is always commutative; and we express the determinant and the latent roots of $f(\phi)$ in terms of the latent roots of ϕ. In § 224 we determine all those rational integral equations (including the rational integral equation of lowest degree) which are satisfied by a given square matrix. Finally § 225 contains some algebraic lemmas with the aid of which particular solutions of the matrix equation $\psi^2 = \phi$ can be found when ϕ is any given undegenerate square matrix.]

§ 222. The latent roots and the characteristic matrix of a square matrix whose elements are constants.

The latent roots of a square matrix $\phi = [a]_m^m$ whose elements are constants are quantities which occur when we attempt to solve the equation

$$[a]_m^m \overline{\underline{x}}_m = \lambda \overline{\underline{x}}_m, \quad \text{or} \quad \{[a]_m^m - \lambda [1]_m^m\} \overline{\underline{x}}_m = 0, \quad \text{.........(1)}$$

where λ is an unspecified scalar number. When we put

$$[a]_m^m = \phi, \quad [1]_m^m = I, \quad \text{..........................(2)}$$

the equation (1) can be written in the form

$$(\phi - \lambda I)\overline{\underline{x}}_m = \begin{bmatrix} a_{11} - \lambda, & a_{12}, & \dots & a_{1m} \\ a_{21}, & a_{22} - \lambda, & \dots & a_{2m} \\ \dots & \dots & \dots & \dots \\ a_{m1}, & a_{m2}, & \dots & a_{mm} - \lambda \end{bmatrix} \begin{bmatrix} x_1 \\ x_2 \\ \vdots \\ x_m \end{bmatrix} = 0. \quad \text{...(3)}$$

If we determine a scalar quantity λ and scalar quantities $x_1, x_2, \dots x_m$, not all zero, such that the equation (3) is satisfied, then λ is called a *latent root* of the square matrix $\phi = [a]_m^m$, and $\overline{\underline{x}}_m$ is called a *solution* of the equation (3) or (1), or a *pole* of ϕ *corresponding to the latent root* λ. The poles of ϕ will be considered in greater detail in Exs. xi and xii of § 236, where the total number of unconnected poles is found.

Since the equation (3) admits of non-zero solutions for $\lceil x \rfloor_m$ when and only when the prefactor on the left is degenerate, we see that the latent roots of the square matrix $\phi = [a]_m^m$ are the roots in λ of the scalar equation

$$\det(\phi - \lambda I) = \begin{vmatrix} a_{11} - \lambda, & a_{12}, & \dots & a_{1m} \\ a_{21}, & a_{22} - \lambda, & \dots & a_{2m} \\ \dots & \dots & \dots & \dots \\ a_{m1}, & a_{m2}, & \dots & a_{mm} - \lambda \end{vmatrix} = 0. \quad \dots\dots\dots(4)$$

Again since the equation (4) has degree m in λ, the square matrix ϕ of order m has exactly m latent roots. These are all finite; but they are not necessarily all distinct, as some of them may be repeated roots of (4).

When λ is regarded as a scalar variable or an arbitrary parameter, the square matrix

$$\phi(\lambda) = \phi - \lambda I = [a]_m^m = \lambda [1]_m^m, \quad \dots\dots\dots\dots\dots\dots\dots(5)$$

which occurs as the prefactor on the left in (3), is called the *characteristic matrix* of $[a]_m^m$, and its determinant in (4) is called the *characteristic determinant* of $[a]_m^m$. Clearly $\lambda = \alpha$ is a latent root of the square matrix $\phi = [a]_m^m$ when and only when $\lambda - \alpha$ is a linear divisor (or an irresoluble divisor) of its characteristic matrix (5), i.e. when and only when $\lambda - \alpha$ is a factor of the characteristic determinant (5). Whenever we speak of the *characteristic linear divisors* or the *characteristic potent divisors* of a square matrix $\phi = [a]_m^m$ whose elements are constants, we shall mean the linear divisors or the potent divisors of the characteristic matrix $\phi(\lambda)$ of ϕ.

When we expand the determinant $\det(\phi - \lambda I)$ in powers of λ, the equation (4) giving the latent roots of ϕ assumes the form

$$\lambda^m - q_1\lambda^{m-1} + q_2\lambda^{m-2} - \dots + (-1)^{m-r} q_{m-r}\lambda^r + \dots + (-1)^m q_m = 0, \quad \dots(6)$$

where q_i is the sum of the diagonal minor determinants (corranged or affected) of $[a]_m^m$ of order i, and where in particular $q_m = (a)_m^m$. To show this, let

$$\begin{bmatrix} a_{11} - \lambda_1, & a_{12}, & \dots & a_{1m} \\ a_{21}, & a_{22} - \lambda_2, & \dots & a_{2m} \\ \dots & \dots & \dots & \dots \\ a_{m1}, & a_{m2}, & \dots & a_{mm} - \lambda_m \end{bmatrix} = [b]_m^m, \quad \begin{vmatrix} a_{11} - \lambda_1, & a_{12}, & \dots & a_{1m} \\ a_{21}, & a_{22} - \lambda_2, & \dots & a_{2m} \\ \dots & \dots & \dots & \dots \\ a_{m1}, & a_{m2}, & \dots & a_{mm} \end{vmatrix} = (b)_m^m.$$

Then if $[u_1 u_2 \dots u_r]$, $[v_1 v_2 \dots v_{m-r}]$ are two complementary corranged minor sequences of $[1\,2 \dots m]$, the coefficient of $\lambda_{u_1} \lambda_{u_2} \dots \lambda_{u_r}$ in $(b)_m^m$ is the same as its coefficient in $(b_{uu})_r^r (b_{vv})_{m-r}^{m-r}$, and is therefore $(-1)^r (a_{vv})_{m-r}^{m-r}$. Putting $\lambda_1 = \lambda_2 = \ldots = \lambda_m = \lambda$, it follows that the coefficient of λ^r in $\det(\phi - \lambda I)$

$$= (-1)^r \sum_v (a_{vv})_{m-r}^{m-r} = (-1)^r q_{m-r},$$

the summation extending over all distinct values of the corranged minor sequence $[v_1 v_2 \ldots v_{m-r}]$. Accordingly we have

$$\begin{aligned}\det(\phi - \lambda I) &= q_m - q_{m-1}\lambda + q_{m-2}\lambda^2 - \ldots + (-1)^r q_{m-r}\lambda^r + \ldots + (-1)^m \lambda^m \\ &= (-1)^m \{\lambda^m - q_1\lambda^{m-1} + q_2\lambda^{m-2} - \ldots + (-1)^{m-r}\lambda^r + \ldots \\ &\qquad + (-1)^m q_m\}. \quad \ldots\ldots(7)\end{aligned}$$

We shall often for the sake of brevity write

$$\overline{\underline{x}}_m = x; \quad \ldots\ldots(8)$$

and when this is done, the equations (1) assume the forms

$$\phi x = \lambda x, \quad (\phi - \lambda I)x = 0. \quad \ldots\ldots(1')$$

Ex. i. If

$$\Delta = \begin{vmatrix} a_{11}-\lambda & a_{12} & \ldots & a_{1r} & a_{1,r+1} & \ldots & a_{1m} \\ a_{21} & a_{22}-\lambda & \ldots & a_{2r} & a_{2,r+1} & \ldots & a_{2m} \\ \ldots & \ldots & \ldots & \ldots & \ldots & \ldots & \ldots \\ a_{r1} & a_{r2} & \ldots & a_{rr}-\lambda & a_{r,r+1} & \ldots & a_{rm} \\ a_{r+1,1} & a_{r+1,2} & \ldots & a_{r+1,r} & a_{r+1,r+1} & \ldots & a_{r+1,m} \\ \ldots & \ldots & \ldots & \ldots & \ldots & \ldots & \ldots \\ a_{m1} & a_{m2} & \ldots & a_{mr} & a_{m,r+1} & \ldots & a_{mm} \end{vmatrix},$$

we can expand Δ in powers of λ by a method similar to that used in the text in the expansion of the determinant $\det(\phi - \lambda I)$, and we then obtain

$$\Delta = (-1)^r \{Q_0\lambda^r - Q_1\lambda^{r-1} + Q_2\lambda^{r-2} - \ldots + (-1)^r Q_r\}, \quad \ldots\ldots(7')$$

where Q_i is the sum of those corranged diagonal minor determinants of $[a]_m^m$ of order $m-r+i$ which involve the last $m-r$ horizontal and vertical rows of $[a]_m^m$ and i additional horizontal and vertical rows belonging to $[a]_r^r$. In particular we have

$$Q_0 = \begin{pmatrix} r+1, & r+2, & \ldots & m \\ & a & & \\ r+1, & r+2, & \ldots & m \end{pmatrix}, \quad Q_r = (a)_m^m.$$

Ex. ii. Any square matrix and its conjugate have the same latent roots.

Ex. iii. The equation giving the latent roots of the square matrix $[abc]_{123}$ is

$$\begin{vmatrix} a_1-\lambda & b_1 & c_1 \\ a_2 & b_2-\lambda & c_2 \\ a_3 & b_3 & c_3-\lambda \end{vmatrix} = 0,$$

or

$$\lambda^3 - (a_1+b_2+c_3)\lambda^2 + \{(bc)_{23} + (ca)_{31} + (ab)_{12}\}\lambda - (abc)_{123} = 0.$$

Ex. iv. If we use the notations of § 129, the equation giving the latent roots of the symmetric matrix

$$\phi = \begin{bmatrix} a & h & g & u \\ h & b & f & v \\ g & f & c & w \\ u & v & w & d \end{bmatrix}$$

is $$\lambda^4-(a+b+c+d)\lambda^3+(A_1+B_1+C_1+A_2+B_2+C_2)\lambda^2-(A+B+C+D)\lambda+\Delta=0.$$

Ex. v. The square matrix $[a]_m^m$ has the latent root a repeated exactly k times when and only when its characteristic matrix $[a]_m^m-\lambda[1]_m^m$ has the linear divisor $\lambda-a$ repeated exactly k times.

By this we mean that $\lambda-a$ is a factor of the characteristic determinant repeated exactly k times. It does not follow that $\lambda-a$ is a linear divisor of the characteristic matrix of any order less than m, nor that $\lambda-a$ is a repeated potent divisor.

Ex. vi. The square matrix $[a]_m^m$ has a zero latent root when and only when it is degenerate.

Ex. vii. If the square matrix $[a]_m^m$ is degenerate and has rank r, it has at least $m-r$ zero latent roots.

Ex. viii. *Latent roots of the powers of a square matrix.*

The latent roots of the matrix $\phi^2=[a]_m^m.[a]_m^m$ *are the squares of the latent roots of the matrix* $\phi=[a]_m^m$.

This follows from the identities

$$(\phi-\lambda I)(\phi+\lambda I)=\phi^2-\lambda^2 I,\quad \det(\phi-\lambda I).\det(\phi+\lambda I)=\det(\phi^2-\lambda^2 I).$$

More generally if k is any positive integer, the latent roots of the matrix

$$\phi^k=[a]_m^m.[a]_m^m.\ldots.[a]_m^m$$

are the kth powers of the latent roots of the matrix $\phi=[a]_m^m$.

For if $1, \omega_1, \omega_2, \ldots \omega_{k-1}$ are the kth roots of 1, we have the identity

$$(\phi-\lambda I)(\phi-\omega_1\lambda I)(\phi-\omega_2\lambda I)\ldots(\phi-\omega_{k-1}\lambda I)=\phi^k-\lambda^k I.$$

These results are particular cases of the theorem proved in § 223.2.

Ex. ix. *If* $\phi=[a]_m^m$, $I=[1]_m^m$, *and if* $\psi=\phi-aI$, *where a is any scalar quantity, then the latent roots of ψ are the latent roots of ϕ each diminished by a.*

For if $$\det(\phi-\lambda I)=(-1)^m(\lambda-\lambda_1)(\lambda-\lambda_2)\ldots(\lambda-\lambda_m),$$

then $$\det(\psi-\lambda I)=(-1)^m(\lambda+a-\lambda_1)(\lambda+a-\lambda_2)\ldots(\lambda+a-\lambda_m).$$

Hence ϕ has the latent root a repeated k times when and only when ψ has the latent root 0 *repeated k times.*

Again if $\Delta(\lambda)$ is any minor determinant of order i of the characteristic matrix $\phi - \lambda I$ of ϕ, then $\Delta(\lambda + a)$ is the correspondingly formed minor determinant of order i of the characteristic matrix $\psi - \lambda I$ of ψ.

Hence $(\lambda - a)^s$ *is a maximum or potent divisor of* $\phi - \lambda I$ *of order* i *when and only when* λ^s *is a maximum or potent divisor of* $\psi - \lambda I$ *of order* i.

Ex. x. *If* $\phi = [a]_m^m$, $I = [1]_m^m$, *and if* $\psi = p\phi + qI$, *where* p *and* q *are any scalar constants such that* $p \neq 0$, *then*:

(1) ϕ *has the latent root* a *repeated* k *times when and only when* ψ *has the latent root* $pa + q$ *repeated* k *times.*

(2) $(\lambda - a)^s$ *is a maximum or potent divisor of* $\phi - \lambda I$ *of order* i *when and only when* $\{\lambda - (pa + q)\}^s$ *is a maximum or potent divisor of* $\psi - \lambda I$ *of order* i.

For if $\Delta(\lambda)$ is any minor determinant of $\phi - \lambda I$ of order i, then $p^i \Delta\left(\frac{\lambda - q}{p}\right)$ is the correspondingly formed minor determinant of $\psi - \lambda I$ of order i.

Ex. xi. *If the square matrix* $\phi = [a]_m^m$ *has one and only one zero latent root, then* ϕ *must have rank* $m - 1$.

For in the equation (6) we must have $q_m = 0$, $q_{m-1} \neq 0$. Thus ϕ is degenerate, but has a non-vanishing diagonal minor determinant of order $m - 1$.

Ex. xii. *If* a *is an unrepeated latent root of the square matrix* $\phi = [a]_m^m$, *then the matrix* $\phi - aI$ *must have rank* $m - 1$.

This follows from Exs. ix and xi.

Ex. xiii. *When the elements of the square matrix* $\phi = [a]_m^m$ *are arbitrary parameters, its characteristic determinant*

$$\Delta(\lambda) = \det(\phi - \lambda I) = \det\{[a]_m^m - \lambda [1]_m^m\}$$

is an irresoluble function of λ *and the elements of* ϕ.

Clearly $\Delta(\lambda)$ is a homogeneous rational integral function of degree m in λ and the elements of ϕ, and is not divisible by λ. If $\Delta(\lambda)$ were not irresoluble, it would have a factor $P(\lambda)$ which is homogeneous of degree p in λ and the elements of ϕ, where p is some non-zero positive integer less than m, and by putting $\lambda = 0$ it would follow that the determinant $\Delta(0) = (a)_m^m$ has the factor $P(0)$ which is homogeneous of degree p in the elements of ϕ. This is impossible because the determinant $(a)_m^m$ is an irresoluble function of its elements.

Ex. xiv. *If* $\phi = [a]_m^m$ *is a compartite matrix of standard form whose parts are square matrices and whose elements are constants, then:*

(1) *The latent roots of* ϕ *are the latent roots of the several parts of* ϕ.

(2) *The potent divisors of the characteristic matrix of* ϕ *are the potent divisors of the characteristic matrices of the several parts of* ϕ.

This follows from Theorem II *b* of § 207 when we observe that the characteristic matrix of ϕ is a compartite matrix of standard form whose successive parts are the characteristic matrices of the successive parts of ϕ.

§ 223. Rational integral functions of a square matrix.

1. *Definition of a rational integral function of a square matrix.*

Let $\phi = [a]_m^m$ be any square matrix of any order m. Then if $I = [1]_m^m$ is the unit matrix of the same order m as ϕ, the positive integral powers of ϕ are the square matrices of order m defined by the equations

$$\phi^0 = I, \quad \phi^1 = \phi = [a]_m^m, \quad \phi^2 = [a]_m^m \cdot [a]_m^m, \quad \phi^3 = [a]_m^m \cdot [a]_m^m \cdot [a]_m^m, \quad \ldots;$$

and these are such that for all positive integral values of r and s

$$\phi^r \times \phi^s = \phi^s \times \phi^r = \phi^{r+s}. \quad \ldots\ldots\ldots\ldots\ldots\ldots\ldots(1)$$

If $$f(x) = a_0 x^n + a_1 x^{n-1} + \ldots + a_{n-1} x + a_n \ldots\ldots\ldots\ldots\ldots\ldots(2)$$

is any rational integral function of the scalar variable x, then for all positive integral values of the order m of ϕ and for all values of the elements of ϕ we will write

$$f(\phi) = a_0 \phi^n + a_1 \phi^{n-1} + \ldots + a_{n-1}\phi^1 + a_n \phi^0$$

$$= a_0 \phi^n + a_1 \phi^{n-1} + \ldots + a_{n-1}\phi + a_n I, \quad \ldots\ldots\ldots\ldots\ldots(2')$$

and call $f(\phi)$ a *rational integral function of the square matrix* ϕ. Clearly $f(\phi)$ is a square matrix of the same order m as ϕ whose elements are rational integral functions of the elements of ϕ. We call $f(\phi)$ a function of ϕ because in using the notation (2′) we can regard the matrix ϕ as variable; in fact we can regard ϕ as a square matrix of any given order m whose elements are arbitrary parameters; and we can also regard it as a perfectly arbitrary square matrix provided that I always means the unit matrix of the same order as ϕ. The coefficients $a_0, a_1, \ldots a_{n-1}, a_n$ of the various powers of ϕ in (2′) are finite scalar quantities which will usually be considered to be given numerical constants; but they may be arbitrary parameters. When the square matrix ϕ is regarded as variable, we call $f(\phi) = 0$ a *rational integral equation in* ϕ, and any particular finite square matrix ϕ will be said to *satisfy* this equation when its elements have such values that every element of the matrix $f(\phi)$ vanishes.

There is a one-one correspondence between all rational integral functions of the square matrix ϕ and all rational integral functions of the scalar variable x, two such corresponding functions $f(\phi)$ and $f(x)$ always having the forms (2′) and (2). The matrix function $f(\phi)$ is said to *vanish identically*, and the equation $f(\phi) = 0$ to be an *identity in* ϕ, when the corresponding scalar function $f(x)$ vanishes identically, i.e. when the coefficients $a_0, a_1, \ldots a_{n-1}, a_n$ in (2′) all vanish. This is the case when and only when the equation $f(\phi) = 0$ is satisfied by every square matrix ϕ; for when $a_0, a_1, \ldots a_{n-1}, a_n$ do not all vanish, we can choose x so that $f(x) \neq 0$, and if we then put

$\phi = x[1]_m^m$, we have $f(\phi) = f(x) . I \neq 0$. When $f(\phi)$ and $f(x)$ do not vanish identically, the degree of $f(x)$ in x is also the *degree* of $f(\phi)$ in ϕ, and the *degree* of the equation $f(\phi) = 0$ in ϕ. Accordingly a rational integral function of ϕ of degree n can always be expressed in the form (2′), where $a_0 \neq 0$.

Two rational integral functions $f(\phi)$ and $g(\phi)$ of the square matrix ϕ are *identically equal*, and the equation $f(\phi) = g(\phi)$ is an *identity in* ϕ, when the difference $f(\phi) - g(\phi)$ vanishes identically, i.e. when the coefficients of corresponding powers of ϕ in $f(\phi)$ and $g(\phi)$ are equal, or when the corresponding scalar functions $f(x)$ and $g(x)$ are identically equal.

If $f(x)$ and $f(\phi)$ are the corresponding rational integral functions of x and ϕ given by (2) and (2′), and if $\alpha_1, \alpha_2, \ldots \alpha_n$ are scalar numbers, the second of the equations

$$f(x) = a_0(x - \alpha_1)(x - \alpha_2)\ldots(x - \alpha_n), \quad \ldots\ldots\ldots\ldots\ldots\ldots(3)$$

$$f(\phi) = a_0(\phi - \alpha_1 I)(\phi - \alpha_2 I)\ldots(\phi - \alpha_n I) \quad \ldots\ldots\ldots\ldots(3')$$

is an identity in ϕ when and only when the first equation is an identity in x; and in both equations the order of arrangement of the factors on the right is immaterial. More generally if $f(x)$, $f_1(x)$, $f_2(x)$, $\ldots f_r(x)$ and $f(\phi)$, $f_1(\phi)$, $f_2(\phi)$, $\ldots f_r(\phi)$ are corresponding rational integral functions of x and ϕ, the second of the equations

$$f(x) = f_1(x) f_2(x) \ldots f_r(x), \quad \ldots\ldots\ldots\ldots\ldots\ldots(4)$$

$$f(\phi) = f_1(\phi) f_2(\phi) \ldots f_r(\phi) \quad \ldots\ldots\ldots\ldots\ldots\ldots(4')$$

is an identity in ϕ when and only when the first equation is an identity in x; and in both equations the order of arrangement of the factors on the right is immaterial. In particular if p is a positive integer, the second of the equations

$$f(x) = \{f_1(x)\}^p, \quad f(\phi) = \{f_1(\phi)\}^p$$

is an identity in ϕ when and only when the first equation is an identity in x. These considerations enable us to express any given rational integral function $f(\phi)$ of the square matrix as a product of irreducible or irresoluble (i.e. linear) factors. They also show that:

The product of two rational integral functions of a square matrix ϕ is always commutative.

For if $f(\phi)$ and $g(\phi)$ are two such functions, we always have the equations

$$f(x) g(x) = g(x) f(x), \quad f(\phi) g(\phi) = g(\phi) f(\phi)$$

which are identities in x and ϕ respectively.

NOTE 1. *Simplest linear functions of a square matrix* $\phi=[a]_m^m$.

To the linear function $x-\lambda$ of x there corresponds the linear function $\phi-\lambda I$ of ϕ, which by § 222 is degenerate or undegenerate according as λ is or is not a latent root of ϕ. From (3) and (3′) we see that every rational integral function of ϕ with numerical coefficients can be expressed as the product of a constant scalar factor and linear factors of this simplest form in one and only one way.

NOTE 2. *Solutions of the equation* $f(\phi)\,\lceil x \rfloor_m=0$ *where* ϕ *is a given square matrix of order* m.

It has been shown in § 222 that the equation $(\phi-\lambda I)\,\lceil x \rfloor_m=0$, which is the simplest equation of this form, has non-zero solutions for $\lceil x \rfloor_m$ when and only when λ is a latent root of ϕ. The non-zero solutions of the more general equation will be considered in § 235.

NOTE 3. *Solutions of the rational integral equation* $f(\phi)=0$.

Whenever the scalar function $f(x)$, in which the coefficients have given numerical values, does not vanish identically and is not merely a constant, it is always possible to determine finite values of x which satisfy the equation $f(x)=0$. Similarly, as will be shown in §§ 224 and 228, whenever the matrix function $f(\phi)$ does not vanish identically and is not merely a constant, it is always possible to determine finite square matrices $\phi=[a]_m^m$ which satisfy the equation $f(\phi)=0$. There does not exist any one-one correspondence between the values of x which satisfy the equation $f(x)=0$ and the matrices $\phi=[a]_m^m$ which satisfy the equation $f(\phi)=0$; but when the linear factors of $f(x)$ and the corresponding linear factors of $f(\phi)$ are known, it is possible (see Ex. iv of § 228) to determine all those matrices $\phi=[a]_m^m$ which satisfy the equation $f(\phi)=0$.

NOTE 4. *Cases in which the equation* $f(\phi)=0$ *is an identity in the elements of* ϕ.

The coefficients of the function $f(\phi)$ are usually considered to be given numerical constants. When however ϕ is a square matrix of given order m with arbitrary elements, and the coefficients of $f(\phi)$ are rational integral functions of the elements of ϕ, it may happen that the equation $f(\phi)=0$, though not an identity in ϕ, is true for all finite values of the elements of ϕ. In such cases the equation $f(\phi)=0$ will be said to be an *identity in the elements of* ϕ.

NOTE 5. *Negative powers of an undegenerate square matrix.*

First let $\phi=[a]_m^m$ be an undegenerate square matrix of order m whose elements are given numerical constants, let $I=[1]_m^m$, and let p be any positive integer. Then we define ϕ^{-1} to be the inverse matrix of ϕ, which is also an undegenerate square matrix of order m, and is the unique solution of each of the equations

$$\phi^{-1}\phi=I, \quad \phi\phi^{-1}=I; \qquad\qquad (5)$$

and we define ϕ^{-p} to be $(\phi^{-1})^p$, i.e. to be the pth power of the square matrix ϕ^{-1}. With these definitions the equations (1) are true for all positive and negative integral values of r and s so long as ϕ is undegenerate; but the definitions become inadmissible when ϕ is degenerate.

Next let $\phi=[a]_m^m$ be a square matrix of order m whose elements are arbitrary parameters, let $\Delta=(a)_m^m$, and let $\ulcorner\underline{A}\urcorner_m^m$ be the conjugate reciprocal of $[a]_m^m$. Then we define ϕ^{-1} to be the square matrix of order m,

$$\phi^{-1}=\frac{1}{\Delta}\ulcorner\underline{A}\urcorner_m^m, \qquad \text{.........(6)}$$

which is the inverse matrix of ϕ and is the unique solution of each of the equations (5); and we define ϕ^{-p} as before to be the pth power of ϕ^{-1}. With these definitions the equations (1) are true for all positive and negative integral values of r and s. When particular values are ascribed to the elements of ϕ, we see from § 124.6 that ϕ^{-1} is infinite when ϕ has rank $m-1$, and indeterminate when ϕ has rank less than $m-1$. Thus for particular values of the elements of ϕ the matrix ϕ^{-1} defined by (6) has a uniquely determinate finite value when and only when ϕ is undegenerate; and the same is true of the matrix ϕ^{-p}, where p is a non-zero positive integer.

NOTE 6. *Use of the notation* $\dfrac{\phi}{\psi}$ *when* ϕ *and* ψ *are square matrices of the same order* m.

Let $\phi=[a]_m^m$ and $\psi=[b]_m^m$ be two square matrices of the same order m whose elements are given numerical constants, and which are such that

(1) the product $\phi\psi$ is commutative,

(2) the matrix ψ is undegenerate.

Then there exists a square matrix $\chi=[c]_m^m$ which is the unique solution of each of the equations

$$[b]_m^m[c]_m^m=[a]_m^m, \quad [c]_m^m[b]_m^m=[a]_m^m,$$

or

$$\psi\chi=\phi, \qquad \chi\psi=\phi. \qquad \text{.........(7)}$$

For if $\psi^{-1}=\ulcorner\underline{B}\urcorner_m^m$ is the inverse matrix of ψ, the two equations (7) have respectively the unique solutions $\chi=\psi^{-1}\phi$, $\chi=\phi\psi^{-1}$; and since

$$(\psi^{-1}\phi)\psi=\psi^{-1}.\phi\psi=\psi^{-1}.\psi\phi=\phi=\phi.\psi^{-1}\psi=(\phi\psi^{-1})\psi,$$

it follows that

$$\psi^{-1}\phi=\phi\psi^{-1}.$$

When χ is the matrix thus determined, we write

$$\chi=\psi^{-1}\phi=\phi\psi^{-1}=[c]_m^m=\frac{\phi}{\psi}, \qquad \text{.........(8)}$$

and the matrix $\dfrac{\phi}{\psi}$ which can be defined in this way when the conditions (1) and (2) are satisfied is such that

$$\psi\frac{\phi}{\psi}=\frac{\phi}{\psi}\psi=\phi. \qquad \text{.........(9)}$$

It corresponds to the algebraical fraction $\dfrac{x}{y}$ which has a definite meaning when $y\neq 0$.

When $\phi=[a]_m^m$, $I=[1]_m^m$, the product $I\phi$ is always commutative. Hence we can always write

$$\phi^{-1}=\frac{I}{\phi}, \qquad \text{.........(8')}$$

both when ϕ is an undegenerate square matrix with constant elements, and when the elements of ϕ are arbitrary parameters; and we then have

$$\phi\frac{I}{\phi}=\frac{I}{\phi}\phi=I. \qquad \text{.........(9')}$$

NOTE 7. *Rational functions of a square matrix* ϕ.

Let $f(\phi)$ and $g(\phi)$ be given rational integral functions of a square matrix $\phi=[a]_m^m$. Since the product $f(\phi)\,g(\phi)$ is always commutative, it follows from Note 6 that we can write

$$\chi=\{g(\phi)\}^{-1}f(\phi)=f(\phi)\{g(\phi)\}^{-1}=\frac{f(\phi)}{g(\phi)}, \qquad \text{.........(10)}$$

where χ is a square matrix of order m which is finite and uniquely determinate whenever the elements of ϕ have such values that

$$\det g(\phi)\neq 0, \qquad \text{.........(11)}$$

it being then the unique solution of each of the equations

$$g(\phi)\cdot\chi=f(\phi),\quad \chi\cdot g(\phi)=f(\phi).$$

We may regard the matrix $\chi=\dfrac{f(\phi)}{g(\phi)}$ as a *rational function* of the square matrix ϕ corresponding to the rational algebraic fraction $\dfrac{f(x)}{g(x)}$. It possesses the property

$$g(\phi)\cdot\frac{f(\phi)}{g(\phi)}=\frac{f(\phi)}{g(\phi)}\cdot g(\phi)=f(\phi),$$

and is uniquely determinate and finite whenever the condition (11) is satisfied.

2. *Latent roots of any rational integral function of a square matrix.*

In the following theorem $\phi=[a]_m^m$ is any square matrix of order m whose elements are given constants, $F(\phi)$ is any rational integral function of the square matrix ϕ, and $F(x)$ is the corresponding rational integral function of the scalar variable x.

Theorem. *If $\alpha_1, \alpha_2, \dots \alpha_m$ are the m latent roots of the square matrix $\phi=[a]_m^m$, and if $F(\phi)$ is any rational integral function of ϕ, then:*

(1) $$F(\alpha_1)F(\alpha_2)\dots F(\alpha_m)=\det F(\phi).$$

(2) *The latent roots of $F(\phi)$ are $F(\alpha_1), F(\alpha_2), \dots F(\alpha_m)$.*

To prove the first part of the theorem let $I=[1]_m^m$, and let

$$\begin{aligned}f(x)=\det(\phi-xI)&=\det\{[a]_m^m-x[1]_m^m\}\\&=(-1)^m(x-\alpha_1)(x-\alpha_2)\dots(x-\alpha_m);\end{aligned}$$

also let n be the degree of $F(x)$ and $F(\phi)$, and let

$$F(x)=b_0(x-\beta_1)(x-\beta_2)\dots(x-\beta_n),$$

so that $$F(\phi)=b_0(\phi-\beta_1 I)(\phi-\beta_2 I)\dots(\phi-\beta_n I)=b_0[c]_m^m.$$

Then

$$\begin{aligned}\det F(\phi) &= b_0^m . \det(\phi - \beta_1 I) . \det(\phi - \beta_2 I) . \dots \det(\phi - \beta_n I)\\ &= b_0^m f(\beta_1) f(\beta_2) \dots f(\beta_n)\\ &= (-1)^{mn} b_0^m \Pi(\beta_j - \alpha_i), \quad (j = 1, 2, \dots n;\ i = 1, 2, \dots m),\\ &= b_0^m \Pi(\alpha_i - \beta_j)\\ &= F(\alpha_1) F(\alpha_2) \dots F(\alpha_m).\end{aligned}$$

To prove the second part of the theorem let λ be any scalar constant, and let

$$F'(x) = F(x) - \lambda, \quad F'(\phi) = F(\phi) - \lambda I.$$

Then from the first part of the theorem we see that

$$\det F'(\phi) = F'(\alpha_1) F'(\alpha_2) \dots F'(\alpha_m),$$

i.e. $\det\{F(\phi) - \lambda I\} = (-1)^m \{\lambda - F(\alpha_1)\}\{\lambda - F(\alpha_2)\}\dots\{\lambda - F(\alpha_m)\}$.

Since this equation is true for all values of λ, it is an identity in λ when we regard λ as a scalar variable; and therefore $F(\alpha_1), F(\alpha_2), \dots F(\alpha_m)$ are the latent roots of $F(\phi)$.

Some special cases of the above theorem are contained in Exs. viii and ix of § 222.

§ 224. Rational integral equations satisfied by a given square matrix.

As before we will write $\phi = [a]_m^m$, $I = [1]_m^m$, ϕ being a square matrix of order m whose elements are constants. The characteristic matrix of ϕ will be taken to be the undegenerate rational integral x-matrix

$$\phi - xI = [a]_m^m - x[1]_m^m; \quad \dots\dots\dots\dots(1)$$

and the maximum and potent factors of this characteristic matrix will be denoted by $D_1(x), D_2(x), \dots D_m(x)$ and $E_1(x), E_2(x), \dots E_m(x)$, these being rational integral functions of the scalar variable x. It will always be considered that $D_m(x)$ is the characteristic determinant of ϕ, so that

$$D_m(x) = \det(\phi - xI) = \det\{[a]_m^m - x[1]_m^m\}. \quad \dots\dots\dots\dots(2)$$

The conjugate reciprocal of the characteristic matrix $\phi - xI$ will be denoted by $\overline{\underline{A}}_m^m$, this being a rational integral x-matrix of the degree $m-1$; and the inverse of $\phi - xI$ is then the matrix

$$\psi = \frac{1}{D_m(x)} \overline{\underline{A}}_m^m, \quad \dots\dots\dots\dots(3)$$

whose elements are *rational functions* of x. The matrix ψ has a determinate finite value for every finite value of x which is not a latent root of ϕ, i.e. whenever $D_m(x) \neq 0$; and it is then the unique solution of each of the equations

$$(\phi - xI)\psi = I, \quad \psi(\phi - xI) = I. \quad \dots\dots\dots\dots(4)$$

Using Notes 6 and 7 of § 223, we can regard ψ as a *rational function* of ϕ, and use for it the notations

$$\psi = (\phi - xI)^{-1} = \frac{I}{\phi - xI}. \quad \text{.........................(5)}$$

For values of x sufficiently great to render the series on the right convergent, we have

$$\psi = -\frac{\phi^0}{x} - \frac{\phi^1}{x^2} - \frac{\phi^2}{x^3} - \frac{\phi^3}{x^4} - \dots, \quad \text{.....................(6)}$$

where $\phi^0 = I$; for this value of ψ clearly satisfies the equations (4); and we can regard (6) as the expansion of $(\phi - xI)^{-1}$ in descending powers of x.

NOTE 1. *Cases in which the elements of ϕ are arbitrary parameters.*

The elements of ϕ are usually to be regarded as given numerical constants. When some or all of them are arbitrary parameters (or are rational integral functions of certain arbitrary parameters), the elements of the characteristic matrix $\phi - xI$ are rational integral functions of x and those arbitrary parameters. In such cases $D_i(x)$ is the product of all those common factors of the minor determinants of $\phi - xI$ of order i which are rational and integral both in x and the arbitrary parameters, those which do not contain x being non-essential. If *all* the elements of ϕ are arbitrary parameters, it follows from Ex. xiii of § 222 that

$$D_1(x) = D_2(x) = \dots = D_{m-1}(x) = 1, \quad E_m(x) = D_m(x).$$

In the following lemmas and theorems we speak of functions of the elements of ϕ only when *all* the elements of ϕ are to be regarded as arbitrary parameters.

Lemma A. *If ψ is the inverse of the characteristic matrix $\phi - xI$ of the square matrix $\phi = [a]_m^m$, and if $f(x)$ and $f(\phi)$ are corresponding rational integral functions of x and ϕ, the product $f(x)\psi$ is rational and integral in x when and only when ϕ satisfies the equation $f(\phi) = 0$.*

Let $f(x)$ and $f(\phi)$ have degree r in x and ϕ, so that we can write

$$f(x) = c_0 + c_1 x + c_2 x^2 + \dots + c_r x^r,$$
$$f(\phi) = c_0 \phi^0 + c_1 \phi^1 + c_2 \phi^2 + \dots + c_r \phi^r,$$

where $c_0, c_1, \dots c_r$ are finite scalar constants and $c_r \neq 0$, and where $\phi^0 = I$. Then from the equations (4) we can deduce the $r+1$ equations

$$\psi = \psi,$$
$$x\psi = \phi\psi - I,$$
$$x^2\psi = \phi^2\psi - \phi - xI,$$
$$x^3\psi = \phi^3\psi - \phi^2 - x\phi - x^2 I,$$
$$x^4\psi = \phi^4\psi - \phi^3 - x\phi^2 - x^2\phi - x^3 I,$$
$$\dots\dots\dots\dots\dots\dots\dots\dots\dots\dots$$
$$x^r\psi = \phi^r\psi - \phi^{r-1} - x\phi^{r-2} - \dots - x^{r-2}\phi - x^{r-1} I.$$

Here the second equation follows immediately from (4); and the equations after the second are formed in succession by a uniform rule, the ith equation being formed by multiplying the $(i-1)$th equation by x and then substituting for $x\psi$ the value given by the second equation. If we multiply these equations by $c_0, c_1, c_2, c_3, c_4, \ldots c_r$ respectively and then add, we obtain

$$f(x)\psi = f(\phi)\psi + \{\psi_0 + x\psi_1 + x^2\psi_2 + \ldots + x^{r-1}\psi_{r-1}\}, \quad \ldots\ldots\ldots\ldots\ldots\ldots(7)$$

where $$\psi_i = -c_{i+1}I - c_{i+2}\phi^1 - \ldots - c_r\phi^{r-i-1}, \quad \psi_{r-1} = -c_rI \neq 0. \quad \ldots\ldots\ldots\ldots\ldots(8)$$

If $f(\phi)=0$, we see from (7) that $f(x)\psi$ is rational and integral in x. Conversely if $f(x)\psi$ is rational and integral in x, we see from (7) that $f(\phi)\psi$ must be rational and integral in x; and it follows from (6) that $f(\phi)=0$; for if

$$f(\phi)\left\{\frac{\phi^0}{x}+\frac{\phi^1}{x^2}+\frac{\phi^2}{x^3}+\ldots\right\} = \chi_0+\chi_1x+\chi_2x^2+\ldots+\chi_px^p$$

for all values of x whose absolute magnitudes exceed R, where R is real and positive, then

$$z^pf(\phi)\{z\phi^0+z^2\phi^1+z^3\phi^2+\ldots\} = z^p\chi_0+z^{p-1}\chi_1+z^{p-2}\chi_2+\ldots+\chi_p$$

for all values of z whose absolute magnitudes are less than $\frac{1}{R}$, and this is only possible when the coefficients of all the powers of z on both sides vanish.

Thus $f(x)\psi$ is rational and integral in x when and only when $f(\phi)=0$, and is then the rational integral x-matrix given by

$$f(x)\psi = \psi_0+x\psi_1+x^2\psi_2+\ldots+x^{r-1}\psi_{r-1}. \quad \ldots\ldots\ldots\ldots\ldots\ldots\ldots\ldots(9)$$

Lemma B. *If all the elements of ϕ are arbitrary parameters, and if the coefficients of $f(x)$ are rational integral functions of the elements of ϕ, then $f(x)\psi$ is rational and integral both in x and in the elements of ϕ when and only when the equation $f(\phi)=0$ is an identity in the elements of ϕ.*

In this case Lemma A holds good for all particular finite values of the elements of ϕ. Hence when $f(\phi)$ vanishes for all finite values of the elements, the equation (9) is true for all finite values of the elements, and therefore $f(x)\psi$ is rational and integral in the elements of ϕ as well as in x. Conversely when $f(\phi)$ does not vanish for all finite values of the elements of ϕ, then $f(x)\psi$ cannot be integral in x, for there are particular values of the elements for which it is not integral in x.

Theorem I a. *If $\phi = [a]_m^m$ is a square matrix whose elements are constants, and if $D_m(x) = \det\{[a]_m^m - x[1]_m^m\}$ is its characteristic determinant, then ϕ satisfies the rational integral equation*

$$D_m(\phi) = 0. \quad \ldots\ldots\ldots\ldots\ldots\ldots\ldots\ldots\ldots\ldots\ldots(\text{A})$$

Further the equation (A) *is an identity in the elements of ϕ when those elements are all arbitrary.*

From (3) we see that

$$D_m(x)\psi = \overline{A}_m^m. \quad \ldots\ldots\ldots\ldots\ldots\ldots\ldots\ldots\ldots(10)$$

Therefore $D_m(x)\psi$ is rational and integral in x, and it follows from Lemma A that $D_m(\phi)=0$. Since this is true for all finite values of the elements of ϕ, the equation (A) must be an identity in those elements when they are all arbitrary. The second part of the theorem could also be deduced directly from Lemma B.

The equation (A) is known as *Cayley's Equation* or as the *Cayley-Hamiltonian Equation* after the names of its discoverers.

Theorem I b. *If all the elements of the square matrix* $\phi = [a]_m^m$ *are arbitrary parameters, and if* $f(\phi)$ *is a rational integral function of* ϕ *in which the coefficients are rational integral functions of the elements of* ϕ, *then the rational integral equation* $f(\phi) = 0$ *is an identity in the elements of* ϕ *when and only when* $f(\phi)$ *has the form*

$$f(\phi) = g(\phi)\, D_m(\phi), \qquad \text{(A}'\text{)}$$

where $g(\phi)$ *is a rational integral function of* ϕ *in which the coefficients are rational integral functions of the elements of* ϕ.

Consequently the equation (A) *of Theorem I a is the rational integral equation of the lowest degree which is rational and integral in the elements of* ϕ *and is an identity in the elements of* ϕ.

First let the equation $f(\phi) = 0$ be an identity in the elements of ϕ. Then from Lemma B we see that there must exist a matrix $[b]_m^m$ which is rational and integral in x and the elements of ϕ and is such that $f(x)\,\psi = [b]_m^m$; and it follows from (3) that

$$f(x)\, \overline{\underline{A}}{}_m^m = D_m(x)\, [b]_m^m. \qquad \text{(11)}$$

When in the identity (11) we equate the highest common factors (rational and integral in x and the elements of ϕ) of the elements of the matrices on the two sides, we obtain

$$f(x) = g(x)\, D_m(x), \qquad \text{(A}''\text{)}$$

where $g(x)$ is the H.C.F. of the elements of $[b]_m^m$; for by Ex. xiii of § 222 the H.C.F. of the elements of $\overline{\underline{A}}{}_m^m$ is 1. From (A″), which is an identity in x and the elements of ϕ, we deduce the equation (A′) which is an identity in ϕ and the elements of ϕ.

Next let $f(\phi)$ be expressible in the form (A′), so that (A′) is an identity in ϕ and the elements of ϕ. Then by Theorem I *a* the equation $f(\phi) = 0$ is an identity in the elements of ϕ.

The second part of Theorem I *b* is now obviously true.

Ex. i. With the notation used in the equation (6) of § 222 we have

$$D_m(x)=(-1)^m\{x^m-q_1x^{m-1}+q_2x^{m-2}-\dots+(-1)^m q_m\},$$

where q_i is the sum of the (corranged or affected) diagonal minor determinants of ϕ of order i; and (A) can be written in the form

$$\phi^m-q_1\phi^{m-1}+q_2\phi^{m-2}-\dots+(-1)^{m-1}q_{m-1}\phi+(-1)^m q_m I=0. \quad\text{.........}(A_1)$$

Ex. ii. If $\lambda_1, \lambda_2, \dots \lambda_m$ are the m latent roots of ϕ, we have

$$D_m(x)=(-1)^m(x-\lambda_1)(x-\lambda_2)\dots(x-\lambda_m);$$

and therefore (A) can also be written in the form

$$(\phi-\lambda_1 I)(\phi-\lambda_2 I)\dots(\phi-\lambda_m I)=0. \quad\text{..........................}(A_2)$$

Ex. iii. *If* $\phi=\begin{bmatrix} a_1 & b_1 \\ a_2 & b_2 \end{bmatrix}$, $I=\begin{bmatrix} 1 & 0 \\ 0 & 1 \end{bmatrix}$, *the square matrix* ϕ *satisfies the equation*

$$\phi^2-(a_1+b_2)\phi+(a_1b_2-a_2b_1)I=0 \quad\text{.............................}(a_1)$$

which is an identity in the elements of ϕ.

In this case $$\phi-xI=\begin{bmatrix} a_1-x, & b_1 \\ a_2\ , & b_2-x \end{bmatrix},$$

$$D_2(x)=\det(\phi-xI)=x^2-(a_1+b_2)x+(a_1b_2-a_2b_1),$$

$$D_2(\phi)=\phi^2-(a_1+b_2)\phi+(a_1b_2-a_2b_1)I$$

$$=\begin{bmatrix} a_1^2+a_2b_1\ , & a_1b_1+b_1b_2 \\ a_1a_2+a_2b_2, & a_2b_1+b_2^2 \end{bmatrix}-(a_1+b_2)\begin{bmatrix} a_1, & b_1 \\ a_2, & b_2 \end{bmatrix}+(a_1b_2-a_2b_1)\begin{bmatrix} 1, & 0 \\ 0, & 1 \end{bmatrix}$$

$$=0.$$

Again if λ_1 *and* λ_2 *are the latent roots of* ϕ, *i.e. the roots of the equation* $D_2(x)=0$, *then* ϕ *satisfies the equation*

$$(\phi-\lambda_1 I)(\phi-\lambda_2 I)=0. \quad\text{...................................}(a_2)$$

In fact

$$(\phi-\lambda_1 I)(\phi-\lambda_2 I)=\begin{bmatrix} a_1-\lambda_1, & b_1 \\ a_2\ , & b_2-\lambda_1 \end{bmatrix}\begin{bmatrix} a_1-\lambda_2, & b_1 \\ a_2\ , & b_2-\lambda_2 \end{bmatrix}=\begin{bmatrix} c_1, & d_1 \\ c_2, & d_2 \end{bmatrix},$$

where $$c_1=a_1^2+a_2b_1-(\lambda_1+\lambda_2)a_1+\lambda_1\lambda_2, \quad c_2=a_2(a_1+b_2)-(\lambda_1+\lambda_2)a_2,$$

$$d_2=a_2b_1+b_2^2-(\lambda_1+\lambda_2)b_2+\lambda_1\lambda_2, \quad d_1=b_1(a_1+b_2)-(\lambda_1+\lambda_2)b_1.$$

Since $\lambda_1+\lambda_2=a_1+b_2$, $\lambda_1\lambda_2=a_1b_2-a_2b_1$, we see that c_1, c_2, d_1, d_2 are all zero.

Theorem II a. *If* $\phi=[a]_m^m$ *is a square matrix whose elements are constants, and if* $E_m(x)$ *is the potent factor of order* m *of the characteristic matrix* $[a]_m^m-x[1]_m^m$ *of* ϕ, *then* ϕ *satisfies the rational integral equation*

$$E_m(\phi)=0. \quad\text{...............................}(B)$$

Let $D_m(x)$ and $D_{m-1}(x)$ be the potent factors of orders m and $m-1$ of the characteristic matrix of ϕ, $D_m(x)$ being chosen to be the characteristic determinant of ϕ. Then the equation (10) deduced from (3) can be written in the form

$$D_m(x)\psi=D_{m-1}(x)\,\overline{\underline{A'}}_m^m, \quad\text{......................}(12)$$

where $\overline{\underline{A'}}{}^{m}_{m}$ is a matrix whose elements are rational integral functions of x having no common factor other than a non-zero constant. Since

$$D_m(x) = D_{m-1}(x)\, E_m(x),$$

it follows from (12) that

$$E_m(x)\,\psi = \overline{\underline{A'}}{}^{m}_{m}. \quad \text{..........................(13)}$$

Consequently $E_m(x)\,\psi$ must be rational and integral in x; and it follows from Lemma A that ϕ satisfies the equation $f(\phi) = 0$.

Theorem II b. *If $\phi = [a]^{m}_{m}$ is a square matrix whose elements are constants, and if $f(\phi)$ is a rational integral function of ϕ, then ϕ satisfies the rational integral equation $f(\phi) = 0$ when and only when $f(\phi)$ has the form*

$$f(\phi) = g(\phi)\, E_m(\phi), \quad \text{..........................(B')}$$

where $g(\phi)$ is a rational integral function of ϕ.

Consequently the equation (B) *of Theorem II a is the rational integral equation of lowest degree which is satisfied by ϕ.*

First let ϕ satisfy the equation $f(\phi) = 0$. Then from Lemma A we see that there exists a rational integral x-matrix $[b]^{m}_{m}$ such that $f(x)\,\psi = [b]^{m}_{m}$; and it follows from (3) that

$$f(x)\,\overline{\underline{A}}{}^{m}_{m} = D_m(x)\,[b]^{m}_{m}. \quad \text{......................(14)}$$

When in the equation (14) we equate the highest common factors (rational and integral in x) of the elements of the matrices on the two sides, we obtain

$$f(x)\, D_{m-1}(x) = g(x)\, D_m(x), \quad \text{....................(15)}$$

where $g(x)$ is the H.C.F. of the elements of $[b]^{m}_{m}$. Since

$$D_m(x) = D_{m-1}(x)\, E_m(x),$$

it follows from (15) that

$$f(x) = g(x)\, E_m(x). \quad \text{..........................(B'')}$$

From the equation (B''), which is an identity in x, we deduce the equation (B') which is an identity in ϕ.

Next let $f(\phi)$ be expressible in the form (B'), so that (B') is an identity in ϕ. Then by Theorem II a the matrix ϕ satisfies the equation $f(\phi) = 0$.

The second part of Theorem II b is now obviously true.

Other proofs of Theorem II b (which includes Theorems I a, I b and II a) are given in Ex. xv of § 230 and Ex. xiii of § 235.

NOTE 2. *Cases in which the elements of ϕ involve arbitrary parameters.*

In proving Theorems II a and II b it has been implicitly assumed that all the elements of ϕ are given numerical constants. But the theorems remain true and the proofs remain valid when some or all the elements of ϕ are arbitrary parameters, or are rational integral functions of certain arbitrary parameters, provided that a rational integral function of x or ϕ means one whose coefficients are rational integral functions of those arbitrary parameters, and provided also that $D_i(x)$ and $E_i(x)$ are interpreted as in Note 1 to be rational and integral in the arbitrary parameters as well as in x.

In the particular case when all the elements of ϕ are arbitrary, Theorems II a and II b become Theorems I a and I b.

Ex. iv. *The square matrix $\phi=[a]_m^m$ satisfies the rational integral equation $f(\phi)=0$ when and only when the potent factor of order m of its characteristic matrix $\phi-xI$ is a factor of $f(x)$, i.e. when and only when all the potent divisors of order m of $\phi-xI$ are factors of $f(x)$.*

This is an alternative form of Theorem II b which follows from the equation (B″).

Hence all the latent roots of ϕ must be roots of the equation $f(x)=0$.

Ex. v. If $x-a_1, x-a_2, \dots x-a_s$ are the distinct linear divisors of the characteristic matrix $\phi-xI$, i.e. if $a_1, a_2, \dots a_s$ are the distinct or unequal latent roots of ϕ, then in (B″) and (B′) we have

$$E_m(x)=(x-a_1)^{\kappa_1}(x-a_2)^{\kappa_2}\dots(x-a_s)^{\kappa_s},$$

$$E_m(\phi)=(\phi-a_1I)^{\kappa_1}(\phi-a_2I)^{\kappa_2}\dots(\phi-a_sI)^{\kappa_s}, \quad\dots\dots\dots\dots(\mathrm{B}_1)$$

where $\kappa_1, \kappa_2, \dots \kappa_s$ are the potent indices of order m of the linear divisors $x-a_1, x-a_2, \dots x-a_s$ for the matrix $\phi-xI$.

Ex. vi. *If*

$$\phi=\begin{bmatrix}6, & 6, & 8\\ 6, & 11, & 12\\ 8, & 12, & 18\end{bmatrix}, \quad I=\begin{bmatrix}1, & 0, & 0\\ 0, & 1, & 0\\ 0, & 0, & 1\end{bmatrix}, \quad \phi-xI=\begin{bmatrix}6-x, & 6, & 8\\ 6, & 11-x, & 12\\ 8, & 12, & 18-x\end{bmatrix},$$

the rational integral equation of lowest degree satisfied by ϕ is

$$E_3(\phi)=(\phi-2I)(\phi-31I)=\phi^2-33\phi+62I=0. \quad\dots\dots\dots\dots(\mathrm{b})$$

In this case, when non-zero constant factors are disregarded, we can write

$$D_1(x)=1, \quad D_2(x)=x-2, \quad D_3(x)=(x-2)^2(x-31),$$

$$E_1(x)=1, \quad E_2(x)=x-2, \quad E_3(x)=(x-2)(x-31).$$

The latent roots of ϕ, being the roots of the equation $D_3(x)=0$, are 2, 2, 31; and the distinct or unequal latent roots are 2, 31.

The conjugate reciprocal of the characteristic matrix $\phi-xI$ is

$$\overline{\underline{A}}{}_3^3=\begin{bmatrix}x^2-29x+54, & 6x-12, & 8x-16\\ 6x-12, & x^2-24x+44, & 12x-24\\ 8x-16, & 12x-24, & x^2-17x+30\end{bmatrix}=(x-2)\begin{bmatrix}x-27, & 6, & 8\\ 6, & x-22, & 12\\ 8, & 12, & x-15\end{bmatrix};$$

and the inverse matrix of $\phi-xI$ is

$$\psi=-\frac{1}{(x-2)^2(x-31)}\overline{\underline{A}}{}_3^3=-\frac{1}{(x-2)(x-31)}\begin{bmatrix}x-27, & 6, & 8\\ 6, & x-22, & 12\\ 8, & 12, & x-15\end{bmatrix}.$$

§ 225. Frobenius's solutions of the matrix equation $\psi^2=\phi$.

The solutions of the matrix equation $\psi^2=\phi$ in which ϕ is any given *undegenerate square matrix* with constant elements are necessarily undegenerate square matrices of the same order as ϕ, and can be regarded as square roots of ϕ. This equation was first solved by Frobenius, his solutions being obtained by the use of the algebraic lemmas which follow. The original theorem proved by Frobenius will be given at the end of this article. Other methods of solving the equation will be described in later chapters.

Lemma 1. *Let a be a given non-zero constant; let $f(x)$ be any given rational integral function of x which is not divisible by $x-a$; and let n be any given non-zero positive integer. Then it is possible to determine a rational integral function $\phi(x)$ of degree less than n which is such that*

$$\chi(x)=\{\phi(x)f(x)\}^2-x \text{ is divisible by } (x-a)^n. \quad \text{.....................(a)}$$

Further there are two and only two such functions, these being

$$\phi(x)=\pm\sqrt{a}\,\psi(x), \quad \text{.....................................(b)}$$

where $\psi(x)$ is a uniquely determinate rational integral function of x of degree less than n whose coefficients are rational functions of a and the coefficients of $f(x)$.

Let the ith derivatives of $f(x)$, $\phi(x)$, $\chi(x)$ be denoted by $f_i(x)$, $\phi_i(x)$, $\chi_i(x)$ respectively, and let

$$\phi(x)=\sqrt{a}\left\{c_0+c_1(x-a)+\frac{c_2}{2!}(x-a)^2+\dots+\frac{c_{n-1}}{(n-1)!}(x-a)^{n-1}\right\}. \quad \text{.........(1)}$$

Then the lemma can be proved by determining such finite values of the constant coefficients $c_0, c_1, c_2, \dots c_{n-1}$ that

$$\chi(a)=0, \quad \chi_1(a)=0, \quad \chi_2(a)=0, \quad \dots \quad \chi_{n-1}(a)=0. \quad \text{..................(2)}$$

Evaluating the successive derivatives of the product $\phi^2.f^2$ by Leibnitz's Theorem, we see from (a) that

$$\chi_1(x)=2\phi\phi_1f^2+2\phi^2ff_1-1,$$

$$\chi_2(x)=2\{(\phi\phi_2+\phi_1^2)f^2+4\phi\phi_1ff_1+\phi^2(ff_2+f_1^2)\},$$

and that when $i>1$,

$$\chi_i(x)=\Phi_if^2+i\Phi_{i-1}F_1+\binom{i}{2}\Phi_{i-2}F_2+\dots+\binom{i}{2}\Phi_2F_{i-2}+i\Phi_1F_{i-1}+\phi^2F_i, \quad \text{......(3)}$$

where $$\Phi_r=\phi_r\phi+r\phi_{r-1}\phi_1+\binom{r}{2}\phi_{r-2}\phi_2+\dots+\binom{r}{2}\phi_2\phi_{r-2}+r\phi_1\phi_{r-1}+\phi\phi_r,$$

and $$F_r=f_rf+rf_{r-1}f_1+\binom{r}{2}f_{r-2}f_2+\dots+\binom{r}{2}f_2f_{r-2}+rf_1f_{r-1}+ff_r.$$

The first two of the conditions (2) are satisfied in succession when and only when

$$c_0=\pm\frac{1}{f(a)}, \quad \frac{c_1}{c_0}=\frac{f(a)-2af_1(a)}{2af(a)}. \quad \text{.............................(4)}$$

When c_0 and c_1 are chosen in accordance with (4), the remaining $n-2$ of the equations (2) determine in succession unique finite values of the ratios $\frac{c_2}{c_0}, \frac{c_3}{c_0}, \dots \frac{c_{n-1}}{c_0}$; and this establishes the lemma.

The actual evaluation of the successive coefficients of $\phi(x)$ is more easily effected by putting

$$y=\frac{x-a}{2af(a)}=\frac{x-a}{k}, \text{ where } k=2af(a)\neq 0,$$

and replacing (1) by

$$\phi(x)=Q\left\{1+a_1y+\frac{a_2}{2!}y^2+\ldots+\frac{a_{n-1}}{(n-1)!}y^{n-1}\right\}, \quad \ldots\ldots(1')$$

where $Q=\sqrt{a}\,.\,c_0\neq 0,\quad a_i=k^i\frac{c_i}{c_0};\;$ so that $\phi_i(a)=\sqrt{a}\,.\,c_i=\frac{Qa_i}{k^i}.$

The first two of the conditions (2) are satisfied in succession when and only when

$$Q=\pm\frac{\sqrt{a}}{f(a)},\quad a_1=f(a)-2af_1(a). \quad \ldots\ldots(4')$$

Suppose that Q, a_1, a_2, ... a_{i-1} are finite quantities which have been so determined that the first i of the conditions (2) are satisfied, where $i \not< 2$, and for the sake of brevity let

$$f_0,\, f_1,\, \ldots f_i,\, \ldots \text{ now stand for } f(a),\, f_1(a),\, \ldots f_i(a),\, \ldots \quad \ldots\ldots(5)$$

Then from (3) we see that

$$k^i\chi_i(a)=Q^2\left\{A_if_0^2+iA_{i-1}F_1k+\binom{i}{2}A_{i-2}F_2k^2+\ldots\right.$$
$$\left.+\binom{i}{2}A_2F_{i-2}k^{i-2}+iA_1F_{i-1}k^{i-1}+F_ik^i\right\}, \ldots\ldots(3')$$

where $A_r=a_r+ra_{r-1}a_1+\binom{r}{2}a_{r-2}a_2+\ldots+\binom{r}{2}a_2a_{r-2}+ra_1a_{r-1}+a_r,$

and $F_r=f_rf_0+rf_{r-1}f_1+\binom{r}{2}f_{r-2}f_2+\ldots+\binom{r}{2}f_2f_{r-2}+rf_1f_{r-1}+f_0f_r.$

It follows that the $(i+1)$th of the conditions (2) will also be satisfied when and only when a_i has the unique finite value given by

$$2a_if_0^2=-A_i'f_0^2-iA_{i-1}F_1k-\ldots-iA_1F_{i-1}k^{i-1}-F_ik^i, \quad \ldots\ldots(3'')$$

where $A_i'=ia_{i-1}a_1+\binom{i}{2}a_{i-2}a_2+\ldots+\binom{i}{2}a_2a_{i-2}+ia_1a_{i-1}.$

If we put $k=2af_0$ in (3″), and observe that $F_1=2f_0f_1$ and that A_i' is divisible by 2, we can cancel $2f_0^2$ on both sides. Therefore a_i is a rational integral function of $a_1, a_2, \ldots a_{i-1}$, $f_0, f_1, \ldots f_i$ and a in which the coefficients are integers. Moreover it is easily seen by induction that a_i has the following properties:

(1) It is a rational integral function of $f_0, f_1, \ldots f_i$ and a which is homogeneous of degree i in $f_0, f_1, \ldots f_i$, and also has degree i in a.

(2) It can be expressed in the form

$$a_i=P_0+P_1a+P_2a^2+\ldots+P_ia^i,$$

where P_r is a homogeneous rational integral function of degree i of $f_0, f_1, \ldots f_i$ which is isobaric of weight r in the suffixes of $f_0, f_1, \ldots f_i$.

For if we assume that the properties (1) and (2) are true for $a_1, a_2, \ldots a_{i-1}$, then they must also be true for every term on the right in (3″) after the factor $2f_0^2$ has been removed.

Ex. i. With the notation (5) the values of a_2 and a_3 in (3″) are

$$a_2=-f_0^2-4af_0f_1+4a^2(f_0f_2-2f_1^2),$$
$$a_3=3f_0^3+6af_0^2f_1-12a^2f_0(f_0f_2-2f_1^2)-8a^3(f_0^2f_3-6f_0f_1f_2+6f_1^3).$$

Ex. ii. The expressions for the coefficients of $\phi(x)$ which occur in (1) and (1′) are always the same no matter what the value of n may be.

Ex. iii. *Special case when $f(x)=x-b$ and $n=2$.*

If $a \neq 0$ and $a-b \neq 0$, there are two and only two functions $\phi(x)$ which have degree less than 2 and are such that $\{\phi(x).(x-b)\}^2-x$ is divisible by $(x-a)^2$. From (1′) and (4′) we see that they are given by

$$\phi(x)=Q\left\{1+a_1\frac{x-a}{2a(a-b)}\right\}, \text{ where } Q=\pm\frac{\sqrt{a}}{a-b}, \quad a_1=-(a+b),$$

i.e. by
$$\phi(x)=\pm\frac{\sqrt{a}}{2a(a-b)^2}\{(a+b)x-a(3a-b)\}.$$

For these two values of $\phi(x)$ we have

$$\{\phi(x).(x-b)\}^2-x=\frac{(x-a)^2}{4a(a-b)^4}\{(a+b)^2x^2-2(a+b)(2a^2-ab+b^2)x+b^2(3a-b)^2\}.$$

Ex. iv. Those functions $\phi(x)$ which have degree less than 2 and are such that $\{\phi(x).(x+3)\}^2-x$ is divisible by $(x-3)^2$ are the constants

$$\phi(x)=\pm\frac{\sqrt{3}}{6}.$$

This is the particular case of Ex. iii in which $a=3$, $b=-3$.

Lemma 2. *Let $F(x)$ be any given rational integral function of x of degree n which is not divisible by x. Then it is possible to determine rational integral functions $f(x)$ of degree less than n which are such that*

$$\chi(x)=\{f(x)\}^2-x \text{ is divisible by } F(x). \quad \text{.........................(a′)}$$

Further if $F(x)$ has exactly s unequal roots $a, b, c, \ldots$, every such function $f(x)$ must satisfy the conditions

$$f(a)=\pm\sqrt{a}, \quad f(b)=\pm\sqrt{b}, \quad f(c)=\pm\sqrt{c}, \quad \ldots, \quad \text{......................(6)}$$

and it is uniquely determinate when $f(a), f(b), f(c), \ldots$ have any assigned values consistent with these s necessary conditions; consequently there are exactly 2^s such functions $f(x)$; moreover these are given by

$$f(x)=\pm\sqrt{a}\,\psi_1(x)\pm\sqrt{b}\,\psi_2(x)\pm\sqrt{c}\,\psi_3(x)\pm\ldots, \quad \text{......................(b′)}$$

where $\psi_1(x), \psi_2(x), \psi_3(x), \ldots$ are uniquely determinate rational integral functions of x each of which has degree less than n.

We can suppose that

$$F(x)=k(x-a)^\alpha(x-b)^\beta(x-c)^\gamma\ldots, \quad \text{..........................(7)}$$

where k is a known non-zero constant; $a, b, c, \ldots$ are known non-zero constants no two of which are equal; and $\alpha, \beta, \gamma, \ldots$ are known non-zero positive integers such that $\alpha+\beta+\gamma+\ldots=n$; and we will further suppose that $\sqrt{a}, \sqrt{b}, \sqrt{c}, \ldots$ are definitely selected square roots of $a, b, c, \ldots$ respectively.

In the first place let the ith derivatives of $f(x)$ and $\chi(x)$ be denoted by $f_i(x)$ and $\chi_i(x)$ respectively. Then if $f(x)$ is any function having the properties mentioned in the lemma, the necessary equations $\chi(a)=0$, $\chi(b)=0$, $\chi(c)=0, \ldots$ show that the conditions (6) must be satisfied. When the value of $f(a)$ has been fixed, the necessary equations $\chi_1(a)=0$, $\chi_2(a)=0, \ldots \chi_{\alpha-1}(a)=0$ show that the values of $f_1(a), f_2(a), \ldots f_{\alpha-1}(a)$ are uniquely determinate. Similarly when the value of $f(b)$ is fixed, the values of $f_1(b), f_2(b), \ldots f_{\beta-1}(b)$ are uniquely determinate; when the value of $f(c)$ is fixed, the values of $f_1(c), f_2(c), \dots f_{\gamma-1}(c)$ are uniquely determinate; and so on. Hence if there are two functions $f(x)=p(x)$, $f(x)=q(x)$ for which the signs in (6) are the same, then

$p(x)-q(x)$ and its first $\alpha-1$ derivatives must vanish when $x=a$,

$p(x)-q(x)$ and its first $\beta-1$ derivatives must vanish when $x=b$,

$p(x)-q(x)$ and its first $\gamma-1$ derivatives must vanish when $x=c$,

and so on. Therefore $p(x)-q(x)$ must be divisible by $F(x)$. Since $p(x)-q(x)$ has degree less than n in x, this is only possible when $p(x)-q(x)$ vanishes identically, i.e. when $q(x)=p(x)$. Consequently there cannot be two different functions $f(x)$ satisfying the conditions of the lemma for which the signs in (6) are the same.

It remains to show that we can determine a function $f(x)$ having the properties mentioned in the lemma when the signs in (6) are arbitrarily assigned.

Changing the meaning of the suffixes, let $F_1(x)$, $F_2(x)$, $F_3(x)$, ... be the functions obtained by dividing $F(x)$ by $(x-a)^\alpha$, $(x-b)^\beta$, $(x-c)^\gamma$, ... respectively, i.e. by omitting these respective factors in (7); let

$$\phi_1(x)=\sqrt{a}\left\{A_0+A_1(x-a)+\frac{A_2}{2!}(x-a)^2+\dots+\frac{A_{\alpha-1}}{(\alpha-1)!}(x-a)^{\alpha-1}\right\},$$

$$\phi_2(x)=\sqrt{b}\left\{B_0+B_1(x-b)+\frac{B_2}{2!}(x-b)^2+\dots+\frac{B_{\beta-1}}{(\beta-1)!}(x-b)^{\beta-1}\right\},$$

$$\phi_3(x)=\sqrt{c}\left\{C_0+C_1(x-c)+\frac{C_2}{2!}(x-c)^2+\dots+\frac{C_{\gamma-1}}{(\gamma-1)!}(x-c)^{\gamma-1}\right\},$$

....

where $A_0, A_1, A_2, \dots, B_0, B_1, B_2, \dots, C_0, C_1, C_2, \dots, \dots$ are constants to be determined; and let

$$f(x)=\phi_1(x)F_1(x)+\phi_2(x)F_2(x)+\phi_3(x)F_3(x)+\dots, \quad \dots\dots\dots(8)$$

where each term is a rational integral function of x having degree less than n. Then since $F_2(x)$, $F_3(x)$, ... are all divisible by $(x-a)^\alpha$, the function $\chi(x)$ defined in (a′) will be divisible by $(x-a)^\alpha$ if and only if $\{\phi_1(x)F_1(x)\}^2-x$ is divisible by $(x-a)^\alpha$. By Lemma 1 the α constants in $\phi_1(x)$ can be chosen so that this condition is satisfied. Similarly the β constants in $\phi_2(x)$ can be so chosen that $\chi(x)$ is divisible by $(x-b)^\beta$, the γ constants in $\phi_3(x)$ can be so chosen that $\chi(x)$ is divisible by $(x-c)^\gamma$, and so on. The function $f(x)$ defined in (8) will be such that $\chi(x)$ is divisible by $F(x)$ when and only when the constants $A_0, A_1, \dots B_0, B_1, \dots C_0, C_1, \dots, \dots$ are obtained in this way. By Lemma 1 the ratios $A_0 : A_1 : A_2 : \dots : A_{\alpha-1}$, $B_0 : B_1 : B_2 : \dots : B_{\beta-1}$, $C_0 : C_1 : C_2 : \dots : C_{\gamma-1}$, ... are then uniquely determinate, and the constants $A_0, B_0, C_0, \dots$ must be so chosen that

$$A_0F_1(a)=\pm 1,\quad B_0F_2(b)=\pm 1,\quad C_0F_3(c)=\pm 1,\quad \dots, \quad \dots\dots\dots(9)$$

the corresponding values of $f(a), f(b), f(c), \dots$ being

$$f(a)=\pm\sqrt{a},\quad f(b)=\pm\sqrt{b},\quad f(c)=\pm\sqrt{c},\quad \dots \quad \dots\dots\dots(9')$$

The signs in (9) can be chosen arbitrarily, and the signs in (9′) are the same as in (9). Thus corresponding to each choice of the signs in (6) we can determine one (and only one) function $f(x)$ having the properties mentioned in the lemma.

Ex. v. *Determination of all those rational integral functions $f(x)$ of x whose degrees are less than 3 and which are such that*

$$\{f(x)\}^2-x \textit{ is divisible by the function } F(x)=(x-a)^2(x-b)$$

when $a\neq 0,\ b\neq 0,\ a-b\neq 0.$

In this case $$f(x)=\phi_1(x)\,F_1(x)+\phi_2(x)\,F_2(x),$$

where $$F_1(x)=x-b, \qquad \phi_1(x)=\sqrt{a}\{A_0+A_1(x-a)\},$$
$$F_2(x)=(x-a)^2, \quad \phi_2(x)=\sqrt{b}\{B_0\}.$$

Determining the constants A_0 and A_1 so that $\{\phi_1(x)\,F_1(x)\}^2-x$ is divisible by $(x-a)^2$ as in Ex. iii, and determining the constant B_0 so that $\{\phi_2(x)\,F_2(x)\}^2-x$ is divisible by $x-b$, we find that

$$A_0=\pm\frac{1}{a-b}, \quad \frac{A_1}{A_0}=-\frac{a+b}{2a(a-b)}, \quad B_0=\pm\frac{1}{(a-b)^2},$$

and we have therefore

$$f(x)=\frac{1}{2a(a-b)^2}[\sqrt{a}\,.\,\{(a+b)\,x-a(3a-b)\}(x-b)+\sqrt{b}\,.\,2a(x-a)^2], \quad \ldots(10)$$

where $\sqrt{a}$ and $\sqrt{b}$ are respectively either one of the two square roots of a and either one of the two square roots of b.

For all four of these functions $f(x)$ we have

$$\{f(x)\}^2-x=\frac{(x-a)^2(x-b)}{4a(a-b)^4}\{P+\sqrt{a}\sqrt{b}\,Q\}, \quad \ldots\ldots\ldots\ldots\ldots\ldots\ldots(11)$$

where $$P=(a^2+6ab+b^2)\,x-(4a^3+9a^2b-6ab^2+b^3),$$
$$Q=4\{(a+b)\,x-a(3a-b)\},$$

$\sqrt{a}$ and $\sqrt{b}$ being the same square roots of a and b in (11) as in (10).

Ex. vi. *Determination of all those rational integral functions $f(x)$ of x whose degrees are less than* 2 *and which are such that*

$$\{f(x)\}^2-x \textit{ is divisible by the function } F(x)=(x-a)(x-b)$$

when $a\neq 0$, $b\neq 0$, $a-b\neq 0$.

In this case $$f(x)=\frac{1}{a-b}\{\sqrt{a}\,.\,(x-b)+\sqrt{b}\,.\,(x-a)\};$$

and then $$\{f(x)\}^2-x=\frac{(x-a)(x-b)}{(a-b)^2}\{(a+b)+2\sqrt{a}\sqrt{b}\}.$$

Here $\sqrt{a}$ and $\sqrt{b}$ are the same square roots of a and b in both equations.

Theorem. *If $\phi=[a]_m^m$ is an undegenerate square matrix of order m whose elements are given constants, it is possible to determine a rational integral function $f(\phi)$ of ϕ whose degree in ϕ is less than m, and which is such that $\{f(\phi)\}^2=\phi$. The matrix $\psi=f(\phi)=[b]_m^m$ is then a particular solution of the matrix equation $\psi^2=\phi$.*

If ϕ has exactly s unequal roots, there are exactly 2^s particular solutions of the equation $\psi^2=\phi$ which can be determined in this way.

Let $I=[1]_m^m$; let x be a scalar variable; let $c_0, c_1, \ldots c_{m-1}$ be finite scalar constants; and let

$$f(x)=c_0+c_1x+c_2x^2+\ldots+c_{m-1}x^{m-1},$$
$$f(\phi)=c_0I+c_1\phi+c_2\phi^2+\ldots+c_{m-1}\phi^{m-1}.$$

Then we have to show that it is possible to determine the finite constants $c_0, c_1, c_2, \ldots c_{m-1}$ so that the given *undegenerate* square matrix $\phi=[a]_m^m$ satisfies the equation

$$\{f(\phi)\}^2-\phi=0. \quad \ldots\ldots\ldots\ldots\ldots\ldots\ldots\ldots\ldots\ldots\ldots(12)$$

Let $\lambda_1, \lambda_2, \ldots \lambda_s$ be the distinct or unequal latent roots of $[a]_m^m$, which are all different from 0 because $(a)_m^m \neq 0$; and let

$$D_m(x) = (-1)^m (x-\lambda_1)^{d_1} (x-\lambda_2)^{d_2} \ldots (x-\lambda_s)^{d_s},$$

and
$$E_m(x) = (x-\lambda_1)^{e_1} (x-\lambda_2)^{e_2} \ldots (x-\lambda_s)^{e_s},$$

where
$$d_1+d_2+\ldots+d_s = m, \quad e_1+e_2+\ldots+e_s = n \ngtr m,$$

be respectively the determinant and the potent factor of order m of the characteristic matrix $[a]_m^m - x[1]_m^m$. By Theorem II *b* of § 224 the given matrix $\phi = [a]_m^m$ satisfies the equation (12) when and only when there exist identities in ϕ and x respectively of the forms

$$\{f(\phi)\}^2 - \phi = g(\phi) E_m(\phi), \quad \{f(x)\}^2 - x = g(x) E_m(x), \quad \ldots\ldots\ldots\ldots\ldots(13)$$

where $g(\phi)$ and $g(x)$ are corresponding rational integral functions of ϕ and x, i.e. when and only when $\{f(x)\}^2 - x$ is divisible by $E_m(x)$.

Now let $F(x)$ be any rational integral function of x of degree r, where $r \ngtr m$, which is not divisible by x and which contains $E_m(x)$ as a factor. By Lemma 2 we can determine a rational integral function $f(x)$ of degree less than r which is such that $\{f(x)\}^2 - x$ is divisible by $F(x)$, and therefore by $E_m(x)$. There then exist identities in x and ϕ of the forms (13); and since the given matrix $\phi = [a]_m^m$ satisfies the equation $E_m(\phi) = 0$, it also satisfies the equation (12). Thus in the above theorem $f(\phi)$ may be the matrix function which corresponds to the scalar function $f(x)$. A solution obtained in this way will not be in its simplest form when its degree exceeds $n-1$; for it can then be reduced in degree by the equation $E_m(\phi) = 0$.

Again by Lemma 2 there are exactly 2^s functions $f(x)$ having degrees less than n which are such that $\{f(x)\}^2 - x$ is divisible by $E_m(x)$. Therefore there are exactly 2^s functions $f(\phi)$ having degrees less than n which satisfy the equation (12). We obtain them by taking $F(x)$ to be $E_m(x)$ in Lemma 2; and when so obtained, they are in their simplest forms, and are all independent. Since every rational integral function of ϕ can be reduced to one whose degree is less than n by the equation $E_m(\phi) = 0$, it follows that every function $f(\phi)$ satisfying the equation (12) must be identical (when its degree is made less than n) with one of the above 2^s independent functions.

Supposing then that the latent roots of ϕ have been found, we can determine particular solutions of the equation $\psi^2 = \phi$ in either of the following two ways, of which the first is the simpler.

First way of determining solutions of $\psi^2 = \phi$. Taking $F(x)$ to be $E_m(x)$ in Lemma 2 we obtain 2^s independent solutions of the forms

$$\psi = g(\phi) = \pm\sqrt{\lambda_1}\, g_1(\phi) \pm \sqrt{\lambda_2}\, g_2(\phi) \pm \ldots \pm \sqrt{\lambda_s}\, g_s(\phi), \quad \ldots\ldots\ldots\ldots\ldots(14)$$

where $g_1(x), g_2(x), \ldots g_s(x)$ are uniquely determinate rational integral functions of x of degrees less than n whose coefficients are rational functions of $\lambda_1, \lambda_2, \ldots \lambda_s$. The functions $g_1, g_2, \ldots g_s$ and g are the same for all matrices $\phi = [a]_m^m$ whose characteristic matrices have the same potent divisors of order m, i.e. for which $\lambda_1, \lambda_2, \ldots \lambda_s$ and $e_1, e_2, \ldots e_s$ are the same. The function $g_1(x)$ is divisible by $(x-\lambda_2)^{e_1}(x-\lambda_3)^{e_3} \ldots (x-\lambda_s)^{e_s}$; it is such that $\{g_1(x)\}^2 - x$ is divisible by $(x-\lambda_1)^{e_1}$; and such that $g_1(\lambda_1) = 1$. Consequently $g_1(\phi)$ has the latent root 1 repeated d_1 times, and all its other latent roots are 0.

Second way of determining solutions of $\psi^2 = \phi$. Taking $F(x)$ to be $D_m(x)$ in Lemma 2 we obtain 2^s independent solutions of the forms

$$\psi = f(\phi) = \pm\sqrt{\lambda_1}\, f_1(\phi) \pm \sqrt{\lambda_2}\, f_2(\phi) \pm \ldots \pm \sqrt{\lambda_s}\, f_s(\phi), \quad \ldots\ldots\ldots\ldots\ldots(15)$$

where $f_1(x)$, $f_2(x)$, ... $f_s(x)$ are uniquely determinate rational integral functions of x of degrees less than m whose coefficients are rational functions of $\lambda_1, \lambda_2, \dots \lambda_s$. The functions $f_1, f_2, \dots f_s$ and f are the same for all matrices $\phi=[a]_m^m$ which have the same latent roots, i.e. for which $\lambda_1, \lambda_2, \dots \lambda_s$ and $d_1, d_2, \dots d_s$ are the same. The function $f_1(x)$ is divisible by $(x-\lambda_2)^{d_2}(x-\lambda_3)^{d_3}\dots(x-\lambda_s)^{d_s}$; it is such that $\{f_1(x)\}^2-x$ is divisible by $(x-\lambda_1)^{d_1}$; and such that $f_1(\lambda_1)=1$. Consequently $f_1(\phi)$ has the latent root 1 repeated d_1 times, and all its other latent roots are 0. When the solutions (15) are reduced in degree by the equation $E_m(\phi)=0$, they become the solutions (14).

Ex. vii. If $\phi=[a]_3^3$ is an undegenerate square matrix of order 3, and if the determinant and the potent factor of order 3 of its characteristic matrix $\phi-xI=[a]_3^3-x[1]_3^3$ are respectively

$$D_3(x)=-(x-\lambda)^2(x-\mu),\quad E_3(x)=(x-\lambda)(x-\mu),$$

where $\lambda\mu\neq 0$, $\lambda\neq\mu$, then the particular solutions of the equation $\psi^2=\phi$ obtained by taking $F(x)$ to be

$$(x-\lambda)(x-\mu),\quad (x-\lambda)^2(x-\mu),\quad (x-\lambda)(x-\mu)(x-\nu)$$

in Lemma 2, where $\lambda\mu\nu\neq 0$, $(\mu-\nu)(\nu-\lambda)(\lambda-\mu)\neq 0$, are respectively

$$(1)\quad \psi=\frac{1}{\lambda-\mu}\{\pm\sqrt{\lambda}\,.\,(\phi-\mu I)\mp\sqrt{\mu}\,.\,(\phi-\lambda I)\};$$

$$(2)\quad \psi=\frac{1}{2\lambda(\lambda-\mu)^2}\{\mp\sqrt{\lambda}\,.\,[(\lambda+\mu)\phi-\lambda(3\lambda-\mu)I](\phi-\mu I)\pm\sqrt{\mu}\,.\,2\lambda(\phi-\lambda I)^2\}$$

$$=\frac{1}{2\lambda(\lambda-\mu)^2}\{\mp\sqrt{\lambda}\,.\,[(\lambda+\mu)\phi^2-(3\lambda^2+\mu^2)\phi+\lambda\mu(3\lambda-\mu)I]$$
$$\pm\sqrt{\mu}\,.\,2\lambda(\phi^2-2\lambda\phi+\lambda^2 I)\};$$

$$(3)\quad \psi=\pm\frac{\sqrt{\lambda}}{(\lambda-\mu)(\lambda-\nu)}(\phi-\mu I)(\phi-\nu I)\pm\frac{\sqrt{\mu}}{(\mu-\nu)(\mu-\lambda)}(\phi-\nu I)(\phi-\lambda I)$$
$$\pm\frac{\sqrt{\nu}}{(\nu-\lambda)(\nu-\mu)}(\phi-\lambda I)(\phi-\mu I).$$

The solutions (2) and (3) become the solutions (1) when their degrees are reduced by the equation

$$E_3(\phi)=(\phi-\lambda I)(\phi-\mu I)=0,\quad\text{or}\quad \phi^2=(\lambda+\mu)\phi-\lambda\mu I.$$

Ex. viii. If ϕ is the square matrix considered in Ex. vi of § 224, we obtain particular solutions of the equation $\psi^2=\phi$ by putting $\lambda=2$, $\mu=31$ and $\nu=7$ in Ex. vii above, these solutions being

$$(1)\quad \psi=\frac{1}{29}\{\mp\sqrt{2}\,.\,(\phi-31I)\pm\sqrt{31}\,.\,(\phi-2I)\};$$

$$(2)\quad \psi=\frac{1}{4\,.\,29^2}\{\mp\sqrt{2}\,.\,(33\phi+50I)(\phi-31I)\pm\sqrt{31}\,.\,4(\phi-2I)^2\};$$

$$(3)\quad \psi=\pm\frac{\sqrt{2}}{145}(\phi-31I)(\phi-7I)\pm\frac{\sqrt{31}}{696}(\phi-7I)(\phi-2I)\mp\frac{\sqrt{7}}{120}(\phi-2I)(\phi-31I).$$

The solutions (2) and (3) become the solutions (1) when their degrees are reduced by the equation

$$E_3(\phi)=(\phi-2I)(\phi-31I)=0,\quad\text{or}\quad \phi^2=33\phi-62I.$$

NOTE. When ϕ is symmetric, the solutions of the equation $\psi^2=\phi$ obtained in this way are all symmetric. A simple equation of this form was solved in § 162. I. (*e*).

CHAPTER XXVI

EQUIMUTANT TRANSFORMATIONS OF A SQUARE MATRIX WHOSE ELEMENTS ARE CONSTANTS

[In §§ 226—7 we define equimutant transformations, point out their importance in the Theory of Substitutions, and find conditions for the existence of an equimutant transformation between two given square matrices. In §§ 228—9 we describe the canonical form Π to which a square matrix ϕ can be reduced by equimutant transformations and the construction of its parts and super-parts; we define unipotent and unilatent square matrices; and we consider some of the properties of a simple canonical square matrix. In §§ 230—1 we determine the rank of any given rational integral function of a given square matrix, and find conditions that such a function shall be non-extravagant. In §§ 232—4 we define the transmutes of a square matrix, and show that there exists an equimutant transformation between two given square matrices when and only when they have the same transmutes. In § 235 we show how to determine complete sets of unconnected solutions of any equation of the form $f(\phi)\,\underline{\overline{x}}_{m}=0$ when ϕ is a given square matrix of order m; in particular we determine the total number of unconnected poles of ϕ. Finally in § 236 we show that the conjugate reciprocal of the characteristic matrix $\phi-\lambda I$ of a given square matrix ϕ is a rational integral function of ϕ; also we obtain expansions for the inverse matrix $(\phi-\lambda I)^{-1}$, and give the equimutant transformations derived from them by Frobenius which convert ϕ into a compartite matrix whose parts are all unilatent square matrices.]

§ 226. Definition of an equimutant transformation.

If $A=[a]_r^r$ and $B=[b]_m^m$ are square matrices of orders r and m, and if $r \not> m$, a relation of the form

$$[h]_m^r\,[a]_r^r\,\underline{\overline{H}}_r^m=[b]_m^m, \qquad \text{.........................(A)}$$

in which $[h]_m^r$ and $\underline{\overline{H}}_r^m$ are two mutually inverse undegenerate matrices of rank r with constant elements, so that

$$\underline{\overline{H}}_r^m\,[h]_m^r=[1]_r^r \qquad \text{.............................(a)}$$

will be called an *equimutant transformation* converting the square matrix A into the square matrix B.

In particular if $A=[a]_m^m$ and $B=[b]_m^m$ are two square matrices of the same order m, a relation of the form

$$[h]_m^m\,[a]_m^m\,\underline{\overline{H}}_m^m=[b]_m^m \qquad \text{.........................(B)}$$

in which $[h]^{m}_{m}$ and $\overline{\underline{H}}{}^{m}_{m}$ are two mutually inverse undegenerate square matrices with constant elements, so that

$$\overline{\underline{H}}{}^{m}_{m}\,[h]^{m}_{m} = [h]^{m}_{m}\,\overline{\underline{H}}{}^{m}_{m} = [1]^{m}_{m}, \quad \text{......................(b)}$$

will be called an *equimutant transformation* converting the square matrix A into the *similar* square matrix B, and may be distinguished from the more general transformation (A) as being an *isomorphic transformation.* Since all the matrices which occur in (B) are square matrices of the same order m, and $[h]^{m}_{m}$ is undegenerate, we can in this case write $h = [h]^{m}_{m}$, $h^{-1} = \overline{\underline{H}}{}^{m}_{m}$, and replace (B) by

$$hAh^{-1} = B.$$

When there exists an equimutant transformation of the form (A), we can (see Note 7 of § 141) deduce from it another transformation of the form

$$[h]^{m}_{m}\begin{bmatrix} a, & 0 \\ 0, & 0 \end{bmatrix}^{r,\,m-r}_{r,\,m-r}\overline{\underline{H}}{}^{m}_{m} = [b]^{m}_{m} \quad \text{......................(C)}$$

in which $[h]^{m}_{m}$ and $\overline{\underline{H}}{}^{m}_{m}$ are two mutually inverse undegenerate square matrices with constant elements formed respectively by adding $m-r$ final vertical rows to $[h]^{r}_{m}$ and $m-r$ final horizontal rows to $\overline{\underline{H}}{}^{m}_{r}$; and (C) is also an equimutant transformation. Conversely the equimutant transformation (C) leads at once to the equimutant transformation (A). Thus every equimutant transformation such as (A) can be replaced by an equivalent equimutant transformation of the special form (B).

The inverse of (B) is the equimutant transformation

$$\overline{\underline{H}}{}^{m}_{m}\,[b]^{m}_{m}\,[h]^{m}_{m} = [a]^{m}_{m}, \quad \text{........................(B}')$$

converting $[b]^{m}_{m}$ into $[a]^{m}_{m}$; and similarly the inverse of (C) is an equimutant transformation converting the square matrix $[b]^{m}_{m}$ into the similar square matrix $\begin{bmatrix} a, & 0 \\ 0, & 0 \end{bmatrix}^{r,\,m-r}_{r,\,m-r}$; but the inverse of (A), viz. the transformation

$$\overline{\underline{H}}{}^{m}_{r}\,[b]^{m}_{m}\,[h]^{r}_{m} = [a]^{r}_{r}, \quad \text{........................(A}')$$

is not an equimutant transformation except when $r = m$.

Since equimutant transformations are a special class of equigradent transformations, two square matrices such as $[a]^{r}_{r}$ and $[b]^{m}_{m}$ in (A) between which an equimutant transformation exists must have the same rank.

Equimutant transformations play a large part in the Theory of Substitutions. It will be seen from Note 5 that in finding the simplest form to which a given square matrix can be reduced by an equimutant transformation, we are at the same time finding the simplest form to which the scalar equations of a given homogeneous linear substitution can be reduced by a change of variables.

NOTE 1. *Appropriateness of the designation 'equimutant.'*

It is convenient to give a name to transformations of the form (A). The appropriateness of the name 'equimutant' will become clear in the following articles of the present chapter. In §§ 233 and 234 the successive transmutes of a square matrix whose elements are constants are defined, and in § 235 it is shown that two square matrices with constant elements are such that one of them can be converted into the other by an equimutant transformation when and only when the two matrices have the same successive transmutes of all orders starting from the first.

NOTE 2. *Symmetric equimutant transformations.*

The equimutant transformation (A) is *symmetric* when $[h]^r_m$ and $\overline{\underline{H}}^{\,m}_{r}$ are mutually conjugate. Thus the most general symmetric equimutant transformation has the form

$$\overline{\underline{l}}^{\,r}_{m}\,[a]^r_r\,[l]^m_r=[b]^m_m, \qquad \text{.................................(D)}$$

where $[l]^m_r$ is a semi-unit matrix of rank r, so that

$$[l]^m_r\,\overline{\underline{l}}^{\,r}_{m}=[1]^r_r.$$

In particular a symmetric equimutant transformation converting one square matrix into another similar square matrix has the form

$$\overline{\underline{l}}^{\,m}_{m}\,[a]^m_m\,[l]^m_m=[b]^m_m, \qquad \text{.................................(E)}$$

where $[l]^m_m$ is a square semi-unit matrix, so that

$$[l]^m_m\,\overline{\underline{l}}^{\,m}_{m}=\overline{\underline{l}}^{\,m}_{m}\,[l]^m_m=[1]^m_m.$$

We shall also call (D) and (E) *symmetric semi-unit transformations.*

NOTE 3. *Composition of equimutant transformations.*

Any number of equimutant transformations applied in succession to a square matrix are clearly together equivalent to a single resultant equimutant transformation; and if the component transformations are all symmetric, then the resultant transformation is also symmetric.

The resultant transformation is formed in the same manner as the resultant of a number of successive equigradent transformations in Note 2 of § 141.

NOTE 4. *Equimutant transformations of a compartite matrix.*

Let

$$\phi=\begin{bmatrix} a, & 0, & \dots & 0 \\ 0, & b, & \dots & 0 \\ \dots & \dots & \dots & \dots \\ 0, & 0, & \dots & c \end{bmatrix}^{p,\,q,\,\dots\,r}_{p,\,q,\,\dots\,r}, \qquad \psi=\begin{bmatrix} \alpha, & 0, & \dots & 0 \\ 0, & \beta, & \dots & 0 \\ \dots & \dots & \dots & \dots \\ 0, & 0, & \dots & \gamma \end{bmatrix}^{P,\,Q,\,\dots\,R}_{P,\,Q,\,\dots\,R},$$

where $P \nless p$, $Q \nless q$, ... $R \nless r$, be two compartite matrices of standard forms having the same number of parts, all the parts being square matrices. Then if

$$[h]_P^p [a]_p^p \overline{\underline{H}}_p^P = [\alpha]_P^P, \quad [h']_Q^q [b]_q^q \overline{\underline{H'}}_q^Q = [\beta]_Q^Q, \quad \ldots \quad [h'']_R^r [c]_r^r \overline{\underline{H''}}_r^R = [\gamma]_R^R \quad \ldots\ldots(1)$$

are equimutant transformations converting the successive parts of ϕ into the corresponding parts of ψ,

$$\begin{bmatrix} h, & 0, & \ldots & 0 \\ 0, & h', & \ldots & 0 \\ \ldots & \ldots & \ldots & \ldots \\ 0, & 0, & \ldots & h'' \end{bmatrix}_{P, Q, \ldots R}^{p, q, \ldots r} \phi \begin{bmatrix} H, & 0, & \ldots & 0 \\ 0, & H', & \ldots & 0 \\ \ldots & \ldots & \ldots & \ldots \\ 0, & 0, & \ldots & H'' \end{bmatrix}_{p, q, \ldots r}^{P, Q, \ldots R} = \psi \quad \ldots\ldots\ldots\ldots(2)$$

is an equimutant transformation converting ϕ into ψ. Conversely if (2) is an equimutant transformation converting ϕ into ψ, then all the transformations (1) are equimutant.

NOTE 5. *Substitutions whose matrices are equimutant.*

If
$$[a]_m^m = [h]_m^m [b]_m^m \overline{\underline{H}}_m^m$$

is an equimutant transformation, the two ordinary homogeneous linear substitutions (or transformations)

$$\overline{\underline{x'}}_m = [a]_m^m \overline{\underline{x}}_m, \quad \overline{\underline{x'}}_m = [b]_m^m \overline{\underline{x}}_m \quad \ldots\ldots\ldots\ldots\ldots\ldots\ldots\ldots(3)$$

are often called 'similar' substitutions (or transformations), because if the two equivalent equations

$$\overline{\underline{x}}_m = [h]_m^m \overline{\underline{y}}_m, \quad \overline{\underline{y}}_m = \overline{\underline{H}}_m^m \overline{\underline{x}}_m \quad \ldots\ldots\ldots\ldots\ldots\ldots\ldots\ldots(4)$$

define a transformation of independent variables, then the substitution defined by the first of the equations (3) when $x_1, x_2, \ldots x_m$ are independent variables is the same as the substitution defined by the equation

$$\overline{\underline{y'}}_m = [b]_m^m \overline{\underline{y}}_m$$

when $y_1, y_2, \ldots y_m$ are independent variables. To give a more precise form to this statement, let $f_1, f_2, f_3, \ldots$ and $g_1, g_2, g_3, \ldots$ be corresponding rational integral functions of the x's and y's defined by the transformations (4) in the way described in § 187. 1, so that

$$g_i(y_1, y_2, \ldots y_m) = f_i(h_{11}y_1 + h_{12}y_2 + \ldots + h_{1m}y_m, \ldots, \ldots),$$
$$f_i(x_1, x_2, \ldots x_m) = g_i(H_{11}x_1 + H_{21}x_2 + \ldots + H_{m1}x_m, \ldots, \ldots).$$

Then $f_1(x_1', x_2', \ldots x_m') = f_2(x_1, x_2, \ldots x_m)$ is an identity in the x's when

$$\overline{\underline{x'}}_m = [a]_m^m \overline{\underline{x}}_m \quad \ldots\ldots\ldots\ldots\ldots\ldots\ldots\ldots\ldots\ldots\ldots\ldots(5)$$

if and only if $g_1(y_1', y_2', \ldots y_m') = g_2(y_1, y_2, \ldots y_m)$ is an identity in the y's when

$$\overline{\underline{y'}}_m = [b]_m^m \overline{\underline{y}}_m. \quad \ldots\ldots\ldots\ldots\ldots\ldots\ldots\ldots\ldots\ldots(5')$$

If $f_1(x_1', x_2', \ldots x_m') = f_1(x_1, x_2, \ldots x_m)$ is an identity in the x's when $\overline{\underline{x'}}_m = [a]_m^m \overline{\underline{x}}_m$, the function f_1 is called an *invariant* of the substitution (5). This is the case if and only if $g_1(y_1', y_2', \ldots y_m') = g_1(y_1, y_2, \ldots y_m)$ is an identity in the y's when $\overline{\underline{y'}}_m = [b]_m^m \overline{\underline{y}}_m$, i.e. if and only if g_1 is an invariant of the substitution (5′). Thus the transformations (4) convert every invariant of the substitution (5) into an invariant of the substitution (5′) and every invariant of the substitution (5′) into an invariant of the substitution (5).

Ex. i. Every symmetric derangement of a square matrix is an equimutant transformation.

Ex. ii. The equimutant transformation (B) can always be replaced by or derived from the equivalent relation

$$[h]^m_m\,[a]^m_m=[b]^m_m\,[h]^m_m. \qquad \text{(B'')}$$

Thus $[h]^m_m$ must be a commutant (see Chapter XXVII) of B and A.

Ex. iii. *Whenever there exists an equimutant transformation of the form* (B), *the square matrices* $[a]^m_m$ *and* $[b]^m_m$, *if their elements are constants, must have the same latent roots.*

For then

$$[b]^m_m-\lambda\,[1]^m_m=[h]^m_m\,\{[a]^m_m-\lambda\,[1]^m_m\}\,\underline{\overline{H}}^m_m, \qquad (6)$$

this equation being an identity in λ; and when we equate the determinants of both sides, we obtain

$$\det\{[b]^m_m-\lambda\,[1]^m_m\}=\det\{[a]^m_m-\lambda\,[1]^m_m\}.$$

Moreover the characteristic matrices $[a]^m_m-\lambda\,[1]^m_m$ *and* $[b]^m_m-\lambda\,[1]^m_m$ *are equipotent and have the same potent divisors.*

This follows from Theorem II of § 214; for (6) is an equigradent transformation, and therefore an equipotent transformation.

Further if $\underline{\overline{x}}_m$ *is a pole of* $[a]^m_m$ *corresponding to the latent root* ρ, *then* $\underline{\overline{y}}_m=[h]^m_m\,\underline{\overline{x}}_m$ *is a pole of* $[b]^m_m$ *corresponding to the same latent root* ρ.

For if $\quad [a]^m_m\cdot\underline{\overline{x}}_m=\rho\,\underline{\overline{x}}_m$, then $[b]^m_m\cdot[h]^m_m\,\underline{\overline{x}}_m=\rho\,[h]^m_m\,\underline{\overline{x}}_m$.

Ex. iv. *Whenever there exists an equimutant transformation of the form* (A), *the square matrices* $[a]^r_r$ *and* $[b]^m_m$, *if their elements are constants, have the same non-zero latent roots.*

Replacing (A) by (C), we have the identity in λ,

$$[b]^m_m-\lambda\,[1]^m_m=[h]^m_m\left\{\begin{bmatrix}a, & 0\\ 0, & 0\end{bmatrix}^{r,\,m-r}_{r,\,m-r}-\lambda\begin{bmatrix}1, & 0\\ 0, & 1\end{bmatrix}^{r,\,m-r}_{r,\,m-r}\right\}\underline{\overline{H}}^m_m; \qquad (6')$$

and when we equate the determinants of both sides we obtain

$$\det\{[b]^m_m-\lambda\,[1]^m_m\}=(-\lambda)^{m-r}\det\{[a]^r_r-\lambda\,[1]^r_r\}.$$

Thus $[b]^m_m$ has certainly $m-r$ zero latent roots, and its remaining latent roots (each of which may or may not be zero) are the r latent roots of $[a]^r_r$.

Moreover the potent divisors of the characteristic matrix $[b]^m_m-\lambda\,[1]^m_m$ *are the potent divisors of the characteristic matrix* $[a]^r_r-\lambda\,[1]^r_r$ *together with* $m-r$ *additional potent divisors all equal to* λ.

For (6′) is an equipotent transformation, and the middle factor matrix on the right is a compartite matrix whose parts are $[a]_r^r - \lambda[1]_r^r$ and $-\lambda[1]_{m-r}^{m-r}$. Therefore the potent divisors of $[b]_m^m - \lambda[1]_m^m$ are the potent divisors of $[a]_r^r - \lambda[1]_r^r$ and $\lambda[1]_{m-r}^{m-r}$.

Ex. v. *Let ϕ and ψ be two square matrices of the same order m, and let $f(\phi)$ and $f(\psi)$ be corresponding rational integral functions of ϕ and ψ. Then if an equimutant transformation converts ϕ into ψ, the same equimutant transformation converts $f(\phi)$ into $f(\psi)$.*

Let $\phi = [a]_m^m,\quad \psi = [b]_m^m,\quad h = [h]_m^m,\quad H = \overline{\underline{H}}_m^m,\quad I = [1]_m^m,$

and let $h\phi H = \psi,$ (where $hH = Hh = I$), ...(7)

be an equimutant transformation converting ϕ into ψ. Then

$$\psi^0 = I = hH = hIH = h\phi^0 H,\quad \psi^1 = h\phi^1 H,$$

$$\psi^2 = h\phi H h\phi H = h\phi^2 H,\quad \psi^3 = h\phi^3 H,\quad \psi^4 = h\phi^4 H,\ \ldots.$$

If we multiply these equations by scalar constants and add, we see that

$$f(\psi) = hf(\phi)H,\quad \text{i.e.}\quad hf(\phi)H = f(\psi). \quad \text{..........................(7′)}$$

Since h is an undegenerate square matrix, we can put $H = h^{-1}$ and replace (7′) by

$$f(h\phi h^{-1}) = hf(\phi)h^{-1}. \quad \text{..................................(7″)}$$

Ex. vi. *Let a given equimutant transformation convert two square matrices ϕ and ψ of order m into two other square matrices Φ and Ψ of the same order m. Then if ψ is a rational integral function of ϕ, the matrix Ψ must be the same rational integral function of Φ.*

Let h be an undegenerate square matrix of order m with constant elements, and let

$$h\phi h^{-1} = \Phi,\quad h\psi h^{-1} = \Psi.$$

Then if $\psi = f(\phi)$, it follows from Ex. v that $h\psi h^{-1} = f(\Phi)$, i.e. $\Psi = f(\Phi)$. We conclude that:

The matrix Ψ is a rational integral function of Φ when and only when the matrix ψ is a rational integral function of ϕ.

Ex. vii. *If $[a]_m^m$ is any square matrix of order m with constant elements, it is possible to determine equimutant transformations of the form*

$$[h]_m^m [a]_m^m \overline{\underline{H}}_m^m = \begin{bmatrix} a, & u \\ 0, & b \end{bmatrix}_{1,m-1}^{1,m-1}. \quad \text{..............................(8)}$$

Then a is one of the latent roots of $[a]_m^m$, and the latent roots of $[b]_{m-1}^{m-1}$ are the remaining $m-1$ latent roots of $[a]_m^m$.

Let ρ be any latent root of $[a]_m^m$; let $\overline{\underline{x}}_m$ be any non-zero solution of the equation

$$[a]_m^m \overline{\underline{x}}_m = \rho\, \overline{\underline{x}}_m;$$

let $[k]_{m-1}^m$ be any undegenerate matrix of rank $m-1$ satisfying the equation

$$[k]_{m-1}^m \overline{\underline{x}}_m = 0;$$

let

$$[h]_m^m = \begin{bmatrix} h \\ k \end{bmatrix}_{1,m-1}^m$$

be an undegenerate square matrix formed by adding an initial horizontal row to $[k]_{m-1}^{m}$; let $\overline{\underline{H}}_{m}^{m}$ be the inverse of $[h]_{m}^{m}$; and let

$$[b]_{m}^{m} = [h]_{m}^{m} [a]_{m}^{m} \overline{\underline{H}}_{m}^{m}. \quad \text{.....................................(9)}$$

Then because $[h]_{m}^{m}$ is undegenerate, we have

$$[h]_{m}^{m} \overline{\underline{x}}_{m} = \begin{bmatrix} \kappa \\ 0 \end{bmatrix}_{1, m-1}^{1}, \quad \text{where } \kappa \neq 0;$$

and because by Ex. iii

$$[b]_{m}^{m} \; [h]_{m}^{m} \overline{\underline{x}}_{m} = \rho \, [h]_{m}^{m} \overline{\underline{x}}_{m},$$

therefore

$$[b]_{m}^{m} \begin{bmatrix} \kappa \\ 0 \end{bmatrix}_{1, m-1}^{1} = \rho \begin{bmatrix} \kappa \\ 0 \end{bmatrix}_{1, m-1}^{1}$$

The last equation shows that $b_{11} = \rho$, $b_{21} = b_{31} = \ldots = b_{m1} = 0$; therefore after a change in the notation the equation (9) becomes (8), where $a = \rho$.

The latent roots of $[a]_{m}^{m}$ are the latent roots of the matrix on the right in (6), i.e. they are a and the latent roots of $[b]_{m-1}^{m-1}$.

Ex. viii. *If* $[a]_{m}^{m}$ *is any square matrix of order m with constant elements, it is possible to determine equimutant transformations of the form*

$$[h]_{m}^{m} [a]_{m}^{m} \overline{\underline{H}}_{m}^{m} = \begin{bmatrix} b_{11} & b_{12} & b_{13} & \ldots & b_{1m} \\ 0 & b_{22} & b_{23} & \ldots & b_{2m} \\ 0 & 0 & b_{33} & \ldots & b_{3m} \\ \ldots & \ldots & \ldots & \ldots & \ldots \\ 0 & 0 & 0 & \ldots & b_{mm} \end{bmatrix}. \quad \text{.........................(10)}$$

Then $b_{11}, b_{22}, \ldots b_{mm}$ *are the m latent roots of* $[a]_{m}^{m}$.

After obtaining the transformation (8) of Ex. vii, we can treat $[b]_{m-1}^{m-1}$ in the same way, and so obtain an equimutant transformation of the form

$$[k]_{m-1}^{m-1} [b]_{m-1}^{m-1} \overline{\underline{K}}_{m-1}^{m-1} = \begin{bmatrix} \beta, & v \\ 0, & c \end{bmatrix}_{1, m-2}^{1, m-2},$$

from which we can deduce an equimutant transformation of the form

$$\begin{bmatrix} 1, & 0 \\ 0, & k \end{bmatrix}_{1, m-1}^{1, m-1} \begin{bmatrix} a, & u \\ 0, & b \end{bmatrix}_{1, m-1}^{1, m-1} \overline{\underline{\begin{matrix} 1, & 0 \\ 0, & K \end{matrix}}}_{1, m-1}^{1, m-1} = \begin{bmatrix} a, & u', & u'' \\ 0, & \beta, & v' \\ 0, & 0, & c \end{bmatrix}_{1, 1, m-2}^{1, 1, m-2}$$

We can now treat $[c]_{m-2}^{m-2}$ in the same way; and so on. Thus by a succession of equimutant transformations we can convert $[a]_{m}^{m}$ into a square matrix having the form shown on the right in (10); and the matrix on the right has the same latent roots as $[a]_{m}^{m}$.

§ 227. First form of the necessary and sufficient conditions for the existence of an equimutant transformation between two given square matrices.

Before stating these conditions we will prove an important lemma which will be extensively used in later articles. It is a particular case of a more general theorem given in the chapter on matrices which are homogeneous and linear in two scalar variables.

Lemma. *Let* $\phi=[\phi]_m^m=x[a]_m^m+[\alpha]_m^m$, $\psi=[\psi]_m^m=x[b]_m^m+[\beta]_m^m$ *be two square matrices of the same order* m *whose elements are linear functions of the single variable* x, *the coefficients of* x *in both of them, i.e. the matrices* $[a]_m^m$ *and* $[b]_m^m$, *being undegenerate.*

Then ϕ *and* ψ *are equipotent when and only when there exists an equigradent transformation of the form*

$$[h]_m^m[\phi]_m^m[k]_m^m=[\psi]_m^m, \qquad \text{(A)}$$

where $[h]_m^m$ *and* $[k]_m^m$ *are undegenerate square matrices with constant elements.*

For the sake of brevity we will write $I=[1]_m^m$ and

$$p=[p]_m^m,\quad q=[q]_m^m,\quad h=[h]_m^m,\quad k=[k]_m^m,$$

$$u=[u]_m^m,\quad v=[v]_m^m,\quad u'=[u']_m^m,\quad v'=[v']_m^m,$$

each of these letters denoting a square matrix of order m.

First suppose that ϕ *and* ψ *are equipotent.*

Then by § 220 there exists an equipotent transformation of the form

$$p\phi q=\psi, \qquad \text{(B)}$$

where p and q are undegenerate square matrices of order m whose elements are rational integral functions of x, and whose determinants are constants independent of x. Let P and Q be the inverses of p and q, these being matrices having the same properties as p and q and satisfying the equations

$$pP=Pp=I,\quad qQ=Qq=I. \qquad \text{(1)}$$

By Ex. v of § 201 we can write

$$p=\psi u+h,\quad q=v\psi+k, \qquad \text{(2)}$$

$$P=\phi u'+H,\quad Q=v'\phi+K, \qquad \text{(3)}$$

where u, v, u', v' are square matrices of order m whose elements are rational integral functions of x; and where h, k, H, K are square matrices of order m whose elements are constants.

Now from (B) and (1) we deduce the equation

$$P\psi=\phi q$$

which is an identity in x; and when we substitute in this identity the values of P and q given by (2) and (3), we obtain

$$\phi(u'-v)\psi=\phi k-H\psi.$$

Here the degree in x of the matrix on the right cannot exceed 1, and therefore the degree in x of $\phi(u'-v)\psi$ cannot exceed 1. But if $u'-v$ does not vanish identically, we see by successive applications of Ex. iii of § 201 that the degree in x of $\phi(u'-v)$ cannot be less than 1, and that the degree in x of $\phi(u'-v)\psi$ cannot be less than 2. We conclude that $u'-v$ must vanish identically, and that we must have

$$u'=v, \quad H\psi=\phi k.$$

Similarly from the identity $\psi Q=p\phi$ we deduce that

$$v'=u, \quad \psi K=h\phi.$$

Hence we can replace (2) and (3) by the equations

$$p=\psi u+h, \quad q=v\psi+k, \quad P=\phi v+H, \quad Q=u\phi+K, \quad \text{.................(4)}$$

in which

$$H\psi=\phi k, \quad \psi K=h\phi. \quad \text{.......................................(5)}$$

Again from (4) and (5) we see that

$$Pp=I=(\phi v+H)p=\phi vp+H(\psi u+h)=\phi(vp+ku)+Hh,$$

or

$$I-Hh=\phi(vp+ku).$$

Since $I-Hh$ is independent of x, and ϕ has degree 1 in x, it follows from Ex. iv of § 201 that $I-Hh=0$, i.e.

$$hH=Hh=I, \quad vp+ku=0; \quad \text{..................................(6)}$$

and from a consideration of the product qQ we can show in a similar way that

$$Kk=kK=I, \quad qu+vh=0. \quad \text{................................(7)}$$

Thus h, H and k, K are two pairs of mutually inverse *undegenerate* square matrices.

By prefixing the *undegenerate* matrix h in the first of the equations (5), or by postfixing the *undegenerate* matrix k in the second of them, we now obtain the equation

$$h\phi k=\psi \quad \text{..(B')}$$

which is an equigradent transformation of the form (A).

Next suppose that there exists an equigradent transformation of the form (A).

Then by Theorem II of § 214 the matrices ϕ and ψ are necessarily equipotent; and this completes the proof of the lemma.

NOTE. *If ϕ and ψ both lie in a restricted domain of rationality Ω and are equipotent, then the matrices $[h]_m^m$ and $[k]_m^m$ in* (A) *can be so chosen as to lie in Ω.*

For we can then choose p and q in (B) and u, v, u', v', h, k, H, K in (2) and (3) so that they all lie in Ω.

Ex. i. *The square matrices u and v are equipotent.*

For from (6) and (7) we obtain the equipotent transformations

$$kuP=-v, \quad quH=-v.$$

Ex. ii. When ϕ and ψ have any degrees r and s in x, and the coefficient of the highest power of x in each of them is undegenerate, we still have the equations (B), (4) and (5) holding good whenever ϕ and ψ are equipotent, the degrees of h and k in x being now less than s, and the degrees of H and K in x being now less than r.

We will now use the foregoing lemma in proving the two theorems which follow.

Theorem I. *Let* $\phi = [a]_m^m$ *and* $\psi = [b]_m^m$ *be two square matrices of the same order m whose elements are constants lying in any domain of rationality* Ω. *Then there exists an equimutant transformation of the form*

$$[h]_m^m [a]_m^m \overline{\underline{H}}_m^m = [b]_m^m, \quad \left(\textit{where } [h]_m^m \overline{\underline{H}}_m^m = \overline{\underline{H}}_m^m [h]_m^m = [1]_m^m\right), \quad \ldots\text{(C)}$$

when and only when the characteristic matrices

$$\phi(\lambda) = [a]_m^m - \lambda [1]_m^m \quad \textit{and} \quad \psi(\lambda) = [b]_m^m - \lambda [1]_m^m$$

of ϕ *and* ψ *have the same potent divisors, i.e. when and only when* $\phi(\lambda)$ *and* $\psi(\lambda)$ *are equipotent; and when this condition is satisfied, there exists an equimutant transformation in* Ω *of the form* (C).

COROLLARY. *In the particular case when* ϕ *and* ψ *in Theorem I are square matrices whose characteristic matrices have only linear potent divisors, there exists an equimutant transformation of the form* (C) *when and only when* ϕ *and* ψ *have the same latent roots, i.e. when and only when*

$$\det\{[a]_m^m - \lambda [1]_m^m\} = \det\{[b]_m^m - \lambda [1]_m^m\}.$$

Since $\phi(\lambda)$ and $\psi(\lambda)$ are undegenerate square matrices having the same rank m, they are equipotent when and only when they have the same potent divisors.

First suppose that there exists an equimutant transformation of the form (C). Then as shown in Ex. iii of § 226 the characteristic matrices $\phi(\lambda)$ and $\psi(\lambda)$ are equipotent, and have the same potent divisors.

Next suppose that $\phi(\lambda)$ and $\psi(\lambda)$ are equipotent. Then by the lemma just proved and the Note which follows it there exist undegenerate square matrices $[h]_m^m$ and $[k]_m^m$ in Ω with constant elements such that

$$[h]_m^m \{[a]_m^m - \lambda [1]_m^m\} [k]_m^m = [b]_m^m - \lambda [1]_m^m;$$

and because this equation is an identity in λ, we have

$$[h]_m^m [a]_m^m [k]_m^m = [b]_m^m, \quad [h]_m^m [k]_m^m = [1]_m^m.$$

Thus $[h]_m^m$ and $[k]_m^m$ are two mutually inverse undegenerate square matrices lying in Ω; and when we write $[k]_m^m = \overline{\underline{H}}_m^m$, we obtain the equation (C), which is an equimutant transformation in Ω.

To deduce the Corollary, we observe that when the potent divisors of each of the characteristic matrices $\phi(\lambda)$ and $\psi(\lambda)$ are all linear (but not necessarily all different), they are $\lambda - \alpha_1, \lambda - \alpha_2, \ldots \lambda - \alpha_m$ and $\lambda - \beta_1, \lambda - \beta_2, \ldots$

$\lambda-\beta_m$ respectively, where $\alpha_1, \alpha_2, \ldots \alpha_m$ are the m latent roots of ϕ and $\beta_1, \beta_2, \ldots \beta_m$ are the m latent roots of ψ. Consequently $\phi(\lambda)$ and $\psi(\lambda)$ have the same potent divisors when and only when ϕ and ψ have the same latent roots.

Theorem II. *Let $\phi=[a]_r^r$ and $\psi=[b]_m^m$ be two square matrices of orders r and m whose elements are constants lying in any domain of rationality Ω, r being not greater than m. Also let*

$$\phi(\lambda)=[a]_r^r-\lambda[1]_r^r \quad and \quad \psi(\lambda)=[b]_m^m-\lambda[1]_m^m$$

be the characteristic matrices of ϕ and ψ.

Then there exists an equimutant transformation of the form

$$[h]_m^r[a]_r^r\overline{\underline{H}}_r^m=[b]_m^m, \quad \left(where\ \overline{\underline{H}}_r^m[h]_m^r=[1]_r^r\right), \quad \ldots\ldots\ldots(D)$$

when and only when the potent divisors of $\psi(\lambda)$ are the potent divisors of $\phi(\lambda)$ together with $m-r$ additional potent divisors all equal to λ; and when this condition is satisfied, there exists an equimutant transformation in Ω of the form (D).

COROLLARY. *In the particular case when ϕ and ψ in Theorem II are square matrices whose characteristic matrices have only linear potent divisors, there exists an equimutant transformation of the form (D) when and only when the latent roots of ψ are those of ϕ with the addition of $m-r$ zero latent roots.*

Since $\phi(\lambda)$ and $\psi(\lambda)$ are undegenerate square matrices having ranks r and m respectively, it is only possible for them to be equipotent in the special case when $m=r$.

First suppose that there exists an equimutant transformation of the form (D). Then as shown in Ex. iv of § 226 the potent divisors of $\psi(\lambda)$ are those mentioned in the enunciation of the theorem.

Next let the potent divisors of $\psi(\lambda)$ be those mentioned in the enunciation, and let

$$\Phi=\begin{bmatrix}a, & 0\\ 0, & 0\end{bmatrix}_{r,\,m-r}^{r,\,m-r}, \quad \Phi(\lambda)=\begin{bmatrix}a, & 0\\ 0, & 0\end{bmatrix}_{r,\,m-r}^{r,\,m-r}-\lambda\begin{bmatrix}1, & 0\\ 0, & 1\end{bmatrix}_{r,\,m-r}^{r,\,m-r},$$

so that Φ is a square matrix of order m lying in Ω, and $\Phi(\lambda)$ is its characteristic matrix. Then $\psi(\lambda)$ has the same rank and the same potent divisors as the compartite matrix $\Phi(\lambda)$, and is equipotent with $\Phi(\lambda)$. Therefore by Theorem I there exists an equimutant transformation in Ω of the form

$$[h]_m^m\begin{bmatrix}a, & 0\\ 0, & 0\end{bmatrix}_{r,\,m-r}^{r,\,m-r}\overline{\underline{H}}_m^m=[b]_m^m, \quad \left(\text{where } \overline{\underline{H}}_m^m[h]_m^m=[1]_m^m\right), \ldots(D')$$

and from (D′) we deduce the equation (D), which is an equimutant transformation in Ω.

To deduce the Corollary let $\alpha_1, \alpha_2, \dots \alpha_r$ and $\beta_1, \beta_2, \dots \beta_m$ be respectively the r latent roots of ϕ and the m latent roots of ψ. Then in the particular case when the potent divisors of each of the matrices $\phi(\lambda)$ and $\psi(\lambda)$ are all linear, those potent divisors are respectively $\lambda - \alpha_1, \lambda - \alpha_2, \dots \lambda - \alpha_r$ and $\lambda - \beta_1, \lambda - \beta_2, \dots \lambda - \beta_m$. Hence by the theorem there exists an equimutant transformation of the form (D) when and only when r of the latent roots $\beta_1, \beta_2, \dots \beta_m$ are the same as $\alpha_1, \alpha_2, \dots \alpha_r$, and the remaining $m - r$ of the β's are 0's.

A way of proving Theorems I and II without using the lemma of this article is indicated in Notes 2 and 3 of § 230.

Ex. iii. There exists an equimutant transformation of the form (D′), which may be a transformation in Ω, when and only when the characteristic matrices of Φ and ψ are equipotent. In the particular case of the Corollary to Theorem II this condition is satisfied when and only when Φ and ψ have the same latent roots.

Ex. iv. *If* $[a]^m_m$ *and* $\overline{\underline{a}}^{m}_{m}$ *are any two mutually conjugate square matrices with constant elements, each of them can be converted into the other by an equimutant transformation.*

For their characteristic matrices clearly have the same maximum factors of all orders, and are therefore equipotent.

Ex. v. *If* $\phi = [a]^m_m$ *is a square matrix whose characteristic matrix* $\phi(\lambda)$ *has only linear potent divisors, and if* $c_1, c_2, \dots c_m$ *are the m latent roots of* ϕ, *there always exists an equimutant transformation of the form*

$$[h]^m_m [a]^m_m \overline{\underline{H}}^{m}_{m} = {}^1[c]_m, \quad \left(\textit{where } [h]^m_m \overline{\underline{H}}^{m}_{m} = \overline{\underline{H}}^{m}_{m} [h]^m_m = [1]^m_m\right). \quad \dots\dots\dots\text{(E)}$$

For the characteristic matrix ${}^1[c]_m - \lambda [1]^m_m$ of the quasi-scalar matrix ${}^1[c]_m$ has the same potent divisors $\lambda - c_1, \lambda - c_2, \dots \lambda - c_m$ as $\phi(\lambda)$.

Ex. vi. *If* $\phi = [a]^m_m$ *is a square matrix of rank r whose characteristic matrix* $\phi(\lambda)$ *has only linear potent divisors, then it has exactly r non-zero latent roots* $c_1, c_2, \dots c_r$, *and there exists an equimutant transformation of the form*

$$[a]^m_m = \overline{\underline{H}}^{r}_{m}\, {}^1[c]_r [h]^m_r, \quad \left(\textit{where } [h]^m_r \overline{\underline{H}}^{r}_{m} = [1]^r_r\right). \dots\dots\dots\dots\dots\dots\text{(E′)}$$

For ϕ must have $m - r$ zero latent roots, and if its remaining latent roots are $c_1, c_2, \dots c_r$, there exists an equimutant transformation of the form (E) in which the last $m - r$ diagonal elements of ${}^1[c]_m$ are 0's. Since ${}^1[c]_m$ must have rank r, therefore $c_1, c_2, \dots c_r$ must all be different from 0; and from (E) we deduce (E′).

§ 228. Reduction of a square matrix to a canonical square matrix of the same order by equimutant transformations.

Since the determinant of the characteristic matrix $\phi(\lambda) = [a]^m_m - \lambda [1]^m_m$ of a square matrix $\phi = [a]^m_m$ has degree m in λ, the product of all the potent divisors of $\phi(\lambda)$, which differs only by a non-zero constant factor from det $\phi(\lambda)$, must have degree m in λ. The next theorems show that when the elements of ϕ are constants which can be chosen arbitrarily, the linear divisors and corresponding potent divisors of $\phi(\lambda)$ can have any assigned values consistent with this necessary condition.

Theorem I a. *We can construct a square matrix ϕ of order m whose characteristic matrix $\phi(\lambda)$ has an arbitrarily assigned complete set of potent divisors*

$$(\lambda - c_1)^{e_1},\ (\lambda - c_2)^{e_2},\ \dots\ (\lambda - c_s)^{e_s}, \quad \dots\dots\dots\dots\dots(1)$$

provided only that $e_1, e_2, \dots e_s$ are non-zero positive integers such that

$$e_1 + e_2 + \dots + e_s = m. \quad \dots\dots\dots\dots\dots(2)$$

In fact these conditions are satisfied by any compartite matrix $\Pi = [\Pi]_m^m$ of standard form which has exactly s parts $\omega_1, \omega_2, \dots \omega_s$ obtained from the square matrix of order e

$$\omega = [\omega]_e^e = \begin{bmatrix} c & 1 & 0 & 0 & \dots & 0 & 0 \\ 0 & c & 1 & 0 & \dots & 0 & 0 \\ 0 & 0 & c & 1 & \dots & 0 & 0 \\ \dots & & & & & & \\ 0 & 0 & 0 & 0 & \dots & c & 1 \\ 0 & 0 & 0 & 0 & \dots & 0 & c \end{bmatrix} \quad \dots\dots\dots\dots\dots(A)$$

by giving to (e, c) the values $(e_1, c_1), (e_2, c_2), \dots (e_s, c_s)$ respectively.

Theorem I b. *The characteristic matrix $[a]_m^m - \lambda[1]_m^m$ of a square matrix $[a]_m^m$ has (1) as its complete system of potent divisors when and only when there exist equimutant transformations of the form*

$$[a]_m^m = [h]_m^m [\Pi]_m^m \overline{\underline{H}}_m^m, \quad \overline{\underline{H}}_m^m [a]_m^m [h]_m^m = [\Pi]_m^m, \quad \dots\dots\dots(B)$$

where $[\Pi]_m^m$ is any one of the compartite matrices constructed as in Theorem I a, and where $[h]_m^m$ and $\overline{\underline{H}}_m^m$ are mutually inverse undegenerate square matrices of order m with constant elements.

NOTE 1. In these theorems $c_1, c_2, \dots c_s$ are any scalar constants and need not be all different. If there are three potent divisors $(\lambda - c)^\alpha$, $(\lambda - c)^\beta$, $(\lambda - c)^\gamma$ corresponding to a given linear divisor $\lambda - c$, the integers α, β, γ being not necessarily all different, then these three potent divisors are three terms of the series (1), and $[\Pi]_m^m$ has three parts corresponding to them.

NOTE 2. If there is a *linear* potent divisor $\lambda - c$ or $(\lambda - c)^1$ in the series (1), the corresponding part of Π in Theorem II *a* is the matrix with one element

$$\omega = [\omega]_1^1 = [c].$$

To prove Theorem I*a* let Π be any one of the compartite matrices constructed as in the enunciation. Then the characteristic matrix

$$\Pi(\lambda) = [\Pi]_m^m - \lambda [1]_m^m$$

of Π is a compartite matrix in standard form whose parts are the characteristic matrices $\omega_1(\lambda), \omega_2(\lambda), \dots \omega_s(\lambda)$ of the parts $\omega_1, \omega_2, \dots \omega_s$ of Π, and as in Ex. xiv of § 222 the potent divisors of $\Pi(\lambda)$ are the potent divisors of $\omega_1(\lambda)$, $\omega_2(\lambda), \dots \omega_s(\lambda)$. Now the characteristic matrix of the typical part $\omega = [\omega]_e^e$ of Π is the matrix

$$\omega(\lambda) = [\omega]_e^e - \lambda [1]_e^e = \begin{bmatrix} c-\lambda, & 1, & 0, & 0, \dots & 0, & 0 \\ 0, & c-\lambda, & 1, & 0, \dots & 0, & 0 \\ 0, & 0, & c-\lambda, & 1, \dots & 0, & 0 \\ \dots & \dots & \dots & \dots & \dots & \dots \\ 0, & 0, & 0, & 0, \dots & c-\lambda, & 1 \\ 0, & 0, & 0, & 0, \dots & 0, & c-\lambda \end{bmatrix},$$

which by Ex. viii of § 205 has only the one potent divisor $(\lambda - c)^e$. Hence by giving to (e, c) the values $(e_1, c_1), (e_2, c_2), \dots (e_s, c_s)$ in turn we see that $\Pi(\lambda)$ has (1) as its complete system of potent divisors, i.e. Π is such a matrix as it was required to construct.

Theorem I*b* now follows immediately from Theorem I of § 227; for it simply states the necessary and sufficient conditions that $\phi(\lambda)$ and $\Pi(\lambda)$ shall be equipotent. When $[h]_m^m$ is arbitrary and $[\Pi]_m^m$ is fixed, the first of the equations (B) is a formula giving every square matrix $\phi = [a]_m^m$ whose characteristic matrix $\phi(\lambda)$ has (1) as its complete system of potent divisors. When $[\Pi]_m^m$ is given and $\phi = [a]_m^m$ is a given square matrix whose characteristic matrix $\phi(\lambda)$ has (1) as its complete system of potent divisors, the second of the equations (B) is an equimutant transformation reducing ϕ to the form $[\Pi]_m^m$.

A compartite matrix in standard form constructed in the manner described in Theorem I*a*, ($c_1, c_2, \dots c_s$ being any scalar quantities, and $e_1, e_2, \dots e_s$ being any non-zero positive integers), will be called a *canonical square matrix* when we are concerned with equimutant transformations; and if $\phi = [a]_m^m$ is any square matrix of order m with constant elements, a canonical square matrix of order m into which ϕ can be converted (or from which ϕ can be derived) by equimutant transformations will be called the *canonical form* or the *canonical reduced form* of ϕ for equimutant transformations.

When the potent divisors (1) are given, the order in which they are arranged being immaterial, the canonical square matrices Π which we can construct in Theorem I *a* differ from one another only in the orders of arrangement of their parts, the parts themselves being completely and uniquely determinate; and any alteration of the potent divisors (1) leads to a corresponding alteration of the parts of the matrices Π. Accordingly if we consider two canonical square matrices whose parts are the same but are differently arranged to be not distinct from one another, then Theorems I *a* and I *b* lead to the following two theorems:

Theorem II a. *There is one and only one distinct canonical square matrix whose characteristic matrix has a given complete set of potent divisors.*

Theorem II b. *If* $\phi = [a]_m^m$ *is a given square matrix of order m with constant elements, there is one and only one distinct canonical square matrix* $\Pi = [\Pi]_m^m$ *of order m into which* ϕ *can be converted (or from which* ϕ *can be derived) by equimutant transformations,* Π *being the canonical square matrix whose characteristic matrix has the same potent divisors as the characteristic matrix of* ϕ. *Further* Π *is the same for all square matrices of order m with constant elements whose characteristic matrices have the same potent divisors (or are equipotent with) the characteristic matrix of* ϕ.

More direct methods of reducing a given square matrix to its canonical form will be described in Chapters XXVII and XXXVII, (*see* Index). The canonical reduced form itself can be found by the use of Theorem IV *b* of § 233, which enables us to find all the potent divisors of the characteristic matrix.

We shall sometimes speak of *equi-canonical* square matrices, thereby meaning square matrices (necessarily of the same order) which have the same canonical reduced form. Thus two square matrices are equi-canonical when and only when their characteristic matrices are equipotent with one another or have the same potent divisors, or again they are equi-canonical when and only when they have equal orders and are convertible into one another by equimutant transformations. Two square matrices A and B (not necessarily of equal orders) will be said to be *equimutant* with one another under any one of the following equivalent sets of circumstances:

(1) when one of them (the one of smaller order) can be converted into the other by an equimutant transformation;

(2) when they have the same successive transmutes, or have any one first transmute in common (see Theorem II of § 234);

(3) when the potent divisors of their characteristic matrices $A(\lambda)$ and $B(\lambda)$ differ only by potent divisors which are equal to λ (see Theorem II of § 227);

(4) when their canonical reduced forms have the same non-zero parts, and differ (if at all) only by zero parts.

A square matrix with constant elements whose characteristic matrix has only one potent divisor (which is a power of a linear divisor) will be called a *unipotent square matrix* with respect to equimutant transformations. A (unipotent) square matrix constructed in the same manner as each of the simple parts of Π in Theorem I a will be called a *simple canonical square matrix*; the formula (A) of Theorem Ia, in which c can be any scalar constant, gives all simple canonical square matrices of order e. The canonical reduced form of any unipotent square matrix is a simple canonical square matrix of the same order.

A square matrix with constant elements which has only one distinct latent root, i.e. one whose characteristic matrix has only one linear divisor, will be called a *unilatent square matrix*. The matrix ϕ of Theorem I a is a unilatent square matrix when $c_1 = c_2 = \ldots = c_s$.

NOTE 3. *Domain of rationality of the transformations of Theorem I b.*

When a given square matrix $\phi=[a]_m^m$ is reduced to its canonical form $\Pi=[\Pi]_m^m$ as in Theorem I b, it follows from Theorem I of § 227 that the equimutant transformations (B) may lie in any domain of rationality which contains both ϕ and Π, i.e. which contains ϕ and all the latent roots $c_1, c_2, \ldots c_s$ of ϕ. Hence if Ω is any domain in which ϕ lies, and if Ω' is the domain formed from Ω by the adjunction of the latent roots of ϕ, then ϕ can be reduced to its canonical form, i.e. to a canonical square matrix of order m, by an equimutant transformation in Ω'. The domain Ω' is either Ω or a domain immediately over Ω.

NOTE 4. *Square matrices whose canonical reduced forms are quasi-scalar matrices.*

Every quasi-scalar matrix is a canonical square matrix whose characteristic matrix has only linear potent divisors; and conversely every canonical square matrix whose characteristic matrix has only linear potent divisors is a quasi-scalar matrix. Accordingly from Theorem II b we see that:

A given square matrix $\phi=[a]_m^m$ whose elements are constants can be reduced to a quasi-scalar matrix ${}^1[c]_m$ of the same order by an equimutant transformation when and only when the potent divisors of its characteristic matrix $\phi(\lambda)=[a]_m^m-\lambda[1]_m^m$ are all linear, i.e. when and only when the potent factor of $\phi(\lambda)$ of order m is a product of unrepeated linear factors.

In fact this reduction is possible when and only when the complete system of potent divisors of $\phi(\lambda)$ is $\lambda-c_1, \lambda-c_2, \ldots \lambda-c_m$; and then $c_1, c_2, \ldots c_m$ (which are not necessarily all unequal) are the m latent roots of ϕ. We see also that:

A given square matrix of order m whose elements are constants can be derived from a quasi-scalar matrix (of the same or lower order) by an equimutant transformation when and only when the potent divisors of its characteristic matrix are all linear.

For if $r \not> m$, a square matrix $\phi=[a]_m^m$ can be derived from a quasi-scalar matrix ${}^1[c]_r$ of order r by an equimutant transformation when and only when it can by equimutant transformations be derived from and converted into the quasi-scalar matrix ${}^1[c]_m$ whose last $m-r$ diagonal elements are 0's.

NOTE 5. *The characteristic numbers and the characteristic symbol of a square matrix with constant elements.*

Let $\phi=[a]_m^m$ be any square matrix whose elements are constants; let $\phi(\lambda)=[a]_m^m-\lambda[1]_m^m$ be the characteristic matrix of ϕ; and let $\lambda-c_1, \lambda-c_2, \ldots \lambda-c_i, \ldots \lambda-c_n$ be all the *distinct* linear divisors of $\phi(\lambda)$, these being n in number. Further let the indices of those powers of $\lambda-c_i$ which are potent divisors of $\phi(\lambda)$ be $e_{i1}, e_{i2}, e_{i3}, \ldots$, this notation to hold good for all the values $1, 2, \ldots n$ of i. Then the integers composing the n groups in the symbol

$$\{(e_{11}, e_{12}, e_{13}, \ldots), \quad (e_{21}, e_{22}, e_{23}, \ldots), \ldots (e_{n1}, e_{n2}, e_{n3}, \ldots)\} \quad \ldots\ldots\ldots\ldots\text{(C)}$$

will be called the *characteristic numbers* of the square matrix ϕ. The sum of all of them is equal to m, the order of ϕ. When ϕ is given, the number of groups and the integers composing each group are perfectly determinate, but the order of arrangement of the groups, and the order of arrangement of the integers in each group are arbitrary. We may call the expression (C) the *characteristic symbol* of the square matrix ϕ when this arbitrariness has been removed in any specified way. Usually the integers composing each group are arranged in ascending (or descending) order of magnitude, and if $i>j$, a group containing i integers always comes before (or always comes after) one containing j integers.

Two square matrices ϕ and ψ are *equi-characteristic* or have the same characteristic numbers when the symbol (C) is the same for both. This is the case when and only when their characteristic matrices $\phi(\lambda)$ and $\psi(\lambda)$ have the same number of distinct linear divisors, and the two sets of linear divisors can be so arranged that the successive potent divisors of $\phi(\lambda)$ corresponding to its ith linear divisor have the same indices as the successive potent divisors of $\psi(\lambda)$ corresponding to its ith linear divisor. The canonical reduced forms of ϕ and ψ constructed as in Theorem I a then differ only in the particular values assumed by the unequal arbitrary constants $c_1, c_2, c_3, \ldots$ occurring in them. It will be clear that equi-characteristic matrices are not necessarily equi-canonical.

NOTE 6. *The super-parts of a canonical square matrix.*

Let $\Pi=[\Pi]_m^m$ be a canonical square matrix which has exactly n distinct or unequal latent roots $c_1, c_2, \ldots c_n$, and let (C) be its characteristic symbol as defined in Note 5. Also let

$$e_{i1}+e_{i2}+e_{i3}+\ldots=e_i, \qquad (i=1, 2, \ldots n),$$

and let $\omega_{i1}, \omega_{i2}, \omega_{i3}, \ldots$ be the parts of Π which correspond to those potent divisors

$$(\lambda-c_i)^{e_{i1}}, \quad (\lambda-c_i)^{e_{i2}}, \quad (\lambda-c_i)^{e_{i3}}, \ldots$$

of the characteristic matrix $\Pi(\lambda)$ which are powers of $\lambda-c_i$. Then when the arrangement of the parts of Π is that corresponding to the arrangement of the characteristic numbers in (C), we can regard Π as a compartite matrix in standard form which has exactly n successive parts $\varpi_1, \varpi_2, \ldots \varpi_n$, where ϖ_i for each of the values $1, 2, \ldots n$ of i is a square matrix of order e_i and is the compartite matrix in standard form whose successive parts are $\omega_{i1}, \omega_{i2}, \omega_{i3}, \ldots$. This can be shown by the symbolical notations

$$\Pi=[\Pi]_m^m=\begin{bmatrix} \varpi_1, & 0, & \ldots & 0 \\ 0, & \varpi_2, & \ldots & 0 \\ \ldots & \ldots & \ldots & \ldots \\ 0, & 0, & \ldots & \varpi_n \end{bmatrix}_{e_1, e_2, \ldots e_n}^{e_1, e_2, \ldots e_n},$$

$$\varpi_i=[\varpi_i]_{e_i}^{e_i}=\begin{bmatrix} \omega_{i1}, & 0, & 0, & \ldots \\ 0, & \omega_{i2}, & 0, & \ldots \\ 0, & 0, & \omega_{i3}, & \ldots \\ \ldots & \ldots & \ldots & \ldots \end{bmatrix}_{e_{i1}, e_{i2}, e_{i3}, \ldots}^{e_{i1}, e_{i2}, e_{i3}, \ldots} \quad \ldots\ldots\ldots\ldots\text{(D)}$$

When the characteristic symbol (C) is given, we shall usually suppose Π to have this form; and we will call $\varpi_1, \varpi_2, \ldots \varpi_n$ the *super-parts* of Π corresponding respectively to the distinct latent roots $c_1, c_2, \ldots c_n$ of Π, or to the distinct linear divisors $\lambda - c_1, \lambda - c_2, \ldots \lambda - c_n$ of $\Pi(\lambda)$. Each of the super-parts of Π is a *unilatent canonical square matrix*; for the super-part ϖ_i has only the one distinct latent root c_i, and its characteristic matrix $\varpi_i(\lambda)$ has only the one distinct linear divisor $\lambda - c_i$. When the parts of Π are arranged as in (D), we shall call Π a canonical square matrix of standard form or simply a *standard canonical square matrix.*

Ex. i. The canonical reduced form of a square matrix $\phi = [a]_9^9$ of order 9 whose characteristic matrix $\phi(\lambda) = [a]_9^9 - \lambda[1]_9^9$ has

$$\lambda^3, \quad \lambda - 2, \quad (\lambda - 3)^3, \quad (\lambda - 2)^2 \quad \ldots\ldots\ldots\ldots\ldots\ldots\ldots(3)$$

as its complete system of potent divisors can be taken to be either one of the square matrices $\Pi = [\Pi]_9^9$ given by

$$\Pi = \left[\begin{array}{ccc|c|ccc|cc}
0&1&0&0&0&0&0&0&0\\
0&0&1&0&0&0&0&0&0\\
0&0&0&0&0&0&0&0&0\\
\hline
0&0&0&2&0&0&0&0&0\\
\hline
0&0&0&0&3&1&0&0&0\\
0&0&0&0&0&3&1&0&0\\
0&0&0&0&0&0&3&0&0\\
\hline
0&0&0&0&0&0&0&2&1\\
0&0&0&0&0&0&0&0&2
\end{array}\right], \qquad
\Pi = \left[\begin{array}{ccc|ccc|c|cc}
0&1&0&0&0&0&0&0&0\\
0&0&1&0&0&0&0&0&0\\
0&0&0&0&0&0&0&0&0\\
\hline
0&0&0&3&1&0&0&0&0\\
0&0&0&0&3&1&0&0&0\\
0&0&0&0&0&3&0&0&0\\
\hline
0&0&0&0&0&0&2&0&0\\
\hline
0&0&0&0&0&0&0&2&1\\
0&0&0&0&0&0&0&0&2
\end{array}\right].$$

Here the characteristic matrix $\Pi(\lambda) = [\Pi]_9^9 - \lambda[1]_9^9$ has (3) as its complete system of potent divisors; and every square matrix of order 9 whose characteristic matrix has (3) as its complete system of potent divisors can be derived from (and converted into) Π by equimutant transformations. The characteristic symbol of every such matrix corresponding to the standard arrangement $\lambda^3, (\lambda-3)^3, \lambda - 2, (\lambda-2)^2$ of the potent divisors (3) is

$$\{(3), (3), (1, 2)\}.$$

The second of the two matrices given above is the corresponding *standard* canonical reduced form of Π. It has 4 simple parts and 3 super-parts.

Ex. ii. *Properties of a canonical square matrix defined by the potent divisors of its characteristic matrix.*

If $\Pi = [\Pi]_m^m$ is the canonical reduced form of the square matrix $\phi = [a]_m^m$, and if $\phi(\lambda) = [a]_m^m - \lambda[1]_m^m$, $\Pi(\lambda) = [\Pi]_m^m - \lambda[1]_m^m$ are the characteristic matrices of ϕ and Π, we may notice the following facts:

(1) To each potent divisor of $\phi(\lambda)$ or $\Pi(\lambda)$ which has the value λ there corresponds a part $[0]_1^1$ of Π having degeneracy 1.

(2) To each potent divisor of $\phi(\lambda)$ or $\Pi(\lambda)$ which has the value λ^p, where $p>1$, there corresponds a part

$$[\omega]_p^p=\begin{bmatrix}0, & 1\\ 0, & 0\end{bmatrix}_{p-1,\,1}^{1,\,p-1}$$

of Π having degeneracy 1.

(3) To each potent divisor $(\lambda-c)^e$ of $\phi(\lambda)$ or $\Pi(\lambda)$ which is not merely a power of λ, i.e. in which $c\neq 0$, there corresponds a part $[\omega]_e^e$ of Π which is an undegenerate square matrix.

(4) The degeneracy of Π is equal to the number of potent divisors of $\phi(\lambda)$ or $\Pi(\lambda)$ which are powers of λ, i.e. which are divisible by λ; consequently Π is undegenerate when and only when λ is not a linear divisor of $\phi(\lambda)$ or $\Pi(\lambda)$, i.e. when and only when no potent divisor of $\phi(\lambda)$ or $\Pi(\lambda)$ is a power of λ.

(5) If $\phi(\lambda)$ or $\Pi(\lambda)$ has exactly r potent divisors equal to λ, then we can write

$$\Pi=\begin{bmatrix}\pi, & 0\\ 0, & 0\end{bmatrix}_{m-r,\,r}^{m-r,\,r},$$

where $[\pi]_{m-r}^{m-r}$ is the canonical square matrix of order $m-r$ the potent divisors of whose characteristic matrix are the remaining potent divisors of $\phi(\lambda)$ or $\Pi(\lambda)$.

If we omit all reference to ϕ and $\phi(\lambda)$, these are properties of a canonical square matrix of order m which is defined by the potent divisors of its characteristic matrix.

Ex. iii. *The degeneracy of a square matrix* $\phi=[a]_m^m$ *is equal to the total number of those potent divisors of its characteristic matrix* $\phi(\lambda)=\phi-\lambda I=[a]_m^m-\lambda[1]_m^m$ *which are divisible by* λ, *i.e. which are powers of* λ.

This follows from Ex. ii, because the rank and degeneracy of ϕ are equal respectively to the rank and degeneracy of the standard reduced matrix Π.

In particular ϕ is undegenerate when and only when λ is not a linear divisor of its characteristic matrix $\phi(\lambda)$, i.e. when and only when ϕ has no zero latent root.

Hence if a *is a scalar quantity, the total number of those potent divisors of* $\phi(\lambda)$ *which are powers of* $\lambda-a$ *is equal to the degeneracy of the square matrix* $\psi=\phi-aI=[a]_m^m-a[1]_m^m$.

For it was shown in Ex. x of § 222 that $(\lambda-a)^p$ is a potent divisor of $\phi(\lambda)$ of order i when and only when λ^p is a potent divisor of order i of the characteristic matrix $\psi(\lambda)=\psi-\lambda I$ of ψ. Therefore the total number of those potent divisors of $\phi(\lambda)$ which are powers of $\lambda-a$ is equal to the total number of those potent divisors of $\psi(\lambda)$ which are powers of λ, and by the first theorem above this number is equal to the degeneracy of ψ.

In particular $\lambda-a$ is a linear divisor of $\phi(\lambda)$, i.e. a is a latent root of ϕ, when and only when the matrix $\psi=\phi-aI$ is degenerate.

Ex. iv. *Construction of all square matrices* $\phi=[a]_m^m$ *which satisfy a given rational integral equation* $f(\phi)=0$.

Let $I=[1]_m^m$, let x be a scalar variable, and let

$$f(\phi)=c(\phi-aI)^p(\phi-\beta I)^q(\phi-\gamma I)^r\ldots,\quad f(x)=c(x-a)^p(x-\beta)^q(x-\gamma)^r\ldots,$$

where c is a non-zero scalar constant, and where $a, \beta, \gamma, \ldots$ are all different.

By Theorem II *b* or Ex. iv of § 224 the matrix $\phi=[a]_m^m$ satisfies the equation $f(\phi)=0$ when and only when every potent divisor of order m of its characteristic matrix $\phi - xI$ is a factor of $f(x)$. This is the case when and only when every potent divisor of $\phi - xI$ is a power of one of the linear factors $x-a$, $x-\beta$, $x-\gamma$, ... of $f(x)$ having an index not greater than the index of that factor in $f(x)$. Consequently the matrices $\phi=[a]_m^m$ which satisfy the equation $f(\phi)=0$ are those given by equimutant transformations of the form

$$[a]_m^m=[h]_m^m\,[\Pi]_m^m\,\overline{\underline{H}}_m^m,$$

where $[\Pi]_m^m$ is a canonical square matrix constructed as in Theorem II *a* in which each part corresponds to one of the linear divisors $x-a$, $x-\beta$, $x-\gamma$, ... of $\Pi - xI$; parts corresponding to the linear divisors $x-a$, $x-\beta$, $x-\gamma$, ... have orders not greater than $p, q, r, \ldots$ respectively; and the sum of the orders of all the parts is equal to m.

Ex. v. If $I=[1]_5^5$, then matrices $\phi=[a]_5^5$ satisfying the rational integral equation

$$f(\phi)=(\phi-aI)^4\,(\phi-\beta I)^2$$

are given by the equimutant transformation

$$\phi=[a]_5^5=[h]_5^5\begin{bmatrix} a & 1 & 0 & 0 & 0\\ 0 & a & 1 & 0 & 0\\ 0 & 0 & a & 0 & 0\\ 0 & 0 & 0 & a & 0\\ 0 & 0 & 0 & 0 & \beta \end{bmatrix}\overline{\underline{H}}_5^5.$$

For the potent factors of orders 1, 2, 3, 4, 5 of the characteristic matrix $\phi - xI$ of every such matrix are respectively

$$1,\ 1,\ 1,\ x-a,\ (x-a)^3\,(x-\beta);$$

and the potent factor of order 5 is a factor of the function $f(x)=(x-a)^4\,(x-\beta)^2$.

§ 229. Some properties of a simple canonical square matrix.

In the following examples it will be understood that $I=[1]_m^m$, and that

$$\phi=[\omega]_m^m=\begin{bmatrix} c & 1 & 0 & \ldots & 0 & 0\\ 0 & c & 1 & \ldots & 0 & 0\\ \cdots & & & & & \\ 0 & 0 & 0 & \ldots & c & 1\\ 0 & 0 & 0 & \ldots & 0 & c \end{bmatrix} \qquad\qquad(1)$$

is the simple canonical square matrix of order m whose characteristic matrix $\phi(\lambda)=\phi-\lambda I$ has only the one potent divisor $(\lambda-c)^m$. Further k will denote any non-zero positive integer, and when $k \not> m$ we shall write

$$m=sk+q$$

where s and q are positive integers, and where $q \not> k-1$. The characteristic symbol of ϕ is $\{m\}$; and by § 224 the rational integral equation of lowest degree satisfied by ϕ is

$$(\phi-cI)^m=0.$$

Ex. i. *Degeneracy of* ϕ^k.

Case I. If $c \neq 0$, ϕ is undegenerate, and therefore ϕ^k is undegenerate.

In this case the single potent divisor of $\phi(\lambda)$ is not divisible by λ.

Case II. If $c=0$, we have $\phi=\begin{bmatrix}0, & 1\\0, & 0\end{bmatrix}_{m-1,1}^{1,m-1}$, and from the identity

$$\begin{bmatrix}0, & 1\\0, & 0\end{bmatrix}_{m-1,1}^{1,m-1}\begin{bmatrix}x\\a\end{bmatrix}_{1,m-1}^{m}=\begin{bmatrix}a\\0\end{bmatrix}_{m-1,1}^{m}$$

we see that

$$\phi^k=\begin{bmatrix}0, & 1\\0, & 0\end{bmatrix}_{m-k,k}^{k,m-k} \text{ or } [0]_m^m \text{ according as } k \not> m \text{ or } k \not< m.$$

Therefore ϕ^k has degeneracy k when $k \not> m$, and degeneracy m when $k \not< m$.

In this case the single potent divisor of $\phi(\lambda)$ is λ^m.

In both cases the degeneracy of ϕ^k is $\rho_1+\rho_2+\ldots+\rho_k$, where ρ_i is the number of potent divisors of $\phi(\lambda)$ which are divisible by λ^i.

For in Case I we have $\rho_1=\rho_2=\ldots=\rho_k=0$; and in Case II we have $\rho_1=\rho_2=\ldots=\rho_m=1$, and $\rho_i=0$ when $i>m$.

Ex. ii. *Degeneracy of $(\phi-aI)^k$, where a is any scalar constant.*

Case I. If $a\neq c$, $\phi-aI$ is undegenerate, and therefore $(\phi-aI)^k$ is undegenerate.

In this case the single potent divisor of $\phi(\lambda)$ is not divisible by $\lambda-a$.

Case II. If $a=c$, we have $\phi-aI=\phi-cI=\begin{bmatrix}0, & 1\\0, & 0\end{bmatrix}_{m-1,1}^{1,m-1}$; and therefore $(\phi-aI)^k$ has degeneracy k when $k \not> m$, and degeneracy m when $k \not< m$.

In both cases the degeneracy of $(\phi-aI)^k$ is $\rho_1+\rho_2+\ldots+\rho_k$, where ρ_i is the number of potent divisors of $\phi(\lambda)$ which are divisible by $(\lambda-a)^i$.

We can also obtain the last result by applying Ex. i to the matrix $\psi=\phi-aI$, which is the canonical square matrix of order m whose characteristic matrix $\psi(\lambda)=\psi-\lambda I$ has only the one potent divisor $(\lambda+a-c)^m$; for the single potent divisor $(\lambda+a-c)^m$ of $\psi(\lambda)$ becomes λ^m when and only when the single potent divisor $(\lambda-c)^m$ of $\phi(\lambda)$ becomes $(\lambda-a)^m$.

Ex. iii. *Degeneracy of any rational integral function $f(\phi)$ of ϕ.*

Let $$f(\phi)=a(\phi-a_1I)^{k_1}(\phi-a_2I)^{k_2}\ldots(\phi-a_nI)^{k_n},$$

where a is a non-zero scalar constant, and where $a_1, a_2, \ldots a_n$ are scalar constants which are all different. Since only one of the quantities $a_1, a_2, \ldots a_n$ can be equal to c, not more than one of the n factors of $f(\phi)$ can be degenerate; and it follows that:

The degeneracy of $f(\phi)$ is equal to the sum of the degeneracies of its n factors.

The degeneracy of each factor is given by Ex. ii.

Ex. iv. *Value of ϕ^k.*

Let $$(1+x)^k=1+p_1x+p_2x^2+\ldots+p_ix^i+\ldots,$$

so that $$p_i=\binom{k}{i}=\binom{k}{k-i}=p_{k-i} \text{ when } i \not> k, \text{ and } p_i=0 \text{ when } i>k.$$

Then if $$\chi=\begin{bmatrix}0, & 1\\0, & 0\end{bmatrix}_{m-1,1}^{1,m-1}, \text{ so that } \phi=cI+\chi,$$

we have $$\phi^k=c^kI+p_1c^{k-1}\chi+p_2c^{k-2}\chi^2+\ldots+p_ic^{k-i}\chi^i+\ldots+\chi^k.$$

Since, as was shown in Ex. i,

$$\chi^i = \begin{bmatrix} 0, & 1 \\ 0, & 0 \end{bmatrix}_{m-i,\, i}^{i,\, m-i} \quad \text{when } i \not> m, \text{ and } \chi^i = 0 \text{ when } i \not< m,$$

it follows that ϕ^k is the square matrix of order m formed with the first m horizontal rows and the first m vertical rows of the indefinite square matrix

$$M^k(c) = \begin{bmatrix} c^k, & p_1c^{k-1}, & p_2c^{k-2}, & p_3c^{k-3}, & \dots & 1, & 0, & 0, & 0 \dots \\ 0, & c^k, & p_1c^{k-1}, & p_2c^{k-2}, & \dots & p_1c, & 1, & 0, & 0 \dots \\ 0, & 0, & c^k, & p_1c^{k-1}, & \dots & p_2c^2, & p_1c, & 1, & 0 \dots \\ 0, & 0, & 0, & c^k, & \dots & p_3c^3, & p_2c^2, & p_1c, & 1 \dots \\ \dots & & & & & & & & \end{bmatrix}$$

which we can regard as the kth power of the indefinite square matrix

$$M(c) = \begin{bmatrix} c, & 1, & 0, & 0, & \dots \\ 0, & c, & 1, & 0, & \dots \\ 0, & 0, & c, & 1, & \dots \\ 0, & 0, & 0, & c, & \dots \\ \dots & & & & \end{bmatrix}.$$

Ex. v. *Inverse of ϕ when ϕ is undegenerate.*

When $c \neq 0$, it can be at once verified by evaluating the product $\phi\phi^{-1}$ that the inverse of ϕ is the square matrix of order m

$$\phi^{-1} = \begin{bmatrix} c^{-1}, & -c^{-2}, & c^{-3}, & \dots & (-1)^{m-1}c^{-m} \\ 0, & c^{-1}, & -c^{-2}, & \dots & (-1)^{m-2}c^{-m+1} \\ 0, & 0, & c^{-1}, & \dots & (-1)^{m-3}c^{-m+2} \\ \dots & & & & \\ 0, & 0, & 0, & \dots & c^{-1} \end{bmatrix}.$$

Ex. vi. *Value of ϕ^{-k} when ϕ is undegenerate.*

When k is a positive integer let the expansion of $(1+x)^{-k}$ in ascending powers of x be

$$(1+x)^{-k} = 1 - q_1x + q_2x^2 - q_3x^3 + \dots + (-1)^i q_i x^i + \dots,$$

so that

$$q_i = {}^{k-1}H_i = \binom{k+i-1}{i} = \binom{k+i-1}{k-1}.$$

Then when $c \neq 0$, it can be verified by evaluating the product $\phi^k\phi^{-k}$ that ϕ^{-k} is the square matrix of order m formed with the first m horizontal rows and the first m vertical rows of the indefinite square matrix

$$M^{-k}(c) = \begin{bmatrix} c^{-k}, & -q_1c^{-k-1}, & q_2c^{-k-2}, & -q_3c^{-k-3}, & \dots \\ 0, & c^{-k}, & -q_1c^{-k-1}, & q_2c^{-k-2}, & \dots \\ 0, & 0, & c^{-k}, & -q_1c^{-k-1}, & \dots \\ 0, & 0, & 0, & c^{-k}, & \dots \\ \dots & & & & \end{bmatrix},$$

which we can regard as the $-k$th power of the matrix $M(c)$ of Ex. iv.

In fact if we define ϕ^{-k} in this way, and put $\phi^k\phi^{-k} = [a]_m^m$, it will be seen that $a_{ij} = 0$ when $i > j$, and that $a_{i,\,i+s} = a_s c^{-s}$, where a_s is the coefficient of x^s in the expansion of $(1+x)^k(1+x)^{-k}$ in ascending powers of x. Accordingly we have $a_{ii} = 1$, $a_{ij} = 0$ when $i \neq j$.

Ex. vii. *Values of* $(\phi - aI)^k$, $(\phi - aI)^{-k}$.

We obtain the value of $(\phi - aI)^k$ from the value of ϕ^k in Ex. iv by substituting $c - a$ for c.

Similarly when $a \neq c$, i.e. when $\phi - aI$ is undegenerate, we obtain the value of $(\phi - aI)^{-k}$ from the value of ϕ^{-k} in Ex. vi by substituting $c - a$ for c.

Ex. viii. *Potent divisors of the characteristic matrix of* ϕ^k.

Case I. If $c \neq 0$, ϕ^k is an undegenerate square matrix of order m whose characteristic matrix $\phi^k(\lambda) = \phi^k - \lambda I$ has only the one potent divisor $(\lambda - c^k)^m$, which is of order m; for from Ex. iv we see that the minor determinant of $\phi^k(\lambda)$ of order $m - 1$ formed with its first $m - 1$ horizontal rows and its last $m - 1$ vertical rows does not vanish when $\lambda = c^k$, and is therefore not divisible by $\lambda - c^k$.

Case II. If $c = 0$, we must distinguish between two sub-cases.

Sub-case 1. When $k \nless m$, we have $\phi^k \doteq 0$; and therefore $\phi^k(\lambda)$ has m potent divisors of orders 1, 2, ... m, each of them having the value λ.

Sub-case 2. When $k \ngtr m$, we have

$$\phi^k = \begin{bmatrix} 0, & 1 \\ 0, & 0 \end{bmatrix}^{k,\, m-k}_{m-k,\, k}; \quad m = sk + q, \quad \text{where } q \ngtr k - 1.$$

Therefore by Ex. vii of § 210 the matrix $\phi^k(\lambda)$ has k potent divisors, of which $k - q$ are equal to λ^s, and q are equal to λ^{s+1}.

In all cases the degeneracy of ϕ^k is equal to the number of potent divisors of $\phi^k(\lambda)$ which are divisible by λ.

Ex. ix. *Potent divisors of the characteristic matrix of* $(\phi - aI)^k$.

Since $\psi = \phi - aI$ is a canonical square matrix of order m whose characteristic matrix $\psi(\lambda)$ has only the one potent divisor $\lambda - (c - a)$, we can deduce the potent divisors of the characteristic matrix $\psi^k(\lambda)$ of $(\phi - aI)^k$ from Ex. viii.

Case I. If $a \neq c$, then $(\phi - aI)^k$ is undegenerate, and its characteristic matrix has only the one potent divisor $\{\lambda - (c - a)^k\}^m$, which is of order m.

Case II. If $a = c$, we must distinguish between two sub-cases.

Sub-case 1. When $k \nless m$, we have $(\phi - aI)^k = 0$, and the characteristic matrix of $(\phi - aI)^k$ has m potent divisors of orders 1, 2, ... m, each of them having the value λ.

Sub-case 2. When $k \ngtr m$, we have

$$(\phi - aI)^k = \begin{bmatrix} 0, & 1 \\ 0, & 0 \end{bmatrix}^{k,\, m-k}_{m-k,\, k}; \quad m = sk + q, \quad \text{where } q \ngtr k - 1.$$

Therefore the characteristic matrix of $(\phi - aI)^k$ has k potent divisors, of which $k - q$ are equal to λ^s and q are equal to λ^{s+1}.

In all cases the degeneracy of $(\phi - aI)^k$ is equal to the number of potent divisors of its characteristic matrix which are divisible by λ.

Ex. x. *Potent divisors of the characteristic matrices of* ϕ^{-k} *and* $(\phi - aI)^{-k}$.

When ϕ is undegenerate, i.e. when $c \neq 0$, then as in Case I of Ex. viii the characteristic matrix of ϕ^{-k} has only the one potent divisor $(\lambda - c^{-k})^m$, which is of order m.

It follows that when $\phi - aI$ is undegenerate, i.e. when $a \neq c$, the potent divisor of the characteristic matrix of $(\phi - aI)^{-k}$ has only the one potent divisor $\{\lambda - (c-a)^{-k}\}^m$, which is of order m.

Ex. xi. *Canonical reduced forms of* ϕ^k, $(\phi - aI)^k$, ϕ^{-k}, $(\phi - aI)^{-k}$.

These are given by Exs. viii—x; for they are known when the potent divisors of the characteristic matrices are known.

NOTE. *Potent divisors of the characteristic matrix of any power of* Φ *when* $\Phi = [a]_m^m$ *is any given square matrix whose elements are constants.*

Let $\Pi = [\Pi]_m^m$ be the canonical reduced form of Φ; let $\Phi(\lambda)$ and $\Pi(\lambda)$ be the characteristic matrices of Φ and Π; and let the parts of Π corresponding to the various common potent divisors of $\Phi(\lambda)$ and $\Pi(\lambda)$ be the canonical square matrices $\omega_1, \omega_2, \omega_3, \ldots$. Also let

$$h\Pi h^{-1} = \Phi \quad \text{..(2)}$$

be an equimutant transformation converting Π into Φ, where $h = [h]_m^m$ is an undegenerate square matrix with constant elements. Then if k is any positive integer, we deduce from (2) the equimutant transformation

$$h\Pi^k h^{-1} = \Phi^k, \quad \text{..(3)}$$

which shows that the characteristic matrices $\Phi^k(\lambda)$ and $\Pi^k(\lambda)$ of Φ^k and Π^k have the same potent divisors; further if Φ is undegenerate, then by equating the inverses of both sides of (3) we obtain the equimutant transformation

$$h\Pi^{-k}h^{-1} = \Phi^{-k}, \quad \text{..(4)}$$

which shows that the characteristic matrices $\Phi^{-k}(\lambda)$ and $\Pi^{-k}(\lambda)$ of Φ^{-k} and Π^{-k} have the same potent divisors. If $k=0$, we have

$$\Phi^k = \Phi^{-k} = \Phi^0 = [1]_m^m, \quad \Phi^0(\lambda) = (1-\lambda) \cdot [1]_m^m,$$

and $\Phi^0(\lambda)$ has m potent divisors all equal to $\lambda - 1$. Dismissing this trivial case, we have two principal cases to consider.

CASE I. *Potent divisors of* $\Phi^k(\lambda)$ *and canonical reduced form of* Φ^k *when* $k > 0$.

Let $\Omega = [\Omega]_m^m$ be the canonical reduced form of Φ^k. Since Π^k is a compartite matrix of order m in standard form whose parts are $\omega_1^k, \omega_2^k, \omega_3^k, \ldots$ therefore (see Ex. xiv of § 222) the potent divisors of $\Pi^k(\lambda)$, and therefore also the potent divisors of $\Phi^k(\lambda)$, are the potent divisors of the characteristic matrices $\omega_1^k(\lambda), \omega_2^k(\lambda), \omega_3^k(\lambda), \ldots$ of $\omega_1^k, \omega_2^k, \omega_3^k, \ldots$, which are known by Ex. viii, and are found as follows:

First let ω *be a part of* Π *corresponding to a potent divisor* $(\lambda - c)^e$ *of* $\Phi(\lambda)$ *which is not a power of* λ, *so that* $c \neq 0$.

Then $\omega^k(\lambda)$ has only the one potent divisor $(\lambda - c^k)^e$. The canonical square matrix Ω has one corresponding part, this being the simple canonical square matrix of order e whose diagonal elements are equal to c^k.

Next let ω *be a part of* Π *corresponding to a potent divisor* λ^e *of* $\Phi(\lambda)$, *where* $e \ngtr k$.

Then $\omega^k(\lambda)$ has exactly e potent divisors all equal to λ. The canonical square matrix Ω has e corresponding parts all equal to $[0]_1^1$.

Lastly let ω be a part of Π corresponding to a potent divisor λ^e of $\Phi(\lambda)$, where $e \nless k$.

We write $e = sk + q$, where s and q are positive integers, and $q \ngtr k-1$, and then $\omega^k(\lambda)$ has exactly k potent divisors, these being

(1) $k-q$ equal to λ^s and q equal to λ^{s+1} when $q \neq 0$,

(2) k equal to λ^s when $q = 0$.

In sub-case (1) the canonical square matrix Ω has k corresponding parts, these being

$$k-q \text{ equal to } \begin{bmatrix} 0, & 1 \\ 0, & 0 \end{bmatrix}_{s-1,\,1}^{1,\,s-1} \quad \text{and} \quad q \text{ equal to } \begin{bmatrix} 0, & 1 \\ 0, & 0 \end{bmatrix}_{s,\,1}^{1,\,s}.$$

In sub-case (2) the corresponding parts of Ω are

$$k \text{ equal to } \begin{bmatrix} 0, & 1 \\ 0, & 0 \end{bmatrix}_{s-1,\,1}^{1,\,s-1}.$$

CASE II. *Potent divisors of $\Phi^{-k}(\lambda)$ and canonical reduced form of Φ^{-k} when Φ is undegenerate and $k > 0$.*

Let $\Omega = [\Omega]_m^m$ be the canonical reduced form of Φ^{-k}. This case is simpler because $\Phi(\lambda)$ has no potent divisor which is a power of λ. Since Π^{-k} is a compartite matrix of order m in standard form whose parts are $\omega_1^{-k}, \omega_2^{-k}, \omega_3^{-k}, \ldots$, the potent divisors of $\Pi^{-k}(\lambda)$, and therefore also the potent divisors of $\Phi^{-k}(\lambda)$, are the potent divisors of the characteristic matrices $\omega_1^{-k}(\lambda), \omega_2^{-k}(\lambda), \omega_3^{-k}(\lambda), \ldots$, which are known by Ex. x.

In fact if ω is a part of Π corresponding to a potent divisor $(\lambda - c)^e$ of $\Phi(\lambda)$, where c cannot be 0, then $\omega^{-k}(\lambda)$ has only the one potent divisor $(\lambda - c^{-k})^e$. The canonical square matrix Ω has one corresponding part, this being the simple canonical square matrix of order e whose diagonal elements are equal to c^{-k}.

Ex. xii. *Potent divisors of the characteristic matrix of any power of Φ when $\Phi = [a]_m^m$ is any given undegenerate square matrix.*

In the special case when Φ is an *undegenerate* square matrix, let

$$(\lambda - c_1)^{e_1}, \quad (\lambda - c_2)^{e_2}, \quad \ldots \quad (\lambda - c_s)^{e_s}$$

be the potent divisors of each of the characteristic matrices $\Phi(\lambda)$ and $\Pi(\lambda)$. Then if k is any non-zero integer (positive or negative), the potent divisors of the characteristic matrices $\Pi^k(\lambda)$ and $\Phi^k(\lambda)$ of Π^k and Φ^k are

$$(\lambda - c_1^k)^{e_1}, \quad (\lambda - c_2^k)^{e_2}, \quad \ldots \quad (\lambda - c_s^k)^{e_s}.$$

In particular if Φ^{-1} and Π^{-1} are the inverses of Φ and Π, the potent divisors of the characteristic matrices $\Pi^{-1}(\lambda)$ and $\Phi^{-1}(\lambda)$ are

$$(\lambda - c_1^{-1})^{e_1}, \quad (\lambda - c_2^{-1})^{e_2}, \quad \ldots \quad (\lambda - c_s^{-1})^{e_s};$$

and the standard reduced form of Π^{-1} and Φ^{-1} is known.

Ex. xiii. If ϕ and Φ are two mutually inverse undegenerate square matrices, the latent roots of Φ are the reciprocals of the latent roots of ϕ; also the potent divisors of the characteristic matrices $\phi(\lambda)$, $\Phi(\lambda)$ of ϕ, Φ can be associated together in pairs of the form $(\lambda - c)^e$, $\left(\lambda - \frac{1}{c}\right)^e$.

§ 230. Rank of any rational integral function of a given square matrix.

1. *Rank of a function which is expressed as a product of linear factors.*

We shall use the notations $\phi = [a]_m^m$, $I = \phi^0 = [1]_m^m$, ϕ being a given square matrix whose elements are numerical constants. The characteristic matrix of ϕ will be taken to be

$$\phi(x) = \phi - xI = \phi - x\phi^0,$$

x being a scalar variable. More generally if ψ is any square matrix with constant elements, we shall use the notations

$$\psi(\lambda) = \psi - \lambda\psi^0, \quad \psi^k(\lambda) = \psi^k - \lambda\psi^0,$$

where λ is any scalar quantity and ψ^0 is the unit matrix of the same order as ψ, so that $\psi(x)$ and $\psi^k(x)$ are the characteristic matrices of ψ and ψ^k. Further $\Pi = [\Pi]_m^m$ will denote the canonical reduced form of ϕ for equimutant transformations constructed as in Theorem I*a* of § 228; and $f(x)$ and $f(\phi)$ will denote corresponding rational integral functions of the scalar variable x and the square matrix ϕ whose coefficients are numerical constants.

Theorem I. *If $\alpha_1, \alpha_2, \ldots \alpha_r$ are distinct scalar constants, and if a rational integral function $f(\phi)$ of ϕ is expressed in the form*

$$f(\phi) = A(\phi - \alpha_1 I)^{k_1}(\phi - \alpha_2 I)^{k_2} \ldots (\phi - \alpha_r I)^{k_r}, \quad \ldots\ldots\ldots\ldots\text{(A)}$$

where A is a non-zero scalar constant, then the degeneracy of the square matrix $f(\phi)$ is equal to the sum of the degeneracies of its r matrix factors in (A).

Here $k_1, k_2, \ldots k_r$ may be any positive integers, but we will suppose that they are all different from 0. This involves no loss of generality because $(\phi - \alpha I)^0$ is the undegenerate unit matrix I, and the factors in (A) are commutative.

Let $$h\phi H = \Pi, \quad \text{where } hH = Hh = I,$$

be an equimutant transformation which converts ϕ into Π. By Ex. v of § 226 the same equimutant transformation converts $f(\phi)$ into $f(\Pi)$, and $(\phi - \alpha I)^k$ into $(\Pi - \alpha I)^k$. Consequently we have

$$f(\Pi) = A(\Pi - \alpha_1 I)^{k_1}(\Pi - \alpha_2 I)^{k_2} \ldots (\Pi - \alpha_r I)^{k_r}. \quad \ldots\ldots\ldots\text{(A}')$$

Since $f(\Pi)$ and its r successive matrix factors in (A′) have the same ranks as $f(\phi)$ and its r successive matrix factors in (A), the above theorem will be true if and only if it is true for the function $f(\Pi)$ in (A′).

Special Case. *When the characteristic matrix $\phi(x)$ has only one potent divisor $(x - \lambda)^m$.*

In this case Π is the simple canonical square matrix of order m whose characteristic matrix has the single potent divisor $(x - \lambda)^m$.

If no one of the constants $\alpha_1, \alpha_2, \dots \alpha_r$ is equal to λ, all the factors of $f(\Pi)$ in (A′) are undegenerate, and $f(\Pi)$ is undegenerate.

If α_u, one of the constants $\alpha_1, \alpha_2, \dots \alpha_r$, is equal to λ, then $(\Pi - \alpha_u I)^{k_u}$ is the only factor of $f(\Pi)$ which is degenerate, and all the other factors of $f(\Pi)$ are undegenerate square matrices. Since the factors of $f(\Pi)$ are commutative, it follows that $f(\Pi)$ has the same rank and the same degeneracy as $(\Pi - \alpha_u I)^{k_u}$.

Thus the theorem is always true in this special case.

General Case. *When the complete system of potent divisors of $\phi(x)$ is*

$$(x-\lambda_1)^{e_1},\ (x-\lambda_2)^{e_2},\ \dots\ (x-\lambda_v)^{e_v},\ \dots\ (x-\lambda_s)^{e_s}. \quad \dots\dots\dots(1)$$

Here it is to be understood that the potent divisors of $\phi(x)$ are those corresponding to its linear divisors; that $\lambda_1, \lambda_2, \dots \lambda_s$ are not necessarily all different; and that $e_1, e_2, \dots e_s$ are any s given non-zero positive integers satisfying the condition $e_1 + e_2 + \dots + e_s = m$. We shall suppose u to be any one of the integers 1, 2, ... r, and v to be any one of the integers 1, 2, ... s.

Let the parts of Π corresponding to the successive potent divisors (1) be $\omega_1, \omega_2, \dots \omega_v, \dots \omega_s$, so that ω_v is the simple canonical square matrix of order e_v whose characteristic matrix $\omega_v(x)$ has the single potent divisor $(x-\lambda_v)^{e_v}$. Then $f(\Pi)$ is the compartite matrix similar to Π whose successive parts are the matrices $f(\omega_1), f(\omega_2), \dots f(\omega_v), \dots f(\omega_s)$ similar to $\omega_1, \omega_2, \dots \omega_v, \dots \omega_s$ respectively, and we can write

$$f(\Pi) = A\,\{\Pi(\alpha_1)\}^{k_1}\{\Pi(\alpha_2)\}^{k_2}\dots\{\Pi(\alpha_u)\}^{k_u}\dots\{\Pi(\alpha_r)\}^{k_r}, \quad \dots\dots(2)$$

$$f(\omega_v) = A\,\{\omega_v(\alpha_1)\}^{k_1}\{\omega_v(\alpha_2)\}^{k_2}\dots\{\omega_v(\alpha_u)\}^{k_u}\dots\{\omega_v(\alpha_r)\}^{k_r}. \quad \dots\dots(3)$$

By § 100 the degeneracy of $f(\Pi)$ is the sum of the degeneracies of its parts such as $f(\omega_v)$; and by the special case the degeneracy of $f(\omega_v)$ is the sum of the degeneracies of its r matrix factors in (3). Consequently the degeneracy of $f(\Pi)$ is the sum of the degeneracies of the rs matrices such as $\{\omega_v(\alpha_u)\}^{k_u}$.

Again $\{\Pi(\alpha_u)\}^{k_u}$ or $(\Pi - \alpha_u I)^{k_u}$ is the compartite matrix similar to Π whose successive parts are the matrices $\{\omega_1(\alpha_u)\}^{k_u}, \{\omega_2(\alpha_u)\}^{k_u}, \dots \{\omega_v(\alpha_u)\}^{k_u}, \dots \{\omega_s(\alpha_u)\}^{k_u}$ similar to $\omega_1, \omega_2, \dots \omega_v, \dots \omega_s$ respectively. By § 100 the degeneracy of $\{\Pi(\alpha_u)\}^{k_u}$ is the sum of the degeneracies of its parts such as $\{\omega_v(\alpha_u)\}^{k_u}$. Therefore the sum of the degeneracies of all the r matrix factors in (2) or in (A′) is equal to the sum of the degeneracies of the rs matrices such as $\{\omega_v(\alpha_u)\}^{k_u}$.

From these two results we see that the degeneracy of $f(\Pi)$ is equal to the sum of the degeneracies of its r matrix factors in (A′), and this establishes the theorem in the most general case.

NOTE 1. Another proof of Theorem I is given in Note 2 of § 235.

From Theorem I it follows that the degeneracy (or rank) of $f(\phi)$ will be known when the degeneracies of its successive matrix factors in (A) are known. The next two theorems show what the degeneracies of those factors are.

Theorem II a. *If k is any non-zero positive integer, the degeneracy of ϕ^k is*

$$r_k = \rho_1 + \rho_2 + \dots + \rho_i + \dots + \rho_k, \quad \text{.....................(B)}$$

where ρ_i is the number of those potent divisors of the characteristic matrix $\phi - xI$ which are divisible by x^i.

If x^κ is the highest power of x which is a potent divisor of $\phi - xI$, then when $k \nless \kappa$ the degeneracy r_k given by (B) *is the sum of the indices of all potent divisors $\phi - xI$ which are powers of x, and when $k \ngtr \kappa$ it is the sum of the same indices when every index greater than k has been replaced by k.*

Let the complete system of potent divisors of each of the characteristic matrices $\phi - xI$ and $\Pi - xI$ be

$$(x-\lambda_1)^{e_1}, \quad (x-\lambda_2)^{e_2}, \quad (x-\lambda_3)^{e_3}, \quad \dots, \quad \text{..................(4)}$$

where $e_1 + e_2 + e_3 + \dots = m$, and where $\lambda_1, \lambda_2, \lambda_3, \dots$ are not necessarily all different; so that

$$\det(\phi - xI) = \det(\Pi - xI) = (-1)^m (x-\lambda_1)^{e_1}(x-\lambda_2)^{e_2}(x-\lambda_3)^{e_3}. \ \dots(5)$$

Also let the parts of Π corresponding to the successive potent divisors (4) be $\omega_1, \omega_2, \omega_3, \dots$, so that Π^k is a compartite matrix in standard form similar to Π whose successive parts are the square matrices $\omega_1^k, \omega_2^k, \omega_3^k, \dots$ similar respectively to $\omega_1, \omega_2, \omega_3, \dots$. The degeneracies of $\omega_1^k, \omega_2^k, \omega_3^k, \dots$ have been determined in Ex. i of § 229, and utilising those results we see that the degeneracy of ϕ^k, which is equal to the degeneracy of Π^k, i.e. to the sum of the degeneracies of $\omega_1^k, \omega_2^k, \omega_3^k, \dots$, must have the value (B), where ρ_i = total number of potent divisors of $\omega_1(x), \omega_2(x), \omega_3(x), \dots$ divisible by x^i = number of potent divisors of $\Pi(x)$ or $\phi(x)$ divisible by x^i.

Ex. i. If ϕ has no zero latent root, i.e. if no one of the potent divisors (4) is a power of x, then ϕ^k is undegenerate; in fact ϕ is undegenerate.

Ex. ii. If ϕ has one or more zero latent roots, and if x^κ is the highest power of x which is a potent divisor of the characteristic matrix $\phi - xI$, then $\rho_1, \rho_2, \dots \rho_n$ are a series of never increasing non-zero positive integers; and $\rho_{\kappa+1} = \rho_{\kappa+2} = \rho_{\kappa+3} = \dots = 0$. Also $r_\kappa = \rho_1 + \rho_2 + \dots + \rho_\kappa$ is the index of the highest power of x which is a factor of the characteristic determinant $\det(\phi - xI)$, i.e. the total number of zero latent roots of ϕ.

If $r_1, r_2, r_3, \dots$ are as in the text the degeneracies of $\phi^1, \phi^2, \phi^3, \dots$, then $r_1, r_2, r_3, \dots r_\kappa$ are a series of positive integers which constantly increase by amounts which never increase; and $r_\kappa = r_{\kappa+1} = r_{\kappa+2} = \dots$. In fact we have

$$r_1 = \rho_1, \ r_2 - r_1 = \rho_2, \ r_3 - r_2 = \rho_3, \ \dots, \ r_\kappa - r_{\kappa-1} = \rho_\kappa; \ r_{\kappa+1} - r_\kappa = \rho_{\kappa+1} = 0, \ \dots .$$

Thus the ranks $m, m-r_1, m-r_2, m-r_3, \dots$ of the successive powers $\phi^0, \phi^1, \phi^2, \phi^3, \dots$ of ϕ constantly diminish by amounts which never increase till we reach the power ϕ^κ whose rank is $m-r_\kappa$, where r_κ is the total number of zero latent roots of ϕ; and all higher powers of ϕ have the same rank as ϕ^κ.

The numbers r_κ and κ are the maximum and potent indices of order m of the linear divisor x of the characteristic matrix $\phi-xI$.

Ex. iii. The numbers of the potent divisors of the characteristic matrix $\phi-xI$ which are equal to

$$x, \quad x^2, \quad x^3, \dots x^{\kappa-1}, \quad x^\kappa$$

are respectively $\quad \rho_1-\rho_2, \quad \rho_2-\rho_3, \quad \rho_3-\rho_4, \dots \rho_{\kappa-1}-\rho_\kappa, \quad \rho_\kappa.$

Ex. iv. There is some power of ϕ which is equal to 0 when and only when $r_\kappa=m$, i.e. when and only when all the m latent roots of ϕ are 0. In this case the lowest power of ϕ which vanishes is ϕ^κ, where x^κ is the potent divisor of $\phi-xI$ of order m.

Ex. v. The matrix ϕ has a zero latent root, i.e. $\phi-xI$ has the linear divisor x, when and only when ϕ is degenerate. In this case all the potent divisors of $\phi-xI$ corresponding to the linear divisor x are linear (i.e. they are all equal to x) when and only when any one of the following equivalent conditions is satisfied:

(1) $\kappa=1$.

(2) The degeneracy of ϕ is equal to the total number of zero latent roots of ϕ.

(3) ϕ^2 has the same rank as ϕ.

Ex. vi. The matrix ϕ has an unrepeated zero latent root when and only when $\rho_1=1$ and $\rho_2=0$, i.e. when and only when ϕ has rank $m-1$, and ϕ^2 has the same rank as ϕ.

Ex. vii. *Potent divisors of the characteristic matrix of ϕ^k.*

It has been shown in the Note of § 229 that, when k is a non-zero positive integer, the potent divisors of $\phi^k(x)$ correspond to the potent divisors of $\phi(x)$ in the following way:

(1) *To each potent divisor $(x-\lambda)^e$ of $\phi(x)$ which is not a power of x there corresponds a single potent divisor $(x-\lambda^k)^e$ of $\phi^k(x)$ which is not a power of x.*

(2) *To each potent divisor x^e of $\phi(x)$ in which $e \not> k$ there correspond exactly e potent divisors of $\phi^k(x)$ all of which are equal to x.*

(3) *To each potent divisor x^e of $\phi(x)$ in which $e \not< k$ there correspond exactly k potent divisors of $\phi^k(x)$. If we put $e=sk+q$, where s and q are positive integers and $q \not> k-1$, these are*

$$k-q \text{ equal to } x^s \text{ and } q \text{ equal to } x^{s+1} \text{ when } q \neq 0;$$
$$k \text{ equal to } x^s \qquad \text{when } q=0.$$

Theorem II b. *If α is a given scalar constant, and k a non-zero positive integer, the degeneracy of $(\phi-\alpha I)^k$ is*

$$r_k = \rho_1+\rho_2+\dots+\rho_i+\dots+\rho_k, \quad \dots\dots\dots\dots\dots\dots\dots(\text{B}')$$

where ρ_i is the number of those potent divisors of the characteristic matrix $\phi-xI$ which are divisible by $(x-\alpha)^i$.

If $(x-\alpha)^{\kappa}$ is the highest power of $x-\alpha$ which is a potent divisor of $\phi - xI$, then when $k \not> \kappa$ the degeneracy r_k given by (B′) is the sum of the indices of all potent divisors of $\phi - xI$ which are powers of $x-\alpha$, and when $k \not< \kappa$ it is the sum of the same indices when every index greater than k has been replaced by k.

Let $\psi = \phi - \alpha I$, so that ψ is a square matrix of order m the potent divisors of whose characteristic matrix $\psi - xI$ are

$$(x+\alpha-\lambda_1)^{e_1},\quad (x+\alpha-\lambda_2)^{e_2},\quad (x+\alpha-\lambda_3)^{e_3},\quad \ldots\ldots\ldots\ldots\ldots(4')$$

corresponding respectively to the potent divisors (4) of $\phi - xI$. Since each potent divisor of $\psi - xI$ is a power of x when and only when the corresponding potent divisor of $\psi - xI$ is a power of $x-\alpha$, and each potent divisor of $\psi - xI$ is divisible by x^i when and only when the corresponding potent divisor of $\phi - xI$ is divisible by $(x-\alpha)^i$, we can obtain Theorem II *b* by applying Theorem II *a* to determine the degeneracy of ψ^k.

We can also prove Theorem II *b* directly with the aid of Ex. ii of § 229.

Exs. viii—xiv which follow correspond exactly to Exs. i—vii.

Ex. viii. If a is not a latent root of ϕ, i.e. if $x-a$ is not a linear divisor of the characteristic matrix $\phi - xI$, so that no one of the potent divisors (4) is a power of $x-a$, then $(\phi - aI)^k$ is undegenerate; in fact $\phi - aI$ is undegenerate.

Ex. ix. If ϕ has one or more latent roots equal to a, i.e. if $x-a$ is a linear divisor of $\phi - xI$, and if $(x-a)^{\kappa}$ is the highest power of $x-a$ which is a potent divisor of $\phi - xI$, then $\rho_1, \rho_2, \ldots \rho_\kappa$ are a series of never increasing non-zero positive integers; and

$$\rho_{\kappa+1} = \rho_{\kappa+2} = \rho_{\kappa+3} = \ldots = 0.$$

Also $r_\kappa = \rho_1 + \rho_2 + \ldots + \rho_\kappa$ is the index of the highest power of $x-a$ which is a factor of the characteristic determinant $\det(\phi - xI)$, i.e. the total number of latent roots of ϕ which are equal to a.

If $r_1, r_2, r_3, \ldots$ are as in the text the degeneracies of $(\phi - aI)^1$, $(\phi - aI)^2$, $(\phi - aI)^3, \ldots$, then $r_1, r_2, \ldots r_\kappa$ are a series of positive integers which constantly increase by amounts which never increase; and $r_\kappa = r_{\kappa+1} = r_{\kappa+2} = \ldots$. In fact we have

$$r_1 = \rho_1,\quad r_2 - r_1 = \rho_2,\quad r_3 - r_2 = \rho_3,\quad \ldots,\quad r_\kappa - r_{\kappa-1} = \rho_\kappa,\quad r_{\kappa+1} - r_\kappa = \rho_{\kappa+1} = 0,\quad \ldots.$$

Thus the ranks $m, m-r_1, m-r_2, m-r_3, \ldots$ of the successive powers $(\phi - aI)^0$, $(\phi - aI)^1$, $(\phi - aI)^2$, $(\phi - aI)^3, \ldots$ of $\phi - aI$ constantly diminish by amounts which never increase till we reach the power $(\phi - aI)^{\kappa}$ whose rank is $m - r_\kappa$, where r_κ is the total number of latent roots of ϕ which are equal to a; and all higher powers of $\phi - aI$ have the same rank as $(\phi - aI)^{\kappa}$.

The numbers r_κ and κ are the indices of the maximum and potent divisors of order m of the characteristic matrix $\phi - xI$ which correspond to (i.e. are powers of) the linear divisor $x-a$.

Ex. x. The numbers of the potent divisors of the characteristic matrix $\phi - xI$ which are equal to

$$x-a,\quad (x-a)^2,\quad (x-a)^3,\quad \ldots\ (x-a)^{\kappa-1},\quad (x-a)^{\kappa}$$

are respectively $\quad \rho_1 - \rho_2,\quad \rho_2 - \rho_3,\quad \rho_3 - \rho_4,\quad \ldots\ \rho_{\kappa-1} - \rho_\kappa,\quad \rho_\kappa.$

Ex. xi. There is some power of $\phi - aI$ which is equal to 0 when and only when $r_\kappa = m$, i.e. when and only when all the m latent roots of ϕ are equal to a. In this case the lowest power of $\phi - aI$ which vanishes is $(\phi - aI)^\kappa$, where $(x-a)^\kappa$ is the potent divisor of $\phi - aI$ of order m.

Ex. xii. The matrix ϕ has a latent root equal to a, i.e. $\phi - xI$ has the linear divisor $x - a$ when and only when $\phi - aI$ is degenerate. In this case all the potent divisors of $\phi - xI$ corresponding to the linear divisor $x - a$ are linear (i.e. they are all equal to $x - a$) when and only when any one of the following equivalent conditions is satisfied:

(1) $\kappa = 1$.

(2) The degeneracy of $\phi - aI$ is equal to the total number of latent roots of ϕ which are equal to a.

(3) $(\phi - aI)^2$ has the same rank as $\phi - aI$.

Ex. xiii. The matrix ϕ has an unrepeated latent root a when and only when $\rho_1 = 1$ and $\rho_2 = 0$, i.e. when and only when $\phi - aI$ has rank $m-1$ and $(\phi - aI)^2$ has the same rank as $\phi - aI$.

Ex. xiv. *Potent divisors of the characteristic matrix of the power $(\phi - aI)^k$ of the square matrix $\psi = \phi - aI$ when k is a positive integer.*

If $k=0$, the characteristic matrix $\psi^k(x) = \psi^0(x)$ has exactly m potent divisors all equal to $x-1$. We shall dismiss this case, and suppose k to be a non-zero positive integer.

There is a one-one correspondence between the potent divisors of $\psi(x)$ and the potent divisors of $\phi(x)$, any two corresponding potent divisors having the same degree and the same order. To a potent divisor $(x-\lambda)^e$ of $\phi(x)$ there corresponds the potent divisor $(x+a-\lambda)^e$ of $\psi(x)$, and conversely to a potent divisor $(x-\lambda)^e$ of $\psi(x)$ there corresponds the potent divisor $(x-a-\lambda)^e$ of $\phi(x)$. Hence by applying Ex. vii to determine the potent divisors of $\psi^k(x)$ we see that the potent divisors of $\psi^k(x)$ correspond to the potent divisors of $\phi(x)$ in the following way:

(1) *To each potent divisor $(x-\lambda)^e$ of $\phi(x)$ which is not a power of $x-a$ there corresponds a single potent divisor $\{x-(\lambda-a)^k\}^e$ of $\psi^k(x)$ which is not a power of x.*

(2) *To each potent divisor $(x-a)^e$ of $\phi(x)$ in which $e \ngtr k$ there correspond exactly e potent divisors of $\psi^k(x)$ all having the value x.*

(3) *To each potent divisor $(x-a)^e$ of $\phi(x)$ in which $e \nless k$ there correspond exactly k potent divisors of $\psi^k(x)$. If we put $e = sk+q$, where s and q are positive integers and $q \ngtr k-1$, these are:*

$$k-q \text{ equal to } x^s \text{ and } q \text{ equal to } x^{s+1} \quad \text{when } q \neq 0;$$

$$k \text{ equal to } x^s \quad \text{when } q = 0.$$

We can obtain the same results directly by observing that $(\phi - aI)^k$ and $(\Pi - aI)^k$ have equipotent characteristic matrices, and that $(\Pi - aI)^k$ is the compartite matrix whose parts are the kth powers of the simple canonical square matrices $\omega_1 - a\omega_1^0, \omega_2 - a\omega_2^0, \omega_3 - a\omega_3^0, \ldots$.

Ex. xv. **Theorem.** *If $f(\phi) = [b]_m^m$ is a rational integral function of the square matrix $\phi = [a]_m^m$ whose elements are given constants, $f(\phi)$ will have rank 0, i.e. will be equal to 0, when and only when it contains as a factor the rational integral function $E_m(\phi)$ which corresponds to the potent factor $E_m(x)$ of order m of the characteristic matrix $\phi - xI$ of ϕ.*

Consequently the rational integral equation of lowest degree satisfied by ϕ is

$$E_m(\phi)=0.$$

Theorems I and II *b*, which enable us to determine the rank of any given rational integral function of a given square matrix, will enable us to prove the above theorem, which is Theorem II *b* of § 224.

Let $\lambda_1, \lambda_2, \ldots \lambda_n$ be the distinct or unequal latent roots of ϕ, and let

$$D_m(x)=(-1)^m(x-\lambda_1)^{s_1}(x-\lambda_2)^{s_2}\ldots(x-\lambda_n)^{s_n}, \quad E_m(x)=(x-\lambda_1)^{\kappa_1}(x-\lambda_2)^{\kappa_2}\ldots(x-\lambda_n)^{\kappa_n}$$

be respectively the determinant of $\phi-xI$ and the potent factor of order m of $\phi-xI$. When $f(\phi)$ does not vanish identically, we can always write

$$f(\phi)=F(\phi)\,G(\phi),$$

where $$F(\phi)=(\phi-\lambda_1 I)^{k_1}(\phi-\lambda_2 I)^{k_2}\ldots(\phi-\lambda_n I)^{k_n}$$

and $$G(\phi)=A(\phi-\mu_1 I)^{h_1}(\phi-\mu_2 I)^{h_2}\ldots(\phi-\mu_r I)^{h_r},$$

the scalar constants $\lambda_1, \lambda_2, \ldots \lambda_n, \mu_1, \mu_2, \ldots \mu_r$ being all different; the indices $k_1, k_2, \ldots k_n, h_1, h_2, \ldots h_r$ being positive integers, each of which may be 0; and A being a non-zero scalar constant. Because the factors of $G(\phi)$ are all undegenerate, therefore $G(\phi)$ is undegenerate, and $f(\phi)$ has the same rank as $F(\phi)$. The degeneracy of $F(\phi)$ is the sum of the degeneracies of its n factors; and by Theorem II *b* the degeneracy of the factor $(\phi-\lambda_i I)^{k_i}$ can never exceed s_i, and is equal to s_i when and only when $k_i \nless \kappa_i$.

Since the degeneracies of the successive factors of $F(\phi)$ cannot exceed $s_1, s_2, \ldots s_n$ respectively, and since $s_1+s_2+\ldots+s_n=m$, it follows that the degeneracy of $F(\phi)$ is equal to m, i.e. $F(\phi)$ has rank 0, when and only when the degeneracies of the successive factors of $F(\phi)$ are $s_1, s_2, \ldots s_n$ respectively, i.e. when and only when

$$k_1 \nless \kappa_1, \quad k_2 \nless \kappa_2, \quad \ldots \quad k_n \nless \kappa_n. \qquad \text{.................................(6)}$$

Thus $f(\phi)$ has rank 0 when and only when the conditions (6) are satisfied, i.e. when and only when $E_m(\phi)$ is a factor of $f(\phi)$.

2. *Rank of a function which is expressed as a product of irreducible factors.*

When ϕ and $f(\phi)$ lie in a restricted domain of rationality Ω, the elements of the matrix $\phi=[a]_m^m$ and the coefficients of the function $f(\phi)$ being given numerical constants in Ω, we can express the maximum and potent factors of order m of the characteristic matrix $\phi-xI$ in the forms

$$D_m(x)=\{t_1(x)\}^{s_1}\{t_2(x)\}^{s_2}\{t_3(x)\}^{s_3}\ldots, \quad E_m(x)=\{t_1(x)\}^{\kappa_1}\{t_2(x)\}^{\kappa_2}\{t_3(x)\}^{\kappa_3}\ldots,$$

where $t_1(x), t_2(x), t_3(x), \ldots$ are distinct irreducible rational integral functions of x in Ω having degrees $p_1, p_2, p_3, \ldots$ such that

$$p_1 s_1+p_2 s_2+p_3 s_3+\ldots=m;$$

and $t_1(x), t_2(x), t_3(x), \ldots$ are then the irreducible divisors of $\phi-xI$ in Ω.

If in this case we consider the potent divisors of $\phi-xI$ to be those corresponding to its irreducible divisors, so that they are powers of $t_1(x)$,

$t_2(x)$, $t_3(x)$, ..., and if we suppose $t(x)$ and $t(\phi)$ to be any two corresponding irreducible functions in Ω of degree p of the scalar variable x and the square matrix ϕ, then we have the following two theorems:

Theorem III. *If $T_1(x)$, $T_2(x)$, ... $T_r(x)$ are distinct irreducible functions of x in Ω, and if $f(\phi)$ is expressed in the form*

$$f(\phi) = \{T_1(\phi)\}^{k_1}\{T_2(\phi)\}^{k_2}\dots\{T_r(\phi)\}^{k_r}, \quad \text{..............(C)}$$

then the degeneracy of $f(\phi)$ is equal to the suni of the degeneracies of its r factors in (C).

Theorem IV. *If $t(x)$ is any irreducible function of x of degree p and if k is a positive integer, then the degeneracy of $\{t(\phi)\}^k$ is*

$$r_k = p(\rho_1 + \rho_2 + \dots + \rho_i + \dots + \rho_k), \quad \text{.................(D)}$$

where ρ_i is the number of those potent divisors of the characteristic matrix $\phi - xI$ which are divisible by $\{t(x)\}^i$.

In particular if $t(x)$ is not one of the irreducible divisors of $\phi - xI$, i.e. not one of the factors of $D_m(x)$, then $t(\phi)$ and all its powers are undegenerate.

These two theorems follow immediately from Theorems I and II *b* when we use the fact that no two distinct irreducible functions of x can have a linear factor in common, and they enable us to determine the rank of $f(\phi)$ when it is expressed in the form (C).

Ex. xvi. If $\{t(x)\}^s$ and $\{t(x)\}^\kappa$ are the maximum and potent divisors of $\phi - xI$ of order m corresponding to an irreducible divisor $t(x)$ whose degree in x is p, the degeneracies of the successive powers of $t(\phi)$ constantly increase by never increasing amounts until we reach the power $\{t(\phi)\}^\kappa$ whose degeneracy is ps; and all the higher powers of $t(\phi)$ have the same degeneracy as $\{t(\phi)\}^\kappa$.

By an argument similar to that used in Note 1 it follows that $f(\phi)$ has rank 0 when and only when it has $E_m(\phi)$ as a factor.

NOTE 2. *Validity of Theorems III and IV when ϕ and $f(\phi)$ involve arbitrary parameters.*

In the general case when the elements of the matrix ϕ and the coefficients of the function $f(\phi)$ are rational integral functions in Ω of certain arbitrary parameters $a_1, a_2, a_3, \dots$, Ω being any domain of rationality, the elements of the characteristic matrix $\phi - xI$ are rational integral functions in Ω of $x, a_1, a_2, a_3, \dots$. Theorems III and IV and the theorem of Ex. xv will remain true in this case provided that:

(1) we take $D_m(x)$ and $E_m(x)$ to be the maximum and potent factors of order m of $\phi - xI$ when $x, a_1, a_2, a_3, \dots$ are all regarded as independent variables;

(2) we consider $t(x)$, $t_1(x)$, $t_2(x)$, ... $T_1(x)$, $T_2(x)$, ... to be irreducible rational integral functions in Ω of $x, a_1, a_2, a_3, \dots$ which actually contain x;

(3) we cancel or leave out of account in each equation a scalar factor common to both sides which is a rational integral function of $a_1, a_2, a_3, \dots$ only.

This follows from the fact that if $t(x)$, $t_1(x)$, $t_2(x)$, $t_3(x)$, ... are distinct irreducible functions in Ω of $x, a_1, a_2, a_3, \ldots$, then for all ordinary particular values of $a_1, a_2, a_3, \ldots$ in Ω they are distinct irreducible functions of x in Ω. Thus when the above interpretations are adopted, Theorems III and IV are true for all ordinary particular values of $a_1, a_2, a_3, \ldots$ in Ω, and are therefore true when $a_1, a_2, a_3, \ldots$ are arbitrary. We can also deduce Theorems III and IV in this case from Theorems I and II *b*, using the fact that if $t(x)$, $t_1(x)$, $t_2(x)$, $t_3(x)$, ... are distinct irreducible functions in Ω of $x, a_1, a_2, a_3, \ldots$, then for all ordinary particular values of $a_1, a_2, a_3, \ldots$, they are functions of x whose linear factors are all unrepeated and all distinct.

In the particular case when all the elements of ϕ are arbitrary $D_m(x)$ is irreducible and is the same as $E_m(x)$. Consequently the rational integral of lowest degree which is satisfied by ϕ and is an identity in the elements of ϕ is

$$D_m(\phi)=0.$$

§ 231. Non-extravagant rational integral functions of a given square matrix.

Let $\phi=[a]_m^m$ be any given square matrix with constant elements; also let $\phi'=\overline{\underline{a}}\,_m^m$ be the conjugate of ϕ, and let $I=[1]_m^m$. Since the characteristic matrices $\phi-xI$ and $\phi'-xI$ of ϕ and ϕ' are equipotent, we may suppose their common potent factor of order m to be

$$E_m(x)=(x-\lambda_1)^{\kappa_1}(x-\lambda_2)^{\kappa_2}\ldots(x-\lambda_n)^{\kappa_n},$$

where $\lambda_1, \lambda_2, \ldots \lambda_n$ are the distinct latent roots of both ϕ and ϕ', and write

$$E_m(\phi)=(\phi-\lambda_1 I)^{\kappa_1}(\phi-\lambda_2 I)^{\kappa_2}\ldots(\phi-\lambda_n I)^{\kappa_n}, \quad \text{.....................(1)}$$

$$E_m(\phi')=(\phi'-\lambda_1 I)^{\kappa_1}(\phi'-\lambda_2 I)^{\kappa_2}\ldots(\phi'-\lambda_n I)^{\kappa_n}. \quad \text{..................(1')}$$

By Ex. xxi of § 165 any matrix $[b]_m^n$ is horizontally (or vertically) non-extravagant when and only when $[b]_m^n\,\overline{\underline{b}}\,_n^m$ (or $\overline{\underline{b}}\,_n^m\,[b]_m^n$) has the same rank as $[b]_m^n$. Hence if $f(\phi)$ is any rational integral function of ϕ, we have the following results:

(1) *The square matrix $f(\phi)$ is horizontally non-extravagant when and only when $f(\phi)f(\phi')$ has the same rank as $f(\phi)$.*

(2) *The square matrix $f(\phi)$ is vertically non-extravagant when and only when $f(\phi')f(\phi)$ has the same rank as $f(\phi)$.*

When $f(\phi)$ is real, both these conditions are necessarily satisfied.

In applying these tests we can strike out from $f(\phi)$ and $f(\phi')$ all those of their linear factors which are not factors of $E_m(\phi)$ and $E_m(\phi')$ respectively; for all such factors are undegenerate. Hence we can always suppose $f(\phi)$ and $f(\phi')$ to be reduced to the forms

$$f(\phi)=(\phi-\lambda_1 I)^{k_1}(\phi-\lambda_2 I)^{k_2}\ldots(\phi-\lambda_r I)^{k_r}, \quad \text{........................(2)}$$

$$f(\phi')=(\phi'-\lambda_1 I)^{k_1}(\phi'-\lambda_2 I)^{k_2}\ldots(\phi'-\lambda_r I)^{k_r}, \quad \text{....................(2')}$$

where $\lambda_1, \lambda_2, \ldots \lambda_r$ are any r of the latent roots $\lambda_1, \lambda_2, \ldots \lambda_n$, and where $k_1, k_2, \ldots k_r$ are non-zero positive integers. We will consider three important special cases.

CASE I. *When the square matrix ϕ is symmetric.*

In this case $f(\phi)$ is horizontally (and vertically) non-extravagant when and only when every linear factor common to $f(\phi)$ and $E_m(\phi)$ has at least as great an index in $f(\phi)$ as in $E_m(\phi)$. ....(A)

Since $f(\phi')=f(\phi)$, the condition for the non-extravagance of $f(\phi)$ is:

$\{f(\phi)\}^2$ *has the same rank, i.e. the same degeneracy, as* $f(\phi)$. ....(a)

Now when $f(\phi)$ is reduced to the form (2), no factor $(\phi-\lambda_i I)^{2k_i}$ of $\{f(\phi)\}^2$ can have a smaller degeneracy than the corresponding factor $(\phi-\lambda_i I)^{k_i}$ of $f(\phi)$. Therefore by Theorem I of § 231 the condition (a) is satisfied when and only when every factor of $\{f(\phi)\}^2$ has the same rank as the corresponding factor of $f(\phi)$, and by Ex. ix of § 231 this is the case when and only when

$$k_1 \nless \kappa_1, \quad k_2 \nless \kappa_2, \quad \ldots \quad k_r \nless \kappa_r.$$

Hence when the potent divisors of the characteristic matrix of ϕ are all linear, every rational integral function of ϕ is non-extravagant. ....(A′)

The results (A) and (A′) clearly remain true whenever $f(\phi')=\pm f(\phi)$, i.e. whenever $f(\phi)$ is symmetric or skew-symmetric.

CASE II. *When the square matrix ϕ is skew-symmetric.*

In this case $f(\phi)$ is horizontally (and vertically) non-extravagant when and only when the following two conditions are satisfied:

(1) *All linear factors other than ϕ common to $f(\phi)$ and $E_m(\phi)$ occur in pairs of the form $\phi-\lambda I$, $\phi+\lambda I$.*

(2) *Every linear factor common to $f(\phi)$ and $E_m(\phi)$ has at least as great an index in $f(\phi)$ as in $E_m(\phi)$.* ....(B)

We assume the result proved in Chapter XXXII that if $\phi-\lambda I$ is a linear factor of $E_m(\phi)$, then $\phi+\lambda I$ is also a linear factor of $E_m(\phi)$; and in proving the theorem (B) we assume that $f(\phi)$ has been reduced to the form (2). Then if

$$g(\phi)=(\phi+\lambda_1 I)^{k_1}(\phi+\lambda_2 I)^{k_2}\ldots(\phi+\lambda_r I)^{k_r}, \quad \text{..............(3)}$$

all linear factors of $g(\phi)$ being linear factors of $E_m(\phi)$, we have $f(\phi')=\pm g(\phi)$, and therefore the condition for the horizontal (or vertical) extravagance of $f(\phi)$ is

$f(\phi)\,g(\phi)$ *has the same rank as* $f(\phi)$. ....(b)

First suppose that the condition (b) is satisfied. Then if $g(\phi)$ had any linear factor not occurring in $f(\phi)$, it would follow from Theorem I of § 231 that $f(\phi)\,g(\phi)$ has a smaller rank than $f(\phi)$; and if $f(\phi)$ had any linear factor not occurring in $g(\phi)$, it would follow in the same way that $f(\phi)\,g(\phi)$ has a smaller rank than $g(\phi)$, and therefore a smaller rank than $f(\phi')$, i.e. than $f(\phi)$. Consequently the linear factors of $g(\phi)$ are identical with the linear factors of $f(\phi)$, which shows that the condition (1) of the theorem (B) is satisfied. Further it now follows by reasoning similar to that used in Case I that the condition (2) of the theorem is also satisfied.

Next suppose that the conditions (1) and (2) of the theorem are satisfied. Then the condition (b) is obviously satisfied; and this proves the theorem (B).

In particular when the potent divisors of the characteristic matrix of ϕ are all linear, every rational integral function of ϕ satisfying the condition (1) *of the theorem* (B) *is non-extravagant.* ....(B′)

CASE III. *When the square matrix ϕ is a semi-unit matrix.*

In this case $f(\phi)$ is horizontally (and vertically) non-extravagant when and only when the following two conditions are satisfied:

(1) *All linear factors other than $\phi - I$ and $\phi + I$ common to $f(\phi)$ and $E_m(\phi)$ occur in pairs of the form $\phi - \lambda I$, $\phi - \frac{1}{\lambda} I$.*

(2) *Every linear factor common to $f(\phi)$ and $E_m(\phi)$ has at least as great an index in $f(\phi)$ as in $E_m(\phi)$.* ....(C)

We assume the result proved in Chapter XXXII that if $\phi - \lambda I$ is a linear factor of $E_m(\phi)$, then $\phi - \frac{1}{\lambda} I$ is also a linear factor of $E_m(\phi)$; and in proving the theorem (C) we assume that $f(\phi)$ has been reduced to the form (2), where $\lambda_1, \lambda_2, \dots \lambda_r$ are all different from 0, because ϕ is undegenerate. Then if

$$g(\phi) = \left(\phi - \frac{1}{\lambda_1} I\right)^{k_1} \left(\phi - \frac{1}{\lambda_2} I\right)^{k_2} \dots \left(\phi - \frac{1}{\lambda_r} I\right)^{k_r}, \quad \text{..................(4)}$$

all linear factors of $g(\phi)$ being linear factors of $E_m(\phi)$, and if $k = k_1 + k_2 + \dots + k_r$, we see from the equations $\phi\phi' = \phi'\phi = I$ that we can convert $f(\phi')$ into $g(\phi)$ both by prefixing and by postfixing the undegenerate square matrix ϕ^k multiplied by a scalar constant, and therefore the condition for the horizontal (or vertical) non-extravagance of $f(\phi)$ is

$$f(\phi)\, g(\phi) \textit{ has the same rank as } f(\phi). \quad \text{..................(c)}$$

First suppose that the condition (c) is satisfied. Then as in Case II the linear factors of $g(\phi)$ must be identical with the linear factors of $f(\phi)$, and therefore the condition (1) of the theorem (C) is satisfied. Further it follows by reasoning similar to that used in Case I that the condition (2) of the theorem is satisfied.

Next suppose that the conditions (1) and (2) of the theorem are satisfied. Then the condition (c) is obviously satisfied; and this proves the theorem (C).

In particular when the potent divisors of the characteristic matrix of ϕ are all linear, every rational integral function of ϕ satisfying the condition (1) of the theorem (C) is non-extravagant. ....(C′)

§ 232. The first transmutes of a square matrix.

Let ϕ be any square matrix of order m and rank r whose elements are numerical constants. If as in § 154 it is expressed in the form

$$\phi = [a]_m^r \, [b]_r^m, \quad \text{..................(A)}$$

where the factor matrices on the right have rank r, then the matrix

$$\phi_1 = [b]_r^m \, [a]_m^r \quad \text{..................(B)}$$

will be called a *first transmute* or simply a *transmute* of ϕ.

Ex. i. The degeneracy of a first transmute of the square matrix ϕ cannot exceed the degeneracy of ϕ.

For if s is the rank of ϕ_1, it follows from § 133 that

$$s + m \nless 2r, \text{ i.e. } r - s \ngtr m - r.$$

The square matrix ϕ has an infinite number of first transmutes, and the following theorem shows how these are related to one another.

Theorem I. *If ϕ is a square matrix of rank r, the various first transmutes of ϕ are square matrices of order r which are convertible into one another by equimutant transformations (i.e. which have equipotent characteristic matrices).*

If ϕ_1 is any one of the first transmutes of ϕ, then ψ is a first transmute of ϕ when and only when it is a square matrix of the same order as ϕ_1 which can be derived from ϕ_1 by an equimutant transformation.

First let
$$\phi = [\alpha]_m^r [\beta]_r^m, \quad \psi = [\beta]_r^m [\alpha]_m^r$$

be respectively any second representation of ϕ in the form (A) and the corresponding first transmute of ϕ. Then

$$[\alpha]_m^r [\beta]_r^m = [a]_m^r [b]_r^m, \quad \text{.........................(1)}$$

and by § 152 we have

$$[\alpha]_m^r = [a]_m^r \underline{\overline{H}}{}_r^r, \quad [\beta]_r^m = [h]_r^r [b]_r^m, \quad \text{..................(2)}$$

where
$$\underline{\overline{H}}{}_r^r [h]_r^r = [h]_r^r \underline{\overline{H}}{}_r^r = [1]_r^r.$$

Therefore
$$\psi = [\beta]_r^m [\alpha]_m^r = [h]_r^r [b]_r^m [a]_m^r \underline{\overline{H}}{}_r^r = [h]_r^r \phi_1 \underline{\overline{H}}{}_r^r. \quad \text{.........(3)}$$

Thus ψ is derivable from ϕ_1 by an equimutant transformation, and ψ and ϕ_1 have equipotent characteristic matrices.

Next let
$$\psi = [h]_r^r \phi_1 \underline{\overline{H}}{}_r^r, \quad (\text{where } [h]_r^r \underline{\overline{H}}{}_r^r = [1]_r^r),$$

be any square matrix of the same order r as ϕ_1 derived from ϕ_1 by an equimutant transformation. Then the matrices $[\alpha]_m^r$, $[\beta]_r^m$ defined by (2) are matrices of rank r such that

$$\phi = [\alpha]_m^r [\beta]_r^m, \quad \psi = [h]_r^r [b]_r^m [a]_m^r \underline{\overline{H}}{}_r^r = [\beta]_r^m [\alpha]_m^r;$$

and this shows that ψ is a first transmute of ϕ.

Ex. ii. The various first transmutes of a given square matrix all have the same rank as well as the same order.

Ex. iii. If two square matrices have any one first transmute in common, then every first transmute of either matrix is also a first transmute of the other matrix, i.e. the two matrices have the same first transmutes.

Ex. iv. *An undegenerate square matrix is itself one of its first transmutes.*

For if $\phi=[a]_m^m$ is undegenerate, and if we express it in the form $\phi=[a]_m^m[1]_m^m$, we see that one of its first transmutes is

$$\phi_1=[1]_m^m[a]_m^m=[a]_m^m=\phi.$$

It follows that all first transmutes of an undegenerate square matrix of order m are undegenerate square matrices of order m.

Ex. v. *First transmutes of the square matrix* $\phi=[a]_m^m=\begin{bmatrix}0, & 1\\ 0, & 0\end{bmatrix}_{m-1,1}^{1,m-1}$.

Writing
$$\phi=[a]_m^m=\begin{bmatrix}1\\ 0\end{bmatrix}_{m-1,1}^{m-1}[0,\ 1]_{m-1}^{1,m-1},$$

we obtain the first transmute

$$\phi_1=[0,\ 1]_{m-1}^{1,m-1}\begin{bmatrix}1\\ 0\end{bmatrix}_{m-1,1}^{m-1}=\begin{bmatrix}0, & 1, & 0\\ 0, & 0, & 1\end{bmatrix}_{m-2,1}^{1,m-2,1}\begin{bmatrix}1, & 0\\ 0, & 1\\ 0, & 0\end{bmatrix}_{1,m-2,1}^{1,m-2}=\begin{bmatrix}0, & 1\\ 0, & 0\end{bmatrix}_{m-2,1}^{1,m-2}.$$

The various first transmutes of ϕ are those square matrices of the same order $m-1$ as ϕ_1 which can be derived from ϕ_1 by equimutant transformations.

Ex. vi. *If* $[a]_q^q$ *is undegenerate, a first transmute* ϕ_1 *of the square matrix* $\phi=\begin{bmatrix}0, & a\\ 0, & 0\end{bmatrix}_{q,p}^{p,q}$ *is obtained by striking out the first* p *horizontal and vertical rows of* ϕ.

For we can write

$$\phi=\begin{bmatrix}a\\ 0\end{bmatrix}_{q,p}^{q}[0,\ 1]_q^{p,q},\quad \phi_1=[0,\ 1]_q^{p,q}\begin{bmatrix}a\\ 0\end{bmatrix}_{q,p}^{q}.$$

When $p \nless q$, we have $\phi_1=[0]_q^q$.

If
$$\phi=\begin{bmatrix}0, & 1\\ 0, & 0\end{bmatrix}_{q,p}^{p,q}\quad\text{we have}\quad \phi_1=\begin{bmatrix}0, & 1\\ 0, & 0\end{bmatrix}_{q-p,p}^{p,q-p}\quad\text{or}\quad \phi_1=[0]_q^q$$

according as $p \ngtr q$ or $p \nless q$.

Ex. vii. *First transmutes of the square matrix* $\phi=[a]_m^m=\begin{bmatrix}0, & 1\\ 0, & 0\end{bmatrix}_{m-p,p}^{p,m-p}$ *in which* $p \nless 1$ *and* $p \ngtr m$.

If we put $m=np+q$, where n and q are positive integers satisfying the conditions $n \nless 1$, $q \ngtr p-1$, so that

$$\phi=\begin{bmatrix}0, & 1\\ 0, & 0\end{bmatrix}_{(n-1)p+q,p}^{p,(n-1)p+q},$$

we see from Ex. vi that a first transmute ϕ_1 of ϕ is given by

$$\phi_1=\begin{bmatrix}0, & 1\\ 0, & 0\end{bmatrix}_{m-2p,p}^{p,m-2p}=\begin{bmatrix}0, & 1\\ 0, & 0\end{bmatrix}_{(n-2)p+q,p}^{p,(n-2)p+q}\quad\text{when } n-1\neq 0,$$

$$\phi_1=[0]_q^q\qquad\text{when } n-1=0.$$

The various first transmutes of ϕ are those square matrices of the same order $m-p$ as ϕ_1 which can be obtained from ϕ_1 by equimutant transformations.

Ex. viii. *The square matrices* $\Phi = \begin{bmatrix} \phi, & 0 \\ 0, & 0 \end{bmatrix}_{m,s}^{m,s}$, $\phi = [\phi]_m^m$ *have the same first transmutes.*

Let ϕ have rank r, and let $[\phi]_m^m = [a]_m^r [b]_r^m$, $[c]_r^r = [b]_r^m [a]_m^r$.

Then since $\Phi = \begin{bmatrix} a \\ 0 \end{bmatrix}_{m,s}^{r} [b, 0]_r^{m,s}$, we see that a first transmute of Φ is

$$\Phi_1 = [b, 0]_r^{m,s} \begin{bmatrix} a \\ 0 \end{bmatrix}_{m,s}^{r} = [b]_r^m [a]_m^r = [c]_r^r.$$

Thus Φ and ϕ have the common first transmute $[c]_r^r$, and therefore they have the same first transmutes.

Ex. ix. *First transmutes of a compartite matrix in standard form whose parts are all square.*

Theorem. *If ϕ is a compartite matrix in standard form whose successive parts are $\omega_1, \omega_2, \omega_3, \ldots$, these being all square matrices, then any compartite matrix ψ in standard form whose successive parts are first transmutes of $\omega_1, \omega_2, \omega_3, \ldots$ is itself a first transmute of ϕ.*

Let ϕ be a compartite matrix in standard form whose successive parts are

$$A = [a]_l^l, \quad B = [b]_m^m, \quad C = [c]_n^n$$

having ranks λ, μ, ν respectively, and let

$$[a]_l^l = [a]_l^\lambda [a']_\lambda^l, \quad [b]_m^m = [\beta]_m^\mu [\beta']_\mu^m, \quad [c]_n^n = [\gamma]_n^\nu [\gamma']_\nu^n,$$

so that
$$\phi = \begin{bmatrix} a, & 0, & 0 \\ 0, & b, & 0 \\ 0, & 0, & c \end{bmatrix}_{l,m,n}^{l,m,n} = \begin{bmatrix} a, & 0, & 0 \\ 0, & \beta, & 0 \\ 0, & 0, & \gamma \end{bmatrix}_{l,m,n}^{\lambda,\mu,\nu} \begin{bmatrix} a', & 0, & 0 \\ 0, & \beta', & 0 \\ 0, & 0, & \gamma' \end{bmatrix}_{\lambda,\mu,\nu}^{l,m,n}.$$

Then
$$[a']_\lambda^\lambda = [a']_\lambda^l [a]_l^\lambda, \quad [b']_\mu^\mu = [\beta']_\mu^m [\beta]_m^\mu, \quad [c']_\nu^\nu = [\gamma']_\nu^n [\gamma]_n^\nu$$

are first transmutes of A, B, C; and since ϕ has rank $\lambda+\mu+\nu$, the matrix

$$\psi = \begin{bmatrix} a', & 0, & 0 \\ 0, & \beta', & 0 \\ 0, & 0, & \gamma' \end{bmatrix}_{\lambda,\mu,\nu}^{l,m,n} \begin{bmatrix} a, & 0, & 0 \\ 0, & \beta, & 0 \\ 0, & 0, & \gamma \end{bmatrix}_{l,m,n}^{\lambda,\mu,\nu} = \begin{bmatrix} a', & 0, & 0 \\ 0, & b', & 0 \\ 0, & 0, & c' \end{bmatrix}_{\lambda,\mu,\nu}^{\lambda,\mu,\nu}$$

is a first transmute of ϕ.

It has been implicitly assumed that λ, μ, ν are all different from zero. If however any part of ϕ is a zero matrix, or has rank 0, then by Ex. viii we simply omit that part in the formation of ψ.

The next theorem shows that the potent divisors of the characteristic matrices of the first transmutes of a square matrix ϕ are known when the potent divisors of the characteristic matrix of ϕ are known.

Theorem II. *If ϕ and ψ are two square matrices of the same order m which have equipotent characteristic matrices (i.e. are convertible into one another by equimutant transformations), then any two first transmutes of ϕ and ψ are square matrices of the same order which have equipotent characteristic matrices (i.e. are convertible into one another by equimutant transformations). Consequently ϕ and ψ have the same first transmutes.*

As before let ϕ have rank r, and let

$$\phi = [a]_m^r\,[b]_r^m, \quad \phi_1 = [b]_r^m\,[a]_m^r. \quad \text{.....................(4)}$$

Further let ψ be a square matrix of order m derived from ϕ by the equimutant transformation

$$\psi = \overline{\underline{H}}_m^m\,\phi\,[h]_m^m,$$

where $[h]_m^m$ and $\overline{\underline{H}}_m^m$ are mutually inverse undegenerate square matrices. Then ψ must have the same rank r as ϕ, and if we write

$$\psi = [\alpha]_m^r\,[\beta]_r^m, \quad \psi_1 = [\beta]_r^m\,[\alpha]_m^r, \text{.....................(4')}$$

we have
$$[\alpha]_m^r\,[\beta]_r^m = \overline{\underline{H}}_m^m\,[a]_m^r \cdot [b]_r^m\,[h]_m^m.$$

From the last equation it follows by § 152 that

$$[\alpha]_m^r = \overline{\underline{H}}_m^m\,[a]_m^r\,\overline{\underline{K}}_r^r, \quad [\beta]_r^m = [k]_r^r\,[b]_r^m\,[h]_m^m, \quad \text{...........(5)}$$

where $[k]_r^r$ and $\overline{\underline{K}}_r^r$ are two mutually inverse undegenerate square matrices; and from (5) it follows that

$$\psi_1 = [\beta]_r^m\,[\alpha]_m^r = [k]_r^r\,[b]_r^m\,[a]_m^r\,\overline{\underline{K}}_r^r = [k]_r^r\,\phi_1\,\overline{\underline{K}}_r^r.$$

Hence by § 227 the first transmutes ϕ_1 and ψ_1 of ϕ and ψ have equipotent characteristic matrices, and it follows from Theorem I that ϕ and ψ have the same first transmutes.

Ex. x. The first transmutes of a square matrix $\phi=[a]_m^m$ are the same as the first transmutes of its standard reduced form $\Pi=[\Pi]_m^m$ constructed as in Theorem I a of § 228.

Ex. xi. *Canonical reduced form of the first transmutes of a square matrix.*

Let $\phi=[a]_m^m$ be a square matrix whose characteristic matrix is $\phi(\lambda)=[a]_m^m-\lambda\,[1]_m^m$, let $\Pi=[\Pi]_m^m$ be the canonical reduced form of ϕ constructed as in Theorem I a of § 228, and let Π_1 be the canonical reduced form of the first transmutes of ϕ. Then we can obtain Π_1 from Π by

(1) omitting every part of the form $[0]_1^1$ corresponding to a potent divisor λ of $\phi(\lambda)$;

(2) replacing every part of the form $[\omega]_e^e = \begin{bmatrix} 0, & 1 \\ 0, & 0 \end{bmatrix}_{e-1,\,1}^{1,\,e-1}$ corresponding to a potent divisor λ^e of $\phi(\lambda)$ in which $e > 1$ by its first transmute

$$[\omega]_{e-1}^{e-1} = \begin{bmatrix} 0, & 1 \\ 0, & 0 \end{bmatrix}_{e-2,\,1}^{1,\,e-2};$$

(3) leaving all other parts of Π unaltered.

For if $\phi(\lambda)$ has exactly ρ potent divisors equal to λ, we have

$$[\Pi]_m^m = \begin{bmatrix} \pi, & 0 \\ 0, & 0 \end{bmatrix}_{m-\rho,\,\rho}^{m-\rho,\,\rho},$$

where $[\pi]_{m-\rho}^{m-\rho}$ is a compartite matrix in standard form whose parts are those parts of Π which correspond to all potent divisors of $\phi(\lambda)$ except those which are equal to λ. The matrix Π_1 formed in the manner just described is by Ex. ix a first transmute of $[\pi]_{m-\rho}^{m-\rho}$; therefore by Ex. viii it is a first transmute of Π; and by Theorem II it is a first transmute of ϕ. Moreover it is a canonical square matrix.

Conversely we can obtain Π from Π_1 when ρ is known.

The next theorem gives the actual values of the potent divisors of the characteristic matrices of the first transmutes of a square matrix when the potent divisors of the characteristic matrix of ϕ are known.

Theorem III. *Let $\phi = [a]_m^m$ be any square matrix of order m whose elements are constants; let the potent divisors of its characteristic matrix $\phi(\lambda) = [a]_m^m - \lambda[1]_m^m$ corresponding to the linear divisor λ be*

$$\sigma_1 \text{ equal to } \lambda,\quad \sigma_2 \text{ equal to } \lambda^2,\ \ldots\ \sigma_\kappa \text{ equal to } \lambda^\kappa;\ \ldots\ldots\ldots\ldots(6)$$

and let the remaining potent divisors of its characteristic matrix $\phi(\lambda)$ be

$$(\lambda - c_1)^{e_1},\quad (\lambda - c_2)^{e_2},\ \ldots\ (\lambda - c_s)^{e_s},\quad \ldots\ldots\ldots\ldots\ldots\ldots(7)$$

where $c_1, c_2, \ldots c_s$ are not necessarily all unequal, but are all different from 0.

Then if ϕ_1 is any first transmute of ϕ, the potent divisors of its characteristic matrix $\phi_1(\lambda)$ corresponding to the linear divisor λ are

$$\sigma_2 \text{ equal to } \lambda,\quad \sigma_3 \text{ equal to } \lambda^2,\ \ldots\ \sigma_\kappa \text{ equal to } \lambda^{\kappa-1},\ \ldots\ldots\ldots(6')$$

and (7) are the remaining potent divisors of $\phi_1(\lambda)$.

Here $\sigma_1, \sigma_2, \ldots \sigma_\kappa, e_1, e_2, \ldots e_s$ are non-zero positive integers such that

$$\sigma_1 + 2\sigma_2 + \ldots + \kappa\sigma_\kappa + e_1 + e_2 + \ldots + e_s = m.$$

To prove Theorem III let Π and Π_1 be the canonical forms to which ϕ and ϕ_1 can be reduced by equimutant transformations. Then the characteristic matrix of ϕ_1 has the same potent divisors as the characteristic matrix of Π_1, and these are the potent divisors of the various parts of Π_1. Referring to Ex. xi we see that Π_1 has no part corresponding to a potent divisor λ of $\phi(\lambda)$, and that its parts correspond to the other potent divisors of $\phi(\lambda)$ in the following way:

(1) To a potent divisor λ^p of $\phi(\lambda)$, where $p > 1$, there corresponds a part $[\omega]_{p-1}^{p-1} = \begin{bmatrix} 0, & 1 \\ 0, & 0 \end{bmatrix}_{p-2,1}^{1,p-2}$ of Π_1 whose characteristic matrix has the single potent divisor λ^{p-1}.

(2) To a potent divisor $(\lambda - c)^e$ of $\phi(\lambda)$ in which $c \neq 0$ there corresponds a part of Π_1 which is the same as the corresponding part of $[\omega]_e^e$ of Π, i.e. a part whose characteristic matrix has the single potent divisor $(\lambda - c)^e$.

Corollary 1. *If the potent divisors of $\phi_1(\lambda)$ and the degeneracy of ϕ are known, then the potent divisors of $\phi(\lambda)$ are known.*

For $\sigma_1 + \sigma_2 + \ldots + \sigma_\kappa$ is equal to the degeneracy of ϕ (see Ex. iii of § 228), and therefore σ_1 is known.

Corollary 2. *If we know any first transmute ϕ_1 of ϕ and the order m of ϕ, then the potent divisors of $\phi(\lambda)$ are uniquely determinate.*

For the order of ϕ_1 is the rank of ϕ, and therefore the degeneracy of ϕ is known.

Ex. xii. If $p > 1$, the number of potent divisors of $\phi(\lambda)$ divisible by λ^p is equal to the number of potent divisors of $\phi_1(\lambda)$ divisible by λ^{p-1}.

Ex. xiii. *If ρ_1 and ρ_2 are the degeneracies of ϕ and ϕ_1, then $\phi(\lambda)$ has exactly ρ_1 potent divisors divisible by λ, and exactly ρ_2 potent divisors divisible by λ^2. Consequently it has exactly $\rho_1 - \rho_2$ potent divisors equal to λ.*

This follows from Exs. ii and iii of § 228 and Ex. xii above.

Ex. xiv. *If ϕ has the latent root 0, i.e. if ϕ is degenerate, then the potent divisors of $\phi(\lambda)$ corresponding to the linear divisor λ are all linear, i.e. are all equal to λ, when and only when ϕ_1, the first transmute of ϕ, is undegenerate.*

This case occurs when and only when the number of zero latent roots of ϕ is equal to the degeneracy of ϕ, or when and only when the number of potent divisors of $\phi(\lambda)$ which are divisible by λ (i.e. are powers of λ) is equal to the degeneracy of ϕ. A proof independent of Theorem III is given in Ex. xxiii.

Ex. xv. *The matrix ϕ has an unrepeated zero latent root when and only when ϕ has rank $m-1$ and ϕ_1 is undegenerate.*

With the notation of Ex. xiii this case occurs when and only when $\rho_1 = 1$ and $\rho_2 = 0$.

Ex. xvi. *The first transmutes of a real symmetric matrix are undegenerate.*

If $\phi=[a]_m^m$ is a real symmetric matrix of rank r, we know from § 147 that we can express it in the form

$$\phi=[a]_m^m=\overline{\underline{l}}_m^{\,r}\begin{bmatrix} e_1 & 0 & \dots & 0 \\ 0 & e_2 & \dots & 0 \\ \dots & \dots & \dots & \dots \\ 0 & 0 & \dots & e_r \end{bmatrix}\cdot\begin{bmatrix} e_1 & 0 & \dots & 0 \\ 0 & e_2 & \dots & 0 \\ \dots & \dots & \dots & \dots \\ 0 & 0 & \dots & e_r \end{bmatrix}[l]_r^m,$$

where $[l]_r^m$ is real, and $e_1, e_2, \dots e_r$ are non-zero quantities. A first transmute of ϕ is

$$\phi_1=\begin{bmatrix} e_1 & 0 & \dots & 0 \\ 0 & e_2 & \dots & 0 \\ \dots & \dots & \dots & \dots \\ 0 & 0 & \dots & e_r \end{bmatrix}\cdot[l]_r^m\,\overline{\underline{l}}_m^{\,r}\cdot\begin{bmatrix} e_1 & 0 & \dots & 0 \\ 0 & e_2 & \dots & 0 \\ \dots & \dots & \dots & \dots \\ 0 & 0 & \dots & e_r \end{bmatrix}.$$

Since $[l]_r^m$ is real, we know by § 72 that the middle factor on the right has rank r; and therefore ϕ_1 is undegenerate.

Now let $\phi=[a]_m^m$ be any square matrix of order m whose elements are constants, let $I=[1]_m^m$, and let $\phi(\lambda)=\phi-\lambda I$ be the characteristic matrix of ϕ, λ being a scalar variable. Also let α be any scalar quantity, let

$$\psi=\phi-\alpha I=\phi(\alpha), \quad \psi(\lambda)=\psi-\lambda I=\phi(\lambda+\alpha),$$

so that $\psi(\lambda)$ is the characteristic matrix of ψ, let ψ_1 be any first transmute of ψ, and let $\psi_1(\lambda)$ be the characteristic matrix of ψ_1.

By Ex. x of § 222 there is a one-one correspondence between the linear and potent divisors of $\phi(\lambda)$ and the linear and potent divisors of $\psi(\lambda)$ such that

$(\lambda-c)^e$ is a potent divisor of $\phi(\lambda)$ of order i when and only when $(\lambda+\alpha-c)^e$ is a potent divisor of $\psi(\lambda)$ of order i,

or $(\lambda-c)^e$ is a potent divisor of $\psi(\lambda)$ of order i when and only when $(\lambda-\alpha-c)^e$ is a potent divisor of $\phi(\lambda)$ of order i.

In particular $\lambda-\alpha$ is a linear divisor of $\phi(\lambda)$, i.e. α is a latent root of ϕ, when and only when λ is a linear divisor of $\psi(\lambda)$, i.e. 0 is a latent root of ψ, or ψ is degenerate. Thus the investigation of the potent divisors of $\phi(\lambda)$ corresponding to the linear divisor $\lambda-\alpha$ of $\phi(\lambda)$ can be reduced to the investigation of the potent divisors of $\psi(\lambda)$ corresponding to the linear divisor λ; and in this way we can obtain the following generalisation of Theorem III:

Theorem IV. *Let α be any latent root of the square matrix $\phi=[a]_m^m$, let $\phi(\lambda)$ be the characteristic matrix of ϕ, let the potent divisors of $\phi(\lambda)$ corresponding to the linear divisor $\lambda-\alpha$ be*

$$\sigma_1 \text{ equal to } \lambda-\alpha,\quad \sigma_2 \text{ equal to } (\lambda-\alpha)^2,\quad \dots\quad \sigma_\kappa \text{ equal to } (\lambda-\alpha)^\kappa,\ \dots\text{(8)}$$

and let the remaining potent divisors of $\phi(\lambda)$ *be*

$$(\lambda - c_1)^{e_1}, \quad (\lambda - c_2)^{e_2}, \ \dots \ (\lambda - c_s)^{e_s}, \quad \dots\dots\dots\dots\dots\dots(9)$$

where $c_1, c_2, \dots c_s$ *are not necessarily all unequal, but are all different from* α.

Then if $\psi = \phi - \alpha I$, *and if* $\psi(\lambda)$ *is the characteristic matrix of* ψ, *the potent divisors of* $\psi(\lambda)$ *corresponding to the linear divisor* λ *are*

$$\sigma_1 \text{ equal to } \lambda, \quad \sigma_2 \text{ equal to } \lambda^2, \ \dots \ \sigma_\kappa \text{ equal to } \lambda^\kappa, \quad \dots\dots\dots(8')$$

and the remaining potent divisors of $\psi(\lambda)$ *are*

$$(\lambda + \alpha - c_1)^{e_1}, \quad (\lambda + \alpha - c_2)^{e_2}, \ \dots \ (\lambda + \alpha - c_s)^{e_s}. \quad \dots\dots\dots\dots(9')$$

Therefore if ψ_1 *is any first transmute of* ψ, *and if* $\psi_1(\lambda)$ *is the characteristic matrix of* ψ_1, *then the potent divisors of* $\psi_1(\lambda)$ *corresponding to the linear divisor* λ *are*

$$\sigma_2 \text{ equal to } \lambda, \quad \sigma_3 \text{ equal to } \lambda^2, \ \dots \ \sigma_\kappa \text{ equal to } \lambda^{\kappa-1}, \ \dots\dots\dots(8'')$$

and (9′) *are the remaining potent divisors of* $\psi_1(\lambda)$.

The first part of this theorem follows from the correspondences mentioned above; and we deduce the second part by applying Theorem III to the matrix ψ.

Ex. xvii. *If* ρ_1 *and* ρ_2 *are the degeneracies of* ψ *and* ψ_1, *then* $\phi(\lambda)$ *has exactly* ρ_1 *potent divisors divisible by* $\lambda - a$, *and exactly* ρ_2 *potent divisors divisible by* $(\lambda - a)^2$. *Consequently it has exactly* $\rho_1 - \rho_2$ *potent divisors equal to* $\lambda - a$.

This follows from Ex. xiii.

Ex. xviii. *If* a *is a latent root of* ϕ, *i.e. if the matrix* $\psi = \phi - aI$ *is degenerate, then the potent divisors of* $\phi(\lambda)$ *corresponding to the linear divisor* $\lambda - a$ *are all linear, i.e. are all equal to* $\lambda - a$, *when and only when the first transmutes of* ψ *are undegenerate.*

This case occurs when and only when the number of latent roots of ϕ having the value a is equal to the degeneracy of ψ, or when and only when the number of potent divisors of $\phi(\lambda)$ corresponding to the linear divisor $\lambda - a$ is equal to the degeneracy of ψ.

Ex. xix. *The matrix* ϕ *has an unrepeated latent root* a *when and only when the matrix* $\psi = \phi - aI$ *has rank* $m - 1$ *and* ψ_1 *is undegenerate.*

With the notation of Ex. xvii this case occurs when and only when $\rho_1 = 1$ and $\rho_2 = 0$.

Ex. xx. *Square matrices having only linear potent divisors.*

All the potent divisors of the square matrix ϕ are linear when and only when the first transmutes of the matrix $\psi = \phi - aI$ are undegenerate for all latent roots a of ϕ, and therefore for all values of a.

Ex. xxi. *Square matrices having only unrepeated latent roots.*

All the latent roots of the square matrix ϕ are unrepeated when and only when the matrix $\psi = \phi - aI$ has degeneracy 1 for all latent roots a of ϕ, and the first transmutes of ψ are undegenerate for all latent roots a of ϕ, i.e. for all values of a.

Ex. xxii. *Properties of any two square matrices ϕ and ψ which are expressed in the forms*

$$\phi=[a]_m^r\,[b]_r^m,\quad \psi=[b]_r^m\,[a]_m^r.$$

The following properties are true both when ϕ has rank r, so that ψ is a first transmute of ϕ, and when ϕ has rank less than r.

(1) *If s is any non-zero positive integer such that $s \not> m$ and $s \not> r$, the sum of the corranged (or affected) diagonal minor determinants of ϕ of order s is equal to the sum of the corranged (or affected) diagonal minor determinants of ψ of order s.*

For if I_s and J_s are the sums of the corranged diagonal minor determinants of ϕ and ψ of order s, we have

$$I_s=\sum_p \det\,[a_{p1}]_s^r\,[b_{1p}]_r^s=\sum_p\sum_q (a_{pq})_s^s\,(b_{qp})_s^s,$$

$$J_s=\sum_q \det\,[b_{q1}]_s^m\,[a_{1q}]_m^s=\sum_p\sum_q (b_{qp})_s^s\,(a_{pq})_s^s,$$

where $[p]_s$ and $[q]_s$ are corranged minors of the sequences $[1\,2\,\ldots\,m]$ and $[1\,2\ldots r]$, and where $\sum_p$ and $\sum_q$ denote summations for all distinct values of $[p]_s$ and $[q]_s$ respectively.

(2) *The matrices ϕ and ψ have the same non-zero latent roots.*

Suppose that $r \not> m$, and let

$$\phi(\lambda)=[a]_m^r\,[b]_r^m-\lambda\,[1]_m^m,\quad \psi(\lambda)=[b]_r^m\,[a]_m^r-\lambda\,[1]_r^r.$$

Then $$\det\phi(\lambda)=(-1)^m\{\lambda^m-I_1\lambda^{m-1}+\ldots+(-1)^r I_r\lambda^{m-r}\}$$

and $$\det\psi(\lambda)=(-1)^r\{\lambda^r-I_1\lambda^{r-1}+\ldots+(-1)^r I_r\};$$

and therefore $$\det\phi(\lambda)=(-\lambda)^{m-r}\det\psi(\lambda).$$

Thus ϕ has certainly $m-r$ zero latent roots, and its remaining r latent roots are identical with the r latent roots of ψ, each of which may or may not be zero.

Ex. xxiii. *If $\phi=[a]_m^m$ is a square matrix of order m and rank r, and if $\phi(\lambda)=[a]_m^m-\lambda\,[1]_m^m$, then λ is not a common factor of all minor determinants of $\phi(\lambda)$ of any one of the orders $1, 2, \ldots r$; but if $r<m$, then all minor determinants of $\phi(\lambda)$ of orders $r+1, r+2, \ldots s, \ldots m$ are divisible respectively by $\lambda, \lambda^2, \ldots \lambda^{s-r}, \ldots \lambda^{m-r}$.*

This follows from Ex. iii of § 228. We can prove it in a more elementary way by means of the expansion given in Ex. i of § 222.

Let D_s be any corranged minor determinant of $\phi(\lambda)$ of order s, and let it contain exactly σ of the diagonal elements of $\phi(\lambda)$. Then D_s is a derangement of a determinant of the form

$$\Delta_s=\det\left\{\begin{bmatrix} a_{\tau\tau}, & a_{\tau v}\\ a_{u\tau}, & a_{uv}\end{bmatrix}_{\sigma,\,s-\sigma}^{\sigma,\,s-\sigma}-\lambda\begin{bmatrix}1, & 0\\ 0, & 0\end{bmatrix}_{\sigma,\,s-\sigma}^{\sigma,\,s-\sigma}\right\},$$

where $[\tau, u, v]_{\sigma,\,s-\sigma,\,s-\sigma}$ is a minor of the sequence $[1\ 2\ \ldots\ m]$; and σ can have any value consistent with the conditions $\sigma \not< 0$, $\sigma \not> s$, $m+\sigma \not< 2s$.

Expanding Δ_s in powers of λ, we have

$$\Delta_s = (-1)^\sigma \{I_{s-\sigma}\lambda^\sigma - I_{s-\sigma+1}\lambda^{\sigma-1} + \ldots + (-1)^{\sigma-1} I_{s-1}\lambda + (-1)^\sigma I_s\},$$

where $$I_{s-\sigma} = (a_{uv})_{s-\sigma}^{s-\sigma}, \quad I_s = \begin{pmatrix} a_{\tau\tau}, & a_{\tau v} \\ a_{u\tau}, & a_{uv} \end{pmatrix}_{\sigma,\, s-\sigma}^{\sigma,\, s-\sigma},$$

and where $I_{s\ \sigma+k}$ is the sum of those corranged diagonal minor determinants of I_s of order $s-\sigma+k$ which contain $I_{s-\sigma}$, all of which are minor determinants of ϕ. We observe that I_s may be any minor determinant of ϕ of order s.

If $s=r$, we can have $I_s \neq 0$, and then Δ_s is not divisible by λ.

If $s>r$, then $I_s, I_{s-1}, \ldots I_{r+1}$ all vanish, and therefore Δ_s is always divisible by λ^{s-r}.

From Ex. xxii we see that λ^{m-r} is the highest power of λ which is a factor of $\det \phi(\lambda)$ when and only when the first transmutes of ϕ are undegenerate. This shows independently of Theorem III that the potent divisors of $\phi(\lambda)$ corresponding to its linear divisor λ are all linear, i.e. all have the value λ, when and only when the first transmutes of ϕ are undegenerate.

§ 233. The successive transmutes of a square matrix.

Let ϕ be any square matrix whose elements are constants, and let ϕ_1 be a first transmute of ϕ, ϕ_2 a first transmute of ϕ_1, ϕ_3 a first transmute of ϕ_2, and so on, so that in the series of square matrices

$$\phi, \phi_1, \phi_2, \phi_3, \ldots \phi_{i-1}, \phi_i, \phi_{i+1}, \ldots$$

each matrix after the first is a first transmute of the preceding matrix. Then $\phi_1, \phi_2, \phi_3, \ldots \phi_i, \ldots$ will be called *successive transmutes* of ϕ, and ϕ_i will be called an *ith transmute* of ϕ or a transmute of the ith order. An ith transmute of ϕ can be formed in a great variety of ways, but from § 232 we see that the two following theorems are true.

Theorem I. *All the ith transmutes of a given square matrix ϕ are square matrices of the same order and the same rank which have equipotent characteristic matrices, i.e. which are convertible into one another by equimutant transformations. Further if ϕ_i is any one ith transmute of ϕ, then every square matrix of the same order as ϕ_i derived from ϕ_i by an equimutant transformation is an ith transmute of ϕ.*

Thus all the ith transmutes of ϕ have the same standard reduced form for equimutant transformations.

Theorem II. *If ϕ and ϕ' are two square matrices of the same order which have equipotent characteristic matrices, then the ith transmutes of ϕ and the ith transmutes of ϕ' have the same common order and the same common rank, and have equipotent characteristic matrices. Consequently ϕ and ϕ' have the same ith transmutes, where i is any non-zero positive integer, i.e. they have the same successive transmutes of all orders starting from the first.*

Clearly if two square matrices of the same order have any one set of successive transmutes in common, then every set of successive transmutes of either matrix is also a set of successive transmutes of the other matrix. Further if any two square matrices have an ith transmute in common, then they have the same ith, $(i+1)$th, $(i+2)$th, ... transmutes; and if they have a first transmute in common, then they have the same successive transmutes of all orders starting from the first.

NOTE. *The 0th transmutes of a square matrix ϕ.*

For the sake of generality we will regard ϕ itself as a 0th transmute of ϕ. Then every square matrix of the same order as ϕ whose characteristic matrix is equipotent with the characteristic matrix of ϕ must be regarded as a 0th transmute of ϕ, and may be denoted by ϕ_0. We shall generally assume that $\phi_0=\phi$.

Let the order of the square matrix ϕ be $m=m_0$; and let the square matrices $\phi, \phi_1, \phi_2, \phi_3, \ldots$ have ranks $m_1, m_2, m_3, m_4, \ldots$ and degeneracies $\rho_1, \rho_2, \rho_3, \rho_4$; so that

$$\phi_i \text{ has order } m_i, \text{ rank } m_{i+1}, \text{ and degeneracy } \rho_{i+1}.$$

Then we can write

$$\begin{aligned} \phi &= [a]_m^m &&= [\alpha]_m^{m_1}\,[\alpha']_{m_1}^m, \\ \phi_1 &= [\alpha']_{m_1}^m\,[\alpha]_m^{m_1} &&= [\beta]_{m_1}^{m_2}\,[\beta']_{m_2}^{m_1}, \\ \phi_2 &= [\beta']_{m_2}^{m_1}\,[\beta]_{m_1}^{m_2} &&= [\gamma]_{m_2}^{m_3}\,[\gamma']_{m_3}^{m_2}, \\ \phi_3 &= [\gamma']_{m_3}^{m_2}\,[\gamma]_{m_2}^{m_3} &&= [\delta]_{m_3}^{m_4}\,[\delta']_{m_4}^{m_3}, \end{aligned} \quad \ldots\ldots\ldots\ldots(1)$$

and so on, where $\phi=\phi_0$ and $m=m_0$. Since the rank m_{i+1} of ϕ_i cannot exceed its order m_i, and since by Ex. i of § 232 the degeneracy of ϕ_i cannot exceed the degeneracy of ϕ_{i-1}, we have

$$m_{i+1} \not> m_i; \quad \text{and} \quad \rho_{i+1} \not> \rho_i, \quad \text{i.e. } m_i - m_{i+1} \not> m_{i-1} - m_i. \quad \ldots\ldots(2)$$

If ϕ is undegenerate, then all its successive transmutes are undegenerate, and have the same order and rank as ϕ. If ϕ is degenerate, then in forming the successive transmutes $\phi_1, \phi_2, \phi_3, \ldots$ we must ultimately arrive either at one which is undegenerate (in which case all the succeeding transmutes are undegenerate) or at one which vanishes and has zero rank; and if we regard the first transmute of a square matrix of zero rank as an undegenerate square matrix of order 0 and rank 0, we may say that we must ultimately arrive at a transmute which is undegenerate. If then ϕ_k is the first of the matrices $\phi, \phi_1, \phi_2, \phi_3, \ldots$ which is undegenerate (or if ϕ_{k-1} is the last of them which is degenerate), we have

$$m_1 < m, \quad m_2 < m_1, \quad \ldots \quad m_k < m_{k-1}; \quad m_k = m_{k+1} = m_{k+2} = \ldots;$$

i.e. $$\rho_1 > 0, \quad \rho_2 > 0, \quad \ldots \quad \rho_k > 0; \quad \rho_{k+1} = \rho_{k+2} = \ldots = 0. \quad \ldots\ldots\ldots(3)$$

From these results and (2) we see that the ranks of the successive matrices $\phi, \phi_1, \phi_2, \ldots \phi_{k-1}, \phi_k, \phi_{k+1}, \ldots$ constantly diminish by never increasing amounts until we reach ϕ_{k-1}, and that all of these matrices which come after ϕ_{k-1} are undegenerate square matrices of the same order, their common order and rank being equal to the rank m_k of ϕ_{k-1}.

Now let $\phi(\lambda)$, $\phi_1(\lambda)$, $\phi_2(\lambda)$, ... denote the characteristic matrices of $\phi, \phi_1, \phi_2, \ldots$; let the potent divisors of $\phi(\lambda)$ corresponding to the linear divisor λ be

$$\sigma_1 \text{ equal to } \lambda, \quad \sigma_2 \text{ equal to } \lambda^2, \quad \ldots \quad \sigma_\kappa \text{ equal to } \lambda^\kappa; \qquad \text{......(a)}$$

and let the remaining potent divisors of $\phi(\lambda)$ be

$$(\lambda - c_1)^{e_1}, \quad (\lambda - c_2)^{e_2}, \quad \ldots \quad (\lambda - c_s)^{e_s}, \qquad \text{..................(b)}$$

where $c_1, c_2, \ldots c_s$ are all different from 0. Then by successive applications of Theorem III of § 232 we obtain the following results:

If $\kappa > 1$, the potent divisors of $\phi_1(\lambda)$ corresponding to the linear divisor λ are

$$\sigma_2 \textit{ equal to } \lambda, \quad \sigma_3 \textit{ equal to } \lambda^2, \quad \ldots \quad \sigma_\kappa \textit{ equal to } \lambda^{\kappa-1}. \qquad \text{......(a}_1\text{)}$$

If $\kappa > 2$, the potent divisors of $\phi_2(\lambda)$ corresponding to the linear divisor λ are

$$\sigma_3 \textit{ equal to } \lambda, \quad \sigma_4 \textit{ equal to } \lambda^2, \quad \ldots \quad \sigma_\kappa \textit{ equal to } \lambda^{\kappa-2}. \qquad \text{......(a}_2\text{)}$$

....

If $\kappa > i$, the potent divisors of $\phi_i(\lambda)$ corresponding to the linear divisor λ are

$$\sigma_{i+1} \textit{ equal to } \lambda, \quad \sigma_{i+2} \textit{ equal to } \lambda^2, \quad \ldots \quad \sigma_\kappa \textit{ equal to } \lambda^{\kappa-i}. \qquad \text{......(a}_i\text{)}$$

....

If $\kappa > 0$, the potent divisors of $\phi_{\kappa-1}(\lambda)$ corresponding to the linear divisor λ are

$$\sigma_\kappa \textit{ equal to } \lambda. \qquad \text{..........................(a}_{\kappa-1}\text{)}$$

The matrices $\phi_\kappa(\lambda)$, $\phi_{\kappa+1}(\lambda)$, ... have no potent divisors which are powers of λ.

The potent divisors of all the matrices $\phi_1(\lambda)$, $\phi_2(\lambda)$, ... $\phi_{\kappa-1}(\lambda)$, $\phi_\kappa(\lambda)$, $\phi_{\kappa+1}(\lambda)$, ... corresponding to linear divisors other than λ are the same, viz. those given in (b).

Hence if we write $e_1 + e_2 + \ldots + e_s = e$, we have

$$\begin{aligned}
\sigma_1 + 2\sigma_2 + 3\sigma_3 + \ldots + \kappa\sigma_\kappa + e &= m; & \sigma_1 + \sigma_2 + \sigma_3 + \ldots + \sigma_\kappa &= \rho_1;\\
\sigma_2 + 2\sigma_3 + \ldots + (\kappa - 1)\sigma_\kappa + e &= m_1; & \sigma_2 + \sigma_3 + \ldots + \sigma_\kappa &= \rho_2;\\
\sigma_3 + \ldots + (\kappa - 2)\sigma_\kappa + e &= m_2; & \sigma_3 + \ldots + \sigma_\kappa &= \rho_3;\\
&\ldots\ldots & &\ldots\ldots\\
\sigma_\kappa + e &= m_{\kappa-1}; & \sigma_\kappa &= \rho_\kappa;
\end{aligned}$$

$$m_\kappa = m_{\kappa+1} = m_{\kappa+2} = \ldots = e; \quad \rho_{\kappa+1} = \rho_{\kappa+2} = \rho_{\kappa+3} = \ldots = 0. \qquad \text{......(4)}$$

The equations on the left follow from the fact that the product of all the potent divisors of $\phi_i(\lambda)$ must have degree m_i in λ; and the equations on the right follow from those on the left, or can be deduced independently from Ex. iii of § 228. From the equations (4) we see that

$$\sigma_1 = \rho_1 - \rho_2,\quad \sigma_2 = \rho_2 - \rho_3,\quad \dots\quad \sigma_i = \rho_i - \rho_{i+1},\quad \dots\quad \sigma_\kappa = \rho_\kappa, \quad \dots\dots(5)$$

and that ϕ_κ is the first of the matrices $\phi, \phi_1, \phi_2, \dots$ which is undegenerate (or $\phi_{\kappa-1}$ is the last which is degenerate). Accordingly the number k in (3) is κ.

We are thus led to the first three of the following four theorems in which $\phi_1, \phi_2, \phi_3, \dots$ are any set of successive transmutes of the square matrix ϕ, and in which $\phi(\lambda), \phi_1(\lambda), \phi_2(\lambda), \dots$ are the characteristic matrices of $\phi, \phi_1, \phi_2, \dots$.

Theorem III a. *If $\phi = [a]_m^m$ is any square matrix whose elements are numerical constants, and if ϕ_κ is the first of the matrices $\phi, \phi_1, \phi_2, \phi_3, \dots$ which is undegenerate (or $\phi_{\kappa-1}$ the last which is degenerate), then λ^κ is the highest power of λ which is a potent divisor of $\phi(\lambda)$.*

Theorem III b. *Moreover if $\rho_1, \rho_2, \rho_3, \dots \rho_\kappa$ are the degeneracies of $\phi, \phi_1, \phi_2, \dots \phi_{\kappa-1}$, then the potent divisors of $\phi(\lambda)$ corresponding to the linear divisor λ are*

$$\left.\begin{array}{l} \rho_1 \textit{ divisible by } \lambda \\ \rho_2 \textit{ divisible by } \lambda^2 \\ \dots\dots\dots\dots \\ \rho_i \textit{ divisible by } \lambda^i \\ \dots\dots\dots\dots \\ \rho_\kappa \textit{ divisible by } \lambda^\kappa \end{array}\right\} \textit{ or } \left\{\begin{array}{ll} \rho_1 - \rho_2 & \textit{equal to } \lambda \\ \rho_2 - \rho_3 & \textit{equal to } \lambda^2 \\ \dots\dots\dots & \\ \rho_i - \rho_{i+1} & \textit{equal to } \lambda^i \\ \dots\dots\dots & \\ \rho_\kappa & \textit{equal to } \lambda^\kappa. \end{array}\right.$$

Theorem III c. *Further the potent divisors of $\phi(\lambda)$ corresponding to all linear divisors other than λ are identical with the potent divisors of $\phi_\kappa(\lambda)$, which are also the potent divisors of each of the matrices $\phi_{\kappa+1}(\lambda), \phi_{\kappa+2}(\lambda), \dots$.*

Theorem III d. *If k is any non-zero positive integer, then ϕ^k has the same rank m_k as ϕ_{k-1}, and more generally ϕ_i^k has the same rank m_{i+k} as ϕ_{i+k-1}. Consequently if r_k is the degeneracy of ϕ^k, we have*

$$r_k = m - m_k = \rho_1 + \rho_2 + \dots + \rho_k.$$

To prove the fourth theorem, we observe that with the notation (1) we have

$$\phi^k = [a]_m^{m_1} \phi_1^{k-1} [a']_{m_1}^m, \quad \phi_1^k = [\beta]_{m_1}^{m_2} \phi_2^{k-1} [\beta']_{m_2}^{m_1}, \quad \phi_2^k = [\gamma]_{m_2}^{m_3} \phi_3^{k-1} [\gamma']_{m_3}^{m_2}, \quad \dots.$$

Therefore by Theorem II of § 131 the matrices

$$\phi^k,\ \phi_1^k,\ \phi_2^k,\ \dots\ \phi_i^k \text{ have the same ranks as } \phi_1^{k-1},\ \phi_2^{k-1},\ \phi_3^{k-1},\ \dots\ \phi_{i+1}^{k-1};$$

and by repeated applications of this result we see that

$$\phi_i^k \text{ has the same rank as } \phi_{i+1}^{k-1},\ \phi_{i+2}^{k-2},\ \dots\ \phi_{i+r}^{k-r},\ \dots\ \phi_{i+k-2}^2,\ \phi_{i+k-1}.$$

Theorems III*a*—III*c* show that if two square matrices ϕ and ϕ' of the same order have the same successive transmutes, then their characteristic matrices are equipotent, i.e. ϕ and ϕ' are convertible into one another by equimutant transformations; and conversely if they have equipotent characteristic matrices, then by Theorems I and II they have the same successive transmutes. More particularly if ϕ and ϕ' have the same rank, if their successive transmutes have the same ranks, and if they have a transmute of any order in common, then they have equipotent characteristic matrices and are convertible into one another by equimutant transformations.

Further these theorems show that we can determine those potent divisors of the characteristic matrix $\phi(\lambda)$ of a square matrix ϕ which correspond to the linear divisor λ (i.e. which correspond to the zero latent roots of ϕ) by determining the ranks of any set of successive transmutes of ϕ, and also by determining the ranks of the successive powers of ϕ.

Theorem III*d* shows that $\rho_1, \rho_2, \rho_3, \ldots r_1, r_2, r_3, \ldots$ have the same meanings in the above theorems as in Theorem II*a* of § 230; and we see (as in Ex. ii of § 230) that $m_\kappa = m - r_\kappa$ is the lowest rank which a power of ϕ can have, and that ϕ^κ is the lowest power of ϕ which has rank m_κ.

Ex. i. If ϕ is undegenerate, then $\rho_1 = 0$, $\kappa = 0$, and λ is not a linear divisor of $\phi(\lambda)$, i.e. ϕ has no zero latent root.

Ex. ii. *When ϕ is degenerate the potent divisor and the maximum divisor of $\phi(\lambda)$ of order m corresponding to the linear divisor λ are respectively λ^κ and λ^s, where*

$$s = \rho_1 + \rho_2 + \ldots + \rho_\kappa = m - m_\kappa = r_\kappa.$$

Ex. iii. If s is the total number of zero latent roots of ϕ, so that $s = m - m_\kappa = r_\kappa$, $m_\kappa = m - s$, then:

The ranks of $\phi, \phi_1, \phi_2, \phi_3, \ldots$ continually diminish by never increasing amounts till we reach the transmute $\phi_{\kappa-1}$ which has rank $m - s$, and all the following transmutes are undegenerate square matrices of order and rank $m - s$.

This result corresponds exactly to that given in Ex. ii of § 230.

Ex. iv. By repeated applications of Ex. xxii of § 232 we see that

$$\det \phi(\lambda) = (-\lambda)^{m-m_1} \det \phi_1(\lambda) = (-\lambda)^{m-m_2} \det \phi_2(\lambda) = \ldots$$
$$= (-\lambda)^{m-m_i} \det \phi_i(\lambda) = \ldots = (-\lambda)^{m-m_\kappa} \det \phi_\kappa(\lambda).$$

Thus if $i < \kappa$, the total number of zero latent roots of the ith transmute ϕ_i is

$$m_i - m_\kappa = \rho_{i+1} + \rho_{i+2} + \ldots + \rho_\kappa,$$

i.e. it is the sum of the degeneracies of ϕ_i and all the following transmutes.

Ex. v. All the matrices $\phi, \phi_1, \phi_2, \phi_3, \ldots$ have the same non-zero latent roots, and these are the m_κ latent roots of ϕ_κ.

Ex. vi. In the case when ϕ has one or more zero latent roots or is degenerate, the potent divisors of $\phi(\lambda)$ corresponding to the linear divisor λ are all linear (see also Ex. v of § 230) when and only when $\rho_2=0$, i.e. when and only when the first transmutes of ϕ are undegenerate.

Ex. vii. The matrix ϕ has an unrepeated zero latent root (see also Ex. vi of § 230) when and only when ϕ has rank $m-1$ and the first transmutes of ϕ are undegenerate.

Ex. viii. *Conditions that $\phi(\lambda)$ shall have only the one potent divisor λ^m.*

This is the case when and only when $\kappa=m$, and we then have

$$\rho_1=\rho_2=\ldots=\rho_m=1;$$

or it is the case when and only when ϕ and its successive transmutes down to the $(m-1)$th all have degeneracy 1, and we then have $\phi_{m-1}=[0]_1^1$, the mth and following transmutes all having order and rank 0, or being non-existent.

The condition that $\phi(\lambda)$ shall have only one potent divisor which is a power of λ is

$$\rho_1=\rho_2=\ldots=\rho_\kappa=1.$$

Ex. ix. *Successive transmutes of the square matrix* $\phi=[a]_m^m=\begin{bmatrix}0, & 1\\ 0, & 0\end{bmatrix}_{m-1,\,1}^{1,\,m-1}$.

Referring to Ex. vi of § 232 we see that if $i \not> m$, we can write

$$\phi_{i-1}=\begin{bmatrix}0, & 1\\ 0, & 0\end{bmatrix}_{m-i,\,1}^{1,\,m-i}, \quad \ldots \quad \phi_{m-2}=\begin{bmatrix}0, & 1\\ 0, & 0\end{bmatrix}_{1,\,1}^{1,\,1}, \quad \phi_{m-1}=[0]_1^1;$$

so that in this case $\qquad \kappa=m, \quad \rho_1=\rho_2=\ldots=\rho_m=1.$

It follows from Theorem III *b* that the characteristic matrix $\phi(\lambda)$ has only the one potent divisor λ^m, and that this is therefore also true of the square matrix

$$[\omega]_m^m=\begin{bmatrix}\lambda & 1 & 0 & \ldots & 0 & 0\\ 0 & \lambda & 1 & \ldots & 0 & 0\\ \ldots & \ldots & \ldots & \ldots & \ldots & \ldots\\ 0 & 0 & 0 & \ldots & \lambda & 1\\ 0 & 0 & 0 & \ldots & 0 & \lambda\end{bmatrix}=\phi(-\lambda).$$

Ex. x. *Successive transmutes of* $\phi=[a]_m^m=\begin{bmatrix}0, & 1\\ 0, & 0\end{bmatrix}_{m-p,\,p}^{p,\,m-p}$, *where* $p \not< 1$ *and* $p \not> m$.

Writing $m=np+q$, where n and q are positive integers such that $n \not< 1$ and $q \not> p-1$, we see from Ex. vii of § 232 that, if $i \not> n$, we can take the successive transmutes of ϕ to be

$$\phi_{i-1}=\begin{bmatrix}0, & 1\\ 0, & 0\end{bmatrix}_{(n-i)p+q,\,q}^{p,\,(n-i)p+q}, \quad \ldots \quad \phi_{n-1}=\begin{bmatrix}0, & 1\\ 0, & 0\end{bmatrix}_{q,\,p}^{p,\,q}, \quad \phi_n=[0]_q^q$$

when $q\neq0$; and that we have the same series terminating with $\phi_{n-1}=[0]_p^p$ when $q=0$. Accordingly in this case we have

$$\kappa=n+1; \quad \rho_1=\rho_2=\ldots=\rho_n=p, \quad \rho_{n+1}=q \qquad \text{when } q\neq0;$$

$$\kappa=n; \quad \rho_1=\rho_2=\ldots=\rho_n=p \qquad \text{when } q=0.$$

It follows from Theorem III *b* that the potent divisors of the characteristic matrix $\phi(\lambda)$ are p in number, viz.

$$p-q \text{ equal to } \lambda^n \text{ and } q \text{ equal to } \lambda^{n+1} \qquad \text{when } q \neq 0;$$

$$p \text{ equal to } \lambda^n \qquad \text{when } q=0.$$

There are no potent divisors other than those corresponding to the linear divisor λ because in both cases $\phi_{\kappa-1}=0$ and ϕ_κ is non-existent.

These results agree with Ex. viii of § 229 and Ex. vii of § 207.

To determine the potent divisors of $\phi(\lambda)$ corresponding to *any* linear divisor $\lambda-\alpha$ we may use the following more general theorems in which $\psi_1, \psi_2, \psi_3, \ldots$ are any set of successive transmutes of the matrix

$$\psi=\phi-\alpha I=[a]_m^m-\alpha[1]_m^m,$$

and in which $\psi(\lambda), \psi_1(\lambda), \psi_2(\lambda), \ldots$ are the characteristic matrices of $\psi, \psi_1, \psi_2, \ldots$. In these theorems and in the examples which follow $m_1, m_2, m_3, \ldots$ and $\rho_1, \rho_2, \rho_3, \ldots$ are the ranks and degeneracies of $\psi, \psi_1, \psi_2, \ldots$, so that $m, m_1, m_2, \ldots$ are the orders of $\psi, \psi_1, \psi_2, \ldots$; and therefore

$$\psi_i \text{ has order } m_i, \text{ rank } m_{i+1}, \text{ and degeneracy } \rho_{i+1}.$$

Theorem IV a. *If* $\phi=[a]_m^m$, $\psi=\phi-\alpha I=[a]_m^m-\alpha[1]_m^m$, *and if* ψ_κ *is the first of the matrices* $\psi, \psi_1, \psi_2, \psi_3, \ldots$ *which is undegenerate (or* $\psi_{\kappa-1}$ *the last which is degenerate), then* $(\lambda-\alpha)^\kappa$ *is the highest power of* $\lambda-\alpha$ *which is a potent divisor of* $\phi(\lambda)$.

Theorem IV b. *Moreover if* $\rho_1, \rho_2, \rho_3, \ldots \rho_\kappa$ *are the degeneracies of* $\psi, \psi_1, \psi_2, \ldots \psi_{\kappa-1}$, *then the potent divisors of* $\phi(\lambda)$ *corresponding to the linear divisor* $\lambda-\alpha$ *are*

ρ_1 *divisible by* $\lambda-\alpha$			$\rho_1-\rho_2$	*equal to* $\lambda-\alpha$
ρ_2 *divisible by* $(\lambda-\alpha)^2$			$\rho_2-\rho_3$	*equal to* $(\lambda-\alpha)^2$
....			....	
ρ_i *divisible by* $(\lambda-\alpha)^i$	*or*		$\rho_i-\rho_{i+1}$	*equal to* $(\lambda-\alpha)^i$
....			....	
ρ_κ *divisible by* $(\lambda-\alpha)^\kappa$			ρ_κ	*equal to* $(\lambda-\alpha)^\kappa$.

Theorem IV c. *Further if the potent divisors of* $\psi_\kappa(\lambda)$ *are*

$$(\lambda-c_1)^{e_1},\ (\lambda-c_2)^{e_2},\ (\lambda-c_3)^{e_3},\ \ldots,$$

then $c_1, c_2, c_3, \ldots$ *are all different from* 0, *and the potent divisors of* $\phi(\lambda)$ *corresponding to all linear divisors other than* $\lambda-\alpha$ *are*

$$(\lambda-\alpha-c_1)^{e_1},\ (\lambda-\alpha-c_2)^{e_2},\ (\lambda-\alpha-c_3)^{e_3},\ \ldots.$$

Theorem IV d. *If k is any non-zero positive integer, then ψ^k has the same rank m_k as ψ_{k-1}. Consequently if r_k is the degeneracy of the matrix $\psi^k = (\phi - \alpha I)^k$, we have*

$$r_k = m - m_k = \rho_1 + \rho_2 + \ldots + \rho_k.$$

We deduce Theorems IVa—IVd from Theorems IIIa—IIId in the same way as we deduced Theorem IV from Theorem III in § 232, i.e. by using the fact that $(\lambda - c)^e$ is a potent divisor of $\psi(\lambda)$ of order i when and only when $(\lambda - \alpha - c)^e$ is a potent divisor of $\phi(\lambda)$ of order i. These theorems show that we can determine those potent divisors of the characteristic matrix $\phi(\lambda)$ of the square matrix ϕ which correspond to any linear divisor $\lambda - \alpha$ (i.e. which correspond to any latent root α of ϕ) by determining the ranks of any set of successive transmutes of the matrix $\psi = \phi - \alpha I$, and also by determining the ranks of the successive powers of ψ.

Theorem IVd shows that in these theorems $\rho_1, \rho_2, \rho_3, \ldots r_1, r_2, r_3, \ldots$ have the same meanings as in Theorem IIb of § 230; and (as in Ex. ix of § 230) we see that $m_\kappa = m - r_\kappa$ is the lowest rank which a power of the matrix $\psi = \phi - \alpha I$ can have, and that $(\phi - \alpha I)^\kappa$ is the lowest power of $\phi - \alpha I$ which has rank m_κ.

Ex. xi. If the matrix $\psi = \phi - aI$ is undegenerate, then $\rho_1 = 0$, $\kappa = 0$, and $\lambda - a$ is not a linear divisor of $\phi(\lambda)$, i.e. a is not a latent root of ϕ.

Ex. xii. *When the matrix $\psi = \phi - aI$ is degenerate, the potent divisor and the maximum divisor of $\phi(\lambda)$ of order m corresponding to the linear divisor $\lambda - a$ are respectively $(\lambda - a)^\kappa$ and $(\lambda - a)^s$, where*

$$s = \rho_1 + \rho_2 + \ldots + \rho_\kappa = m - m_\kappa = r_\kappa.$$

Ex. xiii. If s is the total number of latent roots of ϕ which are equal to a, so that $s = m - m_\kappa = r_\kappa$, $m_\kappa = m - s$, then:

The ranks of $\psi, \psi_1, \psi_2, \psi_3, \ldots$ continually diminish by never increasing amounts till we reach the transmute $\psi_{\kappa-1}$ of ψ which has rank $m - s$, and all the following transmutes of ψ are undegenerate square matrices of order and rank $m - s$.

This result corresponds to and is equivalent to that given in Ex. ix of § 230.

Ex. xiv. The latent roots of ϕ which are not equal to a are the m_κ latent roots of ψ_κ each increased by a.

Ex. xv. In the case when a is a latent root of ϕ the potent divisors of $\phi(\lambda)$ corresponding to the linear divisor $\lambda - a$ are all linear (see also Ex. xii of § 230) when and only when $\rho_2 = 0$, i.e. when and only when the first transmutes of the matrix $\psi = \phi - aI$ are undegenerate.

Ex. xvi. The matrix ϕ has an unrepeated latent root equal to a (see also Ex. xiii of § 230) when and only when $\rho_1 = 1$ and $\rho_2 = 0$, i.e. when and only when $\phi - aI$ has rank $m - 1$ and the first transmutes of $\phi - aI$ are undegenerate.

Ex. xvii. *Condition that $\phi(\lambda)$ shall have only the one potent divisor $(\lambda - a)^m$.*

This is the case when and only when $\kappa = m$, i.e. when and only when the matrix $\psi = \phi - aI$ and its successive transmutes down to the $(m-1)$th all have degeneracy 1.

Ex. xviii. *Condition that the potent divisors of $\phi(\lambda)$ shall all be linear.*

The first transmutes of the matrix $\psi = \phi - aI$ must be undegenerate for all latent roots a of ϕ, and therefore for all values of the scalar quantity a.

§ 234. Second form of the necessary and sufficient conditions for the existence of an equimutant transformation between two given square matrices.

The results obtained in §§ 232 and 233 enable us to replace Theorem I of § 227 by the following theorem:

Theorem I. *Let $\phi = [a]_m^m$ and $\psi = [b]_m^m$ be two square matrices of the same order m whose elements are numerical constants lying in any domain of rationality Ω. Then there exists an equimutant transformation of the form*

$$[h]_m^m [a]_m^m \overline{\underline{H}}_m^m = [b]_m^m, \quad \left(\text{where } [h]_m^m \overline{\underline{H}}_m^m = \overline{\underline{H}}_m^m [h]_m^m = [1]_m^m\right), \quad \ldots(\text{A})$$

when and only when ϕ and ψ have the same first transmutes (and therefore the same successive transmutes of all orders starting from the first); and when this condition is satisfied, there exists an equimutant transformation in Ω of the form (A).

It follows from Theorem II of § 233 that if ϕ and ψ have any one first transmute in common, then they have the same successive transmutes of all orders starting from the first.

First suppose that there exists an equimutant transformation of the form (A). Then by Theorem II of § 232 the matrices ϕ and ψ have the same first transmutes.

Next suppose that ϕ and ψ have the same first transmutes. Then they necessarily have the same rank, and by Theorem II of § 233 they have the same successive transmutes of all orders starting from the first. It therefore follows from Theorems IIIb and IIIc of § 233 that their characteristic matrices have the same potent divisors, and are equipotent. Therefore by Theorem I of § 227 there exists an equimutant transformation in Ω of the form (A). It follows from (A) that ϕ and ψ also have the same 0th transmutes.

We can further replace Theorem II of § 227 by the following generalisation of the theorem given above:

Theorem II. *Let $\phi = [a]_r^r$ and $\psi = [b]_m^m$ be two square matrices of orders r and m whose elements are numerical constants lying in any domain of rationality Ω, r being not greater than m. Then there exists an equimutant transformation of the form*

$$[h]_m^r [a]_r^r \overline{\underline{H}}_r^m = [b]_m^m, \quad \left(\textit{where } \overline{\underline{H}}_r^m [h]_m^r = [1]_r^r\right), \text{.........(B)}$$

when and only when ϕ and ψ have the same first transmutes (and therefore the same successive transmutes of all orders starting from the first); and when this condition is satisfied, there exists an equimutant transformation in Ω of the form (B).

It follows from Theorem II of § 233 that if ϕ and ψ have any one first transmute in common, then they have the same successive transmutes of all orders starting from the first.

First suppose that there exists an equimutant transformation of the form (B). Then if $[h]_m^m$ and $\overline{\underline{H}}_m^m$ are two mutually inverse undegenerate square matrices formed as in Note 7 of § 141 by adding final vertical rows to $[h]_m^r$ and final horizontal rows to $\overline{\underline{H}}_r^m$, we have

$$[h]_m^m \begin{bmatrix} a, & 0 \\ 0, & 0 \end{bmatrix}_{r,\, m-r}^{r,\, m-r} \overline{\underline{H}}_m^m = [b]_m^m, \quad \left(\text{where } \overline{\underline{H}}_m^m [h]_m^m = [1]_m^m\right). \quad \text{...(B}')$$

Therefore by Theorem I the matrices

$$\Phi = \begin{bmatrix} a, & 0 \\ 0, & 0 \end{bmatrix}_{r,\, m-r}^{r,\, m-r}, \quad \psi = [b]_m^m$$

have the same first transmutes; therefore by Ex. viii of § 232 the matrices ϕ and ψ have the same first transmutes.

Next suppose that ϕ and ψ have the same first transmutes. Then by Ex. viii of § 232 the matrices Φ and ψ have the same first transmutes; therefore by Theorem I there exists an equimutant transformation in Ω of the form (B′), from which we can deduce an equimutant transformation in Ω of the form (B).

These proofs depend ultimately on the lemma of § 227; or, if we omit all reference to Ω, they depend on the possibility of reducing a square matrix to its canonical form by an equimutant transformation.

Ex. i. *Variation in the proof of Theorem II.*

We can avoid the use of Note 7 of § 141 in the first part of the proof of Theorem II in the following way.

Suppose that there exists an equimutant transformation of the form (B). Then ϕ and ψ have the same rank. Let their common rank be s, and let

$$[a]^r_r = [p]^s_r\,[q]^r_s, \quad \text{so that} \quad [b]^m_m = [h]^r_m\,[p]^s_r \cdot [q]^r_s\,\overline{\underline{H}}^m_r .$$

Then ϕ and ψ have the common first transmute

$$[q]^r_s\,\overline{\underline{H}}^m_r\,[h]^r_m\,[p]^s_r = [q]^r_s\,[p]^s_r .$$

Ex. ii. *If*

$$[x]^m_r\,[y]^r_m = [a]^m_r\,[b]^r_m ,$$

where all matrices have rank r, *so that* $r \ngtr m$, *and where all matrices lie in* Ω, *then there exists an equimutant transformation in* Ω *of the form*

$$[y]^r_m\,[x]^m_r = [h]^m_m\,[b]^r_m\,[a]^m_r\,\overline{\underline{H}}^m_m .$$

For $\phi = [b]^r_m\,[a]^m_r$ and $\psi = [y]^r_m\,[x]^m_r$ are two square matrices in Ω which have a common first transmute.

If we could give a direct proof of this result, we could deduce Theorem I from it and § 152.

Ex. iii. *If*

$$[x]^m_s\,[y]^s_m = [a]^r_s\,[b]^s_r ,$$

where $r \ngtr m$, *and where all matrices have rank* s *and lie in* Ω, *then there exists an equimutant transformation in* Ω *of the form*

$$[y]^s_m\,[x]^m_s = [h]^r_m\,[b]^s_r\,[a]^r_s\,\overline{\underline{H}}^m_r .$$

For $\phi = [b]^s_r\,[a]^r_s$ and $\psi = [y]^s_m\,[x]^m_s$ are two square matrices in Ω which have a common first transmute.

§ 235. Solutions of any equation of the form $f(\phi)\,\overline{\underline{x}}_m = 0$ when ϕ is a given square matrix of order m.

1. *Notation.*

It will be understood that $\phi = [a]^m_m$ is a square matrix of order m whose elements are given scalar numbers, and that $f(\phi)$ is some rational integral function of ϕ whose coefficients are given scalar numbers which are not all 0. As usual we will write $I = [1]^m_m$; and for the sake of brevity we will put

$$x = \overline{\underline{x}}_m, \qquad \alpha = \overline{\underline{\alpha}}_m, \qquad \beta = \overline{\underline{\beta}}_m, \ \ldots,$$

$$x_i = \begin{bmatrix} x_{i1} \\ x_{i2} \\ \vdots \\ x_{im} \end{bmatrix}, \quad \alpha_i = \begin{bmatrix} \alpha_{i1} \\ \alpha_{i2} \\ \vdots \\ \alpha_{im} \end{bmatrix}, \quad \beta_i = \begin{bmatrix} \beta_{i1} \\ \beta_{i2} \\ \vdots \\ \beta_{im} \end{bmatrix}, \ \ldots,$$

these letters denoting matrices, whilst $\lambda, \lambda_1, \lambda_2, \dots \mu, \mu_1, \mu_2, \dots$ will denote scalar numbers. The equation to be solved is then

$$f(\phi)\,x = 0.$$

We will regard z as a scalar variable, and take the characteristic matrix of ϕ to be

$$\phi(z) = \phi - zI = [a]_m^m - z\,[1]_m^m.$$

We will suppose that λ is a typical latent root of ϕ, and that $(z-\lambda)^s$ and $(z-\lambda)^\kappa$ are the maximum and potent divisors of $\phi - zI$ of order m corresponding to the linear divisor $z - \lambda$, so that λ is a latent root repeated exactly s times.

We will define m_k and r_k to be the rank and degeneracy of $(\phi - \lambda I)^k$, k being any positive integer, and ρ_k to be the number of potent divisors of $\phi - zI$ which are divisible by $(z-\lambda)^k$. Then m_k and ρ_k are the rank and degeneracy of the $(k-1)$th transmutes of $\phi - \lambda I$; the κth transmutes of $\phi - \lambda I$ are the first which are undegenerate, i.e. the $(\kappa - 1)$th transmutes of $\phi - \lambda I$ are the last which are degenerate; and we have

$$r_k = \rho_1 + \rho_2 + \dots + \rho_k = m - m_k, \quad r_\kappa = s = m - m_\kappa.$$

This notation is the same as that used in Ex. ix of § 230 and in § 233, and the positive integers ρ_k, m_k, r_k satisfy the conditions

$$\begin{array}{ll} \rho_1 \nless \rho_2 \nless \rho_3 \nless \dots \nless \rho_{\kappa-1} \nless \rho_\kappa \nless 1, & \rho_{\kappa+1} = \rho_{\kappa+2} = \rho_{\kappa+3} = \dots = 0; \\ m > m_1 > m_2 > \dots > m_{\kappa-1} > m_\kappa \quad , & m_\kappa = m_{\kappa+1} = m_{\kappa+2} = \dots \quad ; \\ 0 < r_1 < r_2 < \dots < r_{\kappa-1} < r_\kappa \quad , & r_\kappa = r_{\kappa+1} = r_{\kappa+2} = \dots \quad ; \end{array}$$

to which we may add the interpretations

$$m_0 = m, \quad r_0 = 0.$$

Further we will suppose that $\lambda_1, \lambda_2, \dots \lambda_n$ are all the distinct or unequal latent roots of ϕ; and we will define integers

$$s_i, \quad \kappa_i, \quad m_{ik}, \quad r_{ik}, \quad \rho_{ik}$$

for the latent root λ_i in the same way as we have defined the integers

$$s, \quad \kappa, \quad m_k, \quad r_k, \quad \rho_k$$

for the latent root λ; so that

$$s_1 + s_2 + \dots + s_n = m.$$

At the same time we may suppose that $\lambda_1, \lambda_2, \dots \lambda_r$, where $r \ngtr n$, are any r distinct or unequal latent roots of ϕ.

2. *Solutions of the special equation* $(\phi - \lambda I)^k x = 0$, *where* $k \ngtr \kappa$.

Let p be any positive integer. Then if μ is not a latent root of ϕ, the matrix $(\phi - \mu I)^p$ is undegenerate, and the equation

$$(\phi - \mu I)^p\, x = 0 \quad \dots\dots\dots\dots(1)$$

has no non-zero solution, and is therefore equivalent to the equation

$$(\phi - \mu I)^0 x = 0, \quad \text{i.e.} \quad Ix = x = 0.$$

If λ is a latent root of ϕ, the matrix $(\phi - \lambda I)^p$ has degeneracy $r_p = m - m_p$, and the equation

$$(\phi - \lambda I)^p x = 0 \quad \text{.............................(2)}$$

has exactly r_p unconnected non-zero solutions. If q is any positive integer less than p, then every solution of the equation $(\phi - \lambda I)^q x = 0$ is also a solution of (2). Consequently if $p > \kappa$, so that $r_p = r_\kappa$, every complete set of r_κ unconnected non-zero solutions of the equation

$$(\phi - \lambda I)^\kappa x = 0 \quad \text{.............................(3)}$$

is also a complete set of unconnected non-zero solutions of (2), i.e. the solutions of (2) are identical with the solutions of (3), and the equation (2) is equivalent to the equation (3). Thus every equation of the form (1) is reducible to one of the form

$$(\phi - \lambda I)^k x = 0, \quad \text{where } k \not> \kappa, \quad \text{.....................(A)}$$

λ being one of the latent roots of ϕ; and as regards the equations of the form (A) we have established the following theorem:

Theorem I a. *When λ is a latent root of ϕ, and $k \not> \kappa$, the respective equations*

$$(\phi - \lambda I) x = 0, \quad (\phi - \lambda I)^2 x = 0, \quad \dots \quad (\phi - \lambda I)^k x = 0, \quad \dots \quad (\phi - \lambda I)^\kappa x = 0 \quad \text{.........(A')}$$

have exactly $r_1, r_2, \dots r_k, \dots r_\kappa$ unconnected non-zero solutions; and all solutions of any one of these equations are also solutions of the following equations.

Clearly if $k \not> \kappa$, we can determine a complete set of r_k unconnected non-zero solutions of the equation $(\phi - \lambda I)^k x = 0$ of which r_{k-1} are given unconnected non-zero solutions of the equation $(\phi - \lambda I)^{k-1} x = 0$; and the remaining $r_k - r_{k-1}$ solutions are then solutions of $(\phi - \lambda I)^k x = 0$ which are not also solutions of $(\phi - \lambda I)^{k-1} x = 0$. By repeated applications of this result we obtain the following second theorem:

Theorem I b. *If $k \not> \kappa$, we can determine a complete set of r_k unconnected non-zero solutions of the equation* (A) *of which*

r_1 *are a complete set of unconnected solutions of* $(\phi - \lambda I) x = 0$,

$r_2 - r_1$ *are solutions of* $(\phi - \lambda I)^2 x = 0$ *but not of* $(\phi - \lambda I) x = 0$,

$r_3 - r_2$ *are solutions of* $(\phi - \lambda I)^3 x = 0$ *but not of* $(\phi - \lambda I)^2 x = 0$,

....

$r_k - r_{k-1}$ *are solutions of* $(\phi - \lambda I)^k x = 0$ *but not of* $(\phi - \lambda I)^{k-1} x = 0$.

The above two theorems could be deduced from Note 1, which also gives the general solution of the equation (A).

Ex. i. If we regard $x=\underset{m}{\overline{x}}$ as a point in homogeneous space ω_m of rank m, the solutions of the successive equations (A′) form a series of κ spacelets $S_1, S_2, \ldots S_k, \ldots S_\kappa$ of ranks $r_1, r_2, \ldots r_k, \ldots r_\kappa$ each of which is contained in the following.

NOTE 1. *Solution of the equation* $(\phi-\lambda I)^k x=0$ *by the reduction of* ϕ *to canonical form.*

Let $\Pi=[\Pi]_m^m$ be the canonical reduced form of the square matrix $\phi=[a]_m^m$, and let

$$[h]_m^m [\Pi]_m^m \overline{H}_m^m=[a]_m^m \qquad \text{.......(a)}$$

be an equimutant transformation by which ϕ can be derived from Π. Then there is a one-one correspondence between the solutions $x=\underset{m}{\overline{x}}$, $X=\underset{m}{\overline{X}}$ of the equations $(\phi-\lambda I)^k x=0$, $(\Pi-\lambda I)^k X=0$ such that

$$\underset{m}{\overline{x}}=[h]_m^m \underset{m}{\overline{X}}.$$

Therefore if $u=r_k$, the general solution of $(\phi-\lambda I)^k x=0$ can be expressed in the form

$$x=[h]_m^m \overline{X}_m^u \underset{u}{\overline{l}}, \qquad \text{.......(b)}$$

where $l_1, l_2, \ldots l_u$ are arbitrary scalar quantities, and where $[X]_u^m$ is an undegenerate horizontal normal to $(\Pi-\lambda I)^k$, so that

$$(\Pi-\lambda I)^k \overline{X}_m^u=0. \qquad \text{.......(c)}$$

We can choose $\overline{X}_m^u$ to be the matrix formed from $[\Pi]_m^m$ when we strike out all vertical rows which pass through parts of Π corresponding to latent roots of ϕ other than λ and replace each part

$$[\omega]_e^e=\begin{bmatrix}\lambda & 1 & 0 & \ldots & 0\\ 0 & \lambda & 1 & \ldots & 0\\ \ldots & \ldots & \ldots & \ldots & \ldots\\ 0 & 0 & 0 & \ldots & \lambda\end{bmatrix} \qquad \text{.......(d)}$$

of Π which corresponds to the latent root λ by $[1]_e^e$ when $e \ngtr k$, and by $\begin{bmatrix}1\\0\end{bmatrix}_{k,\,e-k}^{k}$ when $e \nless k$; $\overline{X}_m^u$ being then a horizontal derangement of $\begin{bmatrix}1\\0\end{bmatrix}_{u,\,m-u}^{u}$. In the particular case when $k=\kappa$, and when $k \nless \kappa$, we have $u=s$, and in forming $\overline{X}_m^s$ we replace every part of Π such as (d) corresponding to the latent root λ by $[1]_e^e$.

For if $\omega_1, \omega_2, \omega_3, \ldots$ are the parts of Π corresponding to the various potent divisors of the characteristic matrix $\phi(z)=\phi-zI$, then $\Pi-\lambda I$ is composed of similar corresponding parts which we may denote by $P_1, P_2, P_3, \ldots$; $(\Pi-\lambda I)^k$ is composed of similar corresponding parts $P_1^k, P_2^k, P_3^k, \ldots$; and $\overline{X}_m^u$ is a compartite matrix in standard form whose successive parts are the conjugates of horizontal normals to the successive parts of $(\Pi-\lambda I)^k$. If $\omega=[\omega]_e^e$ given by (d) is a part of Π corresponding to a potent divisor $(z-\lambda)^e$ of $\phi(z)$ which is a power of $z-\lambda$, the corresponding part of $\Pi-\lambda I$ is

$$P=\begin{bmatrix}0, & 1\\ 0, & 0\end{bmatrix}_{e-1,\,1}^{1,\,e-1};$$

if $k \not< e$, the corresponding part of $(\Pi - \lambda I)^k$ is $P^k = [0]^{e}_{e}$, and we can take the corresponding part of $\underset{m}{\overline{X}}^{u}$ to be $[1]^{e}_{e}$, since $[0]^{e}_{e}\,[1]^{e}_{e} = 0$; if $k \not> e$, the corresponding part of $(\Pi - \lambda I)^k$ is

$P^k = \begin{bmatrix} 0, & 1 \\ 0, & 0 \end{bmatrix}^{k,\, e-k}_{e-k,\, k}$, and we can take the corresponding part of $\underset{m}{\overline{X}}^{u}$ to be $\begin{bmatrix} 1 \\ 0 \end{bmatrix}^{k}_{k,\, e-k}$, since

$$\begin{bmatrix} 0, & 1 \\ 0, & 0 \end{bmatrix}^{k,\, e-k}_{e-k,\, k} \begin{bmatrix} 1 \\ 0 \end{bmatrix}^{k}_{k,\, e-k} = 0.$$

Again if $\omega = [\omega]^{e}_{e}$ is a part of Π corresponding to a potent divisor of $\phi(z)$ which is not a power of $z - \lambda$, then ω is an undegenerate square matrix, the corresponding parts of $\Pi - \lambda I$ and $(\Pi - \lambda I)^k$ are undegenerate square matrices of order e, and the corresponding part of $\underset{m}{\overline{X}}^{u}$ is the non-existent matrix $[0]^{0}_{e}$.

3. *Reduction of any equation of the form $f(\phi)\, x = 0$.*

By resolving the scalar function $f(z)$ corresponding to $f(\phi)$ into its linear factors, we can always express the given equation in the form

$$\Phi x = 0, \quad \text{......................................(4)}$$

where $$\Phi = (\phi - \mu_1 I)^{p_1} (\phi - \mu_2 I)^{p_2} (\phi - \mu_3 I)^{p_3} \ldots, \quad \text{..............(5)}$$

$\mu_1, \mu_2, \mu_3, \ldots$ being scalar numbers, which are all different, and $p_1, p_2, p_3, \ldots$ being positive integers. Let $(\phi - \mu_i I)^{p_i}$ be any one of the factors of Φ in (5). Since the factors of Φ are commutative, we can replace the equation (4) by

$$(\phi - \mu_i I)^{p_i}\, \Phi_i x = 0, \quad \text{..........................(4')}$$

where Φ_i is the product of the remaining factors of Φ.

First suppose that μ_i is not a latent root of ϕ. Then $(\phi - \mu_i I)^{p_i}$ is an undegenerate square matrix of order m, and the equation (4′) is equivalent to

$$\Phi_i x = 0. \quad \text{...............................(4'')}$$

Thus in this case we can omit the factor $(\phi - \mu_i I)^{p_i}$ in Φ, or replace the index p_i by 0.

Next suppose that μ_i is the latent root λ_i of ϕ and that $p_i > \kappa_i$. Then since the solutions of $(\phi - \lambda_i I)^{p_i} x = 0$ are the same as the solutions of

$$(\phi - \lambda_i I)^{\kappa_i} x = 0,$$

the equation (4′) is satisfied when and only when the equation

$$(\phi - \lambda_i I)^{\kappa_i}\, \Phi_i x = 0 \quad \text{..........................(4''')}$$

is satisfied, and (4‴) is equivalent to (4′). Thus in this case we can replace the index p_i in Φ by κ_i.

Treating every factor of Φ in this way, we obtain the following result:

Theorem II. *Every equation of the form $f(\phi)\, x = 0$ can be reduced to an equivalent equation of the form*

$$(\phi - \lambda_1 I)^{k_1} (\phi - \lambda_2 I)^{k_2} \ldots (\phi - \lambda_r I)^{k_r} x = 0, \quad \text{..............(B)}$$

where $\lambda_1, \lambda_2, \ldots \lambda_r$ *are r distinct or unequal latent roots of* ϕ, *and where* $k_1, k_2, \ldots k_r$ *are positive integers such that*

$$k_1 \not> \kappa_1, \quad k_2 \not> \kappa_2, \quad \ldots \quad k_r \not> \kappa_r.$$

In the particular case when none of the quantities $\mu_1, \mu_2, \mu_3, \ldots$ in (5) are latent roots of ϕ, every one of the indices $p_1, p_2, p_3, \ldots$ can be replaced by 0, and the reduced equation is $x = 0$. In this case Φ is an undegenerate square matrix of order m.

4. *Properties of the prefactor* $(\phi - \mu I)^p$.

Here μ is any scalar quantity, and p is any non-zero positive integer. In the lemmas and examples which follow, λ is any latent root of ϕ, and k is any positive integer not greater than κ and not less than 1, κ being defined as in sub-article 1.

Lemma A. *If* α *is any solution of the equation* $(\phi - \lambda I)\,x = 0$, *then:*

$$(1)\quad (\phi - \mu I)\,\alpha = (\lambda - \mu)\,\alpha; \qquad (2)\quad (\phi - \mu I)^p\,\alpha = (\lambda - \mu)^p\,\alpha.$$

For we have $\phi\alpha = \lambda\alpha$, and therefore

$$(\phi - \mu I)\,\alpha = (\lambda - \mu)\,\alpha,$$
$$(\phi - \mu I)^2\,\alpha = (\lambda - \mu)\,.\,(\phi - \mu I)\,\alpha = (\lambda - \mu)^2\,\alpha,$$
$$(\phi - \mu I)^3\,\alpha = (\lambda - \mu)^2\,.\,(\phi - \mu I)\,\alpha = (\lambda - \mu)^3\,\alpha,$$

and so on.

Lemma B. *If* α *is any solution of the equation* $(\phi - \lambda I)^k\,x = 0$, *then:*

$$(1)\quad (\phi - \mu I)\,\alpha = (\lambda - \mu)\,\alpha + \alpha_1; \qquad (2)\quad (\phi - \mu I)^p\,\alpha = (\lambda - \mu)^p\,\alpha + \alpha';$$

where α_1 *and* α' *are solutions of the equation* $(\phi - \lambda I)^{k-1}\,x = 0$.

By Lemma A both results are true when $k = 1$. We may therefore suppose that $k \not< 2$.

To prove (1) let $(\phi - \lambda I)\,\alpha = \alpha_1$. Then prefixing $(\phi - \lambda I)^{k-1}$ on both sides, we see that $(\phi - \lambda I)^{k-1}\,\alpha_1 = 0$; and we have

$$\phi\alpha = \lambda\alpha + \alpha_1, \quad (\phi - \mu I)\,\alpha = (\lambda - \mu)\,\alpha + \alpha_1.$$

Thus (1) is true generally.

Again from (1) we deduce in succession that

$$(\phi - \mu I)^2\,\alpha = (\lambda - \mu)^2\,\alpha + 2\,(\lambda - \mu)\,\alpha_1 + \alpha_2,$$
$$(\phi - \mu I)^3\,\alpha = (\lambda - \mu)^3\,\alpha + 3\,(\lambda - \mu)^2\,\alpha_1 + 3\,(\lambda - \mu)\,\alpha_2 + \alpha_3,$$

....

$$(\phi - \mu I)^p\,\alpha = (\lambda - \mu)^p\,\alpha + \binom{p}{1}(\lambda - \mu)^{p-1}\,\alpha_1 + \ldots + \binom{p}{i}(\lambda - \mu)^{p-i}\,\alpha_i + \ldots$$
$$+ \binom{p}{1}(\lambda - \mu)\,\alpha_{p-1} + \alpha_p,$$

where α_i is a solution of the equation $(\phi - \lambda I)^{k-i} x = 0$ when $i < k$, and where $\alpha_i = 0$ when $i \nless k$. Since $\alpha_1, \alpha_2, \dots \alpha_p$ are all solutions of the equation $(\phi - \lambda I)^{k-1} x = 0$, we see that (2) is true generally.

Ex. ii. *If* $\mu \neq \lambda$, *the prefactor* $(\phi - \mu I)^p$ *cannot annihilate any non-zero solution of the equation*

$$(\phi - \lambda I)^k x = 0.$$

In fact it converts:

(1) *every non-zero solution* a *of the equation* $(\phi - \lambda I) x = 0$ *into a non-zero solution* β *of the same equation;*

(2) *every solution* a *of the equation* $(\phi - \lambda I)^k x = 0$ *which is not a solution of the equation* $(\phi - \lambda I)^{k-1} x = 0$ *into a solution* β *of the equation* $(\phi - \lambda I)^k x = 0$ *which is not a solution of the equation* $(\phi - \lambda I)^{k-1} x = 0$;

(3) *every non-zero solution* a *of* (A) *into a non-zero solution* β *of* (A).

We will establish Ex. ii by proving the results (1), (2) and (3).

The first result follows immediately from Lemma A; for we have

$$(\phi - \mu I)^p a = (\lambda - \mu)^p a = \beta \neq 0, \quad \text{and} \quad (\phi - \lambda I) \beta = 0.$$

In proving the second result we may now assume that $k \nless 2$. Then by Lemma B

$$(\phi - \mu I)^p a = (\lambda - \mu)^p a + a' = \beta,$$

where a' is a solution of $(\phi - \lambda I)^{k-1} x = 0$. Since a by supposition is not a solution of the last equation, it follows that

$$(\phi - \lambda I)^k \beta = 0, \quad (\phi - \lambda I)^{k-1} \beta = (\lambda - \mu)^p . (\phi - \lambda I)^{k-1} a \neq 0.$$

The third result follows from the properties (1) and (2). If a is a solution of the equation $(\phi - \lambda I) x = 0$, it follows from the property (1) that $\beta = (\phi - \mu I)^p a$ is a non-zero solution of $(\phi - \lambda I) x = 0$, and therefore a non-zero solution of (A). If a is not a solution of the equation $(\phi - \lambda I) x = 0$, there must exist a positive integer h, not less than 2 and not greater than k, such that

$$(\phi - \lambda I)^h a = 0, \quad (\phi - \lambda I)^{h-1} a \neq 0;$$

and it follows from the property (2) that $\beta = (\phi - \mu I)^p a$ is a solution of $(\phi - \lambda I)^h x = 0$ which is not a solution of $(\phi - \lambda I)^{h-1} x = 0$; therefore β is a solution of (A) which is not zero. The third result could also be deduced directly from Lemma B.

Ex. iii. *If* $\mu \neq \lambda$, *the prefactor* $(\phi - \mu I)^p$ *converts every set of* r *unconnected non-zero solutions* $a_1, a_2, \dots a_r$ *of* (A) *into a set of* r *unconnected non-zero solutions* $\beta_1, \beta_2, \dots \beta_r$ *of the same equation* (A).

Let $a_1, a_2, \dots a_r$ be any r unconnected non-zero solutions of the equation (A), and let

$$(\phi - \mu I)^p a_1 = \beta_1, \quad (\phi - \mu I)^p a_2 = \beta_2, \quad \dots \quad (\phi - \mu I)^p a_r = \beta_r,$$

i.e. let

$$(\phi - \mu I)^p \, \lceil a \rfloor^{r}_{m} = \lceil \beta \rfloor^{r}_{m}.$$

By Ex. ii the matrices $\beta_1, \beta_2, \dots \beta_r$ are r non-zero solutions of (A). If there exists a relation between them of the form

$$l_1 \beta_1 + l_2 \beta_2 + \dots + l_r \beta_r = \lceil \beta \rfloor^{r}_{m} \lceil l \rfloor_{r} = 0, \text{ so that } (\phi - \mu I)^p \lceil a \rfloor^{r}_{m} \lceil l \rfloor_{r} = 0,$$

where $l_1, l_2, \dots l_r$ are scalar numbers, then

$$x = l_1 a_1 + l_2 a_2 + \dots + l_r a_r = [a]_m^r\,[l]_r$$

is a solution of the equation (A) which is annihilated by the prefactor $(\phi - \mu I)^p$, and it follows from Ex. ii that $x=0$, i.e.

$$[a]_m^r\,[l]_r = 0.$$

Since $a_1, a_2, \dots a_r$ are unconnected, and $[a]_m^r$ is undegenerate of rank r, this is only possible when $l_1 = l_2 = \dots = l_r = 0$. Consequently $\beta_1, \beta_2, \dots \beta_r$ are solutions of (A) which are unconnected.

If h is any positive integer not greater than k, and if $y = [\beta]_m^r\,[l]_r$ is a solution of $(\phi - \lambda I)^h x = 0$, then $y' = (\phi - \lambda I)^h\,[a]_m^r\,[l]_r$ is a solution of $(\phi - \lambda I)^{k-h} x = 0$ which is annihilated by $(\phi - \mu I)^p$, and must vanish. Therefore $x = [a]_m^r\,[l]_r$ is also a solution of

$$(\phi - \lambda I)^h x = 0.$$

Hence if no solution of $(\phi - \lambda I)^k x = 0$ *connected with* $a_1, a_2, \dots a_r$ *is a solution of* $(\phi - \lambda I)^h x = 0$, *then no solution of* $(\phi - \lambda I)^k x = 0$ *connected with* $\beta_1, \beta_2, \dots \beta_r$ *is a solution of* $(\phi - \lambda I)^h x = 0$.

If we use the notation of Ex. i we see that the prefactor $(\phi - \mu I)^p$ in which $\mu \neq \lambda$ converts each of the spacelets $S_1, S_2, \dots S_h, \dots S_k, \dots S_\kappa$ into itself; it converts every spacelet of rank r lying in S_k into a spacelet of rank r lying in S_k; and it converts every spacelet of rank r which lies in S_k and does not intersect S_h into a spacelet of rank r which lies in S_k and does not intersect S_h. If μ is not one of the latent roots of ϕ, then the prefactor $(\phi - \mu I)^p$ converts every spacelet into itself.

Ex. iv. *If* $\mu = \lambda$ *and* $k \ngtr \kappa$, *the prefactor* $(\phi - \lambda I)^p$ *annihilates every solution of the equation* (A) *when and only when* $p \nless k$.

For if $p < k$, the degeneracy of $(\phi - \lambda I)^k$ is greater than the degeneracy of $(\phi - \lambda I)^p$, and therefore the equation (A) has solutions which are not solutions of the equation

$$(\phi - \lambda I)^p x = 0.$$

In particular the prefactor $(\phi - \lambda I)^p$ *annihilates every solution of the equation* $(\phi - \lambda I)^\kappa x = 0$ *when and only when* $p \nless \kappa$.

When this last condition is satisfied, the prefactor $(\phi - \lambda I)^p$ annihilates every solution of the equation $(\phi - \lambda I)^h x = 0$ whatever positive integral value h may have.

Further if $k > h \nless p$, *the prefactor* $(\phi - \lambda I)^p$ *converts every solution* a *of* $(\phi - \lambda I)^k x = 0$ *which is not a solution of* $(\phi - \lambda I)^h x = 0$ *into a solution* β *of* $(\phi - \lambda I)^{k-p} x = 0$ *which is not a solution of* $(\phi - \lambda I)^{h-p} x = 0$.

For if
$$(\phi - \lambda I)^p a = \beta,$$
then
$$(\phi - \lambda I)^{k-p} \beta = (\phi - \lambda I)^k a = 0, \quad (\phi - \lambda I)^{h-p} \beta = (\phi - \lambda I)^h a \neq 0.$$

In particular the prefactor $(\phi - \lambda I)^p$ *converts every solution of* $(\phi - \lambda I)^k x = 0$ *which is not a solution of* $(\phi - \lambda I)^p x = 0$ *into a non-zero solution of* $(\phi - \lambda I)^{k-p} x = 0$.

Ex. v. *If* $k \ngtr \kappa$, $p < k$, $s = r_p$; *if* $\gamma_1, \gamma_2, \ldots \gamma_s$ *are a complete set of* s *unconnected non-zero solutions of the equation* $(\phi - \lambda I)^p x = 0$; *and if* $\gamma_1, \gamma_2, \ldots \gamma_s, a_1, a_2, \ldots a_r$ *are* $s + r$ *unconnected non-zero solutions of the equation* $(\phi - \lambda I)^k x = 0$; *then the prefactor* $(\phi - \lambda I)^p$ *converts* $a_1, a_2, \ldots a_r$ *into* r *unconnected non-zero solutions* $\beta_1, \beta_2, \ldots \beta_r$ *of the equation* $(\phi - \lambda I)^{k-p} x = 0$.

Let $$(\phi - \lambda I)^p a_1 = \beta_1, \quad (\phi - \lambda I)^p a_2 = \beta_2, \quad \ldots \quad (\phi - \lambda I)^p a_r = \beta_r,$$

i.e. let $$(\phi - \lambda I)^p \, \overline{\underline{a}}^{\,r}_{m} = \overline{\underline{\beta}}^{\,r}_{m}.$$

By Ex. iv the matrices $\beta_1, \beta_2, \ldots \beta_r$ are r non-zero solutions of $(\phi - \lambda I)^{k-p} x = 0$. If they are not unconnected, there must exist relations of the forms

$$\overline{\underline{\beta}}^{\,r}_{m} \, \overline{\underline{l}}_{r} = 0, \quad (\phi - \lambda I)^p \, \overline{\underline{a}}^{\,r}_{m} \, \overline{\underline{l}}_{r} = 0$$

in which $l_1, l_2, \ldots l_r$ are not all 0, and therefore a relation of the form

$$\overline{\underline{a}}^{\,r}_{m} \, \overline{\underline{l}}_{r} = \overline{\underline{\gamma}}^{\,s}_{m} \, \overline{\underline{l'}}_{s}.$$

But this is impossible because $\overline{\underline{\gamma, a}}^{\,s,\,r}_{m}$ is an undegenerate matrix of rank $s + r$. Consequently $\beta_1, \beta_2, \ldots \beta_r$ are unconnected.

If $k > h \nless p$, and if $y = \overline{\underline{\beta}}^{\,r}_{m} \, \overline{\underline{l}}_{r}$ is a solution of $(\phi - \lambda I)^{h-p} x = 0$, then $x = \overline{\underline{a}}^{\,r}_{m} \, \overline{\underline{l}}_{r}$ is a solution of $(\phi - \lambda I)^h x = 0$. It follows that:

If no solution of $(\phi - \lambda I)^k x = 0$ *connected with* $a_1, a_2, \ldots a_r$ *is a solution of* $(\phi - \lambda I)^h x = 0$, *then no solution of* $(\phi - \lambda I)^{k-p} x = 0$ *connected with* $\beta_1, \beta_2, \ldots \beta_r$ *is a solution of* $(\phi - \lambda I)^{h-p} x = 0$.

If we use the notation of Ex. i and suppose that $k \ngtr \kappa$ and $p \ngtr \kappa$, we see that the prefactor $(\phi - \lambda I)^p$ annihilates S_k when $k \ngtr p$, and converts S_k into a spacelet S_{kp} of rank $r_k - r_p$ lying in S_{k-p} when $k \nless p$; if $k > p$, it converts a spacelet of rank r which lies in S_k and does not intersect S_p into a spacelet of rank r lying in S_{k-p}; and if $k > h \nless p$, it converts a spacelet of rank r which lies in S_k and does not intersect S_h into a spacelet of rank r which lies in S_{k-p} and does not intersect S_{h-p}. If $p \nless \kappa$, then $(\phi - \lambda I)^p$ annihilates all the spacelets $S_1, S_2, \ldots S_\kappa$.

5. *Properties of the prefactor*

$$\Phi = f(\phi) = (\phi - \mu_1 I)^{p_1} (\phi - \mu_2 I)^{p_2} \ldots (\phi - \mu_r I)^{p_r}.$$

Here $\mu_1, \mu_2, \ldots \mu_r$ are distinct or unequal scalar numbers; $p_1, p_2, \ldots p_r$ are positive integers; and the order of arrangement of the factors of Φ is immaterial. In the lemmas and examples which follow λ is any latent root of ϕ, and k is any positive integer not greater than κ and not less than 1, κ being defined as in sub-article 1. The equation $\Phi x = 0$ has non-zero solutions when and only when at least one of the quantities $\mu_1, \mu_2, \ldots \mu_r$ is a latent root of ϕ.

Lemma C. *If* α *is any solution of the equation* $(\phi - \lambda I)\, x = 0$, *then*

$$\Phi \alpha = (\lambda - \mu_1)^{p_1} (\lambda - \mu_2)^{p_2} \ldots (\lambda - \mu_r)^{p_r} \, \alpha = f(\lambda)\, \alpha.$$

This is proved by successive applications of Lemma A.

Lemma D. *If α is any solution of the equation $(\phi - \lambda I)^k x = 0$, then*

$$\Phi\alpha = (\lambda - \mu_1)^{p_1}(\lambda - \mu_2)^{p_2} \dots (\lambda - \mu_r)^{p_r}\alpha + \alpha' = f(\lambda)\,\alpha + \alpha',$$

where α' is a solution of the equation $(\phi - \lambda I)^{k-1} x = 0$.

This is proved by successive applications of Lemma B.

Ex. vi. *If no one of the quantities $\mu_1, \mu_2, \dots \mu_r$ is equal to λ, the prefactor Φ cannot annihilate any non-zero solution of the equation*

$$(\phi - \lambda I)^k x = 0. \qquad\qquad (A)$$

In fact it converts:

(1) *every non-zero solution a of the equation $(\phi - \lambda I)x = 0$ into a non-zero solution β of the same equation;*

(2) *every solution a of the equation $(\phi - \lambda I)^k x = 0$ which is not a solution of the equation $(\phi - \lambda I)^{k-1}x = 0$ into a solution β of the equation $(\phi - \lambda I)^k x = 0$ which is not a solution of the equation $(\phi - \lambda I)^{k-1}x = 0$;*

(3) *every non-zero solution a of* (A) *into a non-zero solution β of* (A).

We obtain all these results by successive applications of Ex. ii. We can also deduce Ex. vi from Lemmas C and D in the same way as we deduced Ex. ii from Lemmas A and B.

Ex. vii. *If no one of the quantities $\mu_1, \mu_2, \dots \mu_r$ is equal to λ, the prefactor Φ converts every set of u unconnected non-zero solutions $a_1, a_2, \dots a_u$ of* (A) *into a set of u unconnected non-zero solutions $\beta_1, \beta_2, \dots \beta_u$ of the same equation* (A).

This follows from Ex. iii, and can also be deduced from Ex. vi in the same way as Ex. iii was deduced from Ex. ii.

Since $[\beta]^u_m = \Phi\,[a]^u_m$, we see that if h is any positive integer not greater than k, and if $y = [\beta]^u_m [l]_u$ is a solution of $(\phi - \lambda I)^h x = 0$, then $y' = (\phi - \lambda I)^h [a]^u_m [l]_u$ is a solution of $(\phi - \lambda I)^{k-h} x = 0$ which is annihilated by Φ, and must vanish by Ex. vi; therefore $x = [a]^u_m [l]_u$ is also a solution of $(\phi - \lambda I)^h x = 0$.

Hence if no solution of $(\phi - \lambda I)^k x = 0$ connected with $a_1, a_2, \dots a_u$ is a solution of $(\phi - \lambda I)^h x = 0$, then no solution of $(\phi - \lambda I)^k x = 0$ connected with $\beta_1, \beta_2, \dots \beta_u$ is a solution of $(\phi - \lambda I)^h x = 0$.

If we use the notation of Ex. i, we see that in the present case the prefactor Φ converts each of the spacelets $S_1, S_2, \dots S_h, \dots S_k, \dots S_\kappa$ into itself; it converts any spacelet of rank u lying in S_k into a spacelet of rank u lying in S_k; and it converts every spacelet of rank u which lies in S_k and does not intersect S_h into a spacelet of rank u which lies in S_k and does not intersect S_h. If no one of the quantities $\mu_1, \mu_2, \dots \mu_r$ is a latent root ϕ, then the prefactor Φ converts every spacelet into itself.

Ex. viii. *If $k \not> \kappa$, the prefactor Φ annihilates every solution of the equation* (A) *when and only when for one of the values* 1, 2, ... *r of i we have*

$$\mu_i = \lambda, \quad p_i \not< k.$$

In order that Φ may annihilate some non-zero solutions of the equation (A), it is necessary (see Ex. vi) that one of the quantities $\mu_1, \mu_2, \ldots \mu_r$ shall be equal to λ. We will therefore suppose that $\mu_i = \lambda$; and we will denote by Φ_i the product formed from Φ by striking out the ith factor $(\phi - \mu_i I)^{p_i} = (\phi - \lambda I)^{p_i}$.

Let a be any non-zero solution of the equation (A), and let

$$(\phi - \mu_i I)^{p_i} a = (\phi - \lambda I)^{p_i} a = \beta, \quad \Phi a = \Phi_i \beta = \gamma.$$

If $p_i \nless k$, we have $\beta = 0$, and therefore $\gamma = \Phi_i \beta = 0$. But if $p_i < k$, we can choose the solution a so that $\beta \neq 0$; then β is a non-zero solution of the equation $(\phi - \lambda I)^{k-p_i} x = 0$, and by Ex. vi we have $\gamma = \Phi_i \beta \neq 0$. Thus if $\mu_i = \lambda$, then Φ annihilates every solution a of the equation (A) when and only when $p_i \nless k$.

In particular the prefactor Φ annihilates every solution of the equation $(\phi - \lambda I)^{\kappa} x = 0$, i.e. all solutions of every equation of the form $(\phi - \lambda I)^h x = 0$, when and only when for one of the values 1, 2, ... *r of i we have*

$$\mu_i = \lambda, \quad p_i \nless \kappa.$$

In the case when $\mu_i = \lambda$ and $p_i < k$, it being still supposed that $k \ngtr \kappa$, we deduce from Exs. iv and vi the following further result in which h is any one of the integers $k-1, k-2, \ldots p_i$.

If $\mu_i = \lambda$, and if $k > h \nless p_i$, the prefactor Φ converts every solution a of the equation $(\phi - \lambda I)^k x = 0$ which is not a solution of the equation $(\phi - \lambda I)^h x = 0$ into a solution β of the equation $(\phi - \lambda I)^{k-p_i} x = 0$ which is not a solution of the equation $(\phi - \lambda I)^{h-p_i} x = 0$.

Let $$\Phi_i a = a', \quad \Phi a = (\phi - \lambda I)^{p_i} a' = \beta.$$

From Ex. vi we see that a' is a solution of $(\phi - \lambda I)^k x = 0$ which is not a solution of $(\phi - \lambda I)^h x = 0$; and it follows from Ex. iv that β has the properties mentioned.

In particular, when $\mu_i = \lambda$ and $p_i < k$, the prefactor Φ converts every solution of (A) *which is not a solution of $(\phi - \lambda I)^{p_i} x = 0$ into a non-zero solution of $(\phi - \lambda I)^{k-p_i} x = 0$.*

Ex. ix. *If $\mu_i = \lambda$, $k \ngtr \kappa$, $p_i < k$, $v = r_{p_i}$; if $\gamma_1, \gamma_2, \ldots \gamma_v$ are a complete set of v unconnected non-zero solutions of the equation $(\phi - \lambda I)^{p_i} x = 0$; and if $\gamma_1, \gamma_2, \ldots \gamma_v, a_1, a_2, \ldots a_u$ are $v + u$ unconnected non-zero solutions of the equation $(\phi - \lambda I)^k x = 0$; then the prefactor Φ converts $a_1, a_2, \ldots a_u$ into u unconnected non-zero solutions $\beta_1, \beta_2, \ldots \beta_u$ of the equation*

$$(\phi - \lambda I)^{k-p_i} x = 0.$$

Let $$\Phi_i \left[\gamma, a\right]_m^{v,u} = \left[\gamma', a'\right]_m^{v,u}, \quad \Phi \left[a\right]_m^u = (\phi - \lambda I)^{p_i} \left[a'\right]_m^u = \left[\beta\right]_m^u,$$

where Φ_i is defined as in Ex. viii. From Ex. vii it follows that $\gamma_1', \gamma_2', \ldots \gamma_v'$ are a complete set of v unconnected non-zero solutions of $(\phi - \lambda I)^{p_i} x = 0$, and that $\gamma_1', \gamma_2', \ldots \gamma_v', a_1', a_2', \ldots a_u'$ are $v + u$ unconnected non-zero solutions of $(\phi - \lambda I)^k x = 0$; hence by Ex. v the above theorem is true.

If $k > h \nless p_i$, and if $y = \left[\beta\right]_m^u \left[l\right]_u$ is a solution of $(\phi - \lambda I)^{h-p_i} x = 0$, then

$$y' = (\phi - \lambda I)^h \left[a\right]_m^u \left[l\right]_u$$

is a solution of $(\phi - \lambda I)^{k-h} x = 0$ which is annihilated by Φ_i, and must vanish by Ex. vi; therefore $x = \left[a\right]_m^u \left[l\right]_u$ is a solution of $(\phi - \lambda I)^h x = 0$.

Hence if no solution of $(\phi - \lambda I)^k x = 0$ *connected with* $a_1, a_2, \dots a_u$ *is a solution of* $(\phi - \lambda I)^h x = 0$, *then no solution of* $(\phi - \lambda I)^{k-p_i} x = 0$ *connected with* $\beta_1, \beta_2, \dots \beta_u$ *is a solution of* $(\phi - \lambda I)^{h-p_i} x = 0$.

If we use the notation of Ex. i and suppose that $k \not> \kappa$ and $p_i \not> \kappa$, we see that the prefactor Φ in which $\mu_i = \lambda$ annihilates S_k when $k \not> p_i$, and converts S_k into a spacelet of rank $r_k - r_{p_i}$ lying in S_{k-p_i} when $k \not< p_i$; and if $k > h \not< p_i$, it converts a spacelet of rank u which lies in S_k and does not intersect S_h into a spacelet of the same rank u which lies in S_{k-p_i} and does not intersect S_{h-p_i}. If $p_i \not< \kappa$, it of course annihilates all the spacelets $S_1, S_2, \dots S_\kappa$.

All these results are generalisations of those given in Ex. v.

6. *Solutions of the general equation* $f(\phi)\,x = 0$.

Making use of Theorem II we may suppose without loss of generality that the equation has the form

$$\Phi x = 0, \quad \text{.........(C)}$$

where

$$\Phi = f(\phi) = (\phi - \lambda_1 I)^{k_1}(\phi - \lambda_2 I)^{k_2} \dots (\phi - \lambda_r I)^{k_r}, \quad \text{.........(C')}$$

$\lambda_1, \lambda_2, \dots \lambda_r$ being r distinct latent roots of ϕ, and $k_1, k_2, \dots k_r$ being non-zero positive integers which are respectively not greater than $\kappa_1, \kappa_2, \dots \kappa_r$. In the excluded trivial case when $k_1, k_2, \dots k_r$ are all 0, the only solution of (C) is $x = 0$.

Theorem III. *If* $\lambda_1, \lambda_2, \dots \lambda_r$ *are* r *distinct latent roots of the square matrix* ϕ, *there cannot exist any connection between non-zero solutions* $\alpha_1, \alpha_2, \dots \alpha_r$ *of the successive equations*

$$(\phi - \lambda_1 I)^{k_1} x = 0, \quad (\phi - \lambda_2 I)^{k_2} x = 0, \quad \dots \quad (\phi - \lambda_r I)^{k_r} x = 0. \quad \text{......(6)}$$

Let $\Phi_1, \Phi_2, \dots \Phi_r$ be respectively the products of all the factors of Φ in (C′) except the 1st, 2nd, ... rth; and suppose that there exists a relation of the form

$$l_1\alpha_1 + l_2\alpha_2 + \dots + l_r\alpha_r = \left\lceil \alpha \right\rfloor_m^r \left\lceil l \right\rfloor_r = 0, \quad \text{.................(7)}$$

where $l_1, l_2, \dots l_r$ are scalar constants. Then by Ex. vi

$$\Phi_i \alpha_j = 0 \text{ when } j \neq i, \quad \Phi_i \alpha_i = \beta_i,$$

where β_i is a non-zero solution of the equation $(\phi - \lambda_i I)^{k_i} x = 0$. Hence when we prefix Φ_i on both sides of (7), we obtain

$$l_i \beta_i = 0, \quad \text{i.e. } l_i = 0.$$

Prefixing $\Phi_1, \Phi_2, \dots \Phi_r$ in turn, we see that $l_1 = l_2 = \dots = l_r = 0$. Thus there cannot exist any relation of the form (7) in which $l_1, l_2, \dots l_r$ are not all zero, i.e. $\alpha_1, \alpha_2, \dots \alpha_r$ are unconnected.

Theorem IV. *If* $(\alpha_1, \alpha_2, \alpha_3, \dots)$, $(\beta_1, \beta_2, \beta_3, \dots)$, $(\gamma_1, \gamma_2, \gamma_3, \dots)$, ... *are sets of* $n_1, n_2, n_3, \dots$ *unconnected non-zero solutions of the successive equations* (6), *then* $\alpha_1, \alpha_2, \alpha_3, \dots, \beta_1, \beta_2, \beta_3, \dots, \gamma_1, \gamma_2, \gamma_3, \dots, \dots$ *are all unconnected, and are* $n_1 + n_2 + n_3 + \dots$ *unconnected non-zero solutions of the equation* (C).

Suppose that there exists a relation of the form

$$(A_1\alpha_1 + A_2\alpha_2 + \ldots) + (B_1\beta_1 + B_2\beta_2 + \ldots) + (C_1\gamma_1 + C_2\gamma_2 + \ldots) + \ldots, \quad \ldots(8)$$

where $A_1, A_2, \ldots B_1, B_2, \ldots C_1, C_2, \ldots, \ldots$ are scalar constants; and let

$$\alpha = A_1\alpha_1 + A_2\alpha_2 + \ldots, \quad \beta = B_1\beta_1 + B_2\beta_2 + \ldots, \quad \gamma = C_1\gamma_1 + C_2\gamma_2 + \ldots, \ \ldots,$$

so that $\alpha, \beta, \gamma, \ldots$ are solutions of the 1st, 2nd, 3rd, ... of the equations (6). By Theorem III we can have $\alpha+\beta+\gamma+\ldots=0$ when and only when $\alpha, \beta, \gamma, \ldots$ are all 0; and this is only possible when $A_1, A_2, \ldots, B_1, B_2, \ldots, C_1, C_2, \ldots, \ldots$ are all 0. Thus in every relation of the form (8) the constants $A, B, C, \ldots$ must all vanish, i.e. $\alpha_1, \alpha_2, \ldots, \beta_1, \beta_2, \ldots, \gamma_1, \gamma_2, \ldots, \ldots$ are all unconnected; and every one of them is a solution of (C).

Theorem V. *Any r complete sets of unconnected non-zero solutions of the r equations* (6) *together form a complete set of unconnected non-zero solutions of the equation* (C).

Let $\nu_1, \nu_2, \ldots \nu_r$ be the degeneracies of the factors $(\phi - \lambda_1 I)^{k_1}, (\phi - \lambda_2 I)^{k_2}, \ldots (\phi - \lambda_r I)^{k_r}$ of Φ in (C′); let

$$\nu = \nu_1 + \nu_2 + \ldots + \nu_r;$$

and in Theorem IV take $n_1, n_2, n_3, \ldots$ to be $\nu_1, \nu_2, \nu_3, \ldots$; so that $(\alpha_1, \alpha_2, \ldots)$, $(\beta_1, \beta_2, \ldots)$, $(\gamma_1, \gamma_2, \ldots)$, ... are complete sets of $\nu_1, \nu_2, \nu_3, \ldots$ unconnected non-zero solutions of the 1st, 2nd, 3rd, ... of the equations (6). Then these together form ν unconnected non-zero solutions of the equation (C). But by Theorem I of § 230 we know that ν is the degeneracy of Φ. Therefore these together form a complete set of unconnected non-zero solutions of the equation (C).

The reference to Theorem I of § 230 is however not necessary. It has been shown, as a consequence of Theorem IV, that the equation (C) has ν unconnected non-zero solutions; and it follows that the degeneracy of Φ cannot be less than ν. But by the second restriction of § 133 the degeneracy of Φ cannot be greater than ν. Therefore ν is the degeneracy of Φ. Thus it follows from Theorem IV that Theorem I of § 230 is true for all products of the special form (C′); and Theorem IV suffices for the complete establishment of Theorem V.

NOTE 2. *Alternative proof of Theorem I of* § 230.

Let $F(\phi)$ be any rational integral function whatever of ϕ expressed in the form

$$F(\phi) = A(\phi - \mu_1 I)^{p_1}(\phi - \mu_2 I)^{p_2} \ldots (\phi - \mu_s I)^{p_s},$$

where $\mu_1, \mu_2, \ldots \mu_s$ are distinct scalar quantities, $p_1, p_2, \ldots p_s$ are positive integers, and A is a non-zero scalar constant; and let the equation $F(\phi)x=0$ be reduced in the way described in sub-article 3 to the equation $f(\phi)x=0$, where $f(\phi)$ is a product having the special form (C′). It has been shown to be a consequence of Theorem IV that $f(\phi)$ has degeneracy ν; and because the equations $f(\phi)x=0$, $F(\phi)x=0$ have identical solutions, it follows that $F(\phi)$ has degeneracy ν. But ν, which is the sum of the degeneracies of the r factors of $f(\phi)$, must also be the sum of the degeneracies of the s factors of $F(\phi)$. Consequently Theorem I of § 230 is true for the function $F(\phi)$.

Ex. x. *If $\lambda_1, \lambda_2, \ldots \lambda_n$ are all the distinct latent roots of the given square matrix $\phi=[a]_m^m$, we can always form an undegenerate square matrix $\overline{\underline{a}}\,{}^m_m$ of order m each of whose vertical rows is a solution of one of the equations*

$$(\phi-\lambda_1 I)^{\kappa_1}x=0,\quad (\phi-\lambda_2 I)^{\kappa_2}x=0,\quad \ldots\quad (\phi-\lambda_n I)^{\kappa_n}x=0. \qquad \text{.............(9)}$$

For since $(\phi-\lambda_i I)^{\kappa_i}$ has degeneracy s_i such that $s_1+s_2+\ldots+s_n=m$, we can form a square matrix $\overline{\underline{a}}\,{}^m_m$ in which the first s_1, the next s_2, ... the last s_n vertical rows are complete sets of unconnected non-zero solutions of the successive equations (9); and by Theorem IV every such square matrix is undegenerate.

Ex. xi. *We can form an undegenerate matrix $\overline{\underline{a}}\,{}^t_m$ of rank t each of whose vertical rows is a solution of one of the equations*

$$(\phi-\lambda_1 I)x=0,\quad (\phi-\lambda_2 I)x=0,\quad \ldots\quad (\phi-\lambda_n I)x=0 \qquad \text{.................(10)}$$

when and only when the following equivalent conditions are satisfied:

(1) *$t \not> r_{11}+r_{21}+\ldots+r_{n1}$, where r_{i1} is the degeneracy of $\phi-\lambda_i I$;*

(2) *t is not greater than the total number of potent divisors of the characteristic matrix of ϕ.*

For if $r_{11}+r_{21}+\ldots+r_{n1}=u$, we can form a matrix $\overline{\underline{\beta}}\,{}^u_m$ whose first r_{11}, next r_{21}, ... last r_{n1} vertical rows are complete sets of unconnected non-zero solutions of the successive equations (10), and by Theorem IV this matrix has rank u; and if $t \not> u$, we can form an undegenerate matrix $\overline{\underline{a}}\,{}^t_m$ of rank t whose vertical rows are t of the vertical rows of $\overline{\underline{\beta}}\,{}^u_m$. Again if $\overline{\underline{a}}\,{}^t_m$ is an undegenerate matrix of rank t each of whose vertical rows is a solution of one of the equations (10), all its vertical rows must be connected with the vertical rows of $\overline{\underline{\beta}}\,{}^u_m$, and therefore we must have $t \not> u$.

The equivalence of the conditions (1) and (2) follows from Ex. iii of § 228 or Theorem IV *b* of § 233; for r_{i1} is the total number of those potent divisors of the characteristic matrix $\phi(z)$ which are powers of $z-\lambda_i$.

We see that $u=r_{11}+r_{21}+\ldots+r_{n1}$ is the total number of unconnected poles of the square matrix $\phi=[a]_m^m$, i.e. the total number of unconnected solutions of the equation

$$[a]_m^m\,\overline{\underline{x}}_m=\rho\,\overline{\underline{x}}_m \qquad \text{.......................................(11)}$$

when ρ is an unspecified scalar quantity to which any value can be ascribed, and that exactly r_{i1} of these poles correspond to the latent root λ_i.

Thus the total number of unconnected poles of a square matrix $\phi=[a]_m^m$ is equal to the total number of potent divisors of its characteristic matrix $\phi(z)$; and if exactly p potent divisors of $\phi(z)$ are powers of $z-\lambda$, then there are exactly p unconnected poles (or non-zero solutions of the equation (11)) corresponding to the latent root λ of ϕ.

Ex. xii. *We can form an undegenerate square matrix* $\overline{\underline{a}}{}^m_m$ *each of whose vertical rows is a solution of one of the equations* (10) *when and only when* $\kappa_1=\kappa_2=\ldots=\kappa_n=1$, *i.e. when and only when any one of the following equivalent conditions is satisfied:*

(1) *The potent divisors of the characteristic matrix* $\phi(z)=\phi-zI$ *are all linear.*

(2) $E_m(z)=(z-\lambda_1)(z-\lambda_2)\ldots(z-\lambda_n)$.

(3) *There exists an equimutant transformation converting* ϕ *into a quasi-scalar matrix (i.e. the reduced canonical form of* ϕ *is a quasi-scalar matrix).*

For in Ex. xi we have $r_{i1}=s_i$ when $\kappa_i=1$, $r_{i1}<s_i$ when $\kappa_i>1$; and we always have $s_1+s_2+\ldots+s_n=m$. If $\kappa_1=\kappa_2=\ldots=\kappa_n=1$, then $u=s_1+s_2+\ldots+s_n=m$; but if any one of the κ's is greater than 1, then $u<s_1+s_2+\ldots+s_n$, and therefore $u<m$. In fact ϕ has m unconnected poles when and only when $\phi(z)$ has m potent divisors, and this is the case when and only when every potent divisor is linear.

The equivalence of the condition (3) to the other conditions was proved in Note 4 of § 228. We can show directly that (3) is a necessary and sufficient condition in the following way.

First suppose that there exists an equimutant transformation

$$[h]^m_m\,[a]^m_m\,\overline{\underline{H}}{}^m_m={}^1[\lambda]_m \quad\ldots\ldots\ldots\ldots\ldots\ldots\ldots\ldots(12)$$

converting ϕ into a quasi-scalar matrix. Then the deduced equation

$$[a]^m_m\,\overline{\underline{H}}{}^m_m=\overline{\underline{H}}{}^m_m\,{}^1[\lambda]_m \ldots\ldots\ldots\ldots\ldots\ldots\ldots\ldots(12')$$

shows that the successive vertical rows of $\overline{\underline{H}}{}^m_m$ are m unconnected poles of ϕ corresponding respectively to the latent roots $\lambda_1, \lambda_2, \ldots \lambda_m$.

Next suppose that ϕ has m unconnected poles formed by the successive vertical rows of the undegenerate square matrix $\overline{\underline{H}}{}^m_m$, and let these correspond respectively to the latent roots $\lambda_1, \lambda_2, \ldots \lambda_m$ of ϕ, which are not necessarily all different. Then we have the equation (12′); and if $[h]^m_m$ is the inverse of $\overline{\underline{H}}{}^m_m$, we deduce from (12′) the equimutant transformation (12) converting ϕ into a quasi-scalar matrix.

Ex. xiii. *Additional proofs of Theorems IIa and IIb of* § 224.

First let $E_m(z)$ be the potent factor of order m of the characteristic matrix $\phi(z)=\phi-zI$ of the square matrix $\phi=[a]^m_m$, so that

$$E_m(\phi)=(\phi-\lambda_1 I)^{\kappa_1}(\phi-\lambda_2 I)^{\kappa_2}\ldots(\phi-\lambda_n I)^{\kappa_n}.$$

By Theorem IV any complete sets of $s_1, s_2, \ldots s_n$ unconnected non-zero solutions of the successive equations (9) together form m unconnected non-zero solutions of the equation $E_m(\phi)=0$. Therefore $E_m(\phi)$ must have degeneracy m or rank 0. In fact if $\overline{\underline{a}}{}^m_m$ is an undegenerate square matrix of order m formed as in Ex. x, we have $E_m(\phi)\,\overline{\underline{a}}{}^m_m=0$, and therefore

$$E_m(\phi)=0.$$

Again let $f(\phi)=0$ be any rational integral equation satisfied by the square matrix ϕ. Then the prefactor $f(\phi)$, which has rank 0 or is equal to 0, necessarily annihilates every matrix $\lceil x \rfloor_m$. It therefore annihilates all solutions of every one of the equations (9), and it follows from Ex. viii that $E_m(\phi)$ must be a factor of $f(\phi)$, and that $f(\phi)$ must have the form

$$f(\phi)=g(\phi)\,E_m(\phi).$$

Consequently $E_m(\phi)=0$ is the rational integral equation of lowest degree which is satisfied by ϕ.

Note 3. *Solutions of the equation $f(\phi)\,x=0$ when $f(\phi)$ lies in a given restricted domain of rationality Ω.*

The maximum and potent factors of order m of the characteristic matrix $\phi-zI$ of ϕ can be expressed in the forms

$$D_m(z)=\{t_1(z)\}^{s_1}\{t_2(z)\}^{s_2}\ldots\{t_n(z)\}^{s_n},\quad E_m(z)=\{t_1(z)\}^{\kappa_1}\{t_2(z)\}^{\kappa_2}\ldots\{t_n(z)\}^{\kappa_n},$$

where $t_1(z), t_2(z), \ldots t_n(z)$ are distinct irreducible rational integral functions of z in Ω; and the equation $f(\phi)\,x=0$ can always be reduced to one of the form

$$\Phi x=0,$$

where
$$\Phi=\{t_1(\phi)\}^{k_1}\{t_2(\phi)\}^{k_2}\ldots\{t_r(\phi)\}^{k_r},$$

$k_1, k_2, \ldots k_r$ being positive integers which are respectively not greater than $\kappa_1, \kappa_2, \ldots \kappa_r$. The degeneracy of Φ is equal to the sum of the degeneracies of its factors, and any r complete sets of unconnected non-zero solutions of the successive equations

$$\{t_1(\phi)\}^{k_1}x=0,\quad \{t_2(\phi)\}^{k_2}x=0,\ \ldots\ \{t_r(\phi)\}^{k_r}x=0$$

together form a complete set of unconnected non-zero solutions of the equation $\Phi x=0$.

We obtain these results from those of the text by expressing each irreducible function as a product of linear factors as in § 230. 2.

Note 4. *Solution of the equation $f(\phi)\,x=0$ by the reduction of ϕ to canonical form.*

Let $\Pi=[\Pi]_m^m$ be the canonical reduced form of the square matrix $\phi=[a]_m^m$, and let

$$[h]_m^m\,[\Pi]_m^m\,\lceil H\rfloor_m^m=[a]_m^m \qquad \text{(a')}$$

be an equimutant transformation by which ϕ can be derived from Π. Then there is a one-one correspondence between the solutions $x=\lceil x\rfloor_m$, $X=\lceil X\rfloor_m$ of the equations

$$f(\phi)\,x=(\phi-\lambda_1 I)^{k_1}(\phi-\lambda_2 I)^{k_2}\ldots(\phi-\lambda_r I)^{k_r}\,x=0,$$

$$f(\Pi)\,X=(\Pi-\lambda_1 I)^{k_1}(\Pi-\lambda_2 I)^{k_2}\ldots(\Pi-\lambda_r I)^{k_r}\,X=0$$

such that
$$\lceil x\rfloor_m=[h]_m^m\,\lceil X\rfloor_m.$$

Therefore if u_i is the degeneracy of $(\phi-\lambda_i I)^{k_i}$, and if $u=u_1+u_2+\ldots+u_r$ is the degeneracy of $f(\phi)$, the general solution of the equation $f(\phi)\,x=0$ can be expressed in the form

$$x=[h]_m^m\,\lceil X\rfloor_m^u\,\lceil l\rfloor_u, \qquad \text{(b')}$$

where $l_1, l_2, \ldots l_n$ are arbitrary scalar quantities, and where $[X]_n^m$ is an undegenerate horizontal normal to $f(\Pi)$ so that

$$f(\Pi)\,\overline{\underline{X}}_m^u = 0. \quad \text{...(c')}$$

If ϖ_i is the super-part of Π corresponding to the latent root λ_i, we can write

$$\Pi = \begin{bmatrix} \varpi_1, & 0, & \ldots & 0 \\ 0, & \varpi_2, & \ldots & 0 \\ \cdots & \cdots & \cdots & \cdots \\ 0, & 0, & \ldots & \varpi_n \end{bmatrix}_{s_1, s_2, \ldots s_n}^{s_1, s_2, \ldots s_n}, \quad \overline{\underline{X}}_m^u = \begin{bmatrix} P_1, & 0, & \ldots & 0 \\ 0, & P_2, & \ldots & 0 \\ \cdots & \cdots & \cdots & \cdots \\ 0, & 0, & \ldots & P_r \\ 0, & 0, & \ldots & 0 \\ \cdots & \cdots & \cdots & \cdots \\ 0, & 0, & \ldots & 0 \end{bmatrix}_{s_1, s_2, \ldots s_r, s_{r+1}, \ldots s_n}^{u_1, u_2, \ldots u_r}$$

where the ith vertical minor of $\overline{\underline{X}}_m^u$ is an undegenerate matrix $\overline{\underline{X_i}}_m^{u_i}$ satisfying the equation

$$(\Pi - \lambda_i I)^{k_i}\,\overline{\underline{X_i}}_m^{u_i},$$

and determined as in Note 1. The matrix $\overline{\underline{X}}_m^u$ is a horizontal derangement of the matrix $\begin{bmatrix} 1 \\ 0 \end{bmatrix}_{u,\, m-u}^{u}$. When $k_i = \kappa_i$, P_i is a unit matrix of order s_i.

All the results of this article could be deduced from the known structures of Π and $\overline{\underline{X}}_m^u$.

§ 236. The conjugate reciprocal and inverse of the characteristic matrix of a given square matrix.

Let $\phi = [a]_m^m$ be a square matrix of order m whose elements are given constants, let $I = [1]_m^m$, let λ be a scalar variable, let $D_m(\lambda)$ and $E_m(\lambda)$ be respectively the determinant and the potent factor of order m of the characteristic matrix $\phi - \lambda I$ of ϕ, and let

$$f(\lambda) = \det(\lambda I - \phi) = \lambda^m + c_1\lambda^{m-1} + c_2\lambda^{m-2} + \ldots + c_m, \quad \text{......(1)}$$

so that $$f(\lambda) = (-1)^m \det(\phi - \lambda I) = (-1)^m D_m(\lambda). \quad \text{.................(1')}$$

Also, using the notation of Note 5 in § 223, let $(\lambda I - \phi)^{-1}$ and $(\phi - \lambda I)^{-1}$ be the inverses of $\lambda I - \phi$ and $\phi - \lambda I$, so that

$f(\lambda)\,.\,(\lambda I - \phi)^{-1}$ is the conjugate reciprocal of $\lambda I - \phi$,

$D_m(\lambda)\,.\,(\phi - \lambda I)^{-1}$ is the conjugate reciprocal of $\phi - \lambda I$,

and $$(\phi - \lambda I)^{-1} = -(\lambda I - \phi)^{-1}.$$

Then if α is any scalar quantity, $f(\lambda)-f(\alpha)$ is divisible by $\lambda-\alpha$; and if

$$g_i(\lambda)=\lambda^i+c_1\lambda^{i-1}+c_2\lambda^{i-2}+\ldots+c_{i-2}\lambda^2+c_{i-1}\lambda+c_i, \quad \ldots\ldots\ldots(2)$$

$$g_i(\phi)=\phi^i+c_1\phi^{i-1}+c_2\phi^{i-2}+\ldots+c_{i-2}\phi^2+c_{i-1}\phi+c_iI \quad \ldots\ldots\ldots(2')$$

for the values 1, 2, ... $m-1$ of i, we see by actual division that

$$f(\lambda)-f(\alpha)=(\lambda-\alpha)F(\alpha), \quad \ldots\ldots\ldots\ldots\ldots\ldots\ldots\ldots(3)$$

where
$$F(\alpha)=\lambda^{m-1}+g_1(\alpha)\lambda^{m-2}+\ldots+g_{m-2}(\alpha)\lambda+g_{m-1}(\alpha)$$
$$=\alpha^{m-1}+g_1(\lambda)\alpha^{m-2}+\ldots+g_{m-2}(\lambda)\alpha+g_{m-1}(\lambda). \quad \ldots\ldots(4)$$

Since the last two equations are rational integral identities in α as well as in λ, and since $f(\phi)=0$ by Theorem I *a* of § 224, it follows that

$$f(\lambda).I-f(\phi)=f(\lambda).I=(\lambda I-\phi).F(\phi), \quad \ldots\ldots\ldots\ldots(3')$$

and therefore by prefixing $(\lambda I-\phi)^{-1}$ on both sides of (3′) that

$$f(\lambda).(\lambda I-\phi)^{-1}=F(\phi), \quad \ldots\ldots\ldots\ldots\ldots\ldots\ldots\ldots(A)$$

where
$$F(\phi)=\lambda^{m-1}I+g_1(\phi)\lambda^{m-2}+\ldots+g_{m-2}(\phi)\lambda+g_{m-1}(\phi)$$
$$=\phi^{m-1}+g_1(\lambda)\phi^{m-2}+\ldots+g_{m-2}(\lambda)\phi+g_{m-1}(\lambda)I. \quad \ldots(4')$$

From the equation (A) we see that the rational integral function $F(\phi)$ given by (4′) is the conjugate reciprocal of $\lambda I-\phi$, and that $(-1)^{m-1}F(\phi)$ is the conjugate reciprocal of the characteristic matrix $\phi-\lambda I$ of ϕ. Accordingly we have the following theorem and corollaries:

Theorem I. *The conjugate reciprocal of the characteristic matrix $\phi-\lambda I$ of the square matrix $\phi=[a]_m^m$ is a rational integral function of ϕ of degree $m-1$ whose coefficients are rational integral functions of λ, and it is also a rational integral function of λ of degree $m-1$ whose coefficients are rational integral functions of ϕ.*

Corollary 1. *The inverse of the characteristic matrix $\phi-\lambda I$, i.e. the matrix*

$$(\phi-\lambda I)^{-1}=-(\lambda I-\phi)=-\frac{F(\phi)}{f(\lambda)},$$

is a square matrix of order m which is a rational integral function of ϕ; and it is rational but not integral in λ.

Corollary 2. *The conjugate reciprocal and the inverse of the characteristic matrix $\phi-\lambda I$ are both commutative with all other square matrices of order m which are rational integral functions of ϕ.*

Now let α be any latent root of ϕ, and let $(\lambda-\alpha)^\kappa$ be the highest power of $\lambda-\alpha$ which is a potent divisor of $\phi-\lambda I$ or $\lambda I-\phi$, so that it is the potent divisor of order m of each of those matrices corresponding to the linear divisor $\lambda-\alpha$. Also let d_m and d_{m-1} be the maximum indices of orders m and $m-1$ of the linear divisor $\lambda-\alpha$ for the matrices $\phi-\lambda I$ and $\lambda I-\phi$, so that $\kappa=d_m-d_{m-1}$. Then $(\lambda-\alpha)^{d_{m-1}}$ is the highest power of $\lambda-\alpha$ which is a common factor of all the elements of the square matrix $F(\phi)$, and $(\lambda-\alpha)^{d_m}$ is the highest power of $\lambda-\alpha$ which is a factor of $f(\lambda)$. If we divide both sides of (A) by $f(\lambda)$, and expand $\frac{\lambda^{m-1}}{f(\lambda)}, \frac{\lambda^{m-2}}{f(\lambda)}, \ldots$ or $\frac{g_1(\lambda)}{f(\lambda)}, \frac{g_2(\lambda)}{f(\lambda)}, \ldots$ in ascending powers of $\lambda-\alpha$, we obtain an expansion of $(\lambda I-\phi)^{-1}$ in ascending powers of $\lambda-\alpha$ in which the coefficients of the various powers of $\lambda-\alpha$ are rational integral functions of ϕ. Again if in the equation $(\lambda I-\phi)^{-1}=\frac{F(\phi)}{f(\lambda)}$ we cancel the factor $(\lambda-\alpha)^{d_{m-1}}$ which is common to the numerator and denominator on the right, and then expand the resulting fraction in ascending powers of $\lambda-\alpha$, we see that the expansion of $(\lambda I-\phi)^{-1}$ in ascending powers of $\lambda-\alpha$ has the form

$$(\lambda I-\phi)^{-1}=\frac{A_{\kappa-1}}{(\lambda-\alpha)^{\kappa}}+\frac{A_{\kappa-2}}{(\lambda-\alpha)^{\kappa-1}}+\ldots+\frac{A_i}{(\lambda-\alpha)^{i+1}}+\ldots+\frac{A_0}{\lambda-\alpha}$$
$$+B_1+B_2(\lambda-\alpha)+B_3(\lambda-\alpha)^2+\ldots+B_j(\lambda-\alpha)^{j-1}+\ldots, \quad \ldots\text{(B)}$$

where the A's and B's are square matrices of order m. Since this expansion must be the same as that obtained by the previous method, the A's and B's are rational integral functions of ϕ, and are therefore commutative with one another and with all other square matrices of order m which are rational integral functions of ϕ. Moreover $A_{\kappa-1}$ cannot vanish; for if it did vanish, then $F(\phi)=f(\lambda).(\lambda I-\phi)^{-1}$ would be divisible by a higher power of $\lambda-\alpha$ than $(\lambda-\alpha)^{d_{m-1}}$.

We will proceed to determine the ranks of all the matrices A in (B), following the methods used by *Frobenius* and *Stickelberger*. For values of i greater than $\kappa-1$ we will define A_i to be a square matrix of order m having rank 0, so that $A_i=0$ when $i \nless \kappa$, and we may then supply an indefinite number of terms before the first in (B).

Prefixing $\lambda I-\phi$ on the left in (B) and the same matrix written in the form $(\lambda-\alpha)I-(\phi-\alpha I)$ on the right, and then equating the coefficients of $\lambda-\alpha$ on both sides, we obtain

$$A_i=(\phi-\alpha I)A_{i-1}, \quad B_i=(\phi-\alpha I)B_{i+1}, \quad \text{when } i \nless 1; \quad \ldots\ldots(5)$$

and
$$A_0-I=(\phi-\alpha I)B_1. \quad \ldots\ldots(5')$$

Further by repeated applications of (5) and (5′) and as special cases, we obtain

$$A_i=(\phi-\alpha I)^i A_0, \quad A_0-I=(\phi-\alpha I)^i B_i, \quad \text{when } i \nless 1; \quad \ldots\ldots(6)$$

and
$$(\phi-\alpha I)A_{\kappa-1}=A_\kappa=0, \quad (\phi-\alpha I)^\kappa A_0=A_\kappa=0. \quad \ldots\ldots(6')$$

If any one of the matrices A_0, A_1, A_2, ... vanishes, then by (5) all those which follow it vanish. Hence since $A_{\kappa-1}$ does not vanish, we see that:

No one of the matrices $A_0, A_1, A_2, \ldots A_{\kappa-1}$ *in* (B) *vanishes, or has rank* 0.

Let the maximum and potent indices of order m of the linear divisor $\lambda - \alpha$ of the characteristic matrix $\phi - \lambda I$ be s and κ, so that α is a latent root of ϕ repeated exactly s times, and κ has the same meaning as before. Let m_k and ρ_k be the rank and degeneracy of the $(k-1)$th transmute of $\phi - \alpha I$, so that m_k is also the rank of $(\phi - \alpha I)^k$, and ρ_k is also the total number of those potent divisors of $\phi - \lambda I$ which are divisible by $(\lambda - \alpha)^k$; and let

$$r_k = m - m_k = \rho_1 + \rho_2 + \ldots + \rho_k$$

be the degeneracy of $(\phi - \alpha I)^k$. The integers m_k, ρ_k, r_k satisfy the same conditions as in sub-article 1 of § 235; and in particular we have

$$r_\kappa = m - m_\kappa = s.$$

Further let all the distinct latent roots of ϕ be $\alpha, \alpha_1, \alpha_2, \alpha_3, \ldots$, and let the maximum and potent indices of order m of the linear divisors $\lambda - \alpha_1$, $\lambda - \alpha_2$, $\lambda - \alpha_3$, ... of $\phi - \lambda I$ or $\lambda I - \phi$ be $s_1, s_2, s_3, \ldots$ and $\kappa_1, \kappa_2, \kappa_3, \ldots$, so that

$$D_m(\phi) = (\phi - \alpha I)^s (\phi - \alpha_1 I)^{s_1} (\phi - \alpha_2 I)^{s_2} (\phi - \alpha_3 I)^{s_3} \ldots, \quad \ldots\ldots(7)$$

$$E_m(\phi) = (\phi - \alpha I)^\kappa (\phi - \alpha_1 I)^{\kappa_1} (\phi - \alpha_2 I)^{\kappa_2} (\phi - \alpha_3 I)^{\kappa_3} \ldots. \quad \ldots\ldots(7')$$

Then $$s + s_1 + s_2 + s_3 + \ldots = m,$$

and $E_m(\phi) = 0$ is the rational integral equation of lowest degree satisfied by ϕ.

CASE I. *Let* ϕ *have no latent roots distinct from* α.

In this case $s = m = r_\kappa$, $m_\kappa = 0$, $(\phi - \alpha I)^\kappa = E_m(\phi) = 0$, and the equation $(\phi - \alpha I)^\kappa B_\kappa = A_0 - I$ given by (6) shows that

$$A_0 = I.$$

It then follows from the first of the equations (6) that

$$A_i = (\phi - \alpha I)^i.$$

Therefore A_i has rank $m_i = m_i - m_\kappa = r_\kappa - r_i = \rho_{i+1} + \rho_{i+2} + \ldots + \rho_\kappa$; in particular A_0 has rank m, and $A_{\kappa-1}$ has rank ρ_κ.

CASE II. *Let* ϕ *have at least one latent root distinct from* α.

We will first show that if $A_0' = A_0 - I$, the rational integral equations of lowest degrees satisfied by the square matrices A_0 and A_0' are respectively

$$A_0(A_0 - I) = A_0^2 - A_0 = 0, \quad \ldots\ldots\ldots\ldots\ldots\ldots\ldots(8)$$

$$A_0'(A_0' + I) = A_0'^2 + A_0' = 0. \quad \ldots\ldots\ldots\ldots\ldots\ldots\ldots(8')$$

From (6) we see that when $i \not< 1$

$$A_i B_i = (\phi - \alpha I)^i A_0 B_i = A_0 (\phi - \alpha I)^i B_i = A_0(A_0 - I);$$

and when we put $i=\kappa$, we obtain (8). Thus A_0 always satisfies the equation (8), and therefore A_0' always satisfies the equation (8'). If then $\psi(A_0)=0$ is the rational integral equation of lowest degree satisfied by A_0, $\psi(A_0)$ must, when we disregard a non-zero scalar factor, be either A_0 or A_0-I or $A_0^2-A_0$. It cannot be A_0, because $A_0 \neq 0$; and it cannot be A_0-I, for if it were, we should have $A_0=I$, and it would follow from (6) that $(\phi-\alpha I)^\kappa = A_\kappa = 0$, which is only possible when all the latent roots of ϕ are equal to α. Accordingly we must have $\psi(A_0)=A_0^2-A_0$, i.e. (8) is the rational integral equation of lowest degree satisfied by A_0; and it follows that (8') is the rational integral equation of lowest degree satisfied by A_0'.

From (8) we see that the distinct latent roots of A_0 are 0 and 1, and that the potent divisors of the matrix $A_0-\lambda I$ corresponding to its linear divisors λ and $\lambda-1$ are all linear. Let $A_0=\pi(\phi)$, where $\pi(\phi)$ is a rational integral function of ϕ. Then from the equation

$$\pi(\phi)(\phi-\alpha I)^\kappa = (\phi-\alpha I)^\kappa A_0 = 0$$

and (7') it follows by Theorem II*b* of § 224 that $\pi(\phi)$ must contain the factor $(\phi-\alpha_1 I)^{\kappa_1}(\phi-\alpha_2 I)^{\kappa_2}(\phi-\alpha_3 I)^{\kappa_3}\ldots$; and from the relation

$$\pi(\phi)(\phi-\alpha I)^{\kappa-1} = (\phi-\alpha I)^{\kappa-1} A_0 = A_{\kappa-1} \neq 0$$

it then follows that $\pi(\phi)$ does not contain the factor $\phi-\alpha I$. Consequently $\lambda-\alpha_1$, $\lambda-\alpha_2$, $\lambda-\alpha_3$, ..., but not $\lambda-\alpha$, are linear factors of the algebraic function $\pi(\lambda)$, i.e.

$$\pi(\alpha_1)=0,\quad \pi(\alpha_2)=0,\quad \pi(\alpha_3)=0,\quad \ldots,\quad \text{but}\quad \pi(\alpha)\neq 0.$$

We see therefore from the theorem of § 223.2 and from (7) that

$$\begin{aligned}\det\{\lambda I-A_0\} &= \det\{\lambda I-\pi(\phi)\} = \{\lambda-\pi(\alpha)\}^s.\{\lambda-\pi(\alpha_1)\}^{s_1}.\{\lambda-\pi(\alpha_2)\}^{s_2}\ldots\\ &= \{\lambda-\pi(\alpha)\}^s.\lambda^{m-s},\quad \text{where}\quad \pi(\alpha)\neq 0.\end{aligned}$$

Since the only distinct latent roots of A_0 are 0 and 1, we conclude that

$$\pi(\alpha)=1,\quad \det(\lambda I-A_0)=\lambda^{m-s}(\lambda-1)^s. \quad \ldots\ldots\ldots\ldots\ldots(9)$$

Further since the potent divisors of $\lambda I-A_0$ are all linear, it follows from (9) that $\lambda I-A_0$ has exactly $m-s$ potent divisors divisible by λ, i.e. equal to λ. Therefore by Ex. iii of § 228 or Theorem III*a* of § 233 the degeneracy of A_0 is $m-s$, i.e. A_0 has rank s.

From (8') we see in the same way that the distinct latent roots of A_0' are 0 and -1, and that the potent divisors of $A_0'-\lambda I$ are all linear. Moreover

$$\begin{aligned}\det\{\lambda I-A_0'\} &= \det\{(\lambda+1)I-\pi(\phi)\}\\ &= \{\lambda+1-\pi(\alpha)\}^s.\{\lambda+1-\pi(\alpha_1)\}^{s_1}.\{\lambda+1-\pi(\alpha_2)\}^{s_2}\\ &= \lambda^s(\lambda+1)^{m-s};\quad \ldots\ldots\ldots\ldots\ldots\ldots\ldots\ldots\ldots\ldots(9')\end{aligned}$$

and (9') shows that the degeneracy of A_0' is s, i.e. A_0' has rank $m-s$.

Thus the ranks of A_0 and $A_0 - I$ are respectively s and $m - s$.

Now because $\phi - \alpha I$ is not a factor of $\pi(\phi)$, we conclude from the equation

$$A_i = (\phi - \alpha I)^i A_0 = (\phi - \alpha I)^i \pi(\phi),$$

using Theorem I of § 230, that the degeneracy of A_i is equal to the sum of the degeneracies of $(\phi - \alpha I)^i$ and A_0.

Therefore if R_i is the rank of A_i, we have

$$m - R_i = (m - m_i) + (m - s) = m - (m_i - m_\kappa), \quad R_i = m_i - m_\kappa = r_\kappa - r_i.$$

Summarising the results obtained in Cases I and II, we have the following second theorem:

Theorem II. *If α is a latent root of the square matrix $\phi = [a]_m^m$ repeated s times exactly, and if $(\lambda - \alpha)^\kappa$ is the highest power of $\lambda - \alpha$ which is a potent divisor of the characteristic matrix $\phi - \lambda I$, then the expansion of the inverse of $\phi - \lambda I$ in ascending powers of $\lambda - \alpha$ has the form given by*

$$(\lambda I - \phi)^{-1} = \frac{A_{\kappa-1}}{(\lambda - \alpha)^\kappa} + \frac{A_{\kappa-2}}{(\lambda - \alpha)^{\kappa-1}} + \dots + \frac{A_1}{(\lambda - \alpha)^2} + \frac{A_0}{\lambda - \alpha}$$
$$+ B_1 + B_2(\lambda - \alpha) + B_3(\lambda - \alpha)^2 + \dots, \quad \dots(C)$$

where the A's and B's are square matrices of order m which are rational integral functions of ϕ.

If m_i and ρ_i are the rank and degeneracy of the $(i-1)$th transmute of $\phi - \alpha I$, or if m_i and r_i are the rank and degeneracy of $(\phi - \alpha I)^i$, the rank of A_i is

$$m_i - m_\kappa = r_\kappa - r_i = \rho_{i+1} + \rho_{i+2} + \dots + \rho_\kappa;$$

in particular $A_{\kappa-1}$ has rank ρ_κ, and A_0 has rank $m - m_\kappa = r_\kappa = s$.

Ex. i. The rank of $A_{\kappa-1}$ is equal to the number of potent divisors of $\phi - \lambda I$ which are equal to $(\lambda - a)^\kappa$.

Ex. ii. If ϕ and a both lie in a domain of rationality Ω, then the matrices A and B occurring as coefficients in the expansion (C) all lie in Ω.

Ex. iii. ***Frobenius's reduction of a given square matrix to a compartite matrix in standard form, every part of which is a unilatent square matrix.***

Let $\phi = [a]_m^m$ be the square matrix of the text, its latent roots being $a, a_1, a_2, \dots$ repeated $s, s_1, s_2, \dots$ times exactly. Since in (C) the matrices A_0 and $I - A_0$ have ranks s and $m - s$ respectively, we can put

$$A_0 = [p]_m^s [q]_s^m, \quad I - A_0 = [p']_m^{m-s} [q']_{m-s}^m, \quad \dots\dots(10)$$

where each of the factor matrices on the right has rank equal to its passivity, and when we add the two equations (10), we see that

$$[p,\ p']_m^{s,\ m-s}\begin{bmatrix} q \\ q' \end{bmatrix}_{s,\ m-s}^{m} = [1]_m^m = \begin{bmatrix} q \\ q' \end{bmatrix}_{s,\ m-s}^{m} [p,\ p']_m^{s,\ m-s}, \qquad \text{......(11)}$$

that

$$[q]_s^m\,[p']_m^{m-s}=0, \quad [q']_{m-s}^m\,[p]_m^s=0, \qquad \text{......(11')}$$

and that

$$[h]_m^m = \begin{bmatrix} q \\ q' \end{bmatrix}_{s,\ m-s}^{m}, \quad \overline{\underline{H}}_m^m = [p,\ p']_m^{s,\ m-s} \qquad \text{......(12)}$$

are two mutually inverse undegenerate square matrices.

Using the notation (12), let

$$[h]_m^m\,[a]_m^m\,\overline{\underline{H}}_m^m = \begin{bmatrix} b, & \beta \\ \gamma, & c \end{bmatrix}_{s,\ m-s}^{s,\ m-s}, \qquad \text{......(13)}$$

so that

$$[\beta]_s^{m-s} = [q]_s^m\,[a]_m^m\,[p']_m^{m-s}, \quad [\gamma]_{m-s}^s = [q']_{m-s}^m\,[a]_m^m\,[p]_m^s. \qquad \text{......(14)}$$

Then if we in the one case postfix $[p']_m^{m-s}$ and in the other case prefix $[q']_{m-s}^m$ on both sides of the equation

$$A_0\,[a]_m^m = [a]_m^m\,A_0$$

or

$$[p]_m^s\,[q]_s^m\,[a]_m^m = [a]_m^m\,[p]_m^s\,[q]_s^m,$$

we see from (11′) that $[\beta]_s^{m-s}=0$, $[\gamma]_{m-s}^s=0$. Consequently the equimutant transformation (13) has the form

$$[h]_m^m\,[a]_m^m\,\overline{\underline{H}}_m^m = \begin{bmatrix} b, & 0 \\ 0, & c \end{bmatrix}_{s,\ m-s}^{s,\ m-s}. \qquad \text{......(D)}$$

Now in (D) we have

$$[b]_s^s = [q]_s^m\,[a]_m^m\,[p]_m^s;$$

from this equation we deduce by (10) and (8) the equimutant transformations

$$[p]_m^s\,[b]_s^s\,[q]_s^m = A_0\,[a]_m^m\,A_0 = \phi A_0^2 = \phi A_0 = \phi\pi(\phi),$$

$$\overline{\underline{H}}_m^m \begin{bmatrix} b, & 0 \\ 0, & 0 \end{bmatrix}_{s,\ m-s}^{s,\ m-s} [h]_m^m = \phi\pi(\phi);$$

and when we use the theorem of § 223. 2, it follows that

$$\det\left\{\lambda\,[1]_m^m - \begin{bmatrix} b, & 0 \\ 0, & 0 \end{bmatrix}_{s,\ m-s}^{s,\ m-s}\right\} = \det\{\lambda I - \phi\pi(\phi)\},$$

i.e.

$$\det\{\lambda\,[1]_s^s - [b]_s^s\}\,.\,\lambda^{m-s} = \{\lambda - a\pi(a)\}^s\,.\,\{\lambda - a_1\pi(a_1)\}^{s_1}\,.\,\{\lambda - a_2\pi(a_2)\}^{s_2}\ldots$$
$$= (\lambda - a)^s\,.\,\lambda^{m-s},$$

and therefore

$$\det\{\lambda\,[1]_s^s - [b]_s^s\} = (\lambda - a)^s. \qquad \text{......(15)}$$

Also in (D) we have

$$\det\left\{\lambda\,[1]_m^m - \begin{bmatrix} b, & 0 \\ 0, & c \end{bmatrix}_{s,\ m-s}^{s,\ m-s}\right\} = \det(\lambda I - \phi),$$

i.e.

$$\det\{\lambda\,[1]_s^s - [b]_s^s\}\,.\,\det\{\lambda\,[1]_{m-s}^{m-s} - [c]_{m-s}^{m-s}\} = (\lambda - a)^s\,.\,(\lambda - a_1)^{s_1}\,.\,(\lambda - a_2)^{s_2}\ldots,$$

and therefore

$$\det\{\lambda\,[1]_{m-s}^{m-s} - [c]_{m-s}^{m-s}\} = (\lambda - a_1)^{s_1}\,(\lambda - a_2)^{s_2}\ldots \qquad \text{......(15')}$$

Thus in (D) the part $[b]_s^s$ has only the one distinct latent root a, and the latent roots of the part $[b]_{m-s}^{m-s}$ are the other latent roots of ϕ.

We have now proved the following theorem:

Theorem. *If $\phi=[a]_m^m$ is a square matrix with constant elements which has a latent root a repeated s times exactly, and which has other latent roots $a_1, a_2, \ldots$ repeated $s_1, s_2, \ldots$ times exactly, we can determine an equimutant transformation of the form* (D) *in which*

$$\det\{\lambda[1]_s^s-[b]_s^s\}=(\lambda-a)^s, \quad \det\{\lambda[1]_{m-s}^{m-s}-[c]_{m-s}^{m-s}\}=(\lambda-a_1)^{s_1}(\lambda-a_2)^{s_2}\ldots.$$

We can now treat $[c]_{m-s}^{m-s}$ in the same way as $[a]_m^m$, and by a succession of such steps we can reduce ϕ by equimutant transformations to a similar compartite matrix in standard form whose parts are all unilatent square matrices, there being one part corresponding to each distinct latent root of ϕ. This result can be used as a first step in the reduction of ϕ to its standard canonical form by equimutant transformations.

Ex. iv. *If in Ex.* iii *the square matrix A_0 is symmetric, we can replace* (D) *by a symmetric equimutant transformation of the form*

$$\overline{\underline{l}}\,_m^m\,[a]_m^m\,[l]_m^m=\begin{bmatrix} b, & 0 \\ 0, & c\end{bmatrix}_{s,\,m-s}^{s,\,m-s}, \quad \left(\textit{where } [l]_m^m\,\overline{\underline{l}}\,_m^m=[1]_m^m\right). \qquad \text{............(E)}$$

For in this case we can express A_0 and $I-A_0$ in the forms

$$A_0=\overline{\underline{p}}\,_m^s\,[p]_s^m, \quad I-A_0=\overline{\underline{q}}\,_m^{m-s}\,[q]_{m-s}^m,$$

and then in (D) we have

$$[h]_m^m=\begin{bmatrix} p \\ q\end{bmatrix}_{s,\,m-s}^m=\overline{\underline{l}}\,_m^m, \quad \overline{\underline{H}}\,_m^m=\overline{\underline{p,\,q}}\,_m^{s,\,m-s}=[l]_m^m.$$

Ex. v. *If the square matrix A_0 is symmetric and real, then in the symmetric equimutant transformation* (E) *the square semi-unit matrix $[l]_m^m$ can be so chosen as to be real.*

For in this case we can choose $[p]_m^m$ and $[q]_{m-s}^m$ in Ex. ii so that the elements in each horizontal row are either all real or all purely imaginary (see § 160. 3). Then the elements in each horizontal row of $\overline{\underline{l}}\,_m^m$ are either all real or all purely imaginary, and the equation $\overline{\underline{l}}\,_m^m\,[l]_m^m=[1]_m^m$ would be impossible if the elements in any horizontal row of $\overline{\underline{l}}\,_m^m$ were purely imaginary; consequently $[p]_s^m$, $[q]_{m-s}^m$ and $[l]_m^m$ must then be real.

It follows that when A_0 is a real symmetric matrix, it must be definite and have only positive signants.

Ex. vi. *If $[a]_m^m$ is a square semi-unit matrix, so that $[a]_m^m\,\overline{\underline{a}}\,_m^m=[1]_m^m$, and if $a=-1$, then the matrix A_0 is symmetric, and we can replace* (D) *by a symmetric equimutant transformation of the form* (E).

Let $\phi=[a]_m^m$, $\phi'=\overline{\underline{a}}\,_m^m$, so that $\phi\phi'=\phi'\phi=I$. Then since the inverse of the conjugate of an undegenerate square matrix is the conjugate of its inverse, we see from Theorem II that we can write

$$(\lambda I-\phi)^{-1}=\Sigma\frac{A_i}{(\lambda+1)^{i+1}}+\Sigma B_j(\lambda+1)^{j-1}, \qquad \text{........................(16)}$$

$$(\lambda I-\phi')^{-1}=\Sigma\frac{A_i'}{(\lambda+1)^{i+1}}+\Sigma B_j'(\lambda+1)^{j-1}, \qquad \text{........................(17)}$$

where A_i' and B_j' are the conjugates of A_i and B_j, and where i receives the values 0, 1, 2, 3, ... and j the values 1, 2, 3, ...; and when we replace λ by $\frac{1}{\lambda}$ in (17), we have

$$\left(\frac{1}{\lambda}I-\phi'\right)^{-1}=\Sigma\frac{A_i'\lambda^{i+1}}{(\lambda+1)^{i+1}}+\Sigma B_j'\frac{(\lambda+1)^{j-1}}{\lambda^{j-1}}. \quad\text{.....................(17')}$$

Now when we prefix $(\lambda I-\phi)^{-1}$ and postfix $\left(\frac{1}{\lambda}I-\phi'\right)^{-1}$ on both sides of the identity

$$(\lambda I-\phi)\left(\frac{1}{\lambda}I-\phi'\right)=\frac{1}{\lambda}(\lambda I-\phi)+\lambda\left(\frac{1}{\lambda}I-\phi'\right), \quad\text{.....................(18)}$$

we obtain

$$(\lambda I-\phi)^{-1}+\frac{1}{\lambda^2}\left(\frac{1}{\lambda}I-\phi'\right)^{-1}=\frac{I}{\lambda}. \quad\text{...........................(18')}$$

Substituting the expansions (16) and (17') in (18'), we see that for all sufficiently small values of $\lambda+1$ we have

$$\Sigma\frac{A_i}{(\lambda+1)^{i+1}}+\Sigma B_j(\lambda+1)^{j-1}=\frac{I}{\lambda}-\Sigma\frac{A_i'\lambda^{i-1}}{(\lambda+1)^{i+1}}-\Sigma\frac{B_j'(\lambda+1)^{j-1}}{\lambda^{j+1}}. \quad\text{.........(19)}$$

When we expand the positive and negative powers of λ on the right of (19) in ascending powers of $\lambda+1$, and then equate the coefficients of $\frac{1}{\lambda+1}$ on both sides, we see that

$$A_0=A_0',$$

i.e. A_0 is symmetric.

CHAPTER XXVII

COMMUTANTS

[A solution of the matrix equation $AX = XB$, in which A and B must be square matrices, is called a commutant $X = \{A, B\}$; it is a co-commutant when $B = \pm A$, a contra-commutant when $B = \pm A'$, and a non-singular commutant when its elements are independent of any arbitrary parameters occurring in the elements of A and B. The earlier articles (§§ 237—40) deal with the definitions of commutants and commutantal transformations, and with general principles relating to commutants. It is shown in §§ 241—3 that when A and B are unilatent canonical square matrices, the general commutant $\{A, B\}$ is a general ruled compound slope or a zero matrix according as the latent roots of A and B are equal or unequal; and these facts furnish the complete constructions of § 244 for the general commutant $\{A, B\}$ whenever all the elements of A and B are numerical constants. Special attention is given to the rank of a general commutant, the conditions for the existence of undegenerate symmetric or undegenerate skew-symmetric contra-commutants, and the conditions that the general co-commutant $\{A, A\}$ shall be a rational integral function of A. The concluding articles (§§ 245—8) deal with the reduction and construction of commutantal transformations, and the final theorems with the construction of equigradent commutantal transformations converting one given undegenerate square commutant into another. Particular cases are the expressions for a given undegenerate square matrix as a product of symmetric or skew-symmetric matrices (§ 245), and the construction of symmetric semi-unit transformations converting one given symmetric or skew-symmetric matrix into another (§ 248).]

§ 237. Independent matrices of a given simple class.

A compound matrix in which the constituents form r horizontal and s vertical minors will be said to belong to the *class*

$$M\begin{pmatrix} q_1, & q_2, & \dots & q_s \\ p_1, & p_2, & \dots & p_r \end{pmatrix},$$

when the r successive horizontal simple minors counted from the top contain respectively $p_1, p_2, \dots p_r$ horizontal rows and the s successive vertical simple minors counted from the left contain respectively $q_1, q_2, \dots q_s$ vertical rows. We will call $p_1, p_2, \dots p_r$ and $q_1, q_2, \dots q_r$ the horizontal and vertical *index numbers*. A matrix which is not regarded as compound (one in which $r = s = 1$) is called a *simple matrix* or a matrix belonging to a *simple class*.

If $X_1, X_2, \dots X_i$ are i particular matrices with constant elements of a given simple class $M\begin{pmatrix} n \\ m \end{pmatrix}$, i.e. each containing m horizontal and n vertical rows, they will be said to be mutually *independent* when there exists no relation of the form

$$c_1X_1 + c_2X_2 + \dots + c_iX_i = 0, \qquad \dots\dots\dots\dots\dots\dots(1)$$

where $c_1, c_2, \dots c_i$ are scalar constants which are not all equal to 0. Any particular matrix X of the same class is *dependent* on $X_1, X_2, \dots X_i$ when and only when there exists a relation of the form

$$X = c_1X_1 + c_2X_2 + \dots + c_iX_i, \quad \dots\dots\dots\dots(2)$$

where $c_1, c_2, \dots c_i$ are scalar constants.

Let X_{ij} be the matrix of the class $M\binom{n}{m}$ in which the element common to the ith horizontal and jth vertical rows is 1, and all other elements are 0's; and let $X_0 = [0]_m^n$ be the zero matrix of that class. Then the mn non-zero matrices

$$X_{ij}, \quad (i = 1, 2, \dots m; \; j = 1, 2, \dots n),$$

which are mutually independent, will be called the *mn elementary* independent non-zero matrices of that class; and they together with X_0 are $mn+1$ independent matrices of that class. The general matrix X of the given class $M\binom{n}{m}$ can be expressed in the form

$$X = \Sigma x_{ij}X_{ij}, \quad (i = 1, 2, \dots m; \; j = 1, 2, \dots n), \quad \dots\dots\dots\dots(3)$$

where the mn letters x_{ij} denote independent arbitrary parameters, or if we please in the equivalent form

$$X = \Sigma x_{ij}X_{ij} + X_0. \quad \dots\dots\dots\dots(3')$$

Because every matrix of the given class can be regarded as a particularisation of this general matrix, we see that the given class contains exactly mn (but not more) independent non-zero matrices, and exactly $mn+1$ (but not more) independent matrices.

We shall be concerned in this chapter with matrices X of the simple class $M\binom{n}{m}$ which can be expressed in the form

$$X = x_1X_1 + x_2X_2 + \dots + x_iX_i, \quad (i \not> mn), \quad \dots\dots\dots\dots(4)$$

where $x_1, x_2, \dots x_i$ are *independent scalar parameters*, and $X_1, X_2, \dots X_i$ are particular *independent non-zero matrices* of the same class. When X is any such matrix, its elements are homogeneous linear functions of the i independent scalar variables $x_1, x_2, \dots x_i$, and it has exactly i independent non-zero particularisations. We can either regard X as a homogeneous linear function of the scalar variables $x_1, x_2, \dots x_i$, its coefficients being then independent non-zero matrices of the class $M\binom{n}{m}$, or we can regard it as a homogeneous linear function of the i particular independent non-zero matrices $X_1, X_2, \dots X_i$, its coefficients being then independent scalar variables. If Y is the same matrix expressed as a homogeneous linear function of other independent non-zero matrices $Y_1, Y_2, Y_3, \dots$ of the same class, the number of terms in Y must also be equal to i, because independent non-zero particularisations of

X or Y are also independent non-zero particularisations of Y or X, and we have

$$Y = y_1 Y_1 + y_2 Y_2 + \dots + y_i Y_i, \quad \text{.....................(4')}$$

where $y_1, y_2, \dots y_i$ are algebraic homogeneous linear functions of $x_1, x_2, \dots x_i$ which can be treated as independent scalar variables. We can pass from (4) to (4′) either by expressing $X_1, X_2, \dots X_i$ in terms of any i independent non-zero particularisations of X, or by using any ordinary homogeneous linear transformation to replace the variables $x_1, x_2, \dots x_i$ by $y_1, y_2, \dots y_i$. We can always express X in terms of the elementary independent non-zero matrices of the class $M\binom{n}{m}$ in the form (3); but ordinarily there will then be more than i non-zero terms, and the coefficients will not be *independent* variables.

Ex. i. We can replace the equation (2) by

$$X = x_1 X_1 + x_2 X_2 + \dots + x_i X_i + X_0,$$

where X_0 is the zero matrix $[0]_m^n$, and $X_1, X_2, \dots X_i, X_0$ are $i+1$ particular independent matrices of the class $M\binom{n}{m}$.

Ex. ii. If $A = [a]_m^p$ and $B = [b]_p^n$ are general matrices, we can evaluate the product matrix in the product

$$[a]_m^p [b]_p^n = [x]_m^n \text{ or } AB = X$$

by putting

$$A = \Sigma a_{iu} A_{iu}, \quad (i = 1, 2, \dots m;\ u = 1, 2, \dots p), \text{........................(5)}$$
$$B = \Sigma b_{vj} B_{vj}, \quad (v = 1, 2, \dots p;\ j = 1, 2, \dots n), \text{.....................(5')}$$
$$X = \Sigma x_{ij} X_{ij}, \quad (i = 1, 2, \dots m;\ j = 1, 2, \dots n), \text{.....................(5'')}$$

where A_{iu}, B_{vj}, Y_{ij} are the elementary independent matrices of the classes $M\binom{p}{m}$, $M\binom{n}{p}$, $M\binom{n}{m}$, and using the law of matrix multiplication only for the latter matrices. We then have

$$A_{iu} B_{uj} = X_{ij}; \quad A_{iu} B_{vj} = 0 \text{ when } v \neq u; \text{ ....(6)}$$

and it follows that for a given pair of values of i and j we have

$$x_{ij} X_{ij} = \Sigma (a_{iu} b_{uj} . A_{iu} B_{uj}) = (\Sigma a_{iu} b_{uj}) . X_{ij},$$

or

$$x_{ij} = \Sigma a_{iu} b_{uj}.$$

Some writers regard the matrices A, B, X as defined by the equations (5), (5′), (5″) in which a_{iu}, b_{vj}, x_{ij} are scalar variables, and $A_{iu} B_{vj}$, X_{ij} are 'hyper-numbers' which are subject to the law of multiplication defined by (6).

NOTE 1. *Independent rational integral functional matrices.*

When we are concerned with matrices $X, X_1, X_2, X_3, \dots$ of a given simple class whose elements are rational integral functions of given scalar variables $\lambda, \mu, \nu, \dots$, we may define a particular rational integral functional matrix X to be one whose elements are particular rational integral functions of $\lambda, \mu, \nu, \dots$; and the particular rational integral functional matrices $X_1, X_2, \dots X_i$ to be mutually *independent* when there exists no identity in $\lambda, \mu, \nu, \dots$ of the form (1) in which $c_1, c_2, \dots c_i$ are rational integral functions of $\lambda, \mu, \nu, \dots$ which do not all vanish identically. In this case the matrix X is *dependent* on X_1,

$X_2, \dots X_i$ when there exists an identity of the form (2) in which $c_1, c_2, \dots c_i$ are rational functions of $\lambda, \mu, \nu, \dots$, i.e. when there exists an identity of the form

$$cX = c_1X_1 + c_2X_2 + \dots + c_iX_i$$

in which $c, c_1, c_2, \dots c_i$ are rational integral functions of $\lambda, \mu, \nu, \dots$ and c does not vanish identically.

NOTE 2. *Independent rational functional matrices.*

When the elements of $X, X_1, X_2, X_3, \dots$ are particular rational functions of the scalar variables $\lambda, \mu, \nu, \dots$, we can always determine non-zero rational integral functions $\rho, \rho_1, \rho_2, \rho_3, \dots$ of $\lambda, \mu, \nu, \dots$ such that the elements of the matrices

$$Y = \rho X, \quad Y_1 = \rho_1 X_1, \quad Y_2 = \rho_2 X_2, \quad Y_3 = \rho_3 X_3, \dots$$

are rational integral functions of $\lambda, \mu, \nu, \dots$; and we consider $X_1, X_2, \dots X_i$ to be independent or X to be dependent on $X_1, X_2, \dots X_i$ when and only when $Y_1, Y_2, \dots Y_i$ are independent or Y is dependent on $Y_1, Y_2, \dots Y_i$.

NOTE 3. *Relations of this and the next two chapters to the immediately following chapters.*

There are many inter-connections between Chapters XXVII—XXIX, which deal with commutants and invariant transformands, and Chapters XXX—XXXIX, which deal with matrices having assigned commutantal types represented by symbolic commutants, and more particularly with the properties of simple and compound slopes. The latter chapters are placed after the former because the definitions of symbolic commutants and of matrices having assigned commutantal types are derived from the definition of a true commutant. Definitions given in the latter chapters will often be used in the former chapters; and proofs given in the former chapters will sometimes be abbreviated by references to the latter chapters for the properties of simple and compound slopes, which are discussed in them from a more general standpoint than would have been required in the former chapters.

NOTE 4. *Independent powers of a given matrix.*

If ϕ is a given square matrix of order m, there must exist some smallest positive integer r such that ϕ^r is dependent on the lower positive integral powers of ϕ, which are all independent. This integer r, which has been determined in Chapter XXV, must obviously play an important part in the properties of ϕ. If we include ϕ^0 amongst the positive integral powers of ϕ and interpret it to be the unit matrix $[1]_m^m$, the integer r cannot exceed m; if we do not include ϕ^0 amongst the positive integral powers of ϕ, the integer r cannot exceed $m+1$. In both cases there are exactly r independent positive integral powers of A. Corresponding remarks (see Appendix A) can be made with respect to a given matrix of any given simple class.

§ 238. Commutants defined.

1. *The commutants of a pair of square matrices whose elements are constants.*

If $A = [a]_m^m$ and $B = [b]_n^n$ are any two given square matrices of orders m and n with constant elements forming a matrix-pair (A, B) in which A precedes B, any matrix $X = [x]_m^n$ satisfying the equation

$$[a]_m^m [x]_m^n = [x]_m^n [b]_n^n \quad \text{or} \quad AX = XB \quad \dots\dots\dots\dots\dots\dots\text{(a)}$$

will be called a *commutant* of the arranged matrix-pair (A, B), or more shortly a *commutant* $\{A, B\}$. If the elements of X are rational integral functions of certain independent arbitrary parameters, it is to be understood that the equation (a) must be an identity in those parameters.

The expression $\{A, B\}$ will frequently be used to denote 'solution of the equation $AX = XB$,' where A and B must be *square* matrices if it is understood as usual that AX and XB are standard products in each of which the two factors have equal passivities. The equation $X = \{A, B\}$ will be used to indicate that X is a commutant defined as above, and the expression $\{A, B\}$ will then be called the *commutantal type* of the commutant X.

The equation (a) always has the zero solution $X = [0]_m^n$; and if X is any solution, then ρX is a solution, where ρ is any scalar quantity; conversely if ρX is a solution, and ρ does not vanish identically, then X is a solution. Further if $X_1, X_2, \dots X_r$ are solutions, then

$$\rho_1 X_1 + \rho_2 X_2 + \dots + \rho_r X_r$$

is a solution when $\rho_1, \rho_2, \dots \rho_r$ are any r scalar quantities. In solutions involving arbitrary parameters, the elements will usually be supposed to be rational integral functions of those parameters. Rational solutions are not thereby excluded, since they can always be replaced by equivalent rational integral solutions.

By a *particular commutant* $\{A, B\}$ we shall mean a matrix $X = [x]_m^n$ with constant or numerical elements satisfying the equation (a). When the equation (a) has non-zero solutions, let i be the greatest possible number of independent particular non-zero solutions. Then a *general commutant* $\{A, B\}$ will be defined to be a matrix $X = [x]_m^n$ expressible in the form

$$X = \lambda_1 X_1 + \lambda_2 X_2 + \dots + \lambda_i X_i, \quad \text{.....................(b)}$$

where $X_1, X_2, \dots X_i$ are any i independent particular non-zero commutants $\{A, B\}$, and $\lambda_1, \lambda_2, \dots \lambda_i$ are independent arbitrary parameters. If the equation (a) has no non-zero solution, we have $i = 0$, and in this case the general commutant $\{A, B\}$ is the zero matrix $X_0 = [0]_m^n$, which could have been regarded as an additional term in (b). Thus when there exist non-zero commutants $\{A, B\}$, a general commutant $\{A, B\}$ is a commutant whose elements are homogeneous linear functions of a determinate number i of independent scalar parameters or variables; it can be regarded as a homogeneous linear function of those scalar variables, the coefficients being independent non-zero commutants; or it can be regarded as a homogeneous linear function of i particular non-zero commutants with constant elements, the coefficients being independent scalar variables.

When X is a general commutant $\{A, B\}$ expressed in the form (b), we will call $\lambda_1, \lambda_2, \ldots \lambda_i$ the parameters of X. By a *particularisation* of X we shall mean a matrix derived from X by ascribing particular numerical values to the λ's; and by a *specialisation* of X we shall mean a matrix derived from X by substituting rational integral (or rational) functions of certain parameters $\mu_1, \mu_2, \ldots \mu_r$ for the λ's, such functions being usually homogeneous and linear. Every such matrix is a commutant $\{A, B\}$. Any two general commutants $\{A, B\}$ are mutually equivalent in the sense that each of them is a specialisation of the other. In fact when the equation (a) is replaced by mn equivalent scalar equations, it will become clear that every general commutant $X = \{A, B\}$ is a general solution of the equation (a) of which all other solutions are particularisations or specialisations. We shall often speak of *the* general commutant $\{A, B\}$, thereby meaning some selected general commutant $\{A, B\}$.

A symmetric (or skew-symmetric) solution X of the equation (a) will be called a *general symmetric* (or *general skew-symmetric*) commutant $\{A, B\}$ when all symmetric (or all skew-symmetric) solutions can be regarded as particularisations or specialisations of it. When it is not a zero matrix, it is a homogeneous linear function of the greatest possible number of independent particular non-zero symmetric (or skew-symmetric) solutions, the coefficients being arbitrary scalar parameters.

If A, B, C, D are matrices with constant elements, and if X' is the conjugate of X, we can define particular and general solutions of matrix equations of the forms

$$\Sigma AXB = 0, \quad \Sigma AXB + \Sigma CX'D = 0$$

in the same ways as we have defined those of the equation $AX = XB$, a general solution X being a homogeneous linear function of the greatest possible number of independent particular solutions.

NOTE 1. *Continuantal and alternating commutants.*

The symbols $\{A, B\}, \ \{-A, -B\}, \ \{-A, B\}, \ \{A, -B\}$, ...(1)

in which A and B are always square matrices, mean solutions of the respective equations

$$AX = XB, \quad -AX = -XB, \quad -AX = XB, \quad AX = -XB. \qquad \text{............(1')}$$

When a matrix X is represented by the 1st, 2nd, 3rd, 4th of the symbols (1) to indicate that it is a commutant of the 1st, 2nd, 3rd, 4th of the arranged matrix-pairs

$$(A, B), \quad (-A, -B), \quad (-A, B), \quad (A, -B),$$

we will call it a *continuantal* or *alternating* commutant according as one of the first two or one of the last two symbols is used. The first two symbols are interchangeable with one another, as also the last two symbols; and ordinarily we shall only use the two symbols $\{A, B\}$, $\{A, -B\}$. But in defining and manipulating commutantal equations and products it is convenient to treat all four of the symbols (1) as being different.

An alternating commutant $\{A, -B\}$ can always be regarded as a continuantal commutant $\{A, B'\}$ in which $B' = -B$.

The symbols $\{B, A\}, \ \{-B, -A\}, \ \{-B, A\}, \ \{B, -A\}$, ...(2)

in which B and A are always square matrices, mean solutions of the respective equations

$$BY = YA, \quad -BY = -YA, \quad -BY = YA, \quad BY = -YA, \quad \text{............(2')}$$

and must not be confounded with the symbols (1).

NOTE 2. *The commutants and contra-commutants of a single square matrix.*

The particular cases in which B is the same matrix as A or the conjugate of A will often occur.

If $A = [a]_m^m$ is any given square matrix, any square matrix $X = [x]_m^m$ satisfying the equation

$$[a]_m^m [x]_m^m = [x]_m^m [a]_m^m \quad \text{or} \quad AX = XA \quad \text{...........................(c)}$$

will be called a *commutant* of A (or a co-commutant of A), and we can use the equation $X = \{A, A\}$ to indicate that X is a 'commutant of A.' Again if A' is the conjugate of A, any square matrix $Y = [y]_m^m$ satisfying the equation

$$\overline{\underline{a}}_m^m [y]_m^m = [y]_m^m [a]_m^m \quad \text{or} \quad A'Y = YA \quad \text{..........................(c')}$$

will be called a *contra-commutant* of A, and we shall use the equation $Y = \{A', A\}$ to indicate that Y is a 'contra-commutant of A.'

A commutant $\{A', A'\}$ is a 'commutant of A''; and a commutant $\{A, A'\}$ is a 'contra-commutant of A'.'

Since X is a commutant of A when and only when it is commutative with A, the determination of a general commutant of A is equivalent to the determination of all square matrices which are commutative with A.

NOTE 3. *Co-commutants and contra-commutants.*

The generic term *co-commutant* will be applied to commutants of the types

$$\{\pm A, \pm A\}, \quad \{\pm A', \pm A'\},$$

these being the continuantal and alternating commutants of a pair of equal square matrices.

The generic term *contra-commutant* will be applied to commutants of the types

$$\{\pm A', \pm A\}, \quad \{\pm A, \pm A'\},$$

these being the continuantal and alternating commutants of a pair of mutually conjugate square matrices.

Ex. i. *Cases in which A or B is a zero matrix.*

When $B = [0]_n^n$, a general commutant $\{A, B\}$ is a general solution of the matrix equation

$$[a]_m^m [x]_m^n = 0 \quad \text{or} \quad AX = 0;$$

and when $A = [0]_m^m$, a general commutant $\{A, B\}$ is a general solution of the matrix equation

$$[x]_m^n [b]_n^n = 0 \quad \text{or} \quad XB = 0.$$

When A and B are both zero matrices, a general commutant $\{A, B\}$ is a matrix $X = [x]_m^n$ whose elements are independent arbitrary parameters.

Ex. ii. *Cases in which A and B are both scalar or quasi-scalar matrices (which may be zero matrices).*

If $A=a[1]_m^m$ and $B=b[1]_n^n$ are scalar matrices, the general commutant $\{A, B\}$ is:

(1) the zero matrix $[0]_m^n$ when $a \neq b$;

(2) a matrix $[x]_m^n$ with arbitrary elements when $a=b$.

By using this result in conjunction with Theorem II of § 240 we can construct the general commutant $\{A, B\}$ when A and B are given quasi-scalar matrices.

Ex. iii. *Cases in which the general commutant $\{A, B\}$ is a matrix with arbitrary elements.*

If $A=[a]_m^m$ and $B=[b]_n^n$ are square matrices of orders m and n with constant elements, the general commutant $\{A, B\}$ will be a matrix $[x]_m^n$ whose elements are independent arbitrary parameters when and only when A and B are scalar matrices with equal diagonal elements, so that

$$A=c[1]_m^m, \quad B=c[1]_n^n,$$

where c is any scalar number, which may be 0.

For if $X=[x]_m^n$, the matrix obtained by putting all elements of X equal to 0 except x_{ij} will be a commutant $\{A, B\}$ if and only if

$$a_{pi}=0 \text{ when } p \neq i, \quad b_{jq}=0 \text{ when } q \neq j, \quad a_{ii}=b_{jj};$$

and when these conditions are satisfied for all permissible pairs of values of i and j, A and B must have the forms given above.

Ex. iv. *Cases in which the general commutant $\{A, B\}$ is a zero matrix.*

If A and B are defined as in Ex. iii, it will be shown in § 240. 3 that the general commutant $\{A, B\}$ is a zero matrix when and only when A and B have no latent root in common.

Ex. v. *The general commutant $\{A, A\}$ of a single square matrix A is always an undegenerate square matrix.*

For if m is the order of A, the unit matrix $[1]_m^m$ is always a particular undegenerate commutant $\{A, A\}$. It will be shown in § 244 that the general contra-commutant $\{A', A\}$ is always an undegenerate square matrix.

Ex. vi. *Cases in which the general commutant $\{A, B\}$ is an undegenerate square matrix.*

The condition $n=m$ is of course necessary. Let A and B be two square matrices of the same order m with constant elements; and let h and H be two mutually inverse undegenerate square matrices with constant elements.

Then h is a particular undegenerate square commutant $\{A, B\}$ or H is a particular undegenerate square commutant $\{B, A\}$ when and only when

$$hBH=A$$

is an equimutant (or isomorphic) transformation converting B into A.

Thus there exist undegenerate square commutants $\{A, B\}$ or undegenerate square commutants $\{B, A\}$ when and only when A and B are equi-canonical, i.e. have equipotent characteristic matrices.

Further h or H is a particular undegenerate commutant $\{A, A\}$ of the single square matrix A when and only when

$$hAH=A$$

is an equimutant (or isomorphic) transformation converting A into itself, so that A is an invariant transformand of a transformation by h and H.

Ex. vii. *Rational integral functions of a square matrix regarded as commutants.*

If A is any given square matrix of order m, every rational integral function of A is a commutant $\{A, A\}$, i.e. a commutant of A; for (as in Ex. viii) it is commutative with A.

It will be shown in Theorem II of § 244 that *all* commutants of A will be rational integral functions of A when and only when A satisfies no rational integral equation of degree less than m.

Ex. viii. *If $\Phi = f(A)$ is any rational integral function of a given square matrix A of order m, and if A', Φ' are the conjugates of A, Φ, then:*

(i) *Every co-commutant $\{A, A\}$ is also a co-commutant $\{\Phi, \Phi\}$.*
(ii) *Every co-commutant $\{A', A'\}$ is also a co-commutant $\{\Phi', \Phi'\}$.*
(iii) *Every contra-commutant $\{A', A\}$ is also a contra-commutant $\{\Phi', \Phi\}$.*
(iv) *Every contra-commutant $\{A, A'\}$ is also a contra-commutant $\{\Phi, \Phi'\}$.*

Since $\Phi = f(A)$ when and only when $\Phi' = f(A')$, we need only consider the properties (i) and (iii). Let $I = [1]_m^m$, $X = [x]_m^m$, and suppose that

$$\Phi = c_0 I + c_1 A + c_2 A^2 + \dots + c_p A^p, \quad \Phi' = c_0 I + c_1 A' + c_2 A'^2 + \dots + c_p A'^p.$$

Then if X is any commutant of A, the property (i) follows from the equations

$$AX = XA, \quad A^2X = AXA = XA^2, \quad A^3X = AXA^2 = XA^3, \dots,$$

which lead to
$$\Phi X = X\Phi;$$

and if X is any contra-commutant of A, the property (iii) follows from the equations

$$A'X = XA, \quad A'^2X = A'XA = XA^2, \quad A'^3X = A'XA^2 = XA^3, \dots,$$

which lead to
$$\Phi' X = X\Phi.$$

It will be shown in § 252 that a square matrix Φ of order m which has any one of the properties (i), (ii), (iii), (iv) must be a rational integral function of A.

2. *Determination of commutants by the solution of scalar equations.*

The matrix $X = [x]_m^n$ will be a solution of the matrix equation

$$[a]_m^m [x]_m^n = [x]_m^n [b]_n^n \quad \text{or} \quad AX = XB \quad \dots\dots\dots\dots\dots\dots\text{(a)}$$

when and only when its elements satisfy the mn homogeneous linear scalar equations

$$e_{ij} \equiv \sum_{u=1}^{u=m} a_{iu} x_{uj} = \sum_{v=1}^{v=n} x_{iv} b_{vj} = 0, \quad (i = 1, 2, \dots m;\ j = 1, 2, \dots n), \dots\text{(a}')$$

which are together equivalent to (a). We can therefore determine particular or general commutants $\{A, B\}$ by finding particular or general solutions of the mn scalar equations (a′) in which the variables are the mn elements of X.

After determining the general commutant $\{A, B\}$ by this method in certain particular cases, we shall be able to determine it for all particular matrix-pairs by using the general principles described in § 240.

In all cases the scalar equations (a′) can be replaced by a single matrix equation of the form

$$[\omega]_{mn}^{mn} [\xi]_{mn}^{1} = 0 \quad \text{or} \quad \Omega\xi = 0, \quad \dots\dots\dots\dots\dots\dots\dots\text{(a}'')$$

where ξ is a one-rowed matrix whose elements are the mn elements of X, and where Ω is a square matrix of order mn in which $mn(mn-m-n+1)$ of the m^2n^2 elements are always 0's. The square matrix Ω can be made to assume many elegant forms by suitably arranging the variables such as x_{ij} and the equations (a′), i.e. by suitably arranging its horizontal and vertical rows. If we adopt the same order of arrangement for

$$x_{11}, x_{12}, \ldots x_{ij}, \ldots \text{ and } e_{11}, e_{12}, \ldots e_{ij}, \ldots,$$

it has resemblances with a symmetric matrix; and if we adopt the same order of arrangement for

$$x_{11}, x_{12}, \ldots x_{ij}, \ldots \text{ and } e_{11}, e_{21}, \ldots e_{ji}, \ldots,$$

it has resemblances with a skew-symmetric matrix. The general structure of Ω is most clearly shown when the orders of arrangement are as in Exs. xi and xii. We shall not make much actual use of the equation (a″) except (see § 253) in the determination of the general commutant $\{A, A\}$ when all the elements of A are arbitrary.

Since the determinant $\qquad R = \det \Omega$ ....(3)

is the resultant of the mn scalar equations (a′) or of the mn homogeneous linear functions e_{ij} when all the elements of A and B are arbitrary, we see that:

If A and B are square matrices with arbitrary elements, then for all particular values of their elements the general commutant $\{A, B\}$ is a non-zero matrix, i.e. there exist non-zero commutants $\{A, B\}$, when and only when $R=0$. ....(4)

On this account we will call R the *scalar resultant* of the two square matrices A and B. Although difficulties present themselves when we attempt to obtain a complete expansion of the determinant of Ω, some important properties of R can be immediately deduced from the result stated in Ex. iv. In the first place it follows (see § 240. 3) that:

If A and B are square matrices with arbitrary elements, then:

(1) *For all particular values of their elements the vanishing of R is the necessary and sufficient condition that A and B shall have a latent root in common.*

(2) *The scalar resultant R of A and B is the resultant of the two characteristic determinants*

$$det\, A(\lambda) = det\,\{[a]_m^m - \lambda\,[1]_m^m\}, \quad det\, B(\lambda) = det\,\{[b]_n^n - \lambda\,[1]_n^n\},$$

when these are regarded as functions of λ only, the elements of A and B being arbitrary parameters. ....(5)

From the second of these two results it follows further that:

If A and B are square matrices whose elements are rational integral functions of certain arbitrary parameters $\gamma_1, \gamma_2, \gamma_3, \ldots$, so that their scalar *resultant R is a rational integral function of the γ's, then the identical vanishing of R is the necessary and sufficient condition that the two characteristic determinants det A (λ) and det B (λ) shall have a factor in common which is a rational integral function of λ and the γ's; and it is of course also the necessary and sufficient condition that the equation* (a) *or* (a″) *shall have a non-zero solution.* ....(6)

To prove (6) let $f = \det A(\lambda)$, $g = \det B(\lambda)$. Then if f and g have a rational integral factor in common (which must involve λ), the resultant R necessarily vanishes identically. Again if f and g have no rational integral factor in common, then by Ex. viii of § 188, there must exist rational integral identities in λ and the γ's of the form

$$Pf + Qg = \phi, \quad \text{..............................(7)}$$

where ϕ is a rational integral function of the γ's only which does not vanish identically; consequently we can ascribe such particular values to the γ's that f and g become functions of λ having no common root, so that $R \neq 0$; consequently R cannot vanish identically.

In connection with the last result attention may be drawn to the well-known fact that we can always determine rational integral identities in λ and the γ's of the form

$$Pf + Qg = R, \quad \text{..............................(7')}$$

where the degrees of P and Q in λ are less than n and m respectively. This follows from Ex. iv of § 193, where before evaluating Δ by expansion we can add to the last horizontal row the 1st, 2nd, ... $(m+n-1)$th preceding horizontal rows multiplied by $x^1, x^2, \ldots x^{m+n-1}$ respectively.

Ex. ix. *If* $A' = A - c[1]_m^m$, $B' = B - c[1]_n^n$, *where c is any scalar number, the matrix-pairs* (A', B') *and* (A, B) *have the same commutants.*

For we have $A'X = XB'$ when and only when $AX = XB$. The diagonal elements of A and B only occur in the one term $(a_{ii} - b_{jj})\, x_{ij}$ of e_{ij}; consequently the equations (a′) and (a″) are the same for the two matrix-pairs.

Ex. x. *Reduction to the case in which A and B are square matrices of the same order.*

First suppose that
$$n = m + p > m,$$

and let
$$A' = \begin{bmatrix} a, & 0 \\ 0, & 0 \end{bmatrix}_{m,p}^{m,p}, \quad X' = \begin{bmatrix} x \\ y \end{bmatrix}_{m,p}^{n}, \quad X = [x]_m^n, \quad Y = [y]_p^n.$$

Then the equation $A'X' = X'B$ is equivalent to the two equations $AX = XB$, $YB = 0$. Consequently X' will be a general (or particular) commutant $\{A', B\}$ when and only when X is a general (or particular) commutant $\{A, B\}$ and Y is a general (or particular) solution of the equation $YB = 0$; and after determining the general commutant $\{A', B\}$ we shall know the general commutant $\{A, B\}$. If R' is the scalar resultant of A' and B, we have in this case

$$R' = (\det B)^p \,.\, R.$$

Next suppose that
$$m = n + p > n,$$

and let
$$B' = \begin{bmatrix} b, & 0 \\ 0, & 0 \end{bmatrix}_{n,p}^{n,p}, \quad X' = [x, y]_m^{n,p}, \quad X = [x]_m^n, \quad Y = [y]_m^n.$$

Then X' will be a general (or particular) commutant $\{A, B'\}$ when and only when X is a general (or particular) commutant $\{A, B\}$ and Y is a general (or particular) solution of the equation $AY=0$. If R' is the scalar resultant of A and B', we have in this case

$$R'=(\det A)^p \,.\, R.$$

It follows that in discussing the general properties of commutants $\{A, B\}$ we can confine ourselves to the cases in which A and B are square matrices of the same order.

Ex. xi. If $m=n=2$, and if we put $c_{ij}=a_{ii}-b_{jj}$, the equation (a) can be replaced by either of the equations

$$\begin{bmatrix} e_{11} \\ e_{21} \\ e_{12} \\ e_{22} \end{bmatrix} \equiv \begin{bmatrix} c_{11}, & a_{12}, & -b_{21}, & 0 \\ a_{21}, & c_{21}, & 0, & -b_{21} \\ -b_{12}, & 0, & c_{12}, & a_{12} \\ 0, & -b_{12}, & a_{21}, & c_{22} \end{bmatrix} \begin{bmatrix} x_{11} \\ x_{21} \\ x_{12} \\ x_{22} \end{bmatrix} =0,$$

$$\begin{bmatrix} e_{11} \\ e_{12} \\ e_{21} \\ e_{22} \end{bmatrix} \equiv \begin{bmatrix} c_{11}, & -b_{21}, & a_{12}, & 0 \\ -b_{12}, & c_{12}, & 0, & a_{12} \\ a_{21}, & 0, & c_{21}, & -b_{21} \\ 0, & a_{21}, & -b_{12}, & c_{22} \end{bmatrix} \begin{bmatrix} x_{11} \\ x_{12} \\ x_{21} \\ x_{22} \end{bmatrix} =0,$$

where the prefactors on the right are forms of the square matrix Ω in (a″). In this case we have

$$R=(a_{11}-b_{11})(a_{11}-b_{22})(a_{22}-b_{11})(a_{22}-b_{22})-\alpha\{(b_{11}-a_{11})(b_{11}-a_{22})+(b_{22}-a_{11})(b_{22}-a_{22})\}$$
$$-\beta\{(a_{11}-b_{11})(a_{11}-b_{22})+(a_{22}-b_{11})(a_{22}-b_{22})\}+(\alpha-\beta)^2,$$

where $\qquad \alpha=a_{12}a_{21}, \quad \beta=b_{12}b_{21}.$

If $a_{12}=a_{22}=a_{21}=0$, then in accordance with Ex. x we have

$$R=\det B\,.\,\{(b_{11}-a_{11})(b_{22}-a_{11})-b_{12}b_{21}\},$$

where the co-factor of $\det B$ is the scalar resultant of the two square matrices $[a]_1^1$ and $[b]_2^2$.

Ex. xii. If $m=3$ and $n=4$, and if we put $c_{ij}=a_{ii}-b_{jj}$ the equation (a) can be replaced by either of the equations

$$\left[\begin{array}{c} e_{11} \\ e_{21} \\ e_{31} \\ \hline e_{12} \\ e_{22} \\ e_{32} \\ \hline e_{13} \\ e_{23} \\ e_{33} \\ \hline e_{14} \\ e_{24} \\ e_{34} \end{array}\right] \equiv \left[\begin{array}{ccc|ccc|ccc|ccc} c_{11} & a_{12} & a_{13} & -b_{21} & 0 & 0 & -b_{31} & 0 & 0 & -b_{41} & 0 & 0 \\ a_{21} & c_{21} & a_{23} & 0 & -b_{21} & 0 & 0 & -b_{31} & 0 & 0 & -b_{41} & 0 \\ a_{31} & a_{32} & c_{31} & 0 & 0 & -b_{21} & 0 & 0 & -b_{31} & 0 & 0 & -b_{41} \\ \hline -b_{12} & 0 & 0 & c_{12} & a_{12} & a_{13} & -b_{32} & 0 & 0 & -b_{42} & 0 & 0 \\ 0 & -b_{12} & 0 & a_{21} & c_{22} & a_{23} & 0 & -b_{32} & 0 & 0 & -b_{42} & 0 \\ 0 & 0 & -b_{12} & a_{31} & a_{32} & c_{32} & 0 & 0 & -b_{32} & 0 & 0 & -b_{42} \\ \hline -b_{13} & 0 & 0 & -b_{23} & 0 & 0 & c_{13} & a_{12} & a_{13} & -b_{43} & 0 & 0 \\ 0 & -b_{13} & 0 & 0 & -b_{23} & 0 & a_{21} & c_{23} & a_{23} & 0 & -b_{43} & 0 \\ 0 & 0 & -b_{13} & 0 & 0 & -b_{23} & a_{31} & a_{32} & c_{33} & 0 & 0 & -b_{43} \\ \hline -b_{14} & 0 & 0 & -b_{24} & 0 & 0 & -b_{34} & 0 & 0 & c_{14} & a_{12} & a_{13} \\ 0 & -b_{14} & 0 & 0 & -b_{24} & 0 & 0 & -b_{34} & 0 & a_{21} & c_{24} & a_{23} \\ 0 & 0 & -b_{14} & 0 & 0 & -b_{24} & 0 & 0 & -b_{34} & a_{31} & a_{32} & c_{34} \end{array}\right] \left[\begin{array}{c} x_{11} \\ x_{21} \\ x_{31} \\ \hline x_{12} \\ x_{22} \\ x_{32} \\ \hline x_{13} \\ x_{23} \\ x_{33} \\ \hline x_{14} \\ x_{24} \\ x_{34} \end{array}\right] =0, \qquad \text{.........(8)}$$

$$\begin{bmatrix} e_{11}\\ e_{12}\\ e_{13}\\ e_{14}\\ \hline e_{21}\\ e_{22}\\ e_{23}\\ e_{24}\\ \hline e_{31}\\ e_{32}\\ e_{33}\\ e_{34} \end{bmatrix} \equiv \left[\begin{array}{cccc|cccc|cccc} c_{11} & -b_{21} & -b_{31} & -b_{41} & a_{12} & 0 & 0 & 0 & a_{13} & 0 & 0 & 0\\ -b_{12} & c_{12} & -b_{32} & -b_{42} & 0 & a_{12} & 0 & 0 & 0 & a_{13} & 0 & 0\\ -b_{13} & -b_{23} & c_{13} & -b_{43} & 0 & 0 & a_{12} & 0 & 0 & 0 & a_{13} & 0\\ -b_{14} & -b_{24} & -b_{34} & c_{14} & 0 & 0 & 0 & a_{12} & 0 & 0 & 0 & a_{13}\\ \hline a_{21} & 0 & 0 & 0 & c_{21} & -b_{21} & -b_{31} & -b_{41} & a_{23} & 0 & 0 & 0\\ 0 & a_{21} & 0 & 0 & -b_{12} & c_{22} & -b_{32} & -b_{42} & 0 & a_{23} & 0 & 0\\ 0 & 0 & a_{21} & 0 & -b_{13} & -b_{23} & c_{23} & -b_{43} & 0 & 0 & a_{23} & 0\\ 0 & 0 & 0 & a_{21} & -b_{14} & -b_{24} & -b_{34} & c_{24} & 0 & 0 & 0 & a_{23}\\ \hline a_{31} & 0 & 0 & 0 & a_{32} & 0 & 0 & 0 & c_{31} & -b_{21} & -b_{31} & -b_{41}\\ 0 & a_{31} & 0 & 0 & 0 & a_{32} & 0 & 0 & -b_{12} & c_{32} & -b_{32} & -b_{42}\\ 0 & 0 & a_{31} & 0 & 0 & 0 & a_{32} & 0 & -b_{13} & -b_{23} & c_{33} & -b_{43}\\ 0 & 0 & 0 & a_{31} & 0 & 0 & 0 & a_{32} & -b_{14} & -b_{24} & -b_{34} & c_{34} \end{array}\right] \begin{bmatrix} x_{11}\\ x_{12}\\ x_{13}\\ x_{14}\\ \hline x_{21}\\ x_{22}\\ x_{23}\\ x_{24}\\ \hline x_{31}\\ x_{32}\\ x_{33}\\ x_{34} \end{bmatrix} = 0, \quad \ldots(9)$$

where the prefactors are forms of the square matrix Ω in (a″). Except in special cases the scalar resultant $R = \det \Omega$ is most easily evaluated as in Exs. iv and v of § 193 when it is regarded as the resultant of the two functions $\det A(\lambda)$ and $\det B(\lambda)$.

We have corresponding forms of (a″) for all values of m and n. The form corresponding to (8) can be replaced by n homogeneous linear matrix equations in which the unknowns are the n vertical rows of X. The form corresponding to (9) can be replaced by m homogeneous linear matrix equations in which the unknowns are the m horizontal rows of X. When A and B are simple square ante-slopes both these forms of Ω become quadrate hemipteric matrices in which the diagonal constituents are simple square slopes.

Ex. xiii. If we arrange the elements x_{ij} and the linear functions e_{ij} according to their difference-weights, i.e. according to the values of $j-i$, the arrangements in Ex. xii being

$$x_{31};\ x_{32}, x_{21};\ x_{33}, x_{22}, x_{11};\ x_{34}, x_{23}, x_{12};\ x_{24}, x_{13};\ x_{14};$$
$$e_{31};\ e_{32}, e_{21};\ e_{33}, e_{22}, e_{11};\ e_{34}, e_{23}, e_{12};\ e_{24}, e_{13};\ e_{14},$$

we obtain another form of (a″) which can be replaced by $m+n-1$ homogeneous linear matrix equations in which the unknowns are the successive diagonal lines of X. This form is particularly useful when A and B are simple square ante-slopes, the matrix Ω being then hemipteric.

Ex. xiv. If $m=n=3$, and if we put $c_{ij} = a_{ii} - b_{jj}$, we can use for (a″) the form

$$\begin{bmatrix} e_{11}\\ e_{22}\\ e_{33}\\ \hline e_{32}\\ e_{23}\\ \hline e_{13}\\ e_{31}\\ \hline e_{21}\\ e_{12} \end{bmatrix} \equiv \left[\begin{array}{ccc|cc|cc|cc} c_{11} & 0 & 0 & 0 & 0 & a_{13} & -b_{31} & -b_{21} & a_{12}\\ 0 & c_{22} & 0 & -b_{32} & a_{23} & 0 & 0 & a_{21} & -b_{12}\\ 0 & 0 & c_{33} & a_{32} & -b_{23} & -b_{13} & a_{31} & 0 & 0\\ \hline 0 & a_{32} & -b_{32} & 0 & c_{32} & -b_{12} & 0 & a_{31} & 0\\ 0 & -b_{23} & a_{23} & c_{23} & 0 & 0 & a_{21} & 0 & -b_{13}\\ \hline -b_{13} & 0 & a_{13} & a_{12} & 0 & 0 & c_{13} & -b_{23} & 0\\ a_{31} & 0 & -b_{31} & 0 & -b_{21} & c_{31} & 0 & 0 & a_{32}\\ \hline a_{21} & -b_{21} & 0 & -b_{31} & 0 & a_{23} & 0 & 0 & c_{21}\\ -b_{12} & a_{12} & 0 & 0 & a_{13} & 0 & -b_{32} & c_{12} & 0 \end{array}\right] \begin{bmatrix} x_{11}\\ x_{22}\\ x_{33}\\ \hline x_{23}\\ x_{32}\\ \hline x_{31}\\ x_{13}\\ \hline x_{12}\\ x_{21} \end{bmatrix} = 0,$$

which makes Ω a skew-symmetric matrix when $B=A$. We shall use forms corresponding to this when determining the general commutant $\{A, A\}$ in § 253.

3. *The commutants of a pair of square matrices containing arbitrary elements.*

In the definitions of sub-article 1 it has been assumed that all the elements of A and B have given numerical values. When the elements of two square matrices $A=[a]_m^m$ and $B=[b]_n^n$ are rational integral (or rational) functions of certain independent variables $\gamma_1, \gamma_2, \gamma_3, \ldots$, a matrix $X=[x]_m^n$ will be called a *commutant* of the matrix-pair (A, B) or a *commutant* $\{A, B\}$ when its elements are such rational integral (or rational) functions of the γ's that the equation $AX=XB$ is an identity in those variables. If the elements of X involve arbitrary parameters independent of the γ's, the equation must be an identity in them also. It will be understood that A, B, X can be replaced by any matrices derived from them by multiplying or dividing them by scalar rational integral functions which do not vanish identically. Consequently their elements may always be supposed to be rational *integral* functions of the variables and parameters occurring in them.

A *particular commutant* $X=\{A, B\}$ will be understood to be a commutant in which the elements are particular rational integral functions of the variables $\gamma_1, \gamma_2, \gamma_3, \ldots$ with numerical coefficients. If i is the greatest possible number of independent particular non-zero commutants $\{A, B\}$, and if i is not 0, a *general commutant* $X=\{A, B\}$ is a matrix expressible in the form

$$X=\lambda_1 X_1+\lambda_2 X_2+\ldots+\lambda_i X_i, \qquad \text{.....................(b)}$$

where $X_1, X_2, \ldots X_i$ are any i independent particular non-zero commutants $\{A, B\}$, and the λ's are independent arbitrary parameters independent of the γ's. If $i=0$, the general commutant $\{A, B\}$ is a zero matrix.

When X is any general commutant $\{A, B\}$ expressed in the form (b), we will call $\lambda_1, \lambda_2, \ldots \lambda_i$ the parameters of X or the parameters peculiar to X; and when A and B are given, particularisations and specialisations of X will ordinarily have reference only to the parameters of X, and not to the variables of A and B. Thus a *particularisation* of X will mean a matrix derived from X by substituting particular rational integral (or rational) functions of the γ's for the λ's; and a *specialisation* of X will mean a matrix derived from X by substituting for the λ's rational integral functions of parameters $\mu_1, \mu_2, \mu_3, \ldots$ independent of the γ's in which the coefficients are rational integral (or rational) functions of the γ's and may be constants, such functions being usually homogeneous and linear in the μ's. Ordinarily only rational integral functions of the γ's will occur in X; when this restriction is not observed, we will call X a rational commutant. Every commutant $\{A, B\}$ is a particularisation or specialisation of every given general commutant $\{A, B\}$; and any two general commutants $\{A, B\}$ are mutually equivalent in the sense that each of them is a specialisation of the other.

Commutants in which the elements are independent of (or have degree 0 in) the variables of A and B will be called *non-singular*, other commutants being singular. A *particular non-singular commutant* $X=\{A, B\}$ is a matrix with constant or numerical elements which makes the equation $AX=XB$ an identity in the γ's. A *general non-singular commutant* $X=\{A, B\}$ which is not a zero matrix is a homogeneous linear function of the greatest possible number of independent particular non-zero non-singular commutants, the coefficients being independent scalar parameters independent of the variables occurring in A and B. When A and B are square matrices whose elements are constants, the distinction between non-singular and singular commutants disappears; every particular or general commutant $\{A, B\}$ being also à particular or general non-singular commutant $\{A, B\}$. The existence of singular commutants adds considerably to the difficulty of determining a general commutant $\{A, B\}$ when A and B contain arbitrary elements.

In anticipation of § 244 it may be mentioned here that:

(1) *A general commutant $\{A, B\}$ is ordinarily a zero matrix; being a non-zero matrix when and only when the characteristic determinants of A and B have a factor in common.*

(2) *A general (or general non-singular) co-commutant $\{A, A\}$ is always an undegenerate square matrix.*

(3) *A general contra-commutant $\{A', A\}$ is always an undegenerate square matrix.*

To establish (2), it is sufficient to observe that the unit matrix of the same order as A is a commutant $\{A, A\}$. From Exs. xix and xx it will be seen that a general non-singular contra-commutant $\{A', A\}$ can only be undegenerate under special circumstances, and is usually a zero matrix.

If A, B, C, D are matrices whose elements are rational integral (or rational) functions of certain scalar variables $\gamma_1, \gamma_2, \gamma_3, \ldots$, and if X' is the conjugate of X, we define particular and general solutions or non-singular solutions X of matrix equations of the forms

$$\Sigma AXB=0, \quad \Sigma AXB+\Sigma CX'D=0,$$

in the same ways as we have defined those of the equation $AX=XB$.

The commutants occurring in the following examples of this article are determined by solving the scalar equations corresponding to (a′).

NOTE 4. *The commutants of a particularised or specialised matrix-pair.*

Let X be a general commutant $\{A, B\}$, where A and B are square matrices whose elements are rational integral functions of the variables $\gamma_1, \gamma_2, \gamma_3, \ldots$; let $\lambda_1, \lambda_2, \lambda_3, \ldots$ be the parameters of X; and let A_0, B_0, X_0 be matrices derived from A, B, X by particularising or specialising the γ's.

Then in all cases X_0 will be a commutant $\{A_0, B_0\}$, and will therefore be a particularisation or specialisation of every general commutant $\{A_0, B_0\}$.

For in the most general case we may suppose A_0, B_0, X_0 to be derived from A, B, X by substituting for the γ's rational integral functions of certain independent scalar parameters c_1, c_2, c_3, ... which are independent of the λ's. Then because the equation $AX=XB$ is an identity in the λ's and γ's the equation $A_0X=X_0B_0$ is an identity in the c's and λ's, i.e. X_0 is a commutant $\{A_0, B_0\}$.

Moreover if A_0, B_0 are particularisations of A and B, then for ordinary particular values of the γ's the matrix X_0 will be a general commutant $\{A_0, B_0\}$.

For when the γ's receive particular values which do not satisfy certain equations, the process of solving the non-scalar equations (a′) will be exactly the same as when the γ's are independent variables. Alternatively for ordinary particular values of the γ's the matrix Ω in (a″) will become particularised into a matrix Ω_0 of the same rank, and independent particular commutants $\{A, B\}$ will become particularised into independent particular commutants $\{A_0, B_0\}$; consequently every complete set of independent commutants $\{A, B\}$ will become a complete set of independent commutants $\{A_0, B_0\}$. This principle will enable us to pass from theorems concerning the commutants of square matrices whose elements are constants to theorems concerning the commutants of functional square matrices.

When X is the general *non-singular* commutant $\{A, B\}$, which is independent of the γ's, X_0 is the same matrix as X, and the identical equation $A_0X=XB_0$ shows that X is a non-singular commutant $\{A_0, B_0\}$.

Thus every general non-singular commutant $\{A, B\}$ is a particularisation or specialisation of every general non-singular commutant $\{A_0, B_0\}$.

It follows that as the square matrices A and B become more and more specialised, the general non-singular commutant $\{A, B\}$ tends to become less and less specialised. It is most specialised when the elements of A and B are all arbitrary, being then a zero matrix; and it is least specialised when A and B are zero matrices, being then a matrix with arbitrary elements.

As regards the general commutants $\{A, B\}$ and $\{A_0, B_0\}$ we can show that:

If the general commutant $\{A_0, B_0\}$ is a zero matrix, then the general commutant $\{A, B\}$ is also a zero matrix.

For the scalar resultant $R_0=\det \Omega_0$ of A_0 and B_0 is a particularisation or specialisation of the scalar resultant $R=\det \Omega$ of A and B. If the general commutant $\{A_0, B_0\}$ is a zero matrix, R_0 does not vanish identically; therefore R does not vanish identically; therefore the general commutant $\{A, B\}$ is a zero matrix. The corresponding more general principle is that:

The total number of independent non-zero commutants $\{A, B\}$ cannot exceed the total number of independent non-zero commutants $\{A_0, B_0\}$.

Thus as the square matrices A and B become more and more specialised, the total number of independent commutants $\{A, B\}$ tends to become greater and greater.

If the variables of A and B can be divided into two sets a_1, a_2, a_3, ... and β_1, β_2, β_3, ..., and the β's only are particularised or specialised, being replaced in the latter more general case by rational integral functions of variables b_1, b_2, b_3, ... which are independent of the a's and λ's, all the above theorems remain true except that the third must be interpreted to mean that:

Every commutant $\{A, B\}$ independent of the β's is a commutant $\{A_0, B_0\}$ independent of the b's.

When the β's are particularised, the b's are absent, and every commutant $\{A, B\}$ independent of the β's is a commutant $\{A_0, B_0\}$.

Ex. xv. In each of the following two cases non-zero commutants exist because the characteristic determinants of A and B have a factor in common. The letters t and ρ denote independent arbitrary parameters, and X and X' are respectively the general and the general non-singular commutant $\{A, B\}$. In both cases the scalar resultant R vanishes identically, and Ω has degeneracy 1.

$$(1)\quad A=\begin{bmatrix} t\ , & 2t+1 \\ t+3, & 2 \end{bmatrix},\quad B=\begin{bmatrix} t+3, & 2t+4 \\ 2t-2, & t-3 \end{bmatrix},\quad X=\rho\begin{bmatrix} t-1\ , & -(t+2) \\ -(t-1), & t+2 \end{bmatrix},\quad X'=[0]_2^2\ ;$$

$$\det A(\lambda)=(\lambda+t+1)(\lambda-2t+3),\quad \det B(\lambda)=(\lambda+t+1)(\lambda-3t+1).$$

$$(2)\quad A=\begin{bmatrix} t\ , & 2t+1 \\ t+3, & 2 \end{bmatrix},\quad B=\begin{bmatrix} 2t+1, & -t \\ 3t+2, & -(2t+1) \end{bmatrix};\ X=Y=\rho\begin{bmatrix} 1\ , & -1 \\ -1, & 1 \end{bmatrix};$$

$$\det A(\lambda)=(\lambda+t+1)(\lambda-2t+3),\quad \det B(\lambda)=(\lambda+t+1)(\lambda-t-1).$$

Ex. xvi. We give below three simple illustrations of Note 4. The letters a, b denote given scalar numbers such that $b\neq a$; the letters λ, μ, x, y denote independent arbitrary parameters; the matrices A_0, B_0 are particularisations or specialisations of A, B; the matrices X and X' are respectively general and general non-singular commutants $\{A, B\}$; and the matrices Y and Y' are respectively general and general non-singular commutants $\{A_0, B_0\}$. In (1) and (2) we use the specialisation $\mu=\lambda$; and in (3) we use the particularisations $\lambda=a$, $\mu=1$.

$$(1)\quad A=\begin{bmatrix} a, & \lambda \\ 0, & b \end{bmatrix},\quad B=\begin{bmatrix} a, & \mu \\ 0, & b \end{bmatrix};\quad X=\begin{bmatrix} (a-b)x, & \mu x-\lambda y \\ 0\ , & (a-b)y \end{bmatrix},\quad X'=\begin{bmatrix} 0, & 0 \\ 0, & 0 \end{bmatrix};$$

$$A_0=B_0=\begin{bmatrix} a, & \lambda \\ 0, & b \end{bmatrix};\quad Y=\begin{bmatrix} (a-b)x, & \lambda(x-y) \\ 0\ , & (a-b)y \end{bmatrix},\quad Y'=\begin{bmatrix} x, & 0 \\ 0, & x \end{bmatrix}.$$

$$(2)\quad A=\begin{bmatrix} a, & \lambda \\ 0, & a \end{bmatrix},\quad B=\begin{bmatrix} a, & \mu \\ 0, & a \end{bmatrix};\quad X=\begin{bmatrix} \lambda x, & y \\ 0, & \mu x \end{bmatrix},\quad X'=\begin{bmatrix} 0, & y \\ 0, & 0 \end{bmatrix};$$

$$A_0=B_0=\begin{bmatrix} a, & \lambda \\ 0, & a \end{bmatrix};\quad Y=Y'=\begin{bmatrix} x, & y \\ 0, & x \end{bmatrix}.$$

$$(3)\quad A=B=\begin{bmatrix} \lambda, & \mu \\ 0, & \lambda \end{bmatrix};\quad X=X'=\begin{bmatrix} x, & y \\ 0, & x \end{bmatrix};$$

$$A_0=B_0=\begin{bmatrix} a, & 1 \\ 0, & a \end{bmatrix};\quad Y=Y'=\begin{bmatrix} x, & y \\ 0, & x \end{bmatrix}.$$

In all three cases X' is a specialisation of Y', and X_0 (defined as in Note 4) is a specialisation of Y, being in fact the same as Y.

Ex. xvii. If α and β are variables or arbitrary parameters, and if $A(\lambda)$, $B(\lambda)$ are the characteristic matrices of the square matrices

$$A=\begin{bmatrix} -\beta\ , & 2\alpha+2\beta, & \alpha+2\beta \\ 3\alpha+\beta\ , & 3\alpha+\beta, & \alpha+\beta \\ \alpha-3\beta, & \alpha-\beta\ , & -\alpha+\beta \end{bmatrix},\quad B=\begin{bmatrix} -3\beta, & \alpha+3\beta, & 2\alpha+\beta \\ \alpha\ , & -2\beta\ , & 2\alpha-\beta \\ 2\alpha-\beta, & -\alpha+\beta, & -\beta \end{bmatrix},$$

we have
$$\det A(\lambda)=-(\lambda+\alpha+\beta)\{\lambda^2-(3\alpha+2\beta)\lambda-(8\alpha^2+2\alpha\beta-6\beta^2)\},$$
$$\det B(\lambda)=-(\lambda+\alpha+\beta)\{\lambda^2-(\alpha-5\beta)\lambda-2(\alpha^2+5\alpha\beta-4\beta^2)\}.$$

Since the characteristic determinants have the common factor $\lambda+\alpha+\beta$, we know that there must exist non-zero commutants $\{A, B\}$; and it can be shown that the general commutant $\{A, B\}$ differs only by an arbitrary scalar factor from the matrix

$$X=\begin{bmatrix} x, & y, & z \\ -x, & -y, & -z \\ x, & y, & z \end{bmatrix}=\begin{bmatrix} -(\alpha-\beta), & (\alpha+\beta), & -2\beta \\ \alpha-\beta\ , & -(\alpha+\beta), & 2\beta \\ -(\alpha-\beta), & (\alpha+\beta), & -2\beta \end{bmatrix}\ . \qquad \text{............(10)}$$

Even in this comparatively simple case the direct solution of the scalar equations (a′) is troublesome; but it can be simplified by the following considerations.

If we put $A = \alpha A_1 + \beta A_2$, $B = \alpha B_1 + \beta B_2$, it will be found without difficulty that the general commutants $X_1 = \{A_1, B_1\}$, $X_2 = \{A_2, B_2\}$ are

$$X_1 = \rho \begin{bmatrix} -1, & 1, & 0 \\ 1, & -1, & 0 \\ -1, & 1, & 0 \end{bmatrix}, \quad X_2 = \sigma \begin{bmatrix} -1, & -1, & 2 \\ 1, & 1, & -2 \\ -1, & -1, & 2 \end{bmatrix},$$

where ρ and σ are arbitrary parameters, there being only one independent non-zero commutant in each case. Since a non-singular commutant $X = \{A, B\}$ would satisfy both the equations $A_1 X = X B_1$, $A_2 X = X B_2$, we see in the first place that every non-zero commutant $\{A, B\}$ must be singular. Again if the matrix Ω for A and B in (a″) becomes Ω_1 when $\alpha=1$, $\beta=0$ and Ω_2 when $\alpha=0$, $\beta=1$, then because the degeneracies of Ω_1 and Ω_2 must be exactly equal to 1, we see in the second place that the degeneracy of Ω cannot be less than 1, and must be exactly equal to one. We conclude that there is one and only one independent non-zero commutant $X = \{A, B\}$, and that it must be singular. Its elements are proportional to the simple minor determinants of any undegenerate horizontal simple minor of Ω, and it must be homogeneous and linear in α and β, the coefficients of the highest powers of α and β differing only by scalar factors from the matrices X_1 and X_2.

The rule for determining the rank of any general commutant which is given in Note 3 of § 244 shows that in the case considered above the rank of the general commutant $X = \{A, B\}$ must be equal to 1. If we make use of this fact and note the forms of X_1 and X_2, we see that X must have the form shown on the left in (10); and there is then no difficulty in solving the scalar equations (a′), which reduce to three.

Ex. xviii. *If A and B are both simple square ante-slopes or quasi-scalar matrices, the general commutant $\{A, B\}$ is a non-zero matrix when and only when one of the diagonal elements of A is identically equal to one of the diagonal elements of B, i.e. when and only when A and B have a common diagonal element.*

A simple square ante-slope is a square matrix in which all elements lying on one side of the diagonal are equal to 0. Whenever A and B are simple square ante-slopes (which may be quasi-scalar matrices), we see from Ex. xii or xiii that R is the product of the mn differences

$$a_{ii} - b_{jj}, \qquad (i=1, 2, \ldots m \,;\, j=1, 2, \ldots n) \,;$$

and in order that R shall vanish identically, i.e. in order that there shall be a non-zero commutant, it is necessary and sufficient that one of these differences shall vanish identically.

We conclude as in the fourth theorem of Note 4 that:

The general commutant $\{A, B\}$ is a zero matrix whenever A and B can be particularised into simple square ante-slopes (or quasi-scalar matrices) having no common diagonal element.

It is therefore a zero matrix when A and B are square matrices whose elements are independent arbitrary parameters.

Ex. xix. *If $A = [a]_m^m$ is a square matrix whose elements are independent arbitrary parameters, the general non-singular commutant $\{A, A\}$ is a scalar matrix.*

For when we equate the coefficients of a_{ij} on both sides of the equation $AX = XA$, in which $X=[x]_m^m$ is a matrix whose elements are *independent of the elements of* A, we obtain the scalar equations

$$x_{pi}=0 \text{ when } p\neq i, \quad x_{jq}=0 \text{ when } q\neq j, \quad x_{ii}=x_{jj}.$$

Under the same circumstances the general non-singular contra-commutant $\{A', A\}$ *is a zero matrix.*

For when we equate the coefficient of a_{ii} on both sides of the equation $A'X=XA$, we see that we must have

$$x_{pi}=0 \text{ when } p\neq i, \quad x_{iq}=0 \text{ when } q\neq i,$$

i.e. all the non-diagonal elements of X must be 0's; and the equation $A'X=XA$ then becomes equivalent to the $\frac{1}{2}m(m-1)$ scalar equations

$$x_{ii}a_{ij}=x_{jj}a_{ji}, \qquad (j\neq i), \quad \text{..............................(11)}$$

which cannot be identities in the elements of A unless X is a zero matrix.

The general commutant $\{A, A\}$ will be determined in § 253.

Ex. xx. *If* $A=[a]_m^m$ *is an arbitrary simple square ante-slope of order* m, *the general non-singular commutant* $\{A, A\}$ *is a scalar matrix, and the general non-singular contra-commutant* $\{A', A\}$ *is a zero matrix.*

For if a_{ij} is always 0 when $j<i$, and an arbitrary parameter when $j\nless i$, all the results of Ex. xix remain true except that the equations (11) become

$$x_{ii}a_{ij}=0, \qquad (j\nless i).$$

Ex. xxi. *If* $A=[a]_m^m$ *is a quasi-scalar matrix whose successive diagonal elements* $a_1, a_2, \ldots a_m$ *are independent arbitrary parameters, the general non-singular commutant* $\{A, A\}$ *or contra-commutant* $\{A', A\}$, *which are the same, is a quasi-scalar matrix whose diagonal elements are independent arbitrary parameters.*

For if $X=[x]_m^m$ is a matrix whose elements are independent of the elements of A, the matrix equation $AX=XA$ is equivalent to the $\frac{1}{2}m(m-1)$ scalar equations

$$(a_i-a_j)x_{ij}=0, \qquad (j\neq i).$$

4. *The conjugate of a commutant; the inverse and conjugate reciprocal of an undegenerate square commutant.*

First let $A=[a]_m^m$, $B=[b]_n^n$, $X=[x]_m^n$ be matrices whose conjugates are A', B', X'. Then each of the equations $AX=XB$, $B'X'=X'A'$ is deducible from the other by equating the conjugates of both sides, and it follows that:

The conjugate of a particular or general commutant $X=\{A, B\}$ *is a particular or general commutant* $X'=\{B', A'\}$.

Hence when

$$X=\{A, A\}, \quad X=\{A', A'\}, \quad X=\{A', A\}, \quad X=\{A, A'\},$$

we have

$$X'=\{A', A'\}, \quad X'=\{A, A\}, \quad X'=\{A', A\}, \quad X'=\{A, A'\},$$

where the 1st, 2nd, 3rd, 4th representations are to be taken in both series.

If then X is a particular or general contra-commutant of A, its conjugate X' is also a particular or general contra-commutant of A; consequently

$X + X'$ and $X - X'$ are respectively a symmetric and a skew-symmetric contra-commutant of A. In the special case when A is symmetric, there are no distinctions between the commutants of A, the commutants of A', the contra-commutants of A and the contra-commutants of A'.

Putting $n = m$, so that A and B are square matrices of the same order m, we see that:

A particular or most general symmetric (or skew-symmetric) commutant $X = \{A, B\}$ *is also a particular or most general symmetric (or skew-symmetric) commutant* $X = \{B', A'\}$.

Next let A and B be two square matrices of the same order m; let the square matrix X of order m be any undegenerate commutant $\{A, B\}$; and let X^{-1} and $\dot{X}$ be respectively the inverse and conjugate reciprocal of X. Then by prefixing or postfixing X^{-1} or $\dot{X}$ on both sides of the equation $AX = XB$ we see that

$$BX^{-1} = X^{-1}A, \quad B\dot{X} = \dot{X}A.$$

Thus the inverse X^{-1} *and the conjugate reciprocal* $\dot{X}$ *of an undegenerate square commutant* $X = \{A, B\}$, *as well as all matrices differing from them only by non-zero scalar factors, are undegenerate commutants* $\{B, A\}$.

In particular if A' is the conjugate of A, then according as

$$X = \{A, A\}, \qquad X = \{A', A'\}, \qquad X = \{A', A\}, \qquad X = \{A, A'\},$$

we have

$$X^{-1} = \{A, A\}, \quad X^{-1} = \{A', A'\}, \quad X^{-1} = \{A, A'\}, \quad X^{-1} = \{A', A\}.$$

If X is degenerate and has rank less than $m - 1$, we still have $\dot{X} = \{B, A\}$, because $\dot{X} = 0$; but (see Ex. xxii) this is not in general true when X has rank $m - 1$, so that $\dot{X}$ has rank 1.

Ex. xxii. The general commutant $X = \{A, B\}$ and its conjugate reciprocal $\dot{X}$ for the square matrices

$$A = \begin{bmatrix} 3, & 1 \\ 0, & 2 \end{bmatrix}, \quad B = \begin{bmatrix} 3, & 5 \\ 0, & 2 \end{bmatrix} \text{ are } X = \begin{bmatrix} x, & 5x - y \\ 0, & y \end{bmatrix}, \quad \dot{X} = \begin{bmatrix} y, & y - 5x \\ 0, & x \end{bmatrix},$$

where x and y are arbitrary scalar parameters, and in this case we have $B\dot{X} = \dot{X}A$, so that we can put $\dot{X} = \{B, A\}$.

The general commutant $X = \{A, B\}$ and its conjugate reciprocal $\dot{X}$ for the square matrices

$$A = \begin{bmatrix} 3, & 1 \\ 0, & 2 \end{bmatrix}, \quad B = \begin{bmatrix} 3, & 4 \\ 0, & 1 \end{bmatrix} \text{ are } X = \begin{bmatrix} x, & 2x \\ 0, & 0 \end{bmatrix}, \quad \dot{X} = \begin{bmatrix} 0, & -2x \\ 0, & x \end{bmatrix},$$

where x is an arbitrary scalar parameter; and in this case we have $B\dot{X} \neq \dot{X}A$.

§ 239. Commutantal products of true commutants; commutantal equations and transformations.

1. *Definitions.*

If $A = [a]_m^m$, $B = [b]_p^p$, $C = [c]_n^n$ are any three given square matrices, and if $H = [h]_m^p$, $K = [k]_p^n$ are commutants of the arranged matrix-pairs

$$(A, B), \quad (B, C),$$

then the product matrix $X = [x]_m^n$ in the product

$$[h]_m^p [k]_p^n = [x]_m^n \quad \text{or} \quad HK = X \quad \dots\dots\dots\dots\dots\dots(1)$$

is always a commutant of the arranged matrix-pair (A, C). For from the equations

$$[a]_m^m [h]_m^p = [h]_m^p [b]_p^p, \quad [b]_p^p [k]_p^n = [k]_p^n [c]_n^n$$

it follows that

$$[a]_m^m \cdot [h]_m^p [k]_p^n = [h]_m^p [b]_p^p [k]_p^n = [h]_m^p [k]_p^n \cdot [c]_n^n.$$

On this account we will call the product HK on the left in (1) a commutantal product of type $\{A, B\}\{B, C\}$, and the equation (1) a commutantal equation of type

$$\{A, B\}\{B, C\} = \{A, C\} \dots\dots\dots\dots\dots\dots(1')$$

We may formally define a product of two successive commutants of given types $\{A, P\}$, $\{Q, B\}$ to be *commutantal* when $Q = P$.

If $A, P_1, P_2, \dots P_i, B$ are any $i+2$ given square matrices, where $i \nless 1$, and if $H, X_1, X_2, \dots X_{i-1}, K$ are commutants of the $i+1$ arranged matrix-pairs

$$(A, P_1), (P_1, P_2), (P_2, P_3), \dots (P_{i-1}, P_i), (P_i, B),$$

it follows by repeated applications of the result just proved that in the product

$$HX_1X_2 \dots X_{i-1}K = X \quad \dots\dots\dots\dots\dots\dots(2)$$

the product matrix X is a commutant of the arranged matrix-pair (A, B). This fact we will represent symbolically by the equation

$$\{A, P_1\}\{P_1, P_2\}\{P_2, P_3\} \dots \{P_{i-1}, P_i\}\{P_i, B\} = \{A, B\}, \quad \dots\dots(2')$$

which means that any continued product of commutants of the $i+1$ successive matrix-pairs shown above is a commutant of the matrix-pair (A, B). We will call the product on the left in (2) a *commutantal product* of the *type* defined by the product on the left in (2′), and the equation (2) a *commutantal equation* of the *type* (2′). We may formally define a product of any number of successive commutants to be commutantal when the product formed by every two successive factors is commutantal. We may prefix negative signs throughout to any number of the matrices $A, P_1, P_2, \dots P_i, B$, and then the product matrix X in (2) is a continuantal or alternating commutant according as the total number of factor matrices which are alternating commutants is even or odd.

More generally an equation formed by equating two commutantal products whose product matrices are commutants of the same type will be called a *commutantal equation*. For example such an equation is commutantal if it becomes

$$\{A, P\}\{P, Q\}\{Q, R\}\{R, B\} = \{A, S\}\{S, T\}\{T, B\}$$

or

$$\{A, -P\}\{-P, -Q\}\{-Q, R\}\{R, -B\} = \{A, S\}\{S, -T\}\{-T, -B\},$$

when each commutant factor is replaced by the expression representing its type.

If $A=[a]_m^m$, $B=[b]_n^n$ and $P=[p]_u^u$, $Q=[q]_v^v$ are two pairs of given square matrices, a commutantal equation

$$[h]_m^u [x]_u^v [k]_v^n = [y]_m^n \quad \text{or} \quad HXK = Y \text{ ....(3)}$$

of type

$$\{A, P\}\{P, Q\}\{Q, B\} = \{A, B\} \text{ ....(3')}$$

will be called a *commutantal transformation* converting the commutant X of the matrix-pair (P, Q) into the commutant Y of the matrix-pair (A, B), the transformation being equigradent when H and K are matrices with constant elements having ranks u and v.

NOTE 1. In the definitions of commutantal products and commutantal equations the two types $\{P, Q\}$, $\{-P, -Q\}$ and also the two types $\{P, -Q\}$, $\{-P, Q\}$ are treated as different. If we were to regard them as interchangeable, which is of course allowable, and were only to use types such as $\{P, Q\}$, $\{P, -Q\}$, the most general representative of a commutantal equation (2) would be

$$\{A, \pm P_1\}\{P_1, \pm P_2\}\{P_2, \pm P_3\}\dots\{P_{i-1}, \pm P_i\}\{P_i, \pm B\} = \{A, \pm B\},$$

where the sign of B on the right is + or − according as the sign − occurs an even or odd number of times on the left, and may not be the same as the sign of B on the left.

NOTE 2. The (true) commutantal products defined above, in which the factor matrices are true commutants, must be distinguished from the (conventional) commutantal products defined in Chapter XXX, where the factor matrices have assigned commutantal types represented by symbolic commutants.

2. *Some special equigradent commutantal transformations.*

We will consider *equigradent* commutantal transformations (3) in which $u=m$ and $v=n$, so that the two pairs of given square matrices are

$$A=[a]_m^m,\ B=[b]_n^n \quad \text{and} \quad P=[p]_m^m,\ Q=[q]_n^n.$$

In the first place let h, k be any particular *undegenerate* square commutants $\{A, P\}$, $\{Q, B\}$ *with constant elements,* so that by Ex. xxi of § 238 their inverses H, K are particular *undegenerate* square commutants $\{P, A\}$, $\{Q, B\}$ *with constant elements.* Such commutants can only exist when certain conditions are satisfied by the given square matrices. If the elements of the given square matrices are all constants, the necessary and sufficient conditions are that A and P shall be equicanonical and B and Q shall be equicanonical. If the elements of the given square matrices are rational integral functions of certain scalar variables, the identities

$$Ah=hP,\ A=hPH,\ P=HAh \quad \text{and} \quad Qk=kB,\ B=KQk,\ Q=kBK$$

show that P must involve exactly the same variables as and be equipotent with A, and Q must involve exactly the same variables as and be equipotent with B. Moreover the two matrices in each of the pairs (A, P) and

(B, Q) must be expressible as sums of corresponding terms involving the same powers of the variables as factors, the coefficients of corresponding terms being equicanonical square matrices. We will suppose that these conditions for the existence of the commutants h, k and H, K are satisfied, as well as any other conditions which may be necessary.

With every equigradent transformation of the form

$$[h]_m^m [x]_m^n [k]_n^n = [y]_m^n \quad \text{or} \quad hXk = Y \quad \text{.................(4)}$$

we will associate the inverse equigradent transformation

$$[H]_m^m [y]_m^n [K]_n^n = [x]_m^n \quad \text{or} \quad HYK = X. \quad \text{..............(5)}$$

These two transformations are mutually equivalent, each of them being deducible from the other. In the most general case we may suppose that the elements of the given square matrices A, B, P, Q are rational integral functions of certain scalar variables $\gamma_1, \gamma_2, \gamma_3, \ldots$; and the equations (4) and (5) are identities in the γ's as well as in any additional variables or arbitrary parameters which may occur in X and Y. The matrix Y must involve exactly the same variables as X; it must have the same partial and total degrees in those variables as X; and it must be equipotent with X. In the special case when the elements of the given square matrices are all constants, the γ's are absent.

With respect to such equigradent transformations we have the following theorem, which remains true when 'commutant' is interpreted, so far as X and Y are concerned, to mean 'non-singular commutant' or to mean 'commutant independent of certain specified variables.'

Theorem. *If* (4) *is an equigradent transformation in which h and k are particular undegenerate square commutants $\{A, P\}$ and $\{Q, B\}$ with constant elements, the matrix Y will be a commutant or a particular or general commutant $\{A, B\}$ when and only when the matrix X is a commutant or a particular or general commutant $\{P, Q\}$.*

The equations (4) and (5) taken together show that X will be a commutant $\{P, Q\}$ or (4) will be a commutantal transformation of the type

$$\{A, P\}\{P, Q\}\{Q, B\} = \{A, B\} \quad \text{....................(4')}$$

when and only when Y is a commutant $\{A, B\}$ or (5) is a commutantal transformation of the type

$$\{P, A\}\{A, B\}\{B, Q\} = \{P, Q\}; \quad \text{....................(5')}$$

and because Y must contain exactly the same variables or arbitrary parameters as X, it follows that Y will be a *particular* commutant $\{A, B\}$ when and only when X is a *particular* commutant $\{P, Q\}$.

Now let

$X_1 = HY_1K,\ X_2 = HY_2K, \ldots$ be particular commutants $\{P, Q\}$, so that

$$Y_1 = hX_1k, \quad Y_2 = hX_2k, \quad \dots \text{ are particular commutants } \{A, B\}.$$

Then if $c_1, c_2, \dots$ are scalar quantities which are either constants or rational integral functions of the γ's, we have

$$c_1X_1 + c_2X_2 + \dots = 0 \quad \text{if and only if} \quad c_1Y_1 + c_2Y_2 + \dots = 0. \quad \dots\dots(6)$$

Consequently $Y_1, Y_2, \dots$ will be independent particular commutants $\{A, B\}$ when and only when $X_1, X_2, \dots$ are independent particular commutants $\{P, Q\}$; and it follows that

$$Y = hXk = \lambda_1 Y_1 + \lambda_2 Y_2 + \dots + \lambda_i Y_i \quad \text{will be a } \textit{general} \text{ commutant } \{A, B\}$$

when and only when

$$X = HYK = \lambda_1 X_1 + \lambda_2 X_2 + \dots + \lambda_i X_i \quad \text{is a } \textit{general} \text{ commutant } \{P, Q\}.$$

This result remains true when the commutants $X, X_1, X_2, \dots$ and $Y, Y_1, Y_2, \dots$ are restricted to be independent of certain of the γ's, or to be independent of all the γ's, i.e. to be non-singular commutants. In such cases we may without loss of generality suppose $c_1, c_2, \dots$ in (6) to be functions of only those of the γ's which occur in the commutants.

The last part of the theorem can also be proved in the following way.

Let $X = \lambda_1 X_1 + \lambda_2 X_2 + \dots + \lambda_i X_i$ be a general commutant $\{P, Q\}$, the λ's being arbitrary parameters independent of the γ's; let $Y = hXk$; and let Y_0 be any commutant $\{A, B\}$. Then $X_0 = HY_0K$ is a commutant $\{P, Q\}$ and must be a specialisation of X; therefore $Y_0 = hX_0k$ must be a specialisation of Y. Thus Y is a commutant $\{A, B\}$ of which every commutant $\{A, B\}$ is a specialisation, i.e. Y is a general commutant $\{A, B\}$. Similarly if Y is a general commutant $\{A, B\}$, then the matrix $X = HYK$ is a general commutant $\{P, Q\}$.

All the theorems and subsidiary results of § 240.1, including those relating to derangements of commutants, are particular cases of the general theorem of this article.

Ex. i. *The commutants* $\{A, \epsilon B\}, \{A', \epsilon B'\}, \{A', \epsilon B\}, \{A, \epsilon B'\}$, *where* $\epsilon = \pm 1$.

Let $A = [a]_m^m$, $B = [b]_n^n$ be two square matrices whose conjugates are A', B'; let $U = [U]_m^m$, $V = [V]_n^n$ be particular undegenerate contra-commutants $\{A', A\}, \{B, B'\}$ with constant elements; and let $X = [x]_m^n$, $\epsilon = \pm 1$.

Then $X' = UXV$ *is a particular or general commutant* $\{A', \epsilon B'\}$,

$Y = UX$ *is a particular or general commutant* $\{A', \epsilon B\}$,

$Y' = XV$ *is a particular or general commutant* $\{A, \epsilon B'\}$,

if and only if X *is a particular or general commutant* $\{A, \epsilon B\}$.

Since the unit matrix $[1]_m^m$ is a particular commutant of each of the types $\{A, A\}$, $\{A', A'\}$, and the unit matrix $[1]_n^n$ a particular commutant of each of the types $\{\epsilon B, \epsilon B\}$, $\{\epsilon B', \epsilon B'\}$, and since commutants $\{B, B'\}$ are the same as commutants $\{\epsilon B, \epsilon B'\}$, these results are particular cases of the theorem of the text. They follow directly from the respective pairs of equations

$$A'.UXV=U.AX.V, \quad UXV.B'=U.XB.V,$$
$$A'.UX=U.AX, \quad UX.B=U.XB,$$
$$A.XV=AX.V, \quad XV.B'=XB.V,$$

which are consequences of the equations $A'U=UA$, $BV=VB'$.

When A and B are matrices with constant elements (see Ex. vii of § 240), there always exist such undegenerate contra-commutants U and V with constant elements; moreover they can always be so chosen as to be symmetric matrices. Whenever they exist, we can always form four general commutants X, X', Y, Y' of the types mentioned which are connected by the relations

$$X'=UXV, \quad Y=UX, \quad Y'=XV, \quad \text{.........(7)}$$

all four of them being known when any one of them is known. These three relations are then equigradent commutantal transformations by which X', Y, Y' can be derived from X. By prefixing and postfixing inverse matrices we can derive from them three other sets of three equigradent commutantal transformations by which the remaining three general commutants can be derived from X' or from Y or from Y'.

If A and B contain arbitrary elements, the above results remain true when 'commutant' is interpreted to mean 'non-singular commutant' or 'commutant independent of certain specified variables,' provided that there exist contra-commutants U and V with constant elements.

Ex. ii. *The commutants* $\{A, \epsilon A\}$, $\{A', \epsilon A'\}$, $\{A', \epsilon A\}$, $\{A, \epsilon A'\}$, *where* $\epsilon=\pm 1$.

The results of Ex. i remain true when $B=A$, $B'=A'$. Thus whenever there exist undegenerate contra-commutants U, V of types $\{A', A\}$, $\{A, A'\}$ with constant elements, it is possible to form four general commutants

$$X, X', Y, Y' \quad \text{of types} \quad \{A, \epsilon A\}, \{A', \epsilon A'\}, \{A', \epsilon A\}, \{A, \epsilon A'\}$$

which are connected by the relations

$$X'=UXV, \quad Y=UX, \quad Y'=XV=U^{-1}YV,$$

all four of them being known when any one of them is known. In such cases it will be obvious that we can always choose U and V to be mutually inverse, so that

$$X'=UXU^{-1}=V^{-1}XV, \quad Y'=U^{-1}YU^{-1}=VYV. \quad \text{.........(8)}$$

When A is a square matrix *with constant elements* (see Ex. vii of § 240), we can always choose U and V to be symmetric and at the same time mutually inverse; and it follows that in this case:

If V *is any given particular undegenerate symmetric contra-commutant* $\{A, A'\}$, *it is always possible to form two general symmetric or two general skew-symmetric contra-commutants* Y, Y' *of types* $\{A', \epsilon A\}$, $\{A, \epsilon A'\}$ *which are connected by the symmetric relation*

$$Y'=VYV,$$

each of them being uniquely determinate when the other is given.

§ 240. Theorems facilitating the determination of commutants.

1. *The commutants of equicanonical pairs of square matrices.*

Let $A=[a]_m^m$, $B=[b]_n^n$ be two given square matrices of orders m and n with constant (or numerical) elements; let $\alpha=[\alpha]_m^m$, $\beta=[\beta]_n^n$ be two square matrices of orders m and n equicanonical with A, B; and let

$$[a]_m^m = [h]_m^m [\alpha]_m^m [H]_m^m \quad \text{or} \quad A = h\alpha H, \quad \dots\dots\dots(1)$$

$$[b]_n^n = [k]_n^n [\beta]_n^n [K]_n^n \quad \text{or} \quad B = k\beta K, \quad \dots\dots\dots(2)$$

where $$Hh = hH = [1]_m^m, \quad Kk = kK = [1]_n^n,$$

be any two given equimutant (or isomorphic) transformations by which A, B can be derived from α, β. Alternatively let A, B be square matrices whose elements are rational integral functions of certain scalar variables $\gamma_1, \gamma_2, \gamma_3, \dots$, and let (1) and (2) be given equimutant transformations which are identities in those variables. Also let

$$\epsilon = \pm 1.$$

Then we have the following theorem which remains true when 'commutant' is restricted to mean 'non-singular commutant' or to mean 'commutant independent of certain specified scalar variables.'

Theorem I. *If* $X = [x]_m^n$ *and* $\xi = [\xi]_m^n$ *are two matrices such that*

$$[x]_m^n = [h]_m^m [\xi]_m^n [K]_n^n \quad \text{or} \quad X = h\xi K, \quad \dots\dots\dots(\text{A})$$

then X is a commutant $\{A, \epsilon B\}$ or a particular or general commutant $\{A, \epsilon B\}$ when and only when ξ is a commutant $\{\alpha, \epsilon\beta\}$ or a particular or general commutant $\{\alpha, \epsilon\beta\}$.

Here h, k, H, K are undegenerate square matrices with constant elements, and it is to be understood that the equation (A) is an identity in the scalar variables (if any) of A and B as well as in any other variables which may occur in X and ξ. Theorem I is included in the more general theorem of § 239, and follows from the identical equations

$$AX = [h]_m^m \,.\, \alpha\xi \,.\, [K]_n^n, \quad XB = [h]_m^m \,.\, \xi\beta \,.\, [K]_n^n, \quad \dots\dots\dots(\text{A}_1)$$

which show that $AX = XB$ when and only when $\alpha\xi = \xi\beta$. The commutants mentioned in it are both continuantal or both alternating according as $\epsilon = 1$ or $\epsilon = -1$.

The equation (A) is an equigradent transformation which establishes a one-one correspondence between all commutants $\{A, \epsilon B\}$ and all commutants $\{\alpha, \epsilon\beta\}$; X being a commutant $\{A, \epsilon B\}$ when and only when ξ is a commutant $\{\alpha, \epsilon\beta\}$; X having the same rank and involving the same variables or arbitrary parameters as ξ; X being a particular or general commutant or restricted commutant $\{A, \epsilon B\}$ when and only when ξ is a particular or general commutant or similarly restricted commutant $\{\alpha, \epsilon\beta\}$; and X having the same number of independent non-zero particularisations as ξ.

In most applications of the theorem A and B will be square matrices with constant elements, and α and β will be standard canonical square matrices.

Ex. i. *Other forms of Theorem I.*

Let A', B', α', β', h', k', H', K' be the conjugates of A, B, a, β, h, k, H, K; so that A', B' can be derived from α', β' by the isomorphic transformations

$$\overline{\underline{a}}{}^{m}_{m} = \overline{\underline{H}}{}^{m}_{m}\ \overline{\underline{a}}{}^{m}_{m}\ \overline{\underline{h}}{}^{m}_{m} \quad \text{or} \quad A' = H'a'h', \text{ ....(3)}$$

$$\overline{\underline{b}}{}^{n}_{n} = \overline{\underline{K}}{}^{n}_{n}\ \overline{\underline{\beta}}{}^{n}_{n}\ \overline{\underline{k}}{}^{n}_{n} \quad \text{or} \quad B' = K'\beta'k'. \text{ ....(4)}$$

Also let X, X', Y, Y' and ξ, ξ', η, η' be matrices with m horizontal and n vertical rows which are connected by the identical equations

$$[x]^{n}_{m} = [h]^{m}_{m}[\xi]^{n}_{m}[K]^{n}_{n} \quad \text{or} \quad X = h\xi K, \text{ ....(A)}$$

$$[x']^{n}_{m} = \overline{\underline{H}}{}^{m}_{m}\ [\xi']^{n}_{m}\ \overline{\underline{k}}{}^{n}_{n} \quad \text{or} \quad X' = H'\xi'k', \text{ ....(A')}$$

$$[y]^{n}_{m} = \overline{\underline{H}}{}^{m}_{m}\ [\eta]^{n}_{m}[K]^{n}_{n} \quad \text{or} \quad Y = H'\eta K, \text{ ....(B)}$$

$$[y']^{n}_{m} = [h]^{m}_{m}[\eta']^{n}_{m}\ \overline{\underline{k}}{}^{n}_{n} \quad \text{or} \quad Y' = h\eta' k'. \text{ ....(B')}$$

Then if $\epsilon = \pm 1$, *the matrices* X, X', Y, Y' *will be commutants or particular or general commutants of the respective types*

$$X = \{A, \epsilon B\}, \quad X' = \{A', \epsilon B'\}, \quad Y = \{A', \epsilon B\}, \quad Y' = \{A, \epsilon B'\} \text{ ....(5)}$$

when and only when the matrices ξ, ξ', η, η' *are commutants or particular or general commutants of the respective types*

$$\xi = \{a, \epsilon\beta\}, \quad \xi' = \{a', \epsilon\beta'\}, \quad \eta = \{a', \epsilon\beta\}, \quad \eta' = \{a, \epsilon\beta'\}. \text{ ....(5')}.$$

These four results are merely four different forms of Theorem I obtained respectively by using the two isomorphic transformations (1) and (2), (3) and (4), (3) and (2), (1) and (4). In all of them the term 'commutant' may have the restricted meanings mentioned in connection with Theorem I. The equations (A), (A'), (B), (B') establish one-one correspondences analogous to those described in connection with the equation (A). They will usually be regarded as formulae enabling us to construct general commutants X, X', Y, Y' of the types shown in (5) when general commutants ξ, ξ', η, η' of the types shown in (5') are known.

Ex. ii. Whenever ξ, ξ', η, η' are given commutants of the types shown in (5'), the equations (A), (A'), (B), (B') are equigradent commutantal transformations of the types

$$\{A, \epsilon B\} = \{A, a\}\{a, \epsilon\beta\}\{\epsilon\beta, \epsilon B\},$$
$$\{A', \epsilon B'\} = \{A', a'\}\{a', \epsilon\beta'\}\{\epsilon\beta', \epsilon B'\},$$
$$\{A', \epsilon B\} = \{A', a'\}\{a', \epsilon\beta\}\{\epsilon\beta, \epsilon B\},$$
$$\{A, \epsilon B'\} = \{A, a\}\{a, \epsilon\beta'\}\{\epsilon\beta', \epsilon B'\}$$

determining the corresponding commutants X, X', Y, Y' of the types shown in (5).

Ex. iii. *Derangements of commutants.*

When the equimutant (or isomorphic) transformations (1) and (2) are the symmetric derangements

$$[a]^{m}_{m} = [h]^{m}_{m}[a]^{m}_{m}\ \overline{\underline{h}}{}^{m}_{m} \quad \text{or} \quad A = hah', \text{ ....(1')}$$

$$[b]^{n}_{n} = [k]^{n}_{n}\ [\beta]^{n}_{n}\ \overline{\underline{k}}{}^{n}_{n} \quad \text{or} \quad B = k\beta k', \text{ ....(2')}$$

so that $\qquad hh' = h'h = [1]^{m}_{m}, \quad kk' = k'k = [1]^{n}_{n},$

the formulae (A), (A′), (B), (B′) of Ex. i become

$$X = h\xi k', \quad X' = h\xi' k', \quad Y = h\eta k', \quad Y' = h\eta' k'.$$

In this case general commutants ξ, ξ', η, η' of the types shown in (5′) are converted into general commutants X, X', Y, Y' of the types shown in (5) by applying to their horizontal rows the derangements applied to the horizontal rows of a, a' in forming A, A', and applying to their vertical rows the derangements applied to the vertical rows of β, β' in forming B, B'.

We have corresponding results whenever (1) and (2) are *symmetric* equimutant (or semi-unit) transformations.

The matrices B, β, k can be taken to be the same as the matrices A, α, h. We then have $n = m$; and A, A' are mutually conjugate square matrices of order m derived from the mutually conjugate square matrices α, α' of order m by the equivalent isomorphic transformations

$$A = h\alpha H, \quad A' = H'\alpha' h'. \qquad \text{.........................(6)}$$

In this case the formulae (A), (A′), (B), (B′) assume the special forms (a), (a′), (b), (b′) shown below. Accordingly by using the isomorphic transformations (6) we obtain four particular cases of Theorem I which constitute Theorem I *a* relating to co-commutants and Theorem I *b* relating to contra-commutants. In both these theorems we can ascribe to the term 'commutant' the same restricted meanings as in Theorem I.

Theorem I a. *If X, X' and ξ, ξ' are square matrices of order m connected by the identical equimutant (or isomorphic) relations*

$$[x]^m_m = [h]^m_m [\xi]^m_m [H]^m_m \quad \text{or} \quad X = h\xi H, \qquad \text{.................(a)}$$

$$[x']^m_m = \ulcorner H \urcorner^m_m [\xi']^m_m \ulcorner h \urcorner^m_m \quad \text{or} \quad X' = H'\xi' h', \qquad \text{..............(a′)}$$

then X, X' are co-commutants or particular or general co-commutants of the respective types

$$X = \{A, \epsilon A\}, \quad X' = \{A', \epsilon A'\} \qquad \text{.....................(7)}$$

when and only when ξ, ξ' are co-commutants or particular or general co-commutants of the respective types

$$\xi = \{\alpha, \epsilon\alpha\}, \quad \xi' = \{\alpha', \epsilon\alpha'\}. \qquad \text{........................(7′)}$$

Moreover the two corresponding co-commutants X and ξ or X' and ξ' satisfy the same rational integral equations, and are equicanonical when their elements are constants.

The theorem could be deduced from the two pairs of identical equations

$$AX = h \,.\, \alpha\xi \,.\, H, \quad XA = h \,.\, \xi\alpha \,.\, H; \qquad \text{...........(a}_1\text{)}$$

$$A'X' = H' \,.\, \alpha'\xi' \,.\, h', \quad X'A' = H' \,.\, \xi'\alpha' \,.\, h'. \qquad \text{...........(a}_1'\text{)}$$

It will chiefly be used in the construction of most general co-commutants satisfying a given rational integral equation. We can always choose ξ' and X' to be the conjugates of ξ and X.

Theorem Ib. *If Y, Y' and η, η' are square matrices of order m connected by the identical symmetric equigradent relations*

$$[y]_m^m = \overline{\underline{H}}{}_m^m [\eta]_m^m [H]_m^m \quad \text{or} \quad Y = H'\eta H, \quad \ldots\ldots\ldots\ldots\ldots\text{(b)}$$

$$[y']_m^m = [h]_m^m [\eta']_m^m \overline{\underline{h}}{}_m^m \quad \text{or} \quad Y' = h\eta' h', \quad \ldots\ldots\ldots\ldots\ldots\text{(b}')$$

then Y, Y' are contra-commutants or particular or general contra-commutants of the respective types

$$Y = \{A', \epsilon A\}, \quad Y' = \{A, \epsilon A'\} \quad \ldots\ldots\ldots\ldots\ldots\ldots\ldots\text{(8)}$$

when and only when η, η' are contra-commutants or particular or general contra-commutants of the respective types

$$\eta = \{\alpha', \epsilon\alpha\}, \quad \eta' = \{\alpha, \epsilon\alpha'\}. \quad \ldots\ldots\ldots\ldots\ldots\ldots\ldots\ldots\text{(8}')$$

Moreover Y, Y' are particular or general symmetric or skew-symmetric contra-commutants of the types shown in (8) *when and only when η, η' are particular or general symmetric or skew-symmetric contra-commutants of the types shown in* (8′).

The theorem could be deduced directly from the two pairs of identical equations

$$A'Y = H'.\alpha'\eta.H, \quad YA = H'.\eta\alpha\ .H; \quad \ldots\ldots\ldots\ldots\ldots(\text{b}_1)$$

$$AY' = h\ .\alpha\eta'.h', \quad Y'A' = h\ .\eta'\alpha'.h'. \quad \ldots\ldots\ldots\ldots\ldots(\text{b}_1')$$

It will chiefly be used in the construction of general symmetric or general skew-symmetric contra-commutants. Whenever η and Y are undegenerate, we can choose η' and Y' to be their inverses.

Ex. iv. The equations (b_1), (b_1') show that in Theorem I*b* the first, second, third, fourth of the products

$$YA, \quad A'Y, \quad Y'A', \quad AY'$$

will be symmetric (or skew-symmetric) when and only when the first, second, third, fourth of the corresponding products

$$\eta\alpha, \quad \alpha'\eta, \quad \eta'\alpha', \quad \alpha\eta'$$

is symmetric (or skew-symmetric).

Ex. v. From the definitions of contra-commutants it follows that:

(1) A square matrix Y of order m will be a general symmetric (or skew-symmetric) contra-commutant $\{A', A\}$ when and only when it is a most general symmetric (or skew-symmetric) matrix such that the product YA or $A'Y$, whichever we please, is symmetric (or skew-symmetric).

(2) A square matrix Y of order m will be a general symmetric (or skew-symmetric) contra-commutant $\{A', -A\}$ when and only when it is a most general symmetric (or skew-symmetric) matrix such that the product YA, or $A'Y$, whichever we please, is skew-symmetric (or symmetric).

In these results we can of course interchange A and A', and also replace Y by Y'. Also we can replace A, A', Y, Y' by α, α', η, η'.

Ex. vi. *Symmetric derangements of co-commutants and contra-commutants.*

When the equimutant transformations (6) are the symmetric derangements

$$A = hah', \quad A' = ha'h', \text{ ....(6')}$$

so that $$hh' = h'h = [1]_m^m,$$

the formulae (a), (a′), (b), (b′) of Theorems I *a* and I *b* become

$$X = h\xi h', \quad X' = h\xi' h', \quad Y = h\eta h', \quad Y' = h\eta' h'.$$

Thus when there exists a symmetric derangement converting a into A and therefore a' into A', the same symmetric derangement converts general commutants ξ, ξ', η, η' of the types shown in (7′) and (8′) into general commutants X, X', Y, Y' of the types shown in (7) and (8).

We have corresponding results whenever the equimutant transformations (6) are symmetric, i.e. whenever they are symmetric semi-unit transformations.

Ex. vii. *Undegenerate symmetric contra-commutants* $\{A', A\}$, $\{A, A'\}$.

If A in Theorem I *b* is a square matrix with constant elements whose characteristic potent divisors are

$$(\lambda - c_1)^{e_1}, \quad (\lambda - c_2)^{e_2}, \ \dots \ (\lambda - c_s)^{e_s},$$

we can take a to be a canonical square matrix of the class

$$M\begin{pmatrix} e_1, & e_2, & \dots, & e_s \\ e_1, & e_2, & \dots, & e_s \end{pmatrix} \text{ ....(9)}$$

The part-reversant J of this class defined in § 263 (see also Note 1) is clearly an undegenerate symmetric contra-commutant of each of the types $\{a', a\}$, $\{a, a'\}$; for we have

$$JaJ = a', \text{ and therefore } a'J = Ja, \ aJ = Ja'.$$

It follows that the two mutually inverse square matrices

$$U = H'JH = \{A', A\}, \quad V = hJh' = \{A, A'\}$$

are undegenerate symmetric contra-commutants of the types shown in which all the elements are constants.

NOTE 1. A *simple reversant* is a square matrix in which every element of the counter-diagonal (sloping downwards from right to left) is equal to 1, and all other elements are 0′s. The *part-reversant* J of the class (9) is a standard compartite matrix of that class in which each of the diagonal constituents is a simple reversant, and all other constituents are zero matrices. It is symmetric and inverse to itself.

We shall use the notation $[j]_r^r$ for the simple reversant of order r.

2. *The commutants of a pair of standard compartite square matrices.*

Let $A = [a]_m^m$ and $B = [b]_n^n$ be two given compartite square matrices expressed in the standard forms

$$A = \begin{bmatrix} A_1, & 0, & \dots & 0 \\ 0, & A_2, & \dots & 0 \\ \dots & \dots & \dots & \dots \\ 0, & 0, & \dots & A_r \end{bmatrix}_{\alpha_1, \alpha_2, \dots \alpha_r}^{\alpha_1, \alpha_2, \dots \alpha_r}, \quad B = \begin{bmatrix} B_1, & 0, & \dots & 0 \\ 0, & B_2, & \dots & 0 \\ \dots & \dots & \dots & \dots \\ 0, & 0, & \dots & B_s \end{bmatrix}_{\beta_1, \beta_2, \dots \beta_s}^{\beta_1, \beta_2, \dots \beta_s},$$

their successive parts being *all square*; the ith part A_i of A being a square matrix of order α_i; the jth part B_j of B being a square matrix of order β_j;

and $\alpha_1, \alpha_2, \dots \alpha_r$ and $\beta_1, \beta_2, \dots \beta_s$ being two sets of non-zero positive integers whose sums are respectively m and n. The parts of A and B may be either matrices with constant elements, or their elements may be rational integral functions of certain scalar variables. Also let

$$\epsilon = \pm 1.$$

Then the matrix $X = [x]_m^n$ will be a commutant $\{A, \epsilon B\}$ when and only when

$$[a]_m^m [x]_m^n = \epsilon \, . \, [x]_m^n [b]_n^n . \quad \text{.......................(10)}$$

If we express X as a compound matrix in the form

$$X = \begin{bmatrix} X_{11}, & X_{12}, & \dots & X_{1s} \\ X_{21}, & X_{22}, & \dots & X_{2s} \\ \dots & \dots & \dots & \dots \\ X_{r1}, & X_{r2}, & \dots & X_{rs} \end{bmatrix}_{\alpha_1, \alpha_2, \dots \alpha_r}^{\beta_1, \beta_2, \dots \beta_s}, \quad \text{..................(C)}$$

where the constituent X_{ij} is a matrix with α_i horizontal and β_j vertical rows, the equation (10) is equivalent to the rs equations

$$A_i X_{ij} = \epsilon \, . \, X_{ij} B_j, \qquad (i = 1, 2, \dots r, j = 1, 2, \dots s); \text{......(10')}$$

and we have therefore the following theorem which remains true when 'commutant' is interpreted to mean 'non-singular commutant' or to mean 'commutant independent of certain specified variables.'

Theorem II. *The matrix X expressed in the form* (C) *is a commutant or particular commutant $\{A, \epsilon B\}$ if and only if for all permissible values of i and j the constituent X_{ij} of X is a commutant or particular commutant $\{A_i, \epsilon B_j\}$.*

Also X is a general commutant $\{A, \epsilon B\}$ when X_{ij} is always a general commutant $\{A_i, \epsilon B_j\}$, and the arbitrary parameters of the rs general commutants X_{ij} are independent of one another and of the variables (if any) which occur in A and B.

The second part of the theorem is obtained by choosing the independent particular non-zero commutants $\{A, \epsilon B\}$ of the form (C) in such a manner that in each of them only one of the constituents X_{ij} is a non-zero matrix.

3. *Zero and non-zero general commutants.*

First let A and B be two square matrices of orders m and n whose elements are all constants, and let A_0 and B_0 be canonical square matrices having the same characteristic potent divisors and therefore the same latent roots as A and B respectively. By § 227 there exist equimutant (or isomorphic) transformations by which A and B can be derived from A_0 and B_0 respectively; and it follows from Theorem I that the general commutants $\{A, B\}$ and $\{A_0, B_0\}$ have equal ranks, and that each of them is a zero matrix when and only when the other is a zero matrix. But by Ex. xviii of § 238 the general commutant $\{A_0, B_0\}$ is a non-zero matrix when and only when A_0 and B_0 have a latent root in common. Accordingly we have the following theorem:

Theorem III. *If A and B are square matrices with constant elements, then:*

(1) *the general commutant $\{A, B\}$ is a non-zero matrix when and only when A and B have a latent root in common;*

(2) *the general commutant $\{A, -B\}$ is a non-zero matrix when and only when A and $-B$ have a latent root in common, i.e. when and only when there exist latent roots α, β of A, B such that $\alpha+\beta=0$.*

Next let the elements of A and B be independent arbitrary parameters; let $f=\det A(\lambda)$ and $g=\det B(\lambda)$ be the characteristic determinants of A and B; let $R=\det\Omega$ be the scalar resultant of A and B defined in § 238. 2; and let $R(f, g)$ be the resultant of the scalar functions f and g when they are treated as rational integral functions of λ only. Both R and $R(f, g)$ are homogeneous rational integral functions of the elements of A and B of total degree mn. For all particular values of the elements of A and B the vanishing of R is a necessary and sufficient condition for the existence of a non-zero commutant $\{A, B\}$, and therefore by Theorem III a necessary and sufficient condition that A and B shall have a common latent root or that f and g shall have a common root in λ, i.e. a necessary and sufficient condition for the vanishing of $R(f, g)$. Since R and $R(f, g)$ are homogeneous functions of the same variables of the same total degree which do not vanish identically, it follows that they can only differ by a non-zero numerical factor. But in the particular case when A and B are quasi-scalar matrices whose diagonal elements are independent arbitrary parameters we have

$$R=\pm\Pi(a_{ii}-b_{jj}),\quad R(f, g)=\pm\Pi(a_{ii}-b_{jj}).$$

Therefore the numerical factor must be 1 or -1, and we conclude that:

The scalar resultant R of the two square matrices A and B is the same as the resultant $R(f, g)$ of their characteristic determinants $f=\det A(\lambda)$ and $g=\det B(\lambda)$ when these are treated as functions of λ only.

Lastly let the elements of the square matrices A and B be rational integral functions of certain scalar variables $\gamma_1, \gamma_2, \gamma_3, \ldots$, so that f and g are rational integral functions of λ and the γ's, and R is a rational integral function of the γ's. If f and g have a rational integral factor in common (which must involve λ), then the resultant $R=R(f, g)$ vanishes identically. If f and g have no rational integral factor in common, let them be expressed as products of powers of irresoluble factors which are all different. Then we can give such particular values to the γ's that no two of the factors have a common root in λ, in which case A and B will have no common latent root, the general commutant $\{A, B\}$ will be a zero matrix, and R will not vanish;

consequently the resultant $R=R(f, g)$ does not vanish identically. We are thus led again to (6) of § 238 and to the following theorem:

Theorem IV. *If A and B are square matrices whose elements are rational integral functions of certain scalar variables, the general commutant $\{A, B\}$ is a non-zero matrix when and only when the characteristic determinants det $A(\lambda)$ and det $B(\lambda)$ of A and B have an irresoluble factor in common.*

NOTE 2. *Applications of the foregoing theorems.*

We will consider the applications to square matrices A and B whose elements are constants, and we will call (1) and (2) *isomorphic transformations* to indicate that each of them is an equimutant transformation converting a square matrix into a square matrix of the *same order.*

Theorems I and II show that the general commutant $\{A, B\}$ can be constructed when we know:

(1) any particular isomorphic transformations converting A and B into canonical square matrices a and β;

(2) the general commutants of all pairs of simple canonical square matrices, and therefore the general commutant $\{a, \beta\}$.

The existence of isomorphic transformations

$$HAh=a, \quad KBk=\beta \quad \text{or} \quad A=haH, \quad B=k\beta K,$$

where a and β are canonical reduced forms of A and B, has been proved in § 228, and will be proved again in a different way in Chapter XXXVII. The actual determination of such transformations is best effected by the methods of the later chapter, which consist in successive reductions by isomorphic transformations of:

(i) any square matrix with constant elements to a *simple square ante-slope*;

(ii) any simple square ante-slope to a standard compartite matrix whose parts are *unilatent simple square ante-slopes*;

(iii) any unilatent simple square ante-slope to a standard compartite matrix whose parts are *unipotent simple square ante-slopes*;

(iv) any unipotent simple square ante-slope to a *simple canonical square matrix.*

The general commutant of any pair of simple canonical square matrices is determined in § 321, and also in § 242 of the present chapter.

We can choose a and β to be *standard* canonical square matrices, in which the super-parts are *unilatent* canonical square matrices, and the simple parts are *simple* canonical square matrices. Then knowing the general commutants of pairs of simple canonical square matrices, which are simple ante-continuants, we can construct by Theorem II the general commutants of pairs of unilatent canonical square matrices (such as the super-parts of a and β), which are compound ante-continuants; and by a further application of Theorem II we can construct the general commutant $\{a, \beta\}$, which is a standard compartite square matrix whose parts are compound ante-continuants. Finally Theorem I enables us to pass from the general commutant $\{a, \beta\}$ to the general commutant $\{A, B\}$ by the formula

$$\{A, B\}=h\{a, \beta\}K.$$

§ 241. Simple and compound slopes; the greatest common canonical of two square matrices.

1. *Simple slopes; simple continuants and alternants.*

The properties of simple and compound slopes will be discussed systematically in later chapters, but some of them will be used by anticipation in the present chapter. The present article has been inserted in order to reduce the number of references to later chapters.

Square matrices of any order r having the respective forms

$$\begin{bmatrix} x_{11} & x_{12} & \dots & x_{1r} \\ 0 & x_{22} & \dots & x_{2r} \\ \dots & \dots & \dots & \dots \\ 0 & 0 & \dots & x_{rr} \end{bmatrix}, \quad \begin{bmatrix} x_{rr} & \dots & 0 & 0 \\ \dots & \dots & \dots & \dots \\ x_{2r} & \dots & x_{22} & 0 \\ x_{1r} & \dots & x_{12} & x_{11} \end{bmatrix}, \quad \begin{bmatrix} 0 & 0 & \dots & x_{rr} \\ \dots & \dots & \dots & \dots \\ 0 & x_{22} & \dots & x_{2r} \\ x_{11} & x_{12} & \dots & x_{1r} \end{bmatrix}, \quad \begin{bmatrix} x_{1r} & \dots & x_{12} & x_{11} \\ x_{2r} & \dots & x_{22} & 0 \\ \dots & \dots & \dots & \dots \\ x_{rr} & \dots & 0 & 0 \end{bmatrix},$$

where all elements lying on one side of the diagonal or counter-diagonal are 0's, will be called *simple square slopes* of the types

$$\{\pi, \pi\}, \quad \{\pi', \pi'\}, \quad \{\pi', \pi\}, \quad \{\pi, \pi'\}. \qquad \text{.................(a)}$$

In each of these matrices the corner which has been specially marked is the *apex*, and the opposite corner is the *base*, the position of the apex being fixed by the type.

If m and n are any two non-zero positive integers the smaller of which is r, a *simple slope* of the class $M\binom{m}{n}$ and of any one of the types (a) is a matrix which can be formed from a simple square slope of that type by adding

$n - m$ basical vertical rows of 0's if $m = r \ngtr n$,

$m - n$ basical horizontal rows of 0's if $n = r \ngtr m$,

the added rows of 0's being more remote from the apex than the other rows. That element of a simple slope which lies nearest to the apex will be called the *apical element*, and that element which lies nearest to the base will be called the *basical element*, the basical element being always 0 except when $m = n = 1$. Simple slopes of the first two and the last two of the types (a) will be called respectively *ante-slopes* and *counter-slopes*.

Those elements of a simple slope of given class and type which are not 0's by definition will be called the *parametric elements*. If r is the effective (or smaller) order, they form r *parametric diagonal lines* sloping downwards from left to right in an ante-slope or from right to left in a counter-slope. When double suffix notations similar to those shown above are used, the first suffix increasing as we recede from the apex along a vertical row and the second suffix increasing as we approach the apex along a horizontal row, the element denoted by x_{ij} will be said to have *difference-weight* $j - i$. Each parametric diagonal line is formed with all the parametric elements having a given difference-weight. In a *general simple slope* the parametric elements are independent arbitrary parameters.

The forms of simple slopes of all possible types are shown schematically in the figures of Ex. ii, where each constituent is a simple slope. In any simple slope of the class $M\binom{n}{m}$ and of effective order r the rth or last parametric diagonal line counting from the apical element will be called the *paradiagonal* (or the non-major diagonal). The paradiagonal is the only one of the parametric diagonal lines which is a 'diagonal,' i.e. which is drawn through a corner element so as to pass through other elements; and it is the nearer to the apex of those two of the four diagonals which lie between the apex and the base.

We will formally define the horizontal (or vertical) *apical distance* of any element e of a simple slope X of given type to be the number of vertical (or horizontal) rows extending from the one through e to the one through the apical element, both the extreme rows being included; and in speaking of such distances we shall regard the shortest distance between two consecutive horizontal or vertical rows, i.e. the length of a horizontal or vertical 'step,' as the unit of length. The horizontal and vertical apical distances of the apical element are both equal to 1, and the horizontal or vertical apical distance of any other element is equal to its horizontal or vertical distance from the apical element increased by 1. If ξ and η are the horizontal and vertical apical distances of e, and if r is the effective order of X, the integer

$$E = \xi + \eta - 1 - r$$

will be called the *apical excess* of e. It is a measure of the amount by which the distance from the apex of the parallel diagonal line through e exceeds that of the paradiagonal.

Simple continuants of the type (a) are simple slopes of those types in which the elements of each parametric diagonal line are all equal; and *simple alternants* of the respective types

$$\{\pi, -\pi\},\quad \{\pi', -\pi'\},\quad \{\pi', -\pi\},\quad \{\pi, -\pi'\} \quad \text{..............(b)}$$

are simple slopes of the corresponding types (a) in which the elements of each parametric diagonal line differ from one another only by signs which are alternately + and −. A *ruled simple slope* is a simple slope which is either a simple continuant or a simple alternant. A continuant will be called an *ante-continuant* or *counter-continuant* and an alternant an *ante-alternant* or *counter-alternant* according as it is an ante-slope or a counter-slope.

The expressions such as (a) and (b), in which π and π' are not matrices but merely two different letters, will be called *symbolic commutants* when they are used in the way described in Chapter XXX to indicate the assigned commutantal types of matrices; and we will apply the same terminology to them as to the expressions representing the types of true commutants. Each symbolic commutant is formed with two elements, each of which is $\pm\pi$ or $\pm\pi'$. A change in the sign of the first (or second) element of a symbolic commutant will ordinarily be taken to indicate a change in the sign of every element in every alternate horizontal (or vertical) row of the matrix which it represents. For the present it is not necessary to attach any meaning to a change in the sign of both the elements of a symbolic commutant.

Ex. i. *Zero and non-zero elements of a general simple slope.*

Let X be a general (or general ruled) simple slope of any given class $M\binom{n}{m}$ and of any given type whose effective (or smaller) order is r. Then if ξ and η are the horizontal and vertical apical distances of any element e of X, and if E is the apical excess of e, the necessary and sufficient condition:

(1) that e shall lie on the paradiagonal is $\xi+\eta-1=r$, or $E=0$;

(2) that e shall be a non-zero element is $\xi+\eta-1 \not> r$, or $E \not> 0$;

(3) that e shall be a zero element is $\xi+\eta-1>r$, or $E \not< 1$.

We can treat ξ and η as rectangular-coordinates defining the position of e in X when the apex is origin. Provided that m and n are both *non-zero* positive integers, the possible values of ξ, η, $\xi+\eta-1$ for all positions of e are those consistent with the respective sets of conditions

$$\xi \not< 1,\ \xi \not> n;\quad \eta \not< 1,\ \eta \not> m;\quad \xi+\eta-1 \not< 1,\ \xi+\eta-1 \not> m+n-1.$$

The conditions (1), (2), (3) are independent of the notations used for the elements of X. When the double-suffix notations indicated in the text are used, we can replace them by corresponding conditions to be satisfied by the difference-weight κ of e; for we shall have

$$\xi+\eta-1=r-\kappa \quad \text{or} \quad \xi+\eta-1=n-\kappa,$$

according as we choose the element denoted by x_{11} to be the one in which the first apical horizontal row cuts the paradiagonal or the first basical vertical row.

2. *Compound slopes; compound continuants and alternants.*

If θ is any one of the sixteen symbolic commutants such as (a) and (b), we will define a *compound slope* of the class

$$M\begin{pmatrix}\beta_1, & \beta_2, & \dots & \beta_s \\ \alpha_1, & \alpha_2, & \dots & \alpha_r\end{pmatrix}, \qquad (\Sigma\alpha_i=m,\ \Sigma\beta_i=n), \dots\dots\dots\text{(c)}$$

and of type θ to be a compound matrix

$$X=\begin{bmatrix} X_{11}, & X_{12}, & \dots & X_{1s} \\ X_{21}, & X_{22}, & \dots & X_{2s} \\ \dots & \dots & \dots & \dots \\ X_{r1}, & X_{r2}, & \dots & X_{rs}\end{bmatrix}^{\beta_1,\ \beta_2,\ \dots\ \beta_s}_{\alpha_1,\ \alpha_2,\ \dots\ \alpha_r} \qquad \dots\dots\dots\dots\dots\dots\dots\text{(A)}$$

of that class in which every constituent such as X_{ij} is a simple slope of type θ. Such a matrix is a *ruled compound slope* or a *compound continuant* or a *compound alternant* when every constituent is a ruled simple slope or a simple continuant or a simple alternant. In a *general* (or *general ruled*) compound slope the constituents are all general (or general ruled) simple slopes, the parameters occurring in the various constituents being all arbitrary and independent. When a compound slope is expressed in the form (A), we shall ordinarily denote the effective order of the constituent X_{ij} by γ_{ij}, so that γ_{ij} is the smaller of the two integers α_i and β_j.

By a *standardised* compound slope will be meant one in which the horizontal index numbers and the vertical index numbers are arranged in descending orders of magnitude. Every compound slope can be converted into a standardised compound slope of the same type by a 'class-derangement' in which the constituents are moved to new positions by rigid displacements. Whenever we speak of a standardised compound slope M of the class (c) expressed in the form (A) it will be understood that the index numbers $\alpha_1, \alpha_2, \ldots \alpha_r$ and $\beta_1, \beta_2, \ldots \beta_s$ are arranged in descending orders of magnitude. The paradiagonals of the diagonal constituents $X_{11}, X_{22}, \ldots$ will be called the *principal paradiagonals* of M, and the elements composing them will be called the *principal paradiagonal elements* or simply the *principal elements* of M.

By a *quadrate slope* will be meant a compound slope in which the successive vertical index numbers are the same as the successive horizontal index numbers, a necessary condition for this being $s = r$. All its diagonal constituents are simple *square* slopes, and it is necessarily a square matrix. Every canonical square matrix is a ruled quadrate slope of type $\{\pi, \pi\}$, i.e. a quadrate continuant of that type.

We have already defined a *standard* canonical square matrix and its unilatent super-parts in § 228. By analogy with the terminology of compound slopes we will understand a *standardised canonical square matrix* to be a standard canonical square matrix in which each unilatent super-part is a standardised quadrate slope, the orders of its successive simple parts being integers arranged in descending order of magnitude.

By a *quasi-scalaric compound slope* will be meant a compound slope in which the only elements which can be different from 0 are those forming the paradiagonals of the square constituents. In such a matrix every simple constituent which is not square must be a zero matrix.

It will ordinarily be understood that the index numbers in (c) and (A) are different from 0, but the addition of zero index numbers and corresponding 'non-existent' constituents of effective order 0 will be regarded as allowable. In dealing with a standardised compound slope of the class (c) expressed in the form (A) we can always proceed as if we had $s = r$, provided that we interpret α_i to be 0 when $i > r$, and β_i to be 0 when $i > s$.

Ex. ii. The following figures show the parametric diagonal lines of the simple constituents in general compound slopes, compound continuants and compound alternants of all possible types belonging to the class

$$M\begin{pmatrix}12, & 11, & 6, & 5, & 2\\ 12, & 8, & 7, & 4, & 3\end{pmatrix}.$$

Elements not lying on the parametric diagonal lines are all 0's. In a *general compound slope* the elements lying on the parametric diagonal lines are independent arbitrary parameters. In a *general compound continuant* the elements lying on any given parametric diagonal line are all equal to x, where x is an arbitrary parameter, and the parameters of the various parametric diagonal lines are independent. In a *general compound alternant* the elements lying on any given parametric diagonal line are alternately x and $-x$, where x is an arbitrary parameter, and the parameters of the various parametric diagonal lines are independent. In each of the last two cases the total number of independent arbitrary parameters is equal to the total number of parametric diagonal lines in all the simple constituents.

Fig. 1. Types $\pm\{\pi, \pi\}$, $\pm\{\pi, -\pi\}$.

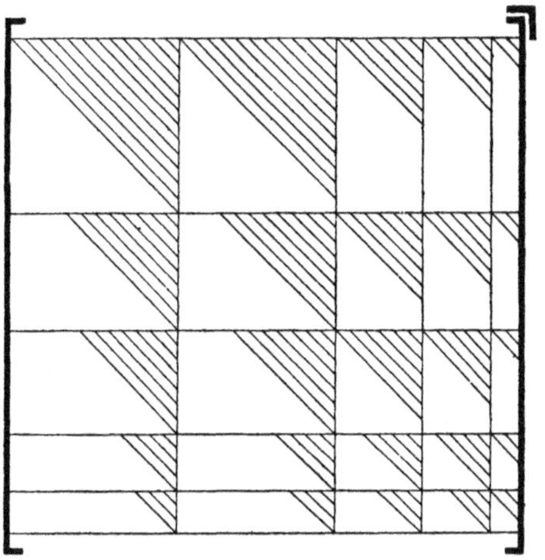

Fig. 2. Types $\pm\{\pi', \pi'\}$, $\pm\{\pi', -\pi'\}$.

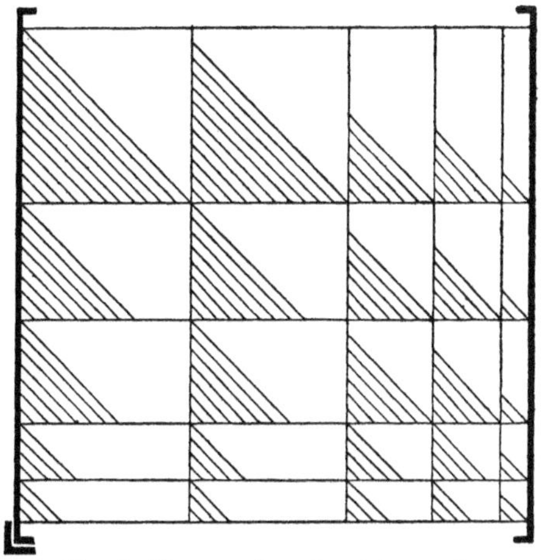

Fig. 3. Types $\pm\{\pi', \pi\}$, $\pm\{\pi', -\pi\}$.

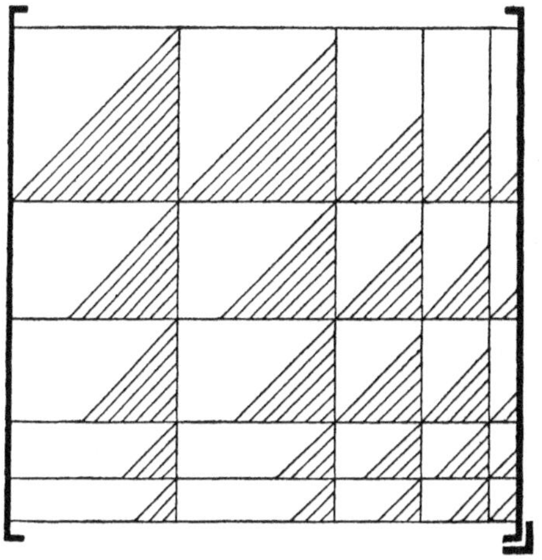

Fig. 4. Types $\pm\{\pi, \pi'\}$, $\pm\{\pi, -\pi'\}$.

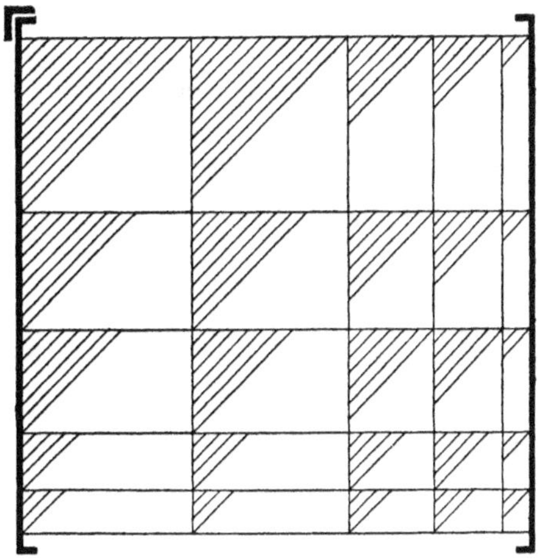

The figures represent standardised compound slopes from which all complete rows of 0's have been struck out. Moreover they illustrate a case in which $s=r$ in (c) and (A). They can be converted into illustrations of other cases by striking out horizontal or vertical minors.

Ex. iii. If we fix our attention on those constituents of a *standardised* general compound slope M which lie in a given horizontal or a given vertical minor, we see that:

(1) If X is a constituent lying in a given *horizontal minor*, the total number of (basical) *horizontal rows of* 0's in X tends constantly to *increase* (or never

diminishes) as X moves *rightward*; on the other hand the total number of (apical) vertical rows of 0's tends constantly to decrease in the same movement.

(2) If X is a constituent lying in a given *vertical minor*, the total number of (apical) *vertical rows of* 0's in X tends constantly to *increase* (or never diminishes) as X moves *downward*; on the other hand the total number of (basical) horizontal rows of 0's tends constantly to decrease in the same movement.

For in the first case the vertical index number of X tends constantly to decrease whilst the horizontal index number remains constant; in the second case the horizontal index number of X tends constantly to decrease, whilst the vertical index number remains constant. We conclude that:

If a horizontal line cuts a constituent X in zero elements only, it cuts all constituents to the right of X in zero elements only.

If a vertical line cuts a constituent X in zero elements only, it cuts all constituents below X in zero elements only.

Ex. iv. Let α be the greatest horizontal and β the greatest vertical index number in (c) and (A). Then a *general* compound slope M of the class (c) and of any type will contain

complete horizontal rows of 0's if and only if $\alpha > \beta$,

complete vertical rows of 0's if and only if $\beta > \alpha$;

for every complete horizontal row of 0's must pass through a constituent having vertical order β and horizontal order greater than β, and every complete vertical row of 0's must pass through a constituent having horizontal order α and vertical order greater than α.

Ex. v. If M is a *standardised* general compound slope, then by striking out all the complete horizontal or vertical rows of 0's (if such exist), we convert it into a *standardised* general compound slope of the same type in which the leading constituent is a square matrix, and the effective orders of all the constituents remain unaltered.

For if γ is the smaller of the two integers α and β of Ex. iv, the class of the new compound slope is derived from that of M by substituting γ for every index number which is greater than γ.

Ex. vi. When M is a *standardised compound slope* of the class (c) expressed in the form (A), let the effective order of a simple constituent X_{ij} be always denoted by γ_{ij}, and let

$$\Omega = \begin{bmatrix} X_{ij}, & X_{ik} \\ X_{hj}, & X_{hk} \end{bmatrix}_{\alpha_i,\ \alpha_h}^{\beta_j,\ \beta_k} = \begin{bmatrix} X_{ij}, & X_{ik} \\ X_{hj}, & X_{hk} \end{bmatrix}_{\alpha,\ \alpha'}^{\beta,\ \beta'}$$

be any corranged minor of M formed by the intersection of two horizontal and two vertical minors, so that

$$i < h, \quad j < k.$$

Then if $$D = (\gamma_{ij} - \gamma_{ik}) - (\gamma_{hj} - \gamma_{hk}) = (\gamma_{ij} - \gamma_{hj}) - (\gamma_{ik} - \gamma_{hk}),$$

we have $$D \not< 0.$$

This is merely a property of a standardised compound slope of a class

$$M\begin{pmatrix} \beta, & \beta' \\ \alpha, & \alpha' \end{pmatrix}$$

with two horizontal and two vertical index numbers. The possible arrangements of all the four index numbers in descending order of magnitude are:

$(\alpha, \alpha', \beta, \beta')$, $(\alpha, \beta, \alpha', \beta')$, $(\alpha, \beta, \beta', \alpha')$ for which $D = 0, \beta - \alpha', \beta - \beta'$;

$(\beta, \beta', \alpha, \alpha')$, $(\beta, \alpha, \beta', \alpha')$, $(\beta, \alpha, \alpha', \beta')$ for which $D = 0, \alpha - \beta', \alpha - \alpha'$.

It will be obvious that the effective order of a simple constituent tends constantly to diminish (i.e. never increases) as it moves downward in a given vertical minor or rightward in a given horizontal minor, and the above result shows that a similar property is possessed by the difference of the effective orders of two constituents lying in the same horizontal minor (the second constituent lying to the right of the first) or of two constituents lying in the same vertical minor (the second constituent lying below the first).

Ex. vii. *Let M be a standardised general compound slope of the class* (c) *expressed in the form* (A)*; let x be any non-zero element of any constituent X, and let the horizontal line through x cut the paradiagonal of a constituent X' which does not lie completely to the right of X in the element x'; also let e and e' be the elements in which the vertical lines through x and x' cut*

any horizontal row of M lying below X or passing through X.

Then if E and E' are the apical excesses of e and e' in the constituents in which they lie, we have $E' \nless E$; consequently if e is a zero element, then e' also is a zero element.

x'	x
e'	e

$$\Omega = \begin{bmatrix} X_{11}, & X_{12} \\ X_{21}, & X_{22} \end{bmatrix}^{\beta_1,\, \beta_2}_{\alpha_1,\, \alpha_2}$$

e	e'
x	x'

.

We will suppose that the constituents in which the elements x, x', e, e' lie are all different. It will be obvious from the proof that the theorem is true in the other special cases.

Since the corranged minor of M formed by the constituents in which x, x', e, e' lie is a standardised general compound slope, there will be no loss of generality in taking it to be the minor Ω shown above, this being merely a matter of notation. We will denote the effective order of any constituent X_{ij} by γ_{ij}. We can take the (horizontal, vertical) apical distances of the elements

$$x, x', e, e' \quad \text{to be} \quad (\xi, \eta), (\xi', \eta), (\xi, \eta'), (\xi', \eta'),$$

in their respective constituents X_{12}, X_{11}, X_{22}, X_{21}, and we are given the relations

$$\xi + \eta \ngtr \gamma_{12}, \quad \xi' + \eta - 1 = \gamma_{11}, \quad \text{involving} \quad \xi' - \xi \nless \gamma_{11} - \gamma_{12}.$$

Since $$E = \xi + \eta' - 1 - \gamma_{22}, \quad E' = \xi' + \eta' - 1 - \gamma_{21},$$

it follows from these relations and Ex. vi that

$$E' - E = (\xi' - \xi) - (\gamma_{21} - \gamma_{22}) \nless (\gamma_{11} - \gamma_{12}) - (\gamma_{21} - \gamma_{22}) \nless 0;$$

and this establishes the theorem. It should be observed that the theorem obviously remains true when the horizontal line through the non-zero element x cuts X' in a *zero element* x'; for then we have $\xi' + \eta - 1 \nless \gamma_{11} + 1$.

By a similar proof it can be shown that:

The theorem remains true in other respects when X' does not lie to the left of X, and e and e' are the elements in which the vertical lines through x and x' cut

any horizontal row of M lying above X or passing through X.

This second theorem also remains true when x' is a zero element instead of a paradiagonal element. Both theorems are generalised in Appendix B.

Ex. viii. *Let M be a standardised general compound slope of the class* (c) *expressed in the form* (A)*; let y be any non-zero element of any constituent Y, and let the vertical line through y cut the paradiagonal of a constituent Y' which does not lie completely below Y in the element y'; also let e and e' be the elements in which the horizontal lines through y and y' cut*

any vertical row of M lying to the right of Y or passing through Y.

Then if E and E' are the apical excesses of e and e' in the constituents in which they lie, we have $E' \nless E$; consequently if e is a zero element, then e' also is a zero element.

y'	e'
y	e

$$\Omega = \begin{bmatrix} X_{11}, & X_{12} \\ X_{21}, & X_{22} \end{bmatrix}^{\beta_1,\, \beta_2}_{\alpha_1,\, \alpha_2}$$

e	y
e'	y'

We will suppose that the constituents in which the elements y, y', e, e' lie are all different. It will be obvious from the proof that the theorem is true in the other special cases.

Since the corranged minor of M formed by the constituents in which y, y', e, e' lie is a standardised general compound slope, there will be no loss of generality in taking it to be the minor Ω shown above, this being merely a matter of notation. We will denote the effective order of any constituent X_{ij} by γ_{ij}. We can take the (horizontal, vertical) apical distances of the elements

$$y, y', e, e' \quad \text{to be} \quad (\xi, \eta), (\xi, \eta'), (\xi', \eta), (\xi', \eta'),$$

in their respective constituents X_{21}, X_{11}, X_{22}, X_{12}, and we are given the relations

$$\xi+\eta-1 \ngtr \gamma_{21}, \quad \xi+\eta'-1=\gamma_{11}, \quad \text{involving} \quad \eta'-\eta \nless \gamma_{11}-\gamma_{21}.$$

Since $$E=\xi'+\eta-1-\gamma_{22}, \quad E'=\xi'+\eta'-1-\gamma_{12},$$

it follows from these relations and Ex. vi that

$$E'-E=(\eta'-\eta)-(\gamma_{12}-\gamma_{22}) \nless (\gamma_{11}-\gamma_{21})-(\gamma_{12}-\gamma_{22}) \nless 0;$$

and this establishes the theorem. It should be observed that the theorem obviously remains true when the vertical line through the non-zero element y cuts Y' in a *zero element* y'; for then we have $\xi+\eta'-1 \nless \gamma_{11}+1$.

By a similar proof it can be shown that:

The theorem remains true in other respects when Y' does not lie completely above Y, and e and e' are the elements in which the horizontal lines through y and y' cut

any vertical row of M lying to the left of Y or passing through Y.

This second theorem also remains true when y' is a zero element instead of a paradiagonal element. Both theorems are generalised in Appendix B.

3. *Rank of a general compound slope.*

Supposing the α's and β's in (c) to be arranged in descending orders of magnitude, we will define

ρ to be the sum of the smaller integers of the pairs $(\alpha_1, \beta_1), (\alpha_2, \beta_2), (\alpha_3, \beta_3), \ldots$ when α_i is interpreted to be 0 if $i > r$, and β_i to be 0 if $i > s$. ....(1)

In such a case ρ is the sum of the effective (or smaller) orders of the diagonal constituents X_{11}, X_{22}, X_{33}, ... of any compound matrix of the class (c) expressed in the form (A), and it is also the total number of principal paradiagonal elements in a (standardised) general compound slope of the class (c). In other cases we will understand ρ to be the integer determined by the rule (1) after the horizontal and vertical index numbers have been arranged in descending orders of magnitude. In every case we may say that

ρ is the sum of the smaller integers in all the pairs of corresponding descendent horizontal and vertical index numbers. ....(1′)

Let X be a *general* (or *general ruled*) compound slope of the class (c). Then if the α's and β's are arranged in descending orders of magnitude, we can form a particularisation of X having rank ρ by putting all elements equal to 0 except the principal paradiagonal elements. It follows that we can obtain a particularisation of X having rank ρ for all orders of arrangement of the α's and β's, and that in all cases

the rank of X cannot be less than ρ. ....(2)

It will be shown that the rank also cannot be greater than ρ, so that the following theorem is true:

Theorem I. *The rank of a general (or general ruled) compound slope of the class* (c) *is always equal to the integer ρ defined in* (1) *or* (1′).

In the particular case when the β's in (c) are the same as the α's, so that

$$s = r \quad \text{and} \quad m = n = \rho,$$

the rank of X necessarily cannot exceed ρ, and Theorem I becomes equivalent to the obvious theorem that:

A general (or general ruled) quadrate slope is always an undegenerate square matrix..(3)

In Vol. I a 'derived product' of order r of any matrix was defined to be a product of r elements of the matrix selected in such a manner that no two of them lie in the same horizontal or vertical row. In a standardised general (or general ruled) compound slope of the class (c) the product of the ρ principal paradiagonal elements is a non-zero derived product of order ρ. Since the parametric elements of a general compound slope are independent and arbitrary, Theorem I is equivalent by virtue of (2) to the statement that:

A compound slope of the class (c) *cannot have any non-zero derived product of order greater than ρ.* ....(4)

There may of course be derived products of order r, where r is greater than ρ, but at least $r - \rho$ of the elements occurring as factors in such a product must be equal to 0.

The most natural proof of Theorem I will be given in Chapter XXXIX of Part II; but Appendix B to the present Part I furnishes the materials for other similar proofs. It will of course be sufficient to prove the theorem for a standardised general compound slope M of the class (c) which is expressed in the form (A).

Referring in the first place to the first part of Appendix B, we see that by moving all the accessary horizontal rows to the bottom and all the accessary vertical rows to the left we can convert M into a compound matrix

$$M_1 = \begin{bmatrix} H, & Y \\ 0\ , & K \end{bmatrix}_{\rho,\ k}^{h,\ \rho},$$

in which Y is a square matrix of order ρ whose diagonal elements are the ρ principal paradiagonal elements of M.

29—2

In the second place by suitably re-arranging the diagonal elements of Y (or the horizontal and vertical rows passing through them) we can convert M_1 into a compound matrix M_2 expressible in the form

$$M_2 = \begin{bmatrix} P, & A, & C \\ 0, & 0, & B \\ 0, & 0, & Q \end{bmatrix}_{p,\,q,\,k}^{h,\,p,\,q}, \qquad \text{where } p+q=\rho, \quad \text{...........(B)}$$

and where A and B are square matrices whose diagonal elements are the principal paradiagonal elements of M. We obtain this form M_2 when the principal elements are arranged:

(1) as in the preclusive groups $g_1, g_2, \ldots g_\gamma$ defined in the second part of Appendix B, the diagonal elements of A and B respectively being in this case the principal elements forming the quasi-terminate and the pleni-terminate preclusive groups;

(2) as in the postclusive groups $g_1, g_2, \ldots g_\gamma$ defined in the third part of Appendix B, the diagonal elements of A and B respectively being in this case the principal elements forming the pleni-terminate and the quasi-terminate postclusive groups;

(3) as in the conclusive groups $g_1, g_2, \ldots g_\gamma$ defined in the fourth part of Appendix B, the diagonal elements of A and B respectively being in this case the principal elements forming the postclusively pleni-terminate and the preclusively pleni-terminate conclusive groups.

Now a non-zero derived product of M_2 cannot contain more than p factors which are elements of the constituents P, A, C; and it cannot contain more than q factors which are elements of the constituents B, Q. Consequently a non-zero derived product of M_2 or M cannot contain more that $p+q$ or ρ elements as factors. Thus the statement (4) is true, and Theorem I is true.

The derangements M_1 and M_2 are most appropriate to a compound slope M of type $\{\pi, \pi\}$. In forming the analogous derangements appropriate to any one of the four types (a) we commence by moving the accessary horizontal rows to the basical side of all the other horizontal rows, and the accessary vertical rows to the basical side of all the other vertical rows.

Ex. ix. *Undegenerate compound slopes.*

From Theorem I we see that a standardised general (or general ruled) compound slope of the class (c) containing m horizontal and n vertical rows will be:

(1) *undegenerate of rank m* when and only when the effective order of every diagonal constituent is its horizontal order, i.e. when and only when

$$a_i \not> \beta_i, \quad (i=1, 2, \ldots r), \quad \text{so that} \quad r \not> s, \quad n \not> m;$$

(2) *undegenerate of rank n* when and only when the effective order of every diagonal constituent is its vertical order, i.e. when and only when

$$\beta_i \not> a_i, \quad (i=1, 2, \ldots s), \quad \text{so that} \quad s \not> r, \quad m \not> n;$$

(3) *an undegenerate square matrix* when and only when it is quadrate, i.e. when and only when

$$s=r,\ n=m \quad \text{and} \quad \beta_i=\alpha_i,\ (i=1,\ 2,\ \dots\ r).$$

Hence whenever a compound slope is an undegenerate square matrix, it must be either a quadrate slope or a class-derangement of a quadrate slope, its horizontal and vertical index numbers being two arrangements of the same integers.

4. *The determinant of a quadrate slope.*

Let X be a quadrate slope of a class

$$M\begin{pmatrix} e_1, & e_2, & \dots & e_s \\ e_1, & e_2, & \dots & e_s \end{pmatrix}, \quad \text{.............................(d)}$$

in which the s index numbers are not necessarily all unequal; and let ϵ be one of the index numbers which is repeated exactly σ times. Then X has exactly σ^2 square constituents (simple square slopes) of order ϵ, exactly σ of them being diagonal constituents. If we select σ^2 correspondingly situated paradiagonal elements of the σ^2 square constituents of order ϵ, they form a corranged square minor

$$Y=[y]_\sigma^\sigma,$$

of X which will be called a *paradiagonal prime minor* of X. The determinant det Y will be called a *prime determinant* of X. It will often be more convenient to regard Y as a corranged square minor of X of order σ in which the diagonal elements are σ correspondingly situated paradiagonal elements of the σ square diagonal constituents of X of order ϵ. It will be proved in Chapter XXXVIII, but will be sufficiently clear from Ex. x below, that the following theorem is true.

Theorem II. *The determinant of any quadrate slope X is equal to the product of all the prime determinants of X.*

Ex. x. If X is either of the quadrate slopes

$$\left[\begin{array}{ccc|ccc|cc|c}
a_{11} & a_{12} & a_{13} & h_{11} & h_{12} & h_{13} & g_{11} & g_{12} & u_{11} \\
0 & a_{22} & a_{23} & 0 & h_{22} & h_{23} & 0 & g_{22} & 0 \\
0 & 0 & a_{33} & 0 & 0 & h_{33} & 0 & 0 & 0 \\
\hline
h_{11}' & h_{12}' & h_{13}' & b_{11} & b_{12} & b_{13} & f_{11} & f_{12} & v_{11} \\
0 & h_{22}' & h_{23}' & 0 & b_{22} & b_{23} & 0 & f_{22} & 0 \\
0 & 0 & h_{33}' & 0 & 0 & b_{33} & 0 & 0 & 0 \\
\hline
0 & g_{12}' & g_{13}' & 0 & f_{12}' & f_{13}' & c_{11} & c_{12} & w_{11} \\
0 & 0 & g_{23}' & 0 & 0 & f_{23}' & 0 & c_{22} & 0 \\
\hline
0 & 0 & u_{13}' & 0 & 0 & v_{13}' & 0 & w_{12}' & d_{11}
\end{array}\right], \quad
\left[\begin{array}{ccc|ccc|cc|c}
a_{13} & a_{12} & a_{11} & h_{13} & h_{12} & h_{11} & g_{12} & g_{11} & u_{11} \\
a_{23} & a_{22} & 0 & h_{23} & h_{22} & 0 & g_{22} & 0 & 0 \\
a_{33} & 0 & 0 & h_{33} & 0 & 0 & 0 & 0 & 0 \\
\hline
h_{13}' & h_{12}' & h_{11}' & b_{13} & b_{12} & b_{11} & f_{12} & f_{11} & v_{11} \\
h_{23}' & h_{22}' & 0 & b_{23} & b_{22} & 0 & f_{22} & 0 & 0 \\
h_{33}' & 0 & 0 & b_{33} & 0 & 0 & 0 & 0 & 0 \\
\hline
g_{13}' & g_{12}' & 0 & f_{13}' & f_{12}' & 0 & c_{12} & c_{11} & w_{11} \\
g_{23}' & 0 & 0 & f_{23}' & 0 & 0 & c_{22} & 0 & 0 \\
\hline
u_{13}' & 0 & 0 & v_{13}' & 0 & 0 & w_{12}' & 0 & d_{11}
\end{array}\right],$$

of the class

$$M\begin{pmatrix} 3, & 3, & 2, & 1 \\ 3, & 3, & 2, & 1 \end{pmatrix},$$

the prime determinants of X are the following six:

$$D_1=\begin{vmatrix} a_{11} & h_{11} \\ h_{11}' & b_{11} \end{vmatrix}, \quad D_2=\begin{vmatrix} a_{22} & h_{22} \\ h_{22}' & b_{22} \end{vmatrix}, \quad D_3=\begin{vmatrix} a_{33} & h_{33} \\ h_{33}' & b_{33} \end{vmatrix}, \quad D_4=c_{11}, \quad D_5=c_{22}, \quad D_6=d_{11}.$$

We have
$$\det X = \pm D_1 . \det X_1,$$
where X_1 is the minor of X complementary to D_1; and X_1 is a quadrate slope of the class
$$M\begin{pmatrix}2, 2, 2, 1\\2, 2, 2, 1\end{pmatrix}$$
whose prime determinants are D_2, D_3, D_4, D_5, D_6. Expanding $\det X_1$ in the same way, we see by repetitions of the same argument that
$$\det X = \pm D_1 D_2 D_3 D_4 D_5 D_6.$$

5. *The greatest common canonical of two square matrices.*

When $A = [a]_m^m$ and $B = [b]_n^n$ are two square matrices of orders m and n with constant elements whose canonical reduced forms are the canonical square matrices $\mathbf{A}$ and $\mathbf{B}$, we will define a *greatest common canonical* of A and B to be a canonical square matrix $\mathbf{C}$ of the greatest possible order which is a diagonal minor of both the matrices $\mathbf{A}$ and $\mathbf{B}$, each of its simple parts being a diagonal minor both of a simple part of $\mathbf{A}$ and of a simple part of $\mathbf{B}$. All canonical square matrices which differ from $\mathbf{C}$ only in the orders of arrangement of their simple parts are greatest common canonicals of A and B, and when we speak of 'the' greatest common canonical of A and B, it will be a matter of indifference which of them is meant. Ordinarily we shall choose $\mathbf{C}$ to be a standard canonical square matrix.

Since the greatest common canonicals of A and B are all square matrices of the same order, the *order of the greatest common canonical* of A and B is a definite positive integer which is independent of the choice of $\mathbf{C}$. Let $A(\lambda) = [a]_m^m - \lambda [1]_m^m$, $B(\lambda) = [b]_n^n - \lambda [1]_n^n$ be the characteristic matrices of A, B; and let
$$f_1(\lambda), f_2(\lambda), \dots f_m(\lambda) \quad \text{and} \quad g_1(\lambda), g_2(\lambda), \dots g_n(\lambda),$$
be those rational integral functions of λ which are respectively the potent factors
$$\text{of orders } 1, 2, \dots m \text{ of } A(\lambda) \quad \text{and} \quad \text{of orders } 1, 2, \dots n \text{ of } B(\lambda).$$
Then from the following examples it will be seen that the order of the greatest common canonical of A and B is the sum of the degrees in λ of the highest common factors of the pairs
$$f_m(\lambda), g_n(\lambda);\ f_{m-1}(\lambda), g_{n-1}(\lambda);\ \dots f_r(\lambda), g_r(\lambda),$$
where r is the smaller of the two integers m and n. Again because the characteristic matrix of $-B$ is $-B(-\lambda)$, the order of the greatest common canonical of A and $-B$ is the sum of the degrees in λ of the highest common factors of the pairs
$$f_m(\lambda), g_n(-\lambda);\ f_{m-1}(\lambda), g_{n-1}(-\lambda);\ \dots f_r(\lambda), g_r(-\lambda).$$

It will be shown in § 244 that the rank of a general commutant $\{A, B\}$ is the order of the greatest common canonical of A and B.

Ex. xi. *Greatest common canonical of two unilatent square matrices.*

Let A and B be unilatent square matrices with constant elements whose latent roots are c and c', and whose characteristic potent divisors are

$$(\lambda-c)^{\alpha_1},\ (\lambda-c)^{\alpha_2},\ \ldots\ (\lambda-c)^{\alpha_r}\ \text{ and }\ (\lambda-c')^{\beta_1},\ (\lambda-c')^{\beta_2},\ \ldots\ (\lambda-c')^{\beta_s},$$

i.e. A and $-B$ (see Ex. x of § 222) be unilatent square matrices whose latent roots are c and $-c'$, and whose characteristic potent divisors are

$$(\lambda-c)^{\alpha_1},\ (\lambda-c)^{\alpha_2},\ \ldots\ (\lambda-c)^{\alpha_r}\ \text{ and }\ (\lambda+c')^{\beta_1},\ (\lambda+c')^{\beta_2},\ \ldots\ (\lambda+c')^{\beta_s},$$

where the α's and β's are *arranged in descending orders of magnitude.* Also let t be the smaller of the two integers r and s; let

$$\gamma_i \text{ be the smaller of the two integers } (\alpha_i,\ \beta_i)$$

for the values $1, 2, \ldots t$ of i; and let

$$\rho=\gamma_1+\gamma_2+\ \ldots\ +\gamma_t, \qquad \text{.....(5)}$$

so that ρ is an integer which could be defined as in (1).

If $c' \neq c$, the greatest common canonical of A and B is a non-existent square matrix of order 0; but if $c'=c$, i.e. if the characteristic potent divisors of A and B are

$$(\lambda-c)^{\alpha_1},\ (\lambda-c)^{\alpha_2},\ \ldots\ (\lambda-c)^{\alpha_r}\ \text{ and }\ (\lambda-c)^{\beta_1},\ (\lambda-c)^{\beta_2},\ \ldots\ (\lambda-c)^{\beta_s}, \qquad \text{.....(6)}$$

the greatest common canonical of A and B can be taken to be the unilatent canonical square matrix $\mathbf{C}$ whose characteristic potent divisors are

$$(\lambda-c)^{\gamma_1},\ (\lambda-c)^{\gamma_2},\ \ldots\ (\lambda-c)^{\gamma_t}.$$

Thus when $c'=c$, the order of the greatest common canonical of A and B is the integer ρ given by (5)*; whilst in other cases it is* 0.

Again if $c' \neq -c$, the greatest common canonical of A and $-B$ is a non-existent square matrix of order 0; but if $c'=-c$, i.e. if the characteristic potent divisors of A and B are

$$(\lambda-c)^{\alpha_1},\ (\lambda-c)^{\alpha_2},\ \ldots\ (\lambda-c)^{\alpha_r}\ \text{ and }\ (\lambda+c)^{\beta_1},\ (\lambda+c)^{\beta_2},\ \ldots\ (\lambda+c)^{\beta_s}, \qquad \text{.....(7)}$$

the greatest common canonical of A and $-B$ can be taken to be the square matrix $\mathbf{C}$ defined above.

Thus when $c'=-c$, the order of the greatest common canonical of A and $-B$ is the integer ρ given by (5)*; whilst in other cases it is* 0.

Ex. xii. *Order of the greatest common canonical of A and B.*

Let A and B be square matrices with constant elements whose distinct or unequal latent roots are

$$c_1,\ c_2,\ \ldots\ c_r,\ a_1,\ a_2,\ \ldots\ \text{ and }\ c_1,\ c_2,\ \ldots\ c_r,\ b_1,\ b_2,\ \ldots, \qquad \text{.....(8)}$$

where the a's, b's and c's are all different; also let $\mathbf{A}$ and $\mathbf{B}$ be standard compartite (or standard canonical) square matrices with unilatent square parts which are equicanonical with A and B respectively, the successive unilatent parts of $\mathbf{A}$ being

$$A_1,\ A_2,\ \ldots\ A_r,\ P_1,\ P_2,\ \ldots\ \text{ with latent roots }\ c_1,\ c_2,\ \ldots\ c_r,\ a_1,\ a_2,\ \ldots,$$

and the successive unilatent parts of $\mathbf{B}$ being

$$B_1,\ B_2,\ \ldots\ B_r,\ Q_1,\ Q_2,\ \ldots\ \text{ with latent roots }\ c_1,\ c_2,\ \ldots\ c_r,\ b_1,\ b_2,\ \ldots.$$

Then the greatest common canonical of A and B can be taken to be the square matrix

$$\mathbf{C}=\begin{bmatrix} C_1, & 0, & \ldots & 0 \\ 0, & C_2, & \ldots & 0 \\ \ldots & \ldots & \ldots & \ldots \\ 0, & 0, & \ldots & C_r \end{bmatrix}_{\rho_1,\ \rho_2,\ \ldots\ \rho_r}^{\rho_1,\ \rho_2,\ \ldots\ \rho_r}$$

in which C_i is the greatest common canonical of A_i and B_i, and $\rho_1, \rho_2, \ldots \rho_r$ are determined by formulae analogous to (5). The order of the greatest common canonical of A and B is the integer

$$\rho=\rho_1+\rho_2+ \ldots +\rho_r, \quad \ldots\ldots(5')$$

which is the sum of the orders of the greatest common canonicals of the unilatent pairs $(A_1, B_1), (A_2, B_2), \ldots (A_r, B_r)$ having the latent roots $(c_1, c_1), (c_2, c_2), \ldots (c_r, c_r)$.

Ex. xiii. *Order of the greatest common canonical of A and* $-B$.

Let A and B be square matrices with constant elements whose distinct or unequal latent roots are

$$c_1, c_2, \ldots c_r, a_1, a_2, \ldots \text{ and } -c_1, -c_2, \ldots -c_r, -b_1, -b_2, \ldots, \quad \ldots\ldots(9)$$

where the a's, b's and c's are all different; also let $\mathbf{A}$ and $\mathbf{B}$ be standard compartite (or standard canonical) square matrices with unilatent square parts, which are equicanonical with A and B respectively, the successive unilatent parts of $\mathbf{A}$ being

$$A_1, A_2, \ldots A_r, P_1, P_2, \ldots \text{ with latent roots } c_1, c_2, \ldots c_r, a_1, a_2, \ldots$$

and the successive unilatent parts of $\mathbf{B}$ being

$$B_1, B_2, \ldots B_r, Q_1, Q_2, \ldots \text{ with latent roots } -c_1, -c_2, \ldots -c_r, -b_1, -b_2, \ldots.$$

Then the greatest common canonical of A and B can be taken to be the square matrix $\mathbf{C}$ of Ex. xii in which C_i is now the greatest common canonical of A_i and $-B_i$, and $\rho_1, \rho_2, \ldots \rho_r$ are integers determined by formulae analogous to (5). The order of the greatest common canonical is the integer ρ given by (5′), which is now the sum of the orders of the greatest common canonicals of the unilatent pairs $(A_1, -B_1), (A_2, -B_2), \ldots (A_r, -B_r)$ having the latent roots $(c_1, c_1), (c_2, c_2), \ldots (c_r, c_r)$.

§ 242. The commutants of a pair of simple canonical square matrices and the commutants correlated with them.

1. *The general commutant of a pair of simple canonical square matrices.*

Let $A=[a]_m^m$ and $B=[b]_n^n$ be the simple canonical square matrices of orders m and n whose latent roots are c and c', so that

$$A=\begin{bmatrix} c & 1 & 0 & \ldots & 0 & 0 \\ 0 & c & 1 & \ldots & 0 & 0 \\ \ldots & \ldots & \ldots & \ldots & \ldots & \ldots \\ 0 & 0 & 0 & \ldots & c & 1 \\ 0 & 0 & 0 & \ldots & 0 & c \end{bmatrix}, \quad B=\begin{bmatrix} c' & 1 & 0 & \ldots & 0 & 0 \\ 0 & c' & 1 & \ldots & 0 & 0 \\ \ldots & \ldots & \ldots & \ldots & \ldots & \ldots \\ 0 & 0 & 0 & \ldots & c' & 1 \\ 0 & 0 & 0 & \ldots & 0 & c' \end{bmatrix}, \quad \ldots\ldots(1)$$

or

$$A=c\,[1]_m^m+\begin{bmatrix} 0, & 1 \\ 0, & 0 \end{bmatrix}_{m-1,\,1}^{1,\,m-1}, \quad B=c'\,[1]_n^n+\begin{bmatrix} 0, & 1 \\ 0, & 0 \end{bmatrix}_{n-1,\,1}^{1,\,n-1}. \quad \ldots\ldots(2)$$

We know from Ex. xviii of § 238 that the general commutant $\{A, B\}$ is a zero matrix when $c' \neq c$; and when $c'=c$, it is easily shown in many ways by the methods of § 238. 2 that the equation $AX=XB$ is satisfied when and only when X is a simple slope of type $\{\pi, \pi\}$ in which the elements of each parametric diagonal line are all equal.

Thus when $c'=c$, *the general commutant* $\{A, B\}$ *is a general simple continuant of type* $\{\pi, \pi\}$; *and when* $c' \neq c$, *it is a zero matrix.*

One direct proof of this result is given in Ex. i below, and another will be found amongst the examples of § 321.

Ex. i. *Direct determination of the general commutant* $X=\{A, B\}$.

If we use the ordinary notations for the elements of the matrix $X=[x]_m^n$ and substitute the forms (2) for A and B, we see that the equation $AX=XB$ can be replaced by the equivalent equation

$$\begin{bmatrix} 0 & x_{11} & x_{12} & \dots & x_{1,n-1} \\ 0 & x_{21} & x_{22} & \dots & x_{2,n-1} \\ \dots & \dots & \dots & \dots & \dots \\ 0 & x_{m-1,1} & x_{m-1,2} & \dots & x_{m-1,n-1} \\ 0 & x_{m1} & x_{m2} & \dots & x_{m,n-1} \end{bmatrix} - \begin{bmatrix} x_{21} & x_{22} & x_{23} & \dots & x_{2n} \\ x_{31} & x_{32} & x_{33} & \dots & x_{3n} \\ \dots & \dots & \dots & \dots & \dots \\ x_{m1} & x_{m2} & x_{m3} & \dots & x_{mn} \\ 0 & 0 & 0 & \dots & 0 \end{bmatrix} = (c-c').[x]_m^n \quad \dots(3)$$

First suppose that $c' \neq c$. Then when we equate in succession the 1st, 2nd, ... nth vertical rows or the mth, $(m-1)$th, ... 1st horizontal rows on both sides of (3), we see that the equation is satisfied when and only when every element vanishes in each of the corresponding rows of X, i.e. when and only when X is a zero matrix.

Next suppose that $c'=c$, so that the right-hand of (3) vanishes. In this case the equation (3) cannot be satisfied unless

$$x_{i+1,j+1}=x_{ij}, \quad (i=1, 2, \dots m-1;\ j=1, 2, \dots n-1), \quad \dots\dots\dots\dots(4)$$

i.e. unless the elements of each line of X parallel to the leading diagonal are all equal. We will suppose that the necessary conditions (4) are satisfied, and distinguish between two sub-cases.

(1) If $m \nless n$, we see by equating in succession the 1st, 2nd, ... nth vertical rows on both sides of (3) that the equation is satisfied if and only if, for each of the values 1, 2, ... n of j, all elements of the jth vertical row of X lying below the jth element of that row (or having difference-weights less than 0) are 0's.

(2) If $n \nless m$, we see by equating in succession the mth, $(m-1)$th, ... 1st horizontal rows on both sides of (3) that the equation is satisfied if and only if, for each of the values 1, 2, ... m of i, all elements of the $(m+1-i)$th horizontal row preceding the last i elements of that row (or having difference-weights less than $n-m$) are 0's.

We have here obtained the mn scalar equations equivalent to the matrix equation $AX=XB$ by equating successive elements of successive vertical or successive horizontal rows, starting always with the basical row and the basical element, where the base of each matrix is interpreted to be the bottom left-hand corner, i.e. the base appropriate to the type $\{\pi, \pi\}$.

Ex. ii. If $c'=c$ and $m=n=r$, the general commutant $X=\{A, B\}$ is the simple square continuant

$$X=[x]_r^r=\begin{bmatrix} x_0 & x_1 & x_2 & x_3 & \dots & x_{r-1} \\ 0 & x_0 & x_1 & x_2 & \dots & x_{r-2} \\ 0 & 0 & x_0 & x_1 & \dots & x_{r-3} \\ 0 & 0 & 0 & x_0 & \dots & x_{r-4} \\ \dots & \dots & \dots & \dots & \dots & \dots \\ 0 & 0 & 0 & 0 & \dots & x_0 \end{bmatrix} \quad \dots\dots\dots\dots(5)$$

of type $\{\pi, \pi\}$ in which $x_0, x_1, x_2, \dots x_{r-1}$ are r arbitrary parameters.

Ex. iii. If $c'=c$ and $m \neq n$, the general commutant $X=\{A, B\}$ is the matrix

$$X=[0, x]_m^{n-m, m} \text{ when } n>m, \quad X=\begin{bmatrix} x \\ 0 \end{bmatrix}_{n, m-n}^n \text{ when } m>n,$$

where the non-zero constituent $[x]_m^m$ or $[x]_n^n$ is a general simple square continuant having the form shown in (5), i.e. X is a general simple continuant of the class $M\binom{n}{m}$ and of type $\{\pi, \pi\}$.

Ex. iv. *If A is a simple canonical square matrix of order r, the general commutant $X=\{A, A\}$ can be regarded as an arbitrary rational integral function of A, or of any simple canonical square matrix of order r.*

If c is the latent root of A, and if $I=[1]_r^r$, the rational integral equation of lowest degree satisfied by A is

$$(A-cI)^r=0,$$

and there are therefore exactly r independent non-zero rational integral functions of A. Moreover there are also exactly r independent particular non-zero commutants $\{A, A\}$ which could be taken to be the coefficients of $x_0, x_1, \dots x_{r-1}$ in the matrix (5). The above theorem follows from the fact that the matrices $I, A, A^2, \dots A^{r-1}$ are a complete set of r independent non-zero rational integral functions, and are also necessarily commutants $\{A, A\}$. In fact from (5) we see that

$$\begin{aligned} X &= x_0 I + x_1(A-cI) + x_2(A-cI)^2 + \dots + x_{r-1}(A-cI)^{r-1} \\ &= \lambda_0 I + \lambda_1 A + \lambda_2 A^2 + \dots + \lambda_{r-1} A^{r-1}, \end{aligned}$$

where the λ's are arbitrary scalar parameters. The alternative form of the theorem is due to the fact that X is the same for all values of c.

2. *The correlated general commutants.*

Let A and B be again the simple canonical square matrices of orders m and n whose latent roots are c and c', and let A' and B' be their conjugates. Also let r be the smaller of the two integers m and n; and (using the notation of § 240, Note 1) let

$$J_m=[j]_m^m, \quad J_n=[j]_n^n,$$

be the simple reversants of orders m and n. It will be obvious that these simple reversants are undegenerate symmetric matrices, and that:

J_m is a contra-commutant of each of the types $\pm\{A', A\}$, $\pm\{A, A'\}$;

J_n is a contra-commutant of each of the types $\pm\{B', B\}$, $\pm\{B, B'\}$.

Therefore if X is a general commutant $\{A, B\}$, then by Ex. i of § 239, or as is otherwise obvious, the square matrices

$$X'=J_m X J_n, \quad Y=J_m X, \quad Y'=XJ_n, \quad \dots\dots\dots\dots\dots\dots(6)$$

are respectively general commutants $\{A', B'\}$, $\{A', B\}$, $\{A, B'\}$. Thus the results of sub-article 1 can be generalised as in the following theorem, in which the symbolic commutants of § 241 are used.

Theorem I a. *The four general continuantal commutants*

$$X=\{A, B\}, \quad X'=\{A', B'\}, \quad Y=\{A', B\}, \quad Y'=\{A, B'\},$$

are all zero matrices when $c' \neq c$. *When* $c' = c$ *(where* c *may be* 0), *they are general simple continuants of the class* $M\binom{n}{m}$ *and of the respective types*

$$\{\pi, \pi\}, \quad \{\pi', \pi'\}, \quad \{\pi', \pi\}, \quad \{\pi, \pi'\},$$

each of them being undegenerate of rank r, *and each of them containing exactly* r *independent arbitrary parameters and having exactly* r *independent non-zero particularisations.*

By changing the signs of corresponding rows in the equations such as $AX = XB$ we can deduce the following second theorem, a formal proof of which will be given in Ex. vii.

Theorem I b. *The four general alternating commutants*

$$\overline{X} = \{A, -B\}, \quad \overline{X}' = \{A', -B'\}, \quad \overline{Y} = \{A', -B\}, \quad \overline{Y}' = \{A, -B'\}$$

are all zero matrices when $c' \neq -c$. *When* $c' = -c$ *(where* c *may be* 0), *they are general simple alternants of the class* $M\binom{n}{m}$ *and of the respective types*

$$\{\pi, -\pi\}, \quad \{\pi', -\pi'\}, \quad \{\pi', -\pi\}, \quad \{\pi, -\pi'\},$$

each of them being undegenerate of rank r, *and each of them containing exactly* r *independent arbitrary parameters and having exactly* r *independent non-zero particularisations.*

Ex. v. *The general commutants when* $m = n = r$.

If $x_0, x_1, \ldots x_{r-1}$ are arbitrary parameters, the four general commutants of Theorem I *a* can be taken to be the four simple square continuants

$$\begin{bmatrix} x_0 & x_1 & \ldots & x_{r-1} \\ 0 & x_0 & \ldots & x_{r-2} \\ \ldots & \ldots & \ldots & \ldots \\ 0 & 0 & \ldots & x_0 \end{bmatrix}, \quad \begin{bmatrix} x_0 & \ldots & 0 & 0 \\ \ldots & \ldots & \ldots & \ldots \\ x_{r-2} & \ldots & x_0 & 0 \\ x_{r-1} & \ldots & x_1 & x_0 \end{bmatrix}, \quad \begin{bmatrix} 0 & 0 & \ldots & x_0 \\ \ldots & \ldots & \ldots & \ldots \\ 0 & x_0 & \ldots & x_{r-2} \\ x_0 & x_1 & \ldots & x_{r-1} \end{bmatrix}, \quad \begin{bmatrix} x_{r-1} & \ldots & x_1 & x_0 \\ x_{r-2} & \ldots & x_0 & 0 \\ \ldots & \ldots & \ldots & \ldots \\ x_0 & \ldots & 0 & 0 \end{bmatrix},$$

which will be denoted by $\qquad [x]_r^r, \; [x']_r^r, \; [y]_r^r, \; [y']_r^r;$

and the four general commutants of Theorem I *b* to be the four simple square alternants

$$\begin{bmatrix} x_0, & -x_1, & \ldots & \pm x_{r-1} \\ 0, & -x_0, & \ldots & \pm x_{r-2} \\ \ldots & \ldots & \ldots & \ldots \\ 0, & 0, & \ldots & \pm x_0 \end{bmatrix}, \quad \begin{bmatrix} \pm x_0, & \ldots & 0, & 0 \\ \ldots & \ldots & \ldots & \ldots \\ \pm x_{r-2}, & \ldots & -x_0, & 0 \\ \pm x_{r-1}, & \ldots & -x_1, & x_0 \end{bmatrix}, \quad \begin{bmatrix} 0, & 0, & \ldots & \pm x_0 \\ \ldots & \ldots & \ldots & \ldots \\ 0, & -x_0, & \ldots & \pm x_{r-2} \\ x_0, & -x_1, & \ldots & \pm x_{r-1} \end{bmatrix}, \quad \begin{bmatrix} \pm x_{r-1}, & \ldots & -x_1, & x_0 \\ \pm x_{r-2}, & \ldots & -x_0, & 0 \\ \ldots & \ldots & \ldots & \ldots \\ \pm x_0, & \ldots & 0, & 0 \end{bmatrix},$$

(all elements in alternate vertical rows having negative signs), which will be denoted by

$$[\bar{x}]_r^r, \; [\bar{x}']_r^r, \; [\bar{y}]_r^r, \; [\bar{y}']_r^r.$$

In each of these matrices we will call x_κ the *parameter of weight* κ.

Ex. vi. *The general commutants for all values of m and n.*

When $m=r \ngtr n$, we can put

$$X=[0, x]_m^{n-m,\, m}, \quad X'=[x', 0]_m^{m,\, n-m}, \quad Y=[0, y]_m^{n-m,\, m}, \quad Y'=[y', 0]_m^{m,\, n-m},$$

$$\bar{X}=[0, \bar{x}]_m^{n-m,\, m}, \quad \bar{X}'=[\bar{x}', 0]_m^{m,\, n-m}, \quad \bar{Y}=[0, \bar{y}]_m^{n-m,\, m}, \quad \bar{Y}'=[\bar{y}', 0]_m^{m,\, n-m},$$

these matrices being formed from the square matrices of Ex. v by the addition of $n-m$ basical vertical rows of 0's.

When $n=r \ngtr m$, we can put

$$X=\begin{bmatrix} x \\ 0 \end{bmatrix}^n_{n,\, m-n}, \quad X'=\begin{bmatrix} 0 \\ x' \end{bmatrix}^n_{m-n,\, n}, \quad Y=\begin{bmatrix} 0 \\ y \end{bmatrix}^n_{m-n,\, n}, \quad Y'=\begin{bmatrix} y' \\ 0 \end{bmatrix}^n_{n,\, m-n},$$

$$\bar{X}=\begin{bmatrix} \bar{x} \\ 0 \end{bmatrix}^n_{n,\, m-n}, \quad \bar{X}'=\begin{bmatrix} 0 \\ \bar{x}' \end{bmatrix}^n_{m-n,\, n}, \quad \bar{Y}=\begin{bmatrix} 0 \\ \bar{y} \end{bmatrix}^n_{m-n,\, n}, \quad \bar{Y}'=\begin{bmatrix} \bar{y}' \\ 0 \end{bmatrix}^n_{n,\, m-n},$$

these matrices being formed from the square matrices of Ex. v by the addition of $m-n$ basical horizontal rows of 0's.

Each of these eight matrices is a ruled simple slope containing r arbitrary parameters. There are no necessary connections between the parameters of different matrices. When the parameters are chosen to be the same in all, the eight matrices are correlated by

simple reversals of the horizontal or vertical rows,

changes of sign in alternate vertical rows.

Ex. vii. *Proof of Theorem I b.*

Let B_1 be the simple canonical square matrix of order n whose latent root is $-c'$; let B_1' be the conjugate of B_1; let X_1, X_1', Y_1, Y_1' be the general commutants

$$X_1=\{A, B_1\}, \quad X_1'=\{A', B_1'\}, \quad Y_1=\{A', B_1\}, \quad Y_1'=\{A, B_1'\}; \quad \text{.........(7)}$$

and let I' be a unit simple square alternant of order n in which all non-diagonal elements are equal to 0 and the successive diagonal elements are alternately 1 and -1. Then from the isomorphic transformation

$$-B=I'B_1I',$$

it follows by Theorem I of § 240 that we can always choose general commutants $\bar{X}, \bar{X}', \bar{Y}, \bar{Y}'$ which are given by the equigradent commutantal transformations

$$\bar{X}=X_1I', \quad \bar{X}'=X_1'I', \quad \bar{Y}=Y_1I', \quad \bar{Y}'=Y_1'I'. \quad \text{..................(8)}$$

After Theorem I a has been proved, we know that X_1, X_1', Y_1, Y_1' are zero matrices when $-c' \neq c$, or general simple continuants of the types shown in Theorem I a when $-c'=c$; and this shows that Theorem I b is true.

It would be sufficient to prove Theorem I b for $\bar{X}$, because by Ex. i of § 239 we can always choose the four general commutants of Theorem I b to be connected by the relations

$$\bar{X}' = J_m\bar{X}J_n, \quad \bar{Y} = J_m\bar{X}, \quad \bar{Y}' = \bar{X}J_n. \quad \text{........................(6')}$$

Ex. viii. The general commutants $\{A, A\}, \{A', A'\}, \{A', A\}, \{A, A'\}$ are general simple square continuants of the types shown in Theorem I a and Ex. v.

Ex. ix. The general commutants $\{A, -A\}, \{A', -A'\}, \{A', -A\}, \{A, -A'\}$ are general simple square alternants of the types shown in Theorem I b and Ex. v.

Ex. x. *If A is a simple canonical square matrix, then all commutants of A are commutative with one another; and if A' is the conjugate of A, then all commutants of A' are commutative with one another.*

These properties follow from Ex. iv; and they are obvious from the forms of the general commutants $X=\{A, A\}$, $X'=\{A', A'\}$, being in fact properties of simple square continuants of the types $\{\pi, \pi\}$, $\{\pi', \pi'\}$.

NOTE 1. *Applications to unipotent square matrices.*

Since the canonical reduced form of a unipotent square matrix with constant elements is a simple canonical square matrix, it follows from Theorem I a that:

If A and B are two unipotent square matrices of orders m and n with constant elements whose latent roots are c and c', the general commutant $X=\{A, B\}$ is a zero matrix when $c' \neq c$; but if $c'=c$, it is an undegenerate matrix of rank r whose elements are homogeneous linear functions of r independent arbitrary parameters, where r is the smaller of the two integers m and n.

In fact if $\mathbf{A}$ and $\mathbf{B}$ are the simple canonical square matrices of orders m and n whose latent roots are c and c', we can put

$$A=h\mathbf{A}H, \quad B=k\mathbf{B}K, \quad X=h\mathbf{X}K$$

where the first two equations are isomorphic transformations, and $\mathbf{X}$ is the general commutant $\{\mathbf{A}, \mathbf{B}\}$ determined by Theorem I a.

Since the general commutant $\{A, A\}$ contains exactly m independent arbitrary parameters, or has exactly m independent non-zero particularisations, we conclude (or could deduce from Ex. x) that:

If A is a unipotent square matrix with constant elements, the general commutant $X=\{A, A\}$ of A is an arbitrary rational integral function of A; and all commutants of A are commutative with one another.

3. *General symmetric and general skew-symmetric commutants.*

Let A and B be the simple canonical square matrices of order r whose latent roots are c and c', and let A' and B' be their conjugates, so that all the general commutants of Theorems I a and I b are ruled simple square slopes of order r. Then by introducing the conditions that these simple slopes may be symmetric or skew-symmetric we obtain the following theorems:

Theorem II a. *The two general symmetric or skew-symmetric commutants*

$$X=\{A, B\}, \quad X'=\{A', B'\},$$

are zero matrices when $c' \neq c$. If however $c'=c$ (where c may be 0), i.e. if $B=A$, then:

(1) *the general symmetric commutants X, X' are undegenerate scalar ante-continuants of rank and order r each containing one arbitrary parameter;*

(2) *the general skew-symmetric commutants X, X' are zero matrices.*

Theorem III a. *The two general symmetric or skew-symmetric commutants*

$$Y=\{A', B\}, \quad Y'=\{A, B'\},$$

are zero matrices when $c' \neq c$. If however $c'=c$ (where c may be 0), i.e. if $B=A$, then:

(1) *the general symmetric commutants Y, Y' are general simple square ante-continuants of types $\{\pi', \pi\}$, $\{\pi, \pi'\}$, each of them being unde-*

generate of rank and order r and containing exactly r arbitrary parameters;

(2) *the general skew-symmetric commutants Y, Y' are zero matrices.*

Theorem II b. *The two general symmetric or skew-symmetric commutants*

$$\bar{X}=\{A,-B\},\quad \bar{X}'=\{A',-B'\}$$

are zero matrices when $c' \neq -c$. If however $c'=-c$ (where c may be 0, so that $B=A$), then:

(1) *the general symmetric commutants $\bar{X}$, $\bar{X}'$ are undegenerate quasi-scalar ante-alternants of rank and order r, each containing one arbitrary parameter;*

(2) *the general skew-symmetric commutants $\bar{X}$, $\bar{X}'$ are zero matrices.*

Theorem III b. *The two general symmetric or skew-symmetric commutants*

$$\bar{Y}=\{A',-B\},\quad \bar{Y}'=\{A,-B'\},$$

are zero matrices when $c' \neq -c$. If however $c'=-c$ (where c may be 0, so that $B=A$), we can distinguish between two cases.

CASE I. *If r is an odd integer and $c'=-c$, then:*

(1) *the general symmetric commutants $\bar{Y}$, $\bar{Y}'$ are undegenerate simple square counter-alternants of rank and order r and of types $\{\pi',-\pi\}$, $\{\pi,-\pi'\}$ in which*

the parameters of odd weights are all equal to 0,
the parameters of even weights are all arbitrary,

the total number of arbitrary parameters in each of them being $\frac{1}{2}(r+1)$;

(2) *the general skew-symmetric commutants $\bar{Y}$, $\bar{Y}'$ are degenerate simple square counter-alternants of order r and rank $r-1$ and of types $\{\pi',-\pi\}$, $\{\pi,-\pi'\}$ in which*

the parameters of even weights are all equal to 0,
the parameters of odd weights are all arbitrary,

the total number of arbitrary parameters in each of them being $\frac{1}{2}(r-1)$.

CASE II. *If r is an even integer and $c'=-c$, then:*

(1') *the general symmetric commutants $\bar{Y}$, $\bar{Y}'$ are degenerate simple square counter-alternants of order r and rank $r-1$ and of types $\{\pi',-\pi\}$, $\{\pi,-\pi'\}$ in which*

the parameters of even weights are all equal to 0,
the parameters of odd weights are all arbitrary,

the total number of arbitrary parameters in each of them being $\frac{1}{2}r$;

(2′) *the general skew-symmetric commutants* $\bar{Y}, \bar{Y}'$ *are undegenerate simple square counter-alternants of rank and order* r *and of types* $\{\pi', -\pi\}$, $\{\pi, -\pi'\}$ *in which*

the parameters of odd weights are all equal to 0,
the parameters of even weights are all arbitrary,

the total number of arbitrary parameters in each of them being $\frac{1}{2}r$.

If J is the simple reversant of order r, the two general symmetric or two general skew-symmetric commutants X, X' or Y, Y' or $\bar{X}, \bar{X}'$ or $\bar{Y}, \bar{Y}'$ can always be so chosen as to be connected by the relations

$$X' = JXJ, \quad Y' = JYJ, \quad \bar{X}' = J\bar{X}J, \quad \bar{Y}' = J\bar{Y}J. \quad \text{.........(9)}$$

NOTE 2. *Applications to a single unipotent square matrix.*

Let A be a single unipotent square matrix of order r with constant elements whose latent root is c; let $\mathbf{A}$ be the simple canonical square matrix of order r whose latent root is c; let

$$A = h\mathbf{A}H$$

be an isomorphic transformation by which A can be derived from $\mathbf{A}$; and let A', $\mathbf{A}'$ be the conjugates of A, $\mathbf{A}$. Then the two general symmetric (or general skew-symmetric) contra-commutants

$$Y = \{A', A\}, \ Y' = \{A, A'\} \quad \text{or} \quad \bar{Y} = \{A', -A\}, \ \bar{Y}' = \{A, -A'\}$$

can be taken to be the square matrices of order r given by the equations

$$Y = H'\mathbf{Y}H, \ Y' = h\mathbf{Y}'h' \quad \text{or} \quad \bar{Y} = H'\bar{\mathbf{Y}}H, \ \bar{Y}' = h\bar{\mathbf{Y}}'h',$$

where $\mathbf{Y}, \mathbf{Y}'$ or $\bar{\mathbf{Y}}, \bar{\mathbf{Y}}'$ are the two symmetric (or two skew-symmetric) commutants determined by Theorem III *a* or III *b*; and we conclude that:

The two general symmetric contra-commutants Y, Y' *are always undegenerate of rank* r, *and each of them contains exactly* r *arbitrary parameters.*

The two general skew-symmetric contra-commutants Y, Y' *are always zero matrices.*

The two general symmetric contra-commutants $\bar{Y}, \bar{Y}'$ *are zero matrices whenever* $c \neq 0$. *If* $c = 0$, *each of them is undegenerate of rank* r *and contains* $\frac{1}{2}(r+1)$ *arbitrary parameters when* r *is odd; and each of them is degenerate of rank* $r-1$ *and contains* $\frac{1}{2}r$ *arbitrary parameters when* r *is even.*

The two general skew-symmetric contra-commutants $\bar{Y}, \bar{Y}'$ *are zero matrices whenever* $c \neq 0$. *If* $c = 0$, *each of them is degenerate of rank* $r-1$ *and contains* $\frac{1}{2}(r-1)$ *arbitrary parameters when* r *is odd; and each of them is undegenerate of rank* r *and contains* $\frac{1}{2}r$ *arbitrary parameters when* r *is even.*

§ 243. The commutants of a pair of unilatent canonical square matrices and the commutants correlated with them.

1. *General commutants.*

Let $A = [a]_m^m$ and $B = [b]_n^n$ be two unilatent canonical square matrices of orders m and n with constant elements whose latent roots are c and c', and whose characteristic potent divisors are

$$(\lambda - c)^{a_1}, \ (\lambda - c)^{a_2}, \ \dots \ (\lambda - c)^{a_r} \ \text{ and } \ (\lambda - c')^{\beta_1}, \ (\lambda - c')^{\beta_2}, \ \dots \ (\lambda - c')^{\beta_s}, \ \text{...(1)}$$

where the indices are non-zero positive integers such that

$$\alpha_1 + \alpha_2 + \dots + \alpha_r = m, \quad \beta_1 + \beta_2 + \dots + \beta_s = n;$$

and let A and B be given in the forms

$$A = \begin{bmatrix} A_1, & 0, & \dots & 0 \\ 0, & A_2, & \dots & 0 \\ \dots & \dots & \dots & \dots \\ 0, & 0, & \dots & A_r \end{bmatrix}_{\alpha_1, \alpha_2, \dots \alpha_r}^{\alpha_1, \alpha_2, \dots \alpha_r}, \quad B = \begin{bmatrix} B_1, & 0, & \dots & 0 \\ 0, & B_2, & \dots & 0 \\ \dots & \dots & \dots & \dots \\ 0, & 0, & \dots & B_s \end{bmatrix}_{\beta_1, \beta_2, \dots \beta_s}^{\beta_1, \beta_2, \dots \beta_s},$$

where A_i, B_j are the simple canonical square matrices of orders α_i, β_j whose latent roots are c, c', i.e. whose single characteristic potent divisors are $(\lambda - c)^{\alpha_i}$, $(\lambda - c')^{\beta_j}$. We will call A and B *standardised* unilatent canonical square matrices when the index numbers $\alpha_1, \alpha_2, \dots \alpha_r$ and $\beta_1, \beta_2, \dots \beta_s$ are arranged in descending orders of magnitude.

Further let J_a and J_b be the part-reversants (see Note 1 and Ex. vii of § 240) of the same classes as the compound matrices A and B, these being undegenerate symmetric contra-commutants of the respective types

$$\pm \{A', A\}, \ \pm \{A, A'\} \ \text{ and } \ \pm \{B', B\}, \ \pm \{B, B'\},$$

where A' and B' are the conjugates of A and B; and let ρ and σ be integers, independent of the orders of arrangement of the α's and β's, such that:

ρ is the sum of the smaller integers of the pairs (α_1, β_1), (α_2, β_2), (α_3, β_3), ... when $\alpha_1, \alpha_2, \dots \alpha_r$ and $\beta_1, \beta_2, \dots \beta_s$ are arranged in descending orders of magnitude and α_i is interpreted to be 0 if $i > r$, β_i to be 0 if $i > s$;

σ is always the sum of the smaller integers of the rs pairs (α_i, β_j).

When the general commutants in question are expressed as compound matrices of the class

$$M \begin{pmatrix} \beta_1, & \beta_2, & \dots & \beta_s \\ \alpha_1, & \alpha_2, & \dots & \alpha_r \end{pmatrix}, \quad \dots\dots\dots\dots\dots\dots\text{(a)}$$

we have the following two theorems, the first parts of which are known to be true by Ex. xviii of § 238 or Theorem III of § 240. The relations (2) and (2′) follow from Ex. i of § 239 (or Theorem I of § 240).

Theorem I a. *The four general continuantal commutants*

$$X = \{A, B\}, \quad X' = \{A', B'\}, \quad Y = \{A', B\}, \quad Y' = \{A, B'\},$$

which can be so chosen as to be connected by the equigradent commutantal relations

$$X' = J_a X J_b, \quad Y = J_a X, \quad Y' = X J_b, \quad \dots\dots\dots\dots\dots\text{(2)}$$

are all zero matrices when $c' \neq c$. If however $c' = c$ (where c may be 0), they are general compound continuants of the class (a) *and of the respective types*

$$\{\pi, \pi\}, \quad \{\pi', \pi'\}, \quad \{\pi', \pi\}, \quad \{\pi, \pi'\},$$

each of them having rank ρ, containing exactly σ arbitrary parameters, and having exactly σ independent non-zero particularisations.

Thus in all cases the rank of each of them is equal to the order of the greatest common canonical of A and B.

Theorem I b. *The four general alternating commutants*

$$\bar{X}=\{A, -B\},\quad \bar{X}'=\{A', -B'\},\quad \bar{Y}=\{A', -B\},\quad \bar{Y}'=\{A, -B'\},$$

which can be so chosen as to be connected by the equigradent commutantal relations

$$\bar{X}'=J_a\bar{X}J_b,\quad \bar{Y}=J_a\bar{X},\quad \bar{Y}'=\bar{X}J_b, \qquad\qquad (2')$$

are all zero matrices when $c' \neq -c$. *If however* $c'=-c$ *(where c may be* 0*), they are general compound alternants of the class* (a) *and of the respective types*

$$\{\pi, -\pi\},\quad \{\pi', -\pi'\},\quad \{\pi', -\pi\},\quad \{\pi, -\pi'\},$$

each of them having rank ρ*, containing exactly* σ *arbitrary parameters, and having exactly* σ *independent non-zero particularisations.*

Thus in all cases the rank of each of them is equal to the order of the greatest common canonical of A and $-B$.

Reference may be made to § 241 for the definitions of compound continuants and compound alternants. There are various considerations which could serve to simplify the proofs of the theorems.

In the first place if the theorems are true when A and B are standardised unilatent canonical square matrices, they must be true in all cases. For A and B can always be derived from standardised unilatent canonical square matrices $\mathbf{A}$ and $\mathbf{B}$ by symmetric class-derangements, which are equivalent to re-arrangements of the simple parts, and by Ex. iii of § 240 the general commutants of Theorems I *a* and I *b* for $\mathbf{A}$ and $\mathbf{B}$ can be converted into general commutants for A and B by corresponding class-derangements, the same for each of the eight commutants, in which the simple constituents receive rigid displacements.

In the second place if Theorem I *a* is true, it follows that Theorem I *b* must also be true. For if I' is a unit compound alternant of the same class as B (in which the diagonal constituents are unit simple alternants and the non-diagonal constituents are zero matrices), and if B_1 is the matrix derived from B by substituting $-c'$ for c', we have

$$-B=I'B_1I',$$

and it follows from § 239 that we can put

$$\bar{X}=X_1I',\quad \bar{X}'=X_1'I',\quad \bar{Y}=Y_1I',\quad \bar{Y}'=Y_1'I',$$

where X_1, X_1', Y_1, Y_1' are general commutants $\{A, B_1\}$, $\{A', B_1'\}$, $\{A', B_1\}$, $\{A, B_1'\}$, the argument being as in Ex. vii of § 242.

In the third place if the theorems are true for X and $\bar{X}$, we see from (2) and (2′) that they must be true for all the commutants. It would therefore be sufficient to prove Theorem I *a* for $\bar{X}$.

We give below a proof which establishes both theorems; the ranks of the general commutants being given by § 241.3 when they are non-zero matrices.

Proof of Theorems I a and I b.

By Theorem II of § 240 the two general commutants $X=\{A, B\}$, $\bar{X}=\{A, -B\}$ are the most general compound matrices of the forms

$$X=\begin{bmatrix} X_{11}, & X_{12}, & \dots & X_{1s} \\ X_{21}, & X_{22}, & \dots & X_{2s} \\ \dots & \dots & \dots & \dots \\ X_{r1}, & X_{r2}, & \dots & X_{rs} \end{bmatrix}_{\alpha_1, \alpha_2, \dots \alpha_r}^{\beta_1, \beta_2, \dots \beta_s}, \quad \bar{X}=\begin{bmatrix} \bar{X}_{11}, & \bar{X}_{12}, & \dots & \bar{X}_{1s} \\ \bar{X}_{21}, & \bar{X}_{22}, & \dots & \bar{X}_{2s} \\ \dots & \dots & \dots & \dots \\ \bar{X}_{r1}, & \bar{X}_{r2}, & \dots & \bar{X}_{rs} \end{bmatrix}_{\alpha_1, \alpha_2, \dots \alpha_r}^{\beta_1, \beta_2, \dots \beta_s}, \quad \dots\dots(3)$$

which can be constructed when the constituent X_{ij} of X is always a commutant $\{A_i, B_j\}$ and the constituent $\bar{X}_{ij}$ of $\bar{X}$ is always a commutant $\{A_i, -B_j\}$.

If $c' \neq c$, it follows from § 242 that all the constituents of X are zero matrices; whilst if $c'=c$, the constituent X_{ij} is a general simple continuant of type $\{\pi, \pi\}$ containing γ_{ij} arbitrary parameters, where γ_{ij} is the smaller of the two integers (α_i, β_j). In the latter case the parameters of the rs constituents of X must be all independent; consequently X contains exactly σ arbitrary parameters; moreover it has exactly σ independent non-zero particularisations, which could be chosen to be the coefficients of those σ parameters.

Again if $c' \neq -c$, it follows from § 242 that all the constituents of $\bar{X}$ are zero matrices; whilst if $c'=-c$, the constituent $\bar{X}_{ij}$ is a general simple alternant of type $\{\pi, -\pi\}$ containing γ_{ij} arbitrary parameters. In the latter case the parameters of the rs constituents of $\bar{X}$ must be all independent; consequently $\bar{X}$ contains exactly σ arbitrary parameters, the coefficients of which form a complete set of independent non-zero particularisations of $\bar{X}$.

The corresponding results for the other general commutants can be deduced from these or obtained directly in similar ways.

It should be observed that the integer σ is the sum of the effective (or smaller) orders of all the rs constituents of any compound matrix of the class (a).

Ex. i. When the successive characteristic potent divisors of A and B are

$$(\lambda-c)^{12}, (\lambda-c)^8, (\lambda-c)^7, (\lambda-c)^4, (\lambda-c)^3 \text{ and } (\lambda-c)^{12}, (\lambda-c)^{11}, (\lambda-c)^6, (\lambda-c)^5, (\lambda-c)^2,$$

the general commutants X, X', Y, Y' are the general compound continuants represented by Figs. 1, 2, 3, 4 in Ex. ii of § 241; and when they are

$$(\lambda-c)^{12}, (\lambda-c)^8, (\lambda-c)^7, (\lambda-c)^4, (\lambda-c)^3 \text{ and } (\lambda+c)^{12}, (\lambda+c)^{11}, (\lambda+c)^6, (\lambda+c)^5, (\lambda+c)^2,$$

the general commutants $\bar{X}, \bar{X}', \bar{Y}, \bar{Y}'$ are the general compound alternants represented by the same figures. In both these cases we have

$$\rho=12+8+6+4+2, \quad \sigma=124,$$

σ being the total number of parametric diagonal lines in each of the figures.

Ex. ii. *Undegenerate commutants.*

The general commutants X, X', Y, Y' are (1) undegenerate of rank m, (2) undegenerate of rank n, (3) undegenerate square matrices when and only when:

(1) the characteristic potent divisors of A of orders $m, m-1, m-2, \dots$ are factors of the characteristic potent divisors of B of orders $n, n-1, n-2, \dots$, which is only possible when $r \ngtr s$, $\rho=m \ngtr n$;

(2) the characteristic potent divisors of B of orders $n, n-1, n-2, \dots$ are factors of the characteristic potent divisors of A of orders $m, m-1, m-2, \dots$, which is only possible when $s \ngtr r$, $\rho=n \ngtr m$;

(3) A and B have the same characteristic potent divisors, which is only possible when $s=r$, $n=m$.

The corresponding results for the general commutants $\bar{X}$, $\bar{X}'$, $\bar{Y}$, $\bar{Y}'$ are obtained by substituting $-B$ for B.

Ex. iii. *Quadrate commutants.*

If A and B have the same successive index numbers $e_1, e_2, \dots e_s$, the general commutants X, X', Y, Y' when $c'=c$ (so that $B=A$), or the general commutants $\bar{X}$, $\bar{X}'$, $\bar{Y}$, $\bar{Y}'$ when $c'=-c$, are ruled quadrate slopes of the class

$$M\begin{pmatrix} e_1, & e_2, & \dots & e_s \\ e_1, & e_2, & \dots & e_s \end{pmatrix}, \text{ where } e_1+e_2+\dots+e_s=m, \quad \dots\dots(b)$$

and are undegenerate square matrices. If $e_1, e_2, \dots e_s$ are arranged in descending order of magnitude, the total number of arbitrary parameters in each of them is the integer

$$\sigma=e_1+3e_2+5e_3+\dots+(2s-1)\,e_s.$$

We can put
$$\sigma=\sigma_0+2\sigma_1,$$
where
$$\sigma_0=e_1+e_2+\dots+e_s=m, \quad \sigma_1=e_2+2e_3+3e_4+\dots+(s-1)\,e_s,$$
σ_0 being the total number of parameters in the diagonal constituents, and σ_1 being the total number of parameters in the non-diagonal constituents lying on one side of the diagonal constituents.

We have $\sigma=m$ when and only when $s=1$, i.e. when and only when A and B are simple canonical square matrices.

Ex. iv. *Interpretations of compound continuants.*

Theorem I*a* shows that the general compound continuants of the class (a) and of the respective symbolic types

$$\pm\{\pi, \pi\}, \ \pm\{\pi', \pi'\}, \ \pm\{\pi', \pi\}, \ \pm\{\pi, \pi'\} \quad \dots\dots(4)$$

can always be regarded as general commutants of the corresponding types

$$\pm\{A, B\}, \ \pm\{A', B'\}, \ \pm\{A', B\}, \ \pm\{A, B'\}, \quad \dots\dots(5)$$

where A and B are unilatent canonical square matrices whose successive characteristic potent divisors are

$$(\lambda-c)^{\alpha_1}, \ (\lambda-c)^{\alpha_2}, \dots (\lambda-c)^{\alpha_r} \text{ and } (\lambda-c)^{\beta_1}, \ (\lambda-c)^{\beta_2}, \dots (\lambda-c)^{\beta_s}, \quad \dots\dots(6)$$

c being any scalar number, which may be 0. When (a) is the quadrate class (b), we have $B=A$.

We can use these interpretations to pass from general properties of commutants such as those given in § 238. 4 to general properties of compound continuants.

Ex. v. *Interpretations of compound alternants.*

Theorem I*b* shows that the general compound alternants of the class (a) and of the respective symbolic types

$$\pm\{\pi, -\pi\}, \ \pm\{\pi', -\pi'\}, \ \pm\{\pi', -\pi\}, \ \pm\{\pi, -\pi'\}$$

can always be regarded as general commutants of the corresponding types

$$\pm\{A, -B\}, \ \pm\{A', -B'\}, \ \pm\{A', -B\}, \ \pm\{A, -B'\},$$

where A and B are unilatent canonical square matrices whose successive characteristic potent divisors are

$$(\lambda-c)^{\alpha_1}, \ (\lambda-c)^{\alpha_2}, \dots (\lambda-c)^{\alpha_r} \text{ and } (\lambda+c)^{\beta_1}, \ (\lambda+c)^{\beta_2}, \dots (\lambda+c)^{\beta_s},$$

c being any scalar number, which may be 0. The square matrix $-B$ is not canonical. When (a) is the quadrate class (b), we have $B=A$ if and only if $c=0$.

We can use these interpretations to pass from general properties of commutants such as those given in § 238. 4 to general properties of compound alternants.

Ex. vi. *The general commutant* $\{A, A\}$ *of a single unilatent canonical square matrix A is a rational integral function of A when and only when A is a simple canonical square matrix.*

Let A be the unilatent canonical square matrix of order m described in Ex. iii, and let e be the greatest of the index numbers $e_1, e_2, \ldots e_s$, so that the rational integral equation of lowest degree satisfied by A is

$$(A-cI)^e=0, \quad \text{where} \quad I=[1]_m^m.$$

The case in which A is a *simple* canonical square matrix has been considered in Ex. iv of § 242. If A is not a simple canonical square matrix, i.e. if $s>1$, we have $\sigma>m$ and $e<m$; thus in this case there are more than m independent non-zero commutants $\{A, A\}$, but only the smaller number e of independent non-zero rational integral functions of A; consequently there must be commutants of A which are not rational integral functions of A.

NOTE 1. *Applications to any two unilatent square matrices A and B with constant elements.*

Let A and B be any two unilatent square matrices of orders m and n with constant elements whose distinct latent roots are c and c' and whose characteristic potent divisors are

$$(\lambda-c)^{\alpha_1}, (\lambda-c)^{\alpha_2}, \ldots (\lambda-c)^{\alpha_r} \text{ and } (\lambda-c')^{\beta_1}, (\lambda-c')^{\beta_2}, \ldots (\lambda-c')^{\beta_s}. \quad \ldots\ldots\ldots(1)$$

Then Theorems Ia and Ib remain true with the new definitions of A and B so far as they relate to the rank and the total number of arbitrary parameters (or independent non-zero particularisations) of each of the general commutants

$$X=\{A, \; B\}, \quad X'=\{A', \; B'\}, \quad Y=\{A', \; B\}, \quad Y'=\{A, \; B'\},$$
$$\bar{X}=\{A, -B\}, \quad \bar{X}'=\{A', -B'\}, \quad \bar{Y}=\{A', -B\}, \quad \bar{Y}'=\{A, -B'\}.$$

For if $\mathbf{A}$ and $\mathbf{B}$ are standard unilatent canonical square matrices which have the same characteristic potent divisors as A and B, and if

$$A=h\mathbf{A}H, \quad B=k\mathbf{B}K$$

are given equimutant transformations (which can certainly be determined), the new general commutants can be regarded as given by the formulae (A) and (B) of § 244. We may in particular take A and B to be unilatent canonical square matrices derived from standard unilatent canonical square matrices $\mathbf{A}$ and $\mathbf{B}$ by symmetric class-derangements.

The conditions that the new general commutants shall be undegenerate are the same as in Ex. ii. In particular the general commutants X, X', Y, Y' are undegenerate square matrices when and only when A and B have the same characteristic potent divisors or are equicanonical; and the general commutants $\bar{X}, \bar{X}', \bar{Y}, \bar{Y}'$ are undegenerate square matrices when and only when A and $-B$ are equicanonical, i.e. when and only when the characteristic potent divisors of A, B can be coupled together in pairs of the form

$$(\lambda-c)^e, \quad (\lambda+c)^e.$$

NOTE 2. *Applications to a single unilatent square matrix A and its conjugate A'.*

Let $A=[a]_m^m$ be a unilatent square matrix with constant elements whose characteristic potent divisors are

$$(\lambda-c)^{e_1}, \; (\lambda-c)^{e_2}, \ldots (\lambda-c)^{e_s},$$

where the indices are arranged in descending order of magnitude; and let A' be the conjugate of A.

Then from the formulae (A′) of § 244 we see that each of the four general commutants

$$X=\{A, A\},\quad X'=\{A', A'\},\quad Y=\{A', A\},\quad Y'=\{A, A'\}$$

is an undegenerate square matrix whose elements are homogeneous linear functions of σ arbitrary parameters, where

$$\sigma=e_1+3e_2+5e_3+\dots+(2s-1)\,e_s \nless m,$$

and that each of them has exactly σ independent non-zero particularisations.

Again from the formulae (B′) of § 244 we see that each of the four general commutants

$$\bar{X}=\{A, -A\},\quad \bar{X}'=\{A', -A'\},\quad \bar{Y}=\{A', -A\},\quad \bar{Y}'=\{A, -A'\}$$

is a zero matrix when $c\neq 0$; whilst when $c=0$, each of them is an undegenerate square matrix whose elements are homogeneous linear functions of σ arbitrary parameters, where σ is the integer given above, and each of them has exactly σ independent non-zero particularisations.

Let $e=e_1$ be the greatest of the indices $e_1, e_2, \dots e_s$, so that the rational integral equation of lowest degree satisfied by A is

$$(A-cI)^e=0,\quad \text{where}\quad I=[1]_m^m.$$

Then if $s=1$, we have $e=\sigma=m$; therefore A has exactly m independent non-zero commutants $\{A, A\}$ which could be taken to be the square matrices

$$I,\ A,\ A^2,\ \dots\ A^{m-1};$$

i.e. the general commutant $\{A, A\}$ is an arbitrary rational integral function of A. But if $s>1$, we have $e<m$ and $\sigma>m$; therefore A has more than m independent non-zero commutants $\{A, A\}$, of which e but not more than e can be rational integral functions of A; consequently A has commutants which are not rational integral functions of A.

Thus the general commutant $X=\{A, A\}$ is a rational integral function of A when and only when A is unipotent or has no characteristic potent divisors of order $m-1$, i.e. when and only when m is the lowest degree of a rational integral equation satisfied by A; and when this condition is satisfied, X is an arbitrary rational integral function of A.

2. *General symmetric and general skew-symmetric commutants.*

Let $A=[a]_m^m$ and $B=[b]_m^m$ be two unilatent canonical square matrices of order m whose latent roots are c and c', and whose successive characteristic potent divisors are

$$(\lambda-c)^{e_1},\ (\lambda-c)^{e_2},\ \dots\ (\lambda-c)^{e_s}\quad\text{and}\quad (\lambda-c')^{e_1},\ (\lambda-c')^{e_2},\ \dots\ (\lambda-c')^{e_s},$$

where

$$e_1+e_2+\dots+e_s=m;$$

and let A and B be given in the forms

$$A=\begin{bmatrix} A_1, & 0, & \dots & 0 \\ 0, & A_2, & \dots & 0 \\ \dots & \dots & \dots & \dots \\ 0, & 0, & \dots & A_s \end{bmatrix}_{e_1, e_2, \dots e_s}^{e_1, e_2, \dots e_s},\qquad B=\begin{bmatrix} B_1, & 0, & \dots & 0 \\ 0, & B_2, & \dots & 0 \\ \dots & \dots & \dots & \dots \\ 0, & 0, & \dots & B_s \end{bmatrix}_{e_1, e_2, \dots e_s}^{e_1, e_2, \dots e_s},$$

where A_i, B_i are the simple canonical square matrices of order e_i whose latent roots are c, c', i.e. whose characteristic potent divisors are $(\lambda-c)^{e_i}$, $(\lambda-c')^{e_i}$. Also let A', B' be the conjugates of A, B; and let J be the part-reversant of the class (b), which makes

$$JAJ=A',\quad A'J=JA,\quad AJ=JA';\qquad JBJ=B',\quad B'J=JB,\quad BJ=JB'.$$

The general commutants of Theorems I *a* and I *b* are in this case either zero matrices or general ruled quadrate slopes of the class (b) as in Ex. iii; and the corresponding general symmetric (or general skew-symmetric) commutants are the most general specialisations of them which can be formed when:

(i) every diagonal constituent, such as X_{ii} in X, is symmetric (or skew-symmetric);

(ii) every two conjugately situated non-diagonal constituents lying on opposite sides of the diagonal constituents, such as X_{ij} and X_{ji} in X when $j \neq i$, are mutually conjugate (or mutually skew-conjugate).

When the conditions (i) and (ii) are satisfied, the diagonal constituents have the characters described in Theorems II *a*, III *a*, II *b*, III *b* of § 242, and the independent parameters in each of the commutants are the parameters of the diagonal constituents together with the parameters of the non-diagonal constituents lying on one side of the diagonal constituents. In the commutants $Y, Y', \bar{Y}, \bar{Y}'$, which are counter-slopes, the conditions (i) and (ii) impose no restrictions on the non-diagonal constituents lying on one side of the diagonal constituents; but in the commutants $X, X', \bar{X}, \bar{X}'$, which are ante-slopes, the conditions can only be satisfied when the only elements which can be different from 0 are those forming the paradiagonals of the square constituents, so that the ante-slopes are *quasi-scalaric* or in particular *scalaric*.

NOTE 3. When the general symmetric or general skew-symmetric commutants thus constructed are not zero matrices, they are general symmetric or general skew-symmetric quadrate slopes, the properties of which are discussed in Chapter XXXVIII. It will there be proved (see also Theorem II of § 241) that the determinant of a quadrate slope is equal to the product of the determinants of all its paradiagonal prime minors. Consequently any given quadrate slope is undegenerate when and only when all its paradiagonal prime minors are undegenerate. The paradiagonal prime minors involve only the elements of the paradiagonals of the square constituents, and in determining them all other elements can be put equal to 0.

NOTE 4. In order to describe the total number of independent arbitrary parameters, in the general symmetric or general skew-symmetric commutants $X, X', \bar{X}, \bar{X}'$, we will define:

k_1 to be the total number of non-diagonal square constituents lying on one side of the (square) diagonal constituents in any compound matrix of the class (b).

If there are exactly r distinct or unequal index numbers $\epsilon_1, \epsilon_2, \ldots \epsilon_r$ in the series $e_1, e_2, \ldots e_s$, and if these are repeated exactly $s_1, s_2, \ldots s_r$ times respectively, so that $s_1+s_2+\ldots+s_r=s$ is the total number of diagonal constituents, we have

$$k_1=\tfrac{1}{2}\{s_1(s_1-1)+s_2(s_2-1)+\ldots+s_r(s_r-1)\},$$
$$s+k_1=\tfrac{1}{2}\{s_1(s_1+1)+s_2(s_2+1)+\ldots+s_r(s_r+1)\},$$
$$s+2k_1=s_1^2+s_2^2+\ldots+s_r^2=k,$$

where k is the total number of square constituents.

NOTE 5. In order to describe the total number of independent arbitrary parameters in the general symmetric or general skew-symmetric commutants Y, Y', $\bar{Y}$, $\bar{Y}'$, we will define:

τ to be the sum of the smallest integers which are respectively $\not< \frac{1}{2}e_1, \frac{1}{2}e_2, \ldots \frac{1}{2}e_s$;

τ' to be the sum of the greatest integers which are respectively $\not> \frac{1}{2}e_1, \frac{1}{2}e_2, \ldots \frac{1}{2}e_s$;

σ_1 to be the sum of the effective orders of all the non-diagonal constituents lying on one side of the diagonal constituents in any compound matrix of the class (b);

so that
$$\tau+\tau'=e_1+e_2+\ldots+e_s=m$$
is the sum of the effective orders of the diagonal constituents. When $e_1, e_2, \ldots e_s$ are arranged in descending order of magnitude, we have
$$\sigma_1=e_2+2e_3+3e_4+\ldots+(s-1)e_s,$$
$$m+\sigma_1=e_1+2e_2+3e_3+\ldots+se_s,$$
$$m+2\sigma_1=e_1+3e_2+5e_3+\ldots+(2s-1)e_s=\sigma,$$
where σ is the sum of the effective orders of all the constituents.

We could also define τ to be one-half the sum of the even integers of the pairs such as (e_i, e_i+1), and τ' to be one-half the sum of the even integers of the pairs such as (e_i, e_i-1).

When the notations of Notes 4 and 5 are used, and reference is made to § 331 or § 335 of Chapter XXXVIII for the properties of the paradiagonal prime minors, the foregoing considerations lead to the following four theorems, the first parts of which were proved in § 240, and also follow from Theorems I *a* and I *b*. The number of independent non-zero particularisations of each commutant is equal to the total number of independent arbitrary parameters occurring in it; for the coefficients of these parameters clearly form a complete set of independent non-zero particularisations, because each of them has non-zero elements peculiar to it. The references to Chapter XXXVIII can easily be avoided by observing the characters of the ϵ_i paradiagonal prime minors
$$D_i=\pm[\alpha]_\mu^\mu, \qquad (i=1, 2, \ldots \epsilon_i),$$
which correspond to an index number ϵ_i repeated exactly μ times. They are all equal in a continuant, and differ only by signs which are alternately + and − in an alternant.

Theorem II a. *The two general symmetric or skew-symmetric commutants*
$$X=\{A, B\}, \quad X'=\{A', B'\},$$
which can be so chosen that $X'=JXJ$, *are zero matrices whenever* $c' \neq c$.

If $c'=c$, *where c may be* 0, *i.e. if* $B=A$, *then:*

(1) *The 'general symmetric' commutants* X, X' *are 'general symmetric' ante-continuants of the class* (b); *they are quasi-scalaric; each of them contains exactly* $s+k_1$ *independent arbitrary parameters; and they are always undegenerate.*

(2) *The 'general skew-symmetric' commutants* X, X' *are 'general skew-symmetric' ante-continuants of the class* (b); *they are quasi-*

scalaric, the diagonal constituents being zero matrices; each of them contains exactly k_1 independent arbitrary parameters; and they are undegenerate when and only when every distinct index number in the series $e_1, e_2, \dots e_s$ is repeated an even number of times.

The paradiagonal prime minors D_i of X and X' are general symmetric matrices of order μ in (1), and general skew-symmetric matrices of order μ in (2).

Theorem III a. *The two general symmetric or skew-symmetric commutants*

$$Y = \{A', B\}, \quad Y' = \{A, B'\},$$

which can be so chosen that $Y' = JYJ$, are zero matrices whenever $c' \neq c$.

If $c' = c$, where c may be 0, i.e. if $B = A$, then:

(1) *The 'general symmetric' commutants Y, Y' are 'general symmetric' counter-continuants of the class* (b) *and of the types $\{\pi', \pi\}$, $\{\pi, \pi'\}$, their diagonal constituents being 'general symmetric' and their non-diagonal constituents 'general' simple counter-continuants; each of them contains exactly $m + \sigma_1$ independent arbitrary parameters; and they are always undegenerate.*

(2) *The 'general skew-symmetric' commutants Y, Y' are 'general skew-symmetric' counter-continuants of the class* (b) *and of the types $\{\pi', \pi\}$, $\{\pi, \pi'\}$, their diagonal constituents being zero matrices, and their non-diagonal constituents being 'general' simple counter-continuants; each of them contains exactly σ_1 arbitrary parameters; and they are undegenerate when and only when every distinct index number in the series $e_1, e_2, \dots e_s$ is repeated an even number of times.*

The paradiagonal prime minors D_i of Y and Y' are general symmetric matrices of order μ in (1), and general skew-symmetric matrices of order μ in (2).

Theorem II b. *The two general symmetric or skew-symmetric commutants*

$$\bar{X} = \{A, -B\}, \quad \bar{X}' = \{A', -B'\},$$

which can be so chosen that $\bar{X}' = J\bar{X}J$, are zero matrices whenever $c' \neq -c$.

If $c' = -c$, where c may be 0 so that $B = A$, then:

(1) *The 'general symmetric' commutants $\bar{X}$, $\bar{X}'$ are 'general symmetric' ante-alternants of the class* (b); *they are quasi-scalaric; each of them contains exactly $s + k_1$ independent arbitrary parameters; and they are always undegenerate.*

(2) *The 'general skew-symmetric' commutants $\overline{X}$, $\overline{X}'$ are 'general skew-symmetric' ante-alternants of the class* (b); *they are quasi-scalaric, their diagonal constituents being zero matrices; each of them contains exactly k_1 independent arbitrary parameters; and they are undegenerate when and only when every distinct index number in the series $e_1, e_2, \ldots e_s$ is repeated an even number of times.*

The paradiagonal prime minors D_i of $\overline{X}$ and $\overline{X}'$ are general symmetric matrices of order μ in (1), and general skew-symmetric matrices of order μ in (2).

Theorem III b. *The two general symmetric or skew-symmetric commutants*

$$\overline{Y} = \{A', -B\}, \quad \overline{Y}' = \{A, -B'\},$$

which can be so chosen that $\overline{Y}' = J\overline{Y}J$, are zero matrices whenever $c' \neq -c$.

If $c' = -c$, where c may be 0 so that $B = A$, then:

(1) *The 'general symmetric' commutants $\overline{Y}$, $\overline{Y}'$ are 'general symmetric' counter-alternants of the class* (b) *and of the types $\{\pi', -\pi\}$, $\{\pi, -\pi'\}$, their diagonal constituents being 'general symmetric' and their non-diagonal constituents 'general' simple counter-alternants; each of them contains exactly $\tau + \sigma_1$ independent arbitrary parameters; and they are undegenerate when and only when every distinct even index number in the series $e_1, e_2, \ldots e_s$ is repeated an even number of times.*

(2) *The 'general skew-symmetric' commutants $\overline{Y}$, $\overline{Y}'$ are 'general skew-symmetric' counter-alternants of the class* (b) *and of the types $\{\pi', -\pi\}$, $\{\pi, -\pi'\}$, their diagonal constituents being 'general skew-symmetric' and their non-diagonal constituents 'general' simple counter-alternants; each of them contains exactly $\tau' + \sigma_1$ independent arbitrary parameters; and they are undegenerate when and only when every distinct odd index number in the series $e_1, e_2, \ldots e_s$ is repeated an even number of times.*

The paradiagonal prime minors D_i of $\overline{Y}$ and $\overline{Y}'$ are general symmetric or general skew-symmetric matrices of order μ according as ϵ_i is odd or even in (1), or according as ϵ_i is even or odd in (2).

Ex. vii. *The sum of a 'general symmetric' and a 'general skew-symmetric' commutant of each of the types considered in the foregoing theorems is always a 'general' commutant of the same type*, provided of course that the *two sets of parameters are independent.*

For the total number of arbitrary parameters in the sum is equal to the total number of arbitrary parameters in the general commutant of the same type, and a non-zero square matrix cannot be both symmetric and skew-symmetric.

NOTE 6. *Applications of Theorems III a and III b to a single unilatent square matrix A and its conjugate A'.*

Let $A=[a]_m^m$ be a unilatent square matrix with constant elements whose characteristic potent divisors are

$$(\lambda-c)^{e_1},\ (\lambda-c)^{e_2},\ \dots\ (\lambda-c)^{e_s},$$

where the indices are arranged in descending order of magnitude; and let A' be the conjugate of A.

Then Theorems III a and III b remain true with the new definition of A in so far as they relate to the rank and the total number of arbitrary parameters or independent non-zero particularisations of the two general symmetric (or skew-symmetric) contra-commutants

$$Y=\{A',\ A\},\ Y'=\{A,\ A'\}\quad or\quad \bar{Y}=\{A',\ -A\},\ \bar{Y}'=\{A,\ -A'\}.$$

Further it is still true that:

The sum of a 'general symmetric' and a 'general skew-symmetric' contra-commutant of any one of the four types mentioned is always a 'general' contra-commutant of that type.

In fact if $\mathbf{A}$ is a standard unilatent canonical square matrix of order m having the same latent root and the same characteristic potent divisors as A, and if

$$A=h\mathbf{A}H$$

is any given isomorphic transformation by which A can be derived from $\mathbf{A}$, we can put

$$Y=H'\mathbf{Y}H,\ Y'=h\mathbf{Y}'h'\quad\text{or}\quad \bar{Y}=H'\bar{\mathbf{Y}}H,\ \bar{Y}'=h\bar{\mathbf{Y}}'h',$$

where $\mathbf{Y}, \mathbf{Y}'$ or $\bar{\mathbf{Y}}, \bar{\mathbf{Y}}'$ are two general symmetric (or skew-symmetric) contra-commutants determined by Theorem III *a* or III *b*.

It should be observed that the contra-commutants $\bar{Y}$, $\bar{Y}'$ defined above are necessarily zero matrices when $c\neq 0$. If $c=0$, then if they are symmetric, they are undegenerate when and only when every distinct potent divisor of the form λ^{2p} is repeated an even number of times; and if they are skew-symmetric, they are undegenerate if and only if every distinct potent divisor of the form λ^{2p+1} is repeated an even number of times.

§ 244. The commutants of any pair of square matrices whose elements are constants.

1. *General commutants.*

Let $A=[a]_m^m$, $B=[b]_n^n$ be any two given square matrices of orders m, n whose elements are constants; and let A', B' be their conjugates. Also let $\mathbf{A}, \mathbf{B}$ be any two canonical square matrices of orders m, n having the same characteristic potent divisors as A, B; and let $\mathbf{A}', \mathbf{B}'$ be their conjugates. From Theorem I of § 240 we see that constructions for the general commutants

$$X=\{A,B\},\quad X'=\{A',B'\},\quad Y=\{A',B\},\quad Y'=\{A,B'\},\ \dots(1)$$
$$\bar{X}=\{A,-B\},\quad \bar{X}'=\{A',-B'\},\quad \bar{Y}=\{A',-B\},\quad \bar{Y}'=\{A,-B'\}\dots(2)$$

can be derived from constructions for the general commutants

$$\mathbf{X}=\{\mathbf{A},\mathbf{B}\},\quad \mathbf{X}'=\{\mathbf{A}',\mathbf{B}'\},\quad \mathbf{Y}=\{\mathbf{A}',\mathbf{B}\},\quad \mathbf{Y}'=\{\mathbf{A},\mathbf{B}'\},\ \dots(1')$$
$$\bar{\mathbf{X}}=\{\mathbf{A},-\mathbf{B}\},\quad \bar{\mathbf{X}}'=\{\mathbf{A}',-\mathbf{B}'\},\quad \bar{\mathbf{Y}}=\{\mathbf{A}',-\mathbf{B}\},\quad \bar{\mathbf{Y}}'=\{\mathbf{A},-\mathbf{B}'\}..(2')$$

by determining particular isomorphic transformations

$$A=h\mathbf{A}H,\quad B=k\mathbf{B}K,\ \dots\dots\dots\dots(a)$$

and using the equigradent commutantal formulae

$$X = h\mathbf{X}K,\quad X' = H'\mathbf{X}'k',\quad Y = H'\mathbf{Y}K,\quad Y' = h\mathbf{Y}'k', \quad \text{......(A)}$$

$$\bar{X} = h\bar{\mathbf{X}}K,\quad \bar{X}' = H'\bar{\mathbf{X}}'k',\quad \bar{Y} = H'\bar{\mathbf{Y}}K,\quad \bar{Y}' = h\bar{\mathbf{Y}}'k'. \quad \text{......(B)}$$

For all choices of **A** and **B**, Theorem II of § 240 and the results of § 242 show that each of the general commutants (1′) and (2′) can be expressed as a known compound matrix in which the non-zero constituents are general ruled simple slopes of a given type. When **A** and **B** are chosen to be *standard* canonical square matrices in which the super-parts are unilatent canonical square matrices, Theorem II of § 240 and the results of § 243 show that each of the general commutants (1′) and (2′) can be expressed as a known compartite matrix in which the non-zero parts are general ruled compound slopes of a given type. In determining the general commutants (1) and (2) we shall usually choose **A** and **B** to be *standard* canonical square matrices on account of the simpler constructions which are thus obtained.

When A and B are themselves canonical square matrices, the transformations (a) are symmetric class-derangements, and the transformations (A) and (B) are class-derangements in which the simple constituents receive rigid displacements.

The formulae (A) and (B) do not in general enable us to pass from constructions for the general symmetric or general skew-symmetric commutants (1′) and (2′) to constructions for the general symmetric or general skew-symmetric commutants (1) and (2); but they do this for the contra-commutants when $k = h$, as can always be the case when $B = A$.

NOTE 1. If U and V are any particular undegenerate square contra-commutants $\{A', A\}$ and $\{B, B'\}$, which by Ex. vii of § 240 always exist and can always be symmetric, we already know that the four general commutants (1) or (2) can always be so chosen as to be connected by the equigradent commutantal transformations

$$X' = UXV,\quad Y = UX,\quad Y' = XV,$$

or

$$\bar{X}' = U\bar{X}V,\quad \bar{Y} = U\bar{X},\quad \bar{Y}' = \bar{X}V,$$

and that all four of them have the same rank, and the same number of independent non-zero particularisations. In particular if the canonical square matrices **A** and **B** are regarded as compound matrices whose diagonal constituents are simple canonical square matrices, and if J_a and J_b are the part-reversants of the same classes as **A** and **B**, the four general commutants (1′) or (2′) can always be so chosen as to be correlated by part-reversals, i.e. connected by the equigradent commutantal transformations

$$\mathbf{X}' = J_a\mathbf{X}J_b,\quad \mathbf{Y} = J_a\mathbf{X},\quad \mathbf{Y}' = \mathbf{X}J_b,$$

or

$$\bar{\mathbf{X}}' = J_a\bar{\mathbf{X}}J_b,\quad \bar{\mathbf{Y}} = J_a\bar{\mathbf{X}},\quad \bar{\mathbf{Y}}' = \bar{\mathbf{X}}J_b.$$

To obtain the most convenient constructions for the general commutants (1), let the distinct latent roots of A and B be so arranged that they are

$$c_1, c_2, \dots c_r, a_1, a_2, a_3, \dots \quad \text{and} \quad c_1, c_2, \dots c_r, b_1, b_2, b_3, \dots,$$

where the a's, b's and c's are all different, and where $c_1, c_2, \dots c_r$ are the distinct common latent roots of A and B. Also let **A** and **B** be standard canonical reduced forms of A and B expressed as standard compartite matrices of classes

$$M\begin{pmatrix}\alpha_1, \alpha_2, \dots \alpha_r, p\\ \alpha_1, \alpha_2, \dots \alpha_r, p\end{pmatrix} \quad \text{and} \quad M\begin{pmatrix}\beta_1, \beta_2, \dots \beta_r, q\\ \beta_1, \beta_2, \dots \beta_r, q\end{pmatrix}$$

in which the successive parts are

$$A_1, A_2, \dots A_r, P \quad \text{and} \quad B_1, B_2, \dots B_r, Q\,;$$

A_i and B_i being unilatent canonical square matrices having the latent root c_i;

P being a compartite square matrix of standard form whose successive parts are unilatent canonical square matrices having the latent roots $a_1, a_2, a_3, \dots$;

Q being a compartite square matrix of standard form whose successive parts are unilatent canonical square matrices having the latent roots $b_1, b_2, b_3, \dots$.

Then by Theorem II of § 240 and the theorems of § 243 we have

$$\mathbf{X} = \begin{bmatrix} X_1, & 0, & \dots & 0, & 0 \\ 0, & X_2, & \dots & 0, & 0 \\ \dots & \dots & \dots & \dots & \dots \\ 0, & 0, & \dots & X_r, & 0 \\ 0, & 0, & \dots & 0, & 0 \end{bmatrix}_{\alpha_1, \alpha_2, \dots \alpha_r, p}^{\beta_1, \beta_2, \dots \beta_r, q},$$

the general commutant $\mathbf{X}$ being a compartite square matrix of standard form in which:

X_i is a general commutant $\{A_i, B_i\}$, i.e. a general compound continuant of the type $\{\pi, \pi\}$;

the parameters of $X_1, X_2, \dots X_r$ are all independent.

The rank of X_i and the number of arbitrary parameters in it are given by Theorem I a of § 243. The rank of $\mathbf{X}$ is the sum of the ranks of its non-zero parts $X_1, X_2, \dots X_r$, and the total number of arbitrary parameters in $\mathbf{X}$ (which is also the total number of independent non-zero particularisations of $\mathbf{X}$) is the sum of the numbers of the arbitrary parameters occurring in $X_1, X_2, \dots X_r$. We can choose A_i and B_i to be always *standardised* unilatent canonical square matrices; and then $X_1, X_2, \dots X_r$ are *standardised* compound continuants.

The four general commutants (1′) are standard compartite matrices of the same class and the same form as $\mathbf{X}$ in which the r non-zero parts are general compound continuants of the respective types

$$\{\pi, \pi\}, \quad \{\pi', \pi'\}, \quad \{\pi', \pi\}, \quad \{\pi, \pi'\}.$$

When isomorphic transformations (a) have been determined, the four general commutants (1) are given by the formulae (A), and they have the same rank, the same arbitrary parameters and the same number of independent non-zero particularisations as the general commutants (1′). From these constructions (see sub-article 5 of § 241) we obtain the following theorem:

Theorem I a. *The rank of each of the four general commutants*

$$X = \{A, B\}, \quad X' = \{A', B'\}, \quad Y = \{A', B\}, \quad Y' = \{A, B'\} \text{......(1)}$$

is equal to the order of the greatest common canonical of A and B.

The total number of independent arbitrary parameters in (or non-zero particularisations of) each of them is equal to the sum of the numbers of the arbitrary parameters occurring in the general commutants of the unilatent pairs $(A_1, B_1), (A_2, B_2), \ldots (A_r, B_r)$ *having the latent roots* $c_1, c_2, \ldots c_r$ *common to A and B.*

Two mutually conjugate square matrices have of course the same canonicals, i.e. the same canonical reduced forms.

NOTE 2. To merely prove Theorem I *a* without aiming at the simplest constructions for the general commutants (1), we could define $\mathbf{A}$ and $\mathbf{B}$ to be compartite square matrices equicanonical with A and B in which the parts are unilatent square matrices (not necessarily canonical) having the same latent roots as before. Then the construction for X would remain valid, the part X_i of $\mathbf{X}$ being a general commutant of the unilatent pair (A_i, B_i), and would include constructions for the other commutants. It would also include constructions for the general commutants (2), and Theorem I *b* would be included in Theorem I *a*.

Ex. i. Each of the four general commutants (1) is a zero matrix when and only when A and B have no latent root in common.

Ex. ii. *Undegenerate commutants* $\{A, B\}, \{A', B'\}, \{A', B\}, \{A, B'\}$.

By Theorem I *a* each of these general commutants is:

(1) *undegenerate of rank m* when and only when the canonicals of A are diagonal minors of the canonicals of B, i.e. when and only when all the characteristic potent divisors $(\lambda - a)^p, (\lambda - b)^q, \ldots$ of A can be associated with corresponding characteristic potent divisors $(\lambda - a)^P, (\lambda - b)^Q, \ldots$ of B in which $P \nless p, Q \nless q, \ldots$;

(2) *undegenerate of rank n* when and only when the canonicals of B are diagonal minors of the canonicals of A, i.e. when and only when all the characteristic potent divisors $(\lambda - a)^p, (\lambda - b)^q, \ldots$ of B can be associated with corresponding characteristic potent divisors $(\lambda - a)^P, (\lambda - b)^Q, \ldots$ of A in which $P \nless p, Q \nless q, \ldots$;

(3) *undegenerate square matrices* when and only when A and B are equicanonical or have the same characteristic potent divisors.

Ex. iii. *The general commutants*

$$\xi = \{\Pi, \Pi\}, \quad \xi' = \{\Pi', \Pi'\}, \quad \eta = \{\Pi', \Pi\}, \quad \eta' = \{\Pi, \Pi'\},$$

when Π *is a standard canonical square matrix whose conjugate is* Π'.

We can use the constructions for the commutants (1′) described in the text by putting $\mathbf{A} = \mathbf{B} = \Pi$.

Let the successive unilatent super-parts of Π be the unilatent canonical square matrices

$$\Pi_1, \Pi_2, \ldots \Pi_r \quad \text{of orders} \quad e_1, e_2, \ldots e_r$$

having latent roots $\qquad c_1, c_2, \ldots c_r,$

where the c's are all different; and let the orders of the successive simple parts of the successive super-parts be

$$e_{11}, e_{12}, \ldots; \; e_{21}, e_{22}, \ldots; \quad \ldots \quad e_{r1}, e_{r2}, \ldots,$$

where $$e_{i1}+e_{i2}+\ldots=e_i, \quad (i=1, 2, \ldots r).$$

Then by Theorem II of § 240 and the theorems of § 243 we have

$$\eta=\begin{bmatrix} Y_1, & 0, & \ldots & 0 \\ 0, & Y_2, & \ldots & 0 \\ \ldots & \ldots & \ldots & \ldots \\ 0, & 0, & \ldots & Y_r \end{bmatrix}_{e_1, e_2, \ldots e_r}^{e_1, e_2, \ldots e_r},$$

where:

Y_i is a general commutant $\{\Pi_i', \Pi_i\}$, i.e. a general quadrate continuant of type $\{\pi', \pi\}$ and of the same class

$$M\binom{e_{i1}, e_{i2}, \ldots}{e_{i1}, e_{i2}, \ldots} \text{ as } \Pi_i;$$

the parameters of $Y_1, Y_2, \ldots Y_r$ are all independent.

The general commutants ξ, ξ', η, η' are compartite matrices of the same class and of the same form as η in which the r parts are general quadrate continuants of the respective types

$$\{\pi, \pi\}, \{\pi', \pi'\}, \{\pi', \pi\}, \{\pi, \pi'\}.$$

They are all undegenerate because by Theorem I a or Note 2 of § 243 the r parts in each of them are undegenerate.

By Ex. iii of § 243 the total number of arbitrary parameters in the part Y_i of η is never less than e_i, and is equal to e_i when and only when Π_i is a simple canonical square matrix. Consequently the total number of arbitrary parameters in each of the general commutants ξ, ξ', η, η' can never be less than m, where m is the order of Π, and is equal to m when and only when all the parts $\Pi_1, \Pi_2, \ldots \Pi_r$ are simple canonical square matrices.

Ex. iv. *The general square commutants*

$$X=\{A, A\}, \quad X'=\{A', A'\}, \quad Y=\{A', A\}, \quad Y'=\{A, A'\}.$$

By Ex. ii each of these general commutants is always an undegenerate square matrix of order m, where m is the order of the square matrix A.

Let the distinct latent roots of A be $c_1, c_2, \ldots c_r$; and for each of the values $1, 2, \ldots r$ of i let the indices of those characteristic potent divisors of A which are powers of $\lambda - c_i$ be

$$e_{i1}, e_{i2}, e_{i3}, \ldots, \quad \text{where} \quad e_{i1}+e_{i2}+e_{i3}+\ldots=e_i.$$

Then if Π is the canonical square matrix described in Ex. iii, and if

$$A=\hbar\Pi H \quad \ldots\ldots\ldots\ldots\ldots\ldots\ldots\ldots\ldots\ldots\ldots\ldots\text{(b)}$$

is any isomorphic transformation by which A can be derived from Π, we can put

$$X=h\mathbf{X}H, \quad X'=H'\mathbf{X}'h', \quad Y=H'\mathbf{Y}H, \quad Y'=h\mathbf{Y}'h', \quad \ldots\ldots\ldots\ldots\text{(A')}$$

where $\mathbf{X}, \mathbf{X}', \mathbf{Y}, \mathbf{Y}'$ are the general commutants ξ, ξ', η, η' of Ex. iii.

The total number of arbitrary parameters in each of the general commutants X, X', Y, Y' can never be less than m, and is equal to m when and only when all the unilatent super-parts $\Pi_1, \Pi_2, \ldots \Pi_r$ of Π are *simple* canonical square matrices, i.e. when and only when A has only one characteristic potent divisor corresponding to each distinct latent root.

To obtain the constructions for the general commutants (2) which are usually the most convenient, let the distinct latent roots of A and B be so arranged that they are

$$c_1, c_2, \dots c_r, a_1, a_2, a_3, \dots \quad \text{and} \quad -c_1, -c_2, \dots -c_r, -b_1, -b_2, -b_3, \dots,$$

where the a's, b's and c's are all different, so that the distinct latent roots of A and $-B$ are

$$c_1, c_2, \dots c_r, a_1, a_2, a_3, \dots \quad \text{and} \quad c_1, c_2, \dots c_r, b_1, b_2, b_3, \dots,$$

and $c_1, c_2, \dots c_r$ are the distinct common latent roots of A and $-B$. Also let $\mathbf{A}$ and $\mathbf{B}$ be standard canonical reduced forms of A and B expressed as standard compartite matrices of classes

$$M\begin{pmatrix}\alpha_1, \alpha_2, \dots \alpha_r, p\\ \alpha_1, \alpha_2, \dots \alpha_r, p\end{pmatrix} \quad \text{and} \quad M\begin{pmatrix}\beta_1, \beta_2, \dots \beta_r, q\\ \beta_1, \beta_2, \dots \beta_r, q\end{pmatrix}$$

in which the successive parts are

$$A_1, A_2, \dots A_r, P \quad \text{and} \quad B_1, B_2, \dots B_r, Q;$$

A_i and B_i being unilatent canonical square matrices having the latent roots $c_i, -c_i$;

P being a compartite square matrix of standard form whose successive parts are unilatent canonical square matrices having the latent roots $a_1, a_2, a_3, \dots$;

Q being a compartite square matrix of standard form whose successive parts are unilatent canonical square matrices having the latent roots $-b_1, -b_2, -b_3, \dots$.

Then by Theorem II of § 240 and the theorems of § 243 we have

$$\bar{\mathbf{X}} = \begin{bmatrix} \bar{X}_1, & 0, & \dots & 0, & 0 \\ 0, & \bar{X}_2, & \dots & 0, & 0 \\ \dots & \dots & \dots & \dots & \dots \\ 0, & 0, & \dots & \bar{X}_r, & 0 \\ 0, & 0, & \dots & 0, & 0 \end{bmatrix}_{\alpha_1, \alpha_2, \dots \alpha_r, p}^{\beta_1, \beta_2, \dots \beta_r, q},$$

the general commutant $\bar{\mathbf{X}}$ being a compartite square matrix of standard form in which:

$\bar{X}$ is a general commutant $\{A_i, -B_i\}$, i.e. a general compound alternant of the type $\{\pi, -\pi\}$;

the parameters of $\bar{X}_1, \bar{X}_2, \dots \bar{X}_r$ are all independent.

The rank of $\bar{X}_i$ and the number of arbitrary parameters in it are given by Theorem I b of § 243. The rank of $\bar{\mathbf{X}}$ is the sum of the ranks of its non-zero parts $\bar{X}_1, \bar{X}_2, \dots \bar{X}_r$, and the total number of arbitrary parameters in $\bar{\mathbf{X}}$ (which is also the total number of independent non-zero particularisations of $\bar{\mathbf{X}}$) is the sum of the numbers of the arbitrary parameters occurring in $\bar{X}_1, \bar{X}_2, \dots \bar{X}_r$. We can choose A_i and B_i to be always *standardised* unilatent canonical square matrices; and then $\bar{X}_1, \bar{X}_2, \ldots \bar{X}_r$ are *standardised* compound alternants.

The four general commutants (2′) are compartite matrices of the same class and the same form as $\bar{\mathbf{X}}$ in which the r non-zero parts are general compound alternants of the respective types

$$\{\pi, -\pi\}, \ \{\pi', -\pi'\}, \ \{\pi', -\pi\}, \ \{\pi, -\pi'\}.$$

When isomorphic transformations (a) have been determined, the four general commutants (2) are given by the formulae (B), and they have the same rank, the same arbitrary parameters and the same number of independent non-zero particularisations as the general commutants (2′). From these constructions (see sub-article 5 of § 241) we obtain the following theorem:

Theorem I b. *The rank of each of the four general commutants*

$$\bar{X}=\{A, -B\}, \quad \bar{X}'=\{A', -B'\}, \quad \bar{Y}=\{A', -B\}, \quad \bar{Y}'=\{A, -B'\} \ldots (2)$$

is equal to the order of the greatest common canonical of A and $-B$.

The total number of independent arbitrary parameters in (or non-zero particularisations of) each of them is equal to the sum of the numbers of the arbitrary parameters occurring in the general commutants of the unilatent pairs $(A_1, -B_1), (A_2, -B_2), \ldots (A_r, -B_r)$ having the latent roots $c_1, c_2, \ldots c_r$ common to A and $-B$.

Ex. v. Each of the four general commutants (2) is a zero matrix when and only when A and $-B$ have no latent root in common, i.e. when and only when there do not exist latent roots $a=c, \beta=-c$ of A, B whose sum is 0.

Ex. vi. *Undegenerate commutants* $\{A, -B\}, \{A', -B'\}, \{A', -B\}, \{A, -B'\}$.

In accordance with Theorem I *b* each of these general commutants is:

(1) *undegenerate of rank m* when and only when the canonicals of A are diagonal minors of the canonicals of $-B$, i.e. when and only when all the characteristic potent divisors $(\lambda-a)^p, (\lambda-b)^q, \ldots$ of A can be associated with corresponding characteristic potent divisors $(\lambda+a)^P, (\lambda+b)^Q, \ldots$ of B in which $P \nless p, Q \nless q, \ldots$;

(2) *undegenerate of rank n* when and only when the canonicals of $-B$ are diagonal minors of the canonicals of A, i.e. when and only when all the characteristic potent divisors $(\lambda+a)^p, (\lambda+b)^q, \ldots$ of B can be associated with corresponding characteristic potent divisors $(\lambda-a)^P, (\lambda-b)^Q, \ldots$ of A in which $P \nless p, Q \nless q, \ldots$;

(3) *undegenerate square matrices* when and only when A and $-B$ are equicanonical, i.e. when and only when the characteristic potent divisors of the square matrices A, B can be coupled together in pairs of the form

$$(\lambda-c)^e, (\lambda+c)^e.$$

If $f_i(\lambda), g_i(\lambda)$ are the potent factors of order i of the characteristic matrices $A(\lambda)$, $B(\lambda)$ of A, B, the third case occurs (see sub-article 5 of § 241) when and only when $n=m$ and we can put

$$g_i(-\lambda)=f_i(\lambda), \qquad (i=1, 2, \ldots m).$$

Ex. vii. *The general commutants*

$$\bar{\xi}=\{\Pi, -\Pi\}, \quad \bar{\xi}'=\{\Pi', -\Pi'\}, \quad \bar{\eta}=\{\Pi', -\Pi\}, \quad \bar{\eta}'=\{\Pi, -\Pi'\},$$

when Π is a standard canonical square matrix whose conjugate is Π'.

To avoid the introduction of two different canonical square matrices **A** and **B** equi-canonical with Π (only one of which could be Π), we will replace the constructions given in the text by others which are simpler in the present special case.

Since a change in the order of arrangement of the super-parts of Π is equivalent to a symmetric class-derangement which merely leads to a corresponding symmetric class-derangement of the general commutants in which their parts receive rigid displacements, we may and will suppose that the successive unilatent super-parts of Π are unilatent canonical square matrices

$$\Pi_0;\ P_1,\ \ Q_1;\ P_2,\ \ Q_2;\ \dots\ P_s,\ \ Q_s;\ \Omega_1,\ \Omega_2,\ \Omega_3,\ \dots,$$

of orders $\quad e_0;\ e_1',\ \ e_1;\ e_2',\ \ e_2;\ \dots\ e_s',\ \ e_s;\ \omega_1,\ \omega_2,\ \omega_3,\ \dots,$

with latent roots $\quad 0;\ c_1,\ -c_1;\ c_2,\ -c_2;\ \dots\ c_s,\ -c_s;\ a_1,\ a_2,\ a_3,\ \dots,$

where the scalar numbers $0,\ c_1,\ -c_1,\ c_2,\ -c_2,\ \dots\ c_s,\ -c_s,\ a_1,\ -a_1,\ a_2,\ -a_2,\ a_3,\ -a_3,\ \dots$ are all different. We may suppose that the orders of the successive simple parts of

$$\Pi_0,\ P_i,\ Q_i,\ \Omega_i \quad \text{are} \quad e_{01},\ e_{02},\ \dots;\ e_{i1}',\ e_{i2}',\ \dots;\ e_{i1},\ e_{i2},\ \dots;\ \omega_{i1},\ \omega_{i2},\ \dots,$$

where $\quad e_{i1}'+e_{i2}'+\ \dots\ =e_i',\quad e_{i1}+e_{i2}+\ \dots\ =e_i,\quad \omega_{i1}+\omega_{i2}+\ \dots\ =\omega_i$

for all permissible values of i. We could simplify the construction by supposing that the index numbers forming every one of these series are arranged in descending order of magnitude, i.e. that all the parts of Π are *standardised* unilatent canonical square matrices, or that Π is a *standardised* canonical square matrix.

By Theorem II of § 240 and the theorems of § 243 we have

$$\bar{\eta}=\begin{bmatrix} Y_0, & 0, & 0, & \dots & 0, & 0, & 0 \\ 0, & 0, & U_1, & \dots & 0, & 0, & 0 \\ 0, & V_1, & 0, & \dots & 0, & 0, & 0 \\ \dots & \dots & \dots & \dots & \dots & \dots & \dots \\ 0, & 0, & 0, & \dots & 0, & U_s, & 0 \\ 0, & 0, & 0, & \dots & V_s, & 0, & 0 \\ 0, & 0, & 0, & \dots & 0, & 0, & 0 \end{bmatrix}\begin{matrix} e_0;\ e_1',\ e_1;\ \dots\ e_s',\ e_s;\ \omega \\ \\ \\ \\ \\ \\ e_0;\ e_1',\ e_1;\ \dots\ e_s',\ e_s;\ \omega \end{matrix},$$

where $\omega=\omega_1+\omega_2+\omega_3+\ \dots$, and:

Y_0 is a general commutant $\{\Pi_0',\ -\Pi_0\}$, i.e. a general quadrate alternant of type $\{\pi',\ -\pi\}$ and of the class $M\begin{pmatrix} e_{01},\ e_{02},\ \dots \\ e_{01},\ e_{02},\ \dots \end{pmatrix}$;

$U_i,\ V_i$ are general commutants $\{P_i',\ -Q_i\},\ \{Q_i',\ -P_i\}$, i.e. general alternants of type $\{\pi',\ -\pi\}$ and of the mutually conjugate classes

$$M\begin{pmatrix} e_{i1},\ e_{i2},\ \dots \\ e_{i1}',\ e_{i2}',\ \dots \end{pmatrix},\quad M\begin{pmatrix} e_{i1}',\ e_{i2}',\ \dots \\ e_{i1},\ e_{i2},\ \dots \end{pmatrix};$$

the parameters of $Y_0,\ U_1,\ V_1,\ \dots\ U_s,\ V_s$ are all independent.

The general commutants $\bar{\xi},\ \bar{\xi}',\ \bar{\eta},\ \bar{\eta}'$ are compartite matrices of the same class and the same form as $\bar{\eta}$ in which the non-zero parts are general alternants of the respective types

$$\{\pi,\ -\pi\},\ \{\pi',\ -\pi'\},\ \{\pi',\ -\pi\},\ \{\pi,\ -\pi'\},$$

all four of them having the same rank and containing the same number of independent arbitrary parameters.

The square matrix $\bar{\eta}$ cannot be undegenerate unless $\omega_1=\omega_2=\omega_3=\ \dots\ =0$; and when this condition is satisfied, it is obviously undegenerate if and only if every one of its $2s+1$ non-zero parts is an undegenerate square matrix. Moreover the part Y_0 is always an undegenerate square matrix; and by Ex. ix of § 231 the parts U_i and V_i are undegenerate square matrices when and only when they are derangements of quadrate alternants, i.e. when and only when $e_{i1}', e_{i2}', \ldots$ and $e_{i1}, e_{i2}, \ldots$ are two arrangements of the same integers. Thus in accordance with Ex. vi the general commutants $\bar{\xi}, \bar{\xi}', \bar{\eta}, \bar{\eta}'$ are undegenerate when and only when Π and $-\Pi$ are equicanonical.

Ex. viii. *The general commutants of Ex.* vii *when* Π *and* $-\Pi$ *are equicanonical and* Π *is in standardised form.*

In this case we may and will suppose that the successive unilatent super-parts of Π are unilatent canonical square matrices

$$\Pi_0;\ P_1,\ Q_1;\ P_2,\ Q_2;\ \ldots\ P_s,\ Q_s$$

of orders $\quad e_0;\ e_1,\ e_1;\ e_2,\ e_2;\ \ldots\ e_s,\ e_s$

with latent roots $\quad 0;\ c_1, -c_1;\ c_2, -c_2;\ \ldots\ c_s, -c_s,$

which are all different, and that the orders of the successive simple parts of the successive super-parts are

$$(e_{01}, e_{02}, \ldots);\ (e_{11}, e_{12}, \ldots), (e_{11}, e_{12}, \ldots);\ (e_{21}, e_{22}, \ldots), (e_{21}, e_{22}, \ldots);\ \ldots\ (e_{s1}, e_{s2}, \ldots), (e_{s1}, e_{s2}, \ldots),$$

where $e_{i1}, e_{i2}, e_{i3}, \ldots$ are integers arranged in descending order of magnitude whose sum is e_i.

The representation of $\bar{\eta}$ becomes

$$\bar{\eta} = \begin{bmatrix} Y_0, & 0, & 0, & \ldots & 0, & 0 \\ 0, & 0, & U_1, & \ldots & 0, & 0 \\ 0, & V_1, & 0, & \ldots & 0, & 0 \\ \ldots & \ldots & \ldots & \ldots & \ldots & \ldots \\ 0, & 0, & 0, & \ldots & 0, & U_s \\ 0, & 0, & 0, & \ldots & V_s, & 0 \end{bmatrix}^{e_0;\ e_1,\ e_1;\ \ldots\ e_s,\ e_s}_{e_0;\ e_1,\ e_1;\ \ldots\ e_s,\ e_s},$$

where

Y_0 is a general commutant $\{\Pi_0', -\Pi_0\}$, i.e. a general quadrate alternant of type $\{\pi', -\pi\}$ and of the class $M\begin{pmatrix} e_{01}, e_{02}, \ldots \\ e_{01}, e_{02}, \ldots \end{pmatrix}$;

U_i, V_i are general commutants $\{P_i', -Q_i\}, \{Q_i', -P_i\}$, i.e. general quadrate alternants of type $\{\pi', -\pi\}$ and of the class $M\begin{pmatrix} e_{i1}, e_{i2}, \ldots \\ e_{i1}, e_{i2}, \ldots \end{pmatrix}$;

the parameters of $Y_0, U_1, V_1, \ldots U_s, V_s$ are all independent.

The general commutants $\bar{\xi}, \bar{\xi}', \bar{\eta}, \bar{\eta}'$ are compartite matrices of the same class and the same form as $\bar{\eta}$ in which the $2s+1$ non-zero parts are general quadrate alternants of the types mentioned in Ex. vii.

Ex. ix. *The general square commutants*

$$\bar{X}=\{A, -A\},\quad \bar{X}'=\{A', -A'\},\quad \bar{Y}=\{A', -A\},\quad \bar{Y}'=\{A, -A'\}.$$

We may suppose the canonical reduced form of A to be the matrix Π described in Ex. vii. Then if

$$A = h\Pi H \quad \text{...(b)}$$

is any isomorphic transformation by which A can be derived from Π, we can put

$$\bar{X}=h\bar{\mathbf{X}}H,\quad \bar{X}'=H'\bar{\mathbf{X}}'h',\quad \bar{Y}=H'\bar{\mathbf{Y}}H,\quad \bar{Y}'=h\bar{\mathbf{Y}}'h', \quad \text{...............(B')}$$

where $\bar{\mathbf{X}}, \bar{\mathbf{X}}', \bar{\mathbf{Y}}, \bar{\mathbf{Y}}'$ are the general commutants $\bar{\xi}, \bar{\xi}', \bar{\eta}, \bar{\eta}'$ of Ex. vii.

By Ex. vi each of these four general commutants is undegenerate when and only when A and $-A$ are equicanonical, i.e. when and only when the characteristic potent divisors of A corresponding to its non-zero latent roots occur in pairs of the form

$$(\lambda - c)^e, \quad (\lambda + c)^e, \qquad \text{where } c \neq 0,$$

so that we can suppose the characteristic potent divisors of A to be

$$\lambda^{t_1}, \lambda^{t_2}, \lambda^{t_3}, \ldots; (\lambda - a)^p, (\lambda + a)^p; (\lambda - b)^q, (\lambda + b)^q; (\lambda - c)^r, (\lambda + c)^r; \ldots,$$

where $a, b, c, \ldots$ are non-zero scalar numbers, which need not be all different. In such a case we may suppose the canonical reduced form of A to be the matrix Π described in Ex. viii, and use the formulae (B′) in which $\bar{\mathbf{X}}, \bar{\mathbf{X}}', \bar{\mathbf{Y}}, \bar{\mathbf{Y}}'$ are the general commutants $\bar{\xi}, \bar{\xi}', \bar{\eta}, \bar{\eta}'$ of Ex. viii.

The final results mentioned in Exs. iii and iv lead to a theorem giving the necessary and sufficient conditions that all commutants $\{A, A\}$ of a square matrix A shall be rational integral functions of A.

Theorem II. *If A is a square matrix of order m with constant elements, the general commutant $X = \{A, A\}$ is a rational integral function of A when and only when A has one of the following mutually equivalent properties:*

(1) *It has only one characteristic potent divisor corresponding to each distinct latent root.*

(2) *Its characteristic matrix $A(\lambda)$ has no potent divisors of order $m-1$, i.e. has only potent divisors of order m.*

(3) *It satisfies no rational integral equation whose degree is less than m.*

When these conditions are satisfied, there are exactly m independent non-zero particular commutants $\{A, A\}$, and the general commutant $X = \{A, A\}$ is an arbitrary rational integral function of A. In all other cases there are more than m independent non-zero particular commutants $\{A, A\}$, and there are commutants $\{A, A\}$ which are not rational integral functions of A.

If $D_i(\lambda)$ and $E_i(\lambda)$ are the maximum and potent factors of order i of the characteristic matrix $A(\lambda)$ of A, the conditions (1), (2), (3) of the theorem are satisfied when and only when we can put

$$D_{m-1}(\lambda) = 1, \quad \text{or} \quad E_{m-1}(\lambda) = 1, \quad \text{(whichever we please)}.$$

Let A be the square matrix of Ex. iv whose distinct latent roots are $c_1, c_2, \ldots c_r$; and for each of the values $1, 2, \ldots r$ of i let ϵ_i be the greatest of the indices $e_{i1}, e_{i2}, e_{i3}, \ldots$ of those characteristic potent divisors of A which are powers of $\lambda - c_i$, so that the rational integral equation of lowest degree satisfied by A is

$$E_m(A) \equiv (A - c_1 I)^{\epsilon_1} (A - c_2 I)^{\epsilon_2} \ldots (A - c_r I)^{\epsilon_r} = 0, \qquad \text{where } I = [1]_m^m,$$

this equation having degree s in A, where

$$s = \epsilon_1 + \epsilon_2 + \ldots + \epsilon_r \ngtr m.$$

First suppose that A has only one characteristic potent divisor corresponding to each distinct latent root, so that

$$s = e_1 + e_2 + \ldots + e_r = m.$$

Then by Exs. iii and iv there are exactly m independent non-zero commutants of A, which could be taken to be the coefficients of the m independent arbitrary parameters of X. Since the square matrices

$$I, A, A^2, \ldots A^{m-1},$$

form a complete set of independent particular non-zero commutants of A, because $E_m(\lambda)$ has degree m in λ, it follows that we could put

$$X = x_0 I + x_1 A + x_2 A^2 + \ldots + x_{m-1} A^{m-1},$$

where the x's are independent arbitrary parameters, i.e. X is an arbitrary rational integral function of A.

Next suppose that A has more than one characteristic potent divisor corresponding to the latent root c_i, so that $\epsilon_i < e_i$, and

$$s < e_1 + e_2 + \ldots + e_m, \quad \text{i.e.} \quad s < m.$$

Then by Ex. iv there are more than m independent non-zero commutants of A, whilst not more than s such commutants can be rational integral functions of A.

2. *General symmetric and general skew-symmetric continuantal contra-commutants.*

Let $A = [a]_m^m$ be a given square matrix with constant elements, and let A' be its conjugate. If V is any particular undegenerate symmetric contra-commutant $\{A, A'\}$, it has been shown in Ex. ii of § 239 that the two general symmetric (or general skew-symmetric) contra-commutants

$$Y = \{A', A\}, \quad Y' = \{A, A'\} \quad \ldots\ldots\ldots\ldots\ldots\ldots\ldots(3)$$

can be so chosen as to be connected by the symmetric equigradent relation

$$Y' = VYV,$$

and that they both have the same rank, and both contain the same number of independent arbitrary parameters. When A is a canonical square matrix, we can choose V to be the part-reversant J of the same class as A.

Let the distinct latent roots of A be $c_1, c_2, \ldots c_r$, and let the canonical reduced form of A be the canonical square matrix Π described in Ex. iii. Then after determining a particular isomorphic transformation

$$A = h\Pi H, \ldots\ldots\ldots\ldots\ldots\ldots\ldots\ldots\ldots\ldots(c)$$

and constructing general symmetric (or general skew-symmetric) contra-commutants

$$\mathbf{Y} = \{\Pi', \Pi\}, \quad \mathbf{Y}' = \{\Pi, \Pi'\}, \quad \ldots\ldots\ldots\ldots\ldots\ldots(3')$$

we can take the general symmetric (or general skew-symmetric) contra-commutants Y, Y' to be the matrices derived from $\mathbf{Y}$, $\mathbf{Y}'$ by the symmetric equigradent commutantal transformations

$$Y = H'\mathbf{Y}H, \quad Y' = h\mathbf{Y}'h'. \quad \ldots\ldots\ldots\ldots\ldots\ldots\ldots(C)$$

We could use the formulae (C) if Π were any square matrix isomorphic with A, but we have so chosen Π that $\mathbf{Y}, \mathbf{Y}'$ are the most general symmetric (or skew-symmetric) specialisations of the matrices η, η' of Ex. iii. Consequently $\mathbf{Y}$ in (C) is what the matrix η of Ex. iii becomes when:

Y_i is a general symmetric (or general skew-symmetric) contra-commutant $\{\Pi_i', \Pi_i\}$, i.e. a general symmetric (or general skew-symmetric) quadrate counter-continuant of type $\{\pi', \pi\}$;

the parameters of $Y_1, Y_2, \dots Y_r$ are all independent;

and $\mathbf{Y}'$ is a matrix of the same form in which the r parts are quadrate counter-continuants of type $\{\pi, \pi'\}$.

The properties of the parts of $\mathbf{Y}$ and $\mathbf{Y}'$ given in Theorem IIIa of § 243 lead to the following theorem, and determine the values of κ and κ' in it. In both parts of the theorem the number of arbitrary parameters in Y or Y' is equal to the number of independent non-zero particularisations of that matrix.

Theorem III a. *The two general symmetric contra-commutants*

$$Y = \{A', A\}, \quad Y' = \{A, A'\}$$

are always undegenerate; and each of them contains exactly κ independent arbitrary parameters, where κ is the sum of the numbers of the arbitrary parameters occurring in the r general symmetric contra-commutants or counter-continuants $Y_1, Y_2, \dots Y_r$ corresponding to the r distinct latent roots

$$c_1, c_2, \dots c_r \quad \textit{of } A.$$

The two general skew-symmetric contra-commutants

$$Y = \{A', A\}, \quad Y' = \{A, A'\}$$

are undegenerate when and only when every distinct characteristic potent divisor of A with given index, such as $(\lambda - c)^e$, is repeated an even number of times. In all cases each of them contains exactly κ' independent arbitrary parameters, where κ' is the sum of the numbers of the arbitrary parameters occurring in the r general skew-symmetric contra-commutants or counter-continuants $Y_1, Y_2, \dots Y_r$ corresponding to the r distinct latent roots

$$c_1, c_2, \dots c_r \quad \textit{of } A.$$

Ex. x. From Ex. vii of § 243 we see that in all cases $\kappa + \kappa'$ is the total number of arbitrary parameters occurring in a general contra-commutant Y or Y'. Since a non-zero square matrix cannot be both symmetric and skew-symmetric, we conclude that:

The sum of a 'general symmetric' and a 'general skew-symmetric' contra-commutant $\{A', A\}$ or $\{A, A'\}$ whose parameters are independent is always a 'general' contra-commutant $\{A', A\}$ or $\{A, A'\}$.

Ex. xi. *Symmetric and skew-symmetric continuantal co-commutants.*

Constructions analogous to those of the text for contra-commutants are only possible when A is a canonical square matrix, and in some other special cases; but it is obvious that:

The general symmetric co-commutants $\{A, A\}$, $\{A', A'\}$ are always undegenerate.

For each of them has the unit matrix $I=[1]_m^m$ as a particularisation.

The general symmetric (or general skew-symmetric) co-commutants $\mathbf{X}=\{\Pi, \Pi\}$, $\mathbf{X}'=\{\Pi', \Pi'\}$ are specialisations of the matrices ξ, ξ' of Ex. iii, and are derived from the matrix η of Ex. iii by taking:

> Y_i to be a general symmetric (or general skew-symmetric) quadrate ante-continuant, which is necessarily quasi-scalaric as described in Theorem IIa of § 243.

But the corresponding co-commutants $X=\{A, A\}$, $X'=\{A', A'\}$ derived from them by the formulae

$$X=h\mathbf{X}H, \quad X'=H'\mathbf{X}'h'$$

are not necessarily symmetric (or skew-symmetric) except in the special case when $H=h'$, i.e. when (c) is a symmetric semi-unit transformation. This special case occurs when A is a canonical square matrix, so that (c) is a symmetric derangement. The above formulae then become symmetric derangements determining the general symmetric (or general skew-symmetric) co-commutants X, X'. Accordingly from Theorem IIa of § 243 we see that:

If A is a canonical square matrix, the general skew-symmetric co-commutants $\{A, A\}$, $\{A', A'\}$ are undegenerate when and only when every distinct characteristic potent divisor of A is repeated an even number of times.

Whenever A is a symmetric matrix, the general symmetric (or general skew-symmetric) co-commutants $\{A, A\}$, $\{A', A'\}$ can be taken to be the general symmetric (or general skew-symmetric) contra-commutants $\{A', A\}$, $\{A, A'\}$ constructed in the text; and whenever A is a skew-symmetric matrix, they can be taken to be the general symmetric (or general skew-symmetric) contra-commutants $\{A, -A'\}$, $\{A', -A\}$ constructed in sub-article 3.

3. *General symmetric and general skew-symmetric alternating contra-commutants.*

Let $A=[a]_m^m$ be a given square matrix with constant elements, and let A' be its conjugate. If V is any particular undegenerate symmetric contra-commutant $\{A, A'\}$, it has been shown in Ex. ii of § 239 that the two general symmetric (or general skew-symmetric) contra-commutants

$$\bar{Y}=\{A', -A\}, \quad \bar{Y}'=\{A, -A'\} \quad \text{.....................(4)}$$

can be so chosen as to be connected by the symmetric equigradent relation

$$\bar{Y}'=V\bar{Y}V,$$

and that they both have the same rank, and both contain the same number of arbitrary parameters. When A is a canonical square matrix, we can choose V to be the part-reversant J of the same class as A.

Let the distinct latent roots of A be so arranged that they are

$$0;\ c_1, -c_1;\ c_2, -c_2;\ \dots\ c_s, -c_s;\ a_1, a_2, a_3, \dots,$$

where the integers $0, c_1, -c_1, c_2, -c_2, \dots c_s, -c_s, a_1, -a_1, a_2, -a_2, a_3, -a_3, \dots$ are all different, and let the canonical reduced form of A be the canonical square matrix Π described in Ex. vii, where the index numbers of the unilatent parts are arranged in descending orders of magnitude. Then after determining a particular isomorphic transformation

$$A = h\Pi H \quad \text{.................................(d)}$$

and constructing general symmetric (or general skew-symmetric) contra-commutants

$$\bar{\mathbf{Y}} = \{\Pi', -\Pi\}, \quad \bar{\mathbf{Y}}' = \{\Pi, -\Pi'\}, \quad \text{.....................(4')}$$

we can take the general symmetric (or general skew-symmetric) contra-commutants $\bar{Y}$, $\bar{Y}'$ to be the matrices derived from $\bar{\mathbf{Y}}$, $\bar{\mathbf{Y}}'$ by the symmetric equigradent commutantal transformations

$$\bar{Y} = H'\bar{\mathbf{Y}}H, \quad \bar{Y}' = h\bar{\mathbf{Y}}'h'. \quad \text{.......................(D)}$$

We could use the formulae (D) if Π were any square matrix isomorphic with A, but we have so chosen Π that $\bar{\mathbf{Y}}$, $\bar{\mathbf{Y}}'$ are the most general symmetric (or skew-symmetric) specialisations of the matrices $\bar{\eta}$, $\bar{\eta}'$ of Ex. vii. Consequently $\bar{\mathbf{Y}}$ in (D) is what the matrix $\bar{\eta}$ of Ex. vii becomes when:

Y_0 is a general symmetric (or general skew-symmetric) contra-commutant $\{\Pi_0', -\Pi_0\}$, i.e. a general symmetric (or skew-symmetric) quadrate counter-alternant of type $\{\pi', -\pi\}$;

U_i is a general contra-commutant $\{P_i', -Q_i\}$, i.e. a general counter-alternant of type $\{\pi', -\pi\}$;

V_i is that general contra-commutant $\{Q_i', -P_i\}$ which is the conjugate U_i' (or the skew-conjugate $-U_i'$) of U_i, being therefore a general counter-alternant of type $\{\pi', -\pi\}$ and of the class conjugate to that of U_i;

the parameters of $Y_0, U_1, U_2, \ldots U_s$ are all independent;

and $\bar{\mathbf{Y}}'$ is a matrix of the same form in which the $2s+1$ non-zero parts are counter-alternants of type $\{\pi, -\pi'\}$. In the particular case when A and $-A$ are equicanonical, Π becomes the canonical square matrix described in Ex. viii, and all the parts of $\bar{\mathbf{Y}}$ and $\bar{\mathbf{Y}}'$ are *quadrate*, the classes of U_i and V_i being the same.

The properties of the parts U_i and Y_0 given respectively in Theorems I *b* and III *b* of § 243 lead to the following theorem, and determine the values of κ and κ' in it. In both parts of the theorem the number of arbitrary parameters in $\bar{Y}$ or $\bar{Y}'$ is equal to the number of independent non-zero particularisations of that matrix.

Theorem III b. *The two general symmetric contra-commutants*

$$\bar{Y} = \{A', -A\}, \quad \bar{Y}' = \{A, -A'\}$$

are undegenerate when and only when every distinct characteristic potent divisor of A not of the form λ^{2p+1} is repeated an even number of times, i.e. when and only when:

(1) *A and $-A$ are equicanonical so that the characteristic potent divisors of A corresponding to its non-zero latent roots occur in pairs of the form*

$$(\lambda - c)^e, \quad (\lambda + c)^e, \qquad \textit{where } c \neq 0;$$

(2) *every distinct characteristic potent divisor of A such as* λ^{2p}, *which is a power of* λ *with even index, is repeated an even number of times;*

and in all cases each of them contains exactly κ *independent arbitrary parameters, where* κ *is the sum of the arbitrary parameters occurring in the 'general symmetric' contra-commutant or counter-alternant* Y_0 *and the 'general' contra-commutants or counter-alternants* $U_1, U_2, \ldots U_s$.

The two general skew-symmetric contra-commutants

$$\bar{Y} = \{A', -A\}, \quad \bar{Y}' = \{A, -A'\}$$

are undegenerate when and only when every distinct characteristic potent divisor of A not of the form λ^{2p} *is repeated an even number of times, i.e. when and only when:*

(1) *A and* $-A$ *are equicanonical, so that the characteristic potent divisors of A corresponding to its non-zero latent roots occur in pairs of the form*

$$(\lambda - c)^e, \quad (\lambda + c)^e, \qquad \text{where } c \neq 0;$$

(2) *every distinct characteristic potent divisor of A such as* λ^{2p+1}, *which is a power of* λ *with odd index, is repeated an even number of times;*

and in all cases each of them contains exactly κ' *independent arbitrary parameters, where* κ' *is the sum of the numbers of the arbitrary parameters occurring in the 'general skew-symmetric' contra-commutant or counter-alternant* Y_0 *and the 'general' contra-commutants or counter-alternants* $U_1, U_2, \ldots U_s$.

Ex. xii. From Ex. vii of § 243 we see that in all cases $\kappa + \kappa'$ is the total number of arbitrary parameters occurring in a general contra-commutant $\bar{Y}$ or $\bar{Y}'$. Since a non-zero square matrix cannot be both symmetric and skew-symmetric, we conclude that:

The sum of a 'general symmetric' and a 'general skew-symmetric' contra-commutant $\{A', -A\}$ *or* $\{A, -A'\}$ *whose parameters are independent is always a 'general' contra-commutant* $\{A', -A\}$ *or* $\{A, -A'\}$.

This theorem and that of Ex. x also follow directly from the fact that a contra-commutant Y of any given type can always be expressed as the sum of a symmetric and a skew-symmetric contra-commutant of that type. In fact the conjugate Y' of Y is a contra-commutant of the same type as Y, and we can put

$$Y = \tfrac{1}{2}(S + T), \quad \text{where} \quad S = Y + Y', \; T = Y - Y'.$$

Ex. xiii. *Symmetric and skew-symmetric alternating co-commutants.*

Constructions analogous to those of the text for the contra-commutants are only possible when A is a canonical square matrix, and in some other special cases.

The general symmetric (or general skew-symmetric) co-commutants $\bar{\mathbf{X}} = \{\Pi, -\Pi\}$, $\bar{\mathbf{X}}' = \{\Pi', -\Pi'\}$ are specialisations of the matrices ξ, ξ' of Ex. vii, and are derived from the matrix $\bar{\eta}$ of Ex. vii by taking:

Y_0 to be a general symmetric (or general skew-symmetric) quadrate ante-alternant, which is necessarily quasi-scalaric;

U_i, V_i to be mutually conjugate (or mutually skew-conjugate) ante-alternants of the same type, which are necessarily quasi-scalaric.

But the corresponding co-commutants $\bar{X}=\{A, -A\}$, $\bar{X}'=\{A', -A'\}$ derived from them by the formulae

$$\bar{X}=h\bar{\mathbf{X}}H, \quad \bar{X}'=H'\bar{\mathbf{X}}'h'$$

are not necessarily symmetric (or skew-symmetric) except in the special case when $H=h'$, i.e. when (d) is a symmetric semi-unit transformation. This special case occurs when A is a canonical square matrix, so that (d) is a symmetric derangement. The above formulae then become symmetric derangements determining the general symmetric (or general skew-symmetric) co-commutants $\bar{X}$, $\bar{X}'$. Accordingly from Ex. vi or Ex. vii of this article and Theorem IIb of § 243, we see that:

If A is a canonical square matrix, then the general symmetric co-commutants $\{A, -A\}$, $\{A', -A'\}$ are undegenerate when and only when A and $-A$ are equicanonical; also the general skew-symmetric co-commutants $\{A, -A\}$, $\{A', -A'\}$ are undegenerate when and only when A and $-A$ are equicanonical, and further every distinct characteristic potent divisor of A which is a power of λ is repeated an even number of times, so that all the characteristic potent divisors of A can be coupled together in pairs of the form

$$(\lambda-c)^e, \quad (\lambda+c)^e.$$

Whenever A is a symmetric matrix, the general symmetric (or general skew-symmetric) co-commutants $\{A, -A\}$, $\{A', -A'\}$ can be taken to be the general symmetric (or general skew-symmetric) contra-commutants $\{A', -A\}$, $\{A, -A'\}$ constructed in the text; and whenever A is a skew-symmetric matrix, they can be taken to be the general symmetric (or general skew-symmetric) contra-commutants $\{A, A'\}$, $\{A', A\}$ constructed in sub-article 2.

Ex. xiv. *If A is a skew-symmetric matrix, every potent divisor of its characteristic matrix $A(\lambda)$ which is not of the form λ^{2p+1} must be repeated an even number of times.*

This follows from Theorem III b, because the general symmetric commutant $\{A, -A'\}$ can be taken to be the general symmetric commutant $\{A, A\}$, and is therefore undegenerate. Alternatively the unit matrix $I=[1]_m^m$ is a particular undegenerate symmetric contra-commutant $\{A, -A'\}$ or $\{A', -A\}$.

NOTE 3. *Rank of any general commutant $\{A, B\}$.*

Let $A=[a]_m^m$ and $B=[b]_n^n$ be square matrices of orders m and n whose elements are rational integral functions of certain scalar variables $\gamma_1, \gamma_2, \gamma_3, \ldots$; and let r be the smaller of the two integers m and n. Also let

$$A(\lambda)=[a]_m^m-\lambda\,.\,[1]_m^m \quad \text{and} \quad B(\lambda)=[b]_n^n-\lambda\,.\,[1]_n^n$$

be the characteristic matrices of A and B; and let

$$f_1(\lambda), f_2(\lambda), \ldots f_m(\lambda) \quad \text{and} \quad g_1(\lambda), g_2(\lambda), \ldots g_n(\lambda)$$

be those rational integral functions of $\lambda, \gamma_1, \gamma_2, \gamma_3, \ldots$ which are the potent factors

of orders 1, 2, ... m of $A(\lambda)$ and of orders 1, 2, ... n of $B(\lambda)$.

These we will call *ascendent* potent factors when they are arranged in the orders shown above, and *descendent* potent factors when they are arranged in the reverse orders. Each of them is a product of powers of irresoluble (or irreducible) divisors of $A(\lambda)$ or $B(\lambda)$, and every such divisor must contain λ; for by Ex. xxiii of § 185 it cannot be a function of the γ's only. Using the principles described in Note 2 of § 238 we can deduce from Theorem Ia that:

The rank ρ of the general commutant $X=\{A, B\}$ is the sum of the degrees in λ of the highest common factors $h_1(\lambda), h_2(\lambda), \ldots h_r(\lambda)$ of the pairs of corresponding descendent potent factors

$$f_m(\lambda), g_n(\lambda)\,;\, f_{m-1}(\lambda), g_{n-1}(\lambda)\,;\, \ldots f_{m+1-r}(\lambda), g_{n+1-r}(\lambda)$$

of the characteristic matrices $A(\lambda)$, $B(\lambda)$. ....(E)

Here $f_1(\lambda)$ or $g_1(\lambda)$ is a member of the last pair of corresponding potent factors according as $r=m \not> n$ or $r=n \not> m$. Moreover there will ordinarily be only a few of the earlier highest common factors which have degrees greater than 0 in λ, all the rest being non-zero constants which can be omitted.

If $T_1, T_2, T_3, \ldots$ are those distinct irresoluble (or irreducible) rational integral functions of $\lambda, \gamma_1, \gamma_2, \gamma_3, \ldots$ which are irresoluble (or irreducible) divisors of both the matrices $A(\lambda)$ and $B(\lambda)$, and if the indices of the highest powers of T_i which are factors of

$$f_1(\lambda), f_2(\lambda), \ldots f_m(\lambda) \quad \text{and} \quad g_1(\lambda), g_2(\lambda), \ldots g_n(\lambda)$$

are

$$a_{i1}, \quad a_{i2}, \ldots \quad a_{im} \quad \text{and} \quad \beta_{i1}, \quad \beta_{i2}, \ldots \beta_{in},$$

the theorem (E) is equivalent to the statement that we can put

$$\rho=\rho_1+\rho_2+\rho_3+\ldots,$$

where ρ_i is the sum of the degrees in λ of the smaller powers of T_i in the pairs

$$(T_i^{a_{im}}, T_i^{\beta_{in}}), \quad (T_i^{a_{i,m-1}}, T_i^{\beta_{i,n-1}}), \ldots.$$

Hence if *corresponding* potent divisors are always powers of the same irresoluble (or irreducible) divisor, we can re-enunciate (E) in the following form:

The rank ρ of the general commutant $X=\{A, B\}$ is the sum of the degrees in λ of the highest common factors of all pairs of corresponding descendent potent divisors of the two characteristic matrices $A(\lambda)$ and $B(\lambda)$. ....(E′)

In this case each pair consists of potent divisors of $A(\lambda)$ and $B(\lambda)$ which are two powers of the same divisor T, and the corresponding highest common factor is that one of the two potent divisors which is the smaller power of T.

We will prove these theorems by taking A_0, B_0, X_0 to be particularisations of A, B, X obtained by ascribing such 'ordinary' particular values to the γ's that:

(1) X_0 is a general commutant $\{A_0, B_0\}$;

(2) X_0 has the same rank as X;

(3) the successive maximum factors of $A(\lambda)$ and $B(\lambda)$ become particularised into the successive maximum factors of $A_0(\lambda)$ and $B_0(\lambda)$;

(4) the common divisors $T_1, T_2, T_3, \ldots$ of $A(\lambda)$ and $B(\lambda)$ become particularised into functions which have the same degrees in λ as $T_1, T_2, T_3, \ldots$, and have no common or repeated linear factors.

These conditions will be secured when certain unspecified rational integral functions of the γ's do not vanish. The successive descendent potent factors of $A(\lambda)$ and $B(\lambda)$ and their highest common factors will then become particularised into the successive descendent potent factors of $A_0(\lambda)$ and $B_0(\lambda)$ and their highest common factors, the degrees in λ being unaltered. Consequently when we regard ρ as the rank of the general commutant $X_0=\{A_0, B_0\}$, it follows from Theorem I a that it has the value ascribed to it in the above theorems.

Ex. xv. *If A and A' are mutually conjugate square matrices whose elements are rational integral functions of certain scalar variables $\gamma_1, \gamma_2, \gamma_3, \ldots$, there always exist undegenerate square contra-commutants*

$$U=\{A', A\}, \quad V=\{A, A'\}.$$

For the characteristic matrices of A and A' have the same potent divisors. It may of course be impossible for such undegenerate contra-commutants to be non-singular, i.e. to be independent of the γ's.

When $A=[a]_2^2$ is a square matrix whose four elements are independent arbitrary parameters, there are two independent particular non-zero contra-commutants $\{A', A\}$, which can be taken to be the square matrices

$$\begin{bmatrix} a_{11}-a_{22}, & a_{12} \\ a_{12} \quad , & 0 \end{bmatrix}, \quad \begin{bmatrix} 0 \;, & a_{21} \\ a_{21}, & a_{22}-a_{11} \end{bmatrix};$$

and there are two independent particular non-zero contra-commutants $\{A, A'\}$, which can be taken to be the square matrices

$$\begin{bmatrix} a_{11}-a_{22}, & a_{21} \\ a_{21} \quad , & 0 \end{bmatrix}, \quad \begin{bmatrix} 0 \;, & a_{12} \\ a_{12}, & a_{22}-a_{11} \end{bmatrix}.$$

NOTE 4. *Condition that the general commutant of any given square matrix A shall be a rational integral function of A.*

By the methods used in Note 3 we can obtain the following generalisation of Theorem II relating to a square matrix $A=[a]_m^m$ whose elements are rational integral functions of scalar variables $\gamma_1, \gamma_2, \gamma_3, \ldots$.

In order that every commutant $\{A, A\}$ or the general commutant $X=\{A, A\}$ shall be a rational integral function of A, it is necessary and sufficient that A shall satisfy no rational integral equation whose degree is less than m.

When this condition is satisfied, there are exactly m independent non-zero particular commutants $\{A, A\}$, and the general commutant $X=\{A, A\}$ is an arbitrary rational integral function of A. In all other cases there are more than m independent particular non-zero commutants $\{A, A\}$, and there are commutants $\{A, A\}$ which are not rational integral functions of A. ....(F)

If $D_i(\lambda)$ and $E_i(\lambda)$ are respectively the maximum and potent factors of $A(\lambda)$ of order i, the condition of the theorem is satisfied when and only when we can put

$$D_{m-1}(\lambda)=1 \quad \text{or} \quad E_{m-1}(\lambda)=1, \qquad \text{(whichever we please)},$$

so that the rational integral equation of lowest degree satisfied by A, viz. the equation $E_m(A)=0$, becomes $D_m(A)=0$.

We will suppose that there are exactly r independent non-zero particular commutants $\{A, A\}$, and that s is the degree of $E_m(\lambda)$ in λ, i.e. the lowest possible degree of a rational integral equation satisfied by A.

We will prove (F) by taking A_0, X_0 to be particularisations of A, X obtained by ascribing such 'ordinary' particular values to the γ's that:

(1) X_0 is a general commutant $\{A_0, B_0\}$;

(2) the highest common factors of the minor determinants of $A(\lambda)$ of orders $m, m-1$ become particularised into the highest common factors of the minor determinants of $A_0(\lambda)$ of orders $m, m-1$;

(3) the irresoluble divisors $T_1, T_2, T_3, \ldots$ of $A(\lambda)$ become particularised into functions which have the same degrees in λ as $T_1, T_2, T_3, \ldots$ and have no common or repeated linear factors.

Under these circumstances there are exactly r independent particular non-zero commutants $\{A_0, A_0\}$, and s is the lowest possible degree of a rational integral equation satisfied by A_0, because the potent factors of $A(\lambda)$ and $A_0(\lambda)$ of order m have the same degree in λ. Therefore by Theorem II we must have either $s=r=m$ or $s<m$, $r>m$.

If $s=r=m$, there are exactly m independent particular non-zero commutants $\{A, A\}$ which can be taken to be the square matrices

$$I, A, A^2, \ldots A^{m-1}, \qquad \text{where } I=[1]_m^m;$$

consequently the general commutant $X=\{A, A\}$ is an arbitrary rational integral function of A.

If $s<m$, $r>m$, there are more than m independent particular non-zero commutants $\{A, A\}$, and there are commutants $\{A, A\}$ which are not rational integral functions of A.

§ 245. Undegenerate square matrix expressed as a product of two square matrices each of which is symmetric or skew-symmetric.

Let $A=[a]_m^m$ be a given undegenerate square matrix with constant elements, and suppose that it is expressed in the form

$$[a]_m^m=[p]_m^m[q]_m^m \quad \text{or} \quad A=pq, \text{.................................(a)}$$

where $p=[p]_m^m$ and $q=[q]_m^m$ are square matrices with constant elements, which are necessarily both undegenerate. Also let $P=[P]_m^m$ and $Q=[Q]_m^m$ be the undegenerate inverses of p and q; and let A', p', q', P', Q' be the conjugates of A, p, q, P, Q. We will investigate the circumstances under which such representations of A are possible in which each of the square matrices p and q is either symmetric or skew-symmetric. The characteristic potent divisors of A will be defined as usual to be the potent divisors of the characteristic matrix

$$A(\lambda)=[a]_m^m-\lambda\,.\,[1]_m^m.$$

CASE I. *When p and q are both symmetric.*

In this case P and Q are both symmetric; and by prefixing P, postfixing Q on both sides we deduce from (a) the equations

$$PA=A'P=A'P'=q, \text{.......................................(1)}$$

$$AQ=QA'=Q'A'=p, \text{.......................................(1')}$$

where the intermediate transformations are due to the facts that q and P are symmetric in (1), p and Q are symmetric in (1'). These equations show that:

P, Q must be undegenerate symmetric contra-commutants $\{A', A\}$, $\{A, A'\}$;

i.e. p, q must be undegenerate symmetric contra-commutants $\{A, A'\}$, $\{A', A\}$.

Conversely if p is any given undegenerate symmetric contra-commutant $\{A, A'\}$, so that its inverse P is an undegenerate symmetric contra-commutant $\{A', A\}$, we have the equations (1) in which q is a known undegenerate symmetric matrix, which by § 239 must be a contra-commutant $\{A', A\}$; and by prefixing p we obtain a representation of A in the form (a). Or if q is any given undegenerate symmetric contra-commutant $\{A', A\}$, so that its inverse Q is an undegenerate symmetric contra-commutant $\{A, A'\}$, we have the equations (1') in which p is a known undegenerate symmetric matrix, which by § 239 must be a contra-commutant $\{A, A'\}$; and by postfixing q we obtain a representation of A in the form (a).

Since by Theorem III a of § 244 or Ex. vii of § 240 there always exist undegenerate symmetric contra-commutants $\{A', A\}$, $\{A, A'\}$, we conclude that:

If A is undegenerate, there always exist equations of the form (a) *in which p and q are both symmetric.*

In every such equation the factors p, q are undegenerate symmetric contra-commutants $\{A, A'\}$, $\{A', A\}$; moreover either factor can be chosen arbitrarily subject to that one of these conditions which is appropriate to it, the other factor being then uniquely determinate.

CASE II. *When p and q are both skew-symmetric.*

In this case P and Q are both skew-symmetric; and by prefixing P, postfixing Q on both sides we deduce from (a) the equations

$$PA=A'P=-A'P'=q, \text{.......................................(2)}$$

$$AQ=QA'=-Q'A'=p, \text{.......................................(2')}$$

where the intermediate transformations are due to the facts that q and P are skew-symmetric in (2), p and Q are skew-symmetric in (2′). These equations show that:

P, Q must be undegenerate skew-symmetric contra-commutants $\{A', A\}$, $\{A, A'\}$;

i.e. p, q must be undegenerate skew-symmetric contra-commutants $\{A, A'\}$, $\{A', A\}$.

Conversely if p is any given undegenerate skew-symmetric contra-commutant $\{A, A'\}$, so that its inverse P is an undegenerate skew-symmetric contra-commutant $\{A', A\}$, we have the equations (2) in which q is a known undegenerate skew-symmetric matrix, which must be a contra-commutant $\{A', A\}$; and by prefixing p we obtain a representation of A in the form (a). Or if q is any given undegenerate skew-symmetric contra-commutant $\{A', A\}$, so that its inverse Q is an undegenerate skew-symmetric contra-commutant $\{A, A'\}$, we have the equations (2′) in which p is a known undegenerate skew-symmetric matrix, which must be a contra-commutant $\{A, A'\}$; and by postfixing q we obtain a representation of A in the form (a).

Referring to Theorem III *a* of § 244 for the necessary and sufficient conditions for the existence of undegenerate skew-symmetric contra-commutants $\{A, A'\}$, $\{A', A\}$, we conclude that:

If A is undegenerate, there exist equations of the form (a) *in which p and q are both skew-symmetric when and only when the characteristic potent divisors of A can be coupled together in pairs of the form*

$$(\lambda - c)^e, \quad (\lambda - c)^e, \qquad (\textit{where } c \neq 0);$$

every distinct characteristic potent divisor being repeated an even number of times.

In every such equation the factors p, q are undegenerate skew-symmetric contra-commutants $\{A, A'\}$, $\{A', A\}$; moreover either factor can be chosen arbitrarily subject to that one of these conditions which is appropriate to it, the other factor being then uniquely determinate.

CASE III. *When p is symmetric and q skew-symmetric.*

In this case P is symmetric and Q skew-symmetric; and by prefixing P, postfixing Q on both sides we deduce from (a) the equations

$$PA = -A'P = -A'P' = q, \quad\quad (3)$$
$$AQ = -QA' = \quad Q'A' = p, \quad\quad (3')$$

where the intermediate transformations are due to the facts that q is skew-symmetric and P symmetric in (3), p is symmetric and Q skew-symmetric in (3′). These equations show that

P must be an undegenerate symmetric contra-commutant $\{-A', A\}$,
Q must be an undegenerate skew-symmetric contra-commutant $\{A, -A'\}$;

i.e. p must be an undegenerate symmetric contra-commutant $\{A, -A'\}$,
q must be an undegenerate skew-symmetric contra-commutant $\{-A', A\}$.

Conversely if p is any given undegenerate symmetric contra-commutant $\{A, -A'\}$, so that its inverse P is an undegenerate symmetric contra-commutant $\{-A', A\}$, we have the equations (3) in which q is a known undegenerate skew-symmetric matrix, which must be a contra-commutant $\{-A', A\}$; and by prefixing p we obtain a representation of A in the form (a). Or if q is any given undegenerate skew-symmetric contra-commutant $\{-A', A\}$, so that its inverse Q is an undegenerate skew-symmetric contra-commutant $\{A, -A'\}$, we have the equations (3′) in which p is a known undegenerate symmetric matrix, which must be a contra-commutant $\{A, -A'\}$; and by postfixing q we obtain a representation of A in the form (a).

Referring to Theorem III *b* of § 244 for the necessary and sufficient conditions for the existence of undegenerate symmetric contra-commutants $\{A, -A'\}$ and undegenerate skew-symmetric contra-commutants $\{-A', A\}$, we conclude that:

If A is undegenerate, there exist equations of the form (a) *in which p is symmetric and q skew-symmetric when and only when the characteristic potent divisors of A can be coupled together in pairs of the form*

$$(\lambda - c)^e, \quad (\lambda + c)^e, \qquad (\textit{where } c \neq 0).$$

In every such equation the factor p is an undegenerate symmetric contra-commutant $\{A, -A'\}$, and the factor q is an undegenerate skew-symmetric contra-commutant $\{-A', A\}$; moreover either factor can be chosen arbitrarily subject to that one of these conditions which is appropriate to it, the other factor being then uniquely determinate.

CASE IV. *When p is skew-symmetric and q symmetric.*

In this case P is skew-symmetric and Q symmetric; and by prefixing P, postfixing Q on both sides we deduce from (a) the equations

$$PA = -A'P = A'P' = q, \quad \text{.................................}(4)$$
$$AQ = -QA' = -Q'A' = p, \quad \text{.................................}(4')$$

where the intermediate transformations are due to the facts that q is symmetric and P skew-symmetric in (4), p is skew-symmetric and Q symmetric in (4′). These equations show that:

P must be an undegenerate skew-symmetric contra-commutant $\{-A', A\}$,
Q must be an undegenerate symmetric contra-commutant $\{A, -A'\}$;
i.e. p must be an undegenerate skew-symmetric contra-commutant $\{A, -A'\}$,
q must be an undegenerate symmetric contra-commutant $\{-A', A\}$.

Arguing as before, and referring to Theorem III *b* of § 244, we conclude that:

If A is undegenerate, there exist equations of the form (a) *in which p is skew-symmetric and q symmetric when and only when the characteristic potent divisors of A can be coupled together in pairs of the form*

$$(\lambda - c)^e, \quad (\lambda + c)^e, \qquad (\textit{where } c \neq 0).$$

In every such equation the factor p is an undegenerate skew-symmetric contra-commutant $\{A, -A'\}$, and the factor q is an undegenerate symmetric contra-commutant $\{-A', A\}$; moreover either factor can be chosen arbitrarily subject to that one of these conditions which is appropriate to it, the other factor being then uniquely determinate.

All the equations occurring in the argument are commutantal; and the equation (a) in

Cases I and II, Cases III and IV

can be regarded as an equigradent commutantal transformation

$$[p]_m^m \, [1]_m^m \, [q]_m^m = [a]_m^m$$

of type

$$\{A, A'\}\{A', A'\}\{A', A\} = \{A, A\}, \quad \{A, -A'\}\{-A', -A'\}\{-A', A\} = \{A, A\}.$$

§ 246. Reduction of a commutantal transformation.

1. *Recapitulatory remarks.*

Every matrix can be regarded as a commutant; for it is always possible to determine two square matrices $A = [a]_m^m$, $B = [b]_n^n$ such that a given matrix $X = [x]_m^n$ is a commutant $\{A, B\}$; in fact this is so when $A = c.[1]_m^m$, $B = c.[1]_n^n$. But it is not always possible to distinguish between two given matrices by representing them as commutants, i.e. the possible representations of one may be exactly the same as the possible representations of the other.

If $A = [a]_m^m$ is any given square matrix with constant elements, and if $\alpha = [\alpha]_m^m$ is a standard canonical reduced form of A or any square matrix equicanonical with A, we know that it is always possible in many ways to determine an equation of the form

$$[a]_m^m = [h]_m^m [\alpha]_m^m [H]_m^m \quad \text{or} \quad A = h\alpha H \text{.................}(1)$$

in which h and H are mutually inverse undegenerate square matrices with constant elements, and every such equation is an *equimutant commutantal transformation* of the type

$$\{A, A\} = \{A, \alpha\}\{\alpha, \alpha\}\{\alpha, A\}; \quad \text{.....................}(1')$$

being in fact an *isomorphic commutantal transformation* of that type, because all the matrices occurring in it are square and of the same order. Further if A and α both lie in a restricted domain of rationality Ω, then (1) can always be a transformation in Ω. The determination of such a transformation is equivalent to the determination of a particular undegenerate commutant $\{A, \alpha\}$ or $\{\alpha, A\}$.

Let $A = [a]_m^m$, $B = [b]_n^n$ be two given square matrices with constant elements, and $\alpha = [\alpha]_m^m$, $\beta = [\beta]_n^n$ be standard canonical reduced forms of A, B or any two square matrices equicanonical with A, B; and let

$$[a]_m^m = [h]_m^m [\alpha]_m^m [H]_m^m, \quad [b]_n^n = [k]_n^n [\beta]_n^n [K]_n^n$$

or

$$A = h\alpha H, \qquad B = k\beta K$$

be two given equimutant (or isomorphic) commutantal transformations of the types

$$\{A, A\} = \{A, \alpha\}\{\alpha, \alpha\}\{\alpha, A\}, \qquad \{B, B\} = \{B, \beta\}\{\beta, \beta\}\{\beta, B\},$$

so that h, k are any two given particular undegenerate commutants $\{A, \alpha\}$, $\{B, \beta\}$. Then if $X = [x]_m^n$ is any given commutant $\{A, B\}$, we know that it is always possible to express X in the form

$$[x]_m^n = [h]_m^m [\xi]_m^n [K]_n^n \quad \text{or} \quad X = h\xi K, \quad \text{...............}(2)$$

where ξ is a uniquely determinate commutant $\{\alpha, \beta\}$; and (2) is an *equigradent commutantal transformation* of the type

$$\{A, B\} = \{A, \alpha\}\{\alpha, \beta\}\{\beta, B\}. \quad \text{.....................}(2')$$

Or if $\xi = [\xi]_m^n$ is any given commutant $\{\alpha, \beta\}$, then (2) is an equigradent commutantal transformation of the type (2′) in which X is a uniquely determinate commutant $\{A, B\}$.

In the following sub-articles we shall use transformations such as (1) and (2) to reduce any commutantal transformation to an equivalent commutantal transformation in which all the matrices are commutants of standard canonical square matrices; and it will be clear that any commutantal equation whatever can be reduced to an equivalent commutantal equation of that special character.

When $B=A$, $\beta=\alpha$, so that $n=m$, we can take (2) to be the *equimutant* (or *isomorphic*) commutantal transformation

$$[x]_m^m=[h]_m^m\,[\xi]_m^m\,[H]_m^m \quad \text{of type} \quad \{A, A\}=\{A, \alpha\}\{\alpha, \alpha\}\{\alpha, A\};$$

and when B, β are the conjugates A', α' of A, α, so that $n=m$, we can take (2) to be the equigradent commutantal transformation

$$[x]_m^m=[h]_m^m\,[\xi]_m^m\,\overline{\underline{h}}_m^m \quad \text{of type} \quad \{A, A'\}=\{A, \alpha\}\{\alpha, \alpha'\}\{\alpha', A'\}$$

which is *symmetric* in form. In both these cases h can be any given particular undegenerate commutant $\{A, \alpha\}$.

NOTE 1. *Generalisation of the transformation* (1); *third form of the conditions for the existence of an equimutant transformation.*

Let $A=[a]_m^m$, $a=[a]_r^r$ be two given square matrices of orders m, r with constant elements, where $r \not> m$.

It has been shown in Theorem II of § 227 that there exists an equimutant transformation

$$[a]_m^m=[h]_m^r\,[a]_r^r\,[H]_r^m \quad \text{or} \quad A=haH, \; \ldots \ldots\ldots\ldots\ldots\ldots\ldots\ldots (1'')$$

in which $Hh=[1]_r^r$, when and only when the canonical reduced forms of a, A can be taken to be

$$[\Pi]_r^r, \quad \begin{bmatrix}\Pi, & 0\\ 0, & 0\end{bmatrix}_{r,\,m-r}^{r,\,m-r},$$

i.e. when and only when the potent divisors of the characteristic matrix $A(\lambda)$ are those of the characteristic matrix $a(\lambda)$ together with $m-r$ others which are all equal to λ. By postfixing h, prefixing H on both sides of (1″) we obtain the equations

$$Ah=ha, \quad aH=HA$$

from which we conclude that every equimutant transformation such as (1″) is an *equimutant commutantal transformation* of the type

$$\{A, A\}=\{A, a\}\{a, a\}\{a, A\}. \; \ldots\ldots\ldots\ldots\ldots\ldots\ldots\ldots\ldots (1')$$

To the two forms of the conditions for the existence of an equimutant transformation (1″) which have been given in §§ 227 and 234, we will here add a third by proving the following theorem:

In order that there shall exist an equimutant transformation (1″) *converting* a *into* A, *it is necessary and sufficient that:*

(1) *there shall exist an undegenerate commutant* $\{A, a\}$ *or* $\{a, A\}$;

(2) A *and* a *shall have equal ranks.*

Moreover when these two conditions are satisfied, we can construct an equimutant transformation (1″) *in which* h *is any given particular undegenerate commutant* $\{A, a\}$, *or* H *any given particular undegenerate commutant* $\{a, A\}$.

The preceding considerations show that the conditions (1) and (2) of the theorem are necessary; and we need only consider their sufficiency.

Suppose that A and a have the same rank t, and let $h=[h]_m^r$ be a particular undegenerate commutant $\{A, a\}$. Then the equation $Ah=ha$ shows that Ah has rank t, and that the equation

$$[a]_m^m [h]_m^r \; \overline{\underline{y}}_r = 0$$

has exactly $r-t$ unconnected non-zero solutions. Consequently exactly $r-t$ unconnected non-zero solutions of the equation

$$[a]_m^m \; \overline{\underline{x}}_m = 0$$

are connected with the vertical rows of h, and we can determine $m-r$ other unconnected non-zero solutions of the equation forming the vertical rows of a matrix $[k]_m^{m-r}$ which is such that the square matrix $[h]_m^m=[h, k]_m^{r,\, m-r}$ has rank m. When $[h]_m^m$ is an undegenerate square matrix thus determined, we have

$$[a]_m^m [h]_m^m = [a]_m^m [h, 0]_m^{r,\, m-r} = [h]_m^r [a, 0]_r^{r,\, m-r};$$

and by postfixing the square matrix $[H]_m^m$ inverse to $[h]_m^m$ we obtain an equimutant transformation (1″) in which h is the given undegenerate commutant $\{A, a\}$. Similarly if A and a have the same rank t, and if $H=[H]_r^m$ is a given undegenerate commutant $\{a, A\}$, we can deduce from the equation $aH=HA$ an equimutant transformation (1″) in which H is the given undegenerate commutant $\{a, A\}$.

We have now shown that the conditions (1) and (2) of the theorem are sufficient as well as necessary, and that the theorem is true in all respects.

We can prove the sufficiency of the conditions (1) and (2) in a simpler way by showing that they are equivalent to the conditions in Theorem II of § 227.

If the condition (1) is satisfied, then by Ex. ii of § 244 we can take the potent divisors of the characteristic matrices $a(\lambda)$ and $A(\lambda)$ to be respectively

$$(\lambda-a_1)^{p_1},\ (\lambda-a_2)^{p_2},\ \dots\ \lambda^{u_1},\ \lambda^{u_2},\ \dots\ \lambda^{u_\rho},$$

and $(\lambda-a_1)^{P_1},\ (\lambda-a_2)^{P_2},\ \dots\ \lambda^{U_1},\ \lambda^{U_2},\ \dots\ \lambda^{U_\rho},\ (\lambda-b_1)^{Q_1},\ (\lambda-b_2)^{Q_2},\ \dots\ \lambda^{V_1},\ \lambda^{V_2},\ \dots\ \lambda^{V_\sigma}$,
where the a's and b's are all different from 0, and where

$$p_1 \not> P_1,\ p_2 \not> P_2,\ \dots\ u_1 \not> U_1,\ u_2 \not> U_2,\ \dots\ u_\rho \not> U_\rho,$$

and

$$r=(p_1+p_2+\dots)+(u_1+u_2+\dots+u_\rho),$$
$$m=(P_1+P_2+\dots)+(U_1+U_2+\dots+U_\rho)+(Q_1+Q_2+\dots)+(V_1+V_2+\dots+V_\sigma).$$

Since the ranks of a and A are respectively $r-\rho$ and $m-(\rho+\sigma)$, the condition (2) will be simultaneously satisfied when and only when $m-r-\sigma=0$, i.e. when and only when

$$(P_1-p_1)+(P_2-p_2)+\dots+(U_1-u_1)+(U_2-u_2)+\dots+Q_1+Q_2+\dots$$
$$+(V_1-1)+(V_2-1)+\dots=0.$$

Because every term on the left is positive, this is only possible when the Q's are all absent, the V's are all equal to 1, and the potent divisors of $A(\lambda)$ are those of $a(\lambda)$ together with $m-r$ others which are all equal to λ, i.e. when and only when there exists an equimutant transformation (1″).

NOTE 2. *Generalisation of the transformation* (2).

Let
$$[a]_m^m=[h]_m^r [a]_r^r [H]_r^m, \quad [b]_n^n=[k]_n^s [\beta]_s^s [K]_s^n$$
or
$$A=haH, \quad B=k\beta K$$

be given equimutant transformations by which the square matrices A, B are derived from the square matrices α, β; so that $r \not> m$ and $s \not> n$.

Then if $$[x]_m^n = [h]_m^r [\xi]_r^s [K]_s^n \quad \text{or} \quad X = h\xi K, \qquad \text{(2'')}$$
it follows from the equations $XB = h.\xi\beta.K$, $AX = h.\alpha\xi.K$ that X is a commutant $\{A, B\}$ when and only when ξ is a commutant $\{\alpha, \beta\}$, and that under those circumstances (2'') is an *equigradent commutantal transformation* of the type
$$\{A, B\} = \{A, \alpha\}\{\alpha, \beta\}\{\beta, B\}. \qquad \text{(2')}$$

The equation (2'') establishes a one-one correspondence between all commutants $\xi = \{\alpha, \beta\}$ and all those commutants $X = \{A, B\}$ which satisfy the equation
$$[x]_m^n = [h]_m^r [H]_r^m . [x]_m^n . [k]_n^s [K]_s^n ;$$
but it does not establish a one-one correspondence between *all* commutants $\{A, B\}$ and all commutants $\{\alpha, \beta\}$, and it does not usually give a *general* commutant $X = \{A, B\}$, when ξ is a general commutant $\{\alpha, \beta\}$. In fact the rank of the general commutant $\{A, B\}$ is ordinarily greater than the rank of the general commutant $\{\alpha, \beta\}$.

2. *Reduction of any commutantal transformation.*

Remembering that an equimutant transformation is called an isomorphic transformation when all the matrices occurring in it are square and of the same order, let

A, B, P, Q be given square matrices of orders r, s, m, n with constant elements, and
$$A = h\mathbf{A}H \quad \text{of type} \quad \{A, A\} = \{A, \mathbf{A}\}\{\mathbf{A}, \mathbf{A}\}\{\mathbf{A}, A\},$$
$$B = k\mathbf{B}K \quad \text{of type} \quad \{B, B\} = \{B, \mathbf{B}\}\{\mathbf{B}, \mathbf{B}\}\{\mathbf{B}, B\},$$
$$P = u\mathbf{P}U \quad \text{of type} \quad \{P, P\} = \{P, \mathbf{P}\}\{\mathbf{P}, \mathbf{P}\}\{\mathbf{P}, P\},$$
$$Q = v\mathbf{Q}V \quad \text{of type} \quad \{Q, Q\} = \{Q, \mathbf{Q}\}\{\mathbf{Q}, \mathbf{Q}\}\{\mathbf{Q}, Q\}$$
be given isomorphic transformations by which they can be derived from given square matrices $\mathbf{A}$, $\mathbf{B}$, $\mathbf{P}$, $\mathbf{Q}$ of the same orders r, s, m, n, which could be taken to be standard canonical square matrices. Whenever the given transformations are isomorphic (or equimutant), they are necessarily commutantal of the types shown. The square matrices h, k, u, v of orders r, s, m, n can be any given particular undegenerate commutants of the types shown; or alternatively their inverses the square matrices H, K, U, V of orders r, s, m, n can be any given particular undegenerate commutants of the types shown. Then we have the following theorem.

Theorem. *Any commutantal transformation of the form*
$$[x]_m^r [e]_r^s [y]_s^n = [E]_m^n \quad \textit{or} \quad XeY = E \qquad \text{(A)}$$
and type $$\{P, A\}\{A, B\}\{B, Q\} = \{P, Q\} \qquad \text{(a)}$$
can be replaced by an equivalent commutantal transformation
$$[\xi]_m^r [\omega]_r^s [\eta]_s^n = [\Omega]_m^n \quad \textit{or} \quad \xi\omega\eta = \Omega \qquad \text{(A')}$$
of type $$\{\mathbf{P}, \mathbf{A}\}\{\mathbf{A}, \mathbf{B}\}\{\mathbf{B}, \mathbf{Q}\} = \{\mathbf{P}, \mathbf{Q}\} \qquad \text{(a')}$$

in which ξ, ω, η, Ω *are the matrices derived from* X, e, Y, E *by the inverses of the equigradent commutantal transformations*

$$X = u\xi H \quad \textit{of type} \quad \{P, A\} = \{P, \mathbf{P}\}\{\mathbf{P}, \mathbf{A}\}\{\mathbf{A}, A\}, \quad \text{.........(i)}$$

$$e = h\omega K \quad \textit{of type} \quad \{A, B\} = \{A, \mathbf{A}\}\{\mathbf{A}, \mathbf{B}\}\{\mathbf{B}, B\}, \quad \text{.........(ii)}$$

$$Y = k\eta V \quad \textit{of type} \quad \{B, Q\} = \{B, \mathbf{B}\}\{\mathbf{B}, \mathbf{Q}\}\{\mathbf{Q}, Q\}, \quad \text{......(iii)}$$

$$E = u\Omega V \quad \textit{of type} \quad \{P, Q\} = \{P, \mathbf{P}\}\{\mathbf{P}, \mathbf{Q}\}\{\mathbf{Q}, Q\} \quad \text{.........(iv)}$$

in which all the transforming factors are square matrices.

The theorem follows immediately from the equations

$$XeY = u\,.\,\xi\omega\eta\,.\,V, \quad E = u\,.\,\Omega\,.\,V, \quad \text{..................(3)}$$

which show that $XeY = E$ when and only when $\xi\omega\eta = \Omega$.

The equigradent substitutions (i), (ii), (iii), (iv) establish a one-one correspondence between all equations (A) and all equations (A′), the equation (A) being true, equigradent, commutantal of type (a) when and only when the corresponding equation (A′) is true, equigradent, commutantal of type (a′); for the matrices X, e, Y, E in (i), (ii), (iii), (iv) have the same ranks as ξ, ω, η, Ω, and they are commutants $\{P, A\}$, $\{A, B\}$, $\{B, Q\}$, $\{P, Q\}$ when and only when ξ, ω, η, Ω are commutants $\{\mathbf{P}, \mathbf{A}\}$, $\{\mathbf{A}, \mathbf{B}\}$, $\{\mathbf{B}, \mathbf{Q}\}$, $\{\mathbf{P}, \mathbf{Q}\}$. When e and E are given commutants of the types shown and the equation (A) is commutantal, we will say that X and Y constitute a commutantal solution of (A); and we have the following corollary.

COROLLARY. *When* e *and* E *are given commutants* $\{A, B\}$ *and* $\{P, Q\}$, *so that* ω *and* Ω *are given commutants* $\{\mathbf{A}, \mathbf{B}\}$ *and* $\{\mathbf{P}, \mathbf{Q}\}$, *the matrices* X *and* Y *constitute a solution or a commutantal solution of the equation* (A) *if and only if the matrices* ξ *and* η *constitute a solution or a commutantal solution of the corresponding equation* (A′).

3. *Reduction of a commutantal transformation converting one commutant into a similar commutant.*

The commutants e and E will be similar matrices when $r = m$ and $s = n$, i.e. when:

- A, P and therefore also $\mathbf{A}$, $\mathbf{P}$, h, u, H, U are square matrices of the same order m;
- B, Q and therefore also $\mathbf{B}$, $\mathbf{Q}$, k, v, K, V are square matrices of the same order n.

In this case the substitutions (i), (ii), (iii), (iv) enable us to replace any commutantal transformation

$$[x]_m^m\,[e]_m^n\,[y]_n^n = [E]_m^n \quad \text{or} \quad XeY = E \quad \text{..............(B)}$$

of type (a) by an equivalent commutantal transformation

$$[\xi]_m^m\,[\omega]_m^n\,[\eta]_n^n = [\Omega]_m^n \quad \text{or} \quad \xi\omega\eta = \Omega \quad \text{..............(B′)}$$

of type (a′). It is possible for a commutantal transformation (B) of type (a) to be *equigradent* when and only when the two square matrices A, P of order m are equicanonical, and also the two square matrices B, Q of order n are equicanonical. When these conditions are satisfied, every commutantal transformation (B) of type (a) can be replaced by an equivalent commutantal transformation (B′) of type

$$\{\mathbf{A}, \mathbf{A}\}\{\mathbf{A}, \mathbf{B}\}\{\mathbf{B}, \mathbf{B}\} = \{\mathbf{A}, \mathbf{B}\}, \quad \text{.....................(b)}$$

where $\mathbf{A}$ is a standard canonical reduced form of A and P, and $\mathbf{B}$ is a standard canonical reduced form of B and Q; the transformation (B) being equigradent when and only when the transformation (B′) is equigradent.

4. *Reduction of a commutantal transformation converting a square commutant into a square commutant.*

The commutants e and E in (A) will be both square matrices when $s=r$ and $n=m$, i.e. when:

A, B and therefore also $\mathbf{A}$, $\mathbf{B}$, h, k, H, K are square matrices of the same order r;

P, Q and therefore also $\mathbf{P}$, $\mathbf{Q}$, u, v, U, V are square matrices of the same order m.

In this case the substitutions (i), (ii), (iii), (iv) enable us to replace any commutantal transformation

$$[x]_m^r\,[e]_r^r\,[y]_r^m = [E]_m^m \quad \text{or} \quad XeY = E \quad \text{..............(C)}$$

of type (a) by an equivalent commutantal transformation

$$[\xi]_m^r\,[\omega]_r^r\,[\eta]_r^m = [\Omega]_m^m \quad \text{or} \quad \xi\omega\eta = \Omega \quad \text{..............(C′)}$$

of type (a′). It is possible for a commutantal transformation (C) of type (a) to be equigradent when and only when the canonicals of A are contained in the canonicals of P, and also the canonicals of B are contained in the canonicals of Q; and this of course necessitates the condition $r \not> m$.

If the conjugates of h, k, H, K, u, v, U, V are denoted by the corresponding dashed letters, the equigradent substitutions (i), (ii), (iii), (iv) will establish a one-one correspondence between all *symmetric* transformations (C) and all *symmetric* transformations (C′) when

$$k = H', \quad K' = h, \quad v = U', \quad V' = u,$$

or

$$hk' = [1]_r^r, \quad uv' = [1]_m^m,$$

i.e. when the given isomorphic transformations are

$$A = h\mathbf{A}H, \quad B = H'\mathbf{B}h', \quad P = u\mathbf{P}U, \quad Q = U'\mathbf{Q}u'.$$

In all such cases the matrices e and E are both symmetric (or both skew-symmetric) when and only when the matrices ω and Ω are both symmetric (or both skew-symmetric). When A, B and P, Q are two pairs of mutually conjugate square matrices, these conditions can always be satisfied by taking the two square matrices $\mathbf{A}$, $\mathbf{B}$ and the two square matrices $\mathbf{P}$, $\mathbf{Q}$ to be mutually conjugate.

The equigradent substitutions (i), (ii), (iii), (iv) will establish a one-one correspondence between all *equimutant* transformations (C) and all *equimutant* transformations (C′) when

$$k = h, \quad v = u,$$

i.e. when the given isomorphic transformations are

$$A = h\mathbf{A}H, \quad B = h\mathbf{B}H, \quad P = u\mathbf{P}U, \quad Q = u\mathbf{Q}U.$$

In all such cases we have

$$e = h\omega H, \quad E = u\Omega U.$$

When $B = A$ and $Q = P$, these conditions can always be satisfied by putting $\mathbf{B} = \mathbf{A}$ and $\mathbf{Q} = \mathbf{P}$.

§ 247. Commutantal transformations of co-commutants and contra-commutants.

1. *General commutantal transformations.*

Let A, P be two given square matrices of orders r, m with constant elements whose conjugates are A', P'; and let

$$[x]_m^r\,[e]_r^r\,[y]_r^m = [E]_m^m \quad \text{or} \quad XeY = E \quad \text{..............(A)}$$

be a commutantal transformation converting

a commutant e of one of the four continuantal types $\{A, A\}$, $\{A', A'\}$, $\{A', A\}$, $\{A, A'\}$

into a commutant E of one of the four continuantal types $\{P, P\}$, $\{P', P'\}$, $\{P', P\}$, $\{P, P'\}$,

so that there are 16 possible types of (A) which can be represented by

$$\{p, a\}\,\{a, b\}\,\{b, q\} = \{p, q\}, \quad \text{........................(a)}$$

where $p = P$ or P', $a = A$ or A', $b = A$ or A', $q = P$ or P'.

Also let $\mathbf{A}$, $\mathbf{P}$ be square matrices of orders r, m equicanonical with A, P (or standard canonical reduced forms of A, P) whose conjugates are $\mathbf{A}'$, $\mathbf{P}'$; and let

$$[\xi]_m^r\,[\omega]_r^r\,[\eta]_r^m = [\Omega]_m^m \quad \text{or} \quad \xi\omega\eta = \Omega \quad \text{..............(A′)}$$

be a commutantal transformation converting

a commutant ω of one of the four continuantal types $\{\mathbf{A}, \mathbf{A}\}$, $\{\mathbf{A}', \mathbf{A}'\}$, $\{\mathbf{A}', \mathbf{A}\}$, $\{\mathbf{A}, \mathbf{A}'\}$

into a commutant Ω of one of the four continuantal types $\{\mathbf{P}, \mathbf{P}\}$, $\{\mathbf{P}', \mathbf{P}'\}$, $\{\mathbf{P}', \mathbf{P}\}$, $\{\mathbf{P}, \mathbf{P}'\}$, so that there are 16 possible types of (A′) which can be represented by

$$\{\mathbf{p}, \mathbf{a}\}\{\mathbf{a}, \mathbf{b}\}\{\mathbf{b}, \mathbf{q}\} = \{\mathbf{p}, \mathbf{q}\}, \quad \text{.......................(a′)}$$

where $\mathbf{p} = \mathbf{P}$ or $\mathbf{P}'$, $\mathbf{a} = \mathbf{A}$ or $\mathbf{A}'$, $\mathbf{b} = \mathbf{A}$ or $\mathbf{A}'$, $\mathbf{q} = \mathbf{P}$ or $\mathbf{P}'$.

Then we have the following theorem which is included in the theorem of § 246.

Theorem I. *A commutantal transformation* (A) *of any one of the* 16 *types* (a) *can always be converted by equigradent commutantal substitutions for* X, e, Y, E *into an equivalent commutantal transformation* (A′) *of any one of the* 16 *types* (a′), *the first transformation being equigradent when and only when the second transformation is equigradent.*

The equivalent transformation will always be supposed to be obtained in the way described in § 246, and the substitutions used to be compounded of those described in Exs. i and ii. The circumstances under which such transformations can be equigradent are known from Ex. ii of § 244.

Ex. i. *Substitutions used in Theorem I: corresponding types.*

Let the type (a′) be said to correspond to the type (a) when it is derived from (a) by substituting $\mathbf{P}$, $\mathbf{P}'$, $\mathbf{A}$, $\mathbf{A}'$ for P, P', A, A'. Also let

$$A = h\mathbf{A}H \quad \text{of type} \quad \{A, A\} = \{A, \mathbf{A}\}\{\mathbf{A}, \mathbf{A}\}\{\mathbf{A}, A\},$$
$$P = u\mathbf{P}U \quad \text{of type} \quad \{P, P\} = \{P, \mathbf{P}\}\{\mathbf{P}, \mathbf{P}\}\{\mathbf{P}, P\}$$

be any given isomorphic (commutantal) transformations by which A, P can be derived from $\mathbf{A}$, $\mathbf{P}$; so that h and H are mutually inverse undegenerate square matrices of order r whose conjugates will be denoted by h' and H', and u and U are mutually inverse undegenerate square matrices of order m whose conjugates will be denoted by u' and U'. We can take h and u to be any particular undegenerate commutants $\{A, \mathbf{A}\}$ and $\{P, \mathbf{P}\}$; and

$$A' = H'\mathbf{A}'h' \quad \text{of type} \quad \{A', A'\} = \{A', \mathbf{A}'\}\{\mathbf{A}', \mathbf{A}'\}\{\mathbf{A}', A'\},$$
$$P' = U'\mathbf{P}'u' \quad \text{of type} \quad \{P', P'\} = \{P', \mathbf{P}'\}\{\mathbf{P}', \mathbf{P}'\}\{\mathbf{P}', P'\}$$

are given isomorphic (commutantal) transformations by which A', P' can be derived from $\mathbf{A}'$, $\mathbf{P}'$.

Then the table at the end shows substitutions for X, e, Y, E which convert a commutantal transformation (A) of any one of the 16 possible types (a) into an equivalent commutantal transformation (A′) of the *corresponding* type (a′).

The transformations of the types shown convert:

a co-commutant into a co-commutant in the four cases I,

a contra-commutant into a contra-commutant in the four cases II,

a co-commutant into a contra-commutant in the four cases III,

a contra-commutant into a co-commutant in the four cases IV.

Symmetric commutantal transformations of the four types II and equimutant commutantal transformations of the four types I will be considered in sub-articles 2 and 3.

Case	*Type of* (A)	X	e	Y	E
I 1	$\{P, A\}\{A, A\}\{A, P\}=\{P, P\}$	$u\xi H$	$h\omega H$	$h\eta U$	$u\Omega U$
I 2	$\{P', A\}\{A, A\}\{A, P'\}=\{P', P'\}$	$U'\xi H$	$h\omega H$	$h\eta u'$	$U'\Omega u'$
I 3	$\{P', A'\}\{A', A'\}\{A', P'\}=\{P', P'\}$	$U'\xi h'$	$H'\omega h'$	$H'\eta u'$	$U'\Omega u'$
I 4	$\{P, A'\}\{A', A'\}\{A', P\}=\{P, P\}$	$u\xi h'$	$H'\omega h'$	$H'\eta U$	$u\Omega U$
II 1	$\{P', A'\}\{A', A\}\{A, P\}=\{P', P\}$	$U'\xi h'$	$H'\omega H$	$h\eta U$	$U'\Omega U$
II 2	$\{P, A'\}\{A', A\}\{A, P'\}=\{P, P'\}$	$u\xi h'$	$H'\omega H$	$h\eta u'$	$u\Omega u'$
II 3	$\{P, A\}\{A, A'\}\{A', P'\}=\{P, P'\}$	$u\xi H$	$h\omega h'$	$H'\eta u'$	$u\Omega u'$
II 4	$\{P', A\}\{A, A'\}\{A', P\}=\{P', P\}$	$U'\xi H$	$h\omega h'$	$H'\eta U$	$U'\Omega U$
III 1	$\{P', A\}\{A, A\}\{A, P\}=\{P', P\}$	$U'\xi H$	$h\omega H$	$h\eta U$	$U'\Omega U$
III 2	$\{P, A\}\{A, A\}\{A, P'\}=\{P, P'\}$	$u\xi H$	$h\omega H$	$h\eta u'$	$u\Omega u'$
III 3	$\{P, A'\}\{A', A'\}\{A', P'\}=\{P, P'\}$	$u\xi h'$	$H'\omega h'$	$H'\eta u'$	$u\Omega u'$
III 4	$\{P', A'\}\{A', A'\}\{A', P\}=\{P', P\}$	$U'\xi h'$	$H'\omega h'$	$H'\eta U$	$U'\Omega U$
IV 1	$\{P, A'\}\{A', A\}\{A, P\}=\{P, P\}$	$u\xi h'$	$H'\omega H$	$h\eta U$	$u\Omega U$
IV 2	$\{P', A'\}\{A', A\}\{A, P'\}=\{P', P'\}$	$U'\xi h'$	$H'\omega H$	$h\eta u'$	$U'\Omega u'$
IV 3	$\{P', A\}\{A, A'\}\{A', P'\}=\{P', P'\}$	$U'\xi H$	$h\omega h'$	$H'\eta u'$	$U'\Omega u'$
IV 4	$\{P, A\}\{A, A'\}\{A', P\}=\{P, P\}$	$u\xi H$	$h\omega h'$	$H'\eta U$	$u\Omega U$

Ex. ii. *Substitutions used in Theorem I: non-corresponding types.*

Substitutions which convert a commutantal transformation (A) of any one of the 16 types (a) into an equivalent commutantal transformation (A) of any other of those types are most easily obtained by the insertion of factors J, J^{-1} and K, K^{-1}, where J and K are particular undegenerate square contra-commutants $\{A', A\}$ and $\{P', P\}$.

Thus let it be required to convert a commutantal transformation

$$X_2 e_2 Y_2 = E_2 \quad \text{of type} \quad \{P', A\}\{A, A\}\{A, P\}=\{P', P\} \quad \ldots\ldots\ldots\ldots\ldots(\mathrm{A}_2)$$

into an equivalent commutantal transformation

$$X_1 e_1 Y_1 = E_1 \quad \text{of type} \quad \{P, A'\}\{A', A'\}\{A', P\}=\{P, P\}. \quad \ldots\ldots\ldots\ldots(\mathrm{A}_1)$$

When we replace (A_1) by the equivalent commutantal equation

$$K\xi J \,.\, J^{-1}\omega J \,.\, J^{-1}\eta = K\Omega \quad \text{of the same type as } (\mathrm{A}_2),$$

we see that the required conversion can be effected by the equigradent commutantal substitutions

$$X_2 = KX_1 J \quad \text{of type} \quad \{P', A\}=\{P', P\}\{P, A'\}\{A', A\},$$
$$e_2 = J^{-1} e_1 J \quad \text{of type} \quad \{A, A\}=\{A, A'\}\{A', A'\}\{A', A\},$$
$$Y_2 = J^{-1} Y_1 \quad \text{of type} \quad \{A, P\}=\{A, A'\}\{A', P\},$$
$$E_2 = KE_1 \quad \text{of type} \quad \{P', P\}=\{P', P\}\{P, P\}.$$

Substitutions which convert a commutantal transformation (A′) of any one of the 16 types (a′) into an equivalent commutantal transformation (A′) of any other of those types can be constructed in similar ways.

By compounding such substitutions with those given in Ex. i we can obtain substitutions converting a commutantal transformation (A) of any one of the 16 types (a) into an equivalent commutantal transformation (A′) of any non-corresponding type (a′). But of course such substitutions can be obtained directly in the same ways as the substitutions described in Ex. i.

NOTE 1. *Commutantal transformations of alternating commutants.*

Theorem I is the first of four corresponding theorems which relate respectively to commutantal transformations converting:

commutants $\{A, A\}, \{A', A'\}, \{A', A\}, \{A, A'\}$
into commutants $\{P, P\}, \{P', P'\}, \{P', P\}, \{P, P'\}$;

commutants $\{A, -A\}, \{A', -A'\}, \{A', -A\}, \{A, -A'\}$
into commutants $\{P, -P\}, \{P', -P'\}, \{P', -P\}, \{P, -P'\}$;

commutants $\{A, A\}, \{A', A'\}, \{A', A\}, \{A, A'\}$
into commutants $\{P, -P\}, \{P', -P'\}, \{P', -P\}, \{P, -P'\}$;

commutants $\{A, -A\}, \{A', -A'\}, \{A', -A\}, \{A, -A'\}$
into commutants $\{P, P\}, \{P', P'\}, \{P', P\}, \{P, P'\}$.

All such transformations are reducible respectively to equivalent commutantal transformations of the types

$\{P, A\}\{A, A\}\{A, P\}=\{P, P\}$, (continuantal into continuantal); ...(i)

$\{P, A\}\{A, -A\}\{-A, -P\}=\{P, -P\}$, (alternating into alternating); ......(ii)

$\{P, A\}\{A, A\}\{A, -P\}=\{P, -P\}$, (continuantal into alternating); ...(iii)

$\{P, A\}\{A, -A\}\{-A, P\}=\{P, P\}$, (alternating into continuantal); ...(iv)

where A and P can be replaced by any square matrices equicanonical with them.

We cannot ordinarily replace a transformation of any one of these four classes by an equivalent transformation of any other of the four classes by the methods which have been described unless there exist undegenerate square commutants $\{A, -A\}$ and $\{P, -P\}$, i.e. unless A and $-A$ are equicanonical with one another, and also P and $-P$ are equicanonical with one another.

2. *Symmetric commutantal transformations of a contra-commutant.*

The square matrices A, $\mathbf{A}$ of order r and P, $\mathbf{P}$ of order m being defined as before, let

$$\overline{\underline{x}}{}^{r}_{m}\,[e]^{r}_{r}\,[x]^{m}_{r}=[E]^{m}_{m} \quad \text{or} \quad X'eX=E \qquad \text{..............(B)}$$

be a *symmetric* commutantal transformation converting

a contra-commutant e of one of the two continuantal types $\{A', A\}, \{A, A'\}$ into

a contra-commutant E of one of the two continuantal types $\{P', P\}, \{P, P'\}$, so that there are four possible types of (B) which can be represented by

$$\{p', a'\}\{a', a\}\{a, p\}=\{p', p\}, \qquad \text{.......................(b)}$$

where $(a', a)=(A', A)$ or (A, A'), $(p', p)=(P', P)$ or (P, P').

Also let

$$\overline{\underline{\xi}}{}^{r}_{m}\,[\omega]^{r}_{r}\,[\xi]^{m}_{r}=[\Omega]^{m}_{m} \quad \text{or} \quad \xi'\omega\xi=\Omega \qquad \text{..............(B')}$$

be a *symmetric* commutantal transformation converting

a contra-commutant ω of one of the two continuantal types $\{\mathbf{A}', \mathbf{A}\}, \{\mathbf{A}, \mathbf{A}'\}$ into

a contra-commutant Ω of one of the two continuantal types $\{\mathbf{P}', \mathbf{P}\}, \{\mathbf{P}, \mathbf{P}'\}$, so that there are four possible types of (B') which can be represented by

$$\{\mathbf{p}', \mathbf{a}'\}\{\mathbf{a}', \mathbf{a}\}\{\mathbf{a}, \mathbf{p}\}=\{\mathbf{p}', \mathbf{p}\}, \qquad \text{....................(b')}$$

where $(\mathbf{a}', \mathbf{a}) = (\mathbf{A}', \mathbf{A})$ or $(\mathbf{A}, \mathbf{A}')$, $(\mathbf{p}', \mathbf{p}) = (\mathbf{P}', \mathbf{P})$ or $(\mathbf{P}, \mathbf{P}')$.

Then from § 246. 4 we obtain the following theorem :

Theorem II. *A symmetric commutantal transformation* (B) *of any one of the four types* (b) *can always be converted by an equigradent commutantal substitution for X and symmetric equigradent commutantal substitutions for e and E into an equivalent symmetric commutantal transformation* (B′) *of any one of the four types* (b′). *The transformation* (B) *is equigradent when and only when the equivalent transformation* (B′) *is equigradent; and the contra-commutants e and E in* (B) *are both symmetric (or both skew-symmetric) when and only when the contra-commutants ω and Ω in* (B′) *are both symmetric (or both skew-symmetric).*

The conversion of (B) into (B′) is supposed to be obtained in the way described in § 246, the substitutions used being compounded of those described in Exs. iii and iv. There exist symmetric commutantal transformations (B) of each of the types (b) in which X is an arbitrary commutant of the prescribed type. By Ex. ii of § 244 such transformations can be equigradent when and only when the canonicals of A are diagonal minors of the canonicals of P.

Ex. iii. *Substitutions used in Theorem II: corresponding types.*

Let
$$A = h\mathbf{A}H, \qquad P = u\mathbf{P}U$$
be the isomorphic transformations described in Ex. i, so that the square matrices h, u of orders r, m are particular undegenerate commutants $\{A, \mathbf{A}\}$, $\{P, \mathbf{P}\}$. Then the following table shows the substitutions for X', e, X, E which convert a symmetric commutantal transformation (B) of any one of the four types (b) into an equivalent symmetric commutantal transformation (B′) of the *corresponding type* (b′) derived from (b) by substituting $\mathbf{A}$, $\mathbf{P}$ for A, P.

Case	*Type of* (B)	X'	e	X	E
1	$\{P', A'\}\{A', A\}\{A, P\} = \{P', P\}$	$U'\xi'h'$	$H'\omega H$	$h\xi U$	$U'\Omega U$
2	$\{P, A'\}\{A', A\}\{A, P'\} = \{P, P'\}$	$u\xi'h'$	$H'\omega H$	$h\xi u'$	$u\Omega u'$
3	$\{P, A\}\{A, A'\}\{A', P'\} = \{P, P'\}$	$u\xi' H$	$h\omega h'$	$H'\xi u'$	$u\Omega u'$
4	$\{P', A\}\{A, A'\}\{A', P\} = \{P', P\}$	$U'\xi' H$	$h\omega h'$	$H'\xi U$	$U'\Omega U$

Ex. iv. *Substitutions used in Theorem II: non-corresponding types.*

Substitutions converting a symmetric commutantal transformation (B) of any one of the four types (b) into an equivalent symmetric commutantal transformation (B) of any other of those types can be most easily constructed in the way described in Ex. ii. Thus if J and K are given particular undegenerate contra-commutants $\{A, A'\}$ and $\{P, P'\}$ whose conjugates are J' and K', then symmetric commutantal transformations

$X_2'e_2X_2 = E_2$ of type $\{P, A'\}\{A', A\}\{A, P'\} = \{P, P'\}$, ...($B_2$)

$X_3'e_3X_3 = E_3$ of type $\{P, A\}\{A, A'\}\{A', P'\} = \{P, P'\}$, ...($B_3$)

$X_4'e_4X_4 = E_4$ of type $\{P', A\}\{A, A'\}\{A', P\} = \{P', P\}$ ....(B_4)

can be converted into equivalent symmetric commutantal transformations

$X_1'e_1X_1 = E_1$ of type $\{P', A'\}\{A', A\}\{A, P\} = \{P', P\}$ ....(B_1)

by the respective sets of equigradent commutantal substitutions

$$X_2' = K'X_1' \quad , \quad e_2 = \quad e_1 \ , \quad X_2 = \quad X_1 K, \quad E_2 = K'E_1 K,$$
$$X_3' = K'X_1'J'^{-1}, \quad e_3 = J'e_1 J, \quad X_3 = J^{-1}X_1 K, \quad E_3 = K'E_1 K,$$
$$X_4' = \quad X_1'J'^{-1}, \quad e_4 = J'e_1 J, \quad X_4 = J^{-1}X_1 \ , \quad E_4 = \quad E_1 \ ,$$

which are most easily obtained by inserting factors in (B_1) to change its type. We can always choose J and K to be symmetric, so that $J' = J$, $K' = K$.

Substitutions converting a symmetric commutantal transformation (B′) of any one of the four types (b′) into an equivalent symmetric commutantal transformation (B′) of any other of those types can be constructed in similar ways.

By compounding such substitutions with those given in Ex. iii we can obtain substitutions converting a symmetric commutantal transformation (B) of any one of the four types (b) into an equivalent symmetric commutantal transformation (B′) of any non-corresponding type (b′).

NOTE 2. *Symmetric commutantal transformations of alternating contra-commutants.*

Theorem II is the first of two corresponding theorems which relate respectively to symmetric commutantal transformations converting:

contra-commutants $\{A', \ A\}$, $\{A, \ A'\}$ into contra-commutants $\{P', \ P\}$, $\{P, \ P'\}$;
contra-commutants $\{A', -A\}$, $\{A, -A'\}$ into contra-commutants $\{P', -P\}$, $\{P, -P'\}$.

All such transformations are reducible respectively to equivalent symmetric commutantal transformations of the types:

$$\{P', A'\}\{A', \ A\}\{\ A, \ P\} = \{P', \ P\}, \quad \text{(continuantal into continuantal)}; \ \text{...(i)}$$
$$\{P', A'\}\{A', -A\}\{-A, -P\} = \{P', -P\}, \quad \text{(alternating into alternating)}; \ \text{......(ii)}$$

where A, P can be replaced by any square matrices equicanonical with them.

We cannot ordinarily replace a symmetric commutantal transformation of any one of these two classes by an equivalent symmetric commutantal transformation of the other class.

3. *Equimutant commutantal transformations of a co-commutant.*

The square matrices A, $\mathbf{A}$ of order r and P, $\mathbf{P}$ of order m being defined as before, let

$$[x]_m^r \, [e]_r^r \, [y]_r^m = [E]_m^m \quad \text{or} \quad XeY = E, \quad \text{..............(C)}$$

where
$$[y]_r^m \, [x]_m^r = [1]_r^r,$$

be an *equimutant* commutantal transformation converting

a co-commutant e of one of the two continuantal types $\{A, A\}$, $\{A', A'\}$ into a co-commutant E of one of the two continuantal types $\{P, P\}$, $\{P', P'\}$, so that there are four possible types of (C) which can be represented by

$$\{p, a\}\{a, a\}\{a, p\} = \{p, p\}, \quad \text{where} \quad a = A \ \text{or} \ A', \quad p = P \ \text{or} \ P'. \ \text{...(c)}$$

Also let
$$[\xi]_m^r \, [\omega]_r^r \, [\eta]_r^m = [\Omega]_m^m \quad \text{or} \quad \xi\omega\eta = \Omega, \quad \text{..............(C')}$$

where
$$[\eta]_r^m \, [\xi]_m^r = [1]_r^r,$$

be an *equimutant* commutantal transformation converting

a co-commutant ω of one of the two continuantal types $\{\mathbf{A}, \mathbf{A}\}$, $\{\mathbf{A}', \mathbf{A}'\}$ into a co-commutant Ω of one of the two continuantal types $\{\mathbf{P}, \mathbf{P}\}$, $\{\mathbf{P}', \mathbf{P}'\}$, so that there are four possible types of (C′) which can be represented by

$$\{\mathbf{p}, \mathbf{a}\}\{\mathbf{a}, \mathbf{a}\}, \{\mathbf{a}, \mathbf{p}\} = \{\mathbf{p}, \mathbf{p}\}, \text{ where } \mathbf{a} = \mathbf{A} \text{ or } \mathbf{A}', \ \mathbf{p} = \mathbf{P} \text{ or } \mathbf{P}'. \quad \ldots(\text{c}')$$

Then from § 246. 4 we obtain the following theorem:

Theorem III. *An equimutant commutantal transformation* (C) *of any one of the four types* (c) *can always be converted by equigradent commutantal substitutions for X and Y and equimutant commutantal substitutions for e and E into an equivalent equimutant commutantal transformation* (C′) *of any one of the four types* (c′). *The co-commutants e, E in* (C) *satisfy the same rational integral equations as the co-commutants ω, Ω in* (C′).

The conversion of (C) into (C′) is supposed to be effected in the way described in § 246, the substitutions used being compounded of those described in Exs. v and vi.

Ex. v. *Substitutions used in Theorem III: corresponding types.*

Let $$A = h\mathbf{A}H, \quad P = u\mathbf{P}U$$

be the isomorphic transformations described in Ex. i, so that the square matrices h, u of orders r, m are particular undegenerate commutants $\{A, \mathbf{A}\}$, $\{P, \mathbf{P}\}$. Then the following table shows the substitutions for X, e, Y, E which convert an equimutant commutantal transformation (C) of any one of the four types (c) into an equivalent equimutant commutantal transformation (C′) of the corresponding type (c′) derived from (c) by substituting $\mathbf{A}$, $\mathbf{P}$ for A, P.

Case	*Type of* (C)	X	e	Y	E
1	$\{P, A\}\{A, A\}\{A, P\}=\{P, P\}$	$u\xi H$	$h\omega H$	$h\eta U$	$u\Omega U$
2	$\{P', A\}\{A, A\}\{A, P'\}=\{P', P'\}$	$U'\xi H$	$h\omega H$	$h\eta u'$	$U'\Omega u'$
3	$\{P', A'\}\{A', A'\}\{A', P'\}=\{P', P'\}$	$U'\xi h'$	$H'\omega h'$	$H'\eta u'$	$U'\Omega u'$
4	$\{P, A'\}\{A', A'\}\{A', P\}=\{P, P\}$	$u\xi h'$	$H'\omega h'$	$H'\eta U$	$u\Omega U$

Ex. vi. *Substitutions used in Theorem III: non-corresponding types.*

Substitutions converting an equimutant commutantal transformation (C) of any one of the four types (c) into an equivalent equimutant commutantal transformation (C) of any other of those types can be most easily constructed in the way described in Ex. ii. Thus if J and K are given particular undegenerate contra-commutants $\{A', A\}$ and $\{P', P\}$, then equimutant commutantal transformations

$$X_2 e_2 Y_2 = E_2 \text{ of type } \{P', A\}\{A, A\}\{A, P'\}=\{P', P'\}, \quad \ldots\ldots\ldots\ldots(\text{C}_2)$$
$$X_3 e_3 Y_3 = E_3 \text{ of type } \{P', A'\}\{A', A'\}\{A', P'\}=\{P', P'\}, \quad \ldots\ldots\ldots\ldots(\text{C}_3)$$
$$X_4 e_4 Y_4 = E_4 \text{ of type } \{P, A'\}\{A', A'\}\{A', P\}=\{P, P\} \quad \ldots\ldots\ldots\ldots(\text{C}_4)$$

can be converted into equivalent equimutant commutantal transformations

$$X_1 e_1 Y_1 = E_1 \text{ of type } \{P, A\}\{A, A\}\{A, P\}=\{P, P\} \quad \ldots\ldots\ldots\ldots(\text{C}_1)$$

by the respective sets of equigradent commutantal transformations

$$X_2 = KX_1 \ , \quad e_2 = e_1 \ , \quad Y_2 = Y_1K^{-1}, \quad E_2 = KE_1K^{-1},$$
$$X_3 = KX_1J^{-1}, \quad e_3 = Je_1J^{-1}, \quad Y_3 = JY_1K^{-1}, \quad E_3 = KE_1K^{-1},$$
$$X_4 = X_1J^{-1}, \quad e_4 = Je_1J^{-1}, \quad Y_4 = JY_1 \ , \quad E_4 = E_1 \ ,$$

which are most easily obtained by inserting factors in (C_1) to change its type.

Substitutions converting an equimutant commutantal transformation (C′) of any one of the four types (c′) into an equivalent equimutant commutantal transformation (C′) of any other of those types can be constructed in similar ways.

By compounding such substitutions with those given in Ex. v we can obtain substitutions converting an equimutant commutantal transformation (C) of any one of the four types (c) into an equimutant commutantal transformation (C′) of any non-corresponding type (c′).

NOTE 3. *Equimutant commutantal transformations of alternating co-commutants.*

Theorem III is the first of two corresponding theorems which relate respectively to equimutant commutantal transformations converting:

co-commutants $\{A, \quad A\}, \{A', \quad A'\}$ into co-commutants $\{P, \quad P\}, \{P', \quad P'\}$;
co-commutants $\{A, \ -A\}, \{A', -A'\}$ into co-commutants $\{P, \ -P\}, \{P', -P'\}$.

All such transformations are reducible respectively to equivalent equimutant commutantal transformations of the types:

$\{P, A\}\{A, \quad A\}\{\ A, \quad P\}=\{P, \quad P\}$, (continuantal into continuantal); ...(i)
$\{P, A\}\{A, -A\}\{-A, -P\}=\{P, -P\}$, (alternating into alternating); ......(ii)

where A, P can be replaced by any square matrices equicanonical with them.

We cannot ordinarily replace an equimutant commutantal transformation of any one of these two classes by an equivalent equimutant commutantal transformation of the other class.

§ 248. Equigradent commutantal transformations converting one given undegenerate square commutant into another of the same order.

1. *General transformations.*

Let A, B and P, Q be two pairs of square matrices with constant elements which are all of the same order m, and let

$$[x]_m^m [e]_m^m [y]_m^m = [E]_m^m \quad \text{or} \quad XeY = E \quad \text{..............(A)}$$

be an equigradent commutantal transformation in which

e, E are undegenerate square commutants $\{A, B\}$, $\{P, Q\}$,

so that (A) has the prescribed type

$$\{P, A\}\{A, B\}\{B, Q\} = \{P, Q\}, \text{.......................(a)}$$

the types of X and Y being fixed by the prescribed types of e and E. By Ex. ii of § 244 there cannot exist undegenerate commutants X, e, Y, E of the prescribed types unless the two square matrices in each of the pairs (P, A), (A, B), (B, Q), (P, Q) are equicanonical.

Hence in order that there may exist equigradent commutantal transformations (A) *of the prescribed type* (a) *in which e and E are undegenerate, it is requisite that all the four square matrices A, B, P, Q shall be equicanonical or have the same characteristic potent divisors.* ....(A_1)

These requisite conditions are also the conditions that there shall exist:

(1) undegenerate square commutants $e = \{A, B\}$ and $E = \{P, Q\}$;
(2) undegenerate square commutants $X = \{P, A\}$ or $Y = \{B, Q\}$;

and they are also the conditions that there shall exist:

(1) undegenerate square commutants $X = \{P, A\}$ and $Y = \{B, Q\}$;

(2) undegenerate square commutants $e = \{A, B\}$ or $E = \{P, Q\}$.

With respect to such transformations we have the following theorem:

Theorem I. *Let the requisite conditions* (A_1) *be satisfied, i.e. let* A, B, P, Q *be equicanonical; and let* e *and* E *be given particular undegenerate square commutants of the prescribed types. Then it is always possible to determine equigradent commutantal transformations* (A) *of the prescribed type* (a) *converting* e *into* E*; moreover either one of the transforming factors* X *and* Y *can be an arbitrarily given particular undegenerate square commutant of the prescribed type, the other transforming factor being then uniquely determinate.*

For if X is any given particular undegenerate square commutant $\{P, A\}$, then by prefixing in succession on both sides of (A) the inverses of X and e, which are undegenerate square commutants $\{A, P\}$ and $\{B, A\}$, we obtain an equivalent equigradent commutantal transformation which determines Y completely and uniquely as an undegenerate square commutant $\{B, Q\}$. Similarly if Y is any given particular undegenerate square commutant $\{B, Q\}$, then by postfixing in succession on both sides of (A) the inverses of Y and e, which are undegenerate square commutants $\{Q, B\}$ and $\{B, A\}$, we obtain an equivalent equigradent commutantal transformation which determines X completely and uniquely as an undegenerate square commutant $\{P, A\}$.

The theorem states that it is possible to determine undegenerate square matrices X and Y which satisfy simultaneously the equations

$$PX = XA, \quad BY = YQ, \quad XeY = E.$$

NOTE 1. When A, B, P, Q are equicanonical, let Π be any one of their canonical reduced forms. Then by § 246 an equigradent commutantal transformation (A) in which e and E are given and undegenerate can be replaced by or derived from a corresponding equivalent equigradent commutantal transformation

$$[\xi]_m^m [\omega]_m^m [\eta]_m^m = [\Omega]_m^m \quad \text{or} \quad \xi\omega\eta = \Omega \quad \text{.........................} (A_0)$$

of type $\{\Pi, \Pi\}\{\Pi, \Pi\}\{\Pi, \Pi\} = \{\Pi, \Pi\}$ .... (a_0)

in which ω and Ω are given undegenerate commutants $\{\Pi, \Pi\}$. Hence Theorem I could be proved by proving the following lemma in which Π is restricted to be a standard canonical square matrix with standardised unilatent super-parts:

Lemma I. *Let* Π *be a standardised canonical square matrix of order* m, *and let* ω *and* Ω *be given particular undegenerate commutants* $\{\Pi, \Pi\}$. *Then it is always possible to construct an equigradent commutantal transformation* (A_0) *of type* (a_0)*; moreover this is possible when either one of the two matrices* ξ *and* η *is a given particular undegenerate commutant* $\{\Pi, \Pi\}$, *the other matrix being then uniquely determinate.*

The proof of Lemma I is of course included in the proof of Theorem I. But we could proceed as in the other lemmas of this article, and replace (A_0) by a number of transformations of the same character in which Π is unilatent, i.e. we could deduce Lemma I from the properties of ruled quadrate slopes.

Ex. i. *If e and E are given particular undegenerate square commutants* $\{A, -B\}$ *and* $\{P, -Q\}$, *the requisite conditions* (A_1) *are that* $A, -B, P, -Q$ *shall be equicanonical, or that:*

A and P shall have the same characteristic potent divisors $(\lambda-c_1)^{e_1}, (\lambda-c_2)^{e_2}, \ldots;$

B and Q shall have the same characteristic potent divisors $(\lambda+c_1)^{e_1}, (\lambda+c_2)^{e_2}, \ldots.$

In this case the prescribed type of (A) is

$$\{P, A\}\{A, -B\}\{-B, -Q\}=\{P, -Q\};$$

and when the requisite conditions are satisfied, we can construct an equigradent commutantal transformation of this type even when X or Y is an arbitrarily given particular undegenerate square commutant of the prescribed type.

Ex. ii. *If e is a given undegenerate square commutant of one of the types*

$$\{A, A\},\ \{A', A'\},\ \{A', A\},\ \{A, A'\}, \qquad \textit{where } A' \textit{ is the conjugate of } A,$$

and E is a given undegenerate square commutant of one of the types

$$\{P, P\},\ \{P', P'\},\ \{P', P\},\ \{P, P'\}, \qquad \textit{where } P' \textit{ is the conjugate of } P,$$

the requisite conditions (A_1) *are that A and P shall be equicanonical.*

In this case the two square matrices A, B and the two square matrices P, Q in (a) are either equal or mutually conjugate.

Ex. iii. *If e is a given undegenerate square commutant of one of the types*

$$\{A, -A\},\ \{A', -A'\},\ \{A', -A\},\ \{A, -A'\}, \qquad \textit{where } A' \textit{ is the conjugate of } A,$$

and E is a given undegenerate square commutant of one of the types

$$\{P, -P\},\ \{P', -P'\},\ \{P', -P\},\ \{P, -P'\}, \qquad \textit{where } P' \textit{ is the conjugate of } P,$$

the requisite conditions (A_1) *are that* $A, -A, P, -P$ *shall be equicanonical, or that A and P shall be equicanonical, their common characteristic potent divisors which are not powers of* λ *being*

$$(\lambda-c_1)^{e_1},\ (\lambda+c_1)^{e_1};\ (\lambda-c_2)^{e_2},\ (\lambda+c_2)^{e_2};\ \ldots\ (\lambda-c_r)^{e_r},\ (\lambda+c_r)^{e_r}.$$

In this case the two square matrices A, B and the two square matrices P, Q of Ex. i are either equal or mutually conjugate.

Ex. iv. The results obtained in § 245 are developments of Theorem I.

2. *Symmetric transformations converting one given undegenerate symmetric or skew-symmetric contra-commutant into another of the same order.*

Let A and P be square matrices of the same order m with constant elements whose conjugates are A' and P', and let

$$\overset{m}{\underset{m}{\overline{x}}}\,[e]_m^m\,[x]_m^m=[E]_m^m \quad \text{or} \quad X'eX=E \quad \ldots\ldots\ldots\ldots\text{(B)}$$

be a symmetric equigradent commutantal transformation in which e and E are two symmetric or two skew-symmetric matrices,

e being an undegenerate contra-commutant $\{A', A\}$ or $\{A, A'\}$,

E being an undegenerate contra-commutant $\{P', P\}$ or $\{P, P'\}$,

so that the possible types of (B) are

$$\{P', A'\}\{A', A\}\{A, P\}=\{P', P\},\quad \{P, A'\}\{A', A\}\{A, P'\}=\{P, P'\},$$
$$\{P, A\}\{A, A'\}\{A', P'\}=\{P, P'\},\quad \{P', A\}\{A, A'\}\{A', P\}=\{P', P\},$$
$$\ldots\ldots\ldots\ldots\text{(b)}$$

the type of (B) being fixed by the prescribed types of e and E. From § 244 we see that:

Symmetric equigradent commutantal transformations (B) in which e and E are both *symmetric* and both undegenerate are possible if and only if A and P are equicanonical. ....(B_1)

Symmetric equigradent commutantal transformations (B) in which e and E are both *skew-symmetric* and both undegenerate are possible if and only if:

(i) A and P are equicanonical;

(ii) every distinct characteristic potent divisor of A (or P) is repeated an even number of times, so that the characteristic potent divisors of A (or P) occur in pairs of the form

$$(\lambda - c)^t, \quad (\lambda - c)^t. \qquad \text{..........................}(B_2)$$

The theorems which will now be enunciated show that if these requisite conditions are satisfied, it is possible to construct transformations (B) when e and E are any two given particular undegenerate contra-commutants of the specified types and characters.

Theorem II a. *Let e, E be given particular undegenerate symmetric contra-commutants $\{A', A\}$ or $\{A, A'\}$, $\{P', P\}$ or $\{P, P'\}$; and let the requisite conditions* (B_1) *be satisfied. Then it is always possible to determine X so that* (B) *is a symmetric equigradent commutantal transformation of the prescribed type converting e into E.*

Theorem II b. *Let e, E be given particular undegenerate skew-symmetric contra-commutants $\{A', A\}$ or $\{A, A'\}$, $\{P', P\}$ or $\{P, P'\}$; and let the requisite conditions* (B_2) *be satisfied. Then it is always possible to determine X so that* (B) *is a symmetric equigradent commutantal transformation of the prescribed type converting e into E.*

In the particular case when $A = P = \Pi$, where Π is a standardised *unilatent canonical square matrix*, these theorems become properties of ruled quadrate slopes which will be proved in Chapter XXXVIII (see Theorem IV of § 342 as applied to quadrate counter-continuants). In the proofs of the theorems which are here given it is assumed by anticipation that they are known to be true in that particular case. In other words the theorems are here shown to be dependent on properties of ruled quadrate slopes which are proved in the appropriate place.

When A and P are equicanonical, let Π be any one of their common canonical reduced forms. Then by § 247 a symmetric equigradent commutantal transformation (B) in which e and E are given particular undegenerate symmetric (or skew-symmetric) contra-commutants of the prescribed types can always be replaced by or derived from an equivalent symmetric equigradent commutantal transformation

$$\underline{\overline{\xi}}_m^m [\omega]_m^m [\xi]_m^m = [\Omega]_m^m \quad \text{or} \quad \xi'\omega\xi = \Omega \qquad \text{..............}(B_0)$$

of type $$\{\Pi', \Pi'\}\{\Pi', \Pi\}\{\Pi, \Pi\} = \{\Pi', \Pi\} \quad \text{.................(b}_0\text{)}$$
in which ω and Ω are given undegenerate symmetric (or skew-symmetric) contra-commutants $\{\Pi', \Pi\}$. Hence Theorems II *a* and II *b* can be completely proved by proving the following lemmas:

Lemma II a. *Let* Π *be a standardised canonical square matrix of order* m *(with standardised unilatent super-parts); and let* ω *and* Ω *be any two given particular undegenerate symmetric contra-commutants* $\{\Pi', \Pi\}$.

Then it is always possible to construct a symmetric equigradent commutantal transformation (B_0) *of type* (b_0) *converting* ω *into* Ω.

Lemma II b. *Let* Π *be a standardised canonical square matrix of order* m *whose characteristic potent divisors occur in pairs of the form*
$$(\lambda - c)^t, \quad (\lambda - c)^t;$$
and let ω *and* Ω *be any two given particular undegenerate skew-symmetric contra-commutants* $\{\Pi', \Pi\}$.

Then it is always possible to construct a symmetric equigradent commutantal transformation (B_0) *of type* (b_0) *converting* ω *into* Ω.

It is shown below that these two lemmas are true, being deducible from properties of ruled quadrate slopes proved in § 342; and it follows that Theorems II *a* and II *b* are true.

Proof of Lemma II a.

Let the successive unilatent super-parts of Π be the square matrices
$$\Pi_1, \Pi_2, \dots \Pi_r \quad \text{of orders} \quad e_1, e_2, \dots e_r$$
with latent roots $\qquad c_1, c_2, \dots c_r \qquad$ which are all different,
the super-part Π_i being a standardised unilatent canonical square matrix of the class
$$M\begin{pmatrix} e_{i1}, & e_{i2}, & \dots \\ e_{i1}, & e_{i2}, & \dots \end{pmatrix} \quad \text{in which} \quad e_{i1}+e_{i2}+\dots=e_i. \quad \text{.....................(1)}$$
Referring to Ex. iii of § 244, we see that in (B_0) we can put
$$\Pi = \begin{bmatrix} \Pi_1, & 0, & \dots & 0 \\ 0, & \Pi_2, & \dots & 0 \\ \dots & \dots & \dots & \dots \\ 0, & 0, & \dots & \Pi_r \end{bmatrix}_{e_1, e_2, \dots e_r}^{e_1, e_2, \dots e_r}, \quad \xi = \begin{bmatrix} \xi_1, & 0, & \dots & 0 \\ 0, & \xi_2, & \dots & 0 \\ \dots & \dots & \dots & \dots \\ 0, & 0, & \dots & \xi_r \end{bmatrix}_{e_1, e_2, \dots e_r}^{e_1, e_2, \dots e_r},$$
$$\omega = \begin{bmatrix} \omega_1, & 0, & \dots & 0 \\ 0, & \omega_2, & \dots & 0 \\ \dots & \dots & \dots & \dots \\ 0, & 0, & \dots & \omega_r \end{bmatrix}_{e_1, e_2, \dots e_r}^{e_1, e_2, \dots e_r}, \quad \Omega = \begin{bmatrix} \Omega_1, & 0, & \dots & 0 \\ 0, & \Omega_2, & \dots & 0 \\ \dots & \dots & \dots & \dots \\ 0, & 0, & \dots & \Omega_r \end{bmatrix}_{e_1, e_2, \dots e_r}^{e_1, e_2, \dots e_r},$$
and then replace (B_0) by the r symmetric equigradent commutantal transformations such as
$$\xi_i' \omega_i \xi_i = \Omega_i \quad \text{of type} \quad \{\Pi_i', \Pi_i'\}\{\Pi_i', \Pi_i\}\{\Pi_i, \Pi_i\} = \{\Pi_i', \Pi_i\}, \quad \text{............(2)}$$
$$(i = 1, 2, \dots r),$$
where ω_i and Ω_i are given undegenerate symmetric commutants. Further we can regard 2) as a symmetric equigradent commutantal transformation
$$\xi_i' \omega_i \xi_i = \Omega_i \quad \text{of type} \quad \{\pi', \pi'\}\{\pi', \pi\}\{\pi, \pi\} = \{\pi', \pi\}, \quad \text{...............(2')}$$

in which all the matrices are ruled quadrate slopes of the class (1) and of the types shown, ω_i and Ω_i being given undegenerate symmetric counter-continuants.

By Theorem IV of § 342 it is possible to determine an undegenerate quadrate ante-continuant ξ_i of the class (1) and of type $\{\pi, \pi\}$ together with its conjugate ξ_i' so that the equation (2′) is satisfied. When $\xi_1, \xi_2, \ldots \xi_r$ have been determined in this way, the equation (B_0) is a commutantal transformation of type (b_0) having the character required in Lemma II *a*.

Proof of Lemma II b.

Proceeding as in the proof of Lemma II *a* we can replace (B_0) by the r equations such as (2) or (2′); but now (for each of the values $1, 2, \ldots r$ of i) every index number in the series $e_{i1}, e_{i2}, \ldots$ is repeated an even number of times, and e_i is an even integer; moreover ω_i and Ω_i are given undegenerate skew-symmetric commutants in (2), and therefore given undegenerate skew-symmetric counter-continuants in (2′).

By Theorem IV of § 342 it is possible to determine an undegenerate quadrate ante-continuant ξ_i of the class (1) and of type $\{\pi, \pi\}$ together with its conjugate ξ_i' so that the equation (2′) is satisfied. When $\xi_1, \xi_2, \ldots \xi_r$ have been determined in this way, the equation (B_0) is a commutantal transformation of type (b_0) having the character required in Lemma II *b*.

3. *Symmetric transformations converting one given undegenerate symmetric or skew-symmetric alternating contra-commutant into another.*

Let A and P be square matrices of the same order m with constant elements whose conjugates are A' and P', and let

$$\overline{\underline{x}}^{\,m}_{\,m}\,[e]^m_m\,[x]^m_m = [E]^m_m \quad \text{or} \quad X'eX = E \text{..................(C)}$$

be a symmetric equigradent commutantal transformation in which e and E are two symmetric or two skew-symmetric matrices,

e being an undegenerate contra-commutant $\{A', -A\}$ or $\{A, -A'\}$,

E being an undegenerate contra-commutant $\{P', -P\}$ or $\{P, -P'\}$,

so that the possible types of (C) are

$$\begin{aligned}
&\{P', A'\}\{A', -A\}\{-A, -P\} = \{P', -P\},\\
&\{P, A\}\{A, -A'\}\{-A', -P'\} = \{P, -P'\},\\
&\{P, A'\}\{A', -A\}\{-A, -P'\} = \{P, -P'\},\\
&\{P', A\}\{A, -A'\}\{-A', -P\} = \{P', -P\}, \quad \text{..............(c)}
\end{aligned}$$

the type of (C) being fixed by the prescribed types of e and E. From § 244 we see that:

Symmetric equigradent commutantal transformations (C) in which e and E are both *symmetric* and both undegenerate are possible if and only if:

(i) A and P are equicanonical;

(ii) the characteristic potent divisors of A (or P) which are not of the form $\lambda^{2\tau+1}$ occur in pairs of the form

$$(\lambda - c)^t, \quad (\lambda + c)^t \text{...........................}(C_1)$$

Symmetric equigradent commutantal transformations (C) in which e and E are both *skew-symmetric* and both undegenerate are possible if and only if:

(i) A and P are equicanonical;

(ii) the characteristic potent divisors of A (or P) which are not of the form $\lambda^{2\tau}$ occur in pairs of the form

$$(\lambda - c)^t, \quad (\lambda + c)^t. \qquad \text{(C}_2\text{)}$$

In (C_1) every distinct characteristic potent divisor which is a power of λ with even index must be repeated an even number of times, but one which is a power of λ with odd index may be repeated either an odd or an even number of times.

In (C_2) every distinct characteristic potent divisor which is a power of λ with odd index must be repeated an even number of times, but one which is a power of λ with even index may be repeated either an odd or an even number of times.

The theorems which will now be enunciated show that if these requisite conditions are satisfied, it is possible to construct transformations (C) when e and E are any two given particular undegenerate contra-commutants of the specified types and characters.

Theorem III a. *Let e, E be given particular undegenerate symmetric contra-commutants $\{A', -A\}$ or $\{A, -A'\}$, $\{P', -P\}$ or $\{P, -P'\}$; and let the requisite conditions* (C_1) *be satisfied. Then it is always possible to determine X so that* (C) *is a symmetric equigradent commutantal transformation of the prescribed type converting e into E.*

Theorem III b. *Let e, E be given particular undegenerate skew-symmetric contra-commutants $\{A', -A\}$ or $\{A, -A'\}$, $\{P', -P\}$ or $\{P, -P'\}$; and let the requisite conditions* (C_2) *be satisfied. Then it is always possible to determine X so that* (C) *is a symmetric equigradent commutantal transformation of the prescribed type converting e into E.*

In the particular case when $A = P = \Pi$, where Π is a standardised *unilatent canonical square matrix*, these theorems become properties of ruled quadrate slopes which will be proved in Chapter XXXVIII (see Theorem IV of § 342 as applied to quadrate counter-alternants). In the proofs of the theorems which are here given it is assumed by anticipation that they are known to be true in that particular case. In other words the theorems are here shown to be dependent on properties of ruled quadrate slopes which are proved in the appropriate place.

When A and P are equicanonical, let Π be any one of their common canonical reduced forms. Then by § 247 a symmetric equigradent commutantal transformation (C) in which e and E are given particular undegenerate symmetric (or skew-symmetric) contra-commutants of the prescribed types can always be replaced by or derived from an equivalent symmetric equigradent commutantal transformation

$$\underset{m}{\overline{\underline{\xi}}}^{m}[\omega]_{m}^{m}[\xi]_{m}^{m}=[\Omega]_{m}^{m} \quad \text{or} \quad \xi'\omega\xi=\Omega \quad \text{..............}(\mathrm{C}_0)$$

of type $\{\Pi', \Pi'\}\{\Pi', -\Pi\}\{-\Pi, -\Pi\}=\{\Pi', -\Pi\}$ ....(c_0)

in which ω and Ω are given undegenerate symmetric (or skew-symmetric) contra-commutants $\{\Pi', -\Pi\}$. Hence Theorems III *a* and III *b* can be completely proved by proving the following lemmas:

Lemma III a. *Let* Π *be a standardised canonical square matrix of order* m *(with standardised unilatent super-parts) whose characteristic potent divisors not of the form* λ^{2r+1} *occur in pairs of the form*

$$(\lambda-c)^t, \quad (\lambda+c)^t;$$

and let ω, Ω *be any two given particular undegenerate symmetric contra-commutants* $\{\Pi', -\Pi\}$.

Then it is always possible to construct a symmetric equigradent commutantal transformation (C_0) *of type* (c_0) *converting* ω *into* Ω.

Lemma III b. *Let* Π *be a standardised canonical square matrix of order* m *whose characteristic potent divisors not of the form* λ^{2r} *occur in pairs of the form*

$$(\lambda-c)^t, \quad (\lambda+c)^t;$$

and let ω, Ω *be any two given particular undegenerate skew-symmetric contra-commutants* $\{\Pi', -\Pi\}$.

Then it is always possible to construct a symmetric equigradent commutantal transformation (C_0) *of type* (c_0) *converting* ω *into* Ω.

It is shown below that these two lemmas are true, being deducible from properties of ruled quadrate slopes proved in § 342; and it follows that Theorems III *a* and III *b* are true.

Proof of Lemma III a.

Let the successive unilatent super-parts of Π be the square matrices

	Π_0;	P_1, Q_1;	P_2, Q_2; ...	P_s, Q_s
of orders	e_0;	e_1, e_1;	e_2, e_2; ...	e_s, e_s
with latent roots	0;	$c_1, -c_1$;	$c_2, -c_2$; ...	$c_s, -c_s$ which are all different;

the super-part Π_0 being a standardised unilatent canonical square matrix of the class

$$M\begin{pmatrix} e_{01}, e_{02}, \dots \\ e_{01}, e_{02}, \dots \end{pmatrix} \quad \text{in which} \quad e_{01}+e_{02}+\dots=e_0, \quad \text{....................}(3)$$

and in which *every even index number is repeated an even number of times*; and the super-parts P_i, Q_i being standardised unilatent canonical square matrices of the class

$$M\begin{pmatrix} e_{i1}, e_{i2}, \dots \\ e_{i1}, e_{i2}, \dots \end{pmatrix} \quad \text{in which} \quad e_{i1}+e_{i2}+\dots=e_i. \quad \text{....................}(4)$$

Referring to Ex. viii of § 244, we see that in (C_0) we can put

$$\Pi=\begin{bmatrix}\Pi_0, & 0, & \dots & 0\\ 0, & \Pi_1, & \dots & 0\\ \dots & \dots & \dots & \dots\\ 0, & 0, & \dots & \Pi_s\end{bmatrix}_{e_0, 2e_1, \dots 2e_s}^{e_0, 2e_1, \dots 2e_s}, \qquad \xi=\begin{bmatrix}\xi_0, & 0, & \dots & 0\\ 0, & \xi_1, & \dots & 0\\ \dots & \dots & \dots & \dots\\ 0, & 0, & \dots & \xi_s\end{bmatrix}_{e_0, 2e_1, \dots 2e_s}^{e_0, 2e_1, \dots 2e_s},$$

$$\omega=\begin{bmatrix}\omega_0, & 0, & \dots & 0\\ 0, & \omega_1, & \dots & 0\\ \dots & \dots & \dots & \dots\\ 0, & 0, & \dots & \omega_s\end{bmatrix}_{e_0, 2e_1, \dots 2e_s}^{e_0, 2e_1, \dots 2e_s}, \qquad \Omega=\begin{bmatrix}\Omega_0 & 0, & \dots & 0\\ 0, & \Omega_1, & \dots & 0\\ \dots & \dots & \dots & \dots\\ 0, & 0, & \dots & \Omega_s\end{bmatrix}_{e_0, 2e_1, \dots 2e_s}^{e_0, 2e_1, \dots 2e_s},$$

where for the values 1, 2, ... s of i the parts Π_i, ξ_i, ω_i, Ω_i are bipartite matrices of the forms

$$\begin{bmatrix}P_i, & 0\\ 0, & Q_i\end{bmatrix}_{ei, ei}^{ei, ei}, \quad \begin{bmatrix}X_i, & 0\\ 0, & Y_i\end{bmatrix}_{ei, ei}^{ei, ei}, \quad \begin{bmatrix}0, & a_i\\ a_i', & 0\end{bmatrix}_{ei, ei}^{ei, ei}, \quad \begin{bmatrix}0, & A_i\\ A_i', & 0\end{bmatrix}_{ei, ei}^{ei, ei}$$

in which the non-zero parts are ruled quadrate slopes of the class (4), and the matrices a_i', A_i' are the conjugates of the matrices a_i, A_i. The matrices Π_0, ξ_0, ω_0, Ω_0 are ruled quadrate slopes of the class (3). Using these forms and denoting the conjugates of Π, ξ, Π_i, ξ_i, P_i, Q_i, X_i, Y_i by the corresponding dashed letters, we can replace (C_0) by the $s+1$ symmetric equigradent commutantal transformations such as the equation

$$\xi_i'\omega_i\xi_i=\Omega_i \quad \text{of type} \quad \{\Pi_i', \Pi_i'\}\{\Pi_i', -\Pi_i\}\{-\Pi_i, -\Pi_i\}=\{\Pi_i', -\Pi_i\}, \quad \dots\dots(5)$$

where i receives the values 0, 1, 2, ... s. But when $i \neq 0$, the equation (5) can be replaced by the two mutually equivalent equations

$$X_i'a_iY_i=A_i, \quad Y_i'a_i'X_i=A_i',$$

and therefore by one only of these equations. Thus finally (C_0) is equivalent to the single symmetric equigradent commutantal transformation

$$\xi_0'\omega_0\xi_0=\Omega_0 \quad \text{of type} \quad \{\Pi_0', \Pi_0'\}\{\Pi_0', -\Pi_0\}\{-\Pi_0, -\Pi_0\}=\{\Pi_0', -\Pi_0\}, \quad \dots\dots(6)$$

in which ω_0 and Ω_0 are given undegenerate symmetric commutants, and the s equigradent commutantal transformations such as

$$X_i'a_iY_i=A_i \quad \text{of type} \quad \{P_i', P_i'\}\{P_i', -Q_i\}\{-Q_i, -Q_i\}=\{P_i', -Q_i\}, \quad \dots\dots(7)$$
$$(i=1, 2, \dots s),$$

where a_i and A_i are given undegenerate commutants. We can regard (6) as a symmetric equigradent commutantal transformation

$$\xi_0'\omega_0\xi_0=\Omega_0 \quad \text{of type} \quad \{\pi', \pi'\}\{\pi', -\pi\}\{-\pi, -\pi\}=\{\pi', -\pi\}, \quad \dots\dots\dots(6')$$

in which all the matrices are ruled quadrate slopes of the class (3) and of the types shown, ω_0 and Ω_0 being given undegenerate symmetric counter-alternants; and we can regard (7) as an equigradent commutantal transformation

$$X_i'a_iY_i=A_i \quad \text{of type} \quad \{\pi', \pi'\}\{\pi', -\pi\}\{-\pi, -\pi\}=\{\pi', -\pi\}, \quad \dots\dots\dots(7')$$

in which all the matrices are ruled quadrate slopes of the class (4) and of the types shown, a_i and A_i being given undegenerate counter-alternants.

By Theorem IV of § 342 it is possible to determine an undegenerate quadrate ante-continuant ξ_0 of the class (3) and of type $\{-\pi, -\pi\}$ or $\{\pi, \pi\}$ together with its conjugate ξ_0' so that the equation (6′) is satisfied. So far as the equation (7′) or (7) is concerned, there is no need to refer to Chapter XXXVIII; for it follows from Theorem I of the present article that it is possible to determine undegenerate commutants X_i' and Y_i of the prescribed types which satisfy the equation (7), X_i being then a commutant $\{P_i, P_i\}$.

When undegenerate commutants $\xi_0, X_1, Y_1, X_2, Y_2, \ldots X_s, Y_s$ have been determined in these ways, the equation (C_0) is a commutantal transformation of type (c_0) having the character required in Lemma III *a*.

Proof of Lemma III b.

The proof is similar to that of Lemma III *a*, but (3) must be a class in which *every odd index number is repeated an even number of times*, and $\Pi_i, \xi_i, \omega_i, \Omega_i$ for the values 1, 2, ... *s* of *i* are bipartite matrices of the forms

$$\begin{bmatrix} P_i, & 0 \\ 0, & Q_i \end{bmatrix}_{e_i, e_i}^{e_i, e_i}, \quad \begin{bmatrix} X_i, & 0 \\ 0, & Y_i \end{bmatrix}_{e_i, e_i}^{e_i, e_i}, \quad \begin{bmatrix} 0, & a_i \\ -a_i', & 0 \end{bmatrix}_{e_i, e_i}^{e_i, e_i}, \quad \begin{bmatrix} 0, & A_i \\ -A_i', & 0 \end{bmatrix}_{e_i, e_i}^{e_i, e_i}.$$

We can replace (C_0) as before by the single symmetric equigradent commutantal transformation (6) or (6′) and the *s* equigradent commutantal transformations such as (7) or (7′); but now ω_0 and Ω_0 are given undegenerate skew-symmetric commutants in (6) or quadrate counter-alternants in (6′).

By Theorem IV of § 342 we can determine an undegenerate quadrate slope ξ_0 of the prescribed type and class which satisfies the new equation (6′); and by Theorem I of the present article we can determine undegenerate commutants X_i and Y_i of the prescribed types which satisfy the new equation (7′). When undegenerate commutants $\xi_0, X_1, Y_1, X_2, Y_2, \ldots X_s, Y_s$ have been determined in these ways, the equation (C_0) is a commutantal transformation of type (c_0) having the character required in Lemma III *b*.

4. *Applications to symmetric semi-unit transformations converting one given symmetric or skew-symmetric matrix into another of the same order.*

Since a symmetric semi-unit transformation is the same thing as a symmetric equimutant transformation, there cannot exist such a transformation converting a square matrix A with constant elements into another square matrix P of the same order unless A and P are equicanonical. When this condition is satisfied, we have the following theorem in which X' is the conjugate of X.

Theorem IV. *Let* $A = [a]_m^m$ *and* $P = [p]_m^m$ *be two given equicanonical square matrices of the same order* m *with constant elements which are either both symmetric or both skew-symmetric. Then it is always possible to determine a square semi-unit matrix* $X = [x]_m^m$ *such that*

$$\overline{\underline{x}}_m^m [a]_m^m [x]_m^m = [p]_m^m \quad \text{or} \quad X'AX = P. \quad \ldots\ldots\ldots\ldots\ldots(D)$$

To do this we have to determine a (necessarily undegenerate) square matrix X of order m which satisfies simultaneously the two equations

$$X'X = I, \quad AX = XP, \qquad \text{where } I = [1]_m^m.$$

If A and P are both symmetric, we can regard I as an undegenerate symmetric contra-commutant of each of the types $\{A', A\}$, $\{A, A'\}$, $\{P', P\}$, $\{P, P'\}$, where A', P' are the conjugates of A, P; consequently the possibility of determining such a matrix X is a particular case of Theorem II *a* obtained by putting $e = E = I$.

If A and P are both skew-symmetric, we can regard I as an undegenerate symmetric contra-commutant of each of the types $\{A', -A\}$, $\{A, -A'\}$, $\{P', -P\}$, $\{P, -P'\}$, and as in Ex. xiv of § 244 the characteristic potent divisors of A (or P) not of the form $\lambda^{2\tau+1}$ occur in pairs of the form $(\lambda - c)^t$, $(\lambda + c)^t$; consequently the possibility of determining such a matrix X is a particular case of Theorem III a obtained by putting $e = E = I$.

NOTE 2. *Determination of the transformations* (D).

The actual determination of X in Theorem IV can of course be reduced to corresponding problems regarding ruled quadrate slopes. Let $\Pi = [\Pi]_m^m$ be a common standard canonical reduced form of A and P with standardised unilatent super-parts, let Π' be the conjugate of Π, and let

$$[a]_m^m = [h]_m^m [\Pi]_m^m [H]_m^m, \quad [p]_m^m = [u]_m^m [\Pi]_m^m [U]_m^m$$

or

$$A = h\Pi H, \qquad P = u\Pi U$$

be given isomorphic transformations by which A and P can be derived from Π, so that h, u are any particular undegenerate commutants $\{A, \Pi\}$, $\{P, \Pi\}$. Then we have to construct a symmetric equigradent commutantal transformation

$$X'IX = \underset{m}{\overset{m}{\overline{\underline{x}}}} [1]_m^m [x]_m^m = [1]_m^m = I$$

of type $\{P', A'\}\{A', A\}\{A, P\} = \{P', P\}$ or $\{P', A'\}\{A', -A\}\{-A, -P\} = \{P', -P\}$, according as A and P are both symmetric or both skew-symmetric, which can be reduced to an equivalent symmetric equigradent commutantal transformation

$$\xi'\omega\xi = \underset{m}{\overset{m}{\overline{\underline{\xi}}}} [\omega]_m^m [\xi]_m^m = [\Omega]_m^m = \Omega$$

of type $\{\Pi', \Pi'\}\{\Pi', \Pi\}\{\Pi, \Pi\} = \{\Pi', \Pi\}$ or $\{\Pi', \Pi'\}\{\Pi', -\Pi\}\{-\Pi, -\Pi\} = \{\Pi', -\Pi\}$, according as A and P are both symmetric or both skew-symmetric, by the equigradent commutantal substitutions

$$X = [x]_m^m = [h]_m^m [\xi]_m^m [U]_m^m = h\xi U,$$

$$I = [1]_m^m = \underset{m}{\overset{m}{\overline{\underline{H}}}} [\omega]_m^m [H]_m^m = H'\omega H,$$

$$I = [1]_m^m = \underset{m}{\overset{m}{\overline{\underline{U}}}} [\Omega]_m^m [U]_m^m = U'\Omega U.$$

Hence if ω and Ω are the given undegenerate symmetric contra-commutants

$$\omega = \underset{m}{\overset{m}{\overline{\underline{h}}}} [h]_m^m = h'h, \quad \Omega = \underset{m}{\overset{m}{\overline{\underline{u}}}} [u]_m^m = u'u$$

which are both of type $\{\Pi', \Pi\}$ when A and P are both symmetric,
or both of type $\{\Pi', -\Pi\}$ when A and P are both skew-symmetric,
and if ξ is any particular undegenerate commutant $\{-\Pi, -\Pi\}$ or $\{\Pi, \Pi\}$ satisfying the symmetric commutantal equation $\xi'\omega\xi = \Omega$, the required conditions are satisfied when

$$X = h\xi U.$$

The matrices ω, Ω, ξ are compartite, their parts being ruled quadrate slopes of types $\{\pi', \pi\}$, $\{\pi', \pi\}$, $\{\pi, \pi\}$ or $\{\pi', -\pi\}$, $\{\pi', -\pi\}$, $\{-\pi, -\pi\}$; and the determination of ξ is effected as in the proof of Lemma II a or Lemma III a.

CHAPTER XXVIII

COMMUTANTS OF COMMUTANTS

[After the commutants of the commutant of a square matrix A have been defined to be those square matrices which are commutative with every commutant of A, it is shown that they could also be defined to be those square matrices (or those commutants of A) which are rational integral functions of A. The rules for constructing a general commutant of the commutant of A lead to special forms of an arbitrary rational integral function of A which are convenient for certain proofs. They are used in deducing Frobenius's Theorem regarding solutions of the matrix equation $\psi^2=\phi$ from the properties of simple square ante-continuants. In the last article (§ 253) the general commutant of a square matrix A whose elements are all arbitrary is determined by a direct method, and is shown to be identifiable with a general commutant of the commutant of A, and with an arbitrary rational integral function of A.]

§ 249. Commutants of commutants defined.

1. *The commutants of the commutant of a square matrix whose elements are constants.*

If $A=[a]_m^m$ is a square matrix of order m whose elements are given constants, a square matrix $\Phi=[\phi]_m^m$ will be called a *commutant of the commutant* of A or a *commutant of the commutants* of A when it is commutative with every particular commutant of A, i.e. when

every commutant $\{A, A\}$ is also a commutant $\{\Phi, \Phi\}$.

If X is a square matrix of order m whose elements can receive any values we please, this is the case if and only if

$$\text{whenever } AX=XA, \quad \text{then also} \quad \Phi X=X\Phi.$$

Let $X=[x]_m^m$ be any general commutant $\{A, A\}$ expressed in the form

$$X=\lambda_1X_1+\lambda_2X_2+\ldots+\lambda_iX_i, \quad \text{.....................(a)}$$

where the λ's are arbitrary scalar parameters, and $X_1, X_2, \ldots X_i$ are a complete set of independent particular non-zero commutants, the integer i being necessarily not less than 1. Then a square matrix $\Phi=[\phi]_m^m$ whose elements are constants is a *particular commutant of the commutant* of A when and only when the equation $X\Phi=\Phi X$ is an identity in the λ's, i.e. when and only when

Φ is a particular non-singular commutant $\{X, X\}$.

Again if j is the greatest possible number of independent particular non-zero commutants of the commutant of A, a *general commutant of the commutant* of A is a square matrix $\Phi=[\phi]_m^m$ expressible in the form

$$\Phi=\mu_1\Phi_1+\mu_2\Phi_2+\ldots+\mu_j\Phi_j, \quad \text{....................(b)}$$

where $\Phi_1, \Phi_2, \ldots \Phi_j$ are j independent particular non-zero commutants of the commutant of A, and the μ's are arbitrary scalar parameters independent of one another and of the λ's occurring in X. In this case the equation $X\Phi = \Phi X$ is an identity in the λ's and μ's, and

$$\Phi \text{ is a general non-singular commutant } \{X, X\}.$$

Every commutant of the commutant of A can be regarded as a particularisation or specialisation of any given general commutant of the commutant of A.

NOTE 1. The commutants of the commutants of A defined above depend on A only, and must be distinguished from the commutants of any specified commutant of A. When X is any general commutant of A, they are the *non-singular* commutants of X, which are independent of the form or choice of X. The singular commutants of X, when they exist, involve the arbitrary parameters of X, and could not properly be called commutants of 'the' commutant of A when X is not specified.

Throughout this chapter 'commutants of commutants' will mean square matrices each of which is a 'commutant of the commutant' of a given square matrix, i.e. square matrices each of which is commutative with all commutants of a given square matrix.

Ex. i. *The general commutant of the commutant of A is always an undegenerate square matrix.*

For the unit matrix of the same order as A is always a commutant of the commutant of A.

Ex. ii. *Every particular rational integral function of A is a commutant of the commutant of A.*

For by Ex. viii of § 238 it is commutative with every particular commutant of A, and is therefore a non-singular commutant of every general commutant $X=\{A, A\}$.

Ex. iii. *The general commutant of the commutant of a scalar or quasi-scalar matrix.*

The general commutant of the commutant of the scalar matrix $A=c.[1]_m^m$, where c may be 0, is the scalar matrix $\Phi=\mu.[1]_m^m$, where μ is an arbitrary parameter. This is shown in Exs. ii and xix of § 238.

If A is a quasi-scalar matrix whose successive diagonal elements are $a_1, a_2, \ldots a_m$, it follows from Ex. ii and Theorem II of § 250 that the general commutant of the commutant of A is a quasi-scalar matrix $\Phi=[\phi]_m^m$ whose successive diagonal elements $\mu_1, \mu_2, \ldots \mu_m$ are arbitrary subject to the conditions that

$$\mu_i=\mu_j \text{ when } a_i=a_j, \quad \text{so that } \mu_i \neq \mu_j \text{ when } a_i \neq a_j.$$

If A contains exactly s unequal diagonal elements $c_1, c_2, \ldots c_s$, then Φ contains exactly s arbitrary parameters $\mu_1, \mu_2, \ldots \mu_s$.

Ex. iv. *If X and Φ are respectively a general commutant of A and a general commutant of the commutant of A, then:*

A is a particularisation of X, A is a particularisation of Φ,

and Φ is a specialisation of X.

These results follow respectively from the facts that A is commutative with A; A is commutative with X; Φ is commutative with A because it is commutative with every commutant of A. By the third theorem in Note 4 of § 238, the last of these three results is a necessary consequence of the first of them; for because A is a particularisation of X, therefore the general non-singular commutant $\Phi=\{X, X\}$ must be a specialisation of the general non-singular commutant $X=\{A, A\}$.

The fact that A is a particularisation of Φ shows that every non-singular commutant $\{\Phi, \Phi\}$ is a commutant $\{A, A\}$; for being commutative with Φ, it must be commutative with A.

Consequently we can regard Φ as the most general square matrix which has X as its general non-singular commutant.

Ex. v. Since every commutant of the commutant of A is a commutant of A, we can say that:

The commutants of the commutant of A are those commutants of A which are commutative with all commutants of A.

Hence when all commutants of A are commutative with one another, every particular or general commutant of A is also a particular or general commutant of the commutant of A; in other cases there exist commutants of A which are not also commutants of the commutant of A.

NOTE 2. *Some anticipatory results.*

If $A=[a]_m^m$ is a square matrix with constant elements, and if s is the lowest possible degree of a rational integral equation satisfied by A, it will be shown in § 252 that:

There are exactly s independent particular non-zero commutants of the commutant of A.

From this result it follows further that:

The general commutant of the commutant of A is an arbitrary rational integral function of A.

The commutants of the commutant of A are those commutants of A (or those square matrices) which are rational integral functions of A.

The general commutant of A is a rational integral function of A when and only when it is a general commutant of the commutant of A, i.e. when and only when all commutants of A are commutative with one another; and this is the case when and only when $s=m$.

Ex. vi. In the following two illustrations a, b, c are given scalar numbers such that $a \neq b$ and $c \neq 0$; $x, y, z, u, v, \lambda, \mu, \nu$ are independent arbitrary parameters; X is a general commutant $\{A, A\}$; Φ is a general non-singular commutant $\{X, X\}$, i.e. a general commutant of the commutant of A; and Ψ is a general commutant $\{X, X\}$.

$$(1) \qquad A=\begin{bmatrix} a & c & 0 \\ 0 & a & 0 \\ 0 & 0 & a \end{bmatrix}, \quad X=\begin{bmatrix} x & z & u \\ 0 & x & 0 \\ 0 & v & y \end{bmatrix}, \quad \Phi=\begin{bmatrix} \lambda & \nu & 0 \\ 0 & \lambda & 0 \\ 0 & 0 & \lambda \end{bmatrix},$$

$$\Psi=\begin{bmatrix} (x-y)\lambda, & (x-y)\nu, & u\mu \\ 0, & (x-y)\lambda, & 0 \\ 0, & v\mu, & (x-y)(\lambda-\mu) \end{bmatrix}.$$

Here there are two independent particular non-zero commutants of the commutant of A, and there are three independent particular non-zero commutants of X, of which only two can be non-singular. The coefficient of μ in Ψ is a singular commutant of X, and is not commutative with all particularisations of X; consequently it is not a commutant of the commutant of A. If $I=[1]_3^3$, we have

$$(A-aI)^2=0, \quad c\Phi=c\lambda . I+\nu . (A-aI).$$

$$(2) \quad A=\begin{bmatrix} a & c & 0 \\ 0 & a & 0 \\ 0 & 0 & b \end{bmatrix}, \quad X=\begin{bmatrix} x & z & 0 \\ 0 & x & 0 \\ 0 & 0 & y \end{bmatrix}, \quad \Phi=\Psi=\begin{bmatrix} \lambda & \nu & 0 \\ 0 & \lambda & 0 \\ 0 & 0 & \mu \end{bmatrix}.$$

Here there are three independent particular non-zero commutants of the commutant of A, and there are three independent particular non-zero commutants of X which are all non-singular, the general commutant $\{X, X\}$ being non-singular. If $I=[1]_3^3$, we have

$$(A-aI)^2(A-bI)=0,$$
$$c(a-b)^2.\Phi=c\lambda.\{b(b-2a)I+2aA-A^2\}+c\mu.\{a^2I-2aA+A^2\}$$
$$+(a-b)\nu.\{abI-(a+b)A+A^2\}.$$

2. *The commutants of the commutant of a square matrix containing arbitrary or variable elements.*

When the elements of a square matrix $A=[a]_m^m$ are rational integral functions of certain scalar variables $\gamma_1, \gamma_2, \gamma_3, \ldots$, we will still define a *commutant of the commutant* of A to be a square matrix which is commutative with every particular commutant of A; but it must be understood that all equations are identities in the γ's.

Let X be a general commutant of A expressed in the form (a), where the elements of $X_1, X_2, \ldots X_i$ are particular rational integral functions of the γ's, and the λ's are arbitrary parameters independent of one another and of the γ's; and let $\Phi=[\phi]_m^m$ be a square matrix whose elements are rational integral functions of the γ's. Then Φ is a *commutant of the commutant* of A when and only when all the equations

$$X_1\Phi=\Phi X_1, \quad X_2\Phi=\Phi X_2, \ldots X_i\Phi=\Phi X_i$$

are identities in the γ's as well as in any arbitrary parameters which may occur in Φ. If we treat all the arbitrary parameters occurring in Φ as independent of the λ's, this is the case when and only when the equation

$$X\Phi=\Phi X$$

is an identity in the γ's, the λ's and the independent arbitrary parameters of Φ. Accordingly we can define Φ to be a *particular commutant of the commutant* of A when its elements are rational integral functions of the γ's only, and it makes the equation $X\Phi=\Phi X$ an identity in the γ's and λ's; and we can define Φ to be a *general commutant of the commutant* of A when it is a homogeneous linear function of the greatest possible number of independent particular non-zero commutants of the commutants of A expressed in the form (b). In the latter case we will call $\mu_1, \mu_2, \ldots \mu_j$ the parameters of (or peculiar to) Φ; they are to be regarded as independent of the parameters $\lambda_1, \lambda_2, \ldots \lambda_i$ of X as well as independent of the scalar variables $\gamma_1, \gamma_2, \gamma_3, \ldots$ of A; and the equation $X\Phi=\Phi X$ is an identity in the γ's, λ's and μ's.

The arguments and results of sub-article 1 can be adapted to the more general case of this sub-article by interpreting non-singular commutants $\{X, X\}$ to mean commutants $\{X, X\}$ which are independent of the λ's; and particularisations or specialisations of X and Φ to be matrices derived from them by particularising or specialising the λ's and μ's in the way described in § 238. 3. When the elements of A, X, Φ are rational functions of the γ's, we can (after multiplying each matrix by a suitable non-zero rational integral scalar function of the γ's) replace them by rational integral functions of the γ's, and all the above remarks then remain applicable.

We shall always regard the general commutant of the commutant of a square matrix A as determined in the way described above, though it can be interpreted (see Note 3) to be an arbitrary rational integral function of A.

Ex. vii. As a square matrix becomes more and more specialised, the commutant of the commutant of A tends in general to become more and more specialised.

Ex. viii. *The conjugates and inverses of commutants of commutants.*

If Φ is a commutant of the commutant of a square matrix A, and if A', Φ' are the conjugates of A, Φ, then Φ' is a commutant of the commutant of A'; moreover when Φ is particular or general, Φ' also is particular or general.

When Φ is undegenerate, its inverse is always another commutant of the commutant of A.

NOTE 3. *Anticipatory results.*

Since the results mentioned in Note 2 must be true for all particularisations of A obtained by ascribing particular values to the γ's, we see by ascribing ordinary particular values to γ's (as at the end of § 252) that they must remain true for the matrix A whose elements are rational integral functions of the γ's. In particular it is still true that:

The general commutant of the commutant of A is an arbitrary rational integral function of A.

Two illustrations are given in Ex. vi, where a, b, c may be independent scalar variables. Other illustrations will be given in § 253.

§ 250. Theorems facilitating the determination of commutants of commutants.

1. *The commutants of the commutants of two equicanonical square matrices.*

Let $A = [a]_m^m$ and $B = [b]_m^m$ be two given equicanonical square matrices of order m with constant elements, and let

$$[a]_m^m = [h]_m^m [b]_m^m [H]_m^m \quad \text{or} \quad A = hBH \quad \ldots\ldots\ldots\ldots\ldots(1)$$

be any given isomorphic transformation by which A can be derived from B. Then we have the following theorem concerning two square matrices Φ and Ψ which are connected by the equation

$$[\phi]_m^m = [h]_m^m [\psi]_m^m [H]_m^m \quad \text{or} \quad \Phi = h\Psi H. \ldots\ldots\ldots\ldots\ldots(A)$$

Theorem I. *The square matrix Φ in (A) is a commutant of the commutant of A when and only when the square matrix Ψ is a commutant of the* *commutant of B; also Φ is a particular or general commutant of the commutant of A when and only when Ψ is a particular or general commutant of the commutant of B.*

As in Theorem I*a* of § 240 we can always select a general commutant $X=\{A, A\}$ and a general commutant $Y=\{B, B\}$ such that

$$[x]_m^m=[h]_m^m [y]_m^m [H]_m^m \quad \text{or} \quad X=hYH. \qquad \text{..............(2)}$$

Then by the same theorem the matrix Φ in (A) is a non-singular commutant $\{X, X\}$ when and only when the matrix Ψ is a non-singular commutant $\{Y, Y\}$; and when Φ and Ψ are non-singular commutants of X and Y respectively, each of them is particular or general when and only when the other is particular or general.

The theorem also follows directly from the pair of equations

$$X\Phi=[h]_m^m . Y\Psi . [H]_m^m, \quad \Phi X=[h]_m^m . \Psi Y . [H]_m^m.$$

Since we can always determine a transformation (1) in which B is a standard canonical square matrix, Theorem I shows that we can always determine the general commutant of the commutant of A when we know the general commutant of the commutant of every standard canonical square matrix.

Ex. i. Theorem I remains true when A and B are square matrices whose elements are rational integral functions of certain scalar variables. The proof will be adapted to this case when we interpret a non-singular commutant of the general commutant X or Y to be a commutant which may involve the scalar variables of A and B, but does not involve the parameters peculiar to X or Y, which are the same for both matrices and are independent of the scalar variables of A and B.

Ex. ii. Taking (1) to be a symmetric derangement $A=hBh'$, we see that:

If there exists a symmetric derangement converting B into A, the same symmetric derangement converts every particular or general commutant of the commutant of B into a particular or general commutant of the commutant of A.

2. *The commutants of the commutant of a compartite square matrix.*

Let

$$A=[a]_m^m=\begin{bmatrix} A_1, & 0, & \dots & 0 \\ 0, & A_2, & \dots & 0 \\ \dots & \dots & \dots & \dots \\ 0, & 0, & \dots & A_r \end{bmatrix}_{a_1, a_2, \dots a_r}^{a_1, a_2, \dots a_r} \qquad \text{..............(3)}$$

be a compartite square matrix in standard form whose parts are all square matrices with constant elements. Also let $X=[x]_m^m$ be a *general* commutant $\{A, A\}$; let $\Phi=[\phi]_m^m$ be *any* commutant of the commutant of A, i.e. any non-singular commutant $\{X, X\}$; and let X and Φ be expressed as compound matrices of the same class as A.

First suppose that no two parts of A have a latent root in common.

When we construct X by Theorem II of § 240, and remember that Φ must be a specialisation of X (because it is commutative with A), we see that we can represent X and Φ as compartite matrices of the same class as A in the forms

$$X = \begin{bmatrix} X_1, & 0 & , \dots & 0 \\ 0 & , X_2, & \dots & 0 \\ \dots & \dots & \dots & \dots \\ 0 & , 0 & , \dots & X_r \end{bmatrix}_{a_1, a_2, \dots a_r}^{a_1, a_2, \dots a_r}, \quad \Phi = \begin{bmatrix} \Phi_1, & 0 & , \dots & 0 \\ 0 & , \Phi_2, & \dots & 0 \\ \dots & \dots & \dots & \dots \\ 0 & , 0 & , \dots & \Phi_r \end{bmatrix}_{a_1, a_2, \dots a_r}^{a_1, a_2, \dots a_r}, \quad \dots\text{(B)}$$

where the part X_i of X is a general commutant $\{A_i, A_i\}$, the parameters of the r parts of X being all independent; and where the part Φ_i of Φ (being a specialisation of X_i) is some commutant $\{A_i, A_i\}$. We proceed to a closer examination of the parts of Φ.

A square matrix Φ with constant elements having the form shown in (B) will be a particular commutant of the commutant of A when and only when the equation $X\Phi = \Phi X$ is an identity in the parameters of X, i.e. when and only when for each of the values 1, 2, ... r of i the equation

$$X_i \Phi_i = \Phi_i X_i$$

is an identity in the arbitrary parameters of i, i.e. when and only when Φ_i is a particular commutant of the commutant of A_i. Since we can choose a complete set of independent particular non-zero commutants of the commutant of A so that in each of them only one of the parts $\Phi_1, \Phi_2, \dots \Phi_r$ is a non-zero matrix, it follows that when Φ is a *general* commutant of the commutant of A, the part Φ_i is a *general* commutant of the commutant of A_i, and the parameters of the r parts are all independent. Accordingly we have the following second theorem:

Theorem II. *When no two parts of the compartite square matrix A shown in* (3) *have a latent root in common, every commutant of the commutant of A is a compartite square matrix Φ of the form shown in* (B) *in which for each of the values* 1, 2, ... *r of i the part Φ_i is a commutant of the commutant of the part A_i of A; moreover Φ will then be a general commutant of the commutant of A when and only when Φ_i is always a general commutant of the commutant of A_i, and the parameters of the r parts $\Phi_1, \Phi_2, \dots \Phi_r$ are all independent.*

When we regard a commutant of the commutant of A as a non-singular commutant $\{X, X\}$, Theorem II becomes merely an application of Theorem II of § 240 after the form of X has been fixed; for every general non-singular commutant $\{X_i, X_i\}$ is a general commutant of the commutant of A_i; and if $j \neq i$, every general non-singular commutant $\{X_i, X_j\}$ is a zero matrix, because it is a specialisation of the general commutant $\{A_i, A_j\}$, which is a zero matrix.

In Theorem II we can take the parts of A to be standardised unilatent canonical square matrices. Then from Theorems I and II it follows that the general commutant of the commutant of every square matrix with constant elements can be constructed when we know the general commutant of the commutant of every standardised unilatent canonical square matrix.

Ex. iii. If the elements of A are rational integral functions of certain scalar variables, we have a corresponding theorem in which the characteristic matrices $A_1(\lambda)$, $A_2(\lambda)$, ... $A_r(\lambda)$ of the parts of A are such that no two of them have an irresoluble divisor in common.

Next let all restrictions on the parts of A be removed, and let X and Φ be expressed in the forms

$$X = \begin{bmatrix} X_{11}, & X_{12}, & \dots & X_{1r} \\ X_{21}, & X_{22}, & \dots & X_{2r} \\ \dots & \dots & \dots & \dots \\ X_{r1}, & X_{r2}, & \dots & X_{rr} \end{bmatrix}_{a_1, a_2, \dots a_r}^{a_1, a_2, \dots a_r}, \quad \Phi = \begin{bmatrix} \Phi_{11}, & \Phi_{12}, & \dots & \Phi_{1r} \\ \Phi_{21}, & \Phi_{22}, & \dots & \Phi_{2r} \\ \dots & \dots & \dots & \dots \\ \Phi_{r1}, & \Phi_{r2}, & \dots & \Phi_{rr} \end{bmatrix}_{a_1, a_2, \dots a_r}^{a_1, a_2, \dots a_r},$$

any parameters which may occur in Φ being regarded as independent of the parameters of X. Every diagonal constituent of X is an undegenerate square matrix; the parameters of the r^2 constituents of X are all independent; and the equation $X\Phi = \Phi X$ is an identity in all the parameters of X.

When we put all parameters of X equal to 0 except those of one of the diagonal constituents, we obtain such equations as

$$X_{ii}\Phi_{ij} = 0, \quad \Phi_{ij}X_{jj} = 0, \quad (j \neq i),$$

which show that all the non-diagonal constituents of Φ must be zero matrices, and that Φ must be a compartite matrix of the same class and form as A. When this condition is satisfied, the identical equation $X\Phi = \Phi X$ becomes equivalent to the identical equations such as

$$X_{ii}\Phi_{ii} = \Phi_{ii}X_{ii}; \quad \Phi_{ii}X_{ij} = X_{ij}\Phi_{ij}, \quad (j \neq i). \quad \dots\dots\dots\dots(4)$$

Thus when Φ is a general commutant of the commutant of A, all its non-diagonal constituents are zero matrices, and its diagonal constituents Φ_{11}, Φ_{22}, ... Φ_{rr} are the most general commutants of the commutants of A_1, A_2, ... A_r which satisfy the second set of relations in (4).

§ 251. The commutants of the commutant of a canonical square matrix.

1. *Simple canonical square matrix.*

Let $A = [a]_m^m$ be the simple canonical square matrix of order m whose latent root is c; let $X = [x]_m^m$ be the general commutant $\{A, A\}$ of A; and let $\Phi = [\phi]_m^m$ be the general commutant of the commutant of A, i.e. a general non-singular commutant $\{X, X\}$ of X. Because A is a particularisation of X, Φ must be a specialisation of X, and we can put

$$X = \begin{bmatrix} x_0 & x_1 & \dots & x_{m-1} \\ 0 & x_0 & \dots & x_{m-2} \\ \dots & \dots & \dots & \dots \\ 0 & 0 & \dots & x_0 \end{bmatrix}, \qquad \Phi = \begin{bmatrix} \rho_0 & \rho_1 & \dots & \rho_{m-1} \\ 0 & \rho_0 & \dots & \rho_{m-2} \\ \dots & \dots & \dots & \dots \\ 0 & 0 & \dots & \rho_0 \end{bmatrix},$$

where the x's are independent arbitrary parameters, and where the ρ's are scalar quantities independent of the x's which have to be so determined that the equation

$$X\Phi = \Phi X \quad \dots\dots(1)$$

is an identity in the x's. Since a matrix Φ having the above form is commutative with X for all values of the ρ's, the ρ's are m independent arbitrary parameters, and we have the following theorem:

Theorem I. *The general commutant of the commutant of the simple canonical square matrix A of order m is a general simple square continuant Φ of order m and of type $\{\pi, \pi\}$ having the form shown above, i.e. having the same form as X.*

It can be regarded as an arbitrary rational integral function of A, or of any simple canonical square matrix of order m.

The last part of the theorem depends on the fact that there are exactly m independent particular non-zero commutants of the commutants of A, the proof being included in Ex. iv of § 242. By Ex. v of § 249 the whole theorem follows from the fact that all commutants of A are commutative with one another, or from the fact that X is a rational integral function of A.

Ex. i. If A' is the conjugate of A, it can be shown in the same way that the general commutant of the commutant of A' is a general simple square continuant Φ' of type $\{\pi', \pi'\}$, and can be regarded as an arbitrary rational integral function of A'. We can clearly take Φ' to be the conjugate of Φ.

Ex. ii. *If Ψ is any undegenerate particularisation of Φ (which may be A when $c \neq 0$), it is possible in two and only two ways to determine such particular values of the m arbitrary parameters of Φ that*

$$\Phi^2 = \Psi. \quad \dots\dots(2)$$

This is a property of simple square ante-continuants of type $\{\pi, \pi\}$ which is proved in § 295. 3 by putting the scalar equations equivalent to (2) into the forms

$$\rho_0^2 = c_0, \quad (c_0 \neq 0); \quad 2\rho_0\rho_\kappa + \sum_{i=1}^{i=\kappa-1} (\rho_i\rho_{\kappa-i}) = c_\kappa, \quad (\kappa = 1, 2, \dots m-1),$$

and solving them for $\rho_0, \rho_1, \dots \rho_{m-1}$ in succession. We have a corresponding theorem relating to Φ'.

2. *Unilatent canonical square matrix.*

Let $A = [a]_m^m$ be a unilatent canonical square matrix, the characteristic potent divisors of whose successive simple parts are

$$(\lambda - c)^{e_1},\ (\lambda - c)^{e_2},\ \dots\ (\lambda - c)^{e_s}, \quad \text{where} \quad e_1 + e_2 + \dots + e_s = m;$$

and let e be the greatest of the integers $e_1, e_2, \dots e_s$. Also let $X = [x]_m^m$ be the general commutant $\{A, A\}$ of A, and $\Phi = [\phi]_m^m$ be the general commutant of the commutant of A, i.e. a general non-singular commutant $\{X, X\}$. Because A is a particularisation of X, Φ must be a specialisation of X, and we can put

$$X = \begin{bmatrix} X_{11}, & X_{12}, & \dots & X_{1s} \\ X_{21}, & X_{22}, & \dots & X_{2s} \\ \dots & \dots & \dots & \dots \\ X_{s1}, & X_{s2}, & \dots & X_{ss} \end{bmatrix}_{e_1,\, e_2,\, \dots\, e_s}^{e_1,\, e_2,\, \dots\, e_s}, \qquad \Phi = \begin{bmatrix} \Phi_{11}, & \Phi_{12}, & \dots & \Phi_{1s} \\ \Phi_{21}, & \Phi_{22}, & \dots & \Phi_{2s} \\ \dots & \dots & \dots & \dots \\ \Phi_{s1}, & \Phi_{s2}, & \dots & \Phi_{ss} \end{bmatrix}_{e_1,\, e_2,\, \dots\, e_s}^{e_1,\, e_2,\, \dots\, e_s},$$

where the constituents of X are general simple continuants of type $\{\pi, \pi\}$ whose parameters are all independent, and where the constituents of Φ are simple continuants of type $\{\pi, \pi\}$ whose parameters are scalar quantities independent of the parameters of X which have to be so determined that the equation

$$X\Phi = \Phi X \quad \dots\dots\dots\dots(3)$$

is an identity in the parameters of X.

By making all constituents of X zero matrices except one of the diagonal constituents, we obtain such equations as

$$X_{ii}\Phi_{ij} = 0, \quad \Phi_{ij}X_{jj} = 0, \qquad (j \neq i);$$

and since these are identities in the parameters of X_{ii} and X_{jj} respectively, they show that all the non-diagonal constituents of Φ must be zero matrices. Accordingly we can put

$$A = \begin{bmatrix} A_1, & 0, & \dots & 0 \\ 0, & A_2, & \dots & 0 \\ \dots & \dots & \dots & \dots \\ 0, & 0, & \dots & A_s \end{bmatrix}_{e_1,\, e_2,\, \dots\, e_s}^{e_1,\, e_2,\, \dots\, e_s}, \qquad \Phi = \begin{bmatrix} \Phi_1, & 0, & \dots & 0 \\ 0, & \Phi_2, & \dots & 0 \\ \dots & \dots & \dots & \dots \\ 0, & 0, & \dots & \Phi_s \end{bmatrix}_{e_1,\, e_2,\, \dots\, e_s}^{e_1,\, e_2,\, \dots\, e_s}, \quad (4)$$

where $\Phi_1, \Phi_2, \dots \Phi_s$ are simple square continuants of type $\{\pi, \pi\}$, which can be regarded as commutants of the commutants of the simple canonical square matrices $A_1, A_2, \dots A_s$. When this form is given to Φ, the equation (1) is equivalent to the equations such as $X_{ii}\Phi_i = \Phi_i X_{ii}$, which are satisfied identically, and the equations such as

$$\Phi_i X_{ij} = X_{ij}\Phi_j, \qquad (j \neq i). \dots\dots\dots\dots (\mathrm{e}_{ij})$$

Now let e_k be any one of the index numbers $e_1, e_2, \dots e_s$ which is equal to e; let e_i be any other of the index numbers; and let $e_i = \epsilon$. Then the equation

$$\Phi_i X_{ik} = X_{ik}\Phi_k, \quad \dots\dots\dots\dots (\mathrm{e}_{ik})$$

which is included in the equations (e_{ij}), has the form

$$[\sigma]_\epsilon^\epsilon\, [0, \xi]_\epsilon^{e-\epsilon,\, \epsilon} = [0, \xi]_\epsilon^{e-\epsilon,\, \epsilon}\, [\rho]_e = [\xi]_e^\epsilon\, [0, \rho]_\epsilon^{e-\epsilon,\, \epsilon}, \quad \dots\dots\dots(5)$$

where the square constituents and matrices are simple square continuants of type $\{\pi, \pi\}$ expressed in standard forms, and where the second product has been transformed by cancelling the zero vertical rows of the first factor with the corresponding horizontal rows of the second factor. The equation (5) is equivalent to the equation

$$[\sigma]_\epsilon^\epsilon [\xi]_\epsilon^\epsilon = [\xi]_\epsilon^\epsilon [\rho]_\epsilon^\epsilon = [\rho]_\epsilon^\epsilon [\xi]_\epsilon^\epsilon, \quad \text{......................(5')}$$

which is an identity in the parameters $\xi_0, \xi_1, \ldots \xi_{\epsilon-1}$ of X_{ik}, and can be replaced by the equation

$$[\sigma]_\epsilon^\epsilon = [\rho]_\epsilon^\epsilon, \quad \text{..................................(5'')}$$

which shows that the ϵ parameters $\sigma_0, \sigma_1, \ldots \sigma_{\epsilon-1}$ of Φ_i must be the same as the first ϵ of the e parameters $\rho_0, \rho_1, \ldots \rho_{e-1}$ of Φ_k. When this condition is satisfied for all the values 1, 2, ... s of i, the equations (e_{ij}) are all satisfied.

Thus Φ is a matrix of the form shown in (4) whose diagonal constituents are the simple square continuants given by

$$\Phi_k = \begin{bmatrix} \rho_0 & \rho_1 & \ldots & \rho_{e-1} \\ 0 & \rho_0 & \ldots & \rho_{e-2} \\ \ldots & \ldots & \ldots & \ldots \\ 0 & 0 & \ldots & \rho_0 \end{bmatrix}; \quad \Phi_i = \begin{bmatrix} \rho_0 & \rho_1 & \ldots & \rho_{\epsilon-1} \\ 0 & \rho_0 & \ldots & \rho_{\epsilon-2} \\ \ldots & \ldots & \ldots & \ldots \\ 0 & 0 & \ldots & \rho_0 \end{bmatrix}, \quad (\epsilon = e_i;\ i = 1, 2, \ldots s),$$

where $\rho_0, \rho_1, \ldots \rho_{e-1}$ are e arbitrary parameters; and we have the following theorem:

Theorem II. *The general commutant of the commutant of the unilatent canonical square matrix A is a general axial ante-continuant of type* $\{\pi, \pi\}$ *and of the class*

$$M\begin{pmatrix} e_1, & e_2, & \ldots & e_s \\ e_1, & e_2, & \ldots & e_s \end{pmatrix}. \quad \text{..............................(a)}$$

It can be regarded as an arbitrary rational integral function of A, or of any unilatent canonical square matrix of the same class as A.

By calling the ante-continuant Φ *axial*, we mean that:

(1) only the paradiagonal constituents can be non-zero matrices (i.e. all the non-diagonal constituents are zero matrices);

(2) in every paradiagonal constituent the parameters of the 1st, 2nd, 3rd, ... ith, ... parametric diagonal lines counted from the paradiagonal are

$$\rho_0, \rho_1, \rho_2, \ldots \rho_i, \ldots,$$

where the ρ's (so far as they occur) are the same for all such constituents.

To indicate the property (2) we shall say that:

All the paradiagonal constituents are formed with the same successive parameters.

The second part of the theorem follows from the fact that the total number of independent particular non-zero commutants of the commutant of A is equal to the lowest possible degree of a rational integral equation satisfied by A. The rational integral equation of lowest degree satisfied by A is

$$(A - cI)^e = 0, \qquad \text{where } I = [1]_m^m;$$

and it is clear that there are exactly e independent particular non-zero commutants of the commutant of A, which could be taken to be the coefficients of $\rho_0, \rho_1, \ldots \rho_{e-1}$ in Φ, but which could also be taken to be the square matrices $I, A, A^2, \ldots A^{e-1}$. In fact we have

$$\begin{aligned}\Phi &= \rho_0 I + \rho_1(A - cI) + \rho_2(A - cI)^2 + \ldots + \rho_{e-1}(A - cI)^{e-1} \\ &= \lambda_0 I + \lambda_1 A + \lambda_2 A^2 + \ldots + \lambda_{e-1} A^{e-1},\end{aligned}$$

where the λ's are arbitrary scalar parameters.

Ex. iii. If A' is the conjugate of A, we could show in the same way that the general commutant of the commutant of A' is a general axial ante-continuant of type $\{\pi', \pi'\}$ and of the class (a); and that it can be regarded as an arbitrary rational integral function of A'. It could be taken to be the conjugate of Φ.

Ex. iv. *If Ψ is any undegenerate particularisation of Φ (which may be A when $c \neq 0$), it is possible in two and only two ways to determine such particular values of the e arbitrary parameters of Φ that*

$$\Phi^2 = \Psi. \qquad \ldots\ldots(6)$$

When Φ is expressed in the form (4), and Ψ in the corresponding form, we can replace (6) by the s equations

$$\Phi_i^2 = \Psi_i, \qquad (i = 1, 2, \ldots s), \qquad \ldots\ldots(2')$$

where Φ_i is a simple square continuant of type $\{\pi, \pi\}$ whose successive parameters are the first e_i of the variables $\rho_0, \rho_1, \ldots \rho_{e-1}$, and where Ψ_i is a simple square continuant of type $\{\pi, \pi\}$ whose successive parameters are the first e_i of certain scalar numbers $\sigma_0, \sigma_1, \ldots \sigma_{e-1}$ of which $\sigma_0 \neq 0$. Supposing that $e_k = e$, it will follow from § 294. 3 (as in Ex. ii) that there are two and only two ways of determining such particular values of $\rho_0, \rho_1, \ldots \rho_{e-1}$ that

$$\Phi_k^2 = \Psi_k,$$

and that then all the equations (2′) are satisfied.

Ex. v. If A is the standardised unilatent canonical square matrix of order 9 whose successive characteristic potent divisors are

$$(\lambda - c)^3, \quad (\lambda - c)^3, \quad (\lambda - c)^2, \quad (\lambda - c),$$

and if Φ is the general commutant of the commutant of A, we can put

$$A = \left[\begin{array}{ccc|ccc|cc|c} c & 1 & 0 & 0 & 0 & 0 & 0 & 0 & 0 \\ 0 & c & 1 & 0 & 0 & 0 & 0 & 0 & 0 \\ 0 & 0 & c & 0 & 0 & 0 & 0 & 0 & 0 \\ \hline 0 & 0 & 0 & c & 1 & 0 & 0 & 0 & 0 \\ 0 & 0 & 0 & 0 & c & 1 & 0 & 0 & 0 \\ 0 & 0 & 0 & 0 & 0 & c & 0 & 0 & 0 \\ \hline 0 & 0 & 0 & 0 & 0 & 0 & c & 1 & 0 \\ 0 & 0 & 0 & 0 & 0 & 0 & 0 & c & 0 \\ \hline 0 & 0 & 0 & 0 & 0 & 0 & 0 & 0 & c \end{array}\right], \quad \Phi = \left[\begin{array}{ccc|ccc|cc|c} \rho_0 & \rho_1 & \rho_2 & 0 & 0 & 0 & 0 & 0 & 0 \\ 0 & \rho_0 & \rho_1 & 0 & 0 & 0 & 0 & 0 & 0 \\ 0 & 0 & \rho_0 & 0 & 0 & 0 & 0 & 0 & 0 \\ \hline 0 & 0 & 0 & \rho_0 & \rho_1 & \rho_2 & 0 & 0 & 0 \\ 0 & 0 & 0 & 0 & \rho_0 & \rho_1 & 0 & 0 & 0 \\ 0 & 0 & 0 & 0 & 0 & \rho_0 & 0 & 0 & 0 \\ \hline 0 & 0 & 0 & 0 & 0 & 0 & \rho_0 & \rho_1 & 0 \\ 0 & 0 & 0 & 0 & 0 & 0 & 0 & \rho_0 & 0 \\ \hline 0 & 0 & 0 & 0 & 0 & 0 & 0 & 0 & \rho_0 \end{array}\right].$$

If A' is the conjugate of A, the general commutants $\{A, A\}$, $\{A', A'\}$, $\{A', A\}$, $\{A, A'\}$ can in this case be taken to be the matrices X, X', Y, Y' shown below, which are general continuants of the respective types $\{\pi, \pi\}$, $\{\pi', \pi'\}$, $\{\pi', \pi\}$, $\{\pi, \pi'\}$ and of the same quadrate class as A, so chosen as to be correlated by part-reversals. When f', g', h', u', v', w' are replaced by f, g, h, u, v, w, the matrices Y and Y' become general symmetric contra-commutants $\{A', A\}$ and $\{A, A'\}$.

$$X = \left[\begin{array}{ccc|ccc|cc|c} a_0 & a_1 & a_2 & h_0 & h_1 & h_2 & g_0 & g_1 & u_0 \\ 0 & a_0 & a_1 & 0 & h_0 & h_1 & 0 & g_0 & 0 \\ 0 & 0 & a_0 & 0 & 0 & h_0 & 0 & 0 & 0 \\ \hline h_0' & h_1' & h_2' & b_0 & b_1 & b_2 & f_0 & f_1 & v_0 \\ 0 & h_0' & h_1' & 0 & b_0 & b_1 & 0 & f_0 & 0 \\ 0 & 0 & h_0' & 0 & 0 & b_0 & 0 & 0 & 0 \\ \hline 0 & g_0' & g_1' & 0 & f_0' & f_1' & c_0 & c_1 & w_0 \\ 0 & 0 & g_0' & 0 & 0 & f_0' & 0 & c_0 & 0 \\ \hline 0 & 0 & u_0' & 0 & 0 & v_0' & 0 & w_0' & d_0 \end{array}\right], \quad X' = \left[\begin{array}{ccc|ccc|cc|c} a_0 & 0 & 0 & h_0 & 0 & 0 & 0 & 0 & 0 \\ a_1 & a_0 & 0 & h_1 & h_0 & 0 & g_0 & 0 & 0 \\ a_2 & a_1 & a_0 & h_2 & h_1 & h_0 & g_1 & g_0 & u_0 \\ \hline h_0' & 0 & 0 & b_0 & 0 & 0 & 0 & 0 & 0 \\ h_1' & h_0' & 0 & b_1 & b_0 & 0 & f_0 & 0 & 0 \\ h_2' & h_1' & h_0' & b_2 & b_1 & b_0 & f_1 & f_0 & v_0 \\ \hline g_0' & 0 & 0 & f_0' & 0 & 0 & c_0 & 0 & 0 \\ g_1' & g_0' & 0 & f_1' & f_0' & 0 & c_1 & c_0 & w_0 \\ \hline u_0' & 0 & 0 & v_0' & 0 & 0 & w_0' & 0 & d_0 \end{array}\right],$$

$$Y = \left[\begin{array}{ccc|ccc|cc|c} 0 & 0 & a_0 & 0 & 0 & h_0 & 0 & 0 & 0 \\ 0 & a_0 & a_1 & 0 & h_0 & h_1 & 0 & g_0 & 0 \\ a_0 & a_1 & a_2 & h_0 & h_1 & h_2 & g_0 & g_1 & u_0 \\ \hline 0 & 0 & h_0' & 0 & 0 & b_0 & 0 & 0 & 0 \\ 0 & h_0' & h_1' & 0 & b_0 & b_1 & 0 & f_0 & 0 \\ h_0' & h_1' & h_2' & b_0 & b_1 & b_2 & f_0 & f_1 & v_0 \\ \hline 0 & 0 & g_0' & 0 & 0 & f_0' & 0 & c_0 & 0 \\ 0 & g_0' & g_1' & 0 & f_0' & f_1' & c_0 & c_1 & w_0 \\ \hline 0 & 0 & u_0' & 0 & 0 & v_0' & 0 & w_0' & d_0 \end{array}\right], \quad Y' = \left[\begin{array}{ccc|ccc|cc|c} a_2 & a_1 & a_0 & h_2 & h_1 & h_0 & g_1 & g_0 & u_0 \\ a_1 & a_0 & 0 & h_1 & h_0 & 0 & g_0 & 0 & 0 \\ a_0 & 0 & 0 & h_0 & 0 & 0 & 0 & 0 & 0 \\ \hline h_2' & h_1' & h_0' & b_2 & b_1 & b_0 & f_1 & f_0 & v_0 \\ h_1' & h_0' & 0 & b_1 & b_0 & 0 & f_0 & 0 & 0 \\ h_0' & 0 & 0 & b_0 & 0 & 0 & 0 & 0 & 0 \\ \hline g_1' & g_0' & 0 & f_1' & f_0' & 0 & c_1 & c_0 & w_0 \\ g_0' & 0 & 0 & f_0' & 0 & 0 & c_0 & 0 & 0 \\ \hline u_0' & 0 & 0 & v_0' & 0 & 0 & w_0' & 0 & d_0 \end{array}\right].$$

3. *Standard canonical square matrix.*

Let $\Pi = [\Pi]_m^m$ be a standard canonical square matrix of order m whose successive unilatent super-parts are the unilatent canonical square matrices

$$\Pi_1, \Pi_2, \ldots \Pi_r \quad \text{of orders} \quad e_1, e_2, \ldots e_r$$

with latent roots $\quad c_1, c_2, \ldots c_r, \quad$ which are all different

the super-part Π_i being a unilatent canonical square matrix of the class

$$M\begin{pmatrix} e_{i1}, & e_{i2}, & e_{i3}, & \ldots \\ e_{i1}, & e_{i2}, & e_{i3}, & \ldots \end{pmatrix}, \quad \text{where} \quad e_{i1} + e_{i2} + e_{i3} + \ldots = e_i, \quad \ldots\ldots(\mathrm{a}_i)$$

so that Π is a compound matrix of the class

$$M\begin{pmatrix} e_{11}, e_{12}, e_{13}, \ldots; & e_{21}, e_{22}, e_{23}, \ldots; & \ldots & e_{r1}, e_{r2}, e_{r3}, \ldots \\ e_{11}, e_{12}, e_{13}, \ldots; & e_{21}, e_{22}, e_{23}; \ldots; & \ldots & e_{r1}, e_{r2}, e_{r3}, \ldots \end{pmatrix}. \quad \ldots\ldots(\mathrm{b})$$

If Φ is the general commutant of the commutant of Π, it follows from Theorem II of § 250 and from sub-article 2 that we can put

$$\Pi = \begin{bmatrix} \Pi_1, & 0\,, & \ldots & 0 \\ 0\,, & \Pi_2, & \ldots & 0 \\ \ldots & \ldots & \ldots & \ldots \\ 0\,, & 0\,, & \ldots & \Pi_r \end{bmatrix}_{e_1, e_2, \ldots e_r}^{e_1, e_2, \ldots e_r}, \qquad \Phi = \begin{bmatrix} \Phi_1, & 0\,, & \ldots & 0 \\ 0\,, & \Phi_2, & \ldots & 0 \\ \ldots & \ldots & \ldots & \ldots \\ 0\,, & 0\,, & \ldots & \Phi_r \end{bmatrix}_{e_1, e_2, \ldots e_r}^{e_1, e_2, \ldots e_r},$$

where Φ_i is the general commutant of the commutant of the unilatent canonical square matrix Π_i, i.e. a general axial continuant of the class (a_i) and of type $\{\pi, \pi\}$, and where the parameters of $\Phi_1, \Phi_2, \ldots \Phi_r$ are all independent. The parts $\Phi_1, \Phi_2, \ldots \Phi_r$ are the same for all values of $c_1, c_2, \ldots c_r$, so long as the r values are all different.

Let ϵ_i be the greatest of the integers $e_{i1}, e_{i2}, e_{i3}, \ldots$ for each of the values $1, 2, \ldots r$ of i, and let

$$s = \epsilon_1 + \epsilon_2 + \ldots + \epsilon_r.$$

Then if $I = [1]_m^m$, the rational integral equation of lowest degree satisfied by Π is the equation

$$(\Pi - c_1 I)^{\epsilon_1} (\Pi - c_2 I)^{\epsilon_2} \ldots (\Pi - c_r I)^{\epsilon} = 0 \qquad \text{of degree } s.$$

Again because Φ_i contains exactly ϵ_i independent arbitrary parameters, there are exactly s independent arbitrary parameters in Φ, and there are exactly s independent non-zero particularisations of Φ which could be taken to be the coefficients of those s parameters, but which could also be taken to be the square matrices

$$I, \Pi, \Pi^2, \ldots \Pi^{s-1},$$

because these are commutants of the commutants of Π, and are all independent. It follows that Φ can also be expressed in the form

$$\Phi = \lambda_0 I + \lambda_1 \Pi + \lambda_2 \Pi^2 + \ldots + \lambda_{s-1} \Pi^{s-1},$$

where the λ's are independent arbitrary scalar parameters.

The results obtained above are summarised in the following theorem:

Theorem III. *The general commutant of the commutant of the standard canonical square matrix* Π *is a standard compartite matrix* Φ *of the same class as* Π *which can be obtained by replacing each unilatent super-part* Π_i *of* Π *by a general axial continuant* Φ_i *of type* $\{\pi, \pi\}$ *and of the same class as* Π_i.

If s *is the lowest possible degree of a rational integral equation satisfied by* Π, *there are exactly* s *independent particular non-zero commutants of the commutants of* Π. *Consequently we can regard* Φ *as an arbitrary rational integral function of* Π; *and every commutant of the commutant of* Π *is a rational integral function of* Π.

We can also regard Φ as an arbitrary rational integral function of any standard canonical square matrix having the same structure as Π, i.e.

belonging to the same class and the same super-class as Π, or having the same characteristic symbol as Π.

Ex. vi. If Π' is the conjugate of Π, we could show in the same way that the general commutant of the commutant of Π' is a standard compartite matrix Φ' of the same class as Φ in which the parts are general axial continuants of type $\{\pi', \pi'\}$; and that it can be regarded as an arbitrary rational integral function of Π'. It could be taken to be the conjugate of Φ.

Ex. vii. *If Φ is the general commutant of the commutant of a standard canonical square matrix Π having exactly r distinct latent roots $c_1, c_2, \dots c_r$, and if Ψ is any undegenerate particularisation of Φ (which may be Π when $c_1, c_2, \dots c_r$ are all different from 0) it is possible in 2^r and only 2^r ways to determine such particular values of the arbitrary parameters of Φ that*

$$\Phi^2=\Psi. \qquad\qquad (7)$$

Supposing Π and Φ to be the matrices of the text, we have

$$\Phi=\begin{bmatrix} \Phi_1, & 0, & \dots & 0 \\ 0, & \Phi_2, & \dots & 0 \\ \dots & \dots & \dots & \dots \\ 0, & 0, & \dots & \Phi_r \end{bmatrix}_{e_1, e_2, \dots e_r}^{e_1, e_2, \dots e_r}, \qquad \Psi=\begin{bmatrix} \Psi_1, & 0, & \dots & 0 \\ 0, & \Psi_2, & \dots & 0 \\ \dots & \dots & \dots & \dots \\ 0, & 0, & \dots & \Psi_r \end{bmatrix}_{e_1, e_2, \dots e_r}^{e_1, e_2, \dots e_r},$$

where Ψ_i is an undegenerate particularisation of the axial ante-continuant Φ_i; and the equation (7) is equivalent to the r equations

$$\Phi_i^2=\Psi_i, \qquad (i=1, 2, \dots r). \qquad\qquad (6')$$

From Ex. iv we see that for each of the values of i there are exactly 2 ways of determining the ϵ_i parameters of Φ_i so that the equation (6′) is satisfied. Consequently there are exactly 2^r ways of determining the $\epsilon_1+\epsilon_2+\dots+\epsilon_r$ parameters of Φ so that all the equations (6′) are satisfied.

Ex. viii. *Let Φ be a general commutant of the commutant of the standard canonical square matrix Π of the text, and let Φ' be the conjugate of Φ. Then if J is the part-reversant of the same class* (b) *as Π and Φ, we have*

$$\Phi'=J\Phi J, \qquad \textit{and therefore} \qquad \Phi'J=J\Phi,\ \Phi J=J\Phi'.$$

When Φ is constructed as in the text, these equations follow from the fact that Φ' can be derived from Φ by reversing the orders of arrangement of the horizontal and vertical rows in every one of the parts. Since they are identities in the parameters of Φ, they remain true when Φ is any particular or general commutant of the commutant of Π, i.e. any particular or general rational integral function of Π, provided of course that Φ' is always the conjugate of Φ. In particular (or by the same argument) we have

$$\Pi'=J\Pi J \qquad \text{and therefore} \qquad \Pi'J=J\Pi,\ \Pi J=J\Pi'.$$

Thus in all cases the part-reversant J is an undegenerate symmetric contra-commutant $\{\Phi', \Phi\}$, $\{\Phi, \Phi'\}$ as well as an undegenerate symmetric contra-commutant $\{\Pi', \Pi\}$, $\{\Pi, \Pi'\}$.

Ex. ix. If Π is a standard canonical square matrix of order m whose conjugate is Π', and if Φ is a square matrix of order m whose conjugate is Φ', then

(i) every co-commutant $\{\Pi, \Pi\}$ is also a co-commutant $\{\Phi, \Phi\}$,

(ii) every co-commutant $\{\Pi', \Pi'\}$ is also a co-commutant $\{\Phi', \Phi'\}$,

(iii) every contra-commutant $\{\Pi', \Pi\}$ is also a contra-commutant $\{\Phi', \Phi\}$,

(iv) every contra-commutant $\{\Pi, \Pi'\}$ is also a contra-commutant $\{\Phi, \Phi'\}$,

when and only when Φ is a commutant of the commutant of Π, i.e. a rational integral function of Π.

Thus the four conditions (i), (ii), (iii), (iv) are mutually equivalent, and the most general square matrix Φ which satisfies any one of them is a general commutant of the commutant of Π, i.e. an arbitrary rational integral function of Π.

We will suppose that Π is the canonical square matrix described in the text, and that J is the part-reversant defined in Ex. viii. Let X, X', Y, Y' be general commutants $\{\Pi, \Pi\}$, $\{\Pi', \Pi'\}$, $\{\Pi', \Pi\}$, $\{\Pi, \Pi'\}$. Then we can prove the above theorem by showing that the equations

$$\Phi X = X\Phi, \quad \Phi' X' = X'\Phi', \quad \Phi' Y = Y\Phi, \quad \Phi Y' = Y'\Phi' \quad \text{.................(8)}$$

are identities in the parameters of X, X', Y, Y' respectively when and only when Φ is a commutant of the commutant of Π; or that when they are regarded as equations to be solved for Φ, the general non-singular solution of each of them is a general commutant of the commutant of Π. Since a commutant of the commutant of Π has been defined to be a non-singular commutant $\{X, X\}$, and has been shown to be the same thing as a rational integral function of Π, it will be sufficient to show that the 2nd, 3rd, 4th of the equations (8) have exactly the same non-singular solutions for Φ as the 1st of these equations.

In dealing with the second of the equations (8) we can take X' to be the conjugate of X, so that the parameters of X' are those of X. Then obviously Φ will make the second equation an identity in the parameters of X' when and only when it makes the first equation an identity in the parameters of X.

In dealing with the third of the equations (8) we can put $Y = JX$, so that the parameters of Y are those of X. Then if Φ is any particular (or general) non-singular solution of the first equation, we have $\Phi' J = J\Phi$ by Ex. viii, and we can construct the successive identities

$$\Phi X = X\Phi, \quad J\Phi X = JX\Phi, \quad \Phi' JX = JX\Phi, \quad \Phi' Y = Y\Phi,$$

starting from the first, which is given; and if Φ is any particular non-singular solution of the third equation, we have $\Phi' J = J\Phi$ because J is a particularisation of Y obtained by putting $X = [1]_m^m$, and we can construct the same successive identities, starting now from the last, which is given. It follows that the third of the equations (8) has exactly the same non-singular solutions as the first equation.

In dealing with the fourth of the equations (8) we can put $Y' = XJ$, so that the parameters of Y' are those of X. Then if Φ is any particular (or general) non-singular solution of the first equation, we have $\Phi J = J\Phi'$ by Ex. viii, and we can construct the successive identities

$$\Phi X = X\Phi, \quad \Phi XJ = X\Phi J, \quad \Phi XJ = XJ\Phi', \quad \Phi Y' = Y'\Phi',$$

starting from the first, which is given; and if Φ is any particular non-singular solution of the fourth equation, we have $\Phi J = J\Phi'$ because J is a particularisation of Y' obtained by putting $X = [1]_m^m$, and we can construct the same successive identities, starting now from the last, which is given. It follows that the fourth of the equations (8) has exactly the same non-singular solutions as the first equation.

The above proof is contained as a particular case in Ex. iv of § 252.

4. *General canonical square matrix.*

If A is any canonical square matrix, it can be derived from a standard canonical square matrix Π by a symmetric class-derangement. Also the general commutants $\{A, A\}$, $\{A', A'\}$, $\{A', A\}$, $\{A, A'\}$ and the general commutants of the commutants of A and A' can be derived respectively from the general commutants $\{\Pi, \Pi\}$, $\{\Pi', \Pi'\}$, $\{\Pi', \Pi\}$, $\{\Pi, \Pi'\}$ and the general commutants of the commutants of Π and Π' by the same symmetric derangement. Thus we can pass immediately from the properties of Π proved in sub-article 3 to the corresponding properties of A; but the most important of the corresponding properties of A are included in the theorems of the next article.

§ 252. The commutants of the commutant of any square matrix whose elements are constants.

Let $A = [a]_m^m$ be any square matrix whose elements are constants, and let Π be a standard canonical reduced form of A. Also let

$$A = h\Pi H \quad \text{.................................(1)}$$

be any given isomorphic transformation by which A can be derived from Π. Then if two square matrices Φ and Ψ of order m are such that

$$\Phi = h\Psi H, \quad \text{.................................(2)}$$

we know from Theorem I of § 250 that Φ is a commutant of the commutant of A or a particular or general commutant of the commutant of A when and only when Ψ is a commutant of the commutant of Π or a particular or general commutant of the commutant of Π. Also we know from Ex. v of § 226 that Φ will be a rational integral function $f(A)$ of A when and only when Ψ is the same rational integral function $f(\Pi)$ of Π. Taking Ψ to be a general commutant of the commutant of Π constructed as in § 251, the formula (2) gives us a construction for the general commutant of the commutant of A, and leads to the following theorem, which must be true generally because it is true in the particular case when A is a standard canonical square matrix.

Theorem I. *The general commutant of the commutant of the square matrix A can be regarded as an arbitrary rational integral function of A; and any square matrix is a commutant of the commutant of A when and only when it is a rational integral function of A.*

In the following examples A', Π' Φ' Ψ' H', h' are the conjugates of A, Π, Φ, Ψ, H, h.

Ex. i. *If s is the lowest possible degree of a rational integral equation satisfied by A, then the general commutant of the commutant of A contains exactly s independent arbitrary parameters, and there are exactly s independent particular non-zero commutants of the commutants of A.*

Though this result is involved in Theorem I, it can be proved otherwise, and then used to prove Theorem I.

For if $f(\Pi)=0$ is the rational integral equation of lowest degree satisfied by Π, then $f(A)=0$ of the same degree is the rational integral equation of lowest degree satisfied by A. Consequently the above result is true because it has been shown in § 251 to be true whenever A is a standard canonical square matrix. It follows that Theorem I must be true because the s non-zero square matrices

$$I=A^0=[1]_m^m,\ A,\ A^2,\ \dots\ A^{m-1}$$

are all independent and are commutants of the commutant of A.

Ex. ii. Since the commutants of the commutant of A are those commutants of A which are rational integral functions of A, we see from Theorem II of § 244 that:

A general commutant $\{A, A\}$ is a general commutant of the commutant of A, i.e. all commutants of A are commutative with one another, when and only when all commutants of A are rational integral functions of A, i.e. when and only when m is the lowest degree of a rational integral equation satisfied by A.

Ex. iii. If Φ is a general commutant of the commutant of A, the unit matrix $I=[1]_m^m$ is always a particular undegenerate symmetric commutant $\{\Phi, \Phi\}$ as well as a particular undegenerate symmetric commutant $\{A, A\}$.

Again if J is the undegenerate symmetric part-reversant of the same class as Π, which is its own inverse, the mutually inverse square matrices

$$U=H'JH, \quad V=hJh'$$

are particular undegenerate symmetric contra-commutants $\{\Phi', \Phi\}$, $\{\Phi, \Phi'\}$ as well as contra-commutants $\{A', A\}$, $\{A, A'\}$. For if $\Phi=h\Psi H$, where Ψ is a general commutant of the commutant Π, it has been shown in Ex. viii of § 251 that J is a contra-commutant $\{\Psi', \Psi\}$, $\{\Psi, \Psi'\}$ as well as a contra-commutant $\{\Pi', \Pi\}$, $\{\Pi, \Pi'\}$.

These results are particular cases of Theorem II.

If Φ is a commutant of the commutant of A, then because it is a rational integral function of A we know (see Ex. viii of § 238) that it has the following four properties:

(i) Every co-commutant $\{A, A\}$ is also a co-commutant $\{\Phi, \Phi\}$.
(ii) Every co-commutant $\{A', A'\}$ is also a co-commutant $\{\Phi', \Phi'\}$.
(iii) Every contra-commutant $\{A', A\}$ is also a contra-commutant $\{\Phi', \Phi\}$.
(iv) Every contra-commutant $\{A, A'\}$ is also a contra-commutant $\{\Phi, \Phi'\}$.

We can now show that conversely every square matrix Φ which has any one of these properties must be a rational integral function of A, and so replace Theorem I by the following more general theorem in which Φ' is always the conjugate of Φ.

Theorem II. *A square matrix Φ has any one of the properties* (i), (ii), (iii), (iv) *when and only when it is a commutant of the commutant of A, i.e. a rational integral function of A; and a most general square matrix Φ having any one of these four properties must be a general commutant of the commutant of A, i.e. an arbitrary rational integral function of A.*

Let X, X', Y, Y' be general commutants $\{A, A\}$, $\{A', A'\}$, $\{A', A\}$, $\{A, A'\}$, and let Φ be a square matrix of order m whose conjugate is Φ'. Then we can prove the theorem by showing that the general non-singular solution Φ of each of the equations

$$\Phi X = X\Phi, \quad \Phi' X' = X'\Phi', \quad \Phi' Y = Y\Phi, \quad \Phi Y' = Y'\Phi' \quad \text{......(3)}$$

is a general commutant of the commutant of A. Now if we put

$$X = h\mathbf{X}H, \quad X' = H'\mathbf{X}'h', \quad Y = H'\mathbf{Y}H, \quad Y' = h\mathbf{Y}'h'$$

and

$$\Phi = h\Psi H, \quad \text{..............................(2)}$$

the matrices $\mathbf{X}, \mathbf{X}', \mathbf{Y}, \mathbf{Y}'$ are general commutants $\{\Pi, \Pi\}$, $\{\Pi', \Pi'\}$, $\{\Pi', \Pi\}$, $\{\Pi, \Pi'\}$, and Φ will be a general non-singular solution of the 1st, 2nd, 3rd,

4th of the equations (3) when and only when Ψ is a general non-singular solution of the 1st, 2nd, 3rd, 4th of the equations

$$\Psi\mathbf{X}=\mathbf{X}\Psi,\quad \Psi'\mathbf{X}'=\mathbf{X}'\Psi',\quad \Psi'\mathbf{Y}=\mathbf{Y}\Psi,\quad \Psi\mathbf{Y}'=\mathbf{Y}'\Psi'\ldots\ldots(3')$$

in which Ψ' is the conjugate of Ψ; moreover Φ will be a general commutant of the commutant of A when and only when Ψ is a general commutant of the commutant of Π. But we know from Ex. ix of § 251 that the general non-singular solution Ψ of each of the equations (3′) is a general commutant of the commutant of Π. It follows that the corresponding general non-singular solution $\Phi=h\Psi H$ of each of the equations (3) is a general commutant of the commutant of A.

Ex. iv. *Alternative proof of Theorem II.*

From Ex. iii we see that if Φ is a general commutant of the commutant of A, it is certainly possible to determine:

a square matrix U which is an undegenerate particular non-singular contra-commutant $\{\Phi', \Phi\}$, and therefore also a contra-commutant $\{A', A\}$;

a square matrix V which is an undegenerate particular non-singular contra-commutant $\{\Phi, \Phi'\}$, and therefore also a contra-commutant $\{A, A'\}$;

and in the present proof we shall suppose U and V to be any two such matrices. In each case the second property is a necessary consequence of the first, because A is a particularisation of Φ; but because Φ is a rational integral function of A, we see that the first property is also a consequence of the second, and that U, V could be any particular undegenerate commutants $\{A', A\}$, $\{A, A'\}$.

Assuming Theorem I to be true, and defining X, X', Y, Y' as before, we will prove Theorem II by showing that the 2nd, 3rd, 4th of the equations (3) have exactly the same non-singular solutions for Φ as the 1st equation.

In dealing with the second of the equations (3) we can take X' to be the conjugate of X, so that the parameters of X' are those of X. Then obviously Φ will make the second equation an identity in the parameters of X' when and only when it makes the first equation an identity in the parameters of X.

In dealing with the third of the equations (3) we can put $Y=UX$, so that the parameters of Y are those of X. Then if Φ is any particular (or general) non-singular solution of the first equation, we have $\Phi'U=U\Phi$ by the definition of U, and we can construct the successive identities

$$\Phi X=X\Phi,\quad U\Phi X=UX\Phi,\quad \Phi'UX=UX\Phi,\quad \Phi'Y=Y\Phi,$$

starting from the first, which is given; and if Φ is any particular non-singular solution of the third equation, we have $\Phi'U=U\Phi$, because U is a particularisation of Y, and we can construct the same successive identities as before, starting now from the last, which is given. It follows that the third of the equations (3) has exactly the same non-singular solutions for Φ as the first equation.

In dealing with the fourth of the equations (3) we can put $Y'=XV$, so that the parameters of Y' are those of X. Then if Φ is any particular (or general) non-singular solution of the first equation, we have $\Phi V=V\Phi'$ by the definition of V, and we can construct the successive identities

$$\Phi X=X\Phi,\quad \Phi XV=X\Phi V,\quad \Phi XV=XV\Phi',\quad \Phi Y'=Y'\Phi',$$

starting from the first, which is given; and if Φ is any particular non-singular solution of the fourth equation, we have $\Phi V=V\Phi'$, because V is a particularisation of Y', and we can construct the same successive identities as before, starting now from the last, which is given. It follows that the fourth of the equations (3) has exactly the same non-singular solutions for Φ as the first equation.

Ex. v. *If A is a square matrix with constant elements, and if Φ is a general commutant of the commutant of A, i.e. an arbitrary rational integral function of A, then the general non-singular commutants*

$$\{\Phi, \Phi\}, \quad \{\Phi', \Phi'\}, \quad \{\Phi', \Phi\}, \quad \{\Phi, \Phi'\} \qquad \text{.............................(4)}$$

can be identified respectively with general commutants

$$X=\{A, A\}, \quad X'=\{A', A'\}, \quad Y=\{A', A\}, \quad Y'=\{A, A'\}; \qquad \text{...............(4')}$$

and Φ is the most general square matrix which can have these four properties or any one of them.

This follows from Theorem II because A is a particularisation of Φ. For example every commutant $\{A', A\}$ is a commutant $\{\Phi', \Phi\}$ because Φ is a rational integral function of A; and every *non-singular* commutant $\{\Phi', \Phi\}$ is a commutant $\{A', A\}$ because A is a particularisation of Φ.

If Φ is a given general commutant of the commutant of A, the general non-singular solutions of the four equations (3) are the four general commutants (4'). If on the other hand X, X', Y, Y' are given general commutants of the types shown in (4'), then the general non-singular solution Φ of each of the equations (3) is a general commutant of the commutant of A.

If Ψ and $\Phi = h\Psi H$ are corresponding commutants of the commutants of Π and A, and if Ψ_0 and $\Phi_0 = h\Psi_0 H$ are corresponding particularisations of Ψ and Φ, we have

$$\Phi^2 - \Phi_0 = h(\Psi^2 - \Psi_0) H. \qquad \text{........................(5)}$$

Hence and from Ex. vii of § 251 we obtain the following third theorem:

Theorem III. *Let A be a square matrix with constant elements having exactly r distinct latent roots. Then if Φ is a general commutant of the commutant of A (or an arbitrary rational integral function of A), and if Φ_0 is any undegenerate particularisation of Φ, it is possible in 2^r and only 2^r ways to determine such particular values of the arbitrary parameters of Φ that*

$$\Phi^2 = \Phi_0. \qquad \text{...................................(6)}$$

Because this theorem has been shown to be true whenever A is a standard canonical square matrix, it must be true generally.

When A is undegenerate, we can take Φ_0 to be A. We thus obtain a corollary which is equivalent to *Frobenius's Theorem* given in § 225.

COROLLARY. *If A is an undegenerate square matrix with constant elements having exactly r distinct latent roots, which are necessarily all different from 0, and if Φ is a general commutant of the commutant of A (or an arbitrary rational integral function of A), it is possible in 2^r and only 2^r ways to determine such particular values of the arbitrary parameters of Φ that*

$$\Phi^2 = A. \qquad \text{...................................(7)}$$

Thus Frobenius's Theorem can be derived from a property of simple square ante-continuants and the form of the general commutant of the commutant of a unilatent canonical square matrix.

Ex. vi. Let $B=[b]_m^m$ be any one of the 2^r undegenerate particularisations of Φ for which the equation (7) of the Corollary is true. Then because every contra-commutant $\{A', A\}$ is a non-singular contra-commutant $\{\Phi', \Phi\}$, and therefore a contra-commutant $\{B', B\}$, it follows that if $Y=[y]_m^m$ is any contra-commutant $\{A', A\}$, we have

$$YA=YB^2=B'YB \quad \text{or} \quad [y]_m^m[a]_m^m=\overline{\underline{b}}_m^m[y]_m^m[b]_m^m. \quad \text{.................(8)}$$

Similarly if $Y=[y]_m^m$ is any contra-commutant $\{A', A\}$, we have

$$AY=B^2Y=BYB' \quad \text{or} \quad [a]_m^m[y]_m^m=[b]_m^m[y]_m^m\overline{\underline{b}}_m^m. \quad \text{.................(8')}$$

Note 1. *Application of Theorem I to a square matrix containing arbitrary elements.*

Let A be a square matrix whose elements are rational integral functions of certain scalar variables $\gamma_1, \gamma_2, \gamma_3, \dots$; and let X and Φ be respectively a general commutant and a general commutant of the commutant of A. Also let s be the lowest possible degree of a rational integral equation satisfied by A.

If we ascribe suitably defined ordinary particular values to the γ's, we shall convert A into a square matrix A_0 with constant elements, and X and Φ into square matrices X_0 and Φ_0 which are respectively a general commutant and a general commutant of the commutant of A_0; moreover s will be the lowest possible degree of a rational integral equation satisfied by A_0, and the parameters peculiar to Φ will be the s parameters of Φ_0. Since there are exactly s independent particular non-zero commutants of the commutant of A, we can conclude that:

Theorem I remains true when A is a square matrix whose elements involve arbitrary parameters.

If $s=m$, X_0 and X contain exactly m arbitrary parameters, and X and Φ are both arbitrary rational integral functions of A. If $s<m$, X_0 and X contain more than m arbitrary parameters, and there are commutants of A which are not rational integral functions of A.

Note 2. *Application of Theorem II to a square matrix containing arbitrary elements.*

In Note 1 above we know from Note 3 of § 244 that there exist undegenerate contra-commutants $\{A', A\}$, $\{A, A'\}$; and because Φ is a rational integral function of A, it follows from Ex. viii of § 238 that these are also undegenerate contra-commutants $U=\{\Phi', \Phi\}$, $V=\{\Phi, \Phi'\}$ which are independent of the parameters peculiar to Φ, but ordinarily not independent of the γ's. The proof of Theorem II given in Ex. iv will remain valid when U and V are two such contra-commutants, and non-singular solutions of the equations (3) are interpreted to be solutions independent of the parameters peculiar to the general commutants X, X', Y, Y'. Hence we can conclude that:

Theorem II remains true when A is a square matrix whose elements involve arbitrary parameters.

Note 3. *Application of Theorem III to a square matrix containing arbitrary elements.*

When A is a square matrix defined as in Note 1, we can solve the equation (6) or (7) by suitably determining the arbitrary parameters peculiar to Φ as functions of the γ's. This can be done in 2^r ways, where r is now the sum of the degrees in λ of the distinct irresoluble divisors of the characteristic matrix $A(\lambda)$ of A. For the ascription of ordinary particular values to the γ's will convert A into a square matrix having exactly r distinct latent roots.

§ 253. The general commutant (or general commutant of the commutant) of a square matrix whose elements are independent arbitrary parameters.

We will define $A = [a]_m^m$ to be a square matrix of order m whose elements are all arbitrary and independent; we will define $X = [x]_m^m$ to be the general commutant $\{A, A\}$, i.e. the general solution of the matrix equation $AX = XB$; and we will determine X by solving the scalar equations which its elements must satisfy. In accordance with previous results it will appear that X can be regarded both as a general commutant of the commutant of A, and as an arbitrary rational integral function of A.

If we put $\quad r = \tfrac{1}{2}m(m-1), \quad \alpha_{ij} = a_{ii} - a_{jj}, \quad \xi_{ij} = x_{ii} - x_{jj},$ ....(1)

so that $\quad \alpha_{ji} = -\alpha_{ij}, \quad \alpha_{ij} = \alpha_{ik} + \alpha_{kj} = \alpha_{kj} - \alpha_{ki},$ ....(2)

and $\quad \xi_{ji} = -\xi_{ij}, \quad \xi_{ij} = \xi_{ik} + \xi_{kj} = \xi_{kj} - \xi_{ki},$ ....(3)

the matrix equation $AX = XB$ is equivalent to m^2 scalar equations, which can be divided into two sets, viz.

(a) the m equations of zero difference-weights obtained by equating corresponding diagonal elements on both sides, these being derived from the equation

$$\Sigma\,(a_{ik}x_{ki} - x_{ik}a_{ki}) = 0, \quad\text{.........................}(e_{ii})$$

(where the summation extends over all the values 1, 2, ... m of k except i), by giving to i the values 1, 2, ... m, and one of them being superfluous;

(b) the $2r$ equations of non-zero difference-weights derived from the equation

$$\alpha_{ij}x_{ij} + \Sigma\,(a_{ik}x_{kj} - x_{ik}a_{kj}) = a_{ij}\xi_{ij}, \quad\text{....................}(e_{ij})$$

(where the summation extends over all the values 1, 2, ... m of k except i and j), by choosing i and j to be two *different* integers of the set 1, 2, ... m.

We will define R to be the resultant of the $2r$ homogeneous linear functions of the x's occurring on the left-hand sides of the equations (b); and in dealing with those equations we will regard the x's on the left as the variables, and the ξ's on the right as having any given values consistent with the conditions (3), so that exactly $m-1$ of them, for example $\xi_{12}, \xi_{13}, \ldots \xi_{1m}$, can be treated as arbitrary parameters in terms of which the others are expressed. It is to be observed that R is not the scalar resultant defined in § 238, which obviously vanishes identically in the present case.

When R is defined as above, the equations (b) have the following properties:

(i) They are correlated by permutations of the suffixes 1, 2, ... m of all the

a's, x's, α's and ξ's in such a way that every permutation which replaces the suffixes i, j by p, q also converts the equation (e_{ij}) into the equation (e_{pq}). Thus all the equations can be derived from any one of them by such permutations.

(ii) They are unaltered in their totality by interchanging the two suffixes in every one of the a's and x's, and leaving the α's and ξ's unaltered, these operations converting each of the two equations (e_{ij}), (e_{ji}) into the equation derived from the other by changing the signs of both sides.

(iii) The resultant R is a determinant of order $2r$ which does not vanish identically. For when A is particularised into a quasi-scalar matrix A_0 whose diagonal elements are independent arbitrary parameters, R becomes particularised into a non-zero determinant R_0 which is a derangement of a quasi-scalar determinant whose diagonal elements are the $2r$ differences $a_{ii} - a_{jj}$, where $j \neq i$. Thus when the ξ's are regarded as given, the equations (b) have a unique finite solution expressing the $2r$ non-diagonal elements x_{ij} of X as homogeneous linear functions of the ξ's in which the coefficients are rational functions of the a's and α's.

(iv) When the equations and their terms are suitably arranged,—the signs of both sides being changed in every equation for greater convenience,—the coefficients of the homogeneous linear functions of the x's occurring on the left form an undegenerate skew-symmetric matrix M of order $2r$. Consequently if Δ is that homogeneous rational integral function of the a's and α's of degree r which is the Pfaffian of M (see Vol. II, Appendix B), and if $\dot{M}$ is the conjugate reciprocal of M, we then have

$$R = \det M = \Delta^2 \neq 0, \quad \dot{M} = \Delta\Omega,$$

where Ω is an undegenerate skew-symmetric matrix of even order $2r$ whose elements are homogeneous rational integral functions of the a's and α's of degree $r-1$. We will call Δ the *Pfaffian discriminant* of A.

(v) The values of the non-diagonal elements x_{ij} of X formed by solving the equations (b) satisfy the equations (a), making them identities in the elements of A and the $m-1$ arbitrary ξ's. This we will proceed to prove.

If $P = [P]_r^1$, $U = [U]_r^1$ are one-rowed matrices whose successive elements are

$$x_{12},\ x_{23},\ x_{34},\ \dots\ x_{13},\ x_{24},\ x_{35},\ \dots,\ \dots,\ x_{1m};$$

$$a_{21}\xi_{12},\ a_{32}\xi_{23},\ a_{43}\xi_{34},\ \dots\ a_{31}\xi_{13},\ a_{42}\xi_{24},\ a_{53}\xi_{35},\ \dots,\ \dots,\ a_{m1}\xi_{1m};$$

and $Q = [Q]_r^1$, $V = [V]_r^1$ are one-rowed matrices whose successive elements are

$$x_{21},\ x_{32},\ x_{43},\ \dots\ x_{31},\ x_{42},\ x_{53},\ \dots,\ \dots,\ x_{m1};$$

$$a_{12}\xi_{12},\ a_{23}\xi_{23},\ a_{34}\xi_{34},\ \dots\ a_{13}\xi_{13},\ a_{24}\xi_{24},\ a_{35}\xi_{35},\ \dots,\ \dots,\ a_{1m}\xi_{1m};$$

the $2r$ scalar equations (b) are equivalent to a matrix equation of the form

$$M.\begin{bmatrix} P \\ Q \end{bmatrix}_{r,r}^1 \equiv \begin{bmatrix} S, & H \\ -K, & T \end{bmatrix}_{r,r}^{r,r} . \begin{bmatrix} P \\ Q \end{bmatrix}_{r,r}^1 = \begin{bmatrix} U \\ -V \end{bmatrix}_{r,r}^1, \qquad \text{(c)}$$

where the prefactor M on the left is an undegenerate skew-symmetric matrix; H and K are mutually conjugate square matrices; S and T are skew-symmetric matrices; and H, S, P, U can be converted into K, T, Q, V by interchanging the two suffixes in every one of the a's and x's, leaving the α's and ξ's unaltered. If Δ is the Pfaffian and $\dot{M}$ the conjugate reciprocal of M, we have

$$\det M = \Delta^2 \neq 0, \quad \dot{M} = \Delta\Omega = \Delta \begin{bmatrix} s, & h \\ -k, & -t \end{bmatrix}_{r,r}^{r,r},$$

where Ω is an undegenerate skew-symmetric matrix, and where h, k, s, t have the same properties as H, K, S, T, this being a consequence of the fact that any symmetric derangement of M must be accompanied by the same symmetric derangement of $\dot{M}$.

When we prefix $\dot{M}$ and cancel Δ on both sides of (c), we obtain the unique solution for the x's of the equations (b) in terms of the ξ's in the form

$$\Delta \begin{bmatrix} P \\ Q \end{bmatrix}_{r,r}^{1} = \begin{bmatrix} s, & h \\ -k, & -t \end{bmatrix}_{r,r}^{r,r} \begin{bmatrix} U \\ -V \end{bmatrix}_{r,r}^{1} \qquad \text{..................(c')}$$

or

$$P = \frac{1}{\Delta}\cdot(sU - hV), \quad Q = \frac{1}{\Delta}\cdot(tV - kU); \qquad \text{..............(c'')}$$

and because of the properties of the matrices occurring on the right in (c''), we see that for all permissible values of i and j such that $j \neq i$ we have

$$\Delta \,.\, x_{ij} = X_{ij}, \qquad \text{.............................}(\mathrm{E}_{ij})$$

where X_{ij} is always a homogeneous linear function of the ξ's with determinate coefficients which are homogeneous rational integral functions of the a's and α's, and where X_{ij} is converted into X_{ji} by interchanging the two suffixes in all the a's, leaving the α's, the ξ's and Δ unaltered. From the last property it follows that the values of the non-diagonal x's given by the equations such as (E_{ij}) satisfy the scalar equations (a) identically when the $m-1$ independent ξ's are arbitrary.

It follows that the equations such as (E_{ij}) constitute the complete solution of the m^2 scalar equations (a) and (b) when we put $\xi_{ij} = x_{ii} - x_{jj}$ and regard the diagonal x's as arbitrary parameters. They are correlated by permutations of the suffixes 1, 2, ... m in all the a's, x's, α's, ξ's and Δ in the same way as the equations such as (e_{ij}); and they are unaltered in their totality when the two suffixes are interchanged in all the a's and x's only, this operation converting the equation (E_{ij}) into the equation (E_{ji}). We conclude that:

If $A = [a]_m^m$ is a square matrix whose elements are independent arbitrary parameters, and if Δ is the Pfaffian discriminant of A, the general commutant $X = \{A, A\}$ can be expressed in the form

$$X = \begin{bmatrix} \Delta x_{11}, & X_{12}, & \dots & X_{1m} \\ X_{21}, & \Delta x_{22}, & \dots & X_{2m} \\ \dots & \dots & \dots & \dots \\ X_{m1}, & X_{m2}, & \dots & \Delta x_{mm} \end{bmatrix}, \quad \dots\dots\dots\dots\dots\dots\text{(d)}$$

where $x_{11}, x_{22}, \dots x_{mm}$ *are new arbitrary parameters, and where* X_{ij} *for every permissible pair of values of* i *and* j *is a uniquely determinate homogeneous linear function of the differences* $x_{ii} - x_{jj}$, *viz. the function occurring on the right in the equation* (E_{ij}). *It can also be expressed as a homogeneous linear function of the independent scalar parameters* $x_{11}, x_{22}, \dots x_{mm}$ *in the form*

$$X = x_{11}\Phi_1 + x_{22}\Phi_2 + \dots + x_{mm}\Phi_m, \dots\dots\dots\dots\dots\dots\text{(e)}$$

where the Φ*'s are completely known independent square matrices.* ....(A)

Consequently there are exactly m *independent particular non-zero commutants of* A *which can be taken to be the uniquely determinate square matrices*

$$\Phi_1, \Phi_2, \dots \Phi_m. \quad \dots\dots\dots\dots\dots\dots\text{(B)}$$

Since m is the lowest degree of a rational integral equation satisfied by A, and the square matrices

$$I, A, A^2, \dots A^{m-1}, \qquad \text{where } I = [1]_m^m,$$

are m independent particular non-zero commutants of A, we can further conclude in succession that:

The general commutant X *described in* (A) *can be regarded as an arbitrary rational integral function of* A, *and each of the matrices* $\Phi_1, \Phi_2, \dots \Phi_m$ *is a determinate rational function of* A; *i.e. we can put*

$$X = \lambda_1 A^{m-1} + \lambda_2 A^{m-2} + \dots + \lambda_{m-1}A + \lambda_m I,$$

where the λ*'s are arbitrary scalar parameters; and each of the square matrices* $\Phi_1, \Phi_2, \dots \Phi_m$ *can be expressed in the same form, where the* λ*'s are uniquely determinate rational integral functions of the elements of* A. ...(C)

All commutants of A *are commutative with one another, and are commutants of the commutant of* A. ....(D)

The general commutant of the commutant of A *is a square matrix* Φ *which is expressible in the same forms as* X. *It can be regarded as an arbitrary homogeneous linear function of* $\Phi_1, \Phi_2, \dots \Phi_m$, *or as an arbitrary rational integral function of* A. ....(E)

Since the equations such as $\Phi_i\Phi_j = \Phi_j\Phi_i$ are identities in the elements of A, and since the conclusions (A) and (B) depend only on the fact that Δ does not vanish identically, we see that:

If the elements of A *are not all arbitrary, but are rational integral functions of certain scalar variables (or are merely constants), all the results* (A), (B), (C), (D), (E) *remain true whenever the Pfaffian discriminant* Δ *does not vanish identically.* ....(F)

In all such cases $\Phi_1, \Phi_2, \dots \Phi_m$ are m independent non-zero rational integral functions of A (or commutants of the commutant of A), and therefore m must be the lowest degree of a rational integral equation satisfied by A. But as is easily verified when $m=2$, m may still be the lowest degree of a rational integral equation satisfied by A when $\Delta=0$. The non-evanescence of Δ is a necessary and sufficient condition that the diagonal elements of a general commutant $\{A, A\}$ shall be independent arbitrary parameters.

Ex. i. *The general commutant of a general square matrix* $A=[a]_2^2$ *of order* 2.

When we put $$a_{12}=a_{11}-a_{22}, \quad \xi_{12}=x_{11}-x_{22},$$

the matrix equation $[a]_2^2[x]_2^2=[x]_2^2[a]_2^2$ is equivalent to the scalar equations

$$a_{12}x_{21}-x_{12}a_{21}=0, \quad \dots\dots(a_1)$$

$$a_{12}x_{12}=a_{12}\xi_{12}, \quad a_{12}x_{21}=a_{21}\xi_{12}. \quad \dots\dots(b_1)$$

The two scalar equations (b_1) are equivalent to the matrix equation

$$M\begin{bmatrix} x_{12} \\ x_{21} \end{bmatrix}=\begin{bmatrix} a_{21}\,\xi_{12} \\ -a_{12}\,\xi_{12} \end{bmatrix}, \quad \text{where } M=\begin{bmatrix} 0, & a_{12} \\ -a_{12}, & 0 \end{bmatrix}, \quad \dots\dots(c_1)$$

and we have $$\det M=\Delta^2=a_{12}^2, \quad \Delta=a_{12}=a_{11}-a_{22}.$$

Since (a_1) is an identity when x_{12} and x_{21} have the values given by (b_1), the general commutant $\{A, A\}$ is the square matrix

$$X=\begin{bmatrix} a_{12}x_{11}, & a_{12}(x_{11}-x_{22}) \\ a_{21}(x_{11}-x_{22}), & a_{12}x_{22} \end{bmatrix}=x_{11}(A-a_{22}I)-x_{22}(A-a_{11}I), \quad \dots\dots(d_1)$$

where I is the unit matrix of order 2, and x_{11} and x_{22} are arbitrary parameters.

The general commutants $\{A', A\}$, $\{A, A'\}$ are shown in Ex. xv of § 244.

Ex. ii. *The general commutants of specialised square matrices of order* 2.

In the following illustrations X is a general commutant of A, Φ is a general commutant of the commutant of A, and the elements of A are arbitrary subject to the restrictions shown or mentioned. The letters x, y, λ, μ denote arbitrary parameters, and I is the unit matrix of order 2.

(1) If $A=\begin{bmatrix} a & c \\ d & b \end{bmatrix}$, then whenever $\Delta=a-b\neq0$, we can put

$$X=\Phi=\begin{bmatrix} (a-b)\,x, & c\,(x-y) \\ d\,(x-y), & (a-b)\,y \end{bmatrix}=x\,(A-bI)-y\,(A-aI).$$

(2) If $A=\begin{bmatrix} a & c \\ d & a \end{bmatrix}$, then whenever c and d are not both 0, we can put

$$X=\Phi=\begin{bmatrix} x & cy \\ dy & x \end{bmatrix}=xI+y\,(A-aI).$$

(3) If $A=\begin{bmatrix} a & 0 \\ 0 & a \end{bmatrix}$, then for all values of a we can put

$$X=\begin{bmatrix} x & \lambda \\ \mu & y \end{bmatrix}, \quad \Phi=\begin{bmatrix} x & 0 \\ 0 & x \end{bmatrix}=xI.$$

Ex. iii. *The general commutant of a general square matrix* $A=[a]_3^3$ *of order* 3.

When we use the notations of the text and replace the matrix equation

$$[a]_3^3[x]_3^3=[x]_3^3[a]_3^3$$

by the nine equivalent scalar equations, the three scalar equations (a) of zero difference-weights are equivalent to the matrix equation

$$\begin{bmatrix} \alpha_{21}, & 0, & \alpha_{31} \\ -\alpha_{21}, & \alpha_{32}, & 0 \\ 0, & -\alpha_{32}, & -\alpha_{31} \end{bmatrix} \begin{bmatrix} x_{12} \\ x_{23} \\ x_{13} \end{bmatrix} = \begin{bmatrix} \alpha_{12}, & 0, & \alpha_{13} \\ -\alpha_{12}, & \alpha_{23}, & 0 \\ 0, & -\alpha_{23}, & -\alpha_{13} \end{bmatrix} \begin{bmatrix} x_{21} \\ x_{32} \\ x_{31} \end{bmatrix}, \quad \text{.........}(\mathrm{a}_2)$$

one of them being superfluous; and the six scalar equations (b) of non-zero difference-weights are equivalent to the matrix equation

$$\left[\begin{array}{ccc|ccc} 0, & \alpha_{31}, & 0 & a_{12}, & 0, & -\alpha_{23} \\ -\alpha_{31}, & 0, & 0 & 0, & a_{23}, & \alpha_{12} \\ 0, & 0, & 0 & -\alpha_{32}, & \alpha_{21}, & a_{13} \\ \hline -a_{12}, & 0, & \alpha_{32} & 0, & -\alpha_{13}, & 0 \\ 0, & -a_{23}, & -\alpha_{21} & \alpha_{13}, & 0, & 0 \\ a_{23}, & -a_{12}, & -a_{13} & 0, & 0, & 0 \end{array}\right] \left[\begin{array}{c} x_{12} \\ x_{23} \\ x_{13} \\ \hline x_{21} \\ x_{32} \\ x_{31} \end{array}\right] = \left[\begin{array}{c} \alpha_{21}\ \xi_{12} \\ \alpha_{32}\ \xi_{23} \\ \alpha_{31}\ \xi_{13} \\ \hline -\alpha_{12}\ \xi_{12} \\ -\alpha_{23}\ \xi_{23} \\ -\alpha_{13}\ \xi_{13} \end{array}\right]. \quad \text{...}(\mathrm{b}_2)$$

If Δ is the Pfaffian and $\dot{M}$ the conjugate reciprocal of the skew-symmetric matrix M on the left in (b_2), we have

$$\Delta = a_{12}a_{23}a_{31} + a_{12}\alpha_{12}\alpha_{21} + a_{23}\alpha_{23}\alpha_{32} + a_{31}\alpha_{31}\alpha_{13};$$
$$\det M = \Delta^2 \neq 0;$$

and

$$\dot{M} = \Delta \, . \left[\begin{array}{ccc|ccc} 0, & -\alpha_{13}a_{31}, & -\alpha_{13}\alpha_{12} & -B_{11}, & -\alpha_{32}\alpha_{12}, & -\alpha_{32}a_{32} \\ \alpha_{13}a_{31}, & 0, & -\alpha_{13}\alpha_{23} & -\alpha_{23}\alpha_{21}, & -B_{22}, & -\alpha_{21}a_{12} \\ \alpha_{13}\alpha_{12}, & \alpha_{13}\alpha_{23}, & 0 & -\alpha_{23}a_{32}, & -\alpha_{12}a_{12}, & B_{33} \\ \hline B_{11}, & \alpha_{23}\alpha_{21}, & \alpha_{23}a_{32} & 0, & \alpha_{31}a_{31}, & \alpha_{31}\alpha_{21} \\ \alpha_{32}\alpha_{12}, & B_{22}, & \alpha_{12}a_{12} & -\alpha_{31}a_{31}, & 0, & \alpha_{31}\alpha_{32} \\ \alpha_{32}a_{32}, & \alpha_{21}a_{12}, & -B_{33} & -\alpha_{31}\alpha_{21}, & -\alpha_{31}\alpha_{32}, & 0 \end{array}\right],$$

where $B_{11} = \alpha_{12}\alpha_{21} + a_{23}a_{31}, \quad B_{22} = \alpha_{23}\alpha_{32} + a_{31}a_{12}, \quad B_{33} = \alpha_{31}\alpha_{13} + a_{12}a_{23}.$

By prefixing $\dot{M}$ on both sides of (b_2) we obtain the unique solution given by the six scalar equations

$$\Delta x_{12} = (\alpha_{12}\alpha_{21} + a_{13}a_{32})\,\alpha_{12}\xi_{12} - (\alpha_{12}\alpha_{31} + \alpha_{32}\alpha_{23})\,\alpha_{13}\xi_{13} - (\alpha_{12}\alpha_{23} + \alpha_{13}a_{13})\,\alpha_{32}\xi_{32},$$
$$\Delta x_{23} = (\alpha_{23}\alpha_{32} + a_{21}a_{13})\,\alpha_{23}\xi_{23} - (\alpha_{23}\alpha_{12} + \alpha_{13}a_{31})\,\alpha_{21}\xi_{21} - (\alpha_{23}\alpha_{31} + \alpha_{21}a_{21})\,\alpha_{13}\xi_{13},$$
$$\Delta x_{31} = (\alpha_{31}\alpha_{13} + a_{32}a_{21})\,\alpha_{31}\xi_{31} - (\alpha_{31}\alpha_{23} + \alpha_{21}\alpha_{12})\,\alpha_{32}\xi_{32} - (\alpha_{31}\alpha_{12} + \alpha_{32}a_{32})\,\alpha_{21}\xi_{21},$$

$$\Delta x_{21} = (\alpha_{21}\alpha_{12} + a_{13}a_{32})\,\alpha_{21}\xi_{12} - (\alpha_{21}\alpha_{13} + \alpha_{23}a_{23})\,\alpha_{31}\xi_{13} - (\alpha_{21}\alpha_{32} + \alpha_{31}a_{13})\,\alpha_{23}\xi_{32},$$
$$\Delta x_{32} = (\alpha_{32}\alpha_{23} + a_{21}a_{13})\,\alpha_{32}\xi_{23} - (\alpha_{32}\alpha_{21} + \alpha_{31}a_{31})\,\alpha_{12}\xi_{21} - (\alpha_{32}\alpha_{13} + \alpha_{12}a_{21})\,\alpha_{31}\xi_{13},$$
$$\Delta x_{13} = (\alpha_{13}\alpha_{31} + a_{32}a_{21})\,\alpha_{13}\xi_{31} - (\alpha_{13}\alpha_{32} + \alpha_{12}a_{12})\,\alpha_{23}\xi_{32} - (\alpha_{13}\alpha_{21} + \alpha_{23}a_{32})\,\alpha_{12}\xi_{21}, \quad \text{...}(\mathrm{c}_2)$$

which are correlated by permutations of the suffixes 1, 2, 3 in all the α's, x's, a's and ξ's, including those occurring in Δ; or given by the three pairs of scalar equations

$$\Delta x_{ij} = (\alpha_{ij}\alpha_{ji} + a_{ik}a_{kj})\,\alpha_{ij}\xi_{ij} - (\alpha_{ij}\alpha_{ki} + \alpha_{kj}a_{jk})\,a_{ik}\xi_{ik} - (\alpha_{ij}\alpha_{jk} + a_{ik}a_{ik})\,\alpha_{kj}\xi_{kj},$$
$$\Delta x_{ji} = (\alpha_{ji}\alpha_{ij} + a_{ik}a_{kj})\,\alpha_{ji}\xi_{ij} - (\alpha_{ji}\alpha_{ik} + \alpha_{jk}a_{jk})\,\alpha_{ki}\xi_{ik} - (\alpha_{ji}\alpha_{kj} + \alpha_{ki}a_{ik})\,\alpha_{jk}\xi_{kj}, \quad \text{......}(\mathrm{c}_2')$$

where i, j, k receive the values 1, 2, 3; 2, 3, 1; 3, 1, 2. The equations (c_2) or (c_2') remain unaltered in their totality when the two suffixes are interchanged in all the α's and x's whilst the a's, the ξ's and Δ remain unaltered.

Since the expression for x_{ji} can be derived from the expression for x_{ij} by interchanging the two suffixes in every one of the a's whilst the α's, the ξ's and Δ remain unaltered, these solutions satisfy the equations (a$_2$) when the ξ's are arbitrary subject to the conditions $\xi_{ji}=-\xi_{ij}$, i.e. they form the complete solutions of the nine scalar equations represented by (a$_2$) and (b$_2$). It follows that the general commutant $\{A, A\}$ is the square matrix

$$X=\begin{bmatrix}\Delta x_{11}, & X_{12}, & X_{13}\\ X_{21}, & \Delta x_{22}, & X_{23}\\ X_{31}, & X_{32}, & \Delta x_{33}\end{bmatrix}, \quad \text{.................................}(\text{d}_2)$$

where x_{11}, x_{22}, x_{33} are arbitrary parameters, and where X_{ij} is the homogeneous linear function of the differences $x_{11}-x_{22}=\xi_{12}$, $x_{22}-x_{33}=\xi_{23}$, $x_{33}-x_{11}=\xi_{31}$ which forms the right-hand side of the equation for Δx_{ij} in (c$_2$).

Ex. iv. *Transformation of the general commutant X of Ex.* iii.

We will transform the expression (d$_2$) for X by writing

$$x,\ y,\ z \ \text{ for }\ x_{11},\ x_{22},\ x_{33};\quad \alpha,\ \beta,\ \gamma\ \text{ for }\ a_{23},\ a_{31},\ a_{12};$$

and putting $\quad P=a_{12}a_{21}-a_{13}a_{31},\quad Q=a_{23}a_{32}-a_{21}a_{12},\quad R=a_{31}a_{13}-a_{32}a_{23};$

$$\omega=a_{12}a_{23}a_{31}-a_{13}a_{32}a_{21},\quad X_{ij}=xA_{ij}+yB_{ij}+zC_{ij};$$

so that
$$\Delta=\alpha\beta\gamma+\alpha a_{23}a_{32}+\beta a_{31}a_{13}+\gamma a_{12}a_{21},$$

and
$$\alpha+\beta+\gamma=0,\quad P+Q+R=0,\quad A_{ij}+B_{ij}+C_{ij}=0.$$

We then have
$$X=x\Phi_1+y\Phi_2+z\Phi_3,$$

where

$$\Phi_1=\begin{bmatrix}\Delta, & A_{12}, & A_{13}\\ A_{21}, & 0, & A_{23}\\ A_{31}, & A_{32}, & 0\end{bmatrix},\quad \Phi_2=\begin{bmatrix}0, & B_{12}, & B_{13}\\ B_{21}, & \Delta, & B_{23}\\ B_{31}, & B_{32}, & 0\end{bmatrix},\quad \Phi_3=\begin{bmatrix}0, & C_{12}, & C_{13}\\ C_{21}, & 0 & C_{23}\\ C_{31}, & C_{32}, & \Delta\end{bmatrix},$$

the non-diagonal elements of Φ_1, Φ_2, Φ_3 being given by the equations

$$A_{23}=a_{23}P-a_{21}a_{13}\alpha,\quad B_{23}=a_{23}(Q+\beta\gamma)-a_{21}a_{13}\beta,\quad C_{23}=a_{23}(R-\beta\gamma)-a_{21}a_{13}\gamma,$$
$$B_{31}=a_{31}Q-a_{32}a_{21}\beta,\quad C_{31}=a_{31}(R+\gamma\alpha)-a_{32}a_{21}\gamma,\quad A_{31}=a_{31}(P-\gamma\alpha)-a_{32}a_{21}\alpha,$$
$$C_{12}=a_{12}R-a_{13}a_{32}\gamma,\quad A_{12}=a_{12}(P+\alpha\beta)-a_{13}a_{32}\alpha,\quad B_{12}=a_{12}(Q-\alpha\beta)-a_{13}a_{32}\beta,$$

and the nine corresponding equations obtained by interchanging the two suffixes in every one of the A's, B's and a's.

The three sets of relations such as

$$A_{23}B_{32}-A_{32}B_{23}=A_{31}B_{13}-A_{13}B_{31}=A_{23}B_{32}-A_{32}B_{23}=\Delta\omega;$$
$$A_{21}B_{13}-B_{21}A_{13}=\ \Delta A_{23},\quad A_{32}B_{21}-B_{32}A_{21}=\ \Delta B_{31},\quad A_{13}B_{32}-B_{13}A_{32}=\ \Delta C_{12};$$
$$A_{31}B_{12}-B_{31}A_{12}=-\Delta A_{32},\quad A_{12}B_{23}-B_{12}A_{23}=-\Delta B_{13},\quad A_{23}B_{31}-B_{23}A_{31}=-\Delta C_{21};$$

in which A, B can be replaced by B, C and C, A on the left, show directly that Φ_1, Φ_2, Φ_3 are commutative with one another and with X, so that the general commutant of the commutant of A is a square matrix Φ of the same form as X.

Again if λ_i, μ_i, ν_i are first determined by equating corresponding diagonal elements on both sides, it will be seen that the equations

$$\Phi_i=\lambda_iA^2+\mu_iA+\nu_iI,\qquad (i=1,\ 2,\ 3),$$

are identities in the elements of A when

$$\lambda_1=a_{33}-a_{22},$$
$$\mu_1=(a_{22}^2-a_{33}^2)+(a_{21}a_{12}-a_{31}a_{13}),$$
$$\nu_1=(a_{33}-a_{22})(a_{22}a_{33}-a_{23}a_{32})+(a_{22}\,.\,a_{31}a_{13}-a_{33}\,.\,a_{21}a_{12}),$$

and the values of λ_2, μ_2, ν_2 and λ_3, μ_3, ν_3 are given by the equations derived from these three by applying cyclical substitutions to all the suffixes. We then have

$$\lambda_1+\lambda_2+\lambda_3=0, \quad \mu_1+\mu_2+\mu_3=0, \quad \nu_1+\nu_2+\nu_3=\Delta.$$

It will be seen that the resultant of the homogeneous linear functions

$$\lambda=\lambda_1x+\lambda_2y+\lambda_3z, \quad \mu=\mu_1x+\mu_2y+\mu_3z, \quad \nu=\nu_1x+\nu_2y+\nu_3z$$

is Δ^2; and because it does not vanish identically, we can put

$$X=\lambda A^2+\mu A+\nu I,$$

where λ, μ, ν are independent arbitrary scalar parameters.

Ex. v. *The general commutants of certain specialised square matrices of order* 3.

In each of the following illustrations A is a given square matrix whose non-zero elements are arbitrary subject to the restrictions shown or mentioned; X is a general commutant of A; and Φ is a general commutant of the commutant of A. The letters x, y, z denote independent arbitrary parameters, and we put

$$\alpha=b-c, \quad \beta=c-a, \quad \gamma=a-b; \quad I=[1]_3^3.$$

$$(1) \quad A=\begin{bmatrix} a & h & g \\ 0 & b & f \\ 0 & 0 & c \end{bmatrix} : X=\Phi=\begin{bmatrix} \Delta x, & h\alpha\beta(x-y), & g\gamma\alpha(z-x)-hf(\alpha x+\beta y+\gamma z) \\ 0, & \Delta y, & f\beta\gamma \\ 0, & 0, & \Delta z \end{bmatrix}$$

whenever $\Delta=(b-c)(c-a)(a-b)\neq 0.$

If $\Phi_1=-\alpha(A-bI)(A-cI), \quad \Phi_2=-\beta(A-cI)(A-aI), \quad \Phi_3=-\gamma(A-aI)(A-bI),$

we have $X=\Phi=x\Phi_1+y\Phi_2+z\Phi_3=\lambda A^2+\mu B^2+\nu I,$

where λ, μ, ν are independent arbitrary parameters.

$$(2) \quad A=\begin{bmatrix} a & d & 0 \\ e & b & 0 \\ 0 & 0 & c \end{bmatrix} : X=\Phi=\Delta_0\begin{bmatrix} (a-b)x, & d(x-y), & 0 \\ e(x-y), & (a-b)y, & 0 \\ 0, & 0 & (a-b)z \end{bmatrix}$$

whenever $\Delta=(a-b)\Delta_0\neq 0,$ where $\Delta_0=(b-c)(c-a)+de.$

The coefficients of x, y, z in X are the square matrices

$$\Phi_1=-\alpha(A-bI)(A-cI)+de(A-cI),$$
$$\Phi_2=-\beta(A-cI)(A-aI)-de(A-cI),$$
$$\Phi_3=-\gamma(A-aI)(A-bI)+de(a-b)I,$$

and we can put $X=\Phi=x\Phi_1+y\Phi_2+z\Phi_3=\lambda A^2+\mu A+\nu I,$

where λ, μ, ν are independent arbitrary scalar parameters.

Ex. vi. *The general commutant of a general square matrix* $A=[a]_4^4$ *of order* 4.

If we use the notations of the text, the four scalar equations (a) are

$$[\;a_{21},\; 0,\; 0,\; a_{31},\; 0,\; a_{41}]P=[\;a_{12},\; 0,\; 0,\; a_{13},\; 0,\; a_{14}]Q,$$
$$[-a_{21},\; a_{32},\; 0,\; 0,\; a_{42},\; 0\;]P=[-a_{12},\; a_{23},\; 0,\; 0,\; a_{24},\; 0\;]Q,$$
$$[\;0,\; -a_{32},\; a_{43},\; -a_{31},\; 0,\; 0\;]P=[\;0,\; -a_{23},\; a_{34},\; -a_{13},\; 0,\; 0\;]Q,$$
$$[\;0,\; 0,\; -a_{43},\; 0,\; -a_{42},\; -a_{41}]P=[\;0,\; 0,\; -a_{34},\; 0,\; -a_{24},\; -a_{14}]Q;$$

...............(a_3)

and the twelve scalar equations (b) are equivalent to the matrix equation

$$\left[\begin{array}{cccccc|cccccc} 0, & a_{31}, & 0, & 0, & a_{41}, & 0 & a_{12}, & 0, & 0, & -a_{23}, & 0, & -a_{24} \\ -a_{31}, & 0, & a_{42}, & 0, & 0, & 0 & 0, & a_{23}, & 0, & a_{12}, & -a_{34}, & 0 \\ 0, & -a_{42}, & 0, & -a_{41}, & 0, & 0 & 0, & 0, & a_{34}, & 0, & a_{23}, & a_{13} \\ 0, & 0, & a_{41}, & 0, & 0, & 0 & -a_{32}, & a_{21}, & 0, & a_{13}, & 0, & -a_{34} \\ -a_{41}, & 0, & 0, & 0, & 0, & 0 & 0, & -a_{43}, & a_{32}, & 0, & a_{24}, & a_{12} \\ 0, & 0, & 0, & 0, & 0, & 0 & -a_{42}, & 0, & a_{31}, & -a_{43}, & a_{21}, & a_{14} \\ \hline -a_{12}, & 0, & 0, & a_{32}, & 0, & a_{42} & 0, & -a_{13}, & 0, & 0, & -a_{14}, & 0 \\ 0, & -a_{23}, & 0, & -a_{21}, & a_{43}, & 0 & a_{13}, & 0, & -a_{24}, & 0, & 0, & 0 \\ 0, & 0, & -a_{34}, & 0, & -a_{32}, & -a_{31} & 0, & a_{24}, & 0, & a_{14}, & 0, & 0 \\ a_{23}, & -a_{12}, & 0, & -a_{13}, & 0, & a_{43} & 0, & 0, & -a_{14}, & 0, & 0, & 0 \\ 0, & a_{34}, & -a_{23}, & 0, & -a_{24}, & -a_{21} & a_{14}, & 0, & 0, & 0, & 0, & 0 \\ a_{24}, & 0, & -a_{13}, & a_{34}, & -a_{12}, & -a_{14} & 0, & 0, & 0, & 0, & 0, & 0 \end{array}\right] \left[\begin{array}{c} x_{12} \\ x_{23} \\ x_{34} \\ x_{13} \\ x_{24} \\ x_{14} \\ \hline x_{21} \\ x_{32} \\ x_{43} \\ x_{31} \\ x_{42} \\ x_{41} \end{array}\right] = \left[\begin{array}{c} a_{21}\xi_{12} \\ a_{32}\xi_{23} \\ a_{43}\xi_{34} \\ a_{31}\xi_{13} \\ a_{42}\xi_{24} \\ a_{41}\xi_{14} \\ \hline -a_{12}\xi_{12} \\ -a_{23}\xi_{23} \\ -a_{34}\xi_{34} \\ -a_{13}\xi_{13} \\ -a_{24}\xi_{24} \\ -a_{14}\xi_{14} \end{array}\right], \qquad \text{..................(b}_3\text{)}$$

or

$$\begin{bmatrix} S, & H \\ -K, & T \end{bmatrix}_{b,\,b}^{b,\,b} \begin{bmatrix} P \\ Q \end{bmatrix}_{b,\,b}^{1} = \begin{bmatrix} V \\ -V \end{bmatrix}_{b,\,b}^{1} \qquad \text{..............................(c}_3\text{)}$$

If we determine the unique solution of the equation (c_3), which necessarily satisfies the equations (a_3), we shall obtain the general commutant $X=\{A, B\}$ in the form corresponding to (d); but we know otherwise that X can also be expressed as an arbitrary rational integral function of A.

CHAPTER XXIX

INVARIANT TRANSFORMANDS

[A solution of the matrix equation $AXB=X$, in which A and B must be square matrices, is called an invariant transformand $X=\text{inv}\ \{A, B\}$. The first three articles (§§ 254—6) deal with definitions and general principles, and with the relations of invariant transformands to commutants, and to bilinear and quadratic scalar invariants. The general invariant transformand of a pair of simple canonical square matrices whose latent roots are both 1 (constructed in §§ 255 and 257) is a simple slope having one and only one arbitrary element in each parametric diagonal line, all the elements being known functions whose coefficients are integers satisfying the equation ${}^iH_\kappa={}^{i-1}H_\kappa+{}^iH_{\kappa-1}$; and that of a pair of unilatent bi-canonical square matrices whose latent roots are c and c' (constructed in § 259) is a completely known compound slope or a zero matrix according as $cc'=1$ or $\neq 1$. The special constructions of §§ 257 and 259 lead to the standard constructions of § 260 for the general (or general symmetric or general skew-symmetric) invariant transformand of any pair of square matrices with constant numerical elements; and these standard constructions show that the independent bilinear and quadratic invariants of homogeneous linear substitutions are always functions having the forms described in § 258.]

§ 254. Invariant transformands defined.

If $A=[a]_m^m$ and $B=[b]_n^n$ are any two given square matrices of orders m and n with constant elements forming a matrix-pair (A, B) in which A precedes B, any matrix $X=[x]_m^n$ satisfying the equation

$$[a]_m^m\,[x]_m^n\,[b]_n^n=[x]_m^n \quad \text{or} \quad AXB=X \quad \text{..............(A)}$$

will be called an *invariant transformand* of the arranged matrix-pair (A, B), or an invariant of a transformation by A and B. More shortly it will be called an invariant $\{A, B\}$; and the equation

$$X=\text{inv}\ \{A, B\}$$

will be used to denote that X is such an invariant, i.e. some solution of the equation $AXB=X$ in which A and B are square matrices. Particular and general solutions of the equation (A) will be called particular and general invariants $\{A, B\}$. Thus a *particular* invariant $\{A, B\}$ is a solution X of the equation (A) in which all the elements of X are constants. If the equation (A) has non-zero solutions, i.e. if there exist non-zero invariants $\{A, B\}$, and if i is the greatest possible number of independent particular non-zero solutions, i.e. of independent particular non-zero invariants $\{A, B\}$, then a *general* invariant $\{A, B\}$ is a matrix X expressible in the form

$$X=\lambda_1X_1+\lambda_2X_2+\ldots+\lambda_iX_i,$$

where $X_1, X_2, \ldots X_i$ are i independent particular non-zero solutions of the equation (A), i.e. i independent particular non-zero invariants $\{A, B\}$, and the λ's are independent arbitrary parameters. If the equation (A) has no non-zero solution, the *general* invariant $\{A, B\}$ is the zero matrix $[0]_m^n$. Every invariant $\{A, B\}$ can be regarded as a particularisation or specialisation of any given general invariant $\{A, B\}$.

An invariant $X = \text{inv}\,\{A, -B\}$ or $X = \text{inv}\,\{-A, B\}$ is of course a solution of the matrix equation

$$AXB = -X.$$

Ex. i. *The invariants of a transformation by two scalar or two quasi-scalar matrices.*

If $A = a[1]_m^m$ and $B = b[1]_n^n$ are two scalar matrices, the general invariant $\{A, B\}$ is the zero matrix $[0]_m^n$ when $ab \neq 1$; and when $ab = 1$, it is a matrix $[x]_m^n$ whose elements are independent scalar parameters.

For when $X = [x]_m^n$, the matrix equation $AXB = X$ is equivalent to the equation

$$(ab - 1)X = 0.$$

This result and Theorem II of the next article enable us to construct the general invariant $\{A, B\}$ when A and B are any two given quasi-scalar matrices.

Ex. ii. *The conjugate of an invariant transformand.*

If A', B' are the conjugates of the square matrices A, B, the matrix X is a particular or general invariant $\{A, B\}$ when and only when its conjugate is a particular or general invariant $\{B', A'\}$.

Ex. iii. *The invariants* $\{A, \epsilon B\}$, $\{A', \epsilon B'\}$, $\{A', \epsilon B\}$, $\{A, \epsilon B'\}$ *when* $\epsilon = \pm 1$.

Let A', B' be the conjugates of A, B; and let U, V be any given particular undegenerate contra-commutants $\{A', A\}$, $\{B, B'\}$, which can always be so chosen as to be symmetric.

Then $X' = UXV$ *is a particular or general invariant* $\{A', \epsilon B'\}$,

$Y = UX$ *is a particular or general invariant* $\{A', \epsilon B\}$,

$Y' = XV$ *is a particular or general invariant* $\{A, \epsilon B'\}$,

if and only if X *is a particular or general invariant* $\{A, \epsilon B\}$.

These three results are particular cases of the theorem of Ex. v, and follow directly from the respective pairs of equations

$$A'X'B' = U.AXB.V, \quad X' = U.X.V;$$
$$A'YB = U.AXB \quad , \quad Y = U.X \quad ;$$
$$AY'B' = \quad AXB.V, \quad Y' = \quad X.V.$$

We can always construct four general invariants X, X', Y, Y' of the four types mentioned above which are connected by the relations

$$X' = UXV, \quad Y = UX, \quad Y' = XV, \quad \text{.............................(1)}$$

all four of them being completely known when any one of them is known.

Ex. iv. *The invariants* $\{A, \epsilon A\}$, $\{A', \epsilon A'\}$, $\{A', \epsilon A\}$, $\{A, \epsilon A'\}$ *when* $\epsilon = \pm 1$.

When $B = A$, $B' = A'$ in Ex. iii, then as shown in Ex. vii of § 240 we can choose U and V to be mutually inverse, and at the same time to be symmetric.

If V is a given particular undegenerate symmetric contra-commutant $\{A, A'\}$, the square matrix Y' given by the symmetric relation

$$Y' = VYV \quad \text{...(2)}$$

is a particular or general symmetric (or skew-symmetric) invariant $\{A, \epsilon A'\}$ when and only when Y is a particular or general symmetric (or skew-symmetric) invariant $\{A', \epsilon A\}$.

Ex. v. Let $A=[a]_m^m$, $B=[b]_n^n$ and $P=[p]_m^m$, $Q=[q]_n^n$ be two pairs of square matrices with constant elements, and let

$$[h]_m^m[x]_m^n[k]_n^n=[y]_m^n \quad \text{or} \quad hXk=Y \quad \text{.........................(3)}$$

be an equigradent transformation in which h, k are particular undegenerate commutants $\{P, A\}$, $\{B, Q\}$, so that their inverses H, K are particular undegenerate commutants $\{A, P\}$, $\{Q, B\}$. Then we have the following theorem :

Theorem. *The matrix X in* (3) *is a particular or general invariant $\{A, B\}$ when and only when the matrix Y is a particular or general invariant $\{P, Q\}$.*

Under these circumstances we can describe (3) as an equigradent transformation of the type

$$\{P, A\}\{A, B\}\{B, Q\}=\{P, Q\} \quad \text{..............................(3')}$$

in which h and k are true commutants, and X and Y are invariant transformands.

Ex. vi. *Cases in which A or B is undegenerate.*

We can identify invariants $\{A, B\}$ with commutants $\{A^{-1}, B\}$ when A is undegenerate, and with commutants $\{A, B^{-1}\}$ when B is undegenerate.

This result is in accordance with the fact that when A and B are two undegenerate square matrices, we can identify commutants $\{A, B\}$ with commutants $\{A^{-1}, B^{-1}\}$.

NOTE. *The invariant transformands of a pair of square matrices whose elements involve arbitrary parameters.*

When A and B are square matrices whose elements are rational integral functions of certain scalar variables $\gamma_1, \gamma_2, \gamma_3, \ldots$, a matrix X whose elements are rational integral functions of the same variables will be an invariant transformand $\{A, B\}$ when it makes the equation $AXB=X$ an identity in those variables. Some general properties of such more general invariant transformands can be derived from the properties of the invariant transformands which we are considering by ascribing ordinary particular values to the γ's.

§ 255. Theorems facilitating the determination of invariant transformands.

1. *Equicanonical pairs of square matrices.*

First let $A=[a]_m^m$, $B=[b]_n^n$ be two given square matrices of orders m, n with constant elements; let $\alpha=[\alpha]_m^m$, $\beta=[\beta]_n^n$ be two square matrices of orders m, n equicanonical with A, B; and let

$$[a]_m^m=[h]_m^m[\alpha]_m^m[H]_m^m \quad \text{or} \quad A=h\alpha H, \quad \text{..............(1)}$$

$$[b]_n^n=[k]_n^n[\beta]_n^n[K]_n^n \quad \text{or} \quad B=k\beta K, \quad \text{..............(2)}$$

where $\quad Hh=hH=[1]_m^m, \quad Kk=kK=[1]_n^n,$

be any two given isomorphic transformations by which A, B can be derived from α, β. Then we have the following theorem in which

$$\epsilon=\pm 1.$$

Theorem I. *If* $X=[x]_m^n$ *and* $\xi=[\xi]_m^n$ *are two matrices such that*

$$[x]_m^n=[h]_m^m[\xi]_m^n[K]_n^n \quad or \quad X=h\xi K, \quad \text{..............(A)}$$

then X *is an invariant* $\{A, \epsilon B\}$ *or a particular or general invariant* $\{A, \epsilon B\}$ *when and only when* ξ *is an invariant* $\{\alpha, \epsilon\beta\}$ *or a particular or general invariant* $\{\alpha, \epsilon\beta\}$.

This is a particular case of the theorem given in Ex. v of § 254, and it follows directly from the two equations

$$AXB=h\,.\,\alpha\xi\beta\,.\,K, \quad X=h\,.\,\xi\,.\,K$$

in which B and β could be replaced by $-B$ and $-\beta$. The theorem shows that we can construct the general invariant $\{A, \epsilon B\}$ when we can construct the general invariant $\{\alpha, \epsilon\beta\}$, where α and β are any square matrices equicanonical with A and B, and that the general invariant $\{A, \epsilon B\}$ has the same rank and contains the same number of independent arbitrary parameters as the general invariant $\{\alpha, \epsilon\beta\}$; moreover the total number of independent particular non-zero invariants $\{A, \epsilon B\}$ is the same as the total number of independent particular non-zero invariants $\{\alpha, \epsilon\beta\}$. In applying the theorem we shall usually choose α and β to be either standard canonical square matrices, or to be compartite square matrices of standard forms whose simple parts are unipotent square matrices of specially simple forms.

Ex. i. *Other forms of Theorem I.*

Let $A', B', a', \beta', h', k', H', K'$ be the conjugates of $A, B, a, \beta, h, k, H, K$; let $\epsilon=\pm 1$; and let X, X', Y, Y' and ξ, ξ', η, η' be matrices with m horizontal and n vertical rows which are connected by the relations

$$X=h\xi K, \quad X'=H'\xi'k', \quad Y=H'\eta K, \quad Y'=h\eta'k'. \quad \text{..................(B)}$$

Then X, X', Y, Y' *will be invariants or particular or general invariants of the respective types*

$$X=inv\,\{A, \epsilon B\}, \quad X'=inv\,\{A', \epsilon B'\}, \quad Y=inv\,\{A', eB\}, \quad Y'=inv\,\{A, \epsilon B'\}$$

when and only when ξ, ξ', η, η' *are invariants or particular or general invariants of the respective types*

$$\xi=inv\,\{a, \epsilon\beta\}, \quad \xi'=inv\,\{a', \epsilon\beta'\}, \quad \eta=inv\,\{a', \epsilon\beta\}, \quad \eta'=inv\,\{a, \epsilon\beta'\}.$$

When ξ, ξ', η, η' are *general* invariants of the types mentioned, we can regard the equations (B) as formulae giving the *general* invariants X, X', Y, Y' of the types mentioned.

Ex. ii. *The invariants* $\{A, \epsilon A\}, \{A', \epsilon A'\}, \{A', \epsilon A\}, \{A, \epsilon A'\}$ *when* $\epsilon=\pm 1$.

When B, β are the conjugates A', a' of A, a, we can take the isomorphic transformations (1) and (2) to be

$$[a]_m^m=[h]_m^m[a]_m^m[H]_m^m \quad \text{or} \quad A=haH, \quad \text{...........................(3)}$$

$$\ulcorner a \lrcorner_m^m=\ulcorner H \lrcorner_m^m \ulcorner a \lrcorner_m^m \ulcorner h \lrcorner_m^m \quad \text{or} \quad A'=H'a'h'. \quad \text{........................(4)}$$

In this case let X, X', Y, Y' and ξ, ξ', η, η' be square matrices of order m connected by the relations

$$X=h\xi H, \quad X'=H'\xi'h', \quad Y=H'\eta H, \quad Y'=h\eta'h'. \quad \text{.....................(C)}$$

Then X, X', Y, Y' will be invariants or particular or general invariants of the respective types

$$X = inv\,\{A, \epsilon A\}, \quad X' = inv\,\{A', \epsilon A'\}, \quad Y = inv\,\{A', \epsilon A\}, \quad Y' = inv\,\{A, \epsilon A'\}$$

when and only when ξ, ξ', η, η' are invariants or particular or general invariants of the respective types

$$\xi = inv\,\{a, \epsilon a\}, \quad \xi' = inv\,\{a', \epsilon a'\}, \quad \eta = inv\,\{a', \epsilon a\}, \quad \eta' = inv\,\{a, \epsilon a'\}.$$

When ξ, ξ', η, η' are *general* invariants of the types mentioned, we can regard the equations (C) as formulae giving the *general* invariants X, X', Y, Y' of the types mentioned.

Since Y and Y' in (C) are symmetric (or skew-symmetric) when and only when η and η' are symmetric (or skew-symmetric), we see that when η and η' are *general symmetric* (or *general skew-symmetric*) invariants of the types mentioned, we can regard the third and fourth of the equations (C) as formulae giving the *general symmetric* (or *general skew-symmetric*) invariants Y and Y' of the types mentioned.

Ex. iii. *Derangements of invariant transformands.*

When the isomorphic transformations (1) and (2) are the symmetric derangements

$$A = ha h', \quad B = k\beta k', \quad \text{where} \quad hh' = [1]_m^m, \quad kk' = [1]_n^n,$$

the equations (B) become

$$X = h\xi k', \quad X' = h\xi' k', \quad Y = h\eta k', \quad Y' = h\eta' k'. \quad \text{......................(B')}.$$

In this case general invariants ξ, ξ', η, η' of the types mentioned are converted into general invariants X, X', Y, Y' of the types mentioned by applying to their horizontal rows the derangement applied to the horizontal rows of a, a' in forming A, A', and to their vertical rows the derangement applied to the vertical rows of β, β' in forming B, B'.

We have corresponding results whenever (1) and (2) are *symmetric* equimutant (i.e. symmetric semi-unit) transformations.

Ex. iv. *Symmetric derangements of the invariant transformands of Ex.* ii.

When the isomorphic transformations (3) and (4) are the symmetric derangements

$$A = hah', \quad A' = ha'h', \quad \text{where} \quad hh' = [1]_m^m,$$

the equations (C) become

$$X = h\xi h', \quad X' = h\xi' h', \quad Y = h\eta h', \quad Y' = h\eta' h'. \quad \text{......................(C')}$$

In this case general invariants ξ, ξ', η, η' of the types mentioned are converted into general invariants X, X', Y, Y' of the types mentioned by applying to them the symmetric derangement which converts a, a' into A, A'.

We have corresponding results whenever (3) and (4) are *symmetric* equimutant (i.e. symmetric semi-unit) transformations.

2. *Compartite square matrices.*

Next let $A = [a]_m^m$, $B = [b]_n^n$ be compartite square matrices with constant elements expressed in the standard forms

$$A = \begin{bmatrix} A_1, & 0, & \dots & 0 \\ 0, & A_2, & \dots & 0 \\ \dots & \dots & \dots & \dots \\ 0, & 0, & \dots & A_r \end{bmatrix}_{a_1, a_2, \dots a_r}^{a_1, a_2, \dots a_r}, \quad B = \begin{bmatrix} B_1, & 0, & \dots & 0 \\ 0, & B_2, & \dots & 0 \\ \dots & \dots & \dots & \dots \\ 0, & 0, & \dots & B_s \end{bmatrix}_{\beta_1, \beta_2, \dots \beta_s}^{\beta_1, \beta_2, \dots \beta_s},$$

where the successive parts are all square matrices; and let $X=[x]_m^n$ be a compound matrix expressed in the form

$$X=\begin{bmatrix} X_{11}, & X_{12}, & \dots & X_{1s} \\ X_{21}, & X_{22}, & \dots & X_{2s} \\ \dots & \dots & \dots & \dots \\ X_{r1}, & X_{r2}, & \dots & X_{rs} \end{bmatrix}_{\alpha_1,\,\alpha_2,\,\dots\,\alpha_r}^{\beta_1,\,\beta_2,\,\dots\,\beta_s} \quad . \quad \dots\dots\dots\dots\dots(\text{D})$$

Then because the matrix equation $AXB=\epsilon X$ is equivalent to the rs matrix equations such as

$$A_i X_{ij} B_j = \epsilon X_{ij},$$

we have the following second theorem:

Theorem II. *If $\epsilon=\pm 1$, the compound matrix X in* (D) *is an invariant $\{A, \epsilon B\}$ when and only when for each of the values* 1, 2, ... r *of i and each of the values* 1, 2, ... s *of j the constituent X_{ij} is an invariant $\{A_i, \epsilon B_j\}$; and X is then a general invariant $\{A, \epsilon B\}$ when and only when X_{ij} is always a general invariant $\{A_i, \epsilon B_j\}$, and the arbitrary parameters of the rs invariants X_{ij} are all independent.*

Theorems I and II reduce the construction of general invariant transformands to the construction of the general invariant transformands of pairs of unipotent square matrices, to which any special forms we please may be ascribed.

Ex. v. It will be obvious that Theorem II remains true when the non-zero elements of A and B are rational integral functions of certain independent scalar variables.

3. *Zero and non-zero general invariants.*

General principles applicable in the determination of the general invariants of pairs of simple square ante-slopes by direct methods will be described in Chapter XXXVII. The results given in Exs. vi and vii below are particular cases of more general theorems obtained in that chapter. It is shown in Ex. vi that if A and B are two unipotent or two unilatent simple square ante-slopes whose latent roots are α and β, in particular if they are two simple or two unilatent canonical square matrices whose latent roots are α and β, the general invariant $\{A, B\}$ is a non-zero matrix when and only when $\alpha\beta=1$; and we are merely expressing this result in another form when we say that the general invariant $\{A, -B\}$ is a non-zero matrix when and only when $\alpha\beta=-1$. From these facts we deduce two useful theorems.

Theorem III a. *If A and B are any two unipotent or any two unilatent square matrices with constant elements whose latent roots are α and β, and if $\epsilon=\pm 1$, then the general invariant $\{A, \epsilon B\}$ is a non-zero matrix when and only when $\alpha\beta=\epsilon$.*

For if **A** and **B** are canonical reduced forms of A and B, it follows from Theorem I that the general invariant $X=\text{inv}\,\{A, \epsilon B\}$ has the same rank as the general invariant $\mathbf{X} = \text{inv}\{\mathbf{A}, \epsilon\mathbf{B}\}$, and is a non-zero matrix when and only when $\mathbf{X}$ is a non-zero matrix, i.e. when and only when $\alpha\beta = \epsilon$.

Theorem III b. *If A and B are any two square matrices with constant elements, and if $\epsilon = \pm 1$, then the general invariant $\{A, \epsilon B\}$ is a non-zero matrix when and only when we can select latent roots α, β of A, B such that $\alpha\beta = \epsilon$.*

From Theorem II and Theorem III *a* we see that this is true whenever A and B are standard canonical square matrices; and it follows from Theorem I that it is true in all cases. The rank of the general invariant is given in Theorem I of § 259 and Ex. i of § 260. As a particular case of Theorem III *b* we see that:

The general invariants $\{A, B\}$ and $\{A, -B\}$ are zero matrices whenever A or B has only zero latent roots.

Ex. vi. *Cases in which the general invariant of two unilatent simple square ante-slopes is a zero or non-zero matrix.*

Let A and B be two given unilatent simple square ante-slopes of type $\{\pi, \pi\}$ and of orders m and n whose latent roots are α and β respectively, so that we can put

$$A = \begin{bmatrix} a_{11} & a_{12} & \dots & a_{1m} \\ 0 & a_{22} & \dots & a_{2m} \\ \dots & \dots & \dots & \dots \\ 0 & 0 & \dots & a_{mm} \end{bmatrix}, \quad B = \begin{bmatrix} b_{11} & b_{12} & \dots & b_{1n} \\ 0 & b_{22} & \dots & b_{2n} \\ \dots & \dots & \dots & \dots \\ 0 & 0 & \dots & b_{nn} \end{bmatrix}, \quad \dots\dots\dots\dots(5)$$

where $$a_{11} = a_{22} = \dots = a_{mm} = \alpha, \quad b_{11} = b_{22} = \dots = b_{nn} = \beta;$$

and with the standard double-suffix notation let $X = [x]_m^n$ be a matrix which satisfies the equation

$$X = AXB. \quad \dots\dots\dots\dots(6)$$

When we equate the element x_{ij} on the left in (6) to the corresponding element of the product matrix on the right, we see that (6) is equivalent to the mn scalar equations such as the equation

$$x_{ij} = [a_{ii} a_{i,i+1} \dots a_{i,m-1} a_{im}] \begin{bmatrix} x_{i1} & x_{i2} & \dots & x_{i,j-1} & x_{ij} \\ x_{i+1,1} & x_{i+1,2} & \dots & x_{i+1,j-1} & x_{i+1,j} \\ \dots & \dots & \dots & \dots & \dots \\ x_{m-1,1} & x_{m-1,2} & \dots & x_{m-1,j-1} & x_{m-1,j} \\ x_{m1} & x_{m2} & \dots & x_{m,j-1} & x_{mj} \end{bmatrix} \begin{bmatrix} b_{1j} \\ b_{2j} \\ \vdots \\ b_{j-1,j} \\ b_{jj} \end{bmatrix} \quad \dots(\text{E})$$

in which $$a_{ii} = \alpha, \quad b_{jj} = \beta.$$

We will call (E) the equation corresponding to the element x_{ij}, and $j - i$ the difference-weight of the element x_{ij}. To facilitate subsequent applications we will divide all the mn equations such as (E) into four classes.

Corresponding to the 'basical' element x_{m1} common to the first vertical row and the last horizontal row of X,—the only element of difference-weight $1 - m$,—we have the single equation

$$(\alpha\beta - 1) \,.\, x_{m1} = 0. \quad \dots\dots\dots\dots(\text{E}_0)$$

Corresponding to any element x_{i1} of X lying in the first vertical row but not in the last horizontal row, we have the equation

$$(\alpha\beta-1).x_{i1}+\beta(a_{i,i+1}x_{i+1,1}+a_{i,i+2}x_{i+2,1}+\ldots+a_{im}x_{m1})=0, \quad\ldots\ldots\ldots\ldots(E_1)$$
$$(i=1, 2, \ldots m-1).$$

Corresponding to any element x_{mj} of X lying in the last horizontal row but not in the first vertical row, we have the equation

$$(\alpha\beta-1).x_{mj}+\alpha(b_{j-1,j}x_{m,j-1}+b_{j-2,j}x_{m,j-2}+\ldots+b_{1j}x_{m1})=0. \quad\ldots\ldots\ldots(E_2)$$
$$(j=2, 3, \ldots n).$$

Corresponding to any element x_{ij} of X lying neither in the first vertical row nor in the last horizontal row, we have the equation

$$(\alpha\beta-1).x_{ij}+\alpha b_{j-1,j}x_{i,j-1}+\beta a_{i,i+1}x_{i+1,j}+f_{ij}=0, \quad\ldots\ldots\ldots\ldots\ldots\ldots(E_3)$$
$$(i=1, 2, \ldots m-1; j=2, 3, \ldots n),$$

where f_{ij} is what the right-hand side of (E) becomes when the three 'apical' elements x_{ij}, $x_{i,j-1}$, $x_{i+1,j}$ are put equal to 0.

We will for the present merely examine the circumstances under which X is necessarily a zero or non-zero matrix; and for this purpose it is sufficient to observe that corresponding to any element x_{ij} of X other than x_{m1} we have an equation of the form

$$(\alpha\beta-1).x_{ij}=c_{ij}, \quad\ldots\ldots\ldots\ldots\ldots\ldots\ldots\ldots\ldots\ldots\ldots\ldots\ldots(7)$$

where c_{ij} is a homogeneous linear function of elements of X having smaller difference-weights than x_{ij}.

Case I. When $\alpha\beta=1$.

If we put all elements of X equal to 0 except x_{1n}, the equation (6) is reduced to the equation

$$x_{1n}=\alpha\beta.x_{1n} \quad \text{or} \quad (\alpha\beta-1).x_{1n}=0,$$

which will be satisfied by all values of x_{1n} when $\alpha\beta=1$. Or when $\alpha\beta=1$, the element x_{1n} does not occur in any of the equations such as (E_0), (E_1), (E_2), (E_3), and can therefore have an arbitrary value in X. We conclude that:

When $\alpha\beta=1$, *the general invariant* $\{A, B\}$ *is a non-zero matrix.*

Case II. When $\alpha\beta\neq 1$.

The equation (E_0) shows that in this case we must have $x_{m1}=0$. Suppose that all the elements of X having difference-weights $1-m$, $2-m$, ... $\kappa-1$ have been proved to be equal to 0, where $\kappa \ngtr n-1$; and let x_{ij} be any element of X having difference-weight κ. Then the equation (7) shows that we must have $x_{ij}=0$. We conclude that:

When $\alpha\beta\neq 1$, *the general invariant* $\{A, B\}$ *is a zero matrix.*

Because a unilatent simple square ante-slope of type $\{\pi', \pi'\}$ can be regarded as a symmetric derangement of a simple square ante-slope of type $\{\pi, \pi\}$ having the same latent root, it follows from Ex. iii that the results obtained in Cases I and II remain true when A and B are any two unilatent simple square ante-slopes whose latent roots are α and β.

Ex. vii. *If A and B are unipotent simple square ante-slopes of type $\{\pi, \pi\}$ having latent roots α and β respectively such that* $\alpha\beta=1$, *the general invariant* $X=inv\{A, B\}$ *is a simple slope.*

We may suppose A and B to be the matrices of Ex. vi in which the elements a_{12}, a_{23}, ... $a_{m-1,m}$ and b_{12}, b_{23}, ... $b_{n-1,n}$ of difference-weight 1 are all different from 0. Then we have to show that in Ex. vi all elements of X having difference-weights less than 0 or less than $n-m$ must be equal to 0. The equations to be satisfied by the elements of X are

$$a_{i,i+1}\, x_{i+1,1}+a_{i,i+2}\, x_{i+2,1}+\ldots+a_{im}\, x_{m1}=0, \quad (i=1, 2, \ldots m-1)\,; \quad \ldots\ldots\ldots\ldots(\mathrm{e}_1)$$
$$b_{j-1,j}\, x_{m,j-1}+b_{j-2,j}\, x_{m,j-2}+\ldots+b_{1j}\, x_{m1}=0, \quad (j=2, 3, \ldots n)\,; \quad \ldots\ldots\ldots\ldots(\mathrm{e}_2)$$
$$ab_{j-1,j}\, x_{i,j-1}+\beta a_{i,i+1}\, x_{i+1,j} \quad +f_{ij} \quad =0, \quad (i=1, 2, \ldots m-1\,; j=2, 3, \ldots n)\,; \ldots(\mathrm{e}_3)$$

where f_{ij} is defined as in (E_3). If all the elements of A and B except those of difference-weights 0 and 1 were equal to 0, we should have

$$f_{ij}=a_{i,i+1}\, b_{j-1,j}\, x_{i+1,j-1}.$$

In the equation (e_3) we can take $x_{i,j-1}$ and $x_{i+1,j}$ to be any two consecutive elements of X having the same difference-weight κ, where the possible values of κ are those consistent with the conditions

$$\kappa \nless 2-m, \quad \kappa \ngtr m-2.$$

The equation shows that if all elements of X having difference-weight $\kappa-1$ and any one element having difference-weight κ are equal to 0, then all elements having difference-weight κ must be equal to 0.

In the equation (e_1) we can take $x_{i+1,1}$ to be any element of the first vertical row of X having difference-weight κ such that

$$\kappa<0, \quad \kappa \nless 1-m,$$

the only excepted element of that row being x_{11}.

In the equation (e_2) we can take $x_{m,j-1}$ to be any element of the last horizontal row of X having difference-weight κ such that

$$\kappa<n-m, \quad \kappa \nless 1-m,$$

the only excepted element of that row being x_{mn}.

By putting $i=m-1$ in (e_1) or $j=2$ in (e_2), we see that we must have

$$x_{m1}=0.$$

First let κ be any integer greater than $1-m$ but less than 0, and suppose that all elements of X having difference-weights $1-m, 2-m, \ldots \kappa-1$ have been proved to be equal to 0. Then the equation (e_1) in which $i=-\kappa$ shows that the element of the first vertical row of X having difference-weight κ is equal to 0; and it follows from (e_3) that all elements of X having difference-weight κ must be equal to 0. We conclude that all elements of X having difference-weights less than 0 must be equal to 0.

Next let κ be any integer greater than $1-m$ but less than $n-m$, and suppose that all elements of X having difference-weights $1-m, 2-m, \ldots \kappa-1$ have been proved to be equal to 0. Then the equation (e_2) in which $j=m+1+\kappa$ shows that the element of the last horizontal row of X having difference-weight κ is equal to 0; and it follows from (e_3) that all elements of X having difference-weight κ must be equal to 0. We conclude that all elements of X having difference-weights less than $n-m$ must be equal to 0.

Ex. viii. *Equations determining the general invariant* $X=inv\,\{A_1, B_1\}$ *when* A_1 *and* B_1 *are the unipotent simple square ante-slopes*

$$A_1=[1]_m^m+\begin{bmatrix}0, & 1\\ 0, & 0\end{bmatrix}_{m-1,1}^{1,m-1}, \quad B_1=[1]_n^n+\begin{bmatrix}0, & 1\\ 0, & 0\end{bmatrix}_{n-1,1}^{1,n-1}.$$

These are the matrices A and B of Ex. vi in which all the elements of difference-weights 0 and 1 are equal to 1, and all other elements are equal to 0. For them the equations (e_1), (e_2), (e_3) of Ex. vii are

$$x_{i+1,1}=0, \quad (i=1, 2, \ldots m-1)\,; \quad x_{m,j-1}=0, \quad (j=2, 3, \ldots n)\,; \ldots\ldots\ldots\ldots(8)$$
$$x_{i,j-1}+x_{i+1,j}+x_{i+1,j-1}=0, \quad (i=1, 2, \ldots m-1\,; j=2, 3, \ldots n). \ldots\ldots\ldots\ldots(9)$$

In this case X has the zero elements given by (8), and its other elements are arbitrary subject to the condition that:

The sum of two consecutive elements having the same difference-weight κ *and the contiguous element of smaller difference-weight* $\kappa-1$ *is always equal to* 0.

We know from Ex. vii, and it would follow from (8) and (9), that X is a simple slope of type $\{\pi, \pi\}$, i.e. we must have

$$x_{ij}=0 \quad \text{whenever} \quad j-i<0 \quad \text{or} \quad j-i<n-m.$$

Consequently X is the most general simple slope of type $\{\pi, \pi\}$ and of the prescribed class which can be constructed when its parametric elements satisfy the condition just mentioned.

In the particular case when $m=n=r$, we have $x_{ij}=0$ whenever $j-i<0$, and the elements with positive difference-weights must be so chosen as to satisfy the equations

$$x_{i+1,j+1}=-x_{ij}-x_{i+1,j}, \qquad (i=1, 2, \dots r-1;\ j=i, i+1, \dots r-1),$$

those in which $j-i$ has a given value κ (where $\kappa=0, 1, 2, \dots r-2$) being obtained by giving to i the values $1, 2, \dots r-1-\kappa$, and those in which $j-i=0$ being

$$x_{i+1,i+1}=-x_{ii}, \qquad (i=1, 2, \dots r-1).$$

Ex. ix. *Derivation of general invariants from general commutants.*

All properties of general invariants can be deduced from properties of general commutants. For when two square matrices A and B with constant elements have been reduced by isomorphic transformations to the respective forms

$$A=\begin{bmatrix}\mathbf{A}, & 0\\ 0, & P\end{bmatrix}_{m,p}^{m,p}, \quad B=\begin{bmatrix}\mathbf{B}, & 0\\ 0, & Q\end{bmatrix}_{n,q}^{n,q},$$

where P and Q are unilatent square matrices with zero latent roots, and $\mathbf{A}$ and $\mathbf{B}$ are undegenerate square matrices (having no zero latent roots), a general invariant $\{A, B\}$ is a matrix X having the form

$$X=\begin{bmatrix}\mathbf{X}, & 0\\ 0, & 0\end{bmatrix}_{m,p}^{n,q},$$

where $\mathbf{X}$ is a general invariant $\{\mathbf{A}, \mathbf{B}\}$, which can also be interpreted to be a general commutant $\{\mathbf{A}^{-1}, \mathbf{B}\}$ or $\{\mathbf{A}, \mathbf{B}^{-1}\}$. The matrix X has the same rank and the same arbitrary parameters as $\mathbf{X}$, its rank being equal to the order of the greatest common canonical of $\mathbf{A}$ and $\mathbf{B}^{-1}$.

4. *Bi-canonical square matrices.*

If α is any *non-zero* scalar number, we will define the square matrix

$$A=\alpha.[1]_m^m+\alpha.\begin{bmatrix}0, & 1\\ 0, & 0\end{bmatrix}_{m-1,1}^{1,m-1}=\begin{bmatrix}\alpha & \alpha & 0 & \dots & 0 & 0\\ 0 & \alpha & \alpha & \dots & 0 & 0\\ 0 & 0 & \alpha & \dots & 0 & 0\\ \dots & & & & & \\ 0 & 0 & 0 & \dots & \alpha & \alpha\\ 0 & 0 & 0 & \dots & 0 & \alpha\end{bmatrix} \quad \dots\dots\dots(10)$$

to be the *simple bi-canonical square matrix* of order m whose latent root is α. It is the simple square ante-slope A of Ex. vi in the special case when

$$a_{11}=a_{22}=\dots=a_{mm}=\alpha, \quad a_{12}=a_{23}=\dots=a_{m-1,m}=\alpha,$$

and all other elements are 0's; or it is the simple square ante-slope of order m and of type $\{\pi, \pi\}$ in which all the elements of difference-weights

0 and 1 are equal to α, and every other element is equal to 0. It is a unipotent square matrix, its characteristic matrix $A(\lambda)$ having the single potent divisor

$$(\lambda - \alpha)^m;$$

and every unipotent square matrix of order m whose latent root is α can be converted into or derived from A by an isomorphic transformation.

More generally we will define a *bi-canonical square matrix* to be a compartite matrix of standard form in which every simple part is a simple bi-canonical or a simple canonical square matrix according as it has a non-zero or a zero latent root. Thus a simple part of order r whose latent root is α will always be the square matrix

$$\alpha . [1]_r^r + \alpha . \begin{bmatrix} 0, & 1 \\ 0, & 0 \end{bmatrix}_{r-1, 1}^{1, r-1} \quad \text{or} \quad \begin{bmatrix} 0, & 1 \\ 0, & 0 \end{bmatrix}_{r-1, 1}^{1, r-1}$$

according as $\alpha \neq 0$ or $\alpha = 0$. A *standard* bi-canonical square matrix is one which is expressed as a compartite matrix of standard form having one unilatent super-part corresponding to each distinct latent root. We know from § 228 that every square matrix with constant numerical elements can be converted into or derived from a bi-canonical square matrix, which may if we please be a standard bi-canonical square matrix, by isomorphic transformations. The simplest methods of actually determining such transformations are described in Chapter XXXVII.

It is implied in the more general definitions that a simple bi-canonical square matrix whose latent root is 0 must be interpreted to be a simple canonical square matrix whose latent root is 0. A simple (or unilatent) bi-canonical square matrix whose latent root is 1 is necessarily a simple (or unilatent) canonical square matrix. A simple bi-canonical square matrix of order 1 whose latent root is α must be interpreted to be the one-element matrix $[\alpha]$ both when $\alpha \neq 0$ and when $\alpha = 0$.

The ordinary constructions for general invariants are derived from those furnished by Exs. vii and viii for the general invariants of simple bi-canonical (or simple canonical) square matrices whose latent roots are 1.

If A and B are the simple bi-canonical square matrices of orders m and n whose latent roots are α and β, we can put

$$A = \alpha A_1, \quad B = \beta B_1, \quad \dots\dots\dots\dots(11)$$

where A_1 and B_1 are the simple bi-canonical (or simple canonical) square matrices of orders m and n whose latent roots are 1. Let A', B' and A_1', B_1' be the conjugates of A, B and A_1, B_1; and let $\epsilon = \pm 1$. We know that the general invariants

$$X = \text{inv}\{A, \epsilon B\}, \quad X' = \text{inv}\{A', \epsilon B'\}, \quad Y = \text{inv}\{A', \epsilon B\}, \quad Y' = \text{inv}\{A, \epsilon B'\}$$

are zero matrices when $\alpha\beta \neq \epsilon$; whilst when $\alpha\beta = \epsilon$, it follows from the equations (11) that for both values of ϵ they are the same as the general invariants

$$\xi = \text{inv}\{A_1, B_1\},\quad \xi' = \text{inv}\{A_1', B_1'\},\quad \eta = \text{inv}\{A_1', B_1\},\quad \eta' = \text{inv}\{A_1, B_1'\}.$$

Again if J_m and J_n are the simple reversants of orders m and n, it will be obvious that A, B or A_1, B_1 can be converted into their conjugates by the symmetric transformations

$$J_m A J_m = A',\quad J_n B J_n = B' \quad\text{or}\quad J_m A_1 J_m = A_1',\quad J_n B_1 J_n = B_1',$$

and that J_m and J_n are respectively undegenerate symmetric contra-commutants

$$\{A', A\},\ \{A_1', A_1\} \quad\text{and}\quad \{B, B'\},\ \{B_1, B_1'\}.$$

Therefore by Ex. iii of this article or Ex. iii of § 254 the general invariants mentioned above can always be so chosen as to be connected by the relations

$$X' = J_m X J_n,\quad Y = J_m X,\quad Y' = X J_n;\quad \xi' = J_m \xi J_n,\quad \eta = J_m \xi,\quad \eta' = \xi J_n;$$

all the four in each set being completely known when any one of them is known. But it has been shown in Exs. vii and viii that ξ is a simple slope of type $\{\pi, \pi\}$. Consequently the four general invariants ξ, ξ', η, η' are simple slopes of the respective types

$$\{\pi, \pi\},\ \{\pi', \pi'\},\ \{\pi', \pi\},\ \{\pi, \pi'\}$$

which can be so chosen as to be correlated by simple reversals. They will all be completely known when ξ has been evaluated by solving the scalar equations of Ex. viii; and they are precisely described in the next two theorems.

Considering first the case in which $m = n = r$, and putting $A_1 = B_1 = \Omega$, we have the following theorem in which ${}^iH_\kappa$ is an integer uniquely determined for all integral values of the arguments i and κ by the definition

$${}^iH_\kappa = \frac{(i+1)(i+2)\dots(i+\kappa)}{1.2.\dots\kappa} = \binom{i+\kappa}{\kappa} \quad\text{when}\quad \kappa > 0;\ \dots(12)$$

and by the interpretations

$${}^iH_0 = 1;\ {}^iH_\kappa = 0 \quad\text{when}\quad \kappa < 0.\ \dots\dots\dots\dots\dots\dots(13)$$

Theorem IV a. *If* $\quad \Omega = [\omega]_r^r = [1]_r^r + \begin{bmatrix} 0, & 1 \\ 0, & 0 \end{bmatrix}_{r-1,1}^{1,r-1}$

is the simple bi-canonical (or simple canonical) square matrix of order r whose latent root is 1, *and if Ω' is its conjugate, the four general invariants*

$$\xi = inv\{\Omega, \Omega\},\quad \xi' = inv\{\Omega', \Omega'\},\quad \eta = inv\{\Omega', \Omega\},\quad \eta' = inv\{\Omega, \Omega'\}$$

can be taken to be the simple square slopes

$$\begin{bmatrix} \xi_{11} & \xi_{12} \dots \xi_{1r} \\ 0 & \xi_{22} \dots \xi_{2r} \\ \dots & \dots \\ 0 & 0 \dots \xi_{rr} \end{bmatrix},\quad \begin{bmatrix} \xi_{rr} \dots 0 & 0 \\ \dots & \dots \\ \xi_{2r} \dots \xi_{22} & 0 \\ \xi_{1r} \dots \xi_{12} & \xi_{11} \end{bmatrix},\quad \begin{bmatrix} 0 & 0 \dots \xi_{rr} \\ \dots & \dots \\ 0 & \xi_{22} \dots \xi_{2r} \\ \xi_{11} & \xi_{12} \dots \xi_{1r} \end{bmatrix},\quad \begin{bmatrix} \xi_{1r} \dots \xi_{12} & \xi_{11} \\ \xi_{2r} \dots \xi_{22} & 0 \\ \dots & \dots \\ \xi_{rr} \dots 0 & 0 \end{bmatrix}$$

in each of which the parametric elements are homogeneous linear functions of r arbitrary parameters $\xi_0, \xi_1, \xi_2, \ldots \xi_{r-1}$, those of difference-weight κ being given by the formula

$$\xi_{ij} = (-1)^{i-1} . \{\xi_\kappa - {}^{i-2}H_1 . \xi_{\kappa-1} + {}^{i-2}H_2 . \xi_{\kappa-2} - \ldots + (-1)^\kappa . {}^{i-2}H_\kappa . \xi_0\}, \quad \ldots\text{(F)}$$
$$(j - i = \kappa;\ i = 1, 2, \ldots r - \kappa;\ \kappa = 0, 1, 2, \ldots r - 1).$$

For the parametric elements of difference-weights 0, 1 the formula (F) becomes

$$\xi_{ii} = (-1)^{i-1} . \xi_0 \qquad , (i = 1, 2, \ldots r - 1); \quad \ldots\ldots(\text{F}_0)$$
$$\xi_{i,i+1} = (-1)^{i-1} . \{\xi_1 - {}^{i-2}H_1 . \xi_0\}, \ (i = 1, 2, \ldots r - 2); \quad \ldots\ldots(\text{F}_1)$$

and the formula includes (when $i = 1$) the equations

$$\xi_{11} = \xi_0, \quad \xi_{12} = \xi_1, \quad \xi_{13} = \xi_2, \ldots \xi_{1r} = \xi_{r-1}.$$

It is not to be implied that the arbitrary parameters are the same in any two different invariants.

By Ex. viii the equations to be satisfied by the parametric elements consist of the $r - 1$ sets

$$\xi_{ij} + \xi_{i+1,j+1} = -\xi_{i+1,j}, \quad (j - i = \kappa;\ i = 1, 2, \ldots r - 1 - \kappa), \quad \ldots(14)$$

in which $\kappa = 0, 1, 2, \ldots r - 2$; the first set in which $\kappa = 0$ being

$$\xi_{ii} + \xi_{i+1,i+1} = 0, \quad (i = 1, 2, \ldots r - 1). \quad \ldots\ldots\ldots\ldots\ldots(15)$$

We can choose one element in each parametric diagonal line to be an arbitrary parameter; and the equations of the sets in which $\kappa = 0, 1, 2, \ldots r - 2$ can then be completely and uniquely solved in succession for the remaining parametric elements of difference-weights $0, 1, 2, \ldots r - 2$. The formula (F) gives the general solution of all the equations when we choose

$$\xi_{11}, \xi_{12}, \ldots \xi_{1r} \text{ to be arbitrary parameters } \xi_0, \xi_1, \ldots \xi_{r-1}.$$

For the formula is obviously true for small difference-weights and for all elements in which $i = 1$; and if we assume that it is true for all parametric elements of difference-weight $\kappa - 1$ and the element ξ_{ij} of difference-weight κ, the equation (14) and the general formula

$${}^iH_{\kappa+1} + {}^{i+1}H_\kappa = {}^{i+1}H_{\kappa+1},$$

which is true for all integral values of i and κ, show that it must also be true for the element $\xi_{i+1,j+1}$ of difference-weight κ.

Passing to the general case in which the two integers m and n are not necessarily equal, we will suppose r to be the smaller of them. Since the general invariant ξ is known by Ex. viii to be a simple slope of type $\{\pi, \pi\}$, we can put:

$$\xi = [0, \xi]_r^{n-r, r} \text{ when } m = r \not> n; \quad \xi = \begin{bmatrix} \xi \\ 0 \end{bmatrix}_{r, m-r}^{r} \text{ when } n = r \not> m;$$

and when these forms of ξ are used, the equation

$$A_1 \xi B_1 = \xi$$

can be reduced by cancellations of zero passive rows with corresponding passive rows and cancellations of corresponding active rows to the equivalent equation

$$[\omega]_r^r [\xi]_r^r [\omega]_r^r = [\xi]_r^r,$$

which shows that the effective constituent $[\xi]_r^r$ of ξ is a general invariant $\{\Omega, \Omega\}$ constructed as in Theorem IV *a*. Accordingly we have the following generalisation of Theorem IV *a*.

Theorem IV b. *If* $A_1 = [1]_m^m + \begin{bmatrix} 0, & 1 \\ 0, & 0 \end{bmatrix}_{m-1,\,1}^{1,\,m-1}$, $B_1 = [1]_n^n + \begin{bmatrix} 0, & 1 \\ 0, & 0 \end{bmatrix}_{n-1,\,1}^{1,\,n-1}$

are the simple bi-canonical (or simple canonical) square matrices of orders m, n whose latent roots are 1, *and if* A_1', B_1' *are their conjugates, the four general invariants*

$$\xi = inv\,\{A_1, B_1\}, \quad \xi' = inv\,\{A_1', B_1'\}, \quad \eta = inv\,\{A_1', B_1\}, \quad \eta' = inv\,\{A_1, B_1'\}$$

can be taken to be the simple slopes

$$[0, \xi]_r^{n-r,\,r}, \quad [\xi', 0]_r^{r,\,n-r}, \quad [0, \eta]_r^{n-r,\,r}, \quad [\eta', 0]_r^{r,\,n-r} \quad \textit{when} \quad m = r \not> n,$$

$$\textit{or} \quad \begin{bmatrix} \xi \\ 0 \end{bmatrix}_{r,\,m-r}^r, \quad \begin{bmatrix} 0 \\ \xi' \end{bmatrix}_{m-r,\,r}^r, \quad \begin{bmatrix} 0 \\ \eta \end{bmatrix}_{m-r,\,r}^r, \quad \begin{bmatrix} \eta' \\ 0 \end{bmatrix}_{r,\,m-r}^r \quad \textit{when} \quad n = r \not> m,$$

in which the effective constituents $[\xi]_r^r$, $[\xi']_r^r$, $[\eta]_r^r$, $[\eta']_r^r$ *are the simple square slopes* ξ, ξ', η, η' *of Theorem IV a.*

Both in Theorem IV *a* and in Theorem IV *b* the general invariants ξ, ξ', η, η' are simple slopes of a known class and of the respective types

$$\{\pi, \pi\}, \ \{\pi', \pi'\}, \ \{\pi', \pi\}, \ \{\pi, \pi'\}$$

whose parametric elements are subject only to the conditions represented by (14), i.e. to the *law of construction* that:

the sum of two consecutive elements of the same difference-weight κ *and the contiguous element of smaller difference-weight* $\kappa - 1$ *is always equal to* 0.

One element in each of the r effective parametric diagonal lines can be regarded as an arbitrary parameter; and the other parametric elements of difference-weights 0, 1, 2, ... $r-1$ can then be determined in succession by the law of construction.

Other forms of these general invariant transformands will be given in Exs. i and ii of § 257.

Ex. x. When $r=6$, the general invariant $\xi = \text{inv}\,\{\Omega, \Omega\}$ of Theorem IV *a* is the simple square slope

$$\xi=\begin{bmatrix} \xi_0, & \xi_1, \xi_2 & , & \xi_3 & , & \xi_4 & , & \xi_5 \\ 0\ , & -\xi_0, \xi_0-\xi_1, & - & \xi_0+\xi_1-\xi_2, & & \xi_0-\ \xi_1+\xi_2-\xi_3, & - & \xi_0+\ \xi_1-\ \xi_2+\xi_3-\xi_4 \\ 0\ , & 0\ , \xi_0 & , & -2\xi_0+\xi_1 & , & 3\xi_0-2\xi_1+\xi_2 & , & -4\xi_0+3\xi_1-2\xi_2+\xi_3 \\ 0\ , & 0\ , 0 & , & -\ \xi_0 & , & 3\xi_0-\ \xi_1 & , & -6\xi_0+3\xi_1-\ \xi_2 \\ 0\ , & 0\ , 0 & , & 0 & , & \xi_0 & , & -4\xi_0+\ \xi_1 \\ 0\ , & 0\ , 0 & , & 0 & , & 0 & , & -\ \xi_0 \end{bmatrix}$$

$$=\begin{bmatrix} 1, & 0, & 0, & 0, & 0, & 0 \\ 0, & -1, & 1, & -1, & 1, & -1 \\ 0, & 0, & 1, & -2, & 3, & -4 \\ 0, & 0, & 0, & -1, & 3, & -6 \\ 0, & 0, & 0, & 0, & 1, & -4 \\ 0, & 0, & 0, & 0, & 0, & -1 \end{bmatrix} \begin{bmatrix} \xi_0 & \xi_1 & \xi_2 & \xi_3 & \xi_4 & \xi_5 \\ 0 & \xi_0 & \xi_1 & \xi_2 & \xi_3 & \xi_4 \\ 0 & 0 & \xi_0 & \xi_1 & \xi_2 & \xi_3 \\ 0 & 0 & 0 & \xi_0 & \xi_1 & \xi_2 \\ 0 & 0 & 0 & 0 & \xi_0 & \xi_1 \\ 0 & 0 & 0 & 0 & 0 & \xi_0 \end{bmatrix} = PX,$$

where P is a particular undegenerate invariant $\{\Omega, \Omega\}$, and X is a general commutant $\{\Omega, \Omega\}$. Every parametric element of ξ or P in a horizontal row below the first is minus the sum of the preceding element in the same horizontal row and the element immediately above that preceding element.

Ex. xi. The general invariant ξ of Theorem IVa can of course be regarded as a general commutant $\{\Omega^{-1}, \Omega\}$ or $\{\Omega, \Omega^{-1}\}$. When $r=6$, we have

$$\Omega=\begin{bmatrix} 1 & 1 & 0 & 0 & 0 & 0 \\ 0 & 1 & 1 & 0 & 0 & 0 \\ 0 & 0 & 1 & 1 & 0 & 0 \\ 0 & 0 & 0 & 1 & 1 & 0 \\ 0 & 0 & 0 & 0 & 1 & 1 \\ 0 & 0 & 0 & 0 & 0 & 1 \end{bmatrix}, \quad \Omega^{-1}=\begin{bmatrix} 1, & -1, & 1, & -1, & 1, & -1 \\ 0, & 1, & -1, & 1, & -1, & 1 \\ 0, & 0, & 1, & -1, & 1, & -1 \\ 0, & 0, & 0, & 1, & -1, & 1 \\ 0, & 0, & 0, & 0, & 1, & -1 \\ 0, & 0, & 0, & 0, & 0, & 1 \end{bmatrix},$$

or
$$\Omega=I+\chi, \quad \Omega^{-1}=I-\chi+\chi^2-\chi^3+\chi^4-\chi^5, \quad \text{where } I=[1]_6^6.$$

Ex. xii. The equation $\xi=PX$ in Ex. x illustrates the following theorem which is generalised in Ex. ii of § 259, and which is included in the general theorem of § 239 when the invariant transformands are interpreted as commutants.

If P, P' are particular undegenerate invariant transformands

$$P=inv\ \{\Omega, \Omega\}, \quad P'=inv\ \{\Omega', \Omega'\},$$

and if X, X', Y, Y' are general commutants $\{\Omega, \Omega\}$, $\{\Omega', \Omega'\}$, $\{\Omega', \Omega\}$, $\{\Omega, \Omega'\}$, then the matrices ξ, ξ', η, η' given by the formulae

$$\xi=PX, \quad \xi'=P'X', \quad \eta=P'Y, \quad \eta'=PY'$$

or
$$\xi=XP, \quad \xi'=X'P', \quad \eta=YP, \quad \eta'=Y'P'$$

are general invariant transformands $\{\Omega, \Omega\}$, $\{\Omega', \Omega'\}$, $\{\Omega', \Omega\}$, $\{\Omega, \Omega'\}$.

We can interpret P, P' to be undegenerate commutants $\{\Omega^{-1}, \Omega\}$, $\{\Omega'^{-1}, \Omega'\}$ or undegenerate commutants $\{\Omega, \Omega^{-1}\}$, $\{\Omega', \Omega'^{-1}\}$; and such matrices can certainly be determined because Ω and Ω^{-1} are equicanonical.

Ex. xiii. *Direct determination of the general invariant $X=inv\ \{A, \epsilon B\}$ when $\epsilon=\pm 1$ and*

$$A=a\,.\,[1]_m^m+a\,.\begin{bmatrix}0, & 1\\ 0, & 0\end{bmatrix}_{m-1,1}^{1,m-1}, \quad B=\beta\,.\,[1]_n^n+\beta\,.\begin{bmatrix}0, & 1\\ 0, & 0\end{bmatrix}_{n-1,1}^{1,n-1}.$$

The equation $AXB=\epsilon X$ in which $X=[x]_m^n$ can be put into the form

$$\alpha\beta.\begin{bmatrix}0 & x_{11} & \dots & x_{1,n-1}\\ \dots & \dots & \dots & \dots\\ 0 & x_{m-1,1} & \dots & x_{m-1,n-1}\\ 0 & x_{m1} & \dots & x_{m,n-1}\end{bmatrix}+\alpha\beta.\begin{bmatrix}x_{21} & x_{22} & \dots & x_{2n}\\ \dots & \dots & \dots & \dots\\ x_{m1} & x_{m2} & \dots & x_{mn}\\ 0 & 0 & \dots & 0\end{bmatrix}+\alpha\beta.\begin{bmatrix}0 & x_{21} & \dots & x_{2,n-1}\\ \dots & \dots & \dots & \dots\\ 0 & x_{m1} & \dots & x_{m,n-1}\\ 0 & 0 & \dots & 0\end{bmatrix}$$

$$=(\epsilon-\alpha\beta)\begin{bmatrix}x_{11} & x_{12} & \dots & x_{1n}\\ \dots & \dots & \dots & \dots\\ x_{m-1,1} & x_{m-1,2} & \dots & x_{m-1,n}\\ x_{m1} & x_{m2} & \dots & x_{mn}\end{bmatrix}.$$

By equating successive elements of successive vertical rows when $m \nless n$ or successive elements of successive horizontal rows when $n \nless m$ on both sides of this equation, starting with the rows through the bottom left-hand element and with the bottom or left-hand elements of the equated rows in each case it will be seen that:

(i) If $\alpha\beta \neq \epsilon$, every element of X must be equal to 0.

(ii) If $\alpha\beta=\epsilon$, all elements of X lying below the diagonal through x_{11} or x_{mn} must be equal to 0.

When $\alpha\beta=\epsilon$, the transformed equation for X is obviously equivalent to the scalar equations

$$x_{i1}=0,\ \ (i=1, 2, \dots m);\ \ x_{mj}=0,\ \ (j=1, 2, \dots n-1);$$

$$x_{ij}+x_{i+1,j+1}+x_{i+1,j}=0,\ \ (i=1, 2, \dots m-1;\ j=1, 2, \dots n-1);$$

which are the equations (8), (9) of Ex. viii, and could be used to prove (ii).

It follows that X is a simple slope of type $\{\pi, \pi\}$ having the same structure as ξ in Theorem IV *b*.

Ex. xiv. If in the matrix ξ of Theorem IV *b* we replace every $-$ sign by a $+$ sign, we obtain the general invariant of the pair of square matrices

$$[1]_m^m+\begin{bmatrix}0, & 1\\ 0, & 0\end{bmatrix}_{m-1,1}^{1,m-1},\quad [1]_n^n-\begin{bmatrix}0, & 1\\ 0, & 0\end{bmatrix}_{n-1,1}^{1,n-1}.$$

NOTE 1. *Properties of* ${}^iH_\kappa$ *when* i *and* κ *are integers.*

The definition (12) and the interpretations (13) are

$${}^iH_\kappa=\frac{(i+1)(i+2)\dots(i+\kappa)}{1.2.\dots\kappa}=\binom{i+\kappa}{\kappa} \text{ when } \kappa>0;\ \dots\dots\dots\dots(12)$$

$${}^iH_0=1 \text{ (when } \kappa=0);\ \ {}^iH_\kappa=0 \text{ when } \kappa<0;\ \dots\dots\dots\dots(13)$$

and they ascribe a unique value to ${}^iH_\kappa$ when i and κ are any two integers, positive or negative. Since (12) and (13) are not symmetric in i and κ, the integers ${}^iH_\kappa$ must be distinguished from the associated integers

$${}^iH'_\kappa={}^\kappa H_i$$

which are such that

$${}^iH'_\kappa=\frac{(\kappa+1)(\kappa+2)\dots(\kappa+i)}{1.2.\dots i} \text{ when } i>0;\ \dots\dots\dots\dots(12')$$

$${}^0H'_\kappa=1 \text{ (when } i=0);\ \ {}^iH'_\kappa=0 \text{ when } i<0.\ \dots\dots\dots\dots(13')$$

The following tables give the values of ${}^iH_\kappa$ and ${}^iH'_\kappa$ for the values of i shown at the top and the values of κ shown on the left. Each integer is the sum of the two adjacent integers to the left of and above it.

Values of ${}^iH_\kappa$.

$i=$	-5,	-4,	-3,	-2,	-1,	0,	1,	2,	3,	4,	5
$\kappa=-5$	0	0	0	0	0	0	0	0	0	0	0
$\kappa=-4$	0	0	0	0	0	0	0	0	0	0	0
$\kappa=-3$	0	0	0	0	0	0	0	0	0	0	0
$\kappa=-2$	0	0	0	0	0	0	0	0	0	0	0
$\kappa=-1$	0	0	0	0	0	0	0	0	0	0	0
$\kappa=\ 0$	1	1	1	1	1	1	1	1	1	1	1
$\kappa=\ 1$	-4	-3	-2	-1	0	1	2	3	4	5	6
$\kappa=\ 2$	6	3	1	0	0	1	3	6	10	15	21
$\kappa=\ 3$	-4	-1	0	0	0	1	4	10	20	35	56
$\kappa=\ 4$	1	0	0	0	0	1	5	15	35	70	126
$\kappa=\ 5$	0	0	0	0	0	1	6	21	56	126	252

Values of ${}^iH'_\kappa$.

$i=$	-5,	-4,	-3,	-2,	-1,	0,	1,	2,	3,	4,	5
$\kappa=-5$	0	0	0	0	0	1	-4	6	-4	1	0
$\kappa=-4$	0	0	0	0	0	1	-3	3	-1	0	0
$\kappa=-3$	0	0	0	0	0	1	-2	1	0	0	0
$\kappa=-2$	0	0	0	0	0	1	-1	0	0	0	0
$\kappa=-1$	0	0	0	0	0	1	0	0	0	0	0
$\kappa=\ 0$	0	0	0	0	0	1	1	1	1	1	1
$\kappa=\ 1$	0	0	0	0	0	1	2	3	4	5	6
$\kappa=\ 2$	0	0	0	0	0	1	3	6	10	15	21
$\kappa=\ 3$	0	0	0	0	0	1	4	10	20	35	56
$\kappa=\ 4$	0	0	0	0	0	1	5	15	35	70	126
$\kappa=\ 5$	0	0	0	0	0	1	6	21	56	126	252

On the supposition that i and κ are positive integers, we have

$${}^iH_\kappa=0 \quad \text{whenever} \quad \kappa<0, \quad \text{or} \quad \kappa+i+1>0>i, \qquad \text{..................(16)}$$

and in no other cases. For ${}^iH_\kappa$ has been interpreted to be 0 when $\kappa<0$; whilst when $\kappa \not< 0$, it is equal to 0 if and only if i satisfies the conditions

$$-1 \not< i \not< -\kappa, \quad \text{which are equivalent to} \quad i<0, \ i+\kappa+1>0,$$

and are incompatible when $\kappa=0$. In particular we have

$${}^{-1}H_\kappa=0 \quad \text{except when} \quad \kappa=0. \qquad \text{..............................(16')}$$

Whenever i and ρ are two integers such that $i+\rho=-(\kappa+1)$, we have

$${}^iH_\kappa=(-1)^\kappa \cdot {}^\rho H_\kappa;$$

for this equation is obviously true when $\kappa>0$, whilst we have

$${}^iH_\kappa={}^\rho H_\kappa=1 \ \text{ when } \ \kappa=0, \quad {}^iH_\kappa={}^\rho H_\kappa=0 \ \text{ when } \ \kappa<0.$$

For positive values of κ each of the two integers ${}^iH_\kappa$, ${}^\rho H_\kappa$ is equal to 0 when and only when

$$i<0, \quad \rho<0.$$

Thus for all integral values of i and κ, positive or negative, we have

$$-{}^iH_\kappa=(-1)^\kappa \,.\, {}^{i-\kappa-1}H_\kappa; \quad \text{.................................(17)}$$

where the two equated integers are different from 0 when and only when κ is positive and the two upper arguments ($-i$, $i-\kappa-1$) have opposite signs, i.e. are not both less than 0.

The definition (12) shows that the functional equation

$${}^{i+1}H_{\kappa+1}={}^{i+1}H_\kappa+{}^iH_{\kappa+1} \quad \text{.................................(18)}$$

is true for all positive and negative integral values of i when $\kappa>0$; and the interpretations (13) can be derived from (18) by putting $\kappa=0, -1, -2, \ldots$ in succession. Consequently (18) is true for all positive and negative integral values of i and κ. From the equation (18), which is symmetric in i and κ, we can deduce the equations

$${}^{i+1}H_{\kappa+r}-{}^{i+1}H_\kappa={}^iH_{\kappa+1}+{}^iH_{\kappa+2}+\ldots+{}^iH_{\kappa+r}, \quad \text{..................(18')}$$

$${}^{i+r}H_{\kappa+1}-{}^iH_{\kappa+1}={}^{i+1}H_\kappa+{}^{i+2}H_\kappa+\ldots+{}^{i+r}H_\kappa, \quad \text{..................(18'')}$$

which are true for all integral values of i and κ when r is any non-zero positive integer.

From (12) and (13) we see that

$${}^iH_\kappa={}^\kappa H_i \quad (\text{or } {}^iH_\kappa={}^iH'_\kappa) \quad \text{.................................(19)}$$

if and only if either $i+\kappa \nless 0$, or both $i<0$ and $\kappa<0$;

i.e.

$${}^iH_\kappa \neq {}^\kappa H_i \quad (\text{or } {}^iH_\kappa \neq {}^iH'_\kappa) \quad \text{.................................(19')}$$

if and only if $i+\kappa<0$, and at the same time either $i \nless 0$ or $\kappa \nless 0$.

We will finally recall the fact that:

When i and κ are both positive integers, the integer ${}^iH_\kappa={}^\kappa H_i$ is the number of terms in a general rational integral function of i variables, or a general homogeneous rational integral function of $i+1$ variables, whose total degree is equal to κ.

NOTE 2. *Solutions of the functional equation* (18) *satisfied by* ${}^iH_\kappa$.

We can regard ${}^iH_\kappa$ as an element $a_{i\kappa}$ of an infinite square matrix A whose middle element is a_{00}. We will here suppose that some of the elements of A are given by definition,—those in which i and κ are both greater than 0 being always given by (12),—and that the others have to be determined in succession by suitable interpretations. It will be evident that we cannot find symmetric interpretations which make (12), (18) and (19) all true universally.

The functional equation (18) makes all elements of A uniquely determinate when we know any one of the following sets of elements:

all elements in which i has a given value x, and one element for every greater value of i, i.e. one for each of the values $x+1, x+2, x+3, \ldots$ of i;

all elements in which κ has a given value y, and one element for every greater value of κ, i.e. one for each of the values $y+1, y+2, y+3, \ldots$ of κ;

all elements in which $i+\kappa$ has a given value w, and one element for every smaller value of $i+\kappa$.

For example the integers ${}^iH_\kappa$ defined in Note 1 form a set of solutions of (18) when ${}^iH_0=1$ for all values of i, and ${}^0H_\kappa=1$ whenever $\kappa>0$; and the integers ${}^iH'_\kappa$ form a set of solutions when ${}^0H_\kappa=1$ for all values of κ, and ${}^iH_0=1$ whenever $i>0$. The mean of these two sets gives a third set of solutions which makes (18) and (19) true universally, but which makes ${}^iH_\kappa$ have half the value given by (12) when $i<0$.

The equations (18) and (19) make ${}^iH_\kappa$ uniquely determinate for all integral values of i and κ if it is known whenever $i=\kappa$. In all symmetric interpretations based on both (18) and (19) we must have

$${}^\tau H_0 = {}^0H_\tau = 1 \qquad \text{when } \tau > 0.$$

If we put ${}^0H_0 = a$, we must also have

$${}^{-1}H_0 = \tfrac{1}{2}a, \quad {}^{-1}H_1 = 1 - a, \quad {}^{-2}H_1 = 1 - \tfrac{3}{2}a;$$

and these values cannot be made to agree with the values

$${}^{-1}H_1 = 0, \quad {}^{-2}H_1 = -1$$

given by (12). Thus we can find symmetric interpretations which make (18) and (19) universally true, and which make (12) true when i and κ are both positive, no matter what values are assigned to ${}^0H_0$, ${}^{-1}H_{-1}$, ...; but we cannot choose them to make (12) true when $i < 0$.

NOTE 3. *Values of ${}^iH_\kappa$ for positive values of i and κ.*

We always have

$${}^iH_\kappa = {}^\kappa H_i = \frac{(i+1)(i+2)\dots(i+\kappa)}{1.2.\dots\kappa} = \frac{(\kappa+1)(\kappa+2)\dots(\kappa+i)}{1.2.\dots i}$$

when i and κ are both greater than 0. It is often convenient to use the interpretations:

$${}^\tau H_0 = {}^0H_\tau = 1 \qquad \text{whenever } \tau \text{ is positive};$$

$${}^iH_\kappa = {}^\kappa H_i = 0 \qquad \text{whenever } i < 0 \text{ or } \kappa < 0;$$

but these interpretations fail to satisfy the equation

$${}^0H_0 = {}^{-1}H_0 + {}^0H_{-1},$$

and therefore fail to make (18) universally true, though it remains true so long as no negative arguments occur. Interesting applications will be found in Vol. II, Chap. XXX of Netto's *Vorlesungen über Algebra.*

NOTE 4. *Values of ${}^iH_\kappa$ for all values of i.*

We may and will regard the definition (12) and the interpretations (13) or the functional equation (18) as rendering ${}^iH_\kappa$ uniquely determinate whenever:

κ is a real integer, positive or negative;

i is any scalar number, which will ordinarily be real and rational.

We still have ${}^iH_\kappa = 0$ whenever $\kappa < 0$; whilst when κ is a positive integer, we have

${}^iH_\kappa = 0$ if and only if i is a real integer such that $i < 0$, $i + \kappa + 1 > 0$.

Subject only to the conditions for convergency the binomial expansion for $(x+y)^n$ is

$$(x+y)^n = {}^nH_0 \,.\, x^n + {}^{n-1}H_1 \,.\, x^{n-1}y + \dots + {}^{n-r}H_r \,.\, x^{n-r}y^r + \dots$$

$$= \binom{n}{0} . x^n + \binom{n}{1} . x^{n-1}y + \dots + \binom{n}{r} . x^{n-r}y^r + \dots, \qquad \text{.................(20)}$$

where $\binom{n}{0} = {}^nH_0 = 1$, and where for every non-zero positive integral value of r

$$\binom{n}{r} = \frac{n(n-1)(n-2)\dots(n-r+1)}{1.2.3.\dots r} = {}^{n-r}H_r.$$

Since ${}^{n-r}H_r = 0$ if and only if n is a positive integer and $r > n$, the series ends at the $(n+1)$th term when n is a positive integer, and is non-terminating with no zero coefficients in all other cases.

The binomial coefficient $\binom{n}{r}$ satisfies the equation

$$\binom{n}{r} = {}^{n-r}H_r$$

whenever r is a non-zero positive integer. If it is defined by that equation for all real integral values of r, we must have:

$$\binom{n}{r} = 0 \text{ if and only if } r < 0, \quad (n \text{ being unrestricted}),$$

$$\text{or } n \text{ is a real positive integer and } r > n;$$

$$\binom{n}{0} = 1; \quad \binom{0}{0} = 1; \quad \binom{0}{r} = 0 \quad \text{when } r \neq 0.$$

§ 256. The corresponding bilinear and quadratic scalar invariants of homogeneous linear substitutions.

1. *Bilinear invariants of any two homogeneous linear substitutions.*

Let $h = [h]_m^m$, $k = [k]_n^n$ be two square matrices with constant elements whose conjugates are h', k'. Also let $x_1, x_2, \ldots x_m$ and $y_1, y_2, \ldots y_n$ be two sets of independent scalar variables; and let $x_1', x_2', \ldots x_m'$ and $y_1', y_2', \ldots y_n'$ be the homogeneous linear functions of the x's and y's respectively determined by the equations

$$\underline{x'}_m = [h]_m^m \underline{x}_m, \quad \underline{y'}_n = [k]_n^n \underline{y}_n, \qquad \text{.............................(1)}$$

which are *ordinary* homogeneous linear transformations of the given variables when h and k are *undegenerate*. Further let f and f' be the bilinear scalar functions given by the equations

$$f = [x]_m [c]_m^n \underline{y}_n, \quad f' = [x']_m [c]_m^n \underline{y'}_n, \qquad \text{......................(1')}$$

where $C = [c]_m^n$ is a matrix with constant elements. The brackets of a one-element matrix such as $[f]$ or $[f']$ can and ordinarily will be omitted.

Then the equation $$f' = f$$
will be an identity in the x's and y's, or f will in this case be a *bilinear scalar invariant* of the substitutions (1), when and only when

$$\underline{h}_m^m [c]_m^n [k]_n^n = [c]_m^n \quad \text{or} \quad h'Ck = C,$$

i.e. when and only when C is an invariant transformand $\{h', k\}$.

It is to be remembered that when a given bilinear function f of the variables x_1, $x_2, \ldots x_m$ and $y_1, y_2, \ldots y_n$ is expressed in the form

$$f = [x]_m [c]_m^n \underline{y}_n = \Sigma c_{ij} x_i y_j,$$

the matrix $C = [c]_m^n$ is uniquely determinate.

Ex. i. When $m = n = 3$, let h and k be the quasi-scalar matrices

$$h = \begin{bmatrix} h_1 & 0 & 0 \\ 0 & h_1 & 0 \\ 0 & 0 & h_2 \end{bmatrix}, \quad k = \begin{bmatrix} k_1 & 0 & 0 \\ 0 & k_2 & 0 \\ 0 & 0 & k_2 \end{bmatrix} \text{ in which } h_2 \neq h_1,\ k_2 \neq k_1.$$

Also let C be the general invariant transformand $\{h', k\}$, and f be the most general bilinear invariant of the substitutions (1), which are now

$$x_1' = h_1 x_1, \quad x_2' = h_1 x_2, \quad x_3' = h_2 x_3; \quad y_1' = k_1 y_1, \quad y_2' = k_2 y_2, \quad y_3' = k_2 y_3.$$

Then in the respective cases when:

$$(1)\ h_1 k_1 = 1,\ h_2 k_2 = 1; \qquad (2)\ h_1 k_1 = 1,\ h_2 k_2 \neq 1\ \ (\text{or } k_2 = 0);$$

we have

$$(1)\ C = \begin{bmatrix} c_{11} & 0 & 0 \\ c_{21} & 0 & 0 \\ 0 & c_{32} & c_{33} \end{bmatrix}; \quad (2)\ C = \begin{bmatrix} c_{11} & 0 & 0 \\ c_{21} & 0 & 0 \\ 0 & 0 & 0 \end{bmatrix};$$

where the c's are independent arbitrary parameters; the corresponding general bilinear invariants being:

$$(1)\ f = c_{11} x_1 y_1 + c_{21} x_2 y_1 + c_{32} x_3 y_2 + c_{33} x_3 y_3;$$

$$(2)\ f = c_{11} x_1 y_1 + c_{21} x_2 y_1.$$

2. *Bilinear invariants of two co-gredient homogeneous linear substitutions.*

In sub-article 1 let $n = m$; and let $x_1', x_2', \ldots x_m'$ and $y_1', y_2', \ldots y_m'$ be the homogeneous linear functions of the x's and y's respectively determined by the equations

$$\overline{\underline{x'}}_m = [h]_m^m \overline{\underline{x}}_m, \qquad \overline{\underline{y'}}_m = [h]_m^m \overline{\underline{y}}_m, \quad \text{.........................(2)}$$

which are *ordinary* co-gredient homogeneous linear transformations of the given variables when h is *undegenerate.* Further let f and f' be the bilinear scalar functions given by the equations

$$f = [x]_m [c]_m^m \overline{\underline{y}}_m, \quad f' = [x']_m [c]_m^m \overline{\underline{y'}}_m, \quad \text{.........................(2')}$$

where $C = [c]_m^m$ is a square matrix with constant elements.

Then the equation $$f' = f$$
will be an identity in the x's and y's, or f will in this sense be a *bilinear scalar invariant* of the co-gredient substitutions (2), when and only when

$$\overline{\underline{h}}_m^m [c]_m^m [h]_m^m = [c]_m^m \quad \text{or} \quad h' C h = C, \quad \text{.........................(2'')}$$

i.e. when and only when C is an invariant transformand $\{h', h\}$.

Ex. ii. When $m = 3$, let h be a quasi-scalar matrix

$$h = \begin{bmatrix} h_1 & 0 & 0 \\ 0 & h_1 & 0 \\ 0 & 0 & h_2 \end{bmatrix} \quad \text{in which } h_2 \neq h_1.$$

Also let C be the general invariant transformand $\{h', h\}$, and f be the most general bilinear invariant of the substitutions (2), which are now

$$x_1' = h_1 x_1, \quad x_2' = h_1 x_2, \quad x_3' = h_2 x_3; \quad y_1' = h_1 y_1, \quad y_2' = h_1 y_2, \quad y_3' = h_2 y_3.$$

Then in the respective cases when:

$$(1)\ h_1 = 1,\ h_2 = -1; \qquad (2)\ h_1^2 = 1,\ h_2^2 \neq 1\ \ (\text{or } h_2 = 0);$$

we have

$$(1)\ C = \begin{bmatrix} c_{11} & c_{12} & 0 \\ c_{21} & c_{22} & 0 \\ 0 & 0 & c_{33} \end{bmatrix}; \quad (2)\ C = \begin{bmatrix} c_{11} & c_{12} & 0 \\ c_{21} & c_{22} & 0 \\ 0 & 0 & 0 \end{bmatrix};$$

where the c's are independent arbitrary parameters; the corresponding general bilinear invariants being:

$$(1)\ f=c_{11}x_1y_1+c_{12}x_1y_2+c_{21}x_2y_1+c_{22}x_2y_2+c_{33}x_3y_3;$$
$$(2)\ f=c_{11}x_1y_1+c_{12}x_1y_2+c_{21}x_2y_1+c_{22}x_2y_2.$$

3. *Bilinear invariants of two contra-gredient homogeneous linear substitutions.*

In sub-article 1 let $n=m$; and let $h=[h]_m^m$ be an *undegenerate* square matrix whose inverse is the square matrix $H=[H]_m^m$. Also let $x_1', x_2', \dots x_m'$ and $y_1', y_2', \dots y_m'$ be the homogeneous linear functions of the x's and y's respectively determined by the equations

$$\overline{\underline{x'}}_m=\overline{\underline{H}}^m_m\,\overline{\underline{x}}_m,\quad \overline{\underline{y'}}_m=[h]_m^m\,\overline{\underline{y}}_m, \qquad \dots\dots(3)$$

which are ordinary contra-gredient homogeneous linear transformations of the given variables. Further let f and f' be the bilinear scalar functions given by the equations

$$f=[x]_m[c]_m^m\,\overline{\underline{y}}_m,\quad f'=[x']_m[c]_m^m\,\overline{\underline{y'}}_m, \qquad \dots\dots(3')$$

where $C=[c]_m^m$ is a square matrix with constant elements.

Then the equation $$f'=f$$
will be an identity in the x's and y's, or f will in this sense be a *bilinear scalar invariant* of the contra-gredient substitutions (3), when and only when

$$[H]_m^m[c]_m^m[h]_m^m=[c]_m^m \quad\text{or}\quad HCh=C,$$

i.e. when and only when C is an invariant transformand $\{H, h\}$, or when and only when C is a commutant $\{h, h\}$ or $\{H, H\}$.

Ex. iii. When $m=3$, let h be a quasi-scalar matrix

$$h=\begin{bmatrix} h_1 & 0 & 0 \\ 0 & h_1 & 0 \\ 0 & 0 & h_2 \end{bmatrix} \quad \text{in which } h_2\neq h_1,\ h_1\neq 0,\ h_2\neq 0.$$

Also let C be the general invariant transformand $\{H, h\}$, and f be the most general bilinear invariant of the substitutions (3), which are now

$$x_1'=h_1x_1,\quad x_2'=h_1x_2,\quad x_3'=h_2x_3;\quad y_1'=h_1^{-1}y_1,\quad y_2'=h_1^{-1}y_2,\quad y_3'=h_2^{-1}y_3.$$

Then in every such case we have

$$C=\begin{bmatrix} c_{11} & c_{12} & 0 \\ c_{21} & c_{22} & 0 \\ 0 & 0 & c_{33} \end{bmatrix},\ f=c_{11}x_1y_1+c_{12}x_1y_2+c_{21}x_2y_1+c_{22}x_2y_2+c_{33}x_3y_3,$$

where the c's are independent arbitrary parameters.

4. *Quadratic invariants of a single homogeneous linear substitution.*

Let $h=[h]_m^m$ be a square matrix with constant elements whose conjugate is h'. Also let $x_1, x_2, \dots x_m$ be m independent scalar variables; and let $x_1', x_2', \dots x_m'$ be the homogeneous linear functions of the x's determined by the equation

$$\overline{\underline{x'}}_m=[h]_m^m\,\overline{\underline{x}}_m, \qquad \dots\dots(4)$$

which is an *ordinary* homogeneous linear transformation of the given variables when h is *undegenerate*. Further let f and f' be the quadratic scalar functions given by the equations

$$f=[x]_m[c]_m^m\,\overline{\underline{x}}_m,\quad f'=[x']_m[c]_m^m\,\overline{\underline{x'}}_m, \qquad \dots\dots(4')$$

where $C=[c]_m^m$ is a *symmetric* matrix with constant elements.

Then the equation $$f' = f$$
will be an identity in the x's, or f will in this sense be a *quadratic scalar invariant* of the substitutions (4), when and only when

$$\overline{\underline{h}}_m^{\,m} [c]_m^m [h]_m^m = [c]_m^m \quad \text{or} \quad h'Ch = C,$$

i.e. when and only when C is a symmetric invariant transformand $\{h', h\}$.

It is to be remembered that when a given quadratic function f of the variables x_1, x_2, ... x_m is expressed in the form

$$f = [x]_m [c]_m^m \overline{\underline{x}}_m = \Sigma c_{ij} x_i x_j, \text{ where } c_{ji} = c_{ij},$$

the symmetric matrix $C = [c]_m^m$ is uniquely determinate. We can of course also express f in many ways in the form

$$f = [x]_m [a]_m^m \overline{\underline{x}}_m = \Sigma a_{ij} x_i x_j,$$

where the square matrix $A = [a]_m^m$ is not symmetric. In fact A can be any square matrix of order m which differs from C by a skew-symmetric matrix.

Ex. iv. When $m = 3$, let h be a quasi-scalar matrix

$$h = \begin{bmatrix} h_1 & 0 & 0 \\ 0 & h_1 & 0 \\ 0 & 0 & h_2 \end{bmatrix} \text{ in which } h_2 \neq h_1.$$

Also let C be a general symmetric invariant transformand $\{h', h\}$, and f be the most general quadratic invariant of the substitution (4), which is now

$$x_1' = h_1 x_1, \quad x_2' = h_1 x_2, \quad x_3' = h_2 x_3.$$

Then in the respective cases when :

(1) $h_1 = 1$, $h_2 = -1$; (2) $h_1^2 = 1$, $h_2^2 \neq 1$ (or $h_2 = 0$);

we have

$$(1)\ C = \begin{bmatrix} c_{11} & c_{12} & 0 \\ c_{12} & c_{22} & 0 \\ 0 & 0 & c_{33} \end{bmatrix}, \qquad (2)\ C = \begin{bmatrix} c_{11} & c_{12} & 0 \\ c_{12} & c_{22} & 0 \\ 0 & 0 & 0 \end{bmatrix},$$

where the c's are independent arbitrary parameters; the corresponding general quadratic invariants being:

$$(1)\ f = c_{11} x_1^2 + 2c_{12} x_1 x_2 + c_{22} x_2^2 + c_{33} x_3^2;$$
$$(2)\ f = c_{11} x_1^2 + 2c_{12} x_1 x_2 + c_{22} x_2^2.$$

5. *General method of determining bilinear and quadratic invariants.*

Let $A = [a]_m^m$, $B = [b]_n^n$ be square matrices of orders m, n with constant elements; let $\alpha = [\alpha]_m^m$, $\beta = [\beta]_n^n$ be square matrices equi-canonical with A, B respectively; and let

$$A = h\alpha H, \quad B = k\beta K$$

be any particular isomorphic transformations by which A, B can be derived from α, β. Also let $x_1, x_2, \ldots x_m$ and $y_1, y_2, \ldots y_n$ be two sets of independent scalar variables, and let $\xi_1, \xi_2, \ldots \xi_m$ and $\eta_1, \eta_2, \ldots \eta_n$ be new variables derived from them by the ordinary linear transformations

$$\overline{\underline{\xi}}_m = H . \overline{\underline{x}}_m, \quad \overline{\underline{\eta}}_n = K . \overline{\underline{y}}_n . \quad \text{..............................(5)}$$

Then $$f = [x]_m [c]_m^n \overline{\underline{y}}_n \quad \text{..(6)}$$

will be a general or particular bilinear invariant of the given substitutions

$$\overline{\underline{x'}}_m = A \cdot \overline{\underline{x}}_m, \quad \overline{\underline{y'}}_n = B \cdot \overline{\underline{y}}_n \quad \text{.............................(7)}$$

when and only when it is expressible in the form

$$f = [\xi]_m [\gamma]_m^n \overline{\underline{\eta}}_n \quad \text{..................................(6')}$$

as a general or particular bilinear invariant of the substitutions

$$\overline{\underline{\xi'}}_m = a \cdot \overline{\underline{\xi}}_m, \quad \overline{\underline{\eta'}}_n = \beta \cdot \overline{\underline{\eta}}_n, \quad \text{...............................(7')}$$

where $C = [c]_m^n$ *and* $\gamma = [\gamma]_m^n$ *are matrices connected by the relation*

$$C = H'\gamma K. \quad \text{..(8)}$$

We know that f expressed in the form (6) will be a general or particular bilinear invariant of the substitutions (7) when and only when C is a general or particular invariant transformand $\{A', B\}$, and that f expressed in the form (6′) will be a general or particular bilinear invariant of the substitutions (7′) when and only when γ is a general or particular invariant transformand $\{a', \beta\}$. The theorem therefore follows from Ex. ii of § 255 and the formula (8).

Let $\xi_1, \xi_2, \dots \xi_m$ and $\eta_1, \eta_2, \dots \eta_n$ be the independent homogeneous linear functions of the variables $x_1, x_2, \dots x_m$ and $y_1, y_2, \dots y_n$ determined by the equations (5). Then the theorem is equivalent to the statement that the second of the two functions

$$f(x_1, x_2, \dots x_m, y_1, y_2, \dots y_n), \quad \text{................................(9)}$$

$$F(x_1, x_2, \dots x_m, y_1, y_2, \dots y_n) = f(\xi_1, \xi_2, \dots \xi_m, \eta_1, \eta_2, \dots \eta_n) \quad \text{...........(9')}$$

is a general or particular bilinear invariant of the given substitutions (7) when and only when the first function is a general or particular bilinear invariant of the substitutions

$$\overline{\underline{x'}}_m = a \cdot \overline{\underline{x}}_m, \quad \overline{\underline{y'}}_n = \beta \cdot \overline{\underline{y}}_n. \quad \text{.............................(7'')}$$

Regarded in this light, it is included as a specially simple case in Note 5 of § 226.

Ordinarily a and β will be chosen to be the bi-canonical square matrices equicanonical with A and B. Then after determining a general invariant transformand $\gamma = \text{inv}\,\{a', \beta\}$, we know the general bilinear invariant (6′) of the substitutions (7′), which a determinate change of variables converts into the general bilinear invariant of the given substitutions (7); or we know the general bilinear invariant (9) of the substitutions (7″), which can be converted into the general bilinear invariant (9′) of the given substitutions (7) by replacing $x_1, x_2, \dots y_1, y_2, \dots$ by the homogeneous linear functions $\xi_1, \xi_2, \dots \eta_1, \eta_2, \dots$.

Again

$$f = [x]_m [c]_m^m \overline{\underline{x}}_m \quad \text{.......................................(10)}$$

will be a general or particular quadratic invariant of the given substitution

$$\overline{\underline{x'}}_m = A \cdot \overline{\underline{x}}_m \quad \text{.......................................(11)}$$

when and only when it is expressible in the form

$$f = [\xi]_m [\gamma]_m^m \overline{\underline{\xi}}_m \quad \text{..................................(10')}$$

as a general or particular quadratic invariant of the substitution

$$\overline{\underline{\xi'}}_m = a \cdot \overline{\underline{\xi}}_m, \quad \text{....................................(11')}$$

where $C = [c]_m^m$ *and* $\gamma = [\gamma]_m^m$ *are symmetric matrices connected by the symmetric relation*

$$C = H'\gamma H. \quad \text{......................................(12)}$$

Let $\xi_1, \xi_2, \dots \xi_m$ be the independent homogeneous linear functions of the variables $x_1, x_2, \dots x_m$ determined by the first of the equations (5). Then this second theorem is equivalent to the statement that the second of the two functions

$$f(x_1, x_2, \dots x_m), \quad \text{.......................................(13)}$$

$$F(x_1, x_2, \dots x_m) = f(\xi_1, \xi_2, \dots \xi_m) \quad \text{..........................(13')}$$

is a general or particular quadratic invariant of the given substitution (11) when and only when the first function is a general or particular quadratic invariant of the substitution

$$\underset{m}{\lceil x' \rfloor} = a \, . \, \underset{m}{\lceil x \rfloor} \quad . \quad \text{..(11'')}$$

Thus the determination of the general quadratic invariant of the given substitution (11) is reducible to the determination of a general symmetric invariant transformand

$$\gamma = \text{inv}\,\{a', a\},$$

where a is a bi-canonical square matrix equicanonical with A.

§ 257. The invariant transformands of a pair of unipotent square matrices.

1. *General invariant transformands.*

Let $A = [a]_m^m$ and $B = [b]_n^n$ be any two unipotent square matrices of orders m and n with constant elements whose conjugates are A' and B', and whose latent roots are α and β, so that the single characteristic potent divisors of A and B are

$$(\lambda - \alpha)^m \quad \text{and} \quad (\lambda - \beta)^n.$$

Also let r be the smaller of the two integers m and n; let U and V be any particular undegenerate contra-commutants $\{A', A\}$ and $\{A, A'\}$; and let

$$\epsilon = \pm 1.$$

Without actually constructing the general invariants mentioned, we can at once enunciate the following theorem:

Theorem I. *The four general invariant transformands*

$$X = inv\,\{A, \epsilon B\}, \quad X' = inv\,\{A', \epsilon B'\}, \quad Y = inv\,\{A', \epsilon B\}, \quad Y' = inv\,\{A, \epsilon B'\}$$

can always be so chosen as to be connected by the relations

$$X' = UXV, \quad Y = UX, \quad Y' = XV;$$

and they are all zero matrices when $\alpha\beta \neq \epsilon$.

When $\alpha\beta = \epsilon$, *each of them is an undegenerate matrix of the class* $M\binom{n}{m}$ *and of rank r containing exactly r independent arbitrary parameters, and having exactly r independent non-zero particularisations.*

In each of them the elements are homogeneous linear functions of the r arbitrary parameters.

The first part of the theorem follows from Ex. iii of § 254 and Theorem III *a* of § 255. The last part of the theorem (in which A and B are necessarily undegenerate) follows from Note 1 of § 242 when we regard X as a general commutant $\{A, \epsilon B^{-1}\}$, because A and B^{-1} are unipotent square matrices whose latent roots are α and β^{-1}, and therefore A and ϵB^{-1} are unipotent square matrices having the same latent root α.

The formulae (B) of § 255 will furnish constructions for the general invariants of Theorem I when $\alpha\beta=\epsilon$ as soon as we have determined the corresponding general invariants for two square matrices A_0 and B_0 equicanonical with A and B respectively. From Ex. vii of § 255 it will be seen that we could choose A_0 and B_0 to be any simple square ante-slopes of type $\{\pi, \pi\}$ equicanonical with A and B respectively; but the simplest constructions are obtained by choosing them to be bi-canonical square matrices.

Accordingly when $\alpha\beta=\epsilon$, we will define

$$A_0=\alpha A_1, \quad B_0=\beta B_1$$

to be the simple bi-canonical square matrices of orders m, n whose latent roots are α, β, and

$$A=hA_0H, \quad B=kB_0K$$

to be any particular isomorphic transformations by which A, B can be derived from A_0, B_0. Then the general invariants

$$\xi=\text{inv}\{A_0, \epsilon B_0\},\ \xi'=\text{inv}\{A_0', \epsilon B_0'\},\ \eta=\text{inv}\{A_0', \epsilon B_0\},\ \eta'=\text{inv}\{A_0, \epsilon B_0'\},$$

or $\xi=\text{inv}\{A_1, B_1\}$, $\xi'=\text{inv}\{A_1', B_1'\}$, $\eta=\text{inv}\{A_1', B_1\}$, $\eta'=\text{inv}\{A_1, B_1'\}$

are the simple slopes described in Theorem IV *b* of § 255; and we can take the general invariants in the second part of Theorem I to be the matrices given by the formulae

$$X=h\xi K, \quad X'=H'\xi'k', \quad Y=H'\eta K, \quad Y'=h\eta'k', \quad \text{.........(A)}$$

in which H', k' are the conjugates of H, k.

Ex. i. *Particularisations of the general invariant* $\xi=inv\{A_1, B_1\}$ *in* (A).

We will suppose that $m=n=r$, so that we can put $A_1=B_1=\Omega$. When $n\neq m$, the following remarks are applicable to the effective constituent $[\xi]_r^r$ of ξ, and the corresponding particularisations of ξ itself are obtained by adding $n-m$ initial vertical or $m-n$ final horizontal rows of 0's according as $m=r \ngtr n$ or $n=r \ngtr m$. From the equation determining ξ it will be at once evident that:

If $[x]_r^r$ *is any matrix which for all positive integral values of r is an undegenerate particularisation of* ξ, *then the matrices*

$$X_\tau=\begin{bmatrix}0, & x\\ 0, & 0\end{bmatrix}_{r-\tau,\ \tau}^{\tau,\ r-\tau} \quad \textit{in which} \quad \tau=0, 1, 2, \ldots r-1$$

form a complete set of r independent non-zero particularisations of ξ *having ranks* $r, r-1, r-2, \ldots 1$; *and we could take the general invariant transformand* ξ *of order r to be the matrix*

$$[\xi]_r^r=\xi_0X_0+\xi_1X_1+\xi_2X_2+\ldots+\xi_{r-1}X_{r-1}, \quad \text{.........................(1)}$$

where the coefficients on the right are r arbitrary scalar parameters.

If $\sigma=(-1)^{r-1}$, two such undegenerate particularisations of ξ are the matrices

$$[\theta]^r_r=\begin{bmatrix} {}^0H_0 & -{}^0H_1 & {}^0H_2 \ldots & -\sigma.{}^0H_{r-2} & \sigma.{}^0H_{r-1} \\ 0 & -{}^1H_0 & {}^1H_1 \ldots & -\sigma.{}^1H_{r-3} & \sigma.{}^1H_{r-2} \\ 0 & 0 & {}^2H_0 \ldots & -\sigma.{}^2H_{r-4} & \sigma.{}^2H_{r-3} \\ \cdots & \cdots & \cdots & \cdots & \cdots \\ 0 & 0 & 0 \ \ldots & -\sigma.{}^{r-2}H_0 & \sigma.{}^{r-2}H_1 \\ 0 & 0 & 0 \ \ldots & 0 & \sigma.{}^{r-1}H_0 \end{bmatrix} = \begin{bmatrix} 1 & -1 & 1 & -1 & 1 & \ldots \\ 0 & -1 & 2 & -3 & 4 & \ldots \\ 0 & 0 & 1 & -3 & 6 & \ldots \\ 0 & 0 & 0 & -1 & 4 & \ldots \\ 0 & 0 & 0 & 0 & 1 & \ldots \\ \cdots & \cdots & \cdots & \cdots & \cdots & \cdots \end{bmatrix}$$

$$[\theta']^r_r=\begin{bmatrix} \sigma.{}^{r-1}H_0 & \sigma.{}^{r-2}H_1 \ldots & \sigma.{}^2H_{r-3} & \sigma.{}^1H_{r-2} & \sigma.{}^0H_{r-1} \\ 0 & -\sigma.{}^{r-2}H_0 \ldots & -\sigma.{}^2H_{r-4} & -\sigma.{}^1H_{r-3} & -\sigma.{}^0H_{r-2} \\ \cdots & \cdots & \cdots & \cdots & \cdots \\ 0 & 0 \ \ldots & {}^2H_0 & {}^1H_1 & {}^0H_2 \\ 0 & 0 \ \ldots & 0 & -{}^1H_0 & -{}^0H_1 \\ 0 & 0 \ \ldots & 0 & 0 & {}^0H_0 \end{bmatrix} = \begin{bmatrix} \cdots & \cdots & \cdots & \cdots & \cdots & \cdots \\ \ldots & 1 & 4 & 6 & 4 & 1 \\ \ldots & 0 & -1 & -3 & -3 & -1 \\ \ldots & 0 & 0 & 1 & 2 & 1 \\ \ldots & 0 & 0 & 0 & -1 & -1 \\ \ldots & 0 & 0 & 0 & 0 & 1 \end{bmatrix}$$

They are the simple square ante-slopes of order r and type $\{\pi, \pi\}$ whose parametric elements are given by the formulae

$$\theta_{ij}=(-1)^{j-1}.{}^{i-1}H_\kappa, \quad \theta'_{ij}=(-1)^{r-i}.{}^{r-j}H_\kappa, \quad (\kappa=j-1), \ \ldots\ldots\ldots\ldots\text{(a)}$$

which involve $\qquad \theta'_{ij}=\theta_{vu}, \quad$ where $\quad u=r+1-i, \quad v=r+1-j.$

If J is the simple reversant of order r, they are connected by the relation

$$[\theta']^r_r=J\,\overline{\underline{\theta}}^{\,r}_{\,r}\,J;$$

and they are connected with the square matrices $[\phi]^r_r$, $[\phi']^r_r$ of Ex. ii by the relations

$$[\theta]^r_r=J.[\phi]^r_r, \quad [\theta']^r_r=J.[\phi']^r_r.$$

Two other such undegenerate particularisations of ξ are the simple square ante-slopes

$$[x]^r_r=[\theta]^r_r+\begin{bmatrix}0, & \theta\\ 0, & 0\end{bmatrix}^{1,r-1}_{r-1,1}=\begin{bmatrix}1, & 0\\ 0, & -\theta\end{bmatrix}^{1,r-1}_{1,r-1};$$

$$[x']^r_r=[\theta']^r_r+\begin{bmatrix}0, & \theta'\\ 0, & 0\end{bmatrix}^{1,r-1}_{r-1,1}=\begin{bmatrix}-\theta', & 0\\ 0, & 1\end{bmatrix}^{r-1,1}_{r-1,1}.$$

For the first of these the formula (1) gives the general invariant ξ described in Theorem IVa of § 255, and is equivalent to the formula

$$[\xi]^r_r=[x]^r_r.\begin{bmatrix}\xi_0 & \xi_1 & \ldots & \xi_{r-1}\\ 0 & \xi_0 & \ldots & \xi_{r-2}\\ \cdots & \cdots & \cdots & \cdots\\ 0 & 0 & \ldots & \xi_0\end{bmatrix}.$$

Ex. ii. *Particularisations of the general invariant* $\eta=inv\,\{A_1', B_1\}$ *in* (A).

We will again suppose that $m=n=r$, so that we can put $A_1=B_1=\Omega$. When $n\neq m$, the following remarks are applicable to the effective constituent $[\eta]^r_r$ of η, and the corresponding particularisations of η itself are obtained by adding $n-m$ initial vertical or $m-n$ initial horizontal rows of 0's according as $m=r \not> n$ or $n=r \not> m$. From the equation determining η it will be at once evident that:

If $[y]_r^r$ *is any matrix which for all positive integral values of* r *is an undegenerate particularisation of* η, *then the matrices*

$$Y_\tau = \begin{bmatrix} 0, & 0 \\ 0, & y \end{bmatrix}_{\tau,\, r-\tau}^{\tau,\, r-\tau} \quad \textit{in which} \quad \tau = 0, 1, 2, \dots r-1$$

form a complete set of r *independent non-zero particularisations of* η *having ranks* $r, r-1, r-2, \dots 1$; *and we could take the general invariant transformand* η *of order* r *to be the matrix*

$$\eta\,(r) = \xi_0 Y_0 + \xi_1 Y_1 + \xi_2 Y_2 + \dots + \xi_{r-1} Y_{r-1}, \quad \text{.........................(2)}$$

where the coefficients on the right are r arbitrary scalar parameters.

If $\sigma = (-1)^{r-1}$, two such undegenerate particularisations of η are the mutually conjugate square matrices

$$[\phi]_r^r = \begin{bmatrix} 0 & 0 & 0 & \cdots & 0 & \sigma.\,{}^{r-1}H_0 \\ 0 & 0 & 0 & \cdots & -\sigma.\,{}^{r-2}H_0 & \sigma.\,{}^{r-2}H_1 \\ \cdots & \cdots & \cdots & \cdots & \cdots & \cdots \\ 0 & 0 & {}^2H_0 & \cdots & -\sigma.\,{}^2H_{r-4} & \sigma.\,{}^2H_{r-3} \\ 0 & -{}^1H_0 & {}^1H_1 & \cdots & -\sigma.\,{}^1H_{r-3} & \sigma.\,{}^1H_{r-2} \\ {}^0H_0 & -{}^0H_1 & {}^0H_2 & \cdots & -\sigma.\,{}^0H_{r-2} & \sigma.\,{}^0H_{r-1} \end{bmatrix} = \begin{bmatrix} \cdots & \cdots & \cdots & \cdots & \cdots & \cdots \\ 0 & 0 & 0 & 0 & 1 & \cdots \\ 0 & 0 & 0 & -1 & 4 & \cdots \\ 0 & 0 & 1 & -3 & 6 & \cdots \\ 0 & -1 & 2 & -3 & 4 & \cdots \\ 1 & -1 & 1 & -1 & 1 & \cdots \end{bmatrix}$$

$$[\phi']_r^r = \begin{bmatrix} 0 & 0 & \cdots & 0 & 0 & {}^0H_0 \\ 0 & 0 & \cdots & 0 & -\,{}^1H_0 & -\,{}^0H_1 \\ 0 & 0 & \cdots & {}^2H_0 & {}^1H_1 & {}^0H_2 \\ \cdots & \cdots & \cdots & \cdots & \cdots & \cdots \\ 0 & -\sigma.\,{}^{r-2}H_0 & \cdots & -\sigma.\,{}^2H_{r-4} & -\sigma.\,{}^1H_{r-3} & -\sigma.\,{}^0H_{r-2} \\ \sigma.\,{}^{r-1}H_0 & \sigma.\,{}^{r-2}H_1 & \cdots & \sigma.\,{}^2H_{r-3} & \sigma.\,{}^1H_{r-2} & \sigma.\,{}^0H_{r-1} \end{bmatrix} = \begin{bmatrix} \cdots & 0 & 0 & 0 & 0 & 1 \\ \cdots & 0 & 0 & 0 & -1 & -1 \\ \cdots & 0 & 0 & 1 & 2 & 1 \\ \cdots & 0 & -1 & -3 & -3 & -1 \\ \cdots & 1 & 4 & 6 & 4 & 1 \\ \cdots & \cdots & \cdots & \cdots & \cdots & \cdots \end{bmatrix}$$

They are the simple square counter-slopes of order r and type $\{\pi', \pi\}$ whose parametric elements are given by the formulae

$$\phi_{ij} = (-1)^{j-1}.\,{}^{i-1}H_\kappa, \quad \phi'_{ij} = (-1)^{r-i}.\,{}^{r-j}H_\kappa, \quad (\kappa = j-i), \quad \text{...............(b)}$$

which involve $\quad \phi'_{ij} = \phi_{vu}, \quad$ where $\quad u = r+1-i, \; v = r+1-j.$

Two other such undegenerate particularisations of η are the mutually conjugate simple square counter-slopes

$$[y]_r^r = [\phi]_r^r + \begin{bmatrix} 0, & 0 \\ 0, & \phi \end{bmatrix}_{1,\, r-1}^{1,\, r-1} = \begin{bmatrix} 0, & -\phi \\ 1, & 0 \end{bmatrix}_{r-1,\, 1}^{1,\, r-1};$$

$$[y']_r^r = [\phi']_r^r + \begin{bmatrix} 0, & 0 \\ 0, & \phi' \end{bmatrix}_{1,\, r-1}^{1,\, r-1} = \begin{bmatrix} 0\,, & 1 \\ -\phi', & 0 \end{bmatrix}_{1,\, r-1}^{r-1,\, 1}.$$

For the first of these the formula (2) gives the general invariant η described in Theorem IVa of § 255, and is equivalent to the formula

$$\eta(r) = [y]_r^r \cdot \begin{bmatrix} \xi_0 & \xi_1 & \cdots & \xi_{r-1} \\ 0 & \xi_0 & \cdots & \xi_{r-2} \\ \cdots & \cdots & \cdots & \cdots \\ 0 & 0 & \cdots & \xi_0 \end{bmatrix}.$$

2. *General symmetric and general skew-symmetric invariant transformands.*

Let $A = [a]_r^r$ be a single unipotent square matrix of order r with constant elements whose latent root is α, so that its single characteristic potent divisor is

$$(\lambda - \alpha)^r;$$

let A' be the conjugate of A; let V be any particular undegenerate symmetric contra-commutant $\{A, A'\}$; and let

$$\epsilon = \pm 1.$$

Also let
$$\Omega = [\omega]_r^r = [1]_r^r + \begin{bmatrix} 0, & 1 \\ 0, & 0 \end{bmatrix}_{r-1,1}^{1,r-1}$$

be the simple bi-canonical (or simple canonical) square matrix of order r whose latent root is 1, and let Ω' be its conjugate. We know from Theorem IV *a* of § 255 that the general symmetric (or general skew-symmetric) invariant transformands $\eta = \text{inv}\,\{\Omega', \Omega\}$, $\eta' = \text{inv}\,\{\Omega, \Omega'\}$ can be taken to be simple square counter-slopes

$$\eta = \begin{bmatrix} 0 & 0 & 0 & \cdots & \xi_{rr} \\ \cdots & \cdots & \cdots & \cdots & \cdots \\ 0 & 0 & \xi_{33} & \cdots & \xi_{3r} \\ 0 & \xi_{22} & \xi_{23} & \cdots & \xi_{2r} \\ \xi_{11} & \xi_{12} & \xi_{13} & \cdots & \xi_{1r} \end{bmatrix}, \quad \eta' = \begin{bmatrix} \xi_{1r} & \cdots & \xi_{13} & \xi_{12} & \xi_{11} \\ \xi_{2r} & \cdots & \xi_{23} & \xi_{22} & 0 \\ \xi_{3r} & \cdots & \xi_{33} & 0 & 0 \\ \cdots & \cdots & \cdots & \cdots & \cdots \\ \xi_{rr} & \cdots & 0 & 0 & 0 \end{bmatrix}, \quad \text{.........(3)}$$

whose parametric elements are given by the formula

$$\xi_{ij} = (-1)^{i-1} \cdot \{\xi_\kappa - {}^{i-2}H_1 \cdot \xi_{\kappa-1} + {}^{i-2}H_2 \cdot \xi_{\kappa-2} - \dots + (-1)^\kappa \cdot {}^{i-2}H_\kappa \cdot \xi_0\}, \quad \text{...(c)}$$
$$(j - i = \kappa;\ i = 1, 2, \dots r - \kappa;\ \kappa = 0, 1, 2, \dots r - 1),$$

where $\xi_0, \xi_1, \xi_2, \dots \xi_{r-1}$ are r parameters which are not all independent, but are arbitrary subject to the conditions that η and η' shall be symmetric (or skew-symmetric) matrices.

With respect to *symmetric* invariant transformands we have the following theorem:

Theorem II. *The two general symmetric invariant transformands*

$$Y = inv\,\{A', \epsilon A\}, \quad Y' = inv\,\{A, \epsilon A'\}$$

can always be so chosen as to be connected by the symmetric relation

$$Y' = VYV;$$

and they are both zero matrices when $\alpha^2 \neq \epsilon$.

If $\alpha^2 = \epsilon$, then :

(1) *when r is odd, they are undegenerate square matrices of order r, each containing exactly $\frac{1}{2}(r+1)$ arbitrary parameters and having exactly $\frac{1}{2}(r+1)$ independent non-zero particularisations ;*

(2) *when r is even, they are degenerate square matrices of order r and rank $r-1$, each containing exactly $\frac{1}{2}r$ arbitrary parameters and having exactly $\frac{1}{2}r$ independent non-zero particularisations.*

The first part of the theorem follows from Ex. iv of § 254 and Theorem III *a* of § 255.

When $\alpha^2 = \epsilon$, we will define

$$A_0 = \alpha\Omega$$

to be the simple bi-canonical square matrix of order r whose latent root is α, and

$$A = hA_0H$$

to be any particular isomorphic transformation by which A can be derived from A_0. Then the *general symmetric* invariants

$$\eta = \text{inv}\{A_0', \epsilon A_0\}, \quad \eta' = \text{inv}\{A_0, \epsilon A_0'\}$$

or

$$\eta = \text{inv}\{\Omega', \Omega\}, \quad \eta' = \text{inv}\{\Omega, \Omega'\}$$

can be taken to be the simple square counter-slopes of order r and of types $\{\pi', \pi\}$, $\{\pi, \pi'\}$ which are completely described in Ex. iii or v; and by Ex. ii of § 255 the *general symmetric* invariant transformands in the second part of Theorem II can be taken to be the square matrices given by the formulae

$$Y = H'\eta H, \quad Y' = h\eta' h' \quad \text{......................(B)}$$

in which h', H' are the conjugates of h, H. The truth of the second part of the theorem follows from (B) and the properties of η and η' described in Ex. iii.

When r is an odd integer $2t+1$, it will be shown that the parametric elements of the simple square counter-slope η or η' are homogeneous linear functions of $t+1$ independent arbitrary parameters

$$\xi_{11} = \xi_0, \quad \xi_{13} = \xi_2, \ \dots \ \xi_{1r} = \xi_{r-1},$$

the elements of the paradiagonal being alternately ξ_0 and $-\xi_0$.

When r is an even integer $2t$, it will be shown that the parametric elements of the simple counter-slope η or η' are homogeneous linear functions of t independent arbitrary parameters

$$\xi_{12} = \xi_1, \ \xi_{14} = \xi_3, \ \dots \ \xi_{1r} = \xi_{r-1},$$

the elements of the paradiagonal being all equal to 0, and the elements of difference-weight 1 being alternately ξ_1 and $-\xi_1$.

In both cases (see Ex. v) we can also regard the parametric elements lying on the median line (or leading diagonal) as the arbitrary parameters of which all other elements are homogeneous linear functions.

If $Y = [y]_r^r$ is a general invariant transformand of one of the types mentioned in Theorem II, it will be obvious that the matrix

$$\mathbf{Y} = \tfrac{1}{2}\{[y]_r^r + \underline{\overline{y}}{}_r^r\}$$

must become a general symmetric invariant transformand of that type when the parameters occurring in its elements have been reduced by substitutions to the smallest number possible. In saying that the elements of a matrix are functions of r *independent* scalar variables, it is ordinarily implied that those variables cannot be reduced by substitutions to a smaller number of variables. Hence we shall not call $\mathbf{Y}$ a *general* symmetric invariant transformand so long as its elements are expressed as functions of the r arbitrary parameters occurring in Y. It is merely a matrix which can be expressed as a general symmetric invariant transformand.

Ex. iii. *The general symmetric invariants* $\eta = inv\ \{\Omega', \Omega\}$, $\eta' = inv\ \{\Omega, \Omega'\}$.

These can be taken to be the simple square counter-slopes (3) whose parametric elements are given by the formula (c), where the r parameters $\xi_0, \xi_1, \dots \xi_{r-1}$ are arbitrary subject to the conditions for symmetry, i.e. the conditions

$$\xi_{ij} = \xi_{JI} \quad \text{whenever} \quad i + I = j + J = r + 1.$$

If we put $$\rho = r - \kappa - 1, \qquad (\kappa = 0, 1, 2, \dots r-1),$$

these conditions are equivalent to the r sets of equations corresponding to the r values of κ such as the sets:

$$2\xi_\kappa - ({}^{\rho-i}H_1 + {}^{i-2}H_1)\,\xi_{\kappa-1} + ({}^{\rho-i}H_2 + {}^{i-2}H_2)\,\xi_{\kappa-2} - \dots + (-1)^\kappa . ({}^{\rho-i}H_\kappa + {}^{i-2}H_\kappa)\,\xi_0 = 0, \quad (\mathrm{a}_\kappa)$$

where $r - \kappa$ is *even*, and $i = 1, 2, \dots \tfrac{1}{2}(r-\kappa)$;

$$0 . \xi_\kappa - ({}^{\rho-i}H_1 - {}^{i-2}H_1)\,\xi_{\kappa-1} + ({}^{\rho-i}H_2 - {}^{i-2}H_2)\,\xi_{\kappa-2} - \dots + (-1)^\kappa . ({}^{\rho-i}H_\kappa - {}^{i-2}H_\kappa)\,\xi_0 = 0, \quad (\mathrm{b}_\kappa)$$

where $r - \kappa$ is *odd*, and $i = 1, 2, \dots \tfrac{1}{2}(r - \kappa + 1)$; the expression on the left reducing to the first term only in each case when $\kappa = 0$.

First suppose that κ is any one of the integers $0, 1, 2, \dots r-1$ which is such that $r - \kappa$ is even, and that (if $\kappa > 0$) all the equations $(\mathrm{b}_{\kappa-1})$ are satisfied. Then by subtracting and using the general formula

$${}^{i+1}H_\kappa - {}^{i}H_\kappa = {}^{i}H_{\kappa-1},$$

it will be seen that every two successive equations (a_κ) are mutually equivalent. Thus all the equations (a_κ) will be satisfied if any one of them (such as the first for which $i = 1$) is satisfied.

Next suppose that κ is any one of the integers $0, 1, 2, \dots r-1$ which is such that $r - \kappa$ is odd, and that (if $\kappa > 0$) all the equations $(\mathrm{a}_{\kappa-1})$ are satisfied. Then by subtracting it will be seen in the same way that every two successive equations (b_κ) are mutually equivalent; moreover the last of these equations is an identity. Thus all the equations (b_κ) are satisfied.

We conclude that the conditions for symmetry will be satisfied when and only when one of the equations (say the first) is satisfied in every such set as (a_κ) corresponding to an even value of $r - \kappa$.

When r is an odd integer $2t + 1$, so that the sets (a_κ), (b_κ) are

$$(b_0), (a_1), (b_2), \dots (a_{r-2}), (b_{r-1}),$$

the conditions for symmetry are equivalent to the t equations such as

$$2\xi_\kappa - {}^{r-\kappa-2}H_1 . \xi_{\kappa-1} + {}^{r-\kappa-2}H_2 . \xi_{\kappa-2} - \dots + (-1)^\kappa . {}^{r-\kappa-2}H_\kappa . \xi_0 = 0, \quad \dots\dots\dots(\mathrm{c}_\kappa)$$

where κ receives the *odd* values 1, 3, 5, ... $r-2$. Hence in the general symmetric invariants η and η' the parameters of even weights, viz. the $t+1$ parameters

$$\xi_0=\xi_{11},\ \xi_2=\xi_{13},\ \dots\ \xi_{r-1}=\xi_{1r},\quad \text{are arbitrary,}$$

whilst the t parameters $\xi_1, \xi_3, \dots \xi_{r-2}$ of odd weights are the homogeneous linear functions of them determined in succession by the equations (c_κ), the first of which (when $r \nless 3$) is

$$2\xi_1 - {}^{r-3}H_1 \,.\, \xi_0 = 0.$$

When r is an even integer $2t$, so that the sets (a_κ), (b_κ) are

$$(a_0),\ (b_1),\ (a_2),\ \dots\ (a_{r-2}),\ (b_{r-1}),$$

the conditions for symmetry are equivalent to the t equations such as

$$2\xi_\kappa - {}^{r-\kappa-2}H_1 \,.\, \xi_{\kappa-1} + {}^{r-\kappa-2}H_2 \,.\, \xi_{\kappa-2} - \dots + (-1)^\kappa \,.\, {}^{r-\kappa-2}H_\kappa \,.\, \xi_0 = 0, \quad \dots\dots\dots (d_\kappa)$$

where κ receives the *even* values 0, 2, 4, ... $r-2$; the expression on the left reducing to the first term only when $\kappa=0$. Hence in the general symmetric invariants η and η' the parameters of odd weights, viz. the t parameters

$$\xi_1=\xi_{12},\ \xi_3=\xi_{14},\ \dots\ \xi_{r-1}=\xi_{1r},\quad \text{are arbitrary,}$$

whilst the t parameters $\xi_0, \xi_2, \dots \xi_{r-2}$ of even weights are the homogeneous linear functions of them determined in succession by the equations (d_κ), the first two of which (when $r \nless 4$) are

$$\xi_0 = 0,$$
$$2\xi_2 - {}^{r-4}H_1 \,.\, \xi_1 = 0.$$

Thus all the elements of η and η' are known homogeneous linear functions of:

the $t+1$ arbitrary parameters $\xi_0, \xi_2, \dots \xi_{r-1}$ of even weights when $r=2t+1$;

the t arbitrary parameters $\xi_1, \xi_3, \dots \xi_{r-1}$ of odd weights when $r=2t$;

and in the latter case we have $\xi_0=0$.

Ex. iv. The conditions for symmetry given by (c_κ) or (d_κ) in Ex. iii are equivalent to the equations

$$\begin{array}{llll}
 & \xi_0=0, & \xi_2=\tfrac{1}{2}\xi_1 & \text{when } r=4; \\
 & \xi_1=\tfrac{3}{2}\xi_0, & \xi_3=\tfrac{1}{2}\xi_2-\tfrac{1}{4}\xi_0 & \text{when } r=5; \\
\xi_0=0, & \xi_2=\tfrac{3}{2}\xi_1, & \xi_4=\tfrac{1}{2}\xi_3-\tfrac{1}{4}\xi_1 & \text{when } r=6; \\
\xi_1=\tfrac{5}{2}\xi_0, & \xi_3=\tfrac{3}{2}\xi_2-\tfrac{5}{2}\xi_0, & \xi_5=\tfrac{1}{2}\xi_4-\tfrac{1}{4}\xi_2+\tfrac{1}{2}\xi_0 & \text{when } r=7; \\
\xi_0=0,\ \ \xi_2=\tfrac{5}{2}\xi_1, & \xi_4=\tfrac{3}{2}\xi_3-\tfrac{5}{2}\xi_1, & \xi_6=\tfrac{1}{2}\xi_5-\tfrac{1}{4}\xi_3+\tfrac{1}{2}\xi_1 & \text{when } r=8.
\end{array}$$

The general symmetric invariant η of order 7 can be taken to be the square matrix

$$\eta(7)=\begin{bmatrix}
0, & 0, & 0, & 0, & 0, & 0, & \xi_0 \\
0, & 0, & 0, & 0, & 0, & -\xi_0, & \tfrac{5}{2}\xi_0 \\
0, & 0, & 0, & 0, & \xi_0, & -\tfrac{3}{2}\xi_0, & \xi_2 \\
0, & 0, & 0, & -\xi_0, & \tfrac{1}{2}\xi_0, & \tfrac{3}{2}\xi_0-\xi_2, & -\tfrac{5}{2}\xi_0+\tfrac{3}{2}\xi_2 \\
0, & 0, & \xi_0, & \tfrac{1}{2}\xi_0, & -2\xi_0+\xi_2, & \xi_0-\tfrac{1}{2}\xi_2, & \xi_4 \\
0, & -\xi_0, & -\tfrac{3}{2}\xi_0, & \tfrac{3}{2}\xi_0-\xi_2, & \xi_0-\tfrac{1}{2}\xi_2, & -\xi_0+\tfrac{1}{2}\xi_2-\xi_4, & \tfrac{1}{2}\xi_0-\tfrac{1}{4}\xi_2+\tfrac{1}{2}\xi_4 \\
\underline{\xi_0}, & \tfrac{5}{2}\xi_0, & \underline{\xi_2}, & -\tfrac{5}{2}\xi_0+\tfrac{3}{2}\xi_2, & \underline{\xi_4}, & \tfrac{1}{2}\xi_0-\tfrac{1}{4}\xi_2+\tfrac{1}{2}\xi_4, & \underline{\xi_6}
\end{bmatrix}$$

in which $\xi_0, \xi_2, \xi_4, \xi_6$ are arbitrary parameters. If the four underlined elements are regarded as given, all the other elements can be determined in succession by the law of construction (described in Ex. v) together with the restrictions imposed by symmetry.

The corresponding invariant $\eta(5)$ can be derived from $\eta(7)$ by putting $\xi_0=0$, striking out the first two horizontal and the first two vertical rows, and substituting ξ_{i-2} for ξ_i. The corresponding invariant $\eta(8)$ can be derived from $\eta(7)$ by adding one initial horizontal and one initial vertical row of 0's and substituting ξ_{i+1} for ξ_i.

Ex. v. *Simpler constructions for the general symmetric invariants* η, η'.

By giving the last instead of the first value to i in (a_κ) we see that the conditions for symmetry in Ex. iii are equivalent to the equations

$$\xi_{i-1,j}=\xi_{i,j+1} \quad \text{whenever} \quad i+j=r+1. \qquad\qquad (4)$$

When an invariant η or η' has been constructed in accordance with the law that

the sum of two consecutive elements of the same difference-weight κ *and the contiguous element of smaller difference-weight* $\kappa-1$ *is always equal to* 0,

it will be symmetric if and only if the conditions for symmetry are satisfied by the elements adjacent to the median line (or leading diagonal); moreover the last mentioned elements are rendered uniquely determinate by the conditions (4) and the law of construction when the parametric elements lying on the median line are given. Consequently a general symmetric invariant η or η' can be regarded as a simple counter-slope of type $\{\pi', \pi\}$ or $\{\pi, \pi'\}$ in which:

(1) the parametric elements lying on the median line are arbitrary parameters;

(2) the parametric elements lying on the two adjacent parallel lines are uniquely determined by the conditions for symmetry and the law of construction;

(3) all the other parametric elements of difference-weights 0, 1, 2, ... are uniquely determined in succession by the law of construction.

For example the general symmetric invariant η of order 7 can be taken to be the square matrix

$$\eta(7)=\tfrac{1}{2}\begin{bmatrix} 0, & 0, & 0, & 0, & 0, & 0, & -2\eta_0 \\ 0, & 0, & 0, & 0, & 0, & 2\eta_0, & -5\eta_0 \\ 0, & 0, & 0, & 0, & -2\eta_0, & 3\eta_0, & -4\eta_0+2\eta_2 \\ 0, & 0, & 0, & \underline{2\eta_0}, & -\eta_0, & \eta_0-2\eta_2, & -\eta_0+3\eta_2 \\ 0, & 0, & -2\eta_0, & -\eta_0, & \underline{2\eta_2}, & -\eta_2, & \eta_2-2\eta_4 \\ 0, & 2\eta_0, & 3\eta_0, & \eta_0-2\eta_2, & -\eta_2, & \underline{2\eta_4}, & -\eta_4 \\ -2\eta_0, & -5\eta_0, & -4\eta_0+2\eta_2, & -\eta_0+3\eta_2, & \eta_2-2\eta_4, & -\eta_4, & \underline{2\eta_6} \end{bmatrix}$$

in which η_0, η_2, η_4, η_6 are arbitrary parameters. If the four underlined elements are regarded as given, all the other elements can be determined uniquely in succession in the way described above. The corresponding invariant $\eta(5)$ can be derived from $\eta(7)$ by striking out the first and last horizontal and vertical rows. The corresponding invariant $\eta(8)$ can be derived from $\eta(7)$ by adding one initial horizontal and one initial vertical row of 0's and substituting η_{i+1} for η_i; or the corresponding invariant $\eta(6)$ by striking out the last horizontal and the last vertical row and substituting η_{i+1} for η_i.

The general symmetric invariant $\eta(r)$ of order r which is constructed in this way is the matrix given by the formula (7) or (7′) of Ex. vi when Y is chosen to be the matrix (9). In fact the matrix (9) is the coefficient of η_0 in $\eta(r)$ when $r=2t+1$. Accordingly we can find a general formula for all the parametric elements ξ_{ij} of $\eta(r)$. If we put

$$y_{ij}(t)=(-1)^{t+1-i}\cdot\tfrac{1}{2}\left({}^{t-j}H_{j-i}+{}^{t+1-j}H_{j-i}\right) \qquad\qquad (5)$$

by evaluating the matrix (9), so that

$$y_{ij}(t)=0 \quad \text{when} \quad j<i \quad \text{or} \quad j>t+1>i, \quad (\text{and also when } j<1 \quad \text{or} \quad i>2t+1),$$

then when $r=2t+1$, the formula is

$$\xi_{ij}=\eta_0\cdot y_{ij}(t)+\eta_2\cdot y_{i,j-2}(t-1)+\ \dots\ +\eta_{r-1}\cdot y_{i,j-r+1}(0); \qquad\qquad (6)$$

and when $r=2t$, the formula is

$$\xi_{ij}=\eta_1 . y_{i,j-1}(t-1)+\eta_3 . y_{i,j-3}(t-2)+ \dots +\eta_{r-1} . y_{i,j-r+1}(0). \quad \dots\dots\dots(6')$$

The series of coefficients of η_0 (or η_1, or η_2, ...) occurring in successive parametric diagonal lines and starting from the median element or an element nearest to the median line are the numbers

$$\tfrac{1}{2}({}^{i-1}H_\kappa+{}^{i}H_\kappa) \quad \text{or} \quad {}^{i-1}H_\kappa+\tfrac{1}{2} . {}^{i}H_\kappa \quad \text{in which} \quad i=0, 1, 2, 3, \dots$$

for the successive values 0, 1, 2, 3, ... of κ.

Ex. vi. *Particularisations of the general symmetric invariant* $\eta=inv\{\Omega', \Omega\}$.

From the matrix equation determining η it will be at once evident that:

If $Y=[y]_{2t+1}^{2t+1}$ *is any symmetric matrix which for all positive integral values of t is a particular undegenerate symmetric invariant* η *of order* $2t+1$, *a complete set of independent non-zero particularisations of the general symmetric invariant* $\eta=\eta(r)$ *of order r is formed by the symmetric matrices:*

$$Y_{2\tau}=\begin{bmatrix} 0, & 0 \\ 0, & y \end{bmatrix}_{2\tau,\, r-2\tau}^{2\tau,\, r-2\tau} \quad \textit{in which } \tau=0, 1, 2, \dots t \quad \textit{when } r=2t+1;$$

$$Y_{2\tau+1}=\begin{bmatrix} 0, & 0 \\ 0, & y \end{bmatrix}_{1+2\tau,\, r-1-2\tau}^{1+2\tau,\, r-1-2\tau} \quad \textit{in which } \tau=0, 1, 2, \dots t-1 \quad \textit{when } r=2t.$$

When any such particular invariant has been determined, we can take the general symmetric invariant $\eta(r)$ to be the matrix:

$$\eta(r)=\eta_0 Y_0+\eta_2 Y_2+ \dots +\eta_{r-1} Y_{r-1} \quad \text{when } r=2t+1; \quad \dots\dots\dots(7)$$

$$\eta(r)=\eta_1 Y_1+\eta_3 Y_3+ \dots +\eta_{r-1} Y_{r-1} \quad \text{when } r=2t; \quad \dots\dots\dots(7')$$

where the coefficients on the right are arbitrary scalar parameters.

If $[\phi]_r^r$, $[\phi']_r^r$ are the two mutually conjugate simple square counter-slopes defined in Ex. ii, we could choose Y to be the symmetric matrix

$$Y=[y]_r^r=\tfrac{1}{2}\{[\phi]_r^r+[\phi']_r^r\}, \quad \text{where } r=2t+1, \quad \dots\dots\dots(8)$$

so that Y is a simple square counter-slope of type $\{\pi', \pi\}$ and of odd order r whose parametric elements are given by the formula

$$2y_{ij}=(-1)^{j-1} . {}^{i-1}H_\kappa+(-1)^{i-1} . {}^{r-j}H_\kappa, \quad \text{where } \kappa=j-i.$$

A more convenient choice is the symmetric simple square counter-slope

$$Y=[\psi]_{2t+1}^{2t+1}=\tfrac{1}{2}\begin{bmatrix} 0, & \phi \\ -\phi', & 0 \end{bmatrix}_{t+1,\, t}^{t,\, t+1}+\tfrac{1}{2}\begin{bmatrix} 0, & -\phi \\ \phi', & 0 \end{bmatrix}_{t,\, t+1}^{t+1,\, t} \quad \dots\dots\dots(9)$$

$$=\begin{bmatrix} 0, & 0, & -\phi \\ 0, & 1, & 0 \\ -\phi', & 0, & 0 \end{bmatrix}_{t,1,t}^{t,1,t}+\tfrac{1}{2}\begin{bmatrix} 0, & 0, & 0 \\ 0, & 0, & -\phi \\ 0, & -\phi', & 0 \end{bmatrix}_{1,t,t}^{1,t,t}, \quad \dots\dots\dots(9')$$

this being that particular symmetric invariant η of order $2t+1$ in which the middle element is equal to 1, and all the other elements lying on the median line are equal to 0. The property (16) of Note 1 in § 255 shows that

$$\phi'_{t+i,\, t+j}(t)=\phi'_{t+1+i,\, t+1+j}(t+1)=-\phi_{ij} \quad \text{when } i \not< 1.$$

Therefore from (9) we see that all the parametric elements of Y are given by the formula

$$\psi_{ij}=(-1)^{t+1-i} . \tfrac{1}{2}\{{}^{t-j}H_\kappa+{}^{t+1-j}H_\kappa\}, \quad (\kappa=j-i), \quad \dots\dots\dots(10)$$

where the first and second terms give the parametric elements of the first and second matrices on the right in (9). We could replace (10) by the equivalent formula

$$\psi_{ij}=(-1)^{t+1-i} . \{{}^{t-j}H_\kappa+\tfrac{1}{2} . {}^{t+1-j}H_{\kappa-1}\}, \quad (\kappa=j-i) \quad \dots\dots\dots(10')$$

corresponding to (9′). But the first and second terms in (10′) do not give separately *all* the parametric elements of the first and second matrices in (9′), which in fact are not separately invariants $\{\Omega', \Omega\}$. Since the integers ψ_{ij} given by (10) or (10′) satisfy the functional equation

$$\psi_{ij}+\psi_{i+1,j+1}+\psi_{i+1,j}=0$$

and also the conditions for symmetry

$$\psi_{ij}=\psi_{JI} \quad \text{whenever} \quad i+I=j+J=2t+2,$$

it would be sufficient to show that they give the elements of the median line correctly.

When we choose Y to be the matrix (9), the formula (7) or (7′) gives the general symmetric invariant η of order r constructed as in Ex. v; for each of the matrices $Y_0, Y_2, \dots$ or $Y_1, Y_3, \dots$ is a symmetric invariant $\{\Omega', \Omega\}$ in which one of the elements of the median line is equal to 1, and all the rest of them are equal to 0.

Ex. vii. The general symmetric invariants $\xi=\text{inv}\{\Omega, \Omega\}$, $\xi'=\text{inv}\{\Omega', \Omega'\}$ are necessarily quasi-scalaric ante-alternants; and they are always zero matrices except when $r=1$ or 2.

With respect to skew-symmetric invariant transformands we have the following theorem:

Theorem III. *For the unipotent square matrix A of order r the two general skew-symmetric invariant transformands*

$$Y=inv\{A', \epsilon A\}, \quad Y'=inv\{A, \epsilon A'\}$$

can always be so chosen as to be connected by the symmetric relation

$$Y'=VYV;$$

and they are both zero matrices when $\alpha^2 \neq \epsilon$.

If $\alpha^2=\epsilon$, then:

(1) *when r is even, they are undegenerate square matrices of order r, each containing exactly $\frac{1}{2}r$ arbitrary parameters and having exactly $\frac{1}{2}r$ independent non-zero particularisations;*

(2) *when r is odd, they are degenerate square matrices of order r and rank $r-1$, each containing exactly $\frac{1}{2}(r-1)$ arbitrary parameters and having exactly $\frac{1}{2}(r-1)$ independent non-zero particularisations.*

The first part of the theorem follows from Ex. iv of § 254 and Theorem III *a* of § 255.

When $\alpha^2=\epsilon$, we will define

$$A_0=\alpha\Omega$$

to be the simple bi-canonical square matrix of order r whose latent root is α, and

$$A=hA_0H$$

to be any particular isomorphic transformation by which A can be derived from A_0. Then the *general skew-symmetric* invariants

$$\eta=\text{inv}\{A_0', \epsilon A_0\}, \quad \eta'=\text{inv}\{A_0, \epsilon A_0'\}$$

or $$\eta=\text{inv}\{\Omega', \Omega\}, \quad \eta'=\text{inv}\{\Omega, \Omega'\}$$

can be taken to be the simple square counter-slopes of order r and types $\{\pi', \pi\}$, $\{\pi, \pi'\}$ which are completely described in Ex. viii or x; and by Ex. ii of § 255 the *general skew-symmetric* invariant transformands in the second part of Theorem III can be taken to be the square matrices given by the formulae

$$Y = H'\eta H, \quad Y' = h\eta' h' \quad \text{......................(C)}$$

in which h', H' are the conjugates of h, H. The truth of the second part of the theorem follows from (C) and the properties of η and η' described in Ex. viii.

When r is an even integer $2t$, it will be shown that the parametric elements of the simple square counter-slope η or η' are homogeneous linear functions of t independent arbitrary parameters

$$\xi_{11} = \xi_0, \ \xi_{13} = \xi_2, \ \dots \ \xi_{1,r-1} = \xi_{r-2},$$

the elements of the paradiagonal being alternately ξ_0 and $-\xi_0$.

When r is an odd integer $2t+1$, it will be shown that the parametric elements of the simple counter-slope η or η' are homogeneous linear functions of t independent arbitrary parameters

$$\xi_{12} = \xi_1, \ \xi_{14} = \xi_3, \ \dots \ \xi_{1,r-1} = \xi_{r-2},$$

the elements of the paradiagonal being all equal to 0, and the elements of difference-weight 1 being alternately ξ_1 and $-\xi_1$.

In both cases (see Ex. x) we can also regard the parametric elements adjacent to and on one side of the median line (or leading diagonal) as the arbitrary parameters of which all other elements are homogeneous linear functions.

If $Y = [y]_r^r$ is a general invariant transformand of one of the types mentioned in Theorem III, it will be obvious that the matrix

$$\mathbf{Y} = \tfrac{1}{2}\{[y]_r^r - \overline{\underline{y}}_r^r\}$$

must become a general skew-symmetric invariant transformand of that type when the parameters occurring in its elements have been reduced by substitutions to the smallest number possible.

Ex. viii. *The general skew-symmetric invariants* $\eta = inv\{\Omega', \Omega\}$, $\eta' = inv\{\Omega, \Omega'\}$.

These can be taken to be the simple square counter-slopes (3) whose parametric elements are given by the formula (c), where the r parameters $\xi_0, \xi_1, \dots \xi_{r-1}$ are arbitrary subject to the conditions for skew-symmetry, i.e. the conditions

$$\xi_{ij} = -\xi_{JI} \quad \text{whenever} \quad i+I=j+J=r+1,$$

including $\qquad \xi_{ij}=0 \quad \text{whenever} \quad i+j=r+1.$

If we put $\qquad \rho = r-\kappa-1, \qquad (\kappa=0, 1, 2, \dots r-1),$

these conditions are equivalent to the r sets of equations corresponding to the r values of κ such as the sets:

$$2\xi_\kappa - ({}^{\rho-i}H_1 + {}^{i-2}H_1)\,\xi_{\kappa-1} + ({}^{\rho-i}H_2 + {}^{i-2}H_2)\,\xi_{\kappa-2} - \dots + (-1)^\kappa . ({}^{\rho-i}H_\kappa + {}^{i-2}H_\kappa)\,\xi_0 = 0, \quad (\mathrm{a}_\kappa')$$

where $r-\kappa$ is *odd*, and $i=1, 2, \dots \frac{1}{2}(r-\kappa+1)$;

$$0.\xi_\kappa-({}^{\rho-i}H_1-{}^{i-2}H_1)\xi_{\kappa-1}+({}^{\rho-i}H_2-{}^{i-2}H_2)\xi_{\kappa-2}-\dots+(-1)^\kappa.({}^{\rho-i}H_\kappa-{}^{i-2}H_\kappa)\xi_0=0, \quad (\mathrm{b}_\kappa')$$

where $r-\kappa$ is *even*, and $i=1, 2, \dots \frac{1}{2}(r-\kappa)$; the expression on the left reducing to the first term only in each case when $\kappa=0$.

By interchanging the words 'even' and 'odd' in the argument of Ex. iii it will be seen that the conditions for skew-symmetry will be satisfied when and only when one of the equations (say the first) in every such set as (a_κ') corresponding to an odd value of $r-\kappa$ is satisfied.

When r is an even integer $2t$, so that the sets (a_κ') and (b_κ') are

$$(b_0'),\ (a_1'),\ (b_2'),\ \dots\ (b'_{r-2}),\ (a'_{r-1}),$$

the conditions for skew-symmetry are equivalent to the t equations such as

$$2\xi_\kappa-{}^{r-\kappa-2}H_1.\xi_{\kappa-1}+{}^{r-\kappa-2}H_2.\xi_{\kappa-2}-\dots+(-1)^\kappa.{}^{r-\kappa-2}H_\kappa.\xi_0=0, \quad \dots\dots(\mathrm{c}_\kappa')$$

where κ receives the *odd* values $1, 3, \dots r-1$. Hence in the general skew-symmetric invariants η and η' the parameters of even weights, viz. the t parameters

$$\xi_0=\xi_{11},\ \xi_2=\xi_{13},\ \dots\ \xi_{r-2}=\xi_{1,r-1}, \quad \text{are arbitrary,}$$

whilst the t parameters $\xi_1, \xi_3, \dots \xi_{r-1}$ of odd weights are the homogeneous linear functions of them determined in succession by the equations (c_κ'), the first of which (when $r \not< 2$) is

$$2\xi_1-{}^{r-3}H_1.\xi_0=0.$$

When r is an odd integer $2t+1$, so that the sets (a_κ') and (b_κ') are

$$(a_0'),\ (b_1'),\ (a_2'),\ \dots\ (b'_{r-2}),\ (a'_{r-1}),$$

the conditions for skew-symmetry are equivalent to the $t+1$ equations such as

$$2\xi_\kappa-{}^{r-\kappa-2}H_1.\xi_{\kappa-1}+{}^{r-\kappa-2}H_2.\xi_{\kappa-2}-\dots+(-1)^\kappa.{}^{r-\kappa-2}H_\kappa.\xi_0=0, \quad \dots\dots(\mathrm{d}_\kappa')$$

where κ receives the *even* values $0, 2, 4, \dots r-2$; the expression on the left reducing to the first term only when $\kappa=0$. Hence in the general skew-symmetric invariants η and η' the parameters of odd weights, viz. the t parameters

$$\xi_1=\xi_{12},\ \xi_3=\xi_{14},\ \dots\ \xi_{r-2}=\xi_{1,r-1}, \quad \text{are arbitrary,}$$

whilst the $t+1$ parameters $\xi_0, \xi_2, \dots \xi_{r-1}$ of even weights are the homogeneous linear functions of them determined in succession by the equations (d_κ'), the first two of which (when $r \not< 3$) are

$$\xi_0=0,$$
$$2\xi_2-{}^{r-4}H_1.\xi_1=0.$$

Thus all the elements of η and η' are known homogeneous linear functions of:

the t arbitrary parameters $\xi_0, \xi_2, \dots \xi_{r-2}$ of even weights when $r=2t$;

the t arbitrary parameters $\xi_1, \xi_3, \dots \xi_{r-2}$ of odd weights when $r=2t+1$;

and in the latter case we have $\xi_0=0$.

Ex. ix. The conditions for skew-symmetry given by (c_κ') or (d_κ') in Ex. viii are equivalent to the equations

$$\begin{array}{llll} \xi_1=\xi_0 & , \xi_3=0 & & \text{when } r=4; \\ \xi_0=0, & \xi_2=\xi_1 & , \xi_4=0 & \text{when } r=5; \\ \xi_1=2\xi_0 & , \xi_3=\xi_2-\xi_0 & , \xi_5=0 & \text{when } r=6; \\ \xi_0=0,\ \xi_2=2\xi_1 & , \xi_4=\xi_3-\xi_1 & , \xi_6=0 & \text{when } r=7; \\ \xi_1=3\xi_0,\ \xi_3=2\xi_2-5\xi_0, & \xi_5=\xi_4-\xi_2+3\xi_0, & \xi_7=0 & \text{when } r=8. \end{array}$$

The general skew-symmetric invariant η of order 8 can be taken to be the square matrix

$$\eta(8)=\begin{bmatrix} 0, & 0, & 0, & 0 & , & 0 & , & 0 & , & 0 & , & -\xi_0 \\ 0, & 0, & 0, & 0 & , & 0 & , & 0 & , & \xi_0 & , & -3\xi_0 \\ 0, & 0, & 0, & 0 & , & 0 & , & -\xi_0 & , & 2\xi_0 & , & -\xi_2 \\ 0, & 0, & 0, & 0 & , & \xi_0 & , & -\xi_0 & , & -2\xi_0+\xi_2 & , & 5\xi_0-2\xi_2 \\ 0, & 0, & 0, & -\xi_0 & , & 0 & , & 3\xi_0-\xi_2 & , & -3\xi_0+\xi_2 & , & -\xi_4 \\ 0, & 0, & \xi_0, & \xi_0 & , & -3\xi_0+\xi_2, & & 0 & , & 3\xi_0-\xi_2+\xi_4, & & -3\xi_0+\xi_2-\xi_4 \\ 0, & -\xi_0, & -2\xi_0, & 2\xi_0-\xi_2 & , & 3\xi_0-\xi_2 & , & -3\xi_0+\xi_2-\xi_4, & & 0 & , & -\xi_6 \\ \underline{\xi_0}, & 3\xi_0, & \underline{\xi_2}, & -5\xi_0+2\xi_2, & & \underline{\xi_4} & , & 3\xi_0-\xi_2+\xi_4 & , & \underline{\xi_6} & , & 0 \end{bmatrix}$$

in which $\xi_0, \xi_2, \xi_4, \xi_6$ are arbitrary parameters. If the four underlined elements are regarded as given, all the other elements can be determined in succession by the law of construction (described in Ex. x) together with the restrictions imposed by symmetry.

The corresponding invariant $\eta(6)$ can be derived from $\eta(8)$ by putting $\xi_0=0$, striking out the first two horizontal and the first two vertical rows, and substituting ξ_{i-2} for ξ_i. The corresponding invariant $\eta(9)$ can be derived from $\eta(8)$ by adding one initial horizontal and one initial vertical row of 0's and substituting ξ_{i+1} for ξ_i.

Ex. x. *Simpler constructions for the general skew-symmetric invariants* η, η'.

By giving the last instead of the first value to i in (a_κ') we see that the conditions for skew-symmetry in Ex. viii are equivalent to the equations

$$\xi_{ij}=0 \quad \text{whenever} \quad i+j=r+1. \qquad\qquad (11)$$

When an invariant η or η' has been constructed in accordance with the law that

the sum of two consecutive elements of the same difference-weight κ *and the contiguous element of smaller difference-weight* $\kappa-1$ *is always equal to* 0,

it will be skew-symmetric if and only if the conditions for skew-symmetry are satisfied by the elements lying on the median line, i.e. if and only if all such elements are equal to 0; moreover the elements adjacent to the median line are rendered uniquely determinate by (11) and the law of construction when those of them which lie on one side of the median line are given. Consequently a general skew-symmetric invariant η or η' can be regarded as a simple counter-slope of type $\{\pi', \pi\}$ or $\{\pi, \pi'\}$ in which:

(i) the elements lying on the median line (or leading diagonal) are all equal to 0;

(ii) the parametric elements adjacent to the median line and lying on one side of it are arbitrary parameters;

(iii) all the other elements of difference-weights 0, 1, 2, ... are uniquely determined in succession by the law of construction.

For example the general skew-symmetric invariant η of order 8 can be taken to be the square matrix

$$\eta(8)=\begin{bmatrix} 0, & 0, & 0, & 0, & 0, & 0, & 0, & \eta_0 \\ 0, & 0, & 0, & 0, & 0, & 0, & -\eta_0, & 3\eta_0 \\ 0, & 0, & 0, & 0, & 0, & \eta_0, & -2\eta_0, & 3\eta_0-\eta_2 \\ 0, & 0, & 0, & 0, & -\eta_0, & \eta_0, & -\eta_0+\eta_2, & \eta_0-2\eta_2 \\ 0, & 0, & 0, & \underline{\eta_0}, & 0, & -\eta_2, & \eta_2, & -\eta_2+\eta_4 \\ 0, & 0, & -\eta_0, & -\eta_0, & \underline{\eta_2}, & 0, & -\eta_4, & \eta_4 \\ 0, & \eta_0, & 2\eta_0, & \eta_0-\eta_2, & -\eta_2, & \underline{\eta_4}, & 0, & -\eta_6 \\ -\eta_0, & -3\eta_0, & -3\eta_0+\eta_2, & -\eta_0+2\eta_2, & \eta_2-\eta_4, & -\eta_4, & \underline{\eta_6}, & 0 \end{bmatrix}$$

in which $\eta_0, \eta_2, \eta_4, \eta_6$ are arbitrary parameters. If the elements on the median line are put equal to 0 and the four underlined elements are regarded as given, all the other parametric elements can be uniquely determined in succession by the law of construction. The corresponding invariant $\eta(6)$ can be derived from $\eta(8)$ by striking out the first and last horizontal and vertical rows. The corresponding invariant $\eta(9)$ can be derived from $\eta(8)$ by adding one initial horizontal and one initial vertical row of 0's, and substituting η_{i+1} for η_i; or the corresponding invariant $\eta(7)$ by striking out the last horizontal and the last vertical row, and substituting η_{i+1} for η_i.

The general skew-symmetric invariant $\eta(r)$ of order r which is constructed in this way is the matrix given by the formula (14) or (14′) of Ex. xi when Y is chosen to be the matrix (16). In fact the matrix (16) is the coefficient of η_0 in $\eta(r)$ when $r=2t$. Accordingly we can find a general formula for all the parametric elements ξ_{ij} of $\eta(r)$. If we put

$$y_{ij}(t)=(-1)^{t-1}\,.\,{}^{t-j}H_{j-i} \quad \text{.................................(12)}$$

by evaluating the matrix (16), so that

$$y_{ij}(t)=0 \quad \text{when} \quad j<i \quad \text{or} \quad j>t>i-1, \quad (\text{and also when } j<1 \text{ or } i>2t),$$

then when $r=2t$, the formula is

$$\xi_{ij}=\eta_0\cdot y_{ij}(t)+\eta_2\cdot y_{i,j-2}(t-1)+\ \dots\ +\eta_{r-2}\cdot y_{i,j-r+2}(1)\,; \quad \text{...........(13)}$$

and when $r=2t+1$, the formula is

$$\xi_{ij}=\eta_1\cdot y_{i,j-1}(t)+\eta_3\cdot y_{i,j-3}(t-1)+\ \dots\ +\eta_{r-2}\cdot y_{i,j-r+2}(1). \quad \text{...........(13′)}$$

The series of coefficients of η_0 (or η_1, or η_2, ...) occurring in successive parametric diagonal lines, and starting from an element adjacent to the median line, are the integers

$${}^iH_\kappa \quad \text{in which} \quad i=0, 1, 2, 3, \dots$$

for the successive values 0, 1, 2, 3, ... of κ.

Ex. xi. *Particularisations of the general skew-symmetric invariant* $\eta=inv\,\{\Omega', \Omega\}$.

From the matrix equation determining η it will be at once evident that:

If $Y=[y]_{2t}^{2t}$ *is any skew-symmetric matrix which for all positive integral values of t is a particular undegenerate skew-symmetric invariant* η *of order 2t, a complete set of independent non-zero particularisations of the general skew-symmetric invariant* $\eta=\eta(r)$ *of order r is formed by the skew-symmetric matrices:*

$$Y_{2\tau}=\begin{bmatrix}0, & 0\\ 0, & y\end{bmatrix}_{2\tau,\ r-2\tau}^{2\tau,\ r-2\tau} \quad \textit{in which } \tau=0, 1, 2, \dots t-1 \qquad \textit{when } r=2t;$$

$$Y_{2\tau+1}=\begin{bmatrix}0, & 0\\ 0, & y\end{bmatrix}_{1+2\tau,\ r-1-2\tau}^{1+2\tau,\ r-1-2\tau} \quad \textit{in which } \tau=0, 1, 2, \dots t-1 \qquad \textit{when } r=2t+1.$$

When any such particular invariant has been determined, we can take the general skew-symmetric invariant $\eta(r)$ to be the matrix:

$$\eta(r)=\eta_0 Y_0+\eta_2 Y_2+\dots+\eta_{r-2} Y_{r-2} \qquad \text{when } r=2t\,; \quad \text{...............(14)}$$

$$\eta(r)=\eta_1 Y_1+\eta_3 Y_3+\dots+\eta_{r-2} Y_{r-2} \qquad \text{when } r=2t+1\,; \quad \text{.........(14′)}$$

where the coefficients on the right are arbitrary scalar parameters.

If $[\phi]_r^r$, $[\phi']_r^r$ are the two mutually conjugate simple square counter-slopes defined in Ex. ii, we could choose Y to be the skew-symmetric matrix

$$Y=[y]_r^r=\tfrac{1}{2}\{[\phi]_r^r-[\phi']_r^r\}, \quad \text{where } r=2t\,; \quad \text{........................(15)}$$

so that Y is a simple square counter-slope of type $\{\pi', \pi\}$ and of even order r whose parametric elements are given by the formula

$$2y_{ij}=(-1)^{j-1}\,.\,{}^{i-1}H_\kappa+(-1)^{i-1}\,.\,{}^{r-j}H_\kappa, \quad \text{where } \kappa=j-i.$$

A more convenient choice is the skew-symmetric simple square counter-slope

$$Y=[\psi]_{2t}^{2t}=\begin{bmatrix}0, & -\phi \\ \phi', & 0\end{bmatrix}_{t,t}^{t,t}, \quad \text{................................(16)}$$

this being that particular skew-symmetric invariant η of order $2t$ in which the first of the parametric elements adjacent to and lying below the median line is equal to 1, and all the rest of them are equal to 0. From (b) of Ex. ii and the property (16) of Note 1 in § 255 it will be seen that all the parametric elements of Y are given by the formula

$$\psi_{ij}=(-1)^{t-i} \cdot {}^{t-j}H_\kappa, \quad \text{where } \kappa=j-i. \quad \text{........................(17)}$$

When we choose Y to be the matrix (16), the formula (14) or (14′) gives the general skew-symmetric invariant η of order r constructed as in Ex. x.

Ex. xii. The general skew-symmetric invariants $\xi=\text{inv}\ \{\Omega, \Omega\}$, $\xi'=\text{inv}\ \{\Omega', \Omega'\}$ are necessarily quasi-scalaric ante-alternants; and they are always zero matrices.

Ex. xiii. *Undegenerate symmetric or skew-symmetric invariant transformands.*

If A is a unipotent square matrix of order r with constant elements, whose latent root is a, and if $\epsilon=\pm 1$, there exist *undegenerate* invariant transformands $\{A', \epsilon A\}$, $\{A, \epsilon A'\}$ which are:

symmetric when and only when $a^2=\epsilon$, and r is odd;

skew-symmetric when and only when $a^2=\epsilon$, and r is even.

Ex. xiv. *If A is a unipotent square matrix, the sum of a 'general symmetric' and a 'general skew-symmetric' invariant $\{A', A\}$ with independent parameters is always a 'general' invariant $\{A', A\}$.*

For when A is the simple bi-canonical square matrix Ω, the sum is a simple counter-slope in which one element of every effective parametric diagonal line can be regarded as arbitrary.

§ 258. Scalar invariants of substitutions by simple bi-canonical square matrices or their conjugates.

1. *Applications to other substitutions.*

The scalar invariants of substitutions by any unipotent square matrices can (see § 256. 3) be derived from the scalar invariants which will be described in this article by a mere change of variables, i.e. by replacing the actual variables $x_1, x_2, \ldots$ and $y_1, y_2, \ldots$ by certain independent homogeneous linear functions $\xi_1, \xi_2, \ldots$ of $x_1, x_2, \ldots$ and $\eta_1, \eta_2, \ldots$ of $y_1, y_2, \ldots$. Moreover the independent bilinear invariants of any two given homogeneous linear substitutions (see Ex. iii of § 260) or the independent quadratic invariants of any one homogeneous linear substitution (see Ex. vi of § 260) can always be taken to consist of sets of functions (of suitably chosen new variables) of the forms described in this article. In fact the forms of all the scalar invariants of substitutions by square matrices A and B are known when the characteristic potent divisors of A and B are known.

2. *Bilinear invariants of two substitutions.*

Let A and B be the simple bi-canonical square matrices of order r whose latent roots are a and β; and let A' and B' be their conjugates. Then if J is the simple reversant of order r, we can take the general bilinear invariants of the pairs of substitutions:

(i) $\overline{\underline{x'}}_r=A\cdot\overline{\underline{x}}_r,\quad \overline{\underline{y'}}_r=B\cdot\overline{\underline{y}}_r$ to be $f=[x]_r\cdot C\cdot\overline{\underline{y}}_r$;

(ii) $\overline{\underline{x'}}_r=A'\cdot\overline{\underline{x}}_r,\quad \overline{\underline{y'}}_r=B'\cdot\overline{\underline{y}}_r$ to be $f'=[x]_r\cdot JCJ\cdot\overline{\underline{y}}_r$;

(iii) $\overline{\underline{x'}}_r=A'\cdot\overline{\underline{x}}_r,\quad \overline{\underline{y'}}_r=B\cdot\overline{\underline{y}}_r$ to be $g=[x]_r\cdot JC\cdot\overline{\underline{y}}_r$;

(iv) $\overline{\underline{x'}}_r=A\cdot\overline{\underline{x}}_r,\quad \overline{\underline{y'}}_r=B'\cdot\overline{\underline{y}}_r$ to be $g'=[x]_r\cdot CJ\cdot\overline{\underline{y}}_r$;

where $C=[c]_r^r$ is a general invariant transformand $\{A', B\}$. When C is a particular invariant transformand $\{A', B\}$, the functions f, f', g, g' are corresponding particular bilinear invariants of the pairs of substitutions (i), (ii), (iii), (iv); and we will call them bilinear invariants of *rank* ρ when the matrix C has rank ρ. We know that C is always a simple square counter-slope of type $\{\pi', \pi\}$.

In describing these bilinear invariants we will use the abbreviated notation

$$f(i+1, i+2, \dots i+s;\ j+1, j+2, \dots j+s)=f(x_{i+1}, x_{i+2}, \dots x_{i+s};\ y_{j+1}, y_{j+2}, \dots y_{j+s}) \quad \dots\dots\dots\dots(1)$$

in which consecutive variables are represented by their suffixes. The function (1) is derived from the function

$$f(1, 2, \dots s;\ 1, 2, \dots s)=f(x_1, x_2, \dots x_s;\ y_1, y_2, \dots y_s)$$

by replacing x_t, y_t by x_{i+t}, y_{i+t}; and the function

$$f(i+s, \dots i+2, i+1;\ j+s, \dots j+2, j+1)=f(x_{i+s}, \dots x_{i+2}, x_{i+1};\ y_{j+s}, \dots y_{j+2}, y_{j+1})$$

is derived from the function (1) by replacing x_{i+t}, y_{i+t} by $x_{i+s+1-t}, y_{j+s+1-t}$.

Since the functions f', g, g' in (ii), (iii), (iv) can be derived from the function f by replacing

$$x_i, y_j \text{ by } x_u, y_v;\quad x_i \text{ by } x_u;\quad y_j \text{ by } y_v;$$

where $$u=r+1-i,\quad v=r+1-j,$$

all bilinear invariants of the four pairs of substitutions will be known when all those of any one of them are known. In fact when we use the notation (1), general or particular bilinear invariants of the pairs of substitutions (i), (ii), (iii), (iv) can always be associated together in sets such as

$$f(1, 2, \dots \overset{*}{r};\ 1, 2, \dots \overset{*}{r}),\ f(r, r-1, \dots \overset{*}{1};\ r, r-1, \dots \overset{*}{1}),$$
$$f(r, r-1, \dots \overset{*}{1};\ 1, 2, \dots \overset{*}{r}),\ f(1, 2, \dots \overset{*}{r};\ r, r-1, \dots \overset{*}{1}),$$

where all four of the functions are known when any one of them is known.

If A and B were the simple bi-canonical square matrices of orders m and n whose latent roots are a and β, where m and n are any two non-zero positive integers the smaller of which is equal to r, and if J_m and J_n are the simple reversants of orders m and n, we could take the general bilinear invariants of the pairs of substitutions:

(i′) $\overline{\underline{x'}}_m = A\,.\,\overline{\underline{x}}_m,\quad \overline{\underline{y'}}_n = B\,.\,\overline{\underline{y}}_n$ to be $[x]_m\,.\,C'\,.\,\overline{\underline{y}}_n$;

(ii′) $\overline{\underline{x'}}_m = A'\,.\,\overline{\underline{x}}_m,\quad \overline{\underline{y'}}_n = B'\,.\,\overline{\underline{y}}_n$ to be $[x]_m\,.\,J_mC'J_n\,.\,\overline{\underline{y}}_n$;

(iii′) $\overline{\underline{x'}}_m = A'\,.\,\overline{\underline{x}}_m,\quad \overline{\underline{y'}}_n = B\,.\,\overline{\underline{y}}_n$ to be $[x]_m\,.\,J_mC'\,.\,\overline{\underline{y}}_n$;

(iv′) $\overline{\underline{x'}}_m = A\,.\,\overline{\underline{x}}_m,\quad \overline{\underline{y'}}_n = B'\,.\,\overline{\underline{y}}_n$ to be $[x]_m\,.\,C'J_n\,.\,\overline{\underline{y}}_n$;

where C' is the simple counter-slope

$$[0, c]_r^{m-r,\,r} \text{ when } m=r \not> n;\quad \begin{bmatrix} 0 \\ c \end{bmatrix}_{m-r,\,r}^{r} \text{ when } n=r \not> m$$

in which $[c]_r^r$ is the same matrix C as in (i), (ii), (iii), (iv). Hence general or particular bilinear invariants of the pairs of substitutions (i′), (ii′), (iii′), (iv′) can always be associated together in sets such as

$$f(1, 2, \dots \overset{*}{r};\ n-r+1, n-r+2, \dots \overset{*}{n}),\ f(r, r-1, \dots \overset{*}{1};\ r, r-1, \dots \overset{*}{1}),$$
$$f(r, r-1, \dots \overset{*}{1};\ n-r+1, n-r+2, \dots \overset{*}{n}),\ f(1, 2, \dots \overset{*}{r};\ r, r-1, \dots \overset{*}{1})$$

when $$m=r \not> n;$$

$$f(m-r+1,\ m-r+2, \ldots \overset{*}{m};\ 1, 2, \ldots \overset{*}{r}),\ f(r, r-1, \ldots \overset{*}{1};\ r, r-1, \ldots \overset{*}{1}),$$
$$f(r, r-1, \ldots \overset{*}{1};\ 1, 2, \ldots \overset{*}{r}),\ f(m-r+1,\ m-r+2, \ldots \overset{*}{m};\ r, r-1, \ldots \overset{*}{1})$$

when $$n=r \not> m;$$

where the functions are the same as before, only the variables having been changed. In all cases only r of the variables $x_1, x_2, \ldots x_m$ and only r of the variables $y_1, y_2, \ldots y_n$ can occur in non-zero bilinear invariants.

The asterisks indicate the 'apical' variables.

It follows that in actual evaluations we can confine ourselves to the pairs of substitutions (i), (ii), (iii), (iv) or to any one of them. We shall direct our chief attention to the substitutions (i) whose scalar equations are

$$x_1' = \alpha(x_1+x_2),\quad x_2' = \alpha(x_2+x_3), \ldots x'_{r-1} = \alpha(x_{r-1}+x_r),\quad x_r' = \alpha x_r;$$
$$y_1' = \beta(y_1+y_2),\quad y_2' = \beta(y_2+y_3), \ldots y'_{r-1} = \beta(y_{r-1}+y_r),\quad y_r' = \beta y_r. \quad \ldots\ldots(a)$$

If $\alpha\beta \neq 1$, the general invariant transformand C is a zero matrix, and there are no non-zero bilinear invariants.

If $\alpha\beta = 1$, we could take the general invariant transformand C to be the simple square counter-slope η described in Theorem IVa of § 255. In this case there are exactly r independent particular non-zero bilinear invariants of each of the pairs of substitutions (i), (ii), (iii), (iv); and the sum of such independent bilinear invariants, each multiplied by an arbitrary scalar parameter, is a general bilinear invariant. After determining any function which is a particular bilinear invariant of maximum rank r for all non-zero positive integral values of r, we can derive from it a complete set of independent particular bilinear invariants by using the following principle, which is immediately deducible from Ex. ii of § 257. The notation defined in (1) is used, i.e. variables are represented by their suffixes.

If (when $\alpha\beta=1$) we have determined functions

$$f_r(1, 2, \ldots \overset{*}{r};\ 1, 2, \ldots \overset{*}{r}),\quad f_r'(\overset{*}{1}, 2, \ldots r;\ \overset{*}{1}, 2, \ldots r),$$
$$g_r(\overset{*}{1}, 2, \ldots r;\ 1, 2, \ldots \overset{*}{r}),\quad g_r'(1, 2, \ldots \overset{*}{r};\ \overset{*}{1}, 2, \ldots r),$$

which are particular non-zero bilinear invariants of maximum rank r of the respective pairs of substitutions (i), (ii), (iii), (iv) *for all non-zero positive integral values of r, then complete sets of independent non-zero bilinear invariants of the substitutions* (i), (ii), (iii), (iv) *having ranks $r, r-1, \ldots r-\rho, \ldots 2, 1$ can be obtained by giving to ρ the values $0, 1, \ldots \rho, \ldots r-2, r-1$ in any one of the four sets of formulae:*

$$f_{r-\rho}(\rho+1, \ldots \overset{*}{r};\ \rho+1, \ldots \overset{*}{r}),\quad f_{r-\rho}(r-\rho, \ldots \overset{*}{1};\ r-\rho, \ldots \overset{*}{1}),$$
$$f_{r-\rho}(r-\rho, \ldots \overset{*}{1};\ \rho+1, \ldots \overset{*}{r}),\quad f_{r-\rho}(\rho+1, \ldots \overset{*}{r};\ r-\rho, \ldots \overset{*}{1});$$
$$f'_{r-\rho}(\overset{*}{r}, \ldots \rho+1;\ \overset{*}{r}, \ldots \rho+1),\quad f'_{r-\rho}(\overset{*}{1}, \ldots r-\rho;\ \overset{*}{1}, \ldots r-\rho),$$
$$f'_{r-\rho}(\overset{*}{1}, \ldots r-\rho;\ \overset{*}{r}, \ldots \rho+1),\quad f'_{r-\rho}(\overset{*}{r}, \ldots \rho+1;\ \overset{*}{1}, \ldots r-\rho);$$
$$g_{r-\rho}(\overset{*}{r}, \ldots \rho+1;\ \rho+1, \ldots \overset{*}{r}),\quad g_{r-\rho}(\overset{*}{1}, \ldots r-\rho;\ r-\rho, \ldots \overset{*}{1}),$$
$$g_{r-\rho}(\overset{*}{1}, \ldots r-\rho;\ \rho+1, \ldots \overset{*}{r}),\quad g_{r-\rho}(\overset{*}{r}, \ldots \rho+1;\ r-\rho, \ldots \overset{*}{1});$$
$$g'_{r-\rho}(\rho+1, \ldots \overset{*}{r};\ \overset{*}{r}, \ldots \rho+1);\quad g'_{r-\rho}(r-\rho, \ldots \overset{*}{1};\ \overset{*}{1}, \ldots r-\rho),$$
$$g'_{r-\rho}(r-\rho, \ldots \overset{*}{1};\ \overset{*}{r}, \ldots \rho+1),\quad g'_{r-\rho}(\rho+1, \ldots \overset{*}{r};\ \overset{*}{1}, \ldots r-\rho).$$
$$\ldots\ldots\ldots\ldots(A)$$

In all cases the asterisks are placed over the integers representing the apical variables. In applying this principle we may and will suppose f_r', g_r, g_r' to be the functions derived from the function f_r by replacing

$$x_i,\ y_j \text{ by } x_{r+1-i},\ y_{r+1-j};\quad x_i \text{ by } x_{r+1-i};\quad y_j \text{ by } y_{r+1-j};$$

so that all four of the functions f_r, f_r', g_r, g_r' are known when any one of them is known. Then the four lines in (A) are four different notations for the same four functions. Possible choices of f_r and the corresponding complete sets of bilinear invariants will be described in the examples.

Every expression for a bilinear invariant given in (A) or in the examples is merely a development of a matrix product

$$[x]_r [e]_r^r \overline{\underline{y}}_r = \Sigma e_{ij} x_i y_j$$

in which $E=[e]_r^r$ is a certain simple square slope. Such developments are only of minor importance, because the properties of the bilinear invariants are most easily seen when they are represented as undeveloped matrix products. In the examples we shall consistently use the third set of notations in (A). In other words we will use the notation

$$[x]_r [e]_r^r \overline{\underline{y}}_r = \phi(1, 2, \dots r;\ 1, 2, \dots r)$$

with two ascending sequences when E is a simple square slope of type $\{\pi, \pi\}$, i.e. when E is an invariant transformand $\{A, B\}$, so that ϕ is a bilinear invariant of the substitutions (iii). The second set of notations, in which functions with two ascending sequences are bilinear invariants of the substitutions (ii), is in many respects the most elegant and suggestive. We can pass from the third to the second notations by reversing all sequences which represent y-variables.

The most easily determined complete sets of bilinear invariants are those derived from the invariant transformands $[\phi]_r^r$, $[\phi']_r^r$ defined in Ex. ii of § 257, or more easily from the corresponding invariant transformands $[\theta]_r^r$, $[\theta']_r^r$ defined in Ex. i of § 257. These are described in Exs. i and iv. The simplest complete sets of bilinear invariants, containing fewer non-zero terms, are those described in Ex. xi, where the bilinear invariants of odd ranks are derived from the symmetric invariant transformand $[\psi]_{2t+1}^{2t+1}$ defined in Ex. vi of § 257, and the bilinear invariants of even ranks are derived from the skew-symmetric invariant transformand $[\psi]_{2t}^{2t}$ defined in Ex. xi of § 257. Each bilinear invariant is expressed as a sum of terms, one corresponding to each non-zero parametric diagonal line in the invariant transformand from which it is derived.

From the fact that the conjugate of an invariant transformand $\{\Omega', \Omega\}$ is always another invariant transformand $\{\Omega', \Omega\}$, or from the forms of the equations (a), it follows that:

The interchange of the letters x and y converts every bilinear invariant of one of the pairs of substitutions (i) *and* (ii) *into another bilinear invariant of the same pair; and converts every bilinear invariant of one of the pairs of substitutions* (iii) *and* (iv) *into a bilinear invariant of the other pair.*

When $\quad \phi(a_1, a_2, \dots a_s;\ b_1, b_2, \dots b_s) = \phi(x_{a_1}, x_{a_2}, \dots x_{a_s};\ y_{b_1}, y_{b_2}, \dots y_{b_s})$ ....(2)

is a given function of s pairs of variables for which the standard abbreviated notation (1) is used, then for the function derived from ϕ by interchanging the letters x and y we will use the notation

$$\phi'(b_1, b_2, \dots b_s;\ a_1, a_2, \dots a_s) = \phi(y_{a_1}, y_{a_2}, \dots y_{a_s};\ x_{b_1}, x_{b_2}, \dots x_{b_s}); \quad \text{......(2')}$$

and we will call (2′) the function ϕ' *conjugate* to ϕ. Further we will call (2) a symmetric function when $\phi' = \phi$, or a skew-symmetric function when $\phi' = -\phi$. The two sequences of ϕ' are formed by interchanging the two sequences of ϕ. We know that ϕ will be a bilinear invariant of the substitutions (i), (ii), (iii), (iv) when and only when ϕ' is a bilinear invariant of the substitutions (i), (ii), (iv), (iii). In deriving the symbol for ϕ' from that for ϕ, asterisks must be moved with the integers to which they are attached; and we can then apply the principle (A) to ϕ' in exactly the same way as to ϕ. The letters f' and g' will always have the meanings ascribed to them in (i), (ii), (iii), (iv), and will not denote the functions conjugate to f and g.

The above considerations show that from any given complete set of bilinear invariants of the substitutions (i) or (ii) we can always deduce a second complete set of bilinear invariants of the same substitutions by interchanging the letters x and y; and that the second set will be different from the first unless the given bilinear invariants are symmetric or skew-symmetric functions.

In the examples which follow we shall temporarily use in conjunction with (1) the still more abbreviated notation

$$f(i+1,\ i+2,\ \dots\ i+s) \quad \dots\dots(3)$$

for any defined function f of the s pairs of variables $(x_{i+1}, y_{i+1}), (x_{i+2}, y_{i+2}), \dots (x_{i+s}, y_{i+s})$ whose suffixes are $i+1, i+2, \dots i+s$; the function (3) being derived from the function

$$f(1,\ 2,\ \dots\ s)$$

of the s pairs of variables $(x_1, y_1), (x_2, y_2), \dots (x_s, y_s)$ by replacing x_t, y_t by x_{i+t}, y_{i+t} for the values $1, 2, \dots s$ of t. Another meaning will be attached to the notation (3) in sub-article 3.

Ex. i. *The complete sets of bilinear invariants derived from the function*

$$[x]_r [\phi]_r^r \overline{\underline{y}}_r = \phi(r, r-1, \dots 2, 1;\ 1, 2, \dots r-1, r).$$

Whenever τ and s are positive integers such that $s \not< 1$, we can and will define a function G_τ of s pairs of variables by the equation

$$G_\tau(1, 2, \dots s;\ 1, 2, \dots s) = \Sigma\{(-1)^i . {}^iH_\tau . x_{1+i}y_{1+i}\}, \qquad (i=0, 1, 2, \dots r-1),$$
$$= {}^0H_\tau . x_1y_1 - {}^1H_\tau . x_2y_2 + {}^2H_\tau . x_3y_3 - \dots + (-1)^{s-1} . {}^{s-1}H_\tau . x_sy_s; \quad \dots\dots(4)$$

so that whenever τ, a, b, s are positive integers and $s \not< 1$, we have:

$$G_\tau(a+s, \dots a+2, a+1;\ b+1, b+2, \dots b+s) = \Sigma\{(-1)^i . {}^iH_\tau . x_{a+s-i}y_{b+1+i}\},$$
$$G_\tau(a+1, a+2, \dots a+s;\ b+s, \dots b+2, b+1) = \Sigma\{(-1)^i . {}^iH_\tau . x_{a+1+i}y_{b+s-i}\},$$
$$G_\tau(a+1, a+2, \dots a+s;\ b+1, b+2, \dots b+s) = \Sigma\{(-1)^i . {}^iH_\tau . x_{a+1+i}y_{b+1+i}\},$$
$$G_\tau(a+s, \dots a+2, a+1;\ b+s, \dots b+2, b+1) = \Sigma\{(-1)^i . {}^iH_\tau . x_{a+s-i}y_{b+s-i}\},$$
$$(i=0, 1, 2, \dots s-1). \quad \dots\dots(4')$$

For any given value of τ the unsigned coefficients of the successive terms of a function G_τ of any number of pairs of variables are always successive integers (starting from the first) of the same series derived from

$${}^iH_\tau \text{ by putting } i=0, 1, 2, 3, \dots.$$

When $\alpha\beta=1$, we can obtain a set of corresponding bilinear invariants of maximum rank r of the pairs of substitutions (i), (ii), (iii), (iv) by taking C to be the matrix $[\phi]_r^r$ defined in Ex. ii of § 257, so that JC is the matrix $[\theta]_r^r$ defined in Ex. i of § 257. By developing the corresponding matrix products it will be seen that these bilinear invariants of the substitutions (i), (ii), (iii), (iv) are the functions

$$f_r(1, 2, \dots r) = \phi(r, r-1, \dots 2, 1;\ 1, 2, \dots r-1, r)$$
$$= \Sigma(-1)^\tau G_\tau(r, r-1, \dots \tau+1;\ \tau+1, \dots r-1, r),$$

$$f_r'(1, 2, \dots r) = \phi(1, 2, \dots r-1, r;\ r, r-1, \dots 2, 1)$$
$$= \Sigma(-1)^\tau G_\tau(1, 2, \dots r-\tau;\ r-\tau, \dots 2, 1),$$
$$g_r(1, 2, \dots r) = \phi(1, 2, \dots r-1, r;\ 1, 2, \dots r-1, r)$$
$$= \Sigma(-1)^\tau G_\tau(1, 2, \dots r-\tau;\ \tau+1, \dots r-1, r),$$
$$g_r'(1, 2, \dots r) = \phi(r, r-1, \dots 2, 1;\ r, r-1, \dots 2, 1)$$
$$= \Sigma(-1)^\tau G_\tau(r, r-1, \dots \tau+1;\ r-\tau, \dots 2, 1),$$
$$(\tau = 0, 1, 2, \dots r-1), \quad \dots\dots\dots\dots\dots(5)$$

where the functions G_τ are defined as above, and where we can and will regard $\phi = \phi_r$ as a completely defined function of r pairs of variables for every non-zero positive integral value of r, the four defining equations in (5) being mutually equivalent. When a suffix is attached to ϕ, it will always be equal to the number of pairs of variables; but as this number is shown by the sequences representing the variables, it will usually be omitted.

Applying the principle (A), it follows from the formulae (5) that when $\alpha\beta = 1$, complete sets of independent non-zero bilinear invariants of the pairs of substitutions (i), (ii), (iii), (iv) having ranks $r, r-1, \dots r-\rho, \dots 2, 1$ can be obtained by putting $\rho = 0, 1, \dots \rho, \dots r-2, r-1$ in the formulae:

$$f_{r-\rho}(\rho+1, \dots r-1, \overset{*}{r};\ \rho+1, \dots r-1, \overset{*}{r})$$
$$= \phi(\overset{*}{r}, r-1, \dots \rho+1;\ \rho+1, \dots r-1, \overset{*}{r})$$
$$= \Sigma(-1)^\tau G_\tau(\overset{*}{r}, r-1, \dots \rho+\tau+1;\ \rho+\tau+1, \dots r-1, \overset{*}{r}),$$
$$f'_{r-\rho}(\overset{*}{1}, 2, \dots r-\rho;\ \overset{*}{1}, 2, \dots r-\rho) = \phi(\overset{*}{1}, 2, \dots r-\rho;\ r-\rho, \dots 2, \overset{*}{1})$$
$$= \Sigma(-1)^\tau G_\tau(\overset{*}{1}, 2, \dots r-\rho-\tau;\ r-\rho-\tau, \dots 2, \overset{*}{1}),$$
$$g_{r-\rho}(\overset{*}{1}, 2, \dots r-\rho;\ \rho+1, \dots r-1, \overset{*}{r}) = \phi(\overset{*}{1}, 2, \dots r-\rho;\ \rho+1, \dots r-1, \overset{*}{r})$$
$$= \Sigma(-1)^\tau G_\tau(\overset{*}{1}, 2, \dots r-\rho-\tau;\ \rho+\tau+1, \dots r-1, \overset{*}{r}),$$
$$g'_{r-\rho}(\rho+1, \dots r-1, \overset{*}{r};\ \overset{*}{1}, 2, \dots r-\rho) = \phi(\overset{*}{r}, r-1, \dots \rho+1;\ r-\rho, \dots 2, \overset{*}{1})$$
$$= \Sigma(-1)^\tau G_\tau(\overset{*}{r}, r-1, \dots \rho+\tau+1;\ r-\rho-\tau, \dots 2, \overset{*}{1}),$$
$$(\tau = 0, 1, 2, \dots r-\rho-1). \quad \dots\dots\dots\dots\dots(5')$$

Here the bilinear invariants of the substitutions (i) correspond to the complete set of independent invariant transformands $\{A', B\}$ described in Ex. ii of § 257 when $[y]_r^r$ is taken to be $[\phi]_r^r$, and the bilinear invariants of the other substitutions can be derived from them by the appropriate reversing substitutions for the variables.

To establish all the formulae (5) and (5′), it is sufficient to obtain for any one of the bilinear invariants f_r, f_r', g_r, g_r' of rank r the development shown in (5). If we put $u = r+1-i$, we have

$$f_r = [x]_r [\phi]_r^r \underset{\smile}{\overset{\frown}{y}}_r = \Sigma \phi_{uj} x_i y_j = \Sigma\{(-1)^{j-1} . {}^{r-i}H_\tau . x_i y_j\},$$

where $$\tau = i+j-r-1,$$

and where the summation extends over all integral values of i and j such that

$$1 \ngtr i \ngtr r, \quad 1 \ngtr j \ngtr r, \quad i+j \nless r+1.$$

The possible values of τ are those consistent with the conditions $\tau \nless 0$, $\tau \ngtr r-1$; and when the terms of the sum are arranged according to their total weights in the suffixes of the x's and y's, we obtain the development

$$f_r = \Sigma(-1)^\tau F_\tau, \qquad (\tau = 0, 1, 2, \dots r-1),$$

where $$F_\tau = \Sigma\{(-1)^\omega . {}^\omega H_\tau . x_{r-\omega} y_{\tau+1+\omega}\}, \qquad (\omega = 0, 1, 2, \dots r-\tau-1),$$
$$= G_\tau(r, r-1, \dots \tau+1;\ \tau+1, \dots r-1, r).$$

In this development of f_r there is one term corresponding to each of the r parametric diagonal lines of the simple square slope $[\phi]_r^r$. In fact we can obtain the development directly by expanding $[\phi]_r^r$ in terms of its successive parametric diagonal lines, i.e. by expressing it as a sum of simple square slopes each having only one non-zero parametric diagonal line. The notations used in (5) and (5′) are most appropriate to the corresponding development of the bilinear invariant

$$g_r = [x]_r [\theta]_r^r \underline{\overline{y}}_r = \phi(1, 2, \dots r;\ 1, 2, \dots r).$$

Ex. ii. *Some properties of the function G_τ.*

In manipulating the functions G_τ we may consider that the identities (4′) are true for all integral values (positive and negative) of τ, a, b, s, provided that we use the natural interpretation $G_\tau = 0$ when $s < 1$.

If G_τ' denotes the function into which G_τ is converted by the substitutions (a), we can verify that the function $f_r(1, 2, \dots r)$ of Ex. i is a bilinear invariant of the substitutions (i) by using the identity

$$G_\tau'(r, r-1, \dots \tau+1;\ \tau+1, \dots r-1, r) - G_\tau(r, r-1, \dots \tau+1;\ \tau+1, \dots r-1, r)$$
$$= Q_{\tau-1}(r, r-1, \dots \tau+2;\ \tau+2, \dots r-1, r) + Q_\tau(r, r-1, \dots \tau+3;\ \tau+3, \dots r-1, r),$$

in which $$Q_\tau = \Sigma\{(-1)^{i+1}\,.\,{}^{i+1}H_\tau\,.\,x_{r-i}\,y_{\tau+3+i}\}, \qquad (i = 0, 1, 2, \dots r-\tau-3),$$
$$= G_\tau(r+1, r, \dots \tau+3;\ \tau+2, \dots r-1, r) \text{ if } x_{r+1} = 0.$$

The identity is true whenever $0 \ngtr \tau \ngtr r-3$; and it is also true for the other values of τ which occur when we use the natural interpretations

$$Q_\tau = 0 \text{ when } \tau < 0;\ Q_{r-2} = Q_{r-1} = 0.$$

Remembering that ${}^{-1}H_\tau = 0$ except when $\tau = 0$, we can use the mutually equivalent identities such as

$$G_\tau(r, \dots \tau+1;\ \tau+1, \dots r) - G_{\tau-1}(r, \dots \tau+1;\ \tau+1, \dots r)$$
$$= {}^{-1}H_\tau x_r y_{\tau+1} - G_\tau(r-1, \dots \tau+1;\ \tau+2, \dots r),$$
$$G_\tau(1, \dots r-\tau;\ \tau+1, \dots r) - G_{\tau-1}(1, \dots r-\tau;\ \tau+1, \dots r)$$
$$= {}^{-1}H_\tau x_1 y_{\tau+1} - G_\tau(2, \dots r-\tau;\ \tau+2, \dots r),$$

to prove the mutually equivalent identities such as

$$\phi_r(r, \dots 1;\ 1, \dots r) + \phi_{r-1}(r, \dots 2;\ 2, \dots r) = x_r y_1 - \phi_{r-1}(r-1, \dots 1;\ 2, \dots r),$$
$$\phi_r(1, \dots r;\ 1, \dots r) + \phi_{r-1}(1, \dots r-1;\ 2, \dots r) = x_1 y_1 - \phi_{r-1}(2, \dots r;\ 2, \dots r).$$

Ex. iii. *The complete sets of bilinear invariants derived from the function*

$$\Phi(r, r-1, \dots 1;\ 1, \dots r-1, r) = \phi(r, r-1, \dots 1;\ 1, \dots r-1, r) + \phi(r, r-1, \dots 2;\ 2, \dots r-1, r)$$
$$= x_r y_1 - \phi(r-1, r-2, \dots 1;\ 2, \dots r-1, r).$$

If when $\alpha\beta = 1$, we take C to be the coefficient of ξ_0 in the invariant transformand η described in Theorem IV a of § 255, i.e. to be the matrix

$$[\Phi]_r^r = [\phi]_r^r + \begin{bmatrix} 0, & 0 \\ 0, & \phi \end{bmatrix}_{1, r-1}^{1, r-1} = \begin{bmatrix} 0, & -\phi \\ 1, & 0 \end{bmatrix}_{r-1, 1}^{1, r-1},$$

we obtain a bilinear invariant

$$f_r = [x]_r [\Phi]_r^r \underline{\overline{y}}_r = \Phi(r, r-1, \dots 1;\ 1, \dots r-1, r)$$
$$= x_r y_1 - \phi(r-1, r-2, \dots 1;\ 2, \dots r-1, r)$$

of maximum rank r of the substitutions (i), the direct development of which is

$$f_r = \Sigma(-1)^\tau F_\tau, \qquad (\tau = 0, 1, 2, \dots r-1),$$

where $$F_\tau = \Sigma \{(-1)^\omega . {}^{\omega-1}H_\tau . x_{r-\omega} y_{\tau+1+\omega}\}$$
$$= {}^{-1}H_\tau . x_r y_{\tau+1} - G_\tau (r-1, r-2, \dots \tau+1; \tau+2, \dots r-1, r),$$
and where of course ${}^{-1}H_\tau = 0$ except when $\tau = 0$.

Consequently a set of corresponding bilinear invariants of maximum rank r of the substitutions (i), (ii), (iii), (iv) is formed, when $\alpha\beta = 1$, by the functions

$$\Phi(\overset{*}{r}, \dots 1; 1, \dots \overset{*}{r}) = \phi(\overset{*}{r}, \dots 1; 1, \dots \overset{*}{r}) + \phi(\overset{*}{r}, \dots 2; 2, \dots \overset{*}{r})$$
$$= x_r y_1 - \phi(r-1, \dots 1; 2, \dots r),$$
$$\Phi(\overset{*}{1}, \dots r; r, \dots \overset{*}{1}) = \phi(\overset{*}{1}, \dots r; r, \dots \overset{*}{1}) + \phi(\overset{*}{1}, \dots r-1; r-1, \dots \overset{*}{1})$$
$$= x_1 y_r - \phi(2, \dots r; r-1, \dots 1),$$
$$\Phi(\overset{*}{1}, \dots r; 1, \dots \overset{*}{r}) = \phi(\overset{*}{1}, \dots r; 1, \dots \overset{*}{r}) + \phi(\overset{*}{1}, \dots r-1; 2, \dots \overset{*}{r})$$
$$= x_1 y_1 - \phi(2, \dots r; 2, \dots r),$$
$$\Phi(\overset{*}{r}, \dots 1; r, \dots \overset{*}{1}) = \phi(\overset{*}{r}, \dots 1; r, \dots \overset{*}{1}) + \phi(\overset{*}{r}, \dots 2; r-1, \dots \overset{*}{1})$$
$$= x_r y_r - \phi(r-1, \dots 1; r-1, \dots 1),$$

which are the functions denoted in Ex. i by

$$f_r + f_{r-1}, \quad f_r' + f'_{r-1}, \quad g_r + g_{r-1}, \quad g_r' + g'_{r-1}.$$

In the derived complete sets of bilinear invariants of ranks r, $r-1, \dots 2, 1$ corresponding to the general invariant transformands described in Theorem IVa of § 255, those of the substitutions (i) are the functions

$$f_r + f_{r-1}, \quad f_{r-1} + f_{r-2}, \quad \dots f_2 + f_1, \quad f_1$$

of Ex. i; and those of the substitutions (ii), (iii), (iv) are formed in similar ways.

Ex. iv. *The complete sets of bilinear invariants derived from the function*

$$[x]_r [\phi']^r_r \overline{\underline{y}}_r = \phi'(1, 2, \dots r-1, r; r, r-1, \dots 2, 1).$$

When $\alpha\beta = 1$, we can obtain a set of corresponding bilinear invariants of maximum rank r of the pairs of substitutions (i), (ii), (iii), (iv) by taking C to be the matrix $[\phi']^r_r$ defined in Ex. ii of § 257, so that JC is the matrix $[\theta']^r_r$ defined in Ex. i of § 257. By developing the corresponding matrix products, it will be seen that these bilinear invariants of the substitutions (i), (ii), (iii), (iv) are the functions

$$f_r\ (1, 2, \dots r) = \phi'(1, 2, \dots r-1, \overset{*}{r}; \overset{*}{r}, r-1, \dots 2, 1)$$
$$= \Sigma (-1)^\tau G_\tau (\tau+1, \dots r-1, \overset{*}{r}; \overset{*}{r}, r-1, \dots \tau+1),$$
$$f_r'\ (1, 2, \dots r) = \phi'(r, r-1, \dots 2, \overset{*}{1}; \overset{*}{1}, 2, \dots r-1, r)$$
$$= \Sigma (-1)^\tau G_\tau (r-\tau, \dots 2, \overset{*}{1}; \overset{*}{1}, 2, \dots r-\tau),$$
$$g_r\ (1, 2, \dots r) = \phi'(r, r-1, \dots 2, \overset{*}{1}; \overset{*}{r}, r-1, \dots 2, 1)$$
$$= \Sigma (-1)^\tau G_\tau (r-\tau, \dots 2, \overset{*}{1}; \overset{*}{r}, r-1, \dots \tau+1),$$
$$g_r'\ (1, 2, \dots r) = \phi'(1, 2, \dots r-1, \overset{*}{r}; \overset{*}{1}, 2, \dots r-1, r)$$
$$= \Sigma (-1)^\tau G_\tau (\tau+1, \dots r-1, \overset{*}{r}; \overset{*}{1}, 2, \dots r-1),$$
$$(\tau = 0, 1, 2, \dots r-1), \quad \dots\dots\dots\dots\dots(6)$$

where the functions G_τ are defined as in Ex. i. The first, second, third, fourth of the functions (6) can be derived from the first, second, fourth, third of the functions (5) of Ex. i by interchanging the two letters x and y, i.e. they are the conjugates of those functions, this being a necessary consequence of the fact that the matrix $[\phi']_r^r$ is the conjugate of the matrix $[\phi]_r^r$. A function $\phi'=\phi_r'$ of r pairs of variables defined as in (6) can always be regarded as the conjugate of a function $\phi=\phi_r$ defined as in (5), where in passing from ϕ' to ϕ the sequences representing the variables are interchanged.

Applying the principle (A), it follows from the formulae (6) that when $\alpha\beta=1$, complete sets of independent non-zero bilinear invariants of the pairs of substitutions (i), (ii), (iii), (iv) having ranks $r, r-1, \dots r-\rho, \dots 2, 1$ can be obtained by putting

$$\rho=0, 1, \dots \rho, \dots r-2, r-1$$

in the formulae:

$$\begin{aligned}
&f_{r-\rho}(\rho+1, \dots r-1, \overset{*}{r}; \rho+1, \dots r-1, \overset{*}{r}) \\
&\qquad = \phi'(\rho+1, \dots r-1, \overset{*}{r}; \overset{*}{r}, r-1, \dots \rho+1) \\
&\qquad = \Sigma(-1)^\tau G_\tau(\rho+\tau+1, \dots r-1, \overset{*}{r}; \overset{*}{r}, r-1, \dots \rho+\tau+1), \\
&f'_{r-\rho}(\overset{*}{1}, 2, \dots r-\rho; \overset{*}{1}, 2, \dots r-\rho) \\
&\qquad = \phi'(r-\rho, \dots 2, \overset{*}{1}; \overset{*}{1}, 2, \dots r-\rho) \\
&\qquad = \Sigma(-1)^\tau G_\tau(r-\rho-\tau, \dots 2, \overset{*}{1}; \overset{*}{1}, 2, \dots r-\rho-\tau), \\
&g_{r-\rho}(\overset{*}{1}, 2, \dots r-\rho; \rho+1, \dots r-1, \overset{*}{r}) \\
&\qquad = \phi'(r-\rho, \dots 2, \overset{*}{1}; \overset{*}{r}, r-1, \dots \rho+1) \\
&\qquad = \Sigma(-1)^\tau G_\tau(r-\rho-\tau, \dots 2, \overset{*}{1}; \overset{*}{r}, r-1, \dots \rho+\tau+1), \\
&g'_{r-\rho}(\rho+1, \dots r-1, \overset{*}{r}; \overset{*}{1}, 2, \dots r-\rho) \\
&\qquad = \phi'(\rho+1, \dots r-1, \overset{*}{r}; \overset{*}{1}, 2, \dots r-\rho) \\
&\qquad = \Sigma(-1)^\tau G_\tau(\rho+\tau+1, \dots r-1, \overset{*}{r}; \overset{*}{1}, 2, \dots r-\rho-\tau), \\
&\qquad\qquad (\tau=0, 1, 2, \dots r-\rho-1). \quad \dots\dots\dots\dots\dots(6')
\end{aligned}$$

Here the bilinear invariants of the substitutions (i) correspond to the complete set of independent invariant transformands $\{A', B\}$ described in Ex. ii of § 257 when $[y]_r^r$ is taken to be $[\phi']_r^r$, and the bilinear invariants of the other substitutions can be derived from them by the appropriate reversing substitutions for the variables. The first, second, third, fourth of the functions (6') can be derived from the first, second, fourth, third of the functions (5') of Ex. i by interchanging the two letters x and y, i.e. they are the conjugates of those functions.

Ex. v. *The complete sets of bilinear invariants derived from the function*

$$\begin{aligned}
&\Phi'(1, \dots r-1, r; r, r-1, \dots 1) \\
&\qquad = \phi'(1, \dots r-1, r; r, r-1, \dots 1) + \phi'(2, \dots r-1, r; r, r-1, \dots 2) \\
&\qquad = x_1 y_r - \phi'(2, \dots r-1, r; r-1, r-2, \dots 1).
\end{aligned}$$

If when $\alpha\beta=1$, we take C to be the coefficient of ξ_0 in the conjugate of the invariant transformand η described in Theorem IV a of § 255, i.e. to be the matrix

$$[\Phi']_r^r = [\phi']_r^r + \begin{bmatrix} 0, & 0 \\ 0, & \phi' \end{bmatrix}_{1, r-1}^{1, r-1} = \begin{bmatrix} 0, & 1 \\ -\phi', & 0 \end{bmatrix}_{1, r-1}^{r-1, 1},$$

we obtain a bilinear invariant

$$\begin{aligned}
f_r = [x]_r [\Phi']_r^r \underset{\smile}{\overset{\frown}{y}}_r &= \Phi'(1, \dots r-1, r; r, r-1, \dots 1) \\
&= x_1 y_r - \phi'(2, \dots r-1, r; r-1, r-2, \dots 1)
\end{aligned}$$

of maximum rank r of the substitutions (i), the direct development of which is

$$f_r = \Sigma (-1)^\tau F_\tau, \qquad (\tau = 0, 1, 2, \dots r-1),$$

where

$$F_\tau = \Sigma \{(-1)^\omega \,.\, {}^{\omega-1}H_\tau \,.\, x_{\tau+1+\omega} y_{r-\omega}\}, \qquad (\omega = 0, 1, 2, \dots r-\tau-1),$$

$$= {}^{-1}H_\tau \,.\, x_{\tau+1} y_r - G_\tau (\tau+2, \dots r-1, r;\ r-1, r-2, \dots \tau+1),$$

and where of course ${}^{-1}H_\tau = 0$ except when $\tau = 0$.

Associated with it we have a set of corresponding bilinear invariants of maximum rank r of the substitutions (i), (ii), (iii), (iv) formed by the functions

$$\Phi'(1, \dots r;\ r, \dots 1) = \phi'(1, \dots r;\ r, \dots 1) + \phi'(2, \dots r;\ r, \dots 2)$$
$$= x_1 y_r - \phi'(2, \dots r;\ r-1, \dots 1),$$

$$\Phi'(r, \dots 1;\ 1, \dots r) = \phi'(r, \dots 1;\ 1, \dots r) + \phi'(r-1, \dots 1;\ 1, \dots r-1)$$
$$= x_r y_1 - \phi'(r-1, \dots 1;\ 2, \dots r),$$

$$\Phi'(r, \dots 1;\ r, \dots 1) = \phi'(r, \dots 1;\ r, \dots 1) + \phi'(r-1, \dots 1;\ r, \dots 2)$$
$$= x_r y_r - \phi'(r-1, \dots 1;\ r-1, \dots 1),$$

$$\Phi'(1, \dots r;\ 1, \dots r) = \phi'(1, \dots r;\ 1, \dots r) + \phi'(2, \dots r;\ 1, \dots r-1)$$
$$= x_1 y_1 - \phi'(2, \dots r;\ 2, \dots r),$$

which are the functions denoted in Ex. iv by

$$f_r + f_{r-1}, \quad f_r' + f'_{r-1}, \quad g_r + g_{r-1}, \quad g_r' + g'_{r-1}.$$

The first, second, third, fourth of them are the conjugates of the first, second, fourth, third of the functions Φ of Ex. iii.

In the derived complete sets of bilinear invariants of ranks $r, r-1, \dots 2, 1$ formed in accordance with the principle (A) those of the substitutions (i) are the functions

$$f_r + f_{r-1}, \quad f_{r-1} + f_{r-2}, \quad \dots \quad f_2 + f_1, \quad f_1$$

of Ex. iv; and those of the substitutions (ii), (iii), (iv) are formed in similar ways. All these functions are derived from the functions of Ex. iii by interchanging the two letters x and y.

Ex. vi. *Illustrations of the formulae of Exs.* i *and* iv.

When $r=5$ and $\alpha\beta=1$, the bilinear invariants of the substitutions (i) described in Ex. iv are the functions:

$$f_5(1, 2, 3, 4, 5) = x_1 y_5 - x_2(y_4 + y_5) + x_3(y_3 + 2y_4 + y_5) - x_4(y_2 + 3y_3 + 3y_4 + y_5)$$
$$+ x_5(y_1 + 4y_2 + 6y_3 + 4y_4 + y_5)$$
$$= (x_1 - x_2 + x_3 - x_4 + x_5)\, y_5 - (x_2 - 2x_3 + 3x_4 - 4x_5)\, y_4 + (x_3 - 3x_4 + 6x_5)\, y_3$$
$$- (x_4 - 4x_5)\, y_2 + x_5 y_1$$
$$= \{x_1 y_5 - x_2 y_4 + x_3 y_3 - x_4 y_2 + x_5 y_1\} - \{x_2 y_5 - 2x_3 y_4 + 3x_4 y_3 - 4x_5 y_2\}$$
$$+ \{x_3 y_5 - 3x_4 y_4 + 6x_5 y_3\} - \{x_4 y_5 - 4x_5 y_4\} + \{x_5 y_5\}$$
$$= G_0(1, \dots 5;\ 5, \dots 1) - G_1(2, \dots 5;\ 5, \dots 2) + G_2(3, 4, 5;\ 5, 4, 3)$$
$$- G_3(4, 5;\ 5, 4) + G_4(5;\ 5);$$

$$f_4(2, 3, 4, 5) = G_0(2, \dots 5;\ 5, \dots 2) - G_1(3, 4, 5;\ 5, 4, 3) + G_2(4, 5;\ 5, 4) - G_3(5;\ 5)$$
$$= \{x_2 y_5 - x_3 y_4 + x_4 y_3 - x_5 y_2\} - \{x_3 y_5 - 2x_4 y_4 + 3x_5 y_3\} + \{x_4 y_5 - 3x_5 y_4\} + \{x_5 y_5\};$$

$$f_3(3, 4, 5) = G_0(3, 4, 5;\ 5, 4, 3) - G_1(4, 5;\ 5, 4) + G_2(5;\ 5)$$
$$= \{x_3 y_5 - x_4 y_4 + x_5 y_3\} - \{x_4 y_5 - 2x_5 y_4\} + \{x_5 y_5\};$$

$$f_2(4, 5) = G_0(4, 5;\ 5, 4) - G_1(5, 5) = \{x_4 y_5 - x_5 y_4\} - \{x_5 y_5\};$$

$$f_1(5) = G_0(5;\ 5) = x_5 y_5.$$

The corresponding bilinear invariants of the substitutions (i) described in Ex. i are derived from these by interchanging the two letters x and y, and also interchanging the two sequences in every function G_τ.

In both cases (see Exs. iii and v) another complete set of bilinear invariants of the substitutions (i) is formed by the functions

$$f_5+f_4,\quad f_4+f_3,\quad f_3+f_2,\quad f_2+f_1,\quad f_1.$$

Ex. vii. *Bilinear invariants of odd ranks derived from the symmetric function*

$$[x]_{2t+1}[\psi]_{2t+1}^{2t+1}\,\underline{\overline{y}}_{2t+1}=\psi\,(2t+1,\,2t,\,\dots\,2,\,1;\;1,\,2,\,\dots\,2t,\,2t+1).$$

Whenever τ and σ are positive integers, we will define a function Γ_τ of

$$s=\tau+1+2\sigma$$

pairs of variables, or rather of the first $\sigma+1$ and last $\sigma+1$ of those pairs, by the equation

$$\begin{aligned}\Gamma_\tau\,(1,\,2,\,\dots\,1+\sigma,\,\dots\,s-\sigma,\,\dots\,s-1,\,s;\;1,\,2,\,\dots\,1+\sigma,\,\dots\,s-\sigma,\,\dots\,s-1,\,s)\\ =\tfrac{1}{2}\Sigma\,(-1)^i\,.\,({}^{i-1}H_\tau+{}^iH_\tau)\,.\,(x_{1+\sigma-i}\,y_{1+\sigma-i}+x_{s-\sigma+i}\,y_{s-\sigma+i}),\\ (i=0,\,1,\,2,\,\dots\,\sigma)\,;\;\dots\dots\dots\dots(7)\end{aligned}$$

so that whenever $\qquad p-a=q-b=\tau+1+2\sigma,$

where τ, σ, a, b are positive integers, we have:

$$\begin{aligned}&\Gamma_\tau\,(p,\,p-1,\,\dots\,a+2,\,a+1;\;b+1,\,b+2,\,\dots\,q-1,\,q)\\ &\quad=\Gamma_\tau\,(a+1,\,a+2,\,\dots\,p-1,\,p;\;q,\,q-1,\,\dots\,b+2,\,b+1)\\ &\quad=\tfrac{1}{2}\Sigma\,(-1)^i\,.\,({}^{i-1}H_\tau+{}^iH_\tau)\,.\,(x_{p-\sigma+i}\,y_{b+1+\sigma-i}+x_{a+1+\sigma-i}\,y_{q-\sigma+i})\,;\\ &\Gamma_\tau\,(a+1,\,a+2,\,\dots\,p-1,\,p;\;b+1,\,b+2,\,\dots\,q-1,\,q)\\ &\quad=\Gamma_\tau\,(p,\,p-1,\,\dots\,a+2,\,a+1;\;q,\,q-1,\,\dots\,b+2,\,b+1)\\ &\quad=\tfrac{1}{2}\Sigma\,(-1)^i\,.\,({}^{i-1}H_\tau+{}^iH_\tau)\,.\,(x_{a+1+\sigma-i}+y_{b+1+\sigma-i}+x_{p-\sigma+i}\,y_{q-\sigma+i})\,;\\ &\qquad\qquad (i=0,\,1,\,2,\,\dots\,\sigma).\;\dots\dots\dots\dots(7')\end{aligned}$$

For any given value of τ the unsigned coefficients of the successive terms of a function Γ_τ of any number of pairs of variables, exceeding τ by an *odd* positive integer, are always successive integers (starting with the first) of the same series derived from

$${}^{i-1}H_\tau+{}^iH_\tau\;\;(\text{or }{}^iH_{\tau-1}+2\,.\,{}^{i-1}H_\tau)\;\;\text{by putting }i=0,\,1,\,2,\,3,\,\dots.$$

When $\alpha\beta=1$ and $r=2t+1$, we can obtain a set of corresponding bilinear invariants of maximum rank $2t+1$ of the pairs of substitutions (i), (ii), (iii), (iv) by taking C to be the symmetric matrix $[\psi]_{2t+1}^{2t+1}$ defined in Exs. v and vi of § 257. By developing the corresponding matrix products (see Ex. viii) we can express these bilinear invariants in the forms

$$\begin{aligned}f_{2t+1}\,(1,\,2,\,\dots\,2t+1)&=\psi\,(\overset{*}{2t}+1,\,2t,\,\dots\,1\;;\;1\;,\,\dots\,2t,\,\overset{*}{2t}+1),\\ &=-x_{t+1}y_{t+1}+\Sigma\,(-1)^\tau\,\Gamma_\tau\,(\overset{*}{2t}+1,\,2t,\,\dots\,\tau+1\,;\;\tau+1,\,\dots\,2t,\,\overset{*}{2t}+1),\\ f'_{2t+1}\,(1,\,2,\,\dots\,2t+1)&=\psi\,(\overset{*}{1},\,2,\,\dots\,2t+1\;;\;2t+1\;,\,\dots\,2,\,\overset{*}{1})\\ &=-x_{t+1}y_{t+1}+\Sigma\,(-1)^\tau\,\Gamma_\tau\,(\overset{*}{1},\,2,\,\dots\,2t+1-\tau\,;\;2t+1-\tau,\,\dots\,2,\,\overset{*}{1}),\\ g_{2t+1}\,(1,\,2,\,\dots\,2t+1)&=\psi\,(\overset{*}{1},\,2,\,\dots\,2t+1\;;\;1\;,\,\dots\,2t,\,\overset{*}{2t}+1)\\ &=-x_{t+1}y_{t+1}+\Sigma\,(-1)^\tau\,\Gamma_\tau\,(\overset{*}{1},\,2,\,\dots\,2t+1-\tau\,;\;\tau+1,\,\dots\,2t,\,\overset{*}{2t}+1),\\ g'_{2t+1}\,(1,\,2,\,\dots\,2t+1)&=\psi\,(\overset{*}{2t}+1,\,2t,\,\dots\,1\;;\;2t+1\;,\,\dots\,2,\,\overset{*}{1})\\ &=-x_{t+1}y_{t+1}+\Sigma\,(-1)^\tau\,\Gamma_\tau\,(\overset{*}{2t}+1,\,2t,\,\dots\,\tau+1\,;\;2t+1-\tau,\,\dots\,2,\,\overset{*}{1}),\\ &\qquad (\tau=0,\,1,\,2,\,\dots\,t),\;\dots\dots\dots\dots(8)\end{aligned}$$

where the functions Γ_τ are defined as above, and where we can and will regard

$$\psi=\psi_{2t+1}$$

as a completely defined function of $2t+1$ pairs of variables for every positive integral value of t, the four defining equations in (8) being mutually equivalent. When a suffix is attached to ψ, it will always be equal to the number of pairs of variables indicated by the sequences, which must be an *odd* integer; but the suffix will usually be omitted as being superfluous. The ψ-functions occurring in the first and second of the formulae (8) are symmetric functions in the sense that each of them remains unaltered when the two letters x and y are interchanged, i.e. when the variables $x_1, x_2, x_3, \ldots$ and $y_1, y_2, y_3, \ldots$ are interchanged.

Whenever $\alpha\beta=1$, we can deduce from the functions (8) sets of independent bilinear invariants of the substitutions (i), (ii), (iii), (iv) of all odd ranks formed in accordance with the principle (A); but we have to distinguish between two cases.

Case I. When $r=2t+1$ and $\alpha\beta=1$, these $t+1$ independent non-zero bilinear invariants of the pairs of substitutions (i), (ii), (iii), (iv) of *odd* ranks $2t+1, 2t-1, \ldots 2t+1-2\rho, \ldots 3, 1$ are the ψ-functions obtained by putting $\rho=0, 1, \ldots \rho, \ldots t-1, t$ in the formulae:

$$f_{2t+1-2\rho}\,(2\rho+1, \ldots \overset{*}{2t}+1;\ 2\rho+1, \ldots \overset{*}{2t}+1)=\psi(\overset{*}{2t}+1, 2t, \ldots 2\rho+1;\ 2\rho+1, \ldots 2t, \overset{*}{2t}+1)$$
$$=-x_{t+1+\rho}\,y_{t+1+\rho}+\Sigma\,(-1)^\tau\,\Gamma_\tau\,(\overset{*}{2t}+1, 2t, \ldots 2\rho+\tau+1;\ 2\rho+\tau+1, \ldots 2t, \overset{*}{2t}+1)$$
$$f'_{2t+1-2\rho}\,(\overset{*}{1}, \ldots 2t+1-2\rho;\ \overset{*}{1}, \ldots 2t+1-2\rho)=\psi\,(\overset{*}{1}, 2, \ldots 2t+1-2\rho;\ 2t+1-2\rho, \ldots 2, \overset{*}{1})$$
$$=-x_{t+1-\rho}\,y_{t+1-\rho}+\Sigma\,(-1)^\tau\,\Gamma_\tau\,(\overset{*}{1}, 2, \ldots 2t+1-2\rho-\tau;\ 2t+1-2\rho-\tau, \ldots 2, \overset{*}{1})$$
$$g_{2t+1-2\rho}\,(\overset{*}{1}, \ldots 2t+1-2\rho;\ 2\rho+1, \ldots \overset{*}{2t}+1)=\psi\,(\overset{*}{1}, 2, \ldots 2t+1-2\rho;\ 2\rho+1, \ldots 2t, \overset{*}{2t}+1)$$
$$=-x_{t+1-\rho}\,y_{t+1+\rho}+\Sigma\,(-1)^\tau\,\Gamma_\tau\,(\overset{*}{1}, 2, \ldots 2t+1-2\rho-\tau;\ 2\rho+\tau+1, \ldots 2t, \overset{*}{2t}+1)$$
$$g'_{2t+1-2\rho}\,(2\rho+1, \ldots \overset{*}{2t}+1;\ \overset{*}{1}, \ldots 2t+1-2\rho)=\psi\,(\overset{*}{2t}+1, 2t, \ldots 2\rho+1;\ 2t+1-2\rho, \ldots 2, \overset{*}{1})$$
$$=-x_{t+1+\rho}\,y_{t+1-\rho}+\Sigma\,(-1)^\tau\,\Gamma_\tau\,(\overset{*}{2t}+1, 2t, \ldots 2\rho+\tau+1;\ 2t+1-2\rho-\tau, \ldots 2, \overset{*}{1}),$$
$$(\tau=0, 1, 2, \ldots t-\rho). \quad\ldots\ldots\ldots\ldots\ldots\ldots(8')$$

Case II. When $r=2t$ (where $t \not< 1$) and $\alpha\beta=1$, these t independent non-zero bilinear invariants of the pairs of substitutions (i), (ii), (iii), (iv) of *odd* ranks

$$2t-1, 2t-3, \ldots 2t+1-2\rho, \ldots 3, 1$$

are the ψ-functions obtained by putting $\rho=1, 2, \ldots \rho, \ldots t-1, t$ in the formulae:

$$f_{2t+1-2\rho}\,(2\rho, \ldots \overset{*}{2t};\ 2\rho, \ldots \overset{*}{2t})=\psi\,(\overset{*}{2t}, 2t-1, \ldots 2\rho;\ 2\rho, \ldots 2t-1, \overset{*}{2t})$$
$$=-x_{t+\rho}\,y_{t+\rho}+\Sigma\,(-1)^\tau\,\Gamma_\tau\,(\overset{*}{2t}, 2t-1, \ldots 2\rho+\tau;\ 2\rho+\tau, \ldots 2t-1, \overset{*}{2t})$$
$$f'_{2t+1-2\rho}\,(\overset{*}{1}, \ldots 2t+1-2\rho;\ \overset{*}{1}, \ldots 2t+1-2\rho)=\psi\,(\overset{*}{1}, 2, \ldots 2t+1-2\rho;\ 2t+1-2\rho, \ldots 2, \overset{*}{1})$$
$$=-x_{t+1-\rho}\,y_{t+1-\rho}+\Sigma\,(-1)^\tau\,\Gamma_\tau\,(\overset{*}{1}, 2, \ldots 2t+1-2\rho-\tau;\ 2t+1-2\rho-\tau, \ldots 2, \overset{*}{1})$$
$$g_{2t+1-2\rho}\,(\overset{*}{1}, \ldots 2t+1-2\rho;\ 2\rho, \ldots \overset{*}{2t})=\psi\,(\overset{*}{1}, 2, \ldots 2t+1-2\rho;\ 2\rho, \ldots 2t-1, \overset{*}{2t})$$
$$=-x_{t+1-\rho}\,y_{t+\rho}+\Sigma\,(-1)^\tau\,\Gamma_\tau\,(\overset{*}{1}, 2, \ldots 2t+1-2\rho-\tau;\ 2\rho+\tau, \ldots 2t-1, \overset{*}{2t}),$$
$$g'_{2t+1-2\rho}\,(2\rho, \ldots \overset{*}{2t};\ \overset{*}{1}, \ldots 2t+1-2\rho)=\psi\,(\overset{*}{2t}, 2t-1, \ldots 2\rho;\ 2t+1-2\rho, \ldots 2, \overset{*}{1})$$
$$=-x_{t+\rho}\,y_{t+1-\rho}+\Sigma\,(-1)^\tau\,\Gamma_\tau\,(\overset{*}{2t}, 2t-1, \ldots 2\rho+\tau;\ 2t+1-2\rho-\tau, \ldots 2, \overset{*}{1}),$$
$$(\tau=0, 1, 2, \ldots t-\rho), \quad\ldots\ldots\ldots\ldots\ldots(8'').$$

Here the bilinear invariants of the substitutions (i) correspond to the set of symmetric invariant transformands $\{A', B\}$ described in Ex. v of § 257; and the bilinear invariants of the substitutions (ii), (iii), (iv) are derived from them by the appropriate reversing substitutions for the variables.

Ex. viii. *Proofs of the formulae of Ex.* vii.

To establish all the formulae (8), (8'), (8'') it is sufficient to obtain the development of the bilinear invariant f_{2t+1} shown in (8). If we put $u=r+1-i$, we have

$$f_{2t+1}=[x]_{2t+1}[\psi]_{2t+1}^{2t+1}\underline{y}_{2t+1}=\Sigma\psi_{uj}x_iy_j$$
$$=\tfrac{1}{2}\Sigma\{(-1)^{t+1-i}({}^{t-j}H_\tau+{}^{t+1-j}H_\tau)\,.\,x_iy_j\}, \quad \text{.....................(9)}$$

where $$\tau=i+j-r-1=i+j-2t-2,$$

and where the summation extends over all integral values of i and j such that

$$1 \not> i \not> r, \quad 1 \not> j \not> r, \quad i+j \not< r+1.$$

The possible values of τ are those consistent with the conditions $\tau \not< 0$, $\tau \not> 2t$; and when τ is given, the possible values of i and j are those consistent with the conditions

$$i+j=2t+2+\tau,$$

and $$\tau+1 \not> j \not> 2t+1 \quad \text{or} \quad 2t+1 \not< i \not< \tau+1,$$

the terms with zero coefficients being those in which

$$t+1<j<t+\tau+1 \quad \text{or} \quad t+\tau+1>i>t+1,$$

and the terms with non-zero coefficients being those in which :

(1) $$\tau+1 \not> j \not> t+1 \qquad \text{or} \quad 2t+1 \not< i \not< t+\tau+1;$$

(2) $$2t+1 \not< j \not< t+\tau+1 \quad \text{or} \quad \tau+1 \not> i \not> t+1.$$

All the terms in which $\tau>t$ have zero coefficients.

By arranging the terms according to their total weights, or by expanding the simple square slope $[\psi]_{2t+1}^{2t+1}$ in terms of its successive parametric diagonal lines, we obtain the development

$$f_{2t+1}=S_0+S_1+\dots+S_\tau+\dots+S_t, \quad \text{..............................(9')}$$

where S_τ is the sum of the terms in which $i+j=2t+2+\tau$. Again by restricting the summation (9) to the two ranges (1) and (2) we see that

$$S_\tau=(-1)^\tau\,.\,\tfrac{1}{2}\Sigma\{(-1)^\omega\,.\,({}^{\omega-1}H_\tau+{}^{\omega}H_\tau)\,.\,(x_{t+1+\tau+\omega}\,y_{t+1-\omega}+x_{t+1-\omega}\,y_{t+1+\tau+\omega})\},$$
$$(\omega=0,\ 1,\ 2,\ \dots\ t-\tau)$$
$$=(-1)^\tau\,\Gamma_\tau\,(2t+1,\ 2t,\ \dots\ \tau+1;\ \tau+1,\ \dots\ 2t,\ 2t+1), \qquad \text{when } \tau=1,\ 2,\ \dots\ t;$$

but for the terms in which $\tau=0$ the corresponding formula is

$$S_0=-x_{t+1}\,y_{t+1}+\Gamma_0\,(2t+1,\ 2t,\ \dots\ 1;\ 1,\ \dots\ 2t,\ 2t+1),$$

because the middle term $x_{t+1}\,y_{t+1}$ is common to the two ranges (1) and (2), and is counted twice in Γ_0. Thus (9′) is the development of f_{2t+1} given in (8).

All the bilinear invariants of Ex. vii can be expressed in terms of the fnnctions defined in Exs. i and iv. In fact the equation (9) in Ex. vi of § 257 leads at once to the formula

$$\psi_{2t+1}\,(2t+1,\ \dots\ 2,\ 1;\ 1,\ 2,\ \dots\ 2t+1)$$
$$=\ \tfrac{1}{2}\{\phi_{t+1}\,(t+1,\ \dots\ 2,\ 1;\ t+1,\ \dots\ 2t,\ 2t+1)+\phi'_{t+1}\,(t+1,\ \dots\ 2t,\ 2t+1;\ t+1,\ \dots\ 2,\ 1\}$$
$$-\tfrac{1}{2}\{\phi_t\ \ (t,\ \ \dots\ 2,\ 1;\ t+2,\ \dots\ 2t,\ 2t+1)+\phi_t'\ \ (t+2,\ \dots\ 2t,\ 2t+1;\ t\ \ ,\ \dots\ 2,\ 1\},$$
$$\text{.....................(9'')}$$

which agrees with the development given in (8) because the formulae (7′) are equivalent to the formulae :

$$\Gamma_\tau\,(p,p-1,\ \dots\ a+2,\ a+1;\ b+1,\ b+2,\ \dots\ q-1,\ q)$$
$$=\Gamma_\tau\,(a+1,\ a+2,\ \dots\ p-1,\ p;\ q,\ q-1,\ \dots\ b+2,\ b+1)$$
$$=\tfrac{1}{2}\,.\,{}^{-1}H_\tau\,.\,(x_{p-\sigma}\,y_{b+1+\sigma}+x_{a+1+\sigma}\,y_{q-\sigma})$$
$$+\tfrac{1}{2}\{G_\tau\,(p-\sigma,\ \dots\ p;\ b+1+\sigma,\ \dots\ b+1)+G_\tau\,(a+1+\sigma,\ \dots\ a+1;\ q-\sigma,\ \dots\ q)\}$$
$$-\tfrac{1}{2}\{G_\tau\,(p-\sigma+1,\ \dots\ p;\ b+\sigma,\ \dots\ b+1)+G_\tau\,(a+\sigma,\ \dots\ a+1;\ q-\sigma+1,\ \dots\ q)\};$$

$$\Gamma_\tau(a+1,\ a+2, \ldots p-1,\ p;\ b+1,\ b+2, \ldots q-1,\ q)$$
$$=\Gamma_\tau(p,\ p-1, \ldots a+2,\ a+1;\ q,\ q-1, \ldots b+2,\ b+1)$$
$$=\tfrac{1}{2}\,.\,{}^{-1}H_\tau\,.\,(x_{a+1+\sigma}y_{b+1+\sigma}+x_{p-\sigma}y_{q-\sigma})$$
$$+\tfrac{1}{2}\{G_\tau(a+1+\sigma, \ldots a+1;\ b+1+\sigma, \ldots b+1)+G_\tau(p-\sigma, \ldots p;\ q-\sigma, \ldots q)\}$$
$$-\tfrac{1}{2}\{G_\tau(a+\sigma, \ldots a+1;\ b+\sigma, \ldots b+1)+G_\tau(p-\sigma+1, \ldots p;\ q-\sigma+1, \ldots q)\};\quad \ldots\ldots(7'')$$

where ${}^{-1}H_\tau=0$ except when $\tau=0$.

Ex. ix. *Bilinear invariants of even ranks derived from the skew-symmetric function*

$$[x]_{2t}^{-}[\psi]_{2t}^{2t}\,\overline{y}_{2t}=\psi(2t,\ 2t-1, \ldots 2,\ 1;\ 1,\ 2, \ldots 2t-1,\ 2t).$$

Whenever τ and σ are positive integers, we will define a function Γ_τ of

$$s=\tau+2+2\sigma$$

pairs of variables, or rather of the first $\sigma+1$ and last $\sigma+1$ of those pairs, by the equation

$$\Gamma_\tau(1,\ 2, \ldots 1+\sigma, \ldots\ldots s-\sigma, \ldots s-1,\ s;\ 1,\ 2, \ldots 1+\sigma, \ldots\ldots s-\sigma, \ldots s-1,\ s)$$
$$=\Sigma(-1)^i\,.\,{}^iH_\tau\,.\,(x_{1+\sigma-i}y_{1+\sigma-i}-x_{s-\sigma+i}y_{s-\sigma+i}),\qquad (i=0,\ 1,\ 2, \ldots \sigma);\ \ldots\ldots(10)$$

so that whenever $\qquad p-a=q-b=\tau+2+2\sigma,$

where τ, σ, a, b are positive integers, we have:

$$\Gamma_\tau(p,\ p-1, \ldots a+2,\ a+1;\ b+1,\ b+2, \ldots q-1,\ q)$$
$$=-\Gamma_\tau(a+1,\ a+2, \ldots p-1,\ p;\ q,\ q-1, \ldots b+2,\ b+1)$$
$$=\Sigma(-1)^i\,.\,{}^iH_\tau\,.\,(x_{p-\sigma+i}y_{b+1+\sigma-i}-x_{a+1+\sigma-i}y_{q-\sigma+i});$$
$$\Gamma_\tau(a+1,\ a+2, \ldots p-1,\ p;\ b+1,\ b+2, \ldots q-1,\ q)$$
$$=-\Gamma_\tau(p,\ p-1, \ldots a+2,\ a+1;\ q,\ q-1, \ldots b+2,\ b+1)$$
$$=\Sigma(-1)^i\,.\,{}^iH_\tau\,.\,(x_{a+1+\sigma-i}y_{b+1+\sigma-i}-x_{p-\sigma+i}y_{q-\sigma+i});$$
$$(i=0,\ 1,\ 2, \ldots \sigma).\ \ldots\ldots(10')$$

For any given value of τ the unsigned coefficients of the successive terms of a function Γ_τ of any number of pairs of variables exceeding τ by an *even* positive integer are always successive integers (starting with the first) of the same series derived from

$${}^iH_\tau \quad \text{by putting} \quad i=0,\ 1,\ 2,\ 3, \ldots.$$

When $\alpha\beta=1$ and $r=2t$ (where $t \not< 1$), we can obtain a set of corresponding bilinear invariants of maximum rank $2t$ of the pairs of substitutions (i), (ii), (iii), (iv) by taking C to be the skew-symmetric matrix $[\psi]_{2t}^{2t}$ defined in Exs. x and xi of § 257. By developing the corresponding matrix products (see Ex. x) we can express these bilinear invariants in the forms

$$f_{2t}(1,\ 2, \ldots 2t)=\psi(2t,\ 2t-1, \ldots 2,\ 1;\ 1,\ 2, \ldots 2t-1,\ 2t)$$
$$=\Sigma(-1)^\tau\,\Gamma_\tau(2t,\ 2t-1, \ldots \tau+1;\ \tau+1, \ldots 2t-1,\ 2t),$$
$$f'_{2t}(1,\ 2, \ldots 2t)=\psi(1,\ 2, \ldots 2t-1,\ 2t;\ 2t,\ 2t-1, \ldots 2,\ 1)$$
$$=\Sigma(-1)^\tau\,\Gamma_\tau(1,\ 2, \ldots 2t-\tau;\ 2t-\tau, \ldots 2,\ 1),$$
$$g_{2t}(1,\ 2, \ldots 2t)=\psi(1,\ 2, \ldots 2t-1,\ 2t;\ 1,\ 2, \ldots 2t-1,\ 2t)$$
$$=\Sigma(-1)^\tau\,\Gamma_\tau(1,\ 2, \ldots 2t-\tau;\ \tau+1, \ldots 2t-1,\ 2t),$$
$$g'_{2t}(1,\ 2, \ldots 2t)=\psi(2t,\ 2t-1, \ldots 2,\ 1;\ 2t,\ 2t-1, \ldots 2,\ 1)$$
$$=\Sigma(-1)^\tau\,\Gamma_\tau(2t,\ 2t-1, \ldots \tau+1;\ 2t-\tau, \ldots 2,\ 1),$$
$$(\tau=0,\ 1,\ 2, \ldots t-1),\ \ldots\ldots(11)$$

where the functions Γ_τ are defined as above, and where we can and will regard

$$\psi=\psi_{2t}$$

as a completely defined function of $2t$ pairs of variables for every non-zero positive integral value of t, the four defining equations in (11) being mutually equivalent. When a suffix is attached to ψ, it will always be equal to the number of pairs of variables indicated by the sequences, which must be an *even* integer; but the suffix will usually be omitted as being superfluous. The ψ-functions occurring in the first and second of the formulae (11) are skew-symmetric functions in the sense that each of them is merely altered in sign when the two letters x and y are interchanged, i.e. when the variables $x_1, x_2, x_3, \dots$ and $y_1, y_2, y_3, \dots$ are interchanged.

Whenever $\alpha\beta=1$, we can deduce from the functions (11) sets of independent bilinear invariants of the substitutions (i), (ii), (iii), (iv) of all even ranks formed in accordance with the principle (A); but we have to distinguish between two cases.

Case I. When $r=2t$ (where $t \nless 1$) and $\alpha\beta=1$, these t independent non-zero bilinear invariants of the substitutions (i), (ii), (iii), (iv) of *even* ranks $2t, 2t-2, \dots 2t+2-2\rho, \dots 4, 2$ are the ψ-functions obtained by putting $\rho=1, 2, \dots \rho, \dots t-1, t$ in the formulae:

$$f_{2t+2-2\rho}\,(2\rho-1, \dots \overset{*}{2t};\ 2\rho-1, \dots \overset{*}{2t})=\psi\,(\overset{*}{2t}, 2t-1, \dots 2\rho-1;\ 2\rho-1, \dots 2t-1, \overset{*}{2t})$$
$$=\Sigma\,(-1)^{\tau}\,\Gamma_\tau\,(\overset{*}{2t}, 2t-1, \dots 2\rho-1+\tau;\ 2\rho-1+\tau, \dots 2t-1, \overset{*}{2t}),$$
$$f'_{2t+2-2\rho}\,(\overset{*}{1}, \dots 2t+2-2\rho;\ \overset{*}{1}, \dots 2t+2-2\rho)=\psi\,(\overset{*}{1}, 2, \dots 2t+2-2\rho;\ 2t+2-2\rho, \dots 2, \overset{*}{1})$$
$$=\Sigma\,(-1)^{\tau}\,\Gamma_\tau\,(\overset{*}{1}, 2, \dots 2t+2-2\rho-\tau;\ 2t+2-2\rho-\tau, \dots 2, \overset{*}{1}),$$
$$g_{2t+2-2\rho}\,(\overset{*}{1}, \dots 2t+2-2\rho;\ 2\rho-1, \dots \overset{*}{2t})=\psi\,(\overset{*}{1}, 2, \dots 2t+2-2\rho;\ 2\rho-1, \dots 2t-1, \overset{*}{2t})$$
$$=\Sigma\,(-1)^{\tau}\,\Gamma_\tau\,(\overset{*}{1}, 2, \dots 2t+2-2\rho-\tau;\ 2\rho-1+\tau, \dots 2t-1, \overset{*}{2t}),$$
$$g'_{2t+2-2\rho}\,(2\rho-1, \dots \overset{*}{2t};\ \overset{*}{1}, \dots 2t+2-2\rho)=\psi\,(\overset{*}{2t}, 2t-1, \dots 2\rho-1;\ 2t+2-2\rho, \dots 2, \overset{*}{1})$$
$$=\Sigma\,(-1)^{\tau}\,\Gamma_\tau\,(\overset{*}{2t}, 2t-1, \dots 2\rho-1+\tau;\ 2t+2-2\rho-\tau, \dots 2, \overset{*}{1}),$$
$$(\tau=0, 1, 2, \dots t-\rho) \quad \dots\dots\dots\dots(11').$$

Case II. When $r=2t+1$ and $\alpha\beta=1$, these t independent non-zero bilinear invariants of the substitutions (i), (ii), (iii), (iv) of *even* ranks $2t, 2t-2, \dots 2t+2-2\rho, \dots 4, 2$ are the ψ-functions obtained by putting $\rho=1, 2, \dots \rho, \dots t-1, t$ in the formulae:

$$f_{2t+2-2\rho}\,(2\rho, \dots \overset{*}{2t}+1;\ 2\rho, \dots \overset{*}{2t}+1)=\psi\,(\overset{*}{2t}+1, 2t, \dots 2\rho;\ 2\rho, \dots 2t, \overset{*}{2t}+1)$$
$$=\Sigma\,(-1)^{\tau}\,\Gamma_\tau\,(\overset{*}{2t}+1, 2t, \dots 2\rho+\tau;\ 2\rho+\tau, \dots 2t, \overset{*}{2t}+1),$$
$$f'_{2t+2-2\rho}\,(\overset{*}{1}, \dots 2t+2-2\rho;\ \overset{*}{1}, \dots 2t+2-2\rho)=\psi\,(\overset{*}{1}, 2, \dots 2t+2-2\rho;\ 2t+2-2\rho, \dots 2, \overset{*}{1})$$
$$=\Sigma\,(-1)^{\tau}\,\Gamma_\tau\,(\overset{*}{1}, 2, \dots 2t+2-2\rho-\tau;\ 2t+2-2\rho-\tau, \dots 2, \overset{*}{1}),$$
$$g_{2t+2-2\rho}\,(\overset{*}{1}, \dots 2t+2-2\rho;\ 2\rho, \dots \overset{*}{2t}+1)=\psi\,(\overset{*}{1}, 2, \dots 2t+2-2\rho;\ 2\rho, \dots 2t, \overset{*}{2t}+1)$$
$$=\Sigma\,(-1)^{\tau}\,\Gamma_\tau\,(\overset{*}{1}, 2, \dots 2t+2-2\rho-\tau;\ 2\rho+\tau, \dots 2t, \overset{*}{2t}+1),$$
$$g'_{2t+2-2\rho}\,(2\rho, \dots \overset{*}{2t}+1;\ \overset{*}{1}, \dots 2t+2-2\rho)=\psi\,(\overset{*}{2t}+1, 2t, \dots 2\rho;\ 2t+2-2\rho, \dots 2, \overset{*}{1})$$
$$=\Sigma\,(-1)^{\tau}\,\Gamma_\tau\,(\overset{*}{2t}+1, 2t, \dots 2\rho+\tau;\ 2t+2-2\rho-\tau, \dots 2, \overset{*}{1}),$$
$$(\tau=0, 1, 2, \dots t-\rho-1) \quad \dots\dots\dots\dots(11'').$$

Here the bilinear invariants of the substitutions (i) correspond to the set of skew-symmetric invariant transformands $\{A', B\}$ described in Ex. x of § 257, and the bilinear invariants of the substitutions (ii), (iii), (iv) can be derived from them by the appropriate reversing substitutions for the variables.

Ex. x. *Proofs of the formulae of Ex.* ix.

To establish all the formulae (11), (11′), (11″) it is sufficient to obtain the development of the bilinear invariant f_{2t} shown in (11). If we put $u=r+1-i$, we have

$$f_{2t}=[x]_{2t}\,[\psi]_{2t}^{2t}\,\overline{\underline{y}}_{2t}=\Sigma\psi_{uj}\,x_i y_j$$
$$=\Sigma\,\{(-1)^{t+1-i}\,.\,{}_{t-j}H_\tau\,.\,x_i\,y_j\}, \quad \dots\dots\dots\dots(12)$$

where $$\tau = i+j-r-1 = i+j-2t-1,$$
and where the summation extends over all integral values of i and j such that
$$1 \ngtr i \ngtr r, \quad 1 \ngtr j \ngtr r, \quad i+j \nless r+1.$$
The possible values of τ are those consistent with the conditions $\tau \nless 0$, $\tau \ngtr 2t-1$; and when τ is given, the possible values of i and j are those consistent with the conditions
$$i+j = 2t+1+\tau,$$
and $$\tau+1 \ngtr j \ngtr 2t \quad \text{or} \quad 2t \nless i \nless \tau+1,$$
the terms with zero coefficients being those in which
$$t < j < t+\tau+1 \quad \text{or} \quad t+\tau+1 > i > t,$$
and the terms with non-zero coefficients being those in which:

(1) $$\tau+1 \ngtr j \ngtr t \quad \text{or} \quad 2t \nless i \nless t+\tau+1;$$
(2) $$2t \nless j \nless t+\tau+1 \quad \text{or} \quad \tau+1 \ngtr i \ngtr t.$$

All the terms in which $\tau > t-1$ have zero coefficients.

By arranging the terms according to their total weights, or by expanding the simple square slope $[\psi]_{2t}^{2t}$ in terms of its successive parametric diagonal lines, we obtain the development
$$f_{2t} = S_0 + S_1 + \ldots + S_\tau + \ldots + S_{t-1}, \quad \text{..............................(12')}$$
where S_τ is the sum of the terms in which $i+j = 2t+1+\tau$. Again by restricting the summation (12) to the two ranges (1) and (2) we see that
$$S_\tau = (-1)^\tau . \Sigma\{(-1)^\omega . {}^\omega H_\tau . (x_{t+1+\tau+\omega}\, y_{t-\omega} - x_{t-\omega}\, y_{t+1+\tau+\omega})\},$$
$$(\omega = 0, 1, 2, \ldots t-\tau-1),$$
$$= (-1)^\tau . \Gamma_\tau (2t, 2t-1, \ldots \tau+1 ; \tau+1, \ldots 2t-1, 2t).$$
Thus (12') is the development of f_{2t} given in (11).

All the bilinear invariants of Ex. ix can be expressed in terms of the functions defined in Exs. i and iv. In fact the equation (16) in Ex. xi of § 257 leads at once to the general formula
$$\psi_{2t} (2t, \ldots 2, 1 ; 1, 2, \ldots 2t)$$
$$= \phi_t' (t+1, \ldots 2t-1, 2t ; t, \ldots 2, 1) - \phi_t (t, \ldots 2, 1 ; t+1, \ldots 2t-1, 2t), \quad \ldots(12'')$$
which agrees with the development given in (11) because the formulae (10') are equivalent to the formulae:
$$\Gamma_\tau (p, p-1, \ldots a+2, a+1 ; b+1, b+2, \ldots q-1, q)$$
$$= -\Gamma_\tau (a+1, a+2, \ldots p-1, p ; q, q-1, \ldots b+2, b+1)$$
$$= G_\tau (p-\sigma, \ldots p ; b+1+\sigma, \ldots b+1) - G_\tau (a+1+\sigma, \ldots a+1 ; q-\sigma, \ldots q);$$
$$\Gamma_\tau (a+1, a+2, \ldots p-1, p ; b+1, b+2, \ldots q-1, q)$$
$$= -\Gamma_\tau (p, p-1, \ldots a+2, a+1 ; q, q-1, \ldots b+2, b+1)$$
$$= G_\tau (a+1+\sigma, \ldots a+1 ; b+1+\sigma, \ldots b+1) - G_\tau (p-\sigma, \ldots p ; q-\sigma, \ldots q). \ldots (10'')$$

Ex. xi. *The complete sets of bilinear invariants derived from the ψ-functions of Exs.* vii *and* ix *when* $\alpha\beta = 1$.

For all non-zero values of r we can obtain a complete set of independent non-zero bilinear invariants of any one of the pairs of substitutions (i), (ii), (iii), (iv) having ranks $r, r-1, \ldots 3, 2, 1$ by forming those of odd ranks as in Ex. vii and those of even ranks as in Ex. ix. A ψ-function of s pairs of variables must be interpreted as in Ex. vii or as in Ex. ix according as s is an odd or even non-zero positive integer; and a function Γ_τ of s pairs of variables must be interpreted as in Ex. vii or as in Ex. ix according as $s-\tau$ is an odd or even non-zero positive integer.

When $r=7$, the complete set of independent non-zero bilinear invariants of the pair of substitutions (i) formed in this way consists of the functions:

$$\begin{aligned} f_7(1, 2, \ldots 7) = &\{x_4y_4-(x_5y_3+x_3y_5)+(x_6y_2+x_2y_6)-(x_7y_1+x_1y_7)\} \\ &-\tfrac{1}{2}\{(x_5y_4+x_4y_5)-3(x_6y_3+x_3y_6)+5(x_7y_2+x_2y_7)\} \\ &+\tfrac{1}{2}\{(x_6y_4+x_4y_6)-4(x_7y_3+x_3y_7)\} \\ &-\tfrac{1}{2}\{x_7y_4+x_4y_7\}; \end{aligned}$$

$$\begin{aligned} f_6(2, 3, \ldots 7) = &\{(x_5y_4-x_4y_5)-(x_6y_3-x_3y_6)+(x_7y_2-x_2y_7)\} \\ &-\{(x_6y_4-x_4y_6)-2(x_7y_3-x_3y_7)\} \\ &+\{x_7y_4-x_4y_7\}; \end{aligned}$$

$$\begin{aligned} f_5(3, 4, \ldots 7) = &\{x_5y_5-(x_6y_4+x_4y_6)+(x_7y_3+x_3y_7)\} \\ &-\tfrac{1}{2}\{(x_6y_5+x_5y_6)-3(x_7y_4+x_4y_7)\} \\ &+\tfrac{1}{2}\{x_7y_5+x_5y_7\}; \end{aligned}$$

$$\begin{aligned} f_4(4, 5, 6, 7) &= \{(x_6y_5-x_5y_6)-(x_7y_4-x_4y_7)\}-\{x_7y_5-x_5y_7\}; \\ f_3(5, 6, 7) &= \{x_6y_6-(x_7y_5+x_5y_7)\}-\tfrac{1}{2}\{x_7y_6+x_6y_7\}; \\ f_2(6, 7) &= \{x_7y_6-x_6y_7\}; \\ f_1(7) &= \{x_7y_7\}. \end{aligned}$$

Ex. xii. *Bilinear invariants of two co-gredient substitutions.*

If A is the simple bi-canonical square matrix of order r whose latent root is a, the only co-gredient substitutions of the kinds considered in this article together with their general bilinear invariants are:

$$\text{(i}''\text{)} \quad \underline{\overline{x'}}_r = A\,.\,\underline{\overline{x}}_r, \quad \underline{\overline{y'}}_r = A\,.\,\underline{\overline{y}}_r \quad \text{for which} \quad f=[x]_r\,.\,C\,.\,\underline{\overline{y}}_r\,;$$

$$\text{(ii}''\text{)} \quad \underline{\overline{x'}}_r = A'\,.\,\underline{\overline{x}}_r, \quad \underline{\overline{y'}}_r = A'\,.\,\underline{\overline{y}}_r \quad \text{for which} \quad f'=[x]_r\,.\,JCJ\,.\,\underline{\overline{y}}_r\,;$$

where $C=[c]_r^r$ is a general invariant transformand $\{A', A\}$.

If $a^2 \neq 1$, C is a zero matrix, and there are no non-zero bilinear invariants. If $a^2=1$, the general bilinear invariants f and f' of (i″) and (ii″) are the same as those of (i) and (ii) when $a\beta=1$.

Ex. xiii. *Bilinear invariants of two contra-gredient substitutions.*

Since the inverse of an undegenerate simple bi-canonical square matrix or its conjugate cannot be a simple bi-canonical square matrix unless its order is equal to 1, the only two contra-gredient substitutions of the kinds considered in this article are

$$x'=a^{-1}x, \quad y'=ay, \quad \text{where } a \neq 0.$$

If c is an arbitrary parameter, the general bilinear invariant of these substitutions is

$$f=cxy.$$

3. *Quadratic invariants of a single substitution.*

Let A be the simple bi-canonical square matrix of order r whose latent root is a; and let A' be its conjugate. Then if J is the simple reversant of order r, we can take the general quadratic invariants of the substitutions:

$$\text{(v)} \quad \underline{\overline{x'}}_r = A\,.\,\underline{\overline{x}}_r \quad \text{to be} \quad f=[x]_r\,.\,C\,.\,\underline{\overline{x}}_r\,;$$

$$\text{(vi)} \quad \underline{\overline{x'}}_r = A'\,.\,\underline{\overline{x}}_r \quad \text{to be} \quad f'=[x]_r\,.\,JCJ\,.\,\underline{\overline{x}}_r\,;$$

where $C=[c]_r^r$ is a general symmetric invariant transformand $\{A', A\}$. When C is a particular symmetric invariant transformand $\{A', A\}$, the functions f and f' are corresponding particular quadratic invariants of the substitutions (v) and (vi); and we will call them quadratic invariants of *rank* ρ when the *symmetric matrix* C has rank ρ.

In describing these quadratic invariants we will use the abbreviated notation

$$f(i+1, i+2, \dots i+s)=f(x_{i+1}, x_{i+2}, \dots x_{i+s}) \quad \text{.....................(13)}$$

in which consecutive variables are represented by their suffixes. The function (13) can be derived from the function

$$f(1, 2, \dots s)=f(x_1, x_2, \dots x_s)$$

by replacing $x_1, x_2, \dots x_s$ by $x_{i+1}, x_{i+2}, \dots x_{i+s}$, and the function

$$f(i+s, \dots i+2, i+1)=f(x_{i+s}, \dots x_{i+2}, x_{i+1})$$

can be derived from the function (13) by replacing x_{i+t} by $x_{i+s+1-t}$ for the values $1, 2, \dots s$ of t.

General or particular quadratic invariants of the substitutions (v), (vi) can always be coupled together in pairs such as

$$f(1, 2, \dots \overset{*}{r}), \quad f(r, r-1, \dots \overset{*}{1}),$$

where both functions are known when one of them is known. Hence in actual evaluations we can confine ourselves to one of the substitutions. General formulae take their simplest forms for the substitution (vi), but we shall direct our chief attention to the substitution (v) whose scalar equations are

$$x_1'=a(x_1+x_2),\ x_2'=a(x_2+x_3),\ \dots\ x'_{r-1}=a(x_{r-1}+x_r),\ x_r'=ax_r. \quad \text{.........(b)}$$

If $a^2 \neq 1$, the general symmetric invariant transformand C is a zero matrix, and there are no non-zero quadratic invariants.

If $a^2=1$, we could take C to be one of the general symmetric invariant transformands η described in Exs. iii and v of § 257. In this case there are exactly $\frac{1}{2}(r+1)$ or exactly $\frac{1}{2}r$ independent particular non-zero quadratic invariants of each of the substitutions (v) and (vi); and the sum of such independent quadratic invariants, each multiplied by an arbitrary scalar parameter, is a general quadratic invariant. After determining any function which is a particular non-zero quadratic invariant of maximum rank r for all odd values of r, we can derive from it a complete set of independent particular quadratic invariants for all values of r by using the following principle, which is immediately deducible from Ex. vi of § 257. The notation defined in (13) is used, i.e. variables are represented by their suffixes.

If $$f_{2t+1}(1, 2, \dots \overset{*}{2t}+1), \quad f'_{2t+1}(\overset{*}{1}, 2, \dots 2t+1)$$

are functions which for all positive integral values of t are particular non-zero quadratic invariants of maximum rank $2t+1$ *of the substitutions* (v), (vi) *whenever* $a^2=1$ *and* $r=2t+1$, *then:*

(1) *when* $a^2=1$ *and* $r=2t+1$, *complete sets of* $t+1$ *independent non-zero quadratic invariants of the substitutions* (v), (vi) *having odd ranks* $2t+1, 2t-1, \dots 2t+1-2\rho, \dots 3, 1$ *can be obtained by giving to* ρ *the values* $0, 1, \dots \rho, \dots t-1, t$ *in either of the two sets of formulae:*

$$f_{2t+1-2\rho}(2\rho+1, \dots 2t, \overset{*}{2t}+1), \quad f_{2t+1-2\rho}(2t+1-2\rho, \dots 2, \overset{*}{1});$$

$$f'_{2t+1-2\rho}(\overset{*}{2t}+1, 2t, \dots 2\rho+1), \quad f'_{2t+1-2\rho}(\overset{*}{1}, 2, \dots 2t+1-2\rho);$$

(2) *when* $a^2=1$ *and* $r=2t$ (*where* $t \not< 1$), *complete sets of independent non-zero quadratic invariants of the substitutions* (v), (vi) *having odd ranks* $2t-1, 2t-3, \dots 2t+1-2\rho, \dots 3, 1$

can be obtained by giving to ρ *the values* 1, 2, ... ρ, ... $t-1$, t *in either of the two sets of formulae:*

$$f_{2t+1-2\rho}(2\rho, \ldots 2t-1, \overset{*}{2t}), \quad f_{2t+1-2\rho}(2t+1-2\rho, \ldots 2, \overset{*}{1});$$
$$f'_{2t+1-2\rho}(\overset{*}{2t}, 2t-1, \ldots 2\rho), \quad f'_{2t+1-2\rho}(\overset{*}{1}, 2, \ldots 2t+1-2\rho). \quad \text{.........(B)}$$

In applying this principle we may and will suppose f_r' to be the function derived from the function f_r by replacing x_i by x_{r+1-i}. Then in each case the two sets of formulae are two different notations for the same functions. We can pass from one notation to the other by reversing the sequences.

The simplest such sets of quadratic invariants are those described in Ex. xiv. They are derived from the sets of corresponding bilinear invariants described in Ex. vii by substituting $x_1, x_2, \ldots x_r$ for $y_1, y_2, \ldots y_r$. From any given bilinear invariant of maximum rank r of one of the pairs of substitutions (i), (ii) we can derive complete sets of quadratic invariants of the substitutions (v), (vi) by the method followed in Ex. xvi.

Ex. xiv. *The complete sets of quadratic invariants derived from the function*

$$[x]_{2t+1}[\psi]_{2t+1}^{2t+1}\ulcorner x \lrcorner_{2t+1} = \psi(2t+1, 2t, \ldots 2, 1).$$

Whenever τ and σ are positive integers, we will define a function Γ_τ of

$$s = \tau + 1 + 2\sigma$$

variables, or rather of the first $\sigma+1$ and last $\sigma+1$ of them, by the mutually equivalent equations

$$\Gamma_\tau(s, s-1, \ldots 2, 1) = \Gamma_\tau(1, 2, \ldots s-1, s)$$
$$= \Sigma\{(-1)^i . ({}^{i-1}H_\tau + {}^iH_\tau) . x_{s-\sigma+i}x_{1+\sigma-i}\}, \quad (i=0, 1, 2, \ldots \sigma), \quad \text{.........(14)}$$

so that whenever

$$p - a = \tau + 1 + 2\sigma,$$

where τ, σ, a are positive integers, we have

$$\Gamma_\tau(p, p-1, \ldots a+2, a+1) = \Gamma_\tau(a+1, a+2, \ldots p-1, p)$$
$$= \Sigma\{(-1)^i . ({}^{i-1}H_\tau + {}^iH_\tau) . x_{p-\sigma+i}x_{a+1+\sigma-i}\}, \quad (i=0, 1, 2, \ldots \sigma). \quad \text{......(14')}$$

We could interpret Γ_τ to be 0 when $p-a=\tau+2+2\sigma$ or $p-a<1$.

When $a^2=1$ and $r=2t+1$, we can obtain quadratic invariants of maximum rank $2t+1$ of the substitutions (v), (vi) by taking C to be the symmetric matrix $[\psi]_{2t+1}^{2t+1}$ defined in Ex. vi of § 257. By developing the corresponding matrix products we can express these quadratic invariants in the forms

$$f_{2t+1}(1, 2, \ldots 2t, \overset{*}{2t}+1) = \psi(\overset{*}{2t}+1, 2t, \ldots 2, 1)$$
$$= -x_{t+1}^2 + \Sigma(-1)^\tau \Gamma_\tau(\overset{*}{2t}+1, 2t, \ldots \tau+1) = -x_{t+1}^2 + \Sigma(-1)^\tau \Gamma_\tau(\tau+1, \ldots 2t, \overset{*}{2t}+1),$$

$$f'_{2t+1}(\overset{*}{1}, 2, \ldots 2t, 2t+1) = \psi(\overset{*}{1}, 2, \ldots 2t, 2t+1)$$
$$= -x_{t+1}^2 + \Sigma(-1)^\tau \Gamma_\tau(\overset{*}{1}, 2, \ldots 2t+1-\tau) = -x_{t+1}^2 + \Sigma(-1)^\tau \Gamma_\tau(2t+1-\tau, \ldots 2, \overset{*}{1}),$$
$$(\tau=0, 1, 2, \ldots t), \ldots(15)$$

where the functions Γ_τ are defined as above, and where we can and will regard $\psi=\psi_{2t+1}$ as a completely defined function of $2t+1$ variables for every positive integral value of t, the two defining equations in (15) being mutually equivalent. In fact the quadratic invariant f_{2t+1} can be derived from the corresponding bilinear invariant f_{2t+1} of Exs. vii and viii by substituting $x_1, x_2, \ldots x_r$ for $y_1, y_2, \ldots y_r$, and the development given in (15) could be obtained directly as in Ex. viii.

In describing the derived complete sets of independent non-zero quadratic invariants of odd ranks formed in accordance with the principle (B) we have to distinguish between two cases.

Case I. When $r=2t+1$ and $a^2=1$, these $t+1$ independent non-zero quadratic invariants of the substitutions (v), (vi) of odd ranks $2t+1,\ 2t-1, \ldots\ 2t+1-2\rho, \ldots\ 3,\ 1$ are the ψ-functions obtained by putting $\rho=0,\ 1, \ldots\ \rho, \ldots\ t-1,\ t$ in the formulae:

$$f_{2t+1-2\rho}(2\rho+1, \ldots\ 2t,\ \overset{*}{2t}+1)=\psi(\overset{*}{2t}+1,\ 2t, \ldots\ 2\rho+1)$$

$$=-x_{t+1+\rho}^2+\Sigma(-1)^\tau\,\Gamma_\tau(\overset{*}{2t}+1,\ 2t, \ldots 2\rho+\tau+1)=-x_{t+1+\rho}^2$$

$$+\Sigma(-1)^\tau\,\Gamma_\tau(2\rho+\tau+1, \ldots 2t,\ \overset{*}{2t}+1);$$

$$f'_{2t+1-2\rho}(\overset{*}{1},\ 2, \ldots\ 2t+1-2\rho)=\psi(\overset{*}{1},\ 2, \ldots\ 2t+1-2\rho)$$

$$=-x_{t+1-\rho}^2+\Sigma(-1)^\tau\,\Gamma_\tau(\overset{*}{1},\ 2, \ldots\ 2t+1-2\rho-\tau)=-x_{t+1-\rho}^2$$

$$+\Sigma(-1)^\tau\,\Gamma_\tau(2t+1-2\rho-\tau, \ldots\ 2,\ \overset{*}{1});$$

$$(\tau=0,\ 1,\ 2, \ldots\ t-\rho). \qquad \text{.........(15')}$$

Case II. When $r=2t$ (where $t \not< 1$) and $a^2=1$, these t independent non-zero quadratic invariants of the substitutions (v), (vi) of odd ranks $2t-1,\ 2t-3, \ldots\ 2t+1-2\rho, \ldots\ 3,\ 1$ are the ψ-functions obtained by putting $\rho=1,\ 2, \ldots\ \rho, \ldots t-1,\ t$ in the formulae:

$$f_{2t+1-2\rho}(2\rho, \ldots\ 2t-1,\ \overset{*}{2t})=\psi(\overset{*}{2t},\ 2t-1, \ldots\ 2\rho)$$

$$=-x_{t+\rho}^2+\Sigma(-1)^\tau\,\Gamma_\tau(\overset{*}{2t}, 2t-1, \ldots 2\rho+\tau)=-x_{t+\rho}^2$$

$$+\Sigma(-1)^\tau\,\Gamma_\tau(2\rho+\tau, \ldots\ 2t-1,\ \overset{*}{2t});$$

$$f'_{2t+1-2\rho}(\overset{*}{1},\ 2, \ldots\ 2t+1-2\rho)=\psi(\overset{*}{1},\ 2, \ldots\ 2t+1-2\rho)$$

$$=-x_{t+1-\rho}^2+\Sigma(-1)^\tau\,\Gamma_\tau(\overset{*}{1}, 2, \ldots 2t+1-2\rho-\tau)=-x_{t+1-\rho}^2$$

$$+\Sigma(-1)^\tau\,\Gamma_\tau(2t+1-2\rho-\tau, \ldots\ 2,\ \overset{*}{1});$$

$$(\tau=0,\ 1,\ 2, \ldots\ t-\rho). \qquad \text{.........(15'')}$$

Ex. xv. *Illustrations of the formulae of Ex.* xiv.

When $r=9$ and $a^2=1$, a complete set of independent non-zero quadratic invariants of the substitution (v) is formed by the functions:

$$f_9(1, 2, \ldots 9)=\{\Gamma_0(1, 2, \ldots 9)-x_5^2\}-\Gamma_1(2, 3, \ldots 9)+\Gamma_2(3, 4, \ldots 9)-\Gamma_3(4, 5, \ldots 9)+\Gamma_4(5, 6, \ldots 9)$$
$$=\{x_5^2-2x_6x_4+2x_7x_3-2x_8x_2+2x_9x_1\}-\{x_6x_5-3x_7x_4+5x_8x_3-7x_9x_2\}$$
$$+\{x_7x_5-4x_8x_4+9x_9x_3\}-\{x_8x_5-5x_9x_4\}+\{x_9x_5\};$$

$$f_7(3, 4, \ldots 9)=\{\Gamma_0(3, 4, \ldots 9)-x_6^2\}-\Gamma_1(4, 5, \ldots 9)+\Gamma_2(5, 6, \ldots 9)-\Gamma_3(6, 7, \ldots 9)$$
$$=\{x_6^2-2x_7x_5+2x_8x_4-2x_9x_3\}-\{x_7x_6-3x_8x_5+5x_9x_4\}$$
$$+\{x_8x_6-4x_9x_5\}-\{x_9x_6\};$$

$$f_5(5, 6, \ldots 9)=\{\Gamma_0(5, 6, \ldots 9)-x_7^2\}-\Gamma_1(6, 7, \ldots 9)+\Gamma_2(7, 8, 9)$$
$$=\{x_7^2-2x_8x_6+2x_9x_5\}-\{x_8x_7-3x_9x_6\}+\{x_9x_7\};$$

$$f_3(7, 8, 9)\ =\{\Gamma_0(7, 8, 9)-x_8^2\}-\Gamma_1(8, 9)=\{x_8^2-2x_9x_7\}-\{x_9x_8\};$$

$$f_1(9)\ =\{\Gamma_0(9)-x_9^2\}=x_9^2;$$

which correspond to the general symmetric invariant transformand $\eta(9)$ described in Ex. v of § 257, the first of them being obtained by taking C to be the symmetric matrix $[\psi]_9^9$ described in Ex. vi of § 257. The corresponding complete set of independent non-zero quadratic invariants of the substitution (vi) is formed by the functions

$$f_9'(1,2,\dots 9)=\{\Gamma_0(1,2,\dots 9)-x_5^2\}-\Gamma_1(1,2,\dots 8)+\Gamma_2(1,2,\dots 7)-\Gamma_3(1,2,\dots 6)+\Gamma_4(1,2,\dots 5),$$
$$f_7'(1,2,\dots 7)=\{\Gamma_0(1,2,\dots 7)-x_4^2\}-\Gamma_1(1,2,\dots 6)+\Gamma_2(1,2,\dots 5)-\Gamma_3(1,2,\dots 4),$$
$$f_5'(1,2,\dots 5)=\{\Gamma_0(1,2,\dots 5)-x_3^2\}-\Gamma_1(1,2,\dots 4)+\Gamma_2(1,2,3),$$
$$f_3'(1,2,3)\quad=\{\Gamma_0(1,2,3)-x_2^2\}-\Gamma_1(1,2),$$
$$f_1'(1)\qquad=\{\Gamma_0(1)-x_1^2\},$$

which are derived from the functions given above by replacing x_i by x_{10-i}.

When $r=8$ and $a^2=1$, a complete set of independent non-zero quadratic invariants of the substitution (vi) is formed by the functions

$$f_7(2,3,\dots 8)=\{\Gamma_0(2,3,\dots 8)-x_5^2\}-\Gamma_1(3,4,\dots 8)+\Gamma_2(4,5,\dots 8)-\Gamma_3(5,6,\dots 8)$$
$$=\{x_5^2-2x_6x_4+2x_7x_3+2x_8x_2\}-\{x_6x_5-3x_7x_4+5x_8x_3\}$$
$$+\{x_7x_5-4x_8x_4\}-\{x_8x_5\};$$
$$f_5(4,5,\dots 8)=\{\Gamma_0(4,5,\dots 8)-x_6^2\}-\Gamma_1(5,6,\dots 8)+\Gamma_2(6,7,8)$$
$$=\{x_6^2-2x_7x_5+2x_8x_4\}-\{x_7x_6-3x_8x_5\}+\{x_8x_6\};$$
$$f_3(6,7,8)\quad=\{\Gamma_0(6,7,8)-x_7^2\}-\Gamma_1(7,8)=\{x_7^2-2x_8x_6\}-\{x_8x_7\};$$
$$f_1(8)\qquad=\{\Gamma_0(8)-x_8^2\}=x_8^2.$$

The corresponding complete set of independent non-zero quadratic invariants of the substitution (vi) is formed by the functions

$$f_7'(1,2,\dots 7)=\{\Gamma_0(1,2,\dots 7)-x_4^2\}-\Gamma_1(1,2,\dots 6)+\Gamma_2(1,2,\dots 5)-\Gamma_3(1,2,\dots 4),$$
$$f_5'(1,2,\dots 5)=\{\Gamma_0(1,2,\dots 5)-x_3^2\}-\Gamma_1(1,2,\dots 4)+\Gamma_2(1,2,3),$$
$$f_3'(1,2,3)\quad=\{\Gamma_0(1,2,3)-x_2^2\}-\Gamma_1(1,2),$$
$$f_1'(1)\qquad=\{\Gamma_0(1)-x_1^2\},$$

which are derived from the functions given above by replacing x_i by x_{9-i}.

Ex. xvi. *The complete sets of quadratic invariants derived from the symmetric invariant transformand*

$$\eta=\tfrac{1}{2}\{[\phi]_r^r+[\phi']_r^r\}.$$

If $[\phi]_r^r$ and $[\phi']_r^r$ are the mutually conjugate square matrices defined in Ex. ii of § 257, and if $a^2=1$, a particular quadratic invariant of the substitution (v) having rank r or $r-1$ according as r is odd or even is the function

$$f(1,2,\dots r)=\tfrac{1}{2}\{[x]_r[\phi]_r^r\,\overline{\underline{x}}_r+[x]_r[\phi']_r^r\,\overline{\underline{x}}_r\}$$
$$=[x]_r[\phi]_r^r\,\overline{\underline{x}}_r=[x]_r[\phi']_r^r\,\overline{\underline{x}}_r$$
$$=\tfrac{1}{2}\Sigma\{(-1)^{j-1}.\,{}^{r-i}H_\tau+(-1)^{i-1}.\,{}^{r-j}H_\tau\}\,x_iy_j,$$

where
$$\tau=i+j-r-1,$$

and where the summation extends over all integral values of i and j such that

$$1\not>i\not>r,\qquad 1\not>j\not>r,\qquad i+j\not<r-1.$$

It is what the bilinear invariant $f_r(1,2,\dots r)$ of Ex. i or iv becomes when $x_1, x_2, \dots x_r$ are substituted for $y_1, y_2, \dots y_r$. If we put

$$s=r-\tau,$$

it can be expressed in the form

$$f(1,2,\dots r)=\Sigma(-1)^\tau F_\tau,\qquad(\tau=0,1,2,\dots r-1),$$

where
$$F_\tau=\tfrac{1}{2}\Sigma(-1)^i.\,\{{}^iH_\tau+(-1)^{s-1}.\,{}^{s-1-i}H_\tau\}\,x_{\tau+1+i}x_{r-i}$$
$$=\Sigma(-1)^i.\,{}^iH_\tau.\,x_{\tau+1+i}x_{r-i},\quad(i=0,1,2,\dots s-1),$$

and where $\quad(-1)^{s-1}=(-1)^\tau$ or $-(-1)^\tau$ according as r is odd or even.

Whenever τ, a, s are positive integers and $s \nless 1$, we can and will define a function G_τ of s variables by the mutually equivalent equations

$$G_\tau(a+1, a+2, \ldots a+s) = G_\tau(a+s, \ldots a+2, a+1)$$
$$= \tfrac{1}{2}\Sigma(-1)^i . \{{}^iH_\tau + (-1)^\tau . {}^{s-1-i}H_\tau\}\, x_{a+1+i}x_{a+s-i}, \quad (i=0, 1, 2, \ldots s-1); \ \ldots(17)$$

so that $G_\tau = 0$ when $\tau+s$ is even. Then when r is odd and $a^2=1$, the functions

$$f_r(1, 2, \ldots \overset{*}{r}) = \Sigma(-1)^\tau G_\tau(\tau+1, \tau+2, \ldots \overset{*}{r}), \quad f_r'(\overset{*}{1}, 2, \ldots r) = \Sigma(-1)^\tau G_\tau(\overset{*}{1}, 2, \ldots r-\tau),$$
$$(\tau=0, 1, 2, \ldots r-1), \ \ldots\ldots\ldots\ldots\ldots(18)$$

are corresponding quadratic invariants of the substitutions (v), (vi) of maximum rank r, the first of them being the function (16); whilst when r is even, they vanish identically.

In describing the derived complete sets of independent non-zero quadratic invariants of odd ranks formed in accordance with the principle (B) we have to distinguish between two cases.

Case I. When r is an odd integer $2t+1$ and $a^2=1$, the corresponding complete sets of independent non-zero quadratic invariants of the substitutions (v), (vi) having odd ranks r, $r-2$, ... $r-2\rho$, ... 3, 1 can be obtained by putting $\rho=0$, 1, ... ρ, ... $t-1$, t in the formulae

$$f_{r-2\rho}(2\rho+1, \ldots r-1, \overset{*}{r}) = \Sigma(-1)^\tau G_\tau(2\rho+1+\tau, \ldots r-1, \overset{*}{r}) = \Sigma(-1)^\tau G_\tau(\overset{*}{r}, r-1, \ldots 2\rho+1+\tau),$$
$$f'_{r-2\rho}(\overset{*}{1}, 2, \ldots r-2\rho) = \Sigma(-1)^\tau G_\tau(\overset{*}{1}, 2, \ldots r-2\rho-\tau) = \Sigma(-1)^\tau G_\tau(r-2\rho-\tau, \ldots 2, \overset{*}{1}),$$
$$(\tau=0, 1, 2, \ldots r-2\rho-1). \ \ldots\ldots\ldots\ldots\ldots(18')$$

Case II. When r is an even integer $2t$ and $a^2=1$, the corresponding complete sets of independent non-zero quadratic invariants of the substitutions (v), (vi) having odd ranks $r-1$, $r-3$, ... $r+1-2\rho$, ... 3, 1 can be obtained by putting $\rho=1$, 2, ... ρ, ... $t-1$, t in the formulae

$$f_{r+1-2\rho}(2\rho, \ldots r-1, \overset{*}{r}) = \Sigma(-1)^\tau G_\tau(2\rho+\tau, \ldots r-1, \overset{*}{r}) = \Sigma(-1)^\tau G_\tau(\overset{*}{r}, r-1, \ldots 2\rho+\tau),$$
$$f'_{r+1-2\rho}(\overset{*}{1}, 2, \ldots r+1-2\rho) = \Sigma(-1)^\tau G_\tau(\overset{*}{1}, 2, \ldots r+1-2\rho-\tau) = \Sigma(-1)^\tau G_\tau(r+1-2\rho-\tau, \ldots 2, \overset{*}{1}),$$
$$(\tau=0, 1, 2, \ldots r-2\rho). \ \ldots\ldots\ldots\ldots\ldots(18'')$$

These quadratic invariants are derived from the corresponding bilinear invariants of Ex. i or iv by substituting $x_1, x_2, \ldots x_r$ for $y_1, y_2, \ldots y_r$.

Ex. xvii. *Illustrations of the formulae of Ex. xvi.*

When $r=9$ and $a^2=1$, a complete set of independent non-zero quadratic invariants of the substitution (v) is formed by the functions

$$\begin{aligned}
f_9(1,2,\ldots 9) &= G_0(1,2,\ldots 9) - G_1(2,3,\ldots 9) + \ldots \quad + G_8(9)\\
&= \{2x_1x_9 - 2x_2x_8 + 2x_3x_7 - 2x_4x_6 + x_5^2\} + \{7x_2x_9 - 5x_3x_8 + 3x_4x_7 - x_5x_6\}\\
&\quad + \{29x_3x_9 - 24x_4x_8 + 21x_5x_7 - 10x_6^2\} + \{55x_4x_9 - 31x_5x_8 + 10x_6x_7\}\\
&\quad + \{71x_5x_9 - 40x_6x_8 + 15x_7^2\} + \{55x_6x_9 - 15x_7x_8\}\\
&\quad + \{29x_7x_9 - 7x_8^2\} + \{7x_8x_9\} + \{x_9^2\};
\end{aligned}$$

$$\begin{aligned}
f_7(3,4,\ldots 9) &= G_0(3,4,\ldots 9) - G_1(4,5,\ldots 9) + \ldots \quad + G_6(9)\\
&= \{2x_3x_9 - 2x_4x_8 + 2x_5x_7 - x_6^2\} + \{5x_4x_9 - 3x_5x_8 + x_6x_7\}\\
&\quad + \{16x_5x_9 - 13x_6x_8 + 6x_7^2\} + \{19x_6x_9 - 6x_7x_8\}\\
&\quad + \{16x_7x_9 - 5x_8^2\} + \{5x_8x_9\} + \{x_9^2\};
\end{aligned}$$

$$\begin{aligned}
f_5(5,6,\ldots 9) &= G_0(5,6,\ldots 9) - G_1(6,7,8,9) + G_2(7,8,9) - G_3(8,9) + G_4(9)\\
&= \{2x_5x_9 - 2x_6x_8 + x_7^2\} + \{3x_6x_9 - x_7x_8\}\\
&\quad + \{7x_7x_9 - 3x_8^2\} + \{3x_8x_9\} + \{x_9^2\};
\end{aligned}$$

$$f_3(7,8,9) = G_0(7,8,9) - G_1(8,9) + G_2(9) = \{2x_7x_9 - x_8^2\} + \{x_8x_9\} + \{x_9^2\}$$

$$f_1(9) = G_0(9) = x_9^2.$$

The corresponding complete set of independent non-zero quadratic invariants of the substitution (vi) can be obtained by replacing x_i by x_{10-i}.

When $r=8$ and $a^2=1$, a complete set of independent non-zero quadratic invariants of the substitution (v) is formed by the functions

$$\begin{aligned} f_7(2,3,\ldots 8) &= G_0(2,3,\ldots 8) - G_1(3,4,\ldots 8) + \ldots + G_6(8) \\ &= \{2x_2x_8 - 2x_3x_7 + 2x_4x_6 - x_5^2\} + \{5x_3x_8 - 3x_4x_7 + x_5x_6\} \\ &\quad + \{16x_4x_8 - 13x_5x_7 + 6x_6^2\} + \{19x_5x_8 - 6x_6x_7\} \\ &\quad + \{16x_6x_8 - 5x_7^2\} + \{5x_7x_8\} + \{x_8^2\}; \end{aligned}$$

$$\begin{aligned} f_5(4,5,\ldots 8) &= G_0(4,5,\ldots 8) - G_1(5,6,7,8) + G_2(6,7,8) - G_3(7,8) + G_4(8) \\ &= \{2x_4x_8 - 2x_5x_7 + x_6^2\} + \{3x_5x_8 - x_6x_7\} \\ &\quad + \{7x_6x_8 - 3x_7^2\} + \{3x_7x_8\} + \{x_8^2\}; \end{aligned}$$

$$f_3(6,7,8) = G_0(6,7,8) - G_1(7,8) + G_2(8) = \{2x_6x_8 - x_7^2\} + \{x_7x_8\} + \{x_8^2\};$$

$$f_1(8) = G_0(8) = x_8^2.$$

The corresponding complete set of independent non-zero quadratic invariants of the substitution (vi) can be obtained by replacing x_i by x_{9-i}.

§ 259. The invariant transformands of a pair of unilatent square matrices.

1. *General invariant transformands.*

Let $A=[a]_m^m$ and $B=[b]_n^n$ be any two unilatent square matrices of orders m and n with constant elements whose latent roots are c and c', and whose characteristic potent divisors are

$$(\lambda-c)^{\alpha_1}, (\lambda-c)^{\alpha_2}, \ldots (\lambda-c)^{\alpha_r} \text{ and } (\lambda-c')^{\beta_1}, (\lambda-c')^{\beta_2}, \ldots (\lambda-c')^{\beta_s},$$

where $\alpha_1, \alpha_2, \ldots \alpha_r$ and $\beta_1, \beta_2, \ldots \beta_s$ are non-zero positive integers arranged in descending orders of magnitude; let A' and B' be their conjugates; and let U and V be any particular undegenerate contra-commutants $\{A', A\}$ and $\{B, B'\}$. Also let

ρ be the sum of the smaller integers of the pairs (α_1, β_1), (α_2, β_2), … (α_i, β_i), …, where α_i is 0 when $i>r$, and β_i is 0 when $i>s$; …………(1)

σ be the sum of the smaller integers of the rs pairs (α_i, β_j); ………(2)

and let $\epsilon = \pm 1$.

Without actually constructing the general invariants mentioned, we can at once enunciate the following theorem:

Theorem I. *The four general invariant transformands*

$$X = inv\{A, \epsilon B\},\quad X' = inv\{A', \epsilon B'\},\quad Y = inv\{A', \epsilon B\},\quad Y' = inv\{A, \epsilon B'\}$$

can always be so chosen as to be connected by the relations

$$X' = UXV, \quad Y = UX, \quad Y' = XV;$$

and they are all zero matrices when $cc' \neq \epsilon$.

When $cc'=\epsilon$, *each of them is a matrix of the class* $M\binom{n}{m}$ *and of rank* ρ *containing exactly* σ *independent arbitrary parameters, and having exactly* σ *independent non-zero particularisations.*

In each of them the elements are homogeneous linear functions of the σ arbitrary parameters.

The first part of the theorem follows from Ex. iii of § 254 and Theorem III a of § 255. The last part of the theorem (in which A and B are necessarily undegenerate) follows from Note 1 of § 243 when we regard X as a general commutant $\{A, \epsilon B^{-1}\}$, because A and B^{-1} are unilatent square matrices whose latent roots are c and $\frac{1}{c'}$, and therefore A and ϵB^{-1} are unilatent square matrices having the same latent root c.

The formulae (B) of § 255 will furnish constructions for the general invariants of Theorem I when $cc' = \epsilon$ as soon as we have determined the corresponding general invariants for two square matrices A_0 and B_0 equicanonical with A and B respectively. The simplest constructions are obtained by choosing A_0 and B_0 to be bi-canonical square matrices.

Accordingly when $cc' = \epsilon$, we will define

$$A_0 = c\mathbf{A}, \quad B_0 = c'\mathbf{B}$$

to be the standardised unilatent bi-canonical square matrices of orders m, n having the same characteristic potent divisors as A, B, and

$$A = hA_0H, \quad B = kB_0K$$

to be any particular isomorphic transformations by which A, B can be derived from A_0, B_0. Then $\mathbf{A}, \mathbf{B}$ are standardised unilatent bi-canonical (or canonical) square matrices having the same latent root 1, and the general invariant transformands in the second part of Theorem I can be taken to be the matrices given by the formulae

$$X = h\xi K, \quad X' = H'\xi'k', \quad Y = H'\eta K, \quad Y' = h\eta'k', \quad \text{.........(A)}$$

where H', k' are the conjugates of H, k, and where ξ, ξ', η, η' are the general invariant transformands

$$\xi = \text{inv}\{A_0, \epsilon B_0\}, \quad \xi' = \text{inv}\{A_0', \epsilon B_0'\}, \quad \eta = \text{inv}\{A_0', \epsilon B_0\}, \quad \eta' = \text{inv}\{A_0, \epsilon B_0'\}$$

or $\quad \xi = \text{inv}\{\mathbf{A}, \mathbf{B}\}, \quad \xi' = \text{inv}\{\mathbf{A}', \mathbf{B}'\}, \quad \eta = \text{inv}\{\mathbf{A}', \mathbf{B}\}, \quad \eta' = \text{inv}\{\mathbf{A}, \mathbf{B}'\}.$

The matrices ξ, ξ', η, η' are the compound slopes described in Ex. i, or again in Ex. ii, and from their properties we deduce the second part of the above theorem.

Ex. i. *Descriptions of the general invariant transformands*

$$\xi = inv\{\mathbf{A}, \mathbf{B}\}, \quad \xi' = inv\{\mathbf{A}', \mathbf{B}'\}, \quad \eta = inv\{\mathbf{A}', \mathbf{B}\}, \quad \eta' = inv\{\mathbf{A}, \mathbf{B}'\}.$$

The unilatent bi-canonical square matrices $\mathbf{A}$ and $\mathbf{B}$ of the text are standard compartite matrices expressible in the forms

$$\mathbf{A} = \begin{bmatrix} A_1, & 0, & \dots & 0 \\ 0, & A_2, & \dots & 0 \\ \dots & \dots & \dots & \dots \\ 0, & 0, & \dots & A_r \end{bmatrix}_{\alpha_1, \alpha_2, \dots \alpha_r}^{\alpha_1, \alpha_2, \dots \alpha_r}, \quad \mathbf{B} = \begin{bmatrix} B_1, & 0, & \dots & 0 \\ 0, & B_2, & \dots & 0 \\ \dots & \dots & \dots & \dots \\ 0, & 0, & \dots & B_s \end{bmatrix}_{\beta_1, \beta_2, \dots \beta_s}^{\beta_1, \beta_2, \dots \beta_s},$$

where every part such as A_i or B_j is a simple bi-canonical (or simple canonical) square matrix whose latent root is 1. If $\mathbf{A}'$, $\mathbf{B}'$ are the conjugates of $\mathbf{A}$, $\mathbf{B}$, and if J_a, J_b are the part-reversants of the same quadrate classes as $\mathbf{A}$, $\mathbf{B}$, the obviously true equations

$$\mathbf{A}'=J_a\mathbf{A}J_a, \quad \mathbf{B}'=J_b\mathbf{B}J_b$$

show that the symmetric matrices J_a and J_b are undegenerate commutants $\{\mathbf{A}', \mathbf{A}\}$ and $\{\mathbf{B}, \mathbf{B}'\}$. Consequently the general invariants ξ, ξ', η, η' can always be so chosen as to be connected by the relations

$$\xi'=J_a\xi J_b, \quad \eta=J_a\xi, \quad \eta'=\xi J_b; \quad \text{.................................(3)}$$

and all four of them are known when any one of them is known.

From Theorem II of § 255 we see that ξ is the most general compound matrix of the form

$$\xi=\begin{bmatrix} \xi_{11}, & \xi_{12}, & \dots & \xi_{1s} \\ \xi_{21}, & \xi_{22}, & \dots & \xi_{2s} \\ \dots & \dots & \dots & \dots \\ \xi_{r1}, & \xi_{r2}, & \dots & \xi_{rs} \end{bmatrix}^{\beta_1, \beta_2, \dots \beta_s}_{\alpha_1, \alpha_2, \dots \alpha_r} \quad \text{.................................(4)}$$

which can be constructed when for all permissible values of i and j the constituent ξ_{ij} is an invariant transformand $\{A_i, B_j\}$. The constituent ξ_{ij} is an undegenerate simple slope of type $\{\pi, \pi\}$ which is completely described in Theorem IV *b* of § 255, the number of arbitrary parameters which it contains (as well as its rank) being the smaller of the two integers (α_i, β_j). The arbitrary parameters of the rs constituents are all independent. Consequently the total number of independent arbitrary parameters in ξ is the sum σ of the effective or smaller orders of all the rs constituents. Further the coefficients of these σ parameters clearly form a complete set of independent non-zero particularisations of ξ. Since $\mathbf{A}$ and $\mathbf{B}$ are undegenerate, we can regard ξ as a general commutant $\{\mathbf{A}, \mathbf{B}^{-1}\}$. Consequently the rank of ξ is the order of the greatest common canonical of $\mathbf{A}$ and $\mathbf{B}^{-1}$, which (because $\mathbf{B}$ and $\mathbf{B}^{-1}$ are equicanonical) is also the order of the greatest common canonical of $\mathbf{A}$ and $\mathbf{B}$, i.e. the integer ρ defined in (1).

The general invariants ξ', η, η' can be constructed in similar ways; but from (3) we see that they can be derived from ξ by part-reversals. We conclude that:

The four general invariant transformands ξ, ξ', η, η' are completely known compound slopes of the standardised class

$$M\begin{pmatrix} \beta_1, & \beta_2, & \dots & \beta_s \\ \alpha_1, & \alpha_2, & \dots & \alpha_r \end{pmatrix} \quad \text{...(a)}$$

and of the respective types

$$\{\pi, \pi\}, \quad \{\pi', \pi'\}, \quad \{\pi', \pi\}, \quad \{\pi, \pi'\}$$

which can always be so chosen as to be correlated by part-reversals. Each of them contains exactly σ independent arbitrary parameters, and each of them has rank ρ.

Ex. ii. *Alternative constructions for the general invariant transformands ξ, ξ', η, η'.*

Let $\mathbf{X}$, $\mathbf{X}'$, $\mathbf{Y}$, $\mathbf{Y}'$ be general commutants $\{\mathbf{A}, \mathbf{B}\}$, $\{\mathbf{A}', \mathbf{B}'\}$, $\{\mathbf{A}', \mathbf{B}\}$, $\{\mathbf{A}, \mathbf{B}'\}$, i.e. general compound continuants of the class (a) and of the respective types

$$\{\pi, \pi\}, \quad \{\pi', \pi'\}, \quad \{\pi', \pi\}, \quad \{\pi, \pi'\}.$$

When we regard the general invariant transformands ξ, ξ', η, η' as general commutants

$$\{\mathbf{A}^{-1}, \mathbf{B}\}, \quad \{\mathbf{A}'^{-1}, \mathbf{B}'\}, \quad \{\mathbf{A}'^{-1}, \mathbf{B}\}, \quad \{\mathbf{A}^{-1}, \mathbf{B}'\},$$

the general theorem of § 239 shows that we can take them to be the matrices given by the formulae

$$\xi=P\mathbf{X}, \quad \xi'=P'\mathbf{X}', \quad \eta=P'\mathbf{Y}, \quad \eta'=P\mathbf{Y}', \quad \text{.........................(5)}$$

where P, P' are any particular *undegenerate* commutants $\{\mathbf{A}^{-1}, \mathbf{A}\}$, $\{\mathbf{A}'^{-1}, \mathbf{A}'\}$ or invariant transformands $\{\mathbf{A}, \mathbf{A}\}$, $\{\mathbf{A}', \mathbf{A}'\}$, which it is certainly possible to determine.

Again when we regard the general invariant transformands ξ, ξ', η, η' as general commutants

$$\{\mathbf{A}, \mathbf{B}^{-1}\}, \quad \{\mathbf{A}', \mathbf{B}'^{-1}\}, \quad \{\mathbf{A}', \mathbf{B}^{-1}\}, \quad \{\mathbf{A}, \mathbf{B}'^{-1}\},$$

the same general theorem shows that we can take them to be the matrices given by the formulae

$$\xi = \mathbf{X}Q, \quad \xi' = \mathbf{X}'Q', \quad \eta = \mathbf{Y}Q, \quad \eta' = \mathbf{Y}'Q', \quad \text{.......................(6)}$$

where Q, Q' are any particular *undegenerate* commutants $\{\mathbf{B}, \mathbf{B}^{-1}\}$, $\{\mathbf{B}', \mathbf{B}'^{-1}\}$ or invariant transformands $\{\mathbf{B}, \mathbf{B}\}$, $\{\mathbf{B}', \mathbf{B}'\}$, which it is certainly possible to determine.

For example if P is a particular *undegenerate* square commutant $\{\mathbf{A}^{-1}, \mathbf{A}\}$, so that $\mathbf{A}^{-1}P = P\mathbf{A}$, and if we put $\xi = P\mathbf{X}$, the equations

$$\mathbf{A}^{-1}\xi = P.\mathbf{AX}, \quad \xi\mathbf{B} = P.\mathbf{XB}$$

show that ξ is a particular or general commutant $\{\mathbf{A}^{-1}, \mathbf{B}\}$ or invariant transformand $\{\mathbf{A}, \mathbf{B}\}$ when and only when $\mathbf{X}$ is a particular or general commutant $\{\mathbf{A}, \mathbf{B}\}$.

The second part of Theorem I follows at once from the formulae (5) or (6). We could give corresponding formulae for the general invariant transformands X, X', Y, Y' themselves in the second part of the theorem.

Ex. iii. *Special case when* $[a_1 a_2 \ldots a_r] = [\beta_1 \beta_2 \ldots \beta_s] = [e_1 e_2 \ldots e_s]$ *in Theorem I.*

If A and B are unilatent square matrices of the same order m whose characteristic potent divisors are

$$(\lambda - c)^{e_1}, (\lambda - c)^{e_2}, \ldots (\lambda - c)^{e_s} \quad \text{and} \quad (\lambda - c')^{e_1}, (\lambda - c')^{e_2}, \ldots (\lambda - c')^{e_s},$$

where the indices are arranged in descending order of magnitude, then in Theorem I we have

$$\rho = m; \quad \sigma = e_1 + 3e_2 + \ldots + (2i-1)e_i + \ldots + (2s-1)e_s.$$

The general invariant transformands X, X', Y, Y' are zero matrices when $cc' \neq \epsilon$; and when $cc' = \epsilon$, they are *undegenerate square matrices* of order m.

Ex. iv. *The four general invariant transformands*

$$X = inv\{A, \epsilon A\}, \quad X' = inv\{A', \epsilon A'\}, \quad Y = inv\{A', \epsilon A\}, \quad Y' = inv\{A, \epsilon A'\}.$$

We can use the notations of Ex. iii, where now $B = A$, $c' = c$.

Then if $c^2 \neq \epsilon$, the four general invariant transformands are all zero matrices; whilst when $c^2 = \epsilon$, each of them is an undegenerate square matrix of order m having exactly σ arbitrary parameters, where σ has the value given in Ex. iii. In the latter case we can use the constructions of the text in which $B_0 = A_0$, $\mathbf{B} = \mathbf{A}$, $k = h$, $K = H$; the matrices ξ, ξ', η, η' being correlated quadrate slopes of the class

$$M\begin{pmatrix} e_1, e_2, \ldots e_s \\ e_1, e_2, \ldots e_s \end{pmatrix}. \quad \text{.....................................(b)}$$

2. *General symmetric and general skew-symmetric invariant transformands.*

Let $A = [a]_r^r$ be a single unilatent square matrix of order r with constant elements whose latent root is c, and whose characteristic potent divisors are

$$(\lambda - c)^{e_1}, \ (\lambda - c)^{e_2}, \ldots (\lambda - c)^{e_s}; \qquad (e_1 + e_2 + \ldots + e_s = r);$$

where the indices are non-zero positive integers (not necessarily all different) arranged in descending order of magnitude; let A' be the conjugate of A; and let V be any particular undegenerate symmetric contra-commutant $\{A, A'\}$. Also let

$$\sigma_1 = e_2 + 2e_3 + 3e_4 + \ldots + (s-1)e_s; \quad \text{.................(7)}$$

τ = sum of the smallest integers which are respectively $\not< \frac{1}{2}e_1, \frac{1}{2}e_2, \ldots \frac{1}{2}e_s$
= half sum of the even integers of the pairs such as (e_i, e_i+1); ...(8)

τ' = sum of the greatest integers which are respectively $\not> \frac{1}{2}e_1, \frac{1}{2}e_2, \ldots \frac{1}{2}e_s$
= half sum of the even integers of the pairs such as (e_i, e_i-1); ...(8′)

and let
$$\epsilon = \pm 1.$$

Then with respect to *symmetric* invariant transformands we have the following theorem:

Theorem II. *The two general symmetric invariant transformands*
$$Y = inv\,\{A', \epsilon A\}, \quad Y' = inv\,\{A, \epsilon A'\}$$
can always be so chosen as to be connected by the symmetric relation
$$Y' = VYV;$$
and they are both zero matrices when $c^2 \neq \epsilon$.

If $c^2 = \epsilon$, *i.e. if* $c = \pm\sqrt{\epsilon}$, *each of them is a symmetric matrix of order* r *containing exactly* $\sigma_1 + \tau$ *arbitrary parameters and having exactly* $\sigma_1 + \tau$ *independent non-zero particularisations; moreover each of them is undegenerate when and only when every distinct characteristic potent divisor* $(\lambda - c)^{2p}$ *with a given even index is repeated an even number of times.*

The first part of the theorem follows from Ex. iv of § 254 and Theorem III *a* of § 255. We will prove the second part of the theorem by constructions based on the formulae (C) of § 255.

When $c^2 = \epsilon$, we will define
$$A_0 = c\mathbf{A}$$
to be the standardised unilatent bi-canonical square matrix of order r having the same characteristic potent divisors as A, and
$$A = hA_0H$$
to be any particular isomorphic transformation by which A can be derived from A_0. Then $\mathbf{A}$ is a standardised unilatent bi-canonical (or canonical) square matrix whose latent root is 1, and the general symmetric invariant transformands in the second part of Theorem II can be taken to be the matrices given by the formulae
$$Y = H'\eta H, \quad Y' = h\eta' h', \quad \text{........................(B)}$$
where H', h' are the conjugates of H, h, and where η, η' are the *general symmetric* invariant transformands
$$\eta = \text{inv}\,\{A_0', \epsilon A_0\}, \quad \eta' = \text{inv}\,\{A_0, \epsilon A_0'\}$$
or
$$\eta = \text{inv}\,\{\mathbf{A}', \mathbf{A}\}, \quad \eta' = \text{inv}\,\{\mathbf{A}, \mathbf{A}'\}.$$
Constructions for η and η' are furnished by Theorem II of § 255; and from the properties of those matrices described in Ex. v we deduce the second part of the above theorem.

The proof of the second part of the theorem could be effected by similar constructions if A_0 were any standard compartite matrix equicanonical with A in which all the parts are unipotent square matrices.

If $Y = [y]_r^r$ is a *general* invariant transformand of one of the types mentioned in Theorem II, it will be obvious that the matrix

$$\mathbf{Y} = \tfrac{1}{2}\left\{[y]_r^r + \overline{\underline{y}}_r^r\right\}$$

will become a *general symmetric* invariant transformand of that type when the arbitrary parameters occurring in its elements have been reduced by substitutions to the smallest number possible.

Ex. v. *Descriptions of the general symmetric invariant transformands*

$$\eta = inv\,\{\mathbf{A}', \mathbf{A}\}, \quad \eta' = inv\,\{\mathbf{A}, \mathbf{A}'\}.$$

It will be obvious that the part-reversant J of the same quadrate class

$$M\begin{pmatrix} e_1, & e_2, & \dots & e_s \\ e_1, & e_2, & \dots & e_s \end{pmatrix} \quad \dots\dots\dots\dots\dots\dots\dots\dots\dots\dots\text{(b)}$$

as $\mathbf{A}$ is an undegenerate symmetric contra-commutant $\{\mathbf{A}', \mathbf{A}\}$, $\{\mathbf{A}, \mathbf{A}'\}$. Consequently η and η' can always be so chosen as to be connected by the symmetric relation

$$\eta' = J\eta J, \quad \dots\dots\dots\dots\dots\dots\dots\dots\dots\dots\dots\dots\text{(9)}$$

and will both be known when one of them is known. Further if we put

$$\mathbf{A} = \begin{bmatrix} A_1, & 0, & \dots & 0 \\ 0, & A_2, & \dots & 0 \\ \dots & \dots & \dots & \dots \\ 0, & 0, & \dots & A_s \end{bmatrix}_{e_1, e_2, \dots e_s}^{e_1, e_2, \dots e_s}, \quad \eta = \begin{bmatrix} \eta_{11}, & \eta_{12}, & \dots & \eta_{1s} \\ \eta_{21}, & \eta_{22}, & \dots & \eta_{2s} \\ \dots & \dots & \dots & \dots \\ \eta_{s1}, & \eta_{s2}, & \dots & \eta_{ss} \end{bmatrix}_{e_1, e_2, \dots e_s}^{e_1, e_2, \dots e_s},$$

where every part of $\mathbf{A}$ such as A_i is a simple bi-canonical (or simple canonical) square matrix whose latent root is 1, we see from Theorem II of § 255 that η (expressed in the form shown above) is the most general *symmetric* compound matrix of the class (b) which can be constructed when for all permissible pairs of values of i and j the constituent η_{ij} is an invariant transformand $\{A_i', A_j\}$.

The diagonal constituent η_{ii} of η is a *symmetric* simple square counter-slope of type $\{\pi', \pi\}$ which has been determined in Ex. iii of § 257 and can be taken to be the symmetric invariant $\eta\,(e_i)$ described in Ex. v of § 257. The total number of arbitrary parameters in the constituent η_{ii} is $\tfrac{1}{2}(e_i+1)$ or $\tfrac{1}{2}e_i$ according as e_i is odd or even, and it has rank e_i or $e_i - 1$ according as e_i is odd or even. If $j > i$, the two non-diagonal constituents η_{ij}, η_{ji} of η are *mutually conjugate*, and η_{ij} is a simple counter-slope of type $\{\pi', \pi\}$ which has been determined in Theorem IV *b* of § 255, other forms of it being described in Ex. ii of § 257. The total number of arbitrary parameters in the non-diagonal constituent η_{ij} (as well as its rank) is the smaller of the two integers (e_i, e_j).

The independent arbitrary parameters of η are those of its diagonal constituents (totalling τ parameters), and those of the constituents lying on one side of the diagonal constituents (totalling σ_1 parameters). Consequently η contains exactly $\sigma_1 + \tau$ independent arbitrary parameters, the coefficients of which clearly form a complete set of independent non-zero particularisations of η.

Because η is a quadrate slope, viz. one of type $\{\pi', \pi\}$, we know by § 241. 4 that det η is the product of the determinants of all the paradiagonal prime minors of η. The paradiagonal prime minors of η involve only the elements of the paradiagonals of the square constituents; and in determining them we may put all other elements of η equal to 0. Corresponding to an index number which is repeated exactly t times in the series $e_1, e_2, \dots e_s$, there are exactly e paradiagonal prime minors each of which is equal to $\pm[y]_t^t$, where $[y]_t^t$ is:

an arbitrary symmetric matrix of order t when e is odd;
an arbitrary skew-symmetric matrix of order t when e is even.

Consequently η is undegenerate, i.e. all the paradiagonal prime minors of η are undegenerate, when and only when every distinct *even* index number e in the series $e_1, e_2, \dots e_s$ is repeated an even number of times, there being no restrictions on the repetitions of odd index numbers.

We can construct η' in a similar way; or it can be derived from η by part-reversals as in (9), i.e. by reversing the orders of arrangement of the horizontal and vertical rows in every constituent.

Thus the two general symmetric invariant transformands η, η' *are completely known symmetric counter-slopes of the quadrate class* (b) *and of the respective types*

$$\{\pi', \pi\}, \quad \{\pi, \pi'\}$$

which can always be so chosen as to be correlated by part-reversals.

All the elements of each of them are homogeneous linear functions of $\sigma_1+\tau$ *independent arbitrary parameters; and each of them is undegenerate when and only when every distinct even index number in the series* $e_1, e_2, \dots e_s$ *is repeated an even number of times.*

Ex. vi. The symmetric matrices η, η' of Ex. v are also the general symmetric invariant transformands

$$\eta = \text{inv}\{A_1', \epsilon B_1\}, \quad \eta' = \text{inv}\{A_1, \epsilon B_1'\},$$

where A_1, B_1 are any unilatent bi-canonical square matrices of the quadrate class (b) having latent roots c, c' such that $cc' = \epsilon$.

Again with respect to skew-symmetric invariant transformands we have the following theorem:

Theorem III. *The two general skew-symmetric invariant transformands*

$$Y = inv\{A', \epsilon A\}, \quad Y' = inv\{A, \epsilon A'\}$$

can always be so chosen as to be connected by the symmetric relation

$$Y' = VYV;$$

and they are both zero matrices when $c^2 \neq \epsilon$.

If $c^2 = \epsilon$, *i.e. if* $c = \pm\sqrt{\epsilon}$, *each of them is a skew-symmetric matrix of order* r *containing exactly* $\sigma_1 + \tau'$ *arbitrary parameters, and having exactly* $\sigma_1 + \tau'$ *independent non-zero particularisations; moreover each of them is undegenerate when and only when every distinct characteristic potent divisor* $(\lambda - c)^{2p+1}$ *with a given odd index is repeated an even number of times.*

The first part of the theorem follows from Ex. iv of § 254 and Theorem III *a* of § 255. We will prove the second part of the theorem by constructions based on the formulae (C) of § 255.

When $c^2 = \epsilon$, we will as before define

$$A_0 = c\mathbf{A}$$

to be the standardised unilatent bi-canonical square matrix of order r having the same characteristic potent divisors as A, and

$$A = hA_0H$$

to be any particular isomorphic transformation by which A can be derived from A_0. Then $\mathbf{A}$ is a standardised unilatent bi-canonical (or canonical) square matrix whose latent root is 1, and the general skew-symmetric invariant transformands in the second part of Theorem III can be taken to be the matrices given by the formulae

$$Y = H'\eta H, \quad Y' = h\eta' h', \quad \text{......................(C)}$$

where H', h' are the conjugates of H, h, and where η, η' are the *general skew-symmetric* invariant transformands

$$\eta = \text{inv}\,\{A_0', \epsilon A_0\}, \quad \eta' = \text{inv}\,\{A_0, \epsilon A_0'\}$$

or

$$\eta = \text{inv}\,\{\mathbf{A}', \mathbf{A}\}, \quad \eta' = \text{inv}\,\{\mathbf{A}, \mathbf{A}'\}.$$

Constructions for η and η' are furnished by Theorem II of § 255, and from the properties of those matrices described in Ex. vii we deduce the second part of the above theorem.

The proof of the second part of the theorem could be effected by similar constructions if A_0 were any standard compartite matrix equicanonical with A in which all the parts are unipotent square matrices.

If $Y = [y]_r^r$ is a *general* invariant transformand of one of the types mentioned in Theorem III, it will be obvious that the matrix

$$\mathbf{Y} = \tfrac{1}{2}\left\{[y]_r^r - \overline{\underline{y}}_r^{\,r}\right\}$$

will become a *general skew-symmetric* invariant transformand of that type when the arbitrary parameters occurring in its elements have been reduced by substitutions to the smallest number possible.

Ex. vii. *Descriptions of the general skew-symmetric invariant transformands*

$$\eta = inv\,\{\mathbf{A}', \mathbf{A}\}, \quad \eta' = inv\,\{\mathbf{A}, \mathbf{A}'\}.$$

If J is the part-reversant of the class (b), η and η' can always be so chosen as to be connected by the symmetric relation

$$\eta' = J\eta J, \quad \text{...(9)}$$

and will both be known when one of them is known. Further if we put

$$\mathbf{A} = \begin{bmatrix} A_1, & 0, & \dots & 0 \\ 0, & A_2, & \dots & 0 \\ \dots & \dots & \dots & \dots \\ 0, & 0, & \dots & A_s \end{bmatrix}_{e_1, e_2, \dots e_s}^{e_1, e_2, \dots e_s}, \quad \eta = \begin{bmatrix} \eta_{11}, & \eta_{12}, & \dots & \eta_{1s} \\ \eta_{21}, & \eta_{22}, & \dots & \eta_{2s} \\ \dots & \dots & \dots & \dots \\ \eta_{s1}, & \eta_{s2}, & \dots & \eta_{ss} \end{bmatrix}_{e_1, e_2, \dots e_s}^{e_1, e_2, \dots e_s},$$

where every part of $\mathbf{A}$ such as A_i is a simple bi-canonical (or simple canonical) square matrix whose latent root is 1, we see from Theorem II of § 255 that η (expressed in the form shown above) is the most general *skew-symmetric* compound matrix of the class (b) which can be constructed when for all permissible pairs of values of i and j the constituent η_{ij} is an invariant transformand $\{A_i', A_j\}$.

The diagonal constituent η_{ii} of η is a *skew-symmetric* simple square counter-slope of type $\{\pi', \pi\}$ which has been determined in Ex. viii of § 257 and can be taken to be the skew-symmetric invariant $\eta(e_i)$ described in Ex. x of § 257. The total number of arbitrary parameters in the constituent η_{ii} is $\frac{1}{2}e_i$ or $\frac{1}{2}(e_i-1)$ according as e_i is even or odd, and it has rank e_i or e_i-1 according as e_i is even or odd. If $j>i$, the two non-diagonal constituents η_{ij}, η_{ji} of η are *mutually skew-conjugate*, and η_{ij} is a simple counter-slope of type $\{\pi', \pi\}$ which has been determined in Theorem IV *b* of § 255, other forms of it being described in Ex. ii of § 257. The total number of arbitrary parameters in the non-diagonal constituent η_{ij} (as well as its rank) is the smaller of the two integers (e_i, e_j).

The independent arbitrary parameters of η are those of its diagonal constituents (totalling τ' parameters), and those of the constituents lying on one side of the diagonal constituents (totalling σ_1 parameters). Consequently η contains exactly $\sigma_1+\tau'$ independent arbitrary parameters, the coefficients of which clearly form a complete set of independent non-zero particularisations of η.

Because η is a quadrate slope, viz. one of type $\{\pi', \pi\}$, we know by § 241. 4 that det η is the product of the determinants of all the paradiagonal prime minors of η. Corresponding to an index number e which is repeated exactly t times in the series $e_1, e_2, \dots e_s$, there are exactly e paradiagonal prime minors each of which is equal to $\pm[y]_t^t$, where $[y]_t^t$ is:

an arbitrary symmetric matrix of order t when e is even;

an arbitrary skew-symmetric matrix of order t when e is odd.

Consequently η is undegenerate, i.e. all the paradiagonal prime minors of η are undegenerate, when and only when every distinct *odd* index number e in the series $e_1, e_2, \dots e_s$ is repeated an even number of times, there being no restrictions on the repetitions of even index numbers.

We can construct η' in a similar way; or it can be derived from η by part-reversals as in (9).

Thus the two general skew-symmetric invariant transformands η, η' are completely known skew-symmetric counter-slopes of the quadrate class (b) *and of the respective types*

$$\{\pi', \pi\}, \quad \{\pi, \pi'\}$$

which can always be so chosen as to be correlated by part-reversals.

All the elements of each of them are homogeneous linear functions of $\sigma_1+\tau'$ independent arbitrary parameters; and each of them is undegenerate when and only when every distinct odd index number in the series $e_1, e_2, \dots e_s$ is repeated an even number of times.

Ex. viii. The skew-symmetric matrices η, η' of Ex. vii are also the general skew-symmetric invariant transformands

$$\eta=\text{inv}\,\{A_1', \epsilon B_1\}, \quad \eta'=\text{inv}\,\{A_1, \epsilon B_1'\},$$

where A_1, B_1 are any unilatent bi-canonical square matrices of the quadrate class (b) having latent roots c, c' such that $cc'=\epsilon$.

Ex. ix. The sum of a 'general symmetric' and a 'general skew-symmetric' invariant transformand (with independent parameters) of one of the types mentioned in Theorem II or III is an invariant transformand of the same type having $2\sigma_1+\tau+\tau'=\sigma$ independent non-zero particularisations, where σ is defined as in (2) and has the value given in Ex. iii; i.e. it is a 'general' invariant transformand of that type, as is otherwise obvious.

§ 260. The invariant transformands of any pair of square matrices whose elements are constants.

1. *The general invariant transformands*

$$X = inv\,\{A,\ \epsilon B\},\quad X' = inv\,\{A',\ \epsilon B'\},\quad Y = inv\,\{A',\ \epsilon B\},\quad Y' = inv\,\{A,\ \epsilon B'\}. \qquad \text{.....................(a)}$$

Let $A = [a]_m^m$ and $B = [b]_n^n$ be any two given square matrices of orders m and n whose elements are constants; let A', B' be their conjugates; and let ϵ be always the same one of the two integers

$$\epsilon = \pm 1.$$

If U and V are any particular undegenerate commutants $\{A',\ A\}$ and $\{B,\ B'\}$, we know (see Ex. iii of § 254) that the four general invariant transformands (a) can always be so chosen as to be connected by the relations

$$X' = UXV, \quad Y = UX, \quad Y' = XV;$$

and that they all have the same rank, and all contain the same number of independent arbitrary parameters, i.e. all have the same number of independent non-zero particularisations.

In order to describe constructions for X, X', Y, Y', we will suppose the distinct latent roots of A and B to be so arranged that they are

$$c_1 = \epsilon c_1'^{-1},\ c_2 = \epsilon c_2'^{-1},\ \ldots\ c_r = \epsilon c_r'^{-1},\ a_1,\ a_2,\ a_3,\ \ldots; \qquad \text{.........(1)}$$

and
$$c_1' = \epsilon c_1^{-1},\ c_2' = \epsilon c_2^{-1},\ \ldots\ c_r' = \epsilon c_r^{-1},\ b_1,\ b_2,\ b_3,\ \ldots; \qquad \text{.........(2)}$$

where
$$c_i c_i' = \epsilon; \ \text{but}\ a_h b_k \neq \epsilon,\ a_h c_i' \neq \epsilon,\ b_k c_i \neq \epsilon.$$

The arrangement is such that λ, μ are (non-zero) latent roots of A, B satisfying the equation $\lambda\mu = \epsilon$ when and only when

$$\lambda = c_i,\ \mu = c_i' \qquad \text{for one of the values } 1, 2, \ldots r \text{ of } i.$$

We could have

$$\text{either}\quad a_h c_i = \epsilon \quad (\text{if } a_h = c_i'), \quad \text{or} \quad b_k c_i' = \epsilon \quad (\text{if } b_k = c_i);$$

and we could have
$$c_i c_j = c_i' c_j' = \epsilon \qquad (\text{if } c_i' = c_j \text{ and } c_j' = c_i);$$

or in particular
$$c_i^2 = c_i'^2 = \epsilon \qquad (\text{if } c_i = c_i' = \pm\sqrt{\epsilon}).$$

We will suppose further that

$$A_0 = \begin{bmatrix} A_1, & 0, \ldots & 0, & 0 \\ 0, & A_2, \ldots & 0, & 0 \\ \cdots & \cdots & \cdots & \cdots \\ 0, & 0, \ldots & A_r, & 0 \\ 0, & 0, \ldots & 0, & P \end{bmatrix}_{\alpha_1,\alpha_2,\ldots\alpha_r,p}^{\alpha_1,\alpha_2,\ldots\alpha_r,p}, \quad B_0 = \begin{bmatrix} B_1, & 0, \ldots & 0, & 0 \\ 0, & B_2, \ldots & 0, & 0 \\ \cdots & \cdots & \cdots & \cdots \\ 0, & 0, \ldots & B_r, & 0 \\ 0, & 0, \ldots & 0, & Q \end{bmatrix}_{\beta_1,\beta_2,\ldots\beta_r,q}^{\beta_1,\beta_2,\ldots\beta_r,q} \qquad \text{.....................(3)}$$

are standard compartite square matrices equicanonical with A, B respectively in which:

A_i is a standardised unilatent bi-canonical square matrix whose single distinct latent root is c_i;

B_i is a standardised unilatent bi-canonical square matrix whose single distinct latent root is c_i';

P is a square matrix whose distinct latent roots are $a_1, a_2, a_3, \ldots$;

Q is a square matrix whose distinct latent roots are $b_1, b_2, b_3, \ldots$;

and that
$$A = hA_0H, \quad B = kB_0K \quad \ldots\ldots(4)$$
are given particular isomorphic transformations by which A, B can be derived from A_0, B_0. We could of course choose A_0 and B_0 to be standard bi-canonical square matrices with standardised unilatent super-parts, one corresponding to each distinct latent root.

If the characteristic potent divisors of A and B corresponding to their latent roots c_i and c_i' are respectively
$$(\lambda - c_i)^{\alpha_{i1}}, (\lambda - c_i)^{\alpha_{i2}}, \ldots \quad \text{and} \quad (\lambda - c_i')^{\beta_{i1}}, (\lambda - c_i')^{\beta_{i2}}, \ldots, \ldots(5)$$
where the indices are arranged in descending orders of magnitude, then A_i and B_i are standard compartite matrices of the quadrate classes
$$M\begin{pmatrix}\alpha_{i1}, & \alpha_{i2}, & \ldots \\ \alpha_{i1}, & \alpha_{i2}, & \ldots\end{pmatrix} \quad \text{and} \quad M\begin{pmatrix}\beta_{i1}, & \beta_{i2}, & \ldots \\ \beta_{i1}, & \beta_{i2}, & \ldots\end{pmatrix}$$
in which all the parts are simple bi-canonical square matrices; and we have
$$\alpha_i = \alpha_{i1} + \alpha_{i2} + \ldots, \quad \beta_i = \beta_{i1} + \beta_{i2} + \ldots.$$

By Ex. i of § 255 the four general invariant transformands (a) can be taken to be the matrices given by the formulae
$$X = h\xi K, \quad X' = H'\xi'k', \quad Y = H'\eta K, \quad Y' = h\eta'k' \quad \ldots\ldots(A)$$
where H', k' are the conjugates of H, k, and where ξ, ξ', η, η' are the general invariant transformands
$$\xi = \text{inv}\{A_0, \epsilon B_0\}, \quad \xi' = \text{inv}\{A_0', \epsilon B_0'\}, \quad \eta = \text{inv}\{A_0', \epsilon B_0\}, \quad \eta' = \text{inv}\{A_0, \epsilon B_0'\},$$
constructions for which are furnished by Theorem II of § 255. In particular (using Theorem III a of § 255) we have
$$\xi = \begin{bmatrix} \xi_1, & 0, & \ldots & 0, & 0 \\ 0, & \xi_2, & \ldots & 0, & 0 \\ \cdots & \cdots & \cdots & \cdots & \cdots \\ 0, & 0, & \ldots & \xi_r, & 0 \\ 0, & 0, & \ldots & 0, & 0 \end{bmatrix}_{\alpha_1, \alpha_2, \ldots \alpha_r, p}^{\beta_1, \beta_2, \ldots \beta_r, q}, \quad \eta = \begin{bmatrix} \eta_1, & 0, & \ldots & 0, & 0 \\ 0, & \eta_2, & \ldots & 0, & 0 \\ \cdots & \cdots & \cdots & \cdots & \cdots \\ 0, & 0, & \ldots & \eta_r, & 0 \\ 0, & 0, & \ldots & 0, & 0 \end{bmatrix}_{\alpha_1, \alpha_2, \ldots \alpha_r, p}^{\beta_1, \beta_2, \ldots \beta_r, q}$$
$$\ldots\ldots(6)$$

where ξ_i, η_i are general invariant transformands
$$\xi_i = \text{inv}\{A_i, \epsilon B_i\}, \quad \eta_i = \text{inv}\{A_i', \epsilon B_i\},$$
these being compound slopes of the class
$$M\begin{pmatrix}\beta_{i1}, & \beta_{i2}, & \ldots \\ \alpha_{i2}, & \alpha_{i2}, & \ldots\end{pmatrix}$$

and of the respective types $\{\pi, \pi\}$, $\{\pi', \pi'\}$ which are completely described in § 259. We can take ξ, ξ', η, η' to be compartite matrices of the same forms as ξ and η in which corresponding parts are correlated compound slopes of the respective types

$$\{\pi, \pi\},\ \{\pi', \pi'\},\ \{\pi', \pi\},\ \{\pi, \pi'\};$$

i.e. if J_a and J_b are the part-reversants of the classes

$$M\begin{pmatrix}\alpha_{11}, & \alpha_{12}, \dots & \alpha_{r1}, & \alpha_{r2}, \dots & p\\ \alpha_{11}, & \alpha_{12}, \dots & \alpha_{r1}, & \alpha_{r2}, \dots & p\end{pmatrix},\quad M\begin{pmatrix}\beta_{11}, & \beta_{12}, \dots & \beta_{r1}, & \beta_{r2}, \dots & q\\ \beta_{11}, & \beta_{12}, \dots & \beta_{r1}, & \beta_{r2}, \dots & q\end{pmatrix},$$

they can be so chosen as to be connected by the relations

$$\xi' = J_a \xi J_b,\quad \eta = J_a \xi,\quad \eta' = \xi J_b.$$

The independent arbitrary parameters of each of the general invariant transformands (a) are the parameters of the parts $\xi_1, \xi_2, \dots \xi_r$ of ξ, which are all independent, and are known by § 259; and the rank of each of them is the sum of the ranks of the parts $\xi_1, \xi_2, \dots \xi_r$ of ξ, which are all known by § 259. Accordingly we see that:

The four general invariant transformands (a) *can be taken to be the matrices given by the formulae* (A). *Each of them has rank* ρ, *and in each of them the elements are homogeneous linear functions of exactly* σ *independent arbitrary parameters, where* ρ *and* σ *are known when the characteristic potent divisors of* A *and* B *are known.*

The above argument remains applicable in the special case when $B = A$; but (1) and (2) are then two different arrangements

$$c_1\ \ ,\ c_2\ \ ,\dots c_r\ \ ,\ a_1, a_2, a_3, \dots, \quad\dots\dots\dots(1')$$
$$c_1' = \epsilon c_1^{-1},\ c_2' = \epsilon c_2^{-1}, \dots c_r' = \epsilon c_r^{-1},\ a_1, a_2, a_3, \dots, \quad\dots\dots\dots(2')$$

of the distinct latent roots of A; and A_0 and B_0 are two different square matrices equicanonical with A. In this special case the construction described in sub-article 2 is more convenient, the places of A_0 and B_0 being taken by a single square matrix Ω equicanonical with A.

We could have taken A_0 and B_0 in (3) to be simply compartite matrices equicanonical with A and B in which the parts are square matrices having the latent roots mentioned, the parts such as A_i and B_i being unilatent square matrices. In particular we could have taken A_0 and B_0 to be standard canonical square matrices equicanonical with A and B. We should still have the formulae (A) and the forms such as (6), the parts of ξ, ξ', η, η' being certain matrices having the same ranks and containing the same numbers of arbitrary parameters as before.

Ex. i. *Rank of the general invariant transformands* (a).

If ρ_i is the rank of the part ξ_i of ξ in (6), the rank of each of the general invariant transformands X, X', Y, Y' is the integer

$$\rho = \rho_1 + \rho_2 + \dots + \rho_r,$$

where ρ_i is the sum of the effective orders of the diagonal constituents of the standardised compound slope ξ_i, i.e. the sum of the smaller integers of the pairs

$$(\alpha_{i1}, \beta_{i1}), (\alpha_{i2}, \beta_{i2}), \ldots$$

formed with the indices of the characteristic potent divisors (5).

Hence if $T_1, T_2, \ldots T_i, \ldots$ are the distinct irresoluble (or irreducible) divisors of the square matrices

$$\mathbf{A}=[a]_m^m-\lambda[1]_m^m, \quad \mathbf{B}=\lambda[b]_n^n-[1]_n^n,$$

and if those potent divisors of $\mathbf{A}$ and $\mathbf{B}$ which are powers of T_i are

$$T_i^{e_{i1}}, T_i^{e_{i2}}, \ldots \quad \text{and} \quad T_i^{e_{i1}'}, T_i^{e_{i2}'}, \ldots,$$

where the indices are arranged in descending orders of magnitude, the rank of the general invariant transformand $\{A, B\}$ is the sum of the degrees in λ of the highest common factors of all such pairs of corresponding descendent potent divisors of $\mathbf{A}$ and $\mathbf{B}$ as

$$(T_i^{e_{i1}}, T_i^{e_{i1}'}), (T_i^{e_{i2}}, T_i^{e_{i2}'}), \ldots.$$

Since the last result must remain true when the elements of A and B involve arbitrary parameters, λ being another arbitrary parameter independent of them, we see that in such a case there exist non-zero invariant transformands $\{A, B\}$ when and only when the two determinants

$$\det\{[a]_m^m-\lambda[1]_m^m\}, \quad \det\{\lambda[b]_n^n-[1]_n^n\}$$

have an irresoluble (or irreducible) factor in common.

Ex. ii. *Invariant transformands which are undegenerate square matrices.*

It follows from Ex. i that each of the general invariant transformands (a) will be an undegenerate square matrix when and only when:

(1) the zero part $[0]_p^q$ of ξ is absent, i.e. the latent roots such as $a_1, a_2, a_3, \ldots$ and $b_1, b_2, \ldots b_3$ in (1) and (2) are all absent;

(2) all the parts of ξ in (6) are quadrate slopes, so that in the characteristic potent divisors such as (5) we always have

$$\beta_{i1}=\alpha_{i1}, \quad \beta_{i2}=\alpha_{i2}, \ldots.$$

This is the case when and only when all the characteristic potent divisors of A and B can be coupled together in pairs such as

$$(\lambda-c)^e, \ (\lambda-\epsilon c^{-1})^e, \qquad \text{where } c\neq 0,$$

there being of course no zero latent roots. Thus, as can be seen otherwise from Ex. ix of § 255:

The general invariant transformands (a) *are undegenerate square matrices when and only when A and B are two undegenerate square matrices of the same order, and moreover are such that*

A and ϵB^{-1} *are equicanonical.*

When these conditions are satisfied, and any undegenerate particularisations X_0, X_0', Y_0, Y_0' of X, X', Y, Y' have been determined, we could take the four general invariant transformands (a) to be the matrices

$$X=PX_0, \ X'=P'X_0', \ Y=P'Y_0, \ Y'=PY_0',$$

where P, P' are general commutants $\{A, A\}, \{A', A'\}$, or to be the matrices

$$X=X_0Q, \ X'=X_0'Q', \ Y=Y_0Q, \ Y'=Y_0'Q',$$

where Q, Q' are general commutants $\{B, B\}, \{B', B'\}$.

Ex. iii. *The bilinear invariants of the substitutions* $\overline{\underline{x'}}_m = A . \overline{\underline{x}}_m, \quad \overline{\underline{y'}}_n = B . \overline{\underline{y}}_n$.

If we put $\epsilon = 1$, and take $u_1, u_2, \ldots u_m$ and $v_1, v_2, \ldots v_n$ to be the independent homogeneous linear functions of $x_1, x_2, \ldots x_m$ and $y_1, y_2, \ldots y_n$ determined by the equations

$$\overline{\underline{u}}_m = H . \overline{\underline{x}}_m, \quad \overline{\underline{v}}_n = K . \overline{\underline{y}}_n,$$

the general bilinear invariant of this pair of substitutions is the function

$$f = [x]_m . Y . \overline{\underline{y}}_n = [u]_m . \eta . \overline{\underline{v}}_n,$$

where η is the general invariant transformand $\eta = \text{inv}\{A_0', B_0\}$ having the form shown in (6). There are exactly σ independent particular non-zero bilinear invariants corresponding one by one to the σ arbitrary parameters in η, the bilinear invariant corresponding to any one parameter being obtained by putting all the other parameters equal to 0.

We can divide the σ independent non-zero bilinear invariants into major sets corresponding one by one to the super-constituents $\eta_1, \eta_2, \ldots \eta_r$ of η; and we can divide the bilinear invariants of the major set corresponding to each compound slope η_i into minor sets corresponding one by one to the simple constituents of η_i. We then have one minor set of non-zero bilinear invariants corresponding to every such pair of characteristic potent divisors of A, B as

$$(\lambda - c_i)^\alpha, \quad (\lambda - c_i^{-1})^\beta, \qquad \text{where } c_i \neq 0.$$

If γ is the smaller of the two integers (α, β), this particular minor set is composed of γ functions which are completely described in sub-article 1 of § 258.

2. *The general invariant transformands*

$$X = inv\{A, \epsilon A\}, \quad X' = inv\{A', \epsilon A'\}, \quad Y = inv\{A', \epsilon A\}, \quad Y' = inv\{A, \epsilon A'\}. \quad \text{......(a')}$$

Let $A = [a]_m^m$ be a single given square matrix of order m whose elements are constants; let A' be the conjugate of A; and as before let ϵ be always the same one of the two integers

$$\epsilon = \pm 1.$$

If S is any particular undegenerate symmetric commutant $\{A, A'\}$, so that S^{-1} is a particular undegenerate symmetric commutant $\{A', A\}$, we know (see Ex. iii of § 254) that the four general invariant transformands (a') can always be so chosen as to be connected by the relations

$$X' = S^{-1}XS, \quad Y = S^{-1}X, \quad Y' = SX = SYS.$$

In order to describe constructions for the general invariant transformands X, X', Y, Y' which are more convenient than those of sub-article 1, we will now suppose the distinct latent roots of A to be so arranged that they are

$$\sqrt{\epsilon}, -\sqrt{\epsilon}; \; c_1, c_1' = \epsilon c_1^{-1}; \; c_2, c_2' = \epsilon c_2^{-1}; \ldots c_r, c_r' = \epsilon c_r^{-1}; \; a_1, a_2, a_3, \ldots; \quad \ldots(7)$$

where $\qquad c_i c_i' = \epsilon$; but $c_i^2 \neq \epsilon, \; a_h^2 \neq \epsilon, \; a_h a_k \neq \epsilon$;

and where one or both of the latent roots $\sqrt{\epsilon}, -\sqrt{\epsilon}$ may be absent. The arrangement is so chosen that if A has two unequal latent roots λ, μ satisfying the equation $\lambda\mu = \epsilon$, then for one of the values $1, 2, \ldots r$ of i we must have either

$$\lambda = c_i, \; \mu = c_i' \quad \text{or} \quad \lambda = c_i', \; \mu = c_i.$$

If A has a latent root λ such that $\lambda^2 = \epsilon$, it must be either $\sqrt{\epsilon}$ or $-\sqrt{\epsilon}$; and A has no latent root λ such that $a_h \lambda = \epsilon$.

We will further suppose that

$$\Omega = \begin{bmatrix} P, & 0, & 0, & 0, & \dots & 0, & 0, & 0 \\ 0, & Q, & 0, & 0, & \dots & 0, & 0, & 0 \\ 0, & 0, & A_1, & 0, & \dots & 0, & 0, & 0 \\ 0, & 0, & 0, & B_1, & \dots & 0, & 0, & 0 \\ \dots & & & & & & & \\ 0, & 0, & 0, & 0, & \dots & A_r, & 0, & 0 \\ 0, & 0, & 0, & 0, & \dots & 0, & B_r, & 0 \\ 0, & 0, & 0, & 0, & \dots & 0, & 0, & C \end{bmatrix}_{p,\, q,\, e_1,\, e_1',\, \dots\, e_r,\, e_r',\, \gamma}^{p,\, q,\, e_1,\, e_1',\, \dots\, e_r,\, e_r',\, \gamma} \quad \text{......(8)}$$

is a square matrix equicanonical with A in which:

P (if it occurs, i.e. if $p \neq 0$) is a standardised unilatent bi-canonical square matrix whose single distinct latent root is $\sqrt{\epsilon}$;

Q (if it occurs, i.e. if $q \neq 0$) is a standardised unilatent bi-canonical square matrix whose single distinct latent root is $-\sqrt{\epsilon}$;

A_i is a standardised unilatent bi-canonical square matrix whose single distinct latent root is c_i;

B_i is a standardised unilatent bi-canonical square matrix whose single distinct latent root is c_i';

C is a square matrix whose distinct latent roots are $a_1, a_2, a_3, \dots$;

and that

$$A = h\Omega H \quad \text{.................................(9)}$$

is a given particular isomorphic transformation by which A can be derived from Ω. Using the definitions of § 255. 4, we could choose Ω to be a standard bi-canonical square matrix having one standardised unilatent superpart corresponding to each distinct latent root of A.

If the characteristic potent divisors of A corresponding to the latent roots c_i and c_i' are

$$(\lambda - c_i)^{e_{i1}},\ (\lambda - c_i)^{e_{i2}},\ \dots \quad \text{and} \quad (\lambda - c_i')^{e_{i1}'},\ (\lambda - c_i')^{e_{i2}'},\ \dots, \quad \text{...(10)}$$

where the indices are arranged in descending orders of magnitude, then A_i and B_i are standard compartite matrices of the quadrate classes

$$M\begin{pmatrix} e_{i1}, & e_{i2}, & \dots \\ e_{i1}, & e_{i2}, & \dots \end{pmatrix} \quad \text{and} \quad M\begin{pmatrix} e_{i1}', & e_{i2}', & \dots \\ e_{i1}', & e_{i2}', & \dots \end{pmatrix},$$

in which all the parts are simple bi-canonical square matrices; and we have

$$e_i = e_{i1} + e_{i2} + \dots, \quad e_i' = e_{i1}' + e_{i2}' + \dots.$$

Again if the characteristic potent divisors of A corresponding to the latent roots $\sqrt{\epsilon}$, $-\sqrt{\epsilon}$ are

$$(\lambda - \sqrt{\epsilon})^{p_1},\ (\lambda - \sqrt{\epsilon})^{p_2},\ \dots \quad \text{and} \quad (\lambda + \sqrt{\epsilon})^{q_1},\ (\lambda + \sqrt{\epsilon})^{q_2},\ \dots \quad \text{...(11)}$$

where the indices are arranged in descending orders of magnitude, then P and Q are standard compartite matrices of the quadrate classes

$$M\begin{pmatrix} p_1, & p_2, & \dots \\ p_1, & p_2, & \dots \end{pmatrix} \quad \text{and} \quad M\begin{pmatrix} q_1, & q_2, & \dots \\ q_1, & q_2, & \dots \end{pmatrix},$$

in which all the parts are simple bi-canonical square matrices; and we have

$$p = p_1 + p_2 + \dots, \quad q = q_1 + q_2 + \dots.$$

By Ex. ii of § 255 the four general invariant transformands (a′) can be taken to be the matrices given by the formulae

$$X = h\xi H, \quad X' = H'\xi'h', \quad Y = H'\eta H, \quad Y' = h\eta'h', \quad \dots\dots(\text{A}')$$

where ξ, ξ', η, η' are the general invariant transformands

$$\xi = \text{inv}\{\Omega, \epsilon\Omega\}, \quad \xi' = \text{inv}\{\Omega', \epsilon\Omega'\}, \quad \eta = \text{inv}\{\Omega', \epsilon\Omega\}, \quad \eta' = \text{inv}\{\Omega, \epsilon\Omega'\},$$

constructions for which are furnished by Theorem II of § 255. In particular (using Theorem III a of § 255) we have

$$\eta = \begin{bmatrix} U, & 0, & 0, & 0, & \dots & 0, & 0, & 0 \\ 0, & V, & 0, & 0, & \dots & 0, & 0, & 0 \\ 0, & 0, & 0, & X_1, & \dots & 0, & 0, & 0 \\ 0, & 0, & Y_1, & 0, & \dots & 0, & 0, & 0 \\ \dots & \dots & \dots & \dots & \dots & \dots & \dots & \dots \\ 0, & 0, & 0, & 0, & \dots & 0, & X_r, & 0 \\ 0, & 0, & 0, & 0, & \dots & Y_r, & 0, & 0 \\ 0, & 0, & 0, & 0, & \dots & 0, & 0, & 0 \end{bmatrix}_{p,\, q,\, e_1,\, e_1',\, \dots\, e_r,\, e_r',\, \gamma}^{p,\, q,\, e_1,\, e_1',\, \dots\, e_r,\, e_r',\, \gamma}, \quad \dots(12)$$

where U, V, X_i, Y_i are general invariant transformands

$$U = \text{inv}\{P', \epsilon P\}, \quad V = \text{inv}\{Q', \epsilon Q\}, \quad X_i = \text{inv}\{A_i', \epsilon B_i\}, \quad Y_i = \text{inv}\{B_i', \epsilon A_i\},$$

these being compound slopes of type $\{\pi', \pi\}$ and of the respective classes

$$M\begin{pmatrix} p_1, & p_2, & \dots \\ p_1, & p_2, & \dots \end{pmatrix}, \quad M\begin{pmatrix} q_1, & q_2, & \dots \\ q_1, & q_2, & \dots \end{pmatrix}, \quad M\begin{pmatrix} e_{i1}', & e_{i2}', & \dots \\ e_{i1}, & e_{i2}, & \dots \end{pmatrix}, \quad M\begin{pmatrix} e_{i1}, & e_{i2}, & \dots \\ e_{i1}', & e_{i2}', & \dots \end{pmatrix},$$

which have been completely described in § 259. We can take ξ, ξ', η, η' to be compartite matrices of the same form as η in which corresponding parts or super-constituents are correlated compound slopes of the respective types

$$\{\pi, \pi\}, \quad \{\pi', \pi'\}, \quad \{\pi', \pi\}, \quad \{\pi, \pi'\};$$

for if $J = \{\Omega, \Omega'\}$ is the part-reversant of the same class

$$M\begin{pmatrix} p_1, p_2, \dots; & q_1, q_2, \dots; & \dots\dots; & e_{i1}, e_{i2}, \dots; & e_{i1}', e_{i2}', \dots; & \dots\dots; & \gamma \\ p_1, p_2, \dots; & q_1, q_2, \dots; & \dots\dots; & e_{i1}, e_{i2}, \dots; & e_{i1}', e_{i2}', \dots; & \dots\dots; & \gamma \end{pmatrix}$$

as the super-slope Ω, we can take them to be matrices connected by the relations

$$\xi' = J\xi J, \quad \eta = J\xi, \quad \eta' = \xi J;$$

or

$$\eta' = J\eta J, \quad \xi = J\eta, \quad \xi' = \eta J.$$

Since the parameters of the various parts of η are all independent, the total number of independent arbitrary parameters in each of the general invariant transformands (a′) is the integer

$$\sigma = u + v + 2\sigma_1 + 2\sigma_2 + \ldots + 2\sigma_r, \quad\ldots\ldots(13)$$

where u, v, σ_i are the numbers of the arbitrary parameters in the parts U, V, X_i of η, which are given in Theorem I and Ex. iii of § 259. Again the rank of each of the general invariant transformands (a′), being the sum of the ranks of the parts of η, is the integer

$$\rho = p + q + 2\rho_1 + 2\rho_2 + \ldots + 2\rho_r, \quad\ldots\ldots(14)$$

where $$p = p_1 + p_2 + \ldots, \quad q = q_1 + q_2 + \ldots;$$

and ρ_i = sum of the smaller integers of the pairs (e_{i1}, e_{i1}'), (e_{i2}, e_{i2}'), ... formed with corresponding descendent indices in (10). We conclude that:

The four general invariant transformands (a′) *can be taken to be the matrices given by the formulae* (A′). *Each of them has rank* ρ, *and in each of them the elements are homogeneous linear functions of exactly* σ *independent arbitrary parameters, where* ρ *and* σ *are the integers given by the equations* (13) *and* (14).

Ex. iv. *Undegenerate invariant transformands.*

In order that the general invariant transformands (a′) shall be undegenerate square matrices, it is necessary and sufficient that the zero part $[0]_\gamma^\gamma$ of the general invariant transformand η corresponding to the latent roots $a_1, a_2, a_3, \ldots$ in (7) shall be absent, and that all the parts of η such as X_i, Y_i shall be undegenerate square matrices, i.e. shall be quadrate slopes, so that

$$e_{i1}' = e_{i1}, \quad e_{i2}' = e_{i2}, \ldots.$$

Thus each of the general invariant transformands (a′) *is an undegenerate square matrix when and only when all the characteristic potent divisors of* A *which are not powers of* $\lambda - \sqrt{\epsilon}$ *or* $\lambda + \sqrt{\epsilon}$ *can be coupled together in pairs of the form*

$$(\lambda - c_i)^e, \quad (\lambda - \epsilon c_i^{-1})^e, \qquad \textit{where } c_i \neq 0;$$

i.e. when and only when A *is an undegenerate square matrix such that*

$$A \textit{ and } \epsilon A^{-1} \textit{ are equicanonical.}$$

If A has a zero latent root, i.e. if it is degenerate, no one of the general invariant transformands (a′) can be undegenerate.

3. *The general symmetric invariant transformands*

$$Y = inv\,\{A', \epsilon A\}, \quad Y' = inv\,\{A, \epsilon A'\}. \quad\ldots\ldots(b)$$

Let $A = [a]_m^m$ be any single given square matrix of order m with constant elements, and let the notations of sub-article 2 be applied to it. We will suppose that the distinct latent roots of A are so arranged that they are

$$\sqrt{\epsilon}, -\sqrt{\epsilon};\ c_1, c_1' = \epsilon c_1^{-1};\ c_2, c_2' = \epsilon c_2^{-1};\ \ldots\ c_r, c_r' = \epsilon c_r^{-1};\ a_1, a_2, a_3, \ldots \quad (7)$$

as in sub-article 2. We will further suppose that a compartite square matrix Ω equicanonical with A has been defined as in (8), and that

$$A = h\Omega H$$

is a particular isomorphic transformation by which A can be derived from Ω. For the characteristic potent divisors of A, which will be regarded as known, we shall use the same notations as in (10) and (11). If S is any particular undegenerate symmetric contra-commutant $\{A, A'\}$, we know (see Ex. iv of § 254) that the two general symmetric invariant transformands (b) can always be so chosen as to be connected by the symmetric relation

$$Y' = SYS.$$

By Ex. ii of § 255 we can take Y and Y' to be the matrices given by the formulae

$$Y = H'\eta H, \quad Y' = h\eta' h', \quad \text{.......................(B)}$$

where H', h' are the conjugates of H, h, and where η, η' are the general symmetric invariant transformands

$$\eta = \text{inv}\,\{\Omega', \epsilon\Omega\}, \quad \eta' = \text{inv}\,\{\Omega, \epsilon\Omega'\},$$

which are the most general symmetric specialisations of the matrices η, η' described in sub-article 2. In particular η can be represented in the form (12), where now:

(1) U and V are symmetric matrices of the respective quadrate classes

$$M\begin{pmatrix} p_1, & p_2, & \dots \\ p_1, & p_2, & \dots \end{pmatrix} \quad \text{and} \quad M\begin{pmatrix} q_1, & q_2, & \dots \\ q_1, & q_2, & \dots \end{pmatrix},$$

viz. 'general symmetric' invariant transformands

$$U = \text{inv}\,\{P', \epsilon P\} \quad \text{and} \quad V = \text{inv}\,\{Q', \epsilon Q\},$$

which are completely known quadrate slopes of type $\{\pi', \pi\}$;

(2) X_i and Y_i are mutually conjugate matrices of the respective classes

$$M\begin{pmatrix} e_{i1}', & e_{i2}', & \dots \\ e_{i1}, & e_{i2}, & \dots \end{pmatrix} \quad \text{and} \quad M\begin{pmatrix} e_{i1}, & e_{i2}, & \dots \\ e_{i1}', & e_{i2}', & \dots \end{pmatrix},$$

viz. 'general' invariant transformands

$$X_i = \text{inv}\,\{A_i', \epsilon B_i\} \quad \text{and} \quad Y_i = \text{inv}\,\{B_i', \epsilon A_i\},$$

which are completely known quadrate slopes of type $\{\pi', \pi\}$;

(3) the parameters of $U, V, X_1, X_2, \dots X_r$ are all independent.

This description of η is correct because the conjugate of a general invariant transformand $\{A_i', \epsilon B_i\}$ is a general invariant transformand $\{B_i', \epsilon A_i\}$.

We can take η' to be a compartite matrix of the same form as η, and corresponding parts or super-constituents of η, η' to be correlated compound slopes of the types

$$\{\pi', \pi\}, \quad \{\pi, \pi'\}.$$

In fact if J is the part-reversant defined in sub-article 2, we can always choose η and η' to be connected by the symmetric relation

$$\eta' = J\eta J.$$

Since the arbitrary parameters of Y_i are now the same as those of X_i, the total number of independent arbitrary parameters in each of the general symmetric invariants (b) is the integer

$$\sigma = u' + v' + \sigma_1 + \sigma_2 + \ldots + \sigma_r, \qquad\ldots\ldots\ldots\ldots(13')$$

where u', v', σ_i are the numbers of the arbitrary parameters in U, V, X_i, which are given in Theorems II and I of § 259. Again because X_i and Y_i have equal ranks, the rank of each of the general symmetric invariant transformands (b) is the integer

$$\rho = p' + q' + 2\rho_1 + 2\rho_2 + \ldots + 2\rho_r, \qquad\ldots\ldots\ldots\ldots(14')$$

where p', q', ρ_i are the ranks of U, V, X_i, so that

$$\rho_i = \text{sum of the smaller integers of the pairs } (e_{i1}, e_{i1}'), (e_{i2}, e_{i2}'), \ldots.$$

We conclude that:

The two general symmetric invariant transformands (b) *can be taken to be the matrices given by the formulae* (B). *Each of them has rank* ρ, *and in each of them the elements are homogeneous linear functions of exactly* σ *independent arbitrary parameters, where* ρ *and* σ *are the integers given by the equations* (13′) *and* (14′).

Ex. v. *Undegenerate symmetric invariant transformands.*

In order that Y and Y' in (b) shall be undegenerate, it is necessary and sufficient that the zero part of the general symmetric invariant transformand η corresponding to the latent roots $a_1, a_2, a_3, \ldots$ in (7) shall be absent, and that all the other parts shall be undegenerate square matrices. The parts $X_1, Y_1, X_2, Y_2, \ldots X_r, Y_r$ will be all undegenerate when and only when they are all *quadrate* slopes, i.e. when and only when in all such pairs as (10) we have

$$e_{i1}' = e_{i1}, \quad e_{i2}' = e_{i2}, \ldots;$$

and the circumstances under which the parts U and V are undegenerate are described in Theorem II of § 259.

Thus the general symmetric invariant transformands Y and Y' in (b) *are undegenerate when and only when:*

(1) *all characteristic potent divisors of A which are not powers of $\lambda - \sqrt{\epsilon}$ or $\lambda + \sqrt{\epsilon}$ can be coupled together in pairs of the form*

$$(\lambda - c)^e, \quad (\lambda - \epsilon c^{-1})^e, \qquad \text{where } c \neq 0,\ c^2 \neq \epsilon;$$

(2) *every distinct characteristic potent divisor $(\lambda - \sqrt{\epsilon})^{2h}$ or $(\lambda + \sqrt{\epsilon})^{2k}$ with a given even index is repeated an even number of times.*

Here the conditions (2) are additional to the necessary conditions that A must be undegenerate and such that

$$A \text{ and } \epsilon A^{-1} \text{ are equicanonical.}$$

Ex. vi. *The quadratic invariants of the substitution* $\overline{x'}_m = A . \overline{x}_m$.

If we put $\epsilon = 1$, and take $u_1, u_2, \ldots u_m$ to be the independent homogeneous linear functions of $x_1, x_2, \ldots x_m$ determined by the equation

$$\overline{u}_m = H . \overline{x}_m,$$

the general quadratic invariant of this substitution is the function

$$f = [x]_m' . Y . \overline{x}_m = [u]_m . \eta . \overline{u}_m,$$

where η is the general symmetric invariant transformand $\eta = \text{inv}\{\Omega', \Omega\}$ expressed in the form (12). There are exactly σ independent particular non-zero quadratic invariants corresponding one by one to the σ arbitrary parameters in η, the quadratic invariant corresponding to any one parameter being obtained by putting all the other parameters equal to 0.

We can divide the σ independent non-zero quadratic invariants into major sets corresponding one by one to the symmetrically placed super-constituents U, V of η and the pairs of conjugately situated super-constituents such as X_i, Y_i; and we can sub-divide the major sets into minor sets corresponding one by one to the symmetrically placed simple constituents (the diagonal constituents of U and V) and the pairs of conjugately situated simple constituents. A minor set corresponding to a symmetrically placed simple constituent is composed of functions which are completely described in sub-article 2 of § 258. A minor set corresponding to a pair of conjugately situated simple constituents is composed of functions which are completely described in sub-article 1 of § 258.

4. *The general skew-symmetric invariant transformands*

$$Y = inv\{A', \epsilon A\}, \quad Y' = inv\{A, \epsilon A'\}. \quad \text{..................(c)}$$

We will suppose A, Ω and the isomorphic transformation (9) to be defined as in sub-articles 2 and 3. If S is any particular undegenerate symmetric commutant $\{A, A'\}$, we know (see Ex. iv of § 254) that the two general skew-symmetric invariants (c) can always be so chosen as to be connected by the symmetric relation

$$Y' = SYS.$$

By Ex. ii of § 255 we can take Y and Y' to be the matrices given by the formulae

$$Y = H'\eta H, \quad Y' = h\eta' h', \quad \text{........................(C)}$$

where H', h' are the conjugates of H, h, and where η, η' are the general skew-symmetric invariant transformands

$$\eta = \text{inv}\{\Omega', \epsilon\Omega\}, \quad \eta' = \text{inv}\{\Omega, \epsilon\Omega'\},$$

which are the most general skew-symmetric specialisations of the matrices η, η' described in sub-article 2. In particular η can be represented in the form (12), where now:

(1) U and V are skew-symmetric matrices of the respective quadrate classes

$$M\begin{pmatrix} p_1, & p_2, & \dots \\ p_1, & p_2, & \dots \end{pmatrix} \quad \text{and} \quad M\begin{pmatrix} q_1, & q_2, & \dots \\ q_1, & q_2, & \dots \end{pmatrix},$$

viz. 'general skew-symmetric' invariant transformands

$$U = \text{inv}\{P', \epsilon P\} \quad \text{and} \quad V = \text{inv}\{Q', \epsilon Q\},$$

which are completely known quadrate slopes of type $\{\pi', \pi\}$;

(2) X_i and Y_i are mutually skew-conjugate matrices of the respective classes

$$M\begin{pmatrix} e_{i1}', & e_{i2}', & \dots \\ e_{i1}, & e_{i2}, & \dots \end{pmatrix} \quad \text{and} \quad M\begin{pmatrix} e_{i1}, & e_{i2}, & \dots \\ e_{i1}', & e_{i2}', & \dots \end{pmatrix}$$

viz. 'general' invariant transformands

$$X_i = \text{inv}\,\{A_i', \epsilon B_i\} \quad \text{and} \quad Y_i = \text{inv}\,\{B_i', \epsilon A_i\},$$

which are completely known compound slopes of type $\{\pi', \pi\}$;

(3) the parameters of $U, V, X_1, X_2, \ldots X_r$ are all independent.

We can take η' to be a compartite matrix of the same form as η, and corresponding parts of η, η' to be correlated compound slopes of the types

$$\{\pi', \pi\}, \ \{\pi, \pi'\}.$$

In fact if J is the part-reversant defined in sub-article 2, we can always choose η and η' to be connected by the symmetric relation

$$\eta' = J\eta J.$$

The total number of independent arbitrary parameters in each of the general skew-symmetric invariant transformands (c) is the integer

$$\sigma = u'' + v'' + \sigma_1 + \sigma_2 + \ldots + \sigma_r, \quad \text{.....................(13'')}$$

where u'', v'', σ_i are the numbers of the arbitrary parameters of U, V, X_i, which are given in Theorems III and I of § 259. Again the rank of each of the general skew-symmetric invariant transformands (b) is the integer

$$\rho = p'' + q'' + 2\rho_1 + 2\rho_2 + \ldots + 2\rho_r, \quad \text{.................(14'')}$$

where p'', q'', ρ_i are the ranks of U, V, X_i, so that

$$\rho_i = \text{sum of the smaller integers of the pairs } (e_{i1}, e_{i1}'), (e_{i2}, e_{i2}'), \ldots.$$

We conclude that:

The two general skew-symmetric invariant transformands (c) *can be taken to be the matrices given by the formulae* (C). *Each of them has rank* ρ, *and in each of them the elements are homogeneous linear functions of exactly* σ *independent arbitrary parameters, where* ρ *and* σ *are the integers given by the equations* (13″) *and* (14″).

Ex. vii. *Undegenerate skew-symmetric invariant transformands.*

In order that Y and Y' in (c) shall be undegenerate, it is necessary and sufficient that the zero part of the general skew-symmetric invariant transformand η corresponding to the latent roots $a_1, a_2, a_3, \ldots$ in (7) shall be absent, and that all the other parts shall be undegenerate square matrices. The parts $X_1, Y_1, X_2, Y_2, \ldots X_r, Y_r$ will be all undegenerate when and only when they are all *quadrate* slopes, i.e. when and only when in all such sets of characteristic potent divisors as (10) we have

$$e_{i1}' = e_{i1}, \quad e_{i2}' = e_{i2}, \ldots;$$

and the circumstances under which the parts U and V are undegenerate are described in Theorem III of § 259. We conclude that:

The general skew-symmetric invariant transformands Y and Y' in (c) *are undegenerate when and only when:*

(1) *all characteristic potent divisors of A which are not powers of $\lambda - \sqrt{\epsilon}$ or $\lambda + \sqrt{\epsilon}$ can be coupled together in pairs of the form*

$$(\lambda - c)^e, \quad (\lambda - \epsilon c^{-1})^e, \qquad \textit{where } c \neq 0,\ c^2 \neq \epsilon;$$

(2) *every distinct characteristic potent divisor $(\lambda - \sqrt{\epsilon})^{2h+1}$ or $(\lambda + \sqrt{\epsilon})^{2k+1}$ with a given odd index is repeated an even number of times.*

Here the conditions (2) are additional to the necessary conditions that A must be undegenerate and such that

$$A \text{ and } \epsilon A^{-1} \text{ are equicanonical.}$$

Ex. viii. *If P and Q are respectively a 'general symmetric' invariant transformand and a 'general skew-symmetric' invariant transformand of the same one of the types*

$$\{A', \epsilon A\}, \quad \{A, \epsilon A'\},$$

and if the parameters of P and Q are all different, then their sum $S = P + Q$ is a 'general' invariant transformand of the same type.

Let S_0 be any particular invariant transformand of the given type, and let S_0' be the conjugate of S. Because S_0' is an invariant transformand of the same type, we can put

$$S_0 = \tfrac{1}{2}(S_0 + S_0') + \tfrac{1}{2}(S_0 - S_0') = P_0 + Q_0,$$

where P_0 and Q_0 are particularisations of P and Q. Thus every particular invariant transformand of the given type is a particularisation of S, i.e. S is a general invariant transformand of the given type.

The same result could be obtained by a comparison of the equations (13′) and (13″) with the equation (13).

APPENDIX A.

(ADDITION TO CHAPTER XXV.)

§ 225 a. Rational integral functions of a matrix which is not square.

First let ϕ be any given *square* matrix of order m. We have obtained a unique interpretation of the 0th power of ϕ by adopting the definition

$$\phi^0 = I, \quad \text{where } I = [1]_m^m. \qquad \text{.........................(1)}$$

This is the only natural interpretation when ϕ is an undegenerate square matrix, but is entirely gratuitous when ϕ is degenerate. Independently of the interpretation (1) we have defined a rational integral function of ϕ to be a square matrix of order m which can be expressed in the form

$$f(\phi) = \lambda_0 I + \lambda_1\phi + \lambda_2\phi^2 + \ldots + \lambda_r\phi^r. \qquad \text{................(2)}$$

It is implied in the definition (2), as well as in the definition (1), that:

the unit matrix I is always to be regarded as a rational integral function of ϕ, even when ϕ is degenerate or a zero matrix. ....(3)

In the particular case when $m = 1$, this implication means that the number 1 is to be regarded as a rational integral function of the number 0. It will be convenient to call $f(\phi)$ a *proper* rational integral function of ϕ when it can be expressed in the form

$$f(\phi) = \lambda_1\phi + \lambda_2\phi^2 + \ldots + \lambda_r\phi^r. \qquad \text{...................(4)}$$

From Cayley's equation we see that the unit matrix I is a proper rational integral function of ϕ when and only when ϕ is undegenerate. If we had avoided the implication (3) by using (4) as the definition of a rational integral function of ϕ, then consistently with the definition (4) we could have retained the interpretation (1) when ϕ is undegenerate whilst interpreting ϕ^0 to be a zero matrix (or leaving it uninterpreted) when ϕ is degenerate; but if Cayley's equation is $g(\phi) = 0$, we could not then have spoken of $g(\phi)$ as a rational integral function of ϕ in the most important case when ϕ is undegenerate.

Now let $A = [a]_m^n$ be any given matrix which is *not square*, so that $n \neq m$. In this case we can only give a unique interpretation to the 0th power of A by using the definition

$$A^0 = [0]_m^n, \qquad \text{..............................(5)}$$

i.e. by interpreting A^0 to be a zero matrix, and we must define a rational integral function of A to be a matrix expressible in the form

$$f(A) = \lambda_1 A + \lambda_2 A^2 + \ldots + \lambda_r A^r. \qquad \text{....................(6)}$$

We will consider separately the two cases $n > m$, $m > n$.

CASE I. When $n > m$, let ϕ be the square matrix of order n formed by adding final horizontal rows of 0's to A, so that

$$A = [a]_m^n, \quad \phi = \begin{bmatrix} a \\ 0 \end{bmatrix}_{m,\, n-m}^n.$$

Then the equations such as

$$A^2 = [a]_m^n [a]_m^n = A\phi, \quad A^3 = [a]_m^n [a]_m^n [a]_m^n = A\phi^2, \dots$$

show that there is a one-one correspondence between all rational integral functions of A and all *proper* rational integral functions of ϕ such that

$$[x]_m^n = \lambda_1 A + \lambda_2 A^2 + \dots + \lambda_r A^r = f(A) \quad \dots\dots\dots\dots(7)$$

is a rational integral function of A when and only when

$$\begin{bmatrix} x \\ 0 \end{bmatrix}_{m,\, n-m}^n = \lambda_1 \phi + \lambda_2 \phi^2 + \dots + \lambda_r \phi^r = f(\phi) \quad \dots\dots\dots\dots(7')$$

is a proper rational integral function of ϕ. Thus all properties of rational integral functions of A can be deduced from corresponding properties of rational integral functions of the square matrix ϕ. In particular $g(A) = 0$ will be a rational integral equation of lowest possible degree s satisfied by A when and only when $g(\phi) = 0$ is a rational integral equation of lowest possible degree s satisfied by the square matrix ϕ; and $g(\phi)$ is necessarily a proper rational integral function of ϕ because ϕ is degenerate.

CASE II. When $m > n$, let ϕ be the square matrix of order m formed by adding final vertical rows of 0's to A, so that

$$A = [a]_m^n, \quad \phi = [a, 0]_m^{n,\, m-n}.$$

Then the equations such as

$$A^2 = [a]_m^n [a]_m^n = \phi A, \quad A^3 = [a]_m^n [a]_m^n [a]_m^n = \phi^2 A, \dots$$

show that there is a one-one correspondence between all rational integral functions of A and all *proper* rational integral functions of ϕ such that

$$[x]_m^n = \lambda_1 A + \lambda_2 A^2 + \dots + \lambda_r A^r = f(A) \quad \dots\dots\dots\dots(8)$$

is a rational integral function of A when and only when

$$[x, 0]_m^{n,\, m-n} = \lambda_1 \phi + \lambda_2 \phi^2 + \dots + \lambda_r \phi^r = f(\phi) \quad \dots\dots\dots(8')$$

is a proper rational integral function of ϕ. Thus all properties of rational integral functions of A can be deduced from corresponding properties of rational integral functions of the square matrix ϕ. In particular $g(A) = 0$ will be a rational integral equation of lowest possible degree s satisfied by A when and only when $g(\phi) = 0$ is a rational integral equation of lowest possible degree s satisfied by the square matrix ϕ; and $g(\phi)$ is necessarily a proper rational integral function of ϕ because ϕ is degenerate.

Rational integral functions of a matrix which is not square do not occur in practical applications of the Calculus of Matrices, because the products involved in them are not standard products. Moreover it will be evident that they do not even possess any special theoretical interest, because the theory of rational integral functions of any matrix can be regarded as co-extensive with, or lying within the range of, the theory of rational integral functions of a square matrix.

This subordination of rectangular matrices to square matrices in one important branch of the Calculus, together with the stress laid upon it by Cayley, seems to have been the chief cause of the long neglect of rectangular matrices which prevailed up to the time of Kronecker.

Ex. i. If
$$A=\begin{bmatrix} 2 & 3 & 1 \\ 1 & 2 & 1 \end{bmatrix}, \quad \phi=\begin{bmatrix} 2 & 3 & 1 \\ 1 & 2 & 1 \\ 0 & 0 & 0 \end{bmatrix},$$
the rational integral equations of lowest degrees satisfied by A and ϕ respectively are
$$A^3-4A^2+A=0, \quad \phi^3-4\phi^2+\phi=0.$$

Ex. ii. If
$$A=\begin{bmatrix} c & 0 & a \\ 0 & c & b \end{bmatrix}, \quad \phi=\begin{bmatrix} c & 0 & a \\ 0 & c & b \\ 0 & 0 & 0 \end{bmatrix},$$
where a, b, c are not all equal to 0, the rational integral equations of lowest degrees satisfied by A and ϕ respectively are
$$A^2-cA=0, \quad \phi^2-c\phi=0.$$

Ex. iii. If
$$A=\begin{bmatrix} 2 & 1 \\ 8 & 4 \\ 2 & 1 \end{bmatrix}, \quad \phi=\begin{bmatrix} 2 & 1 & 0 \\ 8 & 4 & 0 \\ 2 & 1 & 0 \end{bmatrix},$$
the rational integral equations of lowest degrees satisfied by A and ϕ respectively are
$$A^2-6A=0, \quad \phi^2-6\phi=0.$$

APPENDIX B.

(ADDITION TO CHAPTER XXVII.)

241 a. Some properties of a standardised general compound slope.

1. *The accessary horizontal and vertical rows; the dominant accessary elements.*

Throughout this appendix it will be supposed that M is a standardised general compound slope of the class

$$M\begin{pmatrix}\beta_1, & \beta_2, & \ldots & \beta_s\\ \alpha_1, & \alpha_2, & \ldots & \alpha_r\end{pmatrix}, \qquad\qquad \text{(a)}$$

where the α's and β's are arranged in descending orders of magnitude, and that it is expressed in the form

$$M = \begin{bmatrix} X_{11}, & X_{12}, & \ldots & X_{1s}\\ X_{21}, & X_{22}, & \ldots & X_{2s}\\ \cdots & \cdots & \cdots & \cdots\\ X_{r1}, & X_{r2}, & \ldots & X_{rs}\end{bmatrix}_{\alpha_1, \alpha_2, \ldots \alpha_r}^{\beta_1, \beta_2, \ldots \beta_s}, \qquad\qquad \text{(A)}$$

where the constituents such as X_{ij} are general simple slopes. It will be assumed that the index numbers in (a) and (A) are all different from 0, but we shall often use the interpretations

$$\alpha_i = 0 \text{ when } i > r, \quad \beta_i = 0 \text{ when } i > s.$$

It will always be understood that

ρ is the sum of the smaller integers in the pairs of corresponding descendent horizontal and vertical index numbers

as defined in § 241. 3; and the effective order of the constituent X_{ij}, i.e. the smaller of the two integers α_i and β_j, will ordinarily be denoted by γ_{ij}.

By considering separately non-zero elements lying in, below, and to the right of the diagonal constituents it will be seen that if e is a non-zero element of M, either the horizontal or the vertical line through e must cut the paradiagonal of some diagonal constituent, the three possibilities being that there is a principal paradiagonal element:

(1) in the same horizontal and also in the same vertical row as e;

(2) in the same horizontal but not in the same vertical row as e;

(3) in the same vertical but not in the same horizontal row as e.

Every element which lies neither in the same horizontal nor in the same vertical row as a principal diagonal element must be 0. Hence if we strike out the ρ horizontal and ρ vertical rows passing through the ρ principal paradiagonal elements, we strike out all the non-zero elements, and convert M into a zero matrix (which will be non-existent when M is quadrate).

By an *accessary* horizontal or vertical row we shall mean one which does not contain a principal paradiagonal element, i.e. one which does not cut the paradiagonal of any diagonal constituent. The non-zero elements (2) are those which lie in the accessary vertical rows, and the non-zero elements (3) are those which lie in the accessary horizontal rows. The accessary horizontal and vertical rows (which include the complete horizontal or vertical rows of 0's) intersect in zero elements only. They are of course absent when M is quadrate.

If h is the last paradiagonal element which is reached as we pass from above downwards along one of the accessary vertical rows, we will call h the *dominant* accessary element of that row. If H is the constituent in which h lies, the accessary vertical row through h will cut all constituents lying below H in zero elements only, and the horizontal line through h (see Ex. i) will pass through a principal element a of a diagonal constituent A lying completely to the left of H.

If k is the last paradiagonal element which is reached as we pass from left to right along one of the accessary horizontal rows, we will call k the *dominant* accessary element of that row. If K is the constituent in which k lies, the accessary horizontal row through k will cut all constituents lying to the right of K in zero elements only, and the vertical line through k (see Ex. ii) will pass through a principal element a of a diagonal constituent A lying completely above K.

Every accessary horizontal or vertical row will be completely known when its dominant element is specified.

Ex. i. Let a be a principal element of a diagonal constituent A, and let the horizontal line through a cut a constituent E in a non-zero element e, and the paradiagonal of E in the element e_0. If e lies in an accessary vertical row, so does e_0, because it has a greater horizontal apical distance than e in E.

a	...	$e_0\, e$

a	h	h'	h''
0	0	0	0

If E lies to the left of A (or is A), the vertical line through e must cut the paradiagonal of a diagonal constituent lying above E (or the paradiagonal of A), and cannot be one of the accessary vertical rows.

Thus if e or e_0 lies on an accessary vertical row, then E must lie completely to the right of A.

Suppose that e_0 lies on an accessary vertical row which cuts all constituents lying below E in zero elements only. Then by Ex. vii of § 241 the vertical line through a must cut all constituents below A in zero elements only; moreover if there is a constituent E' lying between A and E, and if the horizontal line through a and e_0 cuts the paradiagonal of E' in the element e_0', then the vertical line through e_0' must cut all constituents lying below E' in zero elements only.

Thus if the horizontal line through a cuts any accessary vertical row in a dominant element h, the vertical line through a must cut all constituents lying below A in zero elements only; moreover all such dominant elements h, h', h", ... on the horizontal line through a lie in a series of consecutive constituents situated to the right of A, and starting with the constituent next to A on the right.

Ex. ii. Let a be a principal element of a diagonal constituent A, and let the vertical line through a cut a constituent E in a non-zero element e, and the paradiagonal of E in the element e_0. If e lies in an accessary horizontal row, so does e_0, because it has a greater vertical apical distance than e in E.

$$\begin{array}{|c|} \hline a \\ \vdots \\ e \\ e_0 \\ \hline \end{array} \qquad \begin{array}{c|c} a & 0 \\ k & 0 \\ k' & 0 \\ k'' & 0 \end{array}$$

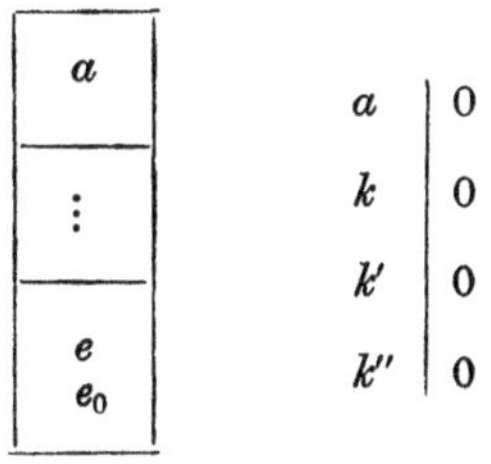

If E lies above A (or is A), the horizontal line through e must cut the paradiagonal of a diagonal constituent lying to the left of E (or the paradiagonal of A), and cannot be one of the accessary horizontal rows.

Thus if e or e_0 lies on an accessary horizontal row, then E must lie completely below A.

Suppose that e_0 lies on an accessary horizontal row which cuts all constituents lying to the right of E in zero elements only. Then by Ex. viii of § 241 the horizontal line through a must cut all constituents to the right of A in zero elements only; moreover if there is a constituent E' lying between A and E, and if the vertical line through a and e_0 cuts the paradiagonal of E' in the element e_0', then the horizontal line through e_0' must cut all constituents to the right of E' in zero elements only.

Thus if the vertical line through a cuts any accessary horizontal row in a dominant element k, the horizontal line through a must cut all constituents lying to the right of A in zero elements only; moreover all such dominant elements k, k', k", ... on the vertical line through a lie in a series of consecutive constituents situated below A, and starting with the constituent next below A.

Ex. iii. *If x and y are non-zero elements lying respectively in an accessary vertical and an accessary horizontal row, and if the paradiagonal of the same diagonal constituent A is cut*

by the horizontal line through x in the principal element x',
by the vertical line through y in the principal element y',

then:

(a) *the four elements common to the vertical rows through x, x' and the horizontal rows through y, y' must be all equal to 0;*

(b) *x' and y' must be different paradiagonal elements of A.*

If $\{x, y\}$ always means the element common to the vertical row through an element x and the horizontal row through an element y, then to prove the first result we have to show that

$$\{x, y\}, \quad \{x', y\}, \quad \{x, y'\}, \quad \{x', y'\} \quad \dots\dots\dots\dots\dots\dots(1)$$

are four zero elements. The second result, which is already known to be true, will follow from the equation $\{x', y'\} = 0$, which shows that x' must have a smaller vertical (or greater horizontal) apical distance in A than y'.

We will denote the constituents in which x and y lie by X and Y. Because X must lie to the right of A and Y below A, it will be sufficient to prove (a) for a standardised compound slope

$$M_1 = \begin{bmatrix} A, & X, & Q \\ P, & X_0, & F \\ Y, & G, & Y_0 \end{bmatrix}_{a,\,h,\,k}^{a,\,p,\,q} \quad \text{or} \quad M_2 = \begin{bmatrix} A, & Q, & X \\ Y, & Y_0, & G \\ P, & F, & X_0 \end{bmatrix}_{a,\,k,\,h}^{a,\,p,\,q}$$

in which X_0 is the diagonal constituent below X, and Y_0 the diagonal constituent to the right of Y. The matrix M_1 or M_2 is derived from the smallest corranged minor of M which contains the constituents mentioned by striking out the complete rows of 0's, this simplification being allowable because no one of the elements x, y, x', y' lies on one of the complete rows of 0's. If X_0 is absent, we can use the form M_2 in which $h=0$; if Y_0 is absent we can use the form M_1 in which $q=0$. Because X_0 must contain vertical and Y_0 horizontal rows of 0's, we must have $p>h$ and $k>q$. Accordingly we have

$$a \nless p > h \nless k > q \text{ in } M_1, \quad a \nless k > q \nless p > h \text{ in } M_2.$$

If we make use of Exs. iii, vii, viii of § 241, the proof is very simple.

In M_1 we have $\{x, y\}=0$ by Ex. iii; therefore also $\{x, y'\}=0$ by Ex. viii; therefore also $\{x', y\}=0$, $\{x', y'\}=0$ by Ex. vii.

In M_2 we have $\{x, y\}=0$ by Ex. iii; therefore also $\{x', y\}=0$ by Ex. vii; therefore also $\{x, y'\}=0$, $\{x', y'\}=0$ by Ex. viii.

A more direct proof can be obtained by supposing the (horizontal, vertical) apical distances of the elements

$$x, y, x', y' \quad \text{to be} \quad (\xi, \lambda),\ (\mu, \eta),\ (\xi', \lambda),\ (\mu, \eta'),$$

in their respective constituents X, Y, A, A of effective orders p, k, a, a, so that those of the four elements (1) in their respective constituents G, X, Y, A are

$$\{\xi, \eta\},\ \{\xi, \eta'\},\ \{\xi', \eta\},\ \{\xi', \eta'\}.$$

Then (see Ex. i of § 241) from the given relations

$$\xi+\lambda-1 \ngtr p,\ \xi'+\lambda=a+1,\ \xi \nless h+1;\quad \mu+\eta-1 \ngtr k,\ \mu+\eta'=a+1,\ \eta \nless q+1,$$

we can deduce the relations

$$\xi+\eta-1-k \nless 1 \text{ in } M_1;\ \xi+\eta-1-p \nless 1 \text{ in } M_2;$$

$$\xi+\eta'-p \nless 1,\ \xi'+\eta-1-k \nless 1,\ \xi'+\eta'-1-a \nless 1 \text{ in } M_1 \text{ and } M_2;$$

and these establish the theorem.

The result (b) is equivalent to the statement that:

If the horizontal line through a principal element a cuts an accessary vertical row in an element x, and the vertical line through a cuts an accessary horizontal row in an element y, then either x or y or both of them must be 0; it is impossible for them to be two non-zero elements.

Ex. iv. *If x and y are non-zero elements lying respectively in an accessary vertical and an accessary horizontal row, and if*

x' is the principal paradiagonal element in the horizontal row through x,

y' is the principal paradiagonal element in the vertical row through y,

then the four elements common to the vertical rows through x, x' and the horizontal rows through y, y' must be all equal to 0.

This theorem is immediately deducible from the general theorem given in Ex. xvi of sub-article 4, but an independent proof will be given. We will again use the notation $\{x, y\}$ for the element of M common to the vertical row through an element x and the horizontal row through an element y, so that the four elements to be proved equal to 0 are

$$\{x, y\},\ \{x', y\},\ \{x, y'\},\ \{x', y'\}. \quad\text{.................................(1)}$$

Let X be the constituent of M in which the element x lies, and let X', X_0 be the diagonal constituents to the left of X, below X, so that X' is the constituent in which x' lies; also let Y be the constituent in which the element y lies, and let Y', Y_0 be the diagonal constituents above Y, to the right of Y,

$$\begin{matrix} X' & X & & Y' & \\ & X_0 & & Y & Y_0 \end{matrix}$$

so that Y' is the constituent in which y' lies. The particular cases in which X_0 or Y_0 is absent are included in the general treatment. In all cases X_0 contains vertical rows of 0's, or has more vertical than horizontal rows; and Y_0 contains horizontal rows of 0's, or has more horizontal than vertical rows. From Ex. iii or from Exs. vii and viii respectively of § 241 we see that:

The vertical rows of M through x and x' cut all horizontal rows which lie below X_0 (or pass through X_0) in zero elements only. ....(2)

The horizontal rows of M through y and y' cut all vertical rows which lie to the right of Y_0 (or pass through Y_0) in zero elements only. ....(2′)

The constituents and elements which have been mentioned all lie in a corranged minor of M which is a standardised general compound slope Ω formed by the intersections of the horizontal and vertical rows passing through the diagonal constituents

$$X',\ X_0,\ Y',\ Y_0, \quad\text{...(3)}$$

and in proving the theorem we can replace M by Ω. We can describe Ω completely by describing the order of arrangement from left to right and from above downwards of the diagonal constituents (3) in M and Ω. Since the two constituents X_0 and Y_0 cannot be the same, and since X' always lies to the left of X_0, and Y' to the left of Y_0, at least three of the four diagonal constituents (3) must be different. If only three of them are different, their possible arrangements in M and Ω are the following four:

$$X'=Y',\ X_0,\ Y_0;\quad X'=Y',\ Y_0,\ X_0;\quad X',\ X_0=Y',\ Y_0;\quad Y',\ Y_0=X',\ X_0.$$

The theorem has been proved for the first two of these arrangements in Ex. iii, where $X'=Y'=A$; and the principles (2) and (2′) respectively suffice to show that it is true for the third and fourth arrangements.

Supposing then that the four diagonal constituents (3) are all different, their possible arrangements in M and Ω are the following six:

(i) X', X_0, Y', Y_0; (ii) X', Y', X_0, Y_0; (iii) X', Y', Y_0, X_0;
(i′) Y', Y_0, X', X_0; (ii′) Y', X', Y_0, X_0; (iii′) Y', X', X_0, Y_0.

The principle (2) applied to horizontal rows of Ω through or below X_0 shows that:

$\{x, y\}, \{x', y\}, \{x, y'\}, \{x', y'\}$ are all 0 in (i);
$\{x, y\}, \{x', y\}$ are both 0 in (ii), (iii′);

the principle (2′) applied to vertical rows of Ω through or to the right of Y_0 shows that:

$\{x, y\}, \{x, y'\}, \{x', y\}, \{x', y'\}$ are all 0 in (i′);
$\{x, y\}, \{x, y'\}$ are both 0 in (ii), (iii′);

from Ex. vii of § 241 applied to horizontal rows below X' and X we see that:

if $\{x, y\}=0$, then $\{x', y\}=0$; if $\{x, y'\}=0$, then $\{x', y'\}=0$ in (i), (ii), (iii);
if $\{x, y\}=0$, then $\{x', y\}=0$ in (ii′), (iii′);

and from Ex. viii of § 241 applied to vertical rows to the right of Y' and Y we see that:

if $\{x, y\}=0$, then $\{x, y'\}=0$; if $\{x', y\}=0$, then $\{x', y'\}=0$ in (i'), (ii'), (iii');

if $\{x, y\}=0$, then $\{x, y'\}=0$ in (ii), (iii).

It follows that in every arrangement all four of the elements (1) are equal to 0. For instance in the arrangement (ii) we have

$\{x, y\}=0$, $\{x', y\}=0$ by (2); $\{x, y'\}=0$ by Ex. viii; $\{x', y'\}=0$ by Ex. vii;

and in the arrangement (iii) we have

$\{x, y\}=0$, $\{x, y'\}=0$ by (2'); $\{x', y\}=0$, $\{x', y'\}=0$ by Ex. vii of § 241.

2. *Preclusive paths and groups.*

Selecting any principal element a_{11} of the leading constituent X_{11} of the standardised general compound slope M, we will construct a zig-zag path by travelling in succession

from a_{11} horizontally rightward to the next paradiagonal element x_{12},
from x_{12} vertically downward to the next principal element a_{22},
from a_{22} horizontally rightward to the next paradiagonal element x_{23},
from x_{23} vertically downward to the next principal element a_{33},

$$\begin{matrix} a_{11} \ldots x_{12} & & \\ & \vdots & \\ & a_{22} \ldots x_{23} & \\ & & \vdots \\ & & a_{33} \end{matrix} \qquad\qquad \begin{matrix} a_{11}' \ldots x_{12}' & & \\ & \vdots & \\ & a_{22}' \ldots x_{23}' & \\ & & \vdots \\ & & a_{33}' \end{matrix}$$

continuing in this way as long as possible, and we will call it a *preclusive path* based on the principal paradiagonals. The paradiagonal elements $a_{11}, x_{12}, a_{22}, x_{23}, a_{33}, \ldots$ lie respectively in the constituents $X_{11}, X_{12}, X_{22}, X_{23}, X_{33}, \ldots$. The suffixes indicate the constituents to which they belong, and not their positions in the constituents. The total number of such paths is equal to the effective order $\gamma=\gamma_{11}$ of the leading diagonal constituent X_{11}, i.e. to the number of elements lying on the paradiagonal of X_{11}. Since a uniquely determinate preclusive path can be drawn backwards from any given principal element of M to a principal element of X_{11}, we see that:

Every principal element a_{ii} of M lies on one and only one of the preclusive paths, viz. one which passes through principal elements $a_{11}, a_{22}, \ldots a_{ii}$ of $X_{11}, X_{22}, \ldots X_{ii}$.

It follows that the principal elements of M can be divided into groups according to the preclusive paths on which they lie. We will call these the *preclusive groups* of the principal elements, and denote by

$$g_1, g_2, \ldots g_\gamma \quad \text{.................................(B)}$$

the groups corresponding to the paths through the principal elements in the 1st, 2nd, ... γth apical horizontal rows of X_{11} (the horizontal row through the apical element being the first apical horizontal row). The elements composing each group will be arranged in the same order from left to right (or from above downwards) as in M.

If X_{ii} is any given diagonal constituent of M, and if u, v, w are the effective orders of X_{ii}, $X_{i,i+1}$, $X_{i+1,i+1}$, so that

$$u \nless v \nless w, \qquad (w \text{ being } 0 \text{ when } X_{i+1,i+1} \text{ is non-existent}),$$

the u preclusive paths which reach the u principal elements of X_{ii} belong to u consecutive groups in (B), and can be divided into three sets of paths belonging to consecutive groups, there being exactly:

$v - w$ preclusive paths 'quasi-terminate in X_{ii},' ....(4)

each of which can be continued from some principal element a_{ii} of X_{ii} to a non-principal paradiagonal element $x_{i,i+1}$ of $X_{i,i+1}$, but cannot be continued any further because the vertical line through $x_{i,i+1}$ cuts all constituents lying below $X_{i,i+1}$ (if such exist) in zero elements only;

w preclusive paths passing through X_{ii} and $X_{i,i+1}$ to $X_{i+1,i+1}$; ...(5)

$u - v$ preclusive paths 'pleni-terminate in X_{ii},' ....(6)

each of which reaches some principal element a_{ii} of X_{ii}, but cannot be continued any further because the horizontal line through a_{ii} cuts all constituents lying to the right of X_{ii} (if such exist) in zero elements only. The w through paths (5) always exist except when X_{ii} is the last diagonal constituent; and they lie between the paths (4) and (6). The *quasi-terminate* paths (4) exist when and only when the effective order of $X_{i+1,i+1}$ is less than those of $X_{i,i+1}$ and X_{ii}, i.e. when and only when $X_{i+1,i+1}$ has a smaller effective order than X_{ii} and also

contains vertical rows of 0's;

and they lie on the *apical horizontal* side of the other paths in X_{ii}. The *pleni-terminate* paths (6) exist when and only when the effective order of X_{ii} is greater than those of $X_{i,i+1}$ and $X_{i+1,i+1}$; and they lie on the *basical horizontal* side of the other paths in X_{ii}.

Every preclusive path must start from some principal element a_{11} of X_{11}, and must terminate in one and only one of the following two ways:

(*a*) It may be *quasi-terminate* at some principal element a_{ii}, i.e. it may proceed horizontally from a_{ii} to a non-principal element $x_{i,i+1}$ from which it cannot be continued vertically. In this case (see Ex. vii of § 241) the vertical lines through the elements a_{ii} and $x_{i,i+1}$ cut all constituents lying below them in zero elements only; $x_{i,i+1}$ is the dominant element of an accessary vertical row; and the diagonal constituent containing a_{ii} must be immediately followed by a diagonal constituent of smaller effective order having vertical rows of 0's, i.e. having more vertical than horizontal rows.

(*b*) It may be *pleni-terminate* at some principal element a_{ii} from which it cannot be continued horizontally. In this case (see Ex. viii of § 241) the horizontal lines through the elements $x_{i-1,i}$ (if $i \neq 1$) and a_{ii} cut all constituents to the right of them in zero elements only;

and the diagonal constituent containing a_{ii} must be immediately followed (if it is not the last) by a diagonal constituent of smaller effective order.

Preclusive paths and groups.

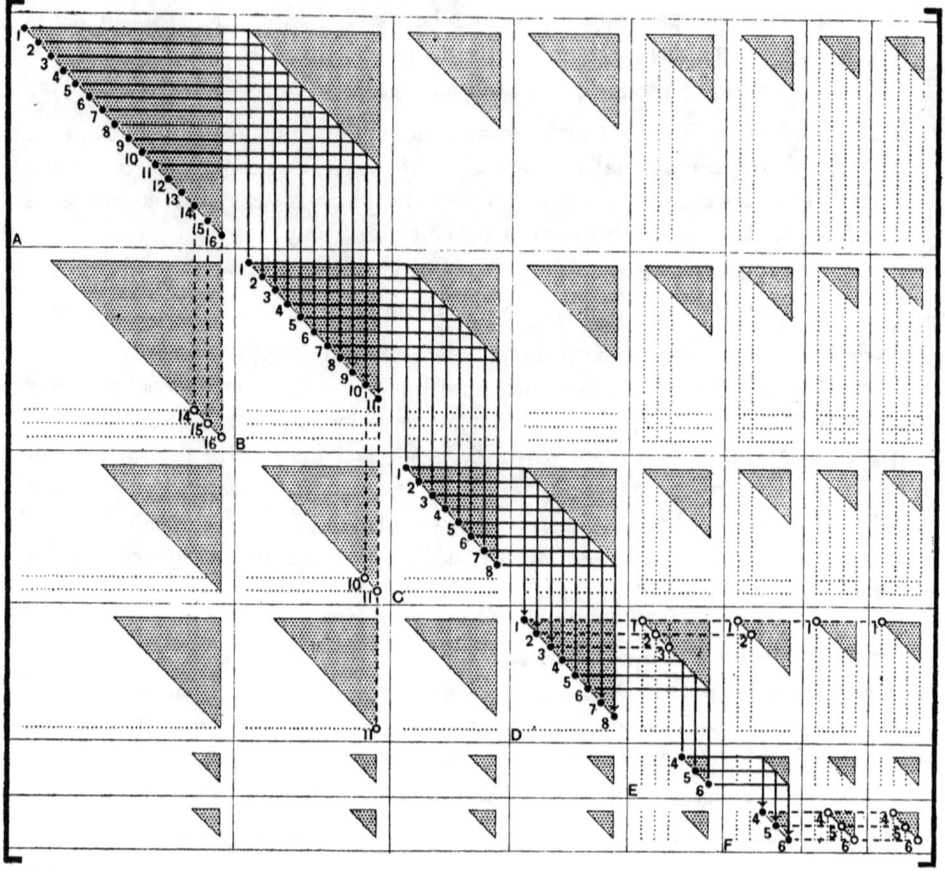

Quasi-terminate groups of principal elements.

$g_1 = (A_1, B_1, C_1, D_1)$,
$g_2 = (A_2, B_2, C_2, D_2)$,
$g_3 = (A_3, B_3, C_3, D_3)$,
$g_4 = (A_4, B_4, C_4, D_4, E_4, F_4)$,
$g_5 = (A_5, B_5, C_5, D_5, E_5, F_5)$,
$g_6 = (A_6, B_6, C_6, D_6, E_6, F_6)$,

Pleni-terminate groups of principal elements.

$g_7 = (A_7, B_7, C_7, D_7)$,
$g_8 = (A_8, B_8, C_8, D_8)$,
$g_9 = (A_9, B_9)$,
$g_{10} = (A_{10}, B_{10})$,
$g_{11} = (A_{11}, B_{11})$,
$g_{12} = A_{12}$,
$g_{13} = A_{13}$,
$g_{14} = A_{14}$,
$g_{15} = A_{15}$,
$g_{16} = A_{16}$.

It will be clear from (4), (5), (6) that the preclusive groups arranged as in (B) are divided naturally into two sets lying on the left and right respectively:

those on the left (the *quasi-terminate* groups) belonging to the paths (a) and containing more and more elements as we pass from left to right;

those on the right (the *pleni-terminate* groups) belonging to the paths (b) and containing fewer and fewer elements as we pass on from left to right.

It can be shown by repeated applications of Exs. vi, vii, viii of § 241 that:

The vertical lines through the principal elements lying on the quasi-terminate preclusive paths (a), *and therefore those through all the elements of any quasi-terminate preclusive group* g, *intersect the accessary horizontal rows in zero elements only.* ....(B_1)

The horizontal lines through the principal elements lying on the pleni-terminate preclusive paths (b), *and therefore those through all the elements of any pleni-terminate preclusive group* g, *intersect the accessary vertical rows in zero elements only.* ....(B_2)

If g *and* g' *are any two preclusive groups such that* g' *occurs later than* g *in the series* (B), *then the vertical rows through the elements of* g *and the horizontal rows through the elements of* g' *intersect one another in zero elements only.* ....(B_3)

The theorems (B_1), (B_2), (B_3) are proved in Exs. vi, v, vii respectively. They remain true when we add to the principal elements of each quasi-terminate group the final dominant accessary element in which the corresponding path terminates, and also any other dominant accessary elements through which the path could be continued horizontally. For (B_1) this follows from the fact that accessary horizontal and vertical rows intersect in zero elements only; for (B_3) it is proved in Ex. viii.

Simpler and more comprehensive proofs of these theorems are given in sub-article 4, (B_1) and (B_2) being included in Ex. xiv, and (B_3) in Exs. xv and xvi.

The figure on p. 642 shows the preclusive paths of a standardised general compound slope of type $\{\pi, \pi\}$ and of the class

$$M\begin{pmatrix}16, & 11, & 8, & 8, & 6, & 5, & 4, & 4\\ 16, & 14, & 10, & 9, & 3, & 3 & & \end{pmatrix},$$

in which the effective orders of the successive diagonal constituents

$$A, B, C, D, E, F, (G, H) \quad \text{are} \quad 16, 11, 8, 8, 3, 3, (0, 0)$$

when the 'non-existent' diagonal constituents G, H of effective order 0 are included. The shaded right-angled triangles represent the effective regions within which and on the boundaries of which the non-zero elements lie, the hypotenuses being the paradiagonals of the simple constituents. The faint dotted lines represent the accessary horizontal and vertical rows; the large dots represent the principal elements; and the small circles represent the dominant accessary elements. The preclusive paths joining the principal paradiagonal elements are shown by the thick continuous lines, the last principal elements on them being marked by arrows; and the extensions of those paths through or to the dominant accessary elements are shown by the thick broken lines. Those principal elements of the 1st, 2nd, 3rd, ... diagonal constituents which lie on the ith *preclusive* path g_i are denoted by A_i, B_i, C_i, ... The occurrence of preclusive groups containing only one principal element is due to the fact that the greatest effective order 16 of a diagonal constituent is not repeated, so that the first diagonal constituent is followed by one of smaller effective order.

Ex. v. *Let a be a principal element lying on a preclusive path p, and let the horizontal line through a cut one of the accessary vertical rows in a non-zero element e.*

Then p cannot terminate at a; and if it is continued through a non-principal paradiagonal element x to another principal element a', the horizontal line through a' must cut the accessary vertical row through e in a non-zero element e'.

$$\begin{matrix} a & x & \dots & e \\ - & a' & \dots & e' \end{matrix}$$

By Ex. i the constituent E containing e must lie completely to the right of the diagonal constituent A containing a; and because the horizontal line through a cuts the paradiagonal of E, it cuts the paradiagonal of the constituent X lying next to A on the right in an element x, i.e. p does not terminate at a, but is continued to x. If E is X, then x (see Ex. i) lies on an accessary vertical row, and in this case p will certainly terminate at x.

If p does not terminate at x, but is continued from x to a principal element a' of a diagonal constituent A' lying immediately below X, then E lies completely to the right of X, and by Ex. viii of § 241 the horizontal line through a' cuts the accessary vertical row through e in a non-zero element e'. By the same argument p cannot terminate at a'.

By repeated applications of this result we see that p cannot terminate at a principal element, i.e. it must be quasi-terminate.

Ex. vi. *Let a be a principal element lying on a preclusive path p, and let the vertical line through a cut one of the accessary horizontal rows in a non-zero element e.*

Then if p does not terminate at a, it is continued through a non-principal paradiagonal element x to another principal element a' which is such that the vertical line through a' cuts the accessary horizontal row through e in a non-zero element e'.

$$\begin{matrix} a & x \\ - & a' \\ \vdots & \vdots \\ e & e' \end{matrix}$$

By Ex. ii the constituent E containing e must lie completely below the diagonal constituent A containing a. If there is no constituent to the right of A, the path certainly terminates at a. Dismissing this case we will denote the constituent next to A on the right by X, and the diagonal constituent immediately below X by A'. If E is the constituent next below A, i.e. if A' is the constituent next to E on the right, the accessary horizontal row through e cuts all constituents to the right of E in zero elements only; therefore by Ex. viii of § 241 the horizontal line through a cuts all constituents to the right of A in zero elements only; therefore p certainly terminates at a.

If p does not terminate at a, but is continued to a paradiagonal element x of X, then there is at least one constituent between A and E. In this case the path p is continued vertically from x to a principal element a' of A'; for otherwise the vertical line through x would cut all constituents below X in zero elements only, and by Ex. vii of § 241 this is only possible when the vertical line through a cuts all constituents below A in zero elements only; moreover because the accessary horizontal row through e lies completely below A', it follows from Ex. vii of § 241 that the vertical line through x cuts it in a non-zero element e'.

By repeated applications of this result we see that p cannot terminate at a non-principal element, i.e. it must be pleni-terminate.

Ex. vii. *Analytic treatment of two preclusive paths.*

Let $a_{11}, a_{22}, \dots a_{qq}$ forming the group g and $a_{11}', a_{22}', \dots a_{pp}'$ forming the group g' be the successive principal elements lying on two different preclusive paths constructed as in the text, and let k be the smaller of the two integers p and q, so that X_{kk} is the last diagonal constituent of M which both paths enter. So far as the paradiagonal elements of these paths are concerned, we may suppose the (horizontal, vertical) apical distances of

$$a_{ii},\ x_{i,i+1},\ a_{ii}',\ x'_{i,i+1} \quad \text{to be} \quad (\lambda_i, \mu_i),\ (\lambda_{i+1}, \mu_i),\ (\lambda_i', \mu_i'),\ (\lambda'_{i+1}, \mu_i')$$

in their respective constituents X_{ii} or $X_{i,i+1}$. Then if the effective order of a constituent X_{ij} is always denoted by γ_{ij}, we have the equations

$$\lambda_{i+1}+\mu_i-1=\gamma_{i,i+1},\ (i=1, 2, \dots q-1);\quad \lambda_i+\mu_i-1=\gamma_{ii},\ (i=1, 2, \dots q);$$
$$\lambda'_{i+1}+\mu_i'-1=\gamma_{i,i+1},\ (i=1, 2, \dots p-1);\quad \lambda_i'+\mu_i'-1=\gamma_{ii},\ (i=1, 2, \dots p);$$

which lead to the equations

$$\lambda_i-\lambda_{i+1}=\gamma_{ii}-\gamma_{i,i+1},\quad \mu_i-\mu_{i+1}=\gamma_{i,i+1}-\gamma_{i+1,i+1},\ (i=1, 2, \dots q-1);$$
$$\lambda_i'-\lambda'_{i+1}=\gamma_{ii}-\gamma_{i,i+1},\quad \mu_i'-\mu'_{i+1}=\gamma_{i,i+1}-\gamma_{i+1,i+1},\ (i=1, 2, \dots p-1);$$

and to the obviously true equations

$$\mu_i'-\mu_i=\lambda_i-\lambda_i'=\tau,\quad (i=1, 2, \dots k),$$

where τ is the amount by which the vertical apical distance of a_{ii}' exceeds that of a_{ii} in the diagonal constituent X_{ii} for the values $1, 2, \dots k$ of i.

Now let $$e_{uv}=\{a_{vv}, a_{uu}'\},\quad (u=1, 2, \dots p;\ v=1, 2, \dots q),$$

be the element of M which is common to the vertical row through a_{vv} and the horizontal row through a_{uu}'; and let E_{uv} be the apical excess of e_{uv} in its constituent X_{uv}. The pq elements such as e_{uv} form a matrix

$$E=[e]_p^q=\begin{bmatrix} e_{11} & e_{12} & \dots & e_{1q} \\ e_{21} & e_{22} & \dots & e_{2q} \\ \dots & \dots & \dots & \dots \\ e_{p1} & e_{p2} & \dots & e_{pq} \end{bmatrix}$$

which is a corranged minor of M; and because the (horizontal, vertical) apical distances of e_{uv} are (λ_v, μ_u'), we have

$$E_{uv}=\lambda_v+\mu_u'-1-\gamma_{uv},\quad (u=1, 2, \dots p;\ v=1, 2, \dots q), \quad \dots\dots\dots\dots(7)$$

and $$E_{11}=E_{22}=\dots=E_{kk}=\tau.$$

By making use of Exs. vii and viii of § 241 as generalised in Ex. xiv we can proceed from this last result to the proof of the theorem (B_3) which is given in Ex. xv.

We can also complete the proof of (B_3) with the aid of Ex. vi of § 241 by observing that the pq integers E_{uv} satisfy the relations such as

$$E_{i,j+1}-E_{ij}=(\gamma_{ij}-\gamma_{i,j+1})-(\lambda_j-\lambda_{j+1})=(\gamma_{ij}-\gamma_{i,j+1})-(\gamma_{jj}-\gamma_{j,j+1})$$
$$\nless 0 \text{ if } j \nless i, \qquad \ngtr 0 \text{ if } j \ngtr i;$$

$$E_{i+1,j}-E_{ij}=(\gamma_{ij}-\gamma_{i+1,j})-(\mu_i'-\mu'_{i+1})=(\gamma_{ij}-\gamma_{i+1,j})-(\gamma_{i,i+1}-\gamma_{i+1,i+1})$$

$$\not<0 \text{ if } i\not<j-1, \qquad \not>0 \text{ if } i\not>j-1;$$

which include $\qquad E_{i,i+1}=\tau, \qquad (i=1, 2, \ldots k-1).$

It follows that the matrix E has the following two properties:

(1) *The apical excess of an element always tends to increase as it moves away from a diagonal element (actual or non-existent) along any horizontal or vertical row.*

(2) *The diagonal elements* $e_{11}, e_{22}, \ldots e_{kk}$ *all have the same apical excess* τ.

Consequently when the non-zero integer τ is positive, all the elements of E are equal to 0; i.e. the theorem (B_3) is true.

Ex. viii. *Augmented preclusive groups.*

If when g in Ex. vii is quasi-terminate we take its elements to be

$$a_{11}, a_{22}, \ldots a_{qq}, \qquad x_{q,q+1}, x_{q,q+2}, \ldots x_{qt},$$

where the added elements are the successive dominant accessary elements in which the horizontal row through a_{qq} cuts the accessary vertical rows, we must replace E by a matrix

$$E'=[e]_p^t$$

formed by adding $t-q$ final vertical rows to E, and we have the additional equations

$$\lambda_{q+i}+\mu_q-1=\gamma_{q,q+i}, \quad \lambda_{q+i}>\gamma_{q+1,q+i}, \quad (i=1, 2, \ldots t-q);$$

leading to

$$\lambda_i-\lambda_{i+1}=\gamma_{qi}-\gamma_{q,i+1}, \qquad (i=q, q+1, \ldots t-1);$$

$$\lambda_q>\gamma_{q+1,q}; \quad \lambda_i>\gamma_{q+1,i}, \qquad (i=1, 2, \ldots t).$$

If there are any new diagonal elements in E' (when $p>q$), they must all be 0's, because the added rows are all accessary and intersect diagonal constituents in zero elements only. For all elements of E and E' we have

$$E_{ij}>0 \quad \text{when} \quad i>q;$$

$$E_{i+1,j}-E_{ij}=(\gamma_{ij}-\gamma_{i+1,j})-(\gamma_{i,i+1}-\gamma_{i+1,i+1}) \qquad \not<0 \text{ if } i\not<j-1, \quad \not>0 \text{ if } i\not>j-1;$$

the first result showing that all horizontal rows of E and E' below the qth (when $p>q$) contain only zero elements; and for the elements added to E in the formation of E' we have the relations

$$E_{i,j+1}-E_{ij}=(\gamma_{ij}-\gamma_{i,j+1})-(\gamma_{qj}-\gamma_{q,j+1}) \qquad \not<0 \text{ if } i\not>q, \quad \not>0 \text{ if } i\not<q,$$

$$(j=q, q+1, \ldots t-1);$$

leading to $\qquad E_{qq}=E_{q,q+1}=E_{q,q+2}= \ldots =E_{qt}=\tau \qquad$ when $p\not<q$;

and $\qquad E_{q+1,q+1}=\tau>0, \quad E_{q+2,q+2}\not>\tau, \ldots \qquad$ when $p>q$.

So far as elements not lying below the qth horizontal row are concerned, E' has the same two properties as E in Ex. vii; and it is therefore a zero matrix whenever τ is positive.

If $p>q$, τ is necessarily positive, because the new diagonal elements of E' are all 0's; and the above relations show that E and E' are zero matrices.

If $k=p\not>q$, E' has exactly the same properties as E, and is a zero matrix whenever τ is positive.

3. *Postclusive paths and groups.*

Selecting any principal element a_{11} of the leading constituent X_{11} of the standardised general compound slope M, we will construct a zig-zag path by travelling in succession

from a_{11} vertically downward to the next paradiagonal element x_{21},
from x_{21} horizontally rightward to the next principal element a_{22},
from a_{22} vertically downward to the next paradiagonal element x_{32},
from x_{32} horizontally rightward to the next principal element a_{33},

$$\begin{array}{lll} a_{11} & & \\ \vdots & & \\ x_{21} \ldots a_{22} & & \\ \qquad\;\; \vdots & & \\ \qquad x_{32} \ldots a_{33} & & \end{array} \qquad\qquad \begin{array}{lll} a_{11}' & & \\ \vdots & & \\ x_{21}' \ldots a_{22}' & & \\ \qquad\;\; \vdots & & \\ \qquad x_{32}' \ldots a_{33}' & & \end{array}$$

continuing in this way as long as possible, and we will call it a *postclusive path* based on the principal paradiagonals. The paradiagonal elements a_{11}, x_{21}, a_{22}, x_{32}, a_{33}, ... lie respectively in the constituents X_{11}, X_{21}, X_{22}, X_{32}, X_{33}, ... The suffixes indicate the constituents to which they belong, and not their positions in the constituents. The total number of such paths is equal to the effective order $\gamma = \gamma_{11}$ of the leading diagonal constituent X_{11}, i.e. to the number of elements lying on the paradiagonal of X_{11}. Since a uniquely determinate postclusive path can be drawn backwards from any given principal element of M to a principal element of X_{11}, we see that:

Every principal element a_{ii} of M lies on one and only one of the postclusive paths, viz. one which passes through principal elements a_{11}, a_{22}, ... a_{ii} of X_{11}, X_{22}, ... X_{ii}.

It follows that the principal elements of M can be divided into groups according to the postclusive paths on which they lie. We will call these the *postclusive groups* of the principal elements, and denote by

$$g_1, g_2, \ldots g_\gamma \quad \ldots\ldots\ldots\ldots\ldots\ldots\ldots\ldots\ldots\ldots\text{(C)}$$

the groups corresponding to the paths through the principal elements in the 1st, 2nd, ... γth apical horizontal rows of X_{11}. The elements composing each group will be arranged in the same order from left to right (or from above downwards) as in M.

If X_{ii} is any given diagonal constituent of M, and if u, v, w are the effective orders of X_{ii}, $X_{i+1,i}$, $X_{i+1,i+1}$, so that

$$u \nless v \nless w, \quad (w \text{ being } 0 \text{ when } X_{i+1,i+1} \text{ is non-existent}),$$

the u postclusive paths which reach the u principal elements of X_{ii} belong to u consecutive groups in (C), and can be divided into three sets of paths belonging to consecutive groups, there being exactly:

$u - v$ postclusive paths 'pleni-terminate in X_{ii},' ....(6′)

each of which reaches some principal element a_{ii} of X_{ii}, but cannot be continued any further because the vertical line through a_{ii} does not cut the paradiagonal of any constituent lying below X_{ii};

w postclusive paths passing through X_{ii} and $X_{i+1,i}$ to $X_{i+1,i+1}$;...(5′)

$v - w$ postclusive paths 'quasi-terminate in X_{ii},' ....(4′)

each of which can be continued from some principal element a_{ii} of X_{ii} to a non-principal paradiagonal element $x_{i+1,i}$ of $X_{i+1,i}$, but cannot be continued further because the horizontal line through $x_{i+1,i}$ does not cut the paradiagonal of any constituent lying to the right of $X_{i+1,i}$. The w through paths (5′) always exist except when X_{ii} is the last diagonal constituent; and they lie between the paths (6′) and (4′). The *pleni-terminate* paths (6′) exist when and only when the effective order of X_{ii} is greater than those of $X_{i+1,i}$ and $X_{i+1,i+1}$; and they lie on the *basical vertical* side of the other paths in X_{ii}. The *quasi-terminate* paths (4′) exist when and only when the effective order of $X_{i+1,i+1}$ is less than those of $X_{i+1,i}$ and X_{ii}, i.e. when and only when $X_{i+1,i+1}$ has a smaller effective order than X_{ii} and also

contains horizontal rows of 0's;

and they lie on the *apical vertical* side of the other paths in X_{ii}.

Every postclusive path must start from some principal element a_{11} of X_{11}, and must terminate in one and only one of the following two ways:

(*a*) It may be *pleni-terminate* at some principal element a_{ii} from which it cannot be continued vertically. In this case (see Ex. vii of § 241) the vertical lines through the elements $x_{i,i-1}$ (if $i \neq 1$) and a_{ii} cut all constituents lying below them in zero elements only; and the diagonal constituent containing a_{ii} must be immediately followed (if it is not the last) by a diagonal constituent of smaller effective order.

(*b*) It may be *quasi-terminate* at some principal element a_{ii}, i.e. it may proceed vertically from a_{ii} to a non-principal element $x_{i+1,i}$ from which it cannot be continued horizontally. In this case (see Ex. viii of § 241) the horizontal lines through the elements a_{ii} and $x_{i+1,i}$ cut all constituents lying to the right of them in zero elements only; $x_{i+1,i}$ is the dominant element of an accessary horizontal row; and the diagonal constituent containing a_{ii} must be immediately followed by a diagonal constituent of smaller effective order having horizontal rows of 0's, i.e. having more horizontal than vertical rows.

It will be clear from (6′), (5′), (4′) that the postclusive groups arranged as in (C) are divided naturally into two sets lying on the left and right respectively:

those on the left (the *pleni-terminate* groups) belonging to the paths (*a*) and containing more and more elements as we pass from left to right;

those on the right (the *quasi-terminate* groups) belonging to the paths (*b*) and containing fewer and fewer elements as we pass on from left to right.

It can be shown by repeated applications of Exs. vi, vii and viii of § 241 that:

The vertical lines through the principal elements lying on the pleni-terminate postclusive paths (a), and therefore those through all the elements of any pleni-terminate postclusive group g, intersect the accessary horizontal rows in zero elements only. ....(C_1)

Postclusive paths and groups.

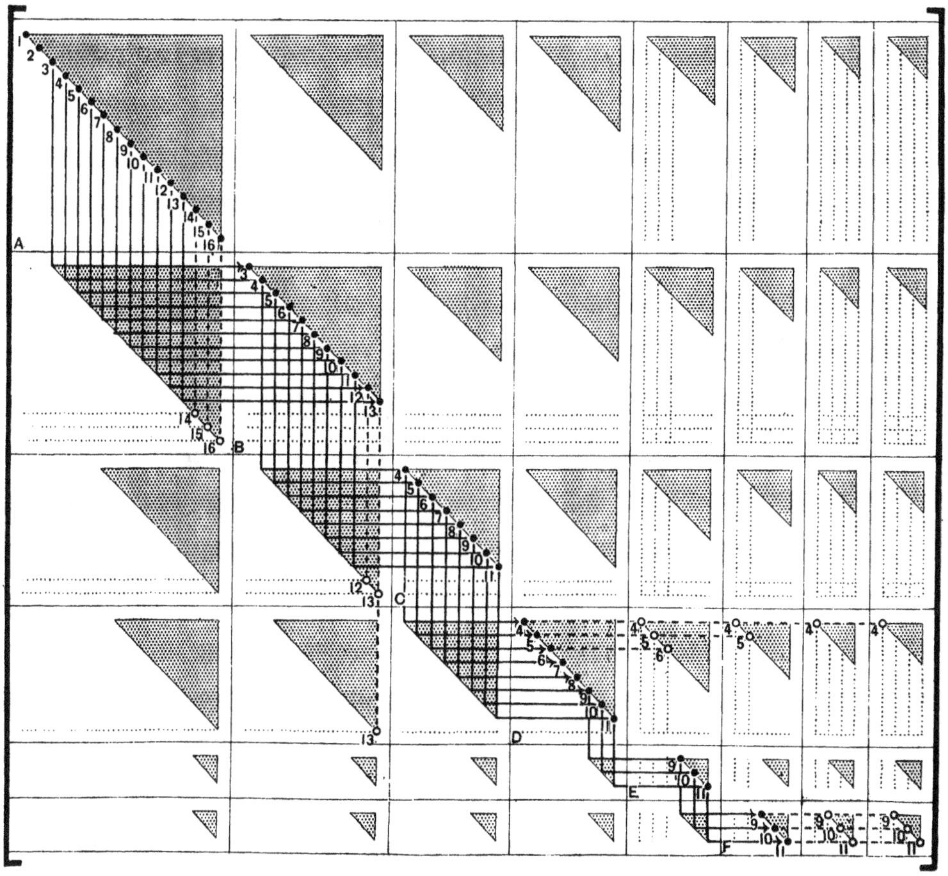

Pleni-terminate groups of principal elements.

$g_1 = A_1$,
$g_2 = A_2$,
$g_3 = (A_3, B_3)$,
$g_4 = (A_4, B_4, C_4, D_4)$,
$g_5 = (A_5, B_5, C_5, D_5)$,
$g_6 = (A_6, B_6, C_6, D_6)$,
$g_7 = (A_7, B_7, C_7, D_7)$,
$g_8 = (A_8, B_8, C_8, D_8)$,
$g_9 = (A_9, B_9, C_9, D_9, E_9, F_9)$,
$g_{10} = (A_{10}, B_{10}, C_{10}, D_{10}, E_{10}, F_{10})$,
$g_{11} = (A_{11}, B_{11}, C_{11}, D_{11}, E_{11}, F_{11})$,

Quasi-terminate groups of principal elements.

$g_{12} = (A_{12}, B_{12})$,
$g_{13} = (A_{13}, B_{13})$,
$g_{14} = A_{14}$,
$g_{15} = A_{15}$,
$g_{16} = A_{16}$.

The horizontal lines through the principal elements lying on the quasi-terminate postclusive paths (*b*), *and therefore those through all the elements of any quasi-terminate postclusive group* g, *intersect the accessary vertical rows in zero elements only.* ....(C_2)

If g *and* g' *are any two postclusive groups such that* g' *occurs later than* g *in the series* (C), *then the vertical rows through the elements of* g *and the horizontal rows through the elements of* g' *intersect one another in zero elements only.* ....(C_3)

These three theorems remain true when we add to the principal elements of each quasi-terminate group the final dominant accessary element in which the corresponding path terminates, and also any other dominant accessary elements through which the path could be continued vertically.

It will be evident that when M, X_{11} are converted into their conjugates M', X_{11}', the preclusive and postclusive paths of M are converted into the postclusive and preclusive paths of M'. In fact if we define $g_i' = g_{\gamma+1-i}$ to be the postclusive group of M which contains the principal element lying in the ith apical *vertical* row of X_{11}, so that the series (C) becomes

$$g_\gamma', \ldots g_2', g_1', \qquad \text{(C')}$$

then g_i' is the preclusive group of M' which contains the principal element lying in the ith apical *horizontal* row of X_{11}'. Thus from any theorem concerning preclusive paths and groups we can deduce a corresponding theorem concerning postclusive paths and groups by merely interchanging the two words 'horizontal' and 'vertical,' as well of course as the two words 'preclusive' and 'postclusive.' In particular the theorems (C_1), (C_2), (C_3) can be deduced from the theorems (B_1), (B_2), (B_3) of sub-article 2 in this way.

A direct proof of (C_3) is given in Exs. ix and x. Simpler and more comprehensive proofs of all three theorems are given in sub-article 4, (C_1) and (C_2) being included in Ex. xiv, and (C_3) in Exs. xv and xvi.

The figure on p. 649 shows the postclusive paths of the standardised general compound slope whose preclusive paths are shown on p. 642. In this case those principal elements of the 1st, 2nd, 3rd, ... diagonal constituents which lie on the ith *postclusive* path g_i are denoted by A_i, B_i, C_i, ... Consequently the allocations of the suffixes are not the same as on p. 642.

Ex. ix. *Analytic treatment of two postclusive paths.*

Let $a_{11}, a_{22}, \ldots a_{qq}$ forming the group g and $a_{11}', a_{22}', \ldots a_{pp}'$ forming the group g' be the successive principal elements lying on two different postclusive paths constructed as in the text, and let k be the smaller of the two integers p and q. So far as the para-diagonal elements of these paths are concerned, we may suppose the (horizontal, vertical) apical distances of

$$a_{ii},\ x_{i+1,i},\ a_{ii}',\ x'_{i+1,i} \quad \text{to be} \quad (\lambda_i, \mu_i),\ (\lambda_i, \mu_{i+1}),\ (\lambda_i', \mu_i'),\ (\lambda'_{i+1}, \mu_i')$$

in their respective constituents X_{ii} or $X_{i+1,i}$. Then if the effective order of a constituent X_{ij} is always denoted by γ_{ij}, we have the equations

$$\lambda_i+\mu_{i+1}-1=\gamma_{i+1,i},\ (i=1,2,\dots q-1);\quad \lambda_i+\mu_i-1=\gamma_{ii},\ (i=1,2,\dots q);$$
$$\lambda_i'+\mu'_{i+1}-1=\gamma_{i+1,i},\ (i=1,2,\dots p-1);\quad \lambda_i'+\mu_i'-1=\gamma_{ii},\ (i=1,2,\dots p);$$

which lead to the equations

$$\lambda_i-\lambda_{i+1}=\gamma_{i+1,i}-\gamma_{i+1,i+1},\quad \mu_i-\mu_{i+1}=\gamma_{ii}-\gamma_{i+1,i},\quad (i=1,2,\dots q-1);$$
$$\lambda_i'-\lambda'_{i+1}=\gamma_{i+1,i}-\gamma_{i+1,i+1},\quad \mu_i'-\mu'_{i+1}=\gamma_{ii}-\gamma_{i+1,i},\quad (i=1,2,\dots p-1);$$

and to the obviously true equations

$$\mu_i'-\mu_i=\lambda_i-\lambda_i'=\tau,\qquad (i=1,2,\dots k),$$

where τ is the amount by which the vertical apical distance of a_{ii}' exceeds that of a_{ii} in the diagonal constituent X_{ii} for the values 1, 2, ... k of i. Defining e_{uv}, E_{uv} and the matrix E as in Ex. vii, we again have

$$E_{11}=E_{22}=\dots=E_{kk}=\tau.$$

By making use of Exs. vii and viii of § 241 as generalised in Ex. xiv we can proceed from this last result to the proof of the theorem (C_3) which is given in Ex. xv.

Alternatively we may observe with the aid of Ex. vi of § 241 that the pq integers E_{uv} satisfy the relations such as

$$E_{i,j+1}-E_{ij}=(\gamma_{ij}-\gamma_{i,j+1})-(\lambda_j-\lambda_{j+1})\ =(\gamma_{ij}-\gamma_{i,j+1})-(\gamma_{j+1,j}-\gamma_{j+1,j+1})$$
$$\not< 0\ \text{ if }\ j\not< i-1,\qquad \not> 0\ \text{ if }\ j\not> i-1;$$
$$E_{i+1,j}-E_{ij}=(\gamma_{ij}-\gamma_{i+1,j})-(\mu_i'-\mu'_{i+1})=(\gamma_{ij}-\gamma_{i+1,j})-(\gamma_{ii}-\gamma_{i+1,i})$$
$$\not< 0\ \text{ if }\ i\not< j,\qquad \not> 0\ \text{ if }\ i\not> j,$$

which include
$$E_{i+1,i}=\tau,\qquad (i=1,2,\dots k-1).$$

It follows that the matrix E has the same general properties as in Ex. vii, so that all its elements will be equal to 0 when τ is positive, i.e. the theorem (C_3) is true.

Ex. x. *Augmented postclusive groups.*

If when g' is quasi-terminate, we take its elements to be

$$a_{11}',\ a_{22}',\ \dots\ a_{pp}',\quad x'_{p+1,p},\ x'_{p+2,p},\ \dots\ x'_{tp},$$

where the added elements are the successive dominant accessary elements in which the vertical row through a_{pp}' cuts the accessary horizontal rows, we must replace E by a matrix

$$E=[e]_t^q,$$

formed by adding $t-p$ final horizontal rows to E, and we have the additional equations

$$\lambda_p'+\mu'_{p+i}-1=\gamma_{p+i,p},\quad \mu'_{p+i}>\gamma_{p+i,p+1},\qquad (i=1,2,\dots t-p);$$

leading to
$$\mu_i'-\mu'_{i+1}=\gamma_{ip}-\gamma_{i+1,p},\qquad (i=p,\ p+1,\dots t);$$
$$\mu_p'>\gamma_{p,p+1};\quad \mu_i'>\gamma_{i,p+1},\qquad (i=1,2,\dots t).$$

If there are any new diagonal elements in E' (when $q>p$), they must all be 0's, because the added rows are all accessary and intersect diagonal constituents in zero elements only. For all elements of E and E' we have

$$E_{ij}>0\quad\text{when}\quad j>p;$$
$$E_{i,j+1}-E_{ij}=(\gamma_{ij}-\gamma_{i,j+1})-(\gamma_{j+1,j}-\gamma_{j+1,j+1})\qquad \not< 0 \text{ if } j\not< i-1,\ \not> 0 \text{ if } j\not> i-1;$$

the first result showing that all vertical rows of E and E' to the right of the pth (when $q\not> p$) contain only zero elements; and for the elements added to E in the formation of E' we have the relations

$$E_{i+1,j}-E_{ij}=(\gamma_{ij}-\gamma_{i+1,j})-(\gamma_{ip}-\gamma_{i+1,p})\qquad \not< 0 \text{ if } j\not> p,\ \not> 0 \text{ if } j\not< p,$$
$$(i=p,\ p+1,\dots t-1);$$

leading to $E_{pp} = E_{p+1,p} = E_{p+2,p} = \ldots = E_{tp} = \tau$ when $q \not< p$;

and $E_{p+1,p+1} = \tau > 0, \quad E_{p+2,p+2} \not> \tau, \ldots$ when $q > p$.

So far as elements not lying to the right of the pth vertical row are concerned, E' has the same two properties as E in Exs. vii and ix; and it is therefore a zero matrix whenever τ is positive.

If $q > p$, τ is necessarily positive, because the new diagonal elements of E' are all 0's, and the above relations show that E and E' are zero matrices.

If $k = q \not> p$, E' has exactly the same properties as E, and is a zero matrix whenever τ is positive.

4. *The conclusive paths and groups of a standardised general compound slope M of the class* (a) *expressed in the form* (A).

We will define a *conclusive path* based on the principal paradiagonal elements to be one which is constructed by drawing in successive preclusive or postclusive portions

from a principal element a_{11} of X_{11} to a principal element a_{22} of X_{22},
from the principal element a_{22} of X_{22} to a principal element a_{33} of X_{33},
from the principal element a_{33} of X_{33} to a principal element a_{44} of X_{44},

and so on in such a manner that:

(1) If $X = X_{ii}$ is a diagonal constituent immediately followed (when it is not the last) by a diagonal constituent $X' = X_{i+1,i+1}$ of smaller effective order, then in its passage from a principal element of X to a principal element of X' the path has a specified character (preclusive or postclusive), being:

preclusive when X' has horizontal rows of 0's, i.e. has more horizontal than vertical rows;

postclusive when X' has vertical rows of 0's, i.e. has more vertical than horizontal rows;

either preclusive or postclusive (the choice between these two alternatives being fixed) when X' is a square matrix.

(2) The character of the path can only change as it passes through a principal element of a diagonal constituent X which is immediately followed by a diagonal constituent X' of smaller effective order which is not square.

The successive principal elements $a_{11}, a_{22}, a_{33}, \ldots$ lying on such a path will be said to form a *conclusive group* of principal elements. Referring to sub-articles 2 and 3, it will be seen that the first two of the conditions under (1) are necessary and sufficient in order that no portion of such a path shall be quasi-terminate, or that every path which reaches a principal element a_{ii} of a diagonal constituent X_{ii} and does not pass through to the next diagonal constituent shall be pleni-terminate in X_{ii}. The third of the conditions under (1) has been added to make the principal elements lying on each path uniquely determinate. The condition under (2) is not so essential, but has been added in order to make the conclusive paths uniquely determinate. It fixes the character of every portion of a conclusive path when M is not quadrate; and when M is quadrate it makes the conclusive paths identical with the preclusive or the postclusive paths according to the one free choice of character. Thus if $\gamma = \gamma_{11}$ is the effective order of X_{11}, there are exactly γ conclusive paths, one drawn through each of the γ principal elements of X_{11}.

Since a uniquely determinate conclusive path can be drawn backwards from any given principal element of M to a principal element of X_{11}, we see that:

Every principal element a_{ii} of M lies on one and only one of the conclusive paths, viz. one which passes through principal elements $a_{11}, a_{22}, \ldots a_{ii}$ of $X_{11}, X_{22}, \ldots X_{ii}$.

It follows that by drawing the γ conclusive paths we can divide all the principal elements of M into γ sets

$$g_1, g_2, \ldots g_\gamma, \qquad \text{(D)}$$

where the set g_i consists of those principal elements which lie on the path drawn through the principal element in the ith apical horizontal, or $(\gamma + 1 - i)$th apical vertical, row of X_{11}. These are the γ conclusive groups of principal elements. The principal elements of the group g_i will always be arranged in the same order from left to right (or from above downwards) as in M.

If we were to discard the condition (2) but retain the conditions (1), there would be a great variety of conclusive paths, but the conclusive groups would still be uniquely determinate and the same as before. The argument would be the same as before if we replace the γ uniquely determinate conclusive paths by any set of γ different conclusive paths drawn through the γ principal elements of X_{11}.

If X_{ii} and $X_{i+1,i+1}$ are two consecutive diagonal constituents having equal effective orders, it will be clear from sub-articles 2 and 3 that every conclusive path which reaches a principal element of X_{ii} can be continued both preclusively and postclusively to one and only one principal element of $X_{i+1,i+1}$. Now let X_{ii} be any diagonal constituent of M of effective order ϵ which is immediately followed by a diagonal constituent $X_{i+1,i+1}$ of *smaller* effective order ϵ', where $\epsilon' = 0$ when X_{ii} is the last diagonal constituent, and consider any set of ϵ different conclusive paths which have reached the ϵ principal elements of X_{ii}. These ϵ paths can always be divided into two sets passing through consecutive principal elements of X_{ii}, there being two cases according as the portion of a path from a principal element of X_{ii} to a principal element of $X_{i+1,i+1}$ is prescribed to be postclusive or preclusive.

Case I. If that portion is *postclusive,* as it always is when $X_{i+1,\,i+1}$ has more vertical than horizontal rows, then those two sets consist respectively of:

(i) $\epsilon - \epsilon'$ paths through the principal elements lying in the first $\epsilon - \epsilon'$ apical horizontal (or last $\epsilon - \epsilon'$ apical vertical) rows of X_{ii}, each of which is '*postclusively pleni-terminate*' at some principal element a_{ii} of X_{ii} because the vertical line through a_{ii} cuts all constituents lying below X_{ii} in zero elements only;

Conclusive paths and groups.

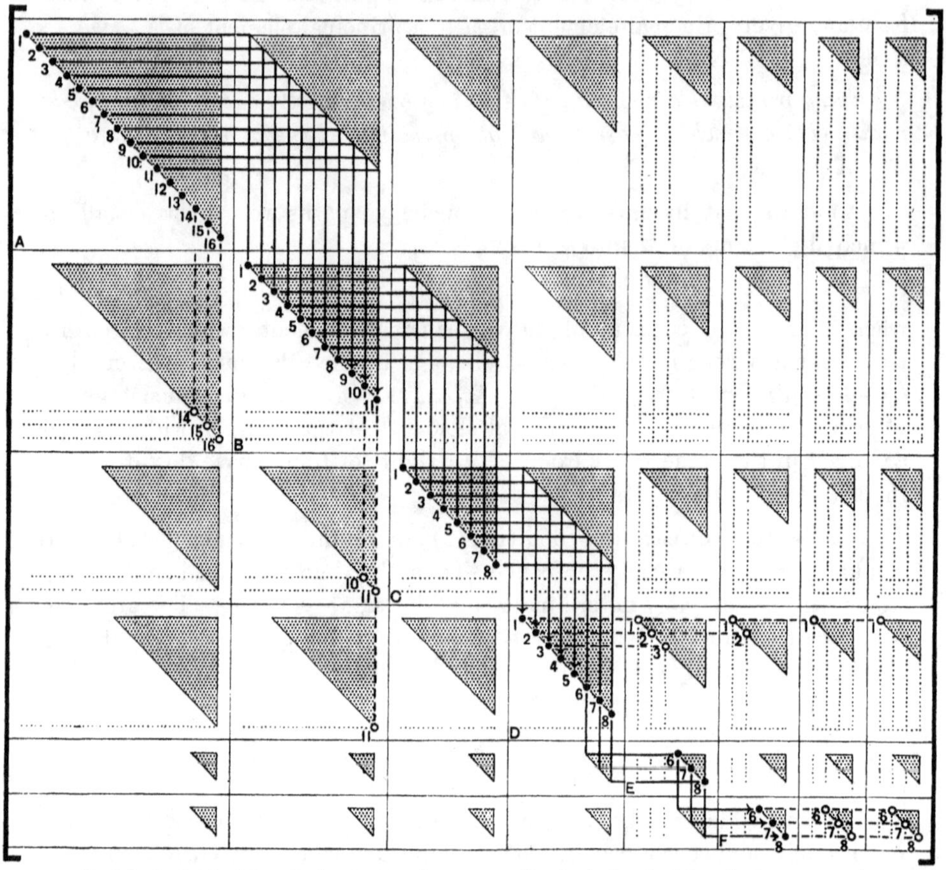

Postclusively pleni-terminate groups of principal elements.

$g_1 = (A_1, B_1, C_1, D_1)$,
$g_2 = (A_2, B_2, C_2, D_2)$,
$g_3 = (A_3, B_3, C_3, D_3)$,
$g_4 = (A_4, B_4, C_4, D_4)$,
$g_5 = (A_5, B_5, C_5, D_5)$,
$g_6 = (A_6, B_6, C_6, D_6, E_6, F_6)$,
$g_7 = (A_7, B_7, C_7, D_7, E_7, F_7)$,
$g_8 = (A_8, B_8, C_8, D_8, E_8, F_8)$,

Preclusively pleni-terminate groups of principal elements.

$g_9 = (A_9, B_9)$,
$g_{10} = (A_{10}, B_{10})$,
$g_{11} = (A_{11}, B_{11})$,
$g_{12} = A_{12}$,
$g_{13} = A_{13}$,
$g_{14} = A_{14}$,
$g_{15} = A_{15}$,
$g_{16} = A_{16}$.

(ii) ϵ' paths through the principal elements lying in the last ϵ' apical horizontal (or first ϵ' apical vertical) rows of X_{ii}, each of which can be continued through $X_{i+1,i}$ to a uniquely determinate principal element of $X_{i+1,i+1}$.

Case II. If that portion is *preclusive*, as it always is when $X_{i+1,i+1}$ has more horizontal than vertical rows, then those two sets consist respectively of:

(i) ϵ' paths through the principal elements lying in the first ϵ' apical horizontal (or last ϵ' apical vertical) rows of X_{ii}, each of which can be continued through $X_{i,i+1}$ to a uniquely determinate principal element of $X_{i+1,i+1}$.

(ii) $\epsilon - \epsilon'$ paths through the principal elements lying in the last $\epsilon - \epsilon'$ apical horizontal (or first $\epsilon - \epsilon'$ apical vertical) rows of X_{ii}, each of which is '*preclusively pleni-terminate*' at some principal element a_{ii} of X_{ii} because the horizontal line through a_{ii} cuts all constituents lying to the right of X_{ii} in zero elements only.

Thus every conclusive path must start from some principal element a_{11} of X_{11}, and terminate in one and only one of the following two ways:

(a) It may be *postclusively terminate* at some principal element a_{ii} from which it cannot be continued vertically.

(b) It may be *preclusively terminate* at some principal element a_{ii} from which it cannot be continued horizontally.

In both cases a_{ii} is a principal element of a diagonal constituent X_{ii} which is immediately followed (if it is not the last) by a diagonal constituent of smaller effective order, the effective order being the horizontal order in (a) or the vertical order in (b).

From Cases I and II above it will be seen that the conclusive groups arranged as in (D) are divided naturally into two sets lying on the left and right respectively:

those on the left (the *postclusively pleni-terminate* groups) belonging to the paths (a) and containing more and more elements as we pass from left to right;

those on the right (the *preclusively pleni-terminate* groups) belonging to the paths (b) and containing fewer and fewer elements as we pass on from left to right.

It can be shown by repeated applications of Exs. vi, vii and viii of § 241 that:

The vertical lines through the principal elements lying on the paths (a), *and therefore those through all the elements of any postclusively pleni-terminate conclusive group* g, *intersect the accessary horizontal rows in zero elements only.* ....(D_1)

The horizontal lines through the principal elements lying on the paths (*b*), *and therefore those through all the elements of any preclusively pleni-terminate conclusive group g, intersect the accessary vertical rows in zero elements only.*

........................(D_2)

If g and g' are any two conclusive groups such that g' occurs later than g in the series (D), *then the vertical rows through the elements of g and the horizontal rows through the elements of g' intersect one another in zero elements only.* ....(D_3)

These three theorems remain true when we add to the principal elements of each conclusive group the final dominant accessary elements (if any) through which the corresponding path could be continued horizontally or vertically from its last principal element.

The theorem (D_1) follows from Ex. xii below and (C_1) of sub-article 3, and the theorem (D_2) from Ex. xiii below and (B_2) of sub-article 2; but both theorems are included in Ex. xiv. The theorem (D_3) is included in Exs. xv and xvi.

The figure on p. 654 shows the conclusive paths of the standardised general compound slope whose preclusive and postclusive paths are shown on p. 642 and p. 649. In this case those principal elements of the 1st, 2nd, 3rd, ... diagonal constituents which lie on the ith *conclusive* path g_i are denoted by A_i, B_i, C_i, ... Consequently the allocations of the suffixes are not the same as on p. 642 or p. 649.

Ex. xi. Let X_{ii} and $X_{i+1,i+1}$ be any two consecutive diagonal constituents of M, and let a_{i+1} be any paradiagonal element of $X_{i+1,i+1}$.

Then if the preclusive and postclusive paths through a_{i+1} cut the paradiagonal of X_{ii} in a_i and a_i' respectively, the vertical apical distance of a_i' in X_{ii} cannot be less than that of a_i.

For if the (horizontal, vertical) apical distances of a_i, a_i', a_{i+1} in X_{ii}, X_{ii}, $X_{i+1,i+1}$ are

$$(\lambda_i, \mu_i),\ (\lambda_i', \mu_i'),\ (\lambda_{i+1}, \mu_{i+1}),$$

we have
$$\lambda_{i+1}+\mu_i-1=\gamma_{i,i+1},\quad \lambda_i'+\mu_{i+1}-1=\gamma_{i+1,i};$$
$$\lambda_i+\mu_i-1=\gamma_{ii},\quad \lambda_i'+\mu_i'-1=\gamma_{ii},\quad \lambda_{i+1}+\mu_{i+1}-1=\gamma_{i+1,i+1};$$

and therefore
$$\mu_i'-\mu_i=\lambda_i-\lambda_i'=(\gamma_{ii}-\gamma_{i,i+1})-(\gamma_{i+1,i}-\gamma_{i+1,i+1})\nless 0.$$

Hence if the conclusive and postclusive paths through a_{i+1} cut the paradiagonal of X_{ii} in a_i and a_i' respectively, the vertical apical distance of a_i' in X_{ii} cannot be less than that of a_i.

For if the conclusive path from a_i to a_{i+1} is not postclusive, it must be preclusive.

Again if the preclusive and conclusive paths through a_{i+1} cut the paradiagonal of X_{ii} in a_i and a_i' respectively, the vertical apical distance of a_i' in X_{ii} cannot be less than that of a_i.

For if the conclusive path from a_i' to a_{i+1} is not preclusive, it must be postclusive.

Ex. xii. *Every principal element of a postclusively pleni-terminate conclusive group g lies on a pleni-terminate postclusive path.*

For the postclusive path through the last principal element of g is necessarily pleni-terminate; and if a_i, a_{i+1} are two consecutive principal elements of g, it follows from the second result of Ex. xi by sub-article 3 that if the postclusive path through a_{i+1} is pleni-terminate, that through a_i is also pleni-terminate.

Ex. xiii. *Every principal element of a preclusively pleni-terminate conclusive group* g *lies on a pleni-terminate preclusive path.*

For the preclusive path through the last principal element of g is necessarily pleni-terminate ; and if a_i, a_{i+1} are two consecutive principal elements of g, it follows from the third result of Ex. xi by sub-article 2 that if the preclusive path through a_{i+1} is pleni-terminate, that through a_i is also pleni-terminate.

Ex. xiv. *Properties of any path which is compounded of preclusive and postclusive paths.*

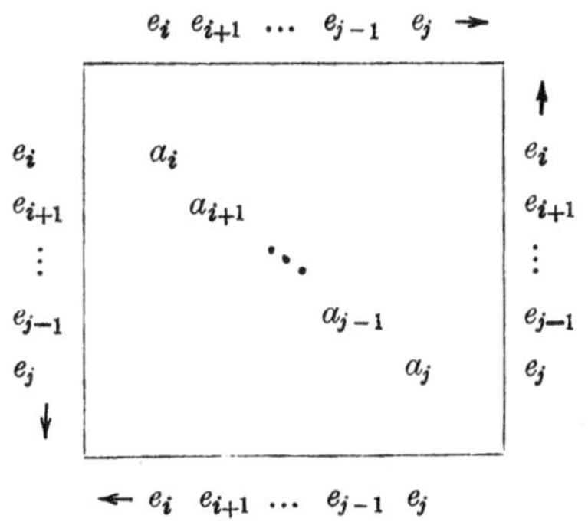

Let a_i always mean a principal element of the diagonal constituent X_{ii}, and let

$$a_i,\ a_{i+1},\ \dots\ a_{j-1},\ a_j,$$

proceeding from left to right, be a series of principal elements of consecutive diagonal constituents such that we can pass from any one (which is not the last) to the next by either a preclusive or a postclusive path. Also let

$$e_i,\ e_{i+1},\ \dots\ e_{j-1},\ e_j$$

be the elements in which :

(1) the vertical lines through $a_i, \dots a_j$ cut any horizontal row of M which lies above or passes through X_{ii} ;

(2) the vertical lines through $a_i, \dots a_j$ cut any horizontal row of M which lies below or passes through X_{jj} ;

(3) the horizontal lines through $a_i, \dots a_j$ cut any vertical row of M which lies to the left of or passes through X_{ii} ;

(4) the horizontal lines through $a_i, \dots a_j$ cut any vertical row of M which lies to the right of or passes through X_{jj}.

Then repeated applications of Ex. vii of § 241 in (1) and (2) or Ex. viii of § 241 in (3) and (4) show that the apical excesses of the elements $e_i, \dots e_j$ or $e_j, \dots e_i$ tend constantly to increase as we pass from one to the next in the direction indicated by the arrow in the figure ; and we conclude that :

If any one of the elements $e_i, \dots e_j$ *or* $e_j, \dots e_i$ *is* 0, *then all those which follow it in the direction of the arrow are* 0's. ....(8)

We can regard a_i, a_{i+1}, ... a_j in (8) as successive elements of a series of successive principal elements

$$a_1,\ a_2,\ \dots\ a_i,\ a_{i+1},\ \dots\ a_{j-1},\ a_j,\ \dots\ a_{p-1},\ a_p$$

lying on a path P which is made up of successive preclusive and postclusive portions constructed as in sub-articles 2 and 3, or which is constructed by drawing in succession

a preclusive or postclusive path from a principal element a_1 of X_{11} to a principal element a_2 of X_{22},

a preclusive or postclusive path from the principal element a_2 of X_{22} to a principal element a_3 of X_{33},

and so on. Without assuming that a_p is the last principal element on P, we will say that the path is

preclusively pleni-terminate at a_p when the horizontal line through a_p cuts all constituents lying to the right of X_{pp} in zero elements only,

postclusively pleni-terminate at a_p when the vertical line through a_p cuts all constituents lying below X_{pp} in zero elements only.

Supposing P to be preclusively pleni-terminate at a_p, let the horizontal lines through $a_1, a_2, \dots a_p$ cut any given vertical row L in elements $e_1, e_2, \dots e_p$ of constituents $E_1, E_2, \dots E_p$. If L passes through the diagonal constituent X_{ii} and if further we have $e_i=0$ (as is necessarily the case when L is an accessary vertical row), then by applying (8) to the series $a_1, a_2, \dots a_i$ and $a_i, a_{i+1}, \dots a_p$ we see that $e_i, e_{i-1}, \dots e_1$ and $e_i, e_{i+1}, \dots e_p$ are all zero elements; whilst if L lies entirely to the right of X_{pp}, we have $e_p=0$, and by applying (8) to the series $a_1, a_2, \dots a_p$ we see that $e_p, e_{p-1}, \dots e_1$ are all zero elements.

Thus if the path P is preclusively pleni-terminate at a_p, the horizontal lines through $a_1, a_2, \dots a_p$ will cut a vertical row L in zero elements only whenever:

(1) *L is an accessary vertical row;*

(2) *L intersects one of those horizontal lines in a zero element lying in the diagonal constituent through which it passes;*

(3) *L lies entirely to the right of all the diagonal constituents containing $a_1, a_2, \dots a_p$.* ....(9)

It can be shown in a similar way that:

If the path P is postclusively pleni-terminate at a_p, the vertical lines through $a_1, a_2, \dots a_p$ will cut a horizontal row L in zero elements only whenever:

(1) *L is an accessary horizontal row;*

(2) *L intersects one of those vertical lines in a zero element lying in the diagonal constituent through which it passes;*

(3) *L lies entirely below all the diagonal constituents containing $a_1, a_2, \dots a_p$.* ...(10)

Ex. xv. *Intersections of the horizontal and vertical rows through the elements of two preclusive or postclusive or conclusive groups.*

Let $a_{11}, a_{22}, \dots a_{qq}$ forming the group g and $a_{11}', a_{22}', \dots a_{pp}'$ forming the group g' be the successive principal elements lying on two different preclusive or two different postclusive or two different conclusive paths based on the principal paradiagonals; and let X_{kk} be the last diagonal constituent which both paths enter, so that k is the smaller of the two integers p and q. Also let

$$e_{uv}=\{a_{vv}, a'_{uu}\}, \quad (u=1, 2, \dots p;\ v=1, 2, \dots q),$$

be the element of M which is common to the vertical row through a_{vv} and the horizontal row through a'_{uu}, so that e_{uv} is an element of the constituent X_{uv}, and let

E_{uv} be the apical excess of e_{uv} in its constituent X_{uv}.

The pq elements such as e_{uv} form a matrix

$$E=[e]_p^q=\begin{bmatrix} e_{11} & e_{12} \dots e_{1q} \\ e_{21} & e_{22} \dots e_{2q} \\ \dots & \dots \\ e_{p1} & e_{p2} \dots e_{pq} \end{bmatrix}=\begin{pmatrix} g \\ g' \end{pmatrix},$$

which is a corranged minor of M. If $i \nless 1$, $i+1 \ngtr k$, we can pass from a_{ii} to $a_{i+1,i+1}$ and from a'_{ii} to $a'_{i+1,i+1}$ either by two preclusive or by two postclusive paths, and in either case (see Exs. vii and ix) we have $E_{ii}=E_{i+1,i+1}$. Consequently we always have

$$E_{11}=E_{22}=\dots=E_{kk}=\tau,$$

where τ is the amount by which the vertical apical distance of a'_{ii} exceeds that of a_{ii} in the diagonal constituent X_{ii} for the values 1, 2, ... k of i. From this result and from Ex. xiv it follows that the matrix E has the following two properties:

(1) *The apical excess of an element always tends to increase as it moves away from a diagonal element (actual or non-existent) along any horizontal or vertical row.*

(2) *The diagonal elements e_{11}, e_{22}, ... e_{kk} all have the same apical excess τ.*

When the non-zero integer τ is positive, all the elements of E must be equal to 0; and this shows that the theorem (D_3) is true.

Ex. xvi. *Intersections of the horizontal and vertical rows through the elements of two augmented preclusive or postclusive or conclusive groups.*

First let g be a preclusive or postclusive or conclusive group whose last principal element is a_{ii}. We will define a corresponding augmented group G.

$$\begin{array}{l|l} a_{ii} & 0 \\ k_{i+1,i} & 0 \\ k_{i+2,i} & 0 \end{array} \qquad \begin{array}{ccc} a_{ii} & h_{i,i+1} & h_{i,i+2} \\ \hline 0 & 0 & 0 \end{array}$$

If the group g is *preclusively pleni-terminate* at a_{ii}, and if there are accessary horizontal rows whose dominant elements $k_{i+1,i}$, $k_{i+2,i}$, ... lie in the same vertical row as a_{ii} (and necessarily in consecutive constituents below X_{ii}), we will take

$$a_{11},\ a_{22},\ \dots\ a_{ii},\ k_{i+1,i},\ k_{i+2,i},\ \dots$$

to be the successive elements of G. This case occurs (see Ex. ii and sub-article 3) when and only when g is *postclusively quasi-terminate* at a_{ii}.

If the group g is *postclusively pleni-terminate* at a_{ii}, and if there are accessary vertical rows whose dominant elements $h_{i,i+1}$, $h_{i,i+2}$, ... lie in the same horizontal row as a_{ii} (and necessarily in consecutive constituents to the right of X_{ii}), we will take

$$a_{11},\ a_{22},\ \dots\ a_{ii},\ h_{i,i+1},\ h_{i,i+2},\ \dots$$

to be the successive elements of G. This case occurs (see Ex. i and sub-article 2) when and only when g is *preclusively quasi-terminate* at a_{ii}.

The augmented groups of g_1, g_2, ... g_γ in (B) or (C) or (D) will be denoted by

$$G_1,\ G_2,\ \dots\ G_\gamma. \quad \dots\dots\dots\dots\dots\dots\dots\dots\text{(D')}$$

Now let G and G' be the augmented groups corresponding to the groups g and g' of Ex. xv. Then the elements of M common to the vertical rows through the elements of G and G' form a matrix

$$E'=[e]_{p'}^{q'}=\begin{pmatrix} G \\ G' \end{pmatrix},$$

which is a corranged minor of M, and which is formed from the matrix E of Ex. xv by adding

$p'-p$ final horizontal rows when g' is preclusively pleni-terminate at a'_{pp},
$q'-q$ final vertical rows when g is postclusively pleni-terminate at a_{qq}.

Because the added rows are all accessary rows which intersect diagonal constituents in zero elements only and intersect one another in zero elements only, the only diagonal elements of E' which can be different from 0 are the diagonal elements $e_{11}, e_{22}, \ldots e_{kk}$ of E; every new diagonal element in E' being necessarily equal to 0. Moreover when we except those elements which are necessarily 0 because of (9) or (10) of Ex. xiv, the matrix E' has the same two properties as the matrix E in Ex. xv.

Hence E' is a zero matrix whenever τ is positive, i.e. whenever G' occurs later than G in the series (D').

A more detailed account of the various possible cases is added.

If g is postclusively and g' preclusively pleni-terminate, it follows from (9) and (10) of Ex. xiv that E' differs from E by zero elements only. Moreover in this case τ is necessarily positive, and E' is a zero matrix.

If g is preclusively and g' postclusively pleni-terminate, E' is the same as E. Moreover in this case τ is necessarily negative, and E or E' is a non-zero matrix.

If g and g' are both postclusively pleni-terminate, E' is formed by adding vertical rows to E. From (10) of Ex. xiv and the properties of dominant accessary elements we see that all horizontal rows of E' below the qth (when $p > q$) contain zero elements only; whilst by applying Exs vii and viii of § 241 (or the argument given in Ex. viii of sub-article 2) to the other added elements we see that with respect to all other elements E' has the same properties as E in Ex. xv. Hence when τ is positive, which is only possible when $q \ngtr p$, E' and E are zero matrices, because all their diagonal elements are 0's. Whenever E' is a non-zero matrix, i.e. whenever τ is negative, we must have $k=p \ngtr q$.

If g and g' are both preclusively pleni-terminate, E' is formed by adding horizontal rows to E. From (9) of Ex. xiv and the properties of dominant accessary elements we see that all vertical rows of E' to the right of the pth (when $q > p$) contain zero elements only; whilst by applying Exs. vii and viii of § 241 (or the argument given in Ex. x of sub-article 3) to the other added elements we see that with respect to all other elements E' has the same properties as E in Ex. xv. Hence when τ is positive, which is only possible when $p \ngtr q$, E' and E are zero matrices, because all their diagonal elements are 0's. Whenever E' is a non-zero matrix, i.e. whenever τ is negative, we must have $k=q \ngtr p$.

Thus whenever E' is a non-zero matrix, it has exactly the same diagonal elements $e_{11}, e_{22}, \ldots e_{kk}$ as E, and has the same two properties as E in Ex. xv.

If then in travelling away from a diagonal element of E' (actual or non-existent) along the horizontal or vertical row through it in either direction we reach a zero element, all the following elements in the same direction are 0's.

The above results are in agreement with the equations

$$E_{pp} = E_{p+1,p} = E_{p+2,p} = \ldots = E_{p'p} = \tau \qquad \text{when } p \ngtr q,$$
$$E_{qq} = E_{q,q+1} = E_{q,q+2} = \ldots = E_{qq'} = \tau \qquad \text{when } q \ngtr p,$$

which show that τ must be positive and E' be a zero matrix when:

there are added horizontal rows in E', and $q > p$;
there are added vertical rows in E', and $p > q$.

5. *Hemipteric derangements of M derived from the preclusive or postclusive or conclusive groups.*

Let $g_1, g_2, \ldots g_\gamma$ be the preclusive groups (B) or the postclusive groups (C) or the conclusive groups (D), and in each case let

$$G_1, G_2, \ldots G_\gamma \quad \ldots\ldots\ldots\ldots\ldots\ldots\ldots\ldots(D')$$

be the corresponding augmented groups defined in Ex. xvi. Every one of the dominant accessary elements occurs in one and only one of the augmented groups. Consequently every horizontal or vertical row of M which is not a complete row of 0's passes through

either a principal element or a dominant accessary element

(but not both) of one and only one of the augmented groups, and can be represented by that element.

Now let all the complete rows of 0's (if such rows occur) be struck out from M, and let the non-zero horizontal and vertical rows be arranged in the orders in which they, i.e. their representative elements, occur in the series (D′). Then M is thereby converted into a compound matrix

$$Y = \begin{bmatrix} Y_{11}, & Y_{12}, & \ldots & Y_{1\gamma} \\ Y_{21}, & Y_{22}, & \ldots & Y_{2\gamma} \\ \cdots & \cdots & \cdots & \cdots \\ Y_{\gamma 1}, & Y_{\gamma 2}, & \ldots & Y_{\gamma\gamma} \end{bmatrix}_{p_1, p_2, \ldots p_\gamma}^{q_1, q_2, \ldots q_\gamma} = \begin{bmatrix} Y_{11}, & Y_{12}, & \ldots & Y_{1\gamma} \\ 0, & Y_{22}, & \ldots & Y_{2\gamma} \\ \cdots & \cdots & \cdots & \cdots \\ 0, & 0, & \ldots & Y_{\gamma\gamma} \end{bmatrix}_{p_1, p_2, \ldots p_\gamma}^{q_1, q_2, \ldots q_\gamma} \quad \ldots(E)$$

in which

$$Y_{ij} = \begin{pmatrix} G_j \\ G_i \end{pmatrix}$$

is the corranged minor of M formed by the intersections of the vertical rows through the elements of G_j and the horizontal rows through the elements of G_i, i.e. one of the corranged minors of M described in Ex. xvi. By restoring the complete rows of 0's we can obtain a hemipteric derangement of M which differs from Y in having initial vertical or final horizontal rows of 0's. By moving the accessary rows only we can obtain a derangement of M having the form (B) of § 241.

Ex. xvii. *Extensions of the compound slope M.*

Suppose in the first place that M has no complete row of 0's. Then to each set of *unaugmented* groups $g_1, g_2, \ldots g_\gamma$ of M there corresponds a way of regarding M as a minor of a quadrate slope $\bar{M}$ of the same type in which all constituents are square matrices (simple square slopes) of order γ, the conversion of M into $\bar{M}$ being carried out in accordance with the following rules:

(1) the groups $\bar{g}_1, \bar{g}_2, \ldots \bar{g}_\gamma$ of $\bar{M}$ and the paths corresponding to them are formed by continuing the groups $g_1, g_2, \ldots g_\gamma$ of M and the paths corresponding to them;

(2) the constituents of $\bar{M}$ through which the paths of $\bar{g}_1, \bar{g}_2, \ldots \bar{g}_\gamma$ pass are *general* simple square slopes;

(3) consecutive paths passing through or terminating in X_{ii} are extended into consecutive paths passing through $\bar{X}_{ii}$.

The process is simplest when the g's are conclusive groups. Ordinarily $\bar{M}$ cannot be a *general* compound slope of its class

$$M\begin{pmatrix}\gamma, \gamma, \dots \gamma\\ \gamma, \gamma, \dots \gamma\end{pmatrix};$$

but when it is made as general as possible, its form as determined by its non-zero elements is unique in each case.

If M has complete rows of 0's, and if c is the greatest of all its index numbers, we can take the extension of M to be a quadrate slope $\bar{M}$ in which all constituents are square matrices of order c. In this case $\bar{M}$ will have $c-\gamma$ initial or final groups which have no elements in common with M according as its complete rows of 0's are vertical or horizontal. Its other groups are formed as before by continuing the groups of M.

Ex. xviii. *Other hemipteric derangements of* M.

Let $\bar{M}$ be the extension of M described in Ex. xvii, and let $\bar{Y}$ be the derangement of $\bar{M}$ which corresponds to (E). Then if we strike out from $\bar{Y}$ those rows which contain no elements of M, we obtain a hemipteric derangement of M (similar in form to that described in the text) in which the horizontal and vertical rows of M are arranged in the orders in which they, i.e. elements lying on them in $\bar{M}$, occur in the successive groups of $\bar{g}_1, \bar{g}_2, \bar{g}_3, \dots$ of $\bar{M}$. Each constituent of this derangement has the same general properties as before, i.e. as we travel away from a diagonal constituent (actual or non-existent) in either horizontal or vertical direction, a zero element can only be followed by zero elements. The final derangement described in Chapter XXXIX is that obtained in this way when the g's are conclusive groups.

Ex. xix. *The paradiagonal prime minors of a standardised quadrate slope.*

When M is quadrate, it has no accessary rows, and none of the groups $g_1, g_2, \dots g_\gamma$ can be augmented. We can divide the group g_i into sub-groups $g_{i1}, g_{i2}, \dots g_{iu}, \dots$, where the sub-group g_{iu} is formed with those principal elements of g_1 which lie in the diagonal constituents having a given effective order ϵ_u. The corranged minor of M formed by the intersections of the horizontal and vertical rows through the principal elements of g_{iu} is one of the paradiagonal prime minors of M mentioned in sub-article 4 of § 241. All the paradiagonal prime minors of M can be constructed in this way, there being one corresponding to each such sub-group as g_{iu}.

APPENDIX C.

WEIERSTRASS'S AND KRONECKER'S REDUCTIONS.

1. *Minimum degrees of connection.*

Let $\phi = [\phi]_m^n$ be a matrix of rank ρ whose elements are rational integral functions of any number of scalar variables.

A rational integral identity in the variables of the form

$$[u]_r^m [\phi]_m^n = 0, \qquad \text{where } r = m - \rho,$$

and where $u = [u]_r^m$ is an undegenerate matrix whose successive horizontal rows have total degrees $\epsilon_1, \epsilon_2, \dots \epsilon_r$ in the variables, represents a complete set of r unconnected connections of degrees $\epsilon_1, \epsilon_2, \dots \epsilon_r$ between the horizontal rows of ϕ. When $\epsilon_1, \epsilon_2, \dots \epsilon_r$ are chosen in succession to be as small as possible, they are the *minimum degrees of horizontal connection* of ϕ.

A rational integral identity in the variables of the form

$$[\phi]_m^n [v]_n^s = 0, \qquad \text{where } s = n - \rho,$$

and where $v = [v]_n^s$ is an undegenerate matrix whose successive vertical rows have total degrees $\eta_1, \eta_2, \dots \eta_s$ in the variables, represents a complete set of s unconnected connections of degrees $\eta_1, \eta_2, \dots \eta_s$ between the vertical rows of ϕ. When $\eta_1, \eta_2, \dots \eta_s$ are chosen in succession to be as small as possible, they are the *minimum degrees of vertical connection* of ϕ.

When ϕ is homogeneous in the variables, we can and always will choose each of the horizontal rows of u and each of the vertical rows of v to be homogeneous in the variables.

The following principle will be used in the next two sub-articles.

If $$[h]_m^m [\phi]_m^n [k]_n^n = [\psi]_m^n \quad \textit{or} \quad h\phi k = \psi$$

is an equigradent transformation converting ϕ into a similar matrix ψ, then the two matrices ϕ and ψ have:

(1) *the same potent divisors;*

(2) *the same minimum degrees of horizontal and vertical connection.* (A)

The first part of the theorem has been proved in Theorem II and Note 1 of § 214. The second part, which was mentioned in § 141, follows from the fact that the first of the equations

$$[u]_r^m \cdot [\psi]_m^n = 0, \quad [u]_r^m [h]_m^m \cdot [\phi]_m^n = 0$$

represents a complete set of unconnected connections of degrees $\epsilon_1, \epsilon_2, \dots \epsilon_r$ between the horizontal rows of ψ when and only when the second equation represents a complete set of unconnected connections of the same degrees $\epsilon_1, \epsilon_2, \dots \epsilon_r$ between the horizontal rows of ϕ, and the fact that the first of the equations

$$[\psi]_m^n \cdot [v]_n^s = 0, \quad [\phi]_m^n \cdot [k]_n^n [v]_n^s = 0$$

represents a complete set of unconnected connections of degrees $\eta_1, \eta_2, \dots \eta_s$ between the vertical rows of ψ when and only when the second equation represents a complete set of unconnected connections of the same degrees $\eta_1, \eta_2, \dots \eta_s$ between the vertical rows of ϕ.

A brief account of the most essential properties of the primitive degrees and the minimum degrees of horizontal and vertical connection of any rational integral functional matrix is contained in a paper by the author published in two parts under the title of 'Primitive Matrices' in the *Bulletin of the Calcutta Mathematical Society*, Vols. VI and VIII (1914–15 and 1916–17).

2. *Weierstrass's reduction of an undegenerate square matrix which is homogeneous and linear in two variables or linear in a single variable.*

Let x and y be two scalar variables, and let

$$X = px + qy, \quad Y = ux + vy \quad \dots\dots\dots\dots(1)$$

be always two distinct non-zero homogeneous linear functions of x and y, so that p, q, u, v are particular scalar numbers subject only to the condition

$$vp - uq \neq 0.$$

Also let
$$\phi = [\phi]_m^m = x\,[a]_m^m + y\,[b]_m^m = xA + yB \quad \dots\dots\dots\dots(2)$$

be an undegenerate square matrix of order m whose elements are homogeneous linear functions of x and y, so that A and B are square matrices of order m whose elements are particular scalar numbers subject only to the condition that the determinant of ϕ does not vanish identically. We may and will suppose that the potent divisors of ϕ corresponding to its linear or irresoluble divisors are

$$T_1 = (\lambda_1 x + \mu_1 y)^{e_1}, \quad T_2 = (\lambda_2 x + \mu_2 y)^{e_2}, \dots \; T_s = (\lambda_s x + \mu_s y)^{e_s}, \quad \dots(3)$$

where
$$e_1 + e_2 + \dots + e_s = m,$$

the last equation following from the fact that the product of all the potent divisors can only differ from det ϕ by a non-zero constant factor.

It was first proved by Weierstrass that it is always possible to reduce ϕ by an equigradent transformation to a standard compartite square matrix

$\psi = [\psi]_m^m$ each of whose parts is a square matrix having the form

$$\omega = [\omega]_e^e = \begin{bmatrix} X & Y & 0 & \dots & 0 & 0 \\ 0 & X & Y & \dots & 0 & 0 \\ \dots & \dots & \dots & \dots & \dots & \dots \\ 0 & 0 & 0 & \dots & X & Y \\ 0 & 0 & 0 & \dots & 0 & X \end{bmatrix} = X \,.\, [1]_e^e + Y \,.\, \begin{bmatrix} 0, & 1 \\ 0, & 0 \end{bmatrix}_{e-1,1}^{1,e-1}, \quad \dots(4)$$

where $e \nless 1$. Since ω is an undegenerate square matrix having the single potent divisor X^e, we see from (A) that there must be exactly s such parts $\omega_1, \omega_2, \dots \omega_s$ corresponding one by one to the potent divisors $T_1, T_2, \dots T_s$; and that in the part ω_i corresponding to T_i we must have

$$e = e_i, \quad p : q = \lambda_i : \mu_i, \quad v\lambda_i - u\mu_i \neq 0.$$

It is easily seen that there are no other restrictions on the values of the coefficients p, q, u, v in the part ω_i, and in that part we can put $X = \lambda_i x + \mu_i y$. Thus we can regard all the parts of ψ as being completely known when the potent divisors (3) are known, though of course they can be chosen in many different ways. In the part corresponding to a potent divisor

$$(x - cy)^e \quad \text{we can put} \quad X = x - cy, \ Y = y;$$

and in the part corresponding to a potent divisor

$$(y - cx)^e \quad \text{we can put} \quad X = y - cx, \ Y = x;$$

where in both cases we may have $c = 0$.

From Weierstrass's reductions of ϕ it follows that:

If $$\phi = x[a]_m^m + y[b]_m^m, \quad \phi' = x[a']_m^m + y[b']_m^m$$

are two undegenerate square matrices of the same order m whose elements are homogeneous linear functions of x and y, it is possible to convert each of them into the other by an equigradent transformation when and only when ϕ and ϕ' have the same potent divisors, i.e. are equipotent. ....(B)

If we put $y = 1$ in the equigradent transformation converting ϕ into the compartite square matrix ψ, we obtain Weierstrass's corresponding reduction of the undegenerate square matrix

$$\Phi = x[a]_m^m + [b]_m^m,$$

whose elements are linear functions of the single variable x, the reduced form of Φ being the compartite square matrix Ψ derived from ψ by putting $y = 1$. To each potent divisor $(x - cy)^e$ of ϕ which is not a power of y, there corresponds a potent divisor $(x - c)^e$ of Φ; and to this potent divisor of Φ there corresponds a part Ω of Ψ of the form (4) in which we can put $X = x - c$, $Y = 1$. To a potent divisor y^e of ϕ which is a power of y, which can only occur when $[a]_m^m$ is degenerate, there corresponds no potent divisor of Φ; but nevertheless there corresponds to it a part Ω of Φ of the form (4) in which we can put $X=1$, $Y=x$. In fact from (B) we see that:

If $$\Phi = x\,[a]_m^m + [b]_m^m, \quad \Phi' = x\,[a']_m^m + [b']_m^m$$

are two undegenerate square matrices of the same order m whose elements are linear functions of the single variable x, it is possible to convert each of them into the other by an equigradent transformation when and only when the two homogenised matrices

$$\phi = x\,[a]_m^m + y\,[b]_m^m, \quad \phi' = x\,[a']_m^m + y\,[b']_m^m$$

have the same potent divisors, i.e. are equipotent. ....(B′)

In the particular case when $[a]_m^m$ and $[a']_m^m$ are both undegenerate, as when they are both equal to $[1]_m^m$, the conditions in (B′) are satisfied when and only when Φ and Φ' are equipotent.

The theorems (B) and (B′) can be proved directly by using the lemma of § 227 and the principles explained in §§ 211 and 212; and in proving them we at the same time prove Weierstrass's reductions which are included in them as particular cases.

The reduction of a square matrix A whose elements are constants to a canonical square matrix by an isomorphic transformation is equivalent to Weierstrass's reduction of the characteristic matrix of A.

3. *Kronecker's reduction of any matrix which is homogeneous and linear in two variables or linear in a single variable.*

Let x and y be two scalar variables, and let X and Y be defined as in (1). Also let

$$\phi = [\phi]_m^n = x\,[a]_m^n + y\,[b]_m^n = xA + yB \quad ....(2')$$

be any matrix of orders m and n whose elements are homogeneous linear functions of x and y. We will suppose that ϕ has rank ρ, and that the potent divisors of ϕ corresponding to its linear or irresoluble divisors are

$$T_1 = (\lambda_1 x + \mu_1 y)^{e_1}, \quad T_2 = (\lambda_2 x + \mu_2 y)^{e_2}, \ \dots \ T_s = (\lambda_s x + \mu_s y)^{e_s}, \quad \dots(3')$$

where of course $$e_1 + e_2 + \dots + e_s \not> \rho.$$

We will further suppose that the minimum degrees of horizontal connection of ϕ consist of

$$0 \text{ repeated } \epsilon_0 \text{ times, and the non-zero integers } \epsilon_1, \epsilon_2, \dots \epsilon_h, \quad \dots(5)$$

and that the minimum degrees of vertical connection of ϕ consist of

$$0 \text{ repeated } \eta_0 \text{ times, and the non-zero integers } \eta_1, \eta_2, \dots \eta_k. \dots(5')$$

Thus ϕ has exactly $\epsilon_0 + h$ unconnected horizontal connections of which exactly ϵ_0 can be of degree 0; it has also exactly $\eta_0 + k$ unconnected vertical connections of which exactly η_0 can be of degree 0; and we have

$$m - \rho = \epsilon_0 + h, \quad n - \rho = \eta_0 + k.$$

It was first proved by Kronecker that it is always possible to reduce ϕ by an equigradent transformation to a standard compartite matrix $\psi = [\psi]_m^n$ each of whose non-zero parts has one of the forms

$$[\omega]_e^e = \begin{bmatrix} X & Y & 0 & \dots & 0 & 0 \\ 0 & X & Y & \dots & 0 & 0 \\ \dots & \dots & \dots & \dots & \dots & \dots \\ 0 & 0 & 0 & \dots & X & Y \\ 0 & 0 & 0 & \dots & 0 & X \end{bmatrix} = X.[1]_e^e + Y.\begin{bmatrix} 0, & 1 \\ 0, & 0 \end{bmatrix}_{e-1,1}^{1,e-1}; \quad \dots\text{(i)}$$

$$[\omega]_{\epsilon+1}^{\epsilon} = \begin{bmatrix} Y & 0 & \dots & 0 \\ X & Y & \dots & 0 \\ 0 & X & \dots & 0 \\ \dots & \dots & \dots & \dots \\ 0 & 0 & \dots & Y \\ 0 & 0 & \dots & X \end{bmatrix} = X.\begin{bmatrix} 0 \\ 1 \end{bmatrix}_{1,\epsilon}^{\epsilon} + Y.\begin{bmatrix} 1 \\ 0 \end{bmatrix}_{\epsilon,1}^{\epsilon}; \quad \dots\dots\dots\dots\text{(ii)}$$

$$[\omega]_{\eta}^{\eta+1} = \begin{bmatrix} X & Y & 0 & \dots & 0 & 0 \\ 0 & X & Y & \dots & 0 & 0 \\ \dots & \dots & \dots & \dots & \dots & \dots \\ 0 & 0 & 0 & \dots & X & Y \end{bmatrix} = X.[1, 0]_{\eta}^{\eta,1} + Y.[0, 1]_{\eta}^{1,\eta}; \quad \dots\text{(iii)}$$

and which in general has in addition a zero part

$$[0]_{\epsilon_0}^{\eta_0}, \quad \dots\dots\dots\dots\dots\dots\dots\dots\dots\text{(iv)}$$

where ϵ_0 and η_0 are positive integers, one or both of which may be 0. The zero part indicates the existence of exactly ϵ_0 horizontal and exactly η_0 vertical complete rows of 0's in ψ. If ϵ_0 only is 0, this part is non-existent, and ψ has η_0 vertical but no horizontal rows of 0's; if η_0 only is 0, this part is non-existent, and ψ has ϵ_0 horizontal but no vertical rows of 0's. In the parts such as (i), (ii), (iii) we always have $e \nless 1$, $\epsilon \nless 1$, $\eta \nless 1$. We could of course interchange X and Y in either of the parts (ii), (iii); that being merely a matter of notation.

On reference to Exs. vi, vii, viii of § 205, it will be seen that:

a part such as (i) has the single potent divisor X^e and no horizontal or vertical connections;

a part such as (ii) has one horizontal connection of minimum degree ϵ, no vertical connection, and no potent divisor;

a part such as (iii) has one vertical connection of minimum degree η, no horizontal connection, and no potent divisor;

the zero part (iv) has exactly ϵ_0 horizontal connections of minimum degree 0, exactly η_0 vertical connections of minimum degree 0, and no potent divisor. Consequently the principle (A) shows that:

there must be exactly s parts such as (i), one corresponding to each of the potent divisors (3′);

there must be exactly h parts such as (ii), one corresponding to each of the non-zero minimum degrees of horizontal connection $\epsilon_1, \epsilon_2, \ldots \epsilon_h$;

there must be exactly k parts such as (iii), one corresponding to each of the non-zero minimum degrees of vertical connection $\eta_1, \eta_2, \ldots \eta_k$;

the integers ϵ_0 and η_0 in the zero part (iv) must be those defined in (5) and (5′), i.e. ϵ_0 is the total number of horizontal and η_0 the total number of vertical connections of ϕ of degree 0.

In the part (i) corresponding to the potent divisor T_i we must have

$$e = e_i, \quad p:q = \lambda_i:\mu_i, \quad v\lambda_i - u\mu_i \neq 0;$$

and p, q, u, v can be any numbers we please satisfying these conditions; in particular we can always put $X = \lambda_i x + \mu_i y$.

In the part (ii) corresponding to the non-zero minimum degree of horizontal connection ϵ_i, we must have $\epsilon = \epsilon_i$; whilst p, q, u, v can be any numbers we please satisfying the condition $vp - uq \neq 0$.

In the part (iii) corresponding to the non-zero minimum degree of vertical connection η_i, we must have $\eta = \eta_i$; whilst p, q, u, v can be any numbers we please satisfying the condition $vp - uq \neq 0$.

Thus we can regard all the parts of ψ as being completely known when the potent divisors and the minimum degrees of horizontal and vertical connection of ϕ are known. We can choose Y to be the same in all the non-zero parts of ψ. It can then be any homogeneous linear function of x and y which is not a linear divisor of ϕ.

From Kronecker's reductions of ϕ it follows that:

If $$\phi = x[a]_m^n + y[b]_m^n, \quad \phi' = x[a']_m^n + y[b']_m^n$$

are any two matrices of the same orders m and n whose elements are homogeneous linear functions of x and y, it is possible to convert each of them into the other by an equigradent transformation when and only when ϕ and ϕ' have:

(1) *the same potent divisors,*

(2) *the same minimum degrees of horizontal and vertical connection.* (C)

If we put $y = 1$ in the equigradent transformation converting ϕ into the compartite matrix ψ, we obtain Kronecker's corresponding reduction of the matrix

$$\Phi = x[a]_m^n + [b]_m^n,$$

whose elements are linear functions of the single variable x, the reduced form of Φ being the compartite square matrix Ψ derived from ψ by putting $y=1$. Hence corresponding to (C) we have the theorem that:

If $$\Phi = x[a]_m^n + [b]_m^n, \quad \Phi' = x[a']_m^n + [b']_m^n$$

are any two similar matrices whose elements are linear functions of the single variable x, it is possible to convert each of them into the other by an equigradent transformation when and only when:

(1) *Φ and Φ' have the same minimum degrees of horizontal and vertical connection;*

(2) *the two homogenised matrices*

$$\phi = x[a]_m^n + y[b]_m^n, \quad \phi' = x[a']_m^n + y[b']_m^n$$

have the same potent divisors. ....(C′)

A direct proof of the theorems (C) and (C′) would establish Kronecker's reductions, which are included in them as particular cases.

Proofs of Weierstrass's and Kronecker's reductions (though not the simplest) will be found in Muth's *Elementarteiler.*

4. *Primitive matrices; the primitive degrees of a matrix.*

If $A = [a]_m^n$ is a matrix of rank r whose elements are rational integral functions of a finite number of scalar variables, there exist uniquely determinate positive integers $\epsilon_1, \epsilon_2, \ldots \epsilon_r$ arranged in ascending order of magnitude which are the lowest possible degrees in all the variables of r one-rowed matrices connected with the horizontal rows of A and forming an undegenerate matrix $H = [h]_r^n$ of rank r, H being a horizontally primitive matrix; also there exist uniquely determinate integers $\eta_1, \eta_2, \ldots \eta_r$ arranged in ascending order of magnitude which are the lowest possible degrees in all the variables of r one-rowed matrices connected with the vertical rows of A and forming an undegenerate matrix $K = [k]_m^r$ of rank r, K being a vertically primitive matrix. The integers $\epsilon_1, \epsilon_2, \ldots \epsilon_r$ and $\eta_1, \eta_2, \ldots \eta_r$ are respectively the *horizontal primitive degrees* and the *vertical primitive degrees* of A.

A matrix $H = [h]_r^n$ whose horizontal rows are unconnected is a *horizontally primitive matrix* when the degrees of its r horizontal rows are its r horizontal primitive degrees; and a matrix $K = [k]_m^r$ whose vertical rows are unconnected is a *vertically primitive matrix* when the degrees of its r vertical rows are its r vertical primitive degrees.

From the definitions of normality given in Vol. II it follows that:

If two matrices are horizontally normal to one another (their horizontal rows being mutually orthogonal), the horizontal primitive degrees of either one of them are the minimum degrees of vertical connection of the other.

If two matrices are vertically normal to one another (their vertical rows being mutually orthogonal), the vertical primitive degrees of either one of them are the minimum degrees of horizontal connection of the other.

Ordinarily a matrix can be undegenerate and impotent without being primitive or primitive without being impotent; but a one-rowed matrix or a matrix whose rows are homogeneous in two variables is undegenerate and impotent when and only when it is primitive; and a primitive matrix whose elements are rational integral functions of a single variable is always impotent.

The reductions of Weierstrass and Kronecker are transformations by matrices which are both primitive and impotent. Transformations of this kind (when not equigradent) will usually increase the total degree of the transformand.

INDEX

The references are to pages. The following is a list of the abbreviations used.

arb.	=arbitrary	equipot.	=equipotent	pot.	=potent
bi-can.	=bi-canonical	equiv.	=equivalent	rat.	=rational
can.	=canonical	eqn.	=equation	recipr.	=reciprocal
charac.	=characteristic	func.	=function	reg.	=regular
coeff.	=coefficient	gen.	=general	sim.	=similar
conj.	=conjugate	homog.	=homogeneous	sq.	=square
const.	=constant	indep.	=independent	st.	=standard
detant.	=determinant	int.	=integral	sym.	=symmetric
detoid.	=determinoid	invar.	=invariant	transfd.	=transformand
el.	=element	irred.	=irreducible	transfn.	=transformation
elem.	=elementary	irresol.	=irresoluble	undeg.	=undegenerate
equican.	=equicanonical	isom.	=isomorphic	unilat.	=unilatent
equigr.	=equigradent	lin.	=linear	unipot.	=unipotent
equim.	=equimutant	max.	=maximum	unit.	=unitary

For EU product safety concerns, contact us at Calle de José Abascal, 56–1°,
28003 Madrid, Spain or eugpsr@cambridge.org.

www.ingramcontent.com/pod-product-compliance
Lightning Source LLC
LaVergne TN
LVHW081522060526
838200LV00044B/1973